POWER SYSTEM RESTORATION

IEEE Press
445 Hoes Lane, P.O. Box 1331
Piscataway, NJ 08855-1331

POWER SYSTEM RESTORATION

Methodologies & Implementation Strategies

Edited by

M. M. Adibi
IRD Corporation

A Selected Reprint Volume

IEEE PRESS

IEEE PRESS SERIES ON POWER ENGINEERING

P. M. Anderson, *Series Editor*

The Institute of Electrical and Electronics Engineers, Inc., New York

This book and other books may be purchased at a discount
from the publisher when ordered in bulk quantities. Contact:

IEEE Press Marketing
Attn: Special Sales
445 Hoes Lane, P.O. Box 1331
Piscataway, NJ 08855-1331
Fax: +1 732 981 9334

For more information on IEEE Press products, visit the
IEEE Online Catalog & Store: http://www.ieee.org/ieeestore.

ISBN 978-0-7803-5397-8

IEEE Order No. PC5830

Library of Congress Cataloging-in-Publication Data

Power system restoration : methodologies & implementation strategies / edited by M.M. Adibi.
p. cm. — (IEEE press series on power engineering)
"A selected reprint volume."
Includes bibliographical references and index.
ISBN 0-7803-5397-8
1. Electric power systems—Maintenance and repair. 2. Electric power failures. 3.
Electric power distribution. 4. Emergency management. I. Adibi. M. M. (Mahmod M.)
1924– II. Series.

TK3091 .P685 2000
621 .31′028′8—dc21

99-058507

BOOKS IN THE IEEE PRESS SERIES ON POWER ENGINEERING

ELECTRIC POWER APPLICATIONS OF FUZZY SYSTEMS
Mohamed E. El-Hawary, *Dalhousie University*
1998 Hardcover 384 pp IEEE Order No. PC5666 ISBN 0-7803-1197-3

RATING OF ELECTRIC POWER CABLES: Ampacity Computations for Transmission, Distribution, and Industrial Applications
George J. Anders, *Ontario Hydro Technologies*
1997 Hardcover 464 pp IEEE Order No. PC5647 ISBN 0-7803-1177-9

ANALYSIS OF FAULTED POWER SYSTEMS, Revised Printing
P. M. Anderson, *Power Math Associates, Inc.*
1995 Hardcover 536 pp IEEE Order No. PC5616 ISBN 0-7803-1145-0

ELECTRIC POWER SYSTEMS: Design and Analysis, Revised Printing
Mohamed E. El-Hawary, *Dalhousie University*
1995 Hardcover 808 pp IEEE Order No. PC5606 ISBN 0-7803-1140-X

POWER SYSTEM STABILITY, Volumes I, II, III
An IEEE Press Classic Reissue Set
Edward Wilson Kimbark, *Iowa State University*
1995 Softcover 1008 pp IEEE Order No. PP5600 ISBN 0-7803-1135-3

ANALYSIS OF ELECTRIC MACHINERY
Paul C. Krause, *Purdue University*
1996 Hardcover 584 pp IEEE Order No. PC4556 ISBN 0-7803-1101-9

SUBSYNCHRONOUS RESONANCE IN POWER SYSTEMS
P. M. Anderson, *Power Math Associates, Inc.*
B. L. Agrawal, *Arizona Public Service Company*
J. E. Van Ness, *Northwestern University*
1990 Softcover 282 pp IEEE Order No. PP2477 ISBN 0-7803-5350-1

POWER SYSTEM PROTECTION
P. M. Anderson, *Power Math Associates, Inc.*
1999 Hardcover 1,344 pp IEEE Order No. PC5389 ISBN 0-7803-3427-2

UNDERSTANDING POWER QUALITY PROBLEMS: Voltage Sags and Interruptions
Math H. J. Bollen, *Chalmers University of Technology, Sweden*
2000 Hardcover 576 pp IEEE Order No. PC5764 ISBN 0-7803-4713-7

POWER AND COMMUNICATION CABLES: Theory and Applications
R. Bartnikas, *University of British Columbia*
K. D. Srivastava, *Institut de Recherche d'Hydro-Quebec*
2000 Hardcover 896 pp IEEE Order No. PC5665 ISBN 0-7803-1196-5

Contents

Preface

To provide a forum to share restoration information and address potential restoration problems, the Power System Restoration Task Force was formed in 1978 as a subgroup of the IEEE Power Engineering Society (PES) System Operation Subcommittee. It was later changed to a working group status and, just recently, Power System Restoration Working Group (PSR-WG) became a subgroup of the Transmission Operations Subcommittee within the PES Power System Operation Committee.

The PSR-WG produced the initial *Power System Restoration,* in order to help provide a source of useful information about restoration. The work was published by the IEEE Power Engineering Society, 93 THO 605-6 PWR, in 1993, and it consisted of 14 reprinted papers written or invited by the IEEE PSR-WG. Demand for the book was heavy, reflecting the wide interest in the subject, and it was soon out of print.

In the interim, a number of excellent papers on the subject have been written. This book responds to the continued interest in restoration by including all the original papers of the first publication along with 73 additional papers. To enhance the organization of the work, the papers are now sorted into seven main parts and, for the sake of continuity, papers in each part are arranged in the order of their original publication date. Also for convenience, a General Discussion and Discussion of Reprints are provided at the beginning of each part.

Part 1 is on Restoration Overview edited by J. J. Ancona. It covers major power system disturbances, power system restoration issues, special considerations in restoration, new approaches in power system restoration, steam plant start-up and control, analytical tools requirements, and a framework for power system restoration. It offers excellent overviews of the power system restoration process, and it provides background and introductory material for those wishing to familiarize themselves with the key aspects of restoration. More importantly, they also include in-depth discussions of the overall restoration process. This helps to illustrate the careful balance that will be required between different objectives, components, constraints, and operations to ensure a successful restoration.

Part 2 is on Restoration Techniques edited by M. M. Adibi. It covers real and reactive power control, sustained and transient overvoltage controls, reactive capability of generators, frequency response of prime movers, protective system issues and line asymmetries, remote cranking of steam electric units, and nuclear plant requirements. It offers various techniques used in planning, simulating, field testing, and verifying different aspects of power system restoration. The models and procedures developed address steady-state, transient, and dynamic behaviors of the power system and its major components. In addition, it provides guidelines and recommendations for development of an efficient restoration plan and its safe execution.

Part 3 is on Restoration Planning edited by L. H. Fink. It covers pre-disturbance conditions and post-disturbance status, post-restoration target system and restoration stages, physical, scheduling, and policy constraints, build-up and build-down tactics and strategies, generic restoration actions and specific building blocks, system modeling, simulation and validation, and estimation of restoration duration. These papers follow the major blackout of 1977 in the northeastern United States. They represent serious reconsideration of system restoration problems and of the need to develop and maintain adequate restoration plans.

Part 4 is on Restoration Training edited by M. Rafian. It covers conventional methods of operator training, development of an operator training simulator (OTS), verification of the restoration plan using OTS, evaluation of restoration tools using OTS, conduct and evaluation of restoration drills, experience using the OTS as a training tool, and restoration training techniques. These papers deal with the use of OTS in system restoration, and they discuss the actual experiences and broad issues surrounding the area of power system restoration training.

Part 5 is on Specific System Restorations edited by R. J. Kafka. It covers the restoration plan for the Pacific Northwest power system, system restoration at Ontario-Hydro, service restoration on the Hydro-Quebec power system, restoration of the French, Italian, Greek and Swedish systems, interactive long-term simulation for restoration, optimum restoration with interactive graphics, and policies for restoration of power system. This group of papers deals with the specific restoration plans of various organizations. Although each plan is different, there are many common characteristics. With respect to restoration planning, we see an evolution from plans based on the experience and expertise of system planning and operations personnel to heuristic trial and error simulations, to computer-guided and optimized simulations.

Part 6 is on Knowledge-Based Systems edited by C. C. Liu. It covers information beyond breaker and relay operations, inconsistent information on the status of breakers and switches, determination of the generator start-up sequence, transmission path findings, determination of the switching sequence, optimal field crew dispatch, and integration of Energy Management System and analytical tools. Power system restoration is one of the important applications for knowledge-based systems (KBSs). This group of papers primarily covers the bulk power system and subtransmission systems, and addresses numerous subproblems due to the complexity and scope of restoration.

Part 7 is on Distribution System Restoration edited by A. J. Monticelli. It covers physically based load models, cold load pickup of large secondary networks, feeder reconfiguration for service restoration, restoration and loss reduction of distribution systems, restoration in subcontrol center, power network restoration support systems, and optimal distribution design for cold load pickup. This group of papers includes models for cold load pickup, transformer thermal behavior, and methods (both combinatorial and analytical) used for determining optimal energization order.

Finally, for quick reference, both an author and a subject index are included at the end of the book.

Though significantly expanded, this edition is not intended as an exhaustive treatment of the subject. Hopefully, however, it will serve as an excellent reference source for those interested in restoration study, planning, and preparation. The PSR-WG welcomes your continued interest and future papers.

M. M. Adibi
IRD Corporation

PART I

Restoration Overview

I. General Discussion

THIS part presents papers that offer excellent overviews of the power system restoration process. They provide background and introductory material for those who want to become familiar with the key aspects of restoration. More importantly, they also include in-depth discussions of the overall restoration process. They help to illustrate the careful balance that will be required between different objectives, components, constraints, and operations to ensure a successful restoration.

II. Discussion of Reprints

1.01 Power System Restoration—A Task Force Report, 1987, p. 3

The first paper offers a comprehensive discussion of the restoration process by first introducing the essential steps in restoration. It then examines restoration considerations for systems with specific types of generation, as well as considerations that are common to all systems such as switching operations timing and cold load inrush. To help address these concerns and plans for restoration, the paper discusses database requirements for generators, particularly with respect to hot restarts. It also provides background on generator mechanical, electrical, and frequency control systems; and how they impact restoration. The paper also discusses the importance of balancing reactive power and controlling voltage during restoration. It goes on to illustrate how load and generation need to be balanced as restoration proceeds, and how controlled islanding and generation isolation may develop. Finally, the paper compares the philosophies of sequential versus parallel restoration in terms of backup power supply problems, blackstart capability, and sectionalizing subsystems.

1.02 Power System Restoration—The Second Task Force Report, 1987, p. 10

This paper investigates specific restoration problems that occurred following major disturbances. These 19 studies identified such problems as faulty synchronization equipment, resynchronization failures, lightning arrester failures due to switching overvoltages, lack of adequate data, and switching errors. The paper then distills these problems into seven general restoration problem categories: (1) reactive power imbalance, (2) load-generation imbalance, (3) lack of load-generation coordination, (4) lack of communication, (5) impediments caused by protective systems, (6) loss of emergency backup power, and (7) lack of restoration plans. Each of these is followed by suggested solutions and then by overall suggestions to help enhance restoration.

1.03 Analytical Tools for Power System Restoration—Conceptual Design, 1988, p. 17

The paper looks at the restoration process as a series of analyses and decisions to meet a set of operating constraints. It proposes a conceptual framework for computer-aided monitoring and assessment during restoration. It also describes a knowledge-based or expert system to help guide restoration. That system would evaluate monitored conditions; suggest an appropriate sequence of actions for assessment; define the problem and select software to be used in an analysis; and then validate the results of the analysis. Based on these results, the system could modify the suggested sequence of actions to help arrive at an improved restoration.

1.04 System Operation Challenges, 1988, p. 24

This paper includes a set of five reports that address various problems that can impact the restoration process. The first involves the general issue of Energy Management System (EMS) software and hardware slowly progressing towards obsolescence, which affects normal operations as well as restoration. The second deals with the difficulty involved in determining EMS requirements. The third raises the concern that EMS alarm systems are designed for normal and limited emergency conditions, and therefore, may need to be modified for restoration conditions. Under these circumstances, only essential alarms should be activated to avoid overwhelming the system. The fourth addresses restoration training concerns such as instructional design, development of performance standards, and inadequacy of resources. Finally, the fifth lists a series of restoration problems in the context of restoration planning, actions during system degradation, and restoration of a stabilized system.

1.05 Power System Restoration Issues, 1991, p. 31

This paper lists major considerations that need to be dealt with during restoration, such as switching transients, remote cranking power, damage assessment versus cause identification, phase angle differences, and generator startup times. The paper then lists the fundamental steps that need to be included in the development of a restoration plan. It goes on to discuss exploration of more advanced methods such as expert systems to aid in restoration and operator training.

1.06 Special Considerations in Power System Restoration, 1992, p. 37

This detailed paper addresses several important restoration issues. It discusses the problem of excessive alarms that may occur, and it suggests that a more confined set of alarms should be used during restoration. It then reviews restoration switching problems and strategies, including the need to consider backup power systems and cold weather problems. In addition, the optimal sequencing of generator startups is discussed in detail. Next, the special problems associated with underground transmission cable are illuminated. Finally, the paper lists the capabilities and limitations of both public and private telecommunication systems during restoration.

1.07 New Approaches in Power System Restoration, 1992, p. 46

This paper looks at how the computer can be used to aid the restoration process. What are the strengths and weaknesses of restoration that is fully computer automated, computed aided, or simply shared cooperatively by both computer and operator? What are the key steps in developing and implementing an expert system? What are the requirements for an Operator Training Simulator (OTS) for use in restoration preparation, and for what typical restoration scenarios would OTS be applicable?

1.08 A Hierarchical Interactive Approach to Electric Power System Restoration, 1992, p. 52

This paper begins with the argument that the use of general guidelines for restoration lacks specificity. It proposes that these guidelines be framed more tightly in terms of multiple, albeit conflicting, objectives; variables that can be controlled; and constraints that need to be honored. The paper recommends a hierarchical interactive control approach such that restoration control actions are decomposed into direct (or localized), optimizing (centralized and global), and adaptive (alternative strategy) layers. The paper then proceeds to explain these concepts to illustrate how they can be used to improve the restoration process.

1.09 Special Consideration in Power System Restoration—The Second Working Group Report, 1994, p. 61

This paper details four restoration issues. The first discusses different types of loads and addresses modeling of cold load pickup both heuristically and through use of physical models. The second looks at variations that may occur in cold load pickup on low-voltage networks. The third discusses the unique role that gas turbines can play in contributing to rapid system restoration. The fourth describes reactive power balancing from the perspectives of transmission line charging, generator capabilities, power-plant auxiliary requirements, and voltage control techniques.

1.10 Steam Plant Startup and Control in System Restoration, 1994, p. 68

This paper indicates the complexity of steam plant restarts following a major system shutdown. It first discusses normal steam unit startups and shutdowns in detail. It then explains how this orderly controlled process differs markedly from a unit that trips and abruptly shuts down following a major disturbance. Finally, it offers methods to improve generator operation during major disturbances through changes in design and operator training.

1.11 An AGC Implementation for System Islanding and Restoration Conditions, 1994, p. 75

This paper begins with an excellent background description of Automatic Generation Control (AGC). From that base, it describes the development of an Island AGC system on the Virginia power system to help improve the restoration process by maintaining a constant frequency. It then shows options that would be available to operators if Island AGC were invoked. It concludes with an analysis of actual test results from the Virginia power system.

1.12 Analytical Tool Requirements for Power System Restoration, 1994, p. 86

This work presents an overview of restoration analytical tools (AT) and serves as a valuable reference because it also addresses practical issues pertaining to restoration. It explains the need for AT in preparing restoration plans and training, but it also implies the need for general restoration guidelines that do not rely on AT. For each type of AT restoration software, the authors indicate: (1) what needs to be studied; (2) specific restoration features needed; (3) recommended preparatory work; (4) study procedures; and (5) why restoration studies are needed.

1.13 A Framework for Power System Restoration Following a Major Power Failure, 1995, p. 96

The last paper in Part I is an organized summary of many of the other restoration papers. As such, it can serve as a general restoration guide for operators, an outline to help prepare a more specific restoration plan, and a guide to help evaluate and improve restoration preparedness. The paper provides a framework for restoration by outlining: (1) recommended goals and objectives of restoration; (2) responses that should be initiated at the onset of abnormal conditions; (3) the sequence of restoration actions that should be taken; and (4) steps that can be taken to enhance restoration preparedness.

POWER SYSTEM RESTORATION - A TASK FORCE REPORT

Contributing Members: M. Adibi (Chairman), P. Clelland, L. Fink, H. Happ, R. Kafka, J. Raine, D. Scheurer, and F. Trefny

Abstract - The IEEE PES System Operation Subcommittee has established the Power System Restoration Task Force to: review operating practices, conduct a literature search, prepare relevant glossaries and bibliographies, and promote information exchange through technical papers. This is the first report of the Task Force.

The problem of bulk power system restoration following a complete or partial collapse is practically as old as the electric utility industry itself. Many electric utilities have developed over the years system restoration schemes that meet the needs of their particular systems. These plans provide a great deal of insight into how the restorative process is viewed by operating and planning personnel and what concerns and constraints any plan must operate under.

The body of the report consists of notes prepared by members of the Task Force. It should not be inferred that a complete reporting on Power System Restoration is undertaken here. The intent is to report upon work of the Task Force to date. The report also reviews several different restoration plans and shows their common concerns and constraints.

I. INTRODUCTION

Today's bulk power systems provide a highly reliable supply of electric power. However, due to a combination of unforeseen circumstances, there is the remote possibility of a system wide outage. It is therefore prudent to be prepared for such an unlikely eventuality by developing an up-to-date, readily accessible, and easily understood power system restoration plan to allow a quick and orderly recovery from a system outage, with resultant minimum impact on the public.

The bulk power supply major disturbances are primarily caused by transient faults and mainly originate in the transimission systems. A very large number of these initiating causes of supply interruptions are due to temporary faults, such as lightning, which are immediately cleared by fast and selective protective relays, leaving the system in an unfaulted condition. In many cases, however, these temporary initiating causes produce subsequent effects which are "permanent," including loss of generation, load and interconnections. These subsequent effects result in a partial to complete collapse of unfaulted power systems. Thus, searching for the originating fault in power failures may be futile, although identifying the status of the collapsed system components would enhance restoration.

86 SM 329-7 A paper recommended and approved by the IEEE Power System Engineering Committee of the IEEE Power Engineering Society for presentation at the IEEE/PES 1986 Summer Meeting, Mexico City, Mexico, July 20 - 25, 1986. Manuscript submitted August 13, 1984; made available for printing May 7, 1986.

The major portion of the initial effort in restoring bulk supply is in restart and reintegration procedures for generation and transmission systems. Load pick-up during these initial phases, i.e. restart and reintegration, is necessary for (a) bringing generators to their stable, minimum generation levels and (b) maintaining satisfactory voltage conditions.

This paper first reviews three different restoration plans. It then describes the power system characteristics relevant to restoration. This is followed by power system considerations. Finally, it provides a bibliography covering 1940 to 1984.

II. REVIEW OF RESTORATION PLANS

1. Thermal Systems(25)

The system restoration plan developed for all thermal systems serving metropolitan areas is based on:

Sectionalization of the system into two or more subsystems and simultaneous reintegration of generation and transmission in each subsystem. Selection of subsystems is based on black-start capability within each subsystem and steam generators with hot restart capability, and any transmission and loads required to interconnect them.

Division of the restoration procedure into restart, reintegration, load pick-up and interconnection phases. The restart phase begins with providing station service for the restart of steam units within each subsystem. In the reintegration phase generating stations are interconnected. During these two phases only the necessary loads are picked-up as dictated by generation and transmission requirements. These two phases end when the subsystems are synchronized and firmly interconnected. During the load pick-up phase loads are picked-up in small increments to avoid excessive underfrequency deviations. Due consideration is given to reactive power balance, voltage conditions and stability of the system. With the start of large steam units and availability of ample reactive absorbing capabilities, the EHV lines are energized and the interconnections are reestablished.

2. Hydro-Thermal Systems (36)

The system restoration plan developed for hydro-thermal systems serving metropolitan areas is based on energizing the entire bulk power transmission system in one step and providing station service to all thermal generating stations. This approach uses the ability of large hydro stations to absorb the charging currents of the complete transmission system, and it is an attempt to avoid the time consuming line sectionalization and switching operations which are normally required in system restoration. The high voltages at the receiving end of lines are avoided by manual operation of hydro's voltage regulators far below normal.

With the availability of large hydros, the interconnection phase can also precede the load pick-up phase.

Reprinted from *IEEE Transactions on Power Systems,* Vol. 2, No. 2, pp. 271–277, May 1987.

3. Primarily Hydro Systems [4]

The system restoration plan developed for all hydro or primarily hydro systems emphasize the switching operation and the response of prime movers to a sudden load pickup. In this plan, due consideratoin is given to the time it takes to isolate and energize the necessary lines by using central control systems to execute the switching operation programs. The off-line dynamic programs are used to determine responses of the system and provide guidelines for load pick-up based on the prevailing generation on-line, transmission configuration and system loading.

4. Common Characteristics of Restoration

Review of the above and several other system restoration plans show a number of common concerns even though their strategies are different:

1. Immediate resupply of station service.

2. Time consuming nature of switching operation.

3. Start-up timings of thermal units.

4. Voltage rise problems of energizing unloaded transmission lines.

5. Frequency response of prime movers to a sudden load pick-up.

6. Cold load inrush, power factors and coincident demand factors.

These system restoration plans generally end when the bulk power system has been "restored" and the pick-up of the next increment of load is based on other criteria than technical requirements of the system.

It is concluded that most bulk power supply systems have certain characteristics in common and behave in a similar manner during the restoration process. It is therefore possible to establish a general procedure and a list of guidelines which would enhance rapid restoration. On the other hand, the detailed plan has to be specifically developed to meet the particular requirements of an individual power system. Thus, it is important to determine and record the relevant characteristics of each power system.

III. POWER SYSTEM CHARACTERISTICS

1. Steam Unit Characteristics

By: R. J. Kafka, Potomac Electric Power Co., Washington, D.C.

The characteristics of a power system relevant to restoration may be divided into three functional groups: those related to reactive power balance, those which affect the load generation balance, and those which primarily deal with the protective and control systems. The characteristics related to load generation are:

- o Number and size of low voltage AC network loads under peak and light load conditions.

- o Generator's maximum and minimum output and rate of loading under normal and emergency conditions.

- o The dynamic response of prime movers to sudden load pick-up.

- o Maximum time intervals for hot restart of drum type of boilers.

- o Minimum elapsed time for hot restart of super-critical once-through boilers.

- o Blackstart capabilities of combustion turbines.

System Restoration entails careful consideration of probable effects of each action on the rest of the system and on the subsequent measures. Transmission line loadings, generator loadings, and phase angle differences between load centers and generators must be evaluated. The time at which units are brought on the line, the rate at which generators can be loaded, the mismatch between load pick-up and generator response, the change in system frequency, and the effect on feeders having under-frequency relays should be considered. Equally important is the recognition and coordination of certain critical time intervals, such as maximum time interval beyond which a unit with a drum type boiler cannot safely be hot restarted, or the minimum time interval after which a super-critical, once-through unit can safely be restarted. While sectionalization of the system into a number of subsystems and simultaneous reintegration of generation in each subsystem could appreciably reduce the duration of an outage, the extent of sectionalization is limited by availability of resources, including the number of system and plant operator teams, as well as the communications available to permit effective and efficient coordination of their efforts.

Hot restart of cycling steam units which are drum type is essential. These units have to be placed back into operation supplying loads within a short time interval following the system outage. Otherwise the restart may not be possible for several hours. Such a time interval then is very critical to system restoration, particularly in those stations where station service has to be supplied from another station. It is important to accurately determine the time intervals between unit trip-outs and hot restart.

2. Combustion Turbine Characteristics for Cold System Start

By: F. Trefny, Houston Lighting and Power Co., Houston, TX

There are two basic sizes and classes of combustion turbines. These are 15MW black plant start and 50MW peaking and/or combined cycle units. Each turbine's starting characteristics are summarized below:

2.1 Characteristics of 15MW Combustion Turbines

These combustion turbines can be started with no external auxiliary power. The units can be brought up from cold start in approximately 30 minutes. However, the 30 minutes include a 15 minute warm-up, but this can be bypassed in an emergency to allow 15 minute start-up time. Units can be restarted immediately from shutdown.

Starting is accomplished by diesel engine via a clutch. No electric motors are used. A second diesel engine driving a direct current generator will supply the necessary auxiliary power. This direct current generator is not normally used unless the plant is blacked out.

The unit may be operated anywhere between 0-15MW. The operator uses MW setpoint control to adjust the turbine to load conditions, and this must be changed as load is increased or decreased.

2.2 Characteristics of 50MW Combustion Turbines

The 50MW combustion turbines require an A.C. source of auxiliary power for starting. The units can be brought from a cold start to full load in about 15 minutes. The same time is required for a hot start. However, before a restart can be attempted, the turbine must roll down to a stop and then be placed on turning gear (this is permissive for starting controls). Roll down requires approximately half an hour. Electrical power must be supplied to a 900HP cranking motor to start the turbine; other auxiliary power requirements are minimal. Often this power can be supplied by a black plant gas turbine which requires approximately 30 minutes to start.

The generator can be operated from zero to full load. MW setpoint can be used to set power. Maximum power is determined by turbine exhaust temperature--power is increased until limit is reached. Governor is set for approximately 4% regulation.

3. The Role of Plant Control System

By: P. Clelland, Philadelphia Electric Co., Philadelphia, PA

3.1 Mechanical Considerations

Many control systems for the "mechanical" portion of the power plant (combustion, feedwater, steam temperature, etc.) are deisgned for use between minimum and maximum load, and are not truly effective during startup or extreme low load operation. Minimum load is probably no lower than 25% of maximum load.

Main and reheat steam temperature controls are often not in control of their associated temperatures at low loads because the steam temperatures are far below their defined setpoints. The controls are taking appropriate action (spray valves closed or tilts fully elevated) but the control objective under these conditions is not correctly defined. The real control objective is to regulate the rate of change of temperatures in order to reduce stresses - not to increase temperature as quickly as possible in an attempt to satisfy some unattainable setpoint. Consequently, during startup the operator must place these temperature control systems on manual and perform his own form of rate control.

Feedwater control systems normally operate in a "three mode" form, where feedwater flow is matched to steam flow and biased to maintain constant drum level as a secondary control objective. During light-off and very low load operation, this control strategy cannot be applied. Before the turbine is started, there is no measured steam flow (though there may actually be steam flow through vents and drains). At extreme low loads, the steam and feedwater flows will be in the lowest range of the associated flow instruments and these measurements will be of questionable accuracy. For this reason, startup feedwater control is performed as a "single mode" operation, with the sole control objective of maintaining drum level at setpoint. As steam flow is initially supplied to the turbine, the single mode control becomes a reaction type controller. It receives no information regarding the outflow from the boiler, but must regulate the inflow to maintain drum level.

It is significantly easier to establish circulation in a boiler which has boiler circulating pumps than it is in a natural circulation unit. In the former unit, circulation can begin before introducing fire to the boiler. This avoids the risk of local hot spots caused by non-uniform firing patterns.

Startup feedwater control is easier to accomplish in a plant which uses feedwater regulating valves for flow control than in a plant which uses turbine-driven feedpumps and speed controllers. This is because the turbine-pump has its own dynamics which must be considered.

Light-off and low load operation is usually not performed with the primary fuel. The process is to warm the unit using igniters and No. 2 oil, transfer to oil torches and No. 6 oil, and finally transfer to pulverized coal. This usually means that we have two fuel supply systems in operation, with one (or none) in automatic and one (or both) in manual. Consequently, regulation of fuel flow is not necessarily accomplished at one work station. Some coordination is lost. Most furnace implosions on pulverized coal units have occurred during this dual-fuel operation.

NFPA codes set minimum air flow requirements to avoid explosive mixtures. This may cause a problem because of the resulting mismatch between minimum air flow and the required fuel flow.

3.2 Electrical Considerations

A wide variety of voltage regulators and associated control schemes exist throughout the power indsutry. This is to control the generator's terminal voltage to some operator-established setpoint. This control is ultimately accomplished by varying the DC field current in the rotor of the generator. The voltage regulator can be considered to act almost instantaneously when compared to other station controls.

In normal operation, the operator will usually establish an initial volt-ampere reactive (VAR) flow from the generator with the voltage regulator on manual. Once the VAR flow is established and the unit has reached a minimum operating point (usually less than 40% load), the voltage regulator is switched to automatic. Any subsequent changes in system voltage levels will result in generator VAR flow changes as the regulator tries to maintain the generator terminal voltage at its setpoint. The major concern of the operator is to keep the field current (and consequently the VAR flow from the generator) high enough to ensure machine stability.

During restart conditions, one should be aware of protective relays which could initiate an undesirable regulator response or generator trip. Some of these relays are:

- Volts per Hertz
- Out-of-Step
- Under-Excitation
- Distance
- Frequency
- Field Forcing

The volts per hertz relay will operate if this ratio becomes excessive (volts too high or frequency too low). Some voltage regulators automatically alter the voltage setpoint for changes in frequency.

The Out-of-Step, Under-Excitation, and Distance relays are impedance-type relays and could operate during short electrical power swings or transients.

The Frequency relay monitors electrical frequency and could operate before turbine speed protection.

The Field Forcing relay will not trip the unit but will establish full field flow during periods of low voltage. Some field rheostats are changed to a new full field position by this relay.

3.3 Frequency Control

Frequency control would probably be inactive during restart conditions and would not be reestablished until the system becomes interconnected.

Units cannot be brought on line with zero load. Some load must be picked up immediately to avoid the possibility of motoring. Thus the generator breaker cannot be closed on a completely cleared bus, but neither can the connected load be excessive.

IV. REACTIVE POWER BALANCE

1. Reactive Power Balance

By: J. G. Raine, Florida Power & Light Co., Miami, FL

The expression "Reactive Power Balance" implies that there is Reactive Generation and Reactive Load and that these must be equal or balanced. Thus it is important to have a clear understanding of what is considered Reactive Generation and what is considered Reactive Load so that we can control their balance to obtain the voltage conditions we desire. Listed below is a possible classification of Generation and Load.

± Generators
± Co-Generators
+ All Circuits (Line Charging)
+ All Shunt Capacitors
- All Shunt Reactors
± Synchronous Condensers
± Static Var Generators
± Customer Demand
+ All Circuits (I^2X) *
+ All Transformers (I^2X)
+ All Transformers (excitation)
+ Generator Auxiliary Equipment
± Interchange

* Series reactors and Series capacitors are treated as changes in the circuit impedance

The first step of actual restoration will involve the energization of some transmission circuit, that is the sudden bringing on line of a Reactive Generator. This will result in an increase in voltage across the system and especially at the "far end" of the circuit energized. To decrease the impact of bringing a large Reactive Generator on line we could at the same time bring on a negative Reactive Generator (a shunt reactor on the far end of the circuit) or a positive Reactive Load (a substation with customer load). The latter would require Real Power Generation to supply the real power load but it would increase the I^2X loss in the system. Care should be taken that no additional Reactive Generators in the form of Shunt capacitors are part of the load. Another approach would be to avoid energizing an EHV circuit or high voltage cable but to energize a lower voltage transmission circuit, that is bring on line a smaller Reactive Generator.

After several steps of transmission restoration have occurred the resulting system may be a "long skinny" system of relative high series reactance so that a high voltage drop occurs from reactive flow through the reactance. To reduce the series reactance, a parallel circuit should be energized and tied through. This energizing of circuits increases the positive reactive generation which in turn requires additional negative reactive generation. If the disturbance has removed most of the real power generators along with their reactive power generation, then these reactive generators must be only reactive generators, that is shunt capacitor, shunt reactors, synchronous condensor or static var generator.

If indeed partial restoration has been accomplished through by-passing of available EHV transmission then the conditions of the above paragraph may be quite difficult. Special switching arrangement may be required to connect an autotransformer and tertiary connected shunt reactors to the EHV circuit prior to energizing the circuit and autotransformers as a unit.

If a high voltage cable is to be energized there should be sufficient negative reactive generation "close" to the cable to balance the added positive reactive generation. Another consideration is whether the cable should be energized raising the system voltage and then shunt reactors be energized to lower the system voltage or should the shunt reactors be energized first to lower the system voltage and then the cable energized to raise the system voltage. Generally, the latter approach is preferred.

In all of these cases the system voltage is monitored to determine the out-of-limit condition but the reactive generation and load plus reactive flows are analyzed to determine the appropriate action. A major requirement of the operating personnel is to know the magnitude and location of the reactive capability of their system and at any moment the magnitude and location of the reactive reserves of their system.

2. Load-Generation Balance

By: M. M. Adibi, IRD Corporation, Bethesda, MD

2.1 System Dynamics During Restoration

The development of a strategy for coordinating power system restorative control actions requires that the behavior of the power system during the periods of outage, startup and reintegration, be thoroughly understood. Unfortunately, the behavior of the system during these periods is uncertain, and the dynamic processes that govern the behavior of relevant system components are highly non-linear and complex: (1) the exact configuration of the system and the operating status of system equipment are not known precisely; (2) load demand changes occur in the course of an outage as a function of several variables, including duration of outage, weather conditions, time of day, customer lifestyle, type of load, and dynamic characteristics of the specific loads involved.

During system restoration, a reasonable balance must be maintained between generation and load to avoid excessive over and under frequency deviations. Load pick-ups at small increments prolong the restoration duration. With large increments there are always risk of frequency decline beyond recovery and recurrence of system outage. The size of load pick-up depends on the response of prime movers. In developing the system restoration plan, it is therefore essential that the response of prime movers be determined.

Unfortunately, the available linearized dynamic models are not suitable nor adequate for evaluating dynamic phenomena such as the variation of system frequency as a function of load-generation mismatch, the effect of underfrequency relays, and the rate at which a generator can be loaded. To consider these phenomena, dynamic behavior of boilers, turbines, combustion controllers, feedwater controllers, temperature controllers, and governor systems must be represented over the operating range of prime movers.

As a substitute for the simulator, a guideline approved by the operating and production personnel may be used. This information could take the form of the largest increment of load increase at one time, which would not lower system frequency more than 0.5 Hz of its previous value. Experience with several systems has shown that a load increment of 5% of each subsystem generation may be satisfactory in that such an increment would reduce the frequency by no more than 0.5 Hz. Also, restoration of smaller and radial loads prior to larger and low voltage AC network load provides a good control of frequency.

2.2 Intentional Islanding Schemes

When two groups of generators begin to lose synchronism, (go out-of-step) it is advisable to end severe voltage fluctuations by separating the two groups of generators, or the two areas. Under such circumstances, it is desirable to select the preplanned locations where: (1) the excess or deficiency of generation in each separated area is minimal, thus resulting in the least disruption of service, and (2) synchronizing equipment is available for reclosing the circuits and restoring the operation back to normal in a minimum time.

In order to ensure satisfactory islanding during emergency conditions, it may be necessary to block some of the relays from tripping where splitting is not desirable, and to trip or allow to trip only those relays where it is desirable. In order to satisfy these requirements, the relay schemes must be able to discriminate between: (1) normal load conditions, (2) swing conditions, (3) out-of-step conditions, and (4) fault conditions.

This would entail extensive transient stability studies of the area(s), installations of appropriate relays, including blocking and transfer tripping, and a thorough relay coordination. These are major undertakings.

2.3 Low Frequency Isolation Scheme

Isolation of generation with matching local loads through relay schemes would avoid startup of the generator(s) from a complete shutdown and consequently would speed up the restoration process.

A dynamic analysis is required to determine the steam units' frequency response under different load and generation mismatch after isolation. The excess generation (i.e., load rejection) may trip the turbine and boiler, and the excess load may trip the generator by the under-frequency relay. The load rejection capabilities can be achieved by proper coordination of turbine controls and boiler interlocks.

Unless, during peak- and light-load conditions, load and generation mismatch are within the limits, the low frequency isolation scheme would not be reliable.

3. SEQUENTIAL VS. PARALLEL RESTORATION

by: D. Scheurer, Philadelphia Electric Company, Philadelphia, PA

The determination of the method to be used in restoring a given system is highly dependent on several factors. These factors include the relative size of the system, the amount of blackstart capacity, the location of the blackstart capacity, and the amount and location of emergency backup power supplies.

The main objective of the early stage of a restoration procedure is to establish a transmission line to all stations and substations in order to reenergize light and power supplies before the integrity of equipment is jeopardized. Small systems with a centralized blackstart capability can probably accomplish these ends using a sequential restoration method. The time required to restore a larger system by this method, however, may be significantly longer, and therefore a parallel restoration method would be required to energize all light and power within a reasonable time frame. The major considerations to be used in determining a restoration method are as follows:

3.1 Backup Power Supplies at Substations, Generating Stations, and Control Centers

A. Telecommuncations equipment

The operability of many batteries used for the backup of supervisory control and data remotes can become questionable within a short time period (1-2 hours). This would result in incomplete data acquisition and the inability to close breakers via supervisory control.

B. Circuit Breakers

Many circuit breakers require an AC feed for compressors and heaters. On a cold day, these breakers may become inoperative within 1-4 hours following interruption due to gas liquefication or loss of pressure. Once rendered inoperative, they may require several hours to recover following the restoration of power.

C. Oil-Static Cable Systems

Cable failure on energizing these systems is a possibility due to oil contamination or loss of insulation quality if no backup AC independent source of pumping is supplied.

3.2 Blackstart Capability

The method of restoration chosen is dependent on the available blackstart capacity and its geographic distribution across the system. Parallel restoration of subsystems requires that sufficient blackstart capability be available within each subsystem to energize all light and power and maintain voltages. Sequential restoration requires that the centralized or main cranking source be capable of carrying sufficient load to hold down system voltages and carry system light and power until main units are on line.

3.3 Criteria for Sectionalizing of Subsystems

The following criteria are applicable for the successful sectionaliziation of subsystems:

A. Each subsystem must have blackstart capability which is sufficient to restore light and power to critical equipment.

B. Each subsystem should have the ability to match generation and load to within prescribed frequency limits.

C. Each subsystem should have adequate voltage controls to maintain a suitable voltage profile. This would include the ability to pick up load, underexcite generating units, change taps on tie transformers or operate synchronous condensers.

D. Each subsystem should be capable of being sufficiently monitored at the system control center in order to ensure its internal security and coordinate switching.

E. In certain areas, subsystems may be formed by isolating steam units along with their auxiliaries and bus loads, either by manually opening transmission lines or via underfrequency relaying.

F. All tie points for subsystems must be capable of swinging generation and measuring synchronization with adjacent subsystems.

In summary, small systems which are capable of restoring light and power to critical points in a relatively short period of time, or large systems with adequate backup power systems in key areas can most likely be successfully restored via the sequential method from a tie point or internal cranking source.

Larger systems, however, are not generally equipped with the extensive resources required to supply emergency AC power at all critical locations. Under these conditions, small self supporting subsystems are the most practical method. This method allows the rapid restoration of station light and power before degradation to equipment becomes a reality, provided that the adequate blackstart capability and criteria listed above can be met.

V. BIBLIOGRAPHY

By: L. H. Fink, Carlson & Fink, Inc., Oakton, VA

1. Adibi, M. M., et al. "Solution Methods for Transient and Dynamic Stability" Proceedings of IEEE, Vol. 62, July 1974.

2. Anderson, Max D., et al. "Power System Operator Training Update," IEEE PES Summer Meeting, Vancouver, B. C., July, 1979, A79-515-8.

3. Arnold, P. F., Discussion to Kafka et al. (1981)

4. Arnold, P. F., "Summary of System Restoration Plan for the Pacific Northwest Power System," Power System Restoration T. F. Mtg., 1984 WPM.

5. Blankenship, G. L., T. A. Trygar, "A Discussion of the Restorative State Control Problem in Electric Power Systems," Proc. EPRI/SIAM Conf. on Electric Power Problems: The Mathematical Challenge, Seattle, March, 1980.

6. Bryson, W. C., and A. P. Hayword, "Restoration of Service on a Metropolitan Power System," Trans. AIEE Vol. 59, 1940.

7. Chaiken, J. M., R. C. Larson, "Methods for Allocating Urban Emergency Units: A Survey," Management Science, 19 (1972), part 2, pp. P-110-130.

8. Chien, K. L., et al., "Dyanmic Analysis of a Boiler," Trans. of the ASME, 80, pp. 1809-1819.

9. Concordia, C. "Equipment Modeling Loads," GE Power System Stability Seminar, September, 1973.

10. Daniels, J. H., et al. "Dynamic Representation of a Larger Boiler-Turbine Unit," 1961, ASME paper No. 61-SA-69.

11. Davidson, D. R., D. N. Ewart and L. K. Kirchmayer, "Long Term Dynamic Response of Power Systems: An Analysis of Major Disturbances," IEEE Trans. on Power Apparatus and Systems, PAS-94 (May/June 1975), pp. 819-826.

12. Dean, S. M., "The Design and Operation of a Metropolitan Electrical System from the Viewpoint of Possible Major Shutdown," Trans. AIEE Vol. 59, 1940.

13. Dryar, H. A. and R. Bailey, "Restoration of Service for Large Metropolitan System After Complete Shutdown," Trans. AIEE, Vol. 59, 1940.

14. EPRI, "Long Term Power System Dynamics," Vol. I Summary and Technical Report, Final Report of EPRI Research Project 90-7, June 1974.

15. Ewart, D. "Whys and Wherefores of Power System Blackouts," IEEE Spectrum, April, 1978, pp. 36-41.

16. Feinstein, J., Discussion of Kafka et al. (1981).

17. Horowitz, S. H., A. Politis, A. F. Gabrielle, "Frequency Actuated Load Shedding and Restoration Part II--Implementation," IEEE Trans. Power Apparatus and Systems, PAS-90 (1971), pp. 541-549.

18. IEEE Committee Report, "Dynamic Models for Steam and Hydro Turbines in Power System Studies," paper T73-089-9, IEEE PES Winter Meeting, Jan. 28-Feb. 2, 1973, New York.

19. IEEE Committee Report, "Survey Report on Current Operational Problems," Paper 81 WM 032-2, IEEE PES Winter Power Meeting, Atlanta, 1981.

20. IEEE Committee Report, "An Updated List of Current Operational Problems," IEEE Trans. on Power Apparatus and Systems, PAS-97, (Jan/Feb 1978), pp. 140-148.

21. IEEE Committee Report, "System Restoration - Deploying the Plan: Current Operational Problems Working Group,: IEEE Trans. Nov., pp. 4263-71, 1982.

22. IEEE Meeting Group on Power Plant Response to Load Changes, "MW Response of Fossil Fueled Steam Units," IEEE Trans. on Power Apparatus and Systems, March/April 1973, Vol. PAS-92, No. 2,

23. Ihara, S., F. C. Schweppe, "Physically Based Modeling of Cold Load Pickup, IEEE/PES Winter Meeting, Atlanta, February, 1981 (81 WM 084-3).

24. Kafka, R. J., D. R. Penders, S. H. Bouchey, and M. M. Adibi, "Role of Interactive and Control Computers in the Development of a System Restoration Plan," 1981 PICA Conf. Proceedings. (Also, IEEE Trans. PAS-101 (January 1983), pp. 43-52).

25. Kafka, R. J., et al. "System Restoration Plan Development for a Metropolitan Electric System," IEEE Trans. - PAS-100 (August 1981), pp. 3703-3713.

26. Knight, U. G., "Aids for the Emergency Control of Power Systems," CIGRE Electra, No. 67, Dec. 1979, pp. 101-134. (Also IEEE PES Winter Power Meeting, New York, 1980, papers A80 002-6, A80 003-4).

27. Knight, U. S., "Remedial Measures to Reduce the Incidence and Effect of Major Disturbances on Power Systems,' CIGRE Study Group 32 Report No. 32-13, August, 1978.

28. Lewis, W. A., ECAR, Chmn.; J. R. Boyle, Sec'y., N. O. Engebretson, "Final Report of System Restoration Working Group" (Established, May, 1972), 1976.

29. Maliszewski, R. M., R. D. Dunlop, G. L. Wilson, "Frequency Actuated Load Shedding and Restoration Part I--Philosophy," IEEE Trans. on Power Apparatus and Systems, PAS-90, (1971), pp. 533-540.

30. Mariani, E., Mastroianni, R., and Romand V. "Field Experiences in Re-energization of Electrical Networks from Thermal and Hydro Units." IEEE 8M WM 093-1.

31. McDonald, J. E., A. M. Bruning and W. R. Mahieu, "Cold Load Pickup," Paper F79-180-1, IEEE PES Winter Power Meeting, New York, 1979.

32. Milne, G. R., "Provisions for Re-energizing with Electric Systems of the Consolidated Edison Company of New York," Trans. AIEE, Vol. 59, 1940.

33. Review of Major Power System Interruptions, National Electric Reliability Council Report, August, 1979.

34. Sakaguchi, T. and K. Matsumoto, "Development of a Knowledge Based System for Power System Restoration," IEEE/PES Summer Meeting 82 SM 351-5, 1982.

35. Scheurer, D., Philadelphia Electric Company System Restoration Plan," presented to PA Electric Association, Pittsburgh, PA, Jan. 25, 1980.

36. Simburger, E. J., Hubert F. J. "Low Voltage Bulk Power System Restoration Simulation," IEEE PES Trans. Vol. PAS-100, Aug., pp. 4479-84.

37. Shultz, R. D. and Mason, G. A., "Blackstart Utilization of Remote Combustion Turbines, Analytical Analysis and Field Test," IEEE 84 WM 092-3.

38. Undrill, J. M., et al. "Interactive Computation in Power Systems Analysis," Proceedings of IEEE, Volume 62, July, 1974.

39. Wilson, G. L. and R. Zarakas, "Anatomy of a Blackout," IEEE Spectrum, pp. 39-46, February, 1978.

40. Wolff, R. F., "Speed Controls Vital During Islanding," Electrical World, November 15, 1979.

41. Zaborszky et al., "A New State Space for Emergency Control in the Interconnected Power System," IEEE Trans., Vol. AC-22, No. 4, August, 1977, pp. 505-516.

42. Zarakas, P., "System Restoration," EPRI Workshop Proc.: Public Policy Aspects of High-Capacity Electric Power Transmission, EPRI WS-79-164, 1979, pp. I. 29-34.

POWER SYSTEM RESTORATION - THE SECOND TASK FORCE REPORT

Contributing Members: M. M. Adibi (Chairman), J. N. Borkoski, and R. J. Kafka

Abstract - In its second report, the Power System Restoration Task Force (PSR-TF) (1) reviews the 1979-83 North American Electric Reliability Council (NERC) Reports on Major Bulk Power System Disturbances (2), identifying restoration problems which have been encountered during that 5-year period.

Generally, these problems fall into three phases of restoration activities: planning for restart, reintegration and restoration of the bulk power supply; control actions during system degradation for saving and retaining critical sources of power; and restoration after a major disturbance when the power system has stabilized.

The report also briefly discusses the applicability of several existing programs to the system restoration process.

The opinions and conclusions expressed in this paper are those of the authors; they are based on the referenced reports and have not been reviewed with NERC or the reporting utilities.

I - INTRODUCTION

In order to reduce the possibility, the extent, and the duration of major failures in bulk power supplies, preventive, corrective, and restorative actions are required, respectively. In the past decade, the industry has undertaken considerable effort in preventive and corrective areas, thus reducing the frequency and intensity of system outages. There appears to be room for additional efforts in the direction of further reducing the duration of an outage.

During the same decade, installation of large and slow starting generating units remote from the load centers, and retirement of small and fast starting generating units within the load centers, have gradually increased the restoration problems, and this trend appears to continue. It is the industry's prediction that, by stretching the existing equipment's capabilities, the restoration difficulties will be further intensified by the early 1990s (4 to 7). Therefore, investigation of ways and means for enhancing power system restoration is a timely endeavor.

In this report, the PSR-TF has reviewed 19 selected major bulk power disturbances in Section II, attempted to identify restoration problems in Section III, and discussed the applicability of several existing computer programs in Section IV.

In a follow-up report, the PSR-TF will present the need for new procedures and explore advanced methodologies.

II - REVIEW OF SELECTED MAJOR BULK POWER DISTURBANCES

The 1979-83 NERC Reports describe 48 selected major bulk power system disturbances out of 149 which had been reported to DOE's Alert Coordination Office (ACO). The DOE ACO's criteria for reporting major bulk power system disturbances are shown in Table I (8).

This review has been limited only to those 19 cases where restoration problems have been reported. As shown in Table II, about 40% of the described major disturbances have had restoration problems.

TABLE I

DOE CRITERIA FOR REPORTING MAJOR BULK POWER SYSTEM DISTURBANCES

FOR PEAK SYSTEM LOAD	$\leq$	3,000 MW
LOSS OF LOAD	$\geq$	15 MINUTES
LOSS OF LOAD	$\geq$	100 MW, OR
LOSS OF LOAD	$\geq$	50% OF TOTAL LOAD
FOR PEAK SYSTEM LOAD	>	3,000 MW
LOSS OF LOAD	$\geq$	15 MINUTES
LOSS OF LOAD	$\geq$	200 MW

TABLE II

MAJOR BULK POWER SYSTEM DISTURBANCES

YEAR	TOTAL REPT'D	DESCRIBED BY NERC	RESTORATION PROBLEMS	%
1979	24	10	3	30
1980	42	12	6	50
1981	43	10	5	50
1982	27	9	3	33
1983	13	7	2	28
TOTAL	149	48	19	40

SUMMARY OF 1979 - 1983 SELECTED RESTORATION PROBLEMS

CASE 1. Restoration could not begin until the ties had been re-established with remote systems to access generation sources. Errors made while switching lines caused delays in the restoration effort. The absence of local var support resulted in low voltage until ties could be re-established.

CASE 2. Restoration was orderly, but delays were encountered while trying to determine the extent of the island. The dispatcher did not have indication of the status of the 345kV breakers. Load restoration is by supervisory control, but each distribution breaker had to be closed separately (80 distribution breakers opened). Since all steps of underfrequency relays operated, restoration was time consuming. The tie had to be established before load could be picked up, since the other 345kV interconnection was locked out by the breaker failure action, and there was insufficient generation on line to carry the local load.

CASE 3. Resynchronizing of the interconnection loop was not accomplished in accordance with the interconnection procedures. The first tie which was closed between the northern and southern islands was a 115kV line, which promptly opened on overload. Schedules were resumed on the 500kV AC intertie before the eastern portion of the loop was closed. All major interconnections between the north and south islands were restored to service and operating normally within 52 minutes. Most customers were restored within 20 minutes and all loads were restored within three hours and 15 minutes.

86 SM 330-5 A paper recommended and approved by the IEEE Power System Engineering Committee of the IEEE Power Engineering Society for presentation at the IEEE/PES 1986 Summer Meeting, Mexico City, Mexico, July 20 – 25, 1986. Manuscript submitted January 31, 1986; made available for printing May 7, 1986.

Reprinted from *IEEE Transactions on Power Systems,* Vol. 2, No. 4, pp. 927–933, November 1987.

CASE 4. The primary restoration method could not be followed due to equipment damaged. Also, some lightning arresters were damaged due to high voltages encountered during restoration. Service to the majority of customers was restored in about six hours with the remaining customers progressively restored over the next seven hours.

CASE 5. Restoration generally proceeded in an orderly and efficient manner considering the magnitude of this disturbance. However, resynchronizing between some islands was delayed due to uncoordinated load restoration. "Black start" capability operated satisfactorily at all major locations with the exception of a power plant where the emergency diesel failed to start due to a dead battery. It has been recommended that member systems should continue to re-emphasize restoration procedures on major transmission systems to all dispatchers and operating personnel. It has also been recommended that the manner in which information is displayed to dispatchers be reviewed to see if improvements could be made to enhance restoration.

CASE 6. A system restoration procedure was developed in 1978 and the system dispatchers were well acquainted with the method proposed for restoration. The procedure was used by the dispatcher in restoring service, and it is concluded that the procedure as written is still a valid document. Essentially, all loads were picked up in about seven and one-half hours. Five steam units were back in service in five and one-half hours. Problems were encountered in closing 15 SF_6 circuit breakers due to low gas pressure caused by low temperatures. From time to time circuits were closed to loads that could not be supported by the transmission facilities, resulting in short periods of restoration to part of the system, followed again by loss of power. Some high voltage conditions were encountered during the restoration process and the need for additional reactive control is being studied.

CASE 7. There were several unsuccessful closures before the island was finally reconnected to the remainder of the interconnection. Efforts need to be made to better coordinate restoration procedures among the many dispatch offices. Switching to return the system to normal was completed in one hour and two minutes following the cascading disturbance.

CASE 8. Customer loads and most of the grid were restored in one hour. The last 500kV circuit was restored to service five hours later when preliminary disturbance investigations were complete. Faulty synchronizing equipment delayed restoration of the grid. It is noted that good cooperation between the various dispatchers following the disturbance permitted quick and efficient restoration of the systems.

CASE 9. Initial load restoration did not begin until about an hour after the start of the disturbance. Reasons for this delay are not known, but it is known that generation in the area was shut down. There was no equipment damaged during the disturbance, but it was approximately four hours before all loads were restored. This extended time was due to the time required to restart generation.

CASE 10. The systems were restored in about three hours and twenty minutes. It is noted that, during restoration, there were several erroneous closings of interconnections due to lack of knowledge of the extent of system separation. The affected systems have recommended that a time-shared computer system be investigated to provide information on the system conditions to dispatch offices with a minimum of input and calling by the dispatcher.

CASE 11. Restoration of most of the load within five minutes is to be commended. The delay in achieving total restoration resulted from the damage to distribution lightning arresters due to high voltage when 161kV transmission lines were reclosed. Insufficient information is available from system records to determine the reason for the length of time before system voltage stabilized.

Interim procedures have been established for re-energizing the 161kV lines to prevent the occurrence of high voltage. Other procedures have been established to facilitate re-establishing subarea generation.

CASE 12. The lack of adequate system data and/or control facilities apparently played a major part in extending the duration of the first interruption. No information is included in the post-disturbance report regarding the existence of a pre-planned restoration procedure. Such procedures appear to have been available three months later, as the duration of that interruption was approximately half that of the first interruption.

CASE 13. Restoration of residential service was complete in three hours and 20 minutes, but industrial load restoration was not complete for approximately 17 hours. The post-disturbance report states that industrial load was increased as generation was brought back on line.

It is noted that a generator which suffered turbine bearing damage when an auxiliary lube oil pump did not start will be out of service for at least four months. As the turbine slowed down a generator hydrogen cooling system leak developed, the hydrogen ignited, and an explosion occurred, blowing the cover off the exciter.

CASE 14. Restoration was delayed due to loss of power at the control center, 69kV line damage on the system, and problems in starting generation on the other system. Neither system had written emergency operating plans prior to the disturbance, but this has been recommended and both systems are considering such plans.

CASE 15. System restoration was hindered by communication, supervisory control, and monitoring equipment failures. This made it difficult to assess the state of the system. The first priority of restoration was to restore the bulk transmission network to normal in order to provide start-up power to the plants. This process was accomplished cautiously with start-up power supplied to the first plant in 36 minutes and the next plant in 90 minutes. Service to the majority of customers was restored within one hour.

Notwithstanding the need for system operators to understand the philosophy of restoration, written procedures should be developed for foreseeable black start conditions.

CASE 16. While load was restored in reasonable time periods (one hour 25 minutes for the first restoration and 25 minutes for the second), complete system restoration required six hours and 44 minutes. Repeated unsatisfactory closures of the 115kV tie and unsuccessful efforts to parallel the other tie indicated sequential restoration procedure problems. Intertie schedules were also resumed before the major loop was restored; this aggravated the disturbance. High voltage was also encountered due to long unloaded 500kV lines terminating on a substation bus.

CASE 17. Load restoration was probably slower than desirable but caution was exercised, since the reasons for some generation loss and the loss of the 345kV line were not fully known. In addition, thunderstorms were in the area and it was probably a wise decision not to restore all the load until the total system conditions were stabilized.

CASE 18. Restoration of the system did not proceed as rapidly as it should have due to lack of data concerning system conditions and lack of supervisory control facilities. There was also a problem in closing the 230kV line which initially caused the disturbance due to a 70-degree phase angle between the buses on each end of the line. This angle was reduced to 50 degrees by altering line loading and increasing generation.

CASE 19. The seven islands were restored to synchronous operation in 24 minutes. Customer loads that had been interrupted were restored within two hours. During restoration, service to an additional 40,000 people was interrupted when dispatchers requested that a 115kV line be opened to prevent conductor damage due to overloading and were unaware that the other lines into the area could not carry the load. It may have been possible to leave the 115kV line in service if the dispatchers had been able to accurately read the line loading and had been provided with emergency short-term overload capability to the line.

The main grid was resynchronized 16 minutes after the disturbance. For a disturbance of this magnitude it must be said that system restoration proceeded at a good pace. The report states that systems should continue to stress the importance of dispatcher training in interconnected system restoration programs.

III. IDENTIFICATION OF RESTORATION PROBLEMS

In this section, an attempt has been made to identify restoration problems. Table III lists the restoration problems encountered in the 19 cases, by year and by category. Table IV shows the same problems by type and frequency of occurrence.

TABLE III

RESTORATION PROBLEMS BY TYPE

Description	79	80	81	81	83	Total
1. REACTIVE POWER BALANCE						
Sustained over Voltage	-	2	1	1	1	5
Sustained under Voltage	-	1	1	-	1	3
Generator under Excitation	-	-	1	-	-	1
Switched Capacitors/Reactors	-	-	1	-	-	1
2. LOAD AND GENERATION BALANCE						
Response/Sudden Increase Load	-	1	1	1	-	3
Underfrequency Load Shedding	-	-	-	-	1	1
3. LOAD AND GENERATION COORDINATION						
Black-start Capability	-	-	-	1	-	1
Steam Unit Start-up Coord.	-	1	1	-	-	2
Switching Operation	1	3	1	1	2	8
Overloads During Restoration	1	-	-	1	-	2
Dispatch Office Coordination	1	-	3	2	1	7
4. MONITORING AND CONTROL						
Inadequate Communication	-	1	-	-	-	1
Inadequate SCADA	2	3	-	-	2	7
Inadequate Displays	-	-	-	1	-	1
System Status Determination	1	1	3	-	1	6
5. PROTECTIVE SYSTEM						
Interlocking Schemes	-	-	-	-	1	1
Synchronization	-	1	2	1	1	5
Standing Angles	1	1	-	-	1	3
6. ENERGY STORAGE	-	1	1	1	-	3
7. SYSTEM RESTORATION PLAN						
No procedure	-	2	-	1	1	4
Procedure Not Followed	-	-	-	2	-	2
Procedure Outdated	2	2	1	1	1	7
No Training	1	1	-	-	-	2

TABLE IV

RESTORATION PROBLEMS BY FREQUENCY OCCURRENCE

	DESCRIPTION	NUMBER*
1	Switching Operation	8
2	Procedure Out-of-Date	7
	SCADA Inadequate	7
	Dispatch offices coordination	7
3	System Status Determination	6
4	Sustained overvoltage	5
	Synchronization location & facilities	5
5	No system restoration plan	4
6	Sustained undervoltage	3
	Response to Sudden Increase in Load	3
	Standing angles	3
	Energy Storage	3
7	Steam Unit Restart	2
	Line Over Load During Restoration	2
	Restoration Procedure not followed	2
	Operator Training	2
8	Generator Under Excitation	1
	Switched Capacitors/Reactors	1
	Underfrequency Load Shedding	1
	Black-Start Capability	1
	Inadequate Communication	1
	Inadequate Displays	1
	Interlocking Schemes	1

*Based on the 48 major disturbance reports

The seven categories of problems listed in Table III are briefly discussed here:

1 - Reactive Power Balance

During early stages of the restoration process, it is necessary to keep system voltages within the allowable range. This is done in several ways: energizing fewer high voltage lines, operating generators at minimum voltage levels, deactivating (over-riding) switched static capacitors, connecting shunt reactors, adjusting transformer taps to appropriate positions, and picking up loads with lagging power factors. The inability to perform these tasks due to the absence of facilities or procedures would cause a reactive power imbalance resulting in, for example, generator self-excitation and runaway voltage rise. Installing minimum excitation level (MEL) relays, particularly on the black start combustion turbine generators, and setting these relays as low as allowable, would help alleviate reactive power imbalance in some cases.

2 - Load and Generation Balance

In the initial phase of restoration, it is necessary to maintain system frequency within the allowable limits by restoring loads in small increments. Smaller and radial loads are restored prior to larger and network loads while maintaining a reasonable constant real to reactive power ratio. Feeders equipped with underfrequency relays are picked up at the subsequent phases of restoration when system frequency has stabilized. Load pick-up in small increments tends to prolong the restoration duration. With large increments there is always the risk of falling into a frequency decline and recurrence of system outage. The size of load pickup depends on the rate of response of prime movers, which more likely are under manual control at this point. A guideline such as allowable load pickup as percentage of generator capability would help maintain load and generation balance at acceptable frequency.

3- Load and Generation Coordination
Generally, the strategies for system restoration consist of restart, reintegration and load restoration. In the restart phase, several load and generation islands are formed. In each island, the objective is to immediately and simultaneously supply station service or cranking power to large thermal plants by black start combustion turbines or, if available, by hydro-electric units. Combustion turbine's behavior is quite different during normal operation when parallelled with a large system and during restoration when they are the prime or the only source of power.

While forming several load and generation islands and simultaneously restarting each island could appreciably reduce the duration of an outage, the number of islands is limited by availability of resources, including the number of operating teams, the number of black start generators within each island, and the available and effective coordination between the various control centers.

In starting steam units, it is important to coordinate certain critical time intervals, such as maximum time interval beyond which certain thermal units cannot safely be restarted, or the minimum time interval required before a thermal unit can be started. Also, the amount of real and reactive generation available at any time, after a system disturbance, is based on the mix of prime movers, their start-up characteristics, and their operating conditions prior to the disturbance.

4- Monitoring and Control
Most major power system disturbances are arrested prior to complete collapse, and some level of interconnection assistance is available. In the event of partial disturbance, it would be necessary to determine the state of the system and enter the restoration procedure at the proper level. Lack of dedicated communications between control centers, power plants and substations, and inadequate monitoring displays and excessive alarms can lengthen the duration of an outage.

5- Protective Systems
The restoration process often has been interrupted or delayed by interlocks preventing, for example, generator breakers from closing on a dead bus, synchronization at locations where there has been no control over frequencies, voltages and phase angles, and lack of procedures for reducing the standing phase angles when closing the ties to the adjacent system or subsystem.

6- Energy Storage
Loss of backup power supplies, such as batteries, battery chargers, on-site generators, and uninterruptable power supplies (UPS), has affected the operations of circuit breakers, DC motor operated disconnect switches, cable pumping plants, telecommunication, SCADA, and control centers, resulting in delays in system restoration.

7- System Restoration Plan
Restoration of power after an outage is a difficult task, complicated by many factors, including unfamiliarity of the operating personnel with the outage situation and the pressure of time. These difficulties have been compounded by installation of large and slow starting generating units remote from the load centers, and installation of extensive underground and overhead high voltage transmission lines between the generating plants and the load centers. Therefore, more than ever before, there is a need for carefully prepared, current, readily accessible and easily understood restoration plans to allow quick and orderly recovery from complete or partial system collapse, with resultant minimum impact on the public and damage to the equipment. Such plans typically are broken down into some logical order. The process provides a discipline whereby there is less risk of overlooking any essential operation.

IV - ENHANCING POWER SYSTEM RESTORATION

Most operating companies are maintaining restoration procedures based on certain operating philosophies, practices, and familiarity with power plant restart and power system operation (3 to 7). In developing these restoration plans, power systems are modeled representing the equipment characteristics and operational data in an interactive computing environment. In some cases, attempts are made to validate these models against the actual operating data, using static state estimator or similar filtering techniques (9 & 10).

In general, the approach is to sectionalize the power system into a number of subsystems and develop a restoration plan for each subsystem by simulating a step-by-step restart and reintegration of generation and load (11). The existing programs most frequently used in these simulations include:

1. Interactive power flow for maintaining reactive power balance without allowing sustained over voltage or generator under excitation and consequential voltage collapse.

2. Dynamic stability for keeping load and generation balance through careful coordination of load pick-up with generators' responses and avoidance of possible risks for system frequency decline.

3. Switching transient for determining how many sections of a high voltage transmission line can be energized at one time without causing transient over voltages detrimental to equipment.

Except for the interactive power flow programs which, with some modifications, are efficient in power system modeling, model validation and simulation (9), the existing programs are large, comprehensive and inefficient when applied to power system restoration. While acceptable for power system planning purposes, dynamic stability and electromagnetic transient programs (EMTP) are not readily applied to system restoration.

During restart and reintegration phases of restoration, a power system often consists of several islands, most of the automatic controls have tripped or are deactivated, and the system is primarily under manual control. During these two restoration phases, wider voltage and frequency ranges are tolerated. Under these conditions, simple quidelines and fast rules based on approximate analysis are more expedient and relevant as the following example on "generation dispatch scenarios" demonstrates.

Generation Dispatch Scenarios (GDS)

In restoring a power system, operators often need to know in advance the amounts of real and reactive generation which would be available. This information could be used to optimize the sequence of the various restoration operations. These operations include energizing lines with high charging currents, which in turn would provide access to larger, but remote, generation capacity. The GDS can allow a proper operating sequence which can significantly reduce the duration of an outage.

The availability of real and reactive generation is based on prime movers' conditions just prior to an outage, and their start-up times. Table V defines nomenclature and lists the types of information required. Figure 1 shows the availability of MW and MVAR absorbing capability of a typical drum type boiler (I = 3), under the hot restart condition (J = 2), with the assumption that there are no network delays, and that only the prime mover's timings are critical. Similar data would need to be compiled for all of the generating units to be considered. At any given time, t, the megawatt and megavar absorbing capability of Unit I, under condition J, can be obtained from the two functions p(I,J,t) and q(I,J,t), respectively. These functions can be derived from individual generator characteristics. The total real and reactive generation available at any time, t, is the sum of the individual p's and q's respectively for all the n generating units within an island.

Figure 2 illustrates the results of a GDS study. In this particular case, generating units with high reactive absorbing characteristics were required to enable the energizing of a transmission line having a large line charging value. The actual requirements were that 500 MVAR of reactive absorbing capability were required within two hours, or the remote generation accessed by the transmission line would become unavailable for timely restart. By bringing on line early in the restoration those units with high rates of response and the proper reactive absorbing capabilities, the restoration process has been shortened by approximately seven hours.

TABLE V

(a) BOILER TURBINE GENERATOR (BTG) CONDITIONS *

AFTER THE OUTAGE, BTGs CAN BE UNDER ONE OF THE FOLLOWING CONDITIONS

J = 1, HOT RESTART
J = 2, HOT BOILER - HOT TURBINE
J = 3, HOT BOILER - COLD TURBINE (1)
J = 4, COLD BOILER - COLD TURBINE (2)
J = 5, VERY COLD

(1) Banking boilers, (2) Water problems
* Conditions are time dependent and overlap

(b) START-UP TIMES *

DEPENDING ON CONDITION (J = 1 to 5), TIMINGS FOR VARIOUS START-UP STAGES INCLUDE:

K = 1, TO PROVIDE CRANKING POWER
K = 2, INTERVAL BETWEEN PARALLELING SUCCESSIVE UNITS IN A PLANT
K = 3, FROM START, "NOTIFICATION", TO PARALLELING
K = 4, FROM PARALLELING TO MINIMUM LOAD
K = 5, SOAKING TIME(1)
K = 6, FROM RELEASE TO MAXIMUM LOAD(2)

(1) Soaking load can be greater than min. load
(2) Release & min. load can be the same
* There are two sets of timings for normal and emergency operations.

(c) BOILER TURBINE GENERATOR TYPES

- o BASE LOAD
- o CYCLING (PEAKING)
- o DRUM-TYPE
- o ONCE THROUGH
- o ONCE THROUGH & SUPER CRITICAL
- o OIL BURNING
- o COAL BURNING

I = 1 to n, identify individual BTG (1)

(1) Even twin units have different timings

(d) BTG CAPABILITIES (IN P.U.) *

	MW	+MVAR(1)	-MVAR(2)
SYNCHRONIZED	.00	.70	-.30
MINIMUM LOAD	.30	.67	-.29
SOAKING LOAD	.40	.65	-.28
90% P.F.	.90	.00(3)	-.12

(1) Max. excitation limit, (2) Min. excitation limit, (3) Max. stator limit.

* There are two sets of MWs for normal & emergency operations.
There are three sets of MVARs depending on H_2 pressures.

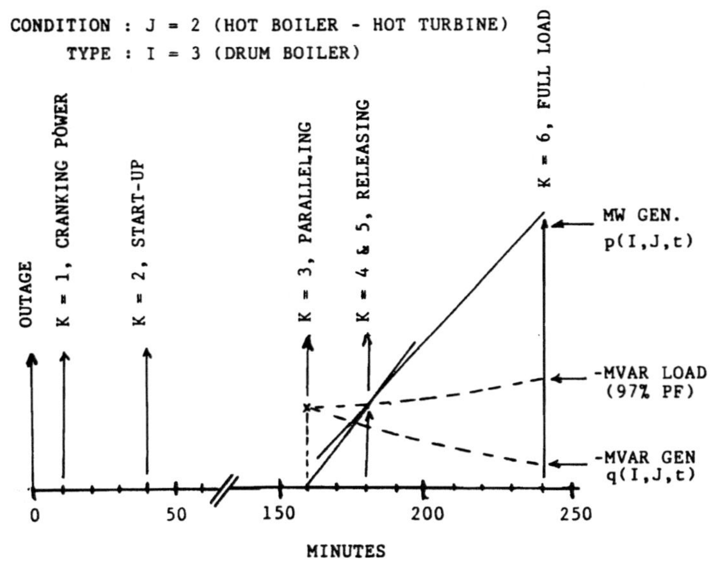

FIGURE 1.

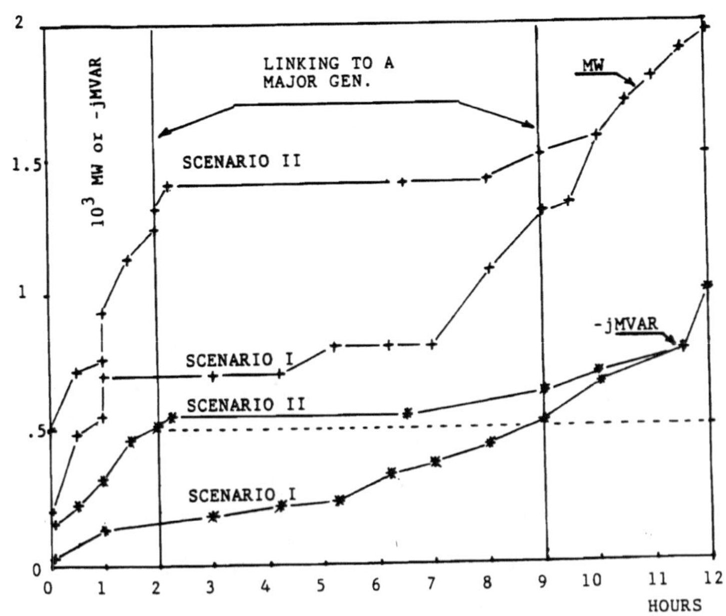

FIGURE 2- REAL POWER & REACTIVE ABSORBING CAPABILITY

The GDS has the potential of being applied in an on-line interactive environment. Readily understood guidelines can be provided with respect to the frequency response of prime movers, cold load pickup, switching transient, and standing angles. These guidelines can be provided by relatively simple and straight forward analysis instead of the more comprehensive power system analysis programs.

V - CONCLUSION

Review of these past major bulk power system disturbances shows that they have occurred at random, each has had somewhat different restoration problems, and in each case the system operator has played a critical role in restoring the system. In all the 19 cases, there have been the following common concerns:

- Identification of the status of the collapsed system, components and equipment.
- Restart and resupply of station service to plants, substations, cable pumping plants, compressed air, etc.
- Coordination of power plant start-up timings with load pick-ups to bring generators to their stable minimum levels and within the range of major analog controllers.
- Energizing large sections of transmission lines within the acceptable transient and sustained over voltages.
- Picking up load in large increments without the risk of frequency decline.
- Reintegration of the skeleton of the bulk power supply with the requisite time-consuming switching operation.
- Deactiviation of automatic load shedding and automatic switched capacitors during initial phases of restoration.
- Maintenance of steady state and transient stability as the system is being restored and when impedances are large.
- Reduction of standing angles when closing loops to firm up transmission paths.
- Probability of successful start-up of thermal units, in particular, combustion turbines.
- Stability of black start and peaking combustion turbines during the time when they constitute a large portion of generation.

Restoring a power system after a partial or complete collapse in a minimal amount of time presents a very complex problem. As a result, most of the existing procedures attempt to solve simpler problems (or subproblems) in a reasonable manner from a limited number of possible initial conditions. Clearly, there is a need for research into advanced methodologies which would allow integration of the operator's heuristic problem solving capabilities, and the applicable analytical tools in the real-time operating environment to provide a general approach to system restoration. Until these advanced methodologies are available, guidelines for restoration can be developed using existing programs and analytic techniques.

VI - REFERENCES

(1) IEEE Committee Report, Adibi, M. M., et al, "System Restoration - A Task Force Report", to be presented at IEEE PES 86 SM.

(2) "Review of Selected Major Electric Bulk Power System Disturbances in North America", NERC 1979-84.

(3) "Kafka, R. J., Penders, D. R., Bouchey, S. H., Adibi, M. M., "System Restoration Plan Development for a Metropolitan Electric System", IEEE Trans. PAS PP. 3703-13, Aug. 1981.

(4) Peach, S., "System Restoration Synthesis - Hydro Quebec", IEEE PES 8th Biennial Workshop on Real-Time Monitoring and Control of Power Systems, Montreal, Oct. 1984.

(5) Scheurer, D., "System Restoration at Philadelphia Electric Company", ibid.

(6) Barrie, D., "System Restoration Following Load and Generation Rejection", ibid.

(7) Arnold, P. F., "Summary of System Restoration Plan for Pacific Northwest Power System", IEEE PSR-TF Mtg., 1984 WPM.

(8) Department of Energy, Alert Coordination Office Publication DOE/EP-0071/1, September 1983.

(9) Kafka, R. J., Penders, D. R., Bouchey, S. H., Adibi, M. M., "Role of Interactive and Control Computers in the development of a System Restoration Plan", IEEE Trans.-PAS PP. 43-52, Jan. '83.

(10) Shultz, R. D. and Mason, G. A., "Blackstart Utilization of Remote Combustion Turbines, Analytical Analysis and Field Test", IEEE Trans.-PAS 84 WM 092-3.

(11) IEEE Committee Report, "System Restoration - Deploying the Plan", Current Operational Problems Working Group, IEEE Trans.-PAS PP. 4263-71, Nov. 1982.

Discussion

David Scheurer (Philadelphia Electric Company, Philadelphia, PA): The problems addressed by the authors are broad in scope; however, it appears that many could be avoided if sufficient information in suitable format were made available to the system operators.

Although most systems incorporate adequate displays into the everyday operating environment, few have developed on-line displays which organize available data in such a way as to be useful during the restoration process. Such displays might include summary tables and diagrams for islands which have been predefined in the restoration procedure. On-line calculations could be performed in order to display such quantities as total island (P,Q) generation and operating reserve. Some additional monitoring might also allow the display of frequency within each individual subsystem.

Other displays might identify critical equipment where rapid degradation or failure is likely during prolonged deenergization. Accumulated information on restoration problems is of little use unless it is presented to those ultimately responsible for startup and integration of the system. The information to be presented is system specific and must be organized in such a way as to allow rapid retrieval by the system operators when required.

Information presented in this format would serve to greatly enhance condition assessment and thereby speed and secure the restoration process.

Operator training is another area which cannot be overemphasized. Since a restoration process is not a routine task, the system operator will in most cases not have the intuitive feel for system response to the controls at his disposal. The operator must be aware of pitfalls and potential areas of concern as well as the details and design criteria of the restoration plan. Training, conducted on a periodic basis including drills utilizing programs such as the interactive load flow and whatever other tools are normally available, can serve to improve personnel response and awareness in an emergency situation.

The use of existing software and data to develop restoration aids as discussed by the authors can greatly reduce the incidence of problems in the short term until more sophisticated on-line analytical tools become available.

Manuscript received August 12, 1986.

M. M. Adibi, J. N. Borkoski, and **R. S. Kafka:** The authors wish to thank Mr. Scheurer for his general comments, which basically support the report by the Task Force.

The first comment is regarding the on-line display of information for restoration. The information needed may be divided into several functional groups, including:

a) physical characteristics which require infrequent updating, such as parameters and topology, equipment ratings, and procedures for sectionalization of subsystems (3);

b) operational characteristics which need to be updated periodically, or as the power system goes through normal and alert operating conditions such as system load and generation, scheduled outages, deratings of equipment and conditions of boiler-turbine-generators;
c) time and temperature sensitive information which change with the elapsed time after a disturbance and need to be updated in real-time such as prevailing and projected availability of real and reactive powers, status of stored energy, SF_6 temperature in circuit breakers, and status of cable pumping plants.

The GDS described in the report is one example of the latter group of information needed for restoration.

The second comment is regarding operator training for restoration. The medium for training, exercising, and guiding power system operators, range from simple instruction manuals, to audio-visual tapes, to highly interactive simulators. The development of the latter encompasses the requirements for all the former media. The available generic simulators can provide procedural training in the early stages of operator training. For exercising and preparing experienced operators to cope with system-specific and time-critical emergency situations, such as restoration, high-fidelity system-referenced simulators are needed, which essentially replicate the control center in terms of functional performance and physical appearance.

The authors are in agreement with the discusser that in both areas there are needs for more sophisticated analytical tools which would allow system-specific simulation and provide timely and relevant information for restoration.

Manuscript received September 2, 1986.

ANALYTICAL TOOLS FOR POWER SYSTEM RESTORATION - CONCEPTUAL DESIGN

Felix F. Wu and A. Monticelli

Department of Electrical Engineering and Computer
Sciences and Electronics Research Laboratory
University of California, Berkeley
Berkeley, CA 94720

Abstract

A conceptual framework for computer-aided monitoring and assessment during system restoration using analytical tools is proposed. The basic structure is similar to the one for security monitoring and assessment. State estimation and the related functions, such as observability analysis, bad data identification, external network modeling, that are used in security analysis, can be modified for application to system restoration monitoring. The work in this area is reported. For restoration assessment, in addition to load flows and optimal power flows that are used in security assessment, a host of analysis/optimization programs is required. These analytical tools are identified and categorized. To synthesize possible control sequences and to select and coordinate analysis procedures for assessing restoration plans is a very complex task. A knowledge-based expert system architecture for this task is suggested. The conceptual design of the knowledge-based system and its interface with the analytical tools are presented.

I. INTRODUCTION

Power systems are operated under two sets of constraints; load constraints and operating constraints [1]. The load constraints impose the requirement that the customer load demand be met, whereas the operating constraints require that the system variables such as line flows, voltages be within acceptable limits. The system is said to be in a *normal state* if both the load constraints and the operating constraints are satisfied. The system is said to be in an *emergency state* if there is a violation of the operating constraints. The system is said to be in a *restorative state* if some load has been lost, i.e., there is service interruption. (See Fig. 1). Since disturbances or contingencies such as lightning strikes on transmission lines and generator failure occur frequently, power systems have been planned and operated so that it has the ability to withstand most contingencies. This is called system *security*. In the last twenty years, great progress has been made in developing analytical tools for security analysis. Sophisticated network analysis software is now installed in modern real-time computer controlled energy management systems (EMS) to perform security monitoring and assessment (Fig.2). These analytical tools contribute to the improvement of system security.

Even for systems designed to be highly secure, unpredictables do happen and cause service interruptions and customer outages. It is therefore imperative to develop strategies to handle service interruption by minimizing its impact and to restore service to customers. Most utilities have system restoration plans [2]. For example, one company has developed system restoration guidelines based on operator-analyst discussions and simulations [3,4]. Their proposed strategy calls for:

(1) sectionalization of power system into islands

(2) restoration of each island

(3) synchronization of islands

The idea behind their proposed strategy is that simultaneous restoration will result in speedy restoration.

In contrast to the rapid advancement in the area of security analysis and control, very little work has been done in providing analytical tools to assist operators during the restoration process. In Japan, an interactive restoration control system has been proposed [5] and an expert system approach has been suggested for restoring a section of a feeder [6]. In a recent survey report on current operational problems in power system [7], it is found that most respondents consider the present approach to system restoration is unreliable, work is needed in problem analysis for system restoration and the new approach should have monitoring/assessment capability. We believe that computer-aided analysis for restorative control can be of great assistance to the operator, the same way as security analysis has demonstrated in normal operation.

In this paper, we

(a) present a conceptual framework for performing monitoring/assessment functions during system restoration

(b) identify analytical tools in system security monitoring that, after necessary modifications, can be transported for use in system restoration monitoring

(c) survey the analytical tools that can be used for system restoration assessment

(d) identify their effective areas of application

(e) propose a knowledge based expert system for coordinating the analytical tools for restoration assessment

Some of the ideas of this paper was outlined in [8] and benefited from [9].

II. A CONCEPTUAL FRAMEWORK FOR SYSTEM RESTORATION

The problem of restoration after service interruption is a complex decision and control problem for the system operator [10]. The problem may be viewed conceptually as a multi-objective, multi-stage, combinatorial, nonlinear constrained optimization problem. The objective of restorative control is the speedy restoration of all customer service,

86 WM 105-1 A paper recommended and approved by the IEEE Power System Engineering Committee of the IEEE Power Engineering Society for presentation at the IEEE/PES 1986 Winter Meeting, New York, New York, February 2 - 7, 1986. Manuscript submitted August 30, 1985; made available for printing November 12, 1985.

Reprinted from *IEEE Transactions on Power Systems*, Vol. 3, No. 1, pp. 10–16, February 1988.

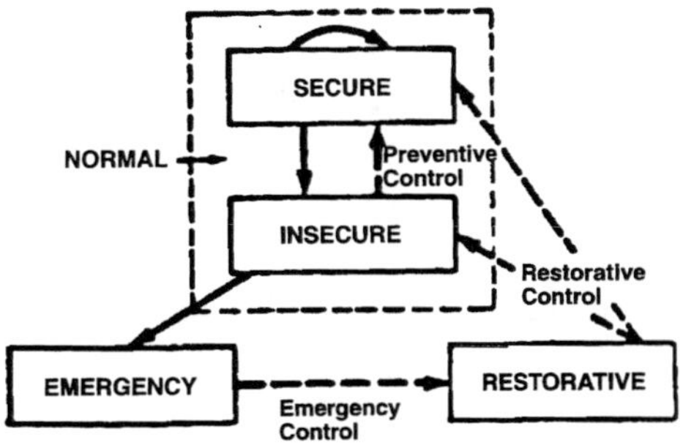

Transition as a result of a contingency
Transition due to a control action

Figure 1. Operating states of a power system.

which involves the minimization of restoration time and the maximization of customer load restoration at each stage. The constraints on the system involved in the restorative control include:

(1) power flow constraints
(power balance between generation and load, line flows and voltage limits)

(2) stability constraints
(transient and dynamic stability of system response, frequency and synchronization considerations)

(3) generator restart constraints
(cold restart or hot restart)

(4) generator load pick-up capability constraints

(5) transmission and tie line switching sequence constraints

The control variables in the restoration problem are the generation schedule of the generators and switching sequences. The decision-maker during restoration is the system operator. The operator's decisions during restoration are based on his knowledge of the

(a) current state of the system

(b) availability of viable alternatives

(c) consequences of each alternative

Computer/communication systems of an energy management system, together with analysis software can assist the operator greatly in the monitoring and assessment functions. Energy management systems have been effective in assisting system operators during normal operation for cost minimization and security enhancement. Additional analytical capability added to the EMS can certainly assist system operators during restoration.

Our proposed conceptual framework for system restoration is centered around an EMS control computer and is shown in Fig. 3. The EMS serves as the interface between the system operator and the power system. The monitoring and assessment functions are divided into three tasks: modeling, analysis/optimization, and synthesis. Here we use the term *modeling* in a more general sense than monitoring. By modeling, we mean the process of assembling from on-line data acquisition and off-line information necessary data regarding the present and future system for use in analysis and assessment. The assessment function is splitted into *analysis/optimization* and *synthesis*. The details of these tasks are described below.

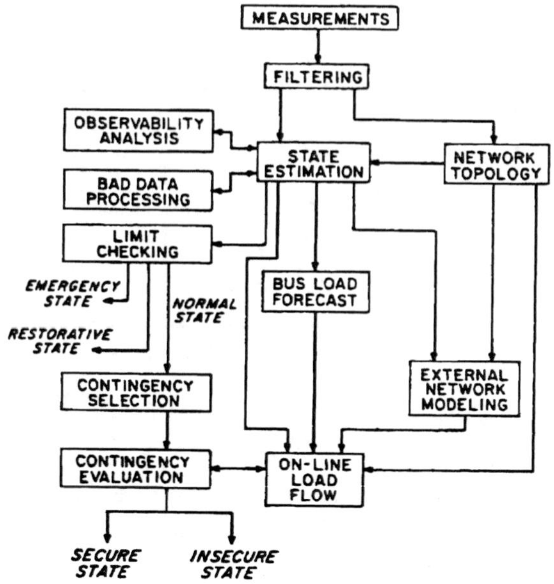

Figure 2: Functional blocks of monitoring and assesment.

III. MODELING

Current capability of EMS is confined to the use of steady-state analysis of power systems using load flows. Recognizing this practical limitation, we propose to formulate the constraints in system restoration as a multi-stage load flow problem. The network configuration and the power flow constraints are represented directly in the load flow model. The stability constraints and the constraints on generator re-start, load pick-up, etc., are transcribed into load flow constraints. The load flow is the workhorse of the monitoring/assessment functions during restoration. The

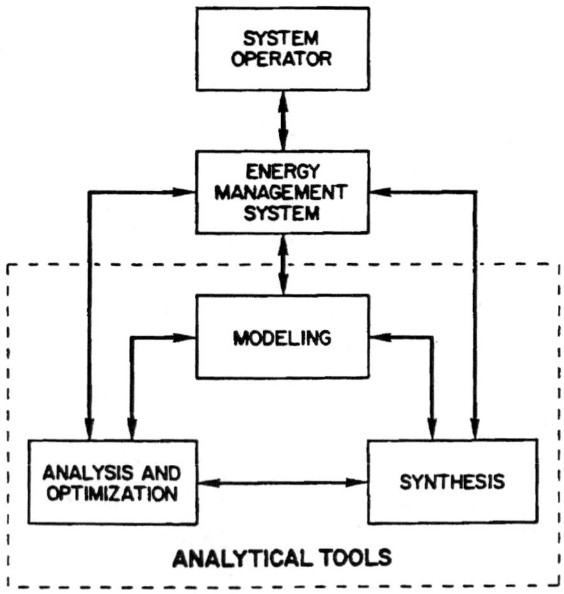

Figure 3: Conceptual framework for system restoration analysis.

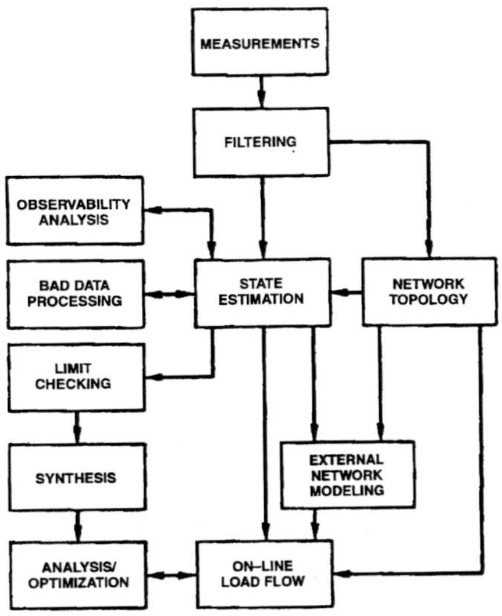

Figure 4: Functional blocks of restoration monitoring and control.

modeling task concerns with the establishment of the load flow model for monitoring/assessment.

Real-time data are obtained through RTU's, SCADA, and communication facilities typically used in EMS. Because there are two networks in the system, one is the electric network and the other is the information (communication) network, we shall carefully distinguish the concepts of *electrical islands* and *observability islands*. During restoration, the system may be splitted into several islands that are electrically unconnected. Depending on the availability of measurements, state estimation can be performed only for a part of the system. This part may contain several islands that are topologically unconnected. The former are called electrical islands and the latter observability islands. For a given system the breakup into electrical islands may not coincident with its observability islands.

The monitoring function during restoration follows closely the same components in the security analysis (Fig. 2) except that the detail requirements are different, as noted below (See Fig.4).

3.1 Observability

We say a network is observable if there are sufficient measurements to make state estimation possible. Communication facilities used during restoration, such as telephone circuits, may be susceptible to overloading during an electrical outage. If the outage is widespread, resulting from natural causes such as severe storms, then certain communication links may be lost to service. Therefore loss of observability is not to be unexpected during restoration. The observability analysis should be able to test observability of the system and in the event it is not observable, to identify *all* observable islands in the system. This is because during restoration it is important to monitor every part that is monitorable. Any observability program that identifies only the largest island is inadequate. We have developed a numerical approach to observability analysis that is capable of simultaneously identifying all observable islands [11-13].

The basic ideas of the multi-island observability analysis method are the following. A network is observable if and only if all measurements are zero implies all line flows are zero. Consider first the case that a network is observable. When all the measurements are set to zero, no matter what reference angle is assigned to the slack bus (ϑ pseudo measurement), the state estimation equation can be solved with a unique solution. This solution should have all angles the same, so the line flows are all zero. When the network is not observable, the fact that all measurements are equal to zero only forces *some* angles to be the same. The result will be several groups of nodes having the same phase angles. Each group of identical phase angle is an observable island. Efficient algorithms have been developed based on these ideas [11-13].

3.2 State Estimation

State estimation processes a set of real-time measurements to give the best estimate of the current state of the system. During restoration, the state estimation is required to handle

- multiple electrical islands
- multiple observability islands

Most state estimation programs can handle multiple electrical islands but not multiple observability islands. The introduction of pseudo-measurements to make unobservable part observable has been suggested, but it may degradate the quality of state estimation results. We have developed a scheme that is capable of performing state estimation for a system with multiple observable islands [11-12]. The process starts from identifying observable islands. The lines flows on the branches crossing two different observable islands will not be observable from the measurements, hence they are unobservable branches. Those injection measurements into the buses that have unobservable branches connected to them are irrelevant in the sense that they are not contributing to the state estimation of the observable part of the network. Once the irrelevant injection measurements are removed and a reference angle is introduced into each observable island, the state estimation program can return the estimated state of all observable islands.

3.3 External Network

The control center receives telemetered data of real-time measurements. The monitored part of the power system that these measurements cover normally consists of one's own system and is usually called the *internal system*. The rest of the interconnection is called the *external system*. Since the division into internal and external systems are for the purpose of state estimation, a better way of defining internal and external systems is via the state-estimation process. The internal system in this context is actually the observable part of the system with respect to the state estimator in one's energy control center. During the restoration process, due to loss and recovery of communication links, the boundary of internal and external systems is constantly changing. Therefore we need an external network model that

(a) is flexible to changing boundary between internal and external systems

(b) does not corrupt the internal system state estimation

We have developed a method that performs internal state estimation and external network modeling simultaneously [14]. It uses one state estimation covering both internal and external systems. The set of pseudo-measurements in the external system is so selected that it makes the external system barely observable. This way the two requirements mentioned above are all satisfied.

3.4 Variable Limit Update

Strictly speaking, system dynamic models are required for analyzing stability and synchronization. With current knowledge and computational capabilities, it is rather impossible to include explicitly the dynamic models for real-time system restorative assessment. In real-time security analysis and control, certain transmission lines loading

Table 1. Subproblems in system restoration

Problem	Number of Islands	Number of Stages	Control Variables	Load Flow Model	Mathematical Problem Formulation	Applications	Relevant References
1. Basic load flow	single	single	continuous (gen/load dispatch)	linear or nonlinear	simultaneous linear or nonlinear equations	testing scenarios compiled by synthesis	interactive load flow
2. Generation/ load scheduling	single	single	continuous (gen/load dispatch)	linear or nonlinear	linear or nonlinear programming	i. max. load restored within an island ii. system voltage maintenance	optimal power flow
3. Generation/ load scheduling with line switching	single	single	continuous (gen/load dispatch) discrete (line switching)	linear	mixed integer programming	network configuration selection	line switching
4. Multi-stage version of 2	single	multiple	continuous	linear	dynamic programming	load pick-up capability	multi-stage generation scheduling
5. Multi-stage version of 3	single	multiple	continuous + discrete	linear	dynamic programming	sequential restoration within an island	none
6. Multiple island versions of 4,5	multiple	multiple	continuous + discrete	linear	dynamic programming	i. defining islands in sectionalization ii. connecting islands during restoration	none
7. Nonlinear versions of 3, 4,5, & 6	single or multiple	single or multiple	continuous + discrete	nonlinear	dynamic programming	comprehensive restoration control	none

limits are established based on off-line stability simulations. Similarly for system restoration, the stability constraints and the constraints on generator re-start, load pick-up, etc. are transcribed into limits on line flows and voltages. These limits are generally functions of the current network topology, generator and load pattern. Therefore during restoration these limits need to be updated from time to time. A viable approach to variable limit update is perhaps through table look-up established based on off-line studies.

IV ANALYSIS/OPTIMIZATION

There are many facets to system restoration. The problem has all the characteristics, and more, of a complex decision and control problem: multi-objective, multi-stage, large-scale, combinatorial, nonlinear, etc. The overall problem defies an analytical solution. However solution techniques are available for some subproblems. Here we categorize the subproblems according to their

- number of stages applicable (single or multiple)
- number of islands applicable (single or multiple)
- control variables (generation rescheduling, load control, line switching)
- linearized (dc) or nonlinear (ac) load flow

Each subproblem can be used naturally as a building block for a more general subproblem (e.g., single-stage for multi-stage) or, when used in a standalone mode, is applied to solve one or more particular aspects of the system restoration problem. We have identified the step-by-step build-up of these subproblems in Table 1. Table 1 also shows the characteristics, the applications areas, and the mathematical formulation of these subproblems.

4.1 Load Flow

The load flow program is the workhorse in the stable of analysis/optimization programs. It is used for checking feasibility of the intermediate steps in a restoration plan and also served as building blocks for other programs.

Approximate models to the nonlinear ac load flow may be used advantageously for restoration analysis because when the case is not feasible the ac load flow will fail to converge without giving the source of the nonconvergence, whereas the approximate models do. The most well-known approximations to the load flow are the dc load flow and the transportation model in which only the Kirchhoff current laws are considered. The solutions to the approximate load flow models can indicate the source of infeasibility. There is a family of approximate models lying between the dc load flow and the transportation model. The dc load flow is equivalent to the optimization problem of minimizing a quadratic function subject to the transportation model constraint [20]. Approximating the quadratic cost by piecewise linear functions gives an approximation to the dc load flow [17].

During restoration the network is frequently splitted into islands. For some studies, the load flow model cannot be applied directly. For example consider the case where one wants to determine the optimal strategy of connecting two islands so that the load in one island can be picked up by increasing the generation in the other island. Transportation model or dc load flow model can be used in this situation by connecting the islands with very high reactance fictitious branches. The resulting sensitivity factors provide the indication of the effectiveness of various connections.

4.2 Other Analysis Programs

The optimal power flow has been used in security analysis for generation/load scheduling. It is also applicable for system restoration analysis. In Europe, the use of line switching as an additional means for security control was proposed [15-18]. We believe that the problem can be more appropriately formulated for system restoration. An LP approach for multi-stage generation scheduling was proposed [19], which is very relevant in restoration analysis. Possible extensions of these methods for restoration analysis/optimization are listed in Table 1.

V SYNTHESIS

During restoration, the operator makes a sequence of decisions concerning

(1) switching sequence

(2) load pick-up sequence

(3) generation/load schedule

The subproblems together with solution techniques identified in Sec. IV form a library of analysis software. In order to make this collection of analysis software useful for the operator, an interface between the analysis software and the operator is needed that can

(a) synthesize appropriate sequences of actions for assessment

(b) select proper subproblems

(c) organize and control the analysis procedure

Such a task can best be accomplished by the employment of a *knowledge-based system*. The knowledge-based, or expert system, is a software consisting of a collection of facts, rules of thumb, and other knowledge about a given field, coupled with methods of applying those rules, to make inferences [21-22]. There are three key components in the construction of an expert system: knowledge base, an inference engine, and a user interface. The knowledge base is the repository of specific knowledge about the problem, usually acquired from an "expert", hence the name "expert system". However this does not have to be so. In fact in the present case, there is no human expert who has the experience serving as an interface between the analysis software and the operator. The knowledge base here is more likely acquired through analysis, heuristics and the understanding of the problem.

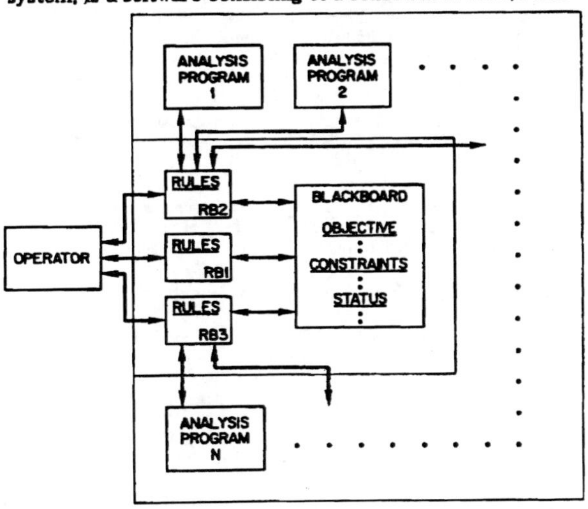

Figure 5: Knowledge-based system architeture for system restoration synthesis.

A knowledge-based system similar to the one for computer-aided control engineering [23] is suggested (Fig. 5). Central to this knowledge-based system is a "list of facts" or a "blackboard." The information in the blackboard may be organized into three categories: objectives, constraints, and status. There are several rule bases in the system. The rules in RB1 suggest appropriate sequences of actions for assessment. The rules in RB2 define an analytical problem and select software in the library for analysis. The rules in RB3 deal with validation of the assessment results through load flow simulations. For example, from RB1, it is suggested that at the present stage to energize a portion of the network. The knowledge-based system writes the following information on the blackboard:

Objective

- to energize the network connecting buses a, b, c, and d.

Constraints

- switching sequence ABCD
- generator ramp constraints on X, Y, Z.

•
•
•

Status

- empty

The problem is then translated into mathematical form using rules in RB2 and appropriate analysis software is selected. This step is recorded on the blackboard:

Objective

- to energize the network connecting buses a, b, c, and d.

Constraints

- switching sequence ABCD
- generator ramp constraints on X, Y, Z.

•
•
•

Status

- use the optimal switching program to determine the switching sequence and simulate the step by step results using load flow.

Suppose that in the load flow simulation of the switching sequence suggested by the optimal switching program, it is found by RB3 that a ramp constraint of the generator is violated. This information is then used to redefine the analytical problem. During the analysis/synthesis process, the current status is always recorded on the blackboard to facilitate the application of the rule bases. An important aspect of the knowledge-based system is its ability to explain the reasoning or inference process. Therefore when the analysis/optimization of a possible control sequence is complete, the knowledge-based system will send the assessment report to the operator.

In the proposed knowledge-based system architecture of Fig. 4, there are two types of programs. The analysis programs perform mainly numerical computations and are coded in an imperative language such as FORTRAN. The decision-oriented programs perform symbolic computations and are coded in a declarative language such as PROLOG. It is reported [24] that research is currently underway in computer science for high performance architectures that support a mixture of numerical and symbolic computations. Our proposed system will profit from any such advance.

VI CONCLUSION

Recent research progress in the development of analytical tools for system security monitoring/assessment has been remarkable. As systems operating closer to their limits and the threat of blackouts increases, system restoration becomes more important and the need for analytical tools assisting the operator for monitoring/assessment during restoration increases.

In this paper we propose a conceptual framework for computer-aided monitoring and assessment during system restoration. The basic structure is rather similar to the one for security monitoring and assessment. State estimation and the related functions, including observability analysis, bad data identification, external network modeling, have been used in security monitoring. They can be modified for application to system monitoring during restoration. For security assessment, the analytical tools used are simply the load flow and the optimal power flow. For system restoration, a host of analysis/optimization programs is required. They are identified and categorized in this paper. The problem of synthesis of possible control sequences and the selection and coordination of analysis procedures for assessing restoration plans is much more complex. A knowledge-based system is suggested to handle this task. To summarize, the same functional diagram for security monitoring/assessment (Fig. 2) can be used for monitoring/assessment during restoration by replacing two blocks. The contingency evaluation block is replaced by a library of analysis/optimization programs and the contingency selection block is replaced by a knowledge-based system (Fig. 4).

Of the components in the proposed framework, the synthesis using the knowledge-based system is the one requires basic research. Currently we are actively working on this problem.

We envisage that the integration of the analytical tools for system restoration into system operation can take place in three levels:

(a) off-line planning studies

(b) operator training simulator

(c) real-time operating environment

VII REFERENCE

[1] T.E. DyLiacco, "System Security: The Computer's Role," *IEEE Spectrum*, vol. 15, June 1978, pp. 43-50.

[2] W.A. Johnson et.al., (Systems Operations Subcommittee Report), "System Restoration - Deploying the Plan," *IEEE Trans. Power Apparatus and Systems*, vol. PAS-101, Nov. 1982, pp. 4263-4271.

[3] R.J. Kafka, D.R. Penders, S.H. Bouchey and M.M. Adibi, "System Restoration Plan Development for a Metropolitan Electric System," *IEEE Trans. Power Apparatus and Systems*, vol. PAS-100, Aug. 1981, pp. 3703-3713.

[4] R.J. Kafka, D.R. Penders, S.H. Bouchey and M.M. Adibi, "Role of Interactive and Control Computers in the Development of a System Restoration Plan", *IEEE Trans. Power Apparatus and Systems*, vol. PAS-101, Jan. 1982, pp. 43-52.

[5] H. Kodama, H. Suzuki, I. Atsumi and K. Ishizuka, "Interactive Restoration Control of Electric Power Systems," *CIGRE-IFAC Symp. Control Applications for Power System Security*, Florence, Sept. 1983, Paper 514-04.

[6] T. Sakaguchi and K. Matsumoto, "Development of a Knowledge-based System for Power System Restoration," *IEEE Trans. on Power Apparatus and Systems*, vol. PAS-102, Feb. 1983, pp. 320-329.

[7] R.P. Schulte et.al., (Systems Operations Subcommittee, Current Operational Problems Working Group Report), "Survey Report on Current Operational Problems," *IEEE Trans. Power Apparatus and Systems*, vol PAS-104, June 1985, pp. 1315-1320.

[8] F.F. Wu, Letter to M.V.F. Pereira, Electric Power Research Institute, April 18, 1984.

[9] M.V.F. Pereira, oral communication.

[10] G.L. Blankenship and T.A. Trygar, "A Discussion of the Restorative State Control Problems in Electric Power Systems," *Proc. of EPRI/SIAM Conference on Electrical Power Problems: the Mathematical Challenge*, Seattle, March, 1980.

[11] A. Monticelli, and F.F. Wu, "Network Observability: Theory," *IEEE Trans Power App. and Systems*, vol. PAS-104, pp. 1042-1048, May 1985.

[12] A. Monticelli and F.F. Wu, "Network Observability: Identification of Observable Islands and Measurement Placement," *IEEE Trans. Power App. and Systems*, vol. PAS-104, pp. 1035-1041, May 1985.

[13] A. Monticelli and F.F. Wu, "Observability Analysis for Orthogonal Transformation Based State Estimation," presented at 1985 PICA Conference, to appear in *IEEE Trans. Power Apparatus and Systems.*

[14] A. Monticelli, and F.F. Wu, "A Method That Combines Internal State Estimation and External Network Modeling," *IEEE Trans Power App. and Systems*, vol. PAS-104, pp. 91-103, Jan. 1985.

[15] J.J. Koglin, and H. Muller, "Corrective Switching: A New Dimension in Optimal Load Flow," *Electrical Power and Energy Systems*, vol. 4, Apr. 1982, pp. 142-149.

[16] H. Kronig and H. Glavitsch, "A Systematic Approach to Corrective Switching in Power Networks," *CIGRE-IFAC Symp. Control Applications for Power System Security*, Paper 206-02, Florence, Sept. 1983.

[17] C.A. Rossier and A. Germond, "Network Topology Optimization for Power System Security Enhancement," *CIGRE-IFAC Symp. Control Appl. for Power System Security*, Paper 206-01, Florence, Sept. 1983.

[18] A.A. Mazi, B.F. Wollenberg, and M.H. Hesse, "Corrective Control of Power System Flows by Line and Bus-bar

Switching," presented at 1985 PICA Conference, to appear in *IEEE Trans. Power Apparatus and Systems*.

[19] B. Krogh, S.H. David and J.H. Chow, "Multi-State Rescheduling of Generation, Load Shedding, and Short-Term Transmission Capacity for Emergency State Control," *IEEE Trans. Power Apparatus and Systems*, vol. PAS-102, May 1983, pp. 1466-1472.

[20] A. Monticelli et.al., "Interactive Transmission Network Planning Using A Least-effort Criterion," *IEEE Trans. Power Apparatus and Systems*, vol. 101, #10, pp. 3919-3925, Oct. 1982.

[21] F. Hayes-Roth, D.A. Waterman and D.B. Lenat, *Building Expert Systems*, Addison-Wesley, 1983.

[22] A. Barr and E.A. Feigenbaum, *The Handbook of Artificial Intelligence*, vol. 2, William Kaufmann, 1981.

[23] J.H. Taylor and D.K. Frederick, "An Expert System Architecture for Computer-Aided Control Engineering," *Proceedings of the IEEE*, vol. 72, No. 12, Dec. 1984, pp. 1795-1805.

[24] C.Y. Cuadrado, and J.L. Cuadrado, "Prolog Goes to Work," *Byte*, Aug. 1985, pp. 151-245.

Discussion

R. B. I. Johnson and B. J. Cory (Imperial College, London, UK): The authors are to be commended on their conceptual design of a knowledge-based system for power system restoration. We are also investigating the use of knowledge-based systems but for the operational planning and real-time operation of power systems. Most expert systems to date have been applied to areas where little causal knowledge exists but in the case of power systems a large number of mathematical models and analytical algorithms are available. The application of knowledge-based techniques to power systems must incorporate these models and we are encouraged to see that the authors have taken an approach broadly similar to ours.

While we agree that knowledge may be acquired from an analysis of the problem we feel that the operator's knowledge about the system's behavior should also be included.

Some objectives which can justify the implementation of an expert system and for which research into expert systems applications in the power system filed should seek, are

i) more complete solutions;
ii) a speed-up in computation;
iii) efficient solutions to search problems;
iv) explanation of how solutions are obtained.

While recognizing the conceptual nature of the paper, we wish to pose the following questions which relate to these objectives.

1) Table 1 of the paper identifies seven algorithms of increasing complexity up to comprehensive restoration control. Would subproblems be selected from these seven algorithms or would a much larger family of algorithms be defined based on model reduction, network equivalenxing, aggregation, etc.?
2) Would rates be included in the knowledge base which assess the reliability of approximate models? This could reduce the need for validation through the load flow simulations.
3) The determination of wwitching sequences is a combinatorial problem for which exhaustive searches must be made. Artificial intelligence techniques of heuristic search may be applied to such problems. If the authors intend to use such techniques, would they be included in a knowledge base or coded as imperative routines?
4) A fundamental problem in man/machine systems is the division of labor between the operator and the computer. Do the authors foresee the operator making decisions based on the computer's assessment report or on a more interactive mode where the operator proposes control sequences and has some control over the search procedures?

Manuscript received February 21, 1986.

G. Morin and H. Horisberger (Hydro-Québec, Montréal, PQ, Canada): The authors have presented a very interesting conceptual framework for computer-aided power system restoration. It should stimulate the development of new analytical tools and their integration with existing security monitoring functions for future energy management systems. For the time being, most utilities have to rely on system restoration guidelines based on operator experience and extensive off-line studies. hydro-Québec has developed a restoration plan [1] which is similar to the one cited in reference [3] of the paper and is based on a three-step strategy (sectionalizing into island, restoration of islands, synchronization of islands).

The objective of the approach is not only to speed up the restoration process but to carefully avoid the risk of equipment damage due to temporary and steady-state overvoltages. These may occur on weak EHV systems with long transmission lines between production and load centers causing large amounts of reactive power to be generated and possibly leading to low-frequency resonance excited by transformer magnetizing in rush currents.

Different restoration plans were validated on our transient network analyzer. For digital simulations, a three-phase harmonic impedance analysis program had to be developed and used together with a conventional loadflow. In the near future we paln to adapt these off-line analysis programs and to integrate them with security monitoring functions fo our energy management system.

Our objective is a relatively simple interactive tool to monitor and predict potentially dangerous system states or switching operations during system restoration.

Manuscript received February 24, 1986.

Reference

[1] G. Morin, "Service Restoration Following a Major Failure on the Hydro-Québec Power System," IEEE PES Winter Meeting, New York, February 2–7, 1986. Paper no. 86 WM 183-8.

F. F. Wu and **A. Monticelli:** We thank the discussers for their valuable comments. We agree with Drs. Johnson and Cory that the operator's knowledge should be included in the rule base. This could be in the form of the selection of appropriate subproblems for analysis, switching sequence, etc. Table 1 presented the results of our preliminary study. A much larger family of algorithms and procedures may be included in restoration control. It seems an excellent idea to develop means to assess the reliability of approximate models and the knowledge be included in the rule-based system. The determination of switching sequence definitely is an area where we see the operator's knowledge can be incorporated directly as part of the knowledge base to reduce the need for a large search. Finally, we believe the development of more interactive software is definitely the right direction to be heading.

We appreciate Messrs. Morin and Horisberger for sharing their experience with us.

Manuscript received September 3, 1987.

SYSTEM OPERATIONS CHALLENGES

A report by the Current Operational Problems Working Group--E. Kenneth Nielsen, Chairman
Contributory authors are: M. M. Adibi, D. Barrie, M. E. Cooper, K. W. Heussner, M. E. Robertson, J. L. Scheidt, and D. Scheurer.

Abstract: This paper is one of a series presented on behalf of the System Operations Subcommittee with the intent of focusing industry attention on current problems faced by electric utilities in the operation of the bulk power supply system.

Five short note reports are presented which direct attention to problem areas being worked on by various Working Groups and Task Forces of the System Operations Subcommittee. Two reports deal with problems associated with the replacement of Energy Management Systems (EMS) and predicting resource requirements. One report describes the problem of making EMS alarming effective. Another focuses on problems in the design, development, and delivery of quality system operator training programs. The final report lists and explains a number of challenges in restoration of the bulk power supply system following a major blackout.

INTRODUCTION

E. Kenneth Nielsen, Member
New England Power Exchange

One of the primary tools of system operators is the Energy Management System (EMS). To be an effective tool, it is essential that the EMS be replaced or upgraded before it becomes obsolete. Yet there are many impediments to keeping EMS facilities up-to-date. One of the most serious problems is predicting the computer resources needed to satisfy the demands which will be placed upon the upgraded system. Many utilities have experienced the frustration of undertaking the massive effort of obtaining authorization for a new EMS and investing years into its implementation, only to find it inadequate. One particularly troublesome EMS function is alarming. The tendency is to provide the system operator with a huge amount of data but not much concise, well organized information, which pinpoints source problems rather than symptoms. These problems are discussed in this report.

Effective control of today's bulk power supply system demands well trained system operators. Training programs must be well conceived. Instructional materials many times provide valuable technical information but do not target on training the system operator to apply the information in performing his job. Another obstacle is the lack of qualified instructors. These problems become more acute as EMS sophistication grows, the power system becomes more complex, and the turnover rate of system operators increases.

87 WM 050-8 A paper recommended and approved by the IEEE Power System Engineering Committee of the IEEE Power Engineering Society for presentation at the IEEE/PES 1987 Winter Meeting, New Orleans, Louisiana, February 1 - 6, 1987. Manuscript submitted August 25, 1986; made available for printing November 13, 1986.

There is little work going on in the area of developing analytical tools to aid in power system restoration. The report presented here identifies a number of the problems faced by the system operator in the hope that it will encourage the development of such tools. While system restoration is not a daily task of the system operator, it is a task he must be prepared and equipped to perform. The threat of power system disturbances is ever present!

THE PROBLEM OF UPGRADING ENERGY MANAGEMENT SYSTEMS

J.L. Scheidt, Senior Member, Florida Power Corporation
M.E. Robertson, Member, Southern California Edison Company

Introduction

Operating personnel of electric utilities have been faced with a significant and growing problem of premature obsolescence in their Energy Management Computer Systems (EMS). Approximately every ten years, the justification of a "new" EMS to replace the existing one becomes a paramount operating necessity but is viewed with increasing skepticism by upper management personnel. This skepticism is, in part, due to the fact that EMS operating personnel are faced with a series of problems associated with the general one stated above, namely:

1. A cost/benefit analysis is difficult due to subjectivity and sparcity of data.

2. Prior justifications may have promised a useful life in excess of that actually being experienced.

3. Actual end-of-life of the existing systems is difficult to establish and more so to prove.

4. The fact that the existing system may, on the surface, be performing adequately merely compounds replacement justifications.

5. Total elapsed time for the complete process of justification, budgeting, specification preparation, bid evaluation, design, manufacturing, testing, and installation is taking approximately five years. From recent experience, more systems are delivered late than on schedule due to first time development problems. Such long lead times tend to compound the limitations experienced today as existing systems are stretched past the point where prudently they should be replaced.

6. Parts of the existing system are replicated in the one being purchased giving rise to the question "Why wholesale replacement?".

The authors co-chair a Task Force comprised of other utility personnel, EMS vendors, and consultants. This group has spent the last three years investigating EMS replacement. The main thrust of the initial work was centered around the specific issues associated with EMS architecture which preclude a logical and phased replacement of EMS subsystems compared to the industry's present practice of an entire EMS replacement. Of necessity, topics con-

Reprinted from *IEEE Transactions on Power Systems,* Vol. 3, No. 1, pp. 118–124, February 1988.

cerning enhancement of existing systems were also evaluated. In order to identify some of the parameters which impact this issue, an EMS architecture problem list was developed. This list is scheduled to be included in its entirety with the report of the Task Force; however, an examination of specifics for various problems observed in different subsystems would illustrate some of the key aspects.

Software Problems

- Software which is not table driven is very difficult to modify.
- Some, if not all, software is machine dependent, thus limiting its transferability and its expandability.
- Operating systems reach a point in time beyond which they are not vendor supported.
- The addition of new system software applications requires major modifications to the data base.
- Improvements in advanced algorithms sometimes cannot be retrofitted in existing programs.

Hardware Problems

- Many systems have finite memory capability thus creating a situation where memory expansion is physically impossible.
- The erosion of hardware computational capabilities for whatever reason (growth, inefficient coding, etc.) slows system response to unacceptable levels.
- Physical constraints limit the number of communications controllers, thus limiting SCADA expansion.
- Vendor specific hardware interfaces preclude expansion from other than the original supplier.
- Original vendor discontinues support of hardware maintenance and/or expansion.

General Problems

- Unrealistically low estimates of ultimate system expansion requirements cause severe compromises late in system life.
- The lack of standard communications protocol has complicated EMS to SCADA and EMS to EMS communications links. This problem is magnified between EMS's that have a wide diversity of capabilities and internal hardware/software design differences.
- Inability to measure the operating system overhead makes it difficult to determine the specific causes of system slowdowns.
- Poor hardware/software documentation results in maintenance and expansion nightmares.

Utilities faced with replacing or upgrading EMS's are concerned with future system architecture that will alleviate many of the aforementioned problems. The current practice of spending 4-6 years designing and procuring an EMS only to have it last 8-12 years has been viewed with justifiable suspicion by utility management. EMS obsolescence has, for the most part, been complete with few, if any, subsystems being candidates for reuse. Some vendors argue the benefits concerning the upgradeability of their specific design while others contend that the unique and specialized interrelationship of hardware and software necessitates that obsolescence and ultimate replacement is a fact of life. The authors question this latter position.

Characteristically conservative by nature, utilities have taken a following role in energy management design. Few companies are willing to expend the dollars needed to support extensive research and development efforts to meet their unique EMS requirements. This, historically, has been left as a vendor prerogative to develop what they market as "standard" offerings. Most standard vendor offerings are subject to limited architectural customization by prospective buyers. On occasion, utilities have attempted to push certain state-of-the-art developments; most of these have been in the more visual portions of the system, namely mapboard display or man-machine interface technology. Many times these state-of-the-art developments have met with less-than-spectacular success. Such experiences have further reduced utility willingness to stray far from standard vendor offerings. There are, however, indications that present day EMS obsolescence may be due to design shortsightedness rather than an inherent problem associated with basic EMS requirements.

Why is This Problem Important

The utility industry has much to gain from participating in the removal of present day EMS design limitations. In the first place, EMS lifecycle costs should drop as individual subsystems become truly upgradeable. This effect will be magnified in many areas such as training, spare parts, etc. Salvageability of operator-familiar subsystems such as logging, alarming methodology, display formats, etc., will greatly facilitate installation of new subsystems. Additionally, vendors can specialize in any combination of subsystem areas which permit design flexibility to cure problems specific to individual utility needs. A great deal of development work is required before mix and match system purchasing becomes a reality; however, a system design which incorporates subsystem replaceability will certainly attract buyers concerned with the unknown software applications of the future. It will, hopefully, also reestablish credibility in the EMS procurement process by upper utility management.

In conclusion, the utility industry, with the cooperation of vendors and consultants, has the opportunity to provide guidelines for the next generation of EMS. Such new systems can, and hopefully will, exhibit much greater flexibility concerning upgradeability. The benefits of such design are not only economic. Utilities will be able to focus on enhanced subsystem capabilities to provide specialized treatment for areas of particular concern. Such tailoring of resources across vendor boundaries to meet specific utility needs is something the industry is hard pressed to now provide.

PROBLEMS ASSOCIATED WITH PREDICTING COMPUTER RESOURCE REQUIREMENTS

David Barrie, Member
Ontario Hydro

What is the Problem

In today's Power System Control Centers, the Energy Management System (EMS) is called upon to process a variety of application programs providing a range of functional requirements to the system operators. The capability of the computer system to carry out this processing is a critical element in determining the overall performance of the system.

The life expectancy of a typical EMS is about 15 years. During this time, the actual requirements will almost certainly differ from those initially specified, given the volatile environment within which utilities must operate. As a result, upgrades, refurbishments, expansions, etc., are virtually unavoidable during the life of an EMS.

This short note focuses on the problem of uncertainty surrounding the planning of computer system upgrades.

Experience in Ontario Hydro

Ontario Hydro's current EMS was installed in 1975. It went into service with a basic suite of application programs running and a more comprehensive suite specified for development in subsequent years. The initial computer system was sized for the initial suite and a plan put in place to upgrade to accommodate the planned growth.

In fact, operating complexity increased far more than originally envisaged, primarily due to inability to build adequate transmission facilities. As a result, increasingly sophisticated application programs and new functional capabilities have been required of the EMS; e.g. much more complex stability limit monitoring software and automatic arming of generation and load rejection schemes. This greatly increased the computing resources required, over and above the initial predictions and triggered two computer upgrades, one in 1978 and another in 1981.

Looking to the future, our ability to predict future requirements may be no better than in the past. We are planning our EMS with the following expectations:

- Continued delay in getting the necessary transmission facilities.
- Power system growth around three percent per annum.
- More tightly integrated operation with interconnected utilities.
- Improved operator training facilities will become essential.

Any or all of the above may not transpire. Furthermore, some new factor, presently unforeseen, may drastically alter our projections. Time alone will tell.

Why Such Uncertainty

The process of planning future computer upgrades involves a sequence of steps, each one fraught with uncertainty. As a result, it is hardly surprising that the overall process is very uncertain. The steps are described below.

Step 1: Identify Application Programs

The requirements for new application programs are driven by the needs of the power system operators. They are a function of the operating complexities imposed by the evolving power system and the extent to which new application programs can assist the operator in his task. Unfortunately, "operating complexity" is largely determined by external events, beyond the control of the electric utility and sometimes unanticipated.

Looking back over the last five to ten years, there have been a number of "external forces" which have had a dramatic effect on the complexity of operation and were largely unanticipated. Specific utilities were affected to a greater or lesser extent by any given factor but a few commonly quoted examples are the following:

- The inability to build adequate transmission facilities as described for Ontario Hydro.
- Major fluctuations in growth rate projections making it very uncertain as to the size of the power system to be monitored by the EMS.
- Increased exports from nonoil-burning utilities. This has led to a more complex transaction process and heavy power flows on key transmission interfaces. The system operator has had to become more aware of system conditions in other utilities for both economic and security issues.
- Proliferation of cogeneration, load management, and other schemes which greatly add to the complexity of generation scheduling and dispatch.

The uncertainty surrounding identification of future application programs is the biggest contributor to the uncertainty of the overall process.

Step 2: Characterize the Programs and Predict Computer Load

Having forecast the functional requirements, the next step is to characterize the application programs. The following questions need to be addressed:

- How frequently will the program be executed?
- How much processing time will it take?
- How much main memory will it occupy?
- How many disk files will it access?

While these are important questions, in a system which has a sequence of interrelated programs, probably the most important question is how the new program will fit into the existing suite of programs.

Knowing the program characteristics, the future computer load can be predicted, however, the questions that need to be addressed are the following:

- How should one express computer loads on a real-time computer system?
- What measurable quantities should one use to express computer loads?
- What power system conditions should one assume in assessing computer loading? It must be recognized that computer loading during power system disturbances is vastly different from normal conditions.

Step 3: Impact on Computer Performance

Having established the computer loading, the next step is to forecast the computer performance. There is a relationship between load and performance, unfortunately, it is not linear.

When the computer is lightly loaded, or in the mid-range, increases in computer loading result in little or no degradation in computer performance. As the computer loading is increased, there is a point at which small increases in computer loading result in major performance degradation. The uncertainty surrounding forecasting computer performance is knowing where the actual loading is, relative to the point where degradation is likely. There is little early warning that major degradation is imminent.

Step 4: Identify Corrective Action

The final step of the process is to identify what corrective actions are appropriate to rectify deficiencies found in the performance analysis. The corrective action can be one or more of the following:

- Optimize existing software to reduce the consumption of resources.
- Install additional hardware or upgrade the existing hardware in the area of deficiency; e.g. disk storage, main memory, or central processing unit.
- Relax the standards of performance.
- Reappraise the requirement for the new program(s).

The major uncertainty in this step is in knowing that the proposed solution will, in fact, solve the problem. In particular, adding or upgrading hardware facilities to solve a specific deficiency may merely move the problem to another area.

EFFECTIVE ALARMING

Kenneth W. Heussner, Member
Houston Lighting and Power Company

Introduction

Effective alarming can be an elusive goal in an Energy Management System (EMS). As systems mature and grow more complex, there is an increasing tendency by system designers to add more information about individual alarms and the conditions that have caused them. The additional information aids the system operator in decision making, provided that the alarms do not overload the ability of the system operator to assimilate information. Under disturbance and emergency conditions, alarms are more numerous and often provide redundant information. Burst alarm processing of 500 to 900 alarms per minute is being incorporated in EMS's being built today.

Since the system operator can be confronted with more alarms than can be handled effectively, alarming should provide processed information that is adequate for system operation, timely, and contains little or no redundant data.

Alarm Conditions

Conditions on the power system and equipment monitored by the EMS can cause the activation of large numbers of alarms. These alarms may be valid or invalid depending upon circumstances. Some examples of conditions which create burst type alarm are as follows:

- System Start-up - When an EMS is initialized or when it is restarted from a failure condition, all telemetry data that does not match a "default" (start-up) condition will appear to the EMS as abnormal situations. As a result, alarms are issued. This will occur during the initial system scans thus causing hundreds or thousands of alarms to be generated in a burst mode.

- Data Set Switch - An EMS requires constant additions and corrections to the data base in order to properly reflect the equipment in the field. Most of the time, the corrections will be minor and will not require a regeneration of the data base. The corrections will be made in an off-line data set and will be switched to replace the on-line data set. Status of points in the new data set which differs from the scan data status will generate alarms.

- Communication System Dropout - The backbone of any EMS is the communication system which ties the system computers with remote terminal units (RTUs). A disturbance which disables communications can cause large numbers of communication errors in addition to a backlog of alarms and state changes that will be detected when the disturbance has cleared. A microwave loop system will exhibit this condition when communications are switched from one direction or path to the other due to a disturbance such as a signal fade. The system will have a communication loss of several seconds but all RTUs using that path will have communication errors. If the path is out for any period of time, RTUs will have alarm state changes queued up.

- System Disturbances - As storms and weather fronts move across the power system, numerous alarms will be received in bursts. The detected alarms can exceed 15 per second. Usually the entire power system is not affected by a storm at the same time. Operators can easily be overloaded with alarms from the portions of the power system being affected by the disturbance.

- Emergency Conditions - Extreme conditions which threaten the integrity of a power system can cause so many alarms that the system operator could not use the EMS. Examples are "network islanding" and operator activation of system voltage reduction on most power system buses when there is insufficient generation to handle the load. Alarms resulting from either of these cases can overwhelm the system operator and mask critical alarms.

Alarm Strategies

The following techniques may be used to insure that a system operator receives optimum information.

- Alarm Acknowledgment - Alarms should not have to be acknowledged by the system operator as the alarms are presented on the CRT. The system operator knows by the CRT display, by output on a printer, or by related events that an alarm has been activated. Having the system operator make a separate alarm acknowledgment is an unnecessary step since each alarm message presented to the system operator requires that he make a decision (ignore, respond, etc.).

- Filtered Alarms - Alarm information should not be redundant. Information which is useful by itself may be either uninformative or not important in certain situations. As an example, an undervoltage alarm is useful information as a single alarm condition but is redundant if status changes indicate loss of power feed.

- Message Buffering - Alarms should be spooled on a large CRT message output buffer. The buffer should consist of multiple pages of messages. This will allow the system operator to go backward in the message output queue on the CRT without being required to hunt back through a printer's output.

- Lost Alarms - Alarms should never be lost due to burst or continuous alarm processing loads. With today's computer systems, there is no reason to ever lose alarms due to computer processing work loads. The system operator may have to wait to see all the alarms, dependent upon CRT output rate, but no alarms should be lost due to the EMS.

- Alarm Categories - Alarms should be separated into categories to allow prioritizing and to allow segmenting by operator position.
- Presentation - Alarm messages should be presented on the CRT using the capabilities of the CRT system (including color, shape, font, screen position, etc.) to enhance alarm identification. Message text should be short and concise.
- Alarm Requirement - Any alarm for an EMS should be thought out and justified as to need, advantage, and dependency. Alarms shown to the system operator should be limited to those alarms necessary for power system operation.
- Preprocessing - Alarm processing decisions should be made where possible by computers and not by the system operator. As a tool, the computer system can do some of the minor decision making thus relieving the system operator of unnecessary work.
- Alarm Priority Types - Alarms should be prioritized by type. Many non-critical alarms can be presented in category type displays while using an operations message display for time critical alarms.

Summary

As more information is being included in the modern EMS, there must be concerted effort to present the system operator with the information that is needed to operate the power system. More preprocessing and filtering of alarms is needed to prevent overloading the system operator. Effective alarming instead of point alarming is required. It is the system operator's actions upon which we must rely to preserve the power system in normal and emergency conditions.

PROBLEMS IN THE DESIGN, DEVELOPMENT, AND DELIVERY OF QUALITY SYSTEM OPERATOR TRAINING PROGRAMS

Maureen E. Cooper, Associate Member
Potomac Electric Power Company

Introduction

The northeast blackouts caused the electric utility industry to focus on the training of Power System Operators (PSOs). Up to that time, most PSOs learned how to operate their system through informal on-the-job training. In the ensuing years, many training programs have been designed, developed, and delivered to PSOs. Although a great deal of effort has gone into their development, most of the programs provide information about electrical systems, but do not train operators how to apply this information in fulfilling their actual job responsibilities. The purpose of this short note is to highlight some of the reasons why little progress has been made in designing, developing, and delivering programs that actually train PSOs to perform their required tasks in a specified manner.

Program Design

There are many problems associated with designing quality PSO training programs. Three of the primary concerns include utilizing qualified instructional designers, developing PSO performance standards, and dedicating adequate resources to the design effort.

It has been well documented that a structured approach to the design of training programs proves to be the most effective in causing specific desired behavior changes. If a structured approach were to be used in designing PSO training programs, an instructional specialist would be assigned the responsibility of designing a training program for a specified group of PSOs. The instructional specialist would work with PSO subject matter experts to determine the performance standards expected of successful trainees; to identify the knowledge, skills, and abilities needed to meet the desired performance standards; and to make recommendations regarding the most effective instructional/learning techniques for ensuring that the trainees acquire the knowledge, skills, and abilities to meet the desired performance standards.

Using this structured approach to training design can take several months or more, depending upon the complexity of the project and the availability of necessary manpower. While this appears to be time-consuming and expensive, this investment is necessary to ensure that the training program will cause desired behavior changes.

Unfortunately, a structured approach to the design of PSO training programs is rarely utilized. More commonly, a senior PSO or an operations engineer is assigned the responsibility of designing a training program on a specific topic. Most electric utilities do not have clearly defined performance standards or procedures. Since they may be unaware of the necessity of determining the desired behavior outcomes and the requisite supporting knowledge, skills, and abilities, the system operator or engineer usually does not make any attempt to link the training program to desired behavior changes. As a result, typical program design consists of an outline of topics to be included in the program without any focus on how the PSO should use the information in performing his job responsibilities.

Program Development

Problems are also encountered in the process of developing quality PSO training programs. Three areas deserving attention are the utilization of qualified program developers, the efficient utilization of available resources, and the ability to support long or expensive program development schedules.

In the ideal situation, an instructional specialist would be assigned the responsibility of overseeing the development of training materials to support the program's approved design. The instructional specialist would be given sufficient time, money, and resource personnel to adequately manage this development effort.

The instructional specialist would oversee the development and testing of all instructional materials, all learning materials, and all evaluation materials. The materials would be made in whatever media was identified in the program design and could include computer simulations, texts, slides, videotapes, role plays, audiotapes, or other materials.

Wherever practical and economical existing training materials would be modified and used. Modifications would be aimed at improving the materials' appropriateness and effectiveness in meeting the new training program's objectives and might include changes such as rewriting or reformatting text to improve its readability, adding relevant illustrations, adding detail to computer simulations, developing company-specific operations exercises to replace generic exercises, or editing material on a videotape. New program materials would be developed where necessary, to meet the program's objectives.

When all materials had been developed, they would be proofed for technical accuracy, field tested, revised where necessary, and then made available for actual training sessions.

The process of developing quality training materials is both time-consuming and expensive. Few electric utilities have instructional specialists with unlimited resources to dedicate to the development of PSO training materials. More frequently, the system operator or engineer who was assigned the responsibility of designing a training program is also assigned the responsibility of developing the related materials in a short time period with a limited budget. As stated earlier, the designated system operator or engineer rarely has any formal training in instructional technology and is, therefore, unaware of the need to consider such things as the amount of practice required to learn how to perform a task, the impact of layout and design on the readability of printed text, or the advantages of using one type of instructional material over another.

Working within these constraints, the program materials tend to be gathered rather than developed, and frequently consist of unmodified, readily available materials such as off-the-shelf videotapes, manufacturers' literature, and copies of operating procedures. Because the program developers are not skilled in making other instructional materials such as videotapes, role-plays, or computer-assisted instruction, the materials developed tend to be oriented towards formal classroom-style training. These materials usually provide technical information, but do not train the PSO how to apply the information in performing his job.

Program Delivery

Given a PSO training program that has been well designed and developed, there are still obstacles to the effective delivery of the program. The lack of qualified instructors, the lack of manhours available for training and the cost associated with using some training programs can prevent good training programs from being used.

The ideal PSO trainer would be not only a system operations subject matter expert, but also a qualified technical instructor. Most PSO trainers are subject matter experts, but have received little or no preparation to assist them in conducting training sessions. If an instructor does not understand the purpose and value of the materials and tests associated with a training program, he is not likely to use them as they were intended.

Given a well designed, well developed program with a qualified instructor, there is still no guarantee the program will be used. Most energy control centers are not staffed to schedule sufficient time for training sessions. In such cases, operators have to work overtime to participate in training sessions, or the training sessions may not be held as frequently as necessary or desirable.

A third problem concerns the cost of some system operator training programs. Although power system simulators and computer-assisted instruction are two of the most effective PSO training tools, their costs have limited their use to a few electric utilities.

Conclusion

The responsibilities of PSOs are in an evolutionary state. Over the next few years there will be significantly greater economic advantages to be gained through better training of PSOs. Three areas which deserve specific attention are:

- Training PSOs to effectively utilize a new generation of energy control systems.
- Training PSOs to operate the electric system with smaller reserve capacity and smaller margin for error.
- Training PSOs to make maximum benefit of energy transactions.

The economic losses possible through incorrect system operations due to inadequate training will be greater in the coming years, and conversely, the economic benefits of correct system operations are best attainable when PSOs have received job-specific training that has been well designed, well developed, and well delivered.

ISSUES AND PROBLEMS IN POWER SYSTEM RESTORATION

M. Mike Adibi, Senior Member, IRD Corporation
David Scheurer, Member, Philadelphia Electric Co.

Introduction

This short note briefly describes some of the problems encountered in the three phases of power system restoration (PSR), with the intent of identifying analytical tools needed for their solutions. The three phases of PSR are the following: planning for restart and reintegration of the bulk power supply, actions during system degradation for saving and retaining critical sources of power, and restoration when the power system has stabilized.

In the planning phase, problems can primarily be addressed by off-line analysis or simulation. During the degradation phase, control problems need solutions in real-time and within the short-time ratings of lines and equipment. In the restoration phase, the problems can be addressed by on-line analysis, presenting the results to the operator for execution.

Restoration Problems

1. Fault location - The major disturbances in the bulk power supply are primarily caused by transient faults and operators' errors, mostly originating in the transmission systems, and resulting in area wide outages. A very large number of initiating causes of interruptions are due to temporary faults, which are immediately cleared, leaving the system in an unfaulted condition. Therefore, identification of the cause of a power failure does not necessarily enhance the restoration, and is not considered as a high priority PSR problem.

2. Interconnection assistance - The development of the system restoration plan from a complete collapse and without interconnection assistance, is relatively straightforward. In practice, due to corrective actions, system disturbances are usually arrested prior to the complete collapse. Furthermore, some level of interconnection assistance would probably be available. The present practice is to activate the EHV interconnections in the system being restored when units with large reactive power absorbing capa-

bilities are in service. There is a need for establishing a simple and quantitative guideline, such as the ratio of available reactive absorbing capability to the charging current of the EHV line, which would allow interconnection activation.

3. Assessment of switching status - Partial system outage and/or reliance on the interconnection necessitates assessments of both system and interconnection switching status. This is a process which uses up critical time. Some utilities automatically open all circuit breakers upon loss of voltage to simplify and speed-up the reintegration process. Others prefer saving the breaker stored energy and minimize breaker operations by judicious selection and operation of breakers. The switching operation has been responsible for slowing many actual restoration processes and demands special attention.

4. Frequency and voltage responses of prime movers - The size of load increments which can be picked-up depends on the frequency and voltage responses of prime movers. Load pick-ups at small increments prolong the PSR. With large increments, there are risks of frequency decline or voltage dip beyond recovery. The challenge is to establish guidelines which would indicate the largest load increment which would keep frequency or voltage to within acceptable limits.

5. Cold load pick-up - It is important to know approximately the amount of real and reactive loads to be picked-up at any time following an outage. As the outage duration lengthens, the coincident factor increases. The load-time variation depends on normal peak and light loads, load power factors, power factor corrections, load type, and the total connected loads. The problem is to use these data to develop real and reactive loads versus time for load pick-up.

6. Optimum sequencing of generating units start-up - In restoring a power system, operators often need to know in advance the amounts of real and reactive generation which would be available at a future time. This information could be used to optimize the sequence of the various restoration operations and allow a significant reduction in the outage duration. The need is for timely identification of critical timings of prime movers and efficient coordination between power plants and power system operations.

7. Switching transient voltages - During the reintegration phase, it is desirable to energize as large a section of high voltage transmission line as transient voltages will allow. Energizing small sections tends to prolong the restoration process. In energizing a large section, there is always the risk of damaging the insulation. The need is for a simple methodology which can readily establish guidelines for energizing transmission lines.

8. Standing phase angles - One of the concerns and constraints in closing the next tie to the adjacent system is the excessive phase-angle difference which may exist. Closing a tie-line on a large phase-angle difference may cause instability. The "standing angle" can be reduced by altering the generation pattern of both interconnecting systems (or subsystems). There is a need for generation rescheduling programs which minimize changes in generation patterns while reducing the specific phase-angle difference.

9. Low frequency isolation scheme - During system degradation, it is desirable to isolate certain generators with "matching" local loads to retain critical sources of power. Under this scheme, the generating units do not have to be restarted from a complete shut-down, thus saving several precious hours. The scheme requires that during peak and light load conditions, load and generation mismatch be within the over-and-under-frequency limits. Furthermore, it requires coordination of plant control, protective systems, and interlock systems to enhance generating units' partial load rejection capabilities.

10. Intentional islanding - When two groups of generators begin to lose synchronism, it is advisable to separate the two groups. It is desirable to select locations for separation where the load-generation mismatch in each group is "minimal", and synchronizing equipment is available for reclosing. To ensure satisfactory islanding, it may be necessary to use "blinders" to block relays from the tripping where splitting is not desirable, and allow relays to trip where it is desirable. The intentional islanding has been used in very few systems. It's implementation is an engineering challenge.

11. Local load shedding - There is a need for an on-line real-time methodology which would relieve the overloaded equipment and prevent the system from further degradation. Such a methodology should in the first order of priority be able to respond within the short-time rating of equipment and in the second order of priority, "minimize" the amount of loads to be curtailed in specific locations.

12. Underfrequency relays and switched capacitors - During the initial phases of restoration, when the system is primarily under manual control, wider frequency and voltage variations are tolerated. Under these conditions, automatically switched capacitors and underfrequency relays, which are responsive to voltage and frequency, respectively, will hinder the restoration process. The problem is whether, and to what extent, these devices should be deactivated during restoration.

Conclusion

Review of the past major power disturbances by the Power System Restoration-Task Force (PSR-TF) shows that these disturbances have occurred at random, each disturbance has had somewhat different restoration problems, and in each case the system operators have played a critical role in restoring the system. It appears to be difficult to analytically formulate generally applicable tools. On the other hand, the knowledge-based expert systems, which would allow operators' heuristic problem solving capabilities, do seem to be suited to the restoration problems. This is especially the case if ways can be found to integrate them with analytical tools outlined in this paper. Until the availability of such methodologies, it is the PSR-TF's position that simple restoration guidelines should be developed based on the identified analytical tools for the short-term implementation.

Power System Restoration Issues

M.M. Adibi[1] *and R.J. Kafka*[2]

This article describes some of the problems encountered in the three phases of power system restoration (PSR). The three phases of PSR are:

- Planning for restart and reintegration of the bulk power supply
- Actions during system degradation for saving and retaining critical sources of power
- Restoration when the power system has stabilized at some degraded level.

We must find ways to integrate knowledge-based expert systems with analytical tools required by PSR.

In the planning phase, problems can be addressed primarily by offline analysis and simulation. During the degradation phase, control problems need solutions in real-time and within the short time ratings of lines and equipment. In the restoration phase, problems can be addressed by online simulation, presenting results to the operator for execution.

Training of operators in restoration techniques and philosophy is of high priority, especially since it should be very rare that operators will have an opportunity to experience a blackout and use it as a learning experience. A training simulator is an effective tool in preparing operators for system restoration.

The knowledge-based, expert-systems approach can be considered as a goal not only for system restoration, but for eventual automatic operation of the power system. Meanwhile, it is prudent to develop detailed restoration plans for study and feasibility purposes. Commissioned plans must be flexible to ensure that they are kept current as the system changes. During development of the plans, it may become apparent that new facilities will be needed to permit implementation of restoration in a timely manner.

Power System Restoration Problems

Review of past major power disturbances, as shown in Figure 1 and reported by the former Federal Power Commission (24 cases in the 1960s) and by the North American Electric Reliability Council (42 cases in the 1980s), indicates that these disturbances have occurred at random, each disturbance has had somewhat different restoration problems, and, in each case, system operators have played a critical role in restoring the system.

These problems may be divided into several functional groups:

- Reactive power balance
- Switching transient voltages
- Interconnection assistance
- Load and generation balance
- Frequency response of prime movers
- Cold load pickup
- Load and generation coordination
- Remote cranking power
- Optimum sequencing of generating units startup
- Fault location
- Assessment of switching status
- Standing phase angles
- Low frequency isolation scheme
- Intentional islanding
- Local load shedding
- Under-frequency and switched capacitor relays.

Reactive power balance

During early stages of the restoration process, it is necessary to keep system voltages within the allowable, usually lower than normal range. This is done in

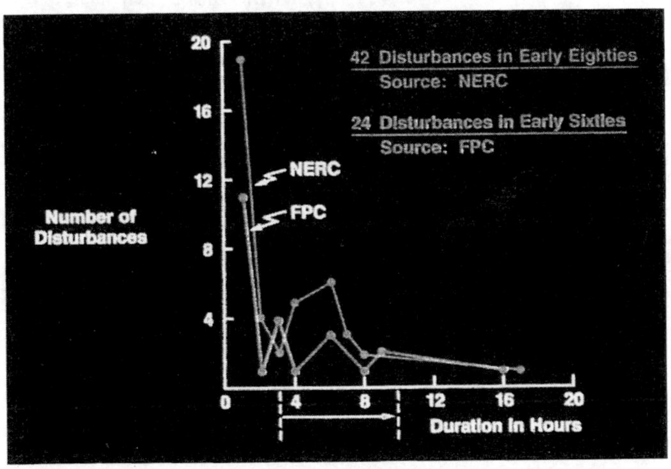

Figure 1. Major bulk power system disturbances

[1] Industrial Research and Development Corporation
[2] Potomac Electric Power Company

Reprinted from *IEEE Computer Applications in Power*, Vol. 4, No. 2, pp. 19–24, April 1991.

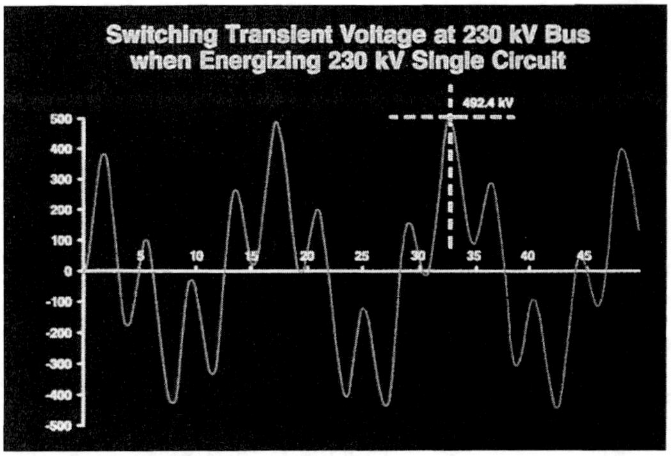

Figure 2. Switching transient voltage

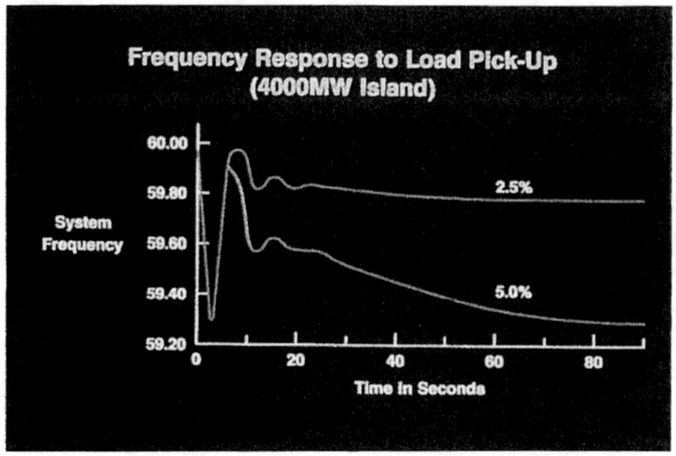

Figure 3. Frequency response to load pickup

several ways: energizing fewer high voltage lines, operating generators at minimum voltage levels, deactivating (over-riding) switched static capacitors, connecting shunt reactors, adjusting transformer taps, and picking up loads with lagging power factors. Inability to perform these tasks due to absence of facilities or procedures would cause a reactive power imbalance resulting in, for example, generator self-excitation and runaway voltage rise.

Installing minimum excitation level relays, particularly on the black-start combustion turbine generators, and setting these relays as low as allowable would help alleviate reactive power imbalance in some cases.

Switching transient voltages

During the reintegration phase, it is desirable to energize as large a section of high-voltage transmission as switching transient voltages would allow. Energizing small sections tends to prolong the restoration process. In energizing a large section, there is the risk of damaging the equipment insulation. Figure 2 shows the plot of a switching transient at the sending end of a 230 kV double-circuit line in response to energizing two sections of the single circuit. The transient voltage of 492 kV exceeds the 470 kV basic impulse level of the transformer at the sending end. Energizing only one section of the single circuit results in a transient voltage of 460 kV, which is below the basic impulse level.

The problem is to determine how to energize a multisection line without exceeding the basic impulse level of the equipment. Some utilities energize lines section by section, one circuit of the double circuit line at one time, and maintain the generator voltage as low as the minimum excitation level allows. This helps reduce the line charging currents at the same time as reducing the transient voltage magnitudes; however, the procedure lengthens the restoration. A simple methodology is needed to readily establish guidelines for energizing transmission lines.

Interconnection assistance

The problem is to determine the point in the restoration procedure at which the tie lines to the adjacent systems can be closed. The present practice is to activate the EHV interconnections in the system being restored when units with large reactive power absorbing capabilities are in service. There is a need for establishing a simple and quantitative guideline, such as the ratio of available reactive absorbing capability to the charging current of the EHV line, that would allow interconnection activation.

Load and generation balance

It is always necessary to maintain system frequency within the allowable limits imposed by turbine resonance, system stability, and the resulting protection settings. This is generally accomplished by picking up loads in increments that can be accommodated by the inertia and response of the restored and synchronized system.

Smaller and radial loads are restored prior to larger and network loads while maintaining a reasonably constant real-to-reactive power ratio. Feeders equipped with under-frequency relays are picked up at the subsequent phases of restoration when system frequency has stabilized.

Load pickup in small increments tends to prolong the restoration duration. With large increments, there is always the risk of falling into a frequency decline and recurrence of the system outage. The size of load pickup depends on the rate of response of prime movers, which more likely are under manual control at this point. A dynamically calculated guideline, such as allowable load pickup as a percentage of generator capability, would help maintain load and generation balance at acceptable frequency.

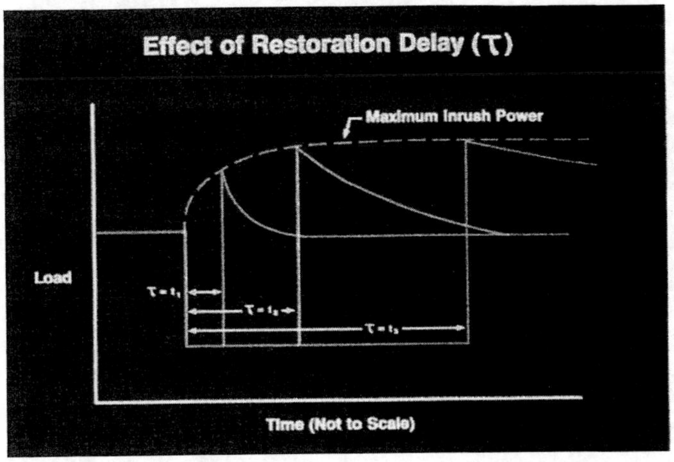

Figure 4. Effect of restoration delay on cold load pickup

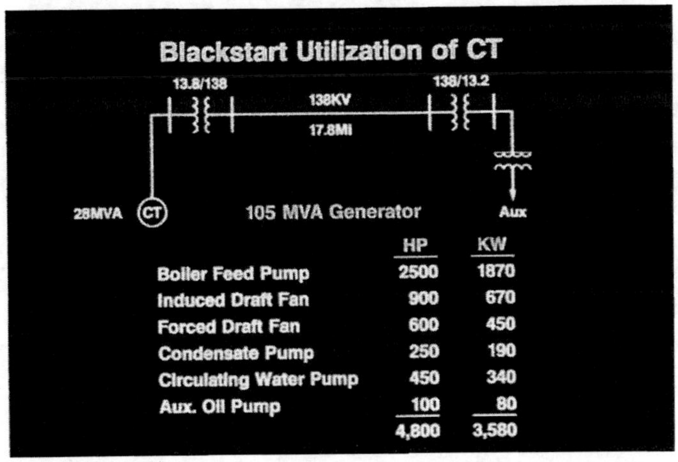

Figure 5. Black-start utilization of a combustion turbine

Frequency response of prime movers

Common practice in the initial stages of restart and reintegration is to rely on combustion turbine, hydro, and drum-type boiler-turbine units. Figure 3 shows the frequency response of a 4,000 MW island to 2.5 percent and 5 percent sudden increases in load. The problem is to establish guidelines that would indicate the largest load increment that would keep the frequency within acceptable limits.

Cold load pickup

At the early stages of restoration, it is important to know the approximate amount of real and reactive power loads to be picked up at any time following an outage. As the outage duration lengthens, the coincident load factor increases. The load-time variation depends on normal peak and light loads, load power factors, power factor corrections, load type, and the total connected loads. The problem is to use this data and to develop real and reactive power loads as a function of elapsed time, as shown in Figure 4.

Load and generation coordination

Generally, the strategies for system restoration consist of restart, reintegration, and load restoration. In the restart phase, one or more load and generation islands are formed. In each island, the objective is to supply station service or cranking power immediately and simultaneously to large thermal plants by black-start combustion turbines or, if available, by hydroelectric units. Combustion turbine behavior is quite different during normal operation when parallelled with a large system as compared to during restoration when they are the prime or the only source of power.

In starting steam units, it is important to coordinate certain critical time intervals, such as the maximum time interval beyond which certain thermal units *cannot* safely be restarted hot, or the minimum time interval required before a thermal unit *can* be started.

Remote cranking power

In restarting steam units, it is often necessary to supply auxiliary power from a remote station. Figure 5 shows the one-line diagram of a remote combustion turbine providing cranking power to a drum-type steam unit. Figure 6 shows a typical plot of the starting frequency and voltage of the induction motor driving the boiler-feed pump. To ensure that these values do not fall below the allowable low limits and, for example, cause locked rotor, combustion turbine frequency and voltage should be maintained at certain values prior to the startup, and proper transformer taps should be selected. These values and tap positions are determined by dynamic simulation of turbine, generator, exciter, transformer, motor, and the specific path selected between the combustion turbine and the steam unit.

Transmission lines are usually represented by their constant impedance and admittance in a nominal π circuit. While such a line model is adequate for small perturbations, it may be inaccurate in the initial phases of restoration where frequency may vary as much as 10 percent, even if this is for a very short time. For this type of simulation, RLC representation is preferred. Also, the selection of a path between the combustion turbine and the steam unit should be limited only to the feasible paths.

Optimum sequencing of generating units startup

In restoring a power system, operators often need to know in advance the amounts of real and reactive generation that will be available at a projected time. This information could be used to optimize the sequence of the various restoration operations and allow a significant increase in restoration of service. The need is for timely identification of critical timings of prime movers, as listed in Table 1, and efficient coordination between power plants and power system operations, as illustrated in Figure 7. It can be seen that although

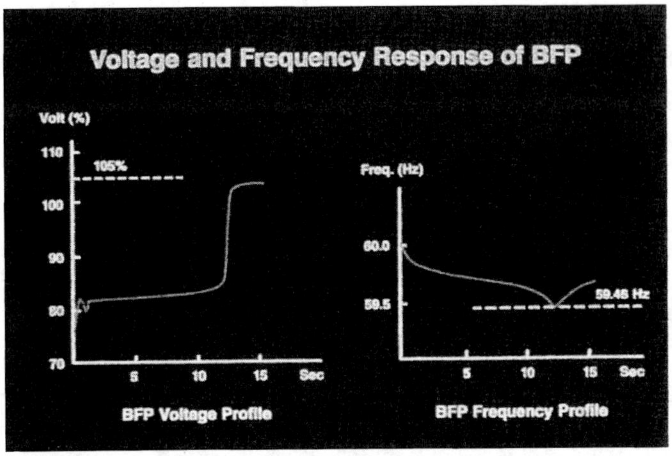

Figure 6. Voltage and frequency response of a boiler-feed pump (source: Illinois Power Company)

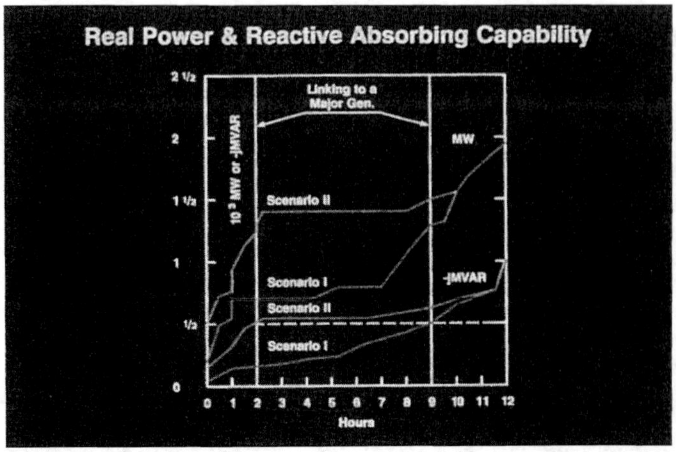

Figure 7. Effect of power system reactive absorbing capability on restoration

both scenarios reach the 1,600 MW load level about 10 hours after the start of restoration, Scenario II serves 17.4 GWh or 42 percent more than the 12.3 GWh for the Scenario I.

Fault location

The major disturbances in bulk power supply are primarily caused by transient faults and operator error, mostly originating in the transmission systems and resulting in area-wide outages. A very large number of initiating causes of interruptions are due to temporary faults, which are cleared immediately, leaving the system in an unfaulted condition. Therefore, identification of the cause of a power failure is not considered to be a high-priority PSR problem.

However, it is recognized that attempting to restore a faulted line or equipment may cause a larger disturbance or extend the restoration time. The knowledge that major transmission paths are unavailable would alter the restoration plan.

Assessment of switching status

Partial system outage and/or reliance on the interconnection necessitate assessments of both system and interconnection switching status. This process uses up critical time and has been responsible for slowing many actual system restorations demanding special attention.

Some utilities open all circuit breakers upon loss of voltage to simplify and speed up the reintegration process. Others prefer saving the breaker stored energy and minimize breaker operations by judicious selection and operation of breakers. The latter approach requires assessment of switching status by dynamic mimic boards and/or automatic path finding through the use of an expert system.

Standing phase angles

One of the concerns and constraints in closing the successive line to the adjacent system is the excessive standing phase-angle difference that may exist. Closing a tie-line on a large standing phase-angle difference could shock the power system and might cause instability. The *standing angle* can be reduced by altering the generation pattern of both sides of the circuit breaker. There is a need for generation rescheduling programs that would minimize changes in generation and load patterns while reducing the specific phase-angle difference.

The range of standing phase-angle differences that a system can withstand depends on the voltage levels and can be determined by steady-state and dynamic simulations. The synchronous relay settings as practiced by one of the northeast interconnections are 60 degrees for 500 kV, 40 degrees for 230 kV, and 20 degrees for 115 kV systems.

Low-frequency isolation scheme

This relaying scheme would be used to isolate certain generators within a plant with *matching* local loads in order to save and retain critical sources of power having large reactive absorbing capabilities. Under such a relaying scheme, the generating units do not have to be restarted from a complete shutdown, thus saving several critical hours. The problem is that both the local load and generation continually change, resulting in varying degrees of mismatch within the isolation scheme. The requirements then for the relay scheme to operate satisfactorily, are two fold:

- Recognition of power system degradation
- Determination of load-generation mismatch within the load rejection capability of the boiler-turbine unit.

In some European power systems, the super-critical, once-through, boiler-turbine units are designed for continuous operation under their own house load (i.e., a load rejection of about 95 percent), and, therefore, there is no need for this type of adaptive relaying scheme. Most coal-fired, steam-generation units currently in service in this country, however, are designed for base load operation with no more than 10 percent by-pass systems. These units trip on a larger load rejection, requiring low frequency isolation schemes to stay in service.

Table 1
Timings for various startup stages of drum-type boiler-turbine-generator

B-T-G Conditions[1]	Hot restart
	Hot boiler, hot turbine
	Hot boiler[2], cold turbine
	Cold boiler, cold turbine
	Very cold[3]
Startup Timing[4]	Time to provide cranking power
	Interval time between paralleling
	Time from start to paralleling
	Time from paralleling to minimum load
	Soaking time[5]
	Time from release to maximum load[6]

[1] Conditions are time dependent and overlap.
[2] Banking boilers.
[3] Water problem.
[4] There are two sets of timings for normal and emergency operations.
[5] Soaking load can be greater than minimum load.
[6] Release and minimum load can be the same.

Intentional islanding

Review of several past major power disturbances shows that relay operations in the initial phases of system separation have lasted for several seconds, for example, approximately 7 seconds in the case of the Northeast and 5 seconds in the case of the PJM major power disturbances of the 1960s. During such a time interval when groups of generators go out of step with one another, it is desirable to end the voltage fluctuations by separating the groups or areas. Under such conditions, it is desirable to select those strategic locations for separation where the load-generation mismatch in each area is minimal, thus resulting in the least disruption of service.

In order to ensure satisfactory system separation, it may be necessary to block some of the relays from tripping where splitting is not desirable and allow tripping only at those locations where it is desirable. This is an online real-time operation and adaptive relaying can play an important role.

Local load shedding

There is a need for an online real-time methodology that would relieve the overloaded equipment and prevent the system from further degradation. Such a methodology should, in the first order of priority, be able to respond within the short-time rating of equipment, and, in the second order of priority, *minimize* the amount of loads to be curtailed in specific locations.

Under-frequency and switched-capacitor relays

During the initial phases of restoration, when the system is primarily under manual control, wider frequency and voltage variations are tolerated. Under these conditions, automatically switched capacitors and under-frequency relays, which are designed for and are responsive to small voltage and frequency perturbation, will hinder the restorative process. The problem is to determine whether, and to what extent, these relays should be deactivated during restoration.

Development of PSR Plan

A review of several PSR plans shows a number of common concerns, even though their strategies are different:

- Identification of the status of the collapsed system, components, and equipment
- Restart and supply of station service to plants, substations, cable pumping plants, compressed air, etc.
- Coordination of power plant startup timings with load pickups to bring generators to their stable minimum levels and within the range of major analog controllers
- Energizing large sections of transmission lines within the acceptable transient and sustained over voltages
- Picking up load in large increments without the risk of frequency decline
- Reintegration of the skeleton of the bulk power supply with the requisite time-consuming switching operation
- Deactivation of automatic load shedding and automatic switched capacitors during initial phases of restoration
- Maintenance of steady-state and transient stability as the system is being restored and when impedances are large
- Reduction of standing angles when closing loops to firm up transmission paths

- Probability of successful startup of thermal units (in particular, combustion turbines)
- Stability of black-start and peaking combustion turbines during the time when they constitute a large portion of generation.

Restoring a power system after a partial or complete collapse in a minimal amount of time presents a very complex problem. As a result, most of the existing procedures attempt to solve simpler problems (or subproblems) in a reasonable manner from a limited number of possible initial conditions. Clearly, there is a need for research into advanced methodologies that will allow integration of applicable analytical tools in the real-time operating environment.

Exploring Advanced Methodologies

During restart and reintegration phases of PSR, a power system often consists of one or two islands, most of the automatic controls have tripped or are deactivated, and the system is primarily under manual control. During these two restoration phases, wide voltage and frequency ranges are tolerated.

Under these conditions, the present dynamic models and simulations, which have been developed primarily for small perturbations of preventive and corrective actions, will not represent accurately the behavior of the power system and its components under a large perturbation of long duration as is required by PSR. It appears to be difficult to analytically formulate generally applicable models for prevention, correction, and restoration of a power system, particularly in view of the limited amount of available data. There is a need for a simulator that can represent power system characteristics relevant to PSR.

Expert systems

Knowledge-based expert systems, which would allow operators heuristic problem-solving capabilities, seem to be suited to the restoration problem. These expert systems are able to:

- Use natural reasoning, i.e., knowledge that experienced operators use
- Diagnose, plan, advise, and, to limited extent, explain their line of reasoning
- Provide very high performance in a relatively narrow field such as PSR
- Expand, continually absorbing new knowledge as it becomes known
- Out-perform human operators under stress, and do so more reliably, especially in those situations that occur infrequently.

The need is to find ways to integrate the capability of knowledge-based expert systems with the analytical tools required by PSR. Such a system would be capable of updating the PSR plan and procedure to allow occurrences of unexpected events and verification of operators's commands by online simulation prior to its implementation. The technology of expert systems is fairly mature, and the time is right to explore their potential use in system restoration.

Operator training

With the increase in power system reliability, major disturbances occur less frequently, and therefore operators receive little experience in restoration. Power system simulators can provide the required opportunity. For the simulator to be an effective training tool and prepare the operators to cope with system-specific and time-critical restoration problems, it must be highly interactive and provide responses to the operator commands similar to those of an actual power system.

PSR training is a high-priority item in the industry and offers ample opportunity for further research.

For Further Reading

R.J. Kafka, *et al*, "System Restoration Plan Development for a Metropolitan Electric System," *IEEE Transactions on Power Apparatus and Systems*, Volume PAS-100, Number 8, pages 3703-3713, August 1981.

M.M. Adibi, *et al*, "System Operations Challenges," *IEEE Transaction on Power Systems*, Volume 3, Number 1, pages 118-126, February 1988.

Y. Kojima, *et al*, "Development of a Guidance Method for Power System Restoration," *IEEE Transaction on Power Systems*, Volume 4, Number 3, pages 1219-1227, August 1989.

D.S. Kirschen and T.L. Volkmann, "Guiding a Power System Restoration with an Expert System," IEEE/PES 90 SM 341-8 PWRS, 1990 Summer Meeting, Minneapolis.

SPECIAL CONSIDERATIONS IN POWER SYSTEM RESTORATION

A report by the Power System Restoration Working Group
M. M. Adibi, Chairman; L. H. Fink, Paper coordinator
Contributing authors are:
C. J. Andrews F. Arsanjani M. W. Lanier J. M. Miller T. A. Volkmann J. Wrubel

INTRODUCTION

This paper is one of a series presented on behalf of the System Operations Subcommittee with the intent of focusing industry attention on power system restoration issues.

Unconventional, somewhat unpredictable new uses of bulk power transmission systems, coming at a time when those systems are operating uncomfortably close to their design limits, are making systems more vulnerable to potential major disturbances, and resulting in renewed attention to the need for prompt and effective restoration following possible major system disturbances. This paper discusses a number of particular problems that require special attention in the development of system restoration plans.

EXCESSIVE ALARMS DURING RESTORATION
C. J. Andrews, ECC, Inc.

Over the years utilities have added a large number of substation alarms to the Energy Management System (EMS) to get better indication of equipment problems. Many of these alarms are activated by zero voltage on the transmission system, and hence flood the EMS during a system blackout. EMS alarm processing performance has not kept up with the increase in alarms, and during a system blackout the EMS may take several minutes, possibly an hour, to process all of the data from the substations. This processing not only delays data pertinent to the restoration efforts, but also affects the performance of the EMS man-machine interface. In order for the EMS to be an effective tool for system operators in monitoring and controling the transmission system during restoration, utilities must recognize and deal with the problem of excessive alarms.

Alarm processing systems in control center computers vary widely in implementation detail, but they all are fundamentally designed to detect abnormal conditions, annunciate these to the system operator, ensure that the alarm message has been recognized (usually by requiring the system operator to acknowledge the alarm), and create a more or less permanent record of the times at which the alarm condition occurred and was acknowledged.

Alarm processing is effective as long as most monitored values are within normal ranges or in normal states, allowing the abnormal values to stand out, but during the restoration process virtually every measured parameter varies outside its normal operating range. A major system collapse may begin with a single initiating event, accompanied by one or more alarms. In the minutes that follow, if the power system cannot be returned to a stable state, additional alarms will follow, indicating loss of generation, frequency decline, undervoltage, automatic underfrequency load shedding, and other protective relay actions. In addition to these primary alarms, secondary alarms based on calculated control center computer values - for example, line flow rates of change - occur. Following shutdown, the rate of new alarms slows, but periodic re-annunciations of persisting abnormal conditions may occur. Finally, as restoration proceeds, return-to-normal alarms occur, further taxing the capabilities of the alarm processing system.

The overall quantities of alarms in a shutdown and restoration scenario can be very high. Moderately sized control center computer installations may have 2,000-5,000 analog measurements, most of which will enter an alarm condition upon shutdown of the power system. Of the 10,000 or more two-state and multi-state measurements in a typical installation, several thousand may enter the alarm state due to relaying actions on generating units, lines, and loads.

Definitions of "Excessive" Alarms

A review of control center user comments related to the alarms that occur during major power system disturbances [1] leads to the conclusion that there are actually two definitions of "excessive" alarms:

1. More alarms than can be handled (either at the computer level or the system operator level) using generally available alarm processing techniques.

2. More alarms than are needed to meet operational objectives.

The problems associated with the two definitions require different solution approaches.

The Alarm Backlog

The problem that must be addressed involves the large number of alarms that were produced during the alert, emergency, and in extremis states [1] leading to the shutdown. These alarms may "clog" the computer control system to the point that the computer system is practically unusable for the first several minutes after electric system shutdown. These alarms typically must be acknowledged, deleted, or otherwise processed by the system operator. The alarms are excessive in the sense of Definition 1, because they stress both the computer system and the system operator. After a certain point in the shutdown, most of the alarms that occurred in the emergency state and shutdown become moot. Some are of value for after-the-fact analysis of the conditions that led to the shutdown, but the only alarms of value to restoration are those that can succinctly point to the cause or extent of the shutdown. Most of the alarms are therefore also excessive in the sense of Definition 2.

To prevent the alarm backlog from hampering the restoration process, some control center installations have implemented special alarm acknowledgement modes that acknowledge all alarms in a single action. The common past practice of making a printed record of alarms as they occur has sometimes become the limiting factor in processing the backlog; the current system approach of keeping only a magnetic permanent record has greatly alleviated this problem.

92 WM 106-5 PWRS A paper recommended and approved by the IEEE Power System Engineering Committee of the IEEE Power Engineering Society for presentation at the IEEE/PES 1992 Winter Meeting, New York, New York, January 26 - 30, 1992. Manuscript submitted August 30, 1991; made available for printing November 25, 1991.

Reprinted from *IEEE Transactions on Power Systems,* Vol. 7, No. 4, pp. 1419–1427, November 1992.

Levels of Alarm Importance

Because system shutdown and restoration represent extraordinary operating conditions, the usual definitions of alarm importance are of limited value for restoration. During restoration, alarms should be associated with one of the following three categories:

1. Currently critical - alarms that must be promptly responded to in order to ensure successful restoration.

2. Currently non-critical - alarms that might, under some conditions, require prompt system operator attention, but during restoration are of secondary importance. This is perhaps the category with the largest number of alarms.

3. Nuisance - alarms that are always of merely nuisance value to the system operator. Most control center support staffs will state that there are none of these, but many system operators would disagree.

The distinction of currently critical versus currently non-critical implies a need for an adaptive alarm processing approach. The alarm system, even if using expert system techniques or other filtering algorithms, must be able to detect (or be told) that the power system is in a restorative state in order to present the most valuable information. Alarms that are currently non-critical may be relegated to a log-only status in which they are only recorded for later analysis.

Automated detection of transitions from the emergency or in extremis states to the restorative state is a significant problem in its own right.

During restoration, much operation continues to be outside of "nominal" ranges; this gives rise to the observation that the alarm system should use a redefined set of limits during restoration in addition to performing filtering based on different alarm type and combination criteria during this phase. To control excessive alarm activity, the redefined limits should be, for the most part, relaxed relative to the non-restorative state limits, but some key monitored values might be monitored more closely during restoration to provide early warning that restoration actions are failing.

These techniques deal largely with the Definition 2 excessive alarms problem. They do not reduce the number of alarms that the control center computer must process, and so do not affect the Definition 1 problem from the computer point of view. They do, though, aid the Definition 1 problem by reducing the number of alarms that the system operator must process.

Critical Alarms to be Expected During Restoration

Of the alarms that may be expected during restoration, the following should be considered currently critical:

Analog alarms:

- Voltage - overvoltage when switching in a line.
- Frequency - frequency excursions due to load-generation mismatch.
- Loading - major transmission line and transformer overloads.

The redefined restoration state limits described earlier should be applied.

Status alarms:

- Switching devices - any uncommanded change of status.
- Relays - most relay actions. Some relay actions (e.g., certain distribution load relays) could be classified as currently non-critical.
- Generating unit status information - only the on-line/off-line status is unequivocally critical. Generating unit tripping and the off-nominal operation that may be required of generating units during restoration also tend to produce a large number of alarms associated with generating unit subsystems, such as auxiliary, governor, or voltage regulator systems. Most of these may be classified as non-critical from the system operator's point of view, on the assumption that the plant operator will be kept aware of these conditions by local monitoring equipment.

Return-to-normal for analog and status alarms should not, in general, be considered currently critical.

Return to Other Power System States

Although, as mentioned earlier, entry to the restorative state is difficult to establish in terms that the control center computer can unambiguously determine, return from the restorative to other states is even more difficult to identify. System operator personnel intervention to indicate the transition appears to be the only practical approach at this time. When this transition occurs, default definitions of criticality may be re-established and use of normal limits may be resumed.

Conclusions

Most of the approaches described in this paper for mitigating the excessive alarms problem from the system operator's point of view can be readily implemented within a typical alarm processing system, once agreement is reached on the criteria for declaring that the system is in the restorative state. The fast alarm acknowledgement scheme and redefined alarm ranges are straightforward extensions of current alarm system features. Defining an alarm as critical or non-critical is also a part of most alarm processing systems; being able to dynamically change this based on a restorative state indicator is not a difficult task. But establishing the values of the redefined alarm limits and performing the final categorization of alarms into currently critical and non-critical categories for the restorative state demand a deep understanding of the electric system and, most likely, detailed analysis and simulation.

A restorative state indicator is not a part of current control systems. No general guides for automatically determining the power system state are in use. The most practical approach is based on the system operator manually indicating the start and end of the restorative state. Some control center installations have already implemented alarm system enhancements that initiate different alarm presentation or acknowledgement strategies based on the rate of alarms. This amounts to a disturbance detector rather than a restorative state detector, but it is useful for preempktive control of alarm backlogs.

The practical capabilities of more comples inference processes, e.g., expert systems, to perform adequately during the stress of system shutdown and restoration remain to be fully demonstrated. The emphasis of most expert system alarm processing work that has been reported to date has been on normal, alert, and emergency state operation [2,3,4]. An extremely large rule base likely will be required for useful restorative state expert system alarm processing. A fruitful initial application of expert system alarm processing for restoration may be analysis of the alarms produced during the alert, emergency, and in-extremis states in an attempt to infer the cause and extent of the shutdown, as this information will guide the restoration process.

SWITCHING DURING RESTORATION

T. A. Volkmann, Northern States Power Co.

One of the first tasks in restarting a power system is to assess the status of the system. It is difficult to reconstruct the sequence of events leading up to a system blackout, and to establish what equipment is in service or out of service. The initial review of the system status may not present the system operator with an accurate indication of equipment serviceability. During the course of the events leading to the system collapse many transmission lines may trip due to out-of-step relay operation, or may temporarily sag into right-of-way obstructions. These facilities can be serviceable, and available for the restoration process. If the initiating event was weather-related, the equipment that appears to the system operator to be in serviceable condition - for example, closed breakers on the system - may indeed be faulted as the result of weather causing problems after the system collapse. Such possibilities mean that the system operator cannot use the status of transmission system breakers after the system collapse as an indication of which equipment is or is not faulted. The effort to ascertain this information ultimately will detract from the restoration activity. However, any positive verifiable failures must be factored into restoration.

Switching Strategies

There are two general switching strategies that may be used in restoration of the transmission system, "all open" and "controlled operation." The "all open" strategy can be accomplished by local substation operators or through EMS programs that open all breakers by substation as selected by the system operator. The "controlled operation" strategy is to control the configuration of the restoration by operating only equipment that is necessary to restore the power system. The advantage of the "all open" strategy is a much cleaner, clearer, and more straight forward configuration for the system operator. The system operator has only to determine what breakers to close, and does not have to be concerned with previously closed breakers. The disadvantage of this strategy is that all breakers have to be opened, creating a greater burden on the stored energy and DC control systems and also leaving several breakers in the open position for an extended period of time.

The advantages of "controlled operation" are minimum burden on the stored energy and DC systems, and breakers not used in the initial operation remaining closed. As breakers become inoperative due to loss of AC, the breaker can continue to be connected in the restoration process at a solid piece of bus. The disadvantages are the continual consideration of what breakers remain closed and the separation of the energized and de-energized systems. In either strategy there is an extensive amount of switching orders, either verbally to field operators or through the EMS. The advent of expert systems may provide a switching system that will greatly reduce the switching time in the restoration process. These tools will develop the switching procedure to accomplish various objectives. The EMS executes the switching procedure bypassing the conventional verification process that normally occurs in EMS SCADA control action, resulting in a 50 to 60% reduction in System Operator switching time.

Operability of Field Equipment

A major issue during system restoration is the operability of the control mechanisms of circuit breakers and motor operator disconnects. The control circuitry for this equipment is normally fed off the battery system and under normal operation is of low power requirements. For motor operated disconnects, however, the opening and closing power generally also comes from the battery system. Operation of these devices places a high power consumption on the station battery system. When motor operated disconnects are used in auto sectionalized schemes, consideration must be given in battery capacity design.

Thus, equipment operation during system restoration must be considered as a major determinent of substation battery size. Generally speaking, the station batteries are sized for a maximum expected operation of the equipment for a creditable substation disturbance followed by a certain period of moderate battery load. To insure adequate battery service in a system utilities must consider an operation of each device in the substation during the restoration process until an AC source is supplied to the station battery. If the substation configuration has a number of motor operated disconnects, consideration must be given to their operation as it affects battery capacity.

In today's transmission operation, utilities have a major dependance on the EMS for operation of substation equipment. In a blackout condition in restarting of a power system it is extremely important that the substation Remote Terminal Units (RTU) and associated communication equipment remain in service. The RTU to EMS communication system should be configured with all DC powered equipment.

Cold Weather Effects

Utilities in cold weather climates have to take additional precautions in their system restoration plans in order to deal effectively with cold weather conditions. Circuit breakers require AC power in two areas, to provide energy for operation of the circuit breaker, and to energize strip heaters to avoid cold weather effects that could cause failure of the operating mechanism.

Cold weather affects circuit breakers in two ways. First, contraction of the physical equipment may result in leaks in the energy storage reservoir used for re-closing of the circuit breakers. Secondly, several high voltage breakers use SF_6 gas as the arc interruption media; in cold weather, loss of AC power to their strip heaters may drop the SF_6 gas pressure to a level such that the breaker may be inoperative, or even may result in liquidation of the SF_6 gas (240-270 PSI). For those circuit breakers that use SF_6 under pressure for arc extension the liquidation point of SF_6 is normally in the range of 20 to 30 degrees Fahrenheit. With the loss of AC heaters to these breakers the time for SF_6 to liquify is on the critical path of the power system restart, and must be factored into the restoration plans. The liquidation of SF_6 is a grave problem, since strip heaters are designed only to maintain the SF_6 temperature, not to regasify liquid SF_6.

Another concern with cold weather is its effect on motor operated disconnects. Under cold weather conditions an increased amount of torque is required to operate these disconnects, resulting in a greater burden on the substation battery system.

In developing system restoration strategies utilities must consider substation equipment problems and the weather conditions in deciding which strategy to use: "all open" or "controlled operation". Both strategies have advantages in restarting of the power system, but utilities must be concerned about the adequacy of the substation equipment and system control center operation.

OPTIMAL SEQUENCING OF STARTING-UP GENERATING UNITS
J. M. Miller, Cajun Electric Power Coop.

While there are various objectives for a system restoration program, one way of stating the overall system restoration objective is to minimize the time required to restore service to 75% - 90% of the system load. There will likely be more units available than will be needed or than can be loaded. The initial objective will be providing some power source to as many plants as possible for equipment protection and preparation for start-up. Where systems have a mix of large and small generating units installed, returning some of the large units to service may be needed before the 75% - 90% system restoration level can be reached. The restoration program needs to be coordinated around the restarting of large units, after restoring enough load, transmission and smaller generating units to secure off-site power to each of the plants.

Procedure for Determining the Start-Up Sequence

The procedure for determining the generating unit start-up sequence should be prescribed ahead of time, before any system emergency actually occurs [5]. It can be developed in various ways, but the important point is for each system to develop its procedure before any severe emergency occurs. The actual sequence will depend upon the particular situation at the time of system restoration, but the procedure for determining the sequence should include certain key steps. The procedure for each system must take into consideration the unique ramifications of the system, but the following key steps should be considered.

1. Poll Plant Availability. Assess the availability of each plant to restart units that tripped off. The assessment should consider any special problems and failures that may have occurred on shut down.

2. Check plant availability with the data base; specifically, the maximum down times that will still permit hot restart, and the cranking power available or that could be made available to each generating unit.

3. Start the load-generation balance projections and tabulations. Also, start the Reactive load-generation balance projections and requirements.

4. Determine how many and which of the slow to restart units will be needed to achieve the system restoration objective of supplying 90% of the load in the least amount of time.

5. Plan the transmission restoration, cranking power provision and area load that must be connected to return the units to service.

6. Determine which of the units that could be restarted quickly would augment the restoration measures.

Supplying 75 to 90% of the load in the least amount of time will most likely require loading the larger, slow-restart units after some of the smaller ones have been reconnected. The system restoration procedure may need to be programmed around these units in order that enough load can be reconnected to permit returning the large units to the stable firing load range early in the start-up. This load range is at least 30% of the units full load capability and 50% of the full load is desirable.

Nearly all thermal units except nuclear units can be restarted hot if cranking power is available. As soon as the turbines start to cool down, the rise in inlet temperatures during restarting must be restricted to limit temperature differentials across the first stage. Steam generators likewise may have thermal stress limits. Fossil-fired steam generating units that can be restarted within ten minutes can be ramped up without waiting for temperature differential restrictions on the turbine and other components. Beyond twenty minutes, depending on the unit size and characteristics, the restart must be slowed to permit temperature equalizations. The maximum time that a dumped unit can be off the line and still be returned on a hot restart is about 30 minutes, but must be prescribed for each unit, or type of unit, and the information made readily accessible in the system restoration data base.

Nuclear units require special treatment. The NRC startup checklists generally do not permit hot restarts and their diesels would not be permitted to supply cranking power to other power stations. Nuclear units that were taken off the line on a controlled shutdown can be returned to service (synchronized) in about 24 hours; more likely 48 hours are required after a scram. While restoring off-site power to nuclear units requires attention, restoring service to nearly all of the load will normally have to be without the benefit of nuclear units.

Full system restoration could easily require more than a day and it is easiest to work with load figures relative to the weekly peak load. It is wise to determine the amount of generation required to supply two levels of the peak load that week: One approximately 90% to represent restoring nearly complete restoration, and another in the range of 75% which may be realizable much faster.

Some plants have black-start capability while most depend upon start-up power from the transmission system. Part of the system restoration procedure should concentrate on restoring enough transmission to provide start-up cranking power to the generating units. Hydroelectric generating units figure very nicely in the restoration plans of the systems fortunate enough to have them available.

Thermal units that are restarted must be loaded to minimum load within a reasonable time. This requires maintaining a load and generation balance sheet throughout the system restoration process. Units that cannot be provided with at least the minimum load required for flame stabilization should not be restarted. More generating units will generally be available for re-start than load. Considering small islands of load and generation, the load-generation balance is critical.

The reactive load requirements and generating unit reactive capabilities can be even more critical in dealing with small islands during system restoration. Voltage excursions on the transmission system risk relay operations that were set for integrated system operations. Reactive savings and voltage excursions imposed on the generators risk generator tripping and consequent load shedding.

The crux of generator sequencing is to determine which units will be needed to supply 75 to 90% of the system load. It is good to project the generating capacity that could be available in eight to 16 hours from the survey of unit availabilities and problems in step 1 above. This will require restarting the larger, slow starting generating units, with enough load and small generating units restored in the island in which each large plant is located that the large units can be held on once they are re-started. Companion papers address the need for assessing transmission availability, whether large units require sub-transmission, transmission or ehv line connections for the large units. Sequencing the generating unit start ups requires projecting load restoration and programming the transmission connections.

Data Base Requirements

Determining the appropriate sequence for returning generating units to service requires that the information that will be needed be readily available in the system restoration data base. Some of the information that should be included are:

1. Generating unit nominal characteristics, minimum load, ramp rate and reactive capabilities.

2. Load reject and black-start capabilities.

3. Normal and start-up transformer system connections. Cranking power MW and MVAr requirements of each unit and the system strength and configuration necessary for start-up.

4. The maximum off-line time permissible before a hot restart. This is generally on the order of half an hour, and may be as short as ten minutes, but the turbines on the 600 MW units in Table I remain hot enough for nearly 24 hours to permit hot restarts

5. The minimum and maximum times required from light off to roll, roll to synchronization, sync. to minimum load and minimum load to full load for hot and cold start-ups. Some systems prefer to define hot, warm and cold restarts; or hot, overnight, weekend and cold restarts; or normal and emergency restarts.

<u>Start-up Definitions</u>

Many of the essential terms are defined in the IEEE Standard Definitions in Power Operations Terminology, some of which are listed below. Others are defined below for convenience.

<u>Load Reject:</u> The action of isolating a generating unit or plant from the system load and leaving the unit to supply only house load.

<u>Black Start Capability:</u> The capability to start-up a generating unit without any power supply from the power system.

<u>Ramp Rate:</u> The rate at which a generating unit increases or decreases its output, usually expressed in megawatts per minute.

<u>Hot Restart:</u> A start-up with hot turbine metal temperatures.

<u>Cold Restart:</u> A slow start-up to limit the turbine metal temperature rise to stay within prescribed limits.

<u>Restart Durations for Unit Types</u>

Although generic start-up times are of limited value when system restoration requires the specific characteristics for the particular system, representative times are shown in Table 1. Similar types of units can require different start-up times. Providing a range of times would be either narrow and exclusive or too broad a range to be meaningful. The characteristics of several particular units are shown in Table I merely to indicate representative times. Some of the data appeared in reference 6.

UNDERGROUND TRANSMISSION SYSTEM CONCERNS

J. N. Wrubel, Public Service Electric & Gas Co.

<u>Physical System Concerns</u>

The primary physical system concern associated with underground transmission system restoration is the integrity of the oil (dielectric fluid) insulation system. The concern is that when all electrical power sources are lost to the pumping plants associated with high pressure oil filled systems, the pressure in this system drops rapidly as the cable system cools down from its previously loaded condition to a condition of no load. The results of this drop in pressure, if no other provision is made, is that fluid pressures at high elevations could decay to low values (approximately 100 psi) where gases would liberate from the fluid, forming gas pockets. It is also possible that a vacuum could develop inside the terminators and the line pipes. Re-energizing with existing gas pockets or vacuum voids could result in immediate electrical failure or damage which could lead to failures in the future.

At PSE&G, these concerns have been addressed in the following manner. First, all pipe cable pumping plants have

<u>TABLE I - STARTUP DURATIONS FOR SELECTED UNITS</u>

Size (1)	MW	20	92	110	235	275	550	500	600	800	936
Type (2)		CT	Drum	Drum	Drum	Drum	Drum	New SCOT	Drum	SCOT	BWR
Fuel		Gas	Coal	Gas	Coal	Oil	Coal	Coal	Coal	Coal	Nuclear
Duty (3)		Cycle	Cycle	Base	Base	Cycle	Base	Cycle	Base	Base	Base
Min Load (4)	MW	5	25	25	50	30	180	150	200	420	400
Req'd Off (5)	Hrs	0	4	N/A	6	4	N/A	N/A	N/A	Hrs	N/A
Hot Sync (6)	Hrs	0.1	2	1.5	6	1.5	4	0.9	4	4	48
Min Load (7)	Hrs	0	0.7	0.1	1	0.7	1	0.5	0.5	1.5	4.5
Cold Sync (8)	Hrs	0.1	6	6	7	7	5	12	12	16	48
Min Load (9)	Hrs	0	2	0.5	1	1.7	1	2	2	3	4.5
Full Load (10)	Hrs	0.1	1.1	1.5	2	1	1.5	2	2	1.5	10

<u>Notes</u>:

(1) Unit size is normal full load gross megawatts
(2) Either Drum type boilers or Super-Critical Once Through (SCOT)
(3) Either base load or peaking operation
(4) Minimum gross load for stable operation
(5) Minimum time required off line due to such things as turbine thermal stress or vibration limits if the unit is not successfully restarted hot
(6) Time required from light-off to synchronization for a hot restart
(7) Time from sync to minimum load on a hot restart
(8) Time from light-off to sync on a cold start-up
(9) Time from sync to minimum load on a cold start
(10) Time from minimum load to full load

provision for maintenance of oil pressure assuming no off-site power source is available. In some locations, an engine-generator set is on the property and that set will be started to provide power to maintain the pumps. At the other locations, a compressed air operated pump is provided which will maintain at least 100 psi pressure at the highest elevation of the cables for a minimum period of six hours.

The next provision for maintaining oil pressure is by installing pressure fittings on the air pumps which can accept a hose connection from a standard air compressor. In the event that it is not expected that off-site power will be restored to the station within the six hour window, arrangements will be made to drive a mobile compressor to the site to continue the air supply to the air pumps.

A third provision for maintaining pipe pressure is via the hydraulic network piping systems that allow circuits in the network to be hydraulically interconnected to other circuits through valving facilities at the pumping plants. Thus it is possible for one pumping plant to pressurize the pipes of other circuits connected to the same hydraulic network.

Should the pressure fall below the minimum allowable operating pressure, for every hour at this reduced pressure there must be a two hour period at the operating pressure before re-energizing the cable.

It should be noted that procedures two and three above require pre-planning and coordination to make sure that the required people and equipment are directed to the various pipe cable terminals in order to dovetail their activities with the electrical restoration of the power system and the pipe-type cables. The whole scheme can fall apart if the communication and coordination procedures are not in place prior to a system breakdown. These procedures are reviewed at PSE&G and made part of the restoration drills on the Pennsylvania, New Jersey, Maryland Interconnection, which are performed on a twice a year basis.

One area of concern which was uncovered during the research for this paper was that there are numerous pieces of underground facilities at many generating stations and substation switchyards which do not show up on System Operator prints. These include transformer, shunt reactor and generator leads. As a result of this effort, operator prints, display media and restoration procedures will be updated to include these facilities.

Power System Concerns

The key concern associated with underground transmission system restoration is the ability of the power system to absorb the charging current reactive power produced by a cable system upon reenergization. There must be enough connected generation, first, to absorb the reactive power and second, to keep voltages low enough to avoid overvoltage on the open end of the cable.

To address the first concern, one must know the charging MVAR of each cable to be energized. For those control center which have a state estimator or an operator's load flow, the charging reactive is part of the model of that element in the power system data base and should be readily available. If not, the System Operators should be provided with a document which shows the charging MVAR of each cable at nominal voltage.

To determine if reactive absorbing capability sufficient to absorb the cable charging is available on the on-line generators, system operators should have the capability curves or an equivalent tabulation for each generator, showing maximum absorptive reactive capability recognizing steady state stability concerns. A rule of thumb used at PSE&G is that there must be at least 2 MW of load being served for each MVAR of charging to be picked up.

If the state estimator/operator's load flow is able to run during the restoration process, a quick study case could be run to examine the feasibility of picking up each cable section. This case would also determine what the voltage would be at the open end of the cable to be energized.

In the case of power systems which have shunt reactors connected to or connectible to the cables via bus switching arrangements, the problem of energizing of the cables can be greatly reduced since only the surplus of cable charging over the amount of the shunt reactors is the net affect on the power system. This could greatly speed up the restoration process.

Before cables may be energized, system operations personnel must be in communications with underground systems personnel to be assured that the integrity of the insulation and oil system has been verified. Also, it greatly assists underground people if system operations identifies, as early as possible in the restoration process, the path or sequence system they intend to use during the restoration procedure. As stated above, underground systems personnel have a logistics problem to address to get knowledgeable people, compressors, and fuel for the compressors to the next two or three cables scheduled for reenergization in order to reestablish oil pressure and insulation integrity prior to reenergization.

Conclusion

As with most other areas of system restoration, the key to rapidly restoring underground transmission systems is to have a well thought out plan in place, good hardware systems which provide flexibility and second or third contingency provisions, and well trained personnel who know what resources are available and how to communicate and coordinate the utilization of these resources.

TELECOMMUNICATION CAPABILITIES AND LIMITATIONS

F. Arsanjani, M. W. Lanier, Virginia Power Co.

Although power system restoration plans may contain detailed technical information on electric operating procedures, they necessarily are critically dependent on the ability to communicate. Unfortunately, too often either little attention is paid to telecommunications availability, or it has been assumed that telecommunication systems will be fully operational during the restoration period, which is usually not the case. This paper will discuss telecommunications system capabilities and limitations during a power system restoration effort.

Public Communications

During widespread emergency conditions, the public switched telephone network (PSTN), including the cellular telephone network, may be inaccessible due to overload caused by public calls. Nevertheless, attempts should be made to establish and maintain communications through the PSTN in order to reduce the demand on any internal communications facilities which might be available. To maximize use of the PSTN during major emergencies, it is recommended that the following items be incorporated into a company's restoration plan:

- Essential Trunk Establishment - Most telephone operating companies have procedures to prioritize service to trunks designated as "essential." Customer local exchange trunks designated as essential can improve their chances for outgoing call capability by being queued ahead of non-essential trunks. Activation of this capability is not automatic and is only implemented when usually high traffic volumes threaten to overload a telephone company central office. It should be understood, however, that this priority to receive dial tone does not provide any assurance that calls can be completed.

o Establishment of PSTN Connections - Since it may be difficult and time consuming to establish PSTN connections, it is suggested that that responsibility be assigned to an individual. Upon going "off-hook," a line may appear dead (no dial tone) which will mean that the telephone company central offices (CO) are overloaded and cannot process any additional calls. It is important that the responsible individual(s) be instructed to remain off-hook and wait for dial tone as long as is needed. Hanging-up the phone and trying again will take the call out of queue with the CO equipment, so it is best to keep the telephone off-hook and wait. A dial tone will be heard as soon as the CO is able to process a call. Once a call is established, it is suggested that it be maintained until it is certain that no further voice contact is necessary. Furthermore, it is suggested that the responsibility for establishing these calls to a predetermined list of stations be distributed to multiple company locations to increase the chances of receiving a dial tone.

Internal Communications

Private Telephone Network. During prolonged electric outages, telephone company equipment may cease to operate. Therefore, internal company telecommunications systems will likely be the only dependable means of communication. More specifically, for companies having their own telephone network via private transmission systems (microwave and fiber optics), the most convenient and reliable form of communications for the restoration coordinator will be "off-premises extensions" OPX's) and/or automatic ringdowns (ARD's) from a private branch exchange (PBX) telephone system. A tie-trunk network can also be used to dial locations which are incorporated in the network directly or assist in contacting remote locations by dialing through network stations to minimize dependence on the telephone company equipment.

Private Transmission Networks. The availability of OPX/ARD circuits and a tie-line network are dependent upon the company's telecommunications transmission (fiber optics and microwave) system and its associated channel equipment. Back-up battery systems and/or generators are usually the limiting factor in terms of time that these systems will be available upon loss of AC; however, there are certain non-power related catastrophic situations which could also impact the operation of telecommunications transmission system.

Therefore, systems should incorporate high reliability and dependability design factors which include:

- redundancy of critical equipment
- eight hour minimum back-up power for backbone sites
- availability of spare equipment for repair
- alarms for major electrical equipment failure, loss of AC/DC power, and extreme environmental conditions
- redundant media transmission (fiber optic and microwave) linking strategic locations
- system configuration to allow alternate routing of circuits

High Frequency Single Side Band Emergency Radio System. If direct communications cannot be established by company or public telephone, a High Frequency Single Side Band (HF SSB) Emergency Radio System can serve as an alternative. The HF SSB system can provide emergency back-up communications to critical power system locations. HF radio systems will communicate over long distances, but it is highly affected by atmospheric conditions. These effects are not specifically predictable for a given frequency, therefore it is recommended that this type of system be a selectable multi-channel system. Back-up power for each radio is supplied by the most dependable means available at each location including generators, uninterruptible power supply, and substations batteries. In addition, the radios can be powered by 12 VDC vehicle batteries.

Two-way FM Land Mobile Radio System. As a last resort, or to locations where no other communications means is available, communications can be established using a two-way FM land mobile radio system by forming a "chain" (voice-relay) of vehicles equipped with radios operating on the same frequency. Personnel in these vehicles will relay the information to the next vehicle. This method of communications will be available as long as radio equipped vehicles (with satisfactory batteries) are available. The spacing of mobile-to-mobile (or portable) radios is dependent upon frequency, terrain, and environmental conditions and appropriate spacing must be determined experimentally. Typically spacing is as follows:

UHF mobile-to-mobile	5-10 miles
VHF mobile-to-mobile	15-20 miles
UHF portable-to-portable	0-1 mile
VHF portable-to-portable	0-3 miles

Conclusion

The capability to communicate under adverse conditions is extremely critical to power restoration efforts. Utility contingency plans should include references to telecommunications systems such as the PSTN and any available internal systems. Audits of these systems, including a complete back-up power check per location, should be completed to note their capabilities and limitations. In addition, non-telecommunications factors such as the availability of adequate personnel and vehicles and the availability of fuel for generators and vehicles will affect a company's ability to communicate and should be considered in restoration planning.

REFERENCES

1. L.H.Fink, K.Carlsen: Operating Under Stress and Strain; IEEE Spectrum, March 1978, pp 48-52

2. W.R.Prince, B.F.Wollenberg, D.B.Bertagnolli: Survey on Excessive Alarms, A Report Prepared for the Power System Control Centers Joint Working Group (WG 73-3), J.L. Scheidt, Chairman; IEEE TRANS. v.PWRS-4, n.3, August 1989, pp 950-956

3. B.F.Wollenberg: Feasibility Study for an Energy Management System Intelligent Alarm Processor; IEEE TRANS. v.PWRS-1, n.2, May 1986, pp 241-247

4. D.B.Tesch, D.C.Yu, L.Fu, K.Vairavan: A Knowledge-Based Alarm Processor for an Energy Management System; IEEE TRANS. v.PWRS-5, n.1, February 1990, pp.492-98.

5. IEEE Committee Report, Adibi, M. M., et al, "Power System Restoration - The Second Task Force Report", IEEE Trans. on Power Systems, Vol. PWRS-3, No. 4, PP 927-933, Nov. 1987.

6. R.J.Kafka, D.R.Penders, S.H.Bouchey, M.M.Adibi: System Restoration Plan Development for a Metropolitan Electric System; IEEE TRANS. v.PAS-100, n.8, pp. 3703-13, Aug. 1981.

DISCUSSION

R.F. CHU (Philadelphia Electric Company, Philadelphia, PA), C.C. LIU and K.L. LIOU (University of Washington, Seattle, WA): We thank the working group and the authors for identifying various restoration issues and possible solutions. Based on the discussions in this paper and findings in our restoration project, we have concluded that:

1. tools are needed to assist operators in restoring the system, and
2. problems associated with many of these issues can be avoided if they are recognized and adequately addressed by the developed tools.

In the on-going EPRI restoration project (RP3104-1), we have developed a restoration expert system (RES) prototype [A] to assist operators in restoring the system. The RES consists of:

1. Generation capacity management - for identifying the start-up sequences based on information on available generation.
2. Path finding algorithm - for generating a feasible path for the start-up sequence.
3. Load restoration procedures.

Many potential restoration problems are formulated as system constraints which must not be violated by the RES generated solution or procedure. Following the RES recommendation, the operators should be confident that many restoration problems can be avoided.

We have applied the RES to provide a feasible start-up sequence for the units given by J.M. Miller. To match the data format for the RES, the data in Table 1 of the paper have been modified and the following assumptions were employed:

1. The nuclear unit is excluded.
2. The hot restart ramping curves are simulated.
3. Between each pair of units, there is at least one transmission path which can be used for cranking.
4. Start-up requirement of a unit is 3% of its capacity.

The modified characteristics of these units are shown in Table A.

Table A. Unit Characteristics

Unit	Type	MW Cap. (MW)	Crank to paral. (hr)	Paral to Min (hr)	Min to Full (hr)	Min load (MW)	Crit. min. int. (hr)	Start req. (MW)
1	CT	20	0:06	[illegible]	0.06	5	N/A	0
2	Drum	92	2:00	0.42	1:06	25	4:00	3
3	Drum	110	1:30	0:36	1:30	25	N/A	3
4	Drum	235	6.00	1:00	2:00	50	6:00	7
5	Drum	275	1:30	0:42	1:00	30	4:00	8
6	Drum	550	4:00	1:00	1:30	180	N/A	17
7	SCOT	500	0:54	0:30	2:00	150	N/A	15
8	Drum	600	4.00	0:30	2:00	200	N/A	18
9	SCOT	800	4:00	1:30	1:30	420	N/A	24

Note that the units, described by J.M. Miller, which, after a certain period, will become "cold" and require a longer time to restart are defined as units with "critical maximum intervals" in Table A. The units which require minimum off-line periods due to turbine thermal stress or vibration limits prior to restart are defined as "critical minimum intervals" units.

The RES start-up sequence is given in Table B. The RES suggests to start the only black-start unit, Unit 1, first. Once Unit 1 can supply sufficient cranking power, Unit 3 and Unit 7 are started. Then, Unit 7 supplies power to crank Units 6, 8, and 9. Units 2 and 5 have critical minimum intervals of 4 hours and Unit 4 has an interval of 6 hours. They will start after the intervals expire and sufficient cranking power can be provided.

Table B. Suggested Generator Start-up Sequence.

UNIT	CRANKING UNIT	START TIME
1	(Black-start)	0:00
3	1	0:10
7	1	0:20
6	7	1:20
8	7	1:30
9	7	1:30
2	7	4:10
5	7	4:10
4	9	6:10

At 1:20, the output of Unit 7 has ramped up to 30 MW which is sufficient to crank either Unit 6, Unit 8, or Unit 9. Under this situation, present RES rule gives priority to the unit which can produce more MWh in the next 60 minutes (this number can be adjusted.) However, if the projected outputs of the units are identical as in this example, RES will give priority to the one (i.e. Unit 6) based on the order of the input data. Note that if RES first selected Unit 9 (the one with the largest capacity as suggested by J.M. Miller), the system generation capability curve is slightly better and reaches the maximum capacity 10 minutes sooner. We have added this suggestion to the RES knowledge base and we thank J.M. Miller for the suggestion.

The two generation capability curves for the two start-up sequences are shown in Figure A. This figure provides a good projection of the available generation during the restoration period. With a dynamic data base, the RES is a very useful tool to assist operators in determining a feasible start-up sequence at the start of the system restoration procedure. As indicated in the paper, information such as Figure A and Table B are valuable to the operators at the start of system restoration.

Figure A. System Generation Capability Curves.

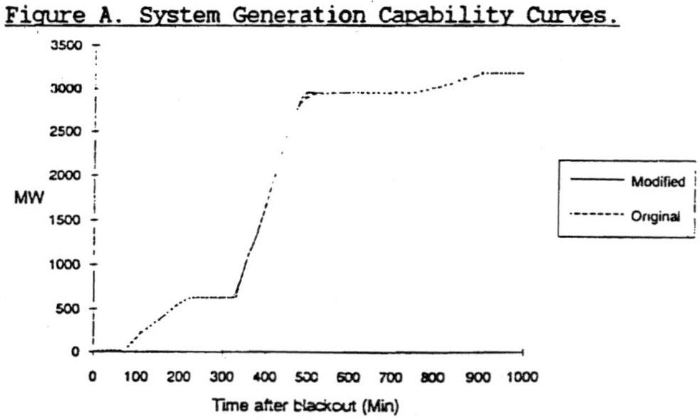

Presently, the system constraints for the RES are heuristic in nature and require little simulation. We are adding to the RES a dynamic data base similar to the one available at the EMS. To anticipate the system response based on a given RES recommendation, power system models (including relay models) and simulation capability (programs to simulate stability, thermal overload, reactive balance, switching transients, load/frequency, etc.) are also being added. These added features will allow the RES to test the next restoration step before making the recommendation.

Referred to C.J. Andrews' discussion on excessive alarms, the final version of the RES will recommend restoration action which will avoid the occurrence of excessive:

1. Analog alarms due to transient overvoltage, frequency excursions, overloads, etc.

2. Status alarms due to relay action and generation tripping due to voltage and frequency problems.

The RES employs the "all open" switching strategy. To cut down the number of feasible paths to be searched by the RES, a default path between any two generating units is included in the RES data base. With reference to J.N. Wrubel's concerns, critical cable systems, if proven to be feasible from off-line simulation, can be incorporated in this default path and the associated time constraints be considered by the RES.

Finally, our main objective is to develop tools to assist operators to avoid as many as possible potential restoration problems. It is critical that we do not overlook any major problems. Hence, we thank the working group and the authors again for providing us a list of these problems and solutions.

Reference

[A] C.C. Liu, K.L. Liou, R.F. Chu and A.T. Holen, "Generation Capability Dispatch for Bulk Power System Restoration," 1992 IEEE PES Winter Meeting Paper no. 92 WM 108-1-PWRS.

Manuscript received February 10, 1992.

M. M. ADIBI, C. J. ANDREWS, F. ARSANJANI, M. W. LANIER, J. M. MILLER, T. A. VOLKMANN, J. N. WRUBEL: The authors wish to thank the discussers for their comments. Messrs. Chu, Liu and Liou point out that system operators should be assisted by an expert system during restoration. They describe such a restoration expert system (RES) in their paper, refer to it in their discussion, and show its capability by finding a start-up sequence from data presented in our Table I. In our paper we have cautioned the reader that the data in Table I are "merely representative timings" and that the actual timings, even for similar units, greatly depend on the specific characteristics of the particular POWER SYSTEM AND ITS PREVAILING operating condition.

We have pointed out five potential problem areas that require special considerations during restoration, whether it be manually directed or computer aided. We agree that the RES would help avoid excessive alarms, and could include special switching configurations. We pointed out that system restoration procedures should not be dependent upon a single communication medium and caution that reliance upon a restoration expert system should not increase that dependence.

We also agree with the discussers that known constraints on underground cables can be included in the determination of a cranking path to non-blackstart units. However, a word of caution: in the event that there is a failure of a pumping plant, or an engine generator fails to start, the constraints known before the blackout are no longer valid. There must be a convenient and rapid way to update the knowledge base of constraints <u>DURING</u> the restoration to update assumed constraints with actual conditions found during restoration.

In general, the RES that was developed by the discussers for Electric Power Research Institute (EPRI) appears to be promising. We encourage the discussers to continue with its development.

Manuscript received April 9, 1992.

NEW APPROACHES IN POWER SYSTEM RESTORATION

A report by the Power System Restoration Working Group
M. M. Adibi, Chairman; L. H. Fink, Paper coordinator
Contributing authors are:
J. Giri D. S. Kirschen S. M. Shahidehpour J. Zaborszky

INTRODUCTION

This paper is one of a series presented on behalf of the System Operations Subcommittee with the intent of focusing industry attention on power system restoration issues.

Renewed, intensive interest in the problem of power system restoration is coming at a time when new technologies are providing powerful new capabilites in areas such as large scale system analysis, communication, data management, artifical intelligence, and allied disciplines. Full advantage should be taken of such an opportune coincidence, but in a responsible manner. The following discussions explore some of the possibilities that are opening up for more power system restoration strategies and practices.

SYSTEM-WIDE COORDINATION AND OVERALL ORGANIZATION OF RESTORATION

John Zaborszky, Washington University

The coordination and overall organization of the several aspects of the restoration process will largely determine the effectiveness of the restoration and its speed of progress. The most basic determinants of this organization are the relative roles played by the operator and the computer. Increasing sophistication in analysis and computational programs, including those derived from artificial intelligence, raises the computer to a significance somewhat comparable to that of the operator. Thus there arises a choice between three basic organizing principles:

1. Automated restoration, computer dominated;
2. Computer aided restoration, operator dominated using interacting programs (possibly artificial intelligence type);
3. Cooperative restoration, an operator-computer partnership with the leading role shifting between them.

Fully automated restoration with the operator playing the role of an observer or supervisor has not been proposed, and its feasibility for any future time is doubtful because of the extensive and time consuming searches (whether conventional or expert system) needed for decision making by the computer, and because of the presence of the operator, the resident expert.

Computer aided restoration in various forms is the state of the art. The operator takes responsibility for the restoration step by step (Fig. 1), and the control loop is closed though him directly. His decision making is helped by interactive computer programs (e.g. load flow) which he can interrogate, and by a restoration manual (which may be stored in the computer) which he studied as part of his training and with which he is therefore expected to be familiar. The restoration proceeds piecewise, with the operator building the system back from those pieces (islands or neighbors) which are still operating.

In cooperative restoration the operator and the computer can be in the loop jointly (Fig. 2), alternating the leadership role between them so that each performs those tasks where it is more efficient. The restoration then may proceed with a broader perspective in planning future steps and then executing them. [1]

In organizing the restoration based on any of the three schemes, the relative strengths and weaknesses of operator and computer must be considered. The operator is the resident expert on his system. His ability to change his system from one operating condition to another is unsurpassable, provided that he knows the current and the target conditions. However, his predictive ability for selecting target conditions is limited in time range and in judging the combined effects of more than one specific change. Furthermore, since the operator can do one thing at a time, his conduct of the restoration is inherently serial. The computer, on the other hand, can effectively predict the combined effects of several changes over a wide section of the system, given the proper database and software. Consequently it can effectively chose target conditions over a relatively long time horizon. Taking the system from one operating condition to a specific target condition, however, requires selecting a sometimes complex sequence of intermediate steps. Such selection may be done more efficiently by the operator using his expert knowledge of his system. For the computer, multiple choice decisions mean searches (directly or organized into expert systems) that are time consuming when the numbers of decisions are large as they are in restoration. Thus operator and computer have compensating strengths and weaknesses which should guide the organization of restoration in whatever scheme, including any of the three organizing principles listed above [1].

Whichever organization principle is followed, the restoration process inherently breaks up into three tasks as indicated in Fig. 3. (Clearly both Fig. 1 and 2 fall into this pattern.) These are:

a) Database building
b1) Target selection
b2) Planning of steps to reach the target
c) Implementation

Most of the necessary database already exists in SCADA-EMS installations, but extensions are needed in order to make intelligent plans for the future. Such items should be added to the existing data base as availability status of currently cold generators and of the crews to start them; loading schedules of these generators for cold start; reenergization time of equipment in maintenance; time requirements to reach an unmanned substation or a manual switch in the yard; estimated load with corrections for the effect of the disturbance; priority rankings of various equipment and load components; schedule for the immediate future of help available from neighbors, and the associated transfer points. All information transfer, as far as possible, should be from computer to computer and available in the database prior to the occurance of any disturbance. A fast and secure communication system,

92 WM 109-9 PWRS A paper recommended and approved by the IEEE Power System Engineering Committee of the IEEE Power Engineering Society for presentation at the IEEE/PES 1992 Winter Meeting, New York, New York, January 26 - 30, 1992. Manuscript submitted August 26, 1991; made available for printing December 11, 1991.

Reprinted from *IEEE Transactions on Power Systems,* Vol. 7, No. 4, pp. 1428–1434, February 1992.

where all transmitters have emergency power supplies (batteries) that would last through the restoration period, is essential to any restoration scheme and certainly for maintaining a reliable database. Database requirements are largely the same regardless of how the planning task is organized.

The planning task (Figure 3) will differ significantly for the different organization principles. However, one necessary element always is choosing for the next restoration stage a target which is judged feasible by operator and computer. The step itself can be narrow or broad. In computer aided restoration (Fig. 1), the operator picks the target for the next move which he feels (based on his expertise, the instruction book and the system status represented in the database) will be feasible, and confirms his judgment by the use of available interacting computer programs. There is an implied time duration for reaching each target. In this type of operation the individual moves will be of limited scope (pick up some loads, reconnect a transmission line, synchronize a machine or an island, cold start a generator for later use). Also the scope and duration of the moves will vary widely between consecutive moves, which will have to be serial at least within the range of activity of one operator. If different operators are in charge of different system segments, coordination needs to be maintained, usually on the telephone, which is time consuming and conducive of errors by misunderstanding even if the telephones are designed for uninterrupted operation during disturbances. These summarize the major overall organizational concerns and their limitations as they apply to target selection in computer aided restoration. Planning the steps to reach the target will be relatively direct and straightforward in this approach, because of the relatively limited scope of the targets. Essentially, the steps to reach the target will be in the operator's head at the time when he selects each target.

In cooperative restoration (Fig. 2) a more direct cooperation between computer and operator allows broader objectives and tighter organization. For instance, a uniform time horizon (say 20 to 40 minutes) can be set for each consecutive restoration stage. The computer can be charged with generating successive system wide targets for these successive time horizons using the prescribed data base. These targets should represent the most complete prioritized restoration possible using all equipment expected to be available at the particular horizon time. This constitutes target selection. Such computations are effectively done by the computer, but it would take excessive computer search time to establish the numerous intermediate steps leading to the target. So the operator then takes over the leading role. As the resident expert he can readily propose these steps if he knows the target condition and trusts the computer that the target will be viable if reached. All these operations can be easily paralleled and staggered in time. The target for the second next horizon time can be computed while the target for the next horizon time is being implemented [1]. Operators or teams of operators can work in parallel. The computer can check steps proposed by the operator, for consistency.

Carrying out the implementation of the target condition can again vary widely depending on the organizational principles used. For effective progress this task should be automated as much as possible. While many steps could be directly commanded by the computer, some will need to be done manually and even after driving to a substation. The computer must monitor the progress of the implementation of the plan to reach the next target, to check for adherence to the schedule and for errors. In case of time lapses or errors, prearranged programs are needed to modify the target in time.

There are numerous algorithmic means possible to generate the software for organizational schemes such as the ones described. Special considerations are indicated however because restoration is a special situation with exceptionally large amounts of software which must be exercised under heavy time pressure and using database information of less than the usual reliability. This indicates the need to develop special algorithmic approaches with limited complexity but with sufficient accuracy [2]. Bulky precision approaches along the lines or in the spirit of optimal load flow have little utility for restoration. Finding algorithmic starting points, which can be established under disrupted conditions and which will help the new algorithms to converge, need special consideration. Naturally all algorithms must take into consideration all applicable limits, including availability limits such as the loading schedules of cold start generators. Priority status of loads must be considered.

Measuring the progress of restoration is of obvious importance, both for tracking its status by the operator and for selecting targets with the computer for effectively pursuing it. Restoration of a large power system is a very complex process, so the quick adoption of some common type of performance measures is probably not the way to go. The measure of progress that is selected should consider such aspects as generation shortfall, shortage of load coverage, and transmission bottlenecks. The priorities of the components in the restored system must be considered. The form of the measure should give a realistic and intuitive feel for the margin to full restoration [1] and should lend itself to the designing of special algorithms to close the margin as fast and with as little computation as possible [2].

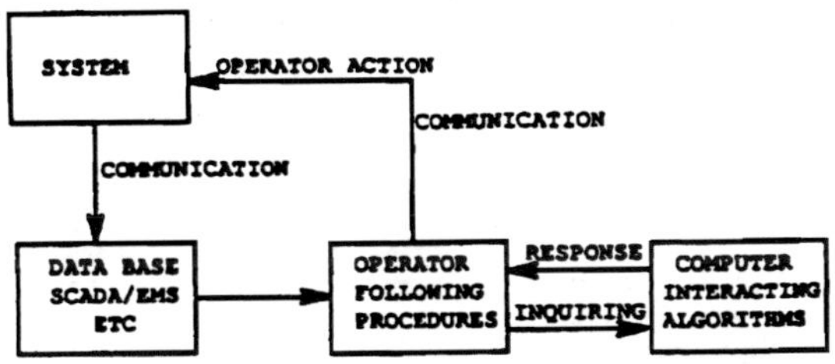

Figure 1. Computer Aided Restoration

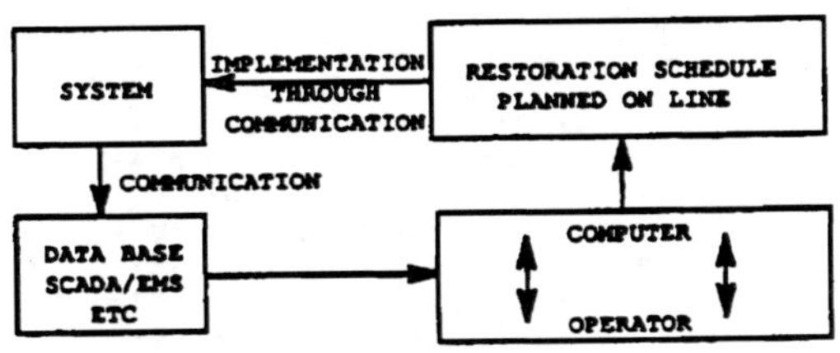

Figure 2. Cooperative Restoration

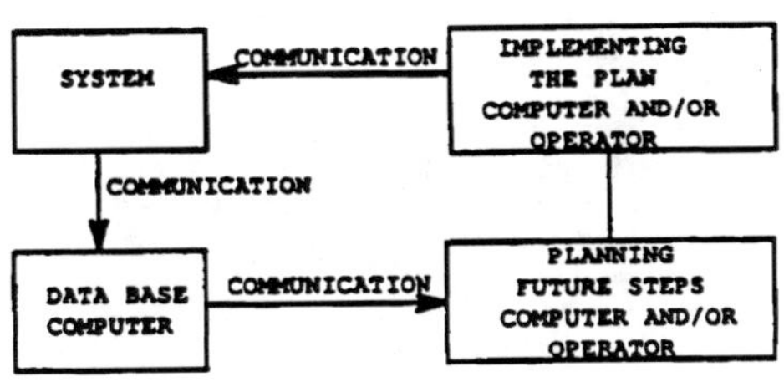

Figure 3. The Three Basic Functions of Restoration

Economics must be considered in any engineering process, and certainly in restoration. Although rare, large blackouts are very expensive in dollars as well as in public goodwill and political repercussions. Designing and developing a computer aided or cooperative arrangement for restoration is a major project, requiring extensive software development and personnel training, so it also would be expensive. Good judgement and analysis of the economics is clearly called for.

EXPERT SYSTEMS FOR POWER SYSTEM RESTORATION

S. M. Shahidehpour, Illinois Institute of Technology
Daniel S. Kirschen, Empros Systems International

Introduction

As power systems continue to grow in size and complexity, the amount of time required to restore power to customers in the aftermath of a blackout reaches a level which produces unacceptable economic and social costs for an industrialized society. Currently, the restoration is performed step-by-step and mostly manually, based on pre-established guidelines and operating procedures. One of the avenues which is being explored with the intent of finding ways to reduce the duration of the blackouts is the application of expert systems designed to help the operators select, schedule and execute restoration actions which best suit the actual status of the power system.

In mathematical terms, the restoration problem may be described as a multi-objective, multi-stage, combinatorial, nonlinear, constrained optimization problem. Unfortunately, the complexity of such a problem currently precludes the development of a general method which would determine a series of restorative control actions for a sufficiently wide range of operating circumstances. A more heuristic approach which would reflect a human operator's approach to the problem thus appears preferable. Such an approach would rely on expert systems and other artificial intelligence techniques and would be part of a package which would help system operators recognize bottlenecks in the operation of the system, minimize further system degradation, and develop strategies for providing a reliable service to the utility customers.

The major steps in the development of an expert system are the compilation of the knowledge required to solve a specific problem, the identification of the process that a human expert undergoes to determine appropriate control actions, the derivation of governing rules for the operation and control of that process, and the implementation and testing of these rules on practical situations. The quality of these rule-based approaches depends, to a great extent, on the flexibility of the model used to represent the actual power system and the closeness with which the rules mimic the problem solving techniques of an expert operator. However, the phenomenal advances in hardware and software which have taken place in recent years have made it possible for expert systems to analyze a much wider range of possibilities in a much shorter time and with much more attention to details than even an experienced operator could ever hope to achieve. Expert systems thus have the potential to identify on-line the restoration strategy best suited to the situation at hand, to quickly develop plans showing how this strategy can be implemented and hence to minimize the time required for system restoration.

The first application of expert systems in power system restoration was introduced by Sakaguchi [3], in which two different types of computations (symbolic and numerical) were considered for a specific case of system restoration. This paper, though limited in its context, was regarded as a landmark by system restoration planners who were searching for a viable approach to assist system operators in critical moments of decision making. Later, Liu [4] extended this approach to distribution systems. The distribution system is generally radial and has fewer constraints, so that the restoration process is more manageable. To restore a maximum number of zones, several steps are followed which include, group restoration, zone restoration, and if necessary, load transfer. Taludkar and others [e.g. 5,6] introduced a systematic approach for developing an intelligent operator's assistant using an expert system. This concept is particularly useful in the power system restoration context. Kojima [7] discussed the realization method for bulk power system restoration. A bulk power system consists of loop and radial systems. In this study, fast, safe, and flexible alternatives are generated automatically without any interactions with the operator. The restoration is decomposed into an energizing phase and a load reconnection phase. Kirschen [8] developed a prototype system for reducing the amount of time required to complete the first phase of restoration. During this process, a skeleton network which links the generating stations and the main substations is reenergized. Nadira [9] formulated the problem as a multi-objective dynamic optimization problem. In order to deal with this problem, a hierarchical time framework is proposed which is basically a temporal time decomposition.

Implementation

Taken individually, none of the tasks which must be performed during the restoration of a power system is fundamentally difficult. However, many constraints must be respected and many conditions must be checked repeatedly as these tasks are executed. It is widely agreed that expert systems can perform these types of activities faster and more reliably than humans, especially when the humans are under stress.

The nature of the restoration problem is such that instead of thinking in terms of a single expert system capable of handling every aspect of the restoration, one should rather design an environment composed of various knowledge-based tools. The operator could ask for advice on a particular question from a specialized expert system or could delegate the execution and the monitoring of a simple task to another system. The expert systems should thus be designed as computerized assistants to the operator. Like all good assistants, they should require as little instructions as possible, stay out of the way of the operator unless called upon, and last but not least, leave the final decision to the operator. The operator should indeed not have to waste time entering data beyond the minimum required to specify unambiguously what is expected from the expert systems. All the digital and analog status data should be obtained directly from the utility's Energy Management System (EMS). Furthermore, integration with the EMS allows the expert systems to base their recommendations on the actual status of the power system and to issue commands directly through the SCADA subsystem of the EMS. The design of the expert systems should be such that the operator should have the ability to ignore their recommendations and bypass their use if desired. This appears simple but it implies that the expert systems must be able independently to track the evolution of the power system.

Most researchers have concentrated their attention so far on the development of expert systems for re-establishing a skeleton network. The remainder of this section will therefore be devoted to a discussion of this problem.

Many utilities have developed written procedures which describe one or more paths along which this skeleton network can be reconstructed. One could thus very easily implement an on-line version of these procedures. However, if unforeseen difficulties arise, this approach would either fail to provide a recommendation or would suggest an unworkable and possibly dangerous course of action. It is thus preferable to design an expert system which can assess the situation and make recommendations based on the actual status of the power system. The fact that such an expert system constructs switching plans from scratch has two important consequences:

- o the system must have the ability to read and manipulate a "road map" of the power system.
- o a method must be devised to avoid a combinatorial explosion in the search for a feasible path.

A "road map" of the power system usually takes the form of an object-oriented description of the various equipments comprising the power system. For efficiency reasons, it is desirable to organize this data in a hierarchical fashion by grouping devices in functional groups. Such a description can be constructed on the basis of the data used for power flow computations.

This hierarchical description of the power system facilitates a two-level approach to the search for feasible paths. At the top level, paths are described in terms of stations and transmission lines. The operator interacts with the expert system at this level to select the direction which will be followed during the reconnection of the skeleton network. Once a decision has been made concerning the next major step in this reconnection, the chosen path is analyzed at the lower level which develops a complete description of the path in terms of switching operations. The operator then has the opportunity to review and possibly modify this detailed plan. This approach allows a separation between strategic and tactical considerations and considerably reduces the amount of searching that must be performed.

A few prototype systems based on these principles have been described in the technical literature [8,10,11]. Significant progress has already been made in the design of object-oriented representations of the power system and in the development of search techniques which provide the ability to adapt to an unforeseen problem such as a faulted line, a disabled station or a blocked breaker. An important benefit of these systems is that they provide a sophisticated graphical representation of the restoration paths. These graphical displays make it possible for the operator to quickly grasp the status of the restoration, both at the system level and at the substation level.

Direction of Future Work

Expert systems develop their recommendations on the basis of logical computations. Numerical computations should also be used to verify that these recommendations will not cause a violation of an operating constraint. An integration of the expert systems with analytical tools would ensure that only feasible recommendations are presented to the operator. For example, the expert system responsible for suggesting reconnection paths should have the ability to call a power flow program to check that a particular path will not cause an overvoltage condition. If the results indicate that the voltage will be excessive, the expert system could then modify the path it recommends to include an inductor or enough load to bring the voltage within the acceptable limits.

The prototypes which have been developed so far have all operated in a standalone fashion. They should be integrated with power system simulators to obtain a more complete verification of their capabilities. This combination of simulator and expert system tools should also be used for restoration drills to evaluate the overall effectiveness of these tools.

While most of the systems described in the literature were designed for implementation in an on-line environment, it should be possible to adapt them for planning or operational planning applications. They provide a convenient vehicle for quickly simulating and evaluating restoration strategies and for verifying that a restoration can actually be performed in a reasonable amount of time given a set of scheduled outages.

Summary

The characteristics of expert systems applied to power system restoration can be summarized as follows:

- o These systems should function as assistants to the operator, leaving the operator responsible for the major decisions.
- o They provide a systematic procedure for making rapid decisions in stressful situations and for a wide range of circumstances.
- o Logical reasoning assisted by algorithmic programs can be used to identify and prioritize control actions.
- o Expert systems integrated within an EMS environment have the potential for significantly reducing the amount of time required to restore service to all of the utility's customers.
- o Expert systems could also be used to plan and test restoration procedures and strategies.

OPERATOR TRAINING SIMULATORS IN POWER SYSTEM RESTORATION

Jay Giri, ESCA Corporation

Introduction

In actual power system operation, severe restoration situations occur infrequently. Therefore, when they do occur, operators have had little first-hand, on-the-job experience. Even so, they are under pressure to restore the system as quickly as possible, while paying close attention to many technical details. They must draw to the utmost from their base of knowledge, experience, and abilities in order to safely, quickly and efficiently restore service.

Operator training simulators provide an effective means by which the operator can be trained to deal with restoration situations in a secure, controllable, offline simulation environment. This training will prepare them to manage a real-life restoration situation in a more prompt and less stressful manner [12].

Operator Training Simulator Overview

Operator training simulators provide a realistic environment for operators and dispatchers to practice normal, everyday operating tasks and procedures, as well as to experience emergency operating situations. The various training activities can be safely and conveniently practiced with the simulator responding in a manner similar to the actual power system [13,14]. The OTS can be used in an investigatory manner to recreate past actual operational scenarios and to formulate system restoration procedures. Scenarios can be created, saved, and re-used. The OTS can be used for evaluating the functionality and performance of new real-time EMS functions, and also for tuning AGC in an offline, secure environment.

A typical operator training simulator consists of the following main subsystems:

1. Energy Control System (ECS): This subsystem emulates the normal EMS functions in the OTS and is the only part of the OTS with which the trainee interacts. It consists of the Supervisory Control and Data Acquisi-

tion (SCADA) system, generation system and all other EMS functions.

2. Power System Dynamic Simulation: This subsystem simulates the dynamic behavior of the power system. System frequency is simulated using a long-term dynamics system model, where frequency of all units is assumed to be the same. The prime mover dynamics are represented by models of the units, turbines, governors, boilers, and boiler auxiliaries. The network flows and states (bus voltages and angles, topology, transformer taps, etc.) are calculated at periodic intervals. Relays are modelled, and they emulate the behaviour of the actual devices in the field.

3. Instructional System: This subsystem includes capabilities to start, stop, re-start, and control simulation. It also includes making and retrieving savecases, re-initializating to a new time, initializing to a specific real-time situation, etc. This subsystem is also used to define event schedules. Events are associated with both the power system simulation and the ECS functions. Events could be deterministic (occur at a pre-defined time), conditional (based on a pre-defined set of power system conditions being met), or probabilistic (occur at random).

A Typical Restoration Scenario

A typical problem is initiated by a series of cascading trips and switching events, which eventually lead to blackout of a portion of the system. A typical sequence of restoration operations includes the following steps:

- Critically study the available real-time information (alarms, SCADA displays, etc.) in order to assess the extent and severity of the blackout.
- Open the system circuit breakers in the blacked-out area, in preparation for implementing the restoration procedure.
- Cold-start a hydro unit or gas turbine in the blacked-out area.
- Energize a transmission path from the blackstart unit to a neighboring unit (possibly a fossil unit) and start up and synchronize this next unit by providing the cranking power from the blackstart unit.
- Gradually increase generation and pick up customer load incrementally, while ensuring that frequency, voltages and flows stay within acceptable limits and do not cause unnecessary relay trips.
- Gradually energize additional transmission paths, pick up more loads, and bring more units on line.
- Continually monitor voltages, frequency, active and reactive power balance, and flows to ensure that they stay within acceptable limits.
- When appropriate, re-synchronize the area with the rest of the live interconnection.
- Repeat above steps until all customer loads have been restored, all units are back in service, and power system conditions are back to normal.

The paramount objective of restoration is to restore customer service and power system normalcy as quickly and safely as possible.

OTS Requirements for Power System Restoration Scenarios

Restoration scenarios involve severely abnormal and off-nominal operating conditions. For effective use in restoration scenarios, the OTS must meet the following simulation requirements:

- Response of the EMS functions and the power system simulation in the OTS must be very similar to the response of the real-time EMS.
- The simulation must accurately represent one's own EMS and one's own power system.
- Simulation fidelity must be maintained over a wide range of voltages and frequencies. This includes the simulation of power system components (generating units, transformers, lines, relays, etc.) and prime-movers. Simulation algorithms must be numerically robust in order to solve severely off-nominal conditions.
- The external power system must be modeled to a greater level of detail for restoration scenarios than for normal operational scenarios. This is to ensure a more accurate simulation of the dynamic behaviour of the interconnection when large disjoint portions of the system have been blacked out.
- The OTS user interface must be able to clearly demonstrate the power system dynamic phenomena that are occurring. This means being able to pause the simulation at any time and use instructor functions and displays to gain a better understanding of the generic system behaviour. Although, in real life, restoration will be performed using a formal restoration plan which uses the SCADA and EMS functions, the use of instructor displays and functions provides an invaluable means of illustrating and explaining the underlying system phenomena (such as effects of transient stability or cold load pickup). This generic understanding of the system is particularly beneficial when the system emergency deviates from the formal documented restoration procedure. Here, the operator has to rely on his own judgement and experience to extrapolate from the plan and make decisions on-the-fly.
- Easy access and transfer of OTS scenario data to other offline planning functions is highly desirable. These functions include transient stability, transient overvoltage simulation, and short circuit calculations. The ability to initialize these functions from the OTS and to conduct studies is invaluable for analyzing system restoration phenomena. Integration of the OTS with a transient stability function, in particular, is highly desirable. The stability function will ascertain whether restoration switching operations will cause inter-unit oscillations that result in relay trips [15].

An operator training simulator must be able to simulate the following:

- Instantaneous trips (and re-connections) of generation and load, large generation-to-load imbalances, and the resulting dynamic behavior of power system frequency.
- Steady state instability phenomena caused by large MW power transfers, which results in abnormally large power angle differences.
- Gradual voltage collapse phenomena caused by lack of generator or supplementary reactive support, which results in significant voltage degradation.
- System breakup into multiple electrical islands some of which may be dead, partial or total system blackout, and re-connection of electrical islands.
- Re-synchronization and ramping of units, and supply of cold cranking power and station service load effects.
- Restoration of customer loads, effects of cold load pickup and duration of de-energization, and effects of voltage and frequency dependence.
- Feeder sectionalization and switching of disconnects and shunt capacitor/reactors, to alleviate generation-load imbalances and maintain flows and voltages within acceptable limits.

- o Relay actions and relay malfunctions pertinent to restoration conditions, such as synchro-check, loss of excitation, overexcitation, over-current, automatic reclosures, over/under-voltage, volts/Hertz, and over/under-frequency.
- o Inclusion of a transient stability program that can be automatically initialized from the OTS [15].
- o Multiple, sequential and consequential, cascading disturbances.
- o Automatic frequency and voltage load-shedding, and disabling of automatic load-shedding during restoration.
- o Verbal communication with field crew, substation operators, plant operators and neighboring system operators.
- o Communication failures with RTUs and other devices in the field.
- o Battery discharge effects on communications, RTUs, and operating mechanisms of field devices (batteries discharge in about 8 hours).

Conclusions and Future Work

Operator training simulators provide an invaluable off-line environment for practicing system restoration. Since restoration situations occur infrequently in real life, and because during restoration there is extreme pressure to quickly and safely restore service, the OTS training will prepare operators to deal with the situation in a much more prompt and competent manner.

The potential of expert systems and full graphics to assist in restoration provides a great deal of promise. At each stage of the restoration, the operator could potentially be presented with the recommended "next action." This action could be the result of using the formal restoration procedure, in conjunction with an expert system network function, to quickly analyze the possible options and to suggest one that is "best." The graphical user interface could be used to highlight and clearly depict system conditions if the "next action" were implemented - possibly by visually highlighting the recommended next action on a display (such as changing color and pulsating the line to be energized next) and showing the flows and voltages that would result from implementing the action.

The expert system rules and conceptual procedures from the heuristic scenario builder in the EPRI operator training simulator [4] could be used to create restoration procedures online. The heuristic scenario builder, in order to create a training scenario, creates a sequence of events that results in degradation of normal operations. During restoration, however, the inverse is required: the expert system needs to analyze the degraded state of the power system in order to create a sequence of events which will restore it to normal.

Graphical displays could be developed to show the desirable sequence of generating unit startup, their time since trip, and the time at which they become available for re-synchronization. As each unit is started up, a display could show the geographical extent of the loads which could potentially be picked up if the unit were loaded to its capacity.

Applications of new technologies, such as neural networks, to power system problems, are showing encouraging results. The OTS provides an invaluable environment to more aggressively explore the potential applications and benefits of these technologies.

REFERENCES

1. J.Zaborszky, A.Huang, P.J.Clelland, D.A.Fagnon: Optimal Partnership of Operator and Computer for Power System Restoration; PROC. 11th Triennial World Contress of IFAC, Tallinn, Estonia, USSR, August 1990, v.VI, pp.73-78.
2. J.Zaborszky, G.Huang, P.Huang, J.B.Liu: A Dynamic Power System Model With Graded Precision for Fast Extraction of Pertinent Load Flow Information; IEEE TRANS. v.PAS-104 n.2, 1985, pp.3367-75.
3. T.Sakaguchi, K.Matsumoto: Development of a Knowledge Based System for Power System Restoration; IEEE Trans. v.PAS-102, n.2, pp.320-329, Feb. 1983
4. C.C.Liu, S.J.Lee, S.S.Venkata: An Expert System Operational Aid for Restoration and Loss Reduction of Distribution Systems; IEEE Trans. v.3, n.2, pp.619-626, May 1988
5. S.N.Talukdar, E.Cardozo, T.Perry: The Operator's Assistant - An Intelligent, Expandable Program for Power System Trouble Analysis; IEEE Trans. v.PWRS-1, n.3, pp.182-187, Aug. 1986
6. S.S.Shah, S.M.Shahidehpour: A Heuristic Approach to Load Shedding Scheme: IEEE Trans. v.PWRS-4, n.4, pp. 1414-1420, Oct. 1989
7. Y.Kojima, S.Warashina, S.Nakamura, K.Matsumoto: Development of a Guidance Method for Power System Restoration; IEEE Trans. v.PWRS-4, n.3, pp.1219-1227, August 1989
8. D.S.Kirschen, T.L.Volkmann: Restoring a Power System with an Expert System; IEEE Trans. v.PWRS-6, n.2, pp. 558-566, May 1991.
9. R.Nadira, T.E.DyLiacco, K.A.Loparo: A Hierarchical Interactive Approach to Electric Power System Restoration; IEEE/PES 1991 Winter Power Meeting, (91 WM 200-6 PWRS)
10. E.Mondon et al.: MARS - An Aid for Network Restoration After a Local Disturbance; Proceedings of the 1991 IEEE Power Industry Computer Application Conference, pp. 344-349.
11. K.Shimakura et al.: A Knowledge-based Method for Making Restoration Plan of Bulk Power System; Proceedings of the 1991 IEEE Power Industry Computer Application Conference, pp. 337-343.
12. M.M.Adibi, R.J.Kafka: Power System Restoration Issues; IEEE/PES Computer Application in Power (CAP), April 1991, pp.19-24.
13. R.Podmore, J.Giri, M.Gorenberg, J.P.Britton, N.Peterson: An Advanced Dispatcher Training Simulator; IEEE TRANS. v.PAS-101, January 1982, pp.17-25.
14. J.G.Waight, K.Nodehi, M.Rafian, H.VanMeeteren, A.Bose, R.Wasley, E.Stackfleth, E.Dobrowolski: An Advanced Transportable Operator Training Simulator; IEEE/PES PROC. PICA, Baltimore, Maryland, May 7-10th, 1991, pp.164-170.
15. S.Vadari, K.Demaree, D.Hwang, G.Miller, R.Waldele: An Online Implementation of Transient Stability in a Dispatcher Training Simulator; IEEE/PES Winter Meeting, Atlanta, February 1990 (90 WM 176-8 PWRS).

A HIERARCHICAL INTERACTIVE APPROACH TO ELECTRIC POWER SYSTEM RESTORATION

Ramón Nadira
Member, IEEE
Universidad Simón Bolívar
Caracas, Venezuela

Tomás E. Dy Liacco
Fellow, IEEE
The Dy Liacco Corporation
Cleveland, Ohio

Kenneth A. Loparo
Senior Member, IEEE
Case Western Reserve University
Cleveland, Ohio

Abstract - This paper addresses the control problem of a power system which is operating in a **restorative** state, i.e., a power system that has lost some of its load and/or has separated into electrical islands as a result of a major system disturbance. By means of a **hierarchical interactive approach,** this control problem is decomposed here into three layers or time-hierarchies (i.e., adaptive, optimizing, and direct); restorative control actions are then determined and executed by the appropriate layer according to the time frame available for doing so. The paper describes the restoration control problem in power systems, and provides the conceptual details of the hierarchical interactive approach proposed for its solution.

Keywords - Power System Operation, Restoration Control, Decentralized Control, Decomposition, Optimization.

INTRODUCTION

The operating condition of a power system can be mathematically described in terms of two sets of equations [1]: the equality (or load flow) constraints, **E,** and the inequality (or limit) constraints, **I.** The load flow constraints impose the requirement that the customer load demand be met at all times, while the limit constraints reflect the fact that the system variables (e.g., voltage magnitudes) must always be kept within limits representing the physical limitations of the power system equipment. Power system operation can be described as being in one of three operating "states": **normal, emergency,** or **restorative** [1, 2]. A power system is said to be in a *restorative* state whenever the set of limit constraints is satisfied, but the set of load flow equations is not completely satisfied. In other words, the restorative state is characterized by feasible operation of the power system equipment but with portions of the load not being served and/or with loss of system integrity. The major control objective in this state is to effect a transition from a partial to a maximum "feasible" load condition in the minimum possible amount of time.

Even though today's bulk power systems provide a very reliable supply of electric power, transitions into a restorative state (generally manifested as blackouts or brownouts) are **not** rare events [3]. The restoration of service following a severe disturbance is a very complex process which typically involves all of the components of the power system, including generation, transmission, and distribution. However, it is customary in electric power utility operating practice to develop distinct strategies for restoring the bulk transmission system and the distribution system [4]. This paper is concerned only with the restoration of service in the bulk transmission system, primarily conducted in a centralized manner from a utility's control center.

Electric utilities have traditionally handled the relatively few instances of transitions into the restorative state by developing general system **restoration plans** [5–10]. These plans provide general *guidelines* on what to do in case the power system experiences a severe disturbance resulting in loss of integrity and/or load. In general, however, these restoration guidelines do not take into account many of the possible system structure and/or operating conditions that can be encountered during the restoration process, and therefore are at best suboptimal. Additionally, the guidelines tend to be general and complex, and in many cases are of little help to the power system operators, who in general have little familiarity with operations in a restorative state.

The increased likelihood of modern power systems sojourning into a restorative state, the very high costs involved in operating the power system in such a state, and the lack of effective restoration control strategies, were important motivations for the research described in this paper.

PROBLEM FORMULATION

As indicated in [11], the restoration control problem can be formulated in general terms as follows:

$$\begin{aligned} \min_{x,u_c} \quad & F(x,u_c) \\ \text{s.t.} \quad & \dot{x} = f(x,u_c,u_d) \\ & g(x,u_c,u_d) = 0 \\ & h(x,u_c,u_d) \leq 0, \end{aligned} \tag{1}$$

91 WM 200-6 PWRS A paper recommended and approved by the IEEE Power System Engineering Committee of the IEEE Power Engineering Society for presentation at the IEEE/PES 1991 Winter Meeting, New York, New York, February 3-7, 1991. Manuscript submitted August 31, 1989; made available for printing January 3, 1991.

Reprinted from *IEEE Transactions on Power Systems,* Vol. 7, No. 3, pp. 1123–1131, August 1992.

where F represents the objective criterion, f, g, and h are nonlinear constraint functions, x is the vector of dependent system variables (such as voltage magnitudes at load buses and system frequency), u_d is the vector of **uncontrollable inputs** (which includes the load demands, the interconnection flows, and the system contingencies such as system faults and equipment malfunction), and u_c is the vector of **control inputs** (generated either locally or by the control center); the elements of x, u_c, and u_d are functions which evolve in time. It is typical to assume in security-constrained control strategies (such as the one to be considered here) that the portion of u_d corresponding to the system contingencies is fixed and known [12].

The **objective criterion** of the restoration control problem (i.e., F) is in general a **multi-objective** function, composed of at least the following three separate objectives: (i) to minimize the restoration time, (ii) to maximize the load served at all times, and (iii) to minimize the number and the magnitude of the control actions to be taken. These three objectives are typically in conflict. For instance, connecting too much load at any given point in time (to maximize the amount of load served) can cause a severe frequency decline which may trigger the load shedding relays, thus delaying the overall restoration process.

The vector of **control variables** (u_c) can be partitioned into a set of **discrete** control variables, y, and a set of **continuous** control variables, u; that is, $u_c^T = [y^T, u^T]$. The discrete variables include such controls as changes to the system configuration (e.g., reconnecting system branches), and settings of certain system components (e.g., in-phase controllable transformers). The vector of continuous variables is composed of controls which can be varied continuously (within limits), such as active and reactive power generations.

The **constraints** associated with the problem of Eq. (1) are of two types, namely algebraic and differential. The *algebraic* equations are the nonlinear load flow relationships, and the constraints related to a safe operation of the equipment. The *differential* equations represent the dynamic behavior of some the important system components, such as the generation units.

In summary, the restoration control problem as formulated here corresponds to a multi-objective, combinatorial (i.e., many discrete control actions), dynamic optimization problem. It goes without saying that any attempt at addressing this control problem without taking advantage of its particular structure and characteristics will probably fail (this is more so if the solution is required in real time). The next section presents a hierarchical interactive approach to restoration control which fully exploits particular problem characteristics and makes the control problem significantly more tractable.

THE HIERARCHICAL INTERACTIVE CONTROL APPROACH

The control approach presented here follows closely the ideas behind the temporal (or multi-layer) decomposition procedure originally proposed in [13], and later specialized to the power system control problem in [1, 2]. By means of this approach, the time horizon for control is partitioned into several scales, *according to the time available for determining and executing a control decision.* The interactive feature of the approach is based on the concept that it is the power system operators who will supervise, and in many instances direct, the computations leading to the determination of a given control strategy. In other words, the approach presented here is developed so as to serve as *an interactive tool for aiding the power system operators in making the correct decisions related to the restoration process.*

The Control Approach.

Determining an optimal control $u_c(t)$ which solves the problem formulated in Eq. (1) is indeed a difficult task, since the time evolution of some of the uncontrollable inputs is not known beforehand and adequate models are not readily available, and the computational requirements needed to generate u_c in real time for realistic systems is beyond the current state-of-the-art. Therefore, rather than attempting to generate a control trajectory $u_c(t)$, we will hereafter assume that the time horizon for restoration control will be partitioned into "small" time periods (or stages); additionally, it will be assumed that the uncontrollable inputs (or their models) will be known at the beginning of each stage, and that they will not change during the entire duration of a given stage.

The control problem then reduces to determining a control vector $u_c(K)$ for all stages K; this input is determined and executed at the beginning of each stage, and is assumed to remain constant throughout that stage. This is similar to the quasi-stationarity assumption made in connection with the determination of the majority of the control strategies in the normal and emergency states [12]. Note that most of the restoration plans currently in place reflect these considerations. That is, most plans restore the system in stages, connecting generation and/or transmission facilities, picking-up some load, and then repeating this process again and again until the system is restored.

The set of restorative control actions available at any stage of the restoration procedure can be classified into the following four categories: (i) changes to the generation patterns, (ii) changes to the system structure and/or connectivity, (iii) changes to the load patterns, and (iv) changes to the settings of the continuous control variables, u. A close look at these control actions reveals that some of them require a longer period of time to be determined and executed, while others can be made and executed in a relatively shorter period of time.

A framework for exploiting this "time-hierarchy" of control decisions in the context of power systems was first proposed in [1, 2]. According to this framework, the overall restorative control problem is decomposed into three control layers, namely, *direct, optimizing,* and *adaptive,* according to the time frame available for making and executing a control decision. The direct control level generates high speed and generally localized decisions, while the centralized decisions of the optimizing and adaptive levels take longer times to be generated (with the adaptive decisions taking the longest). Notice that this same hierarchical control approach is used in the everyday economic operation of power systems. That is, automatic generation control is a form of direct control, economic dispatch is a form of optimizing control, and unit commitment is a form of adaptive control.

One of the major advantages of this decomposition approach is that it makes the restoration control problem of Eq. (1) significantly more tractable; that is, it is possible to make specific assumptions at each of the levels which result in simpler and therefore more easily solvable restoration control problems (see Table I). Note from Table I that the direct control level assumes that all non-local variables are not to be considered for the determination of the direct (or localized) restoration strategies; this greatly simplifies the resulting direct control problem. Similarly, the optimizing level assumes that the discrete variables related to changes in structure and generation patterns are fixed and known; again, this simplifies the restoration control problem at the optimizing level.

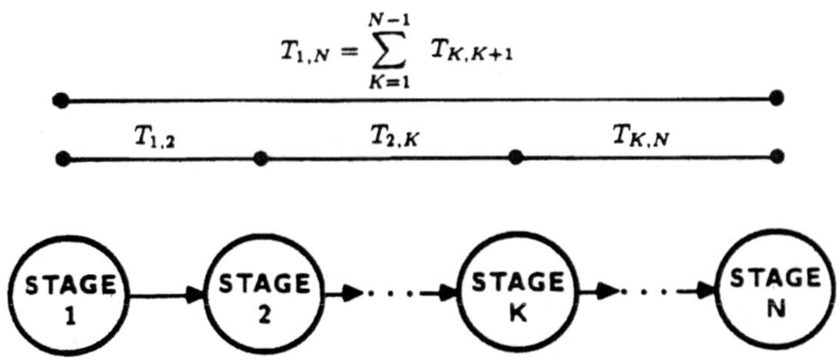

Figure 1. Dynamics of the Restoration Procedure.

The temporal decomposition approach could be implemented as follows. The overall restoration process could be divided into N stages (see Figure 1). Each stage would be characterized by the number of system electrical islands, and by the connectivity and characteristics of the generation-transmission system of each of these islands. Stage 1 of the procedure corresponds to the initial system conditions, once the disintegrating effects of the disturbance that caused loss of integrity and/or load have been halted. Stage N corresponds to those system conditions for which all of the load demand that can be served with the available generation and transmission facilities is actually served. The object of the restoration process is a transition from stage 1 to stage N in minimum time (i.e., to minimize $T_{1,N}$), while at the same time maximizing the amount of load served at each stage and minimizing the control variable changes. Notice that it is important to keep track of the time elapsed between stages 1 and K (i.e., $T_{1,K}$), since some decisions which involve connecting certain power system components (e.g., generation units) are infeasible if that time is either too long or too short.

Table I
Hierarchical Control Problems

CONTROL LEVEL	OBJECTIVE(S)	VARIABLES	CONSTRAINTS	COMMENTS
ADAPTIVE	• Minimize restoration time	• Discrete variables related to changes in structure and generation patterns	• Global connectivity constraints • Global E and I constraints	• Global control problem (for the overall system) • Optimizing variables assumed to be fixed and known
OPTIMIZING	• Maximize load allocation • Minimize control actions	• Continuous control variables • Discrete variables related to load allocations	• E and I for each system island	• Implemented at each observable island • Strategy is basically independent for each of these islands • Adaptive variables assumed to be fixed and known
DIRECT	• Maximize local load allocation • Minimize local control actions	• Local continuous control variables • Local discrete variables related to load allocations and to changes in structure and generation patterns	• Local connectivity constraints • Local E and I constraints	• Implemented at each "important" system bus (local control problem) • Non-local variables assumed to be fixed and known

Transitions between stages K and $K + 1$ will be assumed to occur as a result of either direct, optimizing, or adaptive control actions. In other words, no simultaneous actions from any two control layers will be allowed to take place at any one stage (see Table I). Finally, transitions due to direct control will be more frequent than those due to optimizing or adaptive actions, with the adaptive transitions being the most infrequent.

The overall structure of the proposed hierarchical approach is illustrated in Figure 2 (based on [2]). Notice from Fig. 2 that: (i) the approach assumes the existence of two levels of information processing. Level one processing takes place at a local level (e.g., at each substation), while level two information processing takes place at the control center; (ii) the control strategies generated by the optimizing and adaptive levels are to be actually implemented with the help of both, the relevant application programs and the direct control level, with the supervision of a restoration coordinator (see below); and (iii) the operator is able to monitor and even direct the decision making process

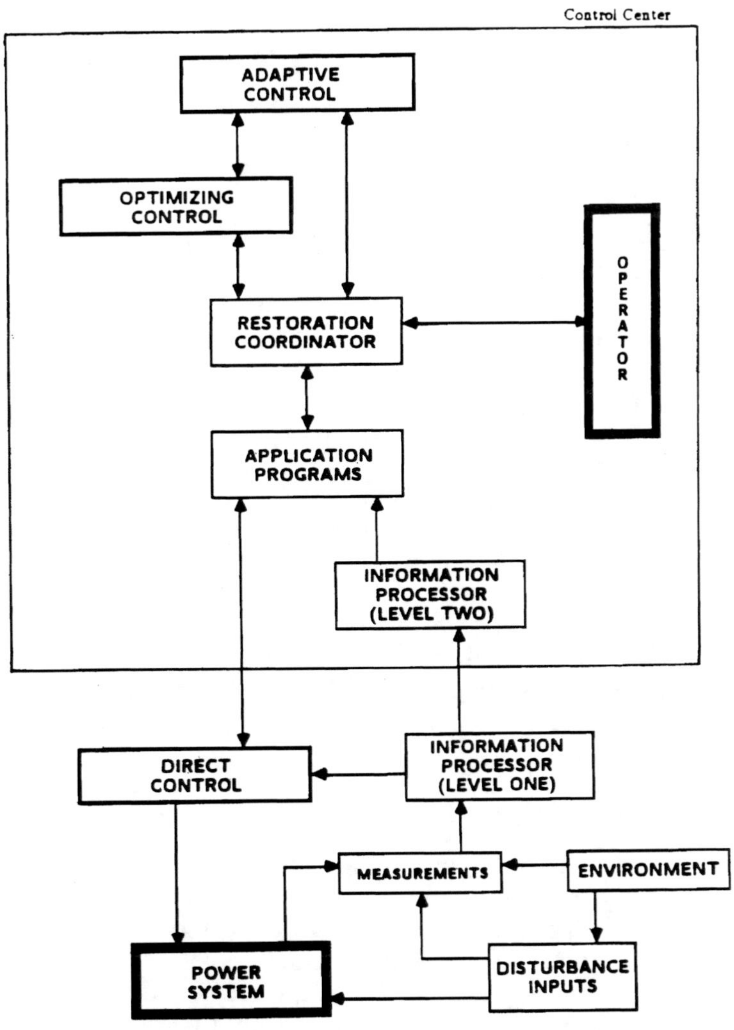

Figure 2. Hierarchical Control Structure.

which takes place at both the optimizing and the adaptive layers of control; this is to be achieved via the interface with the restoration coordinator. In fact, the adaptive level in this approach is not intended to be an entity separated from the system operator, but rather "adaptive control" is generated by a combined effort between the system operator and a computer program (i.e., a man-machine combination).

It is of particular interest to note that the same framework of Fig. 2 could be used in off-line applications geared towards both, the development of operator training programs (with respect to restoration control), and the generation of restoration plans. In those cases, the operator of Fig. 2 is to be replaced by either the trainee or the planner, and the power system is to be replaced by a power system simulation model.

The following subsections elaborate on some of the major components of Fig. 2 including the three layers of control, the restoration coordinator, and the application programs.

The Direct Control Layer. This layer is composed of functions which obtain localized information from the power system, and based upon this and other information obtained from the control center determine and perform automatic high-speed actions on the system based upon simple procedures. These procedures and/or the parameters involved in the procedures may be updated or modified via the interface with the control center. The direct restoration control functions are basically geared towards automatically restoring system integrity and/or load. Therefore, the two basic functions proposed here are:

- a **semi-automatic component reconnection** function, which is to be implemented at *each*, or at least at many, of the bulk transmission buses that are candidates for speedy component reconnection. The objective of this function is to determine the specific (or "micro") control inputs required to execute a general (or "macro") control strategy generated by either the optimizing or the adaptive control levels (this is why we have called it "semi-automatic" rather than "automatic"). Thus, control actions generated by the upper two levels are implemented in a correct and proper way, taking into account both the current system structure and operating conditions, and the connectivity constraints associated with the power system equipment; and

- a **semi-automatic load restoration** function, which is also to be implemented at many of the bulk transmission buses that are candidates for speedy load reconnection (i.e., at high priority buses). The objective of this function is to reconnect **blocks** of load in a semi-automatic manner, whenever the local power system operating conditions permit such a reconnection. The reconnection is semi-automatic since it may take place **only** when the control center enables the bus to do so. The control problem associated with this function follows directly from Eq. (1) by assuming that: (i) all discrete control variables (except those related to changes to the load patterns) are fixed and known, and (ii) all variables which are not local are also fixed and known.

The details of both direct control functions can be found in [11].

The Optimizing Control Layer. This layer is composed of functions which obtain global information about the power system and using this and other stored information compute, in a centralized manner, the set of decision variables which optimizes a suitable objective function subject to system constraints. Since this layer involves solving optimization problems of a generally large scale, it follows that a relatively longer time is required to arrive at the decisions at this level.

The optimizing control function to be considered here is that of **maximal load demand allocation.** For a given power system structure, generation commitment, and tie-line conditions existing in each of the system islands, this function determines the maximum (active and reactive) unserved load that can be connected to each of the observable island buses without violation of any of the relevant power system constraints. The maximal load demand allocation problem follows from Eq. (1) by assuming that all discrete control variables other than changes to the load patterns are fixed and known.

Implementations of the optimizing control layer for restoration control assume that aside from very few coupling constraints (related to the fact that common tie lines will eventually have to be reclosed) the restoration control strategies for each of the system islands are generated **independent** from each other. This follows the accepted restoration practice of sectionalizing the system into islands, restoring each island independently, and then synchronizing each island. The details of the optimizing control function are contained in [11].

The Adaptive Control Layer. The purpose of the adaptive control layer for restoration control as proposed here is to generate alternative strategies for restoring the overall generation-transmission system (considering all system islands simultaneously). These strategies are designed to satisfy important system objectives such as bringing back into service the majority of the available (and necessary) system equipment in minimum time. In contrast with the other two levels, the decisions at this adaptive level are not to be taken automatically, but rather involve man-machine interactions, with the operator playing a vital role [1].

In the context of restoration control, the relevant function is **component restoration.** The objective of this function is to determine which of the available power system component(s) to restore next, in order to minimize the restoration time. The control variables associated with the component restoration function are those related to changes to the generation patterns and changes to the system structure. All other control variables are assumed to be fixed and known. The constraints to be considered by the component restoration function include: (i) feasible connectivity of power system components, (ii) time elapsed since the restoration process started, (iii) minimum start-up times required by those generation units which are in a stand-by condition, (iv) minimum time required to synchronize a unit back into service, and (v) constraints related to the safe resynchronization of any two system islands. These control decisions and constraints are precisely the ones taken into account by the power system planners and operators involved in developing the restoration plans mentioned earlier.

The decisions to be made by the component restoration function at the adaptive level are very system dependent, since component restoration guidelines are likely to be different from system to system. Thus, the particular experience and knowledge that the power system planners and operators have about their own power systems is critical. This suggests the implementation of a knowledge-based approach to power system component restoration.

Since most of the restoration guidelines are generally given in the form of production rules (or IF-THEN rules), a natural way of encoding the experience of the power system planners and operators is by developing and implementing an **expert system** for component restoration [14]. This expert system is viewed here as an interactive assistant to the operator, rather than as an automatic generator of component restoration decisions. In other words, the expert system would be implemented as a decision support system for the power system operators in order to: (i) propose several alternative courses of action for a given operating condition, (ii) explain how it arrived at those alternatives, and (iii) evaluate the feasibility and/or desirability of alternative strategies proposed by the operators. Because of this high degree of interaction between the power system operator and the adaptive layer, and because of the significant computational requirements, it is expected that the control decisions at this level will take a relatively longer time to be determined and executed.

Key to the success of a given expert system development is the availability of experts having the relevant knowledge and experience. For most power system applications, no operator/planner possesses the necessary knowledge to build a complete set of reasoning rules. For instance, no power system operator can determine with reasonable accuracy the maximum amount of load that can be safely connected to a given power system for a specific set of operating conditions, even after decades of familiarity with the system! However, the operator does have valuable knowledge about what adaptive restoration control actions are most effective and desirable; this knowledge can then be effectively used to build a knowledge-based system. Based upon this knowledge, the expert system could propose several alternative adaptive restoration strategies for a given set of operating conditions. These strategies (which could be ranked according to some relevant criteria) could then be reviewed by the operator, who may select one (or several) for further consideration. Recent efforts in restoration control at the bulk transmission level have proposed the use of such knowledge-based systems [15-18].

In conclusion, knowledge-based approaches appear to be excellent candidates for addressing the restoration control problem at the adaptive level. This is in fact the approach proposed here for the *adaptive* component restoration function, and is an area that is currently under investigation.

The Restoration Coordinator. Key to the success of the proposed restoration control approach is the restoration coordinator. This component of the restoration system supervises the overall restoration process. At each stage of the restoration procedure, the responsibilities of the coordinator are: (i) to invoke the proper control layer(s), (ii) to make sure that the proper communication between these layers and the application programs is established (see Fig. 2), (iii) to make sure that the restoration strategies are properly displayed to the operator, and (iv) to direct the testing and evaluation of the alternative control strategies proposed by the operator.

The operations to be performed by the coordinator are very simple and of a procedural nature. These operations could be easily encoded in the form of production rules which could be extracted from a human operator.

The Application Programs. Several application programs are required to support the restoration control procedure presented here. Some of these include: (i) monitoring and assessment programs; (ii) load forecasting programs; (iii) programs for determining the maximum amount of total reconnectable load; and others [11].

In the particular case of control in a restorative state, the **monitoring/assessment** application programs have to be designed in such a way as to be able to handle the typical operating conditions which occur during the restoration process. These conditions include the possible existence of several **electrical islands** (i.e., portions of the system that are electrically isolated), and **observable islands** (i.e., portions of the system whose state can be determined by the state estimator program); in general, the electrical islands and the observable islands are non-coincident. Therefore, the monitoring/assessment programs at the control center must be able to: (i) identify all observable islands, (ii) determine the state of each of these, and (iii) continuously update the models of the internal and external systems for each of the observable islands, since the boundaries of some of these will change as the restoration process progresses. Although the theoretical capabilities for meeting these requirements have already been reported to exist [18], much research and development effort is still needed in order to produce practical and efficient implementations.

Accurate **load forecasts** of the post-disturbance unserved load demand at each of the bulk transmission system buses are essential to the success of any restoration control strategy. In many cases it is not possible to assume that the post-disturbance demand at each of the system buses is equal to the pre-disturbance demand, particularly if a significantly large period of time has elapsed between the outage and the possible reconnection [19]. Application programs which can accurately predict the post-disturbance load demand at most bulk transmission system buses are needed. This is an area that warrants further research.

Finally, it is well known that the restoration control methodology should be designed so that at no point during the restoration process do the restoration transients exceed the settings of the load shedding relays. Otherwise, undesirable load restoration-load shedding cycles may be generated. A practical way of preventing these cycles from occurring is to insure that the amount of load restored per stage (for each system island) is "more than met by the spinning reserve to the extent that the resulting transient [only] involves a small frequency swing" [20]. This **maximum amount of total reconnectable load** for a given available spinning reserve is a system-dependent quantity, which may be determined by performing extensive off-line studies of the transient response of the power system under a variety of structures and operating conditions. Some experiences in this respect are reported in [3] and [20]. Again, this is an area that needs further research and development.

Dynamics of the Restoration Procedure.

A flow chart of the dynamics of the restoration procedure for a given island J at a generic stage K, is illustrated in Figure 3. This flow chart contains the actions to be carried out by the restoration coordinator in order to generate restoration strategies of its own. The restoration coordinator is also able to evaluate alternative restoration strategies suggested by the system operator; this capability is not included in Fig. 3.

In essence, the dynamics of the restoration procedure for each island J are as follows: at the beginning of each restoration stage K, the restoration coordinator finds out whether the current system structure corresponds to a specific "target" system. A target system corresponds to the "ideal" mental image that the operators have of the system configuration for a given set of system components available for reconnection [15, 21]; these target systems are generally determined offline. When the restoration coordinator detects the existence of such a target system, it immediately invokes the optimizing level to determine the maximum amount of load that can be reconnected to each of the island buses (i.e., this is likely to be an optimizing stage). In the unlikely case that the operating conditions for the target system (as computed by the optimizing level) are infeasible, or if the maximum amount of reconnectable load is zero, the restoration coordinator will invoke the adaptive level (resulting in an adaptive stage); this is necessary since neither the optimizing nor the direct levels will be able to correct this infeasible situation. For feasible operating conditions and non-zero allocations, the coordinator will initiate the implementation of the resulting optimizing strategy, provided that the operator has enabled such a procedure. After the implementation of the optimizing strategy, or in the event that the operator disables the implementation of this strategy, the coordinator invokes the adaptive level, since no more optimizing and/or direct control actions can be taken with the current system structure and generation pattern. Observe that in this last case, the next stage will correspond to an adaptive stage.

For initial stage configurations other than a target system, the coordinator determines whether the current operating conditions are infeasible, in which case it invokes the optimizing level to correct the situation. For feasible initial operating conditions, the semi-automatic load restoration function at the direct level is enabled (and the stage will correspond to a direct stage); such a function remains enabled until the island operating conditions become infeasible (due to the direct control actions or some other reason), in which case the optimizing level (or the adaptive level, if necessary) is invoked. This last case requires some clarification. Although the decisions at the adaptive level are made taking the overall system into account, adaptive control actions for a given system island will be implemented only at the beginning of a given (adaptive) restoration stage. Therefore, whenever the adaptive level determines that two given system islands should be synchronized, the sequence of stages for both these islands have to be made concurrent (or coincident) in time, thus forcing one of the islands to stop its direct control procedure short of completion. Notice that even though two islands have been tagged for synchronization, this may not actually happen until several more stages have passed (or in other words, after the synchronization criteria have been met).

CONCLUSION

This paper has presented a hierarchical interactive

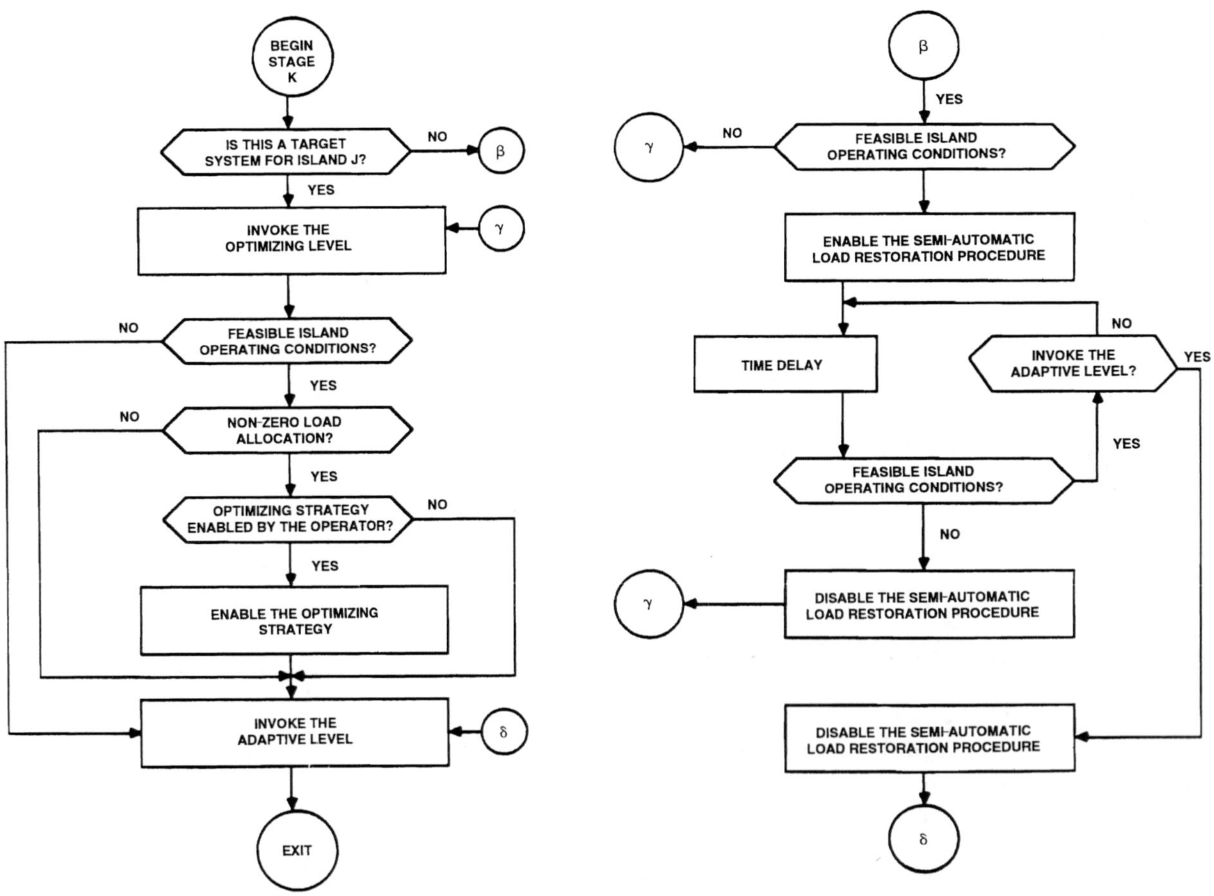

Figure 3. Restoration Procedure.

approach for addressing the security-constrained control problem associated with a power system operating in a restorative state. This approach exploits the inherent "time-hierarchy" associated with the determination and execution of the restorative control actions. Some of the advantages of the temporal decomposition approach include: (i) it makes the restoration control problem more tractable, (ii) it makes it possible to use different and more efficient methodologies at each of the levels of the hierarchy, and (iii) the hierarchical approach is independent of the particular solution methodologies selected to address the control problem at each of the levels, and new and/or improved solution techniques could be easily incorporated, as they become available.

ACKNOWLEDGEMENTS

The work described in this paper was supported in part by the NSF under Award No. ECS 88-16733.

REFERENCES

[1] T. E. Dy Liacco, "Control of Power Systems via the Multi-Level Concept," Ph.D. Dissertion, Case Western Reserve University, Cleveland, Ohio, 1968.

[2] T. E. Dy Liacco, "The Adaptive Reliability Control System," *IEEE Trans. on PAS,* vol. PAS-86, May, 1967, pp. 517-531.

[3] D. N. Ewart, "Whys and Wherefores of Power System Blackouts," *IEEE Spectrum Magazine,* Apr., 1978, pp. 36-41.

[4] G. L. Blankenship, and T.A. Trygar, "A Discussion of the Restorative State Control Problem in Electric Power Systems," in *Electric Power Problems: The Mathematical Challenge,* SIAM, Philadelphia, 1980, pp. 276-294.

[5] J. Gutiérrez, M. Staropolsky, and A. Garcia, "Policies for Restoration of a Power System," *IEEE Trans. on Power Systems,* vol. PWRS-2, May, 1987, pp. 436-442.

[6] IEEE Committee Report, "System Restoration—Deploying the Plan," *IEEE Trans. on PAS,* vol. PAS-101, Nov., 1982, pp. 4263-4271.

[7] IEEE Committee Report, "Power System Restoration—A Task Force Report," *IEEE Trans. on Power Systems,* vol. PWRS-2, May, 1987, pp. 271-277.

[8] R.J. Kafka, D.R. Penders, S.H. Bouchey, and M.M. Adibi, "System Restoration Plan Development for a Metropolitan Electric System," *IEEE Trans. on PAS,* vol. PAS-100, Aug., 1981, pp. 3703-3713.

[9] R. J. Kafka, D. R. Penders, S. H. Bouchey, and M. M. Adibi, "Role of Interactive and Control Computers in the Development of a System Restoration Plan," *IEEE Trans. on PAS,* vol. PAS-101, Jan., 1982, pp. 43-52.

[10] D. Scheurer, "Philadelphia Electric Company System Restoration Plan," presented to the Pennsylvania Electric Association, Pittsburgh, PA, Jan., 1980.

[11] R. Nadira, "A Hierarchical Interactive Approach to Electric Power System Restoration," Ph.D. Dissertion, Case Western Reserve University, Cleveland, Ohio, 1989.

[12] T. E. Dy Liacco, "Mathematical Problems in Electric Power System Operation: An Overview," in *Electric Power Problems: The Mathematical Challenge,* SIAM, Philadelphia, 1980, pp. 21-34.

[13] I. Lefkowitz, "Multi-Level Approach Applied to Control System Design," *Trans. of the ASME,* vol. 88, Series B, June, 1966.

[14] F. Hayes-Roth, D.A. Waterman, and D.B. Lenat, *Building Expert Systems,* Addison-Wesley, 1983.

[15] Y. Kojima, S. Warashina, S. Nakamura, and K. Matsumoto, "Development of a Guidance Method for Power System Restoration." *IEEE Trans. on Power Systems,* vol. 4, Aug., 1989, pp. 1219-1227.

[16] K. Matsumoto, T. Sakaguchi, and K. Uemura, "Heuristic Management of Complex and Large Scale Power Systems in Restorative States," *CIGRE-IFAC Symposium on Control Applications for Power System Security,* Florence, September, 1983, Paper No. 514-01.

[17] T. Sakaguchi, and K. Matsumoto, "Development of a Knowledge-Based System for Power System Restoration," *IEEE Trans. on PAS,* vol. PAS-102, Feb., 1983, pp. 320-329.

[18] F. F. Wu, and A. Monticelli, "Analytical Tools for Power System Restoration–Conceptual Design," *IEEE Trans. on Power Systems,* vol. 3, Feb., 1988, pp. 10-16.

[19] J. E. McDonald, A.M. Bruning, and W.R. Mahieu, "Cold Load Pickup," *IEEE Trans. on PAS,* vol. PAS-98, July/Aug., 1979, pp. 1384-1386.

[20] R. M. Maliszewski, R.D. Dunlop, and G.L. Wilson, "Frequency Actuated Load Shedding and Restoration, Part I–Philosophy," *IEEE Trans. on PAS,* vol. PAS-90, July/Aug., 1971, pp. 1452-1459.

[21] J. Zaborsky, S.Y. Lin, G. Huang, and J.B. Liu, "Restoration of the Large Electric Power System Using a Computer Generated Sequence of Target Systems," *IFAC Symposium on Power Systems and Power Plant Control,* China, 1986, pp. 301-306.

Discussion

D.S. Kirschen, Empros Systems International, Minneapolis, MN: The authors argue convincingly for the adoption of a hierarchical approach to the problem of restoring a power system following a major blackout. A similar decomposition and a prototype expert system have been presented in [A]. This expert system is designed to assist the operator at three levels: strategic, tactical and implementation. At the strategic level it identifies objectives which are reachable from the current state of the power system and evaluates their relative importance with respect to the goal of minimizing the restoration time. Once the operator has selected one of these strategic objectives, the tactical level elaborates a detailed plan of actions leading to this objective. This plan is reviewed by the operator and can then be executed with the assistance of the implementation level of the expert system.

Decentralizing some of the restoration actions to each bus or substation using a direct control level increases the parallelism of the process and might thus theoretically increase the speed of the restoration. However, this approach is not without complications:

- All restoration actions must be carefully coordinated to avoid compromising the stability of the reenergized system.
- Resources (active and reactive power, operator's time) are scarce and must be allocated wisely if the *entire* system is to be restored quickly.

Since most of the control actions can be performed remotely by using the supervisory control functions embedded in modern energy management systems, it is not clear what benefit would be achieved by decentralizing some aspects of the decision making process. The authors comments on this point would be greatly appreciated.

[A] D.S. Kirschen, T.L. Volkmann, "Guiding a Power System Restoration with an Expert System", paper 90 SM 341-8 PWRS presented at the IEEE 1990 Summer Power Meeting.

Manuscript received February 28, 1991.

R. NADIRA, T.E. DY LIACCO AND K.A. LOPARO: Mr. Kirschen raises an interesting point regarding the use of decentralization of restoration actions for electric power systems. The approach discussed in the paper is based on a hierarchical decomposition of the restoration tasks and includes direct, optimizing and adaptive restoration functions. As indicated in the paper the *direct control layer* includes what we referred to as "semi-automatic" component reconnection and load restoration. The terminology semi-automatic was used to indicate that both functions require data and information from the higher levels, i.e. the optimizing and adaptive control layers. In the case of component reconnection, the control inputs required to implement a general strategy developed at a higher level are determined based on local information. For the load restoration problem, decentralization allows a simplification in the optimization problem using only local information. This reduces the computational complexity involved in reaching a solution.

Manuscript received August 12, 1991.

SPECIAL CONSIDERATION IN POWER SYSTEM RESTORATION THE SECOND WORKING GROUP REPORT

A report by the Power System Restoration Working Group,
Chairman: M. M. Adibi, IRD Corporation

Abstract: This paper is one of a series presented on behalf of the System Operation Subcommittee with the intent of focusing industry attention on power system restoration issues. Today's power systems are more than ever vulnerable to potential major disturbances, dictating renewed and closer attention to restoration following a partial or complete collapse. This paper, the second report by the Power System Restoration Working Group, addresses four more issues which require special consideration.

COLD LOAD PICKUP - MODELING

W. W. Price & S. Ihara, GE Industrial & Power System

In performing studies of system restoration, it is necessary to have mathematical models that can be used to predict the load that will be present at various locations on the system. The variation of the load as a function of the length of time that it has been disconnected and other factors, such as weather and voltage, must be considered.

Load Categories

In general, loads can be categorized as: thermostatically-controlled loads, manually-restarted loads, and fixed Loads.

Thermostatically-controlled loads include space heating and cooling, water heating, refrigeration, and some industrial processes, is a major concern for cold load pickup because it tends to increase as the duration of the outage increases. This increase is not due to the increase in power consumption of individual load devices, but rather to the increase in the number of devices that will be in an "on" state following an outage.

Manually-restarted loads include industrial loads which will trip off upon loss of system voltage and will only be restarted manually once voltage is restored. During system restoration, the absence of these loads will represent a decrease in power, partially offsetting the increase due to thermostatically-controlled loads. A related category is loads that switch over to a backup power source during the system outage and are only returned to the system manually after service is restored.

Fixed loads such as lighting, computer/electronic equipment, and some mechanical drive loads, will be essentially the same during restoration as they were prior to the outage, assuming they have not been disconnected during the outage. Loads that switch to backup power during the outage and are automatically reconnected to the system upon restoration of the voltage would also fall into this category.

93 WM 202-2 PWRS A paper recommended and approved by the IEEE Power System Engineering Committee of the IEEE Power Engineering Society for presentation at the IEEE/PES 1993 Winter Meeting, Columbus, OH, January 31 - February 5, 1993. Manuscript submitted August 12, 1992; made available for printing December 14, 1992.

Modeling Approaches

Two approaches have been considered for modeling the behavior of the load during restoration heuristic and physically-base modeling:

Heuristic modeling characterizes the load behavior based on experience and reasoning regarding the overall behavior of load on typical feeders, supported by measured data during system incidents.

Physically-based modeling develops models for the underlying physical phenomena, together with data on the composition of the load.

The heuristic approach is useful for conceptual consideration of load behavior, but it is limited in practical use by a lack of sufficient experience and measured data. The heuristic modeling does not readily lend itself to adjustments called for by varying load compostions and ambient conditions.

The physically-based modeling, while requiring significant development and data collection, has the potential to produce reasonably accurate models for varying load compositions and conditions. Such an approach would involve the following:

1. Classification of loads on each feeder by load components, such as residential space heating, lighting, etc. The variation in load composition with time of day, season, weather, and other factors must be determined in order to characterize the fractional composition of the load at the time of an outage. This type of load characterization based on composition is similar to that required to develop voltage-sensitive load models for load flow and stability analysis. The EPRI LOADSYN program [1,2] has been developed to simplify this process.

2. Each load component must be characterized by its response to a system outage. A first step would be to classify each component as one of the three types discussed above: thermostatic, manually-restarted, or fixed. The impact of manually-restarted and fixed loads are readily apparent. The thermostatic load, which may be a substantial portion of the load, requires further modeling, as discussed below.

Thermostatic Load Modeling

The modeling of thermostatically-controlled loads has received some attention by researchers. The following information is based on reference 3. **Figure 1** shows a schematic block diagram of a residential space heating

Reprinted from *IEEE Transactions on Power Systems,* Vol. 9, No. 1, pp. 15–21, February 1994.

load. The following simple first order dynamic model was shown to produce accurate results of **Figure 2**, for an individual house:

$$Dt/dt = -(1/\tau)(T-T_f -Wt_g),$$

where:

T mean temperature of house,

τ effective thermal time constant,

T_f temperature house will approach with heater off,

T_g temperature gain over T_f attainable with heater on indefinitely, and

W thermostat status (1 or 0).

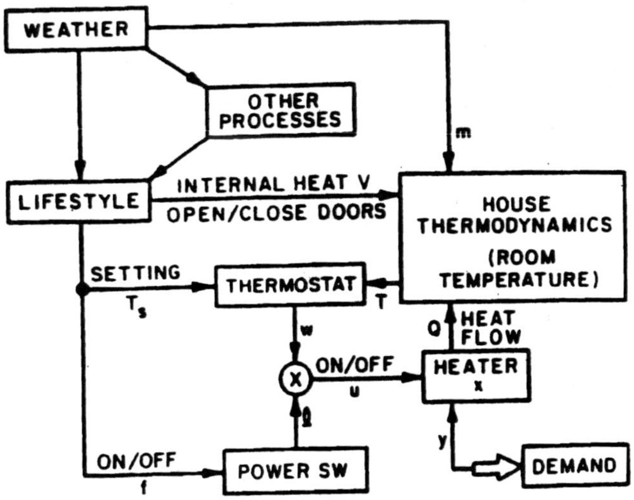

Figure 1. Space heater demand model

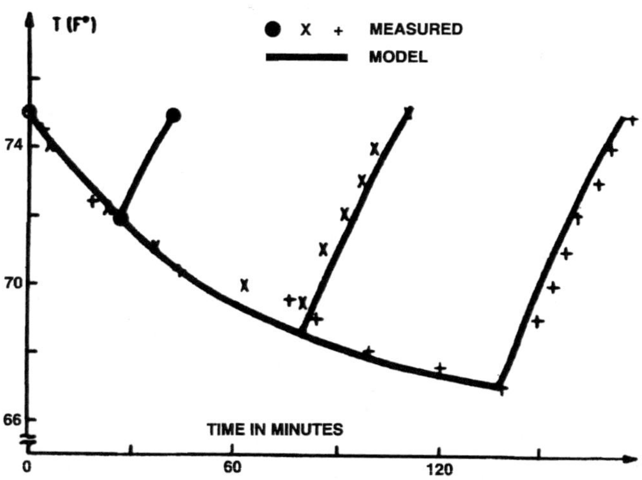

Figure 2. Comparison of model and test results for room temperature vs. time. (Normal thermostat cycle; 80 minimum outage; 140 minimum outage)

The representation of a large number of houses or other thermostatic loads requires analysis of the probability (B) of a given thermostat being on at the time of the outage. For a large number of loads, B is equivalent to the actual load at a particular time divided by the maximum load if all thermostats were on (installed capacity).

Figures 3 & 4 show the simulated response of the heating load (represented by B) for 10,000 homes. **Figure 3** the response following a 30 minute outage for three different initial values of B. **Figure 4** shows the response for a specific initial B for three different outage durations.

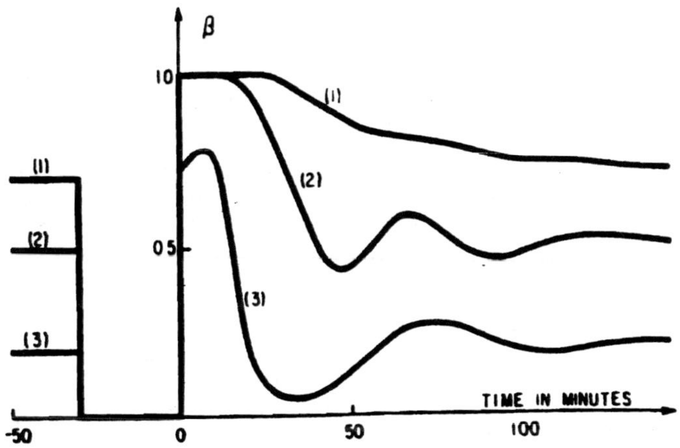

Figure 3. Cold load pickup after a 30 minutes outage effect of initial "B" = actual MW/installed MW - (1) B=0.7, (2) B=0.5, (3) B=0.2

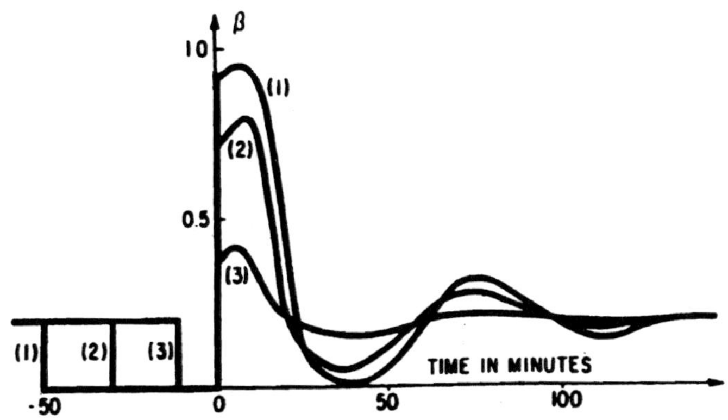

Figure 4. Cold load pickup - effect of outage duration - (1) 50 minutes, (2) 30 minutes, (3) 10 minutes.

This type of modeling could potentially be used to develop aggregate models for the behavior of feeder loads following an outage. The data for such models does not necessarily require detailed modeling of building thermal characteristics, but can be obtained from system diversified demand data and simple measurements of the on and off cycles of thermostats, as described in reference 3. Additional data on the composition of the load, as discussed above, would also be required.

Research Needs

The basic load characteristics that affect cold load pickup response are quite well understood. However, additional research is needed to develop measurement and data collection methods to determine parameters, particularly thermostat on-off cycle statistics for particular feeders at particular times of day, seasons, etc. With such data, simulation studies could be performed to characterize the cold load pickup response of particular feeders or typical feeders. The results of these studies could assist in formulating restoration strategies that properly accounted for the probable or worst-case magnitudes of the load.

COLD LOAD PICK-UP

LOW VOLTAGE NETWORK LOADS

Charles Weliky, Consolidated Edison Co

Over the years a great deal of effort has been expended in concisely forecasting, measuring, and preparing for load in-rush following customer outages. However, another critical consideration following initial restoration is the effect of "Cold Load Pick-Up" which is a sustained increase in load due to the reduction of the normal diversity in the load being served due to the passage of time. Recent experience has demonstrated the importance of being able to know approximately the amount of real and reactive loads which will be picked-up at any time following an outage and also some exceptions to the general rule in the case of lower voltage networks.

"Cold" Cold Load Pick-Up

A dramatic example of consequences of cold load pick-up occurred during the extremely cold weather during Christmas of 1989. Although, in some areas precautions had been taken to limit elected rotating interruptions to 15 minutes, which had been calculated to be maximum time to avoid loss of load diversity, this time was exceeded in hundreds of cases due to forced outages resulting from the very high overloads. As a result, cold load pick-up increased and restoration became very difficult and resulted in further damage to equipment from overloading and extended customer outages. The special nature of the load in this case is of interest. Extremely cold weather in a normally warm climate area will cause more significant increases in load and greater opportunities for cold load pick-up problems. This greater than normal loss of diversity occurs for the following reasons: heating is almost all resistive, insulation and building designs (residential homes on slabs), the lack of alternative heating such as fireplaces, and poor preparation of the population to handle cold weather (availability of warm clothing etc.). These conditions all combine to increase load due to the cold and reduce the probability that customers will turn off heating when requested.

Low Voltage Networks

Low voltage networks consist of multiple high voltage (13 or 27 Kv) distribution feeders supplying transformers which connect to a low voltage grid (240/120 volt) feeding the customers. These networks provide a great deal of redundancy and provide for the most reliable supply of electricity in the United States. However, the nature of these networks, with all distributions feeders emanating from one substation, means that in the very rare case of a substation shutdown, the length of restoration may be extended. In light of these extended outages, the potential for cold load pick-up complications seems to increase. However, experience does not reflect this hypothesis. Experience has demonstrated that long duration outages will not necessarily result in greater load pick-up. Some refrigeration, heating and air conditioning load requires manual intervention to be restarted. A portion of commercial and industrial load also requires manual intervention to start or will be replaced by on-site customer-owned back-up diesels. Based on experience during major area shutdowns, it can be assumed that the load that will be re-energized will be about 80% of the load prior to the shutdown. The following two scenarios demonstrate how these considerations can significantly impact on cold load pick-up.

Short Duration Incident

Two low voltage networks emanating from a single substation, experienced an outage. Since there were no overloads, faults or damaged equipment the networks were re-energized within 35 minutes. Due to the nature of a network, all feeders are energized simultaneously, thereby restoring all the customers to service. This incident occurred on a hot summer day and the networks were experiencing peak load conditions due to the use of air-conditioning. Therefore, based upon normal cold-load pick-up forecasts, the theoretical expectation would be an increase in load. In fact, what was experienced was a reduction in load. This seeming inconsistency is a result of the nature of the load. These networks include many large apartment buildings, commercial businesses, and large hospitals. Many of these customers have standby generators which will be activated to supply essential load. In addition, some of these customers have motors with low voltage protection which require manual resetting prior to restarting. Also, all elevators must be manually restarted as required by housing codes. Another example of the potential for delayed load pick-up is an air-conditioned apartment which is unoccupied and the air conditioner requires a manual reset. These factors combined with high awareness by the customers for the need to reduce load, led to a reduction in the load picked-up immediately after restoration.

Large Complex Re-Energization

This incident demonstrates the dramatic variations which can be experienced with very large-load customers who have load control capabilities. Due to a serious problem in a substation, it was decided to de-energize it for a short period. This station is unusual since it supplies only one customer with more than 100 MW of load. Before the shutdown, the customer was notified and was therefore able to anticipate the loss of power and to start on-site generators.

Once the original problem in the substation was electrically isolated, it was immediately re-energized. However, when the breakers were closed it was observed that there was no load being supplied. The customer strategy had been to completely isolate from the bulk power system and to supply essential load with on-site generators.

Load was then gradually restored by the customer. Had there been problems due to loss of diversity these could have been managed by coordination with the customer.
The high density of the load in an urban center combined with the network design and customer control capabilities significantly modifies cold load pick-up concerns. However, utilities supplying these types of loads must ensure that these mitigating effects are properly managed to optimize their value.

THE ROLE OF GAS TURBINES IN

PSE&G'S RESTORATION PLAN

S. A. Janes & J. N. Wrubel, PSE&G Co.

The current PSE&G restoration plan and philosophy was developed by a Task Force in 1985. The major drivers for the plan were the guidelines developed by PJM, our regional power pool. These guidelines were intended to provide consistency among member companies and facilitate easier integration of the pool following a major disturbance. In 1985, our old restoration plan had been in place for approximately 20 years, essentially untouched. Many changes in the way the bulk power supply system was operated had occurred. Most notably, our system had shutdown in 1967 and system shutdowns had occurred adjacent to us in 1965 and 1977.

Current Operation

The most significant change in PSE&G's system is that of generation supply. Except during peak load period, most of our fossil generating equipment is on reserve shutdown. Only our large nuclear units and load following coal-fired plant are normally on-line. The balance of our load is covered by externally-located jointly-owned units, purchases, or APPs. One of PSE&G's major assets is its sizable gas turbine fleet--in excess of 3000 Ms. Most of these units are quick-start (less than 10 minutes) and boot-strappable.

Restoration Plan

PSE&G's current restoration plan is designed to address this current mode of operation as well as any inherent weaknesses in our equipment or transmission arrangement. The plan relies on the strength of our gas turbines--their locations and capabilities.

A key factor to our plan is a parallel approach. One of the guidelines was to assure that our plan could provide light and power to all generating stations within two hours. As a result of the earlier system shutdowns, small GTs were installed locally at many generating stations. Other large groupings of GT
s have been installed along the transmission system external to the steam power plants. These larger units allow the restoration of the transmission system at the same time that the fossil units are being restarting. They also provide redundancy for light and power sources if local GTs at any generating station should fail to start.

Strengths

Generating Stations

As stated earlier, small (18 MW) GTs have been installed locally at most fossil stations. These units all have bootstrap capability and require minimal switching to provide start-up power. Each station tests their unit monthly to assure readiness and operator familiarity. Typically, each station will also conduct a full steam unit start-up using their isolated GT once a year. Another key designation feature is the ability of each station to synchronize its isolated light and power system to the developing transmission system. This allows fossil station restarts to proceed in parallel with the transmission restoration, thereby not losing valuable time.

System

Large groupings of GTs ($\geq$ 400 MW's) are located in four locations along the transmission system. These GTs provide PSE&G the ability to rapidly restore a transmission core path. These units provide start-up power for those stations not equipped with local GTs or where subsequent synchronization would create a problem, as well as a level of redundancy for the generating stations with local GTs. Selected modules at these four major locations are black start tested every month.

This approach, both in transmission restoration and use of GTs provides PSE&G with the ability to pick up a significant number of megawatts of customer load early, while steam units are being restarting. Restart times for large baseload units and units previously on reserve shutdown can be fairly length (8 to 24 hours).

A major part of our transmission restart plan is to start at one GT location, branching out, establishing a core path among generating stations. This avoids the need to synchronize small islands together at points in the system where personnel are not familiar or lack experience in the process.

Past tests have confirmed that our larger GTs can be black started and held in stand-by mode, and auto-synchronized as needed.

General

Most of our GTs are located with other fossil units. This will assure adequate manpower for restarting at almost all times.

Concerns

Stations

Not all generating stations have small local GTs which can provide light and power for start-up. Larger units, which are available, may cause subsequent synchronization problems. Past testing experienced gained during fossil unit test restarts confirmed difficulties in controlling frequency and load control; especially when starting large loads such as fans and pumps. It should be noted that it is good practice to start the largest motor first, followed, if possible, by motors in decreasing order of size.

System

One significant flaw was discovered in early plan tests of equipment was that unloaded GTs could not be parallelled. This necessitated that the plan provide for small load blocks located near each major GT location, which could be applied quickly to provide load for the GTs.

In addition, all of our GTs are equipped with automatic power factor setting, voltage regulators. To defeat this feature during restoration, each machine must be put in the manual voltage control mode and voltages established as instructed by System Operations. In general, GTs do not provide as much range in voltage regulation, and much less VAR absorption capability, as steam units.

Another major concern is the unproven load regulation and frequency control capabilities of our GTs. How well will they share the load as its picked up? Other than fossil unit test starts, no real test experience is easily obtained without putting firm customer load at risk! One thought on the subject is that good practice would have only the largest single GT (125 MW) on frequency control and the remaining GTs set to hold constant MW output. Under these conditions, the System Operator would have to coordinate the addition of system demand with the synchronization and loading of the smaller GTs to maintain the large GT in its safe operating range.

Finally, the load "blocks" that are used to develop our transmission core do not at present have enough underfrequency load shedding to assure that the loss of the largest unit would be protected by sufficient load shedding to prevent another collapse. This deficiency evolved based on the power pool requirement of only 30% load shedding in three 10% step.

General

Numerous general items of concern about GTs arose during the formulation of our present plan including: battery conditions at GTs; lack of d-c fuel forwarding pumps or inadequate tank head pressure; and possible inadequate system source requirements for fault clearing on the transmission system.

Conclusion

PSE&G has a large fleet of GTs. These units are well suited to provide for fast system restoration, given current typical operating conditions. Our present plan and equipment mix and location provide significant redundancy.

In addition, our plan provides for strong routine and periodic testing, assuring readiness of both equipment and personnel.

PSE&G is currently reviewing and revising its restoration plan, again addressing changes in equipment, practices, operation and mix of generation.

REACTIVE POWER BALANCE

M. M. Adibi, IRD Corporation
M. J. O'Grady, Potomac Electric Power Co.

The characteristics of a power system relevant to restoration may be divided into three functional groups: those related to load-generation balance, those which affect reactive power balance, and those which primarily deal with the protective and control systems. The characteristics affecting the reactive power balance are the subject of this paper [6].

In general, power system operators divide real power into load and generation, referring to what leaves the utility network to the customer as positive load and to what is supplied to the network as positive generation. The reactive power is also divided into reactive load and reactive generation. Since most customers have loads with lagging power factors, the lagging reactive load is termed a positive reactive load and the lagging reactive generation a positive reactive generation [7].

The following tables list positive and negative reactive load and generation:

Reactive Power Generation

± Generators & Co-Generators
+ Lines and Cables (Charging)
+ Shunt Capacitors
± Synchronous Condensers
± Static var Generators
- Shunt Reactors

Reactive Power Load

± Customer Demand
+ Line and Transformer Losses
+ Transformer Excitation
+ Generator Auxiliary Equipment
± Interchanges

During restoration when a high voltage overhead or an underground transmission line is energized, there will be a sudden increase in positive reactive generation (i.e., line charging) which if not balanced with positive reactive load and/or negative reactive generation it will cause sustained overvoltages. With the sustained overvoltages there is always the risk of harmonic resonance generation which may result in damage to equipment [4].

The following table lists typical line charging in MVAR per mile:

Transmission Line Charging

Conductor	Kv	MVAR/MI
954 ACSR,	69	0.03
954 ACSR,	138	0.12
1-795 MCM D,	230	0.29
2-954 MCM R,	345	0.86
3-1127 AAA C,	500	1.74
4-1351 MCM D,	765	4.69

In picking up positive reactive loads one is also concerned with the decline in system frequency below the allowable limit which the sudden increase in associated real load may cause. The negative reactive generation however, can be provided independently of system frequency by synchronous machines, static var generators and shunt reactors. While the amount of reactive generation available from static var generators and shunt reactors are readily known, this is not the case for generators (synchronous machines), because the station and system operating conditions impose certain limitations on their lagging and leading reactive capabilities [8].

Generator VAR Capability

The generator reactive capability curves ABCD of **Figure 5** which are provided by manufacturers, are strictly a function of synchronous machine design parameters at rated terminal voltage. They do not consider generating station and power system operating conditions as limiting factors. These curves show that different generator loads produce greater heating in different parts of generator. The segment AB is limited by rotor heating, BC is limited by armature heating, and CD by armature core end heating.

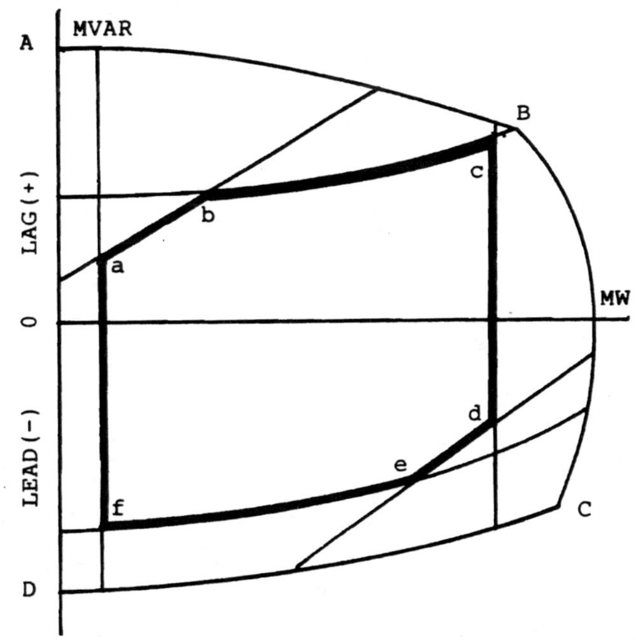

Figure 5. Generator reactive capability Limitation

A typical arrangement for supplying power-plant auxiliaries and services in a generating station is shown in **Figure 6.** The electrical system is composed of many major components and sub-components that operate as a system to supply power to the power system. Each component is designed and rated to operate continuously within certain electrical and mechanical limits specified by the manufacturer. Any component may have operating limits imposed on voltage, current, frequency or a combination of these such as MVA, volt/hertz, and stability. These limitations reduce the reactive capability of a generator to below the designed ranges [9].

In **Figure 5** the cumulative effect of all the operating requirements are superimposed on the manufacturer's design values, i.e., the generator reactive capability curves A-B-C-D. The "true" reactive capability is restricted to the area bounded by the a-b-c-d-e-f as followings:

a-b & d-e	due to auxiliary bus high and low voltage limits;
b-c & e-f	due to generator terminal high and low voltage limits; and
c-d & a-f	due to turbine max. and min. operating limits.

Other possible limitations imposed but not shown are: voltage regulator (VR) max. and min. output settings, volts per hertz limiters, under-excitatio reactive ampere limiting (URAL) relay, and minimum excitation limiting (MEL) relay.

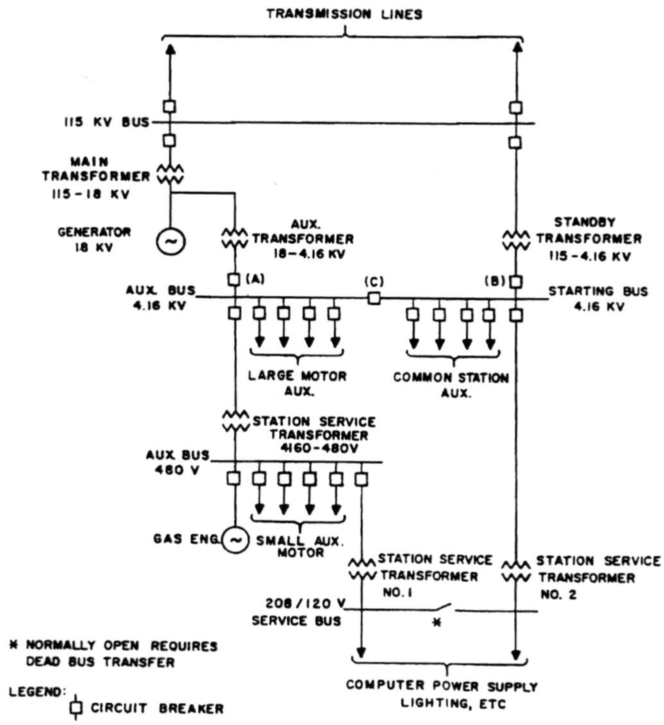

Figure 6. Power distribution to auxiliary equipment

By properly selecting the main and auxiliary transformers tap positions, their two related curves will be moved up or down in parallel, changing the proportion of the lagging and leading reactive capabilities of the generator. However, the sum of generator's lagging and leading reactive capabilities will remain the same unless it is limited by the VR, URAL, MEL or one of the other limiters.

Considerations

1. To ensure availability of adequate reactive absorbing capability it is important that special attention be paid to the main and auxiliary transformer tap positions, voltage regulator and relay settings at the generating plants.

2. In the initial stages of restoration often fast starting combustion turbines are called upon to provide remote cranking power, energize lines or short cables to some local loads. In some installations the URAL or MEL relay is set to unity power factor to protect the turbine during normal parallel operation (apparently from power reversal). Such a relay setting would make the combustion turbine unavailable for leading power factor operation or cranking purposes.

3. In energizing high voltage overhead or underground transmission lines there should be sufficient reactive absorbing (negative reactive generation) capability close to the line to maintain reactive power balance.

4. Generally system voltages are monitored to determine out of limit conditions but the reactive power flows are controlled to maintain acceptable conditions.

5. To avoid excessive charging current when system is lightly loaded, it is advisable to maintain generator voltages as low as the system stability allows, switch off shunt capacitors, and to energize a single circuit in a multiple circuit lines.

6. It is preferable to shed load when low system voltages are observed than to raise system voltages.

7. In order to avoid high sustained overvoltages, in the initial stages of restoration lines with low charging current (sub-transmission and distribution systems) are used. The voltage drop ΔV, across these systems with series impedance of $R+jX$, can be approximated by:

$$\Delta V \approx I\,(R \cos \theta + X \sin \theta),$$

where, I is the current and $\cos \theta$ is the load power factor which is usually kept at about 98%. Since the X/R ratio of these systems is about 5 to 1, therefore the voltage drop due to real and reactive power flows will be equal. This provides a certain rule-of-thumb for guiding reactive power supply and control.

8. The reactive power balance of a system can be facilitated by the use of an interactive power flow program for evaluating generator reactive loading, transformer tap positions, shunt reactor or capacitor switching.

REFERENCES

[1] Price, W.W. et al, "Load Modeling for Power Flow and Transient Stability Computer Studies," IEEE Trans. PWRS-3, pp. 180-7, Feb. 1988.

[2] General Electric Co., "Load Modeling for Power Flow and Transient Stability Computer Studies," Vol. 1-4, EPRI Report EL-5003, 1987.

[3] Ihara, S., F.C. Schweppe, "Physically Based Modeling of Cold Load Pickup," IEEE Trans. PAS-100, pp. 4142-4150, Sept. 1981.

[4] Adibi, M.M Alexander, R.W. Avramovic, B. "Overvoltage Control During Restoration," IEEE Trans. 92WM107-3PWRS.

[5] Alexander, R. W. "Minimum Source Consideration

when Restoring the PJM 500kV System," PEA System Operation Committee, Jan. 1991.

[6] R.J.Kafka et al.:System Restoration Plan Development for a Metropolitan Utility; IEEE Trans. v.PAS-100 n.8, 1981, pp.3703-13.

[7] J.G.Raine: Power System Concept of Reactive Power; The IEEE 8th Biennial Workshop on "Real Time Monitoring and Control of Power Systems," Montreal, Oct. 1984.

[8] D.P.Milanicz :Generator Reactive Capability - Testing and Optimization; Southwestern Electrical Exchange, Alabama, Oct 1989.

[9] M.M.Adibi :Application of Digital Computers in Steam Electric Station; Advanced Development Div. Publication, IBM Corp., May 1961.

STEAM PLANT STARTUP AND CONTROL IN SYSTEM RESTORATION

F. P. de Mello. Fellow IEEE

J. C. Westcott, Fellow ASME

Power Technologies, Inc.
Schenectady, NY

ABSTRACT

The IEEE Working Group on Power System Restoration developed a panel session for the Summer Power Meeting on July 14, 1992 on Special Considerations in Power System Restoration. One of the contributions to this session is presented in this paper dealing with aspects of steam plant startup and control in scenarios of system restoration.

The topics addressed include the complexity of a steam plant, the contrast between normal plant startups and shutdowns and those following major system blackouts including the effects of plant design, automatic controls, bypass valving and operator training.

KEYWORDS

System restoration, steam plant startups and shutdowns, steam plant design, plant controls.

INTRODUCTION

The complexity of system restoration has been well described in other contributions (1,2,3,4) with reference to the electrical plant. Of particular note is the realization that the state of equipment, whether or not ready for restoration, can be affected by the events that led to the system shutdown. These events often admit a large number of combinations and permutations, not all of which have been factored into system design. Equipment failure can be one of the byproducts of a major system shutdown and the result is in extending restoration time periods.

Steam power plants, fossil and nuclear fueled, are a combination of complex thermomechanical equipment as illustrated in Figure 1 showing the elements in a typical coal-fired plant. Operators are concerned with the state of hundreds of variables (pressures, temperatures and flows of water and steam, flue gas temperature and excess oxygen, hydrogen pressure, bearing temperature, vibration, valve positions, etc.) Among these only one and not necessarily the most important variable is the MW output of the generator. The condition of this maze of equipment can be affected significantly by the manner in which events leading to shutdown occurred.

Thermomechanical systems have relatively slow dynamic response relative to electrical phenomena. For this reason, past practice has been to allow operators to manually transfer various plant control systems from automatic to manual control. Traditional control design allows some sub-systems to be manually controlled while others are being automatically controlled. The action of control transfer, either automatic to manual or manual to automatic, alone has caused manyunit trips with resulting equipment damage. Equipment damage due to operator errors is likely considering that the events in question are very rare and in all likelihood operators have not been trained to cope with them. In addition, the robustness of the equipment precludes the occurrence of failure during rapid transients but the rapid transients can cause significant reduction of equipment life expectancy which cannot be overcome by preventative maintenance programs.

93 WM 201-4 PWRS A paper recommended and approved by the IEEE Power System Engineering Committee of the IEEE Power Engineering Society for presentation at the IEEE/PES 1993 Winter Meeting, Columbus, OH, January 31 - February 5, 1993. Manuscript submitted August 27, 1992; made available for printing November 4, 1992.

Aspects of plant startup and control that can play an important role in system restoration are discussed in this paper.

Normal Plant Startups and Shutdowns

Fossil and nuclear fueled steam plants, traditionally, have been classified as base load, cycling or peaking depending on the portion of the load duration curve that they supply. Evidently those plants with the most economic generation capability will be used to supply the base load. Units with a combination of relatively high efficiency and lower fuel costs will be base loaded. These tend to be the larger high pressure and temperature units with correspondingly high cycle efficiencies. The maneuverability of such units is restricted due to temperature gradient limitations affecting stress of thick high pressure and temperature metal parts. This lack of flexibility in load cycling duty is offset by the fact that they normally operate at base load with a small number of startups and shutdowns per year.

The ability of units to load follow during periods of low power requirements is an important aspect particularly in islanded operating conditions. Units with rated steam conditions in the supercritical region, greater than about 3208 psia and 705.5 °F, are the most difficult to operate at low loads and, because of the minimum flow requirements, many units cannot be operated at lower than forty percent of capability. Some supercritical units have been retrofitted to allow them to operate at outputs as low as fifteen to twenty percent and at reduced pressures.

Those units which are used for most of the peaking, cyclic and two-shift-operation duty are generally older and have lower initial pressure and temperature conditions. These can experience startups and shutdowns as often as once a day or once a week. Units as large as 600 mw have been built specifically for two-shift operation.

Figure 2 is a time chart (Ref. 1) showing the startup of a unit after a shutdown for about seven hours. Time from fuel ignition until synchronization is about 85 minutes, 20 percent load is achieved in another 15 minutes and 65 percent load in another 40 minutes. This unit was designed for two-shift operation. It is natural gas fired which helps in attaining this type of startup. A load restoration after a unit trip could be accomplished is less time.

An industry problem has been the fact that changes in costs between fuel types as well as improvements in cycle and equipment efficiencies have, with the passing of time, forced older base loaded units to operate in increasingly cyclic duty. Recognition of this phenomenon should lead to increasing automation and sophistication through revamping of old startup and shutdown controls. Sometimes insufficient attention has been given to this need. An appreciation of the complexity of plant startups can be obtained from reference (2) by the IEEE Steam Station Protection Working Group.

Regardless of how well procedures, manual or automatic, are established for normal startups and shutdowns it should be emphasized that a normal shutdown is quite different from an unanticipated abrupt shutdown.

Reprinted from *IEEE Transactions on Power Systems,* Vol. 9, No. 1, pp. 93–101, February 1994.

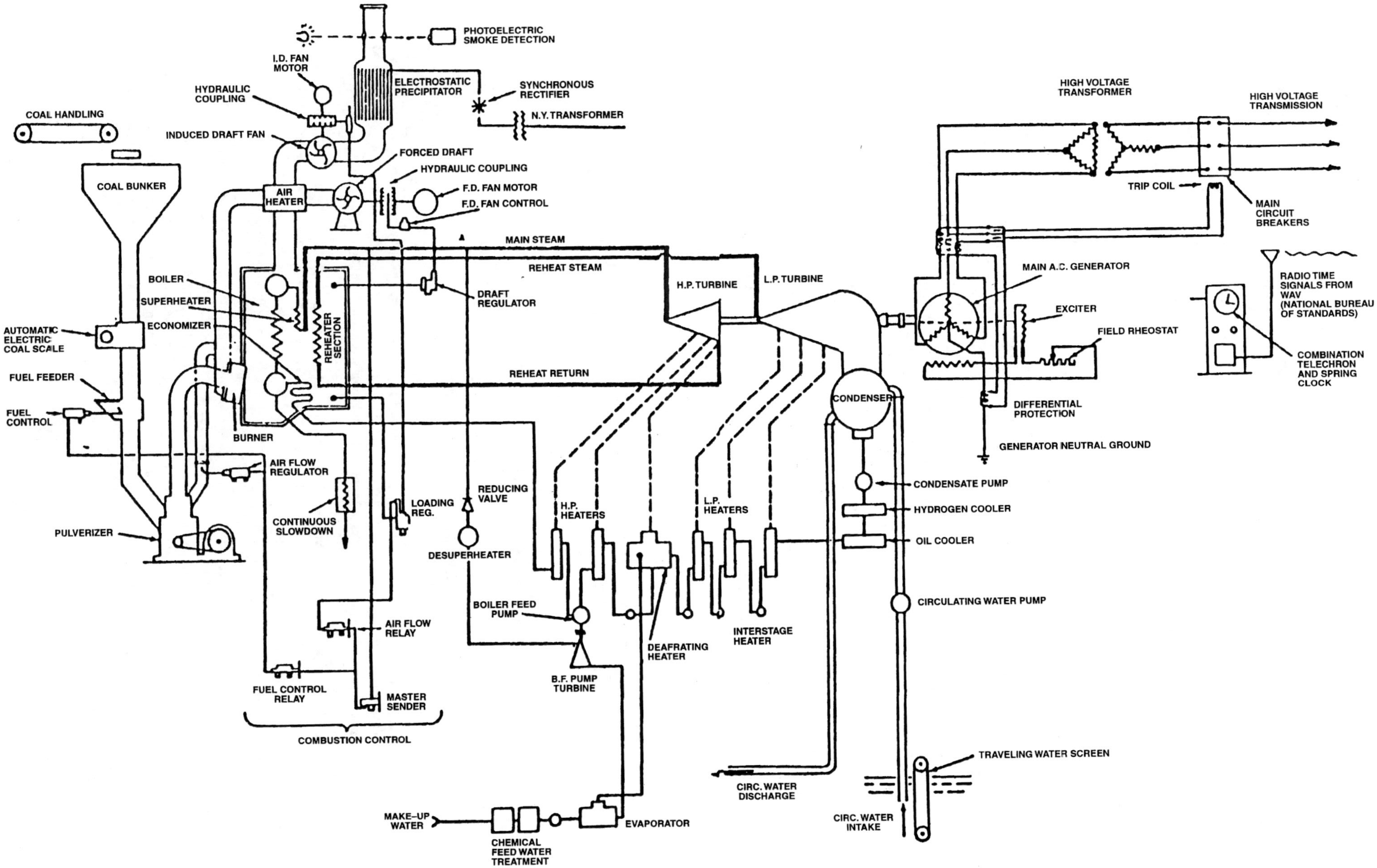

Figure 1*. Elements in Typical Coal-Fired Plant

***Reproduced from Power Engineer's "Power Plant Primer"**

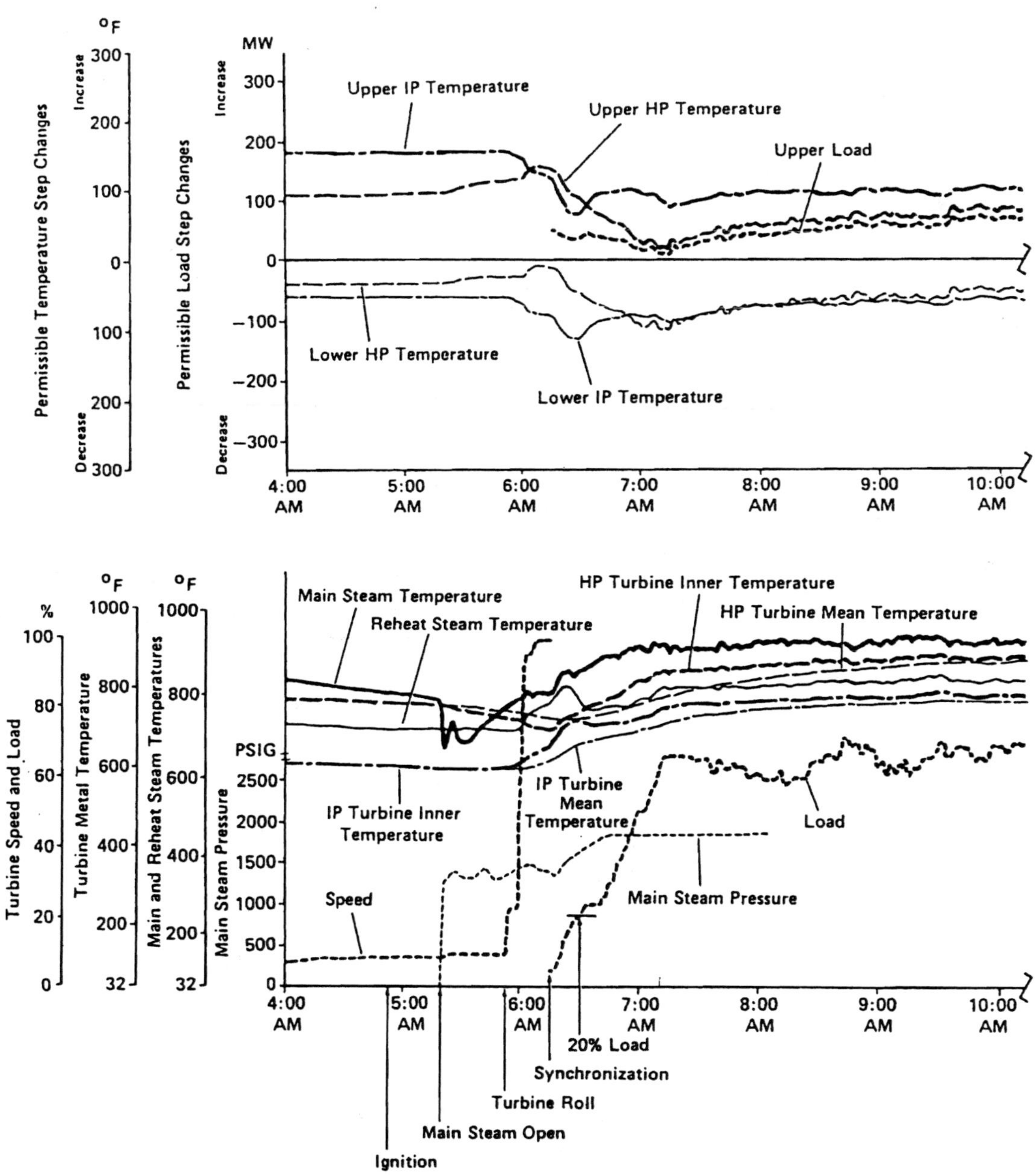

Figure 2. Startup After A Seven Hour Shutdown

A normal shutdown is accomplished at an appropriate controlled rate and, when the unit is finally disconnected from the system, it is done by tripping the turbine with the generator breaker opening after the electrical power has been reduced to zero. There are no steam generator safety valve operations, the fuel combustion in the furnace has been reduced at a safe and desirable rate, and the whole process is slow enough to be handled manually. In a multi-unit control room plant, the normal shutdown procedure is usually handled one unit at a time.

Other aspects of normal shutdowns involve the proper planning of the turbine cool-down and steam generator bottling process depending on the anticipated time before a subsequent startup.

This procedure must be as concerned about rate of change of temperature as during startup. A turbine tripped from full load will not experience any significant rate of change in temperature. A steam generator will experience large temperature and pressure transients during a full load trip and can possibly implode the furnace walls if the combustion is terminated too rapidly.

The procedure of reducing the number of fans, feedwater pumps and other auxiliaries can all be handled manually in a normal shutdown.

The duration of a normal startup can vary from a couple to hours to a day depending on whether it is a cold, warm, or hot start, type of fuel, and depending on the steam supply design, with the once-through design presenting a more difficult process using a complex startup steam generator bypass system which is gradually phased out around 30% load. The following table lists the average startup times after different shut down time periods. These may vary depending on unit conditions and a considered judgement on life expenditure of unit components.

Startup Classification	Time To Roll From Steam Generator Lightoff	Turbine Roll To Synchronization	Loading Rate	Worst Case Turbine Roll To Full Load
Hot Start 8 Hr. Shutdown	30 Minutes	10 Minutes	5 Percent per Minute	30 Minutes
Warm Start 8 to 60 Hour Shutdown (Weekend)	1 to 2 Hours	30 to 40 Minutes	3 Percent per Minute	1½ Hours
Cold Start Greater Than 60 Hours	Up To 24 Hours Depends on Water Chemistry	May Require Heat Soak Holds of up to 2 Hours. 1 Hour for Acceleration	<3 Percent per Minute Depends on temperature rate of change. 1 Hour Heat Soak at 2 Percent Load	5 Hours

The points of these comments on normal startups and shutdowns of steam plants are that:

1. Depending on the design of the plant and controls, the startup and shutdown process can be complicated and lengthy. Contrast this with the push button start, or stop, of a gas turbine or hydro unit.

2. Whether a startup can be considered normal may depend on whether the shutdown preceding it was normal.

3. During a normal shutdown, startup or indeed under normal operation, the power output of the unit, whatever it may be due to plant conditions, can be absorbed by the system which is usually very large relative to the size of the unit. This, in essence, has removed one variable to be controlled when other more important variables (from the perspective of the plant) take precedence.

Shutdowns and Startups Following Major System Blackouts

Unit and plant shutdowns occurring during the unravelling of a major system blackout are almost always initiated by the breakup of the electrical grid through circuit breaker operations. When such breaker operations result in complete isolation of the plant or unit, this creates a condition known as a load rejection. In many cases, a units control capability will not be sufficient to keep the unit from tripping.

A complete load rejection such as a generator trip would be followed by a turbine and steam generator trip within a matter of seconds as the turbine speed controls react to turbine overspeed. Such a shutdown is markedly different from the normal shutdown over many minutes previously discussed. The abrupt transition from loaded to unloaded conditions can be traumatic, particularly when several units experience rejection almost simultaneously.

The scenario in the control room changes abruptly from one of sleep inducing tranquility to a blast of loud bells and whistles, blinking alarm lights, perhaps a throwover to auxiliary lighting, safety valve operation and, much too often, a throwover of many important control functions (drum level for instance) to manual. The integrity of the auxiliary power supply system is often breached as the normal supply from the network fails and throwover schemes malfunction. Those that have witnessed an unanticipated plant shutdown cannot fail to hold in awe the burdens and stress imposed on operators and to marvel at how adrenalin keeps them functioning.

The variations of the type of unanticipated shutdowns are numerous. The simplest case is where a complete load rejection occurs. A more difficult case is where the plant is islanded with the plant capacity being a significant fraction of the island's load. Under this condition the governing performance of the units, which is hardly ever exercised under normal operation, can present problems particularly in cases where steam generator-turbine coordinated control schemes inhibit the usual governor fast response of turbine valves. The plant operators are faced with a condition where the isolated system frequency determines the unit's or plant's MW output. Manual supplementary control of MW output becomes baffling as such control action on a particular unit is negated by system frequency effects acting on the governors of that unit and other parallel units.

Effects of Plant Design

The ability of a steam plant to survive an unanticipated shutdown is very much a function of the plant's design and operator training. Dominant factors are:

- fuel type
- steam conditions and steam generator type
- steam bypass valving and ability to trip to house load
- automatic controls for steam generator, turbine and auxiliaries

Many utilities have upgraded control systems, auxiliaries, and, in some cases, major component modifications to enable units to withstand a generator breaker trip without tripping the turbine or steam generator.

Fuel Type

There is a marked difference in the ability to control heat release in the furnace of a coal-fired unit compared to a gas or oil-fired unit. In the former case, response characteristics of pulverizers can exhibit dead times of up to a minute or more whereas a gas or oil-fired unit has a very predictable short (few seconds) response of the fuel valve. Inherently oil or gas fired units are able to match heat release to steam flow demand transients much better than is possible with coal-fired units. On the other hand this fast response characteristic and low residence time of gas or oil presents other problems such as the possibility of furnace implosions on fuel trips or furnace explosions if

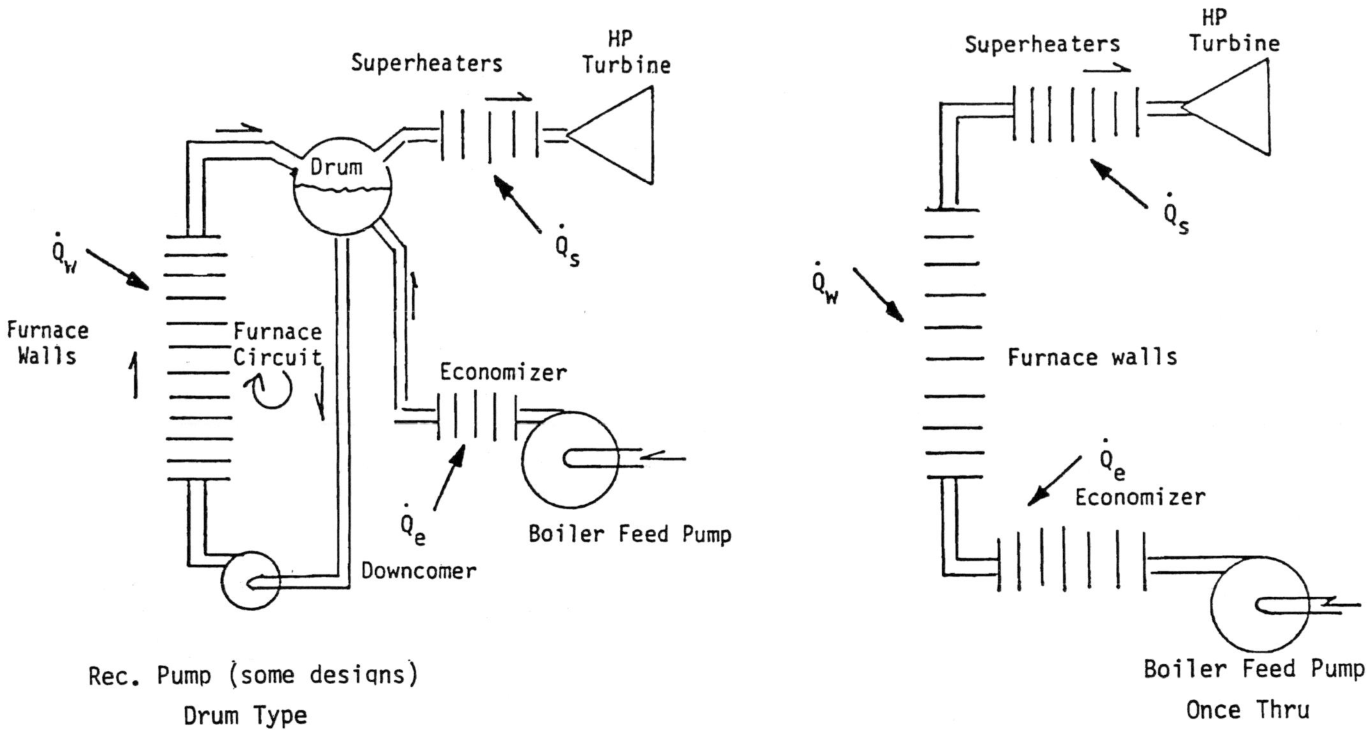

Figure 3. Steam Generator Types

the air supply is not properly coordinated with the fuel flow. A full load trip of a gas or oil fired unit causing a fuel trip is much less traumatic than a similar event on a coal fired unit where fuel shut off results in shutdown of pulverizers loaded with coal.

Steam Conditions and Steam Generator Type

There was a period in the 50's and early 60's when the desire for greater efficiency and economies of scale led to higher and higher steam pressures and temperatures. A significant factor was the introduction of the once-through steam generator design both subcritical and supercritical.

Figure 3 shows schematics of the two basic types of steam generators fluid circuitry on the left is that of the so-called conventional drum-type unit. In this system the pressure level is always below the critical point of steam, 3208.2 psia. The most common maximum is 2400 psig.

Since steam generation occurs in the waterwall sections, the recirculation flow rate around the furnace circuit is five to ten times that of the generated steam flow. Depending on the load the unit is carrying, the waterwall exit flow, for sub-critical designs, is a low quality mixture of steam and water and the furnace circuit remains essentially in a saturated condition at the existing drum pressure. The steam is superheated to rated steam temperature conditions in the superheating sections and passes on into the high pressure turbine.

The fluid circuitry on the right of Figure 3 is that of the once-through steam generator. These units may operate under subcritical and supercritical conditions (i.e., above 3208.2 psi). Note that the flow of fluid through the steam generator above typically 25 to 30% load is continuous, and that as heat is added along the fluid path, the fluid properties change in a continuous manner.

The pressure level in this type unit is basically specified by the flow being forced through the entire boiler circuit by the feed pump. Changes in fuel input will have essentially a transient effect due to the expansion or condensation of the fluid within the steam generator until the new operating conditions are achieved. Since these units operate in the higher pressure ranges, i.e. supercritical, the changes in the fluid specific volume are, relatively, smaller than conventional subcritical units, thus diminishing the effect of heat addition on pressure. In this type unit then, feedwater has a dominant effect on pressure.

An important difference in characteristics and, hence, control requirements of once-through versus drum-type steam generators relates to the fact that the steady state flow rate of fluid in the furnace tubes for once-through steam generators is proportional to load in the normal operating range of 30 to 100%, whereas it is almost independent of load or of feedwater flow in the case of drum-type steam generators. In drum steam generators, changes in feedwater flow or steam flow do not materially affect the fluid flow rate in waterwall tubes and, hence, there is no concern of major imbalance between heat flux to the tubes and heat absorption to the inner fluid. For once-through steam generators, however, it is important to keep a close match between the once-through flow rate, i.e., feedwater flow and fuel flow, since only a few seconds of a major imbalance can result in rapid rise of tube temperatures and burn out. Fuel and feedwater flow must, therefore, be moved in unison and corrections in the ratio of fuel to feedwater should be made within well-defined limits in a slow reset recalibration mode to correct for temperature deviations.

The danger of transient mismatches between heat release and feedwater flow lead to control modes that inhibit the response of the turbine. With inadequately designed steam generator controls, once-through units are much less capable of sustaining islanded operation where governor control of turbine valves is significant.

Another factor, previously mentioned is that these large high pressure and temperature units have much less tolerance to load changes due to fatigue of thick metal parts.

Steam Bypass Valving and Ability to Trip to House Load

Many of the problems of a rapidly imposed load change on steam units can be finessed with the provision of turbine steam bypass valving, a common practice in Europe and particularly other countries where the frequency of system break ups is much greater than in the US, and the interconnected systems are smaller.

The provision of a sufficient capacity bypass system shields the steam generator from rapid changes in turbine steam demand. With such capability it is possible and the usual practice in Europe to be able to trip the unit and have it carry station load. Of course the amount of time that this condition can be supported is limited by temperature induced stresses in the turbine.

Automatic Controls for Steam Generator, Turbine, and Auxiliaries

Perhaps the area where greatest opportunities of improving the reliability and safety of steam plant operation during major contingencies lie in the exploitation of the immense logic and computation power of modern digital control systems.

From the days of pneumatic control to their replacement with analog operational amplifier systems, steam generator control systems were severely limited by the hardware in extending their use through wide excursions in operating conditions. Anytime an additional computation function, such as nonlinear compensation, multiplication or division seemed indicated to improve the control function, this entailed additional hardware with all its problems of calibration, maintenance and reliability. The result was that steam generator control systems were designed for operation under limited operating load conditions. The different requirements through wide excursions or during startup and shutdown were usually handled under manual control.

The transition to Direct Digital Control, DDC, has, for the most part, replaced the old analog controls with a digital implementation of the same limited function.

Although the logic capability available in today's digital systems is literally infinite relative to what is used, extending the function to handle the wide range of contingencies without abdication to manual control requires engineering effort through use of nonlinear multi-variable control design. This involves use of simulation which is again very feasible today with the use of digital computers to an extent never possible in the past with analog computers.

To illustrate, take the schematic of Figure 4 showing the feedwater flow subloop for a once-through steam generator. The flow demand signal generated by another cascade control is shown to be limited so it cannot drop below, say 30%. A controller responds to the error between demand and flow feedback developing a signal to control the speed of the steam generator feed pump through a function generator designed to linearize the speed/flow characteristic of the turbine driven pump. Figure 4 shows the typical nonlinear flow/speed characteristic of a pump supplying the steam generator. There is no flow until the pump speed develops a discharge pressure equal to the steam generator pressure. From that point on, because of the squared law of pressure drop with flow, a very small change in speed produces a very large change in flow and this gain between flow and speed then decreases with increasing flow. A function generator to linearize this characteristic would only be correct at one value of steam generator pressure and supply pressure to the feed pump turbine which can vary appreciably depending on the steam source (extraction from main turbine or an auxiliary steam generator). Evidently it is a simple matter with digital logic to adjust the compensation as a function of these additional variables (steam pressure to feed pump turbine) whereas this would have been discouragingly complicated to attempt with analog implementations.

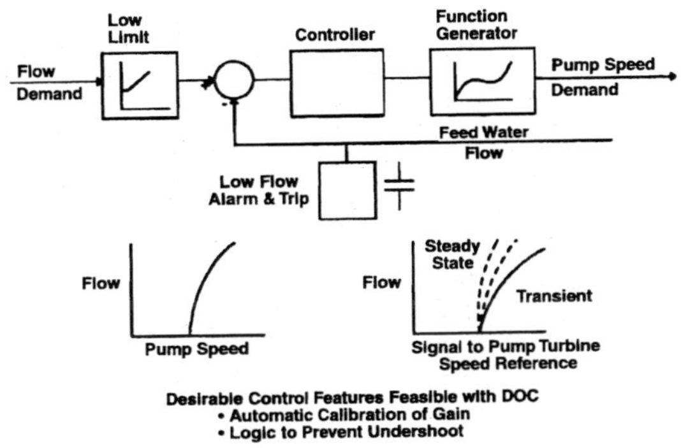

Figure 4. Example of Subloop Control Problem

Another problem in this flow loop is the requirement to prevent flow from dropping below the minimum flow demand level. An alarm and time delay trip is provided to protect the unit from operation at flow rates below minimum. In the configuration of Figure 4 there is no intelligence to the controls to indicate the rate at which the demand signal is approaching the minimum and the particular amount by which the flow is away from minimum. With digital logic it is fairly simple to use the combination of these two items of information to slow down control action in anticipation of the demand signal reaching the low limit, thereby preventing an undershoot in flow.

Another example of an improvement in control that would be of immense value to operators is the problem of drum level control. In many cases these controls trip to manual under upsets such as rapid load changes or load rejections. Even the restoration to automatic usually involves manually bringing the level to setpoint and throwing over to automatic when the error is zeroed. With digital logic incorporating desirable non windup features and appropriate compensation for nonlinearities of pump speed versus flow as function of steam generator pressure this essential control function could easily be provided without any need for manual intervention. One does not achieve these improvements without engineering effort often tailored to the particular plant.

Operator Training

Plants that undergo frequent unanticipated shutdowns provide hands-on experience to operators. The more common case is one where the event is very rare and operators have had little or no prior experience on how to handle the situation.

The concept of training simulators has been well established in the nuclear power plant industry. Similar concepts should also be valid for fossil fired steam plants.

Where control systems are designed to control the plant auxiliaries through wide ranges of upset conditions, the fact that they are not called to perform except in very rare instances, points to the need for periodically checking out the condition of the controls. In the case of digital implementations this does not require checking the computation logic since other normally provided computer diagnostics assure the soundness of the I/O and correct functioning of the CPU. However, the state of actuators, valves and other equipments that are critical to the performance of the controls under these abnormal conditions should be periodically checked out with appropriately designed tests.

CONCLUSIONS

The main points on the problems of steam plant startups in the system restoration process can be summarized as:

(1) The steam plant is a very complex process. Unless the design provides for in the design for the eventuality of unanticipated shutdowns, there is a significant likelihood of equipment damage due to the unusual sequence of events that characterize a system induced plant shutdown.

(2) The ability of a plant to be restarted quickly can be greatly enhanced by basic design features such as bypass valves and tripping to house load as well as resorting to automatic control of auxiliaries through use of the vast control logic power of modern digital systems.

(3) Operator training in unusual shutdown and startup scenario is an important aspect of the restoration problem. The more remote the likelihood of a system induced plant shutdown the more important it is to provide training since the required experience is not developed in day to day normal operating procedures.

F. P. de Mello graduated with a B.Sc and M.Sc degrees in Electrical Engineering from MIT in 1948.

In 1948 he joined Rio Light and Power and Sao Paulo Companies in Brazil and for the next seven years held positions of increasing technical responsibility in system planning and design studies.

From 1955 to 1969, he worked in the General Electric Company Electric Utility Engineering Operation where he specialized in studies of dynamics and electrical machines, excitation control, prime-mover systems, overall power systems and boiler dynamics.

Mr. de Mello joined Power Technologies, Inc. at the time of its formation in August of 1969 as a Principal Engineer, Dynamics and Control. He was appointed Vice President in 1973.

He is presently a Principal Consultant for Power Technologies, Inc.

J. C. Westcott graduated with a B.Sc in Mechanical Engineering from RPI in 1946.

He joined General Electric Company's engineering training program and in 1948 was assigned to the Large Steam Engineering Department working in areas of steam turbine advanced design and development.

In 1960, he joined the Power Generation Sales Division where he eventually became manager of application engineering in power plant automation.

Mr. Westcott joined Power Technologies, Inc. in 1973 as Principal Engineer, Power Generation. He is presently a Principal Consultant for Power Technologies, Inc.

REFERENCES

1) "Special Considerations in Power System Restoration", M. M. Adibi, L. H. Fink, C. J. Andrews, F. Arsanjani, M. W. Lanier, J. M. Miller, T. A. Volkman and J. Wrubel, a report by the Power System Restoration Working Group, 92-WM 106-5 PWRS.

2) "Overvoltage Control During Restoration", M. M. Adibi, R. W. Alexander and B. Avramovic, a report by the Power System Restoration Working Group, 92 W 107-3 PWRS.

3) "New Approaches in Power System Restoration", M. M. Adibi, L. H. Fink, J. Giri, D. S. Kirschen, S. M. Shahidehpour and J. Zaborszky, a report by the Power System Restoration Working Group, 92-109-9 PWRS.

4) "Bulk Power System Restoration Training Techniques", M. E. Cooper, M. M. Adibi, L. H. Fink, R. G. Wasley, R. T. Gonzales, A. B. Storey, F. J. Wood, R. J. Kafka, E. J. Dorowolski, K. Nodehi, J. G. Waight and R. Hoffman.

5) "Two-Shift Handley Station Units #4 and #5", Bill Hoerster, Walter Gorzegno, Heinz Termuehlen, American Power Conference, Chicago, Illinois, April 22-24, 1985.

6) "Steam Station Protection Working Group", Minimum Recommended Protection, Interlocking and Control for Fossil Fuel Unit-Connected Steam Stations, IEEE-Trans. PAS-92, p. 374-401.

AN AGC IMPLEMENTATION FOR SYSTEM ISLANDING AND RESTORATION CONDITIONS

Herman B. Ross

Virginia Power Company

Ning Zhu, *Member IEEE*

Jay Giri, *Sr. Member IEEE*

Barbara Kindel, *Member IEEE*

ESCA Corporation
11120 NE 33rd Pl. Bellevue, WA 98004

ABSTRACT

An Island AGC function has been implemented in a large-scale power system, Virginia Power Company (VP), to assist in system emergency and restoration situations. It is the first time that the conventional AGC function has been enhanced to be utilized, rather than to be suspended, under such circumstances.

The Island AGC function is designed to be implemented in both the real-time and the Dispatcher Training Simulator (DTS) environments. The latter provides a realistic closed-loop simulation environment that mimics the real system behavior, offering an excellent opportunity for operators to be well-trained, aware and prepared for abnormal situations.

This implementation has been tested on the VP system model, which consists of 4400 buses and 775 generators. It has successfully completed custom factory tests, and it is the user's opinion that the Island AGC will provide a valuable additional capability for use during system restoration.

1 INTRODUCTION

AGC has been an important, routine, everyday function of real-time power system operations for many decades. Its overall objectives have been to:

- Match total generation to meet total load.
- Regulate system electrical frequency deviation to zero.
- Modify generation to ensure that the net scheduled area interchange is maintained.
- Allocate generation among the generation sources to minimize operating costs.

The first objective is typically associated with the primary regulation of the governor speed controls; this happens in a few seconds. The next two objectives are attained through the "supplementary control" action of AGC; this occurs in a time frame of tens of seconds. The fourth objective is met by "tertiary control" using the economic dispatch function within AGC; this occurs in a time frame of minutes.

93 SM 519-9 PWRS A paper recommended and approved by the IEEE Power System Engineering Committee of the IEEE Power Engineering Society for presentation at the IEEE/PES 1993 Summer Meeting, Vancouver, B.C., Canada, July 18-22, 1993. Manuscript submitted Sept. 1, 1992; made available for printing April 21, 1993.

PRINTED IN USA

Even during normal operation, the demands on the AGC function are increasing due to the following trends currently facing the electric utility industry:

- Growth. This means that AGC needs to handle more units and tie lines.
- Open transmission access. This is resulting in increased wheeling of power across systems.
- Joint ownership of units. This means having to deal with dispatch and regulation of units that are jointly owned. This also creates situations where one's load and/or generation is physically located in a neighbor's power system.
- Increasing numbers of base-loaded units, such as hydro, nuclear, and non-utility generators. This results in decreasing amounts of generation available for AGC regulation purposes.
- Integration of power plant computers with AGC. This makes real-time plant data available for use in AGC.
- Consideration of environmental and emission constraints during dispatch.
- Consideration of unit fuel constraints and plant fuel contracts during dispatch.
- Impacts of large-non-conforming loads such as steel mills and arc furnaces. These loads swing hundreds of MW in tens of seconds and have a severe impact on AGC.
- Increased emphasis on AGC performance measures.
- Increasing need to include transmission security constraints during dispatch.

Another, more challenging, issue facing utilities is the need to cope with system break-up during emergency and restoration situations.

Typically, AGC systems are designed to suspend control during emergency situations. System islanding is one such emergency situation. AGC is specifically designed to immediately suspend control if an islanding situation is detected. AGC detects an island by monitoring system frequency deviation and/or by comparing system frequency with the frequencies of generating units across the system. If system frequency deviates from normal, or deviates from a unit frequency, by greater than some threshold, it is an indication that, potentially, a portion of the system has been islanded and is not an integral part of the interconnection. AGC systems are typically designed to suspend control under such circumstances since AGC controls units to satisfy interconnected system objectives. AGC is suspended primarily to preclude dangerous and incorrect control of units. Under these conditions, some of the units may actually belong to the island so that controlling them is of no avail, in terms of meeting the interconnection obligations of the AGC function. In fact, under

Reprinted from *IEEE Transactions on Power Systems,* Vol. 9, No. 3, pp. 1399–1410, August 1994.

these circumstances, controlling them would almost definitely result in worsening the island generation/load imbalance and island frequency deviation. This in turn, would result in frequency relay trips of units and loads and eventually lead to a blackout of the island.

This paper describes an Island AGC function. It is an enhancement to the conventional AGC function that is intended to allow AGC to be used effectively during emergency and restoration conditions.

The Island AGC function discussed in this paper is designed to operate in parallel with the primary AGC function on-line just for the island. This enables an islanded part of the system to run on "auto-pilot" while the operator focuses on restoring normalcy across the rest of the system. The current implementation requires the operator to determine that the island exists and which generators are within the island. The generators in the islanded portion are manually assigned to a second AGC function, which has the primary objective of maintaining constant frequency in the island. This selection is done on-line without adversely affecting the primary AGC function. It is highly likely that the primary AGC function would normally have suspended control by now due to the islanding situation. Once the rest of the system is restored, the primary AGC function can be put back on active control. Later, the island can be reconnected to the system and the islanded units put back in the primary AGC; thereby reverting to the normal case of a single AGC function.

The Island AGC function can be implemented in the real-time and in the DTS environments. By implementing this function in the DTS, the operators have the opportunity to experience this emergency situation in a secure, off-line environment without affecting real-time operation. They can study, at their own pace, the behavior of power flows and frequencies in the system and their dynamic characteristics during system breakup, islanding and subsequent reconnection of the island. This permits a better understanding of the power system phenomena that occur during such emergency conditions and, in turn, provides confidence in restoring the system efficiently. By practicing alternate courses of actions in a less stressful, less intimidating and secure environment, operators are better prepared to confidently handle emergency situations if and when, they arise in real life.

The major functions and features of the Island AGC function, as well as test results on the large-scale VP system model, are presented in this paper.

2 THE MOTIVATION

2.1 Why Island AGC?

Utilities are continuing to focus on system breakup emergency scenarios and to address the means to cope with them effectively. As part of their system planning objectives, many utilities are designing their protection schemes to automatically create viable islands during cascading emergencies rather than blacking out portions of the system. The guideline being used is that "a successful island is a much better condition that a blackout." Thus, systems are being designed to survive cascading emergencies by automatically creating viable islands that have a reasonably good generation/load balance. This ensures that, although the power system is operating in an abnormal degraded state, customers are continuing to be served. Once this condition is reached, the operational challenge is to ensure a continual balance between generation and load and to maintain the island frequency at 60Hz. This is where the Island AGC function fits in.

Provision of a second AGC function, which maintains frequency of an island for the short-term, will help the operator focus on issuing manual directives to restore the rest of the system. This feature can be a valuable asset to the operator during this period of restoration. The operator can essentially ignore the island while he addresses the rest of the system. Later this island can be reconnected to the rest of the system. When the reconnection is to be made, the Island AGC will be set to control to the same frequency schedule as the primary AGC function. This will allow the island's phase angle to come in phase with the rest of the system in a smoother manner, significantly reducing the "bump" that might otherwise occur, and helping to insure that the connection will remain intact.

2.2 The Virginia Power System

The Virginia Power system is geographically dispersed with several major generation and load centers. Some portions of the transmission system are more heavily loaded than others. The Virginia Power system is along the east coast of the United States where damage by windstorm and thunderstorms may cause tripping of heavily loaded transmission lines. Under certain circumstances, this might result in additional cascading trips, possibly creating an island within the Virginia Power control areas. Although an islanding situation has not been experienced in real life, engineering studies indicate that, during emergency conditions, the potential for automatic creation of islands within the Virginia Power system exists.

3 THE CONVENTIONAL AGC FUNCTION

The AGC function has played an important role for many years in power system control, monitoring, and operation. A functional diagram, along with a short description of the interfaces and major functions for the conventional AGC system, follow:

3.1 AGC Model

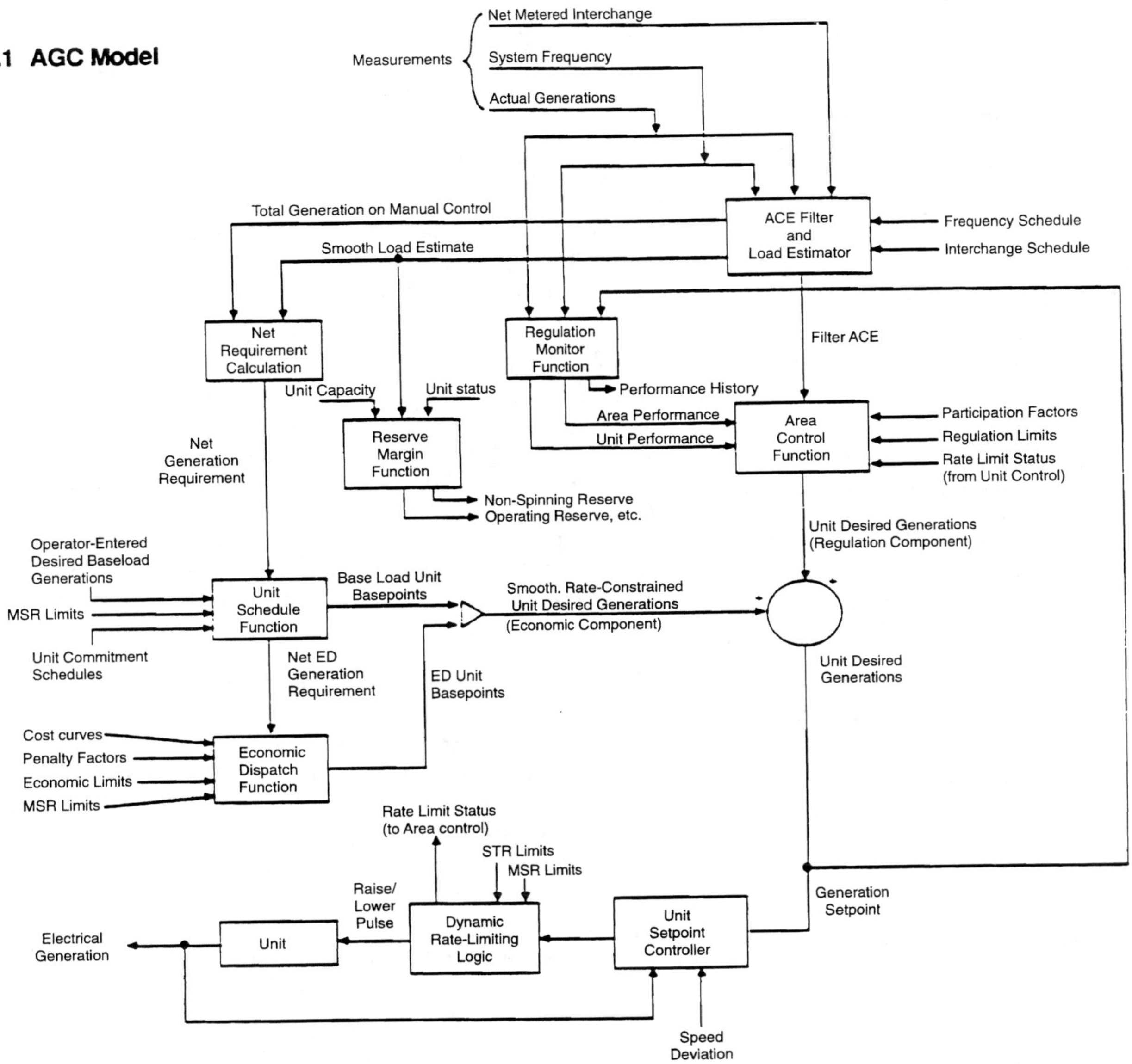

Figure 1. Block Diagram of the Conventional AGC System

3.2 The Interfaces

The AGC function interfaces with the following EMS functions:

- SCADA
- State Estimator
- Security Enhancement
- Load Forecast
- Real-Time Operating Plan

3.3 The Major Functions

The major functions of the Conventional AGC System are:

- Load Frequency Control (LFC)
- Economic Basepoint Calculation (Economic Dispatch)
- Reserve Monitor
- AGC Performance Monitor

4 THE ISLAND AGC FUNCTION

4.1 Overview

The following diagram (Figure 2a.) shows a high-level configuration of the Island AGC function.

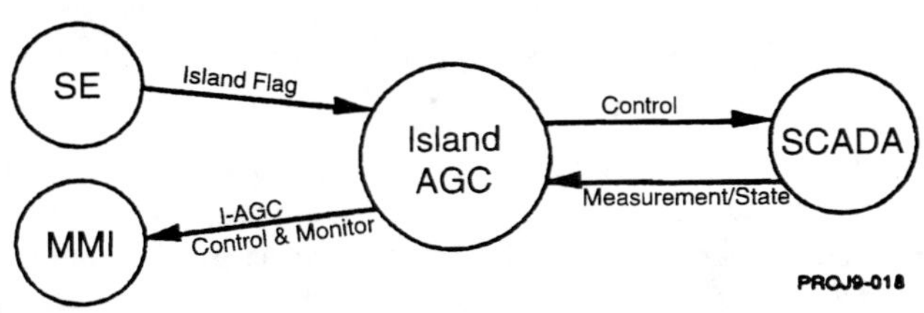

Figure 2a. Island AGC Configuration

The typical sequence of actions in invoking the Island AGC function would be:

- The islanded situation is detected by the State Estimator through network topology processing.

- Once the island is identified, the operator manually specifies the frequency source to be used as well as the units to put under the Island AGC. In the future implementation, this step eventually could be performed by the software, which could automatically assign the available frequency source and units within the islanded area to the Island AGC, according to the islanding information from the State Estimator.
- After activating the Island AGC, the operator can manually adjust the control parameters (such as frequency bias or scheduled frequency) in order to obtain appropriate dynamic performance and to speed up the restoration.
- Eventually, after the rest of the system has been restored to normalcy and the frequencies in both the primary and the island areas are close enough, the island would be reunited with the primary area.

An easy-to-use user interface provides a flexible tool for the operator to modify the Island AGC control parameters. The on/off status of the Island AGC, the desired island frequency source and units, as well as the Island AGC parameters, are all controllable by the system operator through these displays. Figure 2b is a hard copy of the Island State display which shows the current operating information as well as the major AGC parameters of the islanded area.

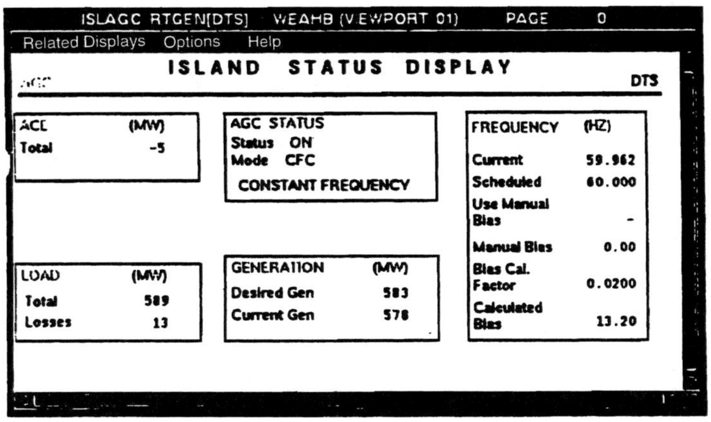

Figure 2b. Island AGC Display

4.2 The Island AGC Features

4.2.1 Working Environment

The main purpose of the Island AGC function is to help with system restoration under the islanded situation. Therefore, the function is designed to supplement real-time control and to run simultaneously with the primary AGC function.

Although the islanding situation occurs rarely in the real world, it can cause severe disturbance and damage to the power system. In order to train system operators to deal with such an unusual situation with greater confidence, the Island AGC function is designed to be implemented in the DTS environment. This offers an excellent opportunity for system operators to simulate the islanding scenario on their own system. They can observe all the events in sequence under different system operating conditions, study the feasible preventive and restoration strategies, manipulate two (primary and island) parallel AGCs, and study the effect of adjusting parameters -as a whole-to prepare for possible islanded situations.

4.2.2 Software Design

Because the islanded situation is not a normal operating condition, the Island AGC function itself is different from the conventional AGC function in the following ways:

4.2.2.1 The Control Mode

The island is isolated from the rest of the system, which means no tie lines exist. The Constant Frequency Control (CFC) mode is used instead of the typical Tie Line Bias Control (TLBC) mode. Also, AGC is specified to be always working in emergency status instead of being dependent on the value of ACE to select the mode (normal, assist, or emergency).

4.2.2.2 The Frequency Bias

In the past, numerous investigations have been conducted to learn how to specify the frequency bias and to select the proper settings [4-6]. Two different ways are provided in this Island AGC implementation. The first is the same as the one used in the conventional AGC function: enter a constant value. The second way is to enter an island frequency bias as a function of the current island operating status. This feature brings the bias closer to the system's nature and offers more flexibility in improving the dynamic performance.

4.2.2.3 The Impacts

The impacts of the Island AGC on the original AGC software are:

- Plant Controller (PLC). Each PLC is under the control of either the Primary AGC of the Island AGC, depending on its designation, and the PLCs in each area are responsible only for the regulations in their own area. Therefore the PLC regulation factors, as well as the PLC control signals, should be calculated separately to correspond to its designated area.
- Dynamic Interchange. An island is an isolated area with no connections to any neighboring areas; it will not participate in dynamic interchanges until it is reconnected with the primary area.
- Reserves. The units in the island can no longer contribute to the reserves of the primary area. They should be excluded from both the operating and the spinning reserve calculations in the primary area.

4.2.3 Operator's Options

The Island AGC function designed here offers operators control and freedom on the following issues:

- Start/Stop the Island AGC function. The system operator has the control of the time and location to start the Island AGC function. The operator also determines when the island is to be integrated with the primary area and when to stop the Island AGC function.
- Define/Adjust control parameters. The major parameters, such as the frequency bias and schedules in the Island AGC function, can be entered and adjusted by the system operator through an easy-to-use user interface. Different values can be used during different phases to adapt and improve the system dynamics.
- Specify the island. The system operator has the full freedom to choose which island to put under the Island AGC control. He can select the frequency source and the units for the specified island. This option gives the system operator more flexibility in handling the system restoration under an islanded situation.

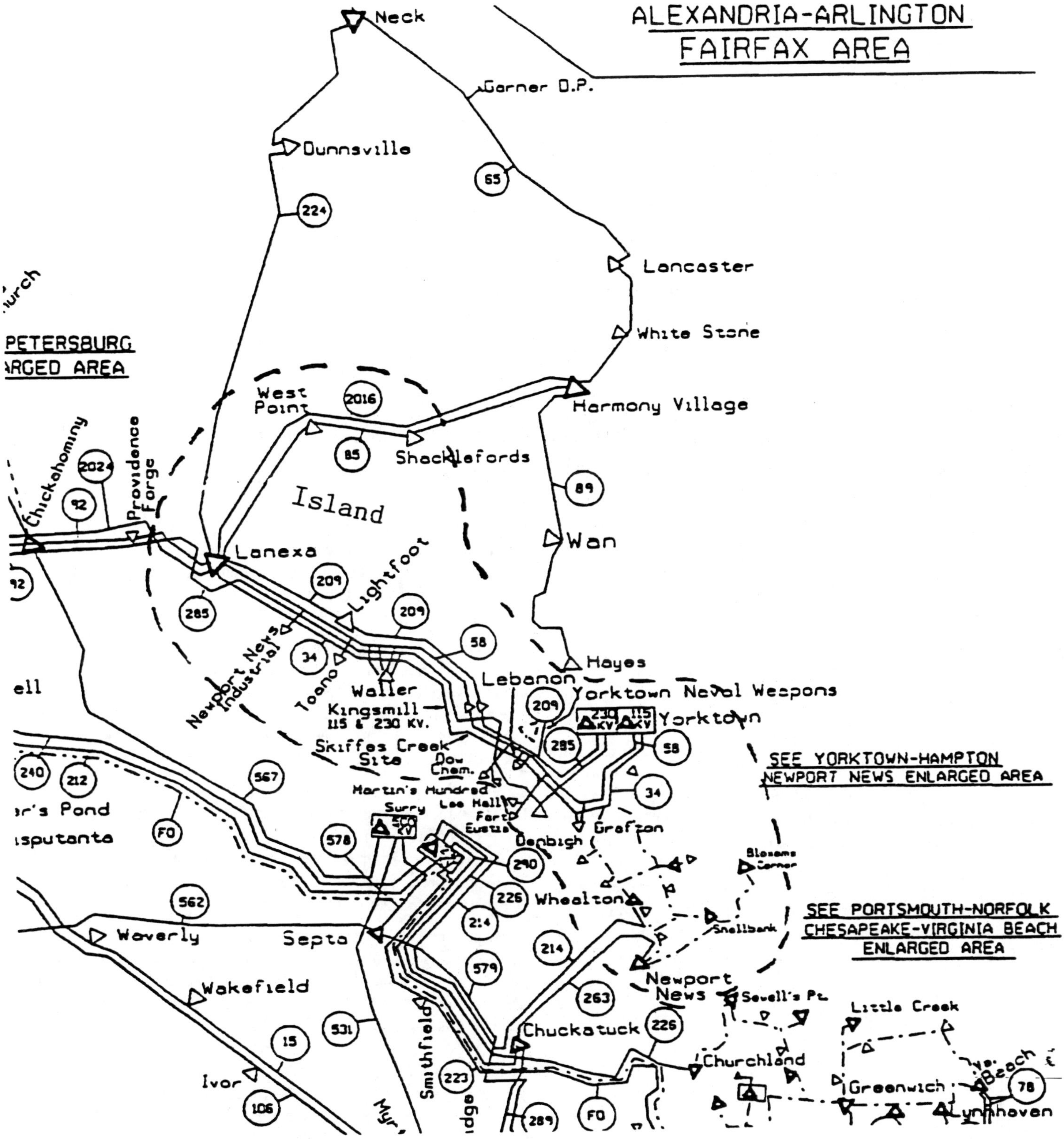

Figure 3. One-line Diagram of the Northeast Portion

5 TESTS AND RESULTS

5.1 Test Scenario

The tests were performed, under the DTS environment, on the Virginia Power system model with 4400 buses, 5000 lines, 2133 loads and 775 generators.

The islanding potential exists in several portions of the system, the northeast portion, shown in Figure 3, was chosen here as our test bed.

The test began with several lines in the north corridor, which is an important transmission link between the northeast and the northwest, being tripped.

- LINE 2024 between the stations LANEXA and CHICKHOMINY.
- LINE 285 between the stations WALLER and CHICKHOMINY.
- LINE 92 between the stations PROVIDENCE FORGE and LANEXA.

This event could occur during a severe windstorm or as a result of the overloaded situation of these lines. The redistribution of the power previously transmitted through these tripped lines might cause overload on the other lines. As a result, the area could be islanded by the consecutive CB openings at the following lines:

- LINE 2016 between the stations LANEXA and HARMONY VILLAGE.
- LINE 85 between the stations SHACKELFORDS and HARMONY VILLAGE.
- LINE 224 between the stations DUNNSVILLE and LANEXA.
- LINE 214 between the stations SURRY and WINCHESTER.
- LINE 263 between the stations NEWPORT NEWS and WINCHESTER.

The northeast islanded area usually appears as a load in the Virginia Power system. At the moment the system was islanded in the test, the primary system had 10961 MW generation, 10540 MW load and 421 MW interchange. The sudden loss of load caused an over-generation situation in the primary area, so the generators tended to speed up (only slightly), while the situation on the island was just the opposite. The mismatch between generation and load created an under-generation case. With a much smaller inertia, the frequency deviation in the island is much bigger than that in the primary area, as observed from Figure 4.

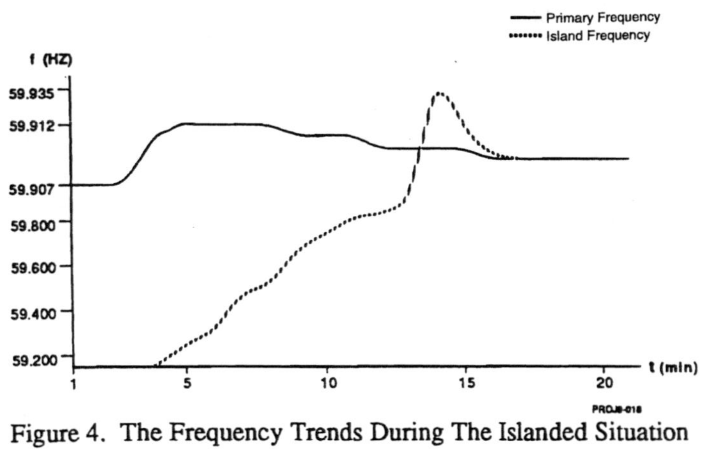

Figure 4. The Frequency Trends During The Islanded Situation

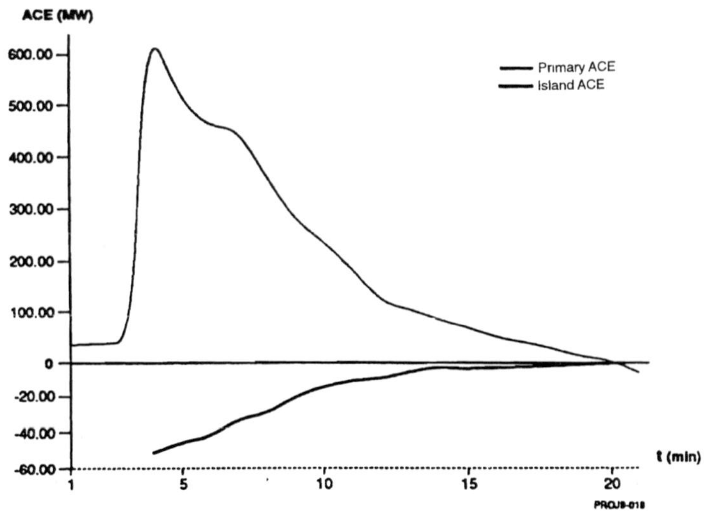

Figure 5. The ACE Trends During The Islanded Situation

5.2 Test Results

Figure 4 displays the frequency profile before, during and after the islanded situation. Figure 5 shows the ACE curves in both the primary and the island areas during the islanded period.

It is apparent, from studying Figure 4, that the frequency of the primary area was hardly disturbed because of the much greater inertia and much smaller relative load disturbance; while a rather big ACE at the islanded moment appears in Figure 5, which was caused by the sudden line tripping. (The real delivered interchange jumped from 421 MW to 980 MW at the islanded moment.)

The islanded area suffered much more than the primary area. The island frequency had a rather big dip at the beginning because of the severe under-generation condition and the much smaller inertia. The advantage of having a smaller inertial is the quicker dynamic response, while the disadvantage is the possible overshoot. These features are displayed in Figure 4.

There is always a trade-off between the speed of the response and the amount of the overshoot. In the islanded frequency restoration process, the bigger the bias factor, the faster the response but at the same time, the larger the overshoot. This can be observed from Figure 6 where curve A uses a smaller bias factor (5%) while curve B uses a greater one (10%).

The dynamic performance can be improved greatly and restoration achieved in less time with the assistance of experienced system operators. Taking the system security and performance into account, control parameters can be adjusted on-line and some restrictions can be relaxed for certain situations (all based on the current operating conditions). For example, the over-generation situation in the primary area during the islanded period can be drastically relieved by manually shutting down some generating units in the Bath County, especially at the beginning stage. The quicker the reduction of the mismatch between the load and generation will greatly improve the system performance so that the ACE in the primary area decreases much faster, as seen from curve B in Figure 7. On the other hand, the under-generation condition in the islanded area can also be improved through the operator's assistance. The recovery of the island frequency would be smoother and faster if the basepoint settings of the islanded units are monitored closely and adjusted properly, especially when the frequency-dependent load takes a fairly large percentage. In addition, putting some fast response units in the manual control mode and ramping up/down at their maximum ramp rate is also helpful. This effect can be observed from Figure 6 by comparing curve C with curve A, where curve C involves the operator's manual control of the islanded units in Yorktown station besides the Island AGC while curve A does not.

Because the two AGC functions are independent, the disturbance of controls only affect the area where they occur. This can be seen by comparing Figure 5 and Figure 7 where the ACE curve in the primary area varies significantly corresponding to different operating conditions, while the island ACE curve is not influenced at all.

The system was integrated when the frequencies in the primary and the island areas were close enough for a smooth reunion, all the CBs opened before were closed and the Island AGC was turned off to allow the Primary AGC function to take care of the whole integrated system.

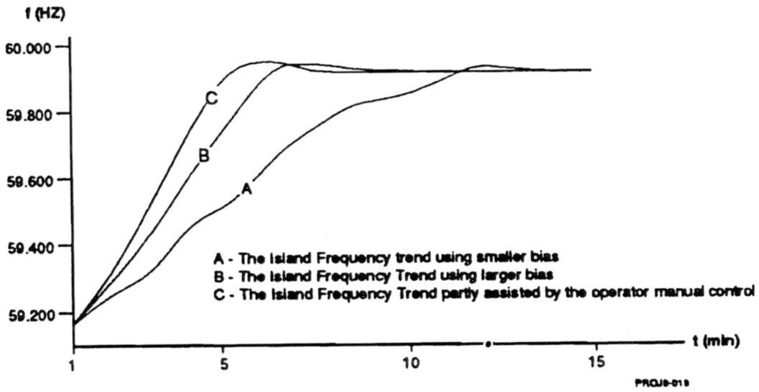

Figure 6. The Island Frequency Trends Under Different Conditions

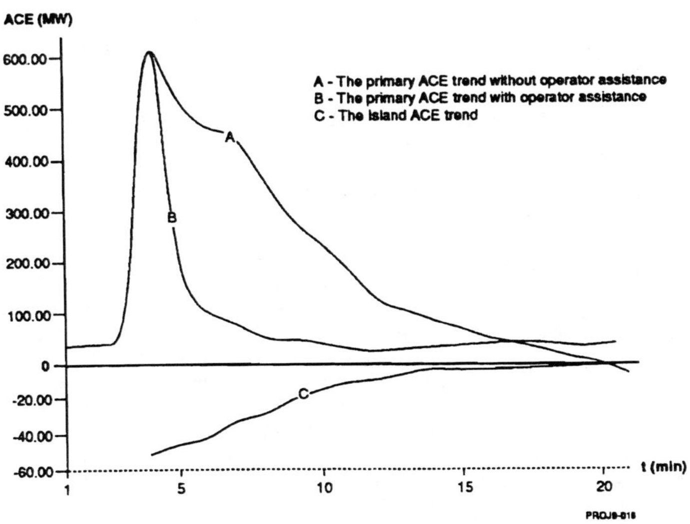

Figure 7. The ACE Trends Under Different Conditions

6 CONCLUSIONS

In the past few years, the power industry has been focusing more closely on system restoration issues.

The Island AGC implementation described here, presents a novel approach for AGC to be used during the islanding condition and system restoration. After islanding, the operator's attention is simultaneously drawn to many diverse and critical issues. This new AGC implementation allows the operator to quickly create a second, short-term AGC function just for the island. The islanded portion of the system can then operate in a self-sustaining, automatic mode so that the operator can focus attention on restoring the rest of the system. At the reunion of the island, the electrical aberration that occurs at the moment of reconnection will be significantly reduced and may prove to be almost "bumpless." This will help insure that the reconnection will hold the first time rather than having the interconnect line(s) relay open, which may then expose the island to enough instability that it could shut down entirely.

The Island AGC function can be implemented in both the real-time and the DTS environments. The DTS provides a realistic closed-loop simulation environment which mimics the real system behavior. It is an excellent environment for development, demonstration and training.

The authors are aware that the Island AGC function is not field proven and that it may not be readily provable until a real situation occurs and someone "steps up to the bar" to try it out. We do, however, believe that an idea such as this must be tested analytically via simulation as a first step in determining the conceptual feasibility of the process. The ESCA simulator is a mature product. The simulator used extensively in the paper to test and validate other real-time EMS functions (such as AGC and the State Estimator) in an offline secure environment, has shown that manually initiating Flat Frequency Control (after the island is identified) of the islanded area can maintain the island frequency at the scheduled level.

The results from the Virginia Power system's simulated islanding demonstrate a successful implementation of the Island AGC function as a supplement to the primary AGC function.

ACKNOWLEDGMENTS

The authors would like to thank Chuck Dubose for his support during the test. Thanks also go to Lindsay Paulson, Andrea Ellis and Daniel Monda for their editorial assistance.

REFERENCES

[1] L. K. Kirchmayer, *Economic Control of Interconnected Systems*. New York, NY: Wiley, 1959.

[2] IEEE Current Operational Problems Working Group, "Current Operating Problems Associated with Automatic Generation Control," IEEE Trans. PAS-98, No.1, 1979.

[3] T. M. Athay, "Generation Scheduling and Control," Proc. of IEEE, Dec. 1987, pp. 1592-1606.

[4] N. Jaleeli, et al., "Understanding Automatic Generation Control," IEEE Trans. on Power Systems, Aug. 1992, pp. 1106-1112.

[5] L. S. VanSlyck, et al., "Implications of Frequency Bias Settings on Interconnected System Operation and Inadvertent Energy Accounting," IEEE TPS, May 1989, pp. 712-723.

[6] T. Kennedy, et al., "Variable, Non-Linear Tie-Line Frequency Bias for Interconnected System Control," IEEE TPS, Aug. 1988, pp. 1244-1253.

[7] *North American Electric Reliability Council Operating Manual*, Guide I. Systems Control, Dec., 1987.

Discussion

M. L. OATTS (Southern Company Services, Inc., Birmingham, AL. 35202): The authors have presented, in an easy to follow paper, an interesting scheme for providing an Island AGC application. Upon review, several implementation issues come to mind. The authors' thoughts on these would be greatly appreciated.

First, the premise for utilizing the proposed Island AGC implementation assumes the existence of what the authors call a "viable" island. Is this implementation only intended for use with islands created via controlled methods such as the protection schemes referred to? If not, several questions come to mind:

- How are uncontrolled (or randomly) created islands presented to the operator so he quickly understands its makeup and can utilize the Island AGC (e.g., select a frequency source, define units in the island, etc.)? Is a special display used, does he rely on island identification on buses from a power flow, other? This is often one of the most difficult problems in an island situation - operator communication.

- If the State Estimator is used as indicated in the paper to notify the operator of an island, is this necessarily sufficient for determining an island? The situation which comes to mind is the continued connection of sub-transmission level lines to the "island" identified by the State Estimator. Often, only bulk power transmission networks are modeled by the State Estimator and in the case mentioned a frequency island would not actually exist.

- How are sufficient frequency source points to be selected and defined in the database to cover the "random" island condition? Off-line simulation of island creation, frequency source at all plants, other?

If instead, the Island AGC is intended for use only with the controlled creation of an island, other questions arise. How would the creation of such an island be ensured by the protection scheme mentioned? How would such a scheme consider the actual dispatch settings of the units and actual "island" load when determining "a reasonably good generation/load balance" as the paper describes? The utilization of unit capabilities is obviously not sufficient since the actual loading may not meet the needs of the island when it is created. For example, units in the island area upon its creation are on minimum but the island load requires near maximum unit output. Since most AGC controllable units can not normally move instantaneously to the desired output of the island, what would the protection scheme do? Unless some intelligent, adaptive protection scheme is utilized, it would seem that a rigid, preset tripping scheme to create the island could cause problems in the island (such as generator trips or load shedding) for system conditions different from that for which the scheme was developed.

Any test results and operator feedback which you have obtained with the Dispatcher Training Simulator that address the preceding questions/issues would be very interesting.

Finally, some general philosophical questions come to mind concerning the idea of using Island AGC. The authors stated that one benefit of the implementation is that it facilitates the phase matching of the island to the rest of the system to reduce the "bump" which would occur upon reconnecting the network. This benefit is understood. Does this not, however, have to be balanced against the risk involved in doing "normal" AGC in the island? Since the determination of the frequency bias factor to be used by the operator in the island will be difficult at best in the "random" island condition, why not let the plants in the island hold frequency via manual control? This is especially true if a "guess" at the frequency bias is used. It does not seem unreasonable to expect that the plant can probably do control as well as an "untuned" AGC. The movement of plant output by plant control would seem to be a less risky alternative under a very tenuous system condition, especially considering that the need to tightly hold a scheduled island frequency is questionable. Several parts of the world control their system frequencies manually and this would seem appropriate in this case. The control of the frequency via plant intervention would seem to minimize the risk of the Island AGC program overcontrolling units or deciding to move them to inappropriate loading points under abnormal system conditions. Would the authors please comment?

Manuscript received August 12, 1993.

LESTER H. FINK, ECC, Inc., Fairfax, Virginia. Suspension of automatic control ("tripping to manual") during emergency conditions has always seemed somewhat incongruous, since those are precisely the conditions under which the operator needs all the help he can get. It is encouraging to see that this anomaly is now being addressed. The authors articulate clearly the need for such capability, and recognize the steps that are required: detection of the islanded situation; identification of the equipment comprising the island; adjustment of control parameters; re-integration of the system. At present, these steps are performed manually, but the potential for automation is pointed out by the authors. Also, and necessarily, this new AGC function has been tested only by simulation using an dispatcher training simulator.

Several aspects of the newly designed function remain unclear. First, it is unclear whether the authors are addressing intentional or unintentional islanding. They cite the development of automatic creation of islands during cascading emergencies as a means of avoiding total blackout of the system. (A prominent example of this development is the "defence plan" now being perfected by Electricité de France [A].) Both intentional and unintentional islanding occur during rapidly deteriorating situations. In the case of intentional islanding, coordinated underfrequency load shedding may be employed to establish a rough load-generation balance. In the case of unintentional islanding, the load-generation imbalance within an island is likely to be large, and any control, to be effective, must be imposed promptly and emphatically. While it is not clearly stated in the paper, the overall impression given is that the authors are dealing with unintentional islanding. In either case, but especially in the latter, it is likely that the present reliance on manual implementation will be unsuccessful, since the rapidity with which such situations evolve will not permit of timely and effective operator action. Thus, effective deployment of the new AGC function will require that it be fully automated throughout the first three steps: island detection, island definition, and parameter adjustment.

The second point that remains unclear is how the frequency bias is calculated "as a function of the current island operating status." Presumably, this is done in some simple, straightforward manner, but several such formulae can be postulated, having various degrees of plausibility. A description of the formula actually used would contribute to assessment of the new function.

The authors are congratulated on presentation of a pioneering concept.

[A] C.Counan et al.: Major Incidents on the French Electric System - Potentiality and Curative Measures; IEEE/PES Summer Meeting, 1992, Seattle (92SM432-5PWRS)

Manuscript received August 16, 1993.

RICHARD P. SCHULZ, American Electric Power Service Corporation, Columbus, Ohio 43215: The authors are to be commended on having put a basically good idea into practice in their System Operating Center (SOC). Tomas DyLiacco has described his experience with use of a local supplementary control within power plants, that is put in service where the plant is islanded [1]. Those applications differ in that they are in smaller, developing utilities, while Virginia Power (VP) operates as part of a large interconnection with an experience record of no islanded operation with local frequency control in recent decades.

The experience of several upsets [2...5] suggests several things about the nature and formation of electrical islands within larger interconnections: 1) usually there is a generation-load imbalance within the island immediately upon its formation; indeed, the power flows across the boundary of the to-be-formed island are usually a factor in the island formation. Then there is consequent rapid change in the island frequency immediately after its formation, with frequency rates up to ±3 Hz/sec; 2) the boundaries for the island are unpredictable in interconnected power systems that rarely experience islanding operations; 3) Where operators are experienced in analyzing in handling one situation, they tend to use the lessons learned in that experience in other situations in which they may not be appropriate. These observations suggest these questions about the operator training on Island AGC with the dispatcher training simulator (DTS):

a) Does the dispatcher training simulator have the capability of simulating any of several scenarios with different island boundaries and different load/generation imbalances within the island?
b) Can the trainer who directs the DTS choose the timing and circumstances of the island formation?

These features will enhance the training of operators by including training in the ability to recognize and act on unexpected conditions.

It also has been observed [6] that power system simulations are unlikely to capture the response of the islanded part of the system, in large part because of poor knowledge about the response of controls and protective devices in these unusual operating conditions. Have the authors considered changing the modeling parameters used within the DTS from time to time so that the operators will experience the unpredictability of an islanded system's response to the islanding? This would enhance the ability of the operator to respond to unexpected consequences of islanding, including complete collapse of the island.

The authors' presentation of the paper described a mechanism within the automatic generation control (AGC) software for using topological network analysis to determine (by software) the boundaries of an island, and to communicate the boundaries to the operator when island(s) occur. They also described the use of several frequency sensors throughout the VP area to provide an independent indication of island boundaries should an island form. This suggests some additional questions. Does the topological islanding detector directly suspend AGC action to the units within the island? Have the authors considered using phasor measuring devices to indicate the islanding of an area? There have been circumstances in which Eastern American Interconnection utilities have created islands during normal switching operations; these islands were characterized by slowly diverging frequencies. The phase displacement from within and outside the island would be a more immediate island indicator than would frequency in those conditions.

Again, the authors are thanked for having described this uncommon improvement in the ability to control the power system during unusual circumstances.

[1] Personal Communication, Tomas E. DyLiacco, July 21, 1993.

[2] Davidson, D.R., Ewart, D.N., and Kirchmayer, L.K., "Long Term Dynamics Response of Power Systems: An Analysis of Major Disturbances", IEEE PAS-94, May/June 1975, pp. 819-826.

[3] Brand, C.W., Usry, R.O., "Gulf Coast Area System Disturbance Final Report", Southern Services, Inc., Birmingham, AL, October 26, 1973. Vol. I, Text and Various Data; Vol. 2, Appendices.

[4] Union Elec. Co., Illinois Power Co., "Islanding of Union Electric Co. and Western Part of Illinois Power Co. Systems, 10:32 a.m., February 13, 1978", Report to MAIN, April 24, 1978.

[5] EPRI, "Simulators: Tough Training for Top Operators", EPRI Journal Vol. 13, No. 4, June, 1988, pp. 22-29.

[6] Schulz, R.P., "Capabilities of System Simulation Tools for Analyzing Severe Upsets", Proceedings, International Symposium on Power System Stability, Ames, Iowa; May, 1985, pp. 209-215.

Manuscript received August 16, 1993.

H. B. Ross, Jr.:

The authors wish to point out that an error was found, after the formal paper had been submitted, on page 6 in section 5.1; the last bullet item should read:

- Line 263 between the stations NEWPORT NEWS and CHUCKATUCK.

This error was commented on at the presentation of the paper. The authors apologize for this oversight and ask for the reader's indulgence.

Author's Discussion :

The authors wish to thank all those who attended the formal presentation of the paper and a special thanks to those who prepared written discussions.

In retrospect, the paper probably should have been titled in such a way that the restoration process would have received the main emphasis rather than the system islanding process. We say this because, indeed, the restoration concerns were the primary motivation for developing the Island AGC function. Under the restoration scenario one is dealing with an already sharply defined geographic area that is without electric power and the load pickup on the available generation is being carefully added manually by field personnel as they reconnect circuits within the island area. With this situation definition, several of the questions voiced by the discussers would probably not have been of concern.

While the primary goals of this functionality are to aid in the system restoration processes and to serve as a training tool for the electric power system operator, by simulation of islanded areas, it is hoped that the functionality will in time come to fruition and do all that is alluded to in the paper and the various discussions. With this in mind, the questions raised by the discussers are addressed in the following paragraphs.

For purposes of clarity we have listed the questions raised by each discusser and followed it with our response.

Discussion by: Mr. Richard P. Schulz
American Electric Power Service Corporation
Columbus, Ohio 43215-2352

Question 1:
Does the dispatcher training simulator (DTS) have the capability of simulating any of several scenarios with different island boundaries and different load/generation imbalances within the island?

Answer 1:
Yes, the simulator is capable of simulating most real world operational configurations. This is so because the scenario is built by using devices in the DTS' real world model of the transmission and sub-transmission equipment, lines, breakers, switches, generators, etc. Whatever island boundaries the trainer or engineer wish to establish, taking into account the real world electrical capabilities, can be set up for the scenario to run and the capability even exists to interactively change the scenario as the simulation progresses.

The load/generation imbalances are not modelled in great detail. Rather, the island area loads will be adjusted during the scenario by a simple load curve model for the load(s). The amount of imbalance can be controlled by modifying the load curve data.

Question 2:
Can the trainer who directs the DTS choose the timing and circumstances of the island formation?

Answer 2:
Yes, please refer to Answer 1 the first paragraph. The trainer can designate specific equipment and the times each item or group of equipment is switched open or closed by the simulator's scenario process.

Question 3:
Have the authors considered changing the modelling parameters used within the DTS from time to time so that the operators will experience the unpredictability of an islanded system's response to the islanding?

Answer 3:
Yes, we have considered this issue. The DTS can simulate not only deterministic events but also conditional events as well as probabilistic events. The modeling parameters, such as the load model and its characteristics, the generator responses, the relay models, probabilities of occurrence, etc., can be changed easily within the DTS.

Question 4:
Does the topological islanding detector directly suspend AGC action to the units within the island?

Answer 4:
If the question is; does the detector suspend AGC to only those units within the island, the answer is no. When an island is first detected, AGC is "tripped" (stopped) for all units in the control area (both inside and outside the islanded area). Additionally, there is program logic for unit frequency checking which will suspend AGC to a unit if that units' frequency differs from the electric power system frequency by more than some threshold since this would indicate that the unit is probably in its own island.

Question 5:
Have the authors considered using phasor measuring devices to indicate the islanding of an area?

Answer 5:
We have not considered using phasor measuring devices to indicate the islanding of an area because our emphasis was not on islanding occurring in the real-time situation but rather what do we do to smooth the restoration process once an island has occurred or whenever the electric system is being pieced back together after a large scale blackout. As such, we have no plans to develop instrumentation schemes that would manage transient conditions during a system break up event.

Discussion by: Mr. Michael L. Oatts
Southern Company Services, Inc.
Birmingham, Alabama

Question 6:
Is this implementation only intended for use with islands created via controlled methods such as the protection schemes referred to?

Answer 6:
No, there is no "controlled creation" of the island capability. Some utilities do design protection schemes this way. The traditional protection schemes are the definitive measures in responding to power system events.

Question 7:
How are uncontrolled (or randomly) created islands presented to the operator so he quickly understands its makeup and can utilize the Island AGC.

Answer 7:
The topology processor will identify an island condition by messaging the electric power system operator. This messaging together with the (EMS) computer actuated circuit breaker indicators on the large mapboard within the System Operation Center provide the system operator with a presentation scheme that will help him to determine that an island exists and where its boundaries lie.

Question 8:
If the State Estimator program is used as indicated in the paper, to notify the operator of an island, is this necessarily sufficient for determining an island?

Answer 8:
Not in many cases due to the possibility of connection of lines at the sub-transmission level that are not monitored by telemetry. However, if the State Estimator program does message the operator that an island exists, the operator should then closely review his mapboard indications and the sub-transmission one-line diagrams shown via his workstation CRT. The operator can also monitor the various frequency transducer readings, from around the control area, for a significant difference with the transducer that is within the general area of question. Virginia Power has installed frequency transducers in five selected power station sites that are located throughout its service area.

Question 9:
Does not the benefit of Island AGC to reduce the electrical transition ("bump") of reconnecting the island into the network have to be balanced against the risk involved in doing "normal" AGC in the island?

Answer 9:
The concern expressed in this question is shared by the authors. For this reason the implementation provides the operator with a means to, (1) observe the manner in which the island reacts to the AGC actions delivered to the generators within the island so that he can (2) manually adjust the AGC frequency error bias factor to dampen the control action if it is causing too much oscillation of the island area's frequency.

One of the purposes of having the island AGC function is to "free" the operators from being occupied by manual duties when they have so many critical tasks to do during the islanded situation.

It is hoped that simulator environment training will help the operators and the engineering support staff to develop some practical guidelines for using this capability in real-world situations.

Discussion by: Mr. Lester H. Fink, ECC, Inc.
4400 Fair Lakes Court
Oakton, Virginia 22033

Question 10:
Are the authors addressing intentional or unintentional islanding?

Answer 10:
Please refer above to the opening discussion where we say that the primary intended uses for this functionality are in the electric power system restoration process and in the training of the electric power system operator. The authors agree that there are many concerns to be dealt with if this concept is to be applied to a dynamic event wherein an island is being created.

Question 11:
Would the authors provide the formula for the automatic calculation of the frequency bias that the functionality would recommend to be used?

Answer 11:
The programming would accept each unit that the operator directs as being in the islanded area and compute the value we refer to as the "calculated island frequency bias" (CIFB) as:

$$\text{CIFB} = \text{K} * \text{SUM (LFCMX of the units under the Island AGC)}$$

where K is enterable/adjustable by the computer system programming analyst.

The recommended value is;

$$\text{K} = \frac{\text{Primary Area Frequency Bias}}{\text{SUM (LFCMX of the units under the Primary Area AGC)}}$$

Manuscript received September 24, 1993.

ANALYTICAL TOOL REQUIREMENTS FOR POWER SYSTEM RESTORATION

M. M. Adibi, LSM
IRD Corporation
Bethesda, Maryland

J. N. Borkoski, SM
Baltimore Gas & Electric Co
Baltimore, Maryland

R. J. Kafka, SM
Potomac Electric Power Co
Washington, D.C.

Abstract- This paper is one of series presented by Power System Restoration Working Group (SRWG) on behalf of the System Operation Subcommittee with the intent of focusing industry attention on power system restoration. In this paper a set of analytical tools is specified which together describe the static, transient and dynamic behavior of a power system during restoration. These tools are identified and described for restoration planning, training and operation. Their applications cover all stages of restoration including pre-disturbance condition, post-disturbance status, post-restoration target system, and minimization of unserved loads. The paper draws on the previous reports by the SRWG.

I. INTRODUCTION

Power system restoration analytical tools (AT) consists of a set of off-line and on-line programs which together describe the static, transient and dynamic behavior of a power system during restoration. The following table lists types and applications of AT. Attempt has been made to cover as many AT and their applications as a single paper allows.

AT meets the requirements of three groups of users: the restoration planning team in an off-line interactive mode, the instructor -trainee team in both off-line interactive and on-line real-time mode, and the shift operators in an on-line real-time mode of operation. AT should be able to assist the above three groups of users in the power system's pre-disturbance conditions, post-disturbance status, post-restoration target system, and minimization of unserved loads. This paper draws on the previous reports by the Power System Restoration Working Group [1,2,3,4].

Pre-disturbance Power System Conditions

The power system operation and equipment condition vary with time and events (such as loads, generation, scheduled and unscheduled maintenance, equipment derating, etc.), from annually to hourly. The planning team should be able to postulate a number of realistic operations and conditions based on their experience and familiarity with their specific power system. They should also be able to validate such operating conditions using AT. The instructors should be able to select one or more of the operating conditions postulated by the planning team as scenarios for their training sessions. The actual operating condition of the power system and equipment should be available to the shift operator, provided that such information are continually updated with the changes during normal operation of power system [5].

94 WM 222-0 PWRS A paper recommended and approved by the IEEE Power System Engineering Committee of the IEEE Power Engineering Society for presentation at the IEEE/PES 1994 Winter Meeting, New York, New York, January 30 - February 3, 1994. Manuscript submitted July 16, 1993; made available for printing January 14, 1994.

Types and Applications of AT

- Power Flow (PF) Program
  - Sustained Overvoltage Control
  - Reactive Power Balance
  - Line/Transformer Thermal Limits
- Transient Stability (TS) Program
  - Subsystem Stability
  - Underfrequency Load Shedding
  - Low Frequency Isolation Scheme
  - Intentional Islanding
- Long-Term Dynamic (LD) Program
  - Frequency Response of Prime Moves
  - Response Reserve
  - Load Frequency Control
- Voltage Transient (VT) Program
  - Switching Surges
  - Insulation Coordination
- Short Circuit (SC) Program
  - Minimum Source Currents
  - Breaker Interruption Ratings
  - Relay Coordination
- Electromagnetic Transient Program (EMTP)
  - Harmonic Resonance
- Standing Phase Angle (SPA) Program
- Cold Load Pickup (CL) Program
  - Heuristic Approach
  - Physical Modeling
- Restoration Coordination Program (CPM)
  - Allocation of Resources
  - Scheduling of Restoration Tasks
  - Estimation of Restoration Duration
  - Evaluation of Restoration Process

Post-disturbance Power System Status

Power system status and equipment condition following a major disturbance can not be predicted accurately, however experienced operators (in the planning team) know in general terms which pre-disturbance condition is likely during a major power system disturbance and which post-disturbance condition would be the subsequent effect. The planning team, using experience, should be able to use the AT and simulate a selected number of pre- and corresponding post-disturbances. The instructors should be able to select one or more of the post-disturbance states postulated and developed by the restoration planning team as scenario(s) for their training sessions. Initially, the shift operator should be able to assess the status of the power system based on his experience, SCADA/EMS displays, communication with power plants, fields and other control centers. In the ultimate use, when AT is integrated with SCADA/EMS, the AT should be able to advise the shift operator of the power system status [6].

Restoration Procedure

Reprinted from *IEEE Transactions on Power Systems,* Vol. 9, No. 3, pp. 1582–1591, August 1994.

Following the assessment of pre-disturbance condition and the recognition of the post-disturbance status, AT should be used in: determination of the post-restoration target system, preparation of power plants and transmission lines, restoration of the bulk power supply, minimization of unserved loads, and interconnection with the adjacent systems. The development of the restoration plan is conducted in an off-line basis and therefore AT should be fully utilized in all the above restoration stages. In restoration training, both the instructor and the trainee should be able to use the tabulated AT results prepared for a set of pre-disturbance conditions and the corresponding set of post-disturbance status. This should enable an on-line simulation and training. In operation however, the need for information may preclude the use of on-line simulation and the tabulated AT results should be presented for use by the shift operator [6,7,8].

Data and parameters necessary to quantify all the unusual phenomena involved, and the non-linear models necessary for analysis, may be difficult to obtain and maintain. In the absence of some data and parameters, particularly those related to thermodynamics of prime movers, representative values should be prepared and used. Although functions specified in the AT would meet requirements of a broad spectrum of different power systems, some of these programs need certain degrees of customization, simplification and adaptation by the users [9,10,11,12].

II. POWER FLOW (PF) PROGRAM

PF Program has a number of specific requirements. First, a great deal of interaction with the program is required as the model of the electric power system is built up from the initial generation-load configuration to the eventual re-connection of the post-restoration target system. The data-base should allow storage of the entire system model; even though most of the elements may be disconnected or out of service. By saving each case after making small changes or additions to the previous case, it will be possible to return to any level of the restoration sequence if the subsequent cases would lead to an unacceptable situation and a new restoration strategy is required. Second, in restoration, each case is built upon the cases which precede it, without the need to examine several alternatives at random. Third, there is a need for graphics output in the form of one-line diagrams showing restored, restoration in progress and unrestored parts of the power system.

PF Program should be able to allow simultaneous solution of several islands, rapid solutions of sequence of cases, and automatic checking of acceptability of the several cases. The post-restoration target system most likely will primarily consist of the bulk power supply, typically requiring a model of no more than several hundred buses [13].

PF Program should be able to start on a flat voltage profile and always converge on the same (unique) solution, i.e., avoid multiple solutions. In the planning mode of operation, the swing bus for each series of power flow cases is selected by the system analyst of the restoration planning team who is experienced in running power flows. In the training and operation modes, if PF Program is to be used, then the swing buses should be selected at generator bus in an island for a sequence of cases.

Finally, in all applications except for Electro-Magnetic Transient Program (EMTP) and Transient Voltage (TV) Program, transmission lines are represented by their constance reactance and susceptance in a nominal π circuit. While such a representation is adequate for off-line use, it may be inaccurate in the on-line (simulator) use, where frequency may vary as much as 10 percent even if this is for a very short time, i.e., several seconds. The simulators in which the representation of inductance and capacitance are independent of system frequency provide a more realistic operating condition in evaluating the reactive power balance, voltage profile and controlling the sustained overvoltages.

In actual restoration, problems associated with equivalency, external system, state estimation convergence, and observability islands (not power system islands) need to be considered [14].

Applications of PF Program

The most important and useful tool in AT is PF Program which should be used for transmission line loadings, reactive power balance, sustained over- and under-voltage conditions, generator loadings, selection of transformer tap positions (voltage and phase angle), shunt capacitor and shunt reactor status, and phase angle differences between load centers and generators (steady state stability). PF Program is the basic requirement for other analytical tools of AT including; transient stability, dynamic stability, transient voltages, AC short circuit, protective relaying, standing phase angles, and cold load pickup [15].

Sustained Overvoltage Control Using The PF Program

During the early stages of restoring high voltage overhead and underground transmission lines, concerns are with the following three related overvoltage problems: sustained power frequency overvoltages, switching transients, and harmonic resonances. Sustained overvoltages should be controlled using the PF Program. The switching transients and harmonic resonances need to be determined using the transient voltage program (VT Program) and the electromagnetic transient program (EMTP) as described in subsequent sections [16].

Sustained power frequency overvoltages are caused by charging currents of lightly loaded transmission lines. If excessive, these currents may cause generator underexcitation, or even self- excitation and instability. Sustained overvoltages also may overexcite transformers, generate harmonic distortions, and cause transformer overheating. The sustained overvoltages should be controlled by absorbing the large charging reactive power of the lightly loaded transmission lines. Otherwise, the imbalance reactive power would result in generator self-excitation and a runaway voltage rise [17].

III. TRANSIENT STABILITY (TS) PROGRAM

TS Program should describe dynamic behavior of a power system for a period of one to five seconds following a perturbation. It is used to determine how certain perturbations may effect the stability of the power system as it is being restored. Typical perturbations in restoration might be a sudden increase in load, loss of a generator, but most probably a fault on a critical line. These perturbations should be introduced into the TS Program systematically for probing the restored subsystem, interconnected subsystems, and interconnected power systems. They can also be introduced by the planning team.

In the initial phases of restoration when a skeleton network linking the generating stations and the load centers is being re-energized, the impedances between generators will be large and generators will be prone to instability with respect to small perturbations. In other words, the stability levels of the power system are lower because of reduced rotating inertia and weak transmission capabilities. Therefore, in the planning mode from the initial restoration phase until a number of generators are firmly interconnected, transient stability of the power system should be examined to avoid recurrence of the outage.

The TS Program determines stability of an operating point with respect to the above perturbations. The PF Program should be used to determine this initial operating point and the TS Program should simulate the behavior of the power system after the initial point has been perturbed.

The models used in TS Program for three phase balanced fault conditions include the positive sequence network model which is similar to that of the PF Program, generator models which include non-salient pole (combustion turbines and steam units) and salient pole (pumped storage and hydro units) rotors, the excitor system, and several types of speed governing systems used in hydro, combustion turbine and steam prime movers.

For unbalanced faults, the fault impedance should be determined from the positive, negative and zero sequence impedances of the power system under consideration. The fault impedance at any point varies with the type of fault including line-to-ground, line-to-line, or two line-to-ground [18].

Applications of TS Program

TS Program should be used to determine the stability of the power system, operation of out-of-step relays, and the automatic underfrequency relays.

The stability of the power system as is being restored can be determined by a sequence of balanced faults placed on critical lines/buses and the automatic checking for each fault (case) of the differences in the angular displacement of generators against the allowable (stable) limits. To allow rapid solutions and automatic checking of the results, TS Program should use simplified load, network and generator models, and the program capacity be dimensioned accordingly.

Out-of-step relay operations and formation of islands can be examined by placing balanced or unbalanced faults at critical locations on the network and checking the apparent impedance trajectories in the vicinity of the operating zone(s) of the transmission relays. To allow the required duration of simulation by relay operation, TS Program should include load, network (including the negative and zero sequence impedance), generator, excitation systems, speed governor systems, and turbine models. Consideration should be given to reducing the solution time(s) by using the dynamic equivalence of coherent generators within the operating plants in the power system of interest [19].

The examination of automatic underfrequency load shedding can be initiated by picking up the next block of load, or by tripping the largest on-line generator and checking the decline in system frequency level (and possibly the frequency rates) against the underfrequency relay setting(s) for the related load shedding schedule(s). Thus, the TS Program is initiated with the sudden excess in load and although the simulation time would allow modeling of generator controls and subtransient effects, the simplified model should be used to conserve the overall simulation time. The simplification does not affect the accuracy of the answers since load shedding schedules are responsive to system frequency decline of 0.7 to 1.5 Hz (i.e., 59.3 to 58.5 Hz), which usually occurs in the first swing or about one second.

IV. LONG-TERM DYNAMIC (LD) PROGRAM

In the course of system restoration, a reasonable balance must be maintained between generation and load, avoiding excessive over- and under-frequency deviation. Load pick-up in small increment tends to lengthen the restoration duration. With large increments there is always the risk of falling into a frequency decline with consequent recurrence of system outage. The size of load pick-up (e.g., low voltage AC network loads) very much depends on the rate of response of prime movers. A related concern in restoration is with the effectiveness of generation reserve in the recovery from loss of a generator in a subsystem (island) or interconnected subsystems. In determining reserve requirements it is important to consider the dynamic capabilities of the sub-system, particularly the responsiveness of prime movers to a large excess load. Another concern is with the inevitable load-generation mismatch within a low frequency isolation (LFI) scheme which may cause excessive frequency deviation.

Since the time scale of prime mover response is typically measured in minutes, the LD Program should incorporate the dynamic phenomena which accounts for fossil fuel and nuclear reactor boiler dynamics which affect pressure and enthalpy at the turbine throttle and reheat energy flow, penstock and conduit dynamics with hydro units, load dynamics, and load-frequency control for multiple islands or subsystems.

The LD Program should be modular and self-initializing. The structure of the power system being restored should be specified by the user and entered in the LD Program interactively, e.g., it can be specified that a hydro unit of certain type (run-of-the river, high head short conduit, or high head long conduit) is the prime mover for particular generator specified at a given bus, etc. Each component of prime mover models should be self contained, defined by a set of parameters supplied by the user of the LD Program, and should initialize itself to attain a meaningful initial total system equilibrium status. The self initialization of LD Program is an important requirements. LD Program should be able to accept the PF Program output in terms of bus voltages and line flows, determine the operating conditions of all the generators, and determine the mechanical torques from the prime movers. These torques should then be used to initialize the governors and turbines associated with each of the prime movers. In the case of fossil fuel and nuclear boilers, the initialization of the turbines should yield the desired pressure, enthalpy, steam flow and reheat energy flow required from the boilers. With hydro units, the gate pressure and flow should be provided. These values should then be used to initialize the penstock/conduit conditions and the fossil fuel and nuclear reactor boilers conditions, producing the power system in equilibrium status ready for LD Program.

The external controls to the LD Program such as relay settings, transformer tap positions, voltage regulator set points, and governor speed changer position (LFC), need to be modeled to represent the behavior of the above modules and the performance of the overall LD Program. It is advantageous to use variable integration step for the numerous differential equations representing behavior of the above modules should be variable. Certain derivatives need to be evaluated more frequently than others because of the wide spectrum of system dynamics inherent in a power system. Both predictor-corrector and simple Euler integration techniques should be used, depending on type of prime mover. Some test should be run to determine which integration technique yields minimum computation time within an integration error tolerance limit [20].

Generally, the models of the fossil fuel boiler, nuclear reactors and steam turbines are probably the most approximate of the models. They should provide only representative dynamic responses associated with these units. The steam turbine and hydro units should be more accurate than those represented in the TS Program for time periods exceeding one or two seconds.

The unit models used in LD Program need to be more comprehensive than those required by TS Program but should not be as detailed or exact as those that are used for designing purposes by manufacturers. Computation time and storage requirements would prohibit utilization of these design models; in addition, getting data for these models would almost be impossible. The approach to be adopted in development of the individual prime mover models should be to group phenomena and represent unit dynamics with relatively few pertinent parameters that are easily understood and not too difficult to obtain for a given unit.

The LD Program should be made more useful by providing a user's manual which incorporates the numerous required data and validity checks for making the LD Program self-explanatory and self-checking [21].

Applications of LD Program

Subsystem Sustaining Capability

The ability of a subsystem to withstand a sudden increase in load (e.g., picking up a large block of load), is a function of its generation response, load droop characteristics, load shedding schedules and protective relay logic. The underfrequency operation of a subsystem results in reduced capacity of auxiliaries and the diminishing supply of energy to fossil fuel boilers and to nuclear units (because of circulation pumps). The overall prime mover dynamics coupled with governor action, auxiliary limitation and underfrequency load shedding will determine whether the system frequency can be restored to normal following the initial decline. In order to determine subsystem's frequency recovery capabilities, simulation should be limited to the prime mover models, load droop characteristics, and load shedding schedules. This would reduce the computational burden by a factor of five to ten. Thus all prime movers within a subsystem should be grouped together, all loads and their drooping characteristics should be lumped together, transmission line losses should be represented by a constant factor, and the underfrequency load shedding schedule should be included in the LD Program for this application.

Response Reserve

Another application of LD Program in restoration of a subsystem is determination of the effectiveness of generation reserve in recovering from loss of the largest unit. The problem is basically a restoration of the energy balance between generation and load. Hence, the dynamics of the primary energy sources play a predominant role in the assessment of frequency restoration capability.

For subsystems with primarily fossil-fuel steam turbine units, boiler response is the determining factor in frequency recovery. Because of the slow dynamics of boilers, system frequency may droop even after governor action has reversed the initial down ward trend. For subsystems of this type, the minimum frequency during boiler recovery is a useful measure of responsiveness.

The frequency recovery characteristic can be affected by changes in the amount of generation reserve and by distribution of generation reserve over prime movers. The importance of the amount of reserve has been recognized. The distribution of reserve which is equally important should be considered. The reserve allocation between prime movers can be determined by simulating the recovery from the loss of the largest generation. However, during restoration it is more desirable to determine an overall relationship between reserve and response, to be used as a simple guideline on the adequacy of a reserve allocation. This overall relationship can be determined on the basis of a number of LD Program simulations.

Restoration Load Frequency Control

In a build-up strategy, the accepted industry approach is to sectionalize the de-energized system into several sub-systems and independently build-up load and generation in each subsystem. This parallel restoration strategy is accepted primarily to take advantage of the hot re-start but time critical capability of re-circulating type steam generators.

The approach calls for one power plant within each subsystem to maintain frequency by manually adjusting the governor speed changer, i.e., flat-frequency control. When two sub-systems are interconnected, and in the absence of automatic generation control system, a power plant within a designated sub-system will maintain the frequency of the interconnected sub-systems, while the other sub-system maintains tie-line flows to within the prescribed range, i.e., flat tie-line control.

Typically, AGC programs are designed to suspend control during emergency operation and particularly when island formation is detected. However, the manual control of frequency and tie-line flow by plant operators should be modeled.

V. TRANSIENT VOLTAGE (TV) PROGRAM

Transient voltages or switching surges are caused by energizing large segments of a transmission line at one time or by switching capacitive elements. These switching transients, which are usually highly damped and of short duration, in conjunction with sustained overvoltages may result in arrester failures. The switching surge overvoltages have become increasingly more important with the design of compact transmission lines and substation facilities. The energy stored in long high voltage lines is large causing significant transient or resonant overvoltages which must carefully be considered during restoration. Generally, if the steady state voltages (power frequency) are less than 1.2 per unit of their nominal values, the switching transients can be managed by typical arresters. A notable exception is energizing transformer terminated lines, which may result in harmonic resonance and damaging overvoltage.

The likelihood and the effect of transient overvoltages should be determined by the analysis of special system conditions using three-phase closure of the breakers at the point when the instantaneous voltage of the source goes through its maximum.

The TV Program should relate the unknown (state) variables of the power system under consideration with the known parameters using current balance at each node and voltage balance in each loop. The state variable should consist of currents through each inductor, and voltage across each capacitor, resulting in a mix of algebraic and ordinary differential equation which should be solved simultaneously using variable step integration [16].

VI. ELECTROMAGNETIC TRANSIENT PROGRAM (EMTP)

Power system equipment are protected against the effect of overvoltages resulting from normal operation by surge arresters which provide a relatively low-impedance path for transient overvoltages. During restoration, however, certain line-transformer configurations when energized generate a combination of sustained and harmonic resonance overvoltages which may cause surge arrester failures and thus system faults, or the subsequent flash-over and damage to equipment.

Harmonic resonance voltages are oscillatory undamped or only weakly damped temporary overvoltages "TOV" of long duration. The TOVs result from several factors that are characteristic of the networks during restoration, including switching operations and equipment non-linearities. The harmonic resonance overvoltage should be analyzed by using EMTP. The EMTP is large, comprehensive and inefficient when applied to restoration training and operation, however, it is indispensable in restoration planning for the analysis of:

- insulation coordination and switching surge,
- harmonic resonance and ferro-resonance, and
- magnetizing transformer inrush overvoltages.

To simplify these analyses, EMTP should be used under several balanced initial conditions on simple systems consisting of lumped RL and C representing transmission lines, non-linear inductances for saturated transformers, circuit breakers, synchronous machines and induction motors. The EMTP solves the algebraic, ordinary and partial differential equations to determine the transient behavior of the power system under study [22].

The harmonic resonance overvoltages should be controlled by:

- connecting sufficient underlying loads at the sending end of a line or by connecting dead load on the transformer to be energized,
- reducing the high source impedance by starting more generators, and
- reducing the reactive power of a lightly loaded system by minimizing the number of unloaded lines to be energized and setting the sending-end transformers at the lowest tap position.

VII. SHORT-CIRCUIT (SC) PROGRAM

Transmission line protection systems are designed with the assumption that the transmission network is intact, with enough generation connected to provide a more than adequate level of fault current(s) to operate the relays. However, during restoration with limited connected generation, the fault level may not be adequate to assure correct operation of transmission line relays. In particular, many schemes employ "fault detectors" which will prevent tripping if the fault level is below the expected load level of the line under normal operating condition. Relaying system selectivity allows location of faults within the "zone of protection" and trips only those circuit breakers which are required to clear the fault. The reduced available fault current during restoration may reduce a relaying system's selectivity. Because of the likelihood of high overvoltages and thus system faults during restoration, it is imperative that the relay protection prevent possible gross damage to equipment.

As the power system restoration progresses, another concern is with short-circuit current(s) exceeding the circuit breaker's interruption ratings. This may occur at some substations due to the prevailing topology being different from that of normal operation for which the circuit breaker rating has been selected and is adequate. The third concern is with the out-of-step separation (islanding) at locations where load-generation mismatch is greater than the islands' sustaining capability (see LD Program).

Transmission line relaying should be modeled because most faults occur on lines due to their vulnerability arising from their exposed lengths. For determining the minimum fault current condition, single-phase-to-ground faults should be considered. Under minimum generation condition, the faults should be placed at one end of line with the adjacent breaker open. The SC Program required should use positive and zero sequence impedances. Usually, the negative and positive sequence impedances are considered to be equal. The double circuit transmission lines which are mutually coupled require mutual reactance representations. The generators should be represented by a fixed voltage back of the transient or sub-transient (on the user's option) reactance. The SC Program should be initialized by the load balance to be provided by the PF Program.

Circuit breaker interruption ratings are based on the symmetrical short-circuit currents generated by three-phase faults. The SC Program should be able to automatically place several single-phase and three-phase faults on the system and check the results. To allow rapid solutions of these cases, the analysis can be simplified by the use of direct current (DC) SC Program. DC analysis provides sufficient accuracy for checking the minimum fault current required for relay operation and the maximum fault current evaluation of the breaker interrupt ratings [23].

Applications of SC Program

The purpose of a restoration short-circuit program is to ensure that: (a) the minimum short-circuit for fault clearing by relays is maintained, (b) the maximum interruption ratings of circuit breakers are not exceeded, and (c) the out-of-step relays (by blocking and transfer tripping) allows minimum of load-generation mismatch.

The first two applications require rapid analysis of sequence of several fault conditions and automatic evaluation of the results. As such they should use a DC short-circuit analysis, which provides sufficiently accurate results. Bus impedance matrices, being more efficient, should be used in these analyses, and faults should be placed on buses or end of open lines. For the three-phase and line-to-ground faults, the SC Program should provide: the total faulted bus currents, contributions in each line connected to the faulted bus, and zero-sequence driving point reactance for the faulted bus.

The distance relay detects an out-of-step condition by the apparent impedance "seen" by the relay at the terminal of the line it protects. The apparent impedance is determined by the ratio of the bus voltage over the line current and as such it requires an AC short-circuit program. The distance relays should include:

- impedance relays, mho relays, reactance relays, and angle-impedance relays (or ohm units) which are used as blinders, and
- pilot relays which provide high-speed fault clearing/reclosing by simultaneously tripping/reclosing circuit breakers at both ends of a line.

VII. STANDING PHASE ANGLE (SPA) REDUCTION PROGRAM

In re-integrating a power system following a major disturbance power system operators often encounter an excessive SPA difference across a breaker which connects the two adjacent stations. These angles may occur across a tie line between two interconnected systems or between two connected subsystems within a system. Closing a power circuit breaker on a large SPA difference could shock the power system due to large circulating currents after breaker closing with possible resultant equipment damage, and the possible re-occurrence of the system collapse due to line tripping, etc.

In order to avoid the inadvertent closing of the circuit breakers on a large SPA, breakers are equipped with synchrocheck relays which prevent their closures on angles greater than the preset values. Under such a condition, the system operator by trial and error attempts to reduce the SPA difference by changing the real power generation in as few power plants as possible and by as little as possible. This time consuming process has shown to prolong the restoration duration.

The range of SPA differences that a system can withstand depends on the nominal voltage level, operating condition and the location of circuit breaker within the network. The range can be determined using the TS Program. Power system operators then can be advised of safe SPA values under which certain breakers can be closed or the synchrocheck relay can be set accordingly.

It is reasonable to assume, that an excessive SPA can be reduced by rescheduling real-power outputs of two or more power plants. However, the selection of the specific power plants which reduce the SPA is limited by both topological and analytical constraints of the system which are imposed by the structure of the transmission network, and by physical limitations of the system components together with network equations (i.e., PF Program). Selection and rescheduling of power plants, i.e. implementation of the SPA Program requires integration of AT with SCADA/EMS. The data requirements of the SPA Program are the same as those required for the PF Program [24].

IX. COLD LOAD PICKUP (CL) PROGRAM

Two approaches are considered in representing the behavior of the load during restoration, heuristic modeling and physical modeling. Heuristic modeling, characterizes the load behavior based on experience and reasoning regarding the overall behavior of load on typical feeders, supported by measured data during system disturbances; for example, load may double after 30 minutes of elapsed outage time. The heuristic approach is useful for conceptual consideration of load behavior, but it is limited in practical use by a lack of sufficient experience and measured data and inability to be adjusted to varying load compositions and ambient conditions.

Physically-based modeling develops models for the underlying physical phenomena, together with data on the composition of the load. Physically-based modeling, while requiring significant development and data collection, has the potential to produce reasonably accurate models for varying load compositions and conditions.

In planning modes of operation, it is necessary to have mathematical models that can be used to predict the load that will be present at various locations on the system. The variation of the load as a function of the length of time that it has been disconnected and other factors, such as weather and voltage, should be considered.

The basic load characteristics that affect cold load pickup response are quite well understood. A first order differential equation relating ambient air temperature and thermal time constant can describe the load adequately. The CL Program is power system specific. It should be developed using parameters for particular feeders at particular day of the week, seasons, etc. Most operating utilities have extensive load data and load forecasting methods based on weather, etc, which can be used in CL Program [25].

X. RESTORATION COORDINATION PROGRAM (CPM)

A functional restoration should entail development of a strategy which meets the primary goal of minimizing the duration of an outage and the unserved customers. Such a strategy typically should include several distinct levels of restoration for various system conditions. Each level in turn should consist of a number of interdependent activities. The entire strategy then should be broken down into some logical order which would provide a discipline whereby there would be less risk of overlooking any essential operation.

For example, while sectionalization of the system into a number of subsystems and simultaneous restoration of service in each subsystem could appreciably reduce the duration of an outage, the extent of sectionalization is limited by availability of resources, including the number of operating teams as well as communications available to permit efficient and effective coordination of their efforts. The coordination of these activities can be evaluated by CPM Program [26].

Some of the questions which CPM Program may answer are:

- What is the estimate of the subsystem restoration duration?
- Among the activities comprising the subsystem, which are likely to affect the subsystem duration?
- What should be the scheduled starting time and finishing time for each of subsystem?
- What are each subsystems maximum and minimum duration for completion?
- What are the probability of successful completion for each subsystem?
- What are the resource requirements in terms of manpower, equipment and information?
- What are the trade offs between subsystem duration and loss of equipment life?
- Because restoration of subsystem never proceeds exactly as planned, how can the deviation be quickly recognized and corrected?

XI. ACKNOWLEDGEMENT

The first author acknowledges support of Baltimore Gas and Electric Company, Potomac Electric Power Company and wishes to thank Mr. L. H. Fink of ECC, Inc. for his valuable suggestions.

XIII. REFERENCES:

[1] Kafka, R. J. et al, "System Restoration Plan Development for a Metropolitan Electric System," IEEE Trans.v. PAS-100 n.8, 1981. pp.3703-13.

[2] IEEE Committee Report, Special Consideration in Power System Restoration, SRWG, M.M. Adibi, chmn, T-PWRS Nov 92 1419 - 1427

[3] IEEE Committee Report, System Operation Challenges, COPWG, E.K. Nielson, chmn, Feb 87, IEEE Trans. v.PWRS-3 n.1, 1987, pp.118 - 26

[4] IEEE Committee Report, System Restoration - Deploying the Plan, COPWG, W.A. Johnson, chmn, IEEE Trans. PAS Nov 1982, 4263 - 71

[5] IEEE Committee Report, Power System Restoration - The Second Task Force Report, SRTF, M.M. Adibi, chmn. T-PWRS Nov 87 927-933

[6] M.M. Adibi and L.H. Fink "Power System Restoration Planning," IEEE/PES 93WM204-8-PWRS

[7] IEEE Committee Report, New Approaches in Power System Restoration, SRWG, M.M. Adibi, chmn. T-PWRS Nov 92 1428-1434

[8] IEEE Committee Report, Power System Restoration - A Task Force Report, SRTF, M.M. Adibi, chmn, T-PWRS May 87 271 - 277

[9] Peach, S., "System Restoration Synthesis - Hydro Quebec," IEEE 8th Biennial Workshop on Real-Time Monitoring and Control of Power Systems, Montreal, Oct. 1984.

[10] Scheurer, D., "System Restoration at Philadelphia Electric Company," IEEE 8th Biennial Workshop on Real-Time Monitoring and Control of Power Systems, Montreal, Oct. 1984.

[11] Barrie, D., " System Restoration Following Load and Generation Rejection - Ontario Hydro, IEEE 8th Biennial Workshop on Real-Time Monitoring and Control of Power Systems, Montreal, Oct. 1984.

[12] Kafka, R. J. et al, "Role of Interactive and Control Computer in the Development of a System Restoration Plan," IEEE Trans. PAS, Vol PAS-101, Aug 1982

[13] Wu, F. F. et al, "Analytical Tools for Power System Restoration - Conceptual Design," IEEE Trans. on Power Systems, Vol. 3, No. 1, pp. 10-16, Feb. 1988.

[14] Schlaepfer, F. M. et al, "An Interactive Load Flow Program," Proc. IEEE/PESW PICA, May 1971, pp 110-117

[15] IEEE Committee Report, "Overvoltage Control During Restoration," SRWG, M.M. Adibi, chmn, T-PWRS, Nov. 92, pp 1464-70

[16] M.M. Adibi and D.P. Milanicz, "Reactive Power Limitation of Synchronous Machines," IEEE/PES 93WM203-0-PWRS.

[17] Adibi, M.M., et al "Solution Methods for Transient and Dynamic Stability," Proceedings of IEEE, vol. 62, July 1974.

[18] Adibi, M. M. "Power System Protective Relaying and Simulation Models," IEEE PES, Power Industry Computer Applications, 1969

[19] Ross, H. B. et al, "An AGC Implementation for System Islanding and Restoration Conditions," IEEE Trans. 93 SM 519-9 PWRS.

[20] IEEE Committee Report,"Dynamic Models for Steam and Hydro Turbines in Power System Studies," T-PAS 73 Nov/Dec pp 1904-1915

[21] Dommel, H. W., et al "Transient Program User's Manual," University of British Columbia Publication, August 1988.

[22] Morshed, Atef S., "Calculating Asymmetrical Fault Currents," IEEE Trans. PAS Aug. 1981, pp. 3785-3790.

[23] Wunderlich, S. et al, "An Approach to Standing Angle Reduction," IEEE/PES 93WM205-5-PWRS.

[24] IEEE Committee Report, "Special Consideration in Power System Restoration - The Second Working Group Report," SRWG, M.M. Adibi, chmn, IEEE/PES 93WP202-2-PWRS

[25] Wiest, J. D., et al, "A Management Guide to PERT/CPM," Englewood Cliffs, N. J.: Prentice-Hall, 1969.

Discussion

G. Krost and **D. Rumpel** (University of Duisburg, Germany): There are several features, which we seemingly do not experience as much under European conditions, possibly due to shorter lines. One is the fear of arrester failures; the other is the fear of harmonic resonance. In any case, we do not use EMTP for harmonic studies, because, despite the detail modeling and computational requirements it remains difficult to represent the damping effects at the harmonic frequencies in the time domain.

Manuscript received February 10, 1994.

W. R. Lachs (University of NSW, Kensington, Australia): The authors are to be commended on their description of an almost exhaustive list of programs associated with power system restoration. This paper, in effect, provides a check list of the studies that should be considered by utility engineers for development of power system restoration plan.

However, there is one gap in their list and it relates to generator negative phase sequence (NSP) protection. A generator's NPS reactance is approximately equal to its subtransient reactance so that some 10% NPS voltage can produce a NPS current equal to the stator rating. When added to normal generator current this can quickly overheat and cause damage to the generator stator and rotor windings and for this reason NPS protection is provided.

Generator NPS protection has produced a major system shut down as hereby described. At a light load period, a utility with an extensive transmission network suffered a single phase fault coupled with a protection failure on the EHV line breaker at zone 1. Because of the long fault duration and the small number of units running, each generator was inflicted with an unusually high NPS voltage. When one generator tripped, the NPS voltages increased on the others, so that all generators were sequentially tripped by their NPS protection, so producing a total blackout.

This is not the direct reason why generator NPS protection must be taken account of in power system restoration. Long EHV lines, particularly if they are not transposed, can have a substantial phase unbalance. By allocating each of the three different phase arrays to the various lines and balancing them at each substation, the phase mismatch can be eliminated without the need to transpose individual lines.

However, due care must be taken in the order of re-energization of line for a power system restoration. If there were to be an excessive phase unbalance by incorrectly energizing lines when only one or two generators are in operation, there would be a high risk of tripping the units by their NPS protection. For this reason, in developing a restoration plan for which a number of long transmission lines need to be energized, care must be taken to minimize the possibility of line phase mismatch. The restoration plan, in these instances, should specify the precise sequence of line re-energization for the guidance of the operators [A].

Reference

[A] Zhu, Y. L., et al. "An Expert System for Power System Restoration," International Power Engineering Conference, Singapore, Mar. 1993.

Manuscript received February 10, 1994.

DENNIS CAUFIELD (Houston Lighting & Power, Houston, TX); The authors have done an excellent job in researching what analytical and expert system tools would be beneficial in power system restoration. The difficult part for anyone interested, however, is justifying the dollars needed to purchase, implement and train personnel on restoration tools such as these when you hope you never have to use them in real-time.

Perhaps information which could be used during normal conditions, alerts, storms, etc. which complement the analytical tools and expert systems which system operators need during actual system restoration could be developed. Tools such as these would be very beneficial. The challenge would be in developing aids which bridge normal operations with system restoration. The one thing system operators do not have during restoration activities is time. Tools such as check-lists, easy to read tables of needed information, graphs of open-ended line voltages versus line length, equipment capabilities, etc could possibly be developed using analytical methods and provided to system operators for their use during normal and abnormal conditions. Support personnel could also use these analytical tools in real-time situations to enhance what information the system operators need during restoration procedures.

Manuscript received February 23, 1994.

R. Paliza and **J. Giri** (ESCA Corporation, Bellveue, WA): The authors are to be complimented on a well-written paper. We would appreciate the author's comments to the following comments:

a. Please clarify the context in which these tools will be utilized. This will enable a clearer definition of the requirements of the tools. For example, will they be used to develop a restoration plan or to assess switching operations.
b. We feel that an essential tool for restoration is a robust state estimator. Please comment on why a state estimator was not mentioned.
c. Please comment on the contingencies used for stability and short circuit—do they change for each stage of restoration?
d. We feel the following requirements should also be included: execution speed, model detail level, and proper scheduling of execution of the tools.

Manuscript received February 23, 1994.

J. J. ANCONA (Niagara Mohawk Power Corp., Syracuse, NY). The author's are to be commended for their well organized comprehensive treatment of power system restoration analytical tools (AT). Besides providing an excellent overview of restoration AT, the paper serves as a valuable reference because it also addresses practical issues pertaining to restoration. It explains the need for restoration AT, while also implying the need for general restoration procedures without AT.

The authors' concise description of each major type of AT restoration software includes: (1) what needs to be studied with each type of AT; (2) software features needed for restoration studies that may not be ordinarily needed; (3) preparatory work that may be needed; (4) how studies should be performed; and (5) why restoration studies need to be conducted. As the authors point out, AT can be used to prepare restoration plans, and can serve as an excellent training tool. It can help operators get a better feel for the interactions of the power system during a major power failure, and give them an insight into restrictions and requirements associated with restoration. Moreover, if a major power failure occurs, efficient and effective AT should allow operators to rebuild a stable electric system and restore larger blocks of unserved energy more quickly and safely with a minimum of operating problems and complications.

Beyond a succinct description of AT, the paper is application oriented from the perspective of power system operators by providing important background information and identifying problems that can be anticipated during restoration. AT can screen major power failure restoration scenarios to help determine if these potential problems are a real concern for a specific power system.

In addressing problems that may occur during restoration, the paper indirectly raises the issue that these concerns need to be resolved whether or not AT can be used. Indeed, even if installed and properly set-up, certain AT may be unavailable or ineffective for even off-line restoration planning and training purposes because: (1) analysts and operators may have insufficient experience with it; (2) resources available for restoration studies may be limited; (3) data base acquisition and modeling for the specific power system may be incomplete [A1, A2]; or (4) field tests needed to verify the AT results have not been conducted [A3].

Furthermore, if a major power failure occurs, some or all AT may be unusable for restoration analyses because: (1) computers are unavailable; (2) necessary SCADA/EMS data acquisition becomes erroneous or inadequate; (3) available operators and support staff are pre-occupied with other problems; or (4) the AT can not produce quick turnaround because: (a) it is inherently cumbersome to use, (b) training sessions and practice drills were inadequate for achieving proficiency in the use of the AT, or (c) appropriate restoration scenarios were not modeled beforehand to help reduce turnaround time.

Conceivably, the shortcomings of AT identified above can be mitigated somewhat. However, if the appropriate AT will not be fully developed, available, or work as intended, a utility needs to rely upon an alternate restoration approach as a backup. In the absence of some or all AT, operators will need to rely on existing restoration plans and procedures, currently available off-line AT study results, and general guidelines. For instance, without the availability of long-term dynamic (LD) software, operators may limit load pick-up to less than 5% of generation to help insure that a subsystem's frequency does not decrease by more than 0.5 Hz [A1, A5]. Alternately, without transient voltage (TV) software, operators may hold sending end voltages to minimum permissible levels to avoid sustained over-voltages when energizing long lightly loaded transmission lines or underground cables [A4, A5]. The paper itself offers restoration guidelines (eg.: harmonic resonant overvoltages can be minimized by connecting dead load on transformers being energized). Thus, lacking some or all AT, operators may need to proceed at a slower pace and work with shorter sections of line and more manageable blocks of load. Obviously this may not be a panacea itself; it may complicate and worsen the restoration process as well. Clearly, being set-up to work with restoration AT is preferred, but it should not be relied upon.

Again, the author's are to be congratulated for having presented an excellent overall description and assessment of restoration AT along with the associated power system problems that need to be anticipated during a major power failure. In explaining the need for restoration AT and implying the need for procedures in the absence of AT, the paper has provided a bridge between the two, and a valuable reference.

References:

[A1] M. M. Adibi et al., "Power System Restoration - A Task Force Report," (IEEE PES 86 SM 329-7), IEEE Trans. Power Syst. Vol. PWRS-2, pp. 271-277, May, 1987.

[A2] M. M. Adibi et al, "Special Considerations in Power System Restoration," IEEE Trans. Power Syst. Vol. PWRS-7, No. 4, pp. 1419-1427, Nov., 1992.

[A3] R. Lindstrom, "Simulation & Field Tests of the Black Start of a Large Coal-Fired Generator Using Small Hydro," IEEE Trans. Power Syst. Vol. 5, No. 12, pp. 162-167, Feb., 1990.

[A4] M. M. Adibi et al, "Overvoltage Control During Restoration," IEEE Trans. Power Syst. Vol. PWRS-7, No. 4, pp. 1464-1470, Nov., 1992.

[A5] R. J. Kafka, "Active and Reactive Power Balance - Potomac Electric Power Co. (PEPCO)," EPRI/ECC Power System Restoration Seminar, Denver, Sep. 1993.

Manuscript received February 24, 1994.

M. M. Adibi, J. N. Borkoski, and R. J. Kafka: The authors wish to thank the discussants for their questions and comments.

Mr. Ancona has correctly identified the point of the paper that general "rule of thumb" based on AT analysis, as well as a conservative restoration approach, are very beneficial to operators during restoration in the absence of on-line AT.

Mr. Caufield is concerned with cost justification of AT for real-time use and recommends that simple guidelines be developed by AT for use by shift operators. As envisaged, AT would serve three groups of users entrusted with restoration tasks; planners, instructor-trainees and shift operators. They use AT for analyzing and assessing pre-disturbance conditions, post-disturbance states, post-restoration targets and restoration procedure to achieving the target systems. Depending on the group of users and their objectives, AT would be used in either off-line, interactive or in real-time modes. For use by the shift operator, to the extent possible, AT would produce tabular results and guidelines for quick reference. Thus the cost of AT will be spread over the three groups of users for three different modes of operations.

The authors wish to express their thanks **Drs. Krost and Rumpel** for their comments based on their experience in Europe.

Dr. Lachs comments on the negative sequence voltage problems caused by the lightly loaded untransposed EHV lines. The authors agree that except for the EMTP and SC Program, the other analytical tools are based on single-phase modeling which assumes that lines are fully transposed and the three-phase voltages and currents are balanced. The present practices in construction and monitoring of EHV lines are to use un-symmetrical conductor spacing, un-transposed lines, and single-element instrumentation.

These practices have caused operational problems during peak-loads because of differences in the three conductor reactances, and during light-loads and restoration because of differences in their charging currents. In particular when providing remote cranking power to a steam unit over EHV lines, high (10 to 15%) negative sequence voltages have been observed which as DR. Lachs comments, could damage the auxiliary motors or impede the restoration process.

Responding to the four questions raised by **Ms. Paliza and Dr. Giri**: (a) AT can be utilized off-line in an interactive mode to develop restoration plans including the assessment of switching operations. In addition, AT can be utilized by instructor-trainee teams in both an off-line interactive mode and simulated "real-time" modes, as well as by shift operators in an on-line "real-time" mode.

(b) The concern with state estimator (SE) is the speed required due to rapidly changing system conditions as a system is being restored. The value of SE will be diminished or even detrimental if SE is not fast enough to keep up with rapidly changing conditions. The other concern is how would the tuning and the subsequent performance of the SE be affected by greatly reduced measurements, and how would the issue of measurement criticality be addressed?

(c) The contingencies used for short circuit and stability studies do change for different stages of restoration. In the beginning of restoration, available fault currents are low due to fewer generators on-line and system impedance to fault may be high due to incomplete transmission system. As restoration progresses more generation is on line and the system topology may be different from normal operation, therefore, exceeding circuit breaker interruption ratings is a major concern. The challenge is to design a short circuit contingency screening technique, which can screen for both "opposing" concerns. Likewise, due to changing voltage profile, system inertia and topology, the contingencies used for stability, will also change depending on the stage of restoration. Again, a screening technique is desirable.

(d) High speed of execution and full detail modeling are conflicting objectives requiring trade-off between speed and detail modeling. The proper scheduling of tools is dependent on the system under study and each system's unique concerns.

Manuscript received April 19, 1994.

A Framework For Power System Restoration Following a Major Power Failure

Jerry J. Ancona, Member, IEEE
Niagara Mohawk Power Corp.
Syracuse, NY 13202 USA

Abstract - **Although a major power failure is a complex unique problem, certain common guidelines for power system restoration have emerged from previous papers. This is a review and organized compilation of those papers. As such, it provides a *framework* to help power system operators limit the extent of major power failures; and if a failure occurs, to help operators rebuild a stable electric system and restore unserved load as safely, smoothly and quickly as possible with minimal adverse impact to the public. These guidelines also offer recommendations to help prepare for a major power failure in advance, thereby further facilitating restoration.**

Keywords: Blackout, Bulk Power System Failure, Emergency Planning, Islands, Major Power Failure, Operational Planning, Power Control, Restoration.

I. Introduction

As an indication of widespread interest within the utility industry, a number of notable papers have been written on specific aspects of power system restoration. This paper is both a comprehensive survey and an organized synthesis of many of those papers. Its intent is to identify the major concerns, interrelationships, and requirements of restoration to serve as: (1) a general restoration guide for operators; (2) a bibliographical reference source for more detailed information; (3) an outline to help prepare an individual restoration plan for a specific power system; and (4) a guide to help evaluate and improve existing restoration preparedness.

This paper focuses on restoration associated with major power failures such as island formations or blackouts resulting from significant loss of generation or bulk power transmission. However, it is also applicable for more standard power failures involving loss of load and regional transmission only. This paper provides a framework for restoration by outlining: (1) recommended goals and objectives of restoration; (2) responses which should be initiated for abnormal conditions; (3) the sequence of restoration actions which should be taken following a major power failure; and (4) steps which can be taken to enhance preparedness.

II. Goals and Objectives of Restoration

The overall goals of restoration should be to *first* rebuild a stable electric system, and *then* restore all remaining unserved load [4, 11, 44]. In the process of attaining these goals, the following objectives need to be achieved as well:

A. Restore Safely

Under the extraordinary circumstances following a major power failure, precautions need to guard against human injury and equipment damage:

1. Special attention needs to be paid to applicable safety procedures to protect people from harm.
2. Operators need to avoid steady-state overloads, and avoid closing ties across excessively large standing phase angles during synchronization [17, 19, 21, 39, 47] to protect equipment from thermal damage.
3. Operators need to take precautions against equipment damage from over-voltages which may occur when lightly loaded long transmission lines, underground cables, or transformers are energized or switched. These problems may take the form of sustained steady-state, transient switching, or harmonic resonant over-voltages [8, 10, 21, 26, 29, 44, 47].

B. Restore Smoothly and Deliberately

To assure a stable rebuilt electric system, several parameters need to be in balance during restoration:

1. Real load needs to be balanced with generator capabilities. Restoring excessively large load blocks can result in unacceptable frequency or voltage excursions; particularly in an islanded power system which is more volatile than a large interconnected system [19, 35, 36, 47]. Cold load pickup should limit frequency drops to 0.5 Hz or less [1, 11, 36, 44].
2. Reactive loads need to be balanced, reactive reserve capabilities need to be maintained, and voltages need to be stabilized to avoid over-voltages, low voltage collapse, and instability [1, 2, 10, 13, 29, 33, 38, 44].
3. Repeated interruptions should be avoided by rebuilding a stable electric system (with respect to voltage, frequency and the ability to pickup load blocks or lose generation blocks) before large amounts of unserved load are restored [4, 11, 44].

C. Minimize Overall Restoration Time

Notwithstanding the need for restoration to be smooth and deliberate, overall restoration time needs to be minimized to accommodate customer needs and equipment limitations:

1. Both minimum and maximum restart times for generators should be observed when deciding unit startup sequences. Priority should be given to restarting tripped

95 WM 143-8 PWRS A paper recommended and approved by the IEEE Power System Engineering Committee of the IEEE Power Engineering Society for presentation at the 1995 IEEE/PES Winter Meeting, January 29, to February 2, 1995, New York, NY. Manuscript submitted July 29, 1994; made available for printing December 2, 1994.

Reprinted from *IEEE Transactions on Power Systems,* Vol. 10, No. 3, pp. 1480–1485, August 1995.

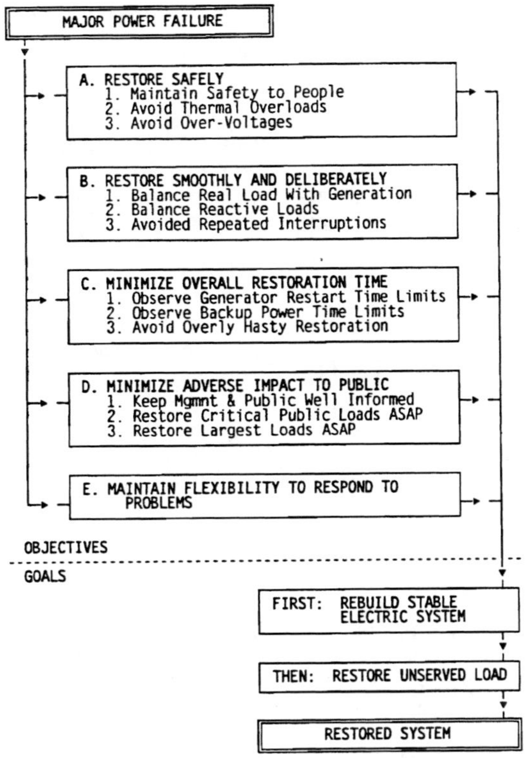

Fig. 1 Restoration Goals and Objectives

drum type steam units (before they cool excessively) and generators that are available for quick startup [1, 11, 17, 21, 25, 31, 32, 37, 44, 47].

2. Time limits should be observed for backup emergency on-site generators or batteries serving critical equipment such as breakers, remote terminal units (RTU's), communication and control systems, and oil filled cable pumps. Station service needs to be re-supplied to these critical loads before their backup power supplies become exhausted; particularly during severe cold weather conditions [14, 17, 24, 26, 45].

3. Obvious restoration actions should be taken immediately. However, overly hasty restoration should be avoided to permit time to analyze and decide upon options before complex actions are taken.

D. Minimize Adverse Impact to the Public

During a major power failure, the restoration process needs to incorporate steps which help minimize detrimental impacts to the public:

1. A separate emergency communication center should be staffed to provide accurate (not overly optimistic) and timely information to management, the media, and the public [4, 46].

2. Loads critical to public safety should be restored as soon as possible [47].

3. Within the limits of the system, the largest possible unserved blocks of load should be restored as rapidly as possible; with plans being developed to restore 75%, 90% and then 100% of total unserved load [25, 47].

E. Maintain Flexibility to Respond to Problems

Because a major power failure is both unique and complex, unusual problems will invariably arise during restoration even with extensive planning and preparation. These may include: excessive and redundant energy management system (EMS) alarms [19, 23]; protective systems which impede restoration [1, 11, 24, 44, 45, 47]; voice or data communication problems [27, 31]; and computer malfunctions. Consequently, to resolve or work around problems as they occur, the restoration process must be flexible and responsive.

III. Initial Responses For Abnormal Conditions

Following a major disturbance, the system should be assessed and stabilized:

A. Assess and Verify Conditions

When load is lost, or the system is unstable or separated, an immediate assessment and verification of conditions should be performed to determine the extent and characteristics of the problem and the overall status of the system [1, 21, 31]. Potential restoration actions would depend upon the type of power system problem encountered:

1. A *standard power failure* involving loss of load and regional transmission only. Load can be restored using straight-forward (though possibly extensive) control actions or system repairs.

2. An *unstable system* with significant degradation in voltage, frequency, or power flow equilibrium signalling an impending major power failure. Emergency control actions including load shedding may help limit further deterioration.

3. A *major power failure island formation* in which a portion of the power system becomes separated from the main synchronized power grid. The island may have been caused by a disturbance which opened ties between two areas due to loss of synchronism. Because islanded generation is less stable, the separated system is more likely to collapse into a blackout.

4. A *major power failure blackout* in which critical generating or bulk transmission facilities are lost resulting in system collapse and significant loss of customer load.

Because a major power failure may have originated with an unacceptable mismatch between load and available supply, and cascading system failures may have followed, customer load can not likely be resupplied using straight-forward methods. A more involved and coordinated restoration procedure will need to be implemented.

B. Limit and Stabilize the Major Power Failure

If part of the system is islanded or blacked-out, immediate steps should be taken to limit further spread of the major power failure and stabilize the remaining system. These actions may include: (1) shedding load; (2) isolating generation; (3) avoiding formation of islands; (4) restoring selected loads to stabilize generation; (5) starting combustion turbines; (6) preparing to restart tripped generation; (7) re-supplying house service to tripped generation; (8) clearing problems such as damaged facilities; and (9) eliminating thermal and voltage violations [9, 11, 19, 21, 25, 42].

IV. Sequence of Restoration Actions

Once a major power failure is assessed, verified and stabilized, achieving the goals and objectives of restoration will require implementation of the following applicable actions:

A. *Start Restoration*

Other operators and management, as well as people needed for additional support should be notified of the failure and current system status. As a follow-up to assessment and without seriously delaying other restoration actions, restoration options should be analyzed in more depth, and a restoration plan should be formulated. Although assessment, analysis and planning can consume critical time, their intent should be to help, not hinder restoration. Ideally, these should proceed in parallel with other restoration actions. Overall, restoration will need to be coordinated, progress will need to be monitored and compared against objectives, and plans may need to be adjusted [17].

B. *Prepare Initial Cranking Source*

If a major power failure blackout occurs, an initial source of cranking power (i.e.: service to house auxiliary loads) for tripped generators must be chosen. Restarting generators "top down" from the bulk power system (BPS) is preferred. Although this may require more coordination and prolong customer outages, it provides a more stable source and more easily allows re-synchronization and reloading of restarted units. Alternating, a "bottoms-up" approach from a black-start or isolated generator will need to be the backup method if obtaining cranking power from the BPS is difficult or infeasible. These black-started or isolated units should pickup local loads in manageable increments until stabilized [4, 5, 7, 9, 11, 20, 25, 34, 43, 49].

C. *Prepare Restoration Path*

In the process of providing a cranking source, a restoration transmission path needs to be prepared either between the BPS and a selected off-line generator, or between a blackstart or isolated generator and the next off-line generator which needs external cranking power. This preparation may include: (1) clearing or bypassing system problems; (2) supplying loads critical to restoration; (3) segmenting portions of the system; (4) matching connected loads with generator pickup capabilities; and (5) taking steps to minimize over-voltages during energization and switching [6, 20, 42, 45].

D. *Build Stable Sub-System*

With a cranking source available, a larger system needs to be built. If a unit is restarted from the BPS, it should be re-synchronized with the BPS and stabilized. Ideally, black-started or isolated generators should be stabilized, and then tied back into the BPS or connected with neighboring islands if a clear interconnection path can be made. Whether or not a tie can successfully be made to an outside system, the following steps, nevertheless, should be taken:

1. Restart, stabilize, and synchronize additional generators via restoration paths so that a number of generators are running in parallel. Automatic Governor Controls (AGC) should be set to maintain constant frequency. In deciding which off-line generators should be restarted from an external source, priority should be given to generators that have: (1) critical restart time limits; (2) good real and reactive power responses; (3) proximity to the cranking source; and (4) proximity to critical loads or other off-line generators requiring cranking power [1, 5, 32, 33, 40].
2. Pickup loads that are critical to the restoration process such as major switching stations, pipe-type underground cable pumps, communication circuits, and loads needed to stabilize generation. General restoration of load should not proceed until a stable integrated system has been rebuilt [1, 24, 26].

E. *Tie Island Sub-System Into BPS*

Because a large interconnected system is more stable and resilient, it is more capable of responding to load changes, surviving disturbances and energizing EHV lines. Thus, a black-started generator, isolated generator, or an islanded system should be interconnected with the BPS as soon as feasible provided both systems are stable and a clear transmission tie path is available [39, 41].

F. *Tie Neighboring Island Sub-Systems Together*

If a tie path to the BPS is unavailable or the BPS is unstable, individual island sub-systems may need to be restarted independently and tied together to form a larger more stable system. This larger sub-system will then need to tie back into the BPS as soon as feasible [39, 41].

G. *Restore Unserved Load*

Once a large, stable, re-integrated system with sufficient generating capability has been rebuilt, overall load restoration should proceed as rapidly as possible taking the following precautions: (1) energize cleared transmission paths to load centers; (2) deactivate or avoid under-frequency relays; (3) restore loads in increments to match generator ramp rates and avoid excessive under-frequency; (4) restore radial lines with lagging power factors to control over-voltages; and (5) restore low voltage AC network large block loads after generating units with adequate frequency response and reactive power capability are on-line [1, 2, 35, 41, 47].

H. *Follow-Up After Restoration*

After load is restored and reintegrated with the BPS, the system should be made secure by restarting additional generation for adequate reserves, energizing and closing parallel transmission paths, and returning to normal operation. Finally, the restoration process should be documented by recording events, actions taken, problems encountered, results obtained, and recommendations for future improvement [16, 47].

V. Enhancing Preparedness

Advanced planning and preparation should improve the effectiveness and efficiency of any restoration process. Moreover, it can benefit day-to-day normal operations. An individual utility needs to decide how much and what type of restoration preparation is cost effective and applicable for its unique characteristics. This could include:

A. *Designate Restoration Chain of Command*

For the extraordinary circumstances of a major power failure, a chain of command authority for both restoration planning and actual execution needs to be decided and agreed

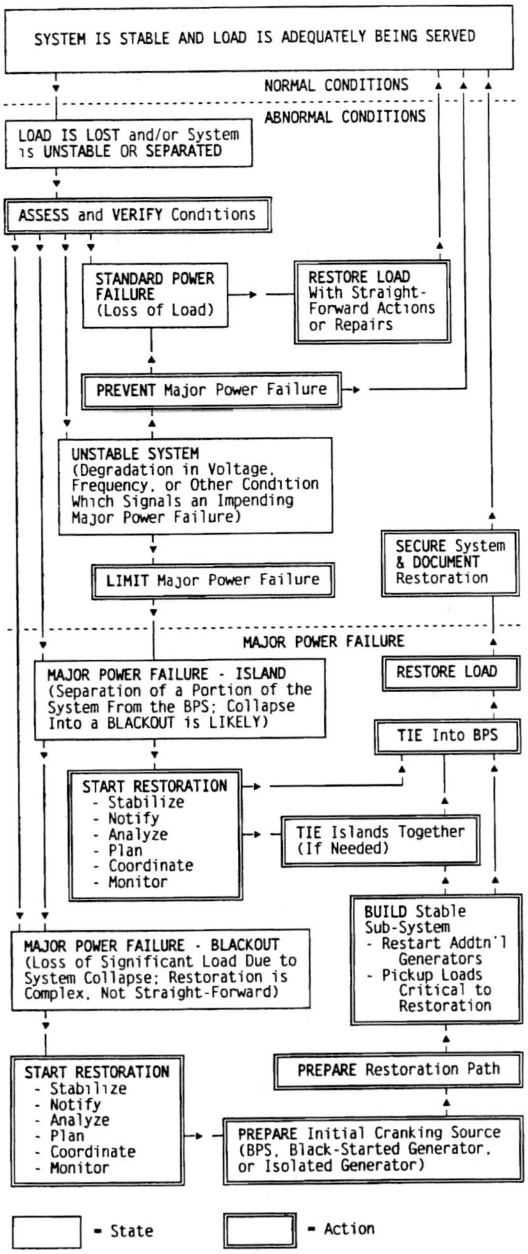

FIG. 2 Power Failure States and Restoration Actions

upon beforehand. It also needs to be clearly understood by everyone potentially involved in restoration to avoid confusion, conflict and lost time [46].

B. Assemble Restoration Team

Due to the complex nature of a major power failure, the expertise and cooperation of numerous departments and disciplines will be essential if restoration is to proceed smoothly. Consequently, individuals that may be needed to help plan or conduct restoration activities should be identified and assigned responsibilities as members of a restoration team [42, 46].

C. Prepare Restoration Data Base

Having organized data available will be extremely valuable during restoration. This could include: (1) generator data (nominal characteristics, blackstart capability or startup requirements, restart time limits, operating capabilities and limitations, and protective relay settings); (2) transmission data (charging currents, reactive power control devices, transformers and phase shifters, underground cables and networks, frequency metering points and synchronizing facilities, under-frequency and other protective relay schemes, and backup power supplies); and (3) load data (loads needed to stabilize restarted generators, critical public loads, low voltage AC networks, and cold load pickup estimates for a variety of conditions) [1, 9, 12, 25, 44].

D. Analyze With Power System Software

Power system software (i.e.: interactive load-flow, short circuit, transient voltage, transient stability, electromagnetic transient, and long term dynamic stability software) can be used to help develop restoration plans. Some of the software may also help guide restoration decision making after a major power failure has occurred [1, 3, 10, 17, 18, 29, 47].

E. Prepare Restoration Plan

Developing a restoration plan consists of three segments:

1. *Formulate and study potential failures*: a list of potential causes of major power failures (eg.: multiple loss of generating or transmission facilities, major disturbances on neighboring systems, malfunctions of protective relays, natural disasters, human error, etc.,.) should be studied to determine their potential impact. From this, a list of conceivable power failures can be generated.

2. *Evaluate restoration options*: given a list of credible major power failures, a corresponding set of restoration options and problem resolutions needs to be formulated and studied to determine their likelihood of success. While an exhaustive study may not be feasible, even a partial analysis offers the advantages of a better understanding of the power system, major power failures, and the general approach that needs to be taken during restoration.

3. *Develop restoration plan*: with some analyses completed, a general restoration plan should be formulated. Portions of the plan may need to be more detailed to accommodate unique problems anticipated for a specific power system. The restoration plan should be well organized and easily understood, practical rather than theoretical, widely distributed to foster cooperation and understanding, and updated and refined regularly [1, 3, 4, 16, 17, 31, 48].

F. Hold Restoration Training Sessions

Regular training sessions should be held to familiarize restoration team members with the unusual conditions of a major power failure, improve coordination and cooperation, help insure people know what to do during a major failure, and avoid counter-productive restoration attempts. Additionally, training should help improve overall understanding of the power system, and may even help prevent a major power failure from occurring [1, 4, 19, 28, 30, 42, 44, 46].

G. Perform Restoration System Tests

System components critical to restoration need to be identified and tested on a regular basis for restoration duty (which may be more rigorous than normal testing). These may include emergency backup power supplies, blackstart sources and paths, generator startup equipment, synchronizing equipment, and computers [16, 30, 42].

H. Conduct Restoration Practice Drills

As a continuation of restoration training, practice drills for major power failure restoration should be developed and regularly held to help assure restoration team members have become familiar with their assigned responsibilities. These practice drills should also help identify restoration problems that need attention, improve the overall restoration process, shorten outage times, and make restoration expectations more realistic [20, 42].

I. Incorporate Recommendations For Improvement

Preparing for a major power failure and planning restoration strategies should be a continual iterative process. Studies, training sessions, system tests, practice drills and restoration team discussions may each reveal weaknesses or problems with the existing system that can and should be corrected. Additionally, they may suggest improvements that can be made in equipment, procedures, software, or administrative policies. Consequently, major power failure restoration plans and preparations need to be reviewed, updated and refined on a regular basis [1, 15, 42, 48].

J. Develop and Implement Preventive Measures

Although this paper has concentrated on restoration, the need for prevention of major power failures should not be overlooked. Thus, restoration preparation should incorporate modifications to procedures or equipment that can help avert major power failures and their associated problems [42, 50].

VI. Conclusions

Major power failures are unfamiliar unique occurrences that make restoration both complex and stressful. Notwithstanding, numerous technical papers have identified aspects which are common to power system restoration in general. This paper draws from those more detailed writings and offers a general restoration guideline that can enable a utility to elaborate upon for its specific power system.

While striving to rebuild a stable electric system, and restore unserved load, the restoration should be conducted as safely, smoothly, and quickly as possible with minimal adverse impact to the public. To help insure objectives are met and avoid counter-productive actions following a major power failure, recommended actions include: assessing; stabilizing; analyzing and planning; preparing a cranking source and restoration path; restarting generators and building a stable sub-system; re-synchronizing with the BPS or neighboring islands; restoring all load; and securing the system. Additionally, recommended preparation and planning steps taken in advance will clearly help improve and expedite the restoration process. Each utility will need to decide what degree of preparation can be justified for its circumstances. Although a major power failure is fairly unlikely for any single utility, its occurrence anywhere and the relative success of the accompanying restoration could have far reaching effects for the entire industry. Therefore, continued discussion and development of general restoration techniques along with sharing of specific restoration plans between utilities is essential.

VII. References

[1] R. J. Kaflka et al, "System Restoration Plan Development For a Metropolitan Electric System," IEEE Trans. v. PAS-100, n.8, Aug 1981, 3703-3713.

[2] E. J. Simburger and F. J. Hubert, "Low Voltage Bulk Power System Restoration Simulation," IEEE Trans. v. PAS-100 n.11, Nov 1981, 4479-4484.

[3] R. J. Kafka, D. R. Penders, S. H. Bouchey, and M. M. Adibi, "Role of Interactive and Control Computers in the Development of a System Restoration Plan", IEEE Trans. v. PAS-101 n.1, Jan 1982, 43-52.

[4] IEEE Committee Report, "System Restoration - Deploying the Plan", COPWG, W. A. Johnson, chmn, IEEE Trans. v. PAS-101, Nov 1982, 4263-4271.

[5] G. Fusco, F. Mazzoldi, A. Possenti, D. Venturini, "Thermal Units Contribution to the Electric Power System Restoration After a Black-Out," (CIGRE 1982, 32-21).

[6] E. Mariani et al, "Field Experience in Reenergization of Electrical Networks from Thermal and Hydro Units," IEEE Trans. v. PAS-103 n.7, Jul 1984, 1707-1713.

[7] R. D. Shultz and G. A. Mason, "Blackstart Utilization of Remote CTs, Analytical Analysis and Field Test," IEEE Trans. v. PAS-103 n.8, Aug 1984, 2186-91.

[8] S. Nishida et al, "Analysis of Overvoltages Caused by Self-Excitation in a Separated Power System With Heavy Load and Large Shunt Capacitance," IEEE Trans. v. PAS-103 n.8, Aug 1984, 2935-2942.

[9] J. L. Lams, M. Monseu, "Operation of Generating Units During System Disturbances," (CIGRE 1986, 39-07).

[10] G. Morin, "Service Restoration Following a Major Failure on the Hydro-Quebec Power System," IEEE Trans. v. PWRD-2, Apr 1987, 454-463.

[11] IEEE Committee Report, "Power System Restoration - A Task Force Report", SRTF, M. M. Adibi, chmn, IEEE Trans. v. PWRS-2, May 1987, 271-277.

[12] PES Task Force Report: Power System Restoration, Part III, Power System Characteristics; IEEE Trans. v. PWRS-2, May 1987, 271-277.

[13] J. G. Raine: Reactive Power Balance (Part IV,1 in PES Task Force Report: Power System Restoration); IEEE Trans. v. PWRS-2, May 1987, 271-277.

[14] M. M. Adibi: Load-Generation Balance (Part IV,2 in PES Task Force Report: Power System Restoration); IEEE Trans. v. PWRS-2, May 1987, 271-277.

[15] R. Kearsley, "Restoration in Sweden and Experience Gained from the Blackout of 1983," IEEE Trans. v. PWRS-2 n.2, May 1987, 422-428.

[16] J. Gutierrez et al., "Policies for Restoration of a Power System," IEEE Trans. v. PWRS-2, May 1987, 436-442.

[17] IEEE Committee Report, "Power System Restoration - The Second Task Force Report", SRTF, M. M. Adibi, chmn, IEEE Trans. v. PWRS-2, n.4, Nov 1987, 927-933.

[18] F. F. Wu and A. Monticelli, "Analytical Tools for Power System Restoration - Conceptual Design" IEEE Trans. v. PWRS-3 n.1, Feb 1988, 10-16.

[19] IEEE Committee Report, "System Operations Challenges", COPWG, E. K. Nielson, chmn, IEEE Trans. v. PWRS-3, Feb 1988, 118-126.

[20] R. Lindstrom, "Simulation & Field Tests of the Black Start of a Large Coal-Fired Generator Using Small Hydro," IEEE Trans. v. PWRS-5, n.12, Feb 1990, 162-167.

[21] M. M. Adibi and R. J. Kafka, "Power System Restoration Issues," IEEE Computer Applications in Power, Vol. 4, No. 2, Apr 1991, 19-24.

[22] IEEE Committee Report, "Special Considerations in Power System Restoration", SRWG, M. M. Adibi, chmn, IEEE Trans. v. PWRS-7, n.4, Nov 1992, 1419-1427.

[23] C. J. Andrews: Excessive Alarms During Restoration (in PES Working Group Report: Special Considerations in Power System Restoration); IEEE Trans. v. PWRS-7, Nov 1992, 1419-1427.

[24] T. A. Volkmann: Switching During Restoration (in PES Working Group Report: Special Considerations in Power System Restoration); IEEE Trans. v. PWRS-7, Nov 1992, 1419-1427.

[25] J. M. Miller: Optimal Sequencing of Starting-Up Generating Units (in PES Working Group Report: Special Considerations in Power System Restoration); IEEE Trans. v. PWRS-7, Nov 1992, 1419-1427.

[26] J. N. Wrubel: Underground Transmission System Concerns (in PES Working Group Report: Special Considerations in Power System Restoration); IEEE Trans. v. PWRS-7, Nov 1992, 1419-1427.

[27] F. Arsanjani and M. W. Lanier: Telecommunication Capabilities and Limitations (in PES Working Group Report: Special Considerations in Power System Restoration); IEEE Trans. v. PWRS-7, Nov 1992, 1419-1427.

[28] IEEE Committee Report, "New Approaches in Power System Restoration", SRWG, M. M. Adibi, chmn, IEEE Trans. v. PWRS-7, n.4, Nov 1992, 1428-1434.

[29] IEEE Committee Report, "Overvoltage Control During Restoration," SRWG, M. M. Adibi, chmn, IEEE Trans. v. PWRS-7, n.4, Nov 1992, 1464-1470.

[30] IEEE Committee Report, "Bulk Power System Restoration Training Techniques", WGOT, M. E. Cooper, chmn and SRWG, M. M. Adibi, chmn, IEEE Trans. v. PWRS, Feb 1993, 191-197.

[31] M. M. Adibi and L. H. Fink, "Power System Restoration Planning", paper 93 WM 204-8-PWRS, presented at IEEE/P-ES 1993 WPM, Columbus, OH Jan 31-Feb 5 1993.

[32] F. P. de Mello and J. C. Westcott, "Steam Plant Startup and Control in System Restoration", paper 93 WM 201-4-PWRS, presented at IEEE/PES 1993 WPM, Columbus, OH, Jan 31-Feb 5, 1993.

[33] M. M. Adibi and D. P. Milanicz, "Reactive Capability Limitation of Synchronous Machines", paper 93 WM 203-0-PWRS, presented at IEEE/PES 1993 WPM, Columbus, OH, Jan 31-Feb 5, 1993.

[34] IEEE Committee Report, "Special Considerations in Power System Restoration", SRWG, M. M. Adibi, chmn, paper 93 WM 202-2-PWRS, presented at IEEE/PES 1993 WPM, Columbus, OH, Jan 31-Feb 5, 1993.

[35] W. W. Price and S. Ihara: Cold Load Pickup - Modeling (in PES Working Group Report: Special Considerations in Power System Restoration); paper 93 WM 202-2-PWRS, presented at IEEE/PES 1993 WPM, Columbus, OH, Jan 31-Feb 5, 1993.

[36] C. Weliky: Cold Load Pickup - Low Voltage Network Loads (in PES Working Group Report: Special Considerations in Power System Restoration); paper 93 WM 202-2-PWRS, presented at IEEE/PES 1993 WPM, Columbus, OH, Jan 31-Feb 5, 1993.

[37] S. A. James and J. N. Wrubel: GT's in PSE&G's Restoration Plan (in PES Working Group Report: Special Considerations in Power System Restoration); paper 93 WM 202-2-PWRS, presented at IEEE/PES 1993 WPM, Columbus, OH, Jan 31-Feb 5, 1993.

[38] M. M. Adibi: Reactive Power Balance (in PES Working Group Report: Special Considerations in Power System Restoration); paper 93 WM 202-2-PWRS, presented at IEEE/PES 1993 WPM, Columbus, OH, Jan 31-Feb 5, 1993.

[39] IEEE Committee Report, "An Approach to Standing Phase Angle Reduction", SRWG, M. M. Adibi, chmn, paper 93 WM 205-5-PWRS, presented at IEEE/PES 1993 WPM, Columbus, OH, Jan 31-Feb 5, 1993.

[40] H. B. Ross et al, "An AGC Implementation for System Islanding and Restoration Conditions," paper 93 SM 519-9 PWRS presented at IEEE/PES 1993 SPM, Vancouver, B. C., July 18 - 22, 1993.

[41] R. M. Hill, "Power System Restoration - Western Area Power Administration," EPRI/ECC Power System Restoration Seminar, Denver, Sep 27-28, 1993.

[42] S. W. Milam, "Black-Start Exercise - Consumers Power", EPRI/ECC Power System Restoration Seminar, Denver, Sep 27-28, 1993.

[43] R. Hamlin, "Initial Black-Start Energy Sources - Con Ed", EPRI/ECC Power System Restoration Seminar, Denver, Sep 27-28, 1993.

[44] R. J. Kafka, "Active and Reactive Power Balance - Potomac Electric Power Co., EPRI/ECC Power System Restoration Seminar, Denver, Sep 27-28, 1993.

[45] T. Volkmann, "Assessment of Switching Status and Switching Operation -- Northern States Power Co.", EPRI/ECC Power System Restoration Seminar, Denver, Sep 27-28, 1993.

[46] S. Lutterodt, "Organization and Human Problems Involved in Restoration", EPRI/ECC Power System Restoration Seminar, Denver, Sep 27-28, 1993.

[47] M. M. Adibi, J. N. Borkowski, and R. J. Kafka, "Analytical Tool Requirements for Power System Restoration", paper 94 WM 222-0-PWRS, presented at IEEE/PES 1994 WPM, New York, NY, Jan 30-Feb 3, 1994.

[48] M. M. Adibi, R. J. Kafka, and D. P. Milanicz, "Expert System Requirements for Power System Restoration", paper 94 WM 223-8-PWRS, presented at IEEE/PES 1994 WPM, New York, NY, Jan 30-Feb 3, 1994.

[49] K.L. Liou, CC. Liu, and R.F. Chu, "Tie Line Utilization During Power System Restoration", paper 94 WM 221-2-PWRS, presented at IEEE/PES 1994 WPM, New York, NY, Jan 30-Feb 3, 1994.

[50] J. A. Juves, R. D. Willoughby, and A. Batenburg. *Utility Survey of Methods for Minimizing the Number and Severity of System Separations*, Palo Alto: EPRI, 1984, EL-3437.

PART II

Restoration Techniques

I. General Discussion

THIS part covers various techniques used in planning, simulating, field testing, and verifying different aspects of power system restoration. The models and procedures developed address steady-state, transient, and dynamic behaviors of the power system and its major components, including frequency response of prime movers, reactive capabilities of generators, asymmetry of transmission lines, overvoltage withstand of system equipment, correct and appropriate operation of protective relays, remote cranking of non-blackstart generating units, startup requirements of stations' large auxiliary induction motors, and maximum demand imposed by cold load pickup. In addition, the following papers provide guidelines and recommendations for developing an efficient restoration plan and its safe execution.

II. Discussion of Reprints

2.01 Cold Load Pickup, 1979, p. 105

A method has been presented for predicting the magnitude and duration of the peak demand following a power outage in cold weather. The method is based on experimental data obtained from electrically heated homes that have thermostatically controlled temperature. It is shown that after an hour of service interruption, the magnitude and duration of demand are high enough to dictate consideration of cold load pickup in both the designs of the power system and the restoration procedures.

2.02 Physically Based Modeling of Cold Load Pickup, 1981, p. 108

A load model for the thermostatically controlled space heating and cooling of a house is described. The model represents the thermal characteristics of the house without modeling details of the appliance loads. The objective has been to provide a quantitative method for predicting the magnitude and duration of the overload following an outage. The paper concludes that to avoid overloads during power system restoration, the cold load pickup phenomena should be carefully considered.

2.03 Blackstart Utilization of Remote Combustion Turbines, Analytical Analysis, and Field Test, 1984, p. 117

Some special operating procedures may be required to blackstart a steam electric station from a remote combustion turbine. This paper presents an analytical method for determining the special procedures needed in a specific case. A blackstart field test has then been conducted on the system, using the procedure determined analytically. The test system voltages and frequency behaviors obtained during the test startups of the large induction motors in the steam plant have compared very well with the predicted analytical results.

2.04 Overvoltage Control During Restoration, 1992, p. 123

Restoration of high-voltage lines may cause three related overvoltages: sustained power frequency overvoltages due to the charging currents of lightly loaded lines, switching transients when energizing large segments of a transmission system, and harmonic resonance voltages when energizing nonlinear equipment such as transformers and iron-core reactors. Power transformers, surge arresters, and circuit breakers are the equipment first affected by these overvoltages. The paper provides three examples describing each type of overvoltage and recommends measures for their control.

2.05 Reactive Capability Limitation of Synchronous Machines, 1994, p. 130

Usually, the generator reactive capability is significantly less than that indicated by manufacturers' curves. This is due to limitations imposed by the plant auxiliaries and the power system itself. The paper develops a method for determining the actual generator reactive power capability within the voltage limits at the generator bus, auxiliary bus, and system bus. The methodology is tested and verified on a large generator. Several recommendations are made for enhancing the generators' reactive power capabilities.

2.06 An Approach to Standing Phase Angle Reduction, 1994, p. 142

The excessive standing phase angles across open circuit breakers cause significant delays in power system restoration. The angle may occur between two connected subsystems. It must be reduced to the allowable limit imposed by the synchrocheck relay before an attempt is made to close the breaker and firm up the interconnection between the two subsystems. This paper describes a method to serve as a guideline for reducing the excessive angle without resorting to the raising and lowering of various generation levels on a trial and error basis.

2.07 Nuclear Plant Requirements During Power System Restoration, 1995, p. 151

The Nuclear Regulatory Commission's "startup checklists" do not permit hot restart of nuclear units following an outage, and the "emergency generators" are not allowed to supply auxiliary power to the other stations. This paper specifies a number of requirements, including provisions for two independent offsite power sources to support all the safety functions for the restart. In order to reduce potential delays in restart activities, the paper recommends that protective measures be taken to improve the reliability of the offsite power supply. The paper concludes that while it is important to provide offsite power sources to nuclear units, the power system load needs to be restored without dependence on the nuclear units.

2.08 Protective System Issues During Restoration, 1995, p. 157

The correct but inappropriate relay operation and relays failing to trip are primarily due to changes in the power system. During restoration, the power system undergoes continual changes, and therefore certain relays are subject to the undesirable operations. This paper identifies ten relay schemes that may affect the restoration duration. It recommends that the location of such relays be identified and their possible inadvertent operations be evaluated under restorative conditions when the power system is going through continual changes. The foreknowledge that certain system operating conditions could cause undesirable operation would help prevent such operating conditions during the development and execution of a restoration plan.

2.09 Remote Cranking of Steam Electric Stations, 1996, p. 163

The paper describes the planning and simulation used in the blackstart of two coal-fired generating units using the remote combustion turbines. A number of concerns and constraints at the cranking source, at the remote station, and in the connecting network path are discussed. It is shown that the combustion turbines should be able to meet the leading reactive power requirements of the connecting line's charging current as well as the lagging reactive power demand of starting the large auxiliary motors within the station. The paper recommends that the present operator training simulators be expanded to accommodate simulation of this basic restoration operation.

2.10 Optimizing Generator Reactive Power Resources, 1999, p. 169

Generators provide a valuable resource in their ability to supply and absorb reactive power. By changing tap positions on the generator step-up and auxiliary transformers, it is possible to vary their lagging and leading operating limits.

The approach described in this paper provides an indication of the range of reactive power outputs that are available for different tap positions. This information can be used to make more effective use of generator reactive power resources.

2.11 Asymmetry Issues in Power System Restoration, 1999, p. 175

The offsite power supply for restarting the large and remote thermal generating units is often provided via the extra-high voltage (EHV) lines. Many of these lines have unsymmetrical conductor spacing without being transposed and consequently generate unacceptable negative sequence currents (NSC) for the startup procedure. This paper evaluates the NSC in typical unsymmetrical and untransposed EHV lines, and it recommends connection of adequate underlying load to reduce the excessive NSC.

2.12 Frequency Response of Prime Movers During Restoration, 1999, p. 181

A simple approach is developed based on an approximate frequency response rate of prime movers for determining the maximum load pickup within the allowable system frequency dip and for determining the amount and distribution of reserve for maintaining firm generation. The paper develops response rates and reserve rates for different types of prime movers to be used in developing several operator guides.

2.13 Energizing High and Extra-High Voltage Lines During Restoration, 1999, p. 187

A simple and approximate approach is presented which evaluates sustained and transient voltages when energizing high and extra-high voltage lines. The sending-end and receiving-end, sustained and transient voltages are determined for lines of different voltage levels and lengths, energized by different sizes of generators, and having different sizes of underlying loads. The models described can readily be used to determine similar steady-state and transient guidelines in other situations and power systems.

COLD LOAD PICKUP

James E. McDonald   Armin M. Bruning, Member, IEEE   William R. Mahieu

A. B. CHANCE COMPANY
210 N. Allen Street, Centralia, Missouri

Abstract - The primary purpose of this paper is to provide a quantitative method for electric power utilities to predict the magnitude and duration of the peak demand following a power outage in cold weather. The study is limited strictly to electrically heated homes. A model has been developed based on experimental data that predicts the power demand of any number of homes following a power outage. The authors believe that this information should influence both the design of the distribution system and the method of power restoration.

INTRODUCTION

With the increasing frequency of large all-electric subdivisions coupled with underground distribution, utilities are finding difficulties in restoring power after a power outage. This paper attempts to quantify the expected overload, magnitude and duration, for an all-electric residential area after a power outage in cold weather. In addition, a brief look is taken at the effects of this type of load on the distribution system.

The method used to analyze cold load pickup was to examine individual electrically heated homes, both experimentally and theoretically, and to extrapolate the results to power outages and restorations involving many homes. The authors have conducted simulated outages on individual homes and then statistically added the effects up for many homes to see what type of power requirements would be expected. To do this, a model has been constructed that predicts the ''average'' inside temperature of a home during the absense of power and the subsequent heatup following power restoration.

EXPERIMENTAL PROCEDURE

Most electrically heated homes have thermostatically controlled power which is fully on when the room air temperature is below the thermostat setting. Following an extended power outage, all homes which experienced the loss of power would require full power simultaneously, with no diversity in the load. One of the major requirements of any model is to predict when the air temperature in any home will recover to the thermostat setting and it will begin cycling the power off periodically. Our tests on individual electrically heated homes measured both air temperatures and electrical power during three distinct time periods: a steady state period where the heating load balanced the home heat loss rate, a cool down period with no heating power on, and a heat up period with maximum power the heating system would consume. All three measurement periods were after sundown to avoid the variable and difficult-to-measure solar input. The predictions based on these experiments should therefore be applicable to the worst case of nighttime power outage with no benefit from solar heating.

F 79 180-1 A paper recommended and approved by the IEEE Transmission and Distribution Committee of the IEEE Power Engineering Society for presentation at the IEEE PES Winter Meeting, New York, NY, February 4-9, 1979.
Manuscript submitted September 5, 1978; made available for printing December 5, 1978.

Temperature of various locations inside and outside of a home were monitored continously. The following schedule was followed with some minor variations in precise times:

TIME	EVENT
Daytime	Temperature Monitoring Started
10:30 p.m.	Power shut off to Heating System
4:00 a.m.	Power was restored
7:00 a.m.	Temperature Monitoring ceased

From the start to 7:00 a.m., power into the home was monitored.

Actual placement of thermocouples was subjective, but the following guidelines were followed as closely as possible:

THERMOCOUPLE	LOCATION
1.	Outside - at least 12 inches from house and three feet off ground.
2.	Inside - halfway between ceiling and floor and at least 12 inches from wall, in central part of house.
3.	Inside - on ceiling in central part of house.
4.	Inside - on floor in central part of house.
5.	Inside - on wall in central part of house.
6.	Inside - on exterior wall halfway between ceiling and floor.
7.	On heat outlet.
8.	On heat intake if it exists.
9. - 24.	Used as needed for any extensions or wings on house upstairs, etc. half way between floor and ceiling and 12 inches from wall.

All thermocouples were located as far as possible from doors, windows, vents, or any other ''perturbations.''

A recording wattmeter was used to monitor input power to the home. To make the test as convenient as possible for the family, in whose home we were testing, no power other than heating was turned off. They were asked not to run any major appliance if that could be avoided. In all cases power consumption was very low during the hours from 10:30 p.m. to 4:00 a.m. The effect of this additional heat was small. An estimate of the magnitude of the effect is made in the theoretical section.

It should be noted that no effort was made to include the effects of wind on a home. In all cases our experimental data was taken when there was little or no wind (<10 mi/h).

Reprinted from *IEEE Transactions on Power Apparatus and Systems*, Vol. 98, No. 4, pp. 1384–1386, July/August 1979.

THEORETICAL RESULTS

An effort was made to keep the mathematical model as simple as possible and still accurately reflect energy consumption. For our purposes the model needs to depend on two parameters, outside temperatures and the length of time power was off. The approach taken was to predict an ''average inside temperature'', which was modeled to follow thermocouple number 2.

A. Steady State

K can be determined from the data collected between the start and 10:30 p.m., when the temperature in the house was essentially constant. K, the thermal conductivity of the house, is given by

$$K = \frac{Q}{t\,(T - T_o)} \tag{1}$$

Where Q is the energy into the house as measured by the wattmeter, t is the length of time over which Q was measured, T_o is outside temperature and T is the inside temperature. In more conventional terminology, this K is equivalent to the product of an overall heat transfer coefficient and the surface area of the home through which the heat is lost.

B. Cool Down

Fifteen minutes after power was shut off all thermocouples in the house recorded temperatures within 2°F of each other. This allows us to use a simple model shown in Figure 1a.

Once power is shut off T is given by

$$T(t) = (T(0) - T_o)\exp\left(-\frac{K}{mC_p}t\right) + T_o \tag{2}$$

Where m is the heated mass of the house and C_p the average specific heat, the product of mC_p can then be determined from the measured time constant of the house.

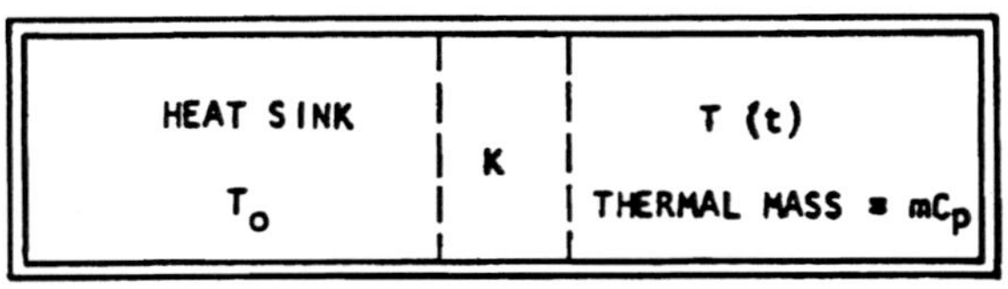

a) COOL DOWN

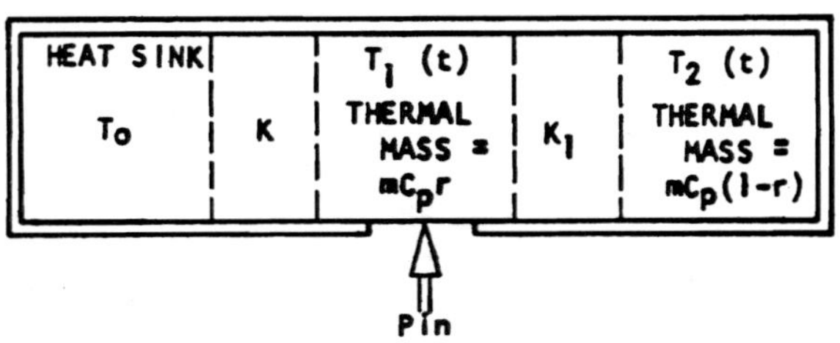

b) HEAT UP
Fig. 1

The size of the error introduced by leaving some power on during the cool down of the house can be estimated by

$$\Delta T = \frac{Q}{mC_p} \tag{3}$$

Where Q the total energy into the house between 10:30 p.m. and 4:00 a.m., it was found in all cases ΔT was less than 5% of total temperature change.

C. Heat Up

A more complicated model is required to accurately predict the power demand of the heating system. One that gave a reasonable fit to the observed response of the house is shown in Figure 1b. In this model, the thermal mass of the house was divided into two parts. The smaller part (mC_pr) represents that portion of the house thermal mass most closely tied to the heating systems (ducts, air mass, etc.). The larger part ($mC_p(1 - r)$) requires an additional heat transfer through K_1 (interior walls, floors, furnishings, etc.). Heat transfer occurs from this total thermal mass through K to the outside air. For this model $T_1(t)$ is matched to thermocouple number 2. The two coupled differential equations that describe this model's behavior are as follows (note that mC_p and K are determined during cool down)

$$rmC_p\frac{dT_1(t)}{dt} = P_{in} - K_1(T_1(t) - T_2(t)) - K(t_1(t) - T_o) \tag{4}$$

$$(1-r)\,mC_p\frac{dT_2(t)}{dt} = K_1(T_1(t) - T_2(t)) \tag{5}$$

Here r is a scaled parameter between 0 and 1, and P_{in} is the input power. By assuming that at t = (when power is restored) $T_1(0) = T_2(0)$ then

$$r = \frac{P_{in} - K(T_1(0) - T_o)}{\frac{dT_1(0)}{dt}\,mC_p} \tag{6}$$

K_1 was determined by an iterative technique so that the model would give the correct time for the room air temperature to recover to the thermostat setting for each test home.

An analytical solution is possible for this system but since we wish to allow P_{in} to turn on or off a digital simulation proved easier than continually matching boundary conditions for each step change in P_{in}. The heat input was set equal to zero when T_1 was 1°F over the set temperature and set back to P_{in} at 1°F lower than set temperature.

The experimental values obtained for these parameters are summarized in Figure 2. These averages and deviations are based on individual tests in six electrically heated homes.

SYSTEM RESPONSE

Using this model a demand curve was constructed by assuming each parameter had a Gaussian distribution with the indicated standard deviation and mean value, then adding the power demand of each house as a function of time while the parameters were statistically varied. Figures 3, 4 and 5 represent a summary of this procedure. Figure 3 gives the average steady state power demand per house. Figure 4 gives the normalized magnitude of the peak demand where 1 corresponds to the condition when all connected heating

PARAMETER	VALUE	STANDARD DEVIATION
Pin	23 kW	4 kW
r	.15	.03
mCp	3.7 kWh/°F	1.2 kWh/°F
T (0)	69 °F	3 °F
K	.14 kW/°F	.02 kW/°F
K1	1.5 kW/°F	.2 kW/°F

Fig. 2 SUMMARY OF PARAMETERS

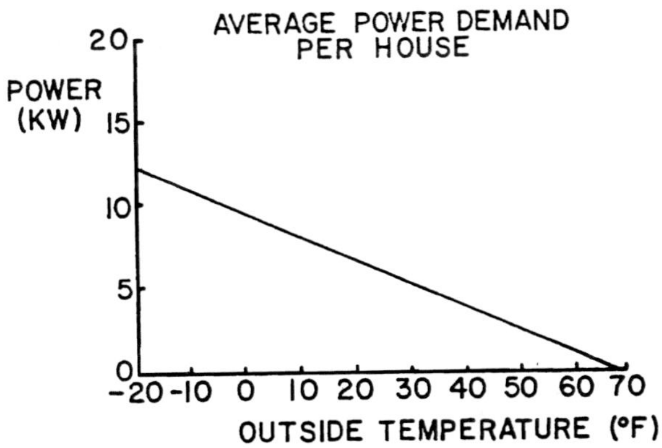

FIG. 3

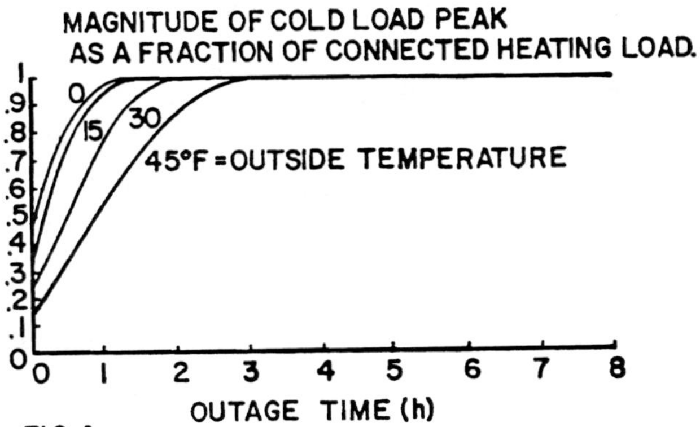

FIG.4

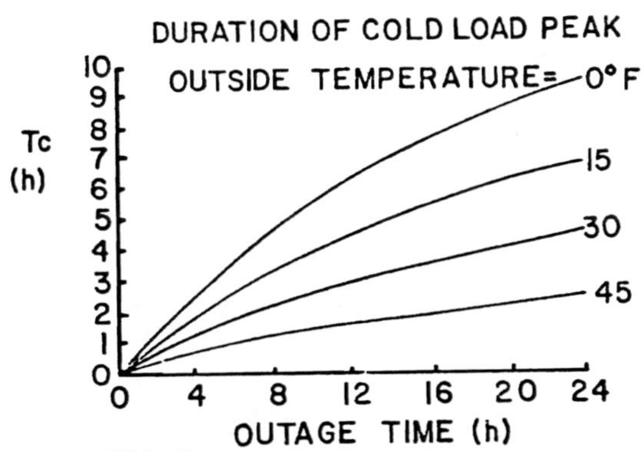

FIG. 5

load is turned on. Figure 5 gives the duration (tc) of the peak power.

CONCLUSIONS

The authors consider this experimental and analytical effort to be part of a first look at the cold load pickup problem to determine its significance.

We found the magnitude and duration of peak power, following outages of a few hours at common winter ambients temperatures, surprisingly high. The realism of these projections needs to be tested by staged outages on actual systems. We are working with utilities to arrange and conduct such tests. If these projections prove to be realistic, utilities will have to consider cold load pickup quantitatively in both the design of their systems and their operating procedures. The implications are that cable sizing, fusing and transformer sizing will be affected.

ACKNOWLEDGEMENTS

The authors would like to thank George Stewart and Mike Cahill of Niagra Mohawk Paper Corporation and Dennis Mastro and Joe Carriere of New York State Electric and Gas Company for the use of their homes. Experiments were also conducted in the homes of several Chance Company employees: Armin Bruning, Carl Yager, George Allen and Ron Azdell. The staged outages on individual homes were conducted by principal author Jim McDonald and technician Ron Azdell. Figures were drawn by Danny Burnett. Manuscript was typed by Lorene Hughes.

PHYSICALLY BASED MODELING OF COLD LOAD PICKUP

Satoru Ihara
Member
General Electric Company
Schenectady, New York

Fred C. Schweppe
Fellow
Massachusetts Institute of Technology
Cambridge, Massachusetts

ABSTRACT

A general, physically based, probabilistic model of power system load has been developed [1]. The model is suitable for short-term prediction of system demand by composition, and accounts for lifestyle dependency and weather dependency as well as effects of other exogenous processes. The conceptual framework and notation applicable to a wide range of load modeling problems have been the key issues in developing a general model.

In contrast to the generality pursued, this paper presents a specific model of the space conditioning load for studying the load behavior during cold load pickup. The objective is to demonstrate how one can use the general load modeling framework to systematically extract the essence of specific modeling problems and to obtain practical models. The key requirements in practical load modeling tasks are the minimum requirement of the data and computation, and the maximum utilization of available data and load modeling experiences.

INTRODUCTION

Restoration of power after a service interruption raises various practical concerns. For example, following an outage on a cold day, a large portion of the installed space heaters will demand power from the system and cause severe overloads on distribution circuits. If an outage involves a large number of customers, the system problems must be avoided through a carefully designed restoration procedure. The objective of modeling cold load pickup is to provide a quantitative method to predict the magnitude and duration of the overload following an outage. Furthermore, the model can be used to study the payback phenomena during load management since loads respond to any power interruptions in the same manner.

Cold load pickup of space conditioning load, specifically heating and cooling load, is of a practical interest because of its major contribution in the total overload. The thermodynamics of individual houses during and after an outage play an important role in determining the magnitude and the duration of the overload; therefore, an efficient modeling of the thermal characteristics of a large number of different houses is critically important. The diversity in the thermal characteristics among houses must be captured through data collection and computation of a reasonable effort.

Physically Based Load Model Structure

Portions of the general model structure [1] will be briefly reviewed to provide a context for the more explicit discussions on the behavior of space conditioning load during cold load pickup. The model structure is based on viewing the total load as seen at any point on the system as being the sum of elemental powers. The power demand in kW and/or kvar of any specific electrical device is called an elemental power. Each elemental power is classified by three subscripts, i, j and k. These subscripts denote classes of elemental power sharing similar characteristics as follows:

Device Class $i = 1,2...,I$: partitioning of the load into sets having similar electrical characteristics (voltage and frequency dependency) e.g., resistance heating, 5-10 hp compressor motors;

Usage Class $j = 1,2...,J$: partitioning of the load into sets having similar usage characteristics, e.g., cooking in surburban homes, cooking in food processing industry;

Billing Entity $k = 1,2...,K$: partitioning of load into sets whose total load is recorded on a single meter, e.g., customers on the system, individually metered loads.

Each elemental power is viewed as belonging to one of the intersections of the three different partitionings. There are $(I \cdot J \cdot K)$ possible intersections.

The dependence of the system-level load on time, economics, weather, voltage, and frequency is modeled by considering the dependence of the elemental power at time t on the exogenous input z:

$$z = (e, d, f, m) \tag{1}$$

where:

- e = vector of economic parameters.
- d = vector of voltage, frequency and device parameters.

81 WM 084-3 A paper recommended and approved by the IEEE Power System Engineering Committee of the IEEE Power Engineering Society for presentation at the IEEE PES Winter Meeting, Atlanta, Georgia, February 1-6, 1981. Manuscript submitted September 2, 1980; made available for printing November 25, 1980.

Reprinted from *IEEE Transactions on Power Apparatus and Systems,* Vol. 100, No. 9, pp. 4142–4150, September 1981.

f = vector of parameters representing customer's lifestyle.

m = vector of meteorological parameters.

Each elemental power y is factored into an elemental <u>potential demand</u> c and an elemental <u>utilization factor</u> u:

$$y_{ijk}(t, z) = c_{ijk}(t, z) \cdot u_{ijk}(t, z) \quad (2)$$

The potential demand is the maximum demand at time t in kW and/or kvar that the customer can achieve by adjusting the controls of the device. The potential demand is further factored into a <u>capital stock</u> x and a <u>demand response term</u> r:

$$c_{ijk}(t, z) = x_{ijk}(t, e) \cdot r_{ijk}(t, d) \quad (3)$$

where the capital stock is the kW and/or kvar rating of the device under nominal conditions, and the demand response term represents the voltage and frequency dependence of the device.

The utilization factor represents the fraction of the potential demand which is actually drawn by the device at time t. The utilization factor is decomposed into a <u>lifestyle term</u> ℓ and a <u>usage response term</u> w:

$$u_{ijk}(t, z) = \ell_{ijk}(t, f) \cdot w_{ijk}(t, m). \quad (4)$$

The lifestyle term principally models the customer's actions on the controls of the device such as a power switch and manual adjustments. On the other hand, the usage response term models the automatic controls of the device such as a thermostat control. In general, both the lifestyle and usage response terms are stochastic, and their statistical properties depend on the exogenous input z, which can also be stochastic.

One of the unavoidable realities of developing a general model structure is that, "everything depends on everything else and often on itself." For example, only the weather dependence of the usage response term w is explicitly shown solely to indicate a subjective interest in its dominant effect on the thermostat behavior. In reality, however, the thermostat setting, which has an equally significant effect on the thermostat behavior, is determined by a customer, whose decision is influenced by economic parameters, lifestyle, and weather. Weather and economic parameters can influence the customer's lifestyle, and the global collection of customers' lifestyles can have impacts on economic parameters.

For the purpose of modeling cold load pickup, the capital stock can be treated as an unknown constant, the demand response term as unity, and the lifestyle term as an unknown function of time. The elemental power model simplifies to:

$$y_{ijk}(t, m) = x_{ijk} \cdot \ell_{ijk}(t) \cdot w_{ijk}(t, m) \quad (5)$$

<u>Elemental Independence</u>

A critical assumption is made concerning the statistical independence of the elemental powers in a conditional sense. (See [1] for the rigorous definition of elemental independence and the significance of the assumption.) For the specialized purpose of modeling cold load pickup presented here, the assumption is interpreted as follows:

It is possible to classify the elemental powers by subscripts i, j, and k such that the expected value of the elemental power is the product of the expected values of the factors; i.e.,

$$\bar{y}_j(t,m) = E\{y_{ijk}(t,m)\} = \chi_j \cdot \alpha_j(t) \cdot \beta_j(t,m) \quad (6)$$

where:

$$\chi_j = E(x_{ijk}) \quad (7)$$

$$\alpha_j(t) = E\{\ell_{ijk}(t)\} \quad (8)$$

$$\beta_j(t,m) = E\{w_{ijk}(t,m)\}. \quad (9)$$

It should be noted that the expected values of the factors do not depend on the device class i nor on the billing entity k reflecting the underlying objective of the usage class definition.

<u>Organization</u>

The thermodynamics of a house as seen at a thermostat is modeled in the next section, and then a usage response model is derived. The subscripts i, j, and k are omitted for simplicity in these sections. A cold load pickup model and an example are presented in the later sections.

The basic modeling procedure is presented in terms of the space heating load where only a single heating unit of an "on-off" type is installed. With a simple sign convention the results are also applicable to the space cooling load.

THERMODYNAMICS AS SEEN AT THERMOSTAT

The mechanism determining the elemental power of a space heater is not trivial (Figure 1). It is not at all easy to find quantitative relationships, even at a global level, between customers' lifestyles and various exogenous processes such as economy and weather. Although the house thermodynamics and the thermostat control seem easier to model, they still need careful consideration [2]. From a purely theoretical point of view, one can use a detailed simulation model which has been developed mainly to study the thermal design of a house. Obviously, this approach is not practical for modeling the system-level heating load because of the enormous amount of data preparation and computation work required for a large number of different houses. One of the most challenging tasks in physically based modeling of the system-level load is to minimize the data and computation requirement.

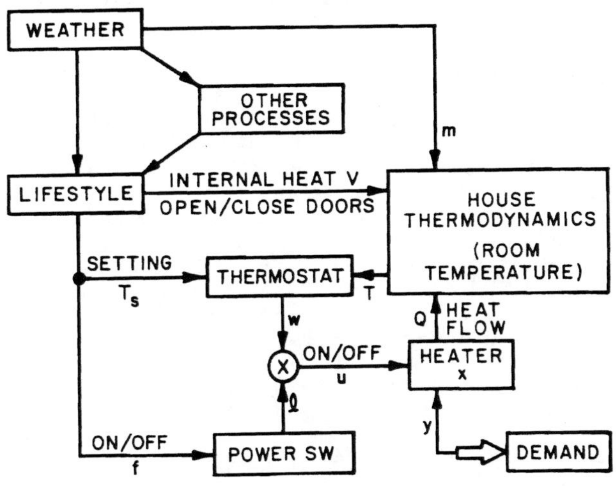

Figure 1. Mechanism Determining Space Heater Demand.

A simple thermodynamic model of a house is derived in this section. The data requirement for the model is minimal, and yet the model is sufficiently accurate for the study of cold load pickup. The model is derived by "condensing" a detailed model. The parameters of the model are very easy to obtain through simple measurements, and still contain physical meaning.

Detailed Thermodynamic Model of a House

A linear dynamic system provides a reasonable approximation of the heat flow to be studied.

$$dX/dt = -AX + Bm + CV + DwQ \tag{10}$$

$$T = H'X \tag{11}$$

where:

- T = room temperature as measured by thermostat
- X = vector (column) of various temperatures
- m = vector of possibly nonlinear functions of meteorological inputs such as outside air temperature, solar intensity and wind velocity
- Q = heat flow of space heater
- w = thermostat status (1 or 0)
- V = vector of other internal heat sources
- A,B,C = constant matrices dependent on house construction
- D,H = constant (column) vectors dependent on house construction (prime on H denotes transpose)

In reality, all the coefficients (A, B, C, D, and H) in Equations (10) and (11) depend on whether doors, windows and/or curtains are open or closed, and therefore they are stochastic. All the inputs (m, V, and Q) are also stochastic. To understand the basic nature of the problem in the simplest setting, all the coefficients and inputs are assumed to be constant at first. Also, the power switch is assumed to be on (ℓ=1). The stochastic nature of the problem will be discussed later.

Since the basic heat transfer has characteristics of exponential decay, the matrix A can be assumed to be positive definite; i.e., all the eigenvalues are positive real.

Thermostat Model

The thermostat is modeled with a flip-flop with a dead-band; i.e., the thermostat status, w, changes from 'off' (w=0) to 'on' (w=1) if the room temperature T becomes lower than or equal to the lower setpoint (T_s-Δ), w changes from 1 to 0 if T becomes higher than or equal to the upper setpoint T_s, and w stays the same if T stays within the dead-band ($T_s - \Delta < T < T_s$).

During an outage, the thermostat remembers whether it was on or off immediately prior to the outage, and the status w does not change as long as the room temperature stays within the dead-band.

Simple Thermodynamic Model of a House

The standard procedure for estimating heat losses (e.g., of a wall) has been based on the overall heat transmission coefficient values of materials used in the wall and the design temperature difference between the air inside and outside of the wall. Conceptually, all the parameters required for the detailed model can be estimated in a similar manner. However, it is very important to recognize that this approach is valid only when insulation material is properly installed without major thermal bridges (locations of heat leakage without passing through insulation). In reality, thermal bridges are abundant due to various reasons such as economics in construction practices, normal aging and differential settling of a house. A report on the heat loss detection using infrared scanner measurements reveals significant thermal bridges [3]. Many other important factors must also be known for the parameter determination. They include house orientations, shade areas, window locations and area. Obviously, it is not practical to determine the parameters of the detailed model when a large number of different houses must be studied.

To avoid unnecessary details, only the thermal characteristics of the house as seen at the thermostat will be sought. The thermostat condenses the thermal characteristics (and the effects of the weather and the lifestyle) into two variables, on-duration and off-duration, where:

<u>on-duration</u> d_1 = minutes required for the room temperature T to rise from the lower setpoint (T_s-Δ) to the upper setpoint T_s.

<u>off-duration</u> d_0 = minutes required for T to descend from T_s to (T_s-Δ).

The mean value of the thermostat status under normal conditions is:

$$\beta^o = d_1/(d_1 + d_0) \ . \tag{12}$$

It should be noted that the mean value of β^o over all the space heaters in the system corresponds to the diversity factor of the space heater load (since ℓ=1 is assumed).

Electric utilities have been conducting load research and synthesis for many years. In the most ideal cases, utilities record time, selected weather variables, and energy consumptions of major devices over a 15 minute period by multi-channel magnetic tape recorders. Customer sampling is usually made such that the utility can determine, principally for forecasting purposes, the system-level demand by adding together the measured device-level demands using appropriate weighting factors. A sufficient amount of data and experience has been accumulated to correlate the diversified demand with time and weather. An example of such data is shown in Figure 2, where the diversified demand of air conditioner load is correlated to the outside air temperature. The outside air temperature is selected only for simplicity. The question of how to actually select the weather variables (e.g., THI, dry bulb, wet bulb, sol-air temperatures, and wind velocity) is not addressed here since the basic modeling procedure is not affected. Instead, the question is addressed on how to use this type of information in order to minimize the additional data required to characterize the cold load pickup. As will become clear later, the statistics of a single parameter (named 'harmonic constant'), describing the thermodynamics of a house seen by the thermostat, is also required.

If all the unnecessary details are ignored and attention is restricted to the on/off-durations of the thermostat, the detailed house model can then be reduced to the following simple model (see Appendix for

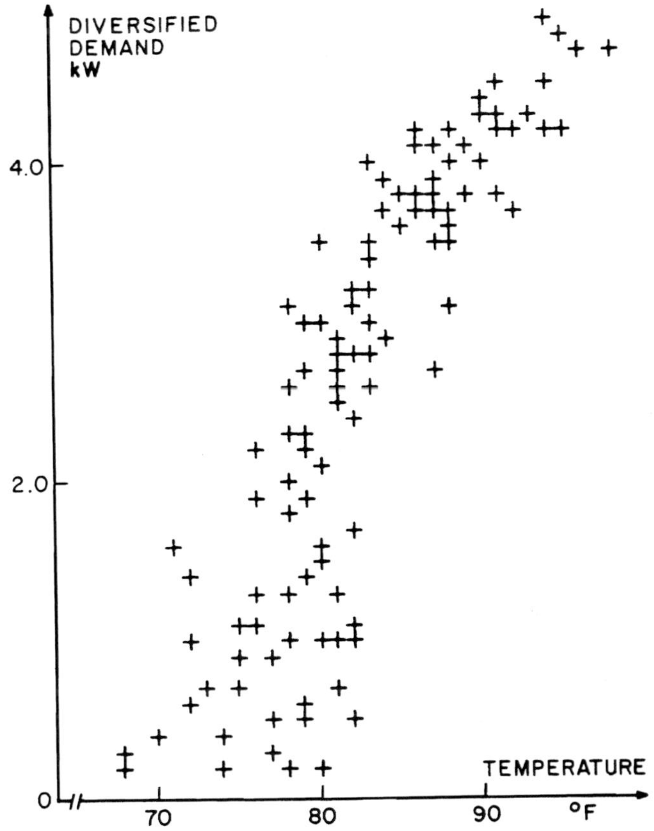

Figure 2. Diversified Demand vs. Temperature.

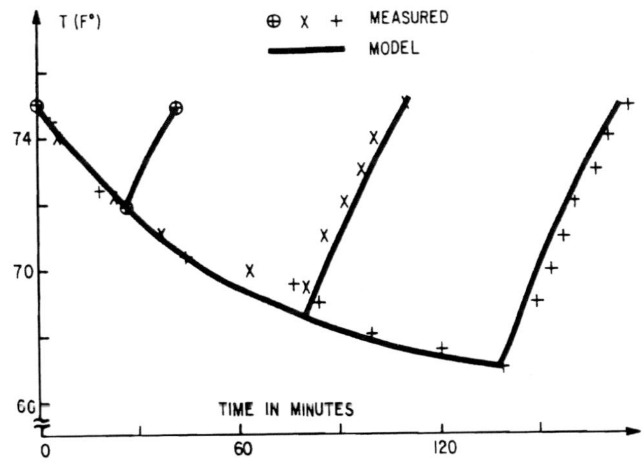

Figure 3. Comparison of Model and Test Results (Room Temperature vs. Time for normal condition, 80 minutes and 140 minutes outages).

the derivation):

$$dT/dt = -(1/\tau)(T - T_f - wT_g) \tag{13}$$

where:

τ = effective thermal time constant (a weighted mean of the time constants of all the parts of a house)

T_f = target temperature which the room temperature T approaches while the heater is off (dependent on weather and internal heat source)

T_g = temperature gain of heater (the heater capacity relative to house heat loss, independent of weather and internal heat source).

A measurement was made to verify the model (Figure 3). Two outages (80 minutes and 140 minutes) were simulated by turning off the heater in the author's house, and the room temperature was measured at the thermostat. The measured temperature excursions were very simple and consistent among the simulated outages. The model of Equation (13) with three parameters reproduced the measured temperature excursions (total of six pieces of curves including the ones for a normal condition) with a satisfactory accuracy. The model parameters were determined simply from the on/off-durations and the maximum temperature drop during the 140 minutes outage:

$$\tau = 64 \text{ minutes} \tag{14}$$

$$T_f = 66\ ^oF \tag{15}$$

$$T_g = 20\ ^oF\ . \tag{16}$$

No optimization techniques were used.

Data Requirement

The on/off-durations completely describe the thermostat behavior under normal conditions. They also describe the thermostat behavior during cold load pickup. For example, the room temperature will descend approximately by 2Δ (twice the dead-band width) during an outage lasting for $2d_0$ (twice the off-duration) requiring the heater to be on during cold load pickup for approximately $2d_1$ (twice the onduration) before the room temperature returns to the original value immediately prior to the outage. The approach pursued here is to determine the on/off-durations under normal conditions by using the diversified demand data together with the minimum of additional data and then to use the on/off-durations to describe the load behavior during cold load pickup. The first question to be answered is what additional data is minimal and practically obtainable.

The on/off-durations are obtained by formally solving Equation (13) for two temperature excursions (with w=1 and w=0) between the setpoints of the thermostat:

$$d_1/\tau = \ln\{1 + g/(1 - s)\} \tag{17}$$

$$d_0/\tau = \ln\{1/(1 - g/s)\} \tag{18}$$

where:

$$s = (T_s - T_f)/T_g \tag{19}$$

$$g = \Delta/T_g\ . \tag{20}$$

It should be noted that the weather dependency and the effect of thermostat setting are contained in a single variable s and that the two parameters τ and g depend only on the thermal characteristics of the house, but not on the weather nor on the thermostat setting. The thermal parameters can be obtained through experiments such as the one of Figure 3. The measured data of Figure 3 yields:

$$g = 0.15 \ . \tag{21}$$

Although this approach is much simpler than determining all the parameters of the detailed model in Equations (10) and (11), it is not sufficiently simple for practical use.

The model of Equations (17) and (18) can be simplified a step further by a linear approximation:

$$d_1 = \theta/(1 - s) \tag{22}$$

$$d_0 = \theta/s \tag{23}$$

where the <u>harmonic constant</u> θ is given by:

$$\theta = \tau g \ . \tag{24}$$

This approximation is very attractive. All the dynamic characteristics of a house are summarized in a single parameter θ. It is stressed that this thermal parameter is independent of weather and internal heat source (within the accuracy of the linear approximation). Determination of the harmonic constant θ is simple. Elimination of s in Equations (22) and (23) yields the harmonic constant as a harmonic sum of the on/off-durations:

$$\theta = d_1 d_0/(d_1 + d_0) \ . \tag{25}$$

Only the on/off-durations are required to find the value of the harmonic constant. For example, the space heater in the experiment of Figure 3 was off for 26 minutes and on for 16 minutes under normal conditions before the experiment. This yields θ = 9.9 minutes, which is slightly larger than, but a good approximation of, the exact value, 9.6 minutes being the product τg in Equations (14) and (21).

Substitution of (22) and (23) into (12) yields:

$$\beta^o = s = (T_s - T_f)/T_g \ . \tag{26}$$

Therefore, the on/off-durations are completely specified by the harmonic constant θ and the thermostat mean status β^o before an outage. The statistics of the harmonic constant θ is easy to obtain. It can be obtained as a harmonic sum of the on/off-durations through simple time measurements (or as a product τg through a detailed simulation model of the house [e.g., 4]. It is the minimum additional data required for the cold load pickup model. The statistics of the thermostat mean status β^o before an outage can be estimated using the diversified demand data (or using a detailed simulation model of the house).

To show the accuracy of the linearization, the approximate relationship of Equation (26) is compared with the true relationship obtained by Equations (17) and (18) for different values of the thermal parameter g (Figure 4). The approximation is accurate if the parameter g is small and becomes less accurate with increasing value of g. It should be noted that the approximation yields pessimistic estimates of overload.

<u>Sign Convention and Normalized Room Temperature</u>

The following sign convention is used:

$T_g > 0$, $\Delta > 0$ for space heater

$T_g < 0$, $\Delta < 0$ for air conditioner.

In order to unify the development in the next section of the elemental power model for both space heater and air conditioner load, the room temperature is normalized by:

$$\xi = (T_s - T)/\Delta \ . \tag{27}$$

The thermostat changes the status from 'on' to 'off' if the normalized room temperature ξ reaches zero (from 'off' to 'on' if ξ reaches unity) for both space heater and air conditioner cases. Due to the sign convention used, increasing ξ corresponds to decreasing T for space heater load and to increasing T for air conditioner load.

USAGE RESPONSE MODEL

The expected value of the thermostat mean status β can be considered as the probability of the thermostat being in the on-status. The behavior of this probability during cold load pickup will be obtained in terms of the on/off-durations and the duration of outage. First, this will be done graphically and then by using the traveling wave concept.

<u>Graphical Model</u>

Immediately prior to an outage, the heater is either on with probability $d_1/(d_1 + d_0)$ or off with probability $d_0/(d_1 + d_0)$ and with an equal probability of the room temperature T being any value within the thermostat dead-band. After an outage of D minutes, the probability of the room temperature being outside of the dead-band is D/d_0 (Figure 5). There are three possible cases at the time (t=0) of the power being restored.

<u>Case (i)</u>: T is outside of the dead-band, and the heater is 'on' at t=0. The probability of this case is:

$$Pr(i) = D/d_0 \ . \tag{28}$$

<u>Case (ii)</u>: T is within the dead-band, the heater is 'on' at t=-D, and remains 'on' at t=0. The probability of this case is:

$$Pr(ii) = \{1 - Pr(i)\} \cdot d_1/(d_1 + d_0) \ . \tag{29}$$

<u>Case (iii)</u>: T is within the dead-band, the heater is 'off' at t=-D, and remains 'off' at t=0. The probability of this case is:

$$Pr(iii) = \{1 - Pr(i)\} \cdot d_0/(d_1 + d_0) \ . \tag{30}$$

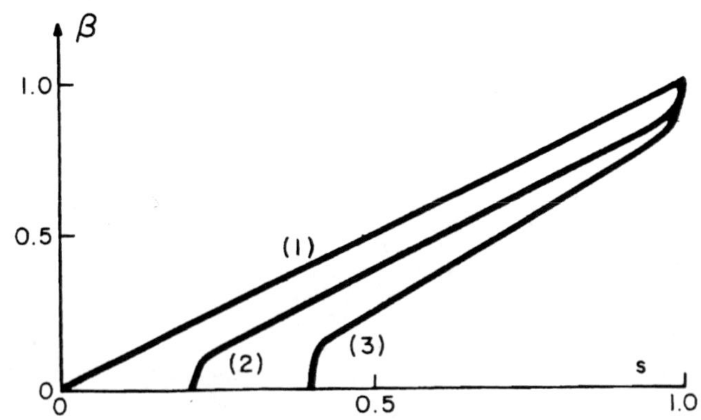

Figure 4. Mean Usage Response Factor β vs. Weather Variable s (1. Linear Approximation; 2. g=0.2; 3. g=0.4; g=thermal gain parameter of house).

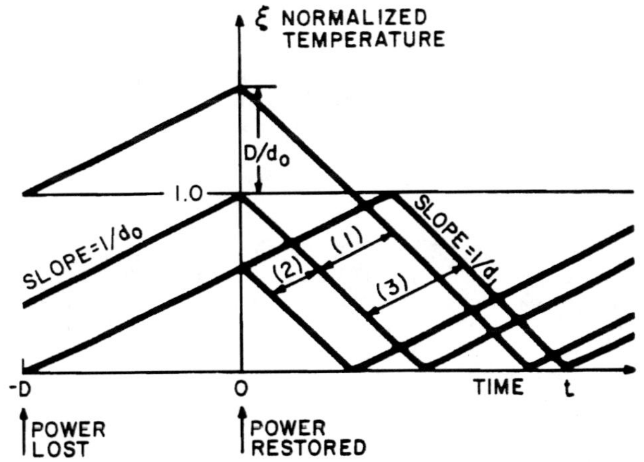

Figure 5. Temperature Behavior for Three Possible Cases (Case 1: $\xi>1$ at $t=0$; Case 2: $\xi<1$ at $t=0$ and heater is on at $t=-D$; Case 3: $\xi<1$ at $t=0$ and heater is off at $t=-D$).

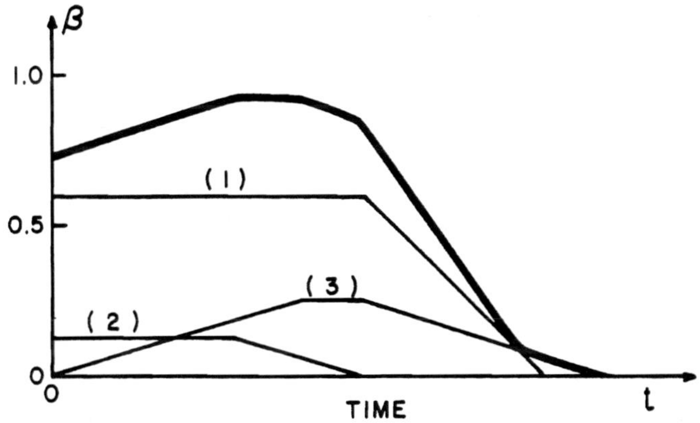

Figure 6. Graphical Model of Cold Load Pickup (three components correspond to three cases of Figure 5).

The total probability $\beta(t)$ of the heater being 'on' during cold load pickup is the sum of the components corresponding to the three possible cases and changes with time as shown in Figure 6. Pr(i) increases as the outage lasts longer. The duration and the magnitude of the overload in terms of $\beta(t)$ can be tabulated according to the duration of outage relative to the on/off-durations before outage (Table 1).

Traveling Wave Type Model

The graphical model can be translated into a simple model of a traveling wave type. Let:

δ = an arbitrarily small time increment ($\delta>0$).

$p(\xi,t)d\xi$ = probability measure of thermostat being 'on' at time t with normalized temperature around ξ.

$q(\xi,t)d\xi$ = probability measure of thermostat being 'off' at time t with normalized temperature around ξ.

The probability $\beta(t)$ of thermostat being 'on' is the integral of $p(\xi,t)$ over ξ; or,

Table 1. Severity of Overload

	$D<d_*$	$d_*<D<d_0$	$d_*<D$
t_1	d_1D/d_0	d_0-D	0
t_2	$\mathrm{Min}\{d_1,d_0-D\}$	d_1D/d_0	d_1D/d_0
β^1	$(D+d_1)/(d_0+d_1)$	$(D+d_1)/(d_0+d_1)$	1
β^2	$D/d_0+d_*/d_1$	1	1

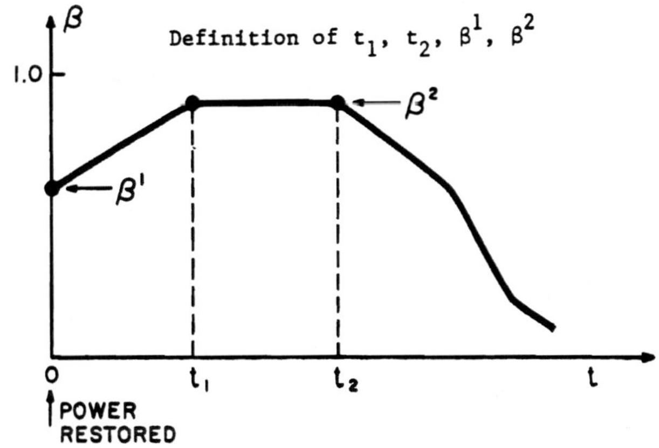

(D = duration of outage; d_0 = off-duration; d_1 = on duration; $d_* = d_1^2/(d_0+d_1)$)

$$\beta(t) = \int p(\xi,t)d\xi \ . \tag{31}$$

Before and after outage, the house with normalized temperature ξ at t will see normalized temperature smaller by δ/d_1 at $(t+\delta)$ if the thermostat is 'on', or higher by δ/d_0 if it is 'off'; i.e.,

$$p(\xi - \delta/d_1, t+\delta) = p(\xi,t) \tag{32}$$

$$q(\xi + \delta/d_0, t+\delta) = q(\xi,t) \tag{33}$$

The thermostat changes the status from 'off' to 'on' if ξ reaches 1:

$$p(1, t+\delta) = q(1, t) \cdot (d_1/d_0) \tag{34}$$

$$q(1, t+\delta) = 0 \tag{35}$$

and from 'on' to 'off' if ξ reaches 0:

$$q(0, t+\delta) = p(0, t) \cdot (d_0/d_1) \tag{36}$$

$$p(0, t+\delta) = 0 \ . \tag{37}$$

Indeed, p and q can be considered waves traveling backward (decreasing ξ direction) and forward, respectively, between the two short-circuited endpoints at $\xi=0$ and $\xi=1$. However, they are different from ordinary traveling waves in that:

- Forward and backward waves have different speed of propagation.
- Backward wave can travel beyond the endpoint at $\xi=1$ during an outage.
- Reflections at the endpoints are selective; i.e., the backward wave $p(\xi,t)$ traveling with ξ larger than 1 (after an outage) passes

through the endpoint at ξ=1 eventlessly, and reflects only after it reaches the other endpoint at ξ=0.

During outage, ξ increases:

$$p(\xi + \delta/d_0, t + \delta) = p(\xi, t) \quad (38)$$

$$q(\xi + \delta/d_0, t + \delta) = q(\xi, t) \quad (39)$$

and the thermostat changes the status only at ξ=1:

$$p(1, t + \delta) = p(1, t) + q(1, t) \quad (40)$$

$$q(1, t + \delta) = 0 . \quad (41)$$

The on/off-durations and the duration of outage together with the initial distributions of p and q complete the usage response model $\beta(t)$ during cold load pickup.

COLD LOAD PICKUP MODEL

The cold load pickup model requires the three factors - the mean capital stock χ, the mean lifestyle $\alpha(t)$, and the mean usage response $\beta(t/m)$ for each usage class j. The mean usage response can be modeled by knowing the value of $\beta^o(m)$ before an outage and the harmonic constant θ. The statistics of the harmonic constant θ is the minimum data required anew. Here, it will be shown that the diversified demand data is sufficient to determine the rest of the parameters - χ, α, and $\beta^o(m)$. Also, an example is provided in the next section to explore some of the basic characteristics of cold load pickup.

Maximum Use of Diversified Demand Data

The stochastic nature of the load behavior during cold load pickup is mostly embedded in the diversified demand data (the rest in the statistics of the harmonic constant). The stochastic nature of the problem can be efficiently captured by working with the diversified demand data and the statistics of the harmonic constant. The following data is typically collected to obtain the diveresity factor:

- o Capital stock data obtained from utility appliance saturation surveys, U.S. census data, appliance sales data, and utility billing data.
- o Demand data obtained by recording 15 minute energy consumptions of sampled appliances with demand recorders.
- o Weather data.

For example, Figure 2 is based on the capital stock data as shown in Figure 7 and on the demand data such as the ones shown in Figure 8. Figure 8b indicates that one of the air conditioners is off (ℓ=0) in the 7 p.m. period. By assuming that all the 22 customers in Figures 7 and 8 belong to the same usage class, the following estimates can be obtained:

χ = 5.54 KW

α = 1.0 , β^o = .72 (2 PM)

α = .96 , β^o = .76 (7 PM)

The assumption of single usage class relates to the elemental independence assumption. For the specific purpose as pursued in this paper (since the higher order statistics are not sought), the assumption implies that there is no correlation among the capital stock, the lifestyle, the usage response and the harmonic constant within a usage class. To test the validity of the assumption, it is necessary to study much larger samples. If correlation is detected, the model can be improved by defining more usage classes.

EXAMPLE

The load behavior of 10,000 customers was calculated to understand the fundamental characteristics of cold load pickup. No actual data was used. Instead, the following assumptions were made for simplicity:

- o The power switches are on for all the customers (α = 1).

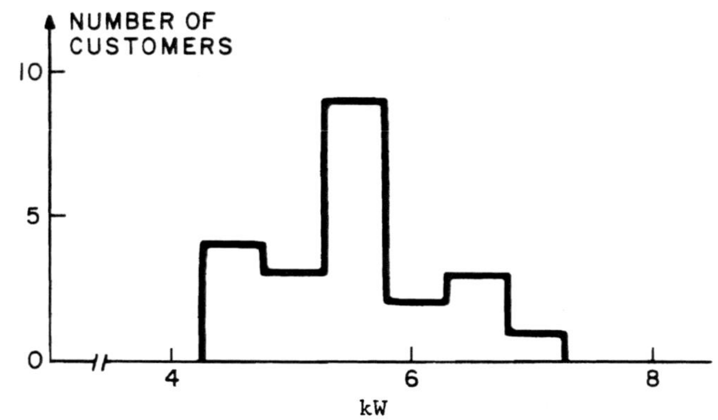

Figure 7. Histogram of Capital Stocks.

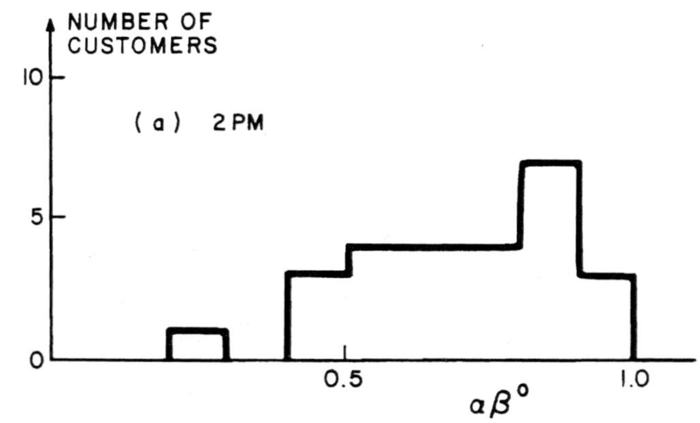

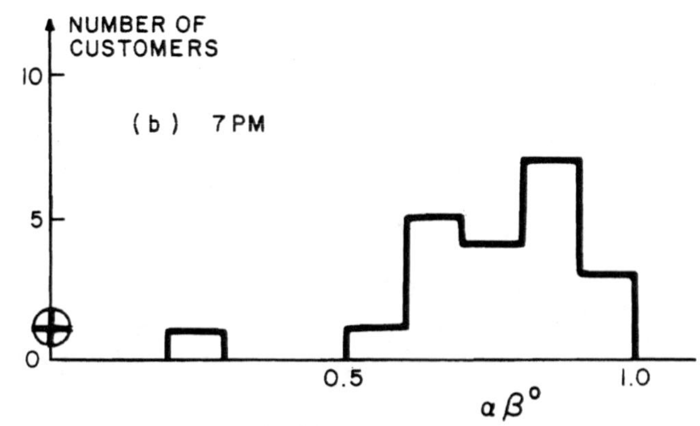

Figure 8. Histograms of Utilization Factors.

- The mean usage response β^o before an outage and the harmonic constant θ are normally distributed.

It should be emphasized that these assumptions are not at all necessary for the use of the model in realistic problems.

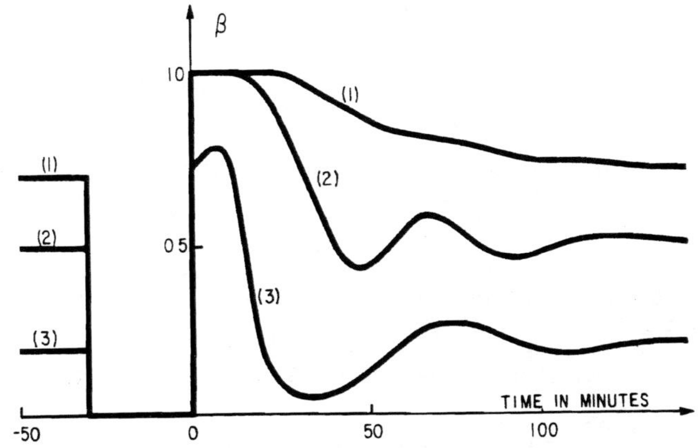

Figure 9. Cold Load Pickup after 30 minutes Outage (β = Actual MW/Installed MW for 10,000 homes; before outage: β=0.7 for (1), β=0.5 for (2), and β=0.2 for (3)).

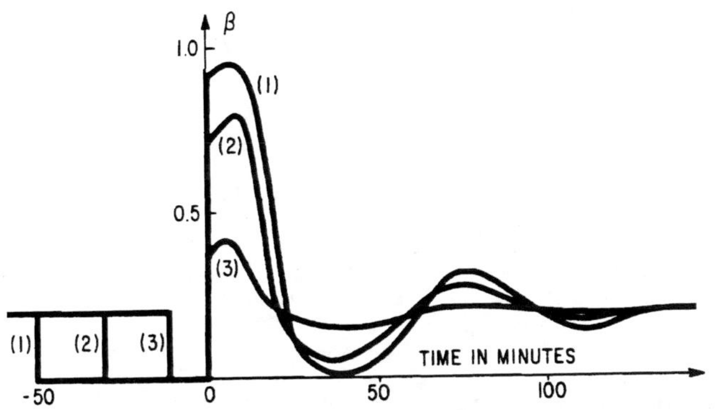

Figure 10. Cold Load Pickup after D minutes Outage (1. D=50; 2. D=30; 3. D=10; β = Actual MW/Installed MW for 10,000 homes).

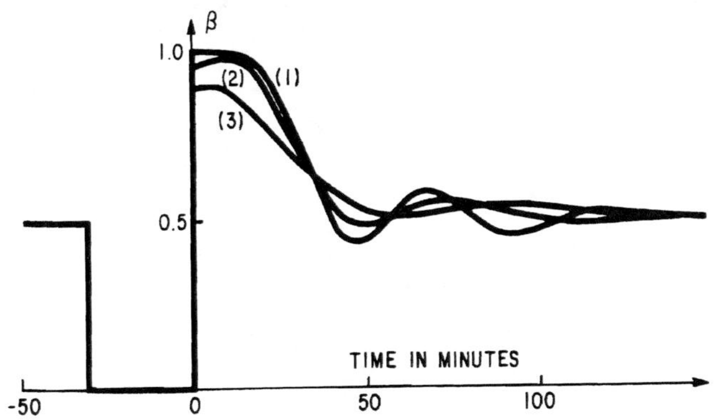

Figure 11. Cold Load Pickup after 30 minutes Outage (β = Actual MW/Installed MW for 10,000 Homes; σ = Standard Deviation of Thermal Parameters; 1. σ=20%, 2. σ=30%, 3. σ=50%).

A total of 121 usage classes were defined by partitioning houses according to the mean usage response before an outage (11 partitions) and to the harmonic constant (11 partitions), and a simple program was developed to compute the evolution of the temperature distribution for each class and the total load behavior during cold load pickup.

The base case chosen is:

Outage duration D = 30 minutes

Mean $\bar{\beta}^o$=.5 , $\bar{\theta}$ = 10 minutes

Standard deviation σ = 30% (always chosen same for both β^o and θ).

Curve (2) of Figure 9 corresponds to the base case. The load behavior during cold load pickup is very dependent on β^o, and therefore dependent on the weather (Figure 9) as well as on the duration of outage (Figure 10). The load behavior becomes less oscillatory with the increasing diversity among customers (Figure 11).

CONCLUDING REMARKS

The model yields a pessimistic estimate of the overload due to the linear approximation used. It should be noted that this approximation implies constant energy recovery (the energy not supplied during an outage is supplied after the outage without any reduction). Some improvements can be achieved by a quadratic approximation of Equations (17) and (18) or by using Equation (13) itself with the flip-flop model of a thermostat. It is conceptually easy to make such improvements, but it is not always easy to justify additional data and computation requirements.

The basic idea presented in this paper is to capture the thermal characteristics of a house at a thermostat without modeling all the details of the house. This idea is not new. In fact, utilities have been collecting diversified demand data for load forecasting purposes and explaining the weather dependencies of load without studying the details of the house. It is not surprising to find that the diversified demand data does not contain all the information required to model cold load pickup. It lacks the information on the house dynamics. Such information is not needed for forecasting. It is pleasantly surprising, however, to find that the house dynamics can be condensed into a single parameter (harmonic constant), and the rest of the required information can be obtained from the diversified demand data.

The behavior of space conditioning load during cold load pickup is strongly dependent on the weather, the duration of outage and the diversity of houses.

ACKNOWLEDGEMENTS

The research described in this paper has been supported by a contract from the Department of Energy, Division of Electric Energy Systems. We wish to thank Dr. T. Trygar of DOE for his helpful comments during the progress of this research.

REFERENCES

[1] General Electric Company and MIT, Systems Engineering for Power V: Load Modeling Methodologies: Interim Report, prepared for the Department of Energy, No. HCP/T5112-01, August, 1979.

[2] J.E. McDonald, A.M. Bruning, and W.R. Mahieu, "Cold Load Pickup." IEEE Transaction on Power Apparatus and Systems, Vol. PAS-98, pp. 1384-1386, July 1979

[3] R.H. Munis, S.J. Marshall, and M.A. Bush, "Detecting Structural Heat Losses with Mobile Infrared Thermography." CRREL Report 76-33, Corps. of Engineers, U.S. Army, Cold Regions Research and Engineering Laboratory, Hanover, N.H., September 1976.

[4] M. Chan, E.N. Marsh, J.Y. Yoon, G.B. Ackerman, and N. Stoughton, "Simulation-Based Load-Synthesis Methodology for Evaluating Load Management Programs." 80 SM 585-0, Paper presented at the IEEE Summer Power Meeting, Minneapolis, Minnesota, 1980.

APPENDIX

Define:

$$F(m, V) = A^{-1} (Bm + CV) \quad (42)$$

$$G(Q) = A^{-1}DQ \quad (43)$$

$$T_f(m, V) = H'F \quad (44)$$

$$T_g(Q) = H'G \quad (45)$$

$$\tau = H'G/H'AG \quad (46)$$

then, Equation (13) approximates Equations (10) and (11) in the behavior of T around the thermostat dead-band. To see this, let:

X_0 = state vector occurring when $T = T_s$ $(=H'X_0)$

X_1 = state vector occurring when $T=T_s-\Delta$ $(=H'X_1)$

Solving Equation (10) for X_0 and X_1 in terms of the on/off-durations yields:

$$X_0 = F+[I-\exp\{-A(d_0+d_1)\}]^{-1}\{I-\exp(-Ad_1)\}G \quad (47)$$

$$X_1 = F+[I-\exp\{-A(d_0+d_1)\}]^{-1} \cdot \{I-\exp(-Ad_1)\}\{\exp(-Ad_0)\}G \quad (48)$$

Now, it is assumed that the on/off-durations are small compared to the time constants of the system. This assumption implies that the following approximation can be used as long as β is not close to zero nor to unity:

$$\exp\{-A(d_0+d_1)\} = I-A(d_0+d_1) \quad (49)$$

Equations (44)-(46) and (11) yield:

$$T_s = T_f + \{d_1/(d_0+d_1)\}T_g \quad (50)$$

$$\Delta = \{d_0d_1/(d_0+d_1)\}T_g/\tau \quad (51)$$

Equation (13) also yields the same equations (50) and (51) with a similar approximation:

$$\exp\{-(d_0+d_1)/\tau\} = 1-(d_0+d_1)/\tau. \quad (52)$$

Therefore, equation (13) approximates Equations (10) and (11) around the dead-band.

Discussion

Robert W. Alford (Texas Power and Light Co., Dallas, TX): Successful prediction in engineering analysis requires careful formulation of problems securely based on an understanding of physical concepts and the exercising of mathematical reasoning. Proper formulation of the problem and its solution technique is heavily dependent upon constructing a mathemical model which incorporates appropriate physical assumptions and mathematical approximations and which approaches the actual situation with sufficient accuracy for the purpose at hand. The authors are to be commended for having devised a technique which requires a minimum of data and computations and which retains the elements of physical meaning while yet being based upon a condensing of the "detailed" model.

It is reassuring to find that the necessary modeling data can be obtained from available sources and that a single parameter properly (and so simply) derived can be used to represent the house thermodynamics.

Mention was made of using the model to study the load management payback phenomena. Have the authors used the model specifically for just that purpose or is that application yet to be implemented? Could the authors comment on special modifications which might have been required in the model to handle the various control strategies which might be employed for space conditioning loads under a typical load management program?

Manuscript received February 26, 1981.

S. Ihara and **F. C. Schweppe:** The authors wish to thank Mr. Robert W. Alford for his kind remarks.

It is our hope to use the model to study the actual load behavior. Such application is yet to be implemented. The model was developed not only for studying the load behavior after service interruptions but also for analyzing the impacts of conservation efforts on hourly load shapes. A customer can reduce the energy consumption for space conditioning in various ways: for example, he can turn off the power switch, adjust the thermostat setting, improve the thermal insulation of his building, and/or replace the space conditioners with ones of higher efficiency. By considering the power switch behavior, α, and the thermostat behavior, β, separately as in equation (6), more information would be extracted from the load research data. Turning off the air conditioner, for example, has variable effects on load shapes. Such effects were observed by some utilities for recent summers in the form of reduced energy sales accompained with sharp peaks. These effects may be detected in α, while the effects of other conservation efforts will appear in β which may be efficiently analyzed by use of equation (26).

Manuscript received May 27, 1981.

Blackstart Utilization of Remote Combustion Turbines, Analytical Analysis and Field Test

R.D. Shultz
Clarkson College of Technology
Potsdam, NY

G. A. Mason
Illinois Power Co.
Decatur, IL

Abstract - This paper illustrates that some special operating procedures may be needed to blackstart a generating station from a remote combustion turbine generator. A method is presented to determine what, if any, special procedures are needed for a specific case. The results of a blackstart field test on an actual system are then compared to the analytical analysis.

Introduction

The ability of a utility to recover from a massive multi-company power outage has been a topic of extreme interest for the past decade. Many companies have responded to this problem by installing gas combustion turbines with backstarting capability. Often these turbines, either due to physical limitations or operational considerations, were located remotely from the generating station which the utility would like to start up under blackstart conditions. This situation results in a combustion turbine generator feeding a load through an isolated transmission line and at least three transformers. The load at the generating station will consist of several large motors which will have to be started. The question which this paper addresses is will there be any voltage or frequency problems as these motors are brought on line, and if so, are there any operational procedures or motor starting orders that will alleviate these problems.

System Studied

The system which was evaluated utilized a 28 MVA combustion turbine generator to blackstart a generating station. The two are electrically connected through 17.8 miles of 138 KV transmission line. The combustion turbine generator operates at 13.8 KV and is connected to the 138 KV system through a step up transformer. The load at the generating station is connected to the 2.4 KV system which is tied to the 138 KV system through a series connection of a 2.4/13.2 KV transformer and a 13.2/138 KV transformer. The load requirements for blackstarting one unit at the generating station were determined to be a 900 hp ID fan, a 600 hp FD fan, a 2500 hp boiler feed pump, a 250 hp condensate pump, a 450 hp circulating pump, and a 100 hp auxillary oil pump. In addition a small miscellaneous load will be present at start up.

Modeling

By neglecting the subtransient response the generator can be modeled using the two-axis model [1]. The corresponding generator equations are:

$$E_d' = V_d + rI_d + X_d' I_q \qquad (1)$$

$$E_q' = V_q + rI_q - X_d' I_d \qquad (2)$$

$$dE_d'/dt = 1/\tau_{q0}' [- E_d + (X_q - X_q') I_q] \qquad (3)$$

$$dE_q'/dt = 1/\tau_{d0}' [- E_{FD} - E_q + (X_d - X_d') I_d] \qquad (4)$$

$$d\omega/dt = 1/2H_G [T_{MG} - D_\omega - T_{EG}] \qquad (5)$$

$$d\delta/dt = \omega - 1 \qquad (6)$$

where
- E_d', E_q' = d and q axis stator voltages
- V_d, V_q = d and q axis terminal voltages
- E_{FD} = d axis stator emf corresponding to field voltage excitation
- I_d, I_q = d and q axis currents
- τ_{d0}', τ_{q0}' = d and q axis open-circuit, transient time constant
- X_d, X_q = d and q axis synchronous reactance
- X_d', X_q' = d and q axis transient reactance
- r = stator resistance
- T_{MG} = mechanical torque on generator
- T_{EG} = electrical torque on generator
- D = damping coefficient of generator
- H_G = inertia constant of generator
- δ = generator rotor angle
- ω = generator speed

The excitation system was represented as an IEEE Type 1 system [2]. The excitor equations, which are coupled to the generator equations through E_{FD} and V_t are:

$$dE_{FD}/dt = 1/\tau_E [V_R - (SE + K_E)] \qquad (7)$$

$$dV_R/dt = 1/\tau_A [-K_A (V_3 + V_1) - V_R + K_A (V_S + V_{REF})] \qquad (8)$$

$$dV_3/dt = 1/\tau_F [-V_3 - K_F/\tau_E (V_R - SE\, E_{FD} - K_E E_{FD})] \qquad (9)$$

$$dV_1/dt = 1/\tau_R [V_t - V_1] \qquad (10)$$

where
- E_{FD} = exciter output voltage
- V_R = regulator output voltage
- V_3 = stabilizing circuit voltage
- V_1 = filtered terminal voltage
- τ_E = exciter time constant
- τ_A = regulator time constant
- τ_F = stabilizing circuit time constant
- τ_R = regulator filter time constant
- K_A = regulator gain

84 WM 092-3 A paper recommended and approved by the IEEE Power System Engineering Committee of the IEEE Power Engineering Society for presentation at the IEEE/PES 1984 Winter Meeting, Dallas, Texas, January 29 - February 3, 1984. Manuscript submitted August 23, 1983; made available for printing November 11, 1983.

Reprinted from *IEEE Transactions on Power Apparatus and Systems*, Vol. 103, No. 8, pp. 2186–2191, August 1984.

K_E = exciter constant

K_F = stabilizing circuit gain

V_t = generator terminal voltage

V_{REF} = reference voltage

V_S = other signals

SE = exciter saturation function

The governor model which was used can be represented mathematically by

$$dP_M/dt = [P_0 + P_\Delta - P_M]/\tau_3 \quad (11)$$

$$dP_\Delta/dt = [(\omega_{ref} - \omega)K_1/\tau_1 - K_1\tau_2/\tau(d\omega/dt) - P_\Delta/\tau_1] \quad (12)$$

$$T_{MG} = P_M/\omega \quad (13)$$

where P_M = mechanical power input to generator

P_0 = Initial mechanical power input to generator

P_Δ = change in power input to generator

K1 = controller power gain

τ_1 = controller lag

ω = generator speed

X_{ref} = initial generator speed

$d\omega/dt$ = generator acceleration

The govenor equations are related to the generator equations through ω and T_{MG}.

Two models are needed for accurate induction motor representation. For motors being brought on line from rest the induction motor equivalent circuit shown in Figure 1 is used [3]. This model allows the impedance representing the load on the system to vary over the full range of speeds of the motor.

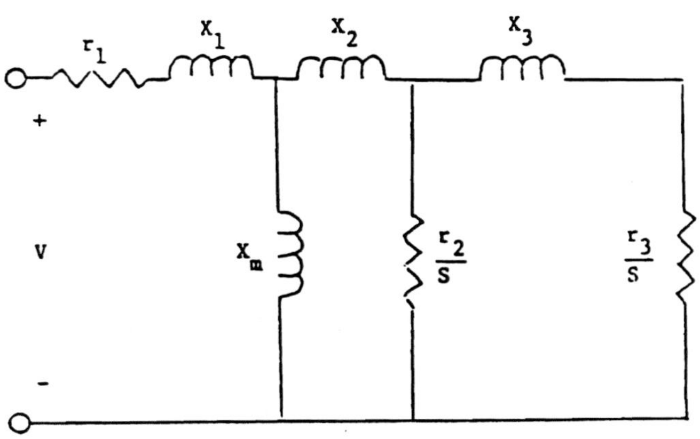

Figure 1. Induction motor equivalent circuit.

The definition of quantities in Figure 1 are:

r_1 = stator resistance

X_1 = stator reactance

r_2, r_3 = rotor resistance referred to stator

X_2, X_3 = rotor reactance referred to stator

X_m = magnetizing reactance

S = rotor slip

A different model is needed to represent the induction motors already on line. This model should represent the motor momentarily going into a generation mode when its terminal voltage suddenly drops. The model used in this paper was the transient equivalent circuit [4], which is shown in Figure 2.

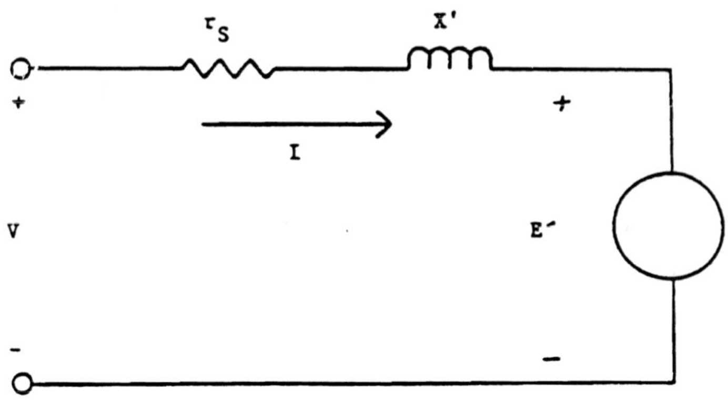

Figure 2. Induction motor transient equivalent circuit.

The electrical and mechanical equations are

$$dE'/dt = -j\omega s E' - 1/\tau'_{do}(E' - j(X - X')I) \quad (16)$$

$$V - E' = (r_1 + jX')I \quad (17)$$

$$dS/dt = 1/2H(T_M - T_e) \quad (18)$$

where E' = voltage behind transient reactance

H = inertia constant

X = input reactance at zero slip

X' = input reactance at unit slip

T'_{do} = rotor time constant, stator open-circuited

The miscellaneous start up load was modeled as a constant power load. The transmission line and transformers are represented by constant impedances.

Method of Approach

Using system data, the appropriate network admittance matrix (Y_{BUS}) is assembled. In order to represent any on-line motors during the load-flow calculations, the induction motor equivalent circuit, shown in Figure 1, with the appropriately rated slip is inserted into Y_{BUS}. With this modified Y_{BUS}, along with any local scheduled loads due to auxiliary equipment, the initial system voltages are calculated by a Newton-Raphson load flow. After the load flow is completed, the on-line motor equivalent impedances are removed from Y_{BUS}. The scheduled loads are replaced by equivalent impedances at the appropriate bus in Y_{BUS}.

The generator and all the on-line induction motors can now be represented by the models proviously presented. As a result, Y_{BUS} is modified by adding one bus to the system for the generator and one bus for each on-line motor and inserting the appropriate impedances. From the bus voltages calculated in the load flow, the appropriate voltages behind impedances can be found.

The pre-startup conditions are used to calculate the initial conditions of the generator, exciter, govenor and on-line motors. The generator, exciter and governor differential equations are solved for the steady-state condition. This is accomplished by equating the derivatives to zero. The mechanical torque of the on-line motors is found from the power being absorbed by the voltage behind the impedance.

The initial slip of the new motor is set equal to one and used in the induction motor equivalent circuit. This circuit is then inserted into Y_{BUS} at the appropriate bus as an impedance to ground. It is assumed that at the instant the new motor is being brought on-line, all the voltages behind impedances of the machine models remain constant. Under these new conditions, the remaining bus voltages are found by solving the network performance equations using an iterative technique [4]. This changes all of the terminal currents of the machines at time $t = 0^{+}$. The rotor current of the new motor is then found with the bus voltage and equivalent circuit. The electrical torque is calculated from the current and slip. The mechanical torque is estimated by using standard load torque curves for various loads which are dependent on the slip of the motor. The angular speed of the motor can then be calculated.

The generator, excitor, on line motor and govenor differential equations are then solved. With the speed of the new motor and the speed of the generator found for the next time step, the slip for the next time step is then calculated. This slip is used to update the impedance of the new motor with the equivalent circuit.

Time is then iterated and tested for a maximum value. If the maximum value is not yet reached, the updated motor impedance is again inserted into Y_{BUS} and another time step set of calculations made. A flow chart illustrating this procedure is given in Figure 3.

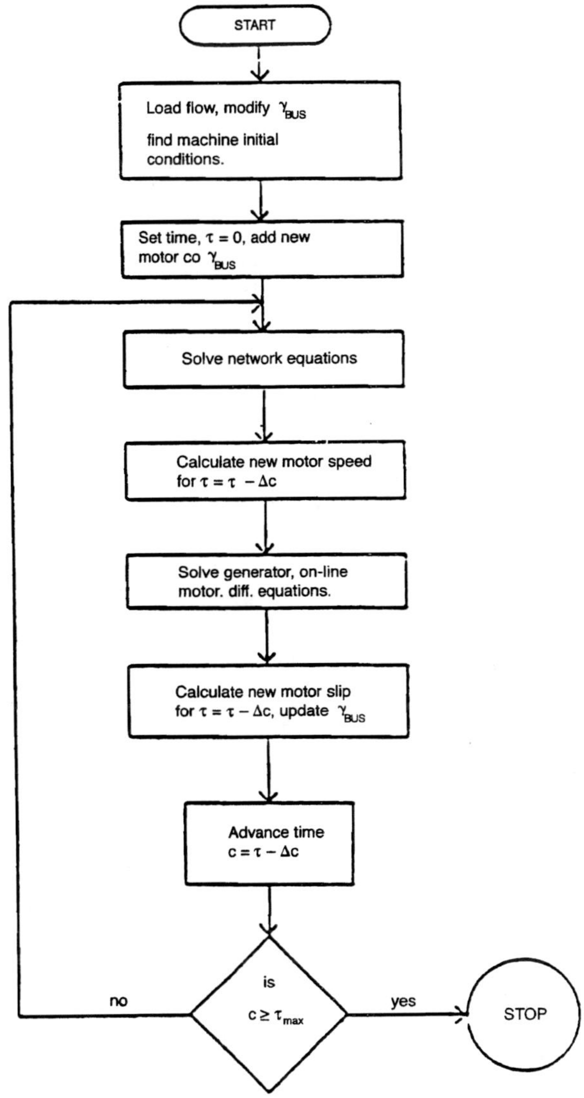

Figure 3. Procedure flow chart.

System Parameters

The parameters of the various elements of the system being studied are given in Tables 1 through 6.

Table 1. Combustion turbine generator data (100 MVA base)

X_d = 6.21 pu	T'_{do} = 5.48 sec
X'_d = 0.828 pu	T'_{qo} = 1.10 sec
X_q = 6.04 pu	D = 2.0 pu
X'_q = 3.436 pu	WR^2 = 101.956 lb-ft^2

Table 2. Exciter Data (100 MVA base)

K_A = 400.	SE(max) = 0.66
K_E = 1.00	SE(75% max) = 0.62
K_F = 0.03	E_{FD}(max) = 4.43 pu
τ_A = 0.02 sec	E_{FD}(min) = 0.0 pu
τ_E = 0.46 sec	V_R(max) = 7.37 pu
τ_F = 1.00 sec	V_R(min) = −6.63 pu
τ_R = 0.005 sec	

Table 3. Govenor Data (100 MVA base)

K_1 = 0.0158	τ_2 = 1.25 sec
τ_1 = 0.5 sec	τ_3 = 0.10 sec

Table 4. Transformer Data

Transformer 1 at load	13.2/138 KV	33 MVA	X = 7.9%
Transformer 2 at load	2.4/13.2 KV	5 MVA	X = 6.8%
Transformer at generator	13.2/138KV	100 MVA	X = 27.47

Table 5. Transmission Line Data (100 MVA base)

R = 0.0183 pu
X = 0.0737 pu
B = 0.0181 pu

Table 6. Induction Motor Data (100 MVA base)

	ID Fan	FD Fan	Boiler Feed Pump	Condensate Pump	Circulating Pump	Aux Oil Pump
r_1(pu)	2.05	3.69	0.30	0	0	0
X_1(pu)	15.73	22.57	4.53	49.24	31.46	123.1
r_2 (pu)	8.21	10.26	2.07	34.47	16.41	61.56
X_2(pu)	7.11	12.31	2.95	29.54	14.23	73.87
r_3(pu)	1.30	1.85	0.33	2.95	2.60	11.08
x_3(pu)	13.68	20.52	4.92	49.24	27.36	123.1
X_m(pu)	484	849	209	999	900	999
$H(\frac{KW\,Sec}{KVA})$	10.0	4.5	3.5	1.0	1.0	1.0
HP (rated)	900	600	2500	250	450	100
RPM (rated)	890	1180	3575	1165	495	3500

Some of the data in Table 6 was unavailable from the manufacturer of the motors. In those cases, typical data was obtained from other manufacturers and consultants. Average values of those typical data were then used, so as to minimize the error introduced.

Blackstart Analytical Evaluation

Studies were done with the large motors started in various orders. In each case, the motors were started one at a time. After each motor was started and the system reached steady state, the frequency at the combustion turbine was adjusted to sixty hertz if needed. In none of the studies did bringing a motor on line cause a severe frequency problem. The largest drop in frequency occurs when the boiler feed pump, the largest motor to be started, is started first. Starting it first results in a minimum frequency of 59.46 hertz. The frequency profile for the boiler feed pump starting first is shown in Figure 4.

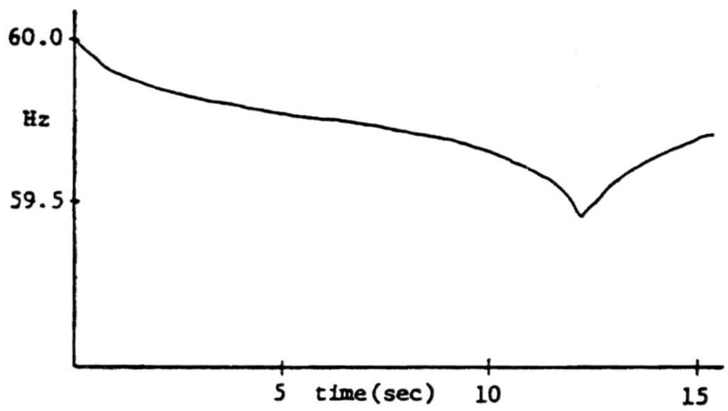

Figure 4. Frequency profile for boiler feed pump.

While changes in frequency appear to cause no problems, the voltage drop from the combustion turbine to the motors' bus can be significant. A lower voltage limit of 0.8 pu. at the load bus was chosen to avoid possible stalling of any on line motors or damage to any other equipment on line. By maintaining the combustion turbine generator bus voltage at 1.027 pu. this criteria is satisfied for starting all the large motors in any order except for the boiler feed pump. The voltage drop associated with starting the boiler feed pump is the smallest when the pump is started first. Figure 5 shows the voltage profile at the load bus while starting the boiler feed pump first with the generator voltage at 1.027 pu.

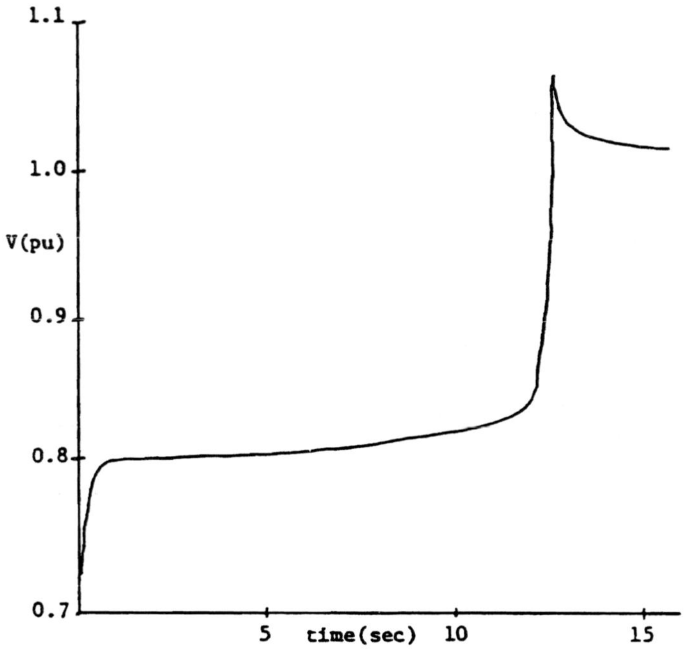

Figure 5. Voltage profile for boiler feed pump with generator voltage of 1.027 pu.

Figure 5 shows that the load voltage is less than or equal to 0.8 pu for 1.7 seconds. To overcome this problem the taps on the 13.2/138 KV transformer were set to a +4%. The voltage profile for starting the boiler feed pump first under these conditions is shown in Figure 6.

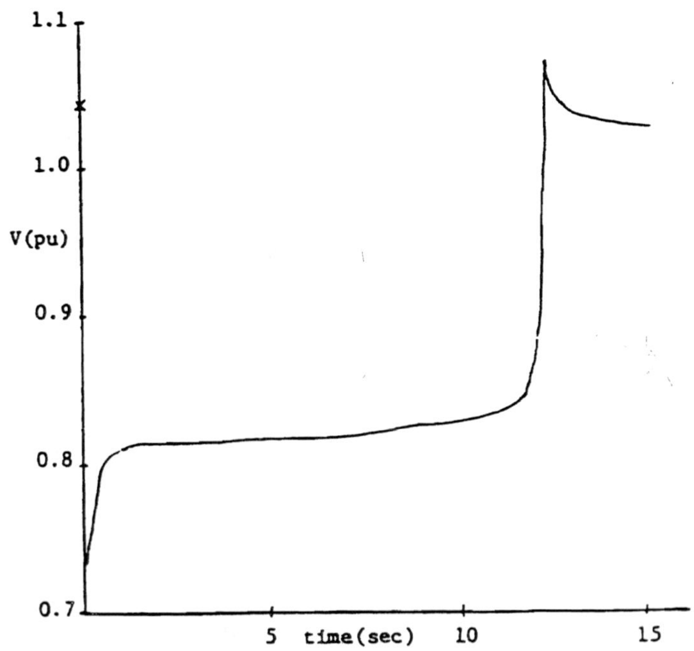

Figure 6. Voltage profile for boiler feed pump with generator voltage of 1.027 pu and transformer taps at +4%.

Figure 6 shows that the load voltage is less than 0.8 pu for only 0.5 seconds. A low voltage condition which is this short lived should not cause severe problems. These results were considered acceptable.

The operational procedure which should be used to blackstart this system consists of three steps:

1. Set the generator voltage to 1.027 pu.
2. Set the taps on the 13.2/138 KV transformer at +4% above nominal voltage.
3. Start the boiler feed pump first.

Blackstart Field Test

The field test was performed by starting the motor loads necessary to run a 105 MVA generator from a combustion turbine source 17.8 miles away. Using operating procedure described in the previous section the blackstart was successfully completed. The electrical path consisted of a 13.2/138 KV transformer at the combustion turbine, 17.8 miles of 138 KV transmission line, 138/13.2 KV and 13.2/2.4 KV transformers at the generating station, and an isolated 2.4 KV reserve bus at the generating station. The motor loads were connected to the isolated reserve bus. The voltage at the combustion turbine was set at 1.027 pu and the taps of the 138/13.2 KV transformer was set to 4% boost. Voltages were recorded at several locations along the electrical path, but the voltage at the load bus was of most interest.

As might be expected from the relative sizes of the motors in Table 6 and the analytical results, the boiler feed punp, ID fan, and FD fan caused the greatest disturbances. For comparison purposes, a copy of line charts of load bus voltage and frequency are shown in Figure 7. The charts in Figure 7 read from right to left. The top chart is voltage, and the bottom is frequency. The major voltage disturbances marked by arrows are the starting of the boiler feed pump, the ID fan, and the FD fan respectively starting from the right. The scales of Figure 7 are 12 seconds per major horizontal division of both chart, 300 volts per major vertical division of the top chart, and 0.5 hertz per major division of the bottom chart. As was predicted by the analytical study, the frequency variation does not cause a problem.

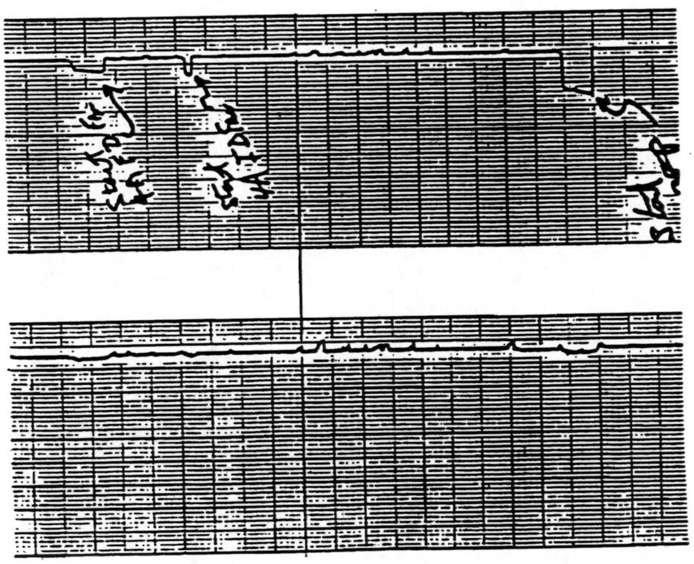

Figure 7. Strip charts of voltage (top) and frequency (bottom) disturbances due to starting boiler feed pump (right), ID fan (center), and FD fan (left). (Scales given in text).

From the field test, as well as the analytical study, it is seen that starting the boiler feed pump causes the largest disturbance. Therefore, it will be used for comparison to the analytical results. Figure 8 shows the voltage at the load bus when the boiler feed pump was started. This chart also reads from right to left. The curvature of the vertical lines are a result of the type of recorder used. The voltage at the load bus did not return to a normal range until after twelve seconds. The minimum voltage recorded was 0.766 pu.

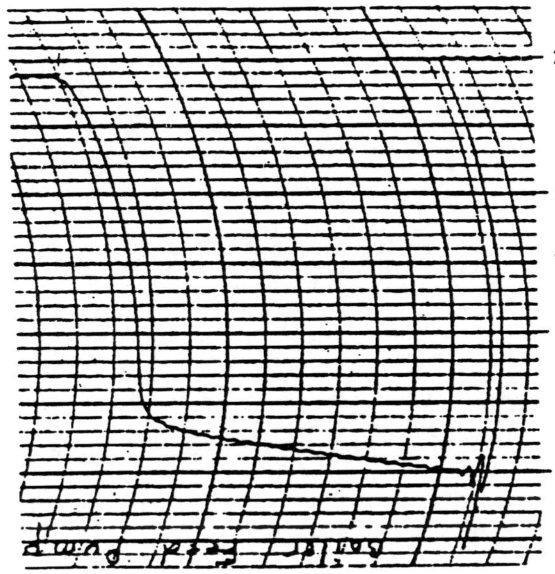

Figure 8. Strip chart of load bus voltage while starting the boiler feed pump.

In Figure 9 the results of Figure 8 are redrawn to the scale of Figure 6. In this manner the results of the analytical analysis and the field test can be compared.

All of the required motors were successfully started. The responses of the motors matched very closely the response predicted by the analytical study. This match can be seen in Figure 9. Some differences are seen in the very fast transient responses. This could be a result of the models used in the analytical study or the response capabilities of the recording system in the field test. However, these differences are of only minor importance for the blackstart conditions being studied.

Conclusions

This paper has shown that special operating procedures may need to be followed to successfully blackstart a generating station from a remote combustion turbine generator. A method was also presented to determine what, if any, special procedures are needed for a specific blackstart situation. A blackstart field test was successfully conducted on a system using a procedure determined by an analytical study. Recorded system voltages and frequency correlated very closely to predicted values from the analytical analysis.

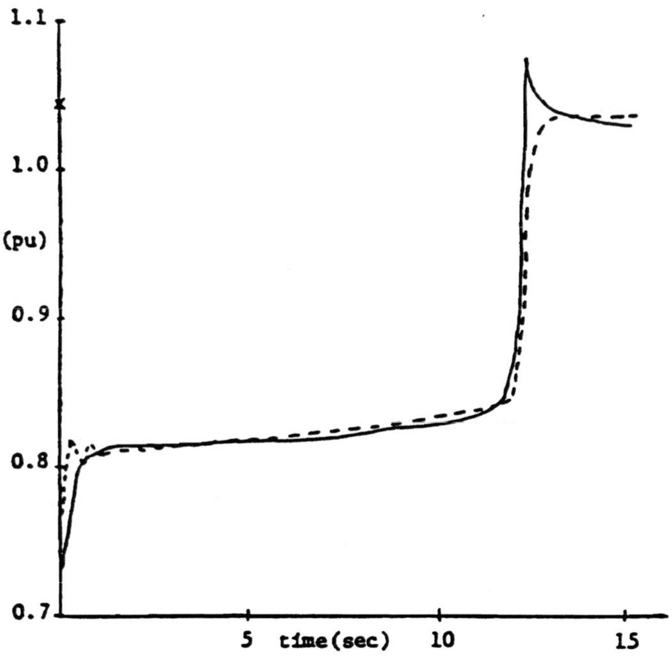

Figure 9. Comparison of load bus voltage calculated (solid line) and from the field test (dashed line).

References

[1] K. Prubhaskankar, W. Janischewskyi, ''Digital simulation of multimachine power systems for stability studies,'' IEEE Transactions on Power Apparatus and Systems, vol. PAS-87, 1968.

[2] IEEE Committee Report, ''Computer representation of excitation systems,'' IEEE Transactions on Power Apparatus and Systems, vol. PAS-99, 1968.

[3] C. P. Arnold, E. J. P. Pacheco, ''Modeling induction motor start-up in a multi-machine transient-stability programme,'' IEEE Transactions on Power Apparatus and Systems, vol. PAS-99, 1980.

[4] G. W. Stagg, A. H. El-Abiad, Computer Methods in Power System Analysis, New York, McGraw-Hill, 1968.

Discussion

K. Hemmaplardh (Engineering Technology Applications, Boeing Computer Services, Tukwila, WA): The authors are to be commended for their work and report. The analysis was clearly présented with the data. The field test results verified the analytical result very well as shown. The paper will serve as a very good reference for more work in the area of blackstarting generating units. In the interest of future work in the area, it would be appreciated if the authors can comment on the following two points:

1. Is there any change in the boiler feed-pump induction motor model that can give simulation result less ''spike'' discrepancy, as shown in Figure 9 of the paper (at time being about 13 seconds)? How does the motor speed-torque curve look for the situations in the Figure?
2. For other blackstart situations, what are the key factors in determining the minimum size of a startup turbine for a given generating plant?

Manuscript received February 23, 1984

R. D. Shultz and G. A. Mason: The authors would like to thank Mr. Hemmaplardh for his interest in our paper. The absence of the transient voltage spikes in the recorded field data is probably due to an inability of the recording equipment used to record transients of such short duration. Because of the short durations of these ''spikes'', they will probably not play an important role in determining the success of a blackstart.

The output torque of an induction motor is a function of the speed of the machine and the magnitude of its terminal voltage. The output torque is directly related to the square of the voltage. For the period during which the startup resulted in reduced voltage, the output torque will be reduced by 35 to 40 percent of what its value would have been had the voltage not decreased.

The work which the authors conducted seemed to indicate that the primary concern in using remote combustion turbines for blackstart was the possibility of unacceptable low voltages at the load bus. The factors that affect the voltage are the size of motors being started, the electrical distance (impedance) between the generator and load, the tap setting of transformers, and the order in which motors are started.

Manuscript received April 12, 1984

Overvoltage Control During Restoration

A report by the Power System Restoration Working Group
Contributing authors:
M. M. Adibi R. W. Alexander B. Avramovic
IRD Corp. PP&L Co. ECC, Inc.

Abstract: This paper is one in a series presented on behalf of the System Operation Subcommittee with the intent of focusing industry attention on power system restoration issues. Restoration of a bulk power supply system presents the operator with unique challenges not normally encountered in daily operations. The initial, and even intermediate transmission system topologies are quite different from the well integrated systems during normal operation. There are several problems that pertain to these non-normal topologies that are of common concern to operators and need to be addressed [1]. One of these, the problem of overvoltages, is the subject of this paper.

Keywords: switching transient, sustained overvoltage, harmonic resonance, surge arrester, generator excitation

Introduction

During the early stages of restoring high voltage overhead and underground transmission lines, one's concerns are with the following three related overvoltage areas: sustained power frequency overvoltages, switching transients, and harmonic resonances.

Sustained power frequency overvoltages are caused by charging currents of lightly loaded transmission lines. If excessive, these currents may cause underexcitation, or even selfexcitation and instability. Sustained overvoltages also overexcite transformers, generate harmonic distortions and cause transformer overheating.

92 WM 107-3 PWRS A paper recommended and approved by the IEEE Power System Engineering Committee of the IEEE Power Engineering Society for presentation at the IEEE/PES 1992 Winter Meeting, New York, New York, January 26 - 30, 1992. Manuscript submitted August 29, 1991; made available for printing December 31, 1991.

Transient voltages or switching surges are caused by energizing large segments of a transmission system or by switching capacitive elements. Switching transients, which are usually highly damped and of short duration, in conjunction with sustained overvoltages may result in arrester failures [2]. Transient overvoltages are not usually a significant factor at transmission voltages below 100 Kv. At higher transmission voltages, overvoltages caused by switching may become significant, because arrester operating voltages are relatively close to normal system voltage and lines are usually long so that energy stored on the lines may be large. In most cases without sustained traveling wave transients, surge arresters have sufficient energy absorbing capability to clamp harmful overvoltages to safe levels without permanent damage. The likelihood and the effect of transient overvoltages are determined by study of special system conditions [3]. Computer aided analysis has proven to be a valuable tool in understanding switching surge overvoltages [4,5].

Harmonic resonance voltages are oscillatory undamped or only weakly damped temporary overvoltages (TOVs) of long duration. They originate from switching operations and equipment non-linearities [2]. They result from several factors that are characteristic of networks during restoration: first, the natural frequency of the series resonance circuit formed by the source inductance and line charging capacitance may, under normal operating conditions, be a low multiple of 60Hz; second, "magnetizing inrush" caused by energizing a transformer produces many harmonics; and finally, during early stages of restoration the lines are lightly loaded, resonance therefore is lightly damped, which in turn means the resulting resonance voltages may be very high [6]. If transformers become overexcited due to power frequency overvoltage, harmonic resonance voltage will be sustained or even grow.

Transmission System Equipment Limitations

Power transformers, surge arresters and circuit breakers will be the equipment earliest affected by overvoltages. Figure 1 shows power frequency voltage withstand vs. time

Reprinted from *IEEE Transactions on Power Systems,* Vol. 7, No. 4, pp. 1464–1470, November 1992.

for typical power transformers and surge arresters as would be applied on a solidly grounded system. As can be seen, a power transformer can survive a seemingly small overvoltage of 1.2 p.u. for only about one minute! both the surge arrester and transformer can withstand 1.4 p.u. voltage for 10 seconds. Above 1.4 p.u. voltage, surge arrester failure will precede transformer damage.

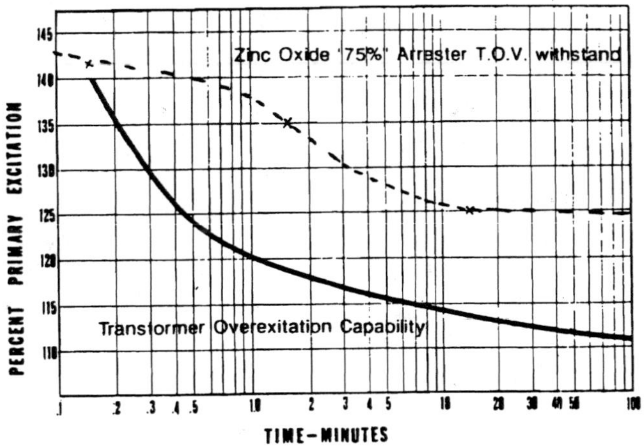

Fig. 1 Temporary Overvoltage (T.O.V.) Capability of Power Transformer and Surge Arrester

Any voltage above 1.1 p.u. will put the transformer into saturation, causing core heating and copious harmonic current generation. Note that the per unit base for the transformer is the particular voltage rating on the tap being energized.

Circuit breakers called upon to operate during periods of high voltage will have reduced interrupting capability (or none at all). At some voltage, even the ability to interrupt line charging current will be lost. This point will vary among specific breaker designs. Although standards require circuit breakers to interrupt against larger than normal recovery voltages with reduced fault currents, there are no overvoltage requirements for line charging current interruption. The authors believe that most HV or EHV circuit breakers could interrupt line charging current at least up to 1.2 p.u. of maximum rating.

Slow clearing of phase - to - ground faults could aggravate an otherwise acceptable overvoltage to the point of damaging surge arresters. Even on well grounded systems the voltage on the un-faulted phases during phase - to - ground faults can be on the order of 1.3 p.u. The longer a fault persists, the higher the likelihood of equipment damage.

In light of the foregoing equipment voltage limitations and fault contingency, it is recommended that no line be energized if by doing so voltage at the remote end would raise to more than 1.2 p.u. of normal or 1.1 x (transformer tap or circuit breaker rated maximum voltage), whichever is lower.

In developing a power system restoration plan the above concerns and constraints can be addressed by analysis and simulation as illustrated by the three sample studies described below.

Sustained Overvoltages

In the following example a load flow program was used to study sustained voltages of the system shown on Figure 2. This 230kV system was energized by a generator connected to bus 1. The objective was to find a sustained voltage profile as a function of the generator's terminal voltage for a no load situation.

Figure 3 shows two voltage profiles of the 220kV System. The upper profile shows voltages to be at an unacceptably high level, i.e., about 130% when the terminal voltages of Generators 1 & 2 is kept at 13.8 KV or 1.0 per unit. The total charging reactive power for the system at this voltage level is about 90 MVAR. The lower profile shows that, to maintain the sustained voltages within the acceptable limits, terminal voltages of Generators 1 & 2 should be kept at about 11.0 KV or 80%, with the resulting reduction in charging reactive power to about 55 MVAR, which is within the reactive absorbing capability of the generators. It is important to note, however, that the extent of the generator's voltage reduction is usually constrained by underexcitation of generators brought about by a number of limiting factors, including generator terminal low voltage limit (as limited by power plant auxiliary equipment), reactive ampere limit relay, minimum excitation limit relay, and rotor core end heating limit [7].

Control of Sustained Overvoltages

Sustained overvoltages can be controlled by absorbing the large charging reactive power of the lightly loaded transmission lines. This can be accomplished by:

- having sufficient under-excitation capability on the generators,

- connecting reactive loads (lagging power factors) to the underlying system including shunt reactors,

- removing all sources of reactive power and switching off shunt capacitors,

- running generators at maximum possible reactive power output to allow margin to adjust for large charging reactive power when lines are switched on,

- operating parallel transformers on different taps to increase circulating currents and reactive power losses,

- energizing only those transmission lines which carry significant load and avoiding the energization of extra lines which will generate unwanted reactive power, and

- maintaining a low voltage profile on the transmission lines, since the charging currents are proportional to the square of the voltage.

Inability to perform the above tasks may cause a serious reactive power imbalance resulting in, for example, generator self-excitation and a runaway voltage rise.

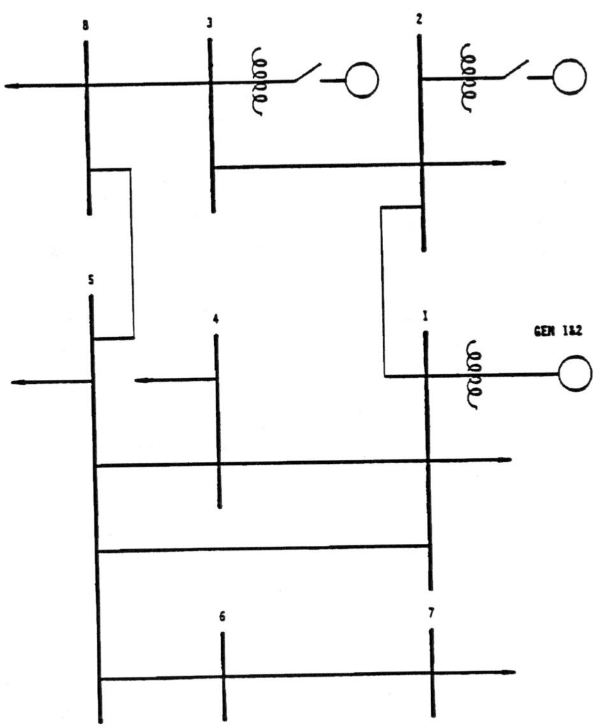

Fig. 2 The 220kV System

Considering the last item, e.g., 100 miles of 500kV line with charging reactive power of 1.8 MVAR per mile generates about 180 MVAR. A 5% voltage reduction will reduce this charging by about 10%, or about 18 MVAR!.

Fig. 3 Sustained Voltages When Energizing The 220k System

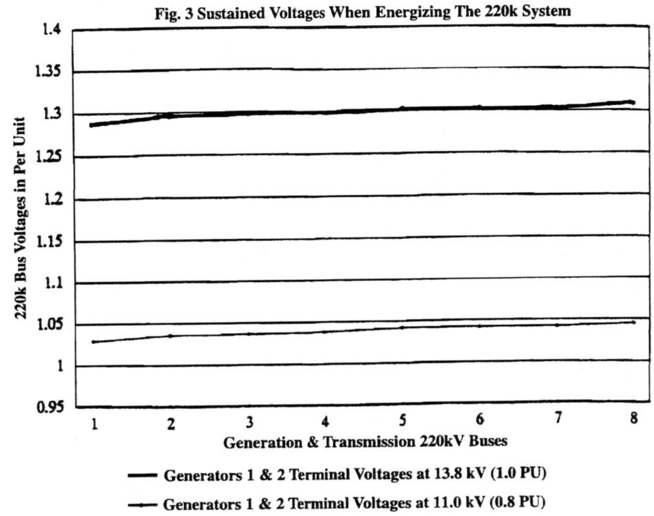

Switching Transients

Severe overvoltages resulting from switching surges may cause flash-over and serious damage to equipment. Switching transients are fast transients that occur in the process of energizing transmission line and bus load capacitances immediately after a power source is connected to the network. Inductances of transmission lines and power sources interact with capacitance to cause very fast oscillations in the process. The main cause of overvoltages is ill-timed closure of the breaker; if it occurs at the point when the instantaneous voltage of the source goes through its maximum a large overvoltage can result. In order to illustrate this phenomenon results of the 220kV System study are shown in Figure 4. The 220kV System of Figure 2 was energized using Generators 1 and 2 at Bus 1, with no surge arresters modeled. When the generators are running at 105% of the rated bus voltage, it can be seen from Figure 4 that the maximum instantaneous phase voltage reaches about 2.3 PU on buses 6 and 7. Typical surge arresters would limit this voltage to 2.1 PU or less while using only 5% of their energy absorbing capability. By reducing terminal voltage on the two generators to 80% of their nominal values, the switching transients are kept well within the arresters ratings.

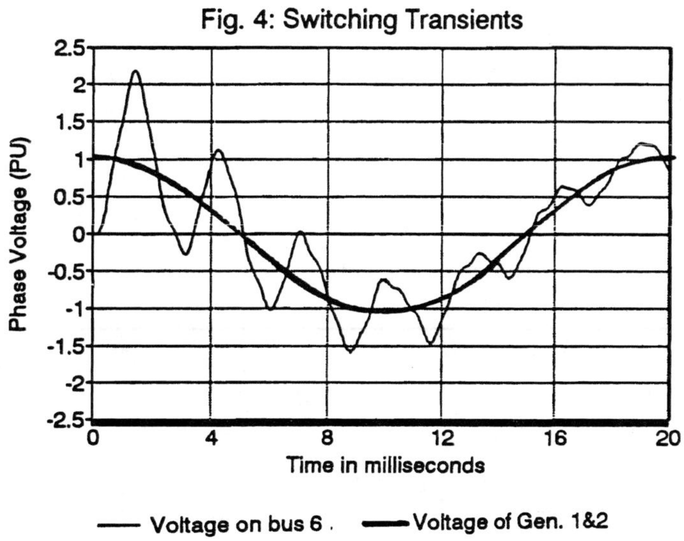

An RLC model of the network is used for this study. Generators are represented as sinusoidal sources; lines and transformers are represented as RLC circuits, with the charging capacitance concentrated at the line ends. Bus loads are converted to equivalent resistances and inductances (for absorbing VAR loads), or capacitances (for generating VAR loads). A model relating unknown (state) variables of the system and known parameters and the input is obtained by observing the basic physical laws of balance of current at each node and balance of voltage in each loop. The state variables consist of currents through each inductor, and voltages across each capacitor. A result is a mix of algebraic and ordinary differential equations, as given in the Appendix. This set of equations is solved by a method that utilizes automatic integration step size adjustment [8].

Control of Switching Transients

Energizing transmission lines or switching capacitive elements causes switching surges of fast front, low energy or slow front, high energy transients. Switching transients are not usually the limiting factor in re-energizing a system. Generally, if the steady state voltages are less than 1.2 per unit of their nominal values, the switching transients can be managed by typical arresters with ease. A notable exception is energizing transformer terminated lines, which may result in harmonic resonance and damaging overvoltages.

Harmonic Resonance

During the reintegration phase, with a "fledgling" system, the capacitive voltage rise due to charging currents will be sufficient to overexcite transformers and generate significant harmonics. If the combination of the system impedance and the line capacitance is adverse, then a harmonic resonance will result. Harmonic distortions produced by transformer saturation will excite these resonances, which can result in damaging overvoltages. Even if transformers are not continuously overexcited (due to sustained overvoltage), harmonics generated by magnetizing inrush on energization is sufficient to excite the resonance. To assure that resonance is damped, sufficient load should be connected to the underlying system at both ends of line to damp the resonance. Analysis for a 500kV line has shown that a load of about three megawatts per mile is adequate.

A major concern with energizing a transmission system is the increase in harmonic overvoltages due to line and transformer configurations different from normal operation. The overall objective of a harmonic analysis study is to determine if the line-transformer configuration resulting from energizing of the line would result in excessive harmonic level.

In the following example, EMTP was used to model a simple system consisting of a 500/230kV transformer at the sending end, a 500kV line, and a 500/230kV transformer at the remote end. The underlying loads were modeled on both transformers. The line length was varied to attain resonance at 3rd, 4th, 5th, 6th and 7th harmonics. Pre-switch voltage was varied from 0.9 to 1.1 p.u. At these resonances, switching angles were varied to produce the maximum sustained overvoltage possible. Lower order resonances produce higher overvoltages. Load levels on both the sending-end and receiving-end transformers were gradually increased until the resonance was controlled by sufficient damping. Approximately, 3 MW load on the underlying 230kV system for every mile of 500kV line connected will damp resonance voltages. Figure 5 shows the one-line diagram of the model and Figures 6 and 7 show sample wave forms.

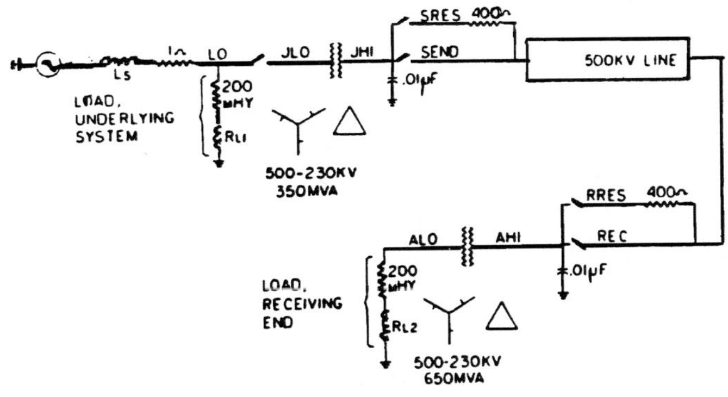

Fig. 5 EMTP Model for Harmonic Resonance Studies

In the case shown, a 650MVA transformer is being energized at the end of an 88 mile 500kV line. B phase has been tuned to 180Hz (3rd harmonic). The pre-switch voltage is 500kV and the transformer being energized is on its 550kV tap. Switching takes place at 0.1 seconds near a voltage zero with no remnant magnetism in the core. The voltage shown is the sending end bus voltage. The receiving end bus voltage is about 8kV higher. In Fig. 6 there is no load. The voltage escalates to 2.0 p.u. and shows no sign of decay during the simulation. Such a sustained resonance voltage would surely cause surge arrester failures and create system faults.

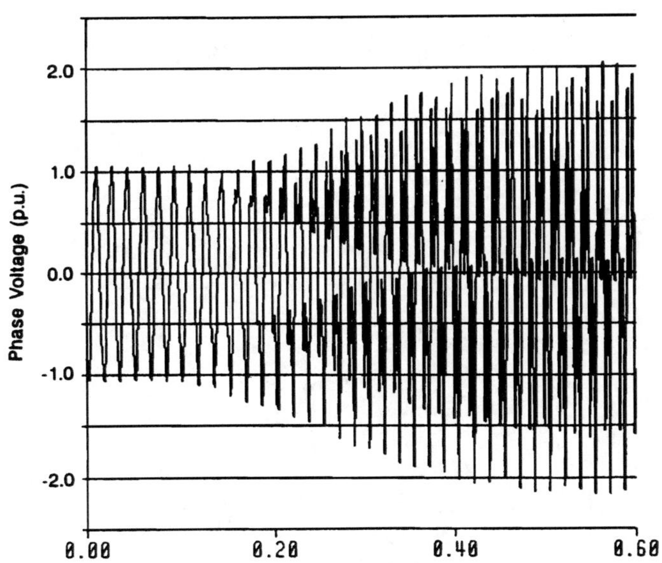

Fig. 6 Harmonic Resonance Excited by Transformer Energization -No Load-

Fig. 7 shows the effect of placing 200 MW of load on the underlying 230kV system at the source end. The resulting voltage is damped to a harmless 1.05 p.u.

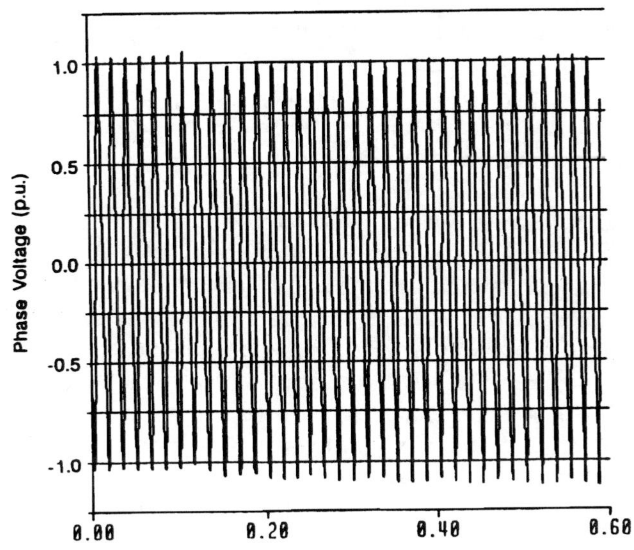

Fig. 7 Case in fig. 6 with 200MW Load on Underlying System

Control of Harmonic Resonance:

- Sustained harmonic overvoltages caused by over excitation of transformers can be controlled by selecting a transformer tap which equals or exceeds the power frequency voltage applied (or lowering system voltage to at or below the tap) before energizing.

- Harmonic resonances can be damped by connecting sufficient underlying loads at the sending end of a line, or by connecting dead load on the transformer to be energized.

- High source impedance can be reduced by starting more generators and connecting underlying loads.

- The reactive power of a lightly loaded system can be reduced by minimizing the number of unloaded lines to be energized and setting the sending-end transformers at the lowest tap position.

Conclusions

Power system equipment is protected against the effect of overvoltages resulting from normal operation by surge arresters which provide a relatively low-impedance path for transient overvoltages. During restoration however, certain line-transformer configurations when energized generate a combination of sustained, transient and harmonic overvoltages which may cause surge arrester failures and thus system faults, or the subsequent flash-over and damage to equipment.

Switching surge overvoltages have become increasingly more important with the design of compact transmission lines and substation facilities. The energy stored in long high voltage lines is large causing significant transient or resonant overvoltages which must carefully be considered during restoration. With the help of analytical tools such as EMTP and those described in this paper it should be possible to establish general procedures and restoration guidelines to avoid overvoltages.

Appendix

The subscripts and notation used in the RLC model of the 220kV System of Figure 2 are:

B - bus
L - line (with terminals i and j)
G - generator

v - instantaneous bus voltage,
i_L - current through line L,
i_G - current through generator G

The model can briefly be described as:

$di_G/dt = 1/L_G\ [\ V \sin(2\pi f) - R_G\ i_G - v_L\ ]$

$di_L/dt = 1/L_L\ (v_i - R_L\ i_L - v_j),$

$\Sigma_L\ c_{BL}\ i_L + \Sigma_G\ c_{BG}\ i_{GL} = i_B,$

$c_{BL} = (1 \text{ if } B = i; -1 \text{ if } B = j; 0 \text{ otherwise})$

$c_{BG} = (1 \text{ if } G \text{ connected to bus } B; 0 \text{ otherwise})$

$dv_B/dt = 1/C_B\ (i_B - v_B/r_B),\ \ r_B \neq 0$

$dv_B/dt = 1/C_B\ i_B,\ \ r_B = 0$

B = 1, 2,..., 8;

L = 1, 2, ..., 9; and

G = 1.

Inductances (in Henries), capacitances (in Farads), and resistances (in Ohms) of the above model are given in the following tables:

Line data

Line L	From (i) To (j)	Inductance L_{ij}	Resistance R_{ij}
1	1 - 2	0.051457	2.6620
2	1 - 4	0.084888	4.0656
3	1 - 5	0.14436	7.5020
4	2 - 3	0.011709	0.3872
5	3 - 8	0.19289	6.3888
6	4 - 5	0.050532	2.4200
7	5 - 6	0.010076	0.50336
8	5 - 8	0.082269	4.2592
9	6 - 7	0.0020182	0.09922

Note: Charging capacitance is lumped at the terminal buses

Source Data

Bus_G	Inductance L_G	Resistance R_G
1	8.0737e-4	2.4757e-3

Bus Data

Bus *B*	Capacitance C_B	Resistance R_B
1	7.2673e-7	0
2	1.584e-7	0
3	5.332e-7	0
4	3.7596e-7	0
5	9.3047e-7	0
6	1.4717e-6	0
7	1.5155e-7	0
8	8.1567e-7	0

References:

[1] Adibi, M. M. and R. J. Kafka, "Power System Restoration Issues," IEEE Computer Application in Power, Vol.4, No.2, April 1991 Issue, pp 19-24.

[2] Rocamora, R. G. (Chairman), "Switching Surge: Part IV - Control and Reduction on AC Transmission Lines," IEEE Trans. on Power Apparatus and Systems, Vol. PAS-101, No.8, pp 2694-2702.

[3] Anderson, J. G. and E. C. Sakshung "Lightning and Surge Protection," McGraw-Hill, Standard Handbook for Electrical Engineers, 11th Edition, 1978.

[4] Natarajan, R. et al, "Adoption of Electromagnetic Transient Program on Personal Computers," IEEE Trans. on Pwr. Sys., Vol.4, No.4, Oct. 1989, pp 1550-1554.

[5] Dommel, H. W., et al, "Transients Program User's Manual," UBC Publication, Aug. 1988.

[6] Alexander, R. W. "Minimum Source Consideration when Restoring the PJM 500kV System," PEA System Operation Committee, Jan. 1991.

[7] Mercer, L. S. (Chairman) "Reactive Output of PJM's Large Generating Units," PJM Reactive Study, July 1983.

[8] Brenan, K.E. et al "Numerical Solution of Initial-Value Problems in Differential Algebraic Equations," North-Holland, 1989.

Reactive Capability Limitation of Synchronous Machines

M. M. Adibi, LM
IRD Corporation
Bethesda, Maryland

D. P. Milanicz, SM
Baltimore Gas & Electric Co.
Baltimore, Maryland

Abstract: Achievable generator reactive capability (GRC) is generally much less than indicated by manufacturers' reactive capability curves, due to constraints imposed by plant auxiliaries and the power system itself. The nature of these constraints is explained and a method for calculating them is provided and verified by field tests on a unit at high and low system voltage levels. Several recommendations are made to enhance the GRC of the tested unit.

Keywords: overexcitation, underexcitation, reactive capability, generator overheating, excitation limitation, reactive requirement.

INTRODUCTION

The generator reactive capability (GRC) curves furnished by manufacturers and used in operation planning typically have a greater range than can be realized during actual operation. Generally, these GRC curves are strictly a function of the synchronous machine design parameters and do not consider plant and system operating conditions as limiting factors. Recently concern over GRC has been heightened by increasing need for reactive power to provide voltage support for large blocks of power transfer.

This paper, describes a method for determining the reactive capability limitation of synchronous machines under plant and system operating conditions. The methodology is tested and verified on a large generator. Then several recommendations are made to enhance the GRC of the unit.

BACKGROUND

1. The Phasor Diagram

The basis for generator reactive capability (GRC) curves is shown in **Figure 1**. To simplify the diagram the effect of saturation and of resistance are neglected, and an unsaturated value of synchronous reactance is used. The generator is assumed to be connected to an infinite bus [1].

For a given terminal voltage V and current I_a at phase angle Φ (i.e., for a given MVA output) the emf E is obtained by adding armature reaction I_aX_d to V at 90° to I_a.

For constant MVA output, I_a is constant and its locus is a circle with center on the end of the voltage phasor V. Constant excitation implies that the locus of E is also a circle with center O.

The angle between E and V is the power angle δ. It can be seen that generator's MW output, $V*I_a*\cos\Phi$, is equal to:

$$(V*E/X_d)*\sin\delta.$$

For δ = 90° the static limit of stability is reached, so that the horizontal line through O represents this limit.

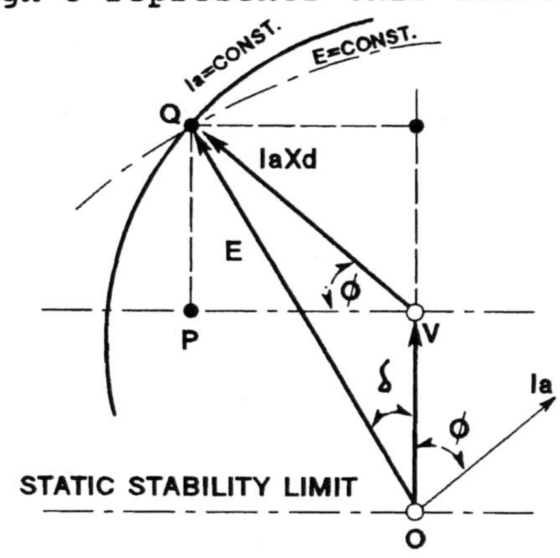

Figure 1. The Phasor Diagram

2. The Rated Gen. Reactive Curves

The GRC curves shown in **Figure 2** are drawn for synchronous reactance X_d = 167 percent. For zero excitation the armature reactance drop (shortcircuit ratio or SCR) is 100/1.67 = 60 percent of full load, representing 60 percent of full MVA scales. The vertical line GO is drawn and scaled to 60 percent MVar, and continued beyond O for lagging MVar. From O a horizontal line, which is similarly scaled, gives power in MW. A circle drawn with O as its center and its radius similarly scaled gives stator current in percent of full load. The radii from O indicate power factor values. Circles of percentage excitation are drawn from center G. A 100 percent excitation (i.e., amperes field no-load or AFNL), corre-

93 WM 203-0 PWRS A paper recommended and approved by the IEEE Power System Engineering Committee of the IEEE Power Engineering Society for presentation at the IEEE/PES 1993 Winter Meeting, Columbus, OH, January 31 - February 5, 1993. Manuscript submitted August 12, 1992; made available for printing December 14, 1992.

Reprinted from *IEEE Transactions on Power Systems,* Vol. 9, No. 1, pp. 29–40, February 1994.

sponds to the fixed terminal voltage GO.

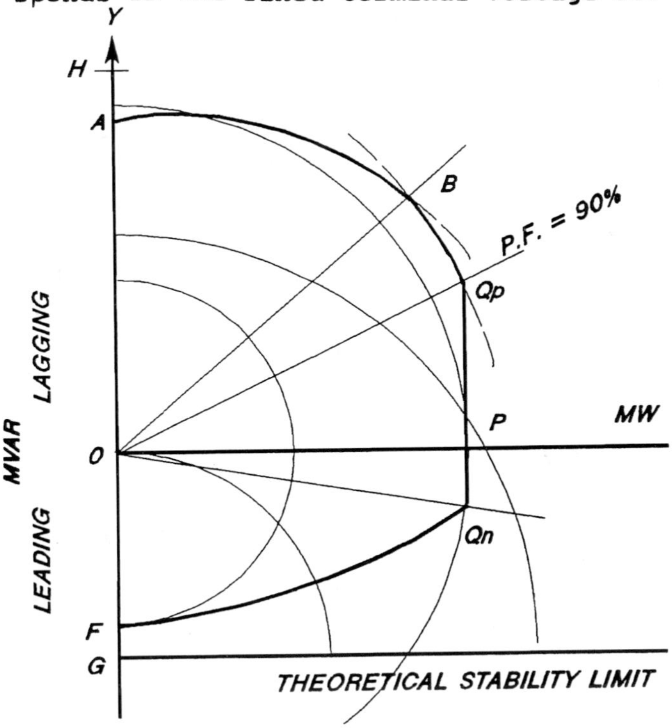

Figure 2. GRC Curves

Taking 100 percent MW as the maximum allowable power (provided by the prime-mover), a vertical limit-line is drawn through P at 100 percent MW. It is assumed that the generator is rated to give 100 percent MW at power factor 0.90 lagging. This determines point Q_P. Limitation of the stator current to the corresponding value requires the limit-line to become circular arc BQ_P about center O. At B a new limit by the exciting current has to be introduced, and it is assumed that the rotor current must not exceed that corresponding to an emf E = 260 percent of terminal voltage V (i.e., GO). The circular arc AB is therefore drawn from center G at this value of excitation. The upper limit ABQ_PP is thus completed.

The line Q_PP cannot be continued downwards to the static limit of stability because of the limitation in the underexcited region by the localized overheating of the stator core ends. This is due to the fact that during weak field operation the retaining rings at the end of the stator are no longer saturated allowing armature end leakage flux to produce localized heating in those areas where the flux is at right angles to the plane of the lamination. The limit imposed by the end heating is also circular with its center at some point H, on the GO line [2 & 3].

The generator reactive capability curves of **Figure 2** which are provided by manufacturers, are strictly a function of synchronous machine design parameters at rated terminal voltage and do not consider generating station and power system operating conditions as limiting factors. These design curves show that different generator loads produce greater heating in different parts of generator. The segment AB is limited by rotor heating due to over-excitation, BQ is limited by armature heating due to large lagging reactive generation, the limit-line Q_PPQ_N by the prime-movers MW output, and Q_NF by armature core end heating due to large leading reactive generation.

GENERATION REACTIVE CAPABILITY MODELING

The operating limits imposed on GRC can be formulated using the design curves. In **Figure 3**, the design limits are described by arcs AB, BC, CD, and DF.

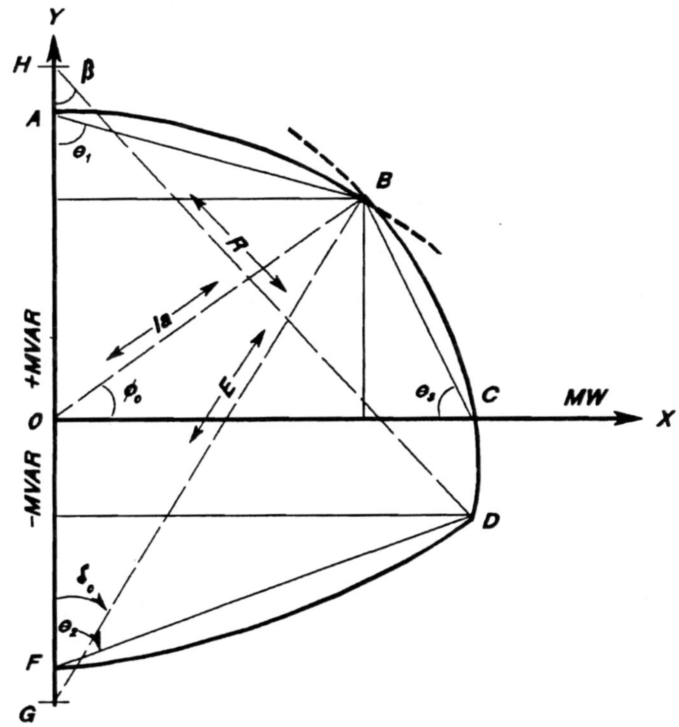

Figure 3. GRC Model

1. The Design Curves

The emf E can be determined from the excitation arc AB which subtends the power angle δ_o at the center G:

Let N_1 be equal to the cord AB, then

$$N_1 = [(A_y - B_y)^2 + B_x^2]^{1/2},$$

$$\Theta_1 = \tan^{-1} [B_x/(A_y - B_y)],$$

$$\cos \Theta_1 = N_1/(2E), \text{ and}$$

$$E = N_1/(2\cos\Theta_1).$$

Consider a point on the excitation arc AB corresponding to the power angle δ, reactive power q_1, and real power p, then

$$\delta = \sin^{-1}(p/E),$$

$$\cos\delta = (q_1+E-A_y)/E, \text{ and}$$

$$q_1 = A_y - E(1 - \cos\delta).$$

The limits imposed by the stator iron end heating during under-excitation operation is a circle with radius R, centered on Y-axis at H. Therefore, the leading reactive power q_2 for a given real power p, can be determined in a manner similar to that of over-excitation operation.

The arc DF subtends the angle β_o at the center H:

Let N_2 be equal to cord DF, then

$$N_2 = [(D_y - F_y)^2 + D_x^2]^{1/2},$$

$$\Theta_2 = \tan^{-1}[D_x/(D_y - F_y)],$$

$$\cos\Theta_2 = N_2/(2R), \text{ and}$$

$$R = N_2/(2\cos\Theta_2).$$

Consider a point on the arc DF corresponding to the angle β, reactive power q_2, and real power p, then

$$\beta = \sin^{-1}(p/R),$$

$$\cos\beta = (q_2+R-F_y)/R, \text{ and}$$

$$q_2 = R(1 - \cos\beta) - F_y.$$

The stator current I_a (assuming 1 pu voltage) can be determined from the arc BC which subtends the phase angle (power factor) ϕ_o at the center O:

$$I_a = (B_y^2 + B_x^2)^{1/2}.$$

Consider a point on the arc BC corresponding to phase angle ϕ, reactive power q_3, and real power p, then

$$q_3 = (I_a^2 - p^2)^{1/2}.$$

For a point on the arc CD,

$$q_4 = -(I_a^2 - p^2)^{1/2}.$$

2. The Plant and System Constraints

A typical arrangement for supplying power-plant auxiliaries and services in a generating station is composed of many major components and sub-components that operate as a system to supply power to the power system. Each component is designed and rated to operate continuously within certain electrical and mechanical limits specified by the manufacturer. Any component may have operating limits imposed on voltage, current, frequency or a combination of these such as MVA, volt/hertz, and stability. These limitations reduce the reactive capability of a generator to below the designed ranges.

The limitations imposed by station equipment on the generator's reactive capability during normal operation can be determined using the system representation of **Figure 4**, where:

T and X are the respective transformer taps and reactances,

g, a, s and as subscripts indicate the generator, auxiliary, system and start-up buses or transformes,

I, V, P and Q are current, voltage, real and reactive powers as indicated, and

mx and mn subscripts indicate the maximum and minimum limits.

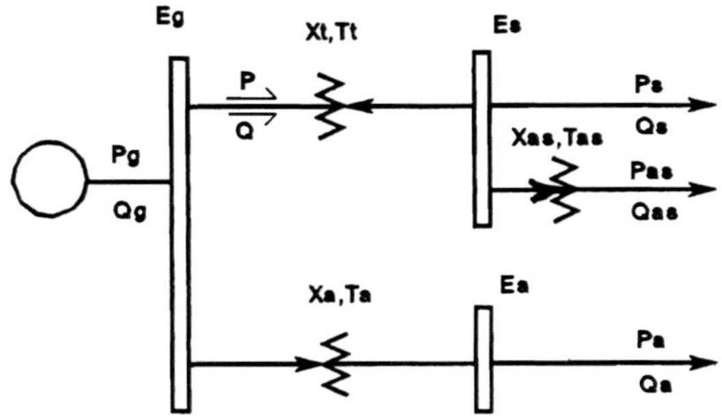

Figure 4. Auxiliary Equipment

In general the main and auxiliary transformers are equipped with fixed taps, whose optimum positions with respect to power system operating conditions need to be determined. Otherwise, they may pose more restrictive limits on either lagging or leading reactive capabilities of the generator.

The auxiliary real and reactive powers, P_a and Q_a, are some functions of generator's output P_g and Q_g, e.g. [5],

$$P_a = P_{amx}(3P_{gmx} + P_g)/(4P_{gmx}), \text{ or}$$

$$Q_a = P_{amx}(3P_{gmx} + P_g)/(4P_{gmx}), \text{ for lagging } Q_g, \text{ and}$$

$$Q_a = 0.85P_{amx}(3P_{gmx} + P_g)/(4P_{gmx}), \text{ for leading } Q_g.$$

Neglecting the effect of the main and auxiliary transformers' resistances, we have:

$$P_g = P_s + P_a, \text{ and}$$

$$Q_g = Q + Q_a + I_a^2 X_a,$$

where:

$$Q = (E_g)^2/X_t - [(E_s E_g/X_t T_t)^2 - P_s^2]^{1/2},$$

$$Q_s = [(E_s E_g/X_t T_t)^2 - P_s^2]^{1/2} - (E_s/T_t)^2/X_t,$$

$$I_a = (P_a + j\ Q_a)^*/E_a, \text{ and}$$

$$E_g = (E_a + j\ I_a X_a) T_a.$$

where:

$(P_a + j\ Q_a)^*$ is the conjugate $(P_a + j\ Q_a)$.

The maximum and minimum auxiliary voltages E_{amx} and E_{amn} reflected to the generator terminal bus are:

$$E_g' = (E_{amx} + j\ I_a * X_a) * T_a, \text{ and}$$

$$E_g'' = (E_{amn} + j\ I_a * X_a) * T_a.$$

If $E_g' > E_{gmx}$, then the E_{gmx} is the overexcitation limiting factor, and if $E_g'' > E_{gmn}$, then the E_{amn} is the underexcitation limiting factor.

Similarly, when E_{g1} and E_{g2} are the maximum and minimum system voltages reflected to the generator terminal bus, then when $E_{g1} > E_{gmx}$, E_{gmx} is the overexcitation limiting factor, and when $E_{g2} > E_{gmn}$, E_{smn} is the underexcitation limiting factor. Thus, the generator reactive power $\pm Q_g$ can be determined as a function of real power output P_g, from minimum generation P_{gmn}, to maximum generation P_{gmn}, for a given main and auxiliary transformer tap positions T_t and T_s, and for system voltage schedule E_s. The above formulation also allows optimization of the two transformer tap positions for a given system voltage schedule under light and peak load conditions.

Figure 5 superimposes the cumulative effect of all the operating requirements in limiting GRC on the manufacturer's design values. The actual reactive capability is restricted to the area bounded by all the operating curves, which are considerably more restrictive than the design curves.

TEST AND VERIFICATION OF GRC MODEL

The operating reactive capability of a 460 MVA generator at Baltimore Gas and Electric Co. was tested in order to verify the GCR methodology and also to determine the extent of improvement which could be made in the generator's reactive capability [4 & 5].

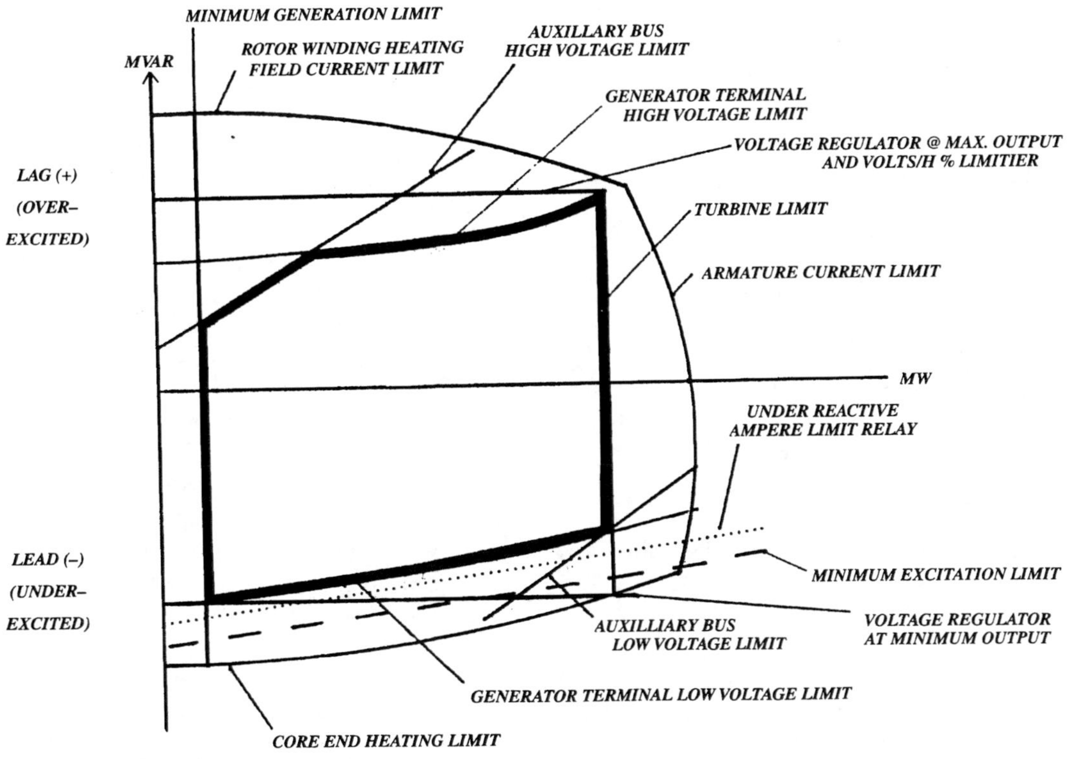

Figure 5. Generator Reactive Capability Limitation

1. Test Conditions and Procedure:

- o Station service load was carried by the unit service transformer,
- o The generator voltage regulator was placed under automatic control,
- o The unit was paralleled and kept at full load for the start of test.
- o Minimum load (14 MW), full load (400 MW), and several intermediate points (i.e. 100, 200, and 300 MW) were selected,
- o System voltage was maintained constant at the desired level (230 and 237 kV) for the duration of test,
- o The generator was kept at each test point for 10 minutes and test data was recorded for each MW loading for both overexcitation and underexcitation, and
- o Tests were repeated for high and low system voltage levels.

In addition to the generators various ratings and plants parameters which were discussed before, the following **"GRC Data Sheet"**, was completed before the start of the test:

2. GRC Test Sheet

Hydrogen Cooling Pressure	60 psi
Gen. Term. Voltage, Hi Limit	25.2 kV
Gen. Term. Voltage, Lo Limit	22.8 kV
Gen. Field Current Limit	3,026 A
Sta.Ser.Bus Voltage, Hi Limit	4576 V
Sta.Ser.Bus Voltage, Lo Limit	3744 V
Turbine-Generator Limit	398 MW
Min. Excitation Limit	Circular
Sta. Ser. Unit Transformer	Low Side
System Voltage Upper Limit	237 kV
System Voltage Lower Limit	230 kV
Field Temperature Limit	80° c

3. Test Results

Table A lists the rated GRC. The simulated and tested GRC are listed in Tables 1 to 4 in the Appendix. GRC for the 237 kV system voltage is shown in **Figure 6.**

ENHANCING GEN. REACTIVE CAPABILITY

Based on the above test results the following changes were recommended to enhance the generator's reactive capability:

- o the step-up transformer tap position to be changed from 235.75 to 241.50 kV to increase the lagging reactive capabilities at 237 kV system voltage,
- o the unit service megawatts to be monitored at the control center to determine the net generation,
- o the voltmeter scale on the unit service transformer to be expanded to read 600 volts AC,
- o the net rather than gross generator MVar to be monitored at the control center, and
- o the local voltage adjuster to be replaced by remote solid state control.

Table A. Rated Machine Reactive Limits

MW	MVar	MAX kV	MVar	MIN kV
0	318	26.21	-251	22.12
20	318	26.21	-257	22.08
50	316	26.21	-263	22.03
75	314	26.20	-266	22.00
100	312	26.18	-268	21.98
125	308	26.15	-269	21.97
150	303	26.12	-268	21.97
175	298	26.09	-266	21.98
200	292	26.05	-263	22.00
225	285	26.00	-258	22.03
250	277	25.94	-251	22.08
275	268	25.87	-243	22.13
300	258	25.80	-234	22.20
335	242	25.69	-217	22.32
370	224	25.55	-196	22.47
400	207	25.43	-175	22.62
430	160	25.09	-150	22.80
460	0	23.94	0	23.94

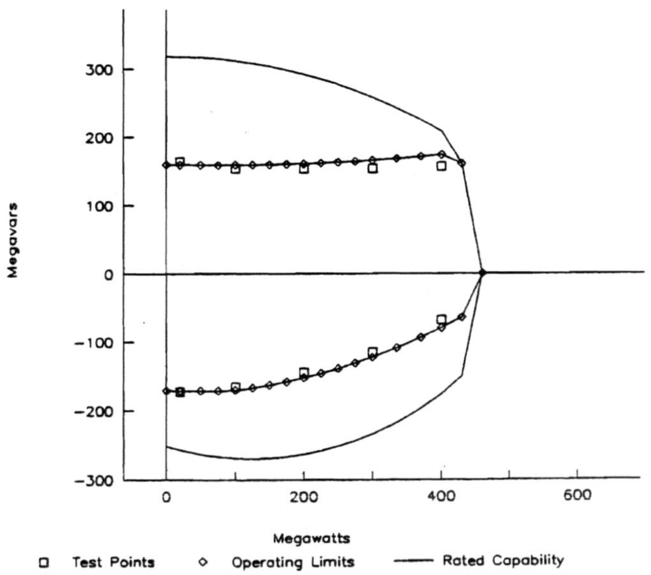

Figure 6. GRC at 237 kV System Voltage

The comparison of GRC simulation results of Tables 1 and 3 with Tables 5 and 6 (for system voltages of 230 kV and 237 kV, respectively), listed in Appendix, show that the implementation of the above recommendations would improve the GRC.

Tables 2 and 4 show the spot tests on the GRC curves, verifying the simulation results as shown in **Figure 6**. **Figure 7** is the plot of MVar lag and lead for the two Tables 3 and 6, showing the GRC improvement for the 237 kV system voltage case.

CONCLUSION

Actually realizable generator reactive capabilities, which are usually significantly less than those indicated by manufacturers' curves, may be calculated according to the expressions developed in this paper. The resulting, verified mathematical model may further be used to identify the step-up and the auxiliary transformers tap positions that will provide the best overall GRC under prevailing high and low system voltage conditions.

The power system reactive capabilities under changing load conditions can be determined by including the GRC models in power system dynamic simulators. This will be the subject of an ensuing effort.

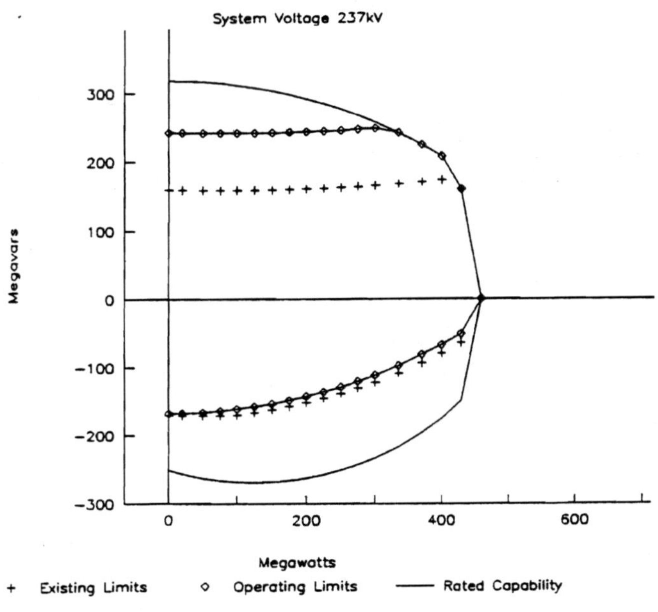

Figure 7. Increased GRC at 237 kV

ACKNOWLEDGEMENT

The authors wish to express their appreciation to Mr. J. E. Siegmund, III, of Electric System Engineering Department of Baltimore Gas & Electric Company and Mr. M. J. O'Grady of System Operation, Potomac Electric Power Company for assisting with GRC Modeling and Programming. The first author acknowledges the partial support of ECC, Inc.

REFERENCES

[1] Say, M.G.: The Performance and Design of Alternating Current Machines; Pitman & Son, Second Edition 1949.

[2] Farnham, S.B. et al.: Field Excitation in Relation to Machine and System Operation; AIEE, Fall GM, Nov 1953, Kansas City.

[3] Nagy, I.: Analysis of Minimum Excitation Limits of Synchronous Machines; IEEE TRANS, VOL PAS-86, NO.12, Jul/Aug 1970, pp.1001-1008.

[4] Prince, W.R.: PJM Generator Reactive Testing; PEA, Hershey, PA, May 1983.

[5] Milanicz, D.P.: Generator Reactive Capability - Testing and Optimization; Southwestern Electrical Exchange, Alabama, Oct 1989.

APPENDIX

Table 1. GRC at 230 kV System Voltage

GROSS MW	LAG MVar	MAX kV	LAG LIMIT	LEAD MVar	MIN kV	LEAD LIMIT
0	262	25.20	3	-77	22.80	8
20	262	25.20	3	-77	22.80	8
50	262	25.20	3	-78	22.80	8
75	262	25.20	3	-78	22.80	8
100	262	25.20	3	-78	22.80	8
125	262	25.20	3	-77	22.80	8
150	262	25.20	3	-77	22.80	8
175	263	25.20	3	-76	22.80	8
200	264	25.20	3	-75	22.80	8
225	265	25.20	3	-74	22.80	8
250	266	25.20	3	-72	22.80	8
275	267	25.20	3	-71	22.80	8
300	258	25.13	1	-69	22.80	8
335	242	25.01	1	-66	22.80	8
370	224	24.87	1	-63	22.80	8
400	207	24.74	1	-60	22.80	8
430	160	24.39	5	-48	22.87	7
460	0	23.20	5	0	23.20	5

Table 2. Spot Tests at 230 kV

20	249	25.2	3	-168	21.8	7
100	255	25.2	3	-159	22.1	7
200	257	25.3	3	-140	22.0	7
300	245	25.1	1	-107	22.9	7
400	195	24.8	1	-57	22.8	7

Table 3. GRC at 237 kV System Voltage

0	159	25.20	3	-170	22.80	8
20	159	25.20	3	-171	22.80	8
50	159	25.20	3	-171	22.80	8
75	158	25.20	3	-171	22.80	8
100	159	25.20	3	-170	22.81	7
125	159	25.20	3	-167	22.83	7
150	159	25.20	3	-163	22.86	7
175	160	25.20	3	-158	22.89	7
200	161	25.20	3	-152	22.93	7
225	161	25.20	3	-146	22.97	7
250	163	25.20	3	-139	23.02	7
275	164	25.20	3	-131	23.07	7
300	165	25.20	3	-122	23.12	7
335	168	25.20	3	-109	23.21	7
370	170	25.20	3	-94	23.30	7
400	173	25.20	3	-79	23.39	7
430	160	25.09	5	-64	23.48	7
460	0	23.94	5	0	23.94	5

Table 4. Spot Tests at 237 kV

GROSS MW	LAG MVar	MAX kV	LAG LIMIT	LEAD MVar	MIN kV	LEAD LIMIT
20	164	25.2	3	-172	22.2	7
100	153	25.2	3	-165	22.1	7
200	153	25.3	3	-144	22.4	7
300	153	25.2	3	-115	22.6	7
400	156	25.1	3	-68	23.1	7

Table 5. Increased Lead GRC at 230 kV

0	318	25.04	1	-158	21.55	7
20	318	25.04	1	-158	21.56	7
50	316	25.03	1	-156	21.57	7
75	314	25.02	1	-154	21.59	7
100	312	25.00	1	-152	21.61	7
125	308	24.98	1	-148	21.64	7
150	303	24.95	1	-143	21.67	7
175	298	24.91	1	-138	21.71	7
200	292	24.86	1	-132	21.75	7
225	285	24.81	1	-125	21.80	7
250	277	24.75	1	-117	21.85	7
275	268	24.68	1	-108	21.91	7
300	258	24.60	1	-99	21.98	7
335	242	24.48	1	0	22.74	6
370	224	24.34	1	0	22.72	6
400	207	24.20	1	0	22.70	6
430	160	23.85	5	0	22.67	6
460	0	22.63	5	0	22.63	5

Table 6. Increased Lag GRC at 237 kV

0	242	25.20	3	-168	22.21	7
20	242	25.20	3	-167	22.22	7
50	242	25.20	3	-166	22.23	7
75	242	25.20	3	-164	22.25	7
100	242	25.20	3	-161	22.27	7
125	242	25.20	3	-158	22.29	7
150	242	25.20	3	-154	22.32	7
175	243	25.20	3	-149	22.36	7
200	244	25.20	3	-143	22.40	7
225	245	25.20	3	-136	22.44	7
250	246	25.20	3	-129	22.49	7
275	247	25.20	3	-121	22.54	7
300	249	25.20	3	-111	22.60	7
335	242	25.14	1	-97	22.69	7
370	224	25.01	1	-81	22.80	6
400	207	24.87	1	-67	22.89	7
430	160	24.53	5	-51	22.99	7
460	0	23.35	5	0	23.35	5

The lag and lead limits identify:

1. Field Heating,
2. Maximum Excitation Limit,
3. Generator Bus Voltage/Voltage Regulator Control at Maximum,
4. Station Service Bus Voltage at Maximum,
5. Stator Heating; Armature Current; Field Current
6. Armature Iron Core End Heating
7. Minimum Excitation Limit
8. Generator Bus Voltage/Voltage Regulator Control at Minimum

9: Station Service Bus Voltage at Minimum

10: Station Instrumentation

Discussion

J. Mercurio and N. E. Nilsson (Ohio Edison Company, Akron, Ohio): The authors are to be congratulated on the development of an excellent paper on the application of generator capability curves. This is a subject of high interest and related papers were presented on this topic at the 1993 IEEE Winter Power Meeting in Columbus, Ohio [1]. The authors clearly demonstrate the dependency of generator reactive capability on the system to which it is connected, including the system voltage regulation schedule, generator step-up transformer (GSU) voltage tap setting and the unit auxiliary transformer (UAT) voltage tap setting.

The discussers have some observations and would like to address some questions to the authors.

Neglecting the effects of armature resistance and saturation may simplify the analysis, but also results in an overly optimistic calculation of overexcited reactive capability. The discussers have found that by including these effects, the discussers can calculate actual rotor field currents within a few percent of measured values.

The loci shown in Figure 2 may need some additional explanation. The thinner lines are not defined. What do they represent? Arc AB should be part of a circle centered on the Reactive Axis at Point G [2]. However, it does not appear to have been constructed in this manner, or even have its center on the Reactive Axis.

The end iron heating limit, as shown on Figure 2, may be circular in shape in some areas, but is not really a circle. Also, the apparent center is not located on the Reactive Axis, but rather somewhere to the right of it (first quadrant). The shape of the end iron heating limit is normally determined by the synchronous generator manufacturer. Accordingly, the manufacturers' capability curve normally looks a little more like Figure D1 in this discussion.

Could the authors discuss in a little more detail the 100 percent MW rating at 0.90 PF lagging? The discussers are more familiar with ratings defined at point B which would make the rating of this synchronous generator more like 0.85 PF. The authors claim that the region BQ in Figure 2 is "limited by armature heating due to large lagging reactive generation." Could the authors provide a little more background on why they believe this is so? The discussers have always been of the opinion that the BQ limit is strictly a stator heating limit occurring because of I^2R losses in the stator.

The discussers have found that the underexcited limit for generator operation is often a result of the steady state stability limit, including some stability margin, as well as coordination of the Minimum Excitation Limiter (MEL) setting with the Underexcitation Alarm relay setting and the Loss-of-Field Tripping relay setting. The limiting characteristics of these devices are illustrated in Figure D2.

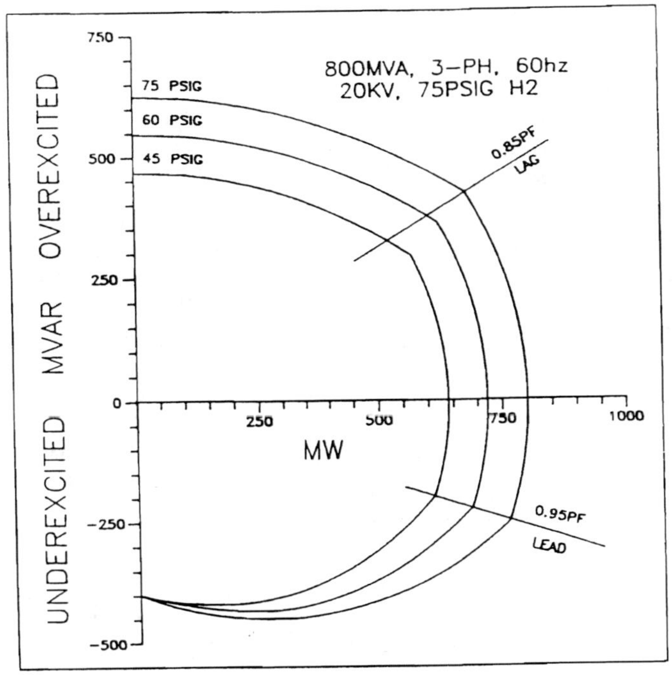

Figure D1: Manufacturer's Capability Curve

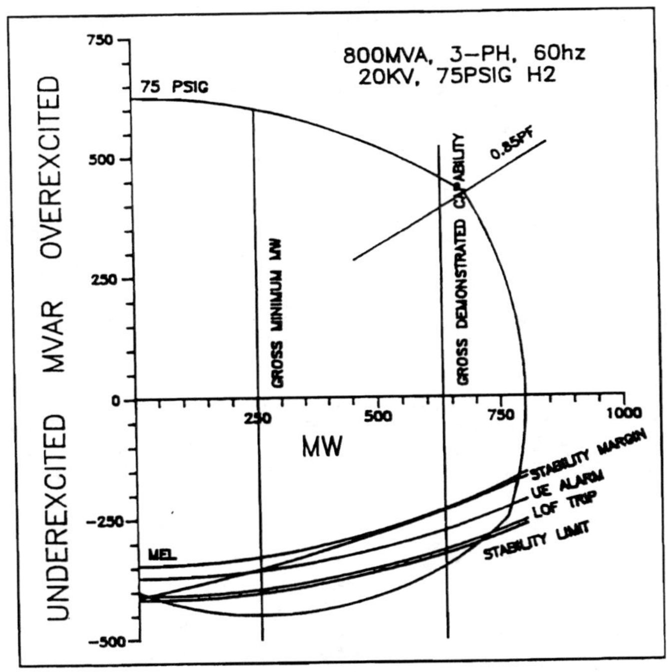

Figure D2: Features of Protective Devices in the Underexcited Region.

It should be noted that the tested limits shown on both Figure 6 and Figure 7 of the paper are only for the case with the high voltage bus at 237 kV. At lower high side bus voltages, the achievable overexcited capability would increase, whereas at higher high side voltages, the achievable underexcited capability would increase. Since the system voltage is regulated between 230 kV and 237 kV, why show the limit of overexcited capability for the maximum 237 kV bus voltage, since it is more likely that MVAR support will be required whenever the high voltage bus is at its minimum levels?

There are perhaps some other significant situations that the authors may want to discuss that impact the application of synchronous generator capability curves. It is not uncommon for synchronous generators to have shorted rotor field turns. Testing and analysis of this phenomenon by the discussers has revealed that rotor field deterioration due to or manifested by turn shorting develops slowly over a period of time [3]. Usually one of three limits will occur which will ultimately require rewinding the rotor field:

1. The bearing vibration increases to unacceptable levels.

2. The exciter will become loaded beyond nameplate when endeavoring to achieve modest levels of exported MVARs.

3. The generator cannot provide rated power in the lagging region which results in low UAT voltages.

The discussers have found it necessary to provide new capability curves to cognizant company personnel showing the loss of over-excited capability on one or two synchronous machines in the company system suffering from shorted turns in the rotor field.

[1] Nilsson, N.E. and J. Mercurio, "Synchronous Generator Capability Curve Testing and Evaluation," paper 93WM042-2PWRD presented on February 4, 1993, at the 1993 Winter Power Meeting in Columbus, Ohio.

[2] Hunt, J. P. "Capability Curves and Excitation Requirements of Saturated Cylindrical Rotor Synchronous Machines," *IEEE Transactions on Power Apparatus and Systems,* Vol. PAS-86, No. 7, July 1967, pp. 855-859.

[3] Nilsson, N.E. and J. Mercurio, "Evaluating the Service Degradation of Large Hydrogen Cooled Generator Rotor Fields," *IEEE Transactions on Power Apparatus and Systems,* Vol. PAS-102, No. 9, September 1983, pp. 2950-2958.

E. S. DEHDASHTI, Pacific Gas and Electric Company, San Francisco, CA; Authors are congratulated for bringing into light an issue which can significantly impact conclusions of reactive power adequacy studies and potential for voltage stability. Can the Authors respond to the following questions:

1. From their experience, is there any evidence leading to the conclusion that the limitations they experienced in their testing apply to many utility grade generators?

2. Regarding the test generator, were the auxiliary equipment supplied by the same manufacturer as the generator itself?

3. Can some of the limitations that were described by the Authors be eliminated by the use of auxiliaries with adequate rating and range of operation to better utilize the reactive capability of synchronous machines?

Manuscript received February 16, 1993.

ALLA PANVINI, Public Service Company of Colorado, Denver, Colorado.

The authors bring attention to an increasingly important issue of reactive power in power systems. The main point of the paper is that the Generator Reactive Capability Curves (GRC), supplied by the manufacturers, can not be used blindly but, rather, the plant house-power equipment limitations as well as the system limitations have to be considered. This is a valid point. The main portion of the paper deals with the machine design limits described by different arcs of the GRC curve. After superimposition of all of the system and equipment limitations on the GRC curve, the authors obtain the boundary of the actual operating area inside of the GRC curve.
A close match of the calculated and tested curves is evidence that the theoretical formulas are correct.

However, the authors' only valuable and practical recommendation for increasing the machine's reactive capability is to change a step-up transformer tap. While this recommendation is a good one, no rigorous field test is required in order to prove that the change is needed.

Public Service Company of Colorado took a more practical approach. Two load points on the generators' GRC curves were spot-checked. This field check revealed the need for numerous changes. Some of the problem areas were step-up and load center transformer taps, temperature alarm settings, calibration of the temperature recorders and electrical meters, maximum excitation limiters and volts per hertz limiters adjustments. The problems were corrected, and classes on reactive power fundamentals were presented to the plant and system operators.

Much more beneficial for the industry would be to have some specific recommendations of how to increase the reactive capability of the generators, as opposed to the methodology on how to calculate the existing limits.

Manuscript received March 1, 1993.

M. M. Adibi and D. P. Milanicz: The authors thank the discussers for their comments and questions, which allow clarification of some points in the paper. We would like to respond to discussers individually.

J. Mercurio & N. E. Nilsson: The authors agree that including the effects of armature resistance and saturation result in more accurate calculations, however the focus of our work was not to attempt to model the generator capability curves but rather identify and predict external system conditions which are typically more limiting than the machine characteristics. In addition, the real limit in this over-excited region is field temperature which is of course directly related to the I^2R losses in the generator field. We have consulted with the generator manufacturers who are typically willing to allow an increase in field current beyond its nameplate rating so long as the field temperature limit is not exceeded.

The discussers are correct in their assumptions related to Figure 2, that is, the thinner arcs are meant to represent loci of constant excitation i.e., arc AB and as such should have their center on the reactive axis located at the per unit value of short circuit ratio (SCR). As for the core end heating characteristic, it may or may not be a circle which may or may not be centered on the reactive axis. Since the graph shown in Figure 2 is not meant to represent any particular case (no values are assigned to the MW & MVAR axis),it was provided for discussion purposes only. Also the steady state stability limit as shown in the figure may or may not be more limiting than the actual machine capability and is shown only to reinforce the point that it must to be considered as a possible limiting factor.

Referring again to Figure 2, the arc BQP is meant to represent that portion of the capability curve limited by stator heating (I^2R losses in the stator) and any limitation imposed by the prime mover. The dashed lines extending upward at point B and downward at point QP are intended to show the continuation of the stator heating limitation just as the dashed line extending downward at point B shows the extension of the field limitation curve. Admittedly, a magnification of this portion of the curve would have improved the discussion. The whole point here is that in this region there are several factors possibly field heating, stator heating and prime mover constraints interacting in a relatively small portion of the capability curve. The selection of the 0.9 PF rating was completely arbitrary and is not meant to represent any particular generating unit on the Baltimore Gas & Electric system.

In order to simplify the discussion of system voltage limitations, please refer to Figures A, B, C and D. For the electric system under consideration, the scheduled system voltages were 230 kV and 237 kV (scheduled in this case means if the voltage falls outside this range, the system operator must take some action to bring it within limits). The task at hand was to attempt to describe the available reactive capability under any allowed system voltage. The approach then, was to consider the extremes of system voltage, determine the capability at these extremes including such parameters as step-up and service transformer taps, voltage limitations on auxiliary equipment, generator terminal voltage limits etc. For system voltages between the scheduled limits the capability could be determined by linear interpolation (not a particularly bad assumption for a small voltage range). A mathematical model was developed to accept the various input data and produce an output of predicted capabilities at the selected high and low voltage levels. Referring to Figures A & B, the outermost curve is the published generator reactive capability from the manufacturer. The inner curve is the predicted reactive capability from the spreadsheet labelled with the appropriate limiting factors. The discrete points on the graphs are actual test points with measured values. Some of the test points in Figure 2 lie outside the predicted locus since the machine low voltage limit was exceeded (of course, machine conditions were closely monitored for these points). The real value of the pretest predictions can now be appreciated. Before any testing is conducted a model of the machine-system can be used to determine which limits can be adjusted or altered (i.e. transformer taps, relay settings, voltage regulator setpoints etc.) to allow optimum reactive capability without the need to retest the unit which prevents additional costs. Figures 3 & 4 illustrate the optimum generator reactive capability with appropriate limiting factors after recommendations were implemented.

The issue of shorted field turns has typically not be a serious problem on any of the machines tested. Bearing vibration is constantly monitored by automatic systems (typically WDPF or similar). The utility involved conducts flux probe testing of generating units to help with the determination of the field condition. Several units have experienced either partial or total field rewinds. Upon occasion, the various generator manufacturers have agreed to modify the nameplate rating of field amps slightly (typically upward) to allow lagging MVAR capability to be reached so long as the field temperature limitations are not violated. Exciter output has never been a problem, presumably due to conservative application and large margins in the original design criterion.

Eddie S. Dehdashti: Based on the testing experiences of 23 generating units, the authors believe the limitations will apply to a large percentage of situations. The following table is a compilation of statistical data related to addressable limiting factors on tested units:

Occurrence of Addressable Limiting
Factors on the Tested Units
"General Limit Areas"

1.Metering Related	52%
2.Excitation System Related	18%
3.Transformer Related	13%
4.Machine Related	10%
5.Procedure Related	7%

The "addressable" refers to limiting conditions which can be possibly modified in some way so as to improve reactive capability:

1. Metering related limitations include; out of calibration, inadequate or improper range, non-existent, improper scales or nomenclature, poor intelligence sources, and defective or not operational.

2. Excitation system related limitations include; MEL or URAL failed to operate or was out of calibration, excitation system range (voltage) incorrect, incorrect limiter and or protective settings, and obsolete or inadequate equipment.

3. Transformer related areas include; station service transformers not on optimum tap setting, generator step-up transformers not on optimum tap, and transformers not correctly applied/sized.

4. Machine related areas include; problems with protective devices, problems with control systems, and artificially imposed heating limitations (stator & field).

5. Procedure related limitations include; operating procedures which do not exploit full reactive capability such as "it has never been done before.

Discussions with several other utilities which have conducted any form of reactive capability testing indicate that many of these problems have occurred on their systems as well.

Regarding the test generator in particular and the other tested units in general, the auxiliary equipment was not supplied by the same manufacturer as the generator.

The discusser is certainly correct in his assumption that the use or purchase of equipment with adequate range and or ratings will improve the reactive capability of the generating units. At the time of installation and based on requirements which existed, the auxiliary equipment and subsystems purchased were probably adequate or in some cases possibly the only alternative. For example, advances in excitation control and protection in the past thirty years allow the operation of machines in areas which were previously avoided. Operating conditions, voltage levels, the need for VAR capability to support power transfers, etc. have all changed dramatically during the period of time many of the operating units were commissioned. To put it simply, conditions have changed but in many cases equipment upgrades or retrofits have not. The purpose and in some cases the success of a reactive capability testing and evaluation program must be judged by these constraints.

Alla Panvini: The authors most definitely agree with the discusser on the point that the manufacturer's designed reactive capability curves are not a true picture of the available machine reactive capability. The whole point of our computer study, which includes input data such as transformer tap settings, maximum and minimum excitation limiter setting, protective device settings, etc., was to be armed with expectation of the test results, possibly correcting any deficiencies before embarking on the expensive (and time consuming) task of generator reactive capability testing. Referring to the previous discussion, certain problems were discovered during testing, a number of them unexpected. This points up the need to actually test the unit under consideration. However, by applying the computer program, we hope to uncover possible problems and solve them before testing if possible, thus augmenting a comprehensive testing program.

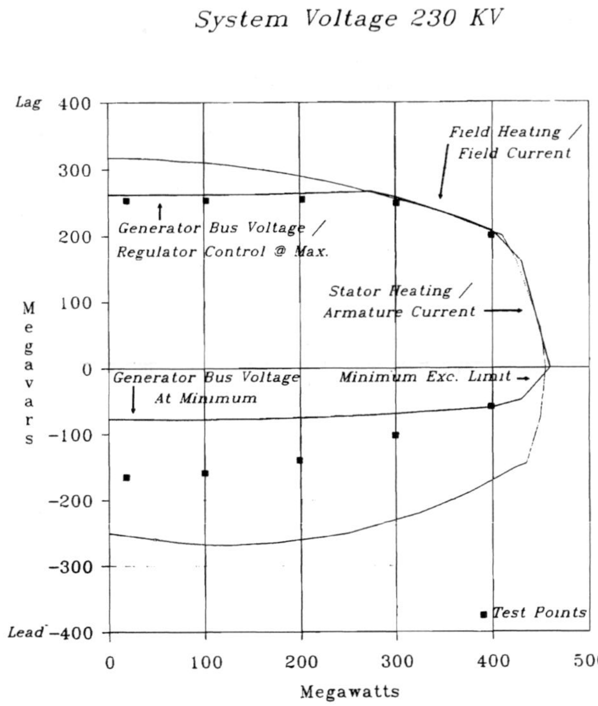

Figure B

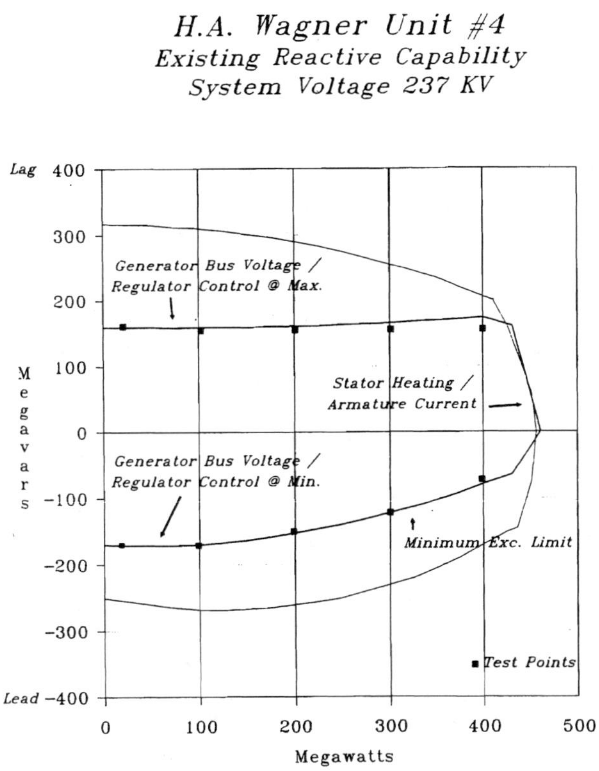

Figure A

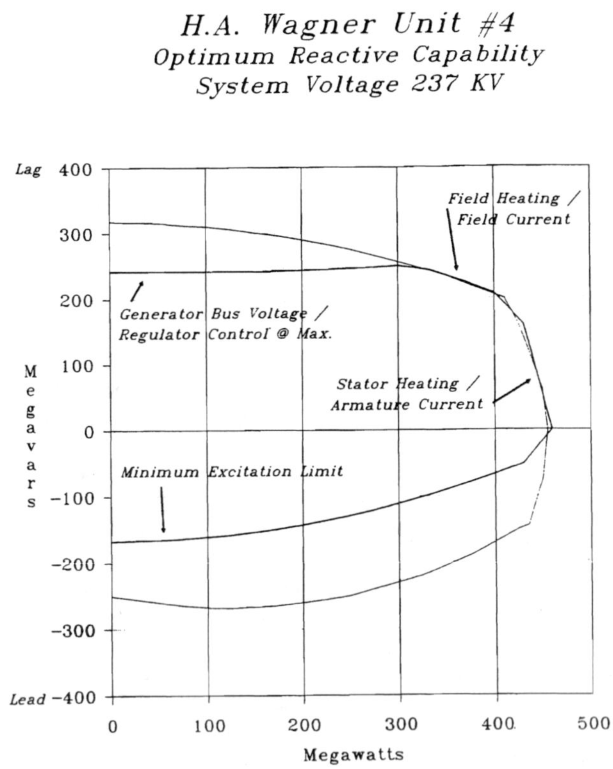

Figure C

System Voltage 230 KV

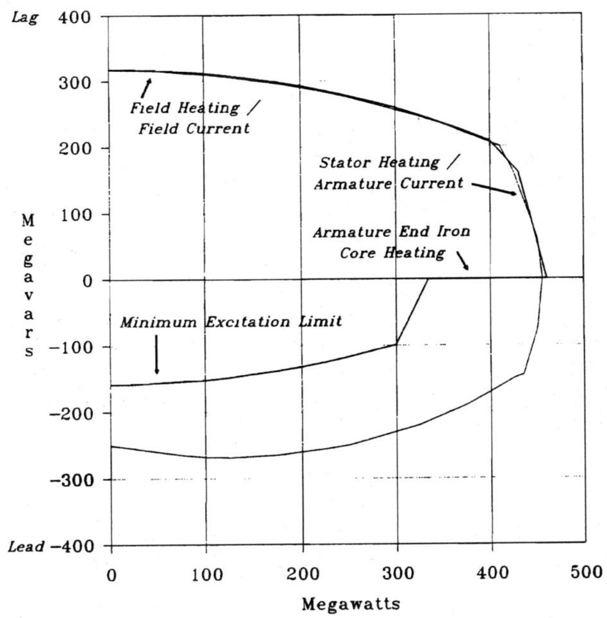

Figure D

Manuscript received March 29, 1993.

An Approach to Standing Phase Angle Reduction

A report by the Power System Restoration Working Group

S. Wunderlich
Drexel University
Philadelphia, PA

M. M. Adibi
IRD Corporation
Bethesda, MD

R. Fischl
Drexel University
Philadelphia, PA

C. O. D. Nwankpa
Drexel University
Philadelphia, PA

Abstract: This paper is part of a series presented on behalf of the System Restoration Subcommittee with the intent of focusing industry attention on power system restoration. The presence of excessive standing phase angle (SPA) differences across open circuit breakers causes significant delays in power system restoration. These angles may occur across a tie line between two systems or between two connected subsystems within a system. They must be within SPA limits before an attempt is made to close breakers to firm up the bulk power transmission system. There has been a need for an efficient methodology to serve as a guideline for reducing excessive SPA differences to allowable limits, without resorting to the raising and lowering of various generation levels on a trial and error basis. This paper describes such a methodology.

Keywords: Standing Phase Angle, Power System Restoration, Generation Schedule, System Disturbance.

1. Introduction

In re-integrating a power system following a major disturbance, power system operators sometimes encounter an excessive standing phase angle (SPA) difference across a line. Closing a power circuit breaker on a large SPA difference can shock the system, causing equipment damage and possible recurrence of the system outage [1].

In order to avoid inadvertent closing on large phase angle differences, breakers are equipped with synchrocheck relays which prevent their closure on angles greater than preset values. Under such a condition, a system operator may attempt to reduce the SPA difference by changing real power generation in as few power plants as possible by trial and error. This time consuming process prolongs the duration of the restoration as illustrated in the following excerpt [2]:

"The restoration of the system did not proceed as rapidly as it should have... . There was also a problem in closing the 230kV line which initially caused the disturbance due to a 70 degree phase angle between buses on each end of the line. This angle was reduced by altering line loading and increasing generation."

The significance of SPA can also be illustrated by a disturbance in another interconnection where the northern and southern parts of a 500kV system nearly separated. A 500kV breaker connecting the two parts was closed on a 40-45 degree SPA difference, shaking a 600 MW generator off its pedestals. There was extensive damage and loss of revenue due to the lengthy generator outage.

The range of SPA differences that a system can withstand depends on the voltage level, operating condition, and location of its circuit breakers. Synchrocheck relay settings as practiced by one northeast interconnection are 20 degrees for 500kV, 40 degrees for 230kV, and 60 degrees for 115kV systems.

This paper describes a method for the systematic reduction of an SPA difference via generation adjustment, with due attention focused upon transmission line thermal security. It is anticipated that the approach will provide the basis for an on-line algorithm designed to help automate the stage-wise restoration of a power system by suggesting a safe sequence of breaker reclosures, generation adjustments and, if necessary, load shedding. Note that the question of islanding, which involves frequency synchronization instead of SPA, is not addressed here.

In the past, a considerable number of studies have been conducted applying mathematical programming techniques to related problems, but none has addressed the issue of rescheduling real power generation in an on-line basis to reduce an excessive SPA difference. An optimal power flow solution was developed using real and reactive power and transformer taps as control variables, under constraints, in [3]. An optimum scheduling of reactive power production for minimization of line losses was described in [4]. The rescheduling of generation and shedding of load was formulated as a nonlinear optimization problem in [5]. Linearized models of power systems were used in [6] to relieve line overloads by active and reactive generation rescheduling and by load shedding. The concept of rotating line overloads for state control was developed in [7].

A scheme was developed in [8] for the control of an underground transmission network at Potomac Electric Power Company. The dual objectives were to: (a) relieve overloaded underground transmission lines within the time limits imposed by equipment short-time ratings, and (b) minimize the amount of load to be curtailed by rescheduling generation and regulating transformer phase-angles. The scheme offered a very fast and reliable algorithm in which accuracy and speed were traded depending on the severity of overloads.

The method presented in this paper first establishes the feasibility of SPA reduction via generator rescheduling, subject to the transmission network thermal limits, without the necessity of load reduction. Once feasibility has been established the alternative rescheduling strategies are then determined. The approach taken is based on the "region-

93 WM 205-5 PWRS A paper recommended and approved by the IEEE Power System Engineering Committee of the IEEE Power Engineering Society for presentation at the IEEE/PES 1993 Winter Meeting, Columbus, OH, January 31 - February 5, 1993. Manuscript submitted September 1, 1992; made available for printing November 30, 1992.

Reprinted from *IEEE Transactions on Power Systems,* Vol. 9, No. 1, pp. 470–478, February 1994.

wise" approach to security assessment, namely the determination of the set of all feasible and secure operating points [9]. This work concentrated on characterizing and identifying steady state security regions dictated by thermal limits, although some work was also done on incorporating voltage limits of ±5% at load buses.

2. Problem Definition

Given a power system with a desired (base case) topology, and the occurrence of one or more sustained outages on that system, an algorithm is to be developed which will recommend an efficient generation reschedule to provide for the safe reclosure of a particular breaker and restoration of the associated line.

Two key aspects of the problem thus stated are *safety* and *efficiency*. *Safety* is ensured by the simultaneous satisfaction of all pertinent constraints, as defined in the operating space of generator real power outputs during each stage of the restoration. Specifically, the constraints to be considered in the formulation of this problem are derived from thermal limits on transmission line flows, high and low limits on power plant outputs, and SPA limits of open breakers prior to their reclosure. (Although the first two of these types of constraints are concerned only with steady-state security, dynamic security can also be enhanced, with little additional complexity, by the inclusion of constraints derived from limitations on angle differences across energized lines.) Note that if all constraints can be satisfied, there will always be some subset of them, called the *binding constraints*, whose satisfaction guarantees the satisfaction of all other constraints in the operating space. The set of all operating points satisfying a set of binding constraints is termed a feasible region. The *efficiency* of a reschedule is related to the amount and number of generation adjustments performed to satisfy the constraints mentioned above. Since these are to be kept to a minimum, a set of preferred generators available for redispatch should be defined.

It is important to note that, in the overall conduct of any restoration scheme, the only time that an operating point should be thermally infeasible is immediately following the occurrence of the outage(s). The first corrective action taken should be to establish thermal feasibility, and all subsequent stages of the restoration effort should maintain that feasibility. It is apparent, then, that any operating point held just prior to the reclosure of an open breaker must satisfy the binding constraints of (i.e., must be contained in) the following three regions:

$S_{TH}(k)$ = The set of operating points satisfying all thermal binding constraints of the system with topology k, the pre-reclosure topology. (The k'th topology corresponds to the k'th *stage* of the restoration: k=0 implies base case topology, k=1 implies post-outage topology, k=2 implies the system topology after the first line is restored, etc.)

$S_{SPA}(k)$ = The set of operating points satisfying the SPA limitation of the breaker to be closed.

$S_{TH}(k+1)$ = The set of operating points satisfying all thermal binding constraints of the system with topology k+1, the post-reclosure system topology.

Consideration of the $S_{TH}(k+1)$ region is necessary because

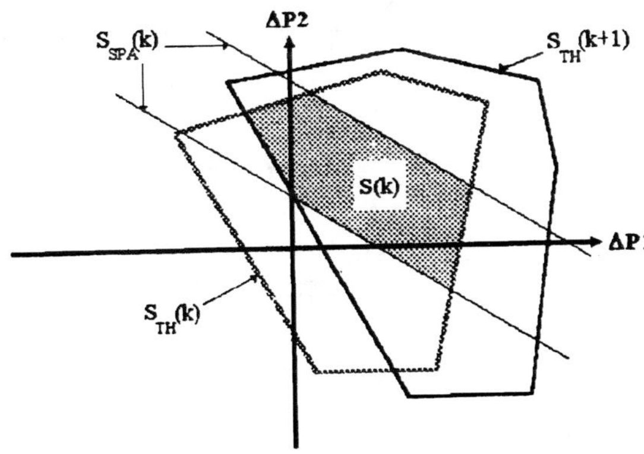

Figure 1: View of a Possible Intersection of Security Regions in ΔP_g-Space

the thermal constraints depend on system topology. It is possible for a line with an initially acceptable power flow to become overloaded after the closure of a breaker. Thus, successful reclosure of any breaker at any stage, k, of the restoration requires the existence of a region S(k) such that

$$S(k) = S_{TH}(k) \cap S_{SPA}(k) \cap S_{TH}(k+1) \neq \emptyset \qquad (1)$$

where $S(k)\neq\emptyset$ means that the region is not empty. An example of the regions involved in (1) is shown in Fig. 1. The first requirement of an efficient reclosure procedure is a fast method of determining if the condition of Eq. (1) is satisfied, i.e., an efficient region finding algorithm.

3. Methodology

The outline of the restoration problem as developed above suggests a two-step approach:

Step 1: Obtain the region S(k) of all feasible generator reschedules defined in Eq. (1) and determine if $S(k)\neq\emptyset$,

Step 2: Select the element in S that minimizes rescheduling effort.

3.1 Identification of Reclosure Feasible Region

The pertinent constraints that define the feasible region for reclosure (i.e., region S(k) of Eq. (1)) are those defining the thermal line and generator limits (denoted by $S_{TH}(k)$) and the SPA limits (denoted by $S_{SPA}(k)$). Mathematically, the sets $S_{TH}(k)$ and $S_{SPA}(k)$ are defined by:

$$S_{TH}(k) = \{\ \Delta \mathbf{P_g} = \mathbf{P_g} - \mathbf{P_g}^0:\ \mathbf{f}(\mathbf{x},\mathbf{P_g},\mathbf{u},\mathbf{Y}^k) = 0;\ |\mathbf{I}| \leq \mathbf{I}^M;\ \mathbf{P_g}^m \leq \mathbf{P_g} \leq \mathbf{P_g}^M\ \} \qquad (2)$$

$$S_{SPA}(k) = \{\ \Delta \mathbf{P_g} = \mathbf{P_g} - \mathbf{P_g}^0:\ \mathbf{f}(\mathbf{x},\mathbf{P_g},\mathbf{u},\mathbf{Y}^k) = 0;\ |\theta_{mn}| \leq \theta_{mn}^M\ \} \qquad (3)$$

where **f** denotes the power balance equations; **x** denotes the vector of state variables (i.e., bus voltages V_i and angles θ_i); $\mathbf{P_g}$ denotes the vector of generator real power injections and $\mathbf{P_g}^0$ denotes the preoutage generation schedule; **u** denotes the preoutage load profile $\mathbf{P_d}^0+j\mathbf{Q_d}^0$, where $\mathbf{P_d}^0$ and $\mathbf{Q_d}^0$ are vectors of real and reactive power demand; $\mathbf{Y}^k$ denotes the bus admittance matrix for the k'th topology; **I** denotes the vector of line current flows, whose limits are given by $\mathbf{I}^M$; $\mathbf{P_g}^m$ and $\mathbf{P_g}^M$ denote minimum and maximum real power generation limits, respectively, on $\mathbf{P_g}$; θ^M_{mn} denotes the maximum allowable limit for SPA θ_{mn} for a breaker on outaged line m to n.

In terms of the above, the region S(k) of Eq. (1) is defined in the space of generation reschedule $\Delta\mathbf{P_g} = \mathbf{P_g} - \mathbf{P_g}^0$. The following should be noted:

Remark 1: If the origin (i.e., $\Delta\mathbf{P_g} = 0$) lies outside S(k) as shown in Fig. 1, rescheduling is needed.

Remark 2: Under certain load conditions the set S(k) may not exist (i.e. S(k) = ∅). In this case, one needs to change the load **u** so as to guarantee the existence of S(k) as indicated in Eq. (1).

Because of the possibility of the above and the existence of many redundant or conflicting constraints, our approach is to first (Step 1) identify the binding constraints of S(k). The procedure for doing this employs a linear programming (LP) algorithm. This means that the constraints which define the regions $S_{TH}(k)$ and $S_{SPA}(k)$ of Eqs. (2) and (3) need to be linearized. Although there are many methods which can do this, the simplest one uses the DC load flow model to derive the appropriate constraints. Thus, based on the DC load flow model, and taking the thermal limits to be based on real power flows ($|\mathbf{P_l}| \le \mathbf{R_l}$, where $\mathbf{P_l}$ is the vector of line flows and $\mathbf{R_l}$ is the vector of line ratings) the regions $S_{TH}(k)$ and $S_{SPA}(k)$ are approximated by [9]

$$\widehat{S}_{TH}(k) = \left\{ \begin{array}{l} \Delta\mathbf{P_g}: \ \mathbf{R^m_l(u)} \le \mathbf{H^k_{TH,g}(k)}\Delta\mathbf{P_g} \le \mathbf{R^M_l(u)}; \\ \mathbf{P^m_g} \le \mathbf{P_g} \le \mathbf{P^M_g}; \ P_D - P^M_{slack} \le \sum_{i=2}^{m} P_{gi} \le P_D - P^m_{slack} \end{array} \right\} \quad (4)$$

$$\widehat{S}_{SPA}(k) = \left\{ \Delta\mathbf{P_g}: \ \mathbf{R^m_{SPA}(u)} \le \mathbf{H^k_{SPA,g}(k)}\Delta\mathbf{P_g} \le \mathbf{R^M_{SPA}(u)} \right\} \quad (5)$$

where

$\mathbf{R^m_l(u)} = -\mathbf{R_l} - \mathbf{H^k_{TH}P^0_{net}}$

$\mathbf{R^M_l(u)} = \mathbf{R_l} - \mathbf{H^k_{TH}P^0_{net}}$

$\mathbf{H^k_{TH}}$, $\mathbf{H^k_{TH,g}}$ are the transfer matrices relating real power line flows, P_{ij}, to net real power bus injection, $\mathbf{P^0_{net}}$, and to the generation reschedule $\Delta\mathbf{P_g}$, respectively, under the k'th topology.

$\mathbf{H^k_{SPA}}$, $\mathbf{H^k_{SPA,g}}$ are the transfer matrices relating bus pair angles, θ_{ij}, to net real power bus injection and to the generation reschedule $\Delta\mathbf{P_g}$, respectively, under the k'th topology.

$\mathbf{R^m_{SPA}(u)} = -\theta^M - \mathbf{H^k_{SPA}(k)P^0_{net}}$

$\mathbf{R^M_{SPA}(u)} = \theta^M - \mathbf{H^k_{SPA}(k)P^0_{net}}$

P_D represents the total system real power demand.

$\mathbf{P^M_{slack}}$, $\mathbf{P^m_{slack}}$ are the maximum and minimum limits, respectively, on slack bus real power injection.

m = # of generators available for adjustment in the system (the slack bus is indexed first)

Note that the notation $\mathbf{R(u)}$ emphasizes the fact that these limits depend on the load profile of the system. The constraints dictated in Eqs. (4) and (5) are combined to form the set defining S(k), a linearized approximation of S(k) in Eq. (1), as follows:

$$\widehat{S}(k) = \left\{ \Delta\mathbf{P_g}: \ [\mathbf{A}]\Delta\mathbf{P_g} \le \mathbf{b} \right\} \quad (6)$$

where

$$[\mathbf{A}]^T = \left[[\mathbf{H^k_{TH,g}}]^T [-\mathbf{H^k_{TH,g}}]^T [\mathbf{H^k_{SPA,g}}]^T [-\mathbf{H^k_{SPA,g}}]^T [\mathbf{H^{k+1}_{TH,g}}]^T [-\mathbf{H^{k+1}_{TH,g}}]^T [\mathbf{I^{m-1}}] [-\mathbf{I^{m-1}}] [1] [-1] \right]$$

$$\mathbf{b}^T = \left[[\mathbf{R^{M,k}_l(u)}]^T [\mathbf{R^{m,k}_l(u)}]^T [\mathbf{R^M_{SPA}(u)}]^T [\mathbf{R^m_{SPA}(u)}]^T [\mathbf{R^{M,k+1}_l(u)}]^T [\mathbf{R^{m,k+1}_l(u)}]^T [\mathbf{P^M_g}]^T [\mathbf{P^m_g}]^T [P_D - P^m_{slack}] [P^M_{slack} - P_D] \right]$$

$\mathbf{I}^{m-1}$ = (m-1) identity matrix

[1] = (m-1) x 1 matrix whose elements are all "1"

[-1] = (m-1) x 1 matrix whose elements are all "-1"

Clearly from Eq. (6), the set $\widehat{S}(k)$ consists of many redundant constraints. Thus, step 1 of our procedure is to identify only the relevant constraints. The technique we use is the one developed in [9]. Essentially, the routine is a linear programming (LP) algorithm followed by a specialized pivoting process aimed at enumerating all binding constraints of the region in question. The algorithm is fast and thus applicable for on-line application. The LP problem solved is

Problem 1 (Identification of the Binding Constraints):
Maximize y subject to the constraints

$$[\mathbf{A} \ \ \mathbf{t}] \begin{bmatrix} \mathbf{x} \\ y \end{bmatrix} \le \mathbf{b} \quad (7)$$
$$\mathbf{x} \ge 0; \ y \ge 0$$

where $\mathbf{x} = \Delta\mathbf{P_g}$; and

$$t_i = \|\mathbf{a}_i\| = \sqrt{\sum_{j=1}^{m-1} a_{ij}^2} \qquad i = 1, \ldots, \text{\# rows of } \mathbf{A} \quad (8)$$

The solution to this problem yields a tableau which is processed via reverse pivoting to identify the set of binding constraints (see Appendix A for more details). This means that the many constraints defining S(k) in (6) are reduced to the minimal set. To emphasize this we use the subscript b to denote that S(k) is only represented by the binding constraints, that is

$$\widehat{S}_b(k) = \left\{ \Delta\mathbf{P_g}: \ [\mathbf{A_b}]\Delta\mathbf{P_g} \le \mathbf{b_b} \right\} \quad (9)$$

where $[\mathbf{A_b}]$ and $\mathbf{b_b}$ are subsets of $[\mathbf{A}]$ and $\mathbf{b}$ representing the binding constraints. Table I illustrates the $[\mathbf{A_b}]$ and $\mathbf{b_b}$ for the 39 bus system (described in Section 4) which (without maximum generation limits) dictates 207 constraints. Note that only 30 constraints are binding.

Another important aspect of the binding constraint identification algorithm is that it allows one to determine if all the constraints are conflicting so that S(k) = ∅. In that case we know that no generation reschedule will alleviate limit violations and the only strategies left are those of changing the load profile $\mathbf{P_d}$ or performing line switching so that condition (1) (i.e. S(k) ≠ ∅) is satisfied [10]. The load shedding can be accomplished by setting up an LP problem in which the cost of load change $\Delta\mathbf{P_d} = \mathbf{P_d} - \mathbf{P_d}^0$ is minimized subject to the constraint of (6) when $\Delta\mathbf{P_g} = 0$. The line switching can be accomplished using line distribution factors and the associated costs of switching lines.

3.2 Generator Rescheduling for Safe Reclosure

Rescheduling for SPA reduction is required prior to reclosure of a breaker whenever the origin is not contained in the S(k)-region associated with that breaker. The rescheduling problem seeks a strategy $\Delta\mathbf{P_g}$ such that

$$\Delta\mathbf{P_g} \in S(k) \quad (10)$$

There are many approaches to find $\Delta\mathbf{P_g}$. They depend on the type of generator control available, i.e., either manual control or Automatic Generation Control (AGC). This section shows that in either case, rescheduling is greatly facilitated by using the solution of Problem 1.

Manual control implies that each generator is controlled one at a time. Thus, manual rescheduling seeks the generator or a set out of the available generators which gives $\Delta \mathbf{P_g}$ that guarantees (9). Note from Fig. 2(b) that if only generator #30 is available for rescheduling then we can never satisfy (9). However, when generator #31 is available, then (9) can be satisfied.

Rescheduling via AGC implies finding a minimum cost strategy which guarantees (10). When the set S(k) is approximated by S_b of (9), then one possible solution is via the following LP problem:

Problem 2(a) (Rescheduling): Find the $\Delta \mathbf{P_g} \in \hat{S}_b$ such that

$$C_1(\Delta \mathbf{P_g}) = \sum_{j=2}^{m} c_j \Delta P_{gj} = \text{minimum}$$

where c_j is the cost of rescheduling unit j.

Note that the manual rescheduling problem is the special case of Problem 2(a) in which $\Delta P_{gk} = 0$ for all k≠j, where the j'th unit is selected for manual scheduling.

Our approach to solving the AGC-based rescheduling is to exploit the information obtained from solving Problem 1. This approach gives the solution to the following simplified rescheduling problem:

Problem 2(b) (Rescheduling): Find the $\Delta \mathbf{P_g} \in \hat{S}_b$ such that

$$C_2(\Delta \mathbf{P_g}) = \sum_{j=2}^{m} c_j (\Delta P_{gj})^2 = \text{minimum}$$

where the cost of rescheduling $c_j = 1$ for all j.

The solution to this problem can be obtained using information provided in the solution of Problem 1. As shown in Appendix A, the same algorithm that identifies the constraints in (9) also identifies the vertices of $S_b(k)$. This information is useful because the optimal reschedule point is either the closest vertex of the $S_b(k)$-polytope, or it is on one of the hyperplanes of the violated constraints defining that vertex, or it is on an 'edge' formed by an intersection of two or more of the constraints, at least one being a violated constraint. Thus, a simple check of the squared distance $\|\Delta \mathbf{P_g}\|^2 = \Sigma(P_{gj})^2$ to these hyperplanes, their intersections, and the feasibility of the closest points on them is all the additional work is required for rescheduling after solving Problem 1. This fact follows directly from the convexity of S(k).

4. Test Results

In order to test the above two-step SPA reduction algorithm, the EPRI 39 bus system shown in Fig. 3 is used. The contingency chosen for restoration operation is a simultaneous outage of lines 2-3, 5-6 and 16-17. Tables B.I. and B.III. show the system loads, generation, and voltages before and after the outage as obtained from an AC load flow program.

4.1 Solution of Problem 1

For Step 1 of the algorithm, it is initially assumed that all ten power plants are available for adjustment, corresponding to a scenario in which AGC is in operation. It is desired to reclose the breaker restoring line 16-17. The control space for this example exists in nine dimensions; one for each non-slack generator. In the formulation of the constraint set for the outage case, there are a total of 207 constraints: 205 for $S_{TH}(k)$ of (4) and 2 for $S_{SPA}(k)$ of (5). The 205 inequalities for Eq. (4) include (2 topologies) x (2 x (49 lines)) = 196 line thermal constraints and 9 $P_{gj} \geq P_{gj}^{m} = 0$ generator constraints (no maximum generation limits or slack generator limits were considered). Only 30 of the 207 constraints were found to be binding. These are given in Table I. Figures 2(b) and 2(c) illustrate two different two-dimensional 'cuts' of the nine-dimensional polytope formed by the binding constraints. Notice, from Fig. 2(b), that the system is found to be insecure immediately following the occurrence of the triple line outage, since the base case operating point violates thermal constraints of lines 6-7 and 4-14. This is corroborated by the AC load flow data given in Table B.II. Also notice that an SPA constraint of the open breaker on line 16-17 is violated, indicating that the SPA across this breaker of interest is greater than the prescribed limit of 10 degrees. These violations indicate that rescheduling is required.

TABLE I. BINDING CONSTRAINTS ($A_b x \leq b_b$)

LINE	G30	G31	G32	G33	G34	G35	G36	G37	G38	b
Thermal										
1-27	-0.25	-0.23	-0.25	-0.28	-0.28	-0.28	-0.28	-0.30	-0 51	5.66
1-39	-0.49	-0.17	-0.18	-0.20	-0.20	-0.20	-0.20	-0.48	-0.45	6.28
2-25	0.34	-0.01	-0.08	-0.09	-0.09	-0.09	-0.09	-0.59	-0.32	6.53
2-30	1.00	0.00	0.00	0.00	0.00	0.00	0.00	0.00	0.00	4.00
6-7	0.03	0.72	0.56	0.37	0.37	0.37	0.37	0.04	0.05	-1.20
6-31	0.00	1.00	0.00	0.00	0.00	0.00	0.00	0.00	0.00	1.05
8-9	-0.10	-0.69	-0.66	-0.63	-0.63	-0.63	-0.63	-0.11	-0.17	5.82
10-32	0.00	0.00	1.00	0.00	0.00	0.00	0.00	0.00	0.00	0.00
13-14	-0.03	0.28	0.44	-0.37	-0.37	-0.37	-0.37	-0.04	-0.05	2.34
15-16	0.00	0.00	0.00	1.00	1.00	1.00	1.00	0.00	0.00	0.93
19-20	0.00	0.00	0.00	0.00	-1.00	0.00	0.00	0.00	0.00	4.75
19-33	0.00	0.00	0.00	1.00	0.00	0.00	0.00	0.00	0.00	0.18
20-34	0.00	0.00	0.00	0.00	1.00	0.00	0.00	0.00	0.00	1.45
21-22	0.00	0.00	0.00	0.00	0.00	0.65	.5244	0.00	0.00	0.09
22-35	0.00	0.00	0.00	0.00	0.00	1.00	0.00	0.00	0.00	0.50
23-36	0.00	0.00	0.00	0.00	0.00	0.00	1.00	0.00	0.00	0.90
25-26	0.34	-0.08	-0.08	-0.09	-0.09	-0.09	-0.09	0.41	-0.32	3.79
25-37	0.00	0.00	0.00	0.00	0.00	0.00	0.00	1.00	0.00	1.10
26-27	0.34	-0.08	-0.08	-0.09	-0.09	-0.09	-0.09	0.41	0.68	2.08
25-37	0.00	0.00	0.00	0.00	0.00	0.00	0.00	0.00	1.00	0.50
SPA										
16-17	-0.00	0.02	0.02	0.05	0.06	0.06	0.06	-0.01	-0.01	-0.03
16-17	0.00	-.02	-0.02	-0.05	-0.06	-0.06	-0.06	0.01	0.01	0.38

Note: The nonnegativity constraints of generators 30, 31, 32, 33, 35, 36, 37, and 38 were also binding.

4.2 Solution of Problem 2(b)

The rescheduling algorithm of Step 2 starts with another result from the region-finding routine: the closest vertex of the polytope, at a distance of 3.290 p.u. from the base case operating point, is formed by an intersection of thermal constraints of lines 6-7, 10-32, 19-33, 21-22, 22-35, 25-37, and 29-38, an SPA constraint of bus pair 16-17, and the nonnegativity constraint of generator 30. As expected, the

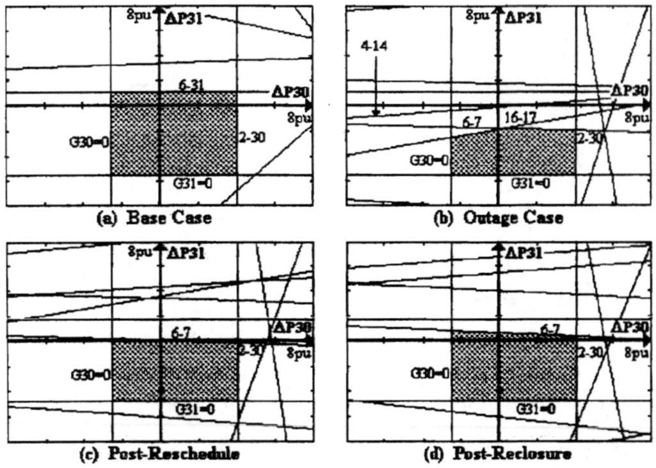

Figure 2: A step-wise view of the outage of lines 2-3, 5-6, 16-17, and the restoration of line 16-17 using all generators for adjustment. Each frame displays a two-dimensional cut of the current security region in the subspace of Gen. 31 vs. 30, centered at the operating point. The shaded regions represent: (a) $S(0) = S_{TH}(0)$; (b) $S(1) = S_{TH}(1) \cap S_{SPA}(1) \cap S_{TH}(2)$; (c) $S(1) = S_{TH}(1) \cap S_{SPA}(1) \cap S_{TH}(2)$; (d) $S(2) = S_{TH}(2)$

TABLE II: THE OPTIMAL RESCHEDULE

GEN BUS #	ΔP (p.u.)	P_{net} (p.u.)	GEN BUS #	ΔP (p.u.)	P_{net} (p.u.)
30	-0.028	2.473	35	-0.320	6.680
31	-0.633	2.817	36	-0.320	5.280
32	-0.488	6.012	37	-0.031	5.369
33	-0.320	6.000	38	-0.046	7.954
34	-0.320	4.732	39*	2.506	4.640

* Gen. 39, the slack bus, is not independently controlled

two violated binding constraints are in this set. The point of optimal reschedule must exist either solely on a hyperplane associated with one of the violated constraints, or on the intersection of one or more of these hyperplanes with those of the satisfied constraints of this set. The vector from the current operating point to the reschedule point must be orthogonal to this hyperplane or intersection of hyperplanes, as the case may be. It was, in fact, found that the point given in Table II., existing solely on a hyperplane associated with line 6-7, satisfies all 30 binding constraints of S(0). This point is the optimal reschedule using nine independently controlled generators. As shown in Fig. 2(c), the SPA constraints of the breaker on line 16-17 are no longer violated, so that safe reclosure is assured. Fig. 2(d) illustrates the thermal security region after reclosure. Notice that the rescheduled operating point remains secure.

Consider next a situation in which the generators must be controlled manually. In this case, it is imperative that as few generators as possible be used to achieve SPA reduction and thermal security. Displays such as the one shown in Fig. 2(b) now become very useful because they show which generators may be individually adjusted to move the operating point inside the security region. For example, it is apparent that trading generation supply between generator 30 and the slack bus can never accomplish the task, whereas decreasing the output of generator 31 by 1.95 p.u. will place the operating point inside of S(0). Of course, the slack generator must then increase its output by 1.95 p.u. to satisfy demand (Eq. (8)).

5. Conclusion

This paper gives a new and efficient method for reducing large standing phase angles (SPA) via generation adjustment subject to the transmission network thermal limits. The method is a two stage procedure in which the first stage establishes the feasibility of SPA reduction via generator rescheduling without the necessity of load reduction and the second stage obtains the optimum rescheduling strategy. Since in each of the above two steps a linear programming algorithm is used to solve the problem, the technique described can be implemented on-line as part of an Energy Management System (EMS). Other features of the method which need mention are:

- If the algorithm determines that there is no feasible generation which can reduce the SPA, then the algorithm is able to determine the amount of load shedding that is necessary to establish the feasibility of generator rescheduling.
- Since in many cases the exact bus load profile is uncertain, the algorithm is able to handle these uncertainties (see [10]).

A number of points needing further investigation include:

- Both the feasibility region finding algorithm and the rescheduling algorithm are based on linear assumptions. The accuracy of the results are then checked using an AC load flow program. We are currently investigating using more accurate models.
- Further work is being done in developing load-shedding strategy in the absence of a feasible generator rescheduling region, so as to minimize the amount of load to be shed.
- Thus far, only SPA, transmission line thermal and generator MW output limits have been considered in the formulation of the problem. Although our procedure can handle other operating limits, such as voltage and VAR limits and voltage stability limits, they have not yet been included.

6. Appendix A

This appendix outlines the region finding algorithm which finds $S_b(k)$ of (9). (For the theoretical basis, see [14].) The first part of the algorithm is a standard LP algorithm which solves Problem 1. The solution to this LP problem yields the "most secure" point; i.e. the point inside the security region that is the maximum possible Euclidean distance away from the closest face of the polytope. More importantly, it yields a tableau, called the *optimal tableau*, which serves as a reliable starting point for a reverse pivoting process. This process then systematically identifies each vertex and creates a special list from which the binding constraints may later be found.

Let us consider Problem 1. The variable y represents the smallest Euclidean distance of the point **x** from the all of the constraints. The solution of Problem 1 produces an optimal tableau giving the optimal values $\{\mathbf{x}^*, \mathbf{y}^*\}$, where $\mathbf{x}^*$ is the "most secure" point inside S(k) and $\mathbf{y}^*$ is the Euclidian distance from $\mathbf{x}^*$ to the closest constraint.

Step two of the region-finding algorithm involves a Simplex pivoting process which progressively reduces the value of y until it can be reduced no further without becoming negative. An ordered list is maintained of those slack variables which are nonbasic in each new tableau obtained in this "backwards Simplex" routine. By keeping this list ordered according to the decreasing values of y for each tableau, duplicate entries are avoided and a path is constructed to guide the process along. Each new tableau is examined for three different types of pivots:

(a) Pivots removing a slack variable s_i from the basis, thereby creating a new set of nonbasic variables to be considered for entry as a new record on the list.

(b) Pivots removing y from the basis, thereby collapsing the higher order polytope onto the original security region. This identifies a new vertex of the security region.

(c) Pivots removing a control variable x_i from the basis. This is similar to (b) above with the added knowledge that $x_i \geq 0$ is binding.

The whole process may be heuristically viewed in three dimensions as first finding the peak of a 'mountain' whose height (y) is a function of horizontal distance from its base boundaries (step one), and then systematically locating spurs with lower and lower elevation (step two). Vertices of the base (i.e., the $S_b(k)$-region) are found at every spur having a slope (i.e., a pivot) going directly down to the base of the mountain.

7. Appendix B

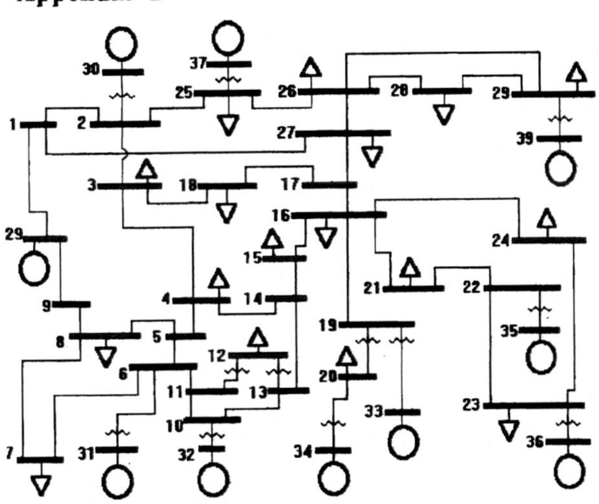

Figure 3: One Line Diagram of 39 Bus System

TABLE B.I. GENERATOR BUS DATA

BUS #	GENERATION			VOLTAGE			
	MW	MVAR		BASE CASE		OUTAGE	
		BASE CASE	OUT-AGE	PU	DEG	PU	DEG
30	250.0	125.1	65.5	1.048	34.9	1.048	38.2
31	545.0	161.4	187.4	0.974	39.8	0.974	42.0
32	650.0	194.0	238.0	0.982	41.6	0.982	43.0
33	632.0	100.0	126.0	0.996	42.0	0.996	46.8
34	505.2	165.9	178.5	1.012	40.5	1.012	45.2
35	700.0	221.2	251.7	1.051	44.8	1.051	49.6
36	560.0	92.7	110.0	1.062	47.0	1.062	51.9
37	540.0	0	5.92	1.031	40.5	1.031	42.9
38	800.0	0	18.2	1.025	44.8	1.025	44.2
39(B)	255.5	-30.0	27.5	1.054	30.0	1.054	30.0
39(O)	264.4						

* Note: (B) = Base Case; (O) = Outage Case

TABLE B.II. LINE IMPEDANCES AND FLOWS

LINE	R	X	RAT-ING	OUTAGE FLOW				FLOW AFTER RE-CLOSURE	
				PRE-RESCHDLE		POST-RESCHDLE			
	(PU)	(PU)	(PU)	P (PU)	Q (PU)	P (PU)	Q (PU)	P (PU)	Q (PU)
1-2	0.004	0.041	6.5	-2.82	-0.27	-2.45	-0.26	-2.58	-0.25
1-27	0.003	0.032	6.5	1.04	0.38	1.71	-0.43	1.32	-0.45
1-39	0.001	0.007	6.5	-0.41	-0.35	-0.90	0.03	-0.64	0.06
1-39	0.002	0.016	6.5	-0.18	-0.18	-0.41	-0.01	-0.29	0.00
1-39	0.001	0.019	6.5	-0.16	-0.77	-0.36	-0.63	-0 26	-0.61
2-3	0.001	0.015	6.5	out	out	out	out	out	out
2-25	0.007	0.009	6.5	-0.24	0.68	0.11	0.82	-0.04	0.75
2-30	0.000	0.018	6.5	-2.50	-0.53	-2.47	-0.58	-2.47	-0.52
3-4	0.001	0.021	6.5	-2.16	0.89	-1.47	0.45	-0.55	0.62
3-18	0.001	0.013	6.5	-0.99	-0.91	-1.65	-0.47	-2.61	-0.64
4-5	0.001	0.013	6.5	-0.88	-0.31	-1.31	-0.33	-1.37	-0.20
4-14	0.001	0.013	6.5	-6.22	-0.54	-5.10	-0.90	-4.13	-0.82
5-6	0.000	0.003	6.5	out	out	out	out	out	out
5-8	0.001	0.011	6.5	-0.84	-0.19	-1.29	-0.22	-1.33	-0.09
6-7	0.001	0.009	6.5	7.52	1.35	6.30	1.42	5.95	1.37
6-11	0.001	0.008	6.5	-2.16	-0.23	-1.56	-0.28	-1.20	-0.30
6-31	0.000	0.025	6.5	-5.43	-1.10	-4.79	-1.13	-4.80	-1.05
7-8	0.000	0.005	6.5	5.15	0.09	3.94	0.31	3.59	0.30
8-9	0.002	0.036	6.5	-0.87	-1.79	-2.53	-1.55	-2.95	-1.41
9-39	0.002	0.018	6.5	-0.50	-0.70	-0.93	-0.69	-1.04	-0.67
9-39	0.002	0.018	6.5	-0.50	-0.70	-0.93	-0.69	-1.04	-0.67
9-39	0.002	0.020	6.5	-0.47	-0.69	-0.87	-0.67	-0.98	-0.66
9-39	0.002	0.020	6.5	-0.47	-0.69	-0.87	-0.67	-0.98	-0.66
10-11	0.000	0.004	6.5	2.29	0.56	1.73	0.57	1.40	0.58
10-13	0.000	0.004	6.5	4.17	0.99	4.25	0.96	4.58	0.87
10-32	0.000	0.020	6.5	-6.47	-1.52	-5.98	-1.51	-5.99	-1.43
11-12	0.002	0.044	6.5	0.14	0.47	0.17	0.47	0.20	0.47
12-13	0.002	0.044	6.5	0.04	-0.41	0.08	-0.41	0.11	-0.41
13-14	0.001	0.010	6.5	4.19	0.56	4.30	0.53	4.67	0.42
14-15	0.002	0.022	6.5	-2.12	-0.38	-0.88	-0.61	0.49	-0.55
15-16	0.001	0.009	6.5	-5.32	-1.65	-4.07	-1.79	-2.69	-1.71
16-17	0.001	0.009	6.5	out	out	out	out	1.39	0.38
16-19	0.002	0.020	6.5	-4.47	-0.92	-3.84	-0.93	-3.83	-1.03
16-21	0.001	0.014	6.5	-3.60	-0.06	-3.22	-0.08	-3.22	-0.15
16-24	0.000	0.006	6.5	-0.59	-1.10	-0.33	-1.11	-0.33	-1.16
17-18	0.001	0.008	6.5	2.56	0.94	3.22	0.56	4.21	0.83
17-27	0.001	0.017	6.5	-2.56	-0.94	-3.22	-0.56	-2.83	-0.32
19-20	0.001	0.014	6.5	1.77	-0.23	2.09	-0.20	2.09	-0.23
19-33	0.001	0.014	6.5	-6.29	-0.73	-5.96	-0.67	-5.96	-0 74
20-34	0.001	0.018	6.5	-5.03	-1.31	-4.71	-1.30	-4.71	-1.33
21-22	0.001	0.014	6.5	-6.34	-1.11	-5.97	-1.09	-5.96	-1.16
22-23	0.001	0.010	6.5	0.61	0.48	0.66	0.46	0.66	0.46
22-35	0.000	0.014	7.5	-6.99	-1.87	-6.66	-1.76	-6.66	-1.84
23-24	0.002	0.035	6.5	3.70	0.18	3.43	0.12	3.43	0.17
23-36	0.001	0.027	6.5	-5.58	-0.35	-5.25	-0.32	-5.25	-0.36
25-26	0.003	0.032	6.5	2.79	-0.10	3.09	0.20	2.96	0.02
25-37	0.001	0.023	6.5	-5.38	0.54	-5.34	0.37	-5.34	0.48
26-27	0.001	0.015	6.5	4.38	0.70	4.59	1.20	4.49	0.91
26-28	0.004	0.047	6.5	-1.26	-0.28	-1.23	-0.42	-1.23	-0.34
26-29	0.006	0.063	6.5	-1.75	-0.32	-1.72	-0.46	-1.72	-0.38
28-29	0.001	0.015	6.5	-3.32	0.23	-3.29	0.08	-3.30	0.17
29-38	0.001	0.016	8.5	-7.95	0.71	-7.89	0.42	-7.90	0.59

TABLE B.III. LOAD BUS DATA

BUS #	LOAD		VOLTAGE			
			BASE CASE		OUTAGE	
	MW	MVAR	PU	DEG	PU	DEG
1	252.7	118.6	1.054	29.8	1.053	29.8
2	0	0	1.053	32.5	1.064	35.8
3	322.0	2.0	1.034	29.8	0.993	24.0
4	500.0	184.0	1.006	29.3	0.976	26.8
5	0	0	1.008	30.6	0.980	27.5
6	0	0	1.009	31.3	0.999	33.4
7	234.0	84.0	1.001	29.2	0.984	29.4
8	522.0	177.0	1.000	28.7	0.982	28.0
9	104.0	125.0	1.046	29.7	1.044	29.6
10	0	0	1.018	33.6	1.006	34.9
11	0	0	1.014	32.8	1.002	34.4
12	9.0	88.0	1.055	32.8	1.041	34.1
13	0	0	1.016	32.9	1.000	33.9
14	0	0	1.014	31.2	0.990	31.5
15	320.0	153.0	1.018	30.8	1.000	34.2
16	329.0	32.0	1.034	32.2	1.020	36.9
17	0	0	1.038	30.9	1.015	25.8
18	158.0	30.0	1.035	30.1	1.005	24.7
19	0	0	1.050	36.7	1.045	41.5
20	680.0	103.0	0.991	35.3	0.988	40.1
21	274.0	115.0	1.033	34.8	1.023	39.6
22	0	0	1.051	39.5	1.045	44.3
23	248.0	85.0	1.045	39.2	1.040	44.0
24	309.0	-92.0	1.039	32.3	1.027	37.1
25	224.0	47.0	1.061	33.7	1.059	36.2
26	139.0	17.0	1.059	32.2	1.048	31.5
27	281.0	76.0	1.047	30.4	1.032	28.2
28	206.0	28.0	1.053	35.4	1.048	34.7
29	284.0	27.0	1.052	38.0	1.048	37.3

9. Acknowledgments

This work was supported in part under National Science Foundation Small Business Initiation Research Grant # ISI-9160779.

8. References

[1] Adibi, M.M. and R.J. Kafka, "Power System Restoration Issues," IEEE Computer Applications in Power, Vol. 4, No. 2, Apr. 1991, pp. 19-24.

[2] "Review of Selected Major Electric Bulk Power System Disturbance in North America,", North American Electric Reliability Council Reports 1979-1983.

[3] Dommel, H.W. and W.F. Tinney, "Optimum Power Flow Solution," IEEE Trans., Power Apparatus and Systems, Vol. PAS-87, No. 10, Oct. 1969, pp. 1866-76.

[4] Peschon, J., et al., "Optimum Control of Reactive Power Flow," IEEE Trans., Power Apparatus and Systems, Vol. PAS-87, No. 1, Jan. 1968, pp. 40-48

[5] Chan, S.M. and F.C. Schweppe, "A Generation Re-Allocation and Load Shedding Algorithm," IEEE Trans., Power Apparatus and Systems, Vol. PAS-98, No. 1, Jan/Feb. 1979, pp. 26-34.

[6] Medicherla, T.K.P, et al., "Generation Rescheduling and Load Shedding to Alleviate Line Overloads," IEEE Trans., Power Apparatus and Systems, Vol. PAS-98, No. 6, Nov/Dec. 1979, pp. 1876-84.

[7] Krogh, B., et al., "Multi-Stage Rescheduling of Generation, Load Shedding and Short-Term Transmission Capacity for Emergency State Control," IEEE Trans., Power Apparatus and Systems, Vol. PAS-102, No. 5, May, 1983,pp. 1466-72.

[8] Adibi, M.M. and D.K. Thorne, "Local Load Shedding," IEEE Trans. PWRS, Vol. 3, No. 3, Aug. 1988, pp. 1220-1229.

[9] Fischl, R. and J.A. DeMaio, "Fast Identification of the Steady-State Security Regions for Power System Security Enhancement," Paper No. A76-076-0, IEEE PES Winter Meeting, Jan., 1976.

[10] Fischl, R., G.C. Ejebe, and J.A. DeMaio, "Identification of Power System Steady-State Security Regions Under Load Uncertainty," Paper No. A76-495-2, IEEE Summer Power Meeting, July 1976.

[11] Taylor, D.G. and L.J. Maahs, "A Reactive Contingency Analysis Algorithm Using MW and MVAR Distribution Factors," 1989 IEEE PES Summer Meeting, Long Beach, CA, July, 1989.

[12] Galiana, F.D., "Bound Estimates on the Severity of Line Outages in Power System Contingency Analysis and Ranking," IEEE Trans. on PAS, Vol. PAS-103, pp. 2612-2624, Sep. 1984.

[13] Schulz, R.P., et al., "Long-Term System Dynamics," Vol. I, Technical Report, EPRI Research Project 90-7-0, Apr. 1974.

[14] Mattheiss, T.H., "An Algorithm for Determining Irrelevant Constraints and All Vertices in Systems of Linear Inequalities," Operations Research, Vol. 21, pp. 247-260, 1973.

DISCUSSION

R. J. KAFKA, Potomac Electric Power Company, Washington, DC; The authors are commended for recognizing a power system restoration problem for which the solution is not trivial. Large standing phase angles may prevent the closing of a breaker needed to strengthen the transmission system or close a transmission loop.

The authors have proposed an interesting solution; in particular, the identification of a feasible region is valuable, for it indicates that a solution exists without further topology changes or load shedding. There are some practical considerations for the generator rescheduling solution, however. During system restoration, automatic control of generation may not be practically realizable, and a solution which identifies a minimum number of generators (perhaps two or three) to be rescheduled is preferred.

Additional constraints should also be placed on the slack bus. While inconsequential in many system studies, the slack bus usually represents a real generator with physical limits in system restoration studies. In the paper's example, most of the burden for the generation reschedule has been placed on the slack bus, Bus 39, since it picks up all the generation reduced by the other generators. It is unfortunate that the solution in the example was to reduce the output of all constrained generators and increase the output of the unconstrained generator which happened to be at the edge of the system.

We recognize that the authors have used a standard system for their example, and also recognize that the graphical representation of Fig. 2 could be useful in identifying the most effective generators to move. Nevertheless, it is likely that system operators familiar with the impedances of their system would have already known this. The example system is not representative of a system during restoration. In fact, it is tightly linked, with multiple parallel lines and many generation busses. No doubt the authors have noticed this, since multiple contingencies were required to produce even a relatively small standing angle.

We would appreciate the authors' comments regarding these issues. The authors' approach is a useful advance which may already be of value in restoration planning. We look forward to additional work in the areas identified by the authors and add a request for solutions using a sequence of network topology changes.

Manuscript received February 16, 1993.

D. P. Milanicz, Baltimore Gas and Electric Company, Baltimore, MD; standing phase angle may occur during restoration when the power system is under manual control or it may occur following a line outage when most of generators can be rescheduled, automatically. Under the former case, the objectives are to determine the minimum number of generators with minimum adjustments to reduce the SPA. The authors have attempted to address the latter case by reducing all the nine generators and picking up the difference of 250.6 Mw from the swing generators (G39). Would the authors please comment on:

(a) How can the former case which is of greater interest in restoration be addressed? In practice the generator pair with maximum phase angle difference is selected, i.e. G39 and G36 of Table B.I having 21.9° angular difference.

(b) Can the latter case be addressed without reducing all the generators and importing the difference (i.e. by G39)?

Manuscript received February 19, 1993.

S. Wunderlich, M. M. Adibi, R. Fischl, and **C. O. D. Nwankpa:** We appreciate both discussions and thank Messrs. Kafka and Milanicz for their interest and insightful comments regarding our paper. Our responses to the issues of concern are categorized below.

1. Slack Bus Generation Problem (mentioned by R. J. Kafka and D. P. Milanicz)

   In the formulation of equation (4), we have included constraints on slack bus generation. These constraints may be tightened by choosing P^M_{slack} and P^m_{slack} to be arbitrarily close to the pre-reclosure operating point. In the limit where $P^M_{slack} = P^m_{slack} = P_{slack}$, we restrict operation to a hyperplane described by $\Sigma_i \Delta P_{gi} = 0$. This means that generation is traded among non-slack generators, and poses no difficulty from a computational standpoint. Since in our paper we were concerned with developing an example to demonstrate the principles of our technique, we did not include tight slack bus constraints (c.f. section 4.1). The use of these constraints would have resulted in much smaller thermal security regions, whereas we wanted large regions for illustrative purposes.

2. Solution for a System Under Manual Control (mentioned by R. J. Kafka and D. P. Milanicz)

   We agree that in most practical situations, AGC is unlikely to be available. Therefore, minimizing the number of generators to be adjusted is of critical importance. We have been investigating the following algorithm as a possible solution. The objective of the algorithm is to (i) minimize k = the number of generators to be adjusted; (ii) select the specific generators to be adjusted; and (iii) determine the amount of adjustment to be made by each generator. In the algorithm given below, we first (Steps 1 and 2) determine which generators are the most likely to move the operating point within the security region. We then (Steps 3 to 7) determine the minimum k and the solution to objective (iii). Specifically:

Step 1: Solve Problem 1 and Problem 2(b) as though AGC were available, to obtain an optimum m-vector of generator real power adjustments. The length of this vector is the minimum possible Euclidean distance to the security region from the current operating point.

Step 2: Rank the generators in descending order of magnitude of real power adjustment in the vector obtained in Step 1. (In our example, this order would be 39, 31, 32, 35, 33, 36, 34, 38, 37, 30 from Table II, where ties were settled by choosing the generator with largest output first, and where we recognize that generator 39 heads the list only because the slack generator was unrestricted.)

Step 3: Set k=2, where k is the number of generators to be considered for rescheduling.

Step 4: Solve Problem 2(b) again, considering only the first k generators from the list obtained in Step 2. (Note that this implies c_j=0 in 2(b) for all generators not under consideation.)

Step 5: If a solution exists in Step 4, record it in a list of solutions.

Step 6: Set k=k+1.

Step 7: If k = m then terminate; otherwise return to Step 4.

This algorithm gives a short list of possible generation reschedules, from which the operator could choose the best based on practical considerations.

We believe the procedure outlined above has several advantages over the current practice, mentioned by Mr. Milanicz, of picking the generator pair with the maximum phase angle difference. The latter approach does not guarantee that the pair of generators chosen has the ability to move the operating point into the security region, and for that matter, the generator selection process does not recognize the possibility that a reclosure security region may not exist. This is why, in our approach, we first determine if the security region exists and then solve the rescheduling problem. In the minimum rescheduling strategy described above, the selection of the generators to be adjusted is based on the knowledge of the region (i.e., the constraints of the network) and is therefore more likely to identify an effective minimum set of generation adjustments.

3. Example System is Tightly Linked (mentioned by R. J. Kafka)

Mr. Kafka is correct in ascertaining our difficulty in producing a significant standing phase angle in our tightly linked, lightly loaded system. The likelihood of encountering large SPA's is much greater in weakly linked systems with heavy loading. In fact, the reclosure security region is less likely to exist in these types of systems and, therefore, our procedure of first establishing the existence of the region becomes even more crucial. In the case of nonexistence of the security region, the only strategy available to the operator is load shedding and/or network switching. An important approach used by the industry is to relax the constraints, i.e., trade-offs between expediting corrective actions and the loss of equipment life. Loss of equipment life is accumulative over 30 to 40 years of equipment life and its formal consideration is also non-trivial. We are currently investigating the solution to this problem, which was not addressed in our paper.

Manuscript received April 13, 1993.

NUCLEAR PLANT REQUIREMENTS DURING POWER SYSTEM RESTORATION

A report by the Power System Restoration Working Group
M. M. Adibi, Chairman; Authors: Gerry Adamski, Ronaldo Jenkins and Paul Gill

Abstract: This paper is one of a series presented on behalf of the System Operation Subcommittee with the intent of focusing industry attention on power system restoration issues. This paper discusses a number of nuclear power plant requirements that require special attention during power system restoration.

POWER SYSTEM REQUIREMENTS

Gerry Adamski, Baltimore G&E Co.

The primary objective of utility companies following an electric system blackout is to restore 100% of customer load in the shortest period of time. For any particular area, this effort essentially becomes a balancing act between customer demand and increasing energy levels as more generating units return to operation and tie line capability is restored. For each utility owning and operating a nuclear facility, the highest priority of any system restoration plan should be the restoration of off-site power sources to the nuclear plant(s). This will ensure that the reactors can be maintained in a safe condition in addition to providing the optimal opportunity to restart the units.

There are obstacles that must be overcome to achieve this objective, in both the generation and the transmission areas. In order to develop a restoration strategy that is technically acceptable and realistic, it becomes imperative that real-time generation status be incorporated into the restoration strategy. Communication needs to be established between the bulk power operations control center and the plant operators at the generating stations as soon as possible following a blackout to gather relevant restoration data.

This effort is particularly important with regard to nuclear plants since they are governed by additional federal regulatory and other license obligations. Stringent guidelines outlined in the plant technical specifications detail the time frames for shutdown and startup and the conditions that must exist for each to proceed. Due to these requirements, the return of nuclear units to the electric grid can take significantly more time than for other types of generating units. This statement is reflected in NERC's Electric System Restoration document dated December 1, 1992:

"Nuclear units require special treatment. Nuclear Regulatory Commission startup checklists generally do not permit hot restarts and the emergency generators do not permit supplying auxiliary power to other stations. Nuclear units that are taken off-line on a controlled shutdown can be restored to service in about 24 hours; more likely 48 hours after a scram. Therefore, while restoring off-site power sources to nuclear units require attention, restoring power to service area load will normally need to be completed without the help of nuclear units."

95 WM 144-6 PWRS A paper recommended and approved by the IEEE Power System Engineering Committee of the IEEE Power Engineering Society for presentation at the 1995 IEEE/PES Winter Meeting, January 29, to February 2, 1995, New York, NY. Manuscript submitted August 12, 1994; made available for printing January 11, 1995.

Therefore, in areas comprised of significant nuclear capacity, full customer restoration may not be achievable for several days following a system blackout.

Initial Stages

Immediately following a system blackout, the first order of business is to ensure that all generating facilities have sufficient energy to safely shutdown. This power will be utilized to operate rotor turning gear, lube oil pumps, hydrogen seal oil pumps, air heaters, etc. At some locations, units may have separated from the system and reverted to supply their own plant station service load in an islandized condition. At other sites, the auxiliary power requirement is satisfied with blackstart combustion turbines, or conventional or pumped hydro units; others have emergency generators; still others have a d.c. source to supply this equipment. In the most extreme cases, some generating facilities have to await the return of off-site power sources for supply to the emergency shutdown loads.

Nuclear plants are equipped with emergency on-site power sources, usually diesel generators, to supply emergency core cooling systems to ensure that the reactors and generators can be safely shutdown following the loss of all off-site power sources. Generally, the nuclear plant can remain in a shutdown mode for a lengthy period of time in this state. Technical specifications direct the plant to proceed to a shutdown mode if off-site power sources are not restored within a specified period of time. Therefore, the mode to which the unit will regress following a blackout is a function of the length of time it takes to restore the off-site sources. Ultimately, this restoration dictates the time it will take for the nuclear unit to heat-up and re-parallel.

Once an assessment is made of the nature and breadth of the system disturbance, the restoration phase begins as each utility implements its system restoration plan. In this phase, the electric system operator begins by determining the status of each generating unit on the system. The system operator should notify the operators at each plant of the extent of the system disturbance and the present system status. Additionally, an indication should be given as to the time frame in which transmission circuits can be energized into each plant.

The plant operator should respond by providing the current status of the plant. Further, the plant operator should detail the cranking power requirements for the unit(s) to return, the time frame to restart and any emergency minimum load level that must be established for stability.

The system operator's primary emphasis at this point is to restart the generating units with enough load to

Reprinted from *IEEE Transactions on Power Systems,* Vol. 10, No. 3, pp. 1486–1491, August 1995.

maintain stability. Units that were on-line at the time of disturbance are targeted first since generally, less time is required to synchronize a unit with a hot boiler than those in an intermediate or cold shutdown mode.

Electric System Requirements

The emphasis of the initial conversation between an electric system operator and the nuclear plant control room operator following a blackout should be to determine that the nuclear plant has successfully shutdown and is in a safe condition. The system operator then should determine the period of time that the nuclear plant can remain in the safe shutdown mode without the establishment of off-site sources. For instance, if the emergency shutdown load is being supplied from diesel generators, the system operator should determine what quantity of on-site reserve fuel is available and for what period of time it will supply the emergency loads. This information will then be used in the overall system restoration strategy as an absolute time frame for which an off-site source needs to be established.

Recall that the highest priority of the system operator is to restore off-site power sources to the nuclear plant. This will guarantee that the reactors can remain in a safe condition. However, if there is any consideration given to utilizing nuclear generation to assist in customer load restoration, rapid off-site restoration is essential. This point becomes the next focus of the discussion between the system and nuclear plant operators.

Typically, the general modes of operation of a nuclear unit are: on-line, hot standby, hot shutdown and cold shutdown. The optimal mode for restart is hot standby. The nuclear control room operator should explicitly detail the conditions that must exist for the unit to remain in a hot standby mode. This description should include the required number of off-site and on-site sources, the length of time the sources need to be established, and the duration that hot standby can be maintained assuming all other conditions are satisfied. Similar discussions should follow for hot and cold shutdown modes.

The nuclear plant operator should then identify the times to restart and synchronize the nuclear units if beginning from each of the three modes. Given that the system operator has a clear understanding of this restart information, an estimate of the time to energize the off-site sources to the nuclear plant should be discussed with the nuclear plant operator. This estimate should establish the mode in which the nuclear plant should be when off-site sources are restored and the additional time needed to restart and synchronize the unit.

This information will then be incorporated into the restoration strategy where it will be used to develop a generation schedule for the course of the restoration. For instance, combustion turbine generation may be brought off-line as nuclear units return in order to prepare for their use during peak periods. Nuclear units return may also allow the restoration of water levels for hydro units. These factors must be identified and included in the restoration strategy.

Throughout the system restoration process, periodic status updates should be shared between the generator and system operations control rooms. No further detailed discussions are needed until prior to energization of the off-site transmission sources to the nuclear plant. When circuit energization is imminent, the system operator should contact the nuclear plant operator and determine the amount of plant load that the circuit will be required to provide during the startup of the unit(s).

The operators should then jointly decide on a procedure to energize the circuit. In this discussion, an understanding needs to be reached on whether to energize the circuit open-ended first or energize directly into the nuclear plant station service buses. The expected voltage rise and other switching transient expectations will guide these decisions.

Finally, energization takes place and plant restart activities should begin once the normal electrical configuration is restored on the plant station service buses and all regulatory hurdles are cleared. Throughout the restart process, nuclear plant operators should periodically convey the synchronization schedule to system operators as it becomes more updated. As parallel draws near, the system operator should be provided with the following generator specific information from the plant operator: the units' emergency minimum load requirement; the unit ramp rate following parallel; the unit overspeed and underfrequency trip settings; the expected leading reactive capability of the unit; the status of the unit voltage regulator; and the expected time to achieve full output. The system operator, in turn, should indicate a specific target bus voltage the unit should maintain or the appropriate reactive output level.

Once the unit is paralleled, periodic communications need to be maintained in order to keep the system operator abreast of unit output changes so that customer load or other generation resources can be shifted to accommodate the change in nuclear unit output.

Transmission Energization

System restoration activities typically follow the "bottoms-up" approach. Generation islands are created as units are brought on-line with enough isolated customer load to maintain stability. Then, the islandized sub-systems are integrated together. Sufficient customer load is restored to maintain adequate load and frequency control. With a solid system established, customer load is restored according to need and priority. Concurrent with or following these activities, interconnection among the utilities begin followed by the restoration of the EHV transmission system.

The "bottoms-up" approach to system restoration provides a strong framework to restore the EHV transmission system. The restoration of the EHV system is extremely important to full customer restoration since it is the viaduct for energy supplied from remote and usually high output generating facilities and from other control areas. Nuclear units fall into this category. Nuclear units are typically high output units, over 600 MWs, that are usually remotely located for environmental or other reasons. Therefore, nuclear units cannot be provided with off-site power sources until the EHV transmission system is restored.

There are a variety of transmission stability factors that must be considered before EHV transmission energization can proceed. Generally, satisfaction of these minimum source requirements for energizing transmission equipment cannot be achieved during the early stages of restoration. These requirements address concerns with:

- Steady state overvoltage caused by excessive var supply from the capacitive effect of EHV lines;
- Transient overvoltage caused by traveling wave phenomena;
- Dynamic overvoltage caused by transformer magnetizing inrush;
- Reduction in proper relaying protection reliability due to insufficient fault current.

It is for these reasons that the unit voltage regulators must be operational and available for automatic operation during the restoration.

Generally, the stronger (i.e., lower source impedance) and more numerous the lines to a particular bus, the less severe the energizing transients become. However, in order to mitigate the capacitive effects of several lightly loaded transmission lines, significant customer load must be locally available to counterbalance the excess reactive. Concurrently, generation must be also be available to supply the customer load and absorb additional reactive from the system.

Summarizing, due to the high customer load and generation capacity requirements, the restoration of the EHV transmission system can take several hours, which, in turn, inhibits a more timely return of nuclear units following a blackout.

Nuclear Unit Integration:

The timing of the generating units' return during restoration typically follows the sequence: hydro units, combustion turbines, fossil-fueled units and nuclear units. A significant portion of hydro, combustion turbine and fossil-fueled units are typically restored within the first 24 hours following a blackout.

The restoration of larger units is not desirable early into a system restoration for two reasons. Larger units typically have higher minimum output levels. Given this fact, it is imprudent to return these units to operation until the electric system can withstand the initial impact of their parallel ad associated load pickup. This shock could cause a frequency deviation resulting in the operation of underfrequency relays or a re-collapse of the fledgling system. Furthermore, given the high output capability, these larger units are typically connected to the higher voltage network and, as outlined earlier, face transmission energization constraints.

For nuclear units, the delay in transmission energization has a detrimental impact on their ability to return to service. For systems comprised of large nuclear capacity, the excessive return time could result in not being able to fully restore all customer load at peak for several days.

In conclusion, an example is provided for a fictitious control area.

Peak Customer Demand:	45,000 MWs
Average Day Demand:	35,000 MWs
Installed Capacity (@ 16% res.):	52,000 MWs
Nuclear Capacity @ 25%	13,000 MWs
Import Capability	3,000 MWs
Generation Outages excl. nuclear:	10,000 MWs
Installed Capacity	52,000 MWs
- Nuclear Unavailability	13,000 MWs
- Other Outages	10,000 MWs
Internal Availability	29,000 MWs
+ Import Capability	3,000 MWs
Total Energy Available	32,000 MWs
- Total Demand	35,000 MWs
Unserved Load	-3,000 MWs

At peak, 3,000 MWs of load would be required to be shed until the return of the nuclear units. Therefore, it is crucial that every effort be made to expeditiously return nuclear units to full operation following a blackout; and that effort begins at the outset with clearly established and effective lines of communication.

NUCLEAR POWER PLANT REQUIREMENTS

R. V. Jenkins and P. S. Gill, NRC

<u>DISCLAIMER:</u> This paper does not represent the official NRC position on the subjects discussed in this document.

Introduction

The Nuclear Regulatory Commission has long recognized the importance of nuclear power plant offsite power supplied by the utility grid and onsite power distribution system to overall nuclear safety. In order to provide for the effective integration of nuclear power plant restoration into an utility's power restoration activities after a system wide outage, several regulatory concerns must be addressed for planning and implementation purposes.

The question of relying on effective restart of nuclear power plants in restoration planning requires that utilities examine several issues both in terms of safety of the nuclear unit, (i.e. Station Blackout, Shutdown Risk Management) and how the plan will interface with regulatory concerns. This paper reviews the regulatory requirements for the offsite and onsite electric power system and discusses a number of problems unique to the nuclear utilities.

Regulatory Requirements for the Offsite and Onsite Power Supply

Nuclear power plant onsite and offsite electrical power system requirements are in Title 10 to Code of Federal Regulations, Appendix A to Part 50 - "General Design Criteria (GDC) for Nuclear Power Plants," Criterion 17. Industry standards and regulatory guides refer to the offsite power system as the "Preferred Power System (PPS)." GDC 17 presents the minimum requirements for the PPS and onsite electric power system in terms of specific system and circuit configurations and functional requirements. GDC 17 requires that:

- An offsite power system and onsite power system shall be provided, each independent of the other and capable of providing power for all safety functions.
- The offsite system shall be comprised of two or more physically independent circuits connecting the transmission network to the onsite safety buses.
- An "immediate access circuit" shall be available within a few seconds to provide power to safety-related equipment following a Loss Of Coolant Accident (LOCA).

- A "delayed access" circuit shall be designed to be available in sufficient time to effect safe shutdown in event of loss of all onsite power and the loss of the other offsite circuit.

- Provisions shall be included to minimize the probability of losing electric power from any of the remaining supplies as a result of, or coincident with, loss of power generated by the nuclear power unit or the transmission network, or the onsite electrical supplies.

- The complete onsite electric power system (Class 1E) must be capable of sustaining a single failure without loss of capability to provide power for the minimum required safety functions.

Although the PPS does not have to meet Class 1E requirements (i.e., redundancy, separation, and single failure criterion), it must have a sound design. IEEE Standard 765-1983, Preferred Power Supply for Nuclear Power Generating Stations, discusses the design criteria for the PPS. This design basis should include consideration for the full range of the transmission and plant distribution steady state and transient conditions.

The Class 1E power system must be capable of providing power under the conditions of the design basis accident. The IEEE Standard 308 provides the criteria for Class 1E power systems for many nuclear power plants. Figures 1(a), (b), and (c) (excerpts from Draft 5 of IEEE Standard 765) illustrate acceptable PPS designs, as well as the division of scope between IEEE Standard 308 and IEEE Standard 765.

Station Blackout Requirements

The Nuclear Regulatory Commission amended its regulations per 10CFR 50.63, "Loss of All Alternating Current Power," to require that each light-water-cooled (i.e. commercial) nuclear power plant be able to withstand and recover from a station blackout (i.e., loss of the offsite electric power system concurrent with reactor trip and unavailability of the onsite emergency ac electric power system) of a specified duration.

The concern about station blackout arose because of the accumulated experience regarding the reliability of ac power supplies. Many operating nuclear plants have experienced a total loss of offsite power (LOOP), and more occurrences are expected in the future. In almost every one of these LOOP events, the onsite emergency ac power supplies were available immediately to supply the power needed by vital safety equipment. However, in some instances, even redundant emergency ac power supplies were unavailable.

The implications of station blackout for a nuclear power plant during a system wide outage may result in a need for additional utility resources to support the shutdown of the plant. Therefore, utility personnel responsible for power system restoration planning should consider contingency plans to provide support as needed during station blackout situations.

Shutdown and Low Power Operating Considerations

Similar to the Station Blackout issue, the NRC staff has become increasingly concerned about the safety of operations during shutdown. Loss of decay heat removal (DHR) during shutdown and refueling has been a continuing problem. More recently, a staff evaluation of the loss of power event at the Vogtle Nuclear Power Plant has emphasized the need for risk management of shutdown operations.

During most of the incidents involving shutdown risk, the availability of electrical power sources to maintain vital equipment has been an issue. It is not expected that a system wide outage would affect plant safety but complications may arise during equipment outages periods. Therefore, as in the case of Station Blackout, utility power system restoration plans should incorporate the plant's shutdown risk philosophy and its contingency plans within the scope of potential activities.

Special Operating Considerations for Nuclear Power Plants

One area of concern during power system restoration which should be considered before realigning the nuclear power plant to the grid, is that of assessing the adequacy of station electric distribution system voltages. Events at operating plants show that adverse effects on the Class 1E equipment can be caused by sustained low and/or high grid voltage conditions when Class 1E buses are connected to the offsite power system. Degraded voltage conditions arising on the Class 1E buses, even with normal grid voltages, can occur due to equipment deficiencies or via starting transients experienced during certain accident events not originally considered in the sizing of these circuits. During the early stages of restoring high voltage overhead and underground transmission lines to service, Class 1E equipment may be subjected to the effects of overvoltage conditions resulting from sustained power frequency overvoltages, switching transients, and harmonic resonances.

One regulatory requirement for nuclear power plants is voltage regulation limits for the offsite power supply which protects vital equipment necessary for safe operation from deleterious undervoltage or overvoltage conditions. The NRC typically requires that the utility analyze the anticipated range of the voltage variations in the offsite power supply to ensure that voltage levels at the safety-related buses remain within the acceptable limits. Therefore, power system restoration plans should consider both the timing of nuclear power plant realignment to the grid and what actions must be taken by the system load dispatcher in order to ensure that overvoltage and/or undervoltage effects will not impact plant equipment operability.

Similar to overvoltage and undervoltage effects on nuclear power plants, the system load dispatcher must also consider to a lesser extent potential underfrequency situations due to load and generation balance problems. Overall, the power system restoration plan should consider any unique constraints which may be part of the operating license for a nuclear unit or temporary licensing commitments limiting the plant electrical power system.

Coordination of Regulatory Oversight During Restoration

Until recently, there was little need for coordination of system restoration activities with regulatory entities. The offsite power system is outside the regulatory scope of the NRC unless it impacts nuclear power plant operations and safety. Since most of the offsite power supply disturbances in the past have resulted in outages lasting only hours, the possibility of an extended system collapse has been considered extremely remote. However, recent trends and events have suggested that there is an increase in the likelihood of a major power outage. The recent trends which could lead to greater power grid vulnerability are (1) a decrease in power generation addition, (2) the retirement and disuse of older, smaller generating capacity and greater reliance on large generating units, and (3) the greater likelihood of unit outages because of the tendency to operate older fossil or less efficient generating units on a load cycle, although these units were originally designed for base load operation.

One recent event which pointed to increased grid vulnerability involved several Northeastern nuclear power plants. As a result of an NRC Electrical Distribution System Functional Inspection (EDSFI), a Northeastern utility performed an engineering analysis which reexamined an earlier Bulk Power analysis using the current 345 kV transmission conditions.

The previous Bulk Power analysis which was performed to support plant licensing efforts determined that the transmission network met all of the basic stability criteria of the Northeast Power Coordination Council (NPCC). These criteria included the assumption that simultaneous ground faults occur on different phases of two transmission lines supported from the same tower. At the utility's station, the multiple nuclear generating units are connected to the transmission network by four shared offsite power lines. Therefore, this NPCC ground fault criterion is especially relevant because the subject lines are arranged in pairs and are routed on separate chain of towers.

The recent reanalysis found that transmission network instability could occur given the postulated NPCC ground fault, full power operation from the units onsite and a transmission network lightly load in the station vicinity coupled with a heavy flow of power in the transmission network from east to west. This design deviation occurred because the utility did not consider the earlier Bulk Power analysis as a living document that requires periodic review and update. Although no significant additions to the 345 kV transmission system had occurred gradual changes in the power system led to the unacceptable condition. These changes included the connections to the transmission network of a large nuclear power unit, a hydroelectric power unit, and several small independently-owned power plants (i.e., co-generation units). Another consequence of the deviation was the possibility that the subject instability could have resulted in a loss of offsite power event at a nearby nuclear power generating unit.

Planning and Implementation of Nuclear Power Plant Restarts

The planning and implementation of nuclear power plant restarts to support a recovery from a system wide outage should begin with the same issues and concerns which would be addressed by any other generating unit. The utility power system restoration plan must also consider the unique regulatory and performance requirements associated with the operation of a nuclear power plant. Any restart of a nuclear power plant following a unit trip, regardless of the reason, necessitates that utility management conduct an event investigation and analysis.

Most existing plant procedures are based either on guidelines generated by the nuclear steam system supplier or on plant specific analysis. As a result of the station blackout rule implementation plan, the nuclear industry has developed station blackout response and ac power restoration procedures and published these in NUMARC 87-00 1988, "Guidelines and Technical Bases for Addressing Initiatives Addressing Station Blackout at Light Water Reactors." Some of these ac power restoration guidlines for operators and load dispatchers personnel for restoring ac power are the following:

- Load dispatchers should give the highest possible priority to restoring power to the nuclear units.
- Should incoming transmission lines to a nuclear unit plant be damaged, high priority should be assigned to repair and restore at least one line.
- Once ac power becomes available, procedures should specify the circuit breaker operations required to restore ac power.

Conclusion

Overall, power system restoration planning which involves nuclear power plant restarts should manage the regulatory requirements for effective implementation by operation and load dispatcher personnel. In order to reduce potential delays in restart activities, utility management should take proactive measures to improve the reliability of the offsite power supply to the nuclear power units.

Although EHV stability problems are not common, because of: (a) national trends in reduced growth of generating capability, (b) the economics of nuclear units versus fossil units leading to retirements of the latter, and (c) cycling of older fossil generating unit designed for base load operation, it is likely that unit outages due to stability may reoccur. This increased potential for a system wide outage leads to the need for coordinating power system restoration plans with the NRC to avoid communication or implementation problems.

REFERENCES

[1] U.S. Code of Federal Regulations, Title 10, "Energy," Part 50.

[2] IEEE Std. 308-1980, IEEE Standard Criteria for Class 1E Electric Systems for Nuclear Power Generating Stations.

[3] IEEE Std. 765-1983, Preferred Power Supply for Nuclear Power Generating Stations.

[4] NUMARC 87-00, REV 1, "Guidelines and Technical Bases for NUMARC Initiatives Addressing Station Blackout at Light Water Reactors," August 1991.

[5] USNRC NUREG 1410, "Loss of Vital AC Power and the Residual Heat Removal System During Mid Loop Operations at Vogtle Unit 1 on March 20, 1990," June 1990.

Figure 1(a)

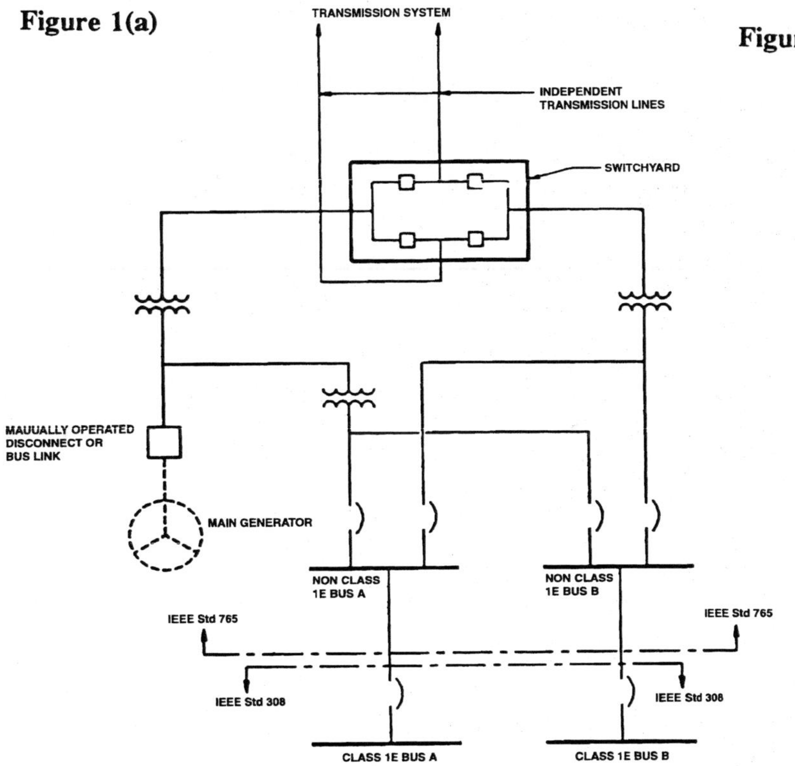

Figure 1(b)

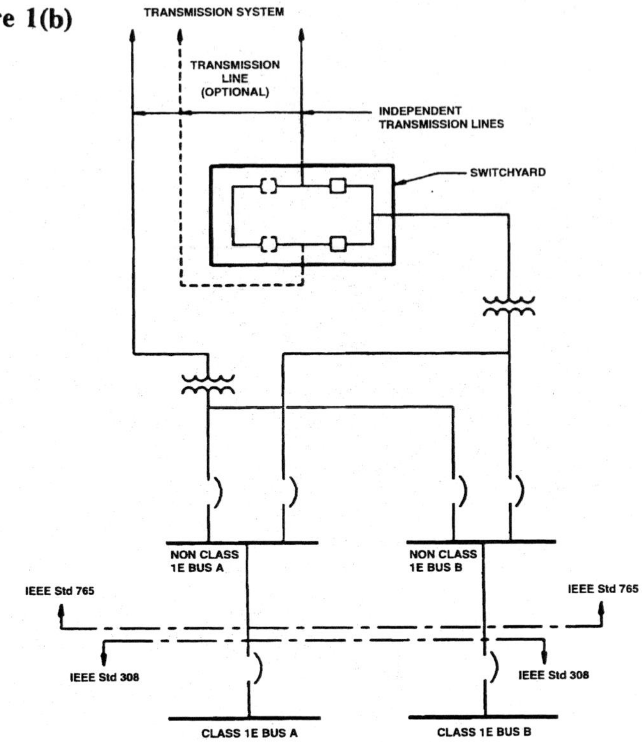

Figure 1(c)

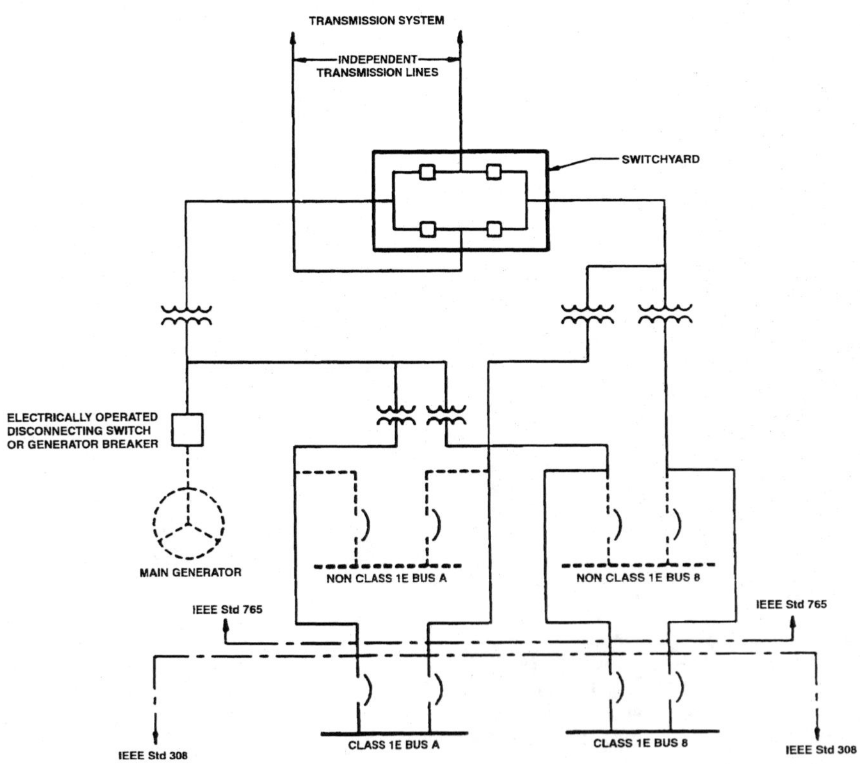

PROTECTIVE SYSTEM ISSUES DURING RESTORATION

M. M. Adibi

IRD Corporation

Bethesda, Maryland

D. P. Milanicz

Baltimore Gas & Electric Co

Baltimore, Maryland

Abstract: This paper is one of a series presented on behalf of the System Operation Subcommittee with the intent of focusing industry attention on power system restoration issues. Performance of protective systems may be measured by the relative percentages of correct and appropriate relay operations, correct but inappropriate operations, wrong tripping, and failure to trip [1]. The primary reason for "correct but inappropriate" operations is attributed to the power system changes. During restoration, the power system undergoes continual changes and therefore it is subject to "correct but inappropriate" relay operations.

Keywords: Protective System, Relaying Scheme, Excitation Relays, Negative Sequence Voltage, Excitation Limitation

INTRODUCTION

This paper reviews protective systems in order to determine the delays in restoration which might be caused by relays and to explore ways for correcting relays or if possible to avoid certain procedures to shorten the delays.

The protective relays which could affect the restoration duration include:

1. Distance Relays/Reclosing Schemes
2. Out-Of-Step Relays,
3. Synchrocheck Relays,
4. Negative Sequence Voltage Relays,
5. Differential Relays Lacking Harmonic Restraint,
6. Volts Per Hertz Relays,
7. Under-Excitation Relays,
8. Generator Excitation Systems
9. Low Frequency Isolation Scheme,
10. Under-frequency and Switched-capacitor Relays

1. Distance Relays/Reclosing Schemes

EHV transmission systems typically use duplicate relaying schemes: the primary relaying which may use solid-state relays and the backup relaying which may use electro-mechanical relays.

The following table shows typical PJM (Pennsylvania-New Jersey-Maryland Interconnection) relay time, signal transmission time, circuit breaker clearing time, and breaker reclosing time for the line voltages used in its bulk power supplies:

Voltage Level:	115	138	230	500 kV
Clearing Time:	8	8	5	3 Cyc
Reclosing Time:	12	12	20	25 Cyc
Relay Time:	2	2	1/4	1/4 Cyc
Comm. Time:	2	2	1/2	1/2 Cyc

95 WM 145-3 PWRS A paper recommended and approved by the IEEE Power System Engineering Committee of the IEEE Power Engineering Society for presentation at the 1995 IEEE/PES Winter Meeting, January 29, to February 2, 1995, New York, NY. Manuscript submitted August 12, 1994; made available for printing January 18, 1995.

The distance relays used generally are of "mho" type using pilot schemes of permissive overreaching and directional comparison scheme. In the following paragraphs these relaying schemes and the fault detector problem will be briefly described [2].

Figure 1(a) shows the mho distance relay characteristics and the fault impedance locus. When used in a "step Distance Scheme" typically, a mho relay has three admittance elements and a timer. Each element is set for a different "reach", protecting three different "zones" and operating at different time intervals from the instance of the fault. The tripping area is inside the circle and above the jX = 0 line.

The impedance "seen" by the relay for a fault during a solid (metallic) line short circuit lies on the line corresponding to the transmission line-impedance p-q and at a distance from the origin p proportional to the distance to the fault. Arc resistance shifts the fault impedance to the right so that the fault-impedance locus of the protected section of transmission line p-q lies in the shaded area of Figure 1(a).

Figure 1(b) shows the permissive overreaching scheme which is typically used on the high voltage transmission system. There are several principal pilot relaying schemes in which current and/or voltage (i.e., impedance, admittance and phase displacement) data are transmitted from each end of the line to the other: current comparison, phase comparison and directional comparison.

Reclosing

Reclosing schemes on EHV circuits are typically "single shot" designs employing dead bus/dead line, dead bus/live line, live bus/dead line or sync check across an open breaker. Of particular interest in system restoration is the live bus/dead line scheme. During the process of isolating/clearing transmission paths in preparation for steam unit start-up, these schemes may undesirably close breakers thus adding to the total reactive (line charging) current required. This additional reactive burden stresses an already precarious electric system in its early stages of restoration. Generally, the automatic reclosing schemes can be defeated remotely (supervisory control) which should be part of the restoration plan. Where this capability does not exist, it should be added and careful consideration should be given to restoration plans involving transmission paths such that the problem is avoided.

Reprinted from *IEEE Transactions on Power Systems,* Vol. 10, No. 3, pp. 1492–1497, August 1995.

Fault Detectors

Considering line p-q in Figure 1(b), with bus voltage Ep, current Ipq and impedance Zpq, then the impedance "seen" by the relay at the p end will be Zp = Ep/Ipq. In order to protect against the loss of voltage source, Ep (e.g., the failure of voltage transformer fuse), an over-current relay (fault detector) is installed in series with the mho relay and set to operate at currents greater than that of the heaviest flow (e.g., under peak load condition). Under this scheme the circuit breaker can only trip when both the mho and the over-current relays operate [3].

During initial phases of restoration when possibility of faults are greater, the short circuit current due to a solid (metallic) fault may not exceed the over-current relay setting due to lack of adequate on-line generation, possibly leading to the re-collapse of the restored system. In order to avoid system re-collapse, the over-current relay fault detector setpoints need to be evaluated for restoration considerations and possibly reset to prevent this occurrence. Preferably, this could be accomplished on mocroprocessor based relays remotely from the control center for "minimum source" operation but manually if necessary[4]. Another approach would be to employ an additional set of fault detector relays calibrated for "minimum source" which could be switched in remotely during system restoration. While certainly more expensive, this approach would save valuable time during system restoration.

Potential Restraint

Consider the universal torque (contact closing torque) equation of a typical electro-mechanical relay:

$$T = K_1 E_p^2 + K_2 I_{pq}^2 + K_3 E_p I_{pq} \cos \Theta - K_4$$

In the above expression, the first term describes the voltage element, the second term describes the current element the third term describes the phase angle element and the constant term represents the control spring constant force tending to hold the contacts open. In the absence of current and voltage the first three terms will be equal to zero. It follows then that torque should be -K_4, that is there should be a contact opening force present preventing trip from being held on the breaker. If however, K_4 approaches zero the contacts could drift closed tripping the breaker and if they remain closed will prevent the breaker from successfully closing later. This points up the need for proper potential restraint to the relay as well as regular maintenance on this type of equipment.

2. Out-of-Step Relays

Figure 1(c) shows characteristics of an Ohm Unit or Angle-Impedance Relay. Ohm relay units are used as "blinders" to control the angle range of tripping on swing conditions. The relay without blinders trips during swings in the range from 90° to 240°, whereas with blinders, the range is limited to 135° to 195°. The out-of-stop blocking is also accomplished by two concentric mho relays as shown in Figure 1(d), thus forming a "buffer area". The apparent impedance passes across the buffer area from normal operating conditions to tripping area gradually for an out-of-step condition, and instantaneously for a fault condition. If the elapsed time, between the crossing of the outer boundary of the buffer area and the inner boundary, is equal to or greater than the pick-up time of the auxiliary relay, then the relay operates, blocking the tripping operation. In other words, there is a timed sequence which distinguishes between faults - power sources and out-of-step conditions. The pick-up time of the auxiliary relay is about 3 to 4 cycles for electro-mechanical relays.

The fact that the loss of synchronism characteristic passes through or near the point representing a three phase short circuit indicates that any distance relay whose range of operation includes this point will have a tendency to operate. Whether the relay will actually complete the operation to trip the breaker depends on the operating speed of the relay as well as the length of time the out of step conditions persist. The effect of changes in system configuration or generating capacities should be taken into account in determining the operating characteristic of distance relays. Such changes will put the position of the "electrical center" within reach of the relay at one location, and later put it within reach of the relays at a different location.

In order to insure satisfactory system separation during out-of-step conditions, it may be necessary to block some of the relays from tripping where splitting is not desirable and to allow trip (transfer tripping) only those relays where it is desirable. To meet these requirements the relay schemes must be able to discriminate between normal load conditions, swing conditions, out-of-step conditions, and fault condition. The out-of-step blocking and transfer tripping is referred to as "controlled islanding" where the deficiency of generation in each separated area does not result in frequency decline beyond the point of no return. Controlled islanding shortens the restoration duration appreciably.

3. Synchro-check Relays

The presence of excessive standing phase angle differences across open circuit breakers causes significant delays in power system restoration. These angles may occur across a tie line between two systems or between two connected subsystems within a system. They must be reduced to within limits before an attempt is made to close breakers to firm up the bulk power transmission system.

The range of standing phase angle differences that a prevailing system or subsystem can withstand depends on voltage levels, operating condition, and location of the circuit breakers. Synchrocheck relay settings as practiced in one interconnection system are set at 20° for 500kV, 40° for 230kV and 60° for 115kV systems. A procedure needs to be established for reducing the standing phase angle differences without resorting to the raising and lowering generation levels on a trial and error basis. This could be accomplished by implementation of an "on line" algorithm designed to automate the process [5].

4. Negative Sequence Voltage Relays

The line conductors in many 500 kV lines are unsymmetrically spaced. The conductors are typically spaced horizontally 35 feet apart. Since lines are not transposed, the inductance and capacitance for phase B conductor which is between the two phases A and C conductors, are respectively less than and greater than the positive-sequence values. These inequalities cause unbalanced voltages and generate negative sequence voltages, particularly under light-load (restoration) conditions. Depending on length of the 500kV line, the negative voltage sequence can be 10 to 15% of the positive sequence voltage [6].

The negative sequence reactance of a typical two pole non-salient generators (i.e., most of the large steam units) is about 10% of its positive reactance. Therefore, a 10% NSV would cause about 100% negative sequence current which together with the positive sequence current over heat the stator and rotor. The generator negative sequence relays are typically set at 4% for alarm and at 10% for tripping.

Negative sequence voltage relays may be applied in power plants to provide protection for undervoltage conditions, single-phasing of motor loads and phase reversal. Such relays may operate undesirably if negative sequence components of sufficient quantity are present in the start-up power supply. This occurrence could delay restart of generating units.

Seriously unbalanced three phase voltages also can damage induction motors due to over heating and stalling. Large induction motors which are usually relayed for the negative sequence voltages will be protected, but the smaller auxiliary induction motors may not be protected. Destruction of these small relatively inexpensive motors can still have a negative impact on restoration efforts.

In providing remote cranking power over 500 kV lines, the extent of NSV should be determined in order: (a) to avoid damage to the auxiliary motors, and (b) to allow synchronization of the incoming generators [7].

5. Differential Relays w/o Harmonic Restraint

Power transformers are equipped with high speed and sensitive differential relays to protect them from internal short circuits of small values. Magnetizing inrush currents are seen by differential relays as internal faults. When transformers are energized magnetizing inrush current occurs to establish steady state flux. The predominant harmonic in the transformer magnetizing inrush current is the second harmonic which may be about 25% of the fundamental values. During energization period, the differential relay is restrained from operation by utilizing the second harmonic currents, i.e., the relays are composed of an operating element and a restraining element, the fundamental (60 Hz) is supplied to the operating element and the second harmonic (120 Hz) is supplied to the restraining element [8].

The generator differential protection is typically not provided with second harmonic restraint Some generator step-up transformers, typically older units, may not use the second harmonic restraint principle. Generally, inrush currents on the these transformers are not as severe since transformer flux is established gradually as the generator voltage is manually raised from zero.

The blackstart units typically are at 4.6 or 13.8 kV levels and usually the path to the steam unit which is to be cranked includes one or two power transformers. The energization of these transformer would cause magnetizing inrush currents which has been known to trip the black-start unit which has no second harmonic restraint. In order to correct these deficiencies it could be possible to supervise tripping on black start by a frequency sensitive element.

6. Volts Per Hertz Relays

Volts/hertz devices sense excessive core flux conditions by measuring generator potential and frequency. The flux is sensed as a ratio of generator voltage to generator frequency. The flux is sensed as a ratio of generator voltage to generator frequency. High volts/hertz condition for either the generator or the generator step-up transformer can occur during shut-down and start-up of the generator. During shut-down, when voltage regulator automatically alters the voltage set point for changes in frequency. During start-up with improper manual voltage adjustments. Many protective schemes employ a "dual level" volts/hertz approach. That is a much higher volts/hertz ratio is tolerated after parallel, than is allowed while the machine is open circuited but still excited. An ideal scheme would result in very fast tripping of the field breaker if the synchronizing breaker is open but inverse time tripping of the synchronizing the field breakers if the machine is paralleled. Most schemes however, utilize only fixed time timers in the trip scheme. This could result in unwarranted tripping during restoration since short term voltage and frequency variations are likely.

7. Under-Excitation Relays

During restoration, generators are operated with leading power factor (low excitation) to provide reactive absorbing capability for the lightly loaded power system. Low excitation if allowed to continue, can cause excessive heating in the stator end iron and can also cause instability between a given generator and the remaining system. The under-excitation relay can alert the operator of the condition which may damage the generator or result in instability. Typically, a directional unit and an impedance unit provide the alarm function. A timer may be included in the alarm circuit to indicate the length of time that the generator has been in operation under low excitation [9].

Loss of Field Relays are typically set to co-ordinate with Minimum Excitation Limiters which are an integral part of the modern excitation system. Usually, the margin between these devices is kept intentionally small in order to make maximum advantage of the machine leading VAR capability. This is especially true if the loss of field device is a mho characteristic device which is supervised by a voltage element. Under normal system conditions, loss of field does not require immediate tripping unless there is an accompanying decrease in terminal voltage that threatens system stability. Under the conditions of restoration however, the electrical system impedance as viewed from the generator terminals appears abnormally high. This set of conditions can result in a loss of co-ordination between the MEL and the loss of field relay causing a unit trip or loss of synchronism before any limiting action can take place. A possible solution to coordination problem is the implementation of a dual under excitation limiter scheme [13].

8. Generator Excitation Systems

The range of voltage adjustment while under automatic control is typically recommended by manufacturers to be ± 7.5% of normal machine terminal voltage, but may be as low as ± 5%. This range may be too restrictive during the restoration process. There is general agreement that the maximum machine terminal voltage should not exceed 110% of rated and the minimum voltage should not exceed 85% of nominal. It is however, necessary to derate the machine MVA in direct proportion to the voltage if operating at lagging power factors e.g. 90% rated MVA @ 90% terminal voltage and by the square of the terminal voltage in the leading in the leading power factor region e.g. 81% MVA @ 90% voltage [10]. In addition, many small peaking units have excitation systems installed which are not equipped with the appropriate limiters/protection which is required of large base load units. This may result in undesired tripping as extreme leading or lagging power factor conditions are reached during the restoration process. Consideration should be given to excitation system retro-fits or control scheme modifications on small generators deemed "black start units"

9. Low Frequency Isolation Scheme

Relay schemes have been theorized which would be used to isolate certain generators within a plant with "matching" local loads in order to save and retain critical sources of power having large reactive absorbing capabilities. Under such relaying schemes, the generating units do not have to be restarted from a complete shutdown, thus saving several critical hours. The problem is that both the local load and generation change, resulting in varying degrees of mismatch within the isolation scheme. The requirements then for the LFI Scheme to operate satisfactory are:

1. Recognition of power system frequency decline below LFI Scheme relay setting,

2. Determination of load-generation mismatch within the load rejection capability of the boiler-turbine unit, and

3. Correction of mismatch by augmenting the LFI Scheme with automatic load shedding relays.

Several LFI Schemes have received serious considerations by utilities and some have been implemented by large, industrial, cogenerating customers (Plant Load Isolation System - PLIS), however they have not gained acceptance by utility companies because of the risk of unintentional operation and complications of the interlocking scheme (switching the LFI Scheme from one unit to another) under normal operation.

Some of the super-critical, once-through, boiler-turbine units are designed or have been retro-fitted for continuous operation under their own house load (i.e., load rejection of about 95%), and, therefore, there is no need for this type of scheme. Most coal-fired steam-generating units, however, are designed for base load operation with no more than 10% by-pass system. These units trip on a larger load rejection, making application of LFI Schemes desirable. Generating units which are good candidates for LFI Schemes are smaller units (200MVA and below) which generate at or near distribution/subtransmission voltage levels.

10. Under-frequency and Switched-capacitor Relays

The under-frequency relays are used to effect load reduction and in some locations to isolate generators. Load reduction is primarily used to prevent system collapse during an under-frequency condition. They are also used for load restoration upon recovery of the system frequency (between 59.3 to 59.9 Hz), automatically or manually by SCADA Systems. In order to prevent damage to steam turbine blades, generators are isolated at frequencies ranging from 58.5 to 57.5 Hz [11].

During the initial phases of power system restoration, when the system is primarily under manual control, wider frequency and voltage variations are tolerated. Under these conditions, automatically switched capacitors and under-frequency relays, which are primarily designed for and are responsive to small voltage and frequency changes, will hinder the restoration process. It is important to determine whether and to what extent, these relays should be deactivated during restoration.

Certain types of underfrequency relays (electro-mechanical devices) exhibit an operating dependence on low temperatures. That is, if the relay is very cold (10° F) the setting tends to drift in the direction which would allow it to operate at higher frequencies than expected. Assuming restoration is attempted under these conditions it is possible these devices could operate and delay the restoration process.

CONCLUSION

The criteria for an efficient protective system includes: (i) dependability - the degree of certainty that the relay or relay scheme will perform correctly, (ii) security - the degree of assurance against incorrect operation due to all extraneous causes, (iii) speed - the minimum fault clearing time and equipment damage, and (iv) selectivity - the minimum removal of equipment from service for isolation of fault or other abnormality. These same criteria should be maintained during restoration [12].

It is important that the performance of relays and relay schemes be evaluated under restorative conditions when the power system is going through continual changes. The foreknowledge that certain system operating conditions could cause "correct and undesired relay operation" allows avoiding such operating condition during development/execution of a restoration plan.

ACKNOWLEDGEMENTS

The first author acknowledges support of PJM Interconnection Association and ECC, Inc.

REFERENCES

[1] "Protective Relaying for Power Systems," IEEE PRESS, Edited by Stanley H. Horowitz, 1980, pp 01370

[2] Adibi, M.M., "Power System Protective Relaying and Simulation Models," proceeding of IEEE Power Industry Computer Application Conference, 1969, Denver.

[3] "Summary Update of Practices on Breaker Failure Protection - Fault Detectors," Report by IEEE Power System Relaying Committee, IEEE Trans. Power App. Syst., vol, PAS-101, no. 3, pp. 555-563, Mar. 1982.

[4] Alexander, R.W., "Minimum Source Considerations When Restoring the PJM 500kV System," Pennsylvania Electric Association, System Operation committee, Jan. 1991.

[5] IEEE Committee Report, "An Approach to Standing Phase Angle Reduction," by S. Wunderlich, M. M. Adibi, R. Fischl, and C.D.O. Nwankpa, paper 93 WM 205-5, presented at IEEE 1993 WPM, Columbus, Ohio, Jan 31 - Feb 5, 1993.

[6] Barkle, J.E. and W.E. Glassburn, "Protection of Generators Against Unbalanced Currents," AIEE Trans. Part III, vol. 72, pp. 282-285, 1953.

[7] Elmore, W.A. and J.L. Blackburn, "Negative-Sequence Directional Ground Relay," AIEE Power App. Syst., pp. 913-920, 1963.

[8] Hayward, C.D., "Harmonic-Current-Restrained Relays for Transformer Differential Protection," AIEE Trans., vol. 60, pp. 377-382, 1941.

[9] Berdy, John, "Loss of Excitation Protection Power App. Syst., vol. PAS-94, pp. 1457-1463, Sept./Oct. 1975.for Modern Synchronous Generators," IEEE Trans.

[10] Milanicz, D. P. "Generator Reactive Capability, Testing and Optimization," Southern Electrical Exchange, Oct 1989

[11] Lockay, H.E. and V. Burtnyk, "Application of Underfrequency Relays for Automatic Load Shedding," IEEE Trans. Power App. Syst., vol. PAS-87, pp. 776-782, Mar. 1968.

[12] Harder, E.L. and W.E. Marter, "Principles and Practices of Relaying in the United States,"AIEE Trans. vol. 67, pp. 1005-1022, 1948

[13] Landgren, G. L. "Extended Use of Generator Reactive Capability by a Dual Under Excitation Limiter", IEEE Trans. vol. PAS-99, July/August 1980.

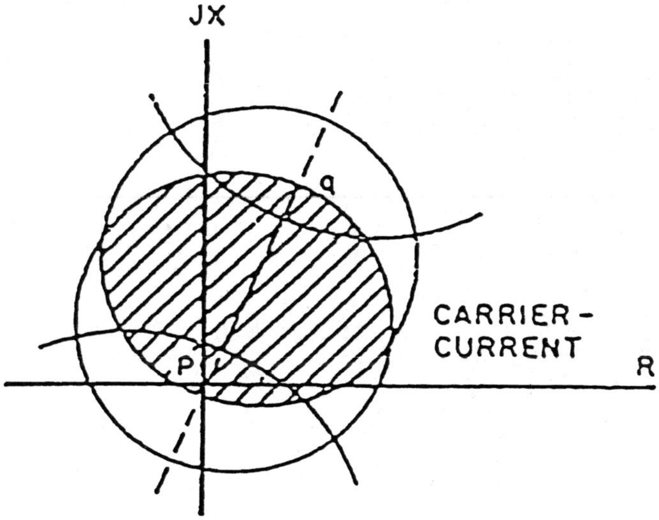

Fig. 1(b) Pilot Relaying

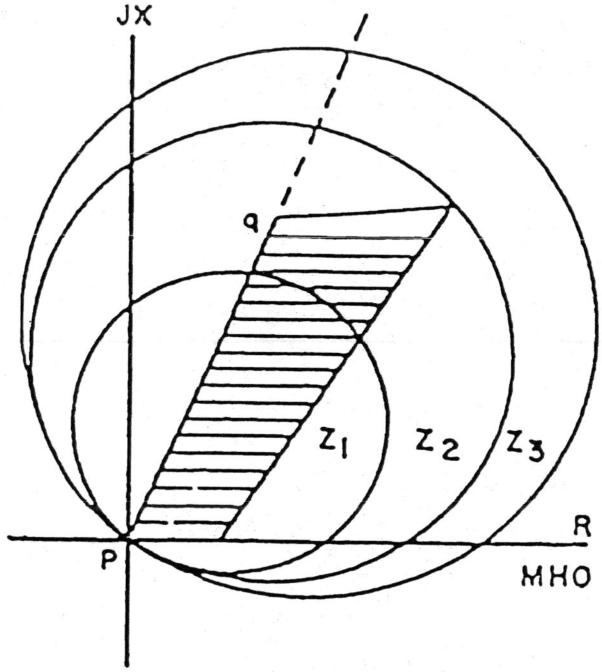

Fig. 1(a) Distance Relays

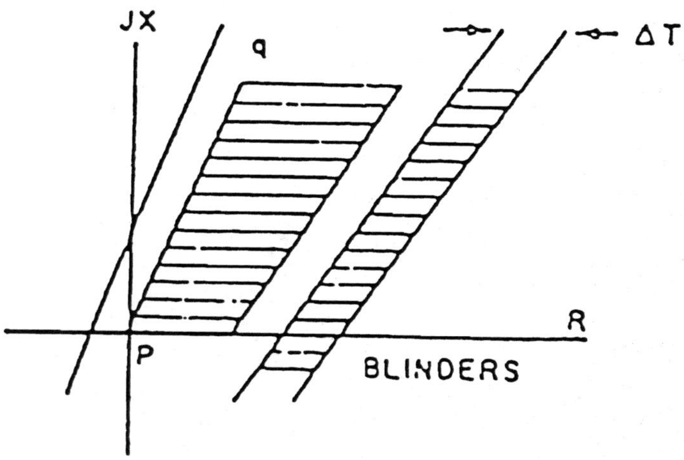

Fig. 1(c) Angle-Impedance Relay

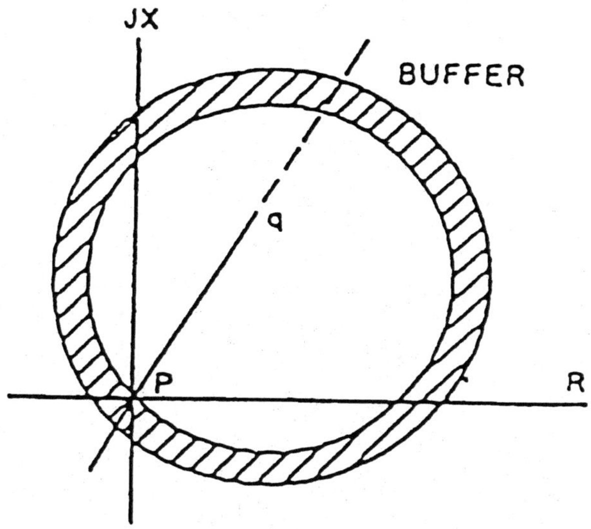

Fig. 1(d) Out-Of-Step Blocking

REMOTE CRANKING OF STEAM ELECTRIC STATIONS

M. M. Adibi, Fellow
IRD Corporation

D. P. Milanicz, SM
Baltimore G & E Co.

T. L. Volkmann, M
Northern States Power Co.

Abstract-This paper is one of series presented on behalf of the System Operation Subcommittee with the intent of focusing industry attention on power system restoration. During the initial phase of restoration, black-start combustion turbines are often considered as remote cranking sources for the start-up of steam electric stations. In general, remote cranking operation is carefully planned, simulated and then verified by field tests. This paper describes the planning and simulation presently used in black-start of two coal-fired generating units utilizing remote combustion turbines. It discusses a number of constraints which have to be considered and simulated in preparation for testing and verification. The paper concludes that, in order to minimize possibility of damage to equipment during testing or actual restoration, it is necessary to undertake extensive simulation of the procedure. It recommends that the realistic generator reactive capability models be made a part of the network models to allow: optimum selections of tap positions for all the transformers within the remote cranking power system, thus providing adequate lead/lag reactive powers and to maintain acceptable voltage profile while proceeding through various stages of operation.

Keywords: Cranking Power, Auxiliary Motors, Reactive Power Requirements, Black-start, Combustion Turbines.

I. INTRODUCTION

Black-start combustion turbines (CT) are often considered for remote cranking of steam electric stations (SES) under partial or complete power system collapse. In general, this type of CT can be started in 5 to 15 minutes, which is well within the 30 minutes critical time interval allowed for the hot restart of drum-type boiler-turbine-generators [1].

In planning for remote cranking power (RCP) operation, a number of constraints need to be considered. These include: frequency and voltage transients of the RCP System, voltage drop at the auxiliary bus in the SES, over and under excitation limits of the CT generators (the cranking source), and the protective relays installed in the RCP System [2].

Past analyses have shown and field tests have verified that the frequency and voltage transients of the RCP System during start-up of auxiliary motors are insignificant. For example, Figures 1 and 2, show the frequency and voltage profiles of a 28 MVA CT used for remote start-up of a 2,500 HP boiler-feed-pump (BFP) motor. As shown in Figure 1, the ½ Hz frequency decline of the CT for the start-up of the 2,500 horse power BFP motor is well within the operating range of the RCP System [3].

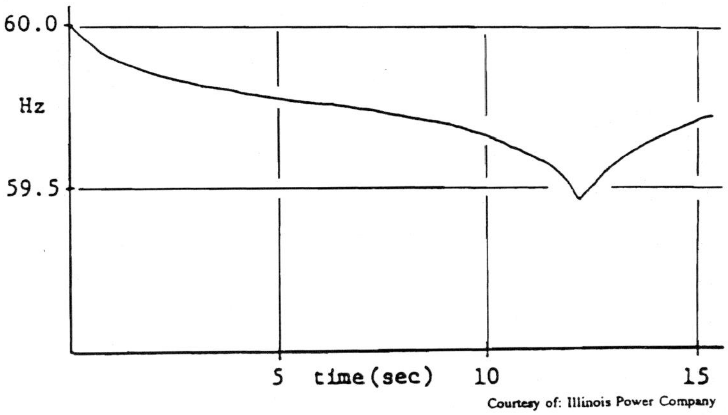

Figure 1. Frequency Response of the Boiler Feed Pump

Figure 2 shows the voltage profile of the auxiliary bus feeding the same BFP motor. It compares the simulated (solid line) with field test (dashed line) results, showing a fairly close match. Although the voltage transients immediately after the start-up and immediately after reaching the running state are insignificant, the magnitude and duration of the "sustained" voltage drop are considerable. The voltage drop is about 20% and lasts over 12 seconds [3].

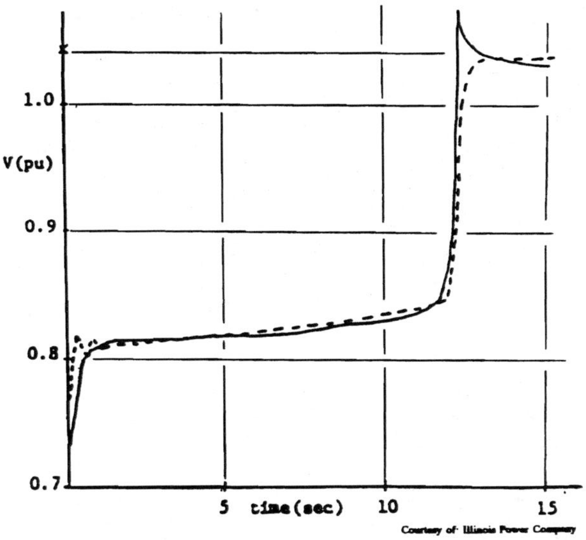

Figure 2. Voltage Profile of the Boiler Feed Pump

96 WM 162-8 PWRS A paper recommended and approved by the IEEE Power System Engineering Committee of the IEEE Power Engineering Society for presentation at the 1996 IEEE/PES Winter Meeting, January 21-25, 1996, Baltimore, MD. Manuscript submitted June 27, 1995; made available for printing November 16, 1995.

Reprinted from *IEEE Transactions on Power Systems,* Vol. 11, No. 3, pp. 1613 1618, August 1996.

The extent and duration of these sustained voltage drops depend on the size and type of combustion-turbine-generators, induction motors and their loads (pumps, fans, etc.), and the interconnecting transformers and transmission lines, thus each RCP operation needs individual and accurate simulation to ensure safe operation.

Comparison of the starting and running currents for the auxiliary induction motors shows a several fold increase, as for example, illustrated in Figure 3. It can be seen that the 4,000 HP induced draft fan (IDF) motor draws about 2,100 amperes during starting as compared to about 300 amperes during running, an increase of about six times. Considering typical values of 90% for running efficiency and 80% for running power factor of the IDF motor, the running reactive power requirement can be determined to be about 2.5 Mvar. The reactive power requirement during start-up of the same IDF motor at typical values of 80% voltage, 20% power factor and 7:1 starting/running current ratio can be estimated to increase to about 23 Mvar, which the cranking source should be able to supply [4].

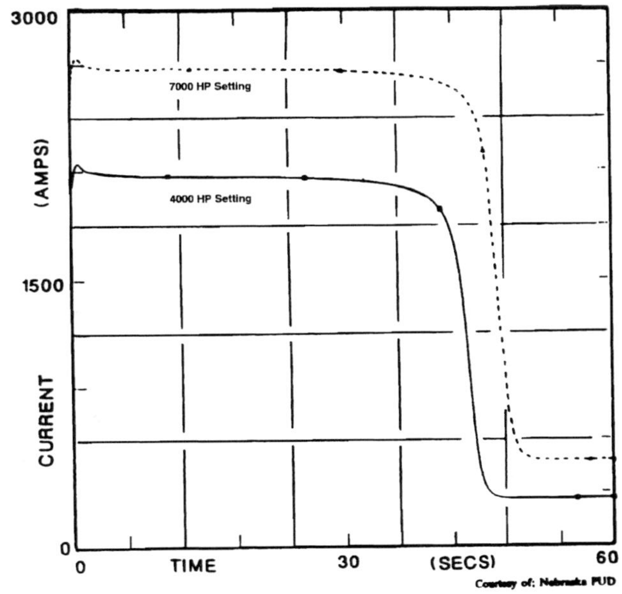

Figure 3. Starting Currents of the Induced Draft Fan

The generator reactive capability (GRC) curves furnished by manufacturers and used in the RCP operation planning typically have a greater range than can be realized during actual operation. Generally, these GRC curves are strictly a function of the synchronous machine design parameters such as rotor heating, stator heating and armature iron core end heating. These curves do not consider power plant and power system operating conditions as the limiting factor. For example, Figure 4, shows the rated capability of a 460 MVA generator and its actual operating limits verified by field tests [5].

In the course of an RCP operation many limiting conditions place severe demands on the reactive capability of the CT generators. For example, one extreme operating condition is during the initial energization of the CT - SES transmission lines when the CT generators absorb the charging currents of the high voltage lines, and the other extreme operating condition is when the CT generators supply the reactive power requirements of starting the largest auxiliary motor in the SES.

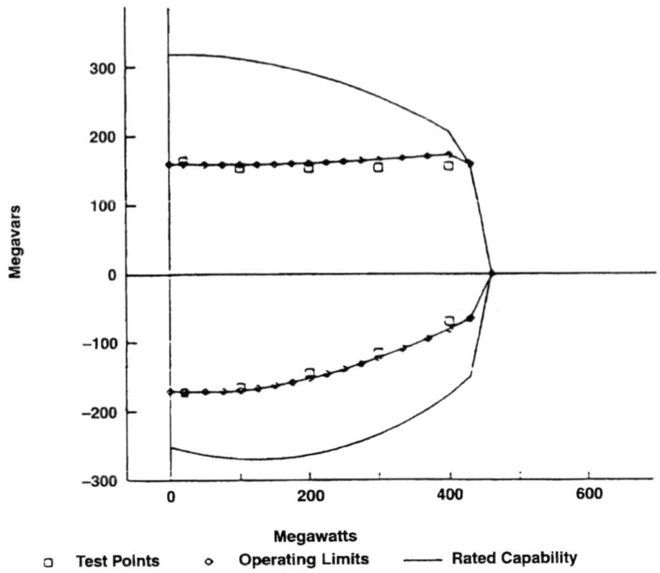

Figure 4. Actual Generator Reactive capability

These severe reactive power demands can be met by proper selection of the CT's unit and auxiliary transformer tap positions while controlling the generator system bus voltage at a correct level [6].

In addition, there are a number of protective relays affecting the RCP operation which should be examined. These include: CT generator excitation system, power reversal (or anti-motoring) relays, under-excitation relays, volts per hertz relays, differential relays lacking harmonic restraints, and negative sequence voltage and current relays. These protective relay issues are described in [7].

Simulation of typical RCP operation is complicated by the following industry practices:

1. The unit transformer and the auxiliary transformer (also referred to as the station service) transformer generally are equipped with no-load (fixed) taps, whose optimum positions need to be determined in the planning phase and prior to the RCP operation, which spans some two dozens of discrete states.

2. The power flow (PF) programs in use either represent the generator reactive power capability by simple high/low hard limits or in the most recent versions by the manufacturers rated parameters, both representations being inadequate, not reflecting the generators actual operating limits [8].

3. The optimal power flow (OPF) programs in use determine a single state of a power system that optimizes a given objective function and satisfies certain set of constraints. The optimal transformer tap position need to satisfy a set of constraints along the entire trajectory of the RCP operation [9].

This paper describes an attempt in simulating an RCP operation for black-start of two coal-fired drum-type units using remote CTs at Northern States Power (NSP) Company. It uses power flow (PF) and generator reactive capability (GRC) programs in an iterative process to determine: the appropriate tap positions for CT's unit and auxiliary transformers, various transformers installed in the CT - SES path, and to control proper voltage level for the generator system bus of the CT plant.

II. RCP SYSTEM DESCRIPTION

Figure 5 shows the one-line diagram of the RCP System. It consists of the CT Plant as the remote cranking source, the SE Plant to be hot restarted and the RCP Transmission System connecting the two plants. Table I lists the auxiliary motors and their horse powers to be started for each SE unit.

A. CT Plant:

It consists of two pairs of generators, each pair rated at 42 MVA, 85% power factor (PF) and 13.8 kV, feeding a common unit transformer rated at 46 MVA, 14.4/118 kV and equipped with 6 no-load taps between 95 and 107.5%. The auxiliary transformer for each pair is rated at 300 kVA, 13.2/0.48 kV, equipped with 5 no-load taps between 95 and 1.05%. The transformer supplying the local load is rated at 42 MVA, 115/34.5 kV, equipped with 16 no-load taps between ±10% on the 34.5 kV side. The power reversal relay for each CT is set at 3.5 MW.

B. SE Plant:

It consists of two 75 MW drum-type boiler-turbine-generators, each unit having a reserve auxiliary transformer rated at 38 MVA, 345/4.16 kV, feeding the two auxiliary buses. These transformers have no taps.

Table I

Steam Electric Station Auxiliary Load Data

Sequence	Load Description	HP
1.	ID Fan	6,000
2.	FD Fan	3,000
3.	Circulating Water Pump	3,000
4.	Primary Air Fan	2,500
5.	Start-up BF Pump	2,500
6.	Boiler Circulating Pump	2,000
7.	Condensate Pump	1,500
8.	River Water MU Pump	900
9.	Aux Cooling Water PU	700
10.	Coal Mill	700
11.	Air compressors	700
12.	Closed Cooling Water	350
13.	Misc Small Motors, etc.	870
Total (rounded)		24,000

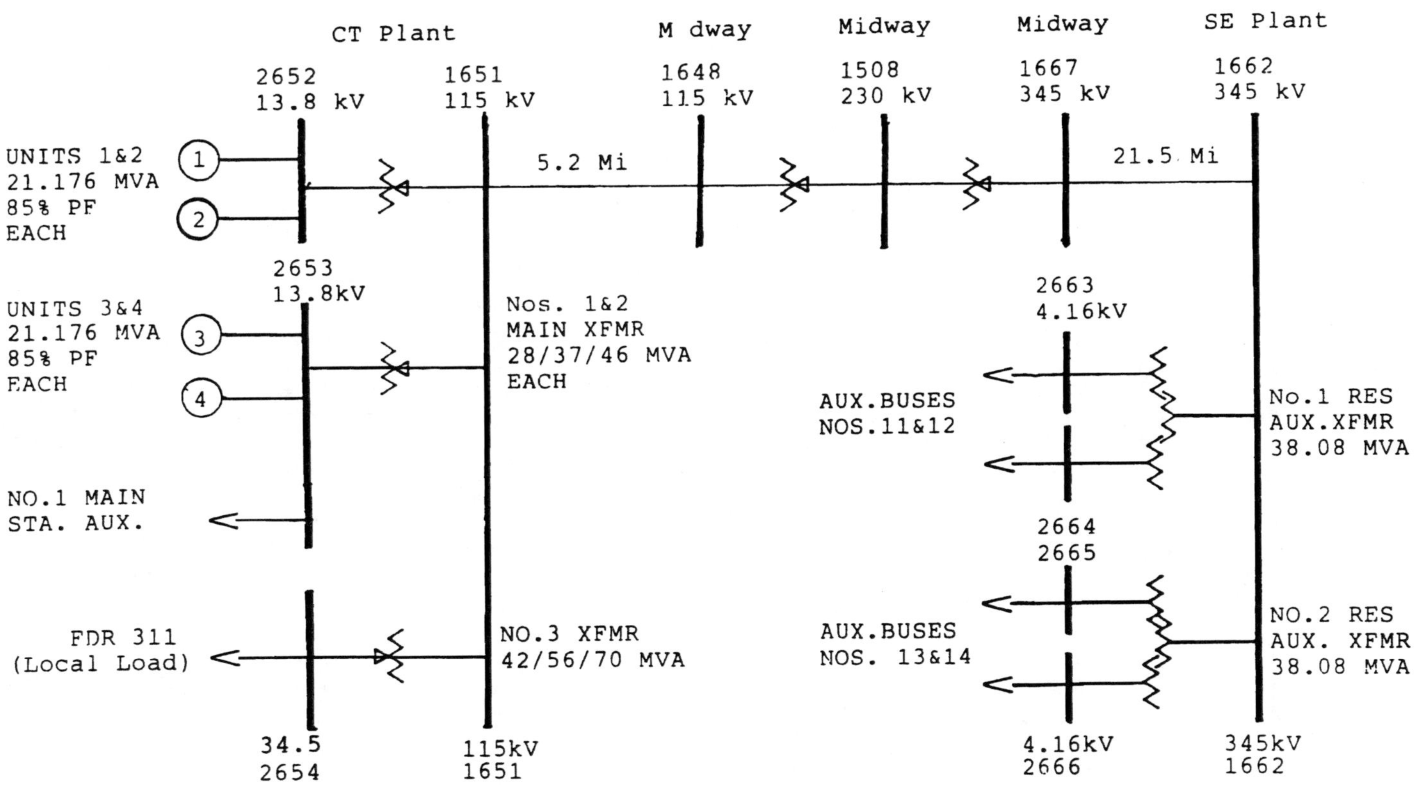

Figure 5. CT Plant - SE Plant One-line Diagram

C. CT Plant - SE Plant Transmission:

The transmission system consists of: 5.2 mi of 115 kV line, a 230/115 kV transformer with 5 no-load taps (i.e., 230 kV ±5%), a 180 MVA, 345/230/13.8 kV autotransformer with 5 on-load or TCUL taps (i.e., 345 kV ±5%), and 21.5 mi of 345 kV line with 18.1 Mvar of line charging.

D. Auxiliary Motor Loads:

The starting and running loads (real and reactive) for all the auxiliary motors were estimated from typical values of 90% for running efficiency, 80% for running power facto, 80% for starting terminal voltage, 20% for starting power factor, 5:1 ratio for starting/running currents, and horse power rating as listed in Table I. Table II lists the cumulative starting and running loads for all the auxiliary motors of the first SE unit and the induced draft fan (IDF) of the second SE unit in the RCP System.

Table II
The Cumulative Starting & Running Reactive Loads

Starting Sequence	Description	Starting and Running MW	Starting and Running MVAR	Running MW	Running MVAR
0	Local Load (FDR 311)	7.5	4.5	7.5	4.5
1	ID Fan	13.7	34.9	12.5	8.2
2	FD Fan	15.6	23.5	15.0	10.1
3	Circulating Water Pump	18.1	25.3	17.4	12.0
4	Primary Air Fan	20.0	24.6	19.5	13.5
5	Startup BF Pump	22.1	26.2	21.6	15.1
6	Boiler Circulating Pump	23.7	25.2	23.2	16.3
7	Condensate Pump	24.8	23.9	24.5	17.2
8	River Water MU Pump	25.4	21.8	25.2	17.8
9	Aux Cooling Water PU	26.0	21.4	25.8	18.2
10	Coal Mill	26.5	21.8	26.4	18.7
11	Air Compressors	27.1	22.2	27.0	19.1
12	Closed Cooling Water	27.3	20.9	27.3	19.3
13	Misc. Small Mtrs., etc.	27.8	19.7	28.0	19.9
14	ID Fan (Unit No.2)	34.2	50.3	33.0	23.6

III. REMOTE CRANKING POWER APPROACH

The objectives were to select the optimum tap positions for the unit and auxiliary transformers at the CT Plant and for the 230/115 kV & 345/230 kV transformers between the CT Plant and the SE Plant, so as to satisfy the following dual constraints under all the above (14 starting and 14 running) discrete loading conditions:

1. the voltage profile of the RCP System stay within the allowable limits, and

2. the leading/lagging reactive power requirements of the RCP System stay within the cranking source capability.

In the absence of a simulator capable of determining the desired optimum tap positions and satisfying the above constraints, numerous power flow and generator reactive capability analyses were conducted by trial and error to arriving at an approximate solution.

A. PF Cases:

The following 14 dual power flow (i.e., starting and running) cases were simulated to verify that the results satisfy the above constraints:

Case 1. After starting the first pair of CTs, 7.5 + j4.5 MVA of local load will be picked up at 34.5 kV bus of the CT Plant. This is to prevent a reverse power relay operation of each CT which is set at 3.5 MW, provide the CTs with lagging reactive load, and also to provide them with damping torques to mitigate frequency decline when in the following case, the CTs are subjected to a sudden load increase. After stabilizing the CTs, the CT Plant - SE Plant transmission system including the two 39 MVA, 4.16 kV reserve auxiliary transformers at the SES will be energized. Generator is checked for under excitation violation and the system voltage profile is checked for voltage violations.

Case 2. The 6,000 HP, IDF motor of Unit No.1 which is one of the largest motor will be started from one of the two 4.16 kV auxiliary buses at the SES Plant. The IDF motor in starting and running states with 6.2 + j30.4 MVA and 5.0 +j3.7 MVA loads respectively, are simulated and the CT generators are checked for over excitation violation. The RCP System voltage profile is also examined for violation.

Cases 3 through 13. The above simulation is repeated for starting each subsequent motor until all the auxiliary motors of the first unit have been started and are running. The total running load for all the auxiliary motors (24,000 HP) of first unit is about 20.5 + j15.4 MVA.

Case 14. While supplying all the auxiliary motors of the first unit, the IDF motor of second unit is started. The total load for running all the auxiliaries of the first unit and the start-up of the second IDF motor is about 26.7 +j45.6 = 52.8 MVA. This is greater than the 42.3 MVA rating of the first pair of generators. In order to crank the second boiler-turbine-generator, the third CT should be started.

B. GRC Cases:

The two CTs, Nos.1 & 2, which are feeding Unit Transformer No.1, are combined as if they are one generator. Table III lists the scaled data representing reactive capability curves furnished by the manufacturer. Table IV lists: generator and auxiliary load limits and various voltage limits furnished by NSP. Other data furnished by NSP and used in the GRC Program are:

Unit transformer: resistance and reactance which are 0.15% and 7.6% on 25 MVA and 118 kV base; and high side, low side and tap position which are 118, 14.4 and 124 kV.

Auxiliary Transformer: reactance which is 4.99% on 0.3 MVA base; and high side, low side and tap position which are 13.2, 0.48 and 13.53 kV.

Miscellaneous: including armature current limit of 1,772 amperes and CT inlet air temperature of 80° F.

Table III
Manufacturer's Reactive Capability Data Scaled for the Combined Generators

Description	Coordinate	MW	MVAR
Gen. Output at Max. Lag,	(Ax,Ay)	0.0	33.4
Gen. Lag at Break Point,	(Bx,By)	36.0	22.0
Peak Gen. Real Power,	(Cx,Cy)	42.2	0.0
Gen. Lead at Beak Point,	(Dx,Dy)	42.2	0.0
Gen. Output at Max. Lead,	(Ex,Ey)	10.0	19.0
Gen. Mvar at Max. MEL,	(Jx,JY)	40.0	3.0
Gen. MW at Max. MEL,	(Kx,Ky)	0.0	18.6

Table IV
The Combined Generators Limits

Description	Unit	Max.	Min.	Nom.
Generator Outputs,	MW	36.00	4.00	------
Auxiliary Loads,	MW	0.15	0.10	------
Generator Bus Voltage,	kV	14.49	13.11	13.80
Auxiliary Bus Voltage,	kV	0.53	0.43	0.48
System Bus Voltage,	kV	120.80	109.30	115.00
Voltage Regulator,	kV	14.84	12.77	------

IV. REMOTE CRANKING POWER RESULTS

Figure 6 shows that with the selections of 103.5 and 102.5% tap positions for the unit and auxiliary transformers and by controlling the generator system bus at 103.0%, the two combined units are capable of absorbing about 12 Mvar and supplying about 32 Mvar. The reactive power absorbing capability of generators is limited by the generators' low terminal voltages between the minimum generation of 4 MW and 20 MW, and by the minimum excitation level relay setting from 20 MW to the maximum generation of 36 MW. The reactive power supply capability of generators is limited by the generators high terminal voltage between the minimum generation of 4 MW and 16 MW, and by the generators field heating from 16 MW to the maximum generation. The reactive power limitations beyond the maximum generation of 36 MW do not apply.

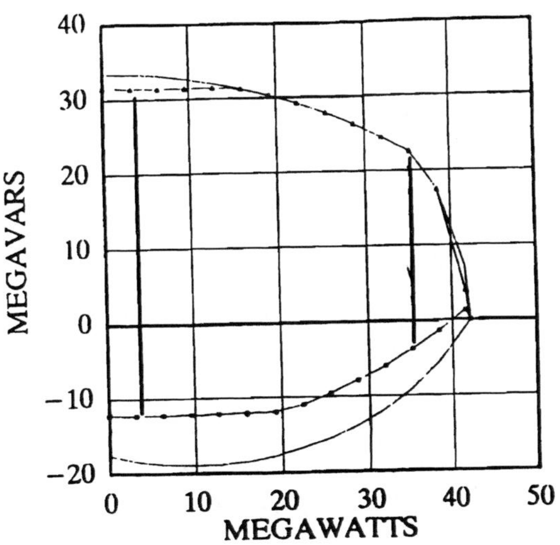

Figure 6. Generator Reactive Capability Curves

The sequence of starting auxiliary motors is determined by the boiler-turbine-generator start-up procedure. However, if there are any options available, it is preferred to start-up the larger motors first. They act as asynchronous generators and help to stablize the stat-up of the smaller motors.

Figure 7 shows the cumulative reactive generation for starting and running of auxiliaries following Table II start-up sequence. The upper curve shows the reactive generation for the start-up of a given auxiliary motor as indicated by its sequence number plus the running reactive generation for all the preceding motors. The lower curve shows the cumulative running reactive generation of the auxiliary motors.

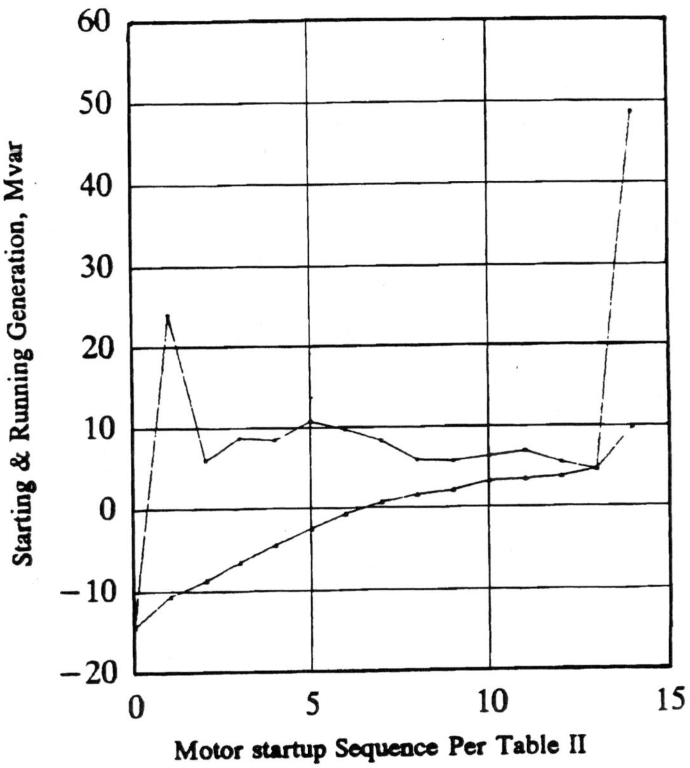

Figure 7. Cumulative Starting and Running Generation

Table V lists the total load, corresponding generation and the 4.16 kV auxiliary bus voltage for the four power flow cases described above. Table VI lists the selected tap positions for all the transformers in the RCP System. These two tables show that throughout the 14 stages of start-up procedure: the auxiliary bus voltages do not exceed the allowable 105.4% and do not drop below 80% limit, and the generator's reactive power absorption and supply stay within its capability. These indicate that the tap positions selected and the controlled system bus voltage of 103.0% provide acceptable results.

Table V
Remote Cranking Power - Significant Cases

Case No.	Auxiliary Load MW + MVAR	Generation MW + MVAR	Bus Voltage %
1	7.5 + j 4.5	7.5 - j 14.5	105.4
2	13.7 + j 34.9	14.0 + j 24.1	86.1
3	27.8 + j 19.7	28.1 + j 4.6	100.0
14*	34.2 + j 50.3	34.7 + j 48.7	83.0

* Three CTs in operation.

Table VI
RCP - Selected Transformer Taps

Description	Tap Position Volts	Auxiliary Tap Position %
Unit Transformer	118,000	103.5
Aux. Transformer	13,530	102.5
Local Transformer	34,500	100.0
230/115 Transformer	224,300	97.5
345/230 Autotransf.	362,300	102.5

V. CONCLUSION:

Considerable portion of the electrical energy is supplied by the base loaded steam electric units which in case of blackout generally need off-site power for start-ups. The installed black-start combustion turbines are used as cycling units to meet the daily peak demands. It is economical and very attractive to be able to use this type of CT for the remote cranking of steam electric units under partial or total power system collapse.

The RCP operation is complicated by the fact that unit and auxiliary transformers at the CT plants are equipped with no-load (fixed) taps whose optimum positions need to be determined by simulation prior to the RCP testing or operation. The present analytical tools (power flow and optimal power flow programs) used for simulation do not adequately represent the actual generators reactive capability to ensure safe operation. Furthermore, these tools usually address only a single state of power system, whereas the RCP procedure covers over two dozens of discrete events representing the various large motor starting and running states.

It is recommended that the present simulators be expanded to accommodate simulation of this basic restoration operation by:

- modeling a realistic reactive capability limits for the generators,
- expanding the present optimal power flow program to address many discrete events and points between such events, and
- integrating the generator reactive capability model with the optimal power flow.

Operator training simulator with the above capabilities would provide guidelines for safe and efficient RCP operation.

VI. REFERENCES

[1] Kafka, R.J., D.R. Penders, S.H. Bouchey and M.M. Adibi, "System restoration plan Development for a Metropolitan Electric System," IEEE Trans. - PAS - 100 (August 1981), pp.3703-3713.

[2] Adibi, M.M., R.J. Kafka and D.P. Milanicz, "Expert System Requirements for Power System Restoration," IEEE Trans. Power System, Vol.9, No.3, Aug. 94, pp.1592-1600.

[3] Shultz, R.D. and Mason G.A. "Black-start Utilization of Remote Combustion Turbines, Analytical Analysis and Field Test," IEEE Trans. PAS 103, Aug. 84, pp. 2186-2191.

[4] Lindstrom, R.R., "Simulation and Field Tests of the Black Start of a Large Coal-Fired Generating Station Utilizing Small Remote Hydro Generation," IEEE Trans. Power System No.1, Feb. 90, pp. 162-168.

[5] Adibi, M.M. and Milanicz, D.P. "Reactive Capability Limitation of Synchronous Machines," IEEE Trans. Power System, Vol.9, No.1, Feb. 94, pp. 29-40.

[6] "Blackstart of Sherburne County SES Utilizing Granite City CTs," an internal report to Northern State Power Company, January 1995.

[7] Adibi, M.M. and Milanicz, D.P. "Protective System Issues During Restoration," 93 WM 145-3 PWRS, New York, New York.

[8] IEEE Committee Report, "Common Format for Exchange of Solved Load Flow Data," Working Group, paper T 72 450-5 presented at the IEEE PES WM, New York, NY Jan. 28 - Feb. 2, 1973

[9] Alsac, O. et al, "Further Developments in LP-based Optimal Power Flow," IEEE Trans. on Power Sys., vol. PWRS-5, pp.697-711, Aug. 1990.

VII. ACKNOWLEDGEMENTS

The first author acknowledges support of NSP and ECC, Inc.

OPTIMIZING GENERATOR REACTIVE POWER RESOURCES

M.M. Adibi, LF — IRD Corporation
D.P. Milanicz, SM — Baltimore G&E Co.
T.L. Volkmann, M — Northern States Power Co.

Abstract: A method is described for selecting generator step-up and auxiliary transformer tap positions to provide the desired reactive power requirements in a coal burning power plant. It is shown how a generator could be made to absorb reactive power, when previously unable to do so, by adjustment of tap positions on the generator step-up and auxiliary transformers. The approach described provides an indication of the range of reactive power outputs that are available for different tap positions. This information would facilitate the alteration of transformer tap positions to make more effective use of generator reactive power resources.

1. INTRODUCTION

The rated reactive power curves furnished by manufacturers, are strictly a function of the generators's design parameters at the rated terminal voltage and hydrogen pressure. These curves do not consider power plant and power system operating conditions as the limiting factors. They show that different generator loads produce greater heating in different parts of generator. **Figure 1** shows a typical generator reactive capability curves. The segment AB is limited by rotor heating, BC is limited by stator heating, and CD by armature core end heating [1].

In **Figure 1,** the limitations due to operating conditions are superimposed on the rated reactive capability curves A-B-C-D. As shown, the reactive capability is restricted to the area bounded by the a-b-c-d-e-f, where, a-b & d-e are the auxiliary bus high/low voltage limits, b-c & e-f are the generator terminal high/low voltage limits, and c-d & a-f are the prime movers max/min operating limits. Other possible limitations to be considered but not shown are: max/min limits for voltage regulator and for excitation. **Table I** lists the rated and operating limits of generator reactive power outputs [2].

PE-732-PWRS-0-06-1997 A paper recommended and approved by the IEEE Power System Operations Committee of the IEEE Power Engineering Society for publication in the IEEE Transactions on Power Systems Manuscript submitted December 4, 1996; made available for printing June 9, 1997.

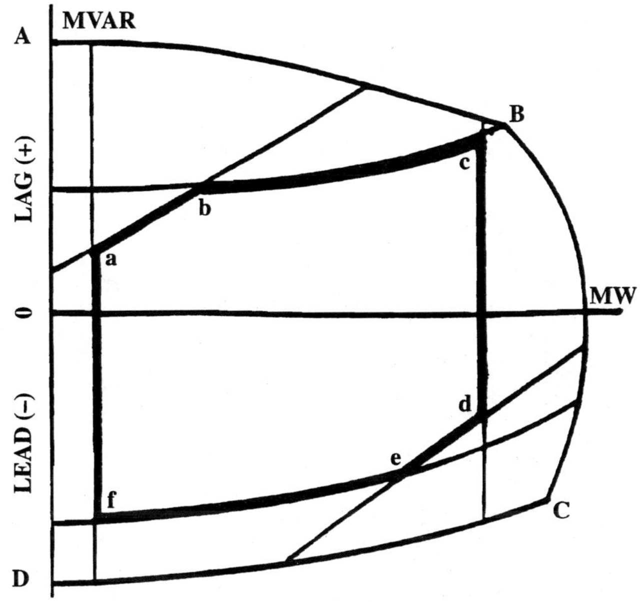

Figure 1. Generator Reactive capability Limitation

Table I. Generator Reactive Power Limitations:

(1) Field Heating

(2) Maximum Excitation Limit

(3) Generator Bus Voltage or Voltage Regulator Control at Maximum

(4) Auxiliary Bus Voltage at Maximum

(5) Stator Heating, Armature Current or Field Current Exceeding Limits

(6) Armature Iron Core End Heating

(7) Minimum Excitation Limit

(8) Generator Bus Voltage or Voltage Regulator Control at Minimum

(9) Auxiliary Bus Voltage at Minimum

The max/min voltage limits for generator terminal, auxiliary bus and generator system bus are interrelated by the tap positions on the generator step-up (GSU) transformer and the auxiliary unit (AUX) transformer. In order to find the GSU and AUX transformer tap positions which would provide the desired proportion of lagging/leading reactive powers, a search is made in the following four sets of operating ranges using the methodology described in Reference [2]:

- prime movers minimum to maximum MW outputs,

Reprinted from *IEEE Transactions on Power Systems,* Vol. 14, No. 1, pp. 319–324, February 1999.

- system bus voltages between peak and light-load conditions,
- GSU transformer range of tap positions, and
- AUX transformer range of tap positions.

In this search the effects of variations in the auxiliary loads as a function of generator output, and voltage drops in both GSU and AUX transformers are considered. Finally, the desired generator reactive capability is modified by the auxiliary load and by the losses in both transformers to determine the reactive powers in and out of generator system bus which is of prime interest.

2. GENERATING UNIT DESCRIPTION

The following data describe the relevant characteristics of the generating used. These data were used to select the correct tap positions for the generator step-up transformer and for the auxiliary transformer to allow an appropriate proportions of over and under-excitation during the entire ranges of system bus voltage conditions and the generator MW outputs.

Generator Ratings:
18 kV, 192 MVA, 85% power factor, 60 and 45 psi hydrogen pressures, generator output limits: max=185 MW and min=50 MW, armature current 6,160 A. Generator terminal voltage limits: ±5%, voltage regulation limits: ±7.5%.

System Bus (Sys Bus) Voltage Range:
Peak load condition 115.6 kV (100.5 %), light-Load condition 119.6 kV (104.0 %), and nominal voltage of 115 kV.

Generator Step-up Transformer (GSU) Ratings:
192 MVA, 115/18 kV nominal ratio, 0.5+j10.8 % impedance based on GSU transformer ratings, five 2.5% taps between 97.5 and 107.5 % of the 115 kV nominal value. The GSU transformer is set at 105% or 120,750 volts.

Auxiliary Transformer (AUX) Ratings:
16 MVA, 18/2.4 kV nominal ratio, j5.75 % reactance based on AUX transformer ratings, five 2.5% taps between 95.0 and 105.0 % of the 18 kV nominal value. The AUX transformer is set at 97.5 % or 17,550 volts.

AUX Bus voltage limits ±5% of 2.4 kV nominal value, and AUX Bus load: max = 10 + j5 MVA and min = 7.5 + j4.5 MVA (coal-fired unit).6.

3. APPROACH

The following iterative process is used to search for the desired tap positions:

1. The Aux transformer tap is kept at its present position (initial value), while: (a) changing the GSU transformer tap position over its entire range, (b) varying Sys Bus voltage over the specified range, and (c) recording the generators maximum lag and lead reactive powers, at the max/min generator MW outputs. The GSU transformer tap position is then selected which provides the required lag/lead reactive powers under the two ranges of Sys Bus voltages and generator MW outputs.

2. The GSU transformer tap is kept at the selected position, while: (a) changing the AUX transformer tap position over its entire range, (b) varying Sys Bus voltage over the same range, and (c) recording the generators maximum lag and lead reactive powers at the same generator MW outputs. The AUX transformer tap position is then selected which provides the required lag/lead reactive powers under the same two ranges of Sys Bus voltages and generator MW outputs.

3. The above procedure is repeated until no changes are necessary in either the GSU or AUX transformer tap positions.

To determine the reactive power capability at the Sys Bus, the effects of auxiliary loads, voltage drops,and reactive power losses in GSU and AUX transformers are considered.

The above approach is illustrated in the following two figures.

Figure 2 shows the generator lagging and leading reactive powers for a range of GSU transformer taps of 97.5, 100, 102.5, 105 & 107.5% as indicated by a, b, c, d & e, and for a range of Sys Bus voltages of 100, 101, 102, 103 & 104% as indicated by the subscripts 1, 2, 3, 4 & 5, all for the present AUX transformer tap position of 97.5.
According to this illustration, GSU tranasformer tap position of 102.5% provides an apprpriate proportion of the lagging and leading reactive powers over the generator output range of 50 to 175 MW. **Figure 2** also shows that for other GSU transformer tap positions the supply or absorption of the reactive powers are reduced at certain Sys Bus voltages. Next the AUX transformer tap position of 97.5% is checked.

Figure 3 shows the generator lagging and leading reactive powers for a range of AUX transformer taps of 95.0, 97.5, 100, 102.5, & 105% as indicated by a, b, c, d & e, and for a range of Sys Bus voltages of 100, 101, 102, 103 & 104% as indicated by the subscript 1, 2, 3, 4 & 5, all for the selected GSU transformer tap position of 102.5%. According to this illustration, AUX transformer tap position of 97.5% provides the appropriate proportion of lagging and leading reactive powers over the generator output range of 50 to 175 MW. **Figure 3** also shows that for other AUX transformer tap positions the supply or absorption of the reactive powers are rduced at certain Sys Bus voltages. The GSU transformer tap position of 102.5% can be checked as the third iteration to ascertain that both transformer taps have been "optimally" selected.

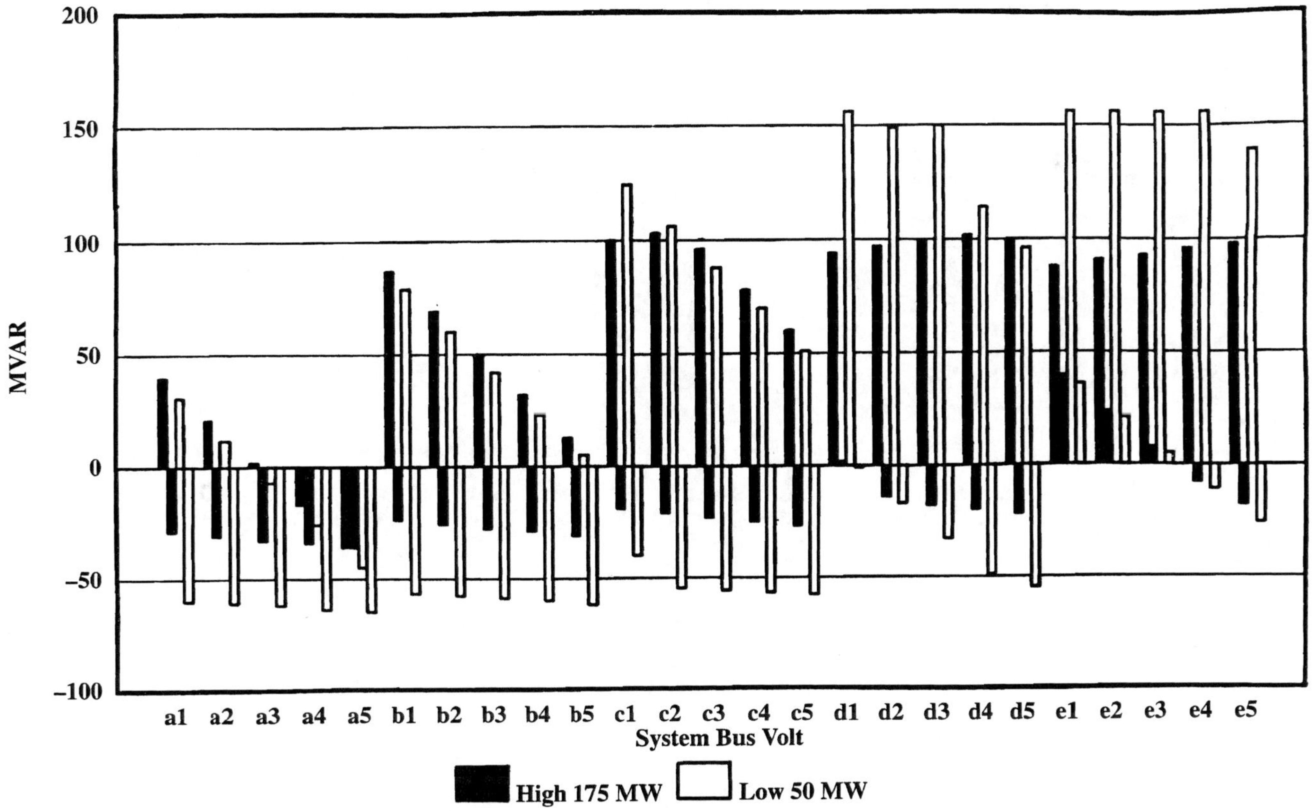

a, b, c, d & e correspond to GSU Xfmr Taps = 97.5, 100, 102.5, 105 & 107.5 %
System Bus Volt Subscripts 1 to 5 correspond to 100, 101, 102, 103 & 104 %

Figure 2. Auxiliary Transformer Tap at 97.5 %

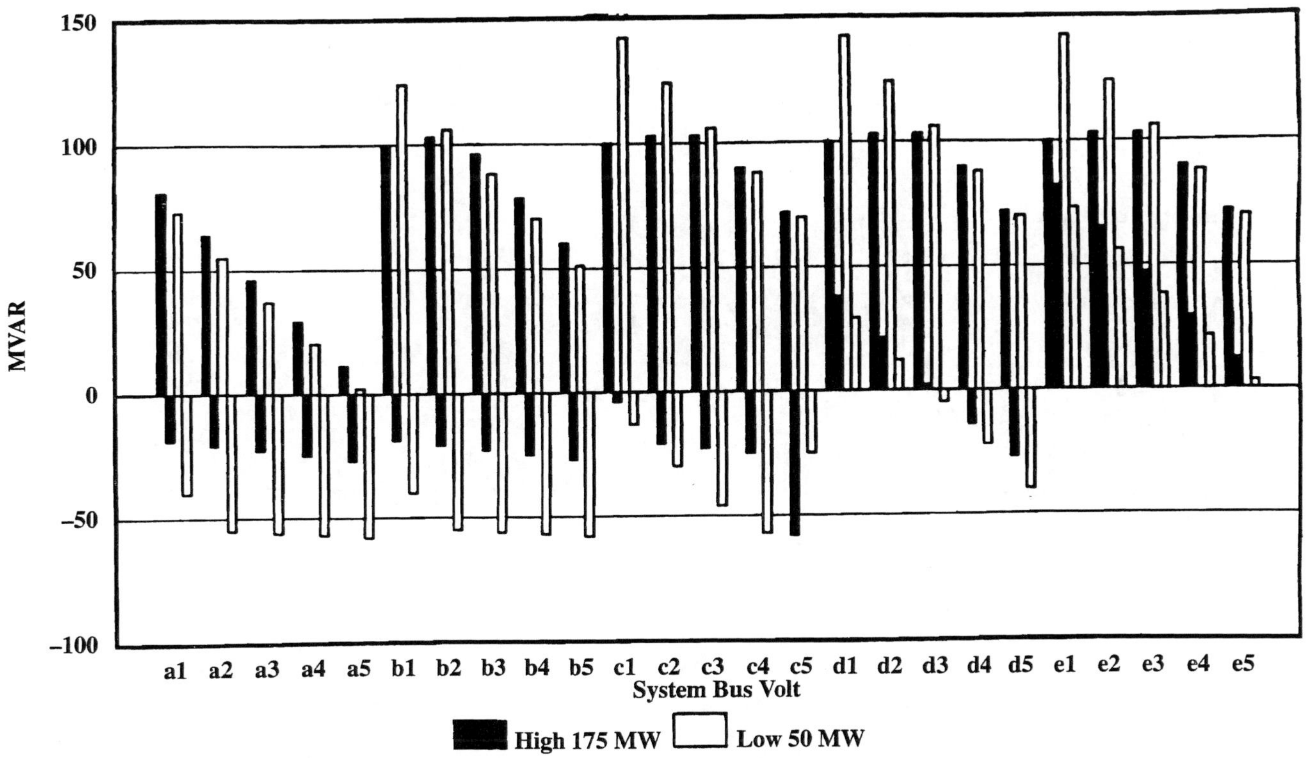

a, b, c, d, & e correspond to GSU Xfmr Taps = 95, 97.5, 100, 102.5, & 105 %
System Bus Volt Subscripts 1 to 5 correspond to 100, 101, 102, 103 & 104%

Figure 3. Generator Step Up Transformer Tap at 102.5 %

Table II. Generator Reactive Capability Under Present Tap Positions

Gross Generator Output = 175 MW

Sys Bus Voltage %	Max Gen Output MVAr (Aux Bus Load = 9.9 + j5.9 MVA)	Gen Bus Voltage %	Aux Bus Voltage %	Rated or Operating Limits	Min Gen Output MVAr (Aux Bus Load = 9.9 + j5.0 MVA)	Gen Bus Voltage %	Aux Bus Voltage %	Operating Limits NB
100.0	94.4	103.3	103.9	(5)	2.4	95.0	95.5	(8)
101.0	96.9	104.0	104.6	(5)	-13.7	95.0	95.5	(8)
102.0	99.4	104.6	105.2	(5)	-18.4	95.7	96.2	(7)
103.0	101.8	105.3	105.9	(5)	-20.2	96.6	97.1	(7)
104.0	100.4	105.0	105.6	(5)	-22.0	97.4	98.0	(7)

Gross Generator Output = 50 MW

Sys Bus Voltage %	Max Gen Output MVAr (Aux Bus Load = 8.2 + j4.9 MVA)	Gen Bus Voltage %	Aux Bus Voltage %	Rated or Operating Limits	Min Gen Output MVAr (Aux Bus Load = 8.2 + j4.2 MVA)	Gen Bus Voltage %	Aux Bus Voltage %	Rated or Operating Limits
100.0	156.4	103.5	104.5	(1)	-0.9	95.0	95.8	(8)
101.0	149.0	104.0	105.0	(4)	-17.0	95.0	95.8	(8)
102.0	131.3	104.0	105.0	(4)	-33.0	95.0	95.8	(8)
103.0	113.7	104.0	105.0	(4)	-49.1	95.0	95.8	(8)
104.0	96.1	104.0	105.0	(4)	-55.4	95.6	96.5	(7)

Table III. Generator Reactive Capability Under Selected Tap Positions

Gross Generator Output = 175 MW

Sys Bus Voltage %	Max Gen Output MVAr (Aux Bus Load = 9.9 + j5.9 MVA)	Gen Bus Voltage %	Aux Bus Voltage %	Rated or Operating Limits	Min Gen Output MVAr (Aux Bus Load = 9.9 + j5.0 MVA)	Gen Bus Voltage %	Aux Bus Voltage %	Operating Limits NB
100.0	100.4	104.9	105.5	(5)	-19.2	96.1	96.6	(7)
101.0	102.9	105.6	106.2	(5)	-21.1	97.0	97.5	(7)
102.0	96.5	104.4	105.0	(4)	-22.9	97.9	98.5	(7)
103.0	78.4	104.4	105.0	(4)	-24.8	98.7	99.4	(7)
104.0	60.3	104.4	105.0	(4)	-26.6	99.6	100.3	(7)

Gross Generator Output = 50 MW

Sys Bus Voltage %	Max Gen Output MVAr (Aux Bus Load = 8.2 + j4.9 MVA)	Gen Bus Voltage %	Aux Bus Voltage %	Operating Limits NB	Min Gen Output MVAr (Aux Bus Load = 8.2 + j4.2 MVA)	Gen Bus Voltage %	Aux Bus Voltage %	Operating Limits NB
100.0	123.6	104.0	105.0	(4)	-40.1	95.0	95.8	(8)
101.0	105.6	104.0	105.0	(4)	-54.7	95.1	95.9	(7)
102.0	87.6	104.0	105.0	(4)	-56.0	96.1	96.9	(7)
103.0	69.5	104.0	105.0	(4)	-57.2	97.0	97.9	(7)
104.0	51.5	104.0	105.0	(4)	-58.4	97.9	98.9	(7)

4. RESULTS

Figure 4 shows the rated and operating generator reactive capability curves for the generator described in **Appendix A**. These curves are drawn for the current GSU and AUX transformer tap positions of 105% and 97.5%, Sys Bus voltage of 100% and generator outputs of 0 to 185 MW. The lagging reactive power (over-excitation) is limited by field and stator limits, and the leading reactive power (under-excitation) is limited by the low generator terminal voltage limit of 17.1 kV or 95%.

Table II lists similar data to those of **Figure 4**, but for a range of Sys Bus voltages of 100 to 104% and for min/max generator output of 50 & 175 MW. The rated or operating limits are identified by numbers (1) to (9) as described in **Table I.**

Figure 5 shows the rated and the operating generator reactive capability curves under 100% system bus voltage condition but with the selected GSU & AUX transformer tap positions of 102.5 % and 97.5%, for 0 to 185 MW generator outputs. The lagging reactive power (over-excitation) is limited by the AUX Bus high voltage limit of 2.52 kV (105%), and the leading reactive power (under-excitation) is limited by the generator terminal voltage limits of 17.1 kV or 95%, between 0 to 125 MW and by the minimum excitation limiter (MEL) relay setting from 125 MW to 185 MW.

Table III lists similar data to those of Figure 5, but for a range of Sys Bus voltages of 100 to 104% and for min/max generator output of 50 & 175 MW. The rated or operating limits are identified by (1) to (9) and described in **Table I.**

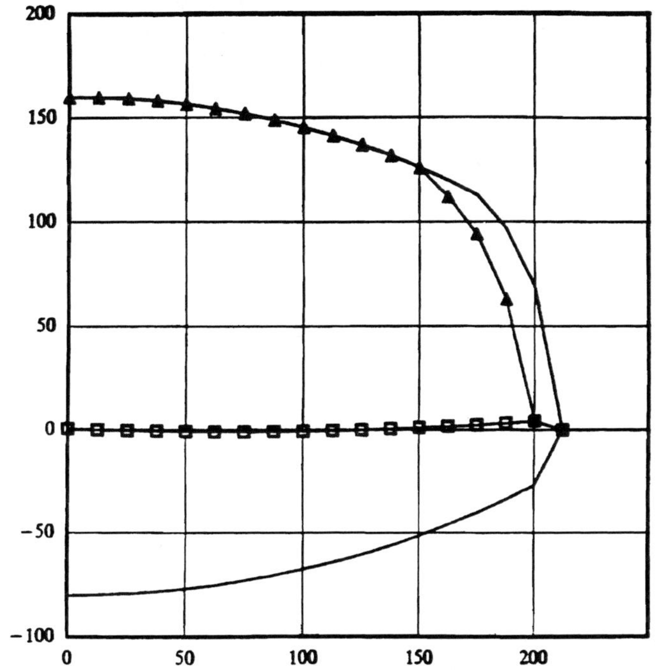

Megavars versus Megawatts

Figure 4.

Generator Step Up Transformer Tap at 105.0 %
Auxiliary Transformer Tap at 97.5 %
System Bus Voltage at 100 %

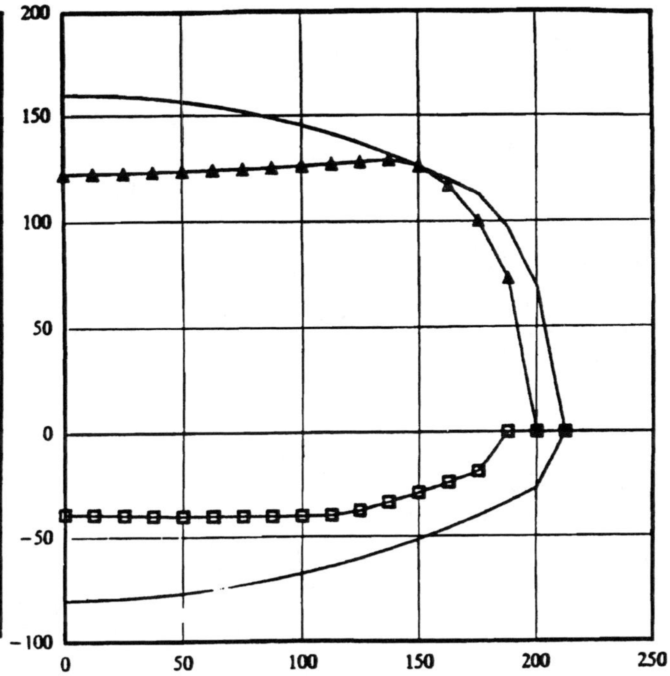

Megavars versus Megawatts

Figure 5.

Generator Step Up Transformer Tap at 102.5 %
Auxiliary Transformer Tap at 97.5 %
System Bus Voltage at 100 %

5. CONCLUSION

Generators provide a valuable resource in their ability to supply and absorb reactive power. The rated reactive power of generators are based on their field, stator and end-iron heating limits. The operating range is further limited by allowable limits of generator terminal and auxiliary bus voltages. By changing tap positions on the generator step-up and auxiliary transformers it is possible to vary the operating limits. For example, under certain operating condition it would be desirable to supply greater amount of reactive power, whereas under other conditions it would be better to absorb more reactive power.

It has been observed that he GSU and AUX transformer tap positions after the initial installations of the generating units are seldom changed. As the power system operating conditions change, it is necessary to check these transformer tap positions and ascertain that adequate over and under-excitation reactive powers are available to meet the needs of the power system under the prevailing operating conditions.

6. REFERENCE

[1] Adibi, M.M. and M.J. O'Grady, Reactive Power Balance During Restoration, in IEEE Committee Report, "Special Consideration in Power System Restoration - The second Working Group Report," paper 93 WM 202-2, presented at IEEE 1993 WPM, Columbus, Ohio Jan 31 - Feb, 1993.

[2] Adibi, M.M. and D.P. Milanicz, "Reactive Capability Limitation of Synchronous Machines," paper 94 WP 203-0 PWRS,T-PWRS presented at IEEE 1994 WPM, New York, February 1994

[3] "IEEE Guide for Transformers Directly Connected to Generators," IEEE C57.116-1989, Approved by American National Standards Institute, January 25, 1990.

[4] Golden, A.W., "Selection of Generator Stepup Transformer Ratings," IEEE Trans on Power Apparatus and Systems, Vol. PAS-100, No. 7 July 1981, pp 3425-3431.

[5] Adibi, M.M., D.P Milanicz & T.L. Volkmann "Simulating Transformer Taps for Remote Cranking Operations," IEEE Computer Application in Power, Vol 9, No 3, July 1996 Issue.

7. ACKNOWLEDGEMENT

The first author acknowledges support of Northern States Power Company for providing the physical and operational data.

ASYMMETRY ISSUES IN POWER SYSTEM RESTORATION

M.M. Adibi, LF
IRD Corporation

D.P. Milanicz, SM
Baltimore G&E Co.

T.L. Volkmann, M
Northern States Power Co.

Abstract: This paper is one of series presented on behalf of the Transmission Operation Subcommittee with the intent of focusing attention on the extra high voltage (EHV) asymmetry issues during restoration. Many existing EHV lines have unsymmetrical (horizontal) conductor spacing without being transposed. These characteristics generate unacceptable negative sequence currents (NSC). Under light-load condition and during restoration, NSC has caused cascade tripping of a number of generators resulting in an area wide blackouts, has prevented synchronization of incoming generators, and has blocked remote cranking of thermal units.

The paper describes a simple and approximate method for assessing the extent of NSC in EHV lines and then determining corrective measures. It recommends that when attempting to provide an off-site source to a thermal power plant the extent of NSC be determined and if unacceptable, either the operation be deferred to after the initial restoration phase, or the appropriate "underlying load" be determined for connection to the receiving end of the EHV lines.

Keywords: untransposed lines, EHV asymmetry, negative sequence current, imbalance voltage, imbalance current.

INTRODUCTION AND BACKGROUND

In monitoring and control of extra high voltage (EHV) systems, it is assumed that three-phase lines are balanced and therefore, only the positive sequence impedances are modeled and single-phase (typically, phase B) measurements are used. Significant imbalance however, can be generated in untransposed long EHV transmission lines. As a result of such asymmetries, the values of individual phase voltages and currents can be higher or lower than the values computed from the balanced analysis.

In building and operating EHV transmission lines, the industry's present practices are to use:

- Positive Sequence (one-line) network models in practically all off-line and on-line analysis.
- Single-phase measurements, consisting of single-phase voltage transformers (VT and CCVT), current transformer (CT), voltage, current, real and reactive powers transducers, etc.
- Horizontal conductor spacing having asymmetrical line parameters.
- Un-transposed lines in almost all EHV transmission lines.

These assumptions and practices under certain operating conditions have caused undesirable operation such as: inadvertent tripping of the protective devices, thereby incurring loss of system components and possibly the start of an area wide outage, producing unacceptable negative sequence voltage levels which have damaged the rotating machineries, and triggering steady state instability or voltage collapse on occurrence of further system events.

The following recent events under light-load conditions and during initial phases of power system restoration have refocussed attention on the EHV asymmetry issues:

1. One west coast utility with a 500 kV, 150 mile, 70% series compensated, untransposed transmission line has observed negative sequence voltages (NSV) during light load in excess of 30% of the positive sequence voltages [1].
2. In a mid-western utility, a large generator could not be synchronized to an energized 345 kV incoming line due to the presence of 11% NSV. Generators are typically set to alarm if NSV levels reach 4% and trip the generator at about 10% [2].
3. In a mid-Atlantic utility, when attempting to provide remote cranking power to a large steam electric station over a 500 kV line, excessive NSV was observed which, if energized, would have tripped or damaged the station's large motors. A 10% (NSV) impressed on a motor or a generator which typically have about 10% negative sequence reactance (i.e., 10% of their positive sequence reactances), produces 100% negative sequence current which together with the positive sequence current overheat the motor or generator.
4. In Australia, during a light-load period, a utility with an extensive 500 kV transmission network has experienced sequential tripping of generators due to excessive NSV producing a total blackout [3].

PE-066-PWRS-0-11-1997 A paper recommended and approved by the IEEE Power System Operations Committee of the IEEE Power Engineering Society for publication in the IEEE Transactions on Power Systems. Manuscript submitted May 10, 1997; made available for printing November 7, 1997.

Reprinted from *IEEE Transactions on Power Systems,* Vol. 14, No, 3, pp. 1085–1091, August 1999.

5. In the mid-south, the asymmetry in a new and long 500 kV line has caused extensive damage to customer rotating machinery.

Several recent papers have drawn attention to the asymmetry in EHV lines. A three-phase load flow method interfacing with the Electromagnetic Transient Program (EMTP) has been developed and applied to an EHV system which effectively utilizes three phase measurements [4]. The combined effects of asymmetry in line parameters and transformer bank impedance mismatch has been corrected using different transformer tap positions in [5]. It has been suggested that the unbalanced system conditions be estimated using a combination of conventional power flow and short circuit (SC) programs in [6].

This paper reviews the effects of unbalanced parameters on restoration of EHV lines, develops a simple and approximate method for their assessment and recommends ways of addressing the problem. In the following sections, first an approximate method is developed for parameter calculations of an EHV line with typical conductor spacing. The methodology then is used to determine the negative sequence currents for several lines of different lengths, under light load conditions. Finally, the approach is validated by field measurements on a 230 kV and 500 kV lines. The paper recommends that under light-load conditions or during initial phases of restoration, prior to providing remote cranking power to a steam plants or providing off-site sources to nuclear plants, appropriate "underlying load" be determined and connected to the receiving end of the EHV lines.

UNSYMMETRICAL LINE PARAMETERS

Figure 1 shows the asymmetrical phase spacing of a typical 500 kV transmission tower. The geometric mean radius (GMR) for the phase conductor of 2300 MCM ACSR 84/19, the self geometric mean distance (D_S) for the two bundle conductors positioned 18 inches apart (d=1.5 ft.), and the mutual geometric mean distance (D_M) for the conductor spacings (D=35 ft.), are:

$$GMR = 0.0607 \text{ ft.},$$

$$D_S = (d\ GMR)^{1/2} = 0.3017 \text{ ft, and}$$

$$D_M = 2^{1/3}\ D = 44.1 \text{ ft.}$$

Neglecting the relatively insignificant coupling between the phases, the inductances and capacitances for phase A, B and C of the untransposed line can be determined from the following formula:

$$L_A = L_C = k_1 \log(2D/D_S) = k_1\ 2.3655,$$

$$L_B = k_1 \log(D/D_S) = k_1\ 2.0645,$$

$$C_A = C_C = k_2/(\log(2D/2^{1/3}D_S)) = k_2 / 2.2652, \text{ and}$$

$$C_B = k_2 / (\log(D/2^{1/3}D_S)) = k_2 / 1.9642.$$

The inductance (L_P) and capacitance (C_P) of the same line when fully transposed are:

$$L_P = k_1 \log(D_M/D_S) = k_1\ 2.1649, \text{ and}$$

$$C_P = (k_2)/(\log(D_M/D_S)) = (k_2)/\ 2.1649.$$

where inductances (L) are in henry/mile, and capacitances (C) are in μf/mile.

The comparison of inductances and capacitances of the fully transposed and the untransposed line shows:

$$L_A/L_P = L_C/L_P = 1.0927,$$

$$C_A/C_P = C_C/C_P = 0.9557,$$

$$L_B/L_P = 0.9536, \text{ and}$$

$$C_B/C_P = 1.1022, \text{ all in per unit.}$$

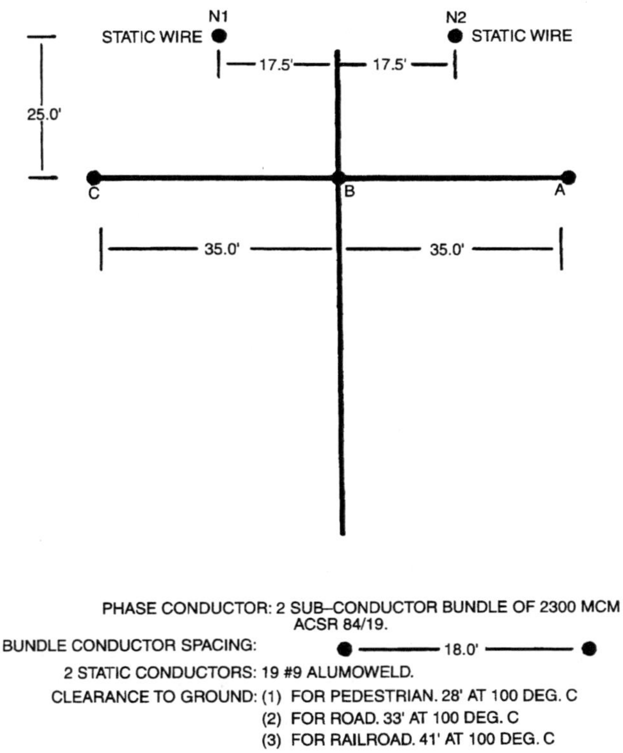

Figure 1. Conductor Spacing - 500 kV Line

The positive-sequence impedance, Z_P, and the total line charging, $3Q_P$, for a 500 kV line, 66.25 mile long with the conductor size and spacing of **Figure 1**, on the basis of 500 kV and 100 MVA are:

$$Z_P = 0.000644 + j\ 0.015513, \text{ per unit, and}$$

$$3Q_P = 119.22, \text{ Mvar.}$$

Using the above L & C ratios, the corresponding impedances and charging currents for phases A and B of the untransposed line can be determined:

$Z_A = 0.000644 + j\ 0.016951$, per unit,

$3Q_A = 113.94$, Mvar,

$Z_B = 0.000644 + j\ 0.014793$, per unit, and

$3Q_B = 131.40$, Mvar.

NEGATIVE SEQUENCE I AND V CALCULATIONS

Consider an unsymmetrical three-phase 500 kV line with symmetrical sending end currents and voltages. **Figure 2** shows the pi-equivalent of one phase of such a line, sectionalized into n sections. It defines the notation used in determining the receiving end current I_n and voltage V_n magnitudes, in per unit.

In the first section of the line, for a given power flow P_i, and a power factor pf, we have:

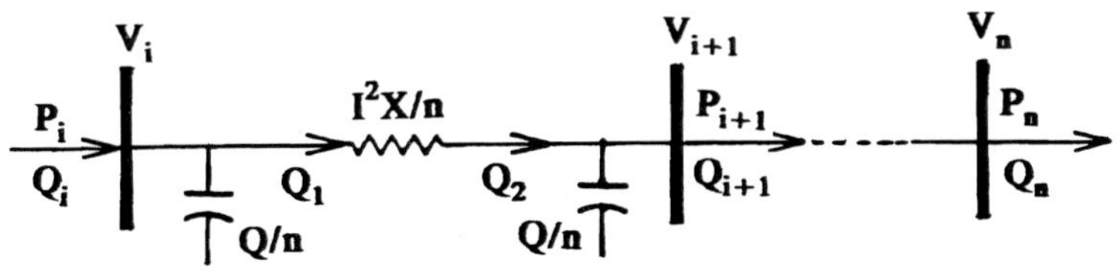

Figure 2. 500 kV Line Pi-Equivalent Model

$P_i = P_{i+1}$, i.e., neglecting line losses, and

$Q_i = P_i \tan(\cos^{-1}\text{pf})$

$Q_1 = Q_i + V_i^2\ Q/n$,

$I = (P_i^2 + Q_1^2)^{1/2}/V_i$,

$Q_2 = Q_1 - I^2\ (X/n)$,

$V_{i+1} = ((V_i - Q_1\ (X/n)/V_i)^2 + (P_i\ (X/n)/V_i)^2)^{1/2}$,

$Q_{i+1} = Q_2 + V_j^2\ Q/n$, and

$I_{i+1} = (P_i^2 + Q_{i+1}^2)^{1/2}/V_{i+1}$.

It can be shown that the differences between the receiving end I_n & V_n for a 100 mile line, sectionalized into n=3 & n=4, are insignificant and therefore in determining I_n & V_n, n=3 has been used.

Using the above pi-equivalent and Phase A line parameters (i.e., substituting Q_A & X_A, proportional to the length of line for Q & X of Figure 2), the Phase A current magnitude I_A and voltage magnitude V_A are determined for a range of P_i power flows. Similarly, using Q_B & X_B, Phase B current magnitude I_B and voltage magnitude V_B, are determined. As shown in Figure 1, Phase A and Phase C are identical, and therefore, I_C & V_C are equal to I_A & V_A, respectively.

It should be noted that under the above operating condition, the receiving end's unequal phase currents and voltages are primarily due to the NSV generated in the untransposed line and as such contain no zero sequence components.

From the symmetrical components we have:

$$I_A = I_1 + I_2 + 0,$$

$$I_B = a^2I_1 + ai_2 + 0, \quad and$$

$$I_C = aI_1 + a^2I_2 + 0.$$

where, I_A, I_B & I_C are the three receiving end phasor currents, I_1 & I_2 are the related positive and negative sequence currents, and a is the operator $e^{j2\pi/3}$ or $\underline{/120°}$.

The *italic I and V* indicate phasor currents and voltages.

Now, $I_1 + a^2I_1 + aI_1 = 0$, *and*

$I_2 + aI_2 + a^2I_2 = 0.$

Therefore,

$$I_A + I_B + I_C = 0,$$

that is, I_A, I_B *and* I_C current phasors form a closed triangle with angles α, ß and γ opposite currents I_A, I_B *and* I_C, respectively.

From the Law of Cosines we have:

$$\alpha = \cos^{-1}[(I_B^2 + I_C^2 + I_A^2)/(2I_BI_C)],$$

$$\beta = \cos^{-1}[(I_A^2 + I_C^2 + I_B^2)/(2I_AI_C)], \text{ and}$$

$$\gamma = \cos^{-1}[(I_B^2 + I_A^2 + I_C^2)/(2I_AI_B)].$$

Let, $I_A = I_Ae^{j\Theta}$,

$I_B = I_Be^{j\Phi}$, and

$I_C = I_Ce^{j\Psi}$.

For, $\Theta = 0.0$, we have,

$\Phi = -(\alpha+\beta)$, and

$\Psi = +(\alpha+\gamma)$.

The positive and negative sequence currents I_1 & I_2 is obtained from the three phase currents:

$$I_1 = (1/3)[I_Ae^{j\Theta} + aI_Be^{j\Phi} + a^2I_Ce^{j\Psi}], \text{ and}$$

$$I_2 = (1/3)[I_Ae^{j\Theta} + a^2I_Be^{j\Phi} + aI_C^{j\Psi}].$$

By similar analysis positive and negative sequence voltages V_1 *and* V_2 can be determined.

Figure 3 shows a plot of negative sequence currents (NSC) at the receiving end of untransposed lines, plotted against MW flows expressed as pu of the surge impedance loading (SIL) of the line under light load conditions. In plotting these curves the following data were used:

Length of lines: 50, 100 and 150 miles,
SIL= $2.5*500^2/1000 = 625$ MW
V_i = 100% or 500 kV
pf = 95 % lead

In Figure 3, the NSC is expressed as % of the positive sequence currents. The NSC curve for the 50 mile 500 kV line shows that to reduce the NSC to below the relay setting of 4%, over 80 MW of underlying load should be connected to the receiving end before energizing the line. The other two curves indicate that with longer lines more underlying loads are required to maintain acceptable NSC.

The generator negative sequence relays are typically set at 4% for alarm and 10% for tripping [7].

Many large steam electric stations and nuclear power stations are connected to load centers via long EHV lines. In providing cranking power to the steam units or providing off-site sources to nuclear plants, the effects of NSC on the large auxiliary motors need to be considered.

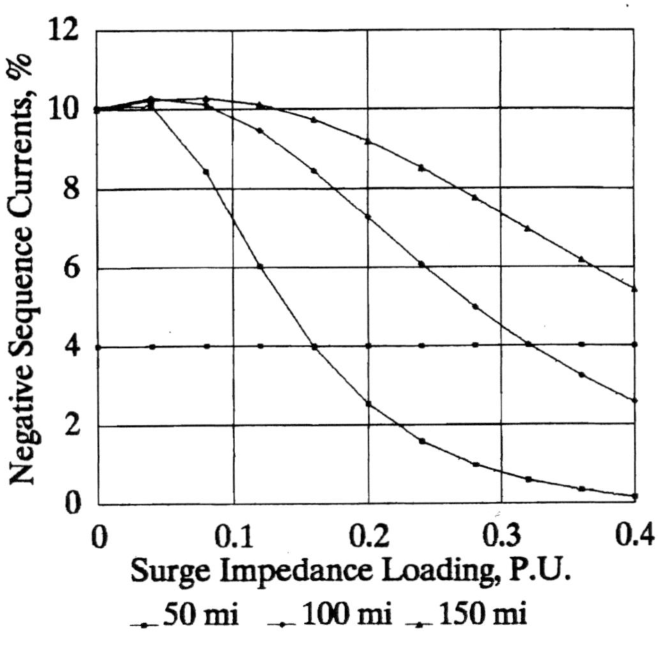

Figure 3. Current Imbalance - 500 kV Lines

TEST AND VERIFICATION

In order to test and show the degrees of asymmetry in the extra high voltage (EHV) lines as compared to the high voltage lines, three phase current measurements were obtained at the receiving end of a 500 kV and a 230 kV lines both being unsymmetrical and untransposed. The current measurements were collected over several days of scanning to ensure adequate ranges of measurements.

The transformer neutral currents for both lines were small and as compared with the phase current magnitudes could be ignored.

The following single scan three phase current measurements illustrate the asymmetry for the 500 kV lines:

$I_A = 510$, $I_B = 589$, $I_C = 522$ in amperes

$\alpha = 54.2°$, $\beta = 69.6°$, $\gamma = 56.2°$,

$\Theta = 0.0°$, $\Phi = 236.2°$, $\Psi = 110.4°$,

$I_1 = 539.1$, and $I_2 = 50.3$ in amperes.

The phase current asymmetries <u>may be defined</u> as :

Ph A = $100(I_A/I_1 - 1)$, in %, etc., therefore,

Ph A =-5.4%, Ph B = 9.3%, Ph C = -3.2%.

Similarly for the 230kV lines we have:

$I_A = 330$ $I_B = 321$ $I_C = 309$ in amperes

$\alpha = 63.1$, $\beta = 60.2°$, $\gamma = 56.7°$,

$\Theta = 0.0°$, $\Phi = 236.7°$, $\Psi = 119.8°$,

$I_1 = 319.9$, $I_2 = 12.1$ in amperes, and

Ph A = 3.2%, Ph B = 0.3%, Ph C = -3.4%.

Figures 4 & 5 show the NSC as a % of positive sequence current (PSC) for the above 230 kV and 500 kV lines. The NSC for the 500 kV line varies from about 85 % at light-load to about 10 % at full load. As shown, the NSC in the 230 kV line is comparatively insignificant.

CONCLUSION

This paper has undertaken a limited analysis of untransposed EHV line based on limited field measurements. It has shown that significant voltage and current imbalances do exist between the three phases.

It has shown that while the symmetry assumption holds for high voltage transmission systems (230 kV and below), the asymmetry is significant in the extra high voltage (345 kV and above) lines and can not be ignored.

The extent of NSC in EHV lines depends on the length and loading of the line. Under light-load conditions, it is the imbalance in line's capacitance which is the dominant factor interfering with the remote cranking operation or causing sequential tripping of the generators.

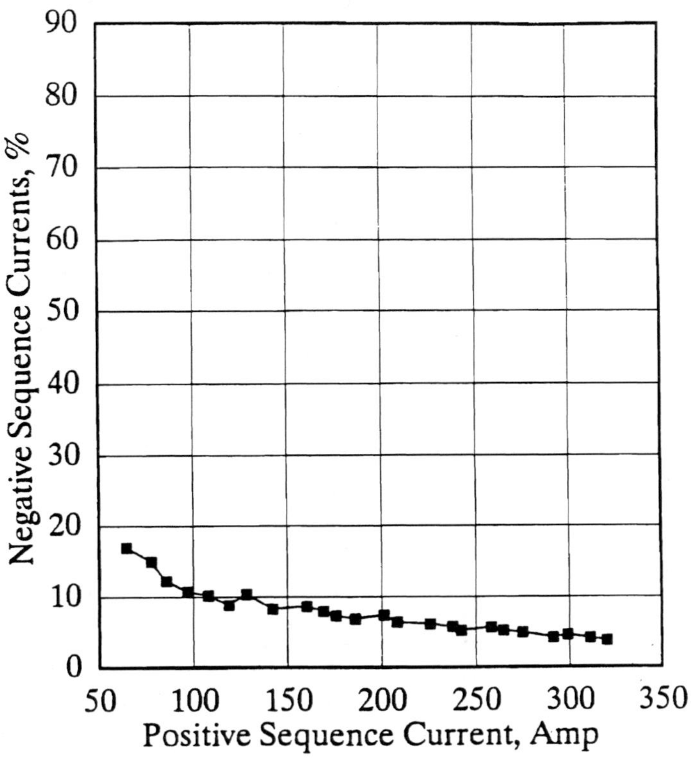

Figure 4. Current Imbalance - 230 kV Line

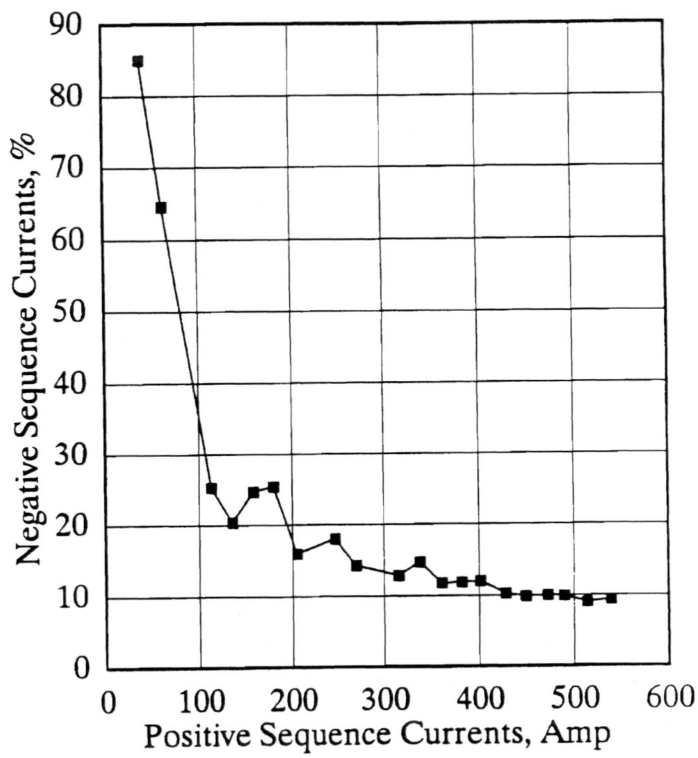

Figure 5. Current Imbalance - 500 kV Line

RECOMMENDATION

In order to alleviate the effect of NSC, it is recommended that, either energization of certain EHV lines be deferred to after the initial phase of restoration, or the adequate underlying load be determined and connected to the receiving end of line before energizing the EHV line.

REFERENCES

[1] IEEE Power Systems Relaying Committee, "EHV Protection Problem," IEEE Trans. Power App. Sys., vol. PAS-100, no.5, pp. 2399-2406, May 1981

[2] Volkmann, T.L., "Assessment of Switching Status and Switching Operation," presented at Seminar on Power System Restoration, Sept 1993, Denver Colorado, EPRI

[3] Lachs, W.R., "Discussion of Analytical Tools Requirements for Power System Restoration," paper 94 WM 222-0-PWRS, presented at IEEE 1994 WPM, New York, Jan 30 - Feb 3, 1994

[4] Allenmong J.J., et al, "Multiphase Power flow Solutions Using EMTP and Newton's Method,' IEEE Trans. on Power Systems, Vol. 8, No. 4, November 1993, pp1455-1462

[5] Gutman, R. and L.B. Wagenaar, "EHV Transformer Bank Unbalance: Practical Issues and Solutions," paper 96 WM 224-6 PWRD, Presented at IEEE 1996 WPM, Baltimore, Maryland.

[6] Reichelt, D. et al, "Estimation of Steady-State Unbalanced System Conditions Combining Conventional Power Flow and Fault Analysis Software," IEEE Trans.on Power Systems, Vol. 11, No. 1, February 1996, pp 422-427.

[7] Adibi, M.M. and D.P. Milanicz, "Protective System Issues During Restoration," IEEE Trans. on Power Systems, Vol 10, No. 3, August 1995, pp 1492-1497.

ACKNOWLEDGEMENT

The author acknowledges support of Baltimore Gas & Electric Company and Potomac Electric Power company for providing physical and operational data.

Discussion

Lester H. Fink: The authors have presented a timely and very helpful discussion of a neglected problem, one which has significance beyond the focus on restoration indicated in the title. In addition to situations refered to in the introduction, for instance, one thinks of potential implications as an unremarked source of uncertainty for operation planning within the new "open access" environment.

A few minor comments may be in order: (1) in the second column of the second page, it might be helpful to explain the use of "3Q"; (2) in the second column of page 3, second paragraph, "no zero sequence components" might better be "negligible zero sequence components"; finally down the same column, the expressions based on the law of cosines should have minus signs preceding the third current within the parentheses.

M.M. Adibi, D.P. Milanicz & T.L. Volkmann: The authors thank Mr. Fink for his comments regarding the potential implications of asymmetry in EHV lines during heavy loadings. A similar analysis was undertaken for the heavy load situations, as illustrated in **Figure A.** It shows a significant voltage difference between Phases B and A at the receiving ends of several 500 kV lines. These voltage unbalances may limit the transfer capabilities of the EHV lines. The presence of NSV may well explain the cause of a state-estimator non-convergence during daily heavy power delivery, as experienced in the southeastern 500 kV interconnection.

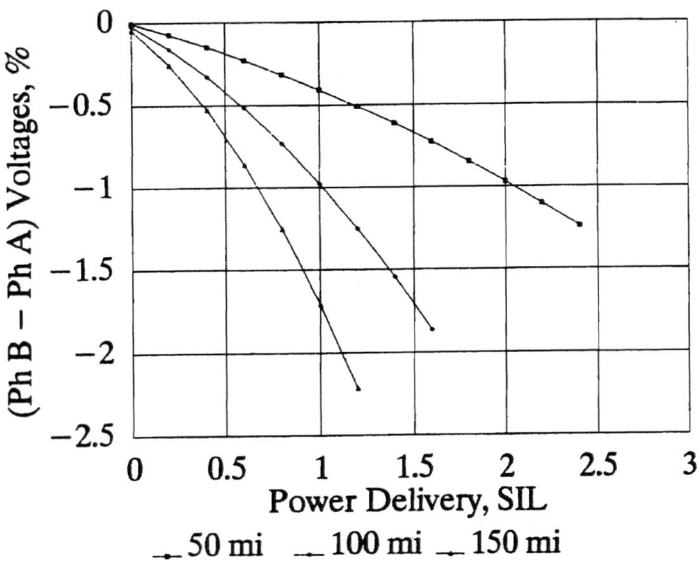

Figure A. Voltage Imbalance In 500 kV Lines
The Sending-End Voltage Kept At 105%
The Receiving-End Voltage Not Less Than 95%

The corrective actions such as: the use of three-phase analytical tools, installation of three-element instrumentations, re-configuration of conductor spacings, and line transpositions, are all cost prohibitive. Since the NSV is a quadratic function of line loadings, it is suggested that for a given line and the range of its loading it should be possible to determine the correction coefficients and correct the measurements used in the single-line analytical tools, i.e.,

$$V_B - V_+ = A + B*P_B + C*P_B^2,$$

where, the correction coefficients A, B, C and measurements P_B and V_B can be used to determine V_+ [a].

Our apologize for the typos "3Q" and the "sign" in the law of cosines. In our analysis, $Z=R+X_L$ and $Q=V^2/X_C$, representing the positive sequence impedance and the charging current of single-line in Pi-equivalents were used. Also, the law of cosines used were:

$$\alpha = \cos^{-1}[(I_B^2 + I_C^2 - I_A^2)/(2I_BI_C)],$$

$$\beta = \cos^{-1}[(I_A^2 + I_C^2 - I_B^2)/(2I_AI_C)], \text{ and}$$

$$c = \cos^{-1}[(I_B^2 + I_A^2 - I_C^2)/(2I_AI_B)].$$

The authors are in agreement with Mr. Fink regarding the presence of both negative and zero sequence current components. Our analysis shows that the NSC and ZSC generated in the untransposed lines are equal in magnitudes. However, the error introduced in determining the NSC% in **Figure 3** of the paper, by neglecting the ZSC, is very small, as shown in **Figure B.** Neglecting the ZSC would allow computations of negative and positive sequence currents and voltages from the available magnitudes of line's three phase measurements which usually are made.

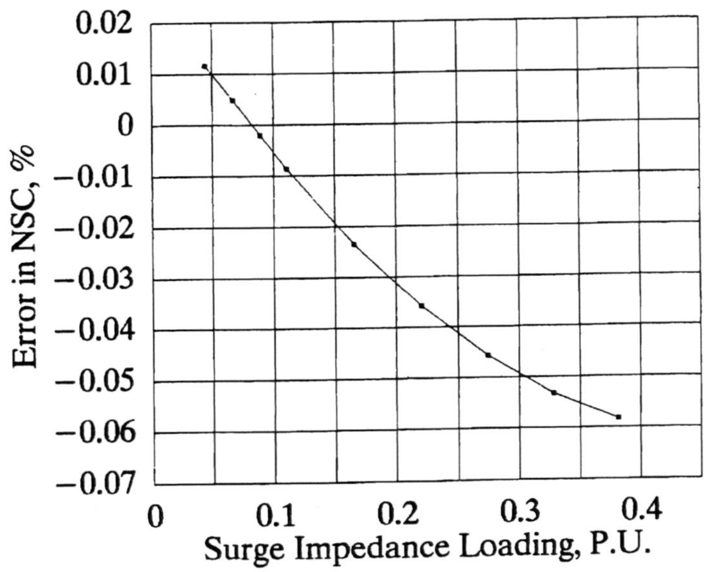

Figure B. NSC Error In A 150 Mile, 500 kV Line
The NSC Error Caused By Neglecting The ZSC
Expressed As % Of The PSC

Reference:

[a] Adibi, M.M. and D.K. Thorne "Remote Measurement Calibration," IEEE Trans. on Power Systems, vol. pwrs-1, no.2, pp.194-203, May 1986

FREQUENCY RESPONSE OF PRIME MOVERS DURING RESTORATION

M.M. Adibi, LF — IRD Corporation
J.N. Borkoski, SM — Baltimore Gas & Elec.
R.J. Kafka, SM — Potomac Elec Power
T.L. Volkmann, SM — Northern States Power

Abstract: In the initial phase of restoration, operators are often concerned with the size of load which can safely be picked-up, and the effectiveness of the generation reserve. There has been a need for a methodology to readily address these concerns by providing simple guidelines to facilitate an orderly power system restoration.

In this paper, a simple approach is developed based on an approximate frequency response rate of prime movers for determining: (a) the maximum load pickup within the allowable system frequency dip, and (b) the amount and distribution of reserve for maintaining firm generation.

Keywords: Frequency Response Rate, Generator Reserve Rate, Reserve Distribution, Load Pickup, Prime Movers,

1. INTRODUCTION

Restoration procedure following a major power system disturbance may span over three periods, depending on the types and availability of prime movers: (a) the initially available prime movers which typically consist of; blackstart combustion turbines (CT), low-head short-conduit hydro-electric (HE), and gas or oil-fired drum-type boiler-turbine steam-electric (SE), (b) the subsequently available prime movers consisting of; large CTs, high-head long-conduit HEs, and combined cycle units, and (c) the finally available prime movers consisting of; large drum-type coal-fired SEs, super-critical once-through units and nuclear plants [1].

During the above three periods and in particular during the initial period, operators are concerned with: the rate at which generators are loaded, the mismatch between load pickup and prime movers' frequency response, the adequacy and distribution of generator reserve, to cope with the loss of the largest unit [2].

In this paper, a simple guideline is developed for: (a) evaluating frequency response rates of prime movers to sudden increase in loads, and (b) determining the amount and distribution of reserve to meet the loss of the largest generator. The frequency responses of a typical CT, HE and SE units are determined for different increments of load pickups. These responses are used to develop guidelines to assist the operators in keeping the frequency above the allowable limits and to maintain adequate reserve. The approach is applied to several case studies and results are verified by simulation. Although these guidelines provide approximate solutions, they do expedite the restoration process.

PE-412-PWRS-0-06-1998 A paper recommended and approved by the IEEE Power System Operations Committee of the IEEE Power Engineering Society for publication in the IEEE Transactions on Power Systems. Manuscript submitted January 5, 1998; made available for printing June 12, 1998.

2. PROCEDURE

2.1. Prime Movers' Block Diagram

Figure 1 shows a simple block diagram for the prime movers under consideration. The $G_1(s)$, $G_2(s)$, and $G_3(s)$ represent the transfer functions for the speed-governor systems, prime movers, and the power system, respectively. The change in system frequency ΔF, is related to the change in load ΔL, by [3]:

$$\Delta F = -F(s)\Delta L/s,$$

where,

$$F(s) = N(s)/D(s),$$
$$N(s) = G_3(s),$$
$$D(s) = 1+G_1(s)G_2(s)G_3(s)/R, \text{ and}$$

R is the governor speed regulation.

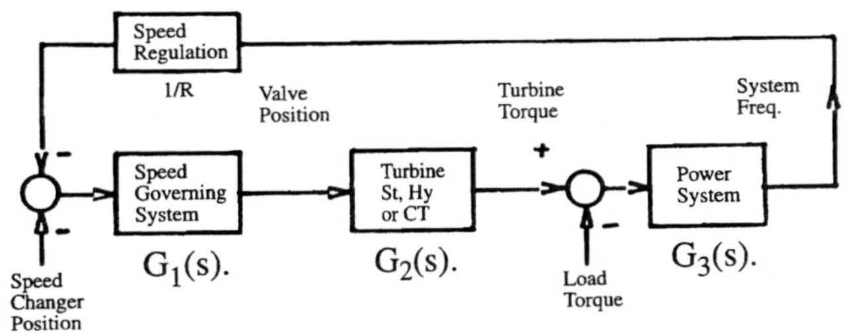

Figure 1 - Prime Movers Block Diagram

2.2 Combustion Turbine Units Frequency Response

The transfer functions and data for a typical blackstart CT are as follows:

$$\Delta F = -F(s)\Delta L/s,$$
$$F(s) = N_1(s)/D_1(s),$$
$$N_1(s) = G_3(s),$$
$$D_1(s) = 1+G_1(s)G_2(s)G_3(s)/R,$$
$$G_1(s) = 1/(1+sT_G),$$
$$G_2(s) = 1/(1+sT_T), \text{ and}$$
$$G_3(s) = (1/D)/(1+sT_P).$$

where,

R , governor speed regulation	=	5.00, %
T_G, governor time constant	=	0.15, sec.
T_T, turbine time constant	=	0.50, sec.
T_P, inertia constant	=	10.00, sec.
D , damping torque coefficient	=	0.75, %.

Figure 2 shows frequency response of a typical CT to sudden increase in loads. The upper curve shows about 0.25 Hz frequency decline in response to a 5% sudden

Reprinted from *IEEE Transactions on Power Systems,* Vol. 14, No. 2, pp. 751–756, May 1999.

increase in load when the CT is under 20% of load. The lower curve shows about 0.5 Hz frequency decline for a 10% sudden load increase under no load condition.

The CT frequency responses for different load pickups have been used to determine the "frequency response rates" of the CT, as shown in **Figure 3**. Accordingly, the frequency response rates for the CT are:

Curves	Load %	ΔF/ΔL Hz/p.u.
CT-1	5.0	-4.81
CT-2	40.0	-4.61
CT-3	75.0	-4.41

From **Figure 3**, it can be seen that: (a) the frequency decline, ΔF is proportional to the load increment, ΔL, and (b) the prime mover load has an insignificant effect on the frequency response rate.

2.3 Steam Electric Units Frequency Response

The transfer functions and data for a typical tandem compound, single reheat SE unit are as follows:

$$\Delta F = -F(s)\Delta L/s,$$
$$F(s) = N_3(s)/D_3(s),$$
$$N_3(s) = G_3(s),$$
$$D_3(s) = 1+G_1(s)G_2(s)G_3(s)/R,$$
$$G_1(s) = 1/(1+sT_G),$$
$$G_3(s) = (1/D)/(1+sT_P), \text{ and}$$
$$G_2(s) = N_2(s)/D_2(s)$$
$$N_2(s) = (1+sT_C)(1+sK_1T_R)+(K_3)$$
$$D_2(s) = (1+sT_S)(1+sT_R)(1+sT_C)$$

where,

T_G, Governor time constant	= 0.20, sec.
T_R, Dashpot time constant	= 5.00, sec.
R , Governor speed regulation	= 3.00, Hz/p.u.
D , Damping torque coefficient	= 3.00, p.u.
T_P, Inertia time constant	= 5.56, sec.
T_S, HP steam box time constant	= 0.50, sec.
T_R, Reheater time constant	= 10.00, sec.
T_C, Crossover (IP-LP) time constant	= 0.50, sec.
K_1, HP turbine Power Fraction	= 0.30, p.u.
K_2, IP turbine power fraction	= 0.40, p.u.
K_3, LP turbine power fraction	= 0.30, p.u.

Figure 4 shows frequency response of a typical drum-type tandem compound, single reheat SE unit to sudden increase in loads. The upper curve shows about 0.4 Hz frequency decline in response to a 5% sudden load increase when the SE unit is under 20% load. The lower curve shows about 0.9 Hz frequency decline for a 10% sudden load increase under no load condition.

The SE frequency responses for different load pickups have been used to determine the "frequency response rates" of the SE, as shown in **Figure 5**. Accordingly, the frequency response rates for the SE unit are:

The frequency response rates for the SE unit are:

Curves	Load %	ΔF/ΔL Hz/p.u.
SE-1	5.0	-8.56
SE-2	40.0	-7.94
SE-3	75.0	-7.33

From **Figure 5**, it can be seen that: (a) the frequency decline, ΔF is proportional to the load increment, ΔL, and (b) the prime mover load has a relatively small damping effect on the frequency response rates (as long as the unit is not loaded near a valve point).

2.4 Hydro Electric Units Frequency Response

The transfer functions and data for a typical run-of-the-river Kaplan turbine are as follows:

$$\Delta F = -F(s)\Delta L/s,$$
$$F(s) = N_4(s)/D_4(s),$$
$$N_4(s) = G_3(s),$$
$$D_4(s) = 1+G_1(s)G_2(s)G_3(s)/R,$$
$$G_1(s) = 1/(1+sT_G)(1+sT_R)/(1+(sT_R)(r/R))$$
$$G_2(s) = (1-sT_W)/(1+sT_W/2), \text{ and}$$
$$G_3(s) = (1/D)/(1+sT_P).$$

where,

R , Permanent speed droop coefficient	= 0.050, p.u.
r , Transient speed droop coefficient	= 0.600, p.u.
T_G, Gate servomotor time constant	= 0.200, sec.
T_R, Dashpot time constant	= 5.000, sec.
T_P, Inertia time constant	= 10.000, sec.
T_W, Water starting time penstock	= 1.000, sec.
D , Damping torque coefficient	= 0.025, p.u.

Figure 6 shows frequency response of a typical Kaplan HE unit to sudden increase in loads. The upper curve shows about 0.4 Hz frequency decline in response to a 5% sudden load increase when the HE unit is under 20% load. The lower curve shows about 0.5 Hz frequency decline for a 10% sudden load increase under no load condition.

Figure 7, is a plot of frequency declines for the HE to sudden load increases. The frequency response rates for the HE unit are:

Curves	Load %	ΔF/ΔL Hz/p.u.
HE-1	5.0	-5.35
HE-2	40.0	-4.61
HE-3	75.0	-1.46

From **Figure 7**, it can be seen that: (a) the frequency decline, ΔF is proportional to the load increment, ΔL, and (b) the HE unit load does affect the frequency response rate.

Figures 3 and 5, show that whereas one can approximate the frequency response rates, ΔF/ΔL, for the CT and the SE units to a constant value over their entire generation outputs, such an assumption does not fully hold for the HE unit as shown in **Figure 7**. However, as the following case study results verify, the above approximation may also be extended to the HE unit.

Accordingly, the frequency response rates for the three

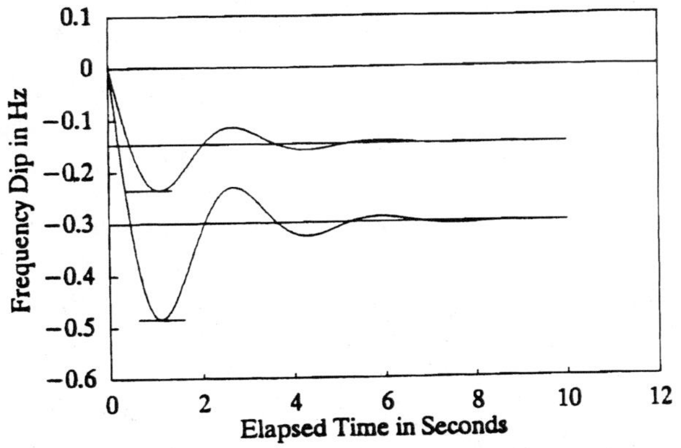

Figure 2. Frequency Response of the CT Unit

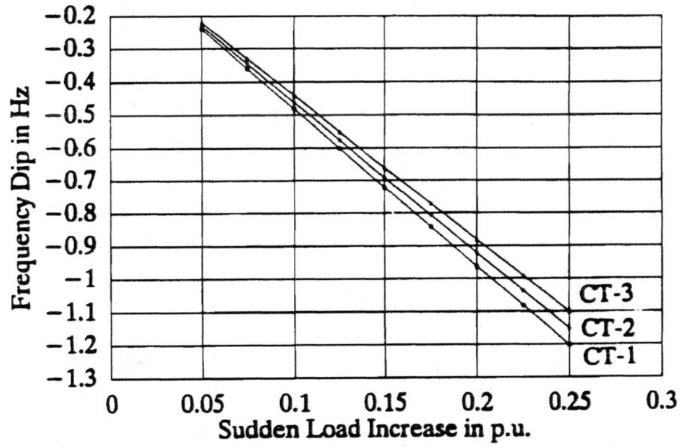

Figure 3. Frequency Response Rates of the CT Unit

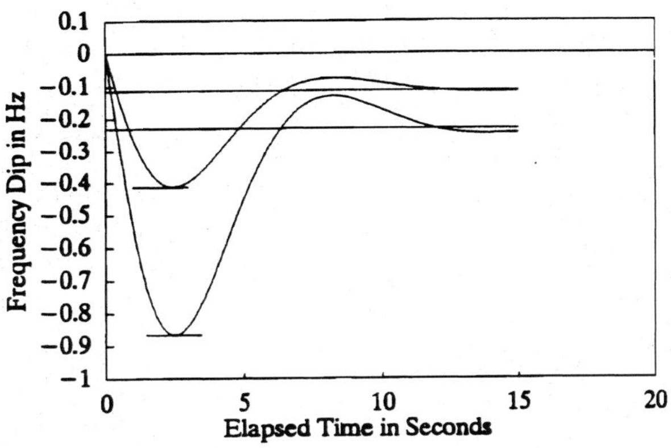

Figure 4. Frequency Response of the SE Unit

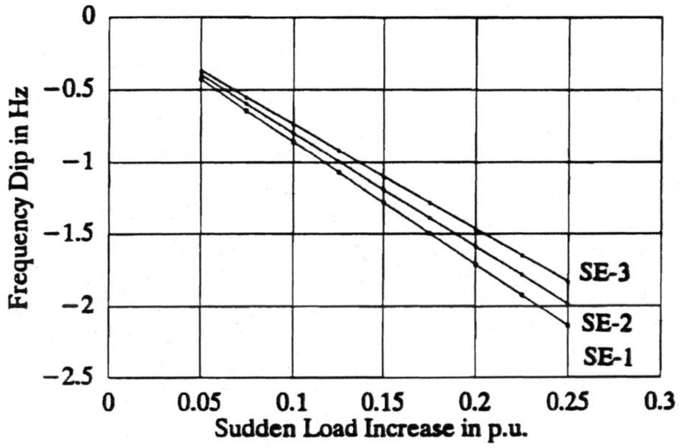

Figure 5. Frequency Response Rates of the SE Unit

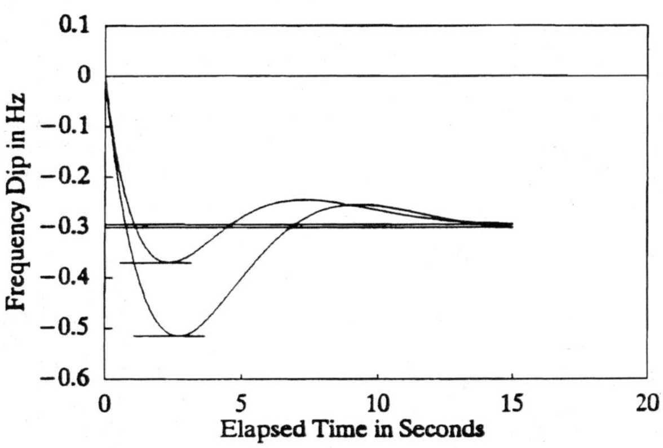

Figure 6. Frequency Response of the HE Unit

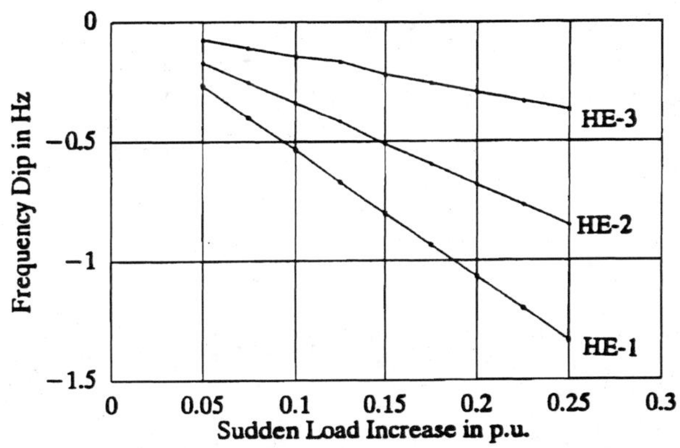

Figure 7. Frequency Response Rates of the HE Unit

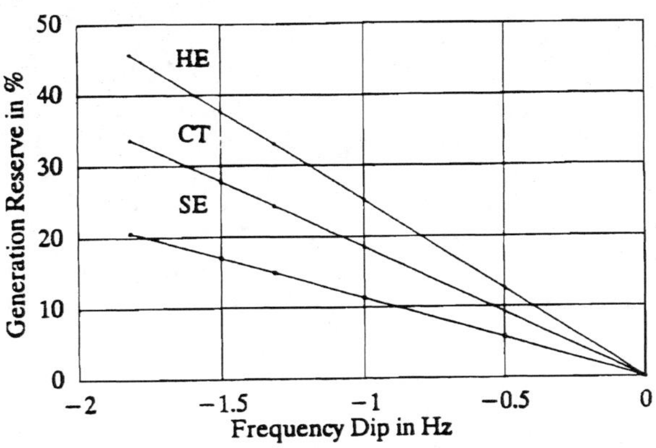

Figure 8. "Optimal" Generation Reserve Distribution

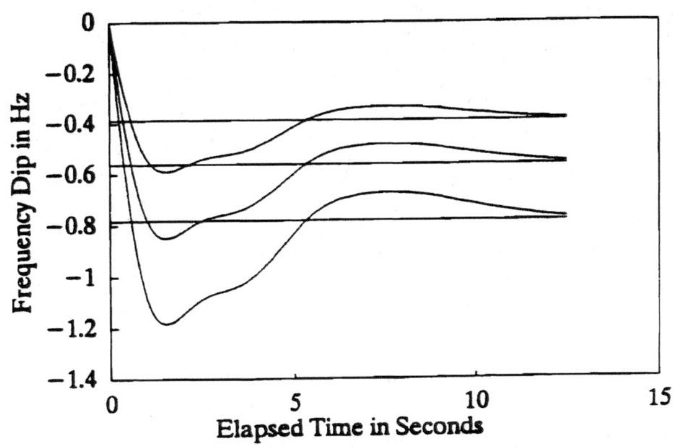

Figure 9. Frequency Response of the Power System

types of prime movers over their entire generation outputs can be approximated as:

Curves	Load %	$\Delta F/\Delta L$ Hz/p.u.
HE-2	40	-3.40
CT-2	40	-4.61
SE-2	40	-7.94

It follows that, in the initial restoration phase, by using the above frequency response rates ($\Delta F/\Delta L$), one can readily determine: (a) the maximum amount of load which can be picked up for a given frequency dip (or vice versa), and (b) the amount and distribution of generator reserve for coping with the loss of the largest generator. These guidelines are illustrated in the following case studies.

3. CASE STUDIES

3.1 Determination ΔL for a Given ΔF'

Table 1. lists the prime movers which may be in operation in the initial phase of a power system restoration. Assuming that the maximum frequency decline allowed is $\Delta F' = -0.75$ Hz, the objective is to find the maximum amount of load which can be picked up. This would guide the operator to, e.g., pickup a low voltage AC (LVAC) network load of an estimated size.

Accordingly, the size of LVAC Network load which can be picked up for $\Delta F' = -0.75$ Hz, is 57 MW or about 13% of the total on-line generation capacity. This assumes that prior to the load pickup: (a) the generation reserve is distributed on the basis of 45, 37 and 18% for the SE, HE and CT units, respectively, and (b) the generation reserve is greater than 57 MW.

3.2 Determination of ΔF for an Estimated ΔL'

In **Table 2.** the size of an LVAC Network load is estimated at about $\Delta L' = 50$ MW, the objective is to find the maximum frequency dip [4].

Accordingly, the maximum frequency dip for a $\Delta L' = 50$ Mw sudden increase in load is $\Delta F = -0.66$ Hz. This again assumes that prior to the load pickup: (a) the generation reserve is distributed on the basis of 45, 37 and 18% for the SE, HE and CT units, respectively, and (b) the generation reserve is greater than 50 MW.

The above two case studies show that: (a) distribution of generation reserve should be based on the frequency response rates of prime movers, and (b) reserve distribution is constant for a given prime mover configuration.

3.3 Amount and Distribution of Generation Reserve

Tables 3, 4 & 5 list prime movers which may be in operation during the initial or subsequent phases of a power system restoration. The objective is to determine the effect of reserve distribution on the frequency dip due to the loss of the largest unit, i.e. 300 MW. Three cases are listed: the uniformly distributed reserve in **Table 3**, the "optimally" distributed reserve for the same frequency dip of -1.32 Hz, in **Table 4**, and the "optimally" distributed reserve for an allowable frequency dip of -0.95 Hz, in **Table 5** [5].

In Table 3, the total load of 782 MW is divided between SE-A, SE-B, HE and CT plants proportional to their capacities, i.e., "uniformly" distributing the generation reserve. After the loss of the largest unit, i.e., SE-A, its load of 270 MW is carried by the remaining three plants, resulting in the overloads of the HE and CT units and their possible trip outs. The frequency dip in case of survival of these two plants would be 1.32 Hz.

In **Table 4**, considering the same frequency dip of 1.32 Hz, the total load of 782 MW is divided between SE-A, SE-B, HE and CT plants proportional to their frequency response rates, i.e., "optimally" distributing the generation reserve. After the loss of the largest unit, i.e., SE-A, its load of 217 MW will be carried by the remaining three plants, resulting in no overloads. It should be noted that after the contingency, all the plants are loaded well below their capacities.

The case studies listed in **Tables 4 and 5** are similar, except that in the latter case, a frequency dip of 0.95 Hz has been allowed. The comparison of the two reserve distributions shows constant reserve rates for a given prime mover configuration, as shown in the following table:

Table 6 - Comparison of the Two Reserve Distributions

Type	Reserve %	Frequency Dip Hz	Reserve Rate %/Hz
From Table 4			
SE-B	14.9	-1.32	-11.3
CT	24.3	-1.32	-18.5
HE	33.0	-1.32	-25.1
From Table 5			
SE-B	10.8	-0.95	-11.3
CT	17.6	-0.95	-18.5
HE	23.8	-0.95	-25.1

Clearly, with no-load on SE-A, there will be no frequency dip after its outage. It can be shown that with the above reserve rates, when SE-A plant carries the full capacity of 300 MW, upon its outage, there will be a frequency dip of 1.82 Hz. Therefore, for a given frequency dip between 0.0 and 1.82 Hz, as long as the generation reserve is distributed on the basis of the above reserve rates, there will be no risk of plant overloads. **Figure 8.** shows the reserve distributions to be used for the CT, HE and SE-B plants to meet the generation contingency for any frequency dip between 0.0 and 1.82 Hz. The reserve distributions are based on reserve rates of the plants.

3.4 Verification

Figure 9. shows the frequency response curves for three of

Table 1. Determination of Load Pickup

GIVEN $\Delta F = -0.75$ Hz, FIND ΔL:

Type	No of Units	Cap. MW	Pi MW	dfi Hz/pu	Pi/dfi MW/Hz	ΔLi MW	Load Li	Reserve %
SE	2	135	270	-7.94	-34.0	25.5	244.5	44.67
HE	3	32	96	-3.40	-28.2	21.2	74.8	37.09
CT	4	16	64	-4.61	-13.9	10.4	53.6	18.24
Total			**430**		**-76.1**	**57.1**	**372.0**	**100.0**

$\Delta L = \Sigma\Delta Li = \Delta F*\Sigma(Pi/dfi) = 57.1$ MW or 13.3 %.

Table 2. Determination of Frequency Dip

GIVEN $\Delta L = 50$ MW, FIND ΔF:

Type	No of Units	Cap. MW	Pi MW	dfi Hz/pu	Pi/dfi MW/Hz	ΔLi MW	Load Li	Reserve %
SE	2	135	270	-7.94	-34.0	22.3	247.7	44.67
HE	3	32	96	-3.40	-28.2	18.5	77.5	37.09
CT	4	16	64	-4.61	-13.9	9.1	54.9	18.24
Total			**430**		**-76.1**	**49.9**	**380.0**	**100.0**

Load Increase = $\Sigma(Pi/dfi) = -76.1$ MW/Hz, and Frequency Dip $\Delta F = 50/\Sigma(Pi/dfi) = -0.66$ Hz.
Accordingly, the frequency dip for picking up a LVAC Network load $\Delta L' = 50$ MW is about $\Delta F" = -66$ Hz.

Table 3. "Uniformly" Distributed Reserve

"Uniform" Reserve Distribution:

Type	No of Units	Pi MW	Li' MW	dfi Hz/pu	Pi/dfi MW/Hz	ΔL MW	Li" MW	Reserve %
SE-A	1	300	217	-7.94	out			27.7
SE-B	2	270	195	-7.94	-34.0	44.7	240	25.0
HE	8	256	185	-3.40	-75.3	99.0	284*	23.7
CT	16	256	185	-4.61	-55.5	73.0	258	23.7
Total		**1082**	**782**		**-164.8**	**217.0**	**782**	**100.0**

$\Sigma Pi = 1082$ MW, $\Sigma Pi - 300 = 782$ MW, $\Sigma(Pi/dfi) = -164.8$ MW/Hz, The Initial Load, $Li' \propto Pi = 72.3\%$,
The Final Load, $Li" = Li' + \Delta L$, and $\Delta F = 217/-164.8 = -1.32$ Hz.
The HE and CT Units Are Overloaded

Table 4. "Optimally" Distributed Reserve

"Optimal" Reserve Distribution Assuming $\Delta F = -1.32$ Hz:

Type	No of Units	Pi MW	dfi Hz/pu	Pi/dfi MW/Hz	ΔL MW	Li MW	Reserve %
SE-A	1	300	-7.94	out		217	27.7
SE-B	2	270	-7.94	-34.0	44.7	225	14.9
HE	8	256	-3.40	-75.3	99.0	157	33.0
CT	16	256	-4.61	-55.5	73.0	183	24.3
Total		**1082**		**-164.8**	**217.0**	**782**	**100.0**

$\Sigma Pi = 1082$ MW, $\Sigma Pi - 300 = 782$ MW, $\Sigma(Pi/dfi) = -164.8$ MW/Hz, $\Delta L = \Delta F*(Pi/dfi)$, and $Li = Pi - \Delta L$.
There are no overloads.

Table 5. "Optimally" Distributed Reserve

"Optimal" Reserve Distribution Assuming $\Delta F = -0.95$ Hz:

Type	No of Units	Pi MW	dfi Hz/pu	Pi/dfi MW/Hz	ΔL MW	Li MW	Reserve %
SE-A	1	300	-7.94	out		157	47.8
SE-B	2	270	-7.94	-34.0	32.3	238	10.8
HE	8	256	-3.40	-75.3	71.5	185	23.8
CT	16	256	-4.61	-55.5	52.8	203	17.6
Total		**1082**		-164.8	**156.6**	**782**	**100.0**

$\Sigma Pi = 1082$ MW, $\Sigma Pi - 300 = 782$ MW, $\Sigma(Pi/dfi) = -164.8$ MW/Hz, $\Delta L = \Delta F*(Pi/dfi)$, and $Li = Pi - \Delta L$.
There are no overloads.

the above case studies. The upper, middle and lower curves show the frequency dips due to: the 50 MW load pickup, the optimal reserve distribution for 0.95 frequency dip, and the "uniform" reserve distribution.

The following table provides a comparison between the simulation results and the results of using the "frequency response rates," $\Delta F/\Delta L$, as a guide:

Table 7 - Simulation & Guideline comparison

ΣPi MW	ΔL MW	ΔF,Hz Sim.	ΔF, Hz Guide	Difference Hz
430	50	-0.59	-0.66	-0.07
782	157	-0.85	-0.95	-0.10
782	217	-1.18	-1.32	-0.14

The differences between the simulation results and the results obtained from the use of "frequency response rates," justify the use of the above approximations in developing the guidelines.

4. CONCLUSION

In this paper, "frequency response rates," and "reserve rates" have been developed for typical CT, SE and HE units. The response and reserve rates have then been used to develop several operator guides for determining:

- the maximum load pickup for a given frequency decline,
- the maximum frequency dip for an estimated Low Voltage AC Network load, and
- the amount and distribution of reserve required for generation contingency.

It has been shown that:

- the distribution of generation reserve should be based on the frequency response rates of prime movers,
- the reserve distribution is constant for a given prime mover configuration, and
- as long as the generation reserve is distributed on the basis of reserve rates, there will be no risk of plant overloads.

The prime mover models described in this paper can readily be used to determine the frequency response rates and reserve rates of the prime movers for developing similar guidelines. It should be emphasized that although these guidelines provide approximate solutions, they do expedite the restoration process.

5. REFERENCES

[1] IEEE Committee Report, "System Operation Challenges)," IEEE TRANS. v.PWRS-3 n.1.1987 pp 118-126

[2] Adibi, M.M. & L.H. Fink, "Power System Restoration Planning," IEEE/PES Winter Meeting Paper No. 93 WM 204-8-PWRS, 1993.

[3] Kirchmayer, L.K., "Economic Control of Interconnected Systems," John Wiley and Sons, New York, 1959.

[4] Borkoski, J.N., "Generation Reserve Disposition at BG&E," Internal Report, Baltimore, MD, June 1985.

[5] Volkmann, T.L., "Governor Evaluation at NSP," Internal Report, Minneapolis, MN, April 1995.

ACKNOWLEDGEMENT

The first author acknowledges the support of Baltimore Gas and Electric Company and Northern States Power Company for providing operational data.

ENERGIZING HIGH AND EXTRA-HIGH VOLTAGE LINES DURING RESTORATION

M. M. Adibi, LF
IRD Corporation

R. W. Alexander, SM
Pennsylvania P&L Co.

D. P. Milanicz, SM
Baltimore G&E Co.

Abstract: In energizing transmission lines, operators are often concerned with: the length of line to be energized, the adequacy of on-line generation, and the size of underlying load. There has been a need for a simple methodology to readily address these issues and provide guidelines for the safe operation of high voltage and extra-high voltage transmission lines.

In this paper, a simple and approximate approach for evaluating sustained and transient voltages is presented. The sending- and receiving-end sustained and transient voltages are determined for energizing lines of different voltage levels and lengths, energized by different sizes of generators, and having different sizes of underlying loads.

Keywords: Sustained Overvoltage, Transient Overvoltage, Source Impedance, Underlying Loads, Temporary overvoltage.

1. INTRODUCTION

In energizing transmission lines during power system restoration, operators are often concerned with: the length of line to be energized, the adequacy of on-line generation, and the size of underlying load [1].

In general, it is desirable to energize as large a sections of lines as the sustained and transient over-voltages will allow. Energizing small sections tends to prolong the restoration process. In energizing a large section however, there is the risk of damaging the equipment insulations [2].

Energizing lines with inadequate sources could result in higher sustained and transient voltages than equipment can withstand. The start-up of more out-of-sequence generators however, would use critical time, and delay the overall restoration process [3].

The underlying loads at both ends of the line tend to reduce the sustained and transient voltages. The operators need to know the minimum size of load which would avoid transient overvoltages. Clearly, there has been a need for a simple methodology to readily address these issues and provide guidelines for the safe operation of high and extra-high voltage transmission lines [4]. Detailed and accurate methods such as the EMTP tools [5,6], can meet the need as described in [7]. However, data unavailability and preparation efforts dictate a simpler and more accessible method.

PE-415-PWRS-0-07-1998 A paper recommended and approved by the IEEE Power System Operations Committee of the IEEE Power Engineering Society for publication in the IEEE Transactions on Power Systems. Manuscript submitted January 5, 1998; made available for printing June 22, 1998.

In this paper, a simple and approximate approach for evaluating sustained and transient voltages is presented. The sending- and receiving-end sustained and transient voltages are determined for energizing lines of different voltage levels and lengths, energized by different sizes of generators, and with different sizes of line-end dead loads. These results are used to provide guidelines to assist operators in energizing high (230 kV) and extra- high (345 & 500 kV) voltage lines. The approach is then applied to the early stages of a power system restoration plan where a hydro-electric-station is called upon to remotely crank a steam-electric-station via a double circuit 230 kV line.

2. PROCEDURE

2.1. The Open-Ended Line Model

Figure 1 shows the one-line diagram of a generator connected to a high or extra-high voltage transmission line. The line is represented by the nominal "π" equivalent with series and shunt impedances Z_3 and $2.Z_2$. The generator step-up transformer impedance is shown as Z_1. For evaluation of switching transient voltages, the generator subtransient reactance Xd", is lumped with Z_1.

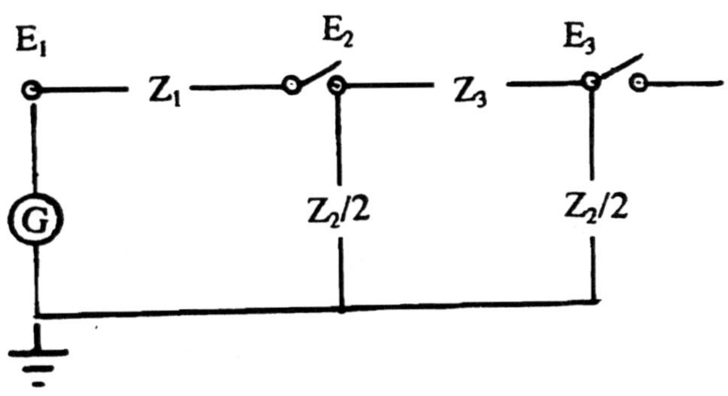

Figure 1 - One-line Diagram of Open-Ended Line (Nominal Pi Equivalent)

The sending-end voltage E_2, and the receiving-end voltage E_3, are related to the generator voltage E_1, by the following two equations:

$$E_2 = (N_1/D_1)E_1, \text{ and}$$

$$E_3 = (N_2/D_1)E_1$$

where

$$N_1 = 2Z_2(2Z_2+Z_3),$$

Reprinted from *IEEE Transactions on Power Systems,* Vol. 14, No. 3, pp. 1121–1126, August 1999.

$N_2 = 4Z_2^2$, and

$D_1 = (Z_1+2Z_2)(2Z_2+Z_3)+(2Z_2Z_1)$.

2.2. The Load-Ended Line Model

For the transformer terminated or load-ended line it is more convenient to represent the line by its nominal "T" equivalent as shown in **Figure 2**. At the receiving end, the step down transformer and loads are lumped together and represented as Z_4.

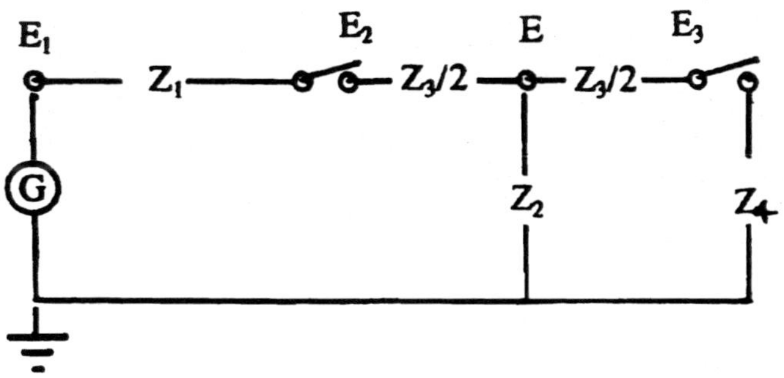

Figure 2 - One-line Diagram of Load-Ended Line (Nominal T Equivalent)

The sending-end voltage E_2, and the receiving-end voltage E_3, are related to the generator voltage E_1, by the following two equations:

$E_2 = (N_1/D_1)E_1$, and

$E_3 = (N_2/D_1)E_1$.

where,

$N_1 = (Z_2+Z_3/2)(Z_4+Z_3/2)+Z_2Z_3/2$,

$N_2 = Z_2Z_4$, and

$D_1 = N_1+Z_1(Z_2+Z_4+Z_3/2)$.

2.3 Evaluating the Sustained Overvoltages

In both of the above models, the following parameters are used:

$Z_1 = R_1 + jwL_1$,

$Z_2 = 1/jwC_2$,

$Z_3 = R_3 + jwL_3$, and

$Z_4 = R_4 + jwL_4$.

Table 1 lists parameters used in evaluating the sustained overvoltages in high and extra-high voltage transmission lines:

Table-1 Transmission System Parameters

Parameters	230 kV	345 kV	500 kV
Resistance (R)	0.0245	0.0047	0.0013
Reactance (wL_3)	0.1520	0.0500	0.0245
Charging (wC)	0.2850	0.8580	1.7400
Xfmr (wL_1)	13.0	15.0	17.0

In the above table:

> Line resistance, reactance and charging currents are in % per mile, based on 100 MVA and their corresponding nominal voltages, and
>
> Transformer reactance is based on transformers' MVA ratings, 13.8/230 kV, 20/345 kV and 20/500 kV.

Figure 3 shows the voltage rise at the receiving end of the open-ended 230, 345 and 500 kV lines with the generator step-up transformer (source) ratings of 50, 100 and 150 MVA, respectively. These source ratings are required to absorb the lines' leading reactive powers [8].

The bottom curve shows that the percentage rise in voltage between the receiving-end and the sending-end is independent of voltage level. It can be seen that the percentage rise between the receiving-end and the sending-end is the same for all voltage levels and is approximately proportional to the square of length of line, i.e., $\Delta E=kx^2$, where x is the length of line, $k=\frac{1}{2}w^2LC$, LC is the propagation constant, and $w=2\pi f$ (this approximation is valid for lines less than about 500 miles in length). This effect is sometimes called Ferranti effect [9].

Figure 4 shows the effect of increase in the source MVA on voltage rise at the receiving end of a 230 kV line. It can be seen that as the generator step-up transformer rating is increased from 25 to 50 and to 75 MVA, the rise in voltage is reduced appreciably.

Figure 5 shows the significant voltage drop at the receiving-end of the 230 kV lines due to 250, 500 and 750 kW underlying loads. Although the magnetizing current of the step-down transformer is sufficient to eliminate the voltage rise, an underlying load needs to be connected to the transformer's secondary to avoid possibility of harmonic resonance.

The harmonic resonance is initiated by the switching transient in combination with the transformer's inrush current. It consists of power frequency and harmonics with energy beyond the capability of surge arresters.

2.4 Evaluating Switching Transient Voltages

In the following models, the coupling between phases and the pole disparity in energizing lines are neglected. In the open-ended model the following parameters are used:

$E_1 - E_2 = R_1I_1 + L_1dI_1/dt$,

$E_2 = 2/C \bullet \int I_2dt$, and

$$E_2 - E_3 = R_3I_3 + L_3dI_3/dt.$$

In the load-ended model the following parameters are used:

$$E_1 - E_2 = R_1I_1 + L_1dI_1/dt,$$

$$E = 1/C \cdot \int I_2dt,$$

$$E - E_3 = (1/2)(R_3I_3 + L_3dI_3/dt), \text{ and}$$

$$E_3 = R_4I_4 + L_4dI_4/dt.$$

where subscript 1, 2, 3 & 4 refer to currents in the branches Z_1. Z_2, Z_3 and Z_4 of the Pi and T one-line diagrams.

Using $E_1(t)=E_1.\cos(wt)$, the transient solutions for $E_2(t)$ and $E_3(t)$ are of the following forms:

(a) Open-Ended

$$E_2(t) = E_1(t)\sum \exp(-\alpha_it)\cos(\beta_it+\delta_i)],$$

$$E_3(t) = E_1(t)\sum \exp(-\alpha_jt)\cos(\beta_jt+\delta_j)],$$

where, i or j = 1, 2 & 3.

In the "π" equivalent model, E_2 and E_3 are related to E_1 by the sum of a 60 cycle sinusoid, i.e., cos(wt), and two high order harmonics which decay with time.

(b) Load-Ended

$$E_2(t) = E_1(t)[\exp(-a_1t+b_1)+ \sum \exp(-\alpha_it)\cos(\beta_it+\delta_i)], \text{ and}$$

$$E_3(t) = E_1(t)[\exp(-a_2t+b_2)+ \sum \exp(-\alpha_jt)\cos(\beta_jt+\delta_j)].$$

where, i or j = 1 & 2.

In the "T" equivalent model, E_2 and E_3 are related to E_1 by the sum of a 60 cycle sinusoid, i.e., cos(wt), a decaying function and one high order harmonics which also decays with time.

Table 2 lists parameters used in evaluating the switching transient overvoltages in high and extra-high voltage transmission lines:

Table-2 Transmission System Parameters

Parameters	230 kV	345 kV	500 kV	Units
Resistance (R)	12.96	5.54	3.15	ohm
Inductance (L)	0.213	0.158	0.162	henry
Capacitance (C)	1.429	1.912	1.846	μ-farad
Xfmr (wL_1)	13	15	17	%
HE Gen (Xd")	25	25	25	%
SE Gen (Xd")	12	23	23	%
CT Gen (Xd")	15	15	15	%

In the above table:

Resistance, inductance and capacitance are for typical 100 miles long lines,

Step-up transformer reactances are in %, based on their MVA ratings and 13.8/230 kV, 20/345 kV & 20/500 kV, and

Generator subtransient reactances for hydro (HE), steam (SE) & combustion turbine (CT) generators' are for typical units.

Figure 6 shows the transient voltages in per unit at the receiving end of the open-ended 230, 345 and 500 kV lines with the generator ratings of 50, 100 and 150 MVA, respectively. These MVA ratings are the minimum required to absorb the charging currents over the different lengths of lines. It can be seen that the switching transient of open-ended lines, for a 230 kV line of 40 mile length, exceeds 2 p.u.

Figure 7 shows the effect of increase in the generator MVA ratings on transient voltage at the receiving-end of a 230 kV line. It can be seen that as the generator ratings are increased from 25 to 50 MVA, the transient voltages are reduced significantly.

In **Figure 8**, the transient voltages are plotted against the receiving-end loads for the 230 and 345 kV lines of 50- and 75-mile lengths, and with 50 and 75 MW generation, respectively. It can be seen that at transient voltage of 1.8 p.u., the minimum load for the 75-mile, 345 kV line is about 25 MW and for the 50-mile, 230 kV line is about 5 MW.

3. CASE STUDY

3.1 Objective

The restoration plan for the power system under study called for using the remote hydro-electric-station (HES) to remotely crank the steam-electric-station (SES) over the 230 kV double circuit transmission line, 30 miles long.

The purpose of this analysis were to examine the transient over-voltages at both, the step-up and step-down transformers with the following objectives: (a) to use as few hydro units as possible, (b) if possible, to energize the two sections of the single circuits of the 230 kV line in one operation, and (c) to determine the minimum required receiving-end load. The primary concern was to limit the switching over-voltages to 2.0 per unit, or 460 kV.

3.2 System Description

The following data describe the relevant characteristics of the HES and the high voltage transmission line to be used for cranking the SES.

The HES consists of five 30 MVA, 13.8 kV generators with subtransient reactance of Xd"=16 % based on the generators' ratings. Two units feed the A-Bus and three units feed the B-Bus.

The leading reactive power capability of each generator at low load is between 7.5 to 10.5 MVAr corresponding to the 230 kV system bus voltages of about 100 to 105%. At

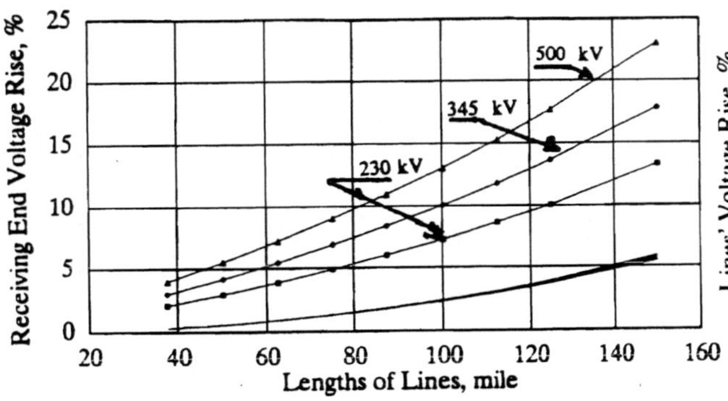

Figure 3. Receiving-End Voltage Rise of the Open-Ended 230. 345 and 500 kV Lines

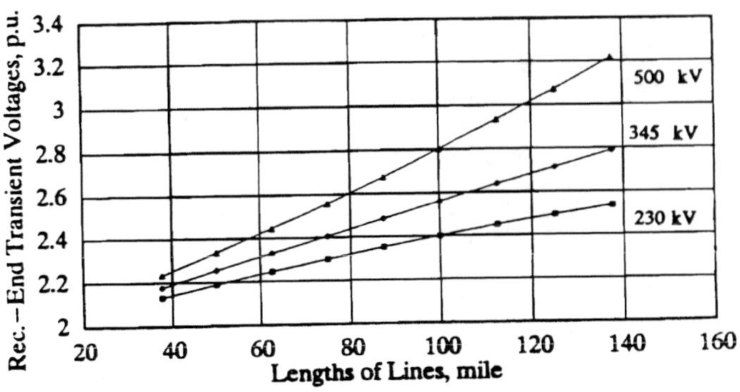

Figure 6 - Receiving-End Transient Voltages of the Open-Ended 230, 345 and 500 kV Lines

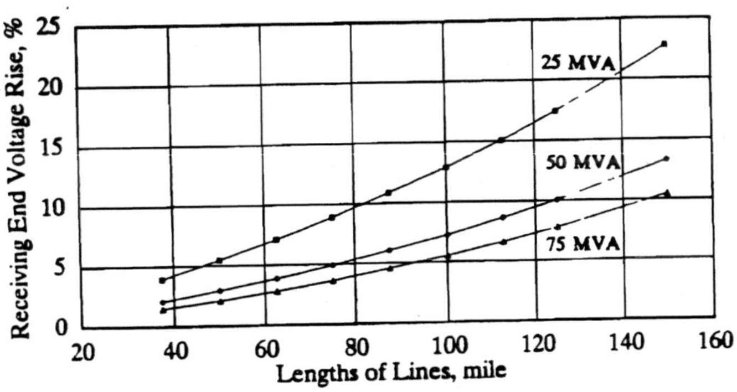

Figure 4 - Effects of Source MVA on the Receiving-End Voltage of the Open-Ended 230 kV Lines

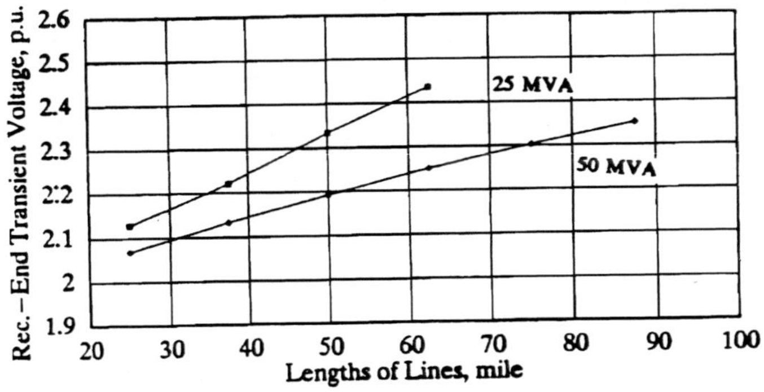

Figure 7 - Effects of Source MVA on the Receiving-End Transient Voltages of the Open-Ended Line

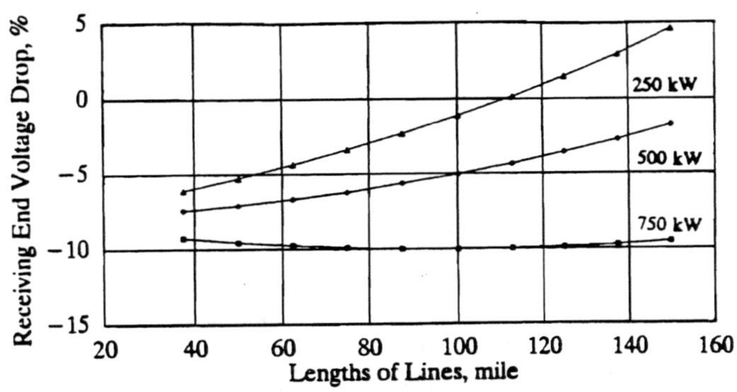

Figure 5 - Effects of Underlying Loads on the Receiving-End Voltage of the 230 kV Lines

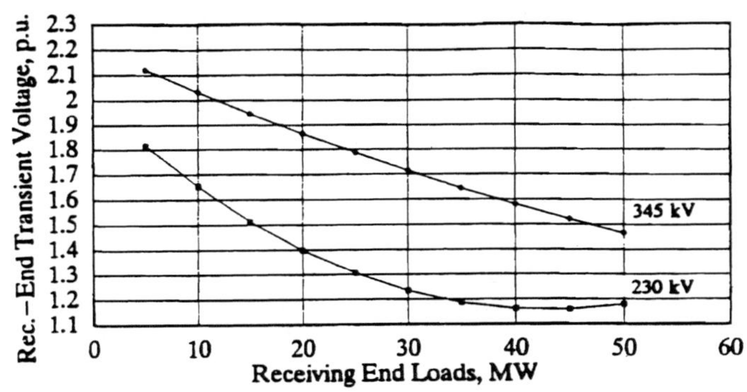

Figure 8 - Effect of Loads on the Receiving-End Transient Voltages of the 230 kV and 345 kV Lines

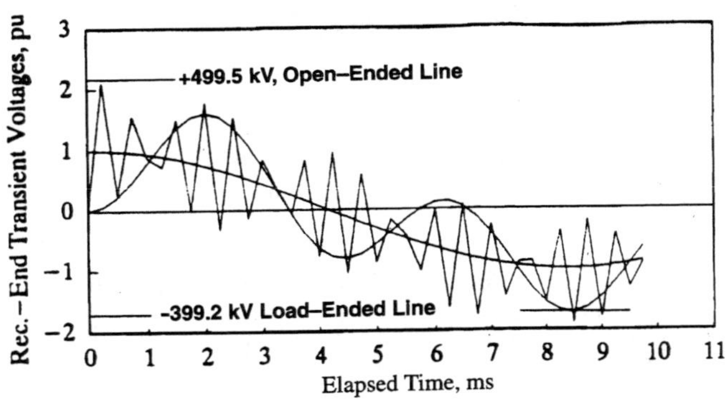

Figure 9 - Receiving-End Transient Voltages of the Open-Ended and Load-Ended 230 kV Line

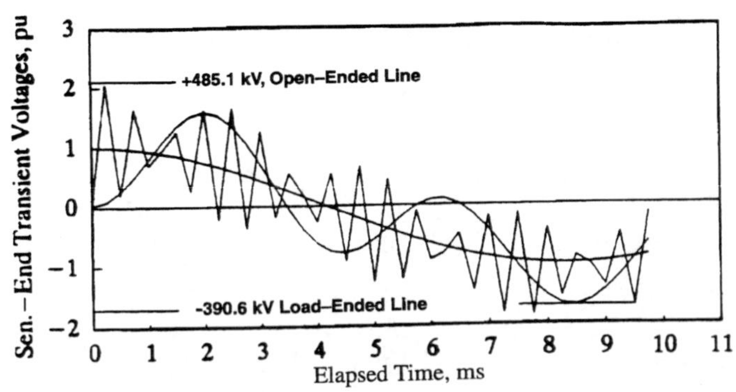

Figure 10 - Sending-End Transient Voltages of the Open-Ended and Load-Ended 230 kV Line

lower system bus voltages, the leading reactive power of the generators reduce appreciably.

The 13.8 kV A-Bus and B-Bus are connected to the 230 kV A-Bus and B-Bus via two step-up transformers of 13.8/230 kV, 28 MVA OA ratings. Each transformer has a resistance of Rt=0.3 % and reactance of Xt=8.2%, based on transformers' ratings. The basic insulation level of these transformers, in rms = 460 kV. As such, the switching surge withstand of these transformers is unusually low, i.e., 2.0 per unit. This level is so low that conventional surge arrester protection will be insufficient.

The transmission line consists of a 230 kV double circuit line sectionalized at mid point, by a 230 kV switching station. The single circuit of each section has a resistance of R_L=0.362 % and reactance of X_L=2.558% on 230 kV and 100 MVA base. The lines charging current is 4.40 MVAr at 230 kV voltage.

The step-down transformer at the receiving end consists of two 30 MVA, 230/13.8 kV ratings, each with no load losses of about 150+j150 kVA.

3.3 Results

Table 3 lists various options for the open-ended line operation. Using the above open-ended line models, for each operating option, the maximum sending-end and receiving-end transient voltages were determined.

Table 3. Open-Ended Line

Case No.	Sending E_2, PU	Receiving E_3, PU
1G,1S	+2.050	+2.085
1G,2S	+2.108	+2.170
2G,1S	+2.000	+2.060
2G,2S	+2.000	+2.120

In the above table; Case No. 1G,1S, refers to one on-line generator, energizing one section of the line, Case No. 2G,2S refers to two on-line generators, energizing two sections of the line, etc.

Table 4 lists various operating options for the transformer-terminated and load-ended line. The load-ended line models were used for each operating option and the maximum sending-end and receiving-end transient voltages were determined.

Table 4. Load-Ended Line

Case No.	Sending E_2, PU	Receiving E_3, PU
1G,T	- 2.054	- 2.089
1G,L	- 1.698	- 1.714
2G,T	- 1.840	- 1.901
2G,L	+1.710	+1.758

In the above table; Case No. 1G,T, refers to one generator and the line-end transformer, Case No, 2G,L refers to two generators and the combined line-end transformer and load, etc.

The step-down transformer excitation and the underlying load used were 150+j150 kVA and 20+j20 MVA, respectively.

Figure 9 compares the sending-end transient voltages of Case No. 1G,2S vs Case No. 1G,L. It shows the significant effects of the underlying load. The transient voltage has been reduced from +5.5% to -15.1% of the transformer switching surge withstand insulation level.

Figure 10 compares the receiving-end transient voltages of the same cases showing a reduction of transient voltages from +8.6% to -14.3%.

3.3 Discussion

The results listed in **Tables 3 & 4** show that there are a number of acceptable operating options. The specific option to be selected is based on the availability of generators, underlying loads and the presence of the switch-man at the mid-point switching station, as discussed below. The concern for the presence of the switch-man is due to the fact that the SCADA System may not be functional in the early stages of restoration.

(a) Open-Ended Line

Under no-load or low-load operating conditions, the leading reactive power capability of each HES generator is between 7.5 to 10.5 MVAr. Therefore, one generator is adequate to absorb the charging currents of two sections of the single circuit. However, as the comparisons in **Table 3** of the two cases 1G,1S & 1G,2S with the two cases 2G,1S & 2G,2S show, two generators are needed to reduce the source impedance and lower the sending-end transient voltages to an acceptable level. Consideration of receiving-end voltage in Case 2G,2S, indicates that even with two on-line generators, the transient voltage is excessive. It can be concluded that the safe switching operation, is to energize the open-ended line one section at a time. This of course requires the presence of the switch-man at the mid-point station.

(b) Load-Ended Line

Review of the four cases in **Table 4** indicates the transient voltages at the receiving end are generally higher than the corresponding values at the sending end. Therefore, unless the insulation levels of the step-up transformers are lower than the step-down transformers, the concern should be primarily with the transient voltages at the receiving. It should be noted the switching surge withstand levels of the transformers in this particular case are unusually low. Normally, switching surge withstand levels would be about 50% higher enabling the transformers to be protected with surge arresters. In such a case the magnitude of switching overvoltage would be irrelevant since the overvoltages would be clamped by the arresters. Then the major overvoltage concerns are with the sustained 60 Hz voltages, and possible harmonic resonance voltages which could fail the arresters and create fault. Harmonic resonance will depend very strongly on the natural frequencies of the system, and can usually be damped with reasonable underlying loads.

The loads' power factors (i.e., R/L ratio) have a significant effect on damping the transient voltages. In general, the load power factor has the tendency of increasing with increase in load, thus increasing the damping effect of the load.

Over a limited range of load and generation, there is a trade-off between increasing generation and increasing load. Comparison of the two cases 1G,T and 1G,L show the effect of increasing load in reducing the transient voltage to the acceptable limit. Comparison of the two cases 1G,T and 2G,L show the effect of increasing generation in reducing the transient voltage.

It can be shown that for the system under consideration, the trade off between the increase in generation and load extends to about 15 MW of load, i.e, 430 kV (93.5 % of Transformer BIL) transient voltage can be obtained by either two generators and 5 MW load or by one generator and 15 MW load.

Based on the availabilities of man-machine resources and results of this approximate analysis it can be concluded that the 30 mile 230 kV single circuit line can be energized with two on-line hydro units and 5 MW of load at the receiving end without exceeding the switching surge withstand level of both the sending-end and the receiving-end transformers.

4. CONCLUSION

The models described in this paper can readily be used to determine similar steady state and transient guidelines in other situations and power systems. Although such guidelines provide approximate solutions, they do expedite the restoration process. They help avoid sustained and transient overvoltages by consideration of:

- the length of high and extra-high voltage lines to be energized,
- the adequacy of energizing sources, and
- the adequacy of the underlying loads at either end of line.

5. REFERENCES

[1] IEEE Committee Report, "System Operation Challenges," IEEE TRANS, v.PWRS-3 n.1.1987 pp 118-126

[2] Alexander, R.W. "Minimum Source Consideration When Restoring PJM 500kV System," PEA System Operation Committee, Jan 1987.

[3] IEEE Committee report, "Power System Restoration - The Second Task Force Report," T-PWRS, Nov. 1987, pp 927-933.

[4] IEEE Committee Report, "Overvoltage Control During Restoration," T-PWRS, Nov 1992 pp 1464-1470.

[5] Natarajan, R. et al, "Adoption of Electromagnetic Transient Program on Personal Computers," IEEE Trans. on Pwr.Sys., Vol.4, No.4, Oct 1989, pp 1550-1554.

[6] Dommel, H.W., et al, "Transmission Program User's Manual," UBC Publication, Aug. 1988.

[7] Morin, G., 'Service Restoration Following a Major Failure on the Hydro-Quebec Power System," 86WM 183-8, T&D.

[8] Adibi, M.M. and D.P. Milanicz, "Reactive Capability Limitation of synchronous Machines," paper 93 WM 201-4, presented at IEEE 1993 WPM, Columbus, Ohio, Jan 31 - Feb 6, 1993.

[9] Cotton, H. and E.W. Golding "The Transmission and Distribution of Electrical Energy," The English University Press Ltd., Feb. 1948, pp 277-278.

ACKNOWLEDGEMENT

The first author acknowledges the support of Baltimore Gas and Electric Company and Pennsylvania Power and Light Company for providing operational data.

PART III

Restoration Planning

I. General Discussion

IN the power system literature, there is a period of nearly forty years, extending from 1940 to the late 1970s, during which no papers on system restoration were published. Following the major blackouts in the northeastern United States in the mid-1960s and early 1970s, industry engineers and operators undertook serious reconsideration of system restoration problems, as well as the need for developing and maintaining adequate restoration plans. The first group of three papers in this section, dating from the early 1980s, represents the early fruits of those efforts. The first two, companion papers report the pioneering work of a major metropolitan utility in developing a comprehensive set of plans for restoration of its system; the first paper deals with the planning process and its initial results, and the second describes the innovative use that was made of computer facilities. The third paper presents multifaceted views of restoration problems from the perspectives of diverse utilities and diverse types of outages.

The second group, comprising four papers dating from the late 1990s, represents the mature results of work done throughout the industry in the intervening years. The first of these presents a summarization and organization of the work of the IEEE/PES Working Group on Power System Restoration that grew out of the work presented in the group of papers. The second describes a software program developed and implemented at Hydro-Quebec. It provides an optimized solution for the control variables (generator voltages, load magnitudes) that will maintain a desired voltage profile as the network is reestablished. The third paper reports work done under an Electric Power Research Institute (EPRI) contract to develop a simulator and artificial intelligence-based program that would guide an engineer through an automated restoration planning process. In particular, procedural practices that had been developed heuristically by a wide variety of utilities worldwide were abstracted to provide generic elements of the planning process. The last paper discusses a matter that has been overlooked, that is, that the duration of the restoration process may be affected significantly by switching operations.

II. Discussion of Reprints

3.01 System Restoration Plan Development for a Metropolitan Electric System, 1981, p. 195

The duration of outages following the major system outages after mid-century led a number of utilities to reexamine their restoration plans. PEPCO, in particular, undertook a significant team effort, involving system analysts, operating personnel, and concurrent use of on-line and off-line computer facilities at the operating center. Among the major benefits was the understanding that was achieved of the importance, during the restoration process, of maintaining reactive power balance, continuous matching of load and generation, coordination of time frames for hot and cold restart of the variety of the system's generating units, and constant awareness of the probable effects of each action on the rest of the system and on subsequent planned actions.

3.02 Role of Interactive and Control Computer in the Development of a System Restoration Plan, 1982, p. 205

This is a companion paper to the previous one, and describes the computer facilities and the power system modeling that were integral to the planning process previously described.

3.03 System Restoration—Deploying the Plan, 1982, p. 214

This paper, a subcommittee report, brings together the contributions of a number of utilities to the consideration of tasks necessary to bridge the gap between a plan for system restoration and its successful implementation under the stressful conditions of an actual system outage. These tasks include making the plan available in formats that will be useful during restoration, training personnel, not only in implementation of the plan, but also in the special operating procedures rarely used during normal operation, and establishing procedures for assembling employees when needed and for enabling necessary extraordinary transportation and communication facilities.

3.04 Power System Restoration Planning, 1994, p. 223

This paper attempts to synthesize the results of a decade of work on power system restoration planning. The system restoration planning process was defined as requiring nine major activities: formation of a planning team, review of system characteristics, formulation of assumptions, establishment of goals, development of strategy and tactics, validation of the draft plan, iteration of the planning process to overcome deficiencies, documentation of the completed plan, and training in deployment of the plan. The system restoration process was characterized in terms of six groups of tasks—system status determination, plant preparation, network preparation, network energization, system rebuilding, and logistics. The overall intent of the paper was to identify, to the extent possible, commonalities in the restoration process across the wide variety of individual utility characteristics.

3.05 A Systematic Method for Power System Restoration Planning, 1995, p. 230

This paper describes a software-aided approach to restoration planning developed within Hydro-Quebec. Hydro-Quebec's practice has been to base restoration on sets of detailed emergency instructions specific to each station and substation. These instructions could be implemented locally with minimal communication between operators and their regional control centers. Networks are energized in stages, each stage comprising the closing of a circuit breaker and the consequent energization of a network component. Each stage presents a new network topology, a new voltage pattern, and a new active/reactive power balance. To satisfy voltage constraints, operators must adjust controls, including generator voltages, transformer taps, shunt reactors, and loads to be picked up. Three types of simulation are required—steady state, harmonic analysis, and switching surge.

Control variables must be optimized across the entire sequence of stages. A software program has been developed around an algorithm to effect such an optimization that satisfies the operating constraints, eliminating the guesswork involved in conventional methods. The problem formulation is that of nonlinear optimization. The software is described and an example is provided.

3.06 From Generic Restoration Actions to Specific Restoration Strategies, 1995, p. 237

This paper grew out of an EPRI project for development of an expert system-based simulator for formulation of restoration plans and, ultimately, operator guidance during the restoration process. In view of these objectives, emphasis was placed on distilling from the plans of numerous types of utilities, generic restoration tasks that could be assembled to meet the system-specific needs of a broad spectrum of users. The general recognition of three stages in the restoration process was clarified by specification of the distinctions between those stages. Generality was achieved by identifying certain conceptual constituent elements of the process, namely, a target system, restoration building blocks (in later practice to be termed *system kernels*), and generic restoration actions.

Since little, if any, information of what has caused the outage is available during the early stages of restoration, an operator's knowledge of what major equipment will be available for restoration enables him to define a target system as the object system to be restored. The restoration building blocks are those minimal elements of the target system that can be restored under local initiative (as in the Hydro-Quebec practice cited in the previous paper) and that can then be assembled under central coordination to achieve the target system. The generic restoration actions resemble the steps taken to achieve the stages in the Hydro-Quebec practice.

The definition of these concepts makes possible the development of generic software that can be utilized not only to assemble the restoration plans of a wide variety of utilities, but at a more fundamental level also to provide guidance for the development of restoration plans.

3.07 Estimating Restoration Duration, 1999, p. 245

In restoration planning, the duration of the process has conventionally been estimated based solely on the availability of the several types of prime movers involved. Specifically, it is assumed that loads can be reenergized as soon as generation becomes available, since—it is assumed—the time required for switching operations to energize successively transformers, lines, and loads is negligible compared to that required to restart generators.

This paper demonstrates that these assumptions are not necessarily correct and that duration should be estimated considering the times required for switching operations as well as for generator restart and initial loading.

The planning strategy that is recommended utilizes the classical Critical Path Method (CPM), which requires only a range of estimated timing for each action—worst case, most likely, and best case. Precedence relations among actions—for example, which action(s) must be completed before others—can be initiated. Use of CPM may dictate revision of a previously adopted overall restoration strategy. A trial application of the plan to the remote cranking operation of a steam turbine-generator indicated that an additional hour beyond the original estimate would be required to complete the operation, which could require scheduling that unit for a cold restart, perhaps requiring three to four hours.

SYSTEM RESTORATION PLAN DEVELOPMENT FOR A METROPOLITAN ELECTRIC SYSTEM

R. J. Kafka, D. R. Penders, and S. H. Bouchey
Potomac Electric Power Company

M. M. Adibi
IRD Corporation

ABSTRACT- The extended duration of recent major power system disturbances have indicated the need for development of system restoration plans (SRP), which would minimize the time required for restoration of service. This paper presents an efficient approach to SRP, from a complete collapse under peak- and light-load conditions, developed at Potomac Electric Power Company (PEPCo). The method is based on: (a) the combined effort of operating personnel and system analysts, (b) the access to data available from the on-line computer facilities, and (c) the use of an interactive load flow program. The SRP development procedure and the results described in this paper are of general interest in that they can be adopted and applied to the other electric systems.

INTRODUCTION

Restoration of power after an area-wide outage is usually a difficult task, complicated by many factors, including unfamiliarity of the operating personnel with the outage situations and the pressure of time. These difficulties are compounded in a metropolitan type of electric system because of extensive underground high voltage transmission lines, the considerable number of large network loads, and generating plants remote from the load centers. Therefore, a carefully prepared, current, readily accessible and easily understood SRP is necessary in order to allow a quick and orderly recovery from a complete or partial system collapse, with resultant minimum impact on the public.

Since the 1965 power system disturbance of the Northeast, the industry has undertaken considerable efforts to develop and implement preventive and corrective measures to reduce the possibility and the extent of system outages. However, relatively little effort has gone into planning system restoration for minimizing the duration of an outage, should one occur.

Most operating companies have maintained restoration procedures (1) based on certain operating philosophies, practices, and familiarity with power plant restart and power system operation. While these procedures have in the past successfully restored power systems following an outage, at PEPCo it was felt that system restoration could significantly be improved by the combined effort of system analysts, operating personnel, and the concurrent use of on-line and off-line computer facilities at the operating center (2).

It was concluded that the results of this development effort were extremely beneficial to PEPCo and should benefit other utilities as well. Of highest significance was the team effort between operating and analytical personnel. This cooperation was facilitated by the ability of the interactive load flow program to provide accurate modeling of the real system and fast turn around of alternate restoration steps suggested by the operators. Another significant benefit was the understanding which the operators obtained about the criticality of:

- maintaining reactive power balance without allowing sustained over voltage or generator under excitation and consequential voltage collapse,
- keeping load and generation balance through careful coordination of network load pick-ups with generator responses and avoidance of possible risks for system frequency decline,
- coordinating time intervals for the start-up and restart of drum type boilers and super-critical once-through units, and
- considering probable effects of each action on the rest of the system and on the subsequent measures.

Figure 1 shows the organization and facilities used in SRP development.

This paper presents a methodological solution to the system restoration problem, using available analytical tools. It summarizes the detailed analysis and applica-

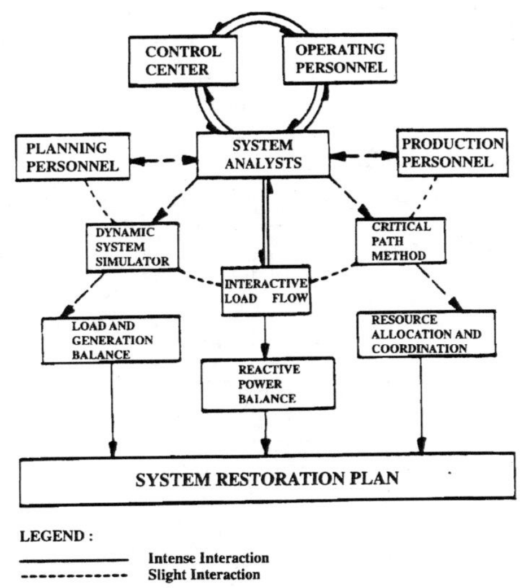

FIGURE 1: Organization and Facilities Used for SRP Development.

81 WM 044-7 A paper recommended and approved by the IEEE Power System Engineering Committee of the IEEE Power Engineering Society for presentation at the IEEE PES Winter Meeting, Atlanta, Georgia, February 1-6, 1981. Manuscript submitted September 2, 1980; made available for printing November 5, 1980.

Reprinted from *IEEE Transactions on Power Apparatus and Systems*, Vol. 100, No. 8, pp. 3703–3713, August 1981.

tion to the PEPCo system. The special problems encountered and the conclusions reached are described.

In a follow-on paper, the authors plan to present the detailed analysis, the role of control computer, and interactive computing in the development of SRP.

Approach and Methodology

Generally, system restoration plans are developed to meet the specific requirements of an individual power system. However, many metropolitan type power systems have certain characteristics in common and, to that extent, the following general approach is useful and applicable:

1. Determination of system characteristics relevant to restoration,
2. Development of a strategy to consist of several distinct levels of restoration, with each level to be comprised of a number of tasks, and each task of several actions,
3. Establishment of guidelines which would allow rapid restoration of the system,
4. Simulation of effects of one or more actions to verify satisfactory operating conditions, and to avoid recurrance of the system outage,
5. Allocation of resources, and coordination of activities to ensure an orderly sequence of actions to a successful completion of restoration in a minimum of time, and finally
6. Preparation of detailed instructions including storage in a convenient media for later access and use by the operating personnel in the event of a system disturbance.

Figure 2 illustrates the iterations which are necessary to make the SRP more effective.

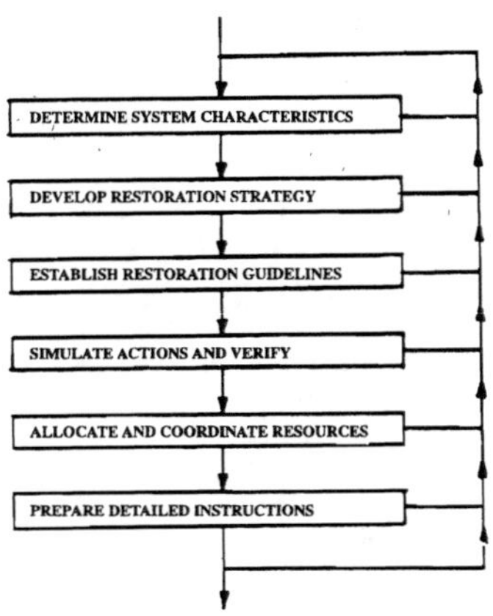

FIGURE 2 : Interactive Process for SRP Development.

System Characteristics

The characteristics of a power system relevant to a restortion plan may be divided into three functional groups: those related to reactive power balance, those which affect the load-generation balance, and those which primarily deal with the protective system. Table 1 lists the characteristic groups.

TABLE I: SYSTEM CHARACTERISTICS

GROUP ONE - RELATED TO REACTIVE POWER BALANCE

1 SHUNT CAPACITORS AND SHUNT REACTORS. THEIR SIZES, VOLTAGE LEVELS, LOCATIONS, AND THEIR MODES OF CONTROL
2. UNDERGROUND CABLES. THEIR VOLTAGE LEVELS. CHARGING CURRENTS, AND THE MAXIMUM SUSTAINED VOLTAGE THEY CAN WITHSTAND.
3. TRANSFORMERS: THEIR REACTANCES. RANGES OF TAP CHANGERS, AND MODES OF CONTROL.
4. SYNCHRONOUS MACHINES. THEIR OVER- AND UNDER-EXCITATION RANGE, AND REACTIVE CAPABILITIES.

GROUP TWO - AFFECTING THE LOAD-GENERATION BALANCE

5. NUMBER AND SIZE OF LOW VOLTAGE AC NETWORK LOADS UNDER PEAK AND LIGHT LOAD CONDITIONS.
6. GENERATOR MAXIMUM AND MINIMUM OUTPUT AND RATE OF LOADING UNDER NORMAL AND EMERGENCY CONDITIONS.
7 THE DYNAMIC RESPONSE OF PRIME MOVERS TO SUDDEN LOAD PICK-UP.
8. MAXIMUM TIME INTERVALS FOR HOT RESTART OF DRUM TYPE OF BOILERS.
9. MINIMUM ELAPSED TIME FOR HOT RESTART OF SUPER-CRITICAL ONCE-THROUGH BOILERS.
10. BLACK-START CAPABILITIES OF COMBUSTION TURBINES

GROUP THREE - DEALING WITH PROTECTIVE SYSTEMS

11. NUMBER AND SIZES OF FEEDERS EQUIPPED WITH UNDER-FREQUENCY RELAYS AND THEIR RELAY SETTINGS
12. NUMBER AND LOCATION OF SCHEMES FOR ISOLATION OF GENERATION WITH LOCAL LOADS.
13. NUMBER AND LOCATION OF SYNCHRONIZING FACILITIES AT OTHER THAN GENERATING PLANTS.
14 NUMBER AND LOCATION OF SYSTEM FREQUENCY METERING POINTS.
15. INTERLOCKING SCHEMES WHICH PREVENT ENERGIZING DEAD BUSES.

Strategy and Guidelines

In practice, due to corrective actions, most system disturbances are arrested prior to complete collapse, and some level of interconnection assistance is available. In the event of a partial disturbance, it would be necessary to determine the state of the system, and enter the restoration procedure at the proper level.

Most power system disturbances are caused by faults which are temporary (3) and are cleared leaving the system in an unfaulted condition. However, there is a high probability of secondary effects causing partial or total collapse. Therefore, an immediate identification of causes of power failure does not appear to enhance restoration of service appreciably, and it is safe to assume that the fault has been cleared by the time the restoration procedure begins.

In this paper, strategy for restoration from a complete collapse and without interconnection assistance is developed. A number of guidelines were established for the study which include:

1. Sectionalization of the system into subsystems, so that each subsystem would contain at least: (a) one drum type boiler with hot restart capabilitiy, and (b) one unit with black-start capability.
2. Simultaneous, immediate, and independent measures to supply station service for the hot restart of the

drum type boilers before elapse of the allowable time interval between shut-down and hot restart. This interval is usually of the order of a half an hour or less.

3. Interconnection of generating stations within each subsystem as soon as possible. Such an interconnection should be carried out after generating units attain the minimum specified generation; but prior to restoration of full load within the sybsystem.

4. Restoration of loads in small increments to avoid excessive under-frequency deviation. The general guide used was to pick-up load when system frequency was high and raise generation when system frequency was low.

5. Maintain steady state stability by having each station supply nearby loads.

6. Restoration of smaller and radial loads prior to larger and low voltage AC network loads, while maintaining a reasonably constant real to reactive power ratio.

7. Maintain system voltages within plus and minus 5% by: (a) maintaining generator voltages at low end of range during early stages of the restoration and (b) adjusting transformer taps to appropriate positions.

8. Maintain reactive balance by: (a) initially energizing only one circuit of double circuit underground transmission lines, and (b) operating generators under-excited to the extent that stability considerations allow.

9. As restoration proceeds give increasing consideration, as reactive limits permit, to firm capability by having: (a) firm transmission in the interconnections between subsystems, and (b) firm generation capacity in each subsystem.

Simulation and Coordination

The SRP entails careful consideration of probable effects of each action on the rest of the system and on the subsequent measures. Transmission line loadings, reactive power balance, over- and under-voltage conditions, generator loadings, transformer tap positions, shunt capacitors and shunt reactors status, and phase angle differences between load centers and generators can be evaluated by means of an interactive load flow program (4).

The time at which units are brought on the line, the rate at which generators can be loaded, the mismatch between load pick-up and generator response, the change in system frequency, and the effect on feeders having under-frequency relays can be evaluated by means of a long-term power system dynamic simulator (5).

Equally important is the recognition and coordination of certain critical time intervals, such as maximum time interval beyond which a unit with a drum type boiler cannot safely be hot restarted, or the minimum time interval after which a super-critical, once-through unit can safely be restarted. While sectionalization of the system into a number of subsystems and simultaneous restoration of service in each subsystem could appreciably reduce the duration of an outage, the extent of sectionalization is limited by availability of resources, including the number of operating teams as well as the communications available to permit effective and efficient coordination of their efforts. These items can be evaluated by the use of the critical path method (6).

Application

The first portion of the general methodology described in the previous section, i.e., the use of an interactive load flow program for simulation, was applied to the PEPCo system. Operator judgement and manual iterative techniques were substituted for the long term dynamic and critical path methods for this initial application.

Comparison of the light-load with the peak-load condition showed sufficient differences to justify development of two separate and detailed procedures. In this paper, emphasis is placed on the light-load SRP. In order to make the presentation of this example more meaningful, the power system, its operating conditions, and the restoration characteristics are briefly described. The description then is followed by the restoration strategy, comparison with peak-load conditions and the special problems encountered.

System Description

The PEPCo power system in 1980 has six generating stations: Stations A, B, C, D, E, and F as shown in Figure 3. Table II, Parts A and B show capabilities of generating units within these stations. To simplifiy SRP, it was convenient to divide the major portion of the system loads into seven load centers: AA, BB, CC, 115, 138, 230, and PC-R as shown in Figure 3. These load centers include 70 low voltage AC network loads of 15 to 32 MW range, radial loads, and large loads under customer control. Briefly, the power system consists of: (a) an outer 230kV system, with strong ties between Stations D, E, and F, (b) an inner system with rather weak links between Station A, B, and C, and (c) 115kV, 138kV, and 230kV overhead lines and underground cables interconnecting the outer and the inner systems as well as supplying the load centers. The 230kV system is tied to PJM, ECAR, and VACAR interconnections at four points.

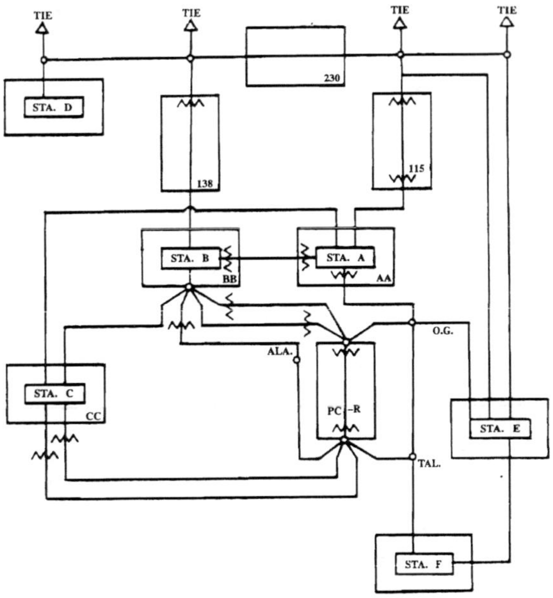

FIGURE 3: Generating Stations and Load Centers.

TABLE II, PART A: COMBUSTION TURBINES AND DIESEL*

		TIME TO MIN. LOAD MIN.	TIME FR. MIN LOAD TO MAX. LOAD	MIN. LOAD (MW)	MAX. LOAD (MW)
	STATION A				
1.	DIESEL 1	n/a	n/a	n/a	0 3
	STATION B				
2.	DIESELS 1&2 E (3)	n/a	n/a	n/a	n/a
3.	CT 1 to 8 E	20	5	56*	150*
4.	DIESELS 1&2 W (5)	n/a	n/a	n/a	n/a
5	CT 1 to 8 W	20	5	56*	160*
	STATION C				
6	DIESEL 1	n/a	n/a	n/a	0.3
	STATION D				
7.	CT 1	2.5	4	5	18**
	STATION E				
8.	CT 1	2.5	5	5	18
9.	CT 2	8.0	12	5	35
	STATION F				
10.	DIESELS 1&2 (11)	n/a	n/a	n/a	n/a
11.	CT 1&2	5	5	7	20
12	CT 3 to 6	10	15	7	

* VALUES ARE FOR BLOCK OF 8 UNITS.
** TRANSFORMER LIMITED TO 13 MW.
n/a NOT APPLICABLE

* Times are in Minutes

NOTES.
(1 & 6) DIESEL GENERATOR PROVIDES EMERGENCY TURNING GEAR END SAFE COOL DOWN POWER
(2, 4 & 10) DIESEL ENGINES PROVIDE BLACK START CAPABILITY FOR COMBUSTION TURBINES AS NOTED
(3 & 5) CT CAN BE STARTED AND OPERATED REMOTELY.
(7 & 8) COMPRESSED AIR PROVIDES BLACK START CAPABILITY
(11) PROVIDES BLACK START CAPABILITY FOR CT 3 TO 6.

TABLE II, PART B: STEAM UNIT CHARACTERISTICS

ITEMS	DESCRIPTION	UNITS	A15&16	B2	B3 to 6	C1&2	C3 to 5	D1 to 3	E1 &2	E3	F1&2
1-	TYPE	DRUM/SCOT	DRUM	DRUM	DRUM	DRUM	DRUM	DRUM	SCOT	DRUM	SCOT
2-	OPERATION	PEAK/BASE	CYCLE	CYCLE	CYCLE	CYCLE	CYCLE	BASE	BASE	CYCLE	BASE
3-	FUEL	OIL/COAL	OIL	OIL	OIL	COAL	COAL	COAL	COAL	OIL	COAL
4-	MIN. OFF TIME	HRS.	4	6	6	4	24	n/a	n/a	4	n/a
5-	MAX. HOT RE-START TIME	HRS.	1/2	1/2	1/2	1/2	1/2	1/2	8	1/2	8
6-	HOT START TIME TO SYNCHRO.	HRS.	1 1/2	2	2	2	2	n/a	n/a	2	n/a
7-	TIME FROM SYNCHRO. TO MIN. LOAD	MIN.	40	10	10	40	50	n/a	n/a	70	n/a
8-	TIME FROM MIN. LOAD TO FULL LOAD	MIN.	60	22	40	65	65	n/a	n/a	60	n/a
9-	MIN. TIME BETWEEN SUCCESSIVE SYNCHRO.	HRS.	1	1/4	1/4	1/2	1/2	n/a	n/a	0	n/a
10-	MIN. GROSS LOAD (EACH)	MW	30	10	10	25	35	75	165	130	190
11-	MAX. (NORMAL) GROSS LOAD (EACH)	MW	275	32	50	92	108	185	340	640	575
12-	EMERGENCY REGULATION RATE	%/MIN	6	1	2	2	2	4	8	36	13
13-	PRIME MOVER RESPONSE	%/HZ	----------ESTIMATED AT 5% PER 1/2 HZ----------								

NOTES:

1. DRUM TYPE BOILER (DRUM) OR SUPER-CRITICAL ONCE THROUGH BOILER (SCOT).
2. STEAM CYCLING OR BASE LOAD.
4. REQUIRED TO AVOID VIBRATION AND ECCENTRICITY PROBLEMS IN ROLLING THE TURBINE UP TO SPEED DUE TO MISMATCH IN STEAM TO TURBINE TEMP.
5. THE INTERVAL BETWEEN UNIT TRIP-OUT AND RE-SYNCHRO.
6. REFERS TO START-UP TIME OF STEAM CYCLING UNITS, HAVING BEEN OFF OVERNIGHT OR LESS THAN 12 HOURS.
7&10. MINIMUM LOAD IS THE LOWEST LOAD LEVEL WITH UNIT UNDER AUTOMATIC CONTROL.
8&11. MAXIMUM (NORMAL) OR FULL LOAD IS THE HIGHEST LEVEL OF LOAD REACHED UNDER AUTOMATIC CONTROL.
9. TIME INTERVAL BETWEEN START-UP OF TWO UNITS WITHIN THE SAME PLANT.
12. REGULATION CAPABIILTY OF STEAM UNITS UNDER EMERGENCY CONDITION.
13. THE RESPONSE OF PRIME MOVERS EXPRESSED IN % LOAD PICK-UP IN ONE TIME TO THE RESULTANT DECLINE IN FREQUENCY.
n/a NOT APPLICABLE.

Operating Conditions and System Characteristics

The SRP for 1980 Summer Light Load was developed using Sunday, April 22, 1979 system conditions, which at 0600 hours represented: 1260 MW of gross generation at leading power factor, the net interchange into the system of about 80 MW at near unity power factor, and 1340 MW of system load. The generating units in operation were: Station F Units 1 and 2, Station E Unit 1, Station D Unit 1 and 2, and Station C Units 4 and 5, all having leading power factors. Table III shows the generating units in service and their outputs. Except for the two combustion turbines under test at Station B, the system is considered as having been under normal conditions. Table IV describes the characteristics of the PEPCo power system relevant to SRP under light-load conditions.

TABLE III: NORMAL LIGHT LOAD CONDITION

GENERATING STATION	GENERATION
STATION A	
STATION B	
*COMBUSTION TURBINE 6E	12.0 + j 0.0
*COMBUSTION TURBINE 4E	12.0 + j 0.0
STATION C	
**STEAM UNIT 4	39.0 − j 2.0
**STEAM UNIT 5	36.0 − j 5.0
STATION D	
**STEAM UNIT 1	78.0 − j 9.0
**STEAM UNIT 2	81.0 − j 6.0
STATION E	
***STEAM UNIT 2	167.0 − j 37.0
STATION F	
***STEAM UNIT 1	549.0 − j 33.0
***STEAM UNIT 2	283.0 − j 60.0
TOTAL GENERATION	1,257.0 − j140.0

NOTES:
* UNDER NORMAL LIGHT LOAD CONDITIONS, CTs WOULD NOT BE IN OPERATION, THESE WERE ON FOR TEST PURPOSES.
** DRUM TYPE OF BOILERS.
*** ONCE-THROUGH SUPER-CRITICAL BOILERS.

TABLE IV. PEPCO SYSTEM CHARACTERISTICS

THE CHARACTERISTICS OF THE POWER SYSTEM RELEVANT TO SRP UNDER LIGHT-LOAD CONDITIONS ARE AS FOLLOWS.

1. ABOUT 1,080 MVAR CONTROLLABLE SHUNT CAPACITOR BANKS ON THE DISTRIBUTION SYSTEM.
2. IN GENERAL, LOADS HAVE A LOWER POWER FACTOR THAN UNDER PEAK LOAD, AND CONSEQUENTLY, RELATIVELY HIGHER REACTIVE POWER, POWER FACTORS AND RATIO OF REAL TO REACTIVE LOADS ARE ABOUT 83% AND 3:2 AS COMPARED TO ABOUT 95% AND 3.1 FOR THE PEAK LOAD CONDITIONS.
3. THERE ARE ABOUT 63 NETWORK LOADS WHICH MUST BE PICKED-UP BY CLOSING GROUP CIRCUIT BREAKERS. THESE NETWORK LOADS ARE ABOUT 25% OF THEIR PEAK VALUES, RANGING FORM 4 TO 8 MW EACH.
4. THE 600 MVAR CHARGING CURRENTS FROM 230, 138, AND 69 KV TRANSMISSION LINES CONSTITUTE A LARGER PERCENTAGE OF SYSTEM MVA. ABOUT 30% AS COMPARED TO 15% FOR THE PEAK-LOAD CONDITIONS.
5. THE UNDERGROUND HIGH VOLTAGE TRANSMISSION LINES MUST BE MAINTAINED AT LESS THAN 105% VOLTAGE AT ALL LOCATIONS.
6. THE COMBUSTION TURBINES AT STATION F AND AT STATION B WITH 420 MW COMBINED PEAK CAPABILITY OPERATE FROM LAGGING TO UNITY POWER FACTOR. I.E., THEY ARE PREVENTED FROM ABSORBING REACTIVE LOADS.
7. STATION F UNITS 1 AND 2 AND STATION E UNIT 1 PROVIDE ABOUT 80% OF SYSTEM GENERATION. THERE THREE UNITS BEING OF SUPER-CRITICAL ONCE-THROUGH TYPE, WOULD NOT BE AVAILABLE AFTER TRIP-OUT FOR SOME EIGHT HOURS.
8. STATION E UNIT 3 HAS 640 MW MAXIMUM CAPACITY, IS A PEAKING UNIT, AND WITH A BLACK-START TIME OF 2 HOURS.
9. MAJOR GENERATING PLANTS D, E, AND F ARE RATHER REMOTE AND CONNECTED TO METROPOLITAN LOAD CENTERS VIA 138 AND 230 KV CABLES.
10. STATION D UNITS 1 AND 2 AND STATION C UNITS 4 AND 5, BEING OF THE DRUM TYPE, WOULD BE AVAILABLE, PROVIDED THEY ARE RESTORED WITHIN ONE HALF AN HOUR AFTER TRIP-OUTS.
11. IN THE ABSENCE OF ANY HYDROGENERATION, THE SYSTEM RESTORATION STRATEGY IS BASED ON BLACK-START CAPABILITIES OF COMBUSTION TURBINES AT STATION F AND STATION B.
12. WITH NO GENERATION AT STATION A, THE 115 LOAD CENTER CAN ONLY BE PICKED-UP FROM THE 230 KV SIDE.

Strategy

The strategy for system restoration under light-load condition is based on immediate and simultaneous supply of station service to Stations C and D, from combustion turbines at Stations B and F, respectively, so that a quick hot restart of Stations C and D drum-type units would be achieved. This strategy is primarily dependent on black-start capabilities of combustion turbines at Stations B and F where multiple unit installations increase the probability of availability. (It should be noted that operating procedures call for the immediate and independent starting of the single combustion turbine at Station D. If this start is successful, it will reduce the restoration time but not change the sequence.) Appendix A and Figures 4, 5, 6, and 7 describe the general approach to SRP under light-load condition in some detail.

Briefly, the SRP calls for four restoration levels: (a) sectionalization of the power system into the inner 69kV and the outer 230kV subsystems, (b) simultaneous restoration of the two subsystems (c) synchronization of the two subsystems and strengthening ties between them, and (d) completion of system restoration. In level four, the remaining loads are restored and the ties are closed in coordination with PJM, ECAR, and VACAR.

Comparison of Light- and Peak-Loads

The strategy for system restoration under peak-load condition differs from the light-load due to the following factors:

1. Low voltage AC network loads are larger, about four times, and thus more difficult to pick-up at early stages of restoration.
2. Those shunt capacitors on 13, 34, and 69 kV busses, which are centrally controlled, can be switched off for control of reactive power balance. (Those on distribution feeders are not under central control.)
3. Cycling steam units at Stations A, B, and E are in operation. These units are all drum type, and can be hot re-started within half an hour after trip-outs.
4. Station B Units 3 or 4 are equipped to be isolated with 33 MW of local loads in the event of system frequency decay. These units do not normally operate during light load periods.

5. With Staton A in operation, the 115 load center can be picked-up from the inner 69 kV subsystem.

6. With cycling steam units in operation and the isolating scheme of Station B available, the restoration strategy is not dependent on black-start capability of combustion turbines.

7. The low voltage AC network loads at all load centers can be picked-up individually after synchronization of the two subsystems.

8. The 138kV double circuit underground cables can be energized after the above synchronization of the two subsystems.

Special Problems

In developing SRP at PEPCo, a number of difficulties were experienced which may be typical of a metropolitan system with predominantly thermal generation. The following measures are under consideration to simplify SRP, to provide flexibility in coping with various system conditions that can result and to enhance rapid restoration. These measures may be applicable in other metropolitan power systems:

1. Additional relay scheme to isolate generation with matching local loads to avoid restoration from a complete collapse. It can be shown for instance, that with incremental load pick-up limited to 5% of the generation within an island, the restoration duration can be halved if restoration begins with double the initial load and generation.

2. Provision of a wider range of voltage regulation for the combustion turbines to assist in maintaining system voltage between plus and minus 5%. The upper limit is set by cables and the lower limit by stability consideration of synchronous generators.

3. In order to strengthen the ties between the inner 69kV and the outer 230kV subsystems, a number of 138 and 230kV cables need to be closed. These cables have about 500 MVAR of charging current at 100% voltage. Shunt reactors are being installed at strategic locations which will offset this charging current and facilitate system restoration.

4. Additional synchronization facilities at strategic locations other than generating stations and interconnection locations, will provide greater flexibility in sectionalization of the system into subsystems and then synchronization of such subsystems. Also, it is essential to measure the frequency of the possible subsystems from the operating center.

5. Hot restart of cycling steam units which are typically drum type is essential. These units have to be placed back into operation supplying loads within a short time interval following the system outtage. Otherwise the re-start may not be possible for several hours. Such a time interval then is very critical to system restoration, particularly in those stations where station service has to be supplied from another station. Again, it is important to accurately determine the time intervals between unit trip-outs and hot restart.

6. In the course of system restoration, a reasonable balance must be maintained between generation and load, avoiding excessive over- and under-frequency deviations. Load pick-up in small increments tends to lengthen the restoration duration. With large increments there is always the risk of falling into a frequency decline and recurrance of system outage. The size of load pick-up very much depends on the rate of response of prime movers. It is essential to determine the rate of response for each individual boiler turbine combinations, or to obtain guidelines from boiler-turbine manufacturers. This information could take the form of the largest increment of load increase at one time, which would not lower system frequency more than 1/2 HZ of its previous value. A long term power system dynamic simulator would be an important tool to determine a balance between load pick-up and response of on-line generators. At PEPCo, in the absence of such a long term dynamic simulator, a load increment of 5% of each subsystem generation was used. It was believed that such an increment would reduce the frequency by no more than one half Hertz.

CONCLUSIONS

1. The nature of system restoration is such that it initially appears that a detailed set of instructions could be prepared for use by the operator in the event of a disturbance. However, upon a closer examination, it becomes clear that the problem is somewhat more complex. For instance, while the development of SRP from a point of complete collapse and without interconnection assistance is relatively straight forward, in practice, due to corrective actions, system disturbances are usually arrested prior to the complete collapse. Furthermore, some level of interconnection assistance would probably be available. However, partial system outages and/or reliance on the interconnection, necessitates assessments of both system and interconnection status, a process which uses up critical time.

Ideally, it would be desirable to develop a master plan covering different system operating conditions and various degrees of system upset. In the event of a system disturbance, after initial assessment of the system status, a proper entry into the master plan would be made, and the specific subroutine selected and followed to a successful restoration. This will be the subject of an ensuing effort.

2. In order to develop an effective SRP, it is essential to have the combined efforts of system analysts, system operators, and production personnel. The plan can then be analyzed and made more efficient by the concurrent use of interactive load flow, long term dynamic simulator, and critical path programs. Unfortunately, the long term dynamic simulator is not generally available.

3. Although power system characteristics under peak- and light-load conditions differ, requiring different SRP, it is important for ease of operation to try, in so far as possible, to develop similar plans.

4. Under light-load conditions, the critical activities include start-up of cycling steam units. Under peak-load conditions, the critical tasks are usually the coordination of network load pick-ups with generation responses and consideration of possible risk for system frequency decline beyond prime movers recovery.

5. A functional SRP entails development of a strategy to minimize the duration of restoration. Such a strategy typically includes several distinct levels of restoration for various phases, parts, and conditions of the system. Each level in turn consists of a number of activities. The entire plan is broken down into some logical order. This process provides a discipline whereby there is less risk of overlooking any essential operation.

6. Most power systems have certain characteristics in common and behave in a similar manner during restoration process. It is therefore possible to establish a general procedure and a list of guidelines, which would enhance rapid restoration. On the other hand, the detailed SRP has to be specifically developed to meet the particular requirements of an individual power system. Thus, it is important to determine and record the relevant characteristics of each power system.

7. Finally, development of a precise and consistent glossary of terms, which could readily be understood and which would be free from ambiguity, becomes more important during restoration than under normal and day-to-day operations.

ACKNOWLEDGEMENT

The authors would like to record their appreciation to Mr. Walter A. Johnson, Potomac Electric Power Company, for his direction in the development of the SRP and helpful suggestions in preparation of this paper.

APPENDIX A

The SRP under light load calls for:

1. Sectionalization of the power system into two subsystems; the inner subsystem and the outer 230kV subsystem.

   a. The inner subsystem consists of three generating stations A, B, and C, and four load centers AA, BB, CC, and 115 (See Figure 4).

   b. The outer 230kV subsystem also consists of three generating stations D, E, and F, and three load centers PC-R, 138 and 230. (See Figure 5).

2. Simultaneous restoration of the two subsystems. In the inner subsystems, the sequence of restoration is as follows (See Figure 4):

   a. Start-up of combustion turbines at Station B, which have black-start capability, and gradual load pick-up at BB load center.

   b. Supply station service to Station C Units 4 and 5 through Station B - Station C 69kV circuits. Start-up of Station C Units 4 and 5 and gradual load pick-up at CC load center.

   c. Interconnection of Stations B and A via the two 34kV circuits and Stations C and A via the 69kV circuit.

   d. Energizing Station A - 115 load center via 115kV single circuit and 115 load center - Station E via 230kV single circuit in preparation for synchronization of the subsystems.

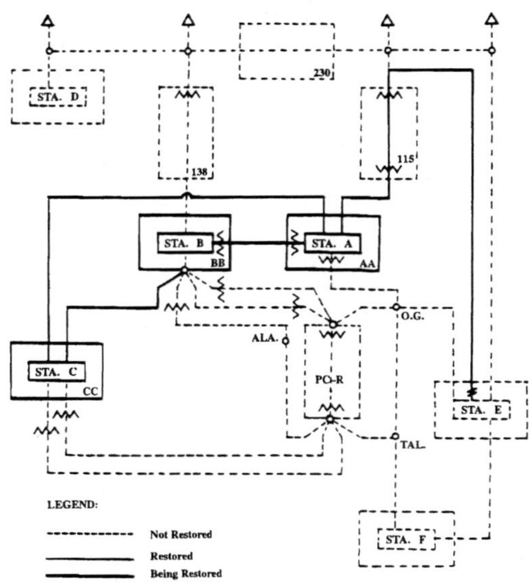

FIGURE 4: Restoration of Inner Subsystem

In the outer 230kV subsystem, the sequence of restoration is as follows (See Figure 5):

a. Start-up of combustion turbines at Station F which have black-start capability. Supply of station service at Station F and to Station E. Gradual local load pick-up at Station E and Station F.

b. Energizing Station F to PC-R load center via a 230kV circuit and gradual load pick-up at PC-R load center.

c. Energizing a single 230kV circuit between Stations E and D. Gradual load pick-up at 230 load center. Supply of station service to Station D and start-up of Units 1 and 2.

d. Closing of a single 230kV circuit between Staton D and 138 load center without closing the 138kV breaker at Station B. Gradual load pick-up at 138 load center.

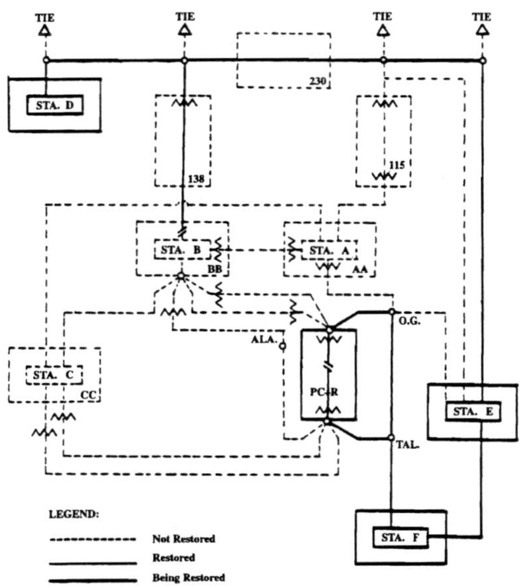

FIGURE 5: Restoration of Outer Subsystems.

3. Synchronization of the two subsystems and the strengthening of the ties between them (See Figure 6):

a. The two subsystems are synchronized at Station E.

b. The two 230kV circuits are tied together through 230/115kV transformers. Gradual load pick-up at 115 load center.

c. Closing of the 138 load center to Station B. Closing of one of the two 230kV circuits between Station C and PC-R load center.

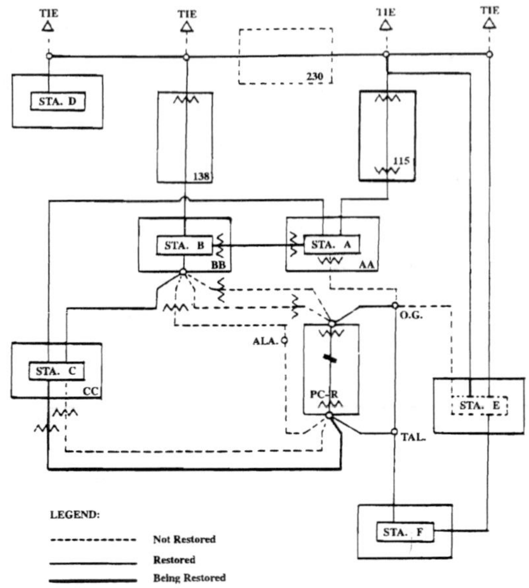

FIGURE 6: Synchronization of the Two Subsystems.

4. Completion of system restoration (See Figure 7): start up of peaking Unit 3 at Station E, interconnection of Station A and Station B to the outer 230kV subsystem and picking up of the 230 load center.

5. Interconnection of PEPCo system with PJM, ECAR, and VACAR interconnections in coordination with interconnection offices.

NOTE: Depending on the availability of Station E, Unit 3, or the supply from PJM Interconnection, the sequence of restoration between above levels 4 and 5 could be reversed.

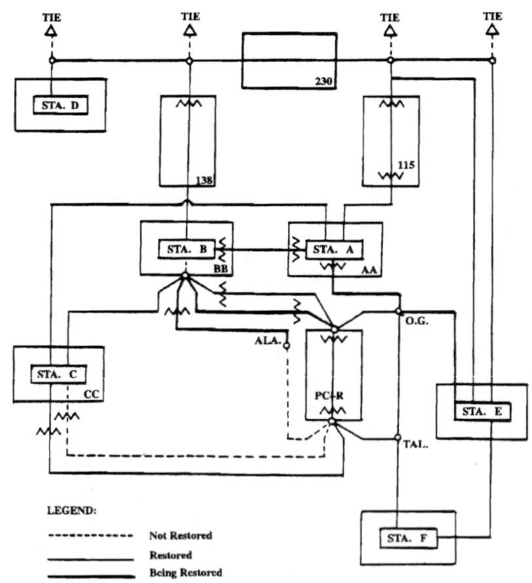

FIGURE 7: Completion of System Restoration.

REFERENCES

1. Scheurer, David, "Philadelphia Electric Company System Restoration Plan", Pennsylvania Electric Association, System Operation Committee, Pittsburgh, Pennsylvania, Jan., 1980.

2. Johnson, Walter A., et al, "The Potomac Electric Power Company's Consolidated Control Center", IEEE PES Summer Meeting Anaheim, CA., July 1974, C74 366-1.

3. "Report of Joint AIEE-EEI Subject Committee on Line Outages", AIEE Committee Report, AIEE Trans. on Power Apparatus and Systems, Jan 1952, pp. 42-56.

4. Undrill, John M., et al, "Interactive Computation in Power System Analysis", Proceedings of IEEE, Vol. 62, July 1974.

5. Adibi, M. M., et al, "Solution Methods for Transient and Dynamic Stability", Proceedings of the IEEE, Vol. 62, July 1974.

6. Wiest, J. D., et al, "A Management Guide to PERT/CPM", Englewood Cliffs, N.J.: Prentice-Hall, 1969.

Discussion

R. O. Usry (Southern Co. Service Inc., Birmingham, Al): The writer's comments are more in the way of observations as opposed to a critical analysis of the paper, but are based upon observations over a large number of years. The writer concurs with the authors in that added emphasis on system restoration plans is a worthwhile endeavor, and additional work is needed in this area. However, it should be recognized that the complexity of the problem compounds tremendously as the size of the system increases.

Accurate and real-time data is essential if a central computer location is to be used to augment restoration procedures. In this regard, reliable back-up power supply systems are a must. Additionally, reliable communications are essential during power failures if data is to be relied upon. This again requires adequate back-up power supply.

Part of the strategy and guidelines noted in the paper is that the drum type boilers would be able to be restarted in one-half an hour or less. A general observation is that from an industry-wide standpoint, this appears optimistic. The writer would be interested in knowing the basis for this ability, and it is presumed that it is based upon actual operating experience.

Additionally, part of the strategy notes that during light load periods, the combustion turbines are immediately available. Normally, the combustion turbines would be disconnected during light load periods, and it appears that a start-up period of ten to fifteen minutes would be more realistic.

The writer would also be interested in knowing if all restoration procedures are anticipated to be handled from one central command point, or does PEPCo have decentralized locations working in concert with the central computer facility.

Manuscript received January 13, 1981.

J. Felinstein (Consolidated Edison Company, New York, NY): Con Edison has a System Restoration Plan (SRP) which includes procedures that would expedite a system restoration effort. It is not surprising that Potomac Electric's SRP methodology is similar to the plan developed by Con Edison since both systems serve metropolitan areas. The only difference between the two systems is the size and complexity of the transmission and generation facilities.

Some differences in the plans are:

- Voltage control of the Con Edison 345kV pipe type cable system will be via load pickup with tie feeders being loaded up to long term emergency ratings (3 hour ratings). Generation will be utilized as it becomes available for both load pickup and voltage control.
- The Con Edison plan assumes restoration will be via seven transmission paths utilizing (assuming they are available) neighboring systems. Re-energization assuming no neighboring systems are available is presently being studied with the intent to pickup large low voltage network load (100-200 MW) via our four major gas turbine sites (2000 MW installed capability).

Another important aspect of Con Edison's SRP is operator training. Once each year, we review the SRP with the entire bulk power operating team (Senior System Operators, System Operators, Power Dispatchers and Transmission Operators for total of 26 shift operators). This is done with groups of 4 to 6 operators over a six week period. Our generating station operating personnel also drill periodically by restarting (following a maintenance outage) a steam generating unit via their black start satellite gas turbine.

Our goal this year is to develop a program for the Control Data System Response Simulator to enable the operators to train in the various restoration stages. A feasibility study is presently underway with a mid-1981 scheduled completion date.

Manuscript received February 26, 1981.

W. H. Underwood (Long Island Lighting Company, Hicksville, N.Y.): The authors have presented in a concise, precise and sound engineering report a guide for helping operators assess the problems of restoring a system or part of a system.

The tools provided here will supply the guidelines of what trouble spots to look for. Since 1965 there have been many disturbances that have been contained by operator action due to their knowledge of what is written in emergency policies. The information now available to him has increased almost 1000 fold since 1965. The operator must still make the decision.

The authors have addressed the last area that requires more detailed study and a determination of what additional tools and guidelines are needed by the operator in the restorative state.

I encourage them to continue the study as planned. My only caution is that they do not become so confident in their ability to analyze energy situations so that the operator can be replaced with a robot. Train the operator, simulate restoration, keep him informed and let him make the decisions on the basis of system conditions as they are not as some usually inadequate model says they should be.

Manuscript received February 26, 1981.

R. J. Kafka, D. R. Penders, S. H. Bouchey, and **M. M. Adibi**: The authors wish to thank the discussors for their comments and questions.

In reply to Mr. Usry, we agree that system restoration complexity increases with size, but point out that restoration complexity can be controlled by judicious selection of subsystems, with the guideline of limiting the complexity of any subsystem to the capabilities of the resources available. It is also true that other items relevant to system restoration, such as back up power supplies, secure communications, and maintenance of critical system components must be addressed before a restoration simulation which assumes the availability of these items can be made. Regarding the start-up times for steam units and combustion turbines, the times listed in the tables have been experienced frequently enough in operation of the electric system to warrant their use in estimating system restoration time. Immediate start of combustion turbines means that the start-up sequence is immediately requested. The combustion turbines would be ready for load pickup 2.5 to 15 minutes afterwards. The one-half hour restart time referenced for drum type boiler units is an estimate, based on operating experience of the maximum allowable time before starting the load pick-up with the unit. If this time is exceeded, the unit must be completely shutdown and may not be available for several hours following initial system collapse. The PEPCo system is operated from one central control center.

Regarding Mr. Feinstein's comments on a Con Ed system restoration plan, we found in our simulation that it is not possible to energize high voltage cables in the initial stages of restoration because the loads necessary to control the voltage could not be picked up with available generation. A significant number of 69kV cables must be energized as they form the bulk of the plant interconnections and substation feeders during the early stages of restoration. Careful selection of feeders to be energized and maintaining minimum voltage on generators were required to control system voltage at nearly all stages of the restoration simulation.

The load pick-up guideline used was to limit each increment to no more than 5 percent of prevailing generation. This would ensure that the frequency would decline by no more than one-half Hertz. The authors caution against picking up 200 or even 100 MW loads with 2000 MW combustion turbine capability as suggested by Mr. Feinstein. A distinction must be made during restoration between capability and actual prevailing generation at the time of the load pick-up. The guidelines and assumptions made prior to the restoration study included that there would be no outside assistance available. While we have not studied a case with interconnection assistance, it is believed that restoration could be significantly enhanced through the use of generation and load sharing.

To date, the training of dispatchers regarding system restoration has not been formalized, but will be included in the normal training for Senior Dispatchers and Transmission Dispatchers. However, the system restoration plan development was a team effort including analysts and dispatchers, and the system operators are already familiar with the restoration plan. We would be interested in any results using the System Response Simulator.

In response to Mr. Underwood, we agree that the system operator must make the decisions based on system conditions in any restoration procedures. In fact, we believe that a system restoration plan for actual use by dispatchers should state goals for each step and the conditions that must be met before proceeding to the next step. Having seen situations in the restoration simulation, and making decisions during this process, the dispatchers have been exercised to cope with the kinds of conditions which may occur in system restoration. The purpose of a detailed simulation is two-fold: to determine the necessary conditions for each step and to assure that each step can be physically accomplished by at least one method. By stating a system restoration plan in this manner, system operators are given guidelines that can be followed with greater confidence than would otherwise be possible.

Manuscript received March 3, 1981.

ROLE OF INTERACTIVE AND CONTROL COMPUTERS IN THE DEVELOPMENT OF A SYSTEM RESTORATION PLAN

R. J. Kafka, D.R. Penders, and S.H. Bouchey
Potomac Electric Power Co., Washington, D.C.

M.M. Adibi
IRD Corporation, Bethesda, Maryland

Abstract. The duration of power system disturbances are minimized by executing a well-maintained system restoration plan (SRP). The SRP is significantly improved by the combined efforts of the operating personnel and systems analysts. The team effort is enhanced by the concurrent use of control computer and interactive computer systems. It is facilitated when both groups and both systems are located at the operating center. Adopting this approach, Potomac Electric Power Company, (PEPCo), has developed an efficient solution to SRP, which has proven extremely beneficial. The approach, and the results presented in this report, should be of interest to other utilities.

INTRODUCTION

During 1979-80, a project team was established at PEPCo to develop plans for restoration of service following a complete system collapse under both light-and peak-load conditions independent of prior plan. The project team consisted of systems analysts and operating personnel, who (a) had access to the on-line, real-time control computer system[1]; and (b) used the off-line interactive computer system.[2] The project team and the control and interactive computer systems were all conveniently located at the system operating center.[3] The project team was assisted by the production personnel, who provided detailed information on the characteristics of generating units, and by the planning personnel, who reviewed the SRP and coordinated the recommendations and requirements of the SRP with their short-term system plans. The results of this operator and analyst team effort and concurrent use of control-interactive computer systems, which were very encouraging, are presented for consideration by other utilities.

Reference is made to the previous paper, "System Restoration Plan Development for a Metropolitan Electric System",[4] which covers methodology, strategies and guidelines, and special restoration problems. This report presents applications of available analytical tools, including Interactive Load Flow (ILF) in the development of the SRP. The need for the development of a long-term power system dynamic simulator,[5] and application of critical path methods[6] are identified and discussed.

COMPUTER SYSTEMS

In order to develop the SRP, it was necessary to model the PEPCo power system by representing the physical and operational data in an interactive computing system. The model then had to be validated against the actual operating data obtained from the control computer system. The project team then formulated a strategy for system restoration. The strategy consisted of sectionalization of the power system into a number of subsystems in accordance with a set of guidelines established by the operating personnel. A restoration plan for each subsystem was then developed by simulating a step-by-step integration of loads and generation. The objectives were to restore the subsystems simultaneously, thus minimizing the restoration duration. This iterative procedure entailed careful consideration of probable effects of each restoration step on the rest of the system and on the subsequent measures. The process was facilitated through the intensive use of ILF. Figure 1 outlines the iterative process involved. A brief description of PEPCo's power system, the control computer system and the interactive computing used in development of the SRP follows.

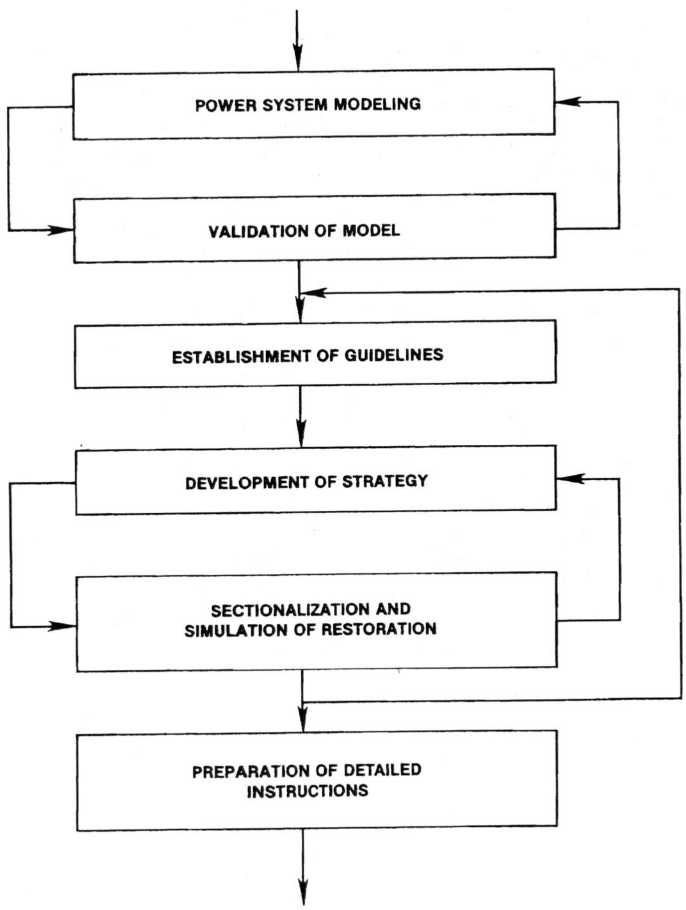

FIGURE 1 — ITERATIVE PROCESS FOR SRP DEVELOPMENT

Reprinted from *IEEE Transactions on Power Apparatus and Systems,* Vol. 101, No. 1, pp. 43–52, January 1982.

Power System. Figure 2 shows the one-line diagram of PEPCo's power system, which at present consists of six generating stations A, B, C, D, E, and F. Stations B, D, E, and F have combustion turbines with black-start capabilities. Total system capacity is about 4400 MW steam and 560 MW combustion turbines. In order to simplify the development of the SRP, the system loads were divided into seven load centers: AA, BB, CC, 115, 138, 230, and PC-R as shown in Figure 2. These load centers include 63 low voltage AC network loads of 15 to 32 MW range, energized with group circuit breakers, radial loads, and large loads under customer control. Briefly, the power system consists of (a) an outer subsystem with strong 230kV ties between Station D, E, and F; (b) an inner subsystem with 69kV links beween Stations A, B, and C; and (c) 115kV, 138kV, and 230kV overhead lines and underground cables interconnecting the outer and the inner subsystems as well as supplying the load centers. The 230kV system is tied to PJM, ECAR, and VACAR interconnections at four points.

Control Computer System. The control computer system had two roles in the development of the SRP. First, it provided system operating data for verfication of the power system steady-state model. Second, it guided, through its man-machine capabilities, how the SRP should be developed for efficient execution. The data acquisition capability of the control computer system provides interfacing for 110 remote terminal units into the system. Each remote terminal unit has the capability of controlling remote equipment, monitoring its status, and digitally telemetering various quantities indicative of power system conditions. The remote terminal units are located at all major generating plants and substations. The generating plant terminals are capable of data acquistion and status monitoring but do not control generator power circuit breakers since full-time operators are available. The substations, on the other hand, are all unmanned and have full remote control as well as data acquisition and status monitoring capabilities. The light pen is the prime tool for calling up displays on CRT's and accomplishing control actions. The man/ machine interface also includes line printers for reports, video copiers for hard copy outputs of system displays, and a display board showing the overall system one-line diagram. Figure 3 shows the control computer system block diagram.

Interactive Computer System. In the course of this project a large number of load flow cases were run. The first group were run during power system modeling process and model validation. The second group were run for development of the SRP, where the power system had to be sectionalized for simultaneous restoration. To enter and run so many cases in the batch load flow mode would have been very inefficient and time-consuming. The ILF program proved to be very efficient while also providing a better understanding of the restoration process.The interactive computer system used for ILF includes: the CRT with standard keyboard for interaction with the load flow program, the printer-plotter to provide hard copy for reports, and storage media for storing and retrieving many cases under study. Figure 4 shows the block diagram of PEPCo's Interactive Computer System.

POWER SYSTEM MODELING

As previously stated, SRP development is an iterative process facilitated by the rapid data changes and solutions available with interactive computing. Three types of models are required to simulate the process of system restoration and reintegration: a steady-state model to represent the electric system at discrete steps, a dynamic model to describe the behavior of the system in transition between the steps, and a model of the operator actions and resulting system changes required to make the transition between steps. Of the three required models, only the load flow (steady-state model) was available to the project team as an interactive computing tool. The load flow enabled evaluation of transmission line loading, reactive power balance, generator loadings, transformer tap positions, shunt capacitor and shunt reactor status, and phase angle between loads and generating stations. Operator actions may dictate the topology and representation of the load flow model and determine the sequence of steps in the restoration procedure. Careful evaluation of expected control actions was critical during the construction of the load flow model, and it was a prime consideration of the system analysts.

Steady State Model. Prior to the investigation of system restoration, PEPCo had been using an interactive load flow as a dispatcher's training and study tool.[8] Experience with the dispatchers load flow data base had been satisfactory, but system restoration studies would require more detailed analysis than had been previously attempted. The study team determined that a formal validation of the existing model was necessary and that additional detail would be required. A one-line diagram was prepared representing the power supply system consisting of generation, transmission, and subtransmission. At the generating stations, the representation included steam and combustion turbine generating units, detailed arrangements for black-start facilities, station service, and switching facilities. Similarly, the substation diagrams included bus sections, circuit breakers, disconnect switches, feeders, and equipment relevant to SRP. During preparation of the one-line diagram, continued comparisons were made to the data acquisition and control displays stored in the control computer system. Figure 5 shows a one-line diagram of a typical substation. As a normal practice in an operating environment, transmission and distribution feeders are identified by circuit numbers. The ILF program, however, identifies circuits by terminal nodes. Tables were constructed relating circuit numbers to load flow terminal nodes for easy conversion between the two representations.

The reactance, resistance, and susceptance of the power system components were obtained from engineering personnel and modified to conform with the operator's representation of the power system. For example, circuit configurations retained specifically to provide redundant supplies to critical loads or to facilitate system restoration, but not normally used in daily operation, were added to the load flow data. Other typical load flow data entered into the ILF included transformer ratios, phase angle ranges, and generating unit characteristics.

Steady State Model Validation. Essentially, validation of the load flow data base consisted of obtaining ''instantaneous'' power system data from the control computer, entering the required information in the form of bus loads, generator outputs and voltages, and transformer ratios and angles into the load flow, and then solving the load flow and comparing the resulting bus voltages and line flows to measured values on the electric system. For the validation phase, operational data were obtained from the hourly logs stored by the control computer. Generating station and tie line flows were obtained within a ten-second scan and all other system data were obtained within three minutes.

The power system was represented fully, although radial systems were modeled by bus loads. Tie line flows were constant MVA models, represented as fixed generators or loads. Since most of PEPCo's transformers have simultaneous tap changer under load (TCUL) control of both voltage ratio and phase angle, it was necessary to enter these transformer values manually based on the telemetered tap positions.

Comparison of the initial ILF solutions with measured and telemetered power system data, not surprisingly, showed significant differences. In order to identify the sources of the inaccuracies, a number of techniques were used. The power system was divided into two subsystems,the outer subsystem, which consisted primarily of the 230kV lines, tie lines, and generating stations D, E, and F as shown in Figure 2; and the inner subsystem, which included the 34, 69, 115, and 138kV lines and generating stations A, B, and C. By making use of existing redundant measurements, comparing parallel lines, performing bus summations, and making minor adjustments in appropriate load flow data, a fair degree of agreement was obtained in the subsystems. The process required further sectionalization into smaller subsystems and many iterations of the adjustment and compare cycle before the subsystems were reintegrated into the complete system model.

During the load flow validation process, a number of observations were made. First, the ILF solutions and measured power system data did not exactly correspond because of inaccuracies in measurement and the time consuming process necessary to fine tune the load flow data base. Second, because of redundancy of measurements and the availability of the IFL for immediate verification, it is possible to locate inaccurate system measurements. Such errors pinpoint instruments that require calibration. Finally, it is possible to obtain a load flow data base which meets the criteria of operating personnel for accuracy. These are: (a) when the computed active and reactive line flows, compared to the measured line flows, produce an error which is a small percentage of the current carrying capacity of the line and (b) when computed voltages for critical busses are close to observed values. Tables I and II show the level of agreement obtained between computed and measured power flows around a 230 kV bus and in a double circuit 230 kV line, respectively.

COMPUTER APPLICATIONS

Interactive Load Flow. Load flow studies for system restoration have a number of requirements that are easily met by an interactive load flow program. First, a great deal of interaction with the data base is required as the model of the electric system is built up from the first generator-load combination to the eventual reconnection of the entire system. The case storage system of the IFL allows the entire system model to be saved, even though most of the elements may be disconnected or out of service. By saving each case, in the form of a disc file, after making small changes or additions to the previous case, it is possible to return to any level of the restoration sequence if the subsequent cases would lead to an unacceptable situation and a new restoration strategy is required. Second, the rapid solution and turnaround time of the ILF allows each case to be built upon the cases which preceeded it, without the need to examine several alternatives at random. With each case leading to the next, efficient use is made of both analytical and computing resources. Third, a menu of output options is available, permitting the selection of detailed and sorted data as required. One of the output options available, but not used during this initial system restoration study, is the graphics output in the form of the one-line diagram. At the start of the study, existing one-line diagram format files were not sufficiently detailed for the configuration of the electric system during restoration. The need for graphic representation became evident by the many hand drawn sketches used to interpret the tabular load flow output. Future system restoration studies will use the graphics output option of the ILF.

Energy Balance. During system restoration, a reasonable balance must be maintained between generation and load to avoid excessive over and under frequency deviations. Load pick-ups at small increments prolong the restoration duration. With large increments there is always the risk of frequency decline beyond recovery and recurrence of system outage. The size of load pick-up depends on the response of prime movers. In developing the SRP it was therefore essential that the rate of response of prime movers be determined.

Unfortunately, the available linearized dynamic models are not suitable nor adequate for evaluating dynamic phenomena such as the variation of system frequency as a function of load-generation mismatch, the effect of underfrequency relays, and the rate at which a generator can be loaded. To consider these phenomena, dynamic behavior of boilers, turbines, combustion controllers, feedwater controllers, temperature controllers, and governor systems must be represented over the operating range of prime movers.

As a substitute for a simulator, a guideline approved by the operating and production personnel was used. This guideline called for an incremental load pick-up equal to or less than 5% of the prevailing generaiton of the subsystem. It was believed that with such a load increment the system frequency would not decline more than 1/2 hertz.

Activity Coordination. In general, system restoration consists of sectionalization of the system into an optimum number of subsystems and simultaneous restoration of each subsystem. The restoration operation included a number of time consuming activities including stripping the generating station busses, closing bus-ties, providing station service and start-up power, energizing transmission lines, closing distribution feeders, adding loads, and stabilizing generation.

The drum type boilers have a maximum elapsed time to hot restart of about 1/2 hour beyond which they would not be available until cold restart, 4 to 6 hours later. The super-critical once-through boilers have a minimum elapsed time after a tripout of several hours before they can be restarted. Loading of generators is also a time-consuming operation, because it involves a number of time dependent characteristics such as maximum hot restart time, hot start time to synchronization, time from synchronization to minimum load, rate of load pick-up and time from minimum load to full load all of which are based on temperature related phenomena inherent in the physical construction of the boiler-turbine-generator complex.

As already discussed, the interactive load flow program basically assists in planning for reactive power balance, and the dynamic simulator when available could help with determination of proper load and generation balance. For coordination of schedules, assignment of resources and the setting of priorities, critical path method (CPM) can be a very helpful tool, providing more precise scheduling and insuring a minimum system restoration duration. The greatest impact of CPM would be in planning, scheduling, and coordination of various restoration tasks for which accurate time estimates cannot be made. With CPM, many parallel and sequential tasks required in system restoration could be planned in an orderly sequence and logical order, from the power system collapse through successful restoration. The process would provide a discipline so that there would be less risk of overlooking any essential operation.

METHODOLOGY

System restoration plans are developed to meet the systems requirements of an individual power system. The development entails evaluation of the relevant power system characteristics, evolution of a strategy for rapid restoration, establishment of guidelines consistent with accepted practices, and preparation of detailed instructions for use by the operating personnel in the event of a system disturbance. The developmental procedure, as shown in Figure 1, is an iterative process. The SRP evolves from simulation of and interaction with the steady-state, dynamic behavior, and scheduling requirements of parts and components of the power system. The characteristics, strategy, and guidelines used in developing the SRP for the PEPCo system are briefly described.

Characteristics. The relevant characteristics of the PEPCo system were divided into three functional groups. The first are related to reactive power balance. The second affect the load and generation balance. The third group deals with feeders equipped with under frequency relays, unit isolation schemes, synchronization facilities at other than generating stations, system frequency metering points, and interlocking schemes that allow or prevent energizing dead busses. Table III lists the PEPCo system characteristics under light-load conditions.

Strategy and Guidelines. The strategy was to (a) sectionalize the system in a manner which would allow each subsystem to contain at least one drum-type boiler with hot restart capability and at least one station with black-start capability, (b) take immediate and simultaneous measures to supply station service for hot restart of the generators with drum type boilers before elapse of the maximum time interval allowed between shut-down and hot restart, (c) interconnect generating stations within each subsystem as soon as possible after the generating units attain the minimum specified generation but before restoration of full load within the subsystem. This strategy for the light load condition divided into four distinct restoration levels, each level consisting of a number of tasks and each task of several actions. Briefly, as shown in Figure 3, the strategy called for sectionalization of the power system into the inner and outer subsystems, synchronization of the two subsystems and strengthening of the ties between them, and completion of system restoration by picking-up the remaining loads and closing of ties in coordination with PJM, ECAR, and VACAR

The guidelines established were the result of several operator-analyst discussions and many power system simulations. Table IV lists the restoration strategy and guidelines used at PEPCo.

RESULTS AND CONCLUSIONS

The operator-analyst team effort and the on-line/off-line computer applications in an operating environment produced the SRP for light- and peak-load conditions which are acceptable to the system operators. This is the most significant result of the project. While it is recognized that the SRP can benefit from further improvements in the future planning efforts, it was the first organized and formal approach to development of SRP at PEPCo.

In the course of the SRP development, a number of difficulties not previously recognized were experienced which are presently under consideration by PEPCo and whose solutions would provide additional flexibility in system restoration. Possible solutions include (a) development of generation isolation schemes to avoid restoration from a complete collapse in each strategic subsystem and thus reduce restoration duration considerably, (b) installation of shunt reactors at strategic locations to offset 138 kV and 230 kV cable charging currents and facilitate rapid restoration, (c) installation of additional synchronization facilities at strategic locations other than generating stations and tie points to provide greater flexibility in sectionalization of the system and subsequent synchronization of the subsystems, (d) measurement and telemetering of system frequency from possible subsystems, (e) accurate determination of prime mover time intervals between shut-down and hot restart or cold restart, timing from restart to sychronization, synchronization to minimum load, and time from minimum load to maximum load, and (f) determination of individual prime mover's frequency response to sudden load changes.

In order to develop an effective SRP, it is essential to combine the efforts of operating personnel and systems analysts. The concurrent use of control computer and interactive computer systems facilitates such a development. While load flow is well established for the steady-state simulation of the system, long-term power system simulators are not available/ adequate for understanding the dynamic behavior of prime movers. Critical path method programs are available and capable of scheduling activities and assigning resources so necessary in system restoration where precise time estimates cannot be made.

With the modification of the power system and changes in its operational procedure, the SRP has to be reviewed and updated annually. Recognizing the power of CPM, the project team intends to fully utilize CPM for the 1981 SRP cycle.

Under light-load conditions, where only the base units are in operation, the critical activities include start-up of cycling steam units. Under peak-load conditions, the critical tasks are usually the coordination of network load pick-up with prime mover's responses. Although power system characteristics under peak- and light-load conditions differ, requiring a different SRP, for ease of execution, it is important that so far as possible, the two SRPs be similar.

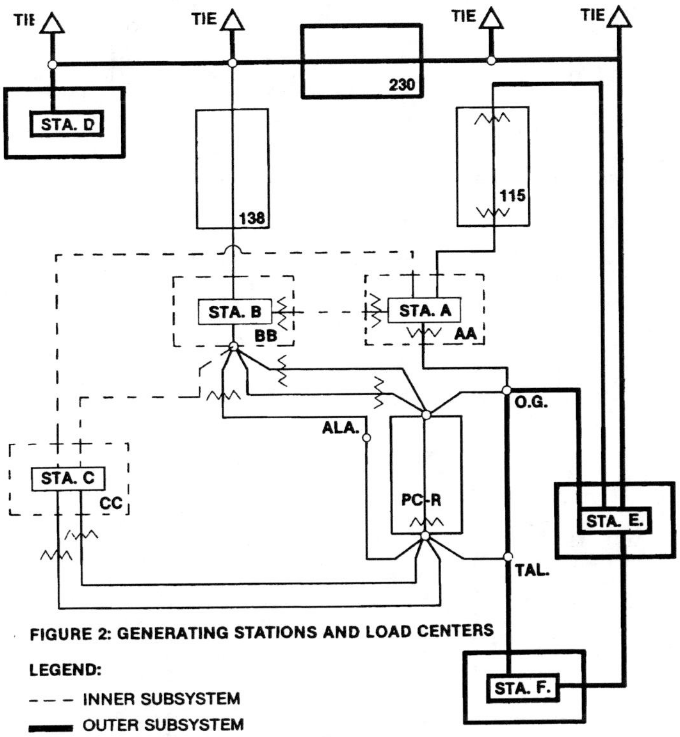

FIGURE 2: GENERATING STATIONS AND LOAD CENTERS

LEGEND:

- - - INNER SUBSYSTEM
▬ OUTER SUBSYSTEM
— 115, 138 AND 230 LINES INTERCONNECTING THE TWO SUBSYSTEMS.

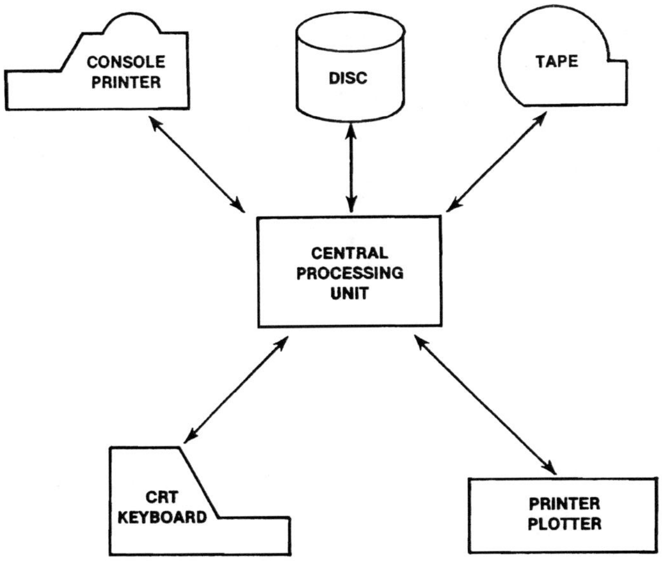

FIGURE 4 — INTERACTIVE COMPUTER SYSTEM

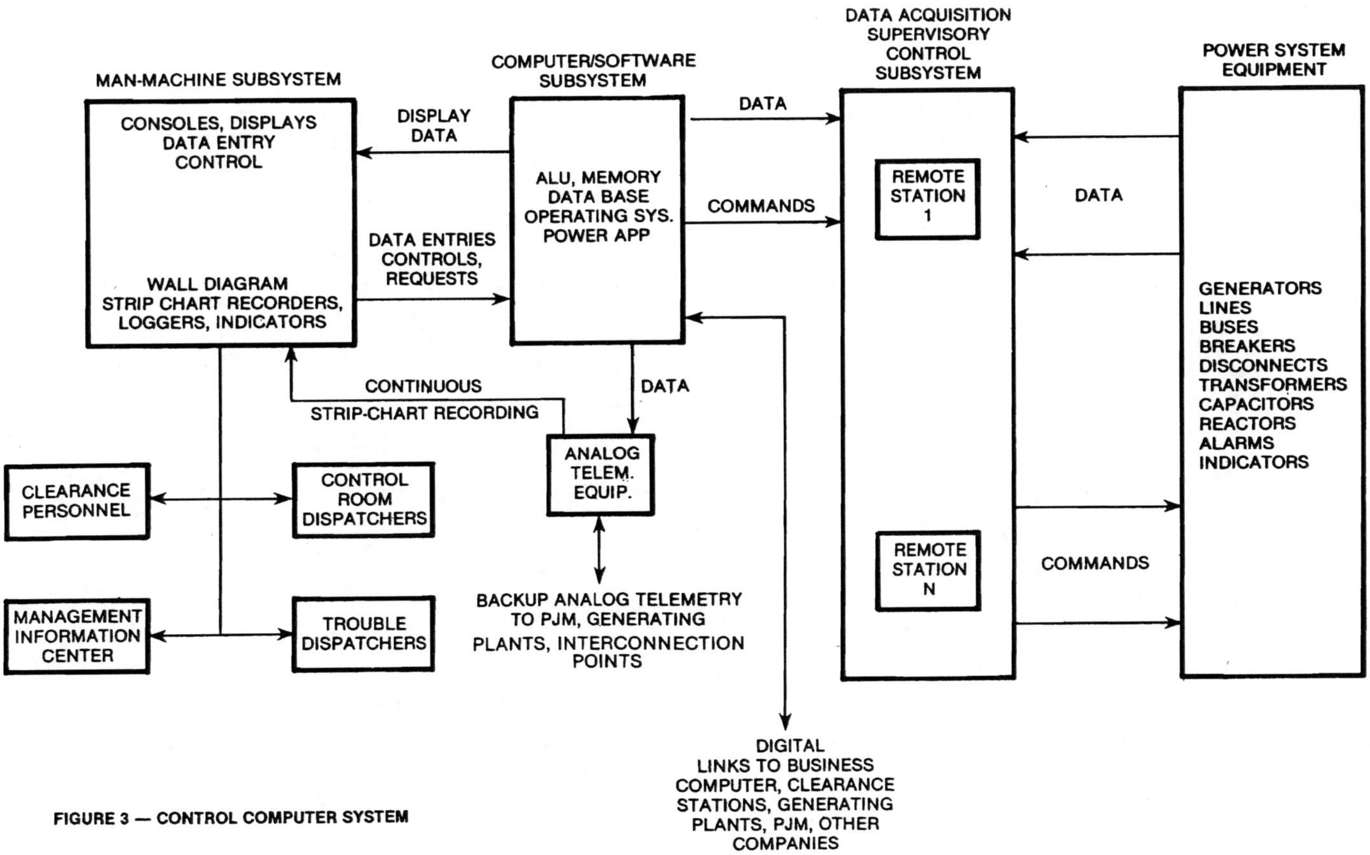

FIGURE 3 — CONTROL COMPUTER SYSTEM

FIGURE 5 — ONE-LINE DIAGRAM OF A TYPICAL SUBSTATION

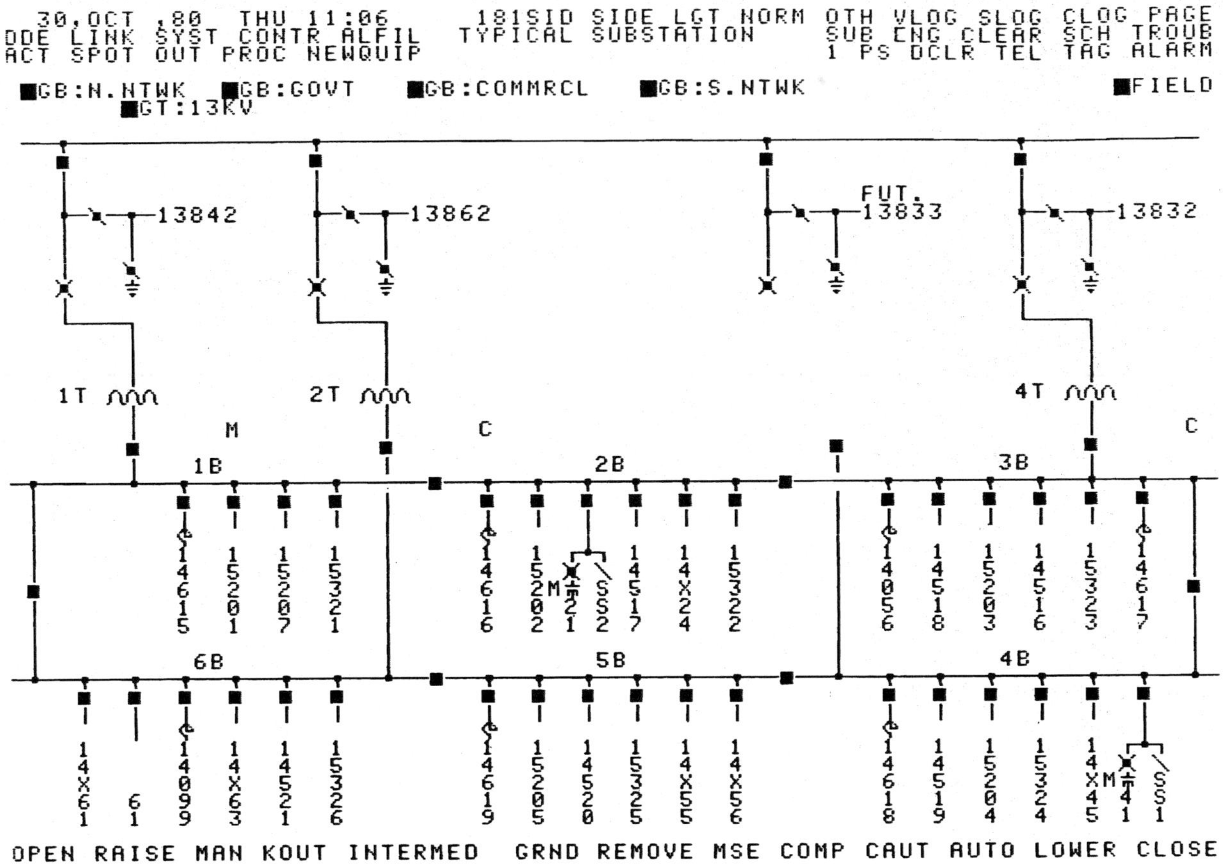

TABLE I. COMPARISON OF COMPUTED AND OBSERVED POWER FLOWS AROUND A 230 kV BUS

			COMPUTED*		OBSERVED**	
FR.	TO BUS	CKT	MW	MVAR	MW	MVAR
554	295	1	− 232.8	− 12.6	− 228	− 5
554	295	2	− 232.8	− 12.5	− 228	− 4
554	390	1	+ 436.8	+ 51.6	+ 428	+ 52
554	390	2	+ 436.8	+ 51.6	+ 427	+ 65
554	574	1	− 127.4	− 23.8	− 118	− 17
554	575	1	− 187.2	− 2.8	− 189	− 25
554	606	1	− 226.3	+ 16.3	− 225	+ 4
554	1341	1	+ 23.0	− 34.1	+ 22	− 30
554	1342	1	+ 209.3	+ 3.0	+ 226	− 11
554	1345	1	+ 122.3	− 62.1	+ 120	+ 37
554	2308	1	− 221.5	+ 25.6	− 229	+ 28
BUS SUMMATION			+ 0.2	+ 0.2	+ 6	+ 42
VOLTAGES			235.24 kV		235.65 kV	

NOTES: POWER FLOWS OUT OF BUS HAVE POSITIVE SIGNS
*INTERACTIVE COMPUTER SYSTEM
**CONTROL COMPUTER SYSTEM kV

TABLE II. COMPARISON OF COMPUTED AND OBSERVED POWER FLOWS IN A DOUBLE CIRCUIT 230 kV LINE

			COMPUTED*		OBSERVED**	
FR.	TO BUS	CKT	MW	MVAR	MW	MVAR
606	295	1	+ 157.5	− 31.3	+ 151	− 18
606	295	2	− 157.5	− 31.3	+ 153	− 22
295	606	1	− 156.8	+ 33.2	− 149	+ 13
295	606	2	− 156.8	+ 33.2	+ 152	+ 18
CKT LOSSES		1	+ 0.7	+ 1.9	+ 2	− 5
CKT LOSSES		1	+ 0.7	+ 1.9	+ 1	− 4
606 BUS VOLTAGE			236.90 kV		238.13 kV	
295 BUS VOLTAGE			238.12 kV		239.62 kV	

NOTES: POWER FLOWS OUT OF BUS HAVE POSITIVE SIGNS
*INTERACTIVE COMPUTER SYSTEM
**CONTROL COMPUTER SYSTEM kV

TABLE III. PEPCO SYSTEM CHARACTERISTICS

The characteristics of the power system relevant to SRP under light-load conditions are as follows:

1. About 1,080 MVAR controllable shunt capacitor banks on the distribution system
2. In general, loads have a lower power factor than underpeak load, and consequently, relatively higher reactive power. Power factors and ratio of real to reactive loads are about 83% and 3:2 as compared to about 95% and 3:1 for the peak load conditions.
3. There are about 63 network loads which must be picked up by closing group circuit breakers. These network loads are about 25% of their peak values, ranging from 4 to 8 MW each.
4. The 600 MVAR charging currents from 230, 138, and 69 KV transmission lines constitute a larger pecentage of system MVA; about 30% as compared to 15% for the peak-load conditions.
5. The underground high voltage transmission lines must be maintained at less than 105% voltage at all locations.
6. The combustion turbines at station F and at station B with 420 MW combined peak capability operate from lagging to unity power factor, i.e., they are prevented from absorbing reactive loads
7. Station F units 1 and 2 and Station E unit 1 provide about 80% of system generation. These three units being of super-critical once-through type, would not be available after trip-out for some eight hours.
8. Station E unit 3 has 640 MW maximum capacity, is a peaking unit, and with a black-start time of 2 hours.
9. Major generating plants D, E, and F are rather remote and connected to metropolitan load centers via 138 and 230 KV cables.
10. Station D units 1 and 2 and station C units 4 and 5, being of the drum type, would be available, provided they are restored within one half an hour after trip-outs.
11. In the absence of any hydro generation, the system restoration strategy is based on black-start capabilities of combustion turbines at station F and station B.
12. With no generation at station A, the 115 load center can only be picked up from the 230 KV side.

TABLE IV: RESTORATION GUIDELINES

1. Sectionalization of the system into subsystems, so that each subsystem would contain at least: (a) one drum type boiler with hot restart capability, and (b) one unit with black-start capability.
2. Simultaneous, immediate, and independent measures to supply station service for the hot restart of the drum type boilers before elapse of the allowable time interval between shut-down and hot restart. This interval is usually of the order of a half an hour or less.
3. Interconnection of generating stations within each subsystem as soon as possible. Such an interconnection should be carried out after generation, but prior to restoration of full load within the subsystem.
4. Restoration of loads in small increments to avoid excessive under-frequency deviation. The general guide used was to pick-up load when system frequency was high and raise generation when system frequency was low.
5. Maintaining steady state stability by having each station supply nearby loads.
6. Restoration of smaller and radial loads prior to larger and low voltage AC network loads, while maintaining a reasonably constant real to reactive power ratio.
7. Maintaining system voltages within plus and minus 5% by: (a) maintaining generator voltages at low end of range during early stages of the restoration and (b) adjusting transformer taps to appropriate positions.
8. Maintaining reactive balance by (a) initially energizing only one circuit of double circuit underground transmission lines, and (b) operating generators under-excited to the extent that stability considerations allow.
9. Consideration of firm capability by having: (a) firm transmission in the interconnections between subsystems, and (b) firm generation capacity in each subsystem.

Future Work. In addition to the need for the development of long-term power system dynamics and introduction of CPM in developing an SRP, the following areas are suggested for furthur study.

First, development of the SRP from a point of complete collapse and without interconnection assistance is relatively straight forward. In practice, however, due to corrective actions, system disturbances are usually arrested prior to complete collapse. Furthermore, some level of interconnection assistance would probably be available. Partial system outages and/or availability of assistance from the interconnection, necessitates assessment of both system and interconnection status prior to restoration, a process which uses up critical time. It would be desirable to develop a master SRP covering different system operating conditions and various degrees of system upset. In the event of a system disturbance, after initial assessment of the system status, a proper entry into the master SRP would be made, and the specific subroutine would be selected and followed to a successful restoration.

Second, is the development of a precise and consistent glossary of terms, which could readily be understood and which would be free from ambiguity. Such a glossary is more important during restoration than under normal, day-to-day operations.

ACKNOWLEDGEMENT

The authors would like to record their appreciation to Mr. Walter A. Johnson, Potomac Electric Power Company, for his direction in the development of the SRP and helpful suggestions in preparation of this paper.

REFERENCES

1. Johnson, Walter A., et al. ''The Potomac Electric Power Company's Consolidated Control Center,'' IEEE PES Summer Meeting Anaheim, California, July 1974, C74 366-1.
2. Undrill, John M., et al. ''Interactive Computation in Power System Analysis,'' Proceedings of IEEE, Volume 62, July 1974.
3. Johnson, Walter A., ''Consolidated Control Center of the Potomac Electric Power Company,'' IFAC, 6th Triennial World Congress, Massachusetts, August 1975.
4. Kafka, Richard J., et al. ''System Restoration Plan Development for a Metropolitan Electric System,'' IEEE PES Winter Meeting, Atlanta, Georgia, February, 1981.
5. Adibi, M. M., et al. ''Solution Methods for Transient and Dynamic Stability'' Proceedings of IEEE, Volume 62, July, 1974.
6. Wiest, J. D., et al. ''A Management Guide to PERT/CPM,'' Englewood Cliffs, New Jersey, Prentice Hall, 1969.
7. Johnson, Walter A., et al. ''Commissioning Experience and Operating Evaluations of a Consolidated Control Center Responsible for the Generation, Transmission, and Distribution Systems Serving the Washington, D. C. Metropolitan Area'' Proceedings of IEE, June 1980.
8. Anderson, Max D., et al. ''Power System Operator Training Update,'' IEEE PES Summer Meeting, Vancouver, B.C., July 1979, A79-515-8.

Discussions

L. H. Fink, (Systems Engineering for Power, Inc., Oakton, VA): This paper and its companion are the first on the subject of system restoration to appear in these Transactions since those presented four decades ago at a session of the AIEE winter convention in New York, January 1940 [1]. It is interesting to note further that in the planning of the special series of articles on Blackouts which appeared in SPECTRUM between February and November 1978, the editors were unable to locate a qualified author willing to write an article on system restoration. The appearance of this pair of papers is a welcome indication that this important topic has re-emerged for substantive discussion.

The authors refer several times to the relevance and importance of transient energy balance considerations, but indicate that "available linearized dynamic models are not suitable nor adequate..." for making such calculations. Did they consider the use of models developed for study of automatic generation control performance? In particular, the "PJM Model" developed at Philadelphia Electric Co. [2] was specifically tailored to reflect the dynamics of load-generation mismatch, and contains low-order, nonlinear models for the various types of plants in use on the PJM System, of which PEPCo is a member company. A related modeling issue is that of estimation and representation of the load demand that will be experienced upon reenergization of distribution circuits. Recent work on this problem [3,4,5] also warrants careful attention.

In addition to the critical times which constrain the restart of both drum-type and once-through units, some consideration should be given to the pressure transients in high pressure oil-pipe type cable circuits [6]. These transients are complex functions of cable current and voltage trajectories, and some energization-deenergization sequences can result in localized pressure drops which vitiate the dielectric integrity of the cable.

The authors are to be commended on a timely and well written paper, as well as on the workmanlike system restoration plan which they describe.

References

[1] Restoration of Service on a Metropolitan Power System, W. C. Bryson, A. P. Hayword; Restoration of Service for Large Metropolitan System After Complete Shutdown, H. A. Dryar, R. Bailey; The Design and Operation of a Metropolitan Electrical System From the Viewpoint of Possible Major Shutdowns, S. M. Dean; Provisions for RE-energizing the Electric System of the Consolidated Edison Company of New York, G. R. Milne, H. C. Otten; TRANSACTIONS AIEE, Vol. 59, 1940; pp. 563-85; combined discussion, ibid, pp. 1150-58.

[2] Use of Computer Model of Interconnected Power System to Assess Generation Control Strategies, L. M. Smith, L. H. Fink, R. P. Schulz; TRANSACTIONS IEEE, Vol. PAS-94, No. 5, Sept./Oct. 1975, pp. 1835-42.

[3] Cold Load Pickup, J. E. McDonald, A. M. Bruning, W. R. Mahien; ibid., Vol. PAS-93, No. 4, July/August 1979, pp. 1384-86.

[4] Physically Based Modeling of Cold Load Pickup, S. Ihara, F. C. Schweppe; IEEE/PES Winter Meeting, Atlanta, February 1981 (81 WM 084-3).

[5] Start-up Response of a Secondary Distribution Network, J. F. Pinto, Z. Zabar; ibid. (81 WM 220-3, 221-1).

[6] Oil Flow and Pressure Calculation for Pipe Type Cable Systems, AIEE Comm. Report; TRANSACTIONS AIEE, Vol. 74, 1955, pp. 251-61.

Manuscript received June 19, 1981.

Anjan Bose, (Control Data Corporation, Minneapolis, MN): The authors are to be congratulated for an interesting paper on the important problem of system restoration, a subject that sorely needs much thought and analysis.

It appears that much of the authors' effort has been spent in simulating the power system behavior under different conditions of failure and restoration. The steady-state simulation has been done by an interactive load flow and the dynamic behavior has been approximated by rules of thumb. I wonder how much simpler the exercise would have been if a dispatcher power flow that ran on the control center computer off the on-line data base would have been used. If the capabilities of a dispatcher training simulator that had both steady-state and dynamic simulation were available, it would probably have been even simpler. Since dispatcher power flows and dispatcher training simulators are getting to be quite common, these questions are quite pertinent. In fact, a more interesting question is whether the restoration procedures should be different if the operator has these sophisticated simulation techniques that can be conveniently initialized to real time conditions. The authors have pointed out the difficulties of setting very explicit restoration procedures because the system may be in a very large number of different conditions at the beginning and during restoration. A convenient on-line simulation capability can provide the capability to formulate the restoration steps "on the fly". This is, of course, a little too optimistic but the answer may lie somewhere in between when the procedure, instead of being completely set apriori, would be modified according to some on-line simulation.

Manuscript received June 1, 1981.

P. R. Arnold, (Bonneville Power Administration, Portland, OR): I compliment the authors for presenting a methodical approach to develop a system restoration plan (SRP). The problem of restoring a power system from a total blackout is complex and multi-faceted. It is agreed that the development of a system restoration plan can best be handled by an iterative process. In fact, the most meaningful iteration may come from experience after the first system collapse.

The authors have made good use of an interactive load flow program. Efforts to validate the model by comparing results to actual system measurements no doubt produced a more accurate model. This comparison also uncovered calibration errors in the system telemetry. Asking for an improved model. However, there is still some question as to whether the model will yield valid results for system restoration studies where single generation-load islands are independently restored following a total blackout. The model would have to be verified by comparison to the actual islanded system. One reason that the model may not compare to the real islanded system is that generator voltage and reactive limits may not be correctly represented in the model. These limits can be controlled by the generation plant operator and must be recognized in the model. These comments are not meant to detract from the fine effort by the authors to produce a valid model. They are meant only as a word of caution and to point out that perhaps just modelling the problem is not good enough.

The authors point out that suitable dynamic models were not available to them to develop the SRP. A similar problem exists at BPA. Dynamic response of generators to line charging and load pickup is a critical and basic element of knowledge that must be factored in to an SRP.

To gain insight into this critical part of restoration, BPA has run four staged system restoration tests in which a block of aluminum industrial load was restored from remote, isolated generator units. During these tests, the frequency response to switching an isolated load was measured. Some of the test results were compared to computer studies. Existing computer models show larger speed deviations that actual tests; in one case, the computer showed 2.1 Hz drop compared to .85 Hz measured. By varying the water starting time constant (Tw) in the governor-turbine model, the computer results come close to actual test results.

Steps will be taken to upgrade governor-turbine models in the off-line transient stability program to give valid dynamic response.

BPA has some evidence using results from both system tests and computer studies that the allowable load increment varies from two to 10 percent. Also, the size of load that can be restored increases as independent islands are connected and regional interties restored. Would the authors explain what is behind the five percent criteria adopted for the Pepco system?

Control computers play a most critical role in the timely restoration of a power system. In addition to sectionalizing into subsystems, individual PCBs must be opened. In the BPA system as well as others, the trend to unman substations places a heavy reliance on SCADA control. BPA assumes that a blackout will occur at a time when all but eight out of 350 substations are unmanned. The task of opening circuit breakers individually by SCADA is time consuming. BPA serves 3,000 MW of aluminum industry load which must be restored within 45 minutes. The time to open the PCBs necessary to restore this load is estimated to take in excess of 80 minutes. For this reason, BPA will implement a single SCADA control point per bus to clear all PCBs. This scheme will be interlocked by 30 undervoltage relays for security.

Will Pepco utilize SCADA to sectionalize their system into subsystems for restoration as indicated by their SRP? How is the problem of individual operation of PCBs for bus clearing addressed?

The overriding assumption here is that SCADA systems will perform adequately during and following a system collapse. But in fact, no one knows how SCADA systems will react. There have been cases of computer systems slowing down during a disturbance to the point where control was severely degraded. Has this problem been experienced on the Pepco system? Is it being addressed as a part of system restoration planning?

Further developments in computational capability both in off-line and on-line computers are necessary to adequately study the dynamics of system restoration. On-line control computers must be capable of providing control during and following system disturbances. Future systems should be designed and tested to handle probably scenarios including total blackout.

Manuscript received June 3, 1981.

S. R. Erwin, (Southern Company Services, Birmingham, AL): The authors have presented an informative paper on a very complex problem. In particular, the need for a team effort of systems analysts, operating personnel, and production personnel and the critical requirement of carefully evaluating expected control actions in preparation of the load flow model are sited. Also the innovative use of Interactive Load Flow and the initial use of operation guidelines to replace the long-term power system dynamics application provides a practical method for other utilities to evaluate system plans.

As noted by the authors, the use of Interactive Load Flow substantially increases the efficiency of power system studies. For other utilities who might use this method of evaluating restoration plans, the writer would be interested in knowing the approximate number of cases required to study this system and if any particular load flow solution problems were found.

Additionally, the authors note the basic assistance the load flow provides is in planning for reactive power balance. The writers experience has been that the actual reactive capability of units in daily operation may be difficult to obtain. Did the authors have a similar problem?

The writer also would be interested in the authors' comments on the use of a state estimator to provide a base case model to develop an SRP and to aid in assessing the status of the power system during a partial outage.

Manuscript received June 15, 1981.

R. J. Kafka, D. R. Penders, S. H. Bouchey, and M. M. Adidi: We thank the discussors for their remarks and technical questions which we will attempt to clarify in our closure. Mr. Erwin asks how many load flow cases were required for the study and whether there were any particular load flow solution problems. The total cases saved on disc numbered approximately 200, and more cases were documented with hard copy output only. Left unrecorded were the many trial cases which did not serve the transistion between two desired or acceptable states represented by the saved cases. Since we were working interactively with the load flow, even the unacceptable cases were seen as steps toward the desired state rather than as rejected cases. The total number of cases run was over 500. No particular load flow problems were encountered, although occasionally bus or line codes were not changed consistently as the system was being restored. One of the features of the interactive load flow allows the analyst to observe the characteristics of the solution progress. When the characteristics indicated that the case was nonconvergent or otherwise unusual the solution would be interrupted and the problem found and corrected by various data base checking activities. The solution would then be resumed.

Messrs. Erwin and Arnold point out that it is difficult to determine the characteristics of generating units for system restoration without testing the units on an islanded system. This is true of both the frequency response and reactive capabilities. The 5 percent of prevailing generation guideline which we used for load increments was chosen as a result of an earlier study[1] of frequency response to load changes and was acceptable to operating personnel based on experience. The reactive capabilities of units depend very much on the electric system to which they are connected. The authors used conservative under- and over-excitation limits obtained during normal operation and from generator tests. It is very helpful to know whether a reactive limit is imposed by the generator or by bus voltage or by a station service voltage requirement or by system stability considerations. Mr. Arnold indicated that hydro-turbine generators had been tested as isolated start up units. The authors would be very interested in seeing published results, even though hydro-turbines are not applicable to our restoration plan, and urge that anyone who has test results for isolated units of any type make them available to the industry through publication.

Messrs. Erwin and Bose ask about the use of on line state estimators and dispatchers load flows, with or without dynamic simulation, in developing the SRP and in carrying out a system restoration. The authors chose historical data at the peak and minimum load levels to use in the restoration study, and manual state estimation techniques were used to insure that the date were valid. If state estimator results were to be saved for historical purposes, then the manual state estimation would not be required. The authors question the use, during restoration, of state estimators and other analytical tools which have been developed for small perturbations under normal system operation. It should be noted that during system restoration the power system is sectionalized into a number of subsystems. However, we do believe there is a need for new sets of analytical tools to determine the status of a power system before attempting restoration, particularly after partial collapse.

The use of a dispatchers load flow, particularly if it had access to historical data, would be helpful because the restoration study would use the same man/machine interface as the actual restoration. This would eliminate the simulation of this interface required in the authors' study. The use of dynamic simulators, while necessary for a restoration study, would not simplify the development of restoration procedures, although it would possibly provide better results. As stated earlier, there is a need for new analytical tools, for example a "restoration load flow" that would calculate the required load at a bus to stabilize line voltages, but we do not believe available simulation techniques could change the restoration procedure appreciably.

It is believed that on-line, real time simulation capability to formulate restoration steps "on the fly" is very difficult to implement. Further, this capability would not prevent the pursuit of a strategy which seems effective at the time but leads to a condition that prevents continued progress or causes recurrence of the outage. The authors' recommendation is to update the SRP periodically and whenever significant changes occur in the power system.

Mr. Fink asks whether the authors considered the use of dynamic models originally developed to study automatic generation control. The authors were aware of the availability of the "PJM Model", but it did not meet our requirements for ease of interaction with the model nor for rapid turn around. Such a model would be useful to verify the entire restoration procedure once it had been developed.

Mr. Fink also points out that proper modeling of the loads is necessary. We agree that careful consideration must be given to inrush on load pickup, loss of diversity and to the general change in load with respect to time. In the initial stages of restoration, when the primary considerations are for generator restart and subsystem intergration, the SRP must specify which loads are acceptable, based on historic data, in terms of proper frequency and voltage control. Later, in the load pick-up phase of restoration, the SRP specifies the load magnitude to be picked-up at certain substations, and it will be up to the system operators to pick-up a sufficient number of feeders to satisfy that requirement.

In reply to Mr. Fink's comments regarding pressure transients in oil filled pipe type cable, we agree that this problem must be considered. PEPCo has tested cables with system restoration in mind, and no undue restrictions were imposed as a result. It is extremely important, however, to maintain oil pressure, by emergency power supplies if necessary, while the electric system is deenergized.

Mr. Arnold asks if PEPCo's implementation of system restoration will involve the use of SCADA type systems for breaker control. PEPCo's switching operations are carried out from the control center through the man/machine interface and the remote terminal units at the substations. It is likely that existing control capabilities used for load shedding can be used for such procedures as bus clearing. PEPCo has addressed the slowing down of the computer system during disturbances by making lower priority functions inactive. It is recommended that the control center computer systems be subjected to vigorous simulation to determine the adequacy of computer speed and storage for emergency operation.

Reference

[1] M. M. Adibi, *et al*, "Solution Methods for Transient and Dynamic Stability", *Proceedings of the IEEE*, Vol. 62, July, 1974.

Manuscript received July 27, 1981.

SYSTEM RESTORATION - DEPLOYING THE PLAN

A report prepared for the System Operations Subcommittee of the Power Systems Engineering Committee by the Current Operational Problems Working Group - Walter A. Johnson, Chairman. Contributing authors are:
F. S. Prabhakara, J. F. Torri, T. Niessink, J. J. Gates, J. P. Stovall, C. N. Whitmire, L. Mills, and J. C. Rodrigue.

ABSTRACT

This paper is one of a series prepared for use in Current Operational Problems (COPs) forums with the goal of focusing the attention of the industry on high priority problems faced by those who are involved in the actual operation of power systems.

The three "short note reports" in this paper provide insights into the problems of deploying a system restoration plan into an operating organization. Many aspects of the deployment problem are discussed and some suggested solutions are provided.

The short note reports presented here are intended to clarify the nature of these operational concerns and problems. It is the objective of COPs forums to solicit discussion and stimulate the creation of responsive solutions.

INTRODUCTION

Walter A. Johnson
Potomac Electric Power Company
Washington, D.C.

Electric system blackouts are not rare events. Each year a number of systems are subjected to widespread outages due to an endlessly changing series of occurrences. The complexity of system restoration efforts and the impact of the blackouts themselves have grown as urban areas have grown. The major blackouts in the northeast United States in the 60s and 70s focused world attention on the social consequences of extended blackouts and industry attention on the problems involved in restoration.

Restoration planners have approached the problem of rapidly reenergizing a complex system in a number of different ways, but in all cases, the resulting restoration plans are complex and have endless alternates. The system operator then faces the challenge of deploying the restoration plan in such a way that it can be effectively executed should a blackout occur.

In an effort to focus on the steps necessary to deploy a restoration plan, a minimum of words are used in describing the details of any given plan or its success in actually restoring a system after a blackout. Instead the concentration is on those aspects of deployment that come between planning and execution. Specific questions and areas of interest include the following:

- How should system operators be trained for its use?
- What format should it be put into to make it usable in an emergency?
- How should computer based security software be configured to fuction after a blackout?
- How should the system operator be instructed to adequately assess the status of his system in enough detail to correctly direct the restoration of his system?
- What special notification and call-in procedures should be instituted to obtain adequately trained dispatchers within the time frame in which they are needed?
- What special team organization should be established for restart? How should the teams be deployed and what special tools do they require?
- Training to handle special restart problems such as hot restart, black start, lightly loaded generator stability, system voltage levels, cold load pickup, island resynchronization.
- Key monitoring indicators the operators should be trained to track during restart.
- Tests, drills, and trials.
- Recordkeeping for restart and after-the-fact evaluations.
- Communications with field, management, media, and government.

The first short note presents a survey of the present industry status in restoration deployment. The second and third short notes provide individual company perspectives.

A PERSPECTIVE OF POWER SYSTEM RESTORATION

F. S. Prabhakara, Senior Member
J. F. Torri, Member
T. Niessink, Senior Member
J. J. Gates, Non-Member
Gilbert/Commonwealth
and
J. P. Stovall, Member
Oak Ridge National Laboratory

INTRODUCTION

Following a major power system event resulting in widespread customer outages, the common objective of electric utilities is to minimize the impact of these outages on electric customers and the public at-large. This paper discusses utility restoration practices used to reenergize electric facilities and restore electric service to customers.

Power system facilities, including generation, transmission, and distribution, are subject to diverse external and internal forces. These forces have caused power fail-

82 WM 196-4 A paper recommended and approved by the IEEE Power System Engineering Committee of the IEEE Power Engineering Society for presentation at the IEEE PES 1982 Winter Meeting, New York, New York, January 31-February 5, 1982. Manuscript submitted October 23, 1981; made available for printing December 8, 1981.

Reprinted from *IEEE Transactions on Power Apparatus and Systems,* Vol. 101, No. 11, pp. 4263–4271, November 1982.

ures resulting in the loss of electric supply. Consequently, individual electric utilities use numerous restoration practices which are collectively referred to as a restoration program. Since there are over 1200 utilities in the United States with varying demographic, physiographic, regulatory and financial characteristics and with varying philosophies of management, restoration programs also vary considerably from utility to utility.

In order to ascertain the differences in the restoration programs a literature survey was conducted. In addition, four utilities (a pool, a large investor owned utility, a large public utility, and a municipal) were interviewed, as part of the survey. No statistical survey was conducted so this discussion is general in nature. The inclusion or the omission of a particular practice in this discussion warrants no consideration of the importance of that practice. Furthermore, this discussion is not intended to establish standards, guidelines or minimum requirements for restoration programs. The purpose of this discussion is to identify the functional requirements of utility restoration programs and to compare the different utility practices adopted to address these functional requirements.

RESTORATION PROGRAMS

Restoration programs were found to consist of both utility and non-utility practices. During restoration, the purpose of these practices is to minimize the socio-economic impact on customers and the public at large until normal electric service is restored. Thus, an effective utility/non-utility interface is required, however, the non-utility practices are not addressed herein.

The diverse restoration practices of numerous electric utilities across the nation have been categorized into eight separate but interdependent utility restoration functions (Figure 1): restoration plan, monitoring, assessment, control, communications, transportation, materials and personnel.

UTILITY RESTORATION FUNCTIONS

The unique nature of the individual utilities requires that a restoration program be tailored to the needs of each. To determine what restoration practices are typically used, a technical survey was made of available literature and four utilities were interviewed.

Restoration Plan

While most utilities have developed portions of a plan, (1,2,3) few utilities, as shown by the literature survey, have a complete formal restoration plan. The portions of complete formal restoration plans are discussed here.

Figure 1. Restoration functions.

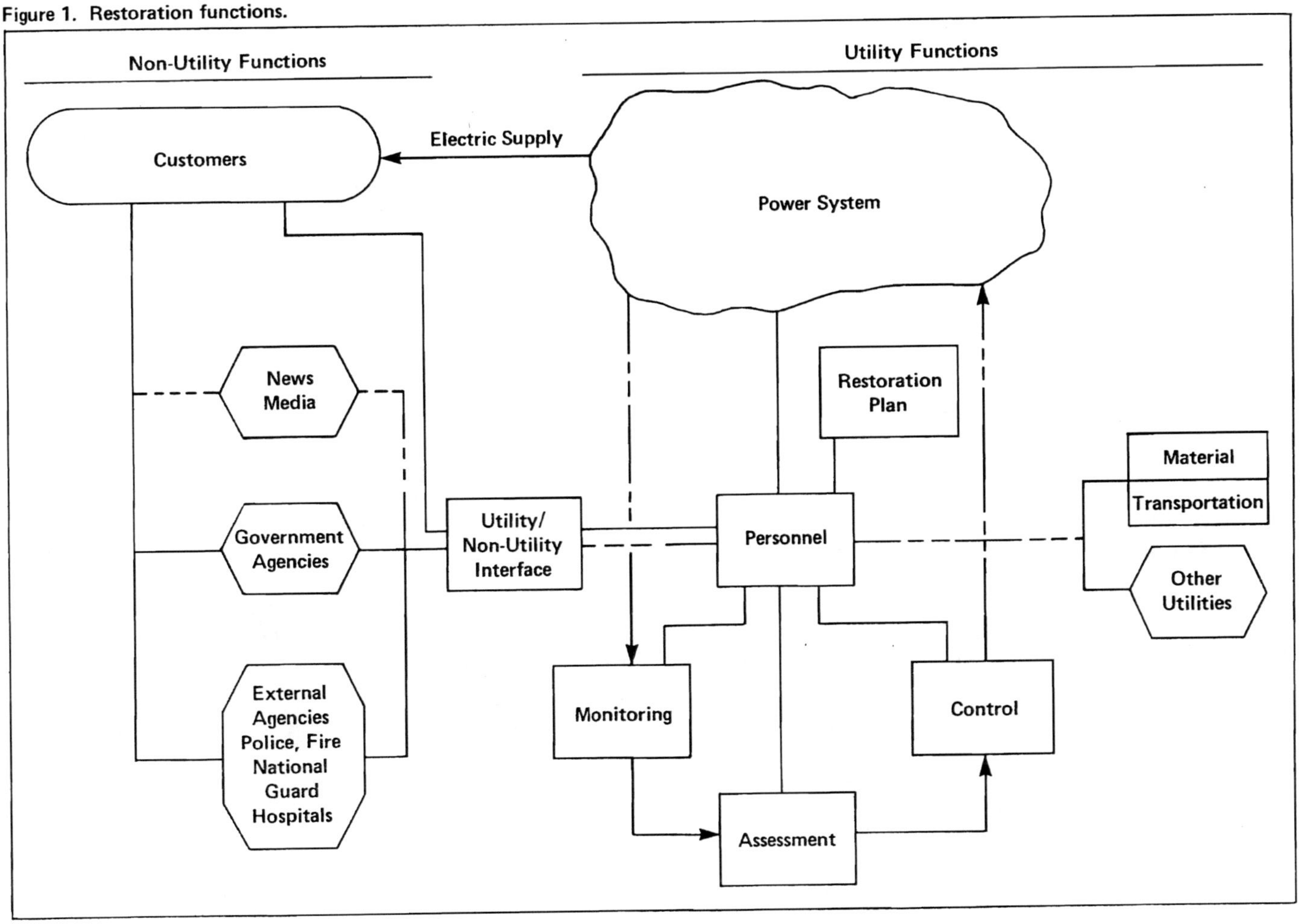

Some restoration plans established priorities of action, taking into account public health, safety and national security. Procedures for communicating and coordinating with other utilities and the non-utility functions are documented in other plans.

Many restoration plans are adaptable for different types of outages to permit proper utilization of manpower, material and equipment, based on the overall severity and extent of a given outage. Restoration plans are also flexible in accommodating new and updated procedures. To this end, restoration plans should be reviewed and revised periodically to incorporate organizational changes, recent experience, new industry practices and equipment and modified system configurations. Plans should also be reviewed as to their compatibility with non-utility efforts, especially the governmental role during major outages.

It is recognized by many utilities that without formal documentation and necessary approvals, the restoration plan does not serve as a useful tool. Some restoration plans result from the utility's commitment to define objectives and policies of the restoration program.

A restoration plan should recognize critical loads and assign them a priority for service restoration. Medical facilities, police and fire stations, communication centers, transportation centers, and water and sewage facilities are some examples of critical loads requiring priority attention, especially if they have no emergency generation facilities. Other loads which may be considered critical are high-rise buildings and customers whose products or equipment might be seriously affected by an electrical supply interruption.

Other restoration plans give major attention to installations that maintain national security, such as military facilities and government offices. The reliability required by some of these facilities may require such things as duplicate and loop transmission and distribution systems. Regardless, priorities for restoration need to be assigned accordingly.

Also included in some restoration plans are priorities of detailed switching procedures to be followed should a major portion of generation and interconnection be lost. Some procedures address generation start-up and synchronization, transmission line reclosing, generation balance and reclosing of interconnection lines.

In the event of widespread extreme damage, damage surveys are conducted in the field to properly establish priorities for service restoration. Some restoration plans contain guidelines for the rapid and specific utilization of damage investigation personnel and for making use of information provided by them. Customer reports of damage and interruptions are considered as part of some utilities' damage surveys.

Communications are recognized as an essential part of a restorative program and are, therefore, specifically addressed in many restoration plans. Procedures can be set forth for gathering and disseminating information. Some restoration plans outline procedures for communicating with other utility segments, particularly neighboring utilities.

Monitoring

A monitoring function is required to provide power system information to the security and restoration programs. Most monitoring functions for restoration programs use facilities implemented for security monitoring.

The primary requirements of monitoring during restoration are to identify the occurrence of a power system interruption, display the performance of the energized portion of the system, and display the status of the de-energized power system facilities. Parameters typically monitored throughout the duration are voltages, substation and generating station load levels, line loadings and network configurations, and facility operating status. These parameters are displayed and utilized in system restoration at load dispatch offices, power pool headquarters, generating stations and substations. Examples of the types of information displays are mimic board, cathode-ray-tubes (CRTs) teleprinter hard copy or meters and strip charts.

Assessment

Assessment is the process of determining the magnitude and type of restoration effort required in the event of a major service interruption. Through assessment, management and system operating personnel can determine the type of utility response necessary for a particular service interruption.

Identification and supplemental studies are two important factors of assessment during the restoration process. Identification determines which facilities have been removed from operation, which ones are operating at critical load levels, and what the potential effect is of removing key facilities from the system. Identification provides information which triggers other factors comprising the total restoration program. Identification may be accomplished by alerting mechanisms and/or communications.

Supplemental studies can be used to determine the priority with which facilities should be returned to service and what switching procedures should be followed. Studies which may be conducted include on-line/off-line load flow and short circuit analyses. Procedures typically used to conduct these studies permit system modeling under a variety of conditions, as the system configuration from the load flow and short circuit viewpoints may be substantially different under outage conditions as opposed to normal operating conditions.

Control

The purpose of control in restoration is to provide the means for preventing equipment damage after the occurrence of a major disruptive event and for intervention to both restore electric service to affected customers and to restore the system to the normal-secure or normal-alert operating states as quickly as possible.

Most control functions used for restoration programs use facilities implemented for security control. Control during restoration may be automatic, remote or manual. Examples of automatic control are generator governor and exciter systems and automatic tap changers on substation transformers. Facilities to permit customer load reductions and remote starting of generation are remote control functions. Remote and manual control functions are performed by restoration personnel depending on the type, magnitude and extent of an outage.

Communications

Communications are essential to the restorative effort within both the utility and non-utility functions. Communications during the restoration program utilize communication facilities typically used in security and may be supplemented with backup and stand-by systems.

Communications are necessary to coordinate and supervise restoration forces, advise neighboring systems of system status, and disseminate timely and accurate information to the non-utility functions. During a major outage, communication channels can become overtaxed or lost to service. Many utilities make provisions for supplemental facilities to overcome these problems.

Communications consist of the system by which information is transmitted, the specific information that is transmitted, and the points of transmission and reception of the information. Communication systems utilized during a restoration effort include telephone, radio, or power line carriers. Information may be audio, video analog or digital. Communication occurs intra-utility, inter-utility and across the utility/non-utility interface.

Intra-utility communication occurs between the system and the control center, between the control center and corporate headquarters and/or between managerial and supervisory personnel and field personnel. Communication between the power system and the control center is largely telemetry, requiring conversion of raw data from the system to radio (or other) signals for transmission. This function is generally common to all states of system operation, and the volume of information is constant except in report-by-exception schemes. The information transmitted is required input to monitoring, assessment, and control functions.

Communication between the control center and utility management is generally over leased telephone lines. During system restoration, large volumes of information are transmitted between these two areas concerning system status, control procedures to be followed, and the availability of reserve generation. This link often becomes overtaxed, and supplemental links are required. Some utilities provide a video monitor (connected to the SCADA system's digital computer) for utility management's use.

The other primary communication link during restoration procedures is by radio. Damage investigators, repair crews, control center personnel and managerial or supervisory personnel must communicate. Essentially, radio is used for this purpose, and normal channels may become overloaded during a restoration. Some utilities maintain stand-by systems specifically used only during restoration.

Communication between utility and non-utility functions is largely by telephone. (The radio and television media are used for dissemination of information to the public, however, the primary contact between the utility and media is telephone.)

Customers out of service are a large user of telephone lines during a major outage, and although this is an important source of information, care is taken by some utilities to reduce transmission of inappropriate or duplicate information.

Other important links which rely on the telephone are contact with external agencies, such as police, fire and national security agencies, and contact with critical load customers.

Telephone circuits are very susceptible to overloading during an electrical outage. If the outage is widespread, resulting from natural causes such as severe storms, then the telephone system itself may be lost to service. Provisions are made by some utilities for supplemental or redundant telephone systems.

Other communication facilities which may be used during restoration include teleprinters, CRTs, microwave, etc. These may or may not be as susceptible to overloading or loss of service as telephone and radio. Backup provisions are made where appropriate.

Transportation

Virtually all utilities have vehicle fleets consisting of passenger vehicles, working trucks and various equipment such as diggers, trenchers, and compressors. During restoration, this fleet becomes essential in transporting personnel, equipment and supplies to and from the areas experiencing service interruptions. Depending on the magnitude of the restorative effort, many utilities can supplement their existing vehicle fleets with those from other organizations.

Transportation is required for all restoration force field personnel. Vehicles such as those utilized by damage investigators are deployed early and may become available for other use after initial field survey work. Supervisory staff, substation crews, generating station crews and repair crews also need transportation. The last three, especially repair crews, may require specialized vehicles such as bucket trucks.

If an outage is extensive and severe, then it may be necessary to supplement existing transportation. Vehicles may be leased or borrowed. An effective way to supplement transportation (and personnel) facilities is through agreements with other utilities. Investor-owned utilities, for example, may participate in the Edison Electric Institute's Mutual Assistance Roster. Detailed checklists for seeking and providing assistance under this roster are published. Municipal utilities may participate in the Municipal Electric Associations.

Air transportation may be desirable to permit rapid patrol of lines, acquisition of material and equipment, and transportation of personnel. Equipment such as airplanes and helicopters may be rented from private companies or may be utility-owned.

For some utilities, an additional source of equipment to perform basic transportation needs, such as trucks, has been the National Guard.

Material

Most utilities have an adequate supply of materials in warehouses, service centers and on utility vehicles, to effectively respond to disruptive events occurring on electric power systems. In some events, the normal material requirements are supplemental from outside sources.

Material over and above normal needs and specifically required by the restorative effort (such as flares, traffic cones, flashlights, etc.) may be stored at strategic locations (substations, generating stations, etc.) throughout the service area and designated as emergency material.

Likewise, provisions are made by some utilities in advance for special equipment such as portable generating sets for major underground systems to support pumps and cooling equipment.

Personnel

Personnel are an integral part of the restoration program, and their proper utilization has been recognized as a key to the successful accomplishment of the program. Few utilities, however, justify a separate restoration force but rely on reallocation of personnel from day-to-day operations to the restoration of generation, transmission and distribution facilities. This reallocation can be pre-established and utilize individuals' talents as applied to the restorative effort. Principal factors that affect the effectiveness and efficiency of personnel are supervision, training and experience.

Supervision is required to assure performance of specifically assigned duties and provide a unified approach to the restorative effort. Trained supervisory staff are especially required at dispatching offices, public relations and customer service offices, generating stations, major switching and substations, and in the field. An organizational chain-of-command should be structured to meet the demands of system restoration work.

Training programs not only address the utilities' routine needs, but also apply to those duties required to perform restoration efforts. Engineers and designers might receive training as damage investigators. Other personnel who may require training are draftspersons and retirees and they might be used in such functions as crew guides for foreign crews, warehouse assistants, data loggers, shuttle drivers, radio and telephone operators, and so forth. Linemen and switchmen, on the other hand, can be trained in their specific restoration duties. Many utilities provide system operators with special training in restoration practices, while customer service representatives receive training in handling voluminous and unique customer communications.

Training may be through courses, seminars, on-the-job training, or on a system simulator.

To a great extent, effective and efficient restoration of service depends on the experience of personnel. Utilities attempt to make maximum utilization of personnel that have assisted during previous restoration efforts.

Other factors considered in formulating the restoration force are safety, physical and mental health, and decision-making ability. Employees are trained to appreciate the need for proper safety procedures and safety considerations become especially important during emergency situations.

Physically and mentally healthy people are required to avoid stress, fatigue and resultant errors. Physical and mental well-being also contribute to the decision-making capability, which can minimize errors and the need for supervision and, thereby increase the efficiency of personnel. Employees of many utilities receive periodic physical checkups to assure peak physical and mental health.

Supplemental personnel may provide direct or indirect assistance to the effort and may include repair crews from neighboring utilities, contractors and police, fire, and National Guard forces. Deployment of supplemental personnel requires a great deal of cooperation, as with the EEI's Mutual Assistance Roster previously discussed, and may be defined in a formal restoration plan.

CONCLUSIONS

The functional requirements of the restoration programs of electric utilities in this country have been categorized into eight separate but interdependent restoration functions.

The literature survey and utility interviews indicate that most utilities have developed portions of restoration plans, however, few utilities have complete and formal restoration plans.

The utility/non-utility interface is an important requirement in coordinating the non-utility restoration functions following major power system outages.

REFERENCES

1. Kakfa, R.J., Penders, D.R., Bouchey, S.H. and Adibi, M.M., "System Restoration Plan Development for a Metropolitan Electric System," IEEE Paper 81-WM-044-7 presented at the 1981 Winter Meeting, Atlanta, Georgia, February 1981.

2. "Emergency Service Restoration Plan," Dallas Power and Light Company, February 1979.

3. "Procedures to be Followed in Severe and Extreme Emergencies," Portland General Electric Company, January 1979.

SYSTEM RESTORATION DEPLOYING THE PLAN

C. N. Whitmire
Florida Power & Light Company

INTRODUCTION

In deploying a "System Restoration Plan" there must be a clear understanding of (1) what the priorities are, (2) what the plan is and, (3) who is to implement the plan. With these three in mind an effective system of directives, supplies and training can be established to deploy the plan.

PRIORITIES

1. Restore off site power to nuclear plants. This also provides transmission out of the plant when the unit returns to service.
2. Restore off site power to all plants.
3. Provide power to pumps for all transmission pressurized oil filled pipe cables.
4. Reenergize the transmission system.
5. Restore service to all customers.

PLAN

1. Determine the boundary of the area that is "blacked out".
2. The System Operator determines the overall plan of restoration and monitors the performance.
3. The Division Dispatchers (the personnel in charge of transmission switching and voltage control) and the plant personnel carry on as much of their work as possible independently and concurrently.
4. Restoration is to proceed from the energized transmission system if possible. If "black start" of portions of the system are required, then synchronizing points of the isolated system to the energized system should be clearly specified.
5. Exercise care in energizing EHV lines and high voltage cables while the system is lightly loaded.
6. Sectionalize lower voltage systems so that the timing of restoration of load and loading on higher voltage circuits can be controlled.

PERSONNEL

1. The initial recognition of condition, establishment of plans, switching for clearing and start of restoration must be done by the normal shift personnel.
2. The additional shift personnel, staff support and regular hour workers who arrive later must be given briefings and work assignment with a minimum of distraction to the personnel already restoring the system.

TRAINING OF SYSTEM OPERATORS (AND DIVISION DISPATCHERS)

Since the conditions requiring restoration are never desired, this means that on-the-job training cannot be used. Most training should be of a classroom type when the personnel are not on regular shift work.

The first training should be in terms of priorities and general guidelines not in terms of specific plans. These guidelines must be in a form that can easily be presented in an oral form that could be memorized.

Second, the operators should be required to devise a plan for initial clearing and restoration following the General Guidelines with no unusual system conditions prior to the "black out".

Third, a well thought out plan of the support staff should be presented for comparison with the operators' plan. The strength and adequacy of each plan should be analyzed.

Fourth, with a number of reasonable but unusual conditions on the system prior to the "black out" require the system operator to derive a plan or plans to restore the system.

This same training should be given to the Division Dispatchers and the Operating Support Staff so that they will be able to understand and implement the System Operators' direction with a minimum of instruction during the emergency.

FORMAT OF THE PLAN

The central feature of restoration is the proper sectionalizing of the transmission grid and the controlled step-by-step restoration of the grid and reenergizing of the generators and loads. Therefore, a special diagram of the transmission grid showing only the necessary controllable switches is a desirable means of planning and controlling the restoration.

The diagram or diagrams can be mounted on separate portable boards, say 18" x 25", which are readily available to the System Operators and Division Dispatchers. The act of selecting the diagrams helps change the mind from normal operation to emergency operation. Also operators at different locations have a common tool for restoration.

ASSESSING THE SYSTEM

Since the system operator must exercise overall control, he should have available to him CRT displays of system overviews which cover as large an area as possible and which indicate open circuit terminals, discontinuity of circuits, MW and MVAR loading on circuits, bus voltage levels and frequency of some busses.

The first assessment the System Operator needs to make is that there is a black out. The boundary of the area that is blacked out must be determined.

Having given instructions as to the plan or plans to be followed in restoration, the System Operator should monitor the system to verify that the steps of restoration are not too large, and after a portion of the system has been restored, that portion is not in distress. By distress, it is meant that some bus voltage is outside emergency limits (too high or too low), some line loading is outside emergency limits or that some area frequency is outside emergency limits (in cases of isolated generation areas).

The System Operator needs the ability to control or stop the restoration work until the distress condition has been eliminated.

SYSTEM RESTORATION DEPLOYING THE PLAN

Laurent Mills
Jean-Claude Rodrigue
Hydro-Quebec
Montreal, Quebec, Canada

INTRODUCTION

At Hydro-Quebec, we saw the need and have had a system restoration plan since 1959. Since that time all of our generating stations have been built far from the major load centers; the latest, in the James Bay area, is almost 700 miles away. Up until 1977, we were restoring our system starting from around the major load centers and reenergizing the system in the direction of the remote generating plants. Because of the natural load growth and because of the many problems associated with energizing long 735 kV lines from a low short circuit source, we have modified our system restoration plan and we are now energizing our long 735 kV lines from our remote generating stations while restoring the part of our system around the load centers. We then synchronize our remote generating stations to the "islanded load centers" just created.

Normally, the operation of the bulk transmission system is centralized. If there is a partial shutdown the restoration process is still centralized. In the case of a total shutdown of the bulk transmission system, we apply a decentralized restoration process; this means that station operators or remote station operators proceed according to a predetermined plan. The System Operator coordinates the energization of the 735 kV lines and the synchronizing of all subsystems.

Since system expansion forces updating of the restoration plan, naturally operator training is required.

FORMAT

Since the restoration plan is decentralized, each power generating station and substation operator is given "emergency instructions" which contain a series of actions he has to take on certain cues. The first cue being when the System Operator informs the Regional System Operator that there is a total shutdown. Other cues are the reenergization of lines at the generating plants or the presence of voltage on the end of a particular line at the substations. All of these instructions have been developed so that a minimum of communications is required between operators.

In order for the System Operator to follow it easily, we have developed a schematic resume of the restoration procedure. He is thus able to take the actions necessary to energize the 735 kV lines, to synchronize the subsystems and then to continue the restoration process.

We have plans to use the computer and CRTs to follow the restoration process as soon as our computerized control center is operational.

OPERATOR TRAINING

Our training program is addressed to the station operators, to the Regional System Operators and to the System Operators. Once a year, the station operator is given an overall view of the restoration procedure and specific instruction on the actions he has to take at the station level. The Regional System Operator is also given an overall view of the restoration procedure and specific instruction pertinent to his region. As for the System Operator, he is given specific instruction on the restoration procedure of strategic stations. Lectures on the restoration procedure are repeated at least once a year with the help of narrated slide projections.

The System Operator is also informed and instructed on all studies made in the area of system stability, voltage control (particularly on the reactions of static compensators and synchronous condensers), and in the area of frequency control by means of automatic load shedding and generation rejection. These instructions, we believe, serve as the basis for evaluating the situation and to control the restoration process.

CALL-IN PROCEDURES

As soon as a partial or total shutdown is confirmed, the Senior System Operator on shift steps back and acts as a coordinator, directs the restoration process via the assistant System Operator and the Relief Operator (normally a System Operator in training). He then calls in the next shift which is at the control center within 30 minutes. The next shift usually keeps record of all events and helps to prepare the short report. At the Regional System Operator level, we recall one operator in the strategic locations of the bulk transmission system.

COMMUNICATIONS

We have recently created an emergency communications center outside the system control center. We operate this center only in the time of major system disturbance. The objectives behind the creation of this center are:

- To limit the number of personnel in the system control center at time of system disturbance.
- To inform the hierarchy on all actions taken and inform the public of the same through our Public Relations Department.
- To group together off-line technical personnel that could be of assistance to the System Operator.

This communications center is now equipped with temporary telephone lines and mobile CRTs. We are satisfied with the results on limiting the number of bodies in the control center but we still have work to do on the other two objectives.

ASSESSING THE SITUATION

Starting from the system configuration and the system status just before the event and recognizing its configuration and status after the event, the system operator normally has a very good idea in what area of the system the original disturbance occurred. The information available today at the system control center is verified with the Regional System Operators concerned and the restoration procedure is initiated.

At Hydro-Quebec, we try to make available to the System Operator on a mimic board or CRTs, field information concerning the status of the major transmission lines, voltage and frequency at strategic points of the system, loading on the major voltage control equipment and the status of all our automatic load shedding and generation rejection systems. With the basic knowledge he has and with all this information, we believe that our System Operator is able to correctly assess the situation.

TEAM ORGANIZATION

By providing an emergency communications center, we have also provided the operator with technical support. When a major system disturbance occurs, selected technical personnel from System Operation, System Performance and System Protection assemble at the center. With information coming from event recorders and oscillographs installed at strategic points of the system, this team confirms the first analysis of the disturbance and is able to guide and/or advise the System Operators on the restoration process.

We found that we are lacking first-hand information in the center and we are planning to telemeter data to that room.

COMPUTER BASED SECURITY SOFTWARE

We are in the process of commissioning our new control center. Our first goal is to obtain maximum relevant information from the field and all of the basic software to operate the system during normal conditions. When that has been achieved and we are satisfied with the performance of the new operating center, we will then address the need for specialized software to be used during a system disturbance or during the restoration process. We have to keep in mind here that our system is mostly affected by stability and voltage control problems during the restoration process and we have developed limit tables which will be introduced in our computerized system. These limits will provide alarms and messages to the system operator in normal and disturbed system operating conditions.

RECORD KEEPING AND AFTER-THE-FACT EVALUATIONS

Normally, within a week after a system disturbance has occurred, there is a meeting held with the System Operators and all Regional System Operators who were on shift at that time. The whole restoration process is reviewed and problems pointed out. These problems vary from lack of data to problems with the restoration plan itself. Specific actions are decided and acted upon as soon as possible.

Concurrently, similar meetings are held by technical staff to evaluate the performance of protection schemes, equipment, load shedding and generation rejection schemes and finally the performance of the system itself Here too, corrections are made as soon as possible.

Valuable information is obtained by these after-the-fact evaluations. That information has permitted us to improve our restoration plan and procedures and to improve the quality and pertinence of our system studies.

CONCLUSION

The experience we have had with system restoration has taught us that in order to have an efficient restoration plan, we need to:

- Continuously improve the plan and procedures considering the system configuration changes. For instance, we have cut our restoration time from 7 hours to around 3 hours by reenergizing the system from our remote generating stations.
- Inform and instruct System Operators of the system studies that are performed; the System Operator is then better equipped to assess what is happening or what has happened.
- System Operators, Regional System Operators and station operators should be instructed at least once a year on the updated system restoration plan and procedures; these lectures are given in early fall when all the major additions to the system have been completed.
- Relevant and readily available information and technical back up are also of major importance for fast and efficient restoration. The commissioning of our new computerized system control center will assure us of readily available information.
- After-the-fact evaluation is probably the area where the most important data is collected; it is not every day that our electrical system and our overall operating procedures are so tested. It is in

this activity that flaws in either our operating procedures or our protection schemes or our system performance evaluations are detected and corrected.

- In order to technically assist the System Operator and to keep management and the public informed of the situation, a fully equipped emergency communications center and the concept of "team organization" is an absolute necessity.

Power System Restoration Planning

M. M. Adibi
IRD Corporation

L. H. Fink
ECC, Inc.

Abstract

System restoration, as an extraordinary mode of system operation, requires careful planning and operator training. The generic tasks of restoration include determination of system and equipment status, preparation of plants and network for systematic restoration, reenergization of the network, and system rebuilding. The procedures for developing an effective restoration plan include formation of a qualified planning team, review of relevant system characteristics, formulation of assumptions regarding blackout scenarios, agreement on restoration goals, development of strategy and tactics, validation of the plan, training, and documentation.

Introduction

Increasing exposure of power systems to extensive blackouts is a consequence of heavier system loading and recent, almost revolutionary changes in industry structure. The impact of prolonged blackouts on the public, on the economy, and on the power system itself makes rapid effective restoration very important. An effective system restoration plan reduces the impact of an outage on customers and on the economy of the affected area while reducing the probability of damage to equipment [1,2].

Power system restoration is one of five states (modes) of power system operation. Operating tasks arise from the interaction of the current state of the system and the operating objectives pertaining to that state [3]. Strategies may then be formulated for carrying out these tasks. The major aspects of system operation in the Restorative State are as follows:

System status is abnormal, with (hopefully decreasing) widespread unserved load, disconnected equipment, disrupted communications, etc.

Restoration objectives are to restore the system to normal operation as quickly as may be consistent with system security, minimizing the amount of unserved kilowatt-hours of energy.

Restoration tasks include ascertaining the detailed state of the system and its elements, preparing the equipment for restoration to service, reintegrating the system, and balancing generation and load as they are brought, in a controlled manner, back to their normal level.

Restoration strategies determine the manner in which the various tasks are coordinated, and the sequence in which they are executed.

This discussion pertains to restoration of the bulk power generation and transmission system. Distribution system restoration has its own problems and solutions, which should be studied and planned for separately.

93 WM 204-8 PWRS A paper recommended and approved by the IEEE Power System Engineering Committee of the IEEE Power Engineering Society for presentation at the IEEE/PES 1993 Winter Meeting, Columbus, OH, January 31 - February 5, 1993. Manuscript submitted August 17, 1992; made available for printing November 30, 1992.

Problematic Aspects of Restoration

The restorative state has some characteristics that make it particularly "user unfriendly." For one, it is extraordinary: widespread blackouts are rare, which is fortunate, but the consequence is that when they do occur, they pose severe and unusual problems. The ordinary task of system operation is that of guiding an ongoing, complex, highly automated process. In restoration, the very different task is that of reassembling the elements of that complex process, and getting it going again. The system may be largely dismembered, with many or most of its elements disconnected. Many of its major elements, the plants, may be in various states of disarray, with abnormal temperatures and pressures outside the range of effective control of plant control loops, or incompatible with rapid return to service.

Operators must work with limited information. Much information that would be helpful in carrying out the restoration process will be unavailable to the operators, either because (a) communications are impaired due to the emergency situation, or (b) the data is not needed in ordinary operation and therefore is not provided. SCADA systems and EMS application programs at the disposal of power system operators were not designed for and are not well adapted to the state of the system that pertains to restoration.

Among the many complicating factors that make restoration of a power system a difficult task is the pressure and stress of completing that task in a minimal amount of time, aggravated by unavailability of many resources, and unfamiliarity of operating personnel with the unusual conditions that predominate. Despite these constraints, priority loads must be restored quickly, and in adverse weather conditions almost all load may be of high priority.

Restoration Tasks and Related Concerns

System restoration involves a broad diversity of tasks, which, depending on the system involved and the circumstances, will not always be carried out in the same sequence. In what follows, we discuss some of the major concerns and problems that may be associated with each of the major groups of tasks.

System Status Determination

The magnitude and actual requirements of system status determination will depend very much on the circumstances and extent of the blackout. In general, this task involves identifying the boundaries of energized areas, and gathering information on frequencies and voltages within energized areas, the status of generating plants, equipment overloads and trouble, and loads interrupted by direct tripping or under-frequency load shedding. Knowledge of the status of stored energy (compressed air, batteries) required for remote switching can be critical.

SCADA system capabilities may be used to the extent possible to assess the status of breakers, but in the absence of effective automatic alarm processing, this may be infeasible due to alarm saturation. The overall quantities of alarms in a shutdown and restoration scenario can be very high. Of the 10,000 or more two-state and multi-state measurements in a typical installation, several thousand may enter the alarm state due to relaying actions on generating units, lines, and loads [4].

Reprinted from *IEEE Transactions on Power Systems,* Vol. 9, No. 1, pp. 22–28, Februay 1994.

Most major disturbances are initiated by transient faults (usually cleared immediately), misoperation of protective equipment, or operator error; in such cases, fault location is not a concern of restoration, but rather of post-disturbance analysis. However, if the disturbance was initiated by a sustained fault, and/or occurred during a major storm, attempts to restore a faulted line or equipment could aggravate the disturbance and extend the time required for restoration. In such cases, knowledge that major transmission paths are unavailable would affect the restoration process.

Plant Preparation

For restoring predominantly thermal power systems, operators need to consider time-critical boiler-turbine start-up characteristics. These include a maximum time interval beyond which a unit with a drum type boiler *cannot* safely be hot restarted, and the minimum time interval after which a super-critical, once-through unit *can* safely be restarted. Drum type boilers have a maximum elapsed time to hot restart of about 1/2 hour beyond which they are not available until cold restart four to six hours later. Super-critical once-through boilers have a minimum elapsed time after a trip-out of several hours before they can be restarted. Other constrained intervals include hot start time to synchronization, time from synchronization to minimum load, and time from minimum load to full load (rate of load pick-up), all of which are based on temperature related phenomena inherent in the physical construction of the boiler-turbine-generator complex [5].

Many thermal plant control systems, designed for normal operation, are not operational during start-up, or at low (<25%) load. Also, protective relays on a unit could initiate an undesirable voltage regulator response or generator trip; e.g. out-of-step, under-excitation, and distance relays could operate during electric power swings or transients [5]. Such conditions make heavy demands on manpower for manual operation, and necessitate carefully coordinated scheduling.

Network Preparation

Along with plant preparation, the network must be prepared for reenergization [6]. De-energized buses must be cleared. Global opening of all breakers following loss of voltage simplifies and may speed up the restoration process, but may exhaust some local supplies of stored energy. The alternative of selected breaker operation will require more data and path analysis. On many systems, in the case of widespread blackouts, it is nearly always advantageous to sectionalize a system into subsystems to allow parallel restoration of islands, and thus reduce restoration duration.

During the initial stages of restoration, when the system is primarily under manual control, wider than normal frequency and voltage variations must be tolerated. Under these conditions, automatically switched capacitors and under-frequency relays responsive to smaller perturbations may hinder the restoration process [7,8].

A feasible start-up sequence for generating units requires scheduling and provision of remote cranking power, with consideration of the effects of energizing particular lines (with awareness of the importance of reactive absorbing capability) and of picking up cold loads [33]. High pressure underground cable is subject to loss of insulation integrity upon loss of pressure, requiring rapid restoration of power to pumping plants in the absence of adequate local energy [9]. Requirements of nuclear units and of critical loads must be accommodated. All of these requirements involve delivery of remote power and in turn require identification of suitable intervening transmission paths. Imperfect knowledge of network status implies in turn a need to provide alternative path plans.

Network Energization

Reintegration of the skeleton of the bulk power system involves time-consuming switching operations. Many phenomena rarely seen during normal system operation may be encountered during restoration. Energization of high voltage lines may incur sustained power frequency overvoltages, switching transients, and harmonic resonance. The first is due to reactive power generation (charging current) of unloaded lines, especially long EHV lines, or of underground cable (oil-filled or high pressure pipe type). Sustained overvoltages may overexcite transformers, generating harmonic distortions and overheating, and may cause generator underexcitation, or even self-excitation and instability. Switching transients on long HV lines, although of short duration, may cause arrestor failures, particularly if coupled with sustained overvoltage conditions. Harmonic resonance on lightly loaded lines may result in very high voltages that may be amplified by transformer overexcitation [10]. In any event, the reactive absorption (under-excitation) capability of generators to be synchronized is of considerable import.

Maintenance of stability of black-start and peaking combustion turbines, during the time when they constitute a large portion of generation and may be feeding into a weak network, may be a difficult problem. Combustion turbines in particular often do not have the reactive absorbing capability required to operate into a long, unloaded transmission line. Another basic task is maintenance of active and reactive power balance as generation is restored [11,12,13].

Data necessary to quantify these phenomena, and the nonlinear models necessary to analyze the data, are not normally available with even the best equipped EMS.

System Rebuilding

If the system is being restored by building up islanded subsystems in parallel, interconnection of generating stations within each subsystem is carried out after generating units attain the minimum specified generation, but prior to restoration of full load within the subsystem. However, certain industrial and municipal loads such as traffic lights are time critical. In each subsystem one plant must be responsible for keeping the frequency near 60 Hertz.

Load restoration and generator loading must be coordinated, considering the frequency response of prime movers to sudden load imposition and cold load pickup characteristics. The former is a critical consideration. In normal operation, plant operators are able to synchronize a generator and take the unit through the above start-up phases with little or no coordination with the power system operators; the incoming unit shares load with the other parallel generators in service. During restoration, however, the various start-up phases have to be coordinated with load pick-up, therefore the activity coordination between power plant, power system, and field operators becomes very important. Closing in on blocks of load larger than can be absorbed by generators on line can cause unit trips and set-back the restoration process.

The amount of latent real and reactive load demand varies with the elapsed time after an outage. Loads restored after some period of outage may be considerably greater than they were at the time of interruption (which is referred to as "cold load pickup"). Numerous studies have shown how coincident load demand may grow during a power outage, and analytic models for estimating this increase have been developed [14,15]. Experience on some systems, however, has indicated that these effects may be counterbalanced by manual or automatic disconnection of load incident upon or during a power failure [8]. Thus, load/time variation and resultant "cold load pickup" magnitudes can be system specific and difficult to predict.

One concern involved in closing line breakers within a partially reestablished network is that of an excessive phase angle difference across the breaker. Closure across such a "standing phase angle" difference could shock the system and possibly cause instability or equipment damage. The magnitude of such phase angle differences that can be tolerated depends on the system voltage; with tolerable angles being greater at lower voltages, where electrical impedances are greater.

Once subsystems have been synchronized, the system will be placed under flat frequency or tie-line/frequency-bias control, depending on circumstances. Full restoration of load and generation may proceed in successive steps, with care that no block of load be picked up at any stage that exceeds the available generation absorption capability.

Logistics

Implementation of the above groups of tasks requires close and continual coordination between power system, power plant, and field operators [9]. The plant and network preparation tasks proceed in parallel, and are mutually dependent in many respects. Unit restart scheduling, for instance, depends not only on coordination of manpower within plants, but also on scheduling of transmission paths to provide cranking and station service to non-black-start units. Also and importantly, the wider and longer the disturbance, the more important it is that neighboring utilities, local governmental authorities, and customers be kept as informed as possible about cause and extent of the disturbance, and the progress of the restoration process. Thus communication policies and facilities are critical and must be coordinated and used as effectively as possible [16].

In a widespread disturbance, or one accompanied by severe storm damage, public communication facilities may be impaired, or inaccessible due to overload caused by public demand, and the utility will have to depend heavily on its own communication facilities. Nevertheless, public communication facilities must be used to the extent they are available in order to minimize the demand on company (internal) communication facilities. Back-up battery systems and/or generators are usually the limiting factors of company telephone networks and emergency radio systems, in terms of how long these systems will be available upon loss of AC power [16].

Criteria for requesting interconnection assistance by a system being restored include (a) the adjacent system being firm and tied to other systems and having adequate reactive absorbing capability, or (b) the system being restored not having black start capability. A criterion for providing assistance to the adjacent system is (c) whether the system being restored is firm and stabilized. Once a tie is closed, the interconnection is responsible for controlling the tie flow and the system being helped should control its frequency.

Return to Normal

The restoration process is considered completed when the final configuration of the target power system has been achieved. Economic Dispatch may be reinstituted and the next load pick-up increment will be based on considerations other than technical requirements of bulk power restoration.

Development of a Restoration Plan

Considering the scope, diversity, and unusual nature of the tasks involved in restoration, and all the things that might go wrong, development of a good plan is a considerable undertaking [17]. Although power system characteristics under different load levels and operating conditions requiring different restoration plans, it is important for the ease of implementation to try, in so far as possible, to develop one master plan or several similar plans. Major elements in the procedure are discussed in the following sections.

Formation of a Planning Team

The first step toward development of an effective plan is formation of a team to provide the combined experience of personnel familiar with production, protection, operations, instrumentation, system analysis, and communications. Because of the unusual system conditions and phenomena that may be encountered during the restoration process, proposed restorative actions will have to be simulated in order to ensure that the system will not inadvertently be thrown back into collapse. This requires the participation of knowledgeable system analysts. Operator-analyst cooperative efforts facilitate accurate modeling and interactive simulation of the real system to provide fast-turn-around analysis of alternate restoration steps suggested by the operators.

Review of System Characteristics

Characteristics of power systems relevant to restoration may be divided into two functional groups: those related to system structure, and those related to system control dynamics. Structural characteristics may differ according to system size (load and generation), the type of service territory (metropolitan or rural, compact or extensive, or combinations), generation mix, transmission voltage levels, type(s) of distribution networks, types and sizes of loads, voltage regulation practices, and availability of interconnection assistance [18]. Prime mover types differ widely in their black-start capabilities or restart restrictions; distribution network configurations affect the degree of control that may be effected over the size of load blocks to be picked up; voltage control capabilities are an important factor in maintaining viable system conditions during the restoration process.

Dynamic characteristics affect reactive and real power balance, and protection and control systems [17]. Characteristics related to reactive power balance include: (a) generators' and synchronous condensers' reactive capabilities (over- and under-excitation) under different system configurations and operating conditions; (b) shunt capacitor and reactor sizes, locations, voltage levels and modes of control; (c) underground cable voltage levels, charging currents and maximum sustainable voltages, (d) transformer reactances, tap ranges and modes of controls; (e) EHV and HV overhead lines charging currents, number of sections per line, and number of circuits per line; etc.

Characteristics related to real power balance include: (a) number and sizes of low voltage AC network loads under peak and light load conditions; (b) generator maximum and minimum outputs under normal and emergency conditions; (c) prime mover time constraints, frequency response to sudden increase in loads, types (drum, supercritical once-through, CT, hydro, nuclear, pumped storage); (d) black start capabilities; (e) cold load pickup of radial feeders and network loads; etc.

Characteristics related to protection and control systems include: (a) number and sizes of feeders with underfrequency and switched capacitor relays and their settings; (b) number sizes and locations of low frequency isolation schemes; (c) number and locations of synchronizing facilities (at other than generating plants); (d) number and locations of frequency metering points; (e) interlocking schemes which prevent energizing dead buses; (f) number and locations of breakers with synchrocheck relays; etc.

Formulation of Assumptions

Restoration plans are usually developed in consideration of a worst case scenario for a major power system disturbance and the most probable load and generating condition under which such a disturbance might occur. It is important (for training as well as for implementation) that restoration plans be as uniform as possible across differing system load conditions. However, constraints affecting restoration may differ markedly, not only between peak and light load conditions, but under other circumstances, all requiring some variations in restoration plans, or even separate plans [17]. Under light-load conditions, critical activities include start-up of cycling steam units. Under peak-load conditions, critical tasks usually include the coordination of network load pick-up with generation response, and consideration of possible risks for system frequency decline beyond prime mover recovery capability.

Conditions for black-start of a system may be more difficult during a weekend at time of system minimum load than on peak. Many stations that would be suitable for hot-restart may not have been on-line at the time of the blackout. On-line generators will be primarily base units; cycling and peaking units will be cold, with no hot re-start capability available for 4 or more hours. Reliance will be on hydro (if available) or on combustion turbines having limited real and reactive power capabilities.

Nevertheless, the occasion often considered is a complete collapse under a given year's off-peak load. From the annual load duration curve, the level of load with the longest duration (hours per year), and the generating units serving that load can be determined. This for example could be an off-peak load condition of about 90% of the annual peak load over about 90% of the time, or 7000 hours per year. Next the worst scenario for a major power system disturbance should be determined using the power system operators' experience and judgment. This for example could be splitting of the power system into subsystems. The power system may either break into the subsystems as the result of a major disturbance, or after a complete blackout be sectionalized by the operator into the subsystems for parallel restoration.

Establishment of Goals

Despite the physical similarity of utility systems, local conditions, whether specific generation mixes, uncommon loads, or perhaps contractual obligations, may mandate establishment of system specific goals for the restoration plan. Strategic, system specific, priority restoration goals are crucial to achieving a restoration that is optimal with respect to maximum restored energy and minimum elapsed time.

In each power system there are certain immediate objectives to be achieved, after which the remaining restoration becomes a more generic (i.e. less system-specific) operation. Depending on the local system characteristics and operating conditions, these objectives could be reenergizing the bulk power network, or reaching a certain level of generation (reactive power absorbing or real power) capability, or picking up a downtown network load, or supplying time-critical industrial loads, or energizing extra high voltage lines to a remote and large generating unit. More specific examples might be: providing cranking power to drum type units within 30 minutes; restoration of the high voltage transmission lines to large units such as super-critical once-through or nuclear units within 3 hours; restoration of 90% of load within 6 hours.

Beyond the earliest attainment of such critical objectives, a good restoration plan will be based on a strategy to minimize the total amount of unserved energy as well as the duration of the restoration.

Development of Strategy and Tactics

Development of a coherent, practicable restoration plan requires postulation of one or more collapse scenarios to be considered [19], and choice of a basic restoration strategy reflecting local system constraints and characteristics. Allocation of resources and coordination of activities to ensure an orderly sequence of actions to successful completion of restoration in a minimum of time is an important consideration.

Generic strategies. In preparing the network for restoration, most utilities open all breakers after a system collapse (an "all-open" strategy) to simplify determination of network status and implementation of the restoration process. Other utilities, particularly in cold climates, follow a "selected operation" strategy, opening and closing breakers as needed by the restoration process in order to conserve stored energy. The latter requires exact knowledge of the status of breakers and disconnect switches.

Development of a restoration plan from a point of complete collapse and without interconnection assistance may be relatively straightforward. Often, however, due to emergency control actions, a system disturbance may be arrested prior to complete collapse. Furthermore, some level of interconnection assistance often will be available. It is thus desirable for a restoration plan to cover different system operating conditions and various degrees of system upset. After initial assessment of the system status, a proper entry into the plan may be made, and the appropriate procedure selected and followed to a successful restoration.

The basic strategic choice is between a "build-down" strategy of reenergizing the bulk power network before resynchonizing most generators, or a "build-up" strategy of restoring islands that will then be mutually interconnected. In comparing the two strategies, it initially appears that the build-down strategy is simpler. However, upon closer examination, it becomes clear that restoration of partial outages and reliance on the interconnection necessitates assessments of both the system being restored and the interconnection status. These assessments use critical time and make the build-down strategy more tasking. On the other hand, while simultaneous restoration of service in subsystems under the build-up strategy can appreciably reduce the duration of an outage, the extent of sectionalization is limited by availability of resources, including the number of operating teams as well as the communications available to permit effective and efficient coordination of their efforts.

The build-down strategy. The build-down restoration strategy, based on reenergization of the bulk power network as a first step, is followed by balanced, step-by-step restoration of loads and generation [11,20]. A major difficulty with this strategy is the reactive power produced by the unloaded high voltage transmission plant, which often exceeds the absorption capability of the generators that are energizing the network, and which at the same time results in unacceptable over-voltages at remote locations [10,20]. This strategy is usually more appropriate for smaller systems without long high voltage lines [21], for predominantly hydro systems with high reactive absorption capability, or for large systems with very compact service territories. On other systems, it usually is selected only when interconnection assistance is available, or after a partial outage. One utility with a large amnount of hydro generation studied a "low voltage" variation on the build-down strategy, wherein the network was reenergized at a minimum perunit voltage, minimizing both the amount of reactive produced and the magnitude of over-voltages [22].

In any case, the build-down strategy includes: (a) assessment of power system status (breaker and disconnect positions), power plant conditions (hot, bottled, cycling, cold start, peaking, and base load), and local energy storage status; (b) energization of HV or EHV lines avoiding sustained overvoltages and harmonic resonances; (c) supply of cranking power to non-black-start plants, and/or supply of station service for the hot restart of drum type boilers before elapse of the maximum critical time; (d) synchronization of units as they become available for service; and (e) full restoration of load and return to normal operation.

The build-up strategy. The more common build-up restoration strategy [17,23,24] is usually selected in conjunction with a scenario involving a complete system collapse and lacking any interconnection assistance. The strategy for this situation includes: (a) assessment of power system status (as above); (b) sectionalization of the system into subsystems, so that each subsystem contains at least one drum type boiler and one unit with black-start capability; (c) supply of cranking power to non-black-start plants, and/or supply of station service for the hot restart of drum type boilers before elapse of the maximum critical time; d) interconnection of generating stations within each subsystem (after generators attain minimum generation but prior to restoration of full load within the subsystem); (e) synchronization of subsystems and load pick up to stabilize all units; (f) start-up and synchronization of large super-critical once through units; and (g) closing of ties with the interconnection when adequate reactive absorbing capability is reached.

Special considerations and tactics. Development of a restoration plan entails careful consideration of the probable effects of each action on the system and on subsequent restoration measures. It is important to consider both system response to control actions, and operator interactions at the control center, generating plants and the load centers, recognizing and coordinating critical time intervals. Especially during the early stages of restoration, firm capability should be emphasized, considering firm transmission in the interconnections between subsystems, and firm generation capacity in each subsystem.

Deactivation of automatic load shedding and automatic switched capacitors may be necessary during the initial stages of restoration, when frequency and voltage excursions may be wider than in normal operation. An alternative approach, in order to avoid the need for global deactivation, is to avoid use of feeders equipped with such devices, but this obviously imposes severe restrictions on the restoration process. Dynamic and transient stability (as the system is being restored and impedances between generators are large) may be enhanced by having each station supply nearby loads.

Load pickup capability of the system increases with the increase in system generation. Initial restoration of loads should be in small increments; restoration of smaller and radial loads should precede that of larger and low voltage AC network loads, while maintaining a reasonably constant real to reactive power ratio. As restoration proceeds, load pick-up should be in as large increments as possible without the risk of dangerous frequency decline.

Reactive balance may be improved by initially energizing only one circuit of double circuit underground transmission lines, and operating generators under-excited to the extent that stability considerations allow. System voltages may be maintained within acceptable bounds (e.g. plus and minus 5%) by such means as keeping generators at the low end of their voltage range during early stages of the restoration, and adjusting system transformer taps to appropriate positions. The general guide used by some operators is to pick-up load when system frequency is high and raise generation when system frequency is low.

In view of the importance of communication during extensive restoration, and the potential unavailability of normal communication facilities, plans will be needed for the provision and use of company owned or other alternate facilities. Moreover, company spokesmen will be needed for communication with local governmental bodies and the public. All these arrangements are necessary elements of an adequate restoration plan.

General principles. It is essential that the system restoration plan adhere to basic restoration principles. Strategies must allow for chance events during the restoration process, including the probability of unsuccessful operation of each task, whether the start-up of a combustion turbine, energization of a line section, or closing of a power circuit breaker. Provision should be made for alternate transmission paths for remote cranking power, to accommodate line outages.

During development of a restoration plan it may become apparent that new facilities or capabilities, such as under-frequency isolation, controlled islanding, under-frequency load shedding capabilities, relay blocking/transfer tripping, and shunt reactors are needed to permit implementation of restoration in a timely manner.

Validation

In order to develop a restoration plan, it is necessary to model the steady state and dynamic behavior of the power system by representing the physical and operational data in one or more interactive computing systems. (More than one model is almost always necessary because of the wide range of dynamic phenomena that must be considered.)

During the restoration process, abnormally wide swings in frequency and voltage and other transient phenomena may be encountered (may even be prominent) that are not germane during normal operation [10,25]. This will require the informed use of a variety of system analysis and simulation tools, including the concurrent use of steady state power flows, transient and dynamic stability, and electromagnetic transient simulations, to validate the feasibility and security of proposed actions.

Transmission line loadings, reactive power balance, over- and under-voltage conditions, generator loadings, transformer tap positions, shunt capacitor and shunt reactor status, and phase angle differences between load centers and generators can be evaluated by means of power flows. Standing phase angles may appear across open switches which will have to be closed to firm up transmission paths, or to complete the reintegration of an island. Reduction of such phase angles before switch closure is a critically important task that may require optimal load flow analysis.

Transient voltages caused by energizing large segments of a transmission system, and the harmonic disturbances produced by line charging and transformer magnetizing currents which may cause damaging resonance overvoltages, can be evaluated by electro-magnetic transient simulation. The time at which units may be bought on line, the rates at which generators can be loaded, mismatch between load pick-up and generator response, changes in system frequency, and effects on feeders having under-frequency relays can be evaluated by means of long-term dynamic simulation. This simulation is particularly needed to evaluate the pick-up of low voltage AC network loads equipped with group breakers.

Simulations require careful preparation of load data, generation data, and other power system characteristics. Data necessary to quantify all the unusual phenomena involved, and the nonlinear models necessary for analysis, may be difficult to obtain.

Iteration

Restoration planning involves two iterative processes, one between power system modeling and model validation, and the other between strategy and tactics. The models have to be validated against measured and telemetered operating data obtained from the control computer system. This iterative process is continued until an acceptable degree of agreement is obtained between the models and the actual power system outputs.

After a strategy for restoration is formulated, and a candidate restoration plan (tactics) is developed, the restoration plan is checked against the initial strategy for conformity and feasibility, and in case of significant departure, the strategy and/or plan may have to be modified [17].

The above iterative procedures entail careful consideration of the probable effects of each restoration step on the rest of the system and on planned subsequent measures.

Documentation

Another important element in restoration planning is effective documentation. Three levels of documentation are required: for reference, for training, and for use during actual restoration.

Reference means a complete record, not just of the plan as finally approved and adopted, but also of the development of the plan, including full documentation of the simulation testing and its results, showing control actions that were rejected because of adverse system response as well as control actions that were incorporated in the plan. Completeness of this reference documentation will greatly facilitate later revisions of the plan.

Training documentation should be developed to incorporate not only the plan itself, but pedagogically effective training procedures.

Convenient *use during actual restoration* implies embodiment of the plan in guidelines for operators. These should abstract the plan itself in such form as to be readily accessible to operators during the restoration process, enabling them to check their procedures against the plan without having to search through one or more encyclopedic volumes.

A system restoration plan should be readily accessible and easily understood; its final effectiveness requires that the information and guidance it provides be readily and easily available to those who will need to make use of it during the restoration process. This imposes requirements in turn upon the manner in which the plan is documented, stored, and made available for retrieval [26]. It should be stored in a convenient media for quick access, and possibly in alternative or complementary media. It would be convenient to be able to access urgently needed information by a light pen on the operator's EMS display.

At all three levels, the restoration plan needs to be documented with utmost care. It should be written clearly using precise terms for ease of understanding, and these should be incorporated in a definitive glossary or dictionary. Furthermore, it should be illustrated with familiar diagrams and charts. The documents should be organized in a logical order so that specific information can be located quickly.

Training

No restoration plan is complete until it has been incorporated in the capabilities and even the reflexes of the operators who will have to put it into action under stressful circumstances. This requires effective, and ongoing, training [27,28].

A wide variety of techniques are available for operator training [27]: simple instruction manuals or audio-visual tapes for independent study, classroom instruction, interactive PCs, restoration drills, and highly interactive operator training simulators augmented for restoration training. Simple methods such as independent study are most appropriate for learning "facts" about system restoration; computer-based instruction can be very useful for inculcating understanding of basic concepts. Role playing and simple simulations are good tools for drill and practice of well-defined scenarios; more sophisticated high fidelity simulators are extremely useful in training system operators to respond to time-critical emergency situations.

In view of the difficulties in developing a generally applicable restoration plan, particularly in the absence of timely information and the presence of many ambiguous alternatives, the operator's heuristic problem solving capabilities should be fully exploited. Training sessions not only enable operators to understand the plan and become proficient in its execution, but also provide an opportunity for them to suggest alternative restoration steps for the improvement of the restoration procedure.

Audits and Updates

A functional system restoration plan is necessarily complex, and despite the best of care and discipline in its development, there will be a risk of overlooking some essential operation or condition. The penalty for any such oversight could be serious. As is natural, because of the close proximity and involvement of the project team to the system restoration plan, no matter how experienced they may be, certain flaws may escape their detection. Therefore the task of technical auditing should be assigned to a qualified person outside the team.

Even an audited restoration plan, to be valid, must be up-to-date. The documents must be regularly revised to reflect the current power system physical characteristics and operating conditions. The plan should be updated on the occasion of changes in the system, changes in the EMS/SCADA installation or major plant controls, changes in other available tools, and at least annually.

Conclusions and Future Needs

Most operating companies maintain restoration plans based on their restoration objectives, operating philosophies and practices; and familiarity with the characteristics of their power plant restart capabilities and their power system reintegration peculiarities. While these procedures have successfully restored power systems in the past, they can be improved significantly by simulations of steady state, transient and dynamic behavior of the power system under various restorative operating conditions, and by engineering and operating judgment reflecting many factors not readily modeled.

Most power systems have certain characteristics in common and behave in a similar manner during restoration process. It is therefore possible to establish a general procedure and a list of guidelines to enhance rapid restoration. A detailed plan, however, has to be developed specifically to meet the particular requirements of an individual power system. Thus, it is important to determine and record the relevant characteristics of each power system.

An on-line restoration *guidance system*, capable of guiding the operator in making decisions as to what steps to take and when to take them, would go a long way toward minimizing the duration of blackouts, and consequently their impact. Preliminary results with some coordinated restoration schemes has indicated a possibility of substantial reduction in unserved load during blackout [1]. The nature of the problem, involving as it does incomplete information and many ambiguous alternatives, suggests a need for an integrated tool, consisting of both an expert system and appropriate analytical tools, as a platform to develop a guidance system for restoration [29-32]. Such a tool would serve as an advisor to bring to the on-line environment many of the tools and capabilities planners and operators use in planning for restoration, while it will still be subject to operator approval of all major decisions.

References

1. PES Task Force Report: Power System Restoration - the Second Task Force Report; *IEEE TRANS.* v.PWRS-3, n.1, 1988, pp.118-126.

2. M.M.Adibi: Discussion of D.S.Kirschen, T.L.Volkmann: Restoring a Power System with an Expert System; *IEEE TRANS.* v.6 n.2, 1991, p.565.

3. L.H.Fink: Power System Operation and Control; *POWER SYSTEMS ANALYSIS AND PLANNING* (ed. A.H.ElAbiad), McGraw Hill, 1983, pp.159-86.

4. C.J.Andrews: Excessive Alarms During Restoration (in PES Working Group: Special Considerations in Power System Restoration; 92WM106-5PWRS)

5. P.J.Clelland: The Role of Plant Control System (Part III,3 in PES Task Force Report: Power System Restoration); *IEEE TRANS.* v.PWRS-2 n.2, May 1987, pp.271-77.

6. T.A.Volkmann: Switching During Restoration (in PES Working Group Report: Special Considerations in Power System Restoration; 92WM106-5PWRS)

7. M.M.Adibi: Issues and Problems in Power System Restoration (in PES Task Force Report: System Operations Challenges); *IEEE TRANS.* v.PWRS-3 n.1, 1987, pp.118-26.

8. J.Feinstein: discussion of PES Task Force Report: System Operations Challenges; *IEEE TRANS.* v.PWRS-3 n.1, 1987, pp.118-26.

9. J.N.Wrubel: Underground Transmission System Concerns; (in PES Working Group Report: Special Considerations in Power System Restoration; 92WM106-5PWRS)

10. M.M.Adibi *et al.*: Overvoltage Control During Restoration; 92WM107-3PWRS

11. R.Kearsley: Restoration in Sweden and Experience Gained from the Blackout of 1983; *IEEE TRANS.* v.PWRS-2 n.2, 1987, pp.422-28.

12. R.D.Shultz, G.A.Mason: Blackstart Utilization of Remote Combustion Turbines, Analytical Analysis and Field Tests; *IEEE TRANS.* v.PAS-103 n.8, 1984, pp.2186-91.

13. J.G.Raine: Reactive Power Balance (Part IV,1 in PES Task Force Report: Power System Restoration); *IEEE TRANS.* v.PWRS-2 n.2, May 1987, pp.271-77.

14. M.H.Yolcu *et al.*: Adaptation of the Simplex Algorithm to Modeling of Cold Load Pickup of a Large Secondary Network Distribution System; *IEEE TRANS.* v.PAS-102 n.7, 1983, pp.2064-68.

15. C-Y.Chong, R.P.Malhami: Statistical Synthesis of Physically Based Load Models with Application to Cold Load Pickup; *IEEE TRANS.* v.PAS-103 n.7, 1984, pp.1621-28.

16. F.Arsanjani, M.W.Lanier: Telecommunication Capabilities and Limitations (in PES Working Group Report: Special Considerations in Power System Restoration; 92WM106-5PWRS)

17. R.J.Kafka *et al.*: System Restoration Plan Development for a Metropolitan Electric System; (81WM 044-7) *IEEE TRANS.* v.PAS-100 n.8, 1981, pp.3703-13.

18. PES Task Force Report: Power System Restoration, Parts III, Power System Characteristics; *IEEE TRANS.* v.PWRS-2 n.2, May 1987, pp.271-77.

19. PES Task Force Report: Power System Restoration, Parts II, IV.3, (Strategies); *IEEE TRANS.* v.PWRS-2 n.2, May 1987, pp.271-77.

20. G.Morin: Service Restoration Following a Major Failure on the Hydro-Québec Power System; (86WM 183-8) (T&D)

21. D.Scheurer: Sequential vs. Parallel Restoration (Part IV,3 in PES Task Force Report: Power System Restoration); *IEEE TRANS.* v.PWRS-2 n.2, May 1987, pp.271-77.

22. E.J.Simburger, F.J.Hubert: Low Voltage Bulk Power System Restoration Simulation; *IEEE TRANS.* v.PAS-100 n.11, 1981, pp.4479-84.

23. E.Mariani *et al.*: Field Experience in Reenergization of Electrical Networks from Thermal and Hydro Units; *IEEE TRANS.* v.PAS-103 n.7, 1984, pp.1707-13.

24. J.Gutierrez *et al.*: Policies for Restoration of a Power System; *IEEE TRANS.* v.PWRS-2 n.2, May 1987, pp.436-42.

25. S.Nishida *et al.*: Analysis of Overvoltages Caused by Self-Excitation in a Separated Power System With Heavy Load and Large Shunt Capacitance; *IEEE TRANS.* v.PAS-103 n.8, 1984, pp.2935-42.

26. S.Mullen: Emergency Plans: Revise, Rewrite, or Leave Them Alone?; *ELECTRICAL WORLD*, July 1991, p.17.

27. PES Working Group: Bulk Power System Restoration Training Techniques; 92WM147-9PWRS

28. J.P.Smith *et al.*: Development of Operator Training Curricula Using the Instructional Systems Development Approach; *IEEE TRANS.* v.PAS-104 n.12, 1985, pp.3446-52.

29. Y.Kojima, *et al.*: Development of a Guidance Method for Power System Restoration; *IEEE TRANS.* v.PWRS-4 n.3, 1989, pp.1219-27.

30. Y.Kojima *et al.*: The Development of Power System Restoration Method for a Bulk Power System by Applying Knowledge Engineering Techniques; *IEEE TRANS.* v.PWRS-4 n.3, August 1989, pp.1228-35.

31. D.S.Kirschen, T.L.Volkmann: Guiding a Power System Restoration with an Expert System; *IEEE TRANS.* v.PWRS-6 n.2, May 1991, pp.558-66.

32. C-C.Liu *et al.*: Generation Capability Dispatch for Bulk Power System Restoration: A Knowledge-Based Approach; 92WM108-1PWRS

33. J.M.Miller: Optimal Sequencing of Starting-Up Generating Units; (in PES Working Group Report: Special Considerations in Power System Restoration; 92WM106-5PWRS)

A SYSTEMATIC METHOD FOR POWER SYSTEM RESTORATION PLANNING

J.A. Huang, L. Audette, Member, IEEE
Institut de recherche d'Hydro-Québec (IREQ)
1800 montée Ste-Julie, Varennes, Québec
Canada J3X 1S1

S. Harrison
Expertise de réseaux, Division Transport
Hydro-Québec, Montréal, Québec
Canada H5B 1H7

ABSTRACT: This paper describes a systematic method for power system restoration planning. System restoration at Hydro-Québec is performed according to a set of detailed emergency instructions. The basis for these instructions is a restoration plan that is prepared off-line by operation planners. This plan includes a switching scenario with a set of values for the control variables – generator voltages, shunt reactors and load levels – that satisfy the operating constraints. Finding a suitable set of control variables usually requires many load-flow runs in a trial-and-error process. However, the problem can be formulated as a nonlinear optimization problem. The optimization algorithm which solves that problem provides values for the control variables automatically. This algorithm is linked to a user interface which constitutes a powerful tool for system restoration planning.

I INTRODUCTION

Power system restoration after a shutdown is a complex decision and control problem for both system operators and operation planners. Many electrical utilities have developed system restoration schemes to meet their particular needs. Most of these schemes contain general guidelines [1-3] used to define the required restoration plans. The latter are usually developed using interactive load-flow programs [4-5] and involve trial-and-error processes. In recent years, expert systems have shown potential, since heuristic knowledge plays an important role in power system restoration [6-8]. However, it seems that little has been done to represent and solve the restoration process as a mathematical problem [9-11].

This paper first describes a systematic method for power system restoration planning. This method was used to design new software that incorporates an optimization algorithm and an advanced user interface. The algorithm formulates the restoration process as a nonlinear optimization problem and provides a way to find a set of values for control variables that satisfies steady-state operating constraints, a situation usually difficult to attain with conventional heuristic approaches. The new approach eliminates the trial-and-error process, thereby saving time in restoration planning. The proposed method is used at Hydro-Québec and is formalized in a general way so that it can be used on other systems (radial or meshed networks).

94 SM 548-8 PWRS A paper recommended and approved by the IEEE Power System Engineering Committee of the IEEE Power Engineering Society for presentation at the IEEE/PES 1994 Summer Meeting, San Francisco, CA, July 24 - 28, 1994. Manuscript submitted December 30, 1993; made available for printing June 10, 1994.

The user interface provides tools that can be used to define restoration scenarios and display and analyze the simulation results.

II SYSTEM RESTORATION AT HYDRO-QUÉBEC

At Hydro-Québec, system restoration is performed according to a set of detailed emergency instructions written for each generating station and substation involved. The instructions are prepared in such a way that the operations can be executed with a minimum of communication between operators and their regional control center. At any given moment, there is only one valid set of instructions for each station or substation.

Responsibility for activities linked to system restoration is shared. The operation planning department prepares and modifies restoration plans according to directions given by the operation department. These directions may include addition of new equipment or removal of equipment for maintenance. The restoration plans constitute the basis for the detailed emergency instructions prepared and distributed by the operations department.

When a complete shutdown of the system occurs, decision-making and execution are transferred from the Hydro-Québec control center (CCR) to dispatchers in regional control centers (CER) and operators at facilities involved in system restoration. The CCR is responsible for monitoring the implementation of the emergency procedures and modifying them in the case of a major equipment failure. An emergency team is also present at the control center for decision support.

The structure of the Hydro-Québec system is quasi-radial: two remote production sites in the northwest and east of Québec are connected to the load in the southwest by long transmission lines. The main problems hampering system restoration are the various overvoltage conditions resulting from line charging, harmonics and switching surges [12].

Hydro-Québec applies the following principles to restore its main grid, which for restoration purposes is divided into five networks, energized independently and simultaneously:

- energize the networks as quickly as possible without necessarily restoring load at this stage;
- minimize the number of control variable adjustments from stage to stage;
- synchronize the networks after energization;
- restore loads in small increments.

A network is reenergized in stages, each stage corresponding to the closing of a circuit breaker and the

Reprinted from *IEEE Transactions on Power Systems,* Vol. 10, No. 2, pp. 869–875, May 1995.

consequent energization of a network component such as a transformer, a line section, a shunt reactor or a load. A complete sequence of switching operations is called a restoration scenario. Each closing of a breaker creates a new network topology, a new set of voltages and a new active/reactive power balance. At each stage, it is essential that voltage and other limits be respected. However, during the early stages of system restoration, maintaining voltage within the acceptable limits, both for transient and steady-state conditions, is difficult because the network is unloaded or lightly loaded, so that the HV transmission lines generate an excess amount of reactive power which creates high voltages [12-13]. This problem is particularly acute on the Hydro-Québec power system. To satisfy voltage constraints, the operator can adjust the generator voltages, transformer taps, switchable shunt reactors, and the amount of load restored. These quantities, i.e. the control variables mentioned above, are estimated by operation planners.

A restoration plan undergoes three types of simulation:

- in steady-state, a switching scenario is defined: the problem involves finding a set of values for that scenario which will keep voltages within limits and keep readjustments of control variables throughout the scenario to a minimum. This work is usually conducted using an interactive load-flow program and involves a time-consuming trial-and-error process;
- harmonic analysis: if a transformer is energized, then harmonic analysis is used to check for any harmful harmonics and prevent equipment damage;
- switching-surge study: the EMTP program is used to check for excessive switching overvoltages.

Most studies involve the modification of existing restoration plans to take into account new equipment or equipment unavailabilities due to maintenance or failure. These modifications usually do not require harmonic or EMTP studies, since they do not affect the frequency response of the system. The work involved, then, is limited to the definition of an alternate switching scenario (if necessary) and the optimization of control variables in the steady-state, which is the focus of this paper.

III CONVENTIONAL RESTORATION PLANNING

Given a network and a switching scenario, finding a set of adequate values for the control variables at any stage of the scenario is not an easy task. It usually involves running a load-flow program in a trial-and-error process until all operating constraints are satisfied. The process is illustrated in Fig. 1.

At stage k, the network topology is N_k and the voltage values are controlled by adjusting vector VG_k of the generator voltages and vector CV_k of other control variables (shunt reactors, loads, transformer taps).

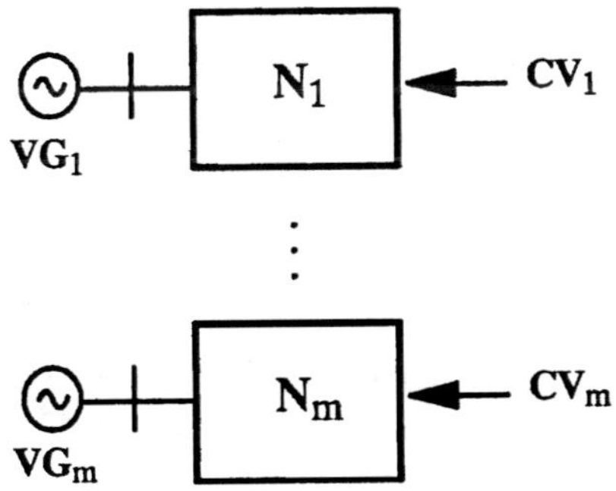

Fig. 1 Finding control variables at each stage

At stage k+1, the topology may change to N_{k+1} (adding load-flow equations), or the dimensions of vectors **VG** and **CV** may be increased by the addition of one control variable in the network. To maintain all voltages within operating limits, values in vectors **VG** and **CV** that were adjusted at the previous stage may need readjustment. Vectors VG_k and CV_k are replaced by VG_{k+1} and CV_{k+1}.

With M stages in the switching scenario, the problem is finding a solution for U, expressed as follows:

$$U = [\,VG_1\,, \dots\,, VG_m\,,\;\; CV_1\,,\;\; \dots\,,\;\; CV_m] \qquad (1)$$

Since, during restoration, the operator is under stress and time is precious, it is preferable to keep the number of readjustments of vectors **VG** and **CV** to a minimum. An optimal solution, shown in Fig. 2, is one where the control variables set at one stage are kept unchanged during all subsequent stages. Thus, at the most, only one control variable needs to be adjusted at each stage. The total adjustments of the control variables are reduced to:

$$U = [\,VG_m\,,\quad CV_m\,] \qquad (2)$$

Finding adequate values for the control variables at one stage is not an easy task, but finding an optimal solution is a even more difficult and time-consuming. However, this problem can be formulated as a non-linear program and be solved systematically.

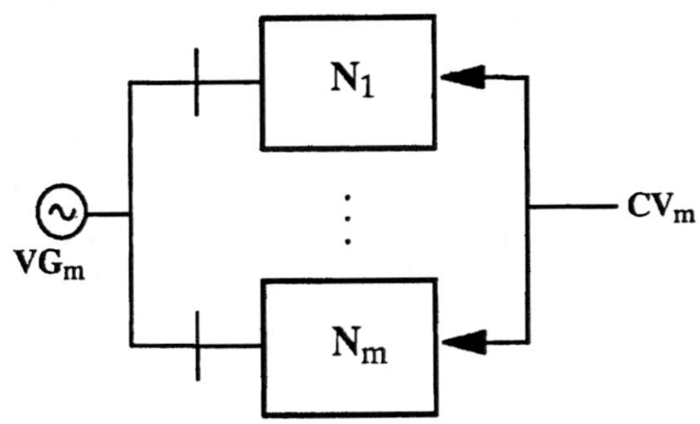

Fig. 2 Optimal solution for the control variables

IV SYSTEMATIC METHOD FOR SYSTEM RESTORATION PLANNING

To help operation engineers find a solution for the control variable problem, Hydro-Québec has developed a software program, REC (standing for Remise En Charge which means load restoration in French), built around an algorithm that eliminates the guesswork involved in conventional methods. The software provides an optimized solution for the control variables that satisfies the operating constraints. The algorithm formulates the problem as a nonlinear optimization problem.

Power flow constraints

At each stage k of the switching scenario, the network equilibrium equations (or power flow equations) can be written as:

$$F_k(X_k, U_k) = 0 \qquad \forall\ k \tag{3}$$

where X_k represents the bus voltage magnitudes and the phase angles in stage k, while U_k represents the control variables in stage k.

The dimension or the values of X_k and U_k vary depending on the element switched on. For each line section switched on, X_k increases by two elements, corresponding to the new voltage magnitude and phase angle, and the value of all other variables changes, since there is now a new network topology and a new set of power flow equations (3). In this case, U_k changes in value but not in dimension. If the switching operation involves the switching of a load or a shunt reactor, the dimension of U_k changes with the addition of a new control variable. In this case, X_k changes in value but not in dimension, since the number of power flow equations remains the same. The example below shows the variations of these variables (Table 2).

Operating constraints

The operating constraints, e.g. voltage and power limits, must be satisfied at each stage k of the restoration scenario. They can be expressed as inequality constraints:

$$G_k(X_k, U_k) \geq 0 \qquad \forall\ k \tag{4}$$

Minimum-adjustments constraint

This constraint states that, once a control variable is introduced, it should remain unchanged for subsequent stages. It can be expressed as follows:

$$U_{jk} = U_{ji} \qquad \forall\ k > i \text{ and } \forall\ j \tag{5}$$

(control variable U_{ji} at bus j is set at stage i and kept unchanged for all subsequent stages k>i).

Objective function

The optimization criterion selected for measuring the performance of the restoration process during the various stages essentially uses bus voltage deviations from the prescribed profile. Since voltage deviations are not of equal importance for different buses at different stages, weighting factors are introduced to strengthen the enforcement where it is most needed, such as at load buses and synchronization buses. It has been found that a quadratic function of the voltage deviations from the desired voltage profile gives satisfactory results. The objective function is therefore:

$$\min \sum_{k=1}^{M} (V_k - I_k) * W_k * (V_k - I_k)^T \tag{6}$$

where:

- M = the total number of restoration stages
- W_k = diag (W_{jk}), j=1, ..., N_i: a diagonal matrix for bus weighting factors at stage k
- V_k = [V_{1k}, V_{2k}, ..., $V_{ni\,k}$]: a vector for bus voltage at stage k
- I_k = [1.0, 1.0, ..., 1.0]: a unit vector (the number of elements is N_k)
- N_k = the number of buses at stage k.

The restoration problem can now be expressed as a nonlinear program:

$$\min \sum_{k=1}^{M} (V_k - I_k) * W_k * (V_k - I_k)^T$$

subject to:

$$F_k(X_k, U_k) = 0 \qquad \forall\ k$$
$$G_k(X_k, U_k) \geq 0 \qquad \forall\ k$$
$$U_{jk} = U_{ji} \qquad \forall\ k > i \text{ and } \forall\ j$$

This algorithm was developed in-house for this problem and integrated into the REC program. However, the solution of the nonlinear program is obtained by calling a commercially available package.

Example

Figure 3 shows a 10-bus network forming part of the Hydro-Québec main grid. This network includes 20 generators at Beauharnois (represented by one generator VG), 20 13.8/120-kV step-up transformers (represented by one transformer, T1), four 120-kV transmission lines, two 120/735-kV transformers (represented as one transformer T2) and two 735-kV transmission lines which connect Châteauguay, Chénier and Duvernay substations. Loads PL1, PL2 and PL3 are supplied at each of these stations through 735/315-kV transformers (T3, T4 and T5).

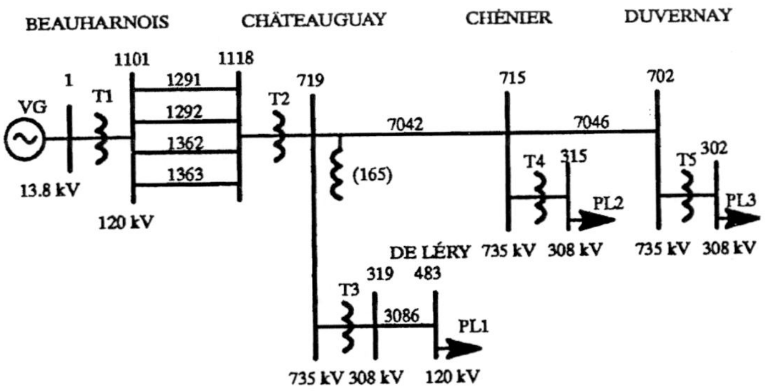

Fig. 3 Example of a Hydro-Québec network

The study calls for the modification of an existing restoration plan to include a 165-Mvar reactor on line 7042. This requires finding values for control variables VG_k, $PL1_k$, $PL2_k$ and $PL3_k$ at each restoration stage k that satisfy the operating constraints. The starting point is at bus 719 and the switching scenario is shown in Table 1. The switching scenario is specified by the user, on the basis of the shortest path available and the number of switching operations.

Table 1 Switching scenario

Stage	Element energized	At bus
1	Transformer T3	719
2	Line 3086	319
3	Load PL1	483
4	Line 7042 with 165-Mvar reactor connected	719
5	Transformer T4	715
6	Load PL2	315
7	Line 7046	715
8	Transformer T5	702
9	Load PL3	302

As mentioned earlier, the network topology, the number of control variables and power-flow equations change at each stage of the switching scenario. This is shown in Table 2.

In the example, there are nine stages and ten network topologies. There are as many as 22 values for the control variables to be determined.

In the proposed optimization method, all power-flow equations are solved simultaneously. This is equivalent to considering the different network topologies produced by the switching operations as one big network. In the example, there are 146 power-flow equations (73 buses) and 22 control variables. One simulation run produces results for all stages of the restoration scenario. By enforcing the minimum adjustments constraint expressed by (5), the solution obtained for control variables is optimal (for example, one value is determined for VG_k instead of 10, and the total number of values for the control variables is 4 instead of 22).

V SOFTWARE DESCRIPTION

The optimization method described above was integrated into the REC software which was designed as a tool for steady-state studies. REC is operated through a graphical user interface (see Fig. 4) designed and programmed with the interface building tool supplied by the manufacturer of the workstations. The interface building tool is a user-friendly graphical environment. It allows users to select interface objects (windows, buttons, scrolling lists, etc.) and to distribute them around the screen. The environment uses a specific language which is automatically translated into C language prior to compilation and linking. The C files generated can then be modified by the programmer to connect interface objects with external modules and data structures. By linking the interface and the external modules, a single executable module is obtained which integrates all operations involved in the preparation of a restoration plan: input/output, definition of the switching scenario, editing network data, editing control-variable constraints and voltage limits, optimization of the scenario, results analysis and report printing.

Table 2 Changes in the control variables and power-flow equations

Stage	No. of power-flow equations	Control variables (U_k)				
		No. of variables	Variations in the values of the control variables			
0	8	1	VG_0			
1	10	1	VG_1			
2	12	1	VG_2			
3	12	2	VG_3	$PL1_3$		
4	14	2	VG_4	$PL1_4$		
5	16	2	VG_5	$PL1_5$		
6	16	3	VG_6	$PL1_6$	$PL2_6$	
7	18	3	VG_7	$PL1_7$	$PL2_7$	
8	20	3	VG_8	$PL1_8$	$PL2_8$	
9	20	4	VG_9	$PL1_9$	$PL2_9$	$PL3_9$
Total	146	22	10	7	4	1

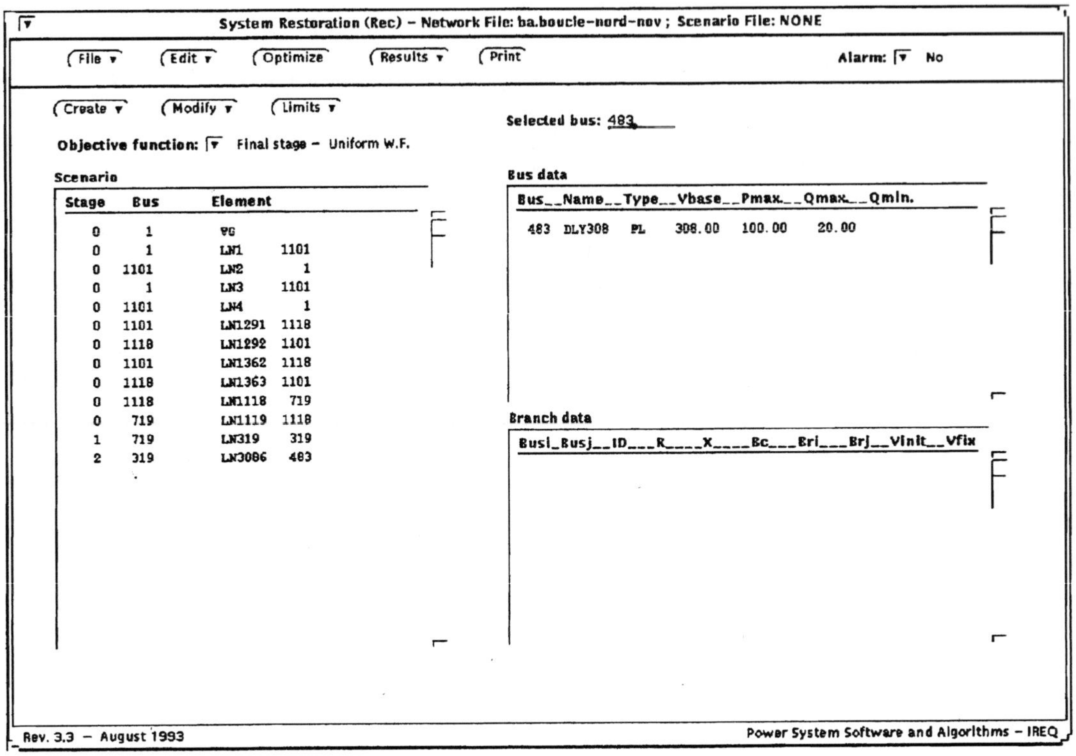

Fig. 4 Editing the switching scenario

The input/output module of REC is compatible with the standard load-flow program used for operations planning, allowing data files used for current simulations to be used for restoration studies, permitting rapid validation of results obtained from REC, and providing data for other types of studies.

The strength of the interface resides in the tools provided for editing the switching scenario and editing the limits for control variables and voltages. The switching scenario is edited by choosing the next element to be connected from two lists of bus and branch elements. A bus element is a load or shunt reactor connected to the bus currently selected. A branch element is a line section or a transformer that can be connected to the bus currently selected. When a branch element is selected, the bus at the other end is automatically selected for the next connection. The switching scenario is displayed in a third list that is automatically updated as stages are added to the current scenario. Most operations are performed by clicking the appropriate mouse button. Limits for the control variables and voltages are edited by selecting the control variable or bus identification in a list and entering modifications in text fields.

Other editing options allow for the modification of bus or branch parameters. In each case, a list is displayed from which the user selects the bus or branch to be modified. All the parameters are then displayed in text fields that can be edited individually.

When editing the scenario, the user can choose from four types of objective functions for the optimization, depending on the weighting factors to be used or on the stages included in the computation of the objective function (all stages or only the last stage). The minimum adjustment constraint can also be relaxed for control variables selected by the user.

Optimization of the scenario is performed by clicking the "Optimize" button (see Fig. 4). Execution of the optimization module is transparent to the user.

Results reports are built automatically. Results include voltage values at all buses for any stage of the scenario, voltage values at one bus for all stages of the scenario, generator output at all stages and control-variable values at all stages. The element or stage for which results are needed can be easily selected in the appropriate list. Information on constraint violations, if any, is also provided for rapid identification of the cause for optimization failure. Results can also be printed.

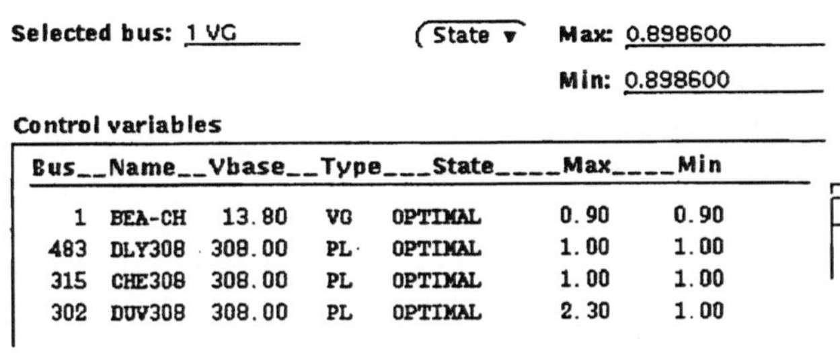

Fig. 5 Editing control variables

VI SAMPLE APPLICATION

The application example was described in section IV. The network includes four control variables with the following constraints:

- generator voltage at Beauharnois: $0.85 \text{ pu} < V_g < 1.0 \text{ pu}$
- load at De Léry: $0 \text{ MW} < PL_1 < 100 \text{ MW}$
- load at Chénier: $0 \text{ MW} < PL_2 < 100 \text{ MW}$
- load at Duvernay: $0 \text{ MW} < PL_3 < 230 \text{ MW}$

Reactive load is described as a ratio of the active load specified by the user and is adjusted automatically. Other voltage limits are set as $0.8 \text{ pu} < V < 1.1 \text{ pu}$.

Using this set of data and constraints, the switching scenario is defined as in Table 1 and a first optimization of that scenario is performed. However optimal, the results obtained at that point cannot be used directly by operators because the precision that can be attained on generator voltages is limited and the load levels that can be set depend on the circuits available and the effective amount of load connected at the moment of shutdown. The values obtained after this first optimization are then fixed by the user of the software to realistic "target values" that take account of an error margin.

A second optimization is performed to make sure that the operating constraints are satisfied. The final results are shown in Fig. 6, 7 and 8.

The example shown may seem simplistic but getting the same results with a simple load-flow program requires far more data manipulation, adjustments and analysis: without REC, the conventional method used before involved getting a valid load flow for the final network (using the usual load-flow program) and proceeding backward to the starting network by removing network elements one at a time, making sure at each step that all operating constraints were satisfied. The complexity of that process grows rapidly with the size of the network under study.

For typical networks, the use of REC to perform a single restoration planning study helps save up to two days of work. Since many equipment unavailabilities are not planned and require restoration plans to be modified rapidly, any time saved is greatly appreciated.

Voltages

#	Bus	Base	V (kV)	Max	Min	V (p.u.)	Max	Min
1	1	13.80	12.40	12.40	12.40	0.8986	0.8986	0.8986
2	1101	120.00	120.79	132.00	96.00	1.0066	1.1000	0.8000
3	1118	120.00	121.49	132.00	96.00	1.0124	1.1000	0.8000
4	719	735.00	709.63	808.50	588.00	0.9655	1.1000	0.8000
5	319	308.00	302.98	338.80	246.40	0.9837	1.1000	0.8000
6	483	308.00	302.83	338.80	246.40	0.9832	1.1000	0.8000
7	715	735.00	716.86	808.50	588.00	0.9753	1.1000	0.8000
8	315	308.00	306.38	338.80	246.40	0.9947	1.1000	0.8000
9	702	735.00	717.42	808.50	588.00	0.9761	1.1000	0.8000
10	302	308.00	306.65	338.80	246.40	0.9956	1.1000	0.8000

Fig. 6 Results of final optimization, voltage values at the end of the switching sequence

Generator output

Stage	ID	Pg	Qg	Pgmax	Qgmax	Qgmin
0		0.000	10.069	640.00	290.00	-290.00
1	LN319	0.000	10.561	640.00	290.00	-290.00
2	LN3086	0.000	8.703	640.00	290.00	-290.00
3	PL	100.214	34.576	640.00	290.00	-290.00
4	LN7042	100.266	-40.248	640.00	290.00	-290.00
5	LN315	100.265	-39.667	640.00	290.00	-290.00
6	PL	200.717	-6.325	640.00	290.00	-290.00
7	LN7046	201.030	-118.674	640.00	290.00	-290.00
8	LN302	201.027	-118.192	640.00	290.00	-290.00
9	PL	301.722	-75.650	640.00	290.00	-290.00

Fig. 7 Results of final optimization, generator output

Control variables

Bus	Name	Vbase	Type	State	Value (pu)
1	BEA-CH	13.80	VG	OPTIMAL	0.90
483	DLY308	308.00	PL	OPTIMAL	1.00
315	CHE308	308.00	PL	OPTIMAL	1.00
302	DUV308	308.00	PL	OPTIMAL	2.30

Fig. 8 Results of final optimization, control variables

VII CONCLUSION

This paper has presented a systematic method for power system restoration planning. This method integrates a powerful optimization algorithm and a user-friendly interface.

By introducing the concept of optimization in system restoration, the process of selecting the control variables can be formulated as a nonlinear program that can be solved by a nonlinear optimization algorithm. This approach removes most of the guesswork found in conventional restoration planning methods and provides a systematic method for finding sets of values for control variables that satisfy operating constraints in the steady state. This method helps operation engineers save considerable time and effort.

The user interface provides a powerful and user-friendly tool for defining the restoration scenarios, and for the display and analysis of the simulation results.

In the future, transformer taps will be added to the list of control variables optimized by REC and the control variables (e.g. loads and shunt reactors) will be discretized automatically (at the present time the model uses continuous variables that do not take account of actual discrete values of equipement available). A path searching algorithm and automatic conversion tools for the harmonic analysis program and the EMTP will be considered. In the long range, links with knowledge-based systems and real-time databases could also be included.

REFERENCES

[1] R.J. Kafka, M.M. Adibi et al., "System Restoration Plan Development for a Metropolitan Electric System," IEEE Transactions on Power Apparatus and System, PAS–100, August 1981, pp. 3703–3711.

[2] M.M. Adibi, et al., "Power System Restoration – A Task Force Report," IEEE Transactions on Power Systems, PWRS–2, No. 2, May 1987, pp. 217–277.

[3] M.M. Adibi et al., "Power System Restoration – The Second Task Force Report," IEEE Transactions on Power Systems, PWRS–2, No. 4, November 1987, pp. 927–933.

[4] R.J Kafka, M.M. Adibi et al., "Role of Interactive and Control Computers in The Development of a System Restoration Plan," IEEE Transactions on Power Apparatus and Systems, PAS–101, January 1982, pp. 43–52.

[5] H. Kodama, H. Suzuki, "Interactive Restoration Control of Electric Power Systems," Report 514–04, CIGRE–IFAC Symposium, Control Applications for Power System Security, Florence, September 26–28, 1983.

[6] T. Sagaguchi, K. Matsumoto, "Development of a Knowledge-Based System for Power System Restoration," IEEE Transactions on Power Apparatus and System, PAS–102, February 1983, pp. 320–329.

[7] E. Modon, B. Heibronn et al., "Mars: an Aid for Network Restoration after a Local Disturbance," PICA'91, Baltimore, Maryland, May 7–10, 1991, pp. 344–349.

[8] K. Matsumoto, T. Sakaguchi, R.A. Kafka, M.M. Adibi, "Knowledge-Based Systems as Operational Aids in Power System Restoration," Proceedings of the IEEE, Vol. 80, No. 5, May 1992, pp. 689–697.

[9] J.A. Huang, F.D. Galiana, G.T. Vuong, "Power System Restoration incorporating interactive Graphics and Optimization," IEEE PICA Conference, Baltimore, Md, May 1991.

[10] R. Nadira, K.A. Loparo, T.E. DyLiacco, "Maximal Load Demand Allocation," IEEE Transactions on Power Systems, Vol. 7, No. 2, May 1992, pp. 528–535.

[11] R. Nadira, T.E. Dy Liacco, K.A. Loparo, "A hierarchical Interactive Approach to Electric Power System Restoration," IEEE Transactions on Power Systems, Vol. 7, No. 3, August 1992, pp. 1123–1131.

[12] G. Morin, "Service Restoration Following a Major Failure on the Hydro-Québec Power System," IEEE Transactions on Power Delivery, PWRD–2, No. 2, April 1987, pp. 454–462.

[13] M.M. Adibi, R.W. Alexander, B. Avramovic, "Overvoltage Control During Restoration," IEEE Transactions on Power Systems, Vol. 7, No. 4, November 1992, pp. 1464–1470.

From Generic Restoration Actions to Specific Restoration Strategies

Lester H. Fink
Life Fellow

ECC, Inc.
Fairfax, Virginia

Kan-Lee Liou
Member

Chen-Ching Liu
Fellow

Department of Electrical Engineering
University of Washington
Seattle, Washington

Abstract—The results presented here reflect work, extending over several years, to identify features underlying restoration processes that are common to almost all utilities and that can serve as guides for preparing and evaluating restoration plans [1], for developing expert system programs that will be widely useful [2], and ultimately for improving the effectiveness of restoration practices. Generality is achieved by identifying constituent elements of the process, namely a "target system," "restoration building blocks (RBBs)," "generic restoration actions (GRAs)," and process constraints, and the factors that distinguish among three clearly demarcated stages. This distillation of the constituent elements of the restoration process provides a basis for development of a comprehensive, knowledge-based restoration expert system appropriate to a broad spectrum of power system utilities.

INTRODUCTION

In reviewing the literature on power system restoration, one conclusion that emerges is that discussions of "restoration strategies" do not lend themselves easily to generalization. "Strategies" that seem straightforward in the context of one company's situation do not conform readily to the situations of most other companies; by reviewing a half-dozen papers it is easy to identify a half-dozen "strategies" that differ in significant aspects. The result has been difficulty in developing guidelines for restoration planning that are meaningful in a general context.

This difficulty became apparent in the course of a project to develop a software tool to support power system restoration efforts (from planning to actual restoration) [3] that would be useful to a wide variety of utilities. This paper reports the results of re-thinking the restoration process in order to resolve this difficulty, and represents a distinct development beyond (but not discontinuity with) earlier views of one of the authors [1].

The other two authors have been working on knowledge-based approaches to power system restoration, especially the generation capability dispatch during restoration [2, 4]. It has been proved that knowledge-based approaches are suitable for restoration applications because the entire restoration process is composed of tasks heavily involving logic reasoning, heuristic search, and/or empirical judgment. However, for some utilities, the first priority goal during the early stage of system restoration is not to maximize the system generation capability. Instead, their initial goal may be energization of a 500 kV ring, security of a nuclear plant, or pickup of critical loads. Therefore, further enhancement of the existing knowledge-based systems is necessary in order to handle the increasing complexity associated with different requirements from different utilities.

In the design and planning areas of Artificial Intelligence (AI), Chandrasekaran proposed that a collection of *generic tasks* can be used to solve a class of design problems [5]. Methods to identify and manipulate these tasks have been developed [6,7]. Motivated by the concept of generic tasks, the authors propose a collection of *generic restoration actions (GRAs)*. The GRAs are the generic tasks that exist in diverse restoration strategies. This paper presents a description of various restoration strategies, a list of GRAs generalized from these strategies, and the use of GRAs to construct RBBs and specific restoration strategies.

94 SM 539-7 PWRS A paper recommended and approved by the IEEE Power System Engineering Committee of the IEEE Power Engineering Society for presentation at the IEEE/PES 1994 Summer Meeting, San Francisco, CA, July 24 - 28, 1994. Manuscript submitted January 4, 1994; made available for printing June 2, 1994.

RESTORATION PROCESS

Review of the literature and discussions with experienced utility operators from a variety of systems reveal that differences in "strategies" are closely related to differences in system characteristics. However, underlying these differences are commonalties in what might be termed more narrowly focused "tactics" that are incorporated into the overall strategies.

Target System

One of the difficulties in attempting to formulate a generally applicable restoration strategy, even for a given company, is the endless multiplicity of postdisturbance initial conditions that might be encountered. Only a small subset of these possibilities can be studied in advance. However, the problem can be reduced significantly if it is viewed as goal-oriented, in terms not of the initial conditions, but of its terminal conditions - i.e. of the postrestoration system that is to be achieved. The postrestoration system that is the *goal* of a particular restoration process is not necessarily identical to the predisturbance system. Some equipment may have been damaged, or may not be available for timely restart; the underlying load demand may have changed

Reprinted from *IEEE Transactions on Power Systems,* Vol. 10, No. 2, pp. 745–751, May 1995.

significantly due to, say, passage of time. Nevertheless, the postrestoration system can be defined with some precision in almost all circumstances, and the steps necessary to achieve that condition can then be considered in logical order.

A *Target System* is that state of the system that is to be achieved by the end of the restoration process (or by some significant earlier stage of the restoration process). The target can be described in terms of units to be in service and loaded, their respective effective capacities, anticipated system load, anticipated power import, export, and/or wheeling levels, and for some systems, a basic (minimum adequate) network. Depending on the status of equipment, this target system may or may not coincide with the predisturbance system.

Given the predisturbance system condition and the postdisturbance status of the system, only a restricted set of postrestoration configurations will be possible, perhaps only one. Definition of the target system is a significant element in restoration planning in that, implicitly, it determines much of the operation that must be carried out during the restoration process.

Restoration Stages

Generally, the restoration process comprises three temporal stages: the first 30 to 60 minutes, the second 3 to 4 hours, and the third extending to completion of the process, perhaps 12 hours. These stages may be termed *Preparation, System Restoration,* and *Load Restoration*. Objectives governing each of these stages differ, as do the circumstances under which they are pursued.

The basic distinction between the first stage and the succeeding stages is that during the first stage time is critical, and many urgent actions must be taken quickly. The basic distinction between the two initial stages and the third is that in initial stages load blocks are control means to maintain stability, whereas in the third stage the restoration of load is the primary objective. This distinction seems to underlie almost all reported strategies.

Stage One: Preparation. In the first stage, the postdisturbance system status is evaluated, the target system for restoration is defined, a strategy for rebuilding the transmission network is selected, and steps are taken to (re)start generation.

In view of the time-critical nature of this stage, it is desirable that, insofar as possible, actions be thoroughly planned for rapid implementation. Given a postdisturbance system status, the operator is in a position, during the time-critical stage of restoration, to define a target system, and, depending on equipment availability, rapidly implement appropriate preplanned elements of the restoration process. Review of system restoration plans of a wide variety of utilities suggests that supporting nuclear units is a stage one task.

Stage Two: System Restoration. During this stage, skeleton transmission paths are energized, islands are resynchronized, and sufficient load is restored to stabilize generation and voltage. Larger, base-load units will be under preparation for restart. The overall goal during this stage is reintegration of the bulk power network, as a means to achieving the goal of load restoration in the next stage.

System Restoration, involving such tasks as energization of higher voltage lines and synchronization of islands, requires either reliable guidelines that have been prepared in advance, or tools for analysis prior to critical switching actions. It is necessary to monitor voltages, VAr balance, stability, voltage and frequency response dynamics, adequacy of unit response capability, fault clearing capability, and to coordinate load-generation trajectories.

Stage Three: Load Restoration. During this final stage, the load itself is restored as fully and rapidly as possible. Once the initial priority objectives have been achieved, and the bulk power system has been reintegrated, restoration of system load now becomes the governing control objective rather than the control means it had been thus far.

Since the overall, and now at this stage the primary, objective of system restoration is to minimize the unserved load, scheduling of load pickup will be based on ability of available generation to assume load (total rate of response capability) rather than on economic dispatch. The effective system response rate, and *pari passu* response reserve, increases with the number of available generators on line that are not against limits. Hence, as the restoration process proceeds load restoration can be effected in increasingly larger steps.

Restoration Constraints

Constraints to which the restoration process is subject may be classified as physical constraints, scheduling constraints, and policy constraints. They are summarized in the following.

Physical Constraints. Most electrical and mechanical constraints are observed directly; other physical constraints are usually observed by way of surrogate time constraints. These constraints appear in the form of minimum or maximum times to restart various types of units, to restore pressure on oil-impregnated underground cables, etc.

The maximum elapsed time for the hot restart of drum-type units is between 30 to 60 minutes, beyond which the units need to be rolled for a cold startup, 2 to 3 hours later. Pumping plants maintain oil pressure in pipe-type cables; the pumping plants need to be re-started within 30 to 45 minutes to maintain the required 12 to 14 atmospheres (about 200 psi) pressure and to prevent generation of vacuum and the release of the dissolved gases, which would result in damage to the cable insulation and pot heads if the cable is re-energized before adequate re-pressurizing. Critical loads such as electrolytic production of aluminum require re-supply of power to the pot lines within 30 to 45 minutes to prevent freezing of the molten aluminum. Substations need stored energy and low voltage AC for heating and preventing SF_6 from liquefying, for cooling fans for transformers, and for operation of relays, breakers, and switches. Many substations rely on the power system to refill the depleting stored energy.

Should an attempt be made to pick up a larger load block than can be accommodated by the response capability of the generators on line, blocks of load already restored that are subject to under-frequency load shedding will be lost. If the frequency swing is severe, one or more generators may trip, precipitating another shut-down. Hence, dynamics of the primary energy sources play a predominant role in the assessment of frequency restoration capability.

Scheduling Constraints. Availability of personnel can impose significant constraints on a utility's ability to restore load quickly. Even with all available personnel mobilized, there will be tasks that will have to wait their turn. An obvious example is within a multi-unit power station, where it is unlikely that enough plant operators will be available to start two or more units simultaneously. Field personnel may be spread thin investigating the status of equipment for which normally transmitted data may be lacking.

Policy Constraints. Policy constraints can be especially utility-specific. Commonly, they will include giving priority to obviously critical loads, such as hospitals and traffic control. Other priorities may be established due to local conditions, e.g., unusual system characteristics, local governmental requirements.

RESTORATION PLANNING

Since the set of possible postdisturbance conditions that might occur is unmanageably large, much of effective planning depends on careful delineation, and analysis, of what conditions will pertain under most conditions. In the preceding sections, three concepts were discussed that provide a framework for such analysis: *Target System*, *Restoration Stages*, and *Restoration Constraints*. In what follows, two concepts are discussed that can shape the planning process: *Restoration Building Blocks* and *Restoration Guidelines*.

Restoration Building Blocks

With regard to the *basic operations* involved in the restoration process, it appears that, common to almost all strategies is the procedure of establishing energized transmission paths to provide cranking power to non-blackstart units. This has been a foundational element of the established Electricité de France (EDF) manual procedure: each non-autonomous (non-blackstart) station has been provided with a set of pre-established "FIDEO sheets" that describe candidate transmission paths and compensation that could be used to provide blackstart power to that unit [8]. Other utilities use very similar concepts adapted to the characteristics of their own systems: Bonneville Power Administration (BPA) "transmission line charging;" Hydro Québec (HQ) "base system restoration procedures" [9]; Ente Nazionale per l'Energia eLettrica (ENEL) "power restoration lines" [10]. Such practices suggest a basis for a *flexible* strategic approach.

Nuclear elements of the restoration process may be identified, which for a given utility are almost invariant, contingent only upon equipment availability. Such elements, referred to as *restoration building blocks,* may be defined, studied, validated, and collected in an RBB library.

A *Restoration Building Block (RBB)* is a minimal configuration of an autonomous stable source of power, together with any necessary associated transmission, adequate to serve as a member of a set of one or more such subsystems that collectively are sufficient to reconstitute the bulk power system.

An RBB must include a source of blackstart power (such as a CT), or an autonomous unit (such as a hydro unit), or an established source of power (such as a thermal unit, protected by low-frequency isolation relaying, that survived the disturbance, or even an energized tie to an external system). If an RBB is dependent on a blackstart source, it may include a non-autonomous boiler-turbine-generator (BTG) unit, which may or may not be adjacent to the source of blackstart power, an electrical path connecting the two, any necessary shunt compensation for voltage control, and any necessary load blocks to meet minimum load requirements of the BTG. On some systems, an RBB may be defined to include a portion of the HV or EHV network such that collectively a set of such RBBs may reconstitute a basic HV or EHV transmission ring or corridor.

An RBB will have associated with it a schedule of switching and other control actions necessary to its establishment, along with estimated times required to effect those actions, and a check list of criteria that must be met for the RBB to be a valid option.

All extensive restorations require the provision of stable sources of generation to re-energize the load that had been lost. The fact that RBBs are minimal configurations required for undertaking this task means that they will be necessary in such cases, and their establishment will constitute sub-processes defining the early stages of the overall restoration process. Accordingly, their definition in the *pre-planning* mode will be a valuable factor in expediting the overall process during its most critical stage.

A system *Island* is a subsystem that may be developed with one or more RBBs as its nucleus, and, in addition to the RBB(s), incorporates re-energized load circuits beyond those required for stable operation of the RBB(s). This distinction between *RBB* and *Island* is necessary in order that a library of unique, relatively invariant RBBs might be established, independently of how they might be used or combined to form islands in particular circumstances. Thus RBBs can be defined in terms of necessary and sufficient elements, together with a defined and validated schedule for its activation, whereas islands can only be defined relative to particular circumstances. An *RBB* may be considered to be the nucleus of an *Island*.

Restoration Guidelines

The development of guidelines in the planning process helps minimize the amount of analysis that must be done on-line during restoration. Reliance on guidelines will entail a more conservative process, but this most likely will be repaid in overall time saving.

Response Reserve. Frequency can exhibit large excursions due to load and generation changes, either planned or unexpected. Generators usually have protective relays or other protection to trip them off line if the frequency exceeds 110 percent or drops below 95 percent. While both frequency limits are important, it is the lower one that causes the greater concern. Frequency decline is due to inadequate system response, following loss of generation or picking up a larger load than can be absorbed by the generators on line.

When a load is restarted or a generator trips, the magnitude of the frequency excursion is determined by the ratio of the amount of the load/generation mismatch to the level of load and generation already restored, and by distribution of generation reserve among prime movers. The importance of the amount of reserve has been recognized. The distribution of reserve which is equally important must also be considered. Distribution is not meant in terms of geographic location of units, but of unit types, since some types of units respond to changes in frequency much more quickly than do other types. For fossil-fuel steam turbine units, boiler response is the determining factor in frequency recovery. Because of the slow dynamics of boilers, system frequency may resume drooping even after governor action has reversed the initial downward trend.

It is desirable to determine an overall relationship between reserve and response, to be used as a guideline for the magnitude of load blocks to be restored. The Pennsylvania-New-Jersey-Maryland Interconnection (PJM) has developed such guidelines in the form of an algorithm for monitoring response ("governor") reserve [11]. *Response Reserve* is the amount of reserve available to preserve system frequency. It consists of two components: (i) reserve on generators available from generator governor action; (ii) system load already restored that is subject to tripping by underfrequency relays.

For illustration, in a more simplified form than is used by PJM, assume that response capability weighting factors have been developed for various types of units. Let these be as follows: steam units, 5%; combustion turbines, 25%; hydro-turbine units, 15%. This means that a steam unit at 60 Hz can respond to a load pick up equal to 5% of its capacity without frequency dropping below, say, 58.5 Hz. Given these assumed capabilities, a system having 300 MW of steam capacity, 400 MW of combustion turbine capacity, and 100 MW of hydro capacity on line, the Response Reserve is calculated as follows: generator response: 5% x 300MW steam = 15; 25% x 400MW CT = 100; 15% x 100MW hydro = 15; for a total of 130MW reserve. This represents the amount of load that could be picked up without violating the low frequency limit.

If security against loss of generation is in question, the contribution of underfrequency load shedding, if any, must be taken into account. Consider that 50 MW of load with under-frequency load shedding has been restored, and assume a discount on under-frequency load shedding, 50% (this discount is a simple way of derating this type of protection due to its coarse nature). The discounted underfrequency load shedding will provide an additional 50% x 50MW = 25MW, yielding a net response reserve of 155MW. However, the response reserve contribution of whatever generator that might be lost would therefore not be available, and must be subtracted from the total; the largest generator should be considered for this purpose. Also, the load protected by under frequency load shedding would be lost.

The response of a steam generator to a sudden change in frequency is affected by characteristics in addition to those of the governor. Various reheat steam cycles or coordination with the boiler controls will affect the frequency transition path before the final frequency defined by the governor speed regulation characteristic is reached. The underfrequency operation of a subsystem results in reduced capacity of auxiliaries and the diminishing supply of energy to fossil fuel boilers and to nuclear units (because of circulation pumps). The overall prime mover dynamics coupled with governor action, auxiliary limitation and underfrequency load shedding will determine whether the system frequency can be restored to normal following the initial decline. These effects need to be taken into account when determining governor reserve guidelines for steam units.

Given reasonably accurate models, response capabilities of all system boiler-turbine-generator units can be simulated at various load levels in order to determine maximum changes in loading that can be accepted without exceeding frequency dip constraints.

Reactive Reserves. Similar to the need for rapidly available real (active) power to limit frequency excursions immediately following restoration of significant blocks of load is the need for VAr absorption capability to limit steady state over-voltages when unloaded high voltage underground or EHV overhead transmission is energized [9,12]. In critical cases, it is necessary to take advantage of all available sources, including shunt compensation, pre-loading, and generator reactive capability. Increasingly, utilities are finding much of the latter capability to have been "bottled up" inadvertently, and that systematic tests to identify and remove limiting conditions can provide significant amounts of additional VAr capability, leading and lagging [13].

Transient Overvoltages. Even generous VAr reserves may not be sufficient to prevent voltage transients. In a comprehensive discussion of risks encountered in restoring an extensive 750kV system, Morin [9] provides a number of guidelines that pertain not only to development of restoration plans, but also may provide valuable guidance during the restoration process. Stated objectives are that steady-state voltages should remain within 0.9 to 1.05 pu, switching overvoltages should be limited to 1.8 pu, and temporary overvoltages (8 cycles) should not exceed 1.5 pu. Results of comprehensive case studies on a given system to determine the best sequences of operations involving particular circuits may be presented in helpful figures for *system specific* on-line guidance.

Synchronization. Early interconnection of islands should provide increased frequency stability and improved security; however, there is a risk that some event could cause the collapse of the entire reconnected subsystem. To assess whether the stage has been reached where advantages outweigh risks, questions such as these, for each subsystem, must be considered and evaluated [11]:

- Are both subsystems ready for synchronization ?
- How stable has frequency been ?
- What is the connected load ?
- How much load is subject to underfrequency tripping, and at what frequencies ?
- What levels of governor and spinning reserve are available ?
- What is the largest hazard or contingency ?
- What is the level of VAr absorption capability ?

- Will the combined system be transiently stable ?

Once a decision to interconnect has been made, additional questions regarding subsequent operation and control must be addressed:

- Where will the interconnection be established?
- When will the interconnection be made, and by whom?
- Will tie flow control be necessary? If so, who will control frequency? Who will control tie-flow? What will be the tie-line schedule?

Final closing of external ties is a milestone in the restoration process. The decisions that are relevant in this case include, but extend beyond, those considered under subsystem reconnection. Additional considerations include:

- What will be the voltage schedule at the tie point?
- Under what conditions will the tie be opened unilaterally?
- What control methods will be used - tie-line/frequency-bias, constant frequency, constant net interchange?

These and other precautions are usually specified in advance by agreements between neighboring systems, and should provide appropriate guidance to the system operator.

EXPERT SYSTEM REALIZATION

Two of the authors developed a knowledge-based system (KBS) for bulk power system restoration [2, 4]. This KBS considers an extensive list of critical constraints and parameters for restoration of the generation and transmission facilities of a power system. Based on these constraints and parameters, a generic goal-driven strategy is developed. Part of the strategy is a procedure to determine the generator starting times, which helps maximize the MW generation capability over the restoration period. This procedure incorporates heuristics related to critical constraints, searches and temporal reasoning. In addition to the internal generation in a power system, this KBS also helps to optimize the use of tie lines from neighboring utilities [4].

Further enhancement is required for this KBS to satisfy all the possible *Targets* during different *Restoration Stages*, considering various *Constraints*. The idea of *Generic Restoration Actions* was inspired by the efforts in the design and planning areas of AI. The main idea is that any restoration *Target* can be achieved with a sequence of restoration actions, each of which is formed by a member from a limited set of GRAs. Therefore, these GRAs can be used to construct the basic *Restoration Building Blocks* while taking advantage of those *Restoration Guidelines*.

The following sections present the design considerations for GRAs, existing restoration strategies for different systems, the identification of a list of GRAs, the relations among GRAs and existing KBS modules, and examples of using GRAs to construct an RBB and a restoration strategy.

Design Considerations

GRAs are identified based on the following design considerations:

1. GRAs are a limited set of actions that operators can take during system restoration.
2. An RBB or a complete restoration plan is a coordinated set of specified GRAs.
3. The modularized framework of our existing KBS is shown in Figure 1. GRAs will be implemented as tools available to the *System Restoration (SR)* module. *SR* can generate sequences of GRAs to achieve system specific RBBs. RBBs and GRAs can be used to build restoration strategies.
4. For restoration planning purposes, the user can try different restoration actions with the help of the KBS.

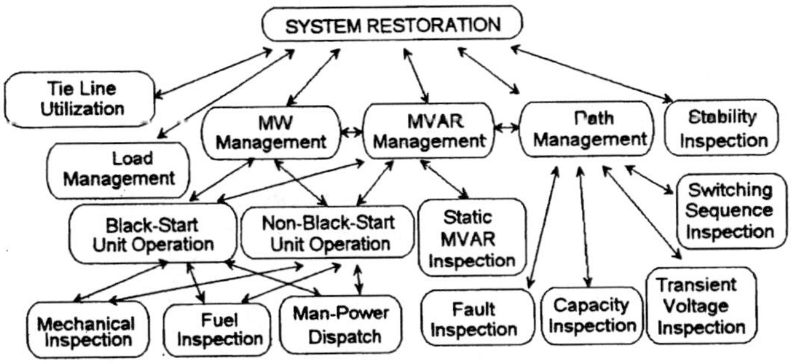

Fig. 1. Existing KBS Modules

Existing Restoration Strategies

Restoration strategies emerge from (i) system specific definition of an RBB library, using GRAs to generate the RBBs, and (ii) assembly of RBBs into a target system using GRAs.

Build-Upward. Many utilities, such as Philadelphia Electric, define islands containing blackstart capabilities, and resynchronize these islands as the restoration process proceeds. The major tasks involved in this restoration strategy are startups of blackstart units, startups of non-blackstart units, restoration of islands, and synchronization of islands.

Build-Downward. Some smaller or closely knit utilities with a lower voltage network can re-energize the network as a means of pooling their blackstart power and providing it to non-blackstart units. The major tasks include startups of blackstart units, energization of the network, and startups of non-blackstart units.

Build-Inward. One of the strategies adopted by Consolidated Edison of New York is to energize its 345 kV corridors from tie lines if tie-line assistance is available. These corridors are built into a central generation station to restart its units, and then from this basic network system restoration will proceed. The major tasks here are energization of 345 kV corridors and startups of non-blackstart units.

Build-Outward. Another strategy adopted by Consolidated Edison is to re-energize a 138 kV ring inside the city if tie-line assistance is not available, and restoration proceeds from this ring outward. The major tasks of this strategy are startups of black-start units, startups of non-blackstart units, and energization of the 138 kV ring.

Build-Together. Hydro Québec is supplied through two long 735 kV corridors from the north. Their strategy is to re-energize these corridors in stages while smaller stations near the load are being black-started. The major tasks here is to energize 735 kV and to start blackstart units to serve local loads.

Serve-Critical. Electricité de France has predefined islands.

Their initial priority is to restore service remotely to their major nuclear units through available corridors from blackstart sources within each island. Pennsylvania-New-Jersey-Maryland has a similar procedure. The major tasks of this strategy include startups of blackstart units and service of critical loads.

Identification of Generic Restoration Actions

Based on the essential tasks for all the strategies discussed in the previous section, a set of GRAs is proposed in this paper. These GRAs can be supported by existing KBS modules. Constraint checking and information request commands are issued by GRAs to the KBS modules.

start_black_start_unit (X). This GRA is to start a specified black-start unit *X* after checking its status by the *Black-Start Unit Operation (BSO)* module, which is supported by the submodules *Fuel Inspection (UI)*, *Mechanical Inspection (MI)*, and *Man-Power Dispatch (MPD)*. The *Stability Inspection (SI)* module is invoked for a scenario which may cause stability problems. The command to execute a GRA is issued by *SR*. For this GRA, the *MW Management (MM)* and *MVAR Management (VM)* modules will be used to monitor the generation capability of unit *X* once it is started.

find_path (X,Y). This GRA is to used to find a transmission path between two generating units, a generating unit and a feeder, or a tie line and a critical load, etc. The path between *X* and *Y* will be searched and its availability and feasibility will be checked by the *Path Management (PM)* module which is supported by the submodules *Fault Inspection (FI)*, *Switching Sequence Inspection (SSI)*, *Capacity Inspection (CI)*, and *Transient Voltage Inspection (TVI)*. It is also necessary to check if the loads connected can be supplied through *MM*, *VM*, and the *Load Management (LM)* module.

energize_line (X). This GRA is to energize a line denoted by *X*. The status and capacity of the line is checked by *PM*. The modules *VM*, *LM* and *Static MVAR Inspection (SMI)* will make sure that the reactive power generated by the line can be absorbed. *MM*, *VM*, and *LM* will check if the loads connected to the line can be supplied.

pick_up_load (X). This GRA is to pick up a specified load *X* as long as its power requirements can be satisfied (*MM, VM, LM*) and the switching actions do not cause any voltage or frequency violation (*SI*).

synchronize(X,Y). This GRA is to synchronize two subsystems, *X* and *Y*, if the prerequisites of synchronization are satisfied. The switching action should not cause any voltage or frequency violation (*SI*), the switching action should not cause excessive transient voltages (*TVI*), and the switching sequence can not violate any regulation (*SSI*).

connect_tie_line (X). This GRA is to connect a tie line *X*. To validate the action, the conditions on tie line connection (e.g., necessary reserves) must be satisfied (*SI*). It is also necessary to plan the use of tie line import (*MM, VM, TM, LM*) or the sources of generation for tie line export (*MM, VM, TM, LM*). Switching actions should not cause any voltage or frequency violation, or excessive voltage transients (*SI, TVI*).

crank_unit (X). This GRA is to crank a specified non-blackstart unit *X*. Conditions to be checked are its status (*Non-Black-Start Unit Operation, NBSO*), its startup prerequisites (such as, energizing a certain line), and its startup requirements (*MM, VM, LM, NBSO*).

energize_busbar (X). This GRA is to energize a specified busbar *X*. The status and capacity (*PM*) of *X* and the loads connected to it must be checked (*MM, VM, LM, SMI*).

The specific restoration strategies mentioned previously have been analyzed to establish the sufficiency of the proposed GRAs. In the analysis, a sequence of GRAs is used to implement each of the six specific restoration strategies.

The following two examples illustrate the construction procedures from GRAs to RBBs, and then to fully extended restoration strategies.

Using GRAs to Construct RBBs

A primary purpose of the proposed GRAs is to construct the RBBs. Note that GRAs are generic but the RBBs may be system-specific. This section uses an example to illustrate how the proposed GRAs can be combined to form various RBBs. Assume an RBB is composed of a blackstart unit, a non-blackstart unit, and a transmission path between them, as shown in Figure 2. RBBs of this type are fundamental in several restoration strategies, such as the build-upward strategy.

The GRAs required to construct this RBB are identified in the following. Note that GRAs here are assumed to be feasible, i.e., all relevant constraints are met.

1. Find a transmission path between BS and NBS:
   find_path (BS,NBS)
2. Start black-start unit BS:
   start_black_start_unit (BS)
3. Build the transmission path from busbar B1 to B3 and pick up loads for stabilization:
   energize_busbar (B1), pick_up_load (C1), energize_line (L1)
   energize_busbar (B2), pick_up_load (C2), energize_line (L2)
   energize_busbar (B3), pick_up_load (C3)
4. Crank non-black-start unit NBS:
   crank_unit (NBS)

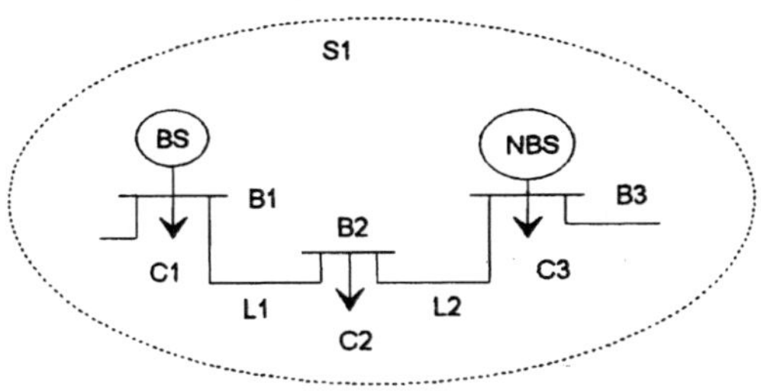

Fig. 2. An RBB

Using RBBs and GRAs to Build Restoration Strategies

The existing restoration knowledge-based system can help optimize and coordinate generator starting times. With the help of RBBs and GRAs, an enhanced expert system can help develop restoration plans that satisfy system-specific requirements. This section uses an example to illustrate the procedure to construct a restoration strategy using GRAs. Assume a build-downward strategy is adopted for a simple power system shown in Figure 3.

To restore this system, an RBB library prepared off-line will be searched to find the applicable RBBs. After the RBBs are built, they will be integrated and a complete restoration plan will be constructed with the assistance of GRAs. The RBBs and GRAs for this restoration strategy are identified in the following construction procedure.

1. Two RBBs are selected:
   RBB1: use unit BS1 to energize bus B1
   RBB2: use unit BS2 to energize bus B2
2. Build RBB1:
   a. Prepare load C1a for unit BS1:
      pick_up_load(C1a)
   b. Start blackstart unit BS1:
      start_black_start_unit (BS1)
   c. Energize bus B1:
      energize_busbar (B1)
3. Build RBB2:
   a. Prepare load C2a for tie line T2:
      pick_up_load(C2a)
   b. Start blackstart unit BS2:
      start_black_start_unit (BS2)
   c. Energize bus B2:
      energize_busbar (B2)
4. Energize the ring:
   energize_line (L1), synchronize (B1,B2), energize_line (L2), energize_busbar (B3), energize_line (L3)
5. restart non-blackstart unit NBS:
   crank_unit (NBS)
6. Pick up loads along the ring:
   pick_up_load (C1b), pick_up_load (C2b), pick_up_load (C3)

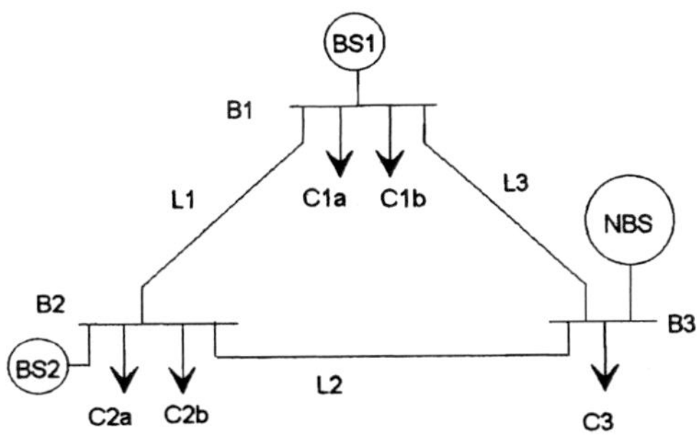

Fig. 3. Build-Downward Strategy

CONCLUSIONS

The diversity of utilities is reflected in the restoration procedures that they require. Utilities are, however, built from the same component elements and operate under the same physical laws. This paper has sought the generic characteristics that must, accordingly, underlie ostensibly diverse restoration strategies. The target system and restoration building block concept, and the categorization of constraints, provide means for system specific customization of basic generic aspects of restoration. Identification of what distinguishes the three stages of restoration, and the concept of generic restoration actions encourage a disciplined approach to restoration planning, and also open up opportunities for development of comprehensive knowledge-based methodologies. Beyond the utilization of GRAs in the development of RBBs, in the preparation stage of restoration and for a variety of current strategies, as presented herein, KBS may be applied during the system restoration stage to the scheduling of generator restarting and of the integration of the RBBs into the target system, and during the load restoration stage to the maximization of loads. The proposed methodology is conceptual; much remains to be done before a practical, generic system restoration software tool can be achieved.

ACKNOWLEDGMENT

The work reported in this paper was partially supported by the Electric Power Research Institute (EPRI) under RP3104-2. The authors affirm their indebtedness and gratitude to M. M. Adibi. The analysis presented herein, although not mutually shared, has grown out of a long association, many extended discussions, and tutelage.

REFERENCES

[1] M.M. Adibi, L.H. Fink, "Power System Restoration Planning;" IEEE/PES Winter Meeting Paper No. 93 WM 204-8-PWRS, 1993.

[2] C.C. Liu, K.L. Liou, *et al.*, "Generation Capability Dispatch for Bulk Power System Restoration - A Knowledge-Based Approach;" IEEE Trans. on Power Systems, Vol. 8, No. 1, Feb. 1993, pp. 316-325.

[3] "Development of Practical Control Center Tools for Power System Restoration," EPRI Project RP3104-2.

[4] K.L. Liou, C.C. Liu, and R.F. Chu, "Tie Line Utilization during Power System Restoration," IEEE/PES Winter Meeting Paper No. 94 WM 221-2-PWRS, 1994.

[5] B. Chandrasekaran, "Toward a Taxonomy of Problem-Solving Types," AI Magazine, Vol. 4, 1983, pp. 9-17.

[6] B. Chandrasekaran, "Generic Tasks in Knowledge-Based Reasoning: High-Level Building Blocks for Expert System," IEEE Expert, Vol. 1, 1986, pp. 23-30.

[7] D. C. Brown and B. Chandrasekaran, *Design Problem Solving*, Morgan Kaufmann Publishers, Inc., 1989.

[8] P. H. Darnault *et al.*, "An Expert System as a Guide for Information Synthesis and Decision Making in the Restoration of Power system After a Blackout," Proc. 2nd Symp. on Expert Systems Application to Power Systems, Seattle, Washington, 1989, pp. 399-404.

[9] G. Morin, "Service Restoration Following a Major Failure on the Hydro-Québec Power System," IEEE Trans. on Power Delivery, Vol. 2, No. 2, April 1987, pp.454-462.

[10] E. Mariani *et al.*, "Field Experience in Reenergization of Electrical Networks from Thermal and Hydro Units," IEEE Trans. on Power Apparatus and Systems, Vol. PAS-103, No. 7, 1984, pp.1707-13.

[11] R. J. Kafka, "Load and Generation Coordination," presented at Seminar on Power System Restoration, April 1992, Arlington, Virginia, and October 1992, Maastricht, Netherlands, ECC, Inc.

[12] IEEE/PES Power System Restoration W. G., "Special Considerations in Power System Restoration - The 2nd W.G. Report," IEEE/PES Winter Meeting Paper No. 93 WM 202-2-PWRS, 1993.

[13] P. B. Johnson *et al.*, "Maximizing the Reactive Capability of AEP Generating Stations," Proc. American Power Conf., Chicago, Illinois, April 1990.

Discussion

R. J. KAFKA, Potomac Electric Power Co., Washington, DC; The authors should be recognized for a significant advancement in formalizing system restoration frameworks. Such a formalization is a necessary step in any significant penetration of knowledge-based or expert system methods into system restoration. Indeed, attempts to arrive at a common set of guidelines for system restoration, or many proposed or existing expert systems for power system restoration have been have been inhibited by the lack of a common method for viewing the problem. The application of Generic Restoration Actions (GRA) and Restoration Building Blocks (RBB) offers the common framework that has been missing.

It must be recognized that while the concept of GRAs and RBBs is a significant step, the unresolved detail is not trivial. An application of the method requires considerable judgement and analysis to define RBB goals or targets and the GRAs of which the RBBs are constituted. Additional analysis of a given GRA is required to assure that constraints are not violated. No implication is made that this is a flaw or even an unreasonable detail yet to be defined. This discusser has gone through a similar process in an essentially manually generated system restoration plan. In an extremely intensive set of power flow simulations and constraint checking activities, the obvious question was whether any of these activities could be sufficiently defined so that they could be assigned to others or even automated. At the time, the question was left unanswered, but it is obvious that repetitive actions can certainly be somehow grouped, defined and packaged for reuse.

The authors implicitly acknowledge this unresolved detail by reference to the expert system (KBS) realization. The KBS is not the main contribution of the paper, but its inclusion invites questions. Figure 1 of the paper shows "existing KBS modules," but the text leaves unresolved the actual state of development of the KBS. Would the authors please comment on the current state of the various modules shown in Figure 1? A future paper or set of papers on the KBS and its application to a real system restoration plan or event would be a noteworthy contribution to the industry.

Manuscript received August 22, 1994.

Kan-Lee Liou (EnergyLine Systems Inc.) and **Chen-Ching Liu** (University of Washington): The authors would like to thank Mr. Kafka for his encouraging comments. His experience in developing restoration plans helps to highlight the need for better software tools for system restoration. We believe that our methodology using GRAs and RBBs provides a foundation for a *generic* restoration operating/ planning/training software tool. We also agree that definition of the specific RBBs and restoration strategies for a utility is a non-trivial task. As a matter of fact, the need for a tool to automate the constraint-checking process was recognized from the beginning of our project. To ensure that each restorative action will meet the relevant constraints, our knowledge-based system (KBS) incorporates many parameter and constraint models, ranging from generator ramping rates to the available loads [2]. However, utility planning engineers still have to specify the values for the constraints, which can be time-consuming. Once the software is coupled to an operator training simulator, it would be possible to generate many parameter values and thresholds automatically.

A software prototype of the proposed KBS illustrated in Figure 1 was developed before the publication of [2]. A simplified model of the Philadelphia electric system was used to test the prototype. Two important enhancements have been completed after the prototyping phase. First of all, the entire Philadelphia electric system was modeled in detail; details of the model include each individual breaker and busbar section at all transmission substations. The KBS software was tested extensively using the detailed power system model. Secondly, the issue of tie line utilization during restoration was addressed through the development of a new KBS module [4]. This new module was integrated into the KBS and tested with the other modules using the Philadelphia system data. The software was transferred to EPRI in 1993 for further development and commercialization.

Finally, we would like to add that research and development of the proposed GRA-RBB methodology is not yet complete. The authors are pursuing a systematic technique to further automate the coordination and constraint-checking procedures involved in the development of specific RBBs and restoration strategies.

Manuscript received November 1, 1994.

ESTIMATING RESTORATION DURATION

M. M. Adibi, LF
IRD Corporation

D. P. Milanicz, SM
Baltimore G and E Co.

Abstract: Restoration duration is typically estimated based on the availability of various prime movers types, sizes and numbers. In these estimates it is assumed that load can be picked up as soon as generation becomes available, and the time required for switching operation to energize transformers, energize lines and pickup loads is much less than the time required to make the generation available.

In this paper a case study is presented, supported by field test, showing that the above assumptions may not necessarily be correct, and restoration duration should be estimated considering both the prime movers timings as well as the switching operations.

The paper recommends that the restoration plans should be reviewed and broken down to its restoration levels, tasks and basic operating actions. Operating actions to include: opening/closing breakers, raising/lowering transformer taps, adjusting voltage and frequency set points, starting auxiliary moters, etc. Then, based on operators' experience the optimistic time, pessimistic time and the most likely time for each action be estimated and assigned. The overall restoration duration then, can be estimated using the Critical Path Method, which in turn may dictate revision of the overall restoration plan.

I. INTRODUCTION

Restoration planning entails development of a strategy which meets the primary goals of minimizing the duration of an outage and maximizing customer service. Such a strategy typically should include several distinct levels of restoration for various system conditions. Each level in turn should consist of a number of interdependent activities. The entire strategy then should be broken down into some logical order to provide a discipline that reduces the risk of overlooking any essential operation [1,2].

PE-353-PWRS-0-12-1998 A paper recommended and approved by the IEEE Power System Operations Committee of the IEEE Power Engineering Society for publication in the IEEE Transactions on Power Systems. Manuscript submitted June 23, 1998; made available for printing December 17, 1998.

The Critical Path Method (CPM) is a technique which can be used in restoration planning, scheduling and evaluation. Its greatest value is due to the fact that precise time estimate does not have to be made for each action. The CPM allows decomposition of the power system into individual restoration activities, specifies the precedence relationships that exists among the activities, estimates the duration of each activity, and schedules the activities to achieve the restoration goals [3,4].

Questions that CPM may address, include:

- What are the estimates of durations of restoration plans?
- Among the activities comprising a specific milestone, which are the ones likely to affect the restoration duration?
- What should be the scheduled starting time and finishing time for each restoration action?
- What are the resource requirements in terms of manpower, equipment and information?
- Because restoration never proceeds exactly as planned, how can the deviations be quickly recognized and corrected?

In this paper a case study is presented, supported by a field test, which by using CPM estimates duration of a cranking operation [5,6].

II. Using Critical Path Method

A major difficulty in the system restoration process is that several actions are involved but certain conditions must be satisfied before some control actions can be taken. In addition, although some actions can be taken simultaneously, others must be performed sequentially. In view of this fact, proper scheduling and coordination between the different actions are required to obtain an effective system restoration plan. The CPM is one technique for applying to such a scheduling problem. The steps involved in using the CPM include:

1. Decompostion of the power systems into individual restoration activities and classification into; fine and coarse activities with appropriate level of details. The creativity and insight required for

Reprinted from *IEEE Transactions on Power Systems,* Vol. 14, No. 4, pp. 1493–1498, August 1999.

decomposing restoration suggests the need for a combined effort by system analyst, plant and system operators.

2. Specification of the precedence relationships that exist among the activities should be established. Usually, some activities cannot begin until others have finished. The restoration planning team should specify the predecessors of each activity. These consist of all those activities that must be completed before the activity under consideration can begin. The list of predecessor activities should be kept as small as possible by excluding all but immediate predecessors and by eliminating redundant and cycling precedence relationships. The successors of each activity are not explicitly specified because this information is implicit in the predecessor list.

3. Estimation of the duration of each activity, i.e., the time that will elapse between its start and finish, must be determined. An activity's duration may be subject to uncontrollable random variation, or subject to controllable variation. The latter variation may depend on the amount of resources in terms of man-hour, material, equipment, and loss of equipment life, allocated to it. An activity duration may be expressed in terms of worst-case, best-case, and most-likely estimates.

4. The planning and scheduling of activities should be developed based on the above information and the use of Critical Path Method (CPM). The planning and scheduling cycle should be repeated until all activities are well coordinated, the peak requirement is kept as low as possible, and the use of resources is as level as possible.

5. Restoration rarely progresses exactly as planned. Once the execution of restoration begins, the plan must undergo frequent revisions as restoration progresses and new information becomes available.

   As the development of restoration plan progresses, CPM should be able to closely monitor the restoration actions, comparing the planned progress to the estimated progress. The monitoring will enable a quick recognition and reaction to unpredictable events that may significantly alter the estimated duration of one or more activities.

6. When an unforeseen event occurs that may significantly alter the estimated duration of one or more activities, CPM should begin to determine the effects of the event. Considering all available information, CPM should revise: the predecessor list, the estimate of every activity duration, and rescheduling activities. At the time of revision there is no uncertainty about the duration of activities that have already finished, and the current information should lead to more accurate estimates of the durations of activities in progress or those not yet started.

III. CPM Methodology

A special requirement of CPM is the treatment of activity durations as a variable, i.e., for each activity three estimation of duration are made: an optimistic time "A", a most likely "M", and a pessimistic time "B". Assuming a beta (ß) distribution of probability of possible duration of activities, the expected time for an activity is approximately

$$t_c = (A + 4*M + B)/6, \text{ and}$$

$$\sigma = (B - A)/6.$$

The expected length of the critical path is

$$T_C = \Sigma t_C \text{ and}$$

$$\sigma_C = (\Sigma\sigma^2)^{1/2},$$

Considering only the longest (critical) path, may lead to an overly optimistic estimate of the probability of completion. The error can be eliminated by averaging many simulated trials of the near-critical paths in the restoration plan.

As an example, restoration activities required for providing cranking power to a drum-type boiler-turbine-generator (BTG) unit by a remote black-start combustion turbine (CT) will be briefly described. The steps involved in a typical CT-BTG restoration include: decomposing the CT-BTG restoration into individual activities with appropriate level of details, specifying the predecessor successor relationships that exists among activities, estimating the time that elapses between the start and finish of each activity (including the time for manual operation), scheduling activities to achieve the CT-BTG restoration goal of 30 to 60 minutes, monitoring progress of activities, and adapting the CT-BTG restoration plan to a new environment when an unforeseen event occurs. The CT-BTG restoration activities include the following steps:

1. Placing the CT under manual control mode of operation.

2. Deactivating dead-bus interlocks and sealed relays at both ends of the path, i.e., at the CT bus and the BTG bus.

3. Adjusting the tap positions on the step-up transformer at the CT end and the standby (step-down) transformer at the BTG end. This action assumes that these transformers are equipped with automatic tap changers.

4. Selecting the CT-BTG path.

5. Selecting loads and their related paths which are needed for stabilization of the CT.

6. Opening all the circuit breakers which are needed in the CT-BTG restoration to disarm all the automatic reclosure relays, including those breakers which connect the path to other parts of the network.

7. Over-riding voltage sensitive switched capacitors on radial feeders which are to be used for load pick ups.

8. Deactivating load shedding under-frequency relays to prevent their undesired and inadvertent operation.

9. Starting the CT, closing the breaker on the dead bus, and energizing the step-up transformer.

10. Energizing the CT-BTG path one section at the time, picking up loads to stabilize the CT and to prevent it from under-excitation.

11. Energizing the BTG standby transformer to supply power to the auxiliary equipment.

12. Picking up BTG auxiliary equipment in a descending order of power requirement, i.e., boiler feed pump, induced draft fan, forced draft fan, condensate pump, circulating water pump, and auxiliary oil pump.

13. Synchronizing the BTG unit to the high voltage bus using the generator stepup transformer, or if the high voltage bus is sectionalized, placing the BTG on the dead bus section.

14. Picking up radial or small loads to reach the BTG's minimum generation capability (about 30% of full load), at which point the BTG can be placed under automatic control, i.e., combustion controller, temperature controller, feed-water controller, etc., and released to the power system operator by the plant operator.

15. Coordinating the rate of load pick up so that the time-load trajectory is enveloped by the minimum and maximum ramp rates of the BTG.

It should be noted that the above steps 13 through 15 are beyond the remote cranking operation.

If for each of the first 12 actions a uniform optimistic time of 2 minutes, most likely time of 3 minutes and pessimistic time of 4 minutes are assumed, then we have:

$$t_c = (2 + 4*3 + 4)/6 = 3 \text{ minutes, and}$$

$$\sigma = (4 - 2)/6 = 1/3 \text{ minute.}$$

The total time for the cranking operation, i.e., activities 1 through 12, is:

$$T_C = \Sigma t_C = 36 \text{ minutes, and}$$

$$\sigma_C = (\Sigma\sigma^2)^{1/2} = \pm\ 1.15 \text{ minutes.}$$

This means that under the above (unrealistic) assumptions, the estimated time for cranking the drum-type BTG with 67% probability is between 34.8 and 37.2 minutes.

In order to estimate a realistic CT-BTG cranking operation, the following factors should be considered:

- The estimated duration for each of the 12 activities should be determined by the power system and power plant operators.
- It is recognized that restoration activities are not performed sequentially but consist of a combination of series and parallel actions.
- The successor activities do not necessarily begin when all the predecessor activities are ended. For example after an incremental load pick up, the CT should be given time to stabilize, or sometimes a BTG should be provided with the soaking time to even out turbine temperature.
- Some activities may begin while the preceding activities have not been completed.
- In some cases when predecessor activities are completed, a specific successor can not begin with certainty, and alternate paths are possible, each with some probability.

The CPM Program will consider all the above factors and estimate the restoration duration by averaging many simulated trials of the restoration plans.

In the initial phase of development, the restoration plan may consist of coarse activities, e.g., each CT-BTG restoration could be considered as one activity with pessimistic, most-likely and optimistic durations. As the development proceeds, activities such as CT-BTG cranking operation can be broken down into finer activities.

IV. Field Test

Figure 1, shows the one-line diagram of the test system for conducting remote cranking operation of the Steam Electric (SE) Generator #2 by the Combustion Turbine (CT) Generator #5. Table I, lists the various actions required by:

> the Bulk Power Operators (BPO), located at the Energy Control Center (ECC),
>
> the Power Plant Operators (PPO), located at the SE and CT sites, and
>
> the System Operator, (SO) out on the electrical system. The SO may be at a number of locations depending on the time of day, problems with various equipment, etc.

The numbers in Table I do not necessarily refer to the sequence of actions, but rather, to the equipment locations. The PPO at the SE Plant conducted activities 16 through 28 and all of this equipment is located within the confines of the SE Plant. Table I, also lists the estimated durations for each action, in A, B & M sequence. It should be noted that the maximum time estimates ignore all possible equipment malfunctions which may render the "time to completion" open ended. In Table II, attempt has been made to indicate the activities which are on the critical path for the cranking operation, these actions do not including the SE #2 startup..

TABLE I - TEST PROCEDURE

1. BPO opens L/S (low side) of OCB on 110-6 Transformer at the CT Plant. This is accomplished remotely (as are all operations of the BPO) from the ECC, (5, 1, 2).
2. BPO opens H/S (high side) motor operated disconnect switch of 110-6 Transformer at the CT Plant (5, 1 2).
3. BPO opens automatic reclosing switch and then opens 110524/110553 Tie OCB at the CT Plant (5, 1, 2).
4. BPO checks circuit for no load and then opens 110524 line air break switch at CT Plant (10, 2, 5).
5. SO opens SE #5 Generator/110-4 Transformer 13 kV Bus OCB's at CT Plant (5, 1, 2).
6. BPO opens 110524 Bus and Tie OCB's and 110532 Bus OCB at Alpha Substation (10, 2, 5).
7. BPO cuts out 115 kV Bus OCB's at Beta Substation, 4 OCB's (10, 2, 5).
8. BPO opens SE #2 Gen/110532 Tie OCB and L/S 110-3 Transformer switch at Beta Substation 5, 1, 2).
9. BPO opens the two OCB on the 230 kV Bus, at SE Plant (5, 1, 2).
10. PPO at CT Plant sets Programmer switches, gives unit a start signal and brings generator up to sync speed/rated voltage. The somewhat lengthy maximum time allows for the fact that all engine pairs may not start on the first command (69, 6, 10).
11. CT Generator #5 Sync OCB's automatically close. Time for this event is included in step above.
12. PPO at CT Plant increases voltage on CT Generator #5 to 13 kV (3, 1, 2).
13. BPO closes H/S disconnect switch on 110-4 Transformer at CT Plant energizing 110524 circuit to Alpha Substation (5, 1, 2).
14. BPO closes 110524/110532 Tie OCB at Alpha Substation energizing 110532 line to Beta Substation (5, 1, 2).
15. BPO operator closes 110532 Bus OCB, energizes Alpha Station #1 115 kV Bus and 230-1 Transformer to SE Plant (5, 1, 2).
16. PPO at SE Plant closes L/S P-13000-2 station service transformer (15, 3, 5).
17. PPO energizes plant lighting bus (5, 1, 3).
18. PPO energizes condensate pump, picking up approximately 1 MW (10, 3, 5).
19. PPO energizes auxiliary plant equipment and an additional condensate pumps (15, 5, 10).
20. PPO energizes make-up water pump (10, 3, 5).
21. PPO energizes several small pieces of plant equipment approximately 0.25 MW (30, 5, 10).
22. PPO energizes #21 boiler feed booster pump, picking up about 1.5 MW (15, 3, 10).
23. PPO energizes #22 boiler feed booster pump, picking up an additional 1.5 MW (15, 3, 10).
24. PPO energizes the start-up boiler feed pump, 3.0 MW additional load (15, 3, 10).
25. PPO starts #21 Induced Draft Fan (15, 3, 10).
26. PPO starts #21 Forced Draft Fan (15, 3, 10).
27. PPO starts #22 Induced Draft Fan (15, 3, 10).
28. PPO starts #22 Forced Draft Fan (15, 3, 10).

The test was terminated at this point, but the estimated time to parallel of SE Generator #2 was an additional 60 minutes. It is unlikely that many of the SE plant activities could be performed in parallel with system activities since these operations cannot be executed until cranking power is available. Activities 18 and 19 however, could be performed simultaneously as could activities 20 and 21 if proper coordination could be maintained in the SE plant control room. This would be necessary in order to prevent

Table II

Activity Number (CPM)	Maximum Time Minute (A)	Minimum Time Minute (B)	Most Likely Minute (M)	Expected Time Minute (t_C)	Standard Deviation Minute (σ)	Variance Minute (v)
1	5	1	2	2.3	0.7	0.4
2	5	1	2	2.3	0.7	0.4
3	5	1	2	2.3	0.7	0.4
4	10	2	5	5.3	1.3	1.8
5	5	1	2	2.3	0.7	0.4
10	60	6	10	17.7	9.0	81.0
12	3	1	2	2.0	0.3	0.1
13	5	1	2	2.3	0.7	0.4
14	5	1	2	2.3	0.7	0.4
15	5	1	2	2.3	0.7	0.4
16	15	3	5	6.3	2.0	4.0

Estimated Duration, Minutes, $T_C = \Sigma t_C = 47.7$

Standard Deviation, Minutes, $\sigma_C = (\Sigma\sigma^2)^{1/2} = 9.5$

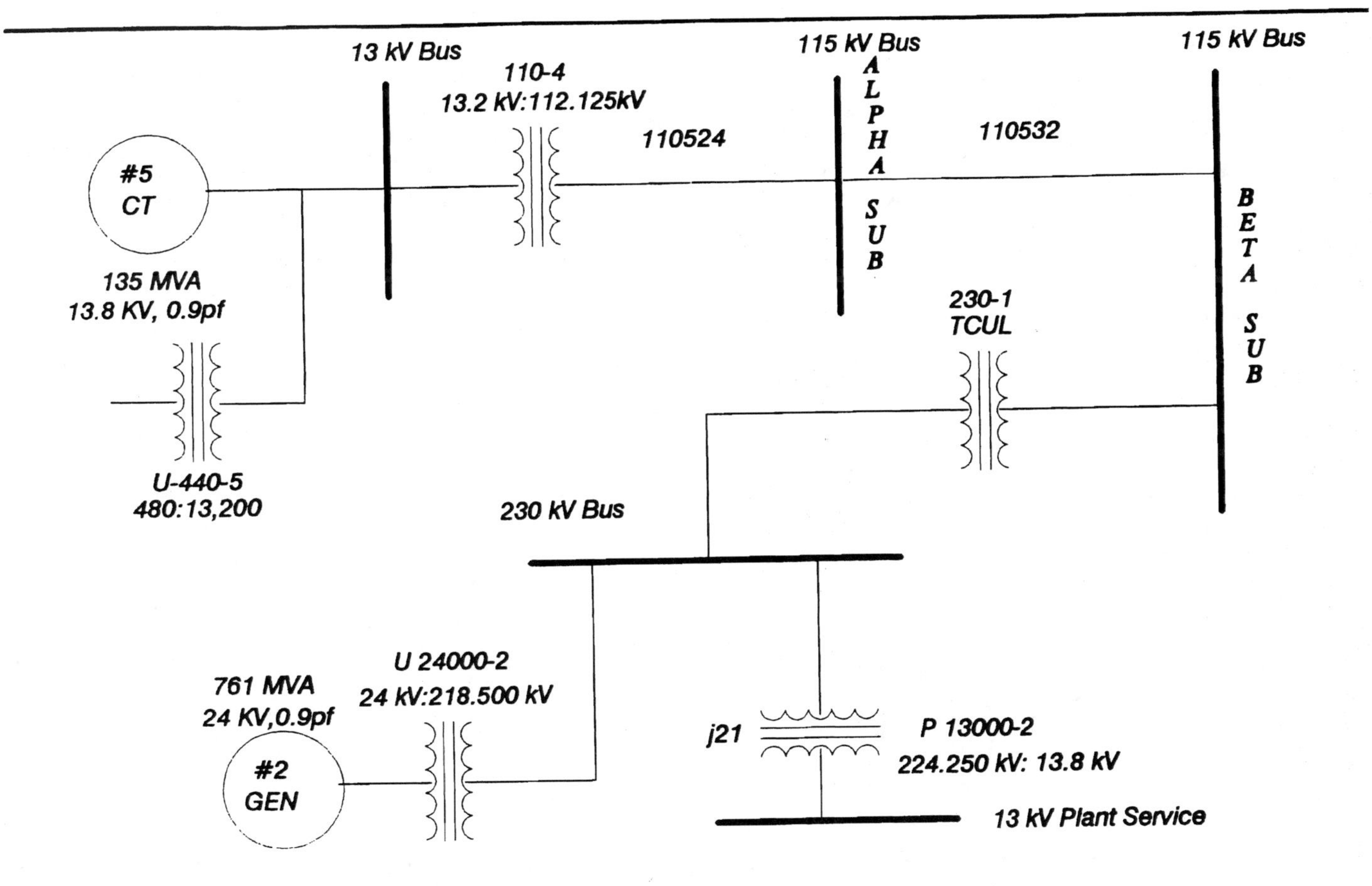

Figure 1. Cranking Steam Electric Unit #2 Using Remote CT #5

two very large pieces of equipment being started simultaneously. Obviously training, including use of the plant simulator, would be required to facilitate these tasks. Activities involving clearing up the transmission path could also possibly be performed in parallel if the initial switching activities at the CT Plant (activities 1 through 5) were handled by a dedicated operator, and activities at Alpha Substation, Beta Station, and SE Switchyard (activities 6 through 9) were handled by a second BPO. As in the plant control room, it would be necessary to maintain close coordination between the operators in the bulk power office in order to prevent details from being overlooked in the switching procedures or actually increasing switching time due to duplication of effort and the ensuing confusion which may result.

The estimated duration for the cranking operation including the startup of SE Generator #2 is $T_C = \Sigma t_C = 200.8$, with the standard deviation of $\sigma_C = (\Sigma\sigma^2)^{1/2} = 12.2$, minutes.

V. Use of CPM in Conducting Drills (or Exercises)

Restoration drills rarely progress exactly as planned. Once the execution of restoration drill begins, the plan must undergo frequent revisions as restoration drill progresses and new information becomes available.

As the restoration drill progresses, the CPM Program should be able to closely monitor the restoration actions, comparing the planned progress to the actual or estimated progress. The monitoring will enable a quick recognition and reaction to unpredictable events that may significantly alter the estimated duration of one or more activities.

When an unforeseen event occurs that may significantly alter the estimated duration of one or more activities, CPM should begin to determine the effects of the event. Considering all available information, CPM should revise: the predecessor list, the estimate of every activity duration, and rescheduling activities. At the time of revision there is no uncertainty about the duration of activities that have already finished, and the current information should lead to more accurate estimates of the durations of activities in progress or those not yet started.

VI. Conclusion

Restoration duration should be estimated with the consideration of both the prime movers timings and activities by the power plant operators as well as the switching operations performed by the electrical system operators.

In a restoration process, some actions can be taken simultaneously while others must be performed sequentially. Proper scheduling of and coordination between the different actions is required to obtain an effective system restoration plan.

The critical path method is a technique which can be used in restoration planning, scheduling and evaluation. Its strength lies in the fact that precise time estimate does not have to be made for each action.

Using the critical path method may dictate revision of the overall restoration strategy. For example, in using a combustion turbine or a hydro unit for the remote cranking of a drum-type boiler-turbine-generator, if the time estimate for the cranking operation is well within 60 minutes, the hot startup of the B-T-G could be planned, if not, the B-T-G should be scheduled for a cold restart which requires an elapse time of 3 to 4 hours after the system collapse.

VII. References

[1] Kafka, R.J., et al "System Restoration Plan Development for a Metropolitan Electric system," IEEE Trans. - PAS - 100, Aug 1981, pp.3703-3713

[2] IEEE Committee Report, "Power System Restoration - The Second Task force Report," T-PWRS, Nov. 1987, pp 927-933

[3] Wiest, J.d., et al, "A Management Guide to PERT/CPM," Englewood Cliffs, N.J. Prentice Hall, 1969

[4] Thronley, G. (editor), "Critical Path Analysis in Practice," Tavistock Publications, London (1968

[5] Levy, F.K., et al "The ABC's of Critical Path Method," Harvard Business Review, Sept-Oct, 1963.

[6] Adibi, M.M. et al, "Remote Cranking of Steam Electric Stations," T-PWRS, Aug 1996, pp. 1613-1618

Acknowledgement

The first author acknowledges the support of Baltimore Gas & Electric Company for providing operational data.

PART IV

Restoration Training

I. General Discussion

THIS group of papers covers a variety of tools used to develop and deliver effective system restoration training programs. The topics addressed include simple methods such as self-study for learning "facts" about system restoration, computer-based instruction for assessing an understanding of basic concepts, role playing and simulation for drills and practices of well-defined scenarios, and use of high-fidelity simulators for training system operators to respond to situations that have not been experienced. Several papers describe restoration guidance systems capable of guiding the operator in making decisions toward minimizing the duration of blackouts, and consequently their impacts. Other papers describe the use of training simulators for testing and verifying the restoration plans. The simulators described include a number of analytical tools and are capable of representing different power systems as well as different control centers.

II. Discussion of Reprints

4.01 Development of a Large-Scale Dispatcher Training Simulator and Training Results, 1986, p. 255

A dispatcher training simulator is described for a large power system with hierarchical control. The power system model includes generator, load, substation controllers, and protective relays. The analytical tools used consist of topology processor, power flow, transient stability and frequency calculation. The dispatching center model includes Supervisory Control and Data Acquisition (SCADA), Energy Management System (EMS), and operator interface functions. The simulator is used for scenario building, training sessions, playback, and evaluation. The playback feature reproduces events that have occurred during a training session. Initially, loads are dispatched among the units through the economic load dispatch function. Voltages are controlled automatically by on-load transformer taps, switched shunt reactors, and capacitors. The load shed due to under-frequency or voltage degradation is automatically restored after frequency or voltage recovery. The hardware configuration, simulator setup, and training experience are also described.

4.02 Dispatcher Training Simulators—Lessons Learned, 1991, p. 264

Five short-notes are presented which address the problems utilities experience in the use of dispatcher training simulators. They report on justification of training simulators, adequacy of power system models, use in restoration drills, staffing criteria, and suitable facilities for training. The paper begins with a description of the design consideration of the EPRI training simulator. This is followed by a focus on the requirements for building and maintaining a real-time database, and on the importance of timely and correct representation of the power system and its components. The third short-note discusses experience with an early version of training simulator in restoration drills. The last two short-notes describe staffing and scheduling for formalized classes, as well as simulation training. The paper concludes that for the training simulator to be effective, it should be consistent with the real-time power system.

4.03 The Use of an Operator Training Simulator for System Restoration, 1991, p. 271

The use of an EPRI training simulator is described for blackstarts of three different restoration scenarios. The restoration procedures, the simulator results and problems, and the adequacy of the analytical tools that are presented. The EPRI simulator is described and consists of power system model, control center model, trainees' consoles, and instructor position. The power system model includes different types of prime movers, network, loads, and protective relays. The control center model includes typical SCADA and EMS functions. The instructor position allows simulator initialization, scenario building, and snapshot reviews. The paper concludes that for the training simulator to be effective, it should include an expert system to guide the operators in restoration of unforeseen events.

4.04 An Advanced Transportable Operator Training Simulator, 1991, p. 278

The design objectives of a transportable training simulator are outlined. Configuration of the simulator are described in terms of power system model, control center model, and instructor position. Modeling technology, instructor capabilities, and training examples are discussed. Implementation and testing of the training simulator at the host utility are reported, and the requirements for its transportability to other utilities are discussed. A special feature of the simulator as described is its heuristic scenario builder which has been applied to voltage collapse, islanding, and blackstart restoration scenarios. The paper concludes that by standardizing the required software and the design techniques, transportability has been achieved.

4.05 Restoration Simulator Prepares Operators for Major Blackouts, 1991, p. 285

A restoration drill is described using a training simulator at an operating company. The training simulator has been used to verify the restoration procedure and then to train the operators to restore part of the system in a simulated environment. The restoration scenarios used are blackstart combustion turbine and hydroelectric units. Several improvements are suggested for the restoration mode of the training simulator.

4.06 Bulk Power System Restoration Training Techniques, 1993, p. 291

The paper presents six short-notes on power system restoration training addressing the restoration issues, performing job task analysis, using conventional methods, applying computer-based systems, supporting restoration by training simulators, and comparing with other industries. The first two short-notes identify the operating conditions to be considered in restoration planning/training programs and analyze how operators should recognize and respond to these conditions. The next three short-notes describe the variety of training methods used at the time, ranging from simple classroom lectures to the use of advanced simulators. The last short-note provides a synopsis of the emergency preparedness in other industries.

4.07 Experiences Using the Dispatcher Training Simulator as a Training Tool, 1993, p. 298

An overview is presented of a training simulator and its use at a large power pool. The paper describes generic configuration of the training simulator and its adaptation for use in the power pool. The simulator includes a number of analytical tools, including cold load pickup and multiple island operation. The dynamic phenomena, such as voltage and frequency transients, are taught via classroom lectures in conjunction with the simulator. The paper reports on a number of simulated restoration exercises in which dispatchers have learned about the frequency responses of prime movers, overvoltage controls, and protective relays. In one case and under an actual emergency condition, operators have been able to use their training and make a more informed decision to save the system. The authors recommend a number of improvements to the training simulator.

4.08 Verification of an Advanced Power System Restoration Support System Using an Operator Training Simulator, 1994, p. 305

The paper describes a restoration support system for subtransmission and local control centers. An operator training simulator is used to test and verify the restoration support system, which consists of a network fault diagnosis and a restoration plan determination. Several criteria for testing and verification of the restoration support model are proposed, including selections of subtransmission systems and support functions. The testing and verification have resulted in improving the relay and circuit breaker tripping models. A special feature of this knowledge-based system is the use of a model-based task shell for diagnosis. The paper concludes that the restoration support system can readily be modified to accommodate changes in the power system due to scheduled maintenance, system expansion, or operating policy changes.

4.09 Evaluating a Restoration Tool Using Consolidated Edison's Training Simulator, 1996, p. 312

A restoration assistant tool is described which enables verification or modification of an existing restoration procedure. The restoration tool reported is an object-oriented knowledge-based software. It is integrated with an operator training simulator to allow restoration plans to be simulated, tested, and verified. Both the restoration tool and the training simulator share the same data and start from a common snapshot of the power system under study. During processing of an existing restoration plan, if certain steps are not feasible they can be modified and used in the training simulator for reevaluation. Based on their field experience, authors believe that the restoration assistant tool can guide the operator through a restoration procedure in a flexible manner. The paper claims that the same tool can be used in the control room environment to capture the system state through a real-time snapshot as a starting case for both the restoration assistant tool and the operator training simulator. This would allow verification of the plan before its implementation. The paper concludes that for a restoration tool to be effective, it should be able to adapt, modify, and replace the procedures if necessary.

4.10 System Restoration Guidelines: How to Set Up, Conduct, and Evaluate a Drill, 1996, p. 319

Periodic drills and exercises are recommended to prepare operators for power system restoration, modify the restoration training program, and improve the restoration plan itself. A basic terminology is provided, together with type of activities involved in a drill. Key participants are identified, along with their role in the drill. The paper describes drill scenarios, their objectives, use of training simulators, and how the drills can be evaluated. An immediate post-exercise evaluation and critiques are recommended, and an evaluation questionnaire and a checklist are provided to help the operating companies design effective drill scenarios. It is concluded that periodic drills can provide a measure of an organization's ability and preparedness to respond to a real power system disturbance.

4.11 Power System Restoration Training Questionnaire Results, 1996, p. 330

The survey results of North American Electric Reliability Council (NERC) utilities is reported by a joint task force to determine the current level of restoration training and additional needs, and to provide a reference. The survey consists of power system profiles, energy management system capabilities, system restoration plans, training programs, training evaluations, training contents, and experience with restoration drills. The paper concludes that most utilities include restoration training as part of their basic training program, use of simulators to help in the training process is growing, operators are receptive to the use of the simulator's advanced capabilities, and restoration training is a valuable module in system training curriculum. The survey also concludes that additional tools are needed for evaluation of the effectiveness of the training programs.

DEVELOPMENT OF A LARGE-SCALE DISPATCHER TRAINING SIMULATOR AND TRAINING RESULTS

I. Susumago M. Suzuki K. Miyama

The Tokyo Electric Power Co., Inc.

Tokyo, Japan

T. Tsuji K. Dan A. Yamanishi

Mitsubishi Electric Corp.

Kobe, Japan

Abstract

The Tokyo Electric Power Co., Inc. (TEPCO) and Mitsubishi Electric Corp. developed a large-scale dispatcher training simulator ELDIS (Electric Load DIspatching Simulator) in February 198 after three years of joint research. This simulator can be used not only for training in normal operating activities, but also for training in the handling of emergency and restorative operations under abnormal conditions. The simulator has been installed at the Dispatcher Training Center in TEPCO where it has been used for four to five days a week for training of about 250 dispatchers in the Central, Area, and Local Load Dispatching Centers. It has contributed greatly to achieving reliable power system operations. This paper gives an overview of the ELDIS and describes the training course, the simulator functions, and the training results.

INTRODUCTION

Power System and Operation Organization

TEPCO is the largest private electric power company in the world, having had a peak demand of 35.7 GW in the summer of 1984. The region, which includes the Tokyo metropolitan area, is served through the 500kV/275kV network which transmits and distributes the electric power generated in nuclear, thermal, and hydroelectric power stations, as shown in Figure 1. The power system is controlled by the hierarchical operation organization, including a Central, three Area, and nineteen Local Load Dispatching Centers, as shown in Figure 2. The Central Load Dispatching Center has the functions of power system operation planning, security monitoring, and automatic generation control (AGC). The Area Load Dispatching Centers monitor and control power flows and bus voltages in the 500kV/275 kV network. The Local Load Dispatching Centers are responsible for the operation of the power system with voltage levels below 154kV, including small power stations, primary transmission substations, distribution substations.

Dispatchers operating a large-scale and highly complicated power system needs to have both a high degree of expertise and various skills to control their power system securely and reliably. However, because of the reduced incidence of breakdowns as a result of recent advances in the reliability of electric power facilities, it is becoming very difficult for dispatchers to gain operating skills under actual operating conditions. A simulator is the most effective means of providing the necessary training opportunities. Training with a simulator allows dispatchers to act with confidence when faced with faults in an actual system. The dispatcher training simulator described here is a large-scale system which provides a realistic power system operating environment, including realistic representation of the power system and exact representation of the dispatching control systems [1].

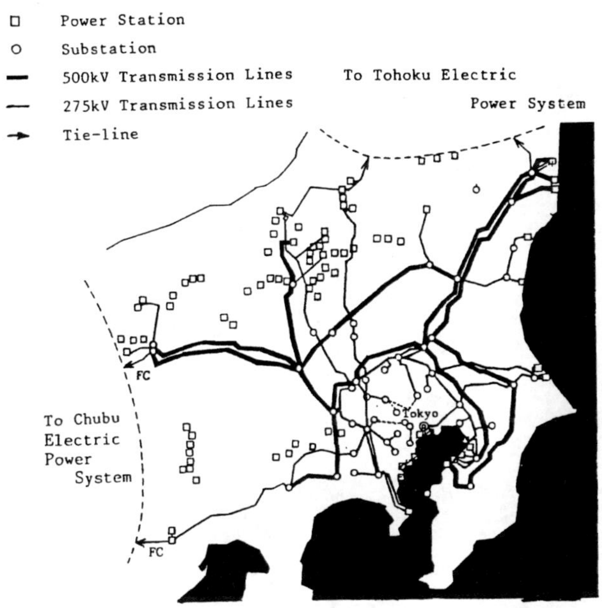

Figure 1 TEPCO's Power System

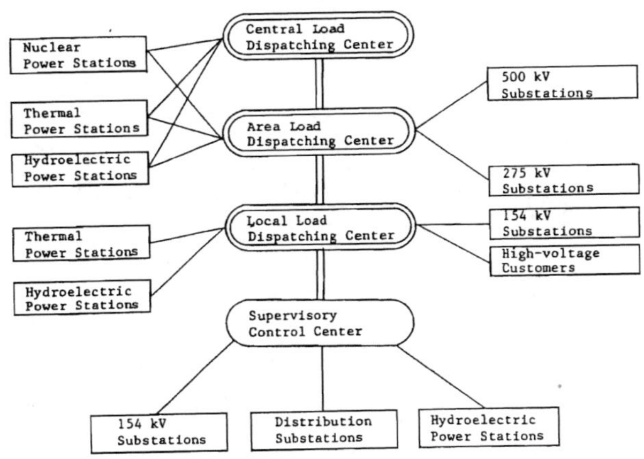

Figure 2 Operation Organization

85 SM 497-3 A paper recommended and approved by the IEEE Power System Engineering Committee of the IEEE Power Engineering Society for presentation at the IEEE/PES 1985 Summer Meeting, Vancouver, B.C., Canada, July 14 - 19, 1985. Manuscript sub-Rmitted September 5, 1984; made available for printing May 17, 1985.

Reprinted from *IEEE Transactions on Power Systems,* Vol. 1, No. 2, pp. 67–75, May 1986.

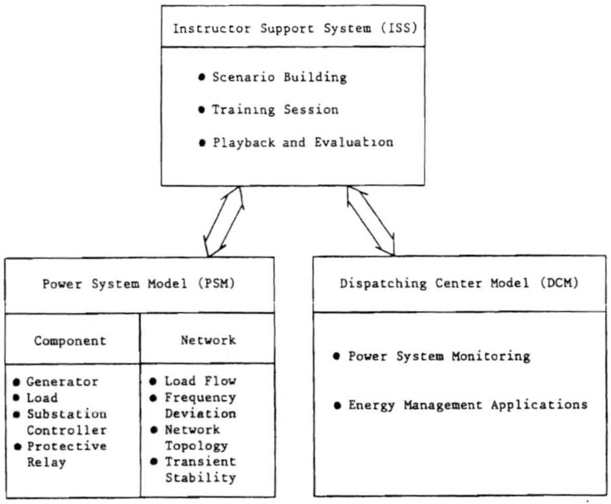

Figure 3 Simulator Function Structure

Basic Simulator Function

The simulator is organized around the functional elements of the Instructor Support System (ISS), the Power System Model (PSM) and the Dispatching Center Model (DCM), as shown in Figure 3.

The ISS consists of instructor support functions of scenario building, training sessions, and playback/evaluation of training sessions.

The PSM realistically represents the dynamic characteristics of the power system, simulating the system frequency deviation and the load flow. It includes the following models :

- Power system component models of generators, loads, substation controllers, protective relays, and so forth
- Power system network models including load flow calculation, frequency deviation calculation, network topology analysis, and transient stability calculation

The DCM is organized around the principal functions of actual dispatching control systems, such as data acquisition, supervisory control, power system monitoring, man-machine interface, and other application programs.

Simulator Plan

The following targets were set for a study of the basic configuration of a simulator system :

- Variations in load flow, voltage, and frequency in the bulk power system to be simulated in real time at three-second intervals, which is about the same sampling period as the data update period in actual dispatching centers.
- The dispatching control functions, the response speed, and the man-machine interface should closely resemble those used in actual dispatching centers, to provide trainees with a realistic power system operating environment. Trainees should be grouped into teams of three as in actual dispatching centers.
- The CRT-based man-machine interface and automatic processing functions should be adequate to allow four instructors to perform a variety of jobs quickly and accurately.

To achieve these goals, the following items had been considered :

- The use of one computer for the PSM which simulates the behavior of the bulk power system at three-second intervals with great accuracy, and the use of an array processor for computation of the load flow.
- For the DCM, the use of the same type of computer and the same software as those in the actual dispatching control system so that three trainees can experience a realistic representation of actual dispatching center work.
- The use of a computer for the ISS since the man-machine interface and automatic processing functions for four instructors require a computing load which exceeds the DCM capabilities.
- Allocation of a power system data base to be shared between the three functions in a shared memory area.

The simulator system is a distributed processing system with each of the three basic functions supported by a computer. This configuration promotes fast processing, simple and compact software structure, and effective use of actual dispatching control software.

We studied whether it was better for this system to be configured independently of the actual dispatching control system or be included in it. We selected the stand-alone system for the following reasons :

- TEPCO has 23 dispatching centers and it is not economically viable to build a separate simulator for each.
- The simulated model power system must be several times larger in scale than the power system which the actual dispatching control system monitors and controls, and therefore the simulated system cannot be run on data input from the actual system only.
- System reliability is crucial to any dispatching control system. But training should be able to proceed regardless of the conditions in the dispatching control system.
- Since there are three trainees in each group, the simulator must be able to independently operate the same types of man-machine interfaces for three trainees as those of the actual control system, even if the computer in the actual control system offers shared processing capabilities.
- Since the Dispatcher Training Center is established independently of the 23 dispatching centers, the stand-alone simulator is the most appropriate for training.

Model Power System

Four model power systems have been prepared, as outlined in Table 1, to train all dispatchers in the Central, Area, and Local Load Dispatching Centers. The trunk power system is the exact model of TEPCO's 500kV/275kV bulk power system, and is used for the training of dispatchers in the Central and the Area Load Dispatching Centers.

Table 1 Model Power Systems

Model Power Systems		Trainee	System Size (Feb., 1984)			
			Generators	Loads	Buses	Branches
Trunk Power System		Dispatchers in Central and Area Load Dispatching Centers	95	140	510	768
Local Power Systems	Urban System	Dispatchers in Local Load Dispatching Centers	0	60	294	386
	Suburban System		10	42	211	332
	Rural System		12	49	259	370

Table 2 Training Courses

Team Training Course		Course for a team of dispatchers to provide training in the restoration from a major system fault.
Personal Training Course	Elementary	Course for elementary dispatchers to teach fundamental network operations under normal conditions and elementary restoration procedures.
	Intermediate	Course for middle-level dispatchers to cover the basics of system restoration procedures ranging from a single line fault to a complicated system fault.
	Advanced	Course for the personnel in the supervisory level to train restoration procedures from an extremely complicated system fault.
License Examination Course	A-grade	Course for the personnel preparing to obtain a supervisor's license.
	B-grade	Course for the personnel preparing to obtain a middle-level operator's license.
General Exercise Course		Course for the personnel other than dispatchers, including operators in power stations or substations, to teach the outline of power system behavior and power system operations.

The local power systems consist of three models, all with voltage levels below 154kV: an Urban System, a Suburban System, and a Rural System. These models are used to train operators in the Local Load Dispatching Centers.

TRAINING COURSE AND TRAINING PROCEDURE

Training Course

Four courses have been prepared, as outlined in Table 2. The Team Training Course provides training for a team of dispatchers not only to increase the expertise of each operator, but also to advance the leadership ability of the supervisory personnel. The Personal Training Course consists of the elementary course, the intermediate course, and the advanced course, and the dispatcher is trained in one of these three courses according to his level of expertise. The License Examination Course is offered at two levels : A-grade for personnel preparing to obtain a supervisor's license, and B-grade for personnel preparing to obtain a middle-level operator's license. The General Exercise Course provides exercise for operators in power stations and substations to learn the outline of power system behavior and power system operations.

Training Procedure

The simulator operates in five different modes : Idle Mode, Base Case Generating Mode, Scenario Building Mode, Training Session Mode, and Playback/Evaluation Mode. The Idle Mode is the mode in which the simulator is on "stand-by" , waiting to be put into one of the other four modes. The Base Case Generating Mode creates the base case data of simulator operating conditions.

The Scenario Building Mode, after copying the base case generated in the previous mode, or selecting a scenario from the library of predefined scenarios, creates the desired scenario by modifying operating conditions and/or fault conditions if the desired initial conditions are not present.

The actual training is carried out in the Training Session Mode, as illustrated in Figure 4. The instructor selects the desired scenario from the scenario library, and then displays the initial power system operating conditions on the mimic-board and the CRT's. After initiating the training simulation, predefined events occur at time intervals relative to the start time of the scenario. The trainee grasps the fault conditions through the mimic board , CRT displays , and telephone inquiries. After mapping out a course of action for the restoration, the trainee uses a telephone to carry out the restoration by dispatching control commands. The instructors execute operations as outside operators would, corresponding to these telephone commands, and, at the same time, provide the training guidance.

The Playback/Evaluation Mode has the capability of reproducing the sequence of events which occur during the training session, including the contents of telephone dialogs. The simulator has the pause/resumption feature for temporarily stopping the playback, the training "switch-over" feature which allows re-entry to the Training Session Mode for re-execution of the training session during the course of playback, and the termination feature which discontinues the playback at any desired point.

INSTRUCTOR SUPPORT SYSTEM

Instructor support functions in the ISS are useful for increasing the efficiency and the effectiveness of the training, and include supports for scenario building, training session, and playback/evaluation.

Scenario Building

The major support functions of load assignment and load flow initialization are described here.

o Load Assignment Support --- This function is effective in the modification of load conditions in the Scenario Building Mode. The load assignment to each bus can be carried out by the following three steps :

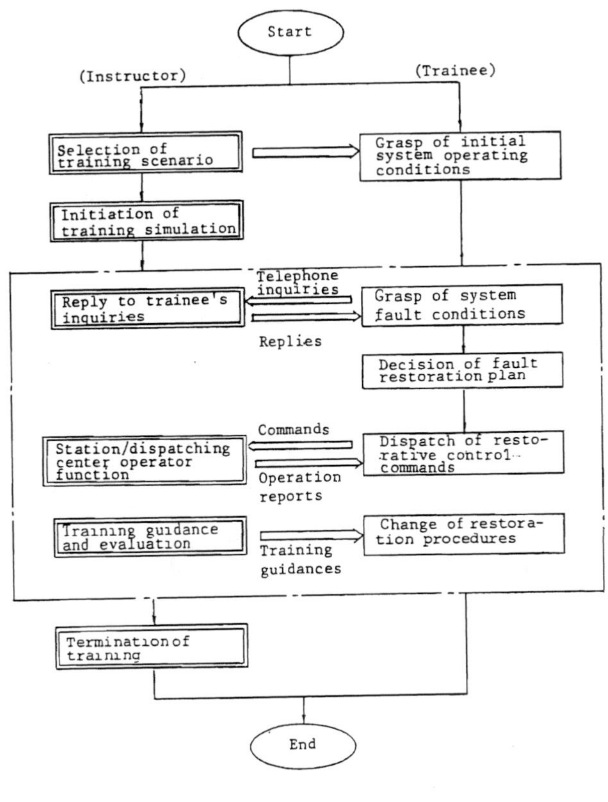

Figure 4 Flow Chart of Training Session

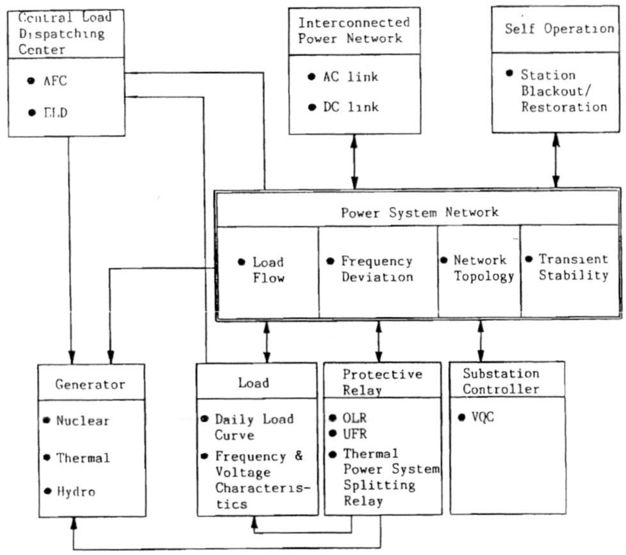

Figure 5 Software Configuration of Power System Simulation

- Selecting a load curve for each bus from normalized daily load curves
- Setting a value corresponding to 100% of the selected load curve
- Setting a start time for the training session and a system peak load

When the system peak load and/or the start time of the training session are changed, the load for each bus is modified automatically.

o Load Flow Initialization Support --- After automatically assigning the load for each bus, the generating power of each generator is also automatically dispatched through the Economic Load Dispatching (ELD) function. As for bus voltages and reactive power flows, the switching in/off of shunt capacitors or reactors and the raising/lowering of Load-Ratio transformer Taps (LRT) are automatically carried out using the voltage Reactive Power Controller (VQC) function.

Training Session

In response to inquiries about fault conditions and control commands from the trainee, the instructor must play the roles of operators in a large number of power stations, substations, neighboring dispatching centers, supervisory control centers. Some of the actions of outside operators are automatically simulated to reduce the burden on the instructor. For example, the switching operations of circuit breakers in case of a substation blackout / restoration are carried out through previously programmed switching sequences without any instructor's intervention, though these operations would be executed by substation operators in an actual power system.

Playback/Evaluation

This function is used to review and analyze the events which occured during the training session. The playback support automatically reproduces either all parts or only selected portions of the training session, including the contents of telephone dialogs. During the playback, it is possible to shift back into the Training Session Mode using the "switch-over" feature.

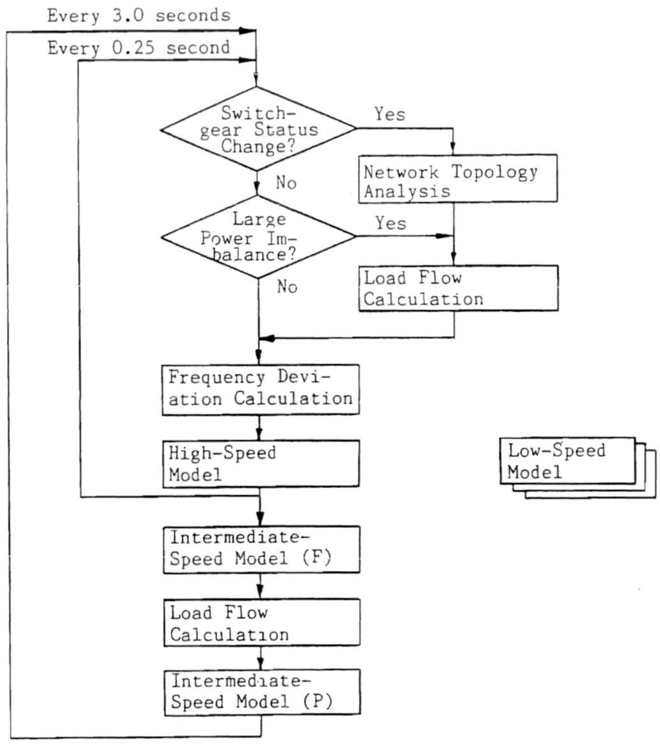

Figure 6 Flow Chart of Power System Simulation

The evaluation support automatically generates records of the following items : interrupted load MW and MWH, number of telephone calls to each station, the start time of system restoration, the time required for restoration, status changes of switchgears and relays, and a telemetering alarm summary.

POWER SYSTEM MODEL

As illustrated in Figure 5, the PSM consists of power system component models of generators, loads, substation controllers, protective relays, and etc., and power system network models including load flow calculation, frequency deviation calculation, network topology analysis, and transient stability calculation. These models are simulated, as shown in the flow chart in Figure 6. The system frequency and the load flow are calculated every 0.25 second and 3.0 seconds, respectively, including the simulation of power system component models. In the event of status changes in switchgears, the network topology analysis and the load flow calculation are executed at once, and in the event of an energy imbalance like load shedding, only the load flow is calculated.

Power System Component Model

o Generator Model --- The generator model is divided into three types : a generating unit model, a power station model, and a scheduled generating model. The generating unit model simulates the detailed characteristics of an individual unit, and is applied to major thermal and nuclear generators. The power station model treats a group of generators in a power station as a single unit, and is applied for hidro-electric power stations. The scheduled generating model is the simplest one having no governor action. The generating power is controlled on the basis of the output pattern over a 24-hour period.

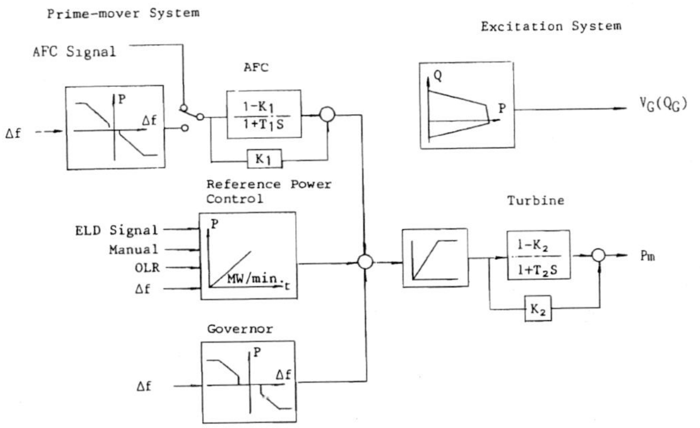

Figure 7 Generator Model

As illustrated in Figure 7, the generating unit model is structured around a prime-mover system and an excitation system. The prime-mover system controls a turbine mechanical power output, and consists of a reference power control block, an Automatic Frequency Control (AFC) block, a governor block, and a turbine block. The reference power control block determines the reference generating power based on the normal speed rate or the emergency speed rate established for each unit. The AFC and the governor blocks regulate the generating power based on the system frequency. The excitation system controls the unit terminal voltage or the unit reactive power according to the operating mode of the unit Automatic Voltage Regulator (AVR).

o Load Model --- All loads are updated every minute according to their daily load curves. The load has voltage and frequency characteristics, and is normally simulated by the combination of a constant power load and a constant impedance load. The load is changed to only the constant impedance characteristics, should the load voltage drop below 85% of the rated voltage. The following conditions are also simulated : load shedding due to frequency drop, automatic load restoration initiated by an instructor, load drop-out during a period of voltage degradation, and automatic load restoration after voltage recovery.

o Interconnected Power Network Model --- Two interconnected power networks are simulated. The network of the Tohoku Electric Power Co., Inc. is simulated by an equivalent generator, having a governor and a turbine, and an equivalent load. It is connected with TEPCO's network through AC lines. The network of the Chubu Electric Power Co., Inc. is connected with TEPCO's network through DC lines using frequency converters (FC) because of the difference in the system frequencies.

o Protective Relay Model --- The following types of relays are simulated : overload relay (OLR), under/overfrequency relay (UFR/OFR), thermal power system splitting relay, undervoltage relay, and loss-of-field relay. An OLR detects an overload in a transmission line or a transformer and sends control commands to power stations to increase/decrease generating power or to trip generating units, and/or also sends control commands to substations to shed loads.

A UFR performs emergency load shedding to insure the supply-demand balance when underfrequency limits are exceeded. A thermal power system splitting relay separates a small 154kV network, including thermal power stations and local loads, from the island, in case of sudden or large frequency deviations. This measure is designed to save the small 154kV network mentioned above in the event of severe disruption.

o Substation Controller Model --- The Voltage Reactive Power Controller (VQC) model controls transformer tap settings, capacitors, and shunt reactors, so that secondary bus voltages and reactive power flows in transformers do not exceed their respective specified ranges.

o Self Operation Model --- This model simulates the self operations that substation or power station operators take, based on their own judgement, in the event of the blackout/restoration.

Power System Network Model

The frequency deviation calculation and the load flow calculation are combined in the following manner. The frequency of generator k is expressed by the following equation :

$$\frac{M_k}{f_K}\dot{f}_k = P_{mk} - P_{ek} \qquad (1)$$

where $f_k = f_0 + \Delta f_k$ and $\Delta f_k \ll f_0$

Thus

$$\frac{M_k}{f_0}\Delta\dot{f}_k = P_{mk} - P_{ek} \qquad (2)$$

Assuming $\Delta f_1 = \Delta f_2 \ldots = \Delta f_{sys}$, equation (2) becomes

$$\frac{M_k}{f_0}\Delta\dot{f}_{sys} = P_{mk} - P_{ek} \qquad (3)$$

Summing up equation (3),

$$\frac{1}{f_0}\cdot M_{sys}\cdot\Delta\dot{f}_{sys} = P_{acc} \qquad (4)$$

where $M_{sys} = \sum_k M_k$ and $P_{acc} = \sum_k P_{mk} - \sum_k P_{ek}$

P_{acc} expresses the accelerating power of the system.

If $\Delta\dot{f}_{sys}$ is eliminated from equations (3) and (4),

$$P_{mk} - P_{ek} - \frac{M_k}{M_{sys}}\cdot P_{acc} = 0 \qquad (5)$$

This equation is the power balance equation under the constraint of equation (6), relating to the reactive power of node k :

$$Q_{sk} - Q_k = 0 \qquad (6)$$

The frequency deviation Δf_{sys} can be calculated by integrating equation (4). P_{acc} can be derived from the load flow calculation using equations (5) and (6).

o Load Flow Calculation --- The load flow is calculated by solving the power balance equations (5) and (6). The imbalance in supply and demand is distributed among the generators in proportion to their inertia constants, as the accelerating power of the system. The solution of the load flow equations is based on the AC method detailed in Refs.[2] and [3]. To enhance the load flow calculation speed, two methods have been adopted.

The first uses voltage phase angles calculated through the DC method for the initial values in the load flow calculation. Using this method, in return for the increased time required for the initial value calculation, a great reduction in the load flow calculation time has been accomplished because of the reduction in the number of iterations required .

The second is the application of an array processor to the load flow calculation. The load flow program for the array processor was developed using the following procedure :

- First, the program was written in FORTRAN, and the correctness of the calculation logic was checked by using a general computer.
- Next, the FORTRAN program was re-written in array processor assembler language. In order to facilitate debugging, the program was written without considering the pipeline architecture.
- Using the array processor debugger, the program was debugged through comparing the results of the program with those of the previously written FORTRAN program.
- Finally, the program was modified for parallel processing considering the pipeline architecture.

Table 3 shows the load flow calculation time for a typical network.

o Frequency Deviation Calculation --- Figure 8 shows the idea of the frequency deviation calculation. The frequency deviation Δf_{sys} is calculated by integrating equation (4) for each island, using the trapezoidal integration rule. P_{acc} is calculated by means of the following method.

The relationship between generating powers and loads is expressed by :

$$\sum_k P_{ek} = \sum_k P_{lk} + P_{loss} \quad (7)$$

Thus,

$$\sum_k (P^z_{ek} + \Delta P_{ek}) = \sum_k (P^z_{lk} + \Delta P_{lk}) + (P^z_{loss} + \Delta P_{loss}) \quad (8)$$

where P^z_{ek}, P^z_{lk}, and P^z_{loss} are the values of P_{ek}, P_{lk}, and P_{loss} in the previous load flow calculation, respectively, and ΔP_{ek}, ΔP_{lk} and ΔP_{loss} are the deviations of P_{ek}, P_{lk}, and P_{loss} from P^z_{ek}, P^z_{lk}, and P^z_{loss}, respectively.

From the previous load flow calculation,

$$\sum_k P^z_{ek} = \sum_k P^z_{lk} + P^z_{loss} \quad (9)$$

Thus, the equation (8) becomes

$$\sum_k \Delta P_{ek} = \sum_k \Delta P_{lk} + \Delta P_{loss} \quad (10)$$

Table 3 Load Flow Calculation Time

(in seconds)

Condition \ Method	Original Method	Method with Initial Phase Angle Calculation	Method with Initial Phase Angle Calculation and an Array Processor
Peak-load Conditions	19.7 (44)	5.9 (7)	0.89 (7)
Middle-load Conditions	8.2 (13)	4.8 (4)	0.83 (4)

Power system consists of 220 nodes and 254 branches.
The calculation accuracy is 1.5MW/1.5MVar in each node mismatch.
( ) shows the number of iterations

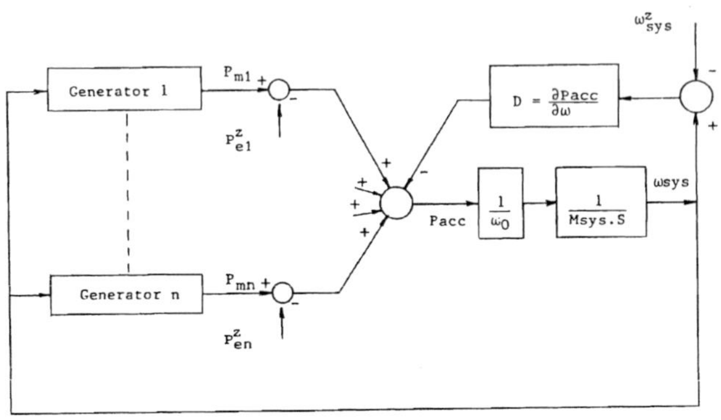

Figure 8 Frequency Deviation Calculation

From $\sum_k P_{lk} >> \Delta P_{loss}$

$$\sum_k \Delta P_{ek} = \sum_k \Delta P_{lk} \quad (11)$$

$\sum_k \Delta P_{lk}$ is expressed by the following equation (12):

$$\sum_k \Delta P_{lk} = \sum_k P^0_{lk} \cdot k_p \cdot \Delta f = D \cdot \Delta f \quad (12)$$

where k_p is the frequency characteristic constant of the load, P^0_{lk} is the active power of load k at the rated frequency, $\Delta f = \Delta f_{sys} - \Delta f^z_{sys}$, and $D = k_p \cdot \sum_k P^0_{lk}$

Thus,

$$\begin{aligned} P_{acc} &= \sum_k P_{mk} - \sum P_{ek} \\ &= \sum_k P_{mk} - (\sum_k P^z_{ek} + \sum_k \Delta P_{ek}) \\ &= \sum_k P_{mk} - (\sum_k P^z_{ek} + D.\Delta f) \end{aligned} \quad (13)$$

o Transient Stability Calculation --- The transient stability calculation faithfully reproduces power system stability conditions after a system fault, including the operations of out-of-step protective relays. This calculation is executed in the Scenario Building Mode, and is used to build a scenario of the fault condition. The calculation results are also displayed on the mimic board to give a realistic representation of the system stability after a fault. The stability calculation uses a generator model with a constant voltage behind xd' and a load model which combines constant power and constant impedance characteristics.

Fault types 1LG, 2LG, 2LS, 3LG, and 3LS can be simulated. The array processor is used for the fast network calculation.

HARDWARE CONFIGURATION

Figures 9 is an block diagram of the hardware system configuration and Figure 10 is an overview of the instructor/trainee room. The simulator is a multi-computer system including three digital computer MELCOM 350-50s, each of which has a main memory of one megabyte and is assigned the ISS, the PSM, or the DCM function, respectively. These three computers are connected to a shared memory of 1.5 megabytes, in order that they may easily communicate with one another. The PSM computer is attached to an array processor, AP-120B, for the fast load flow calculation. Three 300 megabyte disk drives are used for mass data storage.

The instructor/trainee interface includes a rotating type mimic board, four instructor's consoles, three trainee's consoles, two hard copies, two line printers, and several audible alarms. Each console has two CRT display units of a seven color limited graphic type, a light-pen, a joystick, and a group of push buttons. The mimic board consists of nine rotating-type cubicles, each of which has three display panels. Any one of the four model power systems can be selected and displayed by rotating the cubicles.

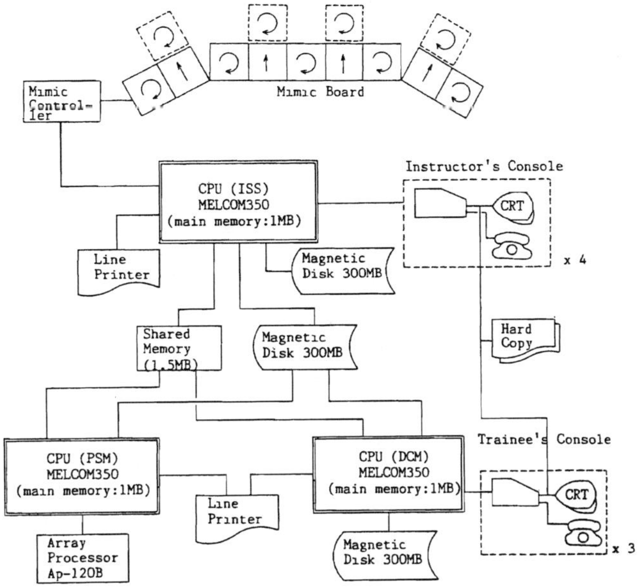

Figure 9 Hardware System Configuration

Figure 10 Overview of Instructor/Trainee Room

TRAINING RESULTS

Training -- mainly team training -- started in April 1984, with yearly training courses. As of the end of August 1984, the total number of training days was 80 and the total number of trainees was about 300. All the dispatching center personnel have attended at least one training course. The trainees could master such the basic dispatcher operating skills as the correct evaluation of fault conditions, the deciding on the best corrective action to restore the system quickly and securely, and the issuing of restorative commands at the proper time.

These training results has pointed out the followings :

- Training is more effective when repeated regularly.
- The larger the scale of the emergency, the greater the chance of oversights in basic operating procedures.
- Correct evaluation of the fault situation at an early stage has a great effect on subsequent restoration.
- The learning effect is increased by including a variety of events in training scenarios.

The above analysis suggests the following points for future training methods :

- Training courses should be run at least three times a year.
- Many training scenarios, including a variety of events, should be prepared.
- On-the-job training should be strengthened using training evaluation results.

CONCLUSIONS

This paper has outlined the dispatcher training simulator developed jointly by TEPCO and Mitsubishi Electric Corp. describing the power system simulation methods and training results. This simulator has the following features :

- The frequency and load flow behaviour of the power system are realistically simulated.
- Four model power systems have been prepared, including the exact model of TEPCO's 500kV/275kV bulk power system.
- Trainees feel they are at work in their actual dispatching control center environment.
- A large number of instructor support functions have been provided.

The electric companies are developing automatic control equipment to operate power systems securely and economically. However emergency and restorative operations, which are difficult to execute completely automatically, will continue to be dependent on the skills and expertise of the dispatchers. Our training results show that training using the simulator will become increasingly important.

NOMENCLATURE

M_k : Unit inertia constant of generator k
f_k : Operating frequency of generator k
$\dot{f}_k$: First differential of f_k
f_0 : Rated frequency
Δf_k : Deviation of f_k from f_0
Δf_{sys} : Deviation of the system frequency from f_0
P_{mk} : Mechanical power of generator k
P_{ek} : Electrical power of generator k
P_{lk} : Active power of load k
P_{loss} : Total transmission loss
Q_{sk} : Specified reactive power at node k
Q_k : Reactive power at node k

REFERENCES

[1] T.E. Dyliacco, M.K. Enns, J.D. Schoettler, et al., "Considerations in Developing and Utilizing Operator Training Simulators", PICA, 1983

[2] S. Iwamoto, Y. Tamura, "A Fast Load Flow Method Retaining Non-Linearity", IEEE, 1978

[3] H. Shiota, Y. Tamenaga, T. Tsuji, K. Dan, "Development of Training Simulator for Power System Operators" , PICA, 1983

Discussion

R. Ramanathan and **H. Ramchandani** (Boeing Computer Services, Energy Control Systems, Tukwila, WA): The authors are to be commended for developing a large-scale dispatcher training simulator. It would be much appreciated to have the following points further clarified by the authors.

1. According to the paper external area is simulated by an equivalent generator and an equivalent load, by this how do the authors justify proper training can be given to the trainee in learning the effect of external area in the function of internal area.
2. What parallel processing algorithm is used for power flow and what modifications where carried out?
3. If this simulator power flow is run every 3 seconds and load is computed only once in a minute. During a minute is the load assumed constant?
4. From Fig. 6 how 0.25 seconds can be achieved if there is a large imbalance in power.
5. What type, and what model have you implemented for nuclear plants?
6. Do you have any backtracking capabilities? That is to restart the DCM and PSM from the playback snapshot/save case snapshot? If yes, please explain.

Manuscript received August 7, 1985.

R. P. Schulte and **W. L. McReynolds** (Bonneville Power Administration, Portland, OR): The authors have accomplished the implementation of a dispatcher training simulator and training program that many utilities can only dream about. Our congratulations.

Could the authors explain the faithfulness of the power system model to simulate actual system responses. For instance, have you recreated actual disturbances? Have you performed before and after checks of actual line or generator outages to a simulation of the same conditions?

The Substation Controller Model appears to perform some functions that may be directed by a dispatcher to a substation operator or SCADA operator. It also is explained that some switching is prearranged in the training session. This reduces instructor workload. Are these two functions the same model?

During a training session, a trainee may take an unexpected course of action. Does the training scenario literally follow unexpected scenario shifts? Do subsequent preprogrammed events get cancelled if they no longer follow the trainee's course of action? Analysis of past wide spread disturbances shows that poor judgment during restoration actually worsened the power system state. Does this simulator and the training session account for these possibilities?

Figure 6 shows a disconnected low speed model. Would the authors explain its relationship to simulation?

BPA has completed an extensive blackout restoration program. Two keys to it's success were stability studies and actual practice in generator black start, paralleling, line charging voltage control, and finally load pickup. The amount of generator voltage bucking and load pickup increments took considerable study and practice. A training simulator would have helped. Is the generator model capable of simulating a similar black start scenario described above?

Could you summarize the staff time required to develop this system? How many people maintain the hardware? The software? What is the size and background of the training staff?

Manuscript received August 8, 1985.

Anjan Bose (Arizona State University, Tempe, AZ): The authors are to be congratulated on developing a stand-alone dispatcher training simulator of high fidelity. The power system, the control system, and the control center environment has not been duplicated in this detail in any one previous training simulator. The instructional capability of scenario building, scenario control, and training session monitoring and review are also significantly enhanced. The following quetions pertain to the power system model which has features specifically developed for their suitability to real time training simulation, and are raised for clarification.

• Generator Model — The excitation system appears to be the usual generator bus model for a power flow with the var limits linearly dependent on the generator power. The dynamics of the excitation system are in the transient range and are normally neglected in the slow dynamic simulation is a training simulator. However, it is known that this fast action of the AVR can affect slow oscillations in the power system but it is not clear how this can be modeled. Was any consideration given to the modeling of excitation system dynamics?

• Load Model — The main difficulty with modeling loads at the high voltage buses is the absence of available empirical data. Load models are often designed for flexibility in case load characteristics become available. Were the load models chosen here influenced by TEPCO load data or were they chosen for convenience? For example, why was a constant current load (linear dependence on voltage) not included? Why was 85% rated voltage chosen as the cut over to all constant impedance load? For frequency characteristics of load, was anything other than system load damping used in the model?

• Interconnected Power Network Model — The Tohoku system is represented very simply as one generator and one load. Does this provide sufficient accuracy in the power flows when a Tokyo system high voltage line is suddenly opened near the Tohoku system?

• Substation Controller Model — Are the VQ controls modeled as usual in the power flow or are their dynamics taken into account?

• Load Flow Calculation — The DC power flow is the same as the first half iteration of the Fast Decoupled power flow. Did the authors try continuing the Fast Decoupled method instead of reverting back to a full Jacobian in rectangular coordinate voltages? In taking advantage of the pipeline structure of the array processor, were the modifications done only to the computer program or also to the solution algorithm itself?

• Frequency Deviation Calculation — The step size used is 0.25 second which is different from the one second size used in many long term dynamic simulation. Was the smaller step size necessary for sufficient accuracy?

• Transient Stability Calculation — The use of a transient stability program to determine the breaker openings after a system fault is meaningful only if the protective relay characteristics are modeled in the program. In addition, the man-machine interface to the program should be such that the scenario building informtion could be easily selected (automatically or easily by the instructor) from the voluminous output. What features does this transient stability program have that are different from the usual program used for off-line studies? If a sequence of events is built from the usual program used for off-line studies? If a sequence of events is built from this transient stability calculation into the middle of a long training scenario, it is possible that certain trainee actions in the early portion of the scenario can make the sequence invalid. Are there any facilities for the instructor to watch out for such conditions and take alternative action?

Manuscript received October 14, 1985.

James G. Waight (Control Data Corp., Minneapolis, MN): The authors are to be congratulated on developing a sophisticated large scale dispatcher training simulator. It is evident from the results that this simulator is the outcome of a major development effort which was planned and executed well. The following questions elicit additional information about the capabilities of the TEPCO simulator, and request classification of why certain design choices were made.

- **System Configuration**
  The authors have designed a configuration of three identical computers (ISS, PSM, DCM) plus peripherals and an array processor. Consoles are attached to the ISS and CCM computer. Separation of the DCM as a separate computer allows replication of the Dispatch Center with high fidelity. Would the authors please comment on why the ISS and PSM models were assigned to two identical computers instead of to a single larger computer.
- **Dispatch Center Model**
  To what extent were the functions of the Central Load Dispatching Center replicated in the DCM. Figure 5 lists the AFC and ELD functions where other application functions such as supervisory control and emergency load shedding are included.
  The organization of TEPCO's operations is hierarchical with central, area and local dispatching centers. The simulator is used for training dispatchers from all three centers. To what extent does the Man-Machine Interface and control functions of the simulator match the central, area and local dispatching centers.
- **Transient Stability**
  The authors have included a transient stability program in the TEPCO simulator. What types of features, Man-Machine Interface and functional interfaces where built into this program to reduce instructor workloads.

- **Playback/Evaluation**
  The TEPCO simulator includes a substantially improved playback/evaluation function. Any further informtion the authors could provide about this feature would be welcome.

Manuscript received November 1, 1985.

I. Susumago, M. Suzuki, K. Miyama, T. Tsuji, K. Dan and **A. Yamanishi**: We are very pleased at the response of so many discussers to our paper and acknowledge their contributions to the subject.

In reply to Prof. Anjan Bose's question

• Generator Model — As the low-frequency oscillation by the fast action of the AVR is negligible in the TEPCO's power system and the main objective of the training simulator is to teach operation skills of system restoration after system faults, the simulation related to the dynamics of the excitation system is considered to be unnecessary.

• Load Model — The load model is determined by combining the constant impedance load and the constant power load, based on actual results in the TEPCO's power system. The constant current load can be simulated by this method. Changing the load characteristics entirely to the constant impedance ones below 85 percent rated voltage is a means of ensuring that the load flow can be calculated even in the severe conditions of the power system, and the training can be continued. The load flow calculation is unlikely to converge below 85 percent rated voltage. For load frequency characteristics, both active power and reactive power are considered.

• Interconnected Power Network Model — The system capacity ratio between the TEPCO and the Tohoku is 5:1, and the TEPCO's system employs flat frequency control and the Tohoku system tie-line bias control. A fault near the Tohoku system is not included in the training items, because it is certain that such a fault does not seriously affect the TEPCO's power system.

• Substation Controller Model — VQC controls substation secondary bus voltages and transformer primary reactive powers within their specified ranges. It integrates the deviations from standard voltage and reactive power values every three seconds, and automatically controls the transformer tap settings, capacitors, or shunt reactors, when the deviations exceed the permitted values.

• Load Flow Calculation — In the training simulator, clearing a severe fault may produce separated systems, the large energy imbalance, and/or unstable power flow conditions. The load flow calculation method using a full Jacobian was adopted to get more reliable solutions under these circumstances. The load flow program was first developed and checked in FORTRAN. Then the program was re-written in array processor assembler language to make best use of the pipeline architecture (add, multiply, data fetch, and data store).

• Frequency Deviation Calculation — TEPCO's UFR is a digital type using microprocessor, with a f/ t function. As the minimum t is specified as 0.5 second, the step size of 0.25 second was adopted for the integration width of the frequency deviation calculation.

• Transient Stability Calculation — The transient stability calculation program includes out-of-step relay models, and is applied only to evaluate a fault at training start. When an instructor inputs a fault case, the program simulates the power system behavior for about two seconds, and determines operated out-of-step relays, tripped CB's, and their trip times. These data are automatically inserted into the training scenario. At the instructor's request, these transient stability data can be displayed on the CRT, and the power system behavior CB/LS conditions, power flows, bus voltages, and frequency can also be displayed on the mimic board at each integration step of the stability calculation.

Regarding the comments and questions from Mr. James G. Waight

• System Configuration — To effectively utilize the operating software and hardware, the same kinds of computers as the DCM computer were adopted for the ISS and the PSM.

• Dispatch Center Model — In the central dispatching center model, the automated computer system functions (power system monitoring, AFC, and ELD) were replicated except demand-supply planning functions. In the area and local dispatching center models, the automated computer functions were replicated exactly. As the man-machine interface, trainee's consoles and telephone consoles can be changed corresponding to each dispatching center model.

• Transient Stability — This point was described in Prof. Anjan Bose's question.

• Playback/Evaluation — All man-machine inputs by instructors and trainees during the training session are stored in the disk, and instructors can make snapshots by designating any points during the training. The playback is performed by executing the same simulation as during the training session from the selected snapshot. The playback, it is possible to shift back into the training session mode by using the "switch-over" feature. Evaluation data collected and stored during the training session can be printed out on the line printer at the instructor's request. There are evaluation data such as interrupted load MW and MWH, number of telephone calls to each station, the start time of system restoration, and the time required for restoration.

In answer to the remarks and questions from Drs. R. P. Shulte and W. L. McReynolds

• Since all system data (P, Q, V and f) are not obtained when a fault occurs accurately, it is very difficult to compare exactly the simulation results of the training simulator with the data during an actual system fault. Therefore, we justified the power system simulation by studying the calculation result of the training simulator in case of past power system faults.

• The Substation controller model performs the VQ control function in the actual power system. The self-operation model performs the self-operations that power station and/or substation operators take independently in the event of blackout/restoration. This model reduces instructors workload very much.

• For the training session, it is necessary to set fault time, fault equipment, operated relays and tripped CB's for the training. Since the power system facilities/equipments are totally modeled, unexpected actions or poor judgment of trainee can be faithfully simualted in the power system.

• In Fig. 6, the low-speed models related to the frequency and power flows/voltages are executed in the positions of the intermediate-speed model (F) and the intermediate-speed model (P), respectively.

• The taining in an extensive blackout restoration is conducted as one of the major training scenarios with great training effect.

• In the TEPCO, the system development period was about four years including the staff time of about fourteen man-years. The training simulator hardware and software are maintained by Mitsubishi Electric, as the manufacturer. The training staff consists of five persons who have experienced as dispatchers in dispatching centers or made operational plans in the head office and branches, and designed automated dispatching computer systems.

In reply to the remarks and questions from Drs. R. Ramanathan and H. Ramchandai:

• The items 1 and 2 were answered in the Prof. Anjan Bose's questions.

• Basic load capacities (P_{LO} and Q_{LO}) are determined as constant values for a minute, and effective loads (P_L and Q_L) are changed every 3 seconds by considering the frequency and bus voltage characteristics.

• When a large power imbalance occurs, the load flow is calculated first by interrupting the frequency deviation calculation. In this load flow calculation, the simulation time is not updated.

• A nuclear plant is modeled by BWR. Since the nuclear plant of an actual system is operated at the constant generating power, the reactor has the simplified model responding the sudden deviation of the system frequency.

• The item 6 was answered in Mr. James G. Waight's question.

Finally, we would like to thank again all the participants for their valuable comments. We hope that the training using the simulator will become increasingly important.

Manuscript received October 14, 1985.

DISPATCHER TRAINING SIMULATORS LESSONS LEARNED

Contributory authors are: J. M. Bucciero, J. A. Dodge, R. N. Gillespie,

R. O. Hinkel, S. L. Mann, S. M. Martin, R. P. Schulte

A report by the Current Operational Problems Working Group — D. Hayward, Chairman

ABSTRACT:

This paper is one of a series presented on behalf of the System Operations Subcommittee with the intent of focusing industry attention on current problems faced by electric utilities in the operation of the bulk power system.

Five short-note reports are presented which direct attention to the problem areas being experienced by utilities in effective utilization of dispatcher training simulators as part of the overall dispatcher training programs. Each report focuses on a different aspect of dispatcher training and utilization of power system simulation in the training curriculum. The first report addresses the justification of a training simulator and focuses on the reasons a utility should incorporate the use of a simulator in the dispatchers' training program. The second report identifies the issues associated with simulation modelling of the real power system versus generic power system modelling. The third report describes the problems experienced with usage of a simulator for power system restoration drills. The fourth report develops a set of criteria for establishing an adequate training staff to support training using a power system simulation tool. The final report explains the challenges in establishing a facility that is suitable for dispatcher training. Each of these reports is descriptive of real experiences and is intended to highlight the issues that must be addressed by the utilities in establishing a dispatcher training program and by the industry in developing the tools needed to effectively train dispatchers using power system simulation techniques. These reports contain representative lessons the industry has learned concerning the use of a dispatcher training simulator.

Keywords - Dispatcher Training Simulator, Operator Training, Training, Simulation, Modelling, Restoration, Staffing, Facilities, Control Center

90 SM 330-1 PWRS A paper recommended and approved by the IEEE Power System Engineering Committee of the IEEE Power Engineering Society for presentation at the IEEE/PES 1990 Summer Meeting, Minneapolis, Minnesota, July 15-19, 1990. Manuscript submitted February 8, 1990: made available for printing July 3, 1990.

INTRODUCTION

Joseph M. Bucciero, Senior Member

ESCA Corporation

The energy control center is strategic to the operation and information management of an electric utility, and well-trained operating personnel are essential to the safe, reliable, and economic operation of the power system. The economic pressures of the electric power marketplace, the constraints of the regulatory agencies, the demands of environmental groups, and heightened customer awareness have made it increasingly more difficult for utilities to meet their service objective while attempting to survive in their radically changing industry. Effective operation of the increasingly complex and tightly-coupled power systems requires extensive training for the dispatchers so that they can react quickly and decisively to routine and emergency conditions.

The application of digital computer technology is prevalent throughout the electric utility industry for monitoring and control of the power system; however, the utilization of computer technology as an integral part of system dispatcher training programs is a more recent phenomenon and, therefore, not as wide-spread in its implementation. Following the 1977 New York black-out, the established training programs in the industry gained a higher degree of visibility, and the concept of utilizing a simulated power system as a dispatcher training tool gained credibility. Traditionally, dispatcher training has consisted of mainly on-the-job instruction and supplementary classroom experience. Few training programs were developed or conducted by education professionals. With today's highly complex power networks, the need for advanced, hands-on training is an essential part of a system dispatcher's training program. A dispatcher training simulator is intended to provide this hands-on, real-time, power system simulation under routine and emergency conditions.

There have been many papers and articles written on the subject of dispatcher training simulators, but little work has been done to report the industry's experience with this training tool. Over the last decade, many dispatcher training simulators have been purchased and installed by electric utility companies, however only a few have been well-integrated into the training program. The reports presented here address some of the practical experiences associated with establishing, maintaining, and investing in a dispatcher training curriculum that contains simulation of the power network. These reports focus on lessons learned by some leading utilities in determining what portions of the power system and the computer-based control system should be simulated, how to apply the simulator to everyday and emergency situations, how large a staff is needed to

Reprinted from *IEEE Transactions on Power Systems,* Vol. 6, No. 2, pp. 594–601, May 1991.

adequately support simulation training, and what type of facility is necessary to establish a realistic training environment.

With sincere appreciation, the Working Group thanks the authors and the Working Group on Operator Training for their time and effort in preparing these reports.

JUSTIFICATION AND REQUIREMENTS

Robert P. Schulte, Member

Consultant

System operator (dispatcher) training simulator development is a problem within a problem. First, the power system operating problems are such that training programs require simulations. Second, the development of training simulators to satisfy dispatcher training for routine and emergency operating problems is in itself a problem area. This short-note presents development and requirement issues that encompass power system operational problems and operator training requirements.

Training Simulator Requirements

Simulators for system operator training became the focus of attention in the aftermath of the 1977 New York black-out. The concern was could a simulator meet system operator training requirements, and if so, how will training needs be met? The focus, however, was on training for emergency conditions. Crisis training on a safe, non-damaging simulator was seen as a necessity. The capabilities to be included were emergency response, islanding, blackstart, and restoration. Later developments added requirements for such capabilities as system stability, voltage collapse, heavy bulk power transfers, resource scarcity, and power transactions.

An EPRI workshop entitled "Power System Operator Training Simulators" was held in New York in September 1978 [1]. Utility representatives, EMS consultants and suppliers, and training personnel from other industries with simulator experience attended. Many issues were discussed, and some of the key issues were:

1. Simulators must be realistic so that what is practiced applies directly to actual operations.
2. With realism, it is believed that the operator will feel the stress of critical situations.
3. Training a crew together or at least simulating the communication between the operational staff is a requirement.
4. Power system size, degrees of freedom, and potential event scenarios make realistic simulation of the power system a very difficult problem.
5. Evaluation of simulator training performance is difficult.

Simulator Development

At the time of the workshop in New York, small generic training simulators were being developed. Even from this early work, it was apparent that simulator training for system operators would be of value. Some EMS manufacturers began to develop and include training simulators in their products.

Power system training simulator development issues were investigated in the EPRI project RP1915-1, "Considerations in Designing and Using Power System Operator Training Simulators" [2]. The major simulator design areas investigated included requirements, uses, configuration, hardware and software guidelines, and instructor position.

EPRI OTS Project

The current EPRI training simulator project, RP1915-4 - "Operator Training Simulator", was undertaken as a result of the EPRI RP1915-1 project report and inquiries from the power industry. The goals of the first phase of this project were to study and specify detailed requirements for a large, effective, realizable operator training simulator suitable for many EPRI members and to prototype major elements of the simulator. The work done in the first phase has been reported in various IEEE/PES papers [3-8].

OTS Project Goals

The EPRI OTS project goals and requirements focused on the following major areas:

1. Realizable: Use readily available hardware and software, including the utility's own hardware and software for the control center model, EPRI software for the power system model and the instructor position, and a full graphics workstation for the instructor position. EPRI designs and requirements for the OTS system, including the power system model, the computer hardware, and the associated data links will be available to its members.
2. Size: Provide the capability to support a major power system model containing 1,000 busses, 2,000 AC transmission lines and transformers, 10 DC lines, and 200 generators including turbine and boiler modelling.
3. Realism: Develop a fast, accurate power system simulation, including high speed topology processor, network solution with dynamic energy source models, five second update time, islanding, and blackstart capability. Provide an exact duplicate of the control center man-machine interface (MMI) and controls, including voice communication facilities.
4. Instructor Position: Develop scenario building tools, including an events scheduler, an expert system based heuristic scenario builder, and transient stability program. Build the instructor's MMI to easily generate, control, monitor, and evaluate training presentations.

OTS Status

The second phase of the EPRI RP1915-4 project is to build, demonstrate, and evaluate the simulator. At this writing, the simulator has been built by Control Data Corporation's Energy Management Systems Division. Subcontractors are Macro Corporation for utility interface work, Seville Division of Flight Safety Training for training requirements, and Arizona State University for model and algorithm development. The EPRI OTS has completed factory tests and does provide the size, realism, and instructor position features described above.

The next step is to demonstrate the OTS in a utility environment. Philadelphia Electric Company (PECO) was chosen as the host utility. The simulator will be utilized by PECO in the training of the system operators, and its effectiveness will be evaluated. The evaluation process will consider the effectiveness of the simulator as a training tool and the effectiveness of simulator training on operator on-the-job performance.

At the conclusion of the evaluation process, two items remain: (1) OTS implementation in more and different utility environments and (2) accumulate simulator training experience throughout the power industry. The OTS software has been made transportable and is readily available for other users. Industry experience in actually implementing this software with other simulators is essential.

As more computer-based simulators are used by utilities as part of their dispatcher training programs, the evolution of training and the types of scenarios and situations used by the utility's training department will be of significant import to the industry at-large.

MODELLING AND DATABASE DEFINITION

Robert O. Hinkel P.E., Member

Pennsylvania Power & Light

This short-note presents a number of Pennsylvania Power and Light (PP&L) Company's experiences with the implementation of a Dispatcher Training Simulator (DTS) for operator training and for system testing associated with the replacement of PP&L's Energy Management System. While PP&L was extremely successful in utilizing the simulator for these purposes, a number of technical and philosophical challenges were encountered in the process and are worthy of sharing with other users. This short-note focuses on the database aspects of simulator setup and maintenance, and discusses areas where improvements can be made to provide better utilization of the DTS as well as better maintainability.

PP&L's Energy Management System is a hierarchical system comprised of 6 Division SCADA Systems (DSS) and a Power Control Center (PCC) connected by a data communications network. PP&L is a part of the Pennsylvania-New Jersey-Maryland (PJM) Interconnection, and data communication links between PJM and PP&L provide data and control signal transfer between the systems.

The models and databases of the PP&L system are developed based on the major application areas of SCADA, Generation, Network Analysis, and Dispatcher Training Simulator. Additional models are provided for Load Forecasting and Loadshed control. Each of the databases developed for real-time control is used in the DTS with additional information provided for such elements as relays, unit dynamics, and training scenarios. With this arrangement, the development of the databases for real-time operation ensures that the DTS environment will present the operator with the same electrical system configuration as the real-time system.

It was PP&L's decision to implement the DTS using the same databases, displays, and power system representation that are used in the real-time control system, rather than implementing a generic model or a subset of the actual system model. It is PP&L's view that the maximum training benefit is gained by allowing the operator to interact with a simulation system that includes the same generation units, transmission representation, displays, and controls as the actual power system. This approach resulted in the need for a number of custom features in the DTS to handle finite limits and boundaries within the simulation; however, the overall value of the simulator for operator training and its long-term maintainability are increased by this approach. The implementation of a generic model or a reduced size model might have resulted in a faster initial implementation, however, the applicability of the DTS to PP&L-specific problems for training cases would have been reduced by this approach. In addition, the on-going maintenance of a separate set of databases, displays, and other parameters for a generic model would be more costly than using the actual system model.

SCADA Database Implications

Since the basic function of the DTS involves periodic calculation of various quantities by means of a Power Flow and transfer of this information to a SCADA processor in lieu of telemetered data, the first database consideration for the DTS is to relate the SCADA and Power Flow databases. This requirement was easily met by PP&L, since the SCADA database needed to be mapped to various network analysis databases for the EMS. Basic rules were established as system databases were constructed to ensure that naming conventions and methods of representing the power system would be compatible with all of the system processes that utilized the data.

Several unique problems were encountered due to the hierarchical structure of the PP&L system. PP&L took the approach that the processing of "raw data" into "information" should be accomplished at the lowest level of the hierarchy. This means that transformer MVA flows, for example, are calculated at a DSS site, if the transformer telemetry is "native" to that site. Any other site utilizing that data will receive the results of the calculation. While this approach has excellent merit from the operational performance perspective, it creates problems for the DTS environment. Since the DTS does not implicitly know the nature of the calculations performed at lower levels of the hierarchy, it cannot reproduce the calculations without additional information. This design has lead to the inclusion of additional calculation functions within the DTS, or, in some cases, the duplication of calculations which are performed at lower levels of the hierarchy. While this problem has occurred at PP&L due to the DSS-PCC hierarchy, it is anticipated that other utilities are or will be experiencing this problem as more functions are distributed to intelligent remotes and front-end processors.

Beyond the basic parallelism of the SCADA system for real-time operation and simulation, the database requirements for SCADA simulation also include a number of additional items that are not required in the real-time environment. These items include relay information, reclosure characteristics, and other miscellaneous control parameters. The biggest problem encountered was the lack of adequate equipment documentation and availability of the data to include in the models.

Generation Modelling

Generation control information for the DTS represents perhaps the single, largest effort required for gathering modelling data beyond the requirements for real-time operation. The data requirements for simulation of unit dynamics go well beyond that required for real-time control and economic dispatch. PP&L took advantage of a number of generic models for various unit types to provide database information for the majority of the modelled units. Specific information for large (greater than 400MW) units on the PP&L system was available and was used in the detailed modelling of these units.

A unique requirement of the PP&L system was the need to provide a simulation of the PJM economic dispatch and unit control algorithms. It was necessary for the DTS to include an overall simulation of the PJM dispatch algorithms so that signals which are normally provided to PP&L from PJM would be developed as a part of the simulation. Meeting these requirements involved the inclusion of specific PJM units in the various DTS databases as well as the inclusion of equivalents for other PJM units and external network units. The simulation uses load forecast information and unit information to perform an economic dispatch for all of PJM's units. This develops the required control parameters for the PP&L portion of the simulation and develops the required unit data for the non-PP&L portion of the simulation. With this data, the simulation can produce realistic generation, load, and interchange patterns for the simulated network.

Tracking Database Changes

Following the initial development of the databases for the DTS, tracking the changes to information in the real-time system and incorporating them in the DTS is a critical consideration. One of the functions of the DTS is to train dispatchers on changes to power system configurations and changes to EMS applications. This means that the simulation must look and feel the same as the real system. Maintenance efforts for the PP&L DTS are greatly reduced based on the approach to model the actual power system. Changes made to a real-time database are implicitly included in the DTS database for the same application. Not all elements of the system, however, can be as easily maintained. Of particular concern is the area of dispatcher maintained information. PP&L has chosen to make many of the specific power system parameters dispatcher enterable. These parameters include line and transformer limits, interface limits, and voltage schedules. When these parameters are altered or updated in the real-time system, no specific mechanism currently exists to provide the automatic update of the parameter in the DTS. This problem is currently being handled by manual procedures; however, a more automated mechanism is needed to ensure that the DTS is always parallel with the real-time system.

Another DTS database maintenance concern is the updating of non-PP&L equipment information such as units, power system elements, and limits. As indicated above, a number of functional areas being simulated are dependent on information associated with non-PP&L equipment. The availability and accuracy of this information as well as the need to perform equivalence operations on the information presents a continuing resource requirement.

In conclusion, the acceptance and utilization of a training simulator is linked closely to the ability to provide timely and correct representations of the power system and its constituent elements. Design decisions in the implementation of the DTS and the real-time database will impact the realism of the simulation as well as the level of support required to maintain the DTS.

RESTORATION DRILL SCENARIOS

J. A. Dodge, Member

Baltimore Gas & Electric

Baltimore Gas & Electric (BG&E) Company has been utilizing a Dispatcher Training Simulator since 1982. This short-note will concentrate on BG&E's usage of its DTS as a training tool for the system operators and as a simulation tool for system restoration drills. It is especially important to note that the version of DTS utilized by BG&E was one of the first prototype simulators developed and does not contain some of the improvements incorporated in later simulators.

Usage of DTS as a Training Tool

The DTS is very useful in providing system operators with training and experience on the basic functionality of the EMS and on basic power flow theory. As a part of an orientation program, new system operators at BG&E are trained on the DTS before they are exposed to the real-time system environment. During this training process, the operators are instructed on the following DTS functions:

1. Supervisory Control (open/close breakers, switches, etc.)
2. Man-Machine Interface (call-up/review of control screens, system summaries, and one-line diagrams)
3. Voltage Control (transformer tap changer & capacitor control)

In addition, system operators are instructed on selected basic event sequences. These event sequences are setup for several different operating conditions, such as transmission faults that recently occurred on the power system, probable fault occurrences, and different maintenance outages. System operators observe and, if necessary, react to these different scenarios, enabling them to learn the general magnitude and direction of power flows across the BG&E system under different operating conditions.

BG&E's use of the DTS is limited by the capacity of the computer equipment, the size of the power system model, and the DTS functional capability. The DTS model has pre-defined size limitations of 100 busses, 185 branches, 150 loads, 65 stations, and 5 islands. As a result, the exact model contained in the on-line EMS cannot be duplicated in the DTS environment. Only a small portion of the power system plus a few selected major bulk power stations can be modelled. Representation of other bulk power stations is greatly reduced or simplified. Distribution substations are not modelled on the DTS.

The DTS does not duplicate the functionality of the on-line EMS. Many of the features of the on-line system are either partially represented or not included in the DTS. Ideally, the DTS should include the capability to simulate and duplicate exactly the functions of the on-line system including: alarm processing, man-machine interface, manual load shedding, voltage reduction, automatic generation control, load management, state estimation, contingency analysis, and network study programs.

In summary, the limitations of the DTS power system model capacity and program functionality have restricted BG&E's use of the DTS to instructing system operators on the elementary features of the EMS and reviewing basic power flow principles.

Usage of DTS as a System Restoration Simulation Tool

BG&E is a member of the Pennsylvania-New Jersey-Maryland (PJM) Interconnection. As part of its system operator training program, PJM conducts semi-annual System Restoration Drills. During these drills, PJM and each of the member companies assume there has been a total system black-out, and then, proceed to simulate the restoration of the power pool (assuming that there are no outside sources of power available).

As a part of BG&E's participation in the drills, the DTS is utilized by engineering/support personnel to simulate the actions of the system operators and to provide them with feedback on their actions. BG&E has been utilizing its DTS, with limited success, to perform this function since 1986.

Due to the limited modelling capabilities of the DTS, it is too awkward for the system operators to directly use the DTS to simulate the system restoration process during the drills. To address this problem, a complete reproduction of all of the on-line EMS substation one-line diagrams was established through the On-Line Display Generator (ODG). Please refer to Figure 1. This set of displays is separate from the displays used in conjunction with the power flow solution program. On each of the duplicate one-line diagrams, all of the on-line breakers and switches are represented. The breaker and switch statuses along with the line and bus voltage values are not linked to the DTS powerflow model, but are linked to a separate ODG file. The operators are able to simulate the opening and closing of breakers and switches through the use of ODG commands (as opposed to the "open" and "close" commands available through the EMS man-machine interface).

During the System Restoration Drills, the system operators "operate" from the duplicate displays: stripping busses, black-starting generators, forming islands, paralleling islands, and re-connecting the power system. As the system operators perform their switching actions, engineering/support personnel review the system operators' switching actions, decipher the general direction of their restoration strategy, attempt to simulate the restoration process on the restricted DTS powerflow model, and provide the system operators with feedback. The voltages from the DTS powerflow solution are manually transferred to the separate ODG file and displayed to the system operators for review and appropriate corrective action.

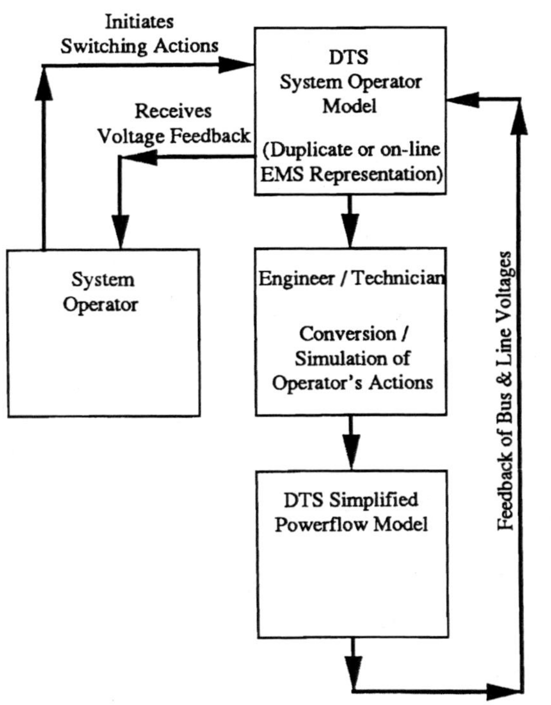

USAGE OF DTS SYSTEM FOR
B.G.&E. SYSTEM RESTORATION DRILL SIMULATION

This simulation process is very manpower intensive. At times, it is very difficult for the engineer to simulate the actions of the system operator, to obtain a convergent powerflow solution, and to provide feedback in a timely manner. Many of the difficulties with the simulation process result from the modelling limitations and program limitations of BG&E's DTS including inappropriate treatment of islands, no black-start combustion turbine capability, insufficient powerflow diagnostics, and lack of a convergence monitor.

Even with these limitations, BG&E has been utilizing its DTS to simulate the System Restoration Drills with reasonable results. The simulator is by far a more valuable tool for restoration drills than any of our other power flow tools.

Conclusion

BG&E has gained valuable experience in utilizing its DTS for System Operator Training and System Restoration Drill Simulation and has included state-of-the-art DTS features in its specification for a new EMS.

DTS STAFFING REQUIREMENTS

Stephen L. Mann, Member and Sharon M. Martin, Member
Carolina Power & Light

The success of power system operations is dependent upon the ability of the power system dispatcher. To properly train the power system dispatcher, many utilities have developed formalized classes. In addition, several utilities have purchased a DTS. To receive dividends from these substantial investments, it is essential that an appropriate number of personnel be dedicated to dispatcher training.

Background

The DTS used by Carolina Power & Light (CP&L) Company was placed in service in 1982. It contains a 500 bus, 1000 line model and provides a nearly identical representation of our real-time EMS.

CP&L has instituted a four-part dispatcher training program consisting of on-the-job, classroom, field work experience (FWE), and DTS training. The training staff is responsible for development and instruction of all classroom material, coordination of the FWE program, and all aspects involved in use of the DTS. This short-note provides information on staff requirements for the DTS portion of the training program, based on experience gained by the training staff at CP&L.

Staff Requirements

When determining the requirements of a Dispatcher Training Staff, the following major points must be addressed:

1. Training Program Goals
2. DTS Scenario Development
3. Instruction/Qualifications/Administration
4. DTS Database
5. Scheduling

Training Program Goals

The goals of the training program, the purpose of the program, and the amount of training material to be covered per year need to be determined. At CP&L, our goals are accomplished through two discrete levels of simulator training. The first level involves instruction of dispatching technicians who are training to become dispatchers. The scenarios used for the technicians focus on the basics of dispatching. One of the most effective uses of the simulator is to implement the material which has been taught in the classroom. This allows the technicians to take the theory from the classroom and apply it to everyday situations. The technicians each receive 5 to 6 one-week sessions of DTS training per year.

The second level of simulator training involves instruction of dispatchers. The scenarios used for dispatchers cover all aspects of dispatching but focus on situations that occur infrequently in the real world. The dispatchers each receive one week of DTS training per year.

Aside from the regular DTS sessions, the simulator is used for classes in system restoration and low frequency operation.

DTS Scenario Development

The time required to prepare new DTS scenarios is approximately eight hours of development for each hour of simulation. This amount of time is used to gather information, program the computer, and prepare for the sessions. For each one week session, six to eight scenarios are prepared, which generally range three to four hours in length. The remainder of the session is spent in review, critique, and instruction. Two dispatchers are trained per session, allowing them each the opportunity to work on generation and transmission. Each set of dispatchers participates in basically the same set of scenarios. However, for diversity, the scenarios are modified (that is, change faulted lines or fault location, trip different generating units, etc.). Changing the basic set of scenarios requires about two hours of modification for each hour of simulation.

Instruction/Qualifications/Administration

The programmed scenarios are only one element of effective simulator training. For successful training, the total control room environment must be duplicated. This requires that the continuous interface with generating plants, neighboring utilities, field personnel, and management be realistic. Our experience has shown the need for a one-to-one relationship between instructors and trainees to create this environment. This allows for simultaneous events to occur on generation and transmission, appropriate dialogue to be maintained, and the simulator to be properly operated.

Possibly the most important part of simulator training is a detailed critique at the end of each session. This is where constructive discussion takes place — questions are answered, suggestions are made to improve performance, explanations are given on why the power system responded in a certain way, procedures are explained in detail, and other various topics are discussed. These critiques often require follow-up time, where the instructors research answers to questions that were not adequately covered during training. The instructors also maintain records of the training sessions. The time requirement for administration and follow-up is approximately 0.75 times the amount of instruction given.

To achieve realistic simulation and to properly critique the sessions, the instructors are required to be familiar with many areas of power system operation and design, in addition to power system dispatching. Some of these areas include:

- Generating plants, substations, and transmission lines (equipment operation, protective relay schemes, typical problems)
- Impact of outages facilities
- Operation within a large interconnection
- Trouble spots in CP&L's system and in the interconnection
- Theoretical understanding of power systems

Obviously, the instructors must have a wide range of knowledge. CP&L has successfully used electrical engineers as instructors.

DTS Database

The time requirement to update the DTS database and associated instructor/dispatcher displays each year will vary greatly depending upon the size of the DTS model and extent of database update. CP&L has found it unnecessary to update the external model every year; every two to four years has been adequate. However, it is imperative to keep the internal model accurate. Any errors or discrepancies in the internal model or displays will be noticed by the dispatchers, and consequently distract from training. We have found it advantageous to put future substations, transmission lines, and forecasted load into the database (six months in advance) to familiarize dispatchers with upcoming peak loads and new facilities. Therefore, we update the internal database at least once each year. Our present 500 bus model takes about 26 person-weeks to complete the update and associated work.

Scheduling

The individuals responsible for training and control room staffing must work very closely to plan and coordinate the dispatchers' training schedules. Since most training staffs work during the day shift, the schedules of the dispatchers (who work on rotating shifts) must be manipulated to provide the desired amount of training. It is important to schedule enough dispatchers during the day shift so neither the control room operations nor the training sessions are disturbed. Scheduling of dispatchers for training requires about three person-weeks.

TRAINING FACILITY REQUIREMENTS

Roger N. Gillespie

Gulf States Utilities

The hardware and facility setup for a training simulator should mimic the equipment used and the environment of the control center as much as possible. This provides for trainee familiarity with the equipment which results in more beneficial training sessions. This short-note will explain in more detail some of the hardware and facility requirements for a training simulator.

Trainee Console(s)

The trainee consoles should be configured to appear and function in the same manner as in the control center. This should include the same type of keyboard and light pen, trackball, or mouse. Definable keys on the trainee consoles should be set up the same as the control center consoles. These consoles should also be provided with the office equipment required in the operations environment.

The number of trainee consoles required is dictated by the operational structure of the control center. That is to say that, if two operators are needed to operate the power system, then there should be two trainee consoles. Training can be done for two operators on operational problems and scenarios while an individual operator may receive training in only a particular area.

Instructor Console

Experience using a training simulator has shown that the displays the instructor uses to control the simulation process should provide as much associated information on the same page as possible. The instructor is usually the busiest person involved in the training session and needs to respond to operator requests in an expedient manner. Thus, the need for related information on the same display is critical.

While one instructor console provides the needed control of the simulator, most training sessions need more than one instructor to effectively administer the training, answer phones, etc. A second instructor console would provide the instructors with quicker response and better coordination of the training session.

Mapboard

The dynamic mapboard is a vital element in the operation of the power system and a mapboard of some degree should also be included in the training area. The mapboard for simulator use should also provide information comparable with the one in the control center. However, due to the size of the training facility, the simulator mapboard should be scaled accordingly. The mapboard should be placed in clear view of the training consoles.

Chart Recorders

If possible, chart recorders should be provided to give the trainees a more realistic training session and to help interpret disturbances in both the internal and external power system. These recorders should be placed in clear view of the training consoles.

Telephones

The trainee console(s) and the instructor console(s) should be equipped with telephone communications equipment equivalent to that used in the control center.

Simulator Location

In the DTS configuration at GSU, the training simulator area is located in the same building but on an upper floor overlooking the operating area. This arrangement has provided dual functionality to the training consoles. When training is not in session, these consoles are utilized for studies, monitoring the generation system, monitoring the power system network, and the other control center functions. A remote location may be beneficial if the training area is susceptible to interruption and noise.

DTS Computer System Configuration

The training simulator should have a computer system designated solely for the purpose of running the simulator, and depending on the quantity of training should have a backup system. The ability to run the simulator should not be reliant upon the availability of all of the computers in the control center.

Summary

Ideally, the hardware configuration for the simulator should replicate the computer system used in the control center in as many aspects as is possible. The number of trainee and instructor consoles for training depends upon the type of training to be done. Mapboard, chart recorders, and telephones should also be provided to give the training environment more realism. A training facility that closely replicates the control center is highly desirable. Finally, the simulator should be run on a computer system that is not dependent on the condition of the on-line computer system for its functionality and availability.

CONCLUSION

J. M. Bucciero, Senior Member

ESCA Corporation

A dispatcher training simulator must be realistic in its simulation of the power system and the real-time control system to be an effective training tool. A simulator can provide dispatchers with a broad range of problem situations that are impossible and inappropriate to stage in a conventional classroom environment and on the actual power system. The effectiveness of the training simulator as a dispatcher training tool will be enhanced by the availability of database and system maintenance tools that will automatically maintain consistency between the real-time system and the training simulator system while allowing the development of training scenarios for specific routine and emergency conditions. The best dispatcher training results will be obtained through the close coordination of the training and operating staffs and with utility investments in training facilities and personnel that can provide the dispatchers with the knowledge to securely and economically operate the power system with confidence.

1. "Workshop Proceedings: Power System Operator Training Simulators," EPRI WS-78-128, August 1979

2. "Considerations in Designing and Using Power System Simulators," EPRI EL-3192, Project 1915-1, Final Report, April 1984

3. V. Koda, A. Bose, M. M. Anderson, "Power Plant Models for Operator Training Simulators, "IEEE/PES Summer Power Meeting, Portland OR, July 1988

4. M. Prais, A. Bose, "A Topology Processor That Tracks Network Modifications Over Time," IEEE/PES Summer Power Meeting, San Francisco, July 1987

5. Y. Chen, A. Bose, "Choosing the Appropriate Boundary for Adaptive Reduction," Conference Papers, PICA 1987, Montreal, Canada, May 1987

6. M. Prais, G. Zhang, Y. Chen, A. Bose, D. Curtice, "Operator Training Simulator: Algorithms and Test Results," IEEE/PES Winter Meeting, New York, February 1989

7. M. Prais, C. Johnson, A. Bose, D. Curtice, "Operator Training Simulator: Component Models," IEEE/PES Winter Meeting, New York, February 1989

8. G. Zhang, A. Bose, "Scenario Building for Operator Training Simulators Using a Transient Stability Program," IEEE/PES Winter Meeting, New York, February 1989

THE USES OF AN OPERATOR TRAINING SIMULATOR FOR SYSTEM RESTORATION

R.F. Chu
Research Division

E.J. Dobrowolski
Engineering Division

E.J. Barr, J. McGeehan, D. Scheurer
System Operation Division
Philadelphia Electric Company
Philadelphia, PA 19101

K. Nodehi
EMPROS Systems International
Plymouth, MN 55441

ABSTRACT

Philadelphia Electric Company (PECo) is the host utility for the EPRI Operator Training Simulator (OTS). One of the OTS functional requirements is to provide system restoration training.

This paper presents some preliminary results from the use of the OTS for the following restoration scenarios: (1) blackstart and resynchronization of two islands, (2) blackstart of a hydro generating station, and (3) blackstart of a combustion turbine for providing cranking power for another generating unit within an isolated island.

The restoration procedure, the system component models, the OTS results, and potential problem areas are also presented. The purpose of the paper is to provide a forum to discuss the capability of existing analytical tools and future research needs to aid system operators in system restoration.

KEYWORDS

Operator Training Simulator, Power System Restoration, Restoration Training

INTRODUCTION

Recent trends in utility generation planning have inadvertently increased both the likelihood of a major blackout and the complication of the restoration efforts. In the sixties and seventies, it was the installation of large, slow starting units remote to load centers and the retirement of small, fast starting units at the load center. In the eighties and nineties, it is the lack of construction of new generating capacity needed to meet the increase of the demand for electricity.

With an exceedingly heavy import of power across an already stressed transmission network, a critical contingency could eventually lead to a major blackout as evident in the 1977 New York incident [1]. To minimize the social and economical impact of a prolonged blackout, a well planned restoration procedure and trained operators familiar with using the procedure to effectively restore the system are needed.

In the restoration procedure, it is necessary that the documentation is well prepared and up-to-date, and the restoration instructions are precise, concise, and easy to understand and execute. As for restoration training, it is very important to have a realistic training environment in which operators: (1) can practice on the execution of the restoration procedure, and (2) will be trained to make appropriate changes on the restoration instructions if some unforseen events prevent the original instructions from being executed.

Since it is inconceivable to wait for a blackout or to shut down the system to establish the necessary restoration training environment, alternative training methods are needed. At Philadelphia Electric Company, the decision was to acquire an Operator Training Simulator where simulation-based restoration training can be conducted.

As a result of the decision, PECo has actively participated in the EPRI research project (RP1915) [2] for developing an Operator Training Simulator [3] and is the host utility for the simulator. Presently, the OTS project is near completion, the simulator is in operation, and two EPRI OTS workshops have been given. In the following sections, the PECo system and operating environment, details of the EPRI OTS, and some preliminary results on using the OTS for restoration training are given.

PHILADELPHIA ELECTRIC COMPANY SYSTEM

Operating Environment

Philadelphia Electric Company is a metropolitan utility serving over 1.45 million customers in approximately 2,475 square miles of Philadelphia and its suburbs. PECo is a summer peaking company with an all time peak of 6,826 MW occurring on August 15, 1988. The total generating capacity (summer rating) is 7,759 MW consisting of nuclear (37.2%), minemouth-coal fired (9.1%), coal fired (8.8%), oil fired (28.3%) and hydro and pumped storage (16.6%) generating units.

PECo is a member of the Pennsylvania-Jersey-Maryland (PJM) Interconnection. The PJM Interconnection schedules units of its member companies on a daily basis and is responsible for pool Automatic Generation Control (AGC) dispatch. PECo receives a steam MW regulation signal, a hydro MW regulation signal, and an economic MW loading signal from PJM every 14-seconds and uses these signals to assign on-line generation of the PECo units.

Reprinted from *IEEE Proceedings of the PICA Conference*, pp. 171–177, May 1991.

The PECo System Automatic Monitoring and Control (SAMAC) System or Energy Management System (EMS) consists of dual Burroughs A10H computers and is located at the company's headquarters in downtown Philadelphia. Supervisory control capabilities are available at SAMAC and at the substations. For system restoration, the SAMAC dispatchers can conduct switching operations: (1) remotely via the available supervisory control capability, or (2) phoning in the switching operations to the substations where the switching is executed by local personnel.

Restoration Procedure and Strategies

Philadelphia Electric Company has maintained a written restoration procedure for many years. The previous edition was written and tested in 1966 with the aid of a DC simulator [4]. This procedure was used to successfully restore the system without any significant problems following the 1967 blackout. Since then, the restoration procedure has been updated annually to incorporate system changes and checked using results from multiple runs of an off-line AC interactive load flow program.

The PECo blackout scenario is that a critical contingency occurs in the PJM transmission system during the summer peak loading period and the contingency leads to a system voltage collapse. Consequently, a complete shutdown of the entire PJM power pool is assumed in the PECo procedure. With no outside generation available from the PJM Interconnection, the restoration instructions are: (1) to blackstart the combustion turbines and hydro and pumped-storage units to provide cranking power to the remaining generating stations, and (2) to re-establish the PECo network.

The basic restoration strategies employed by PECo and many other utilities are:

1. Sectionalization of the system into islands - In the planning stage, each island is designed to have: (1) units with adequate blackstart capability (i.e., hydro units, combustion turbines (CTs), etc.) to provide cranking power to the remaining generating facilities within the island and to supply power to critical loads, and (2) sufficient generation to supply most of the load within the island.

2. Restoration of each island - Immediately after the shutdown, the operators will open almost all system circuit breakers and, then, restart the blackstart units in each island. The next target in the restoration procedure is to establish a path between these units and fossil units which require offsite cranking power. Along these paths, some critical loads are restored for stabilizing the blackstart units. After the fossil units have been restarted, other customers within the island can be reenergized.

3. Synchronization of islands - After each island has been reenergized, the last step in the restoration procedure calls for the resynchronization of these islands.

Since the nuclear generation will not be available for at least 36 hours, PECo can only carry about 60% of its peak summer day load and must get outside power purchases to restore the entire load. Consequently, the procedure "pauses" when a generation and load balance has been reached and "proceeds" when help from the neighboring PJM member utility becomes available.

Restoration Training

The last complete shut-down of the PECo system was in 1967. Some of the operators from that era are still with the company and their experience is valuable input to the restoration plan and training requirement. Since the system constantly undergoes changes, a continuous training effort has been imposed.

Once again, the system cannot be shut down just to provide a training ground for practicing system restoration. However, it is possible to stage a restoration drill in which a shut-down is assumed and the operators are to act through each step of the restoration plan. This is the approach taken by PECo as well as other member utilities in the PJM Interconnection [5].

Specifically, about three to four times a year, PJM organizes a restoration drill in which each company will individually simulate: (1) a complete shut-down with no external help, (2) blackstart of the internal units, (3) re-establish the individual company's system, (4) restoring customers, and finally, (5) re-establishing the transmission system between companies. PJM will assume the pool dispatching position once two or more companies have been reconnected.

Throughout the drill, operators "act" through each restoration step as well as the communications to the PJM pool dispatchers and dispatchers from other companies. The drill normally starts at 0700 hours and ends after 1900 hours on the same day. During the drill, each major restoration step is documented. On the following day, PECo holds a review meeting to discuss the drill results, to identify potential problems, and to make improvements on the existing procedure if needed.

The advantages of holding a post-drill meeting are evident. For the October 25, 1989 drill, the fact that Muddy Run pumped-storage hydro units would have run out of water at the ninth hour (1800 hours) was identified in the post-drill meeting. Assuming that these units are still available, as is the case during the drill, will produce optimistic results on the restoration effort. Consequently, another drill starting at the ninth hour of the October 25th drill was repeated. In this drill, a different shift of operators was involved and the new situation faced by the operators was a load shedding scheme to account for the loss of the pumped-storage hydro generation.

Concerns and Solution Approach

Despite the efforts in generating a well prepared plan and in providing training, there are some concerns:

1. the procedure has not actually been tested and, during the drill, the system provides no feedback to the operators. The operators have to rely on their experience to envision the system behavior resulting from each step of the procedure,

2. the existing procedure does not consider system physical and operational constraints such as the availability of generating capacity, active and reactive balance, cold load pickup, etc., and

3. the training of operators at these drills is not uniform since a different shift of operators is involved in different drills. The experience gained in one drill by one crew of operators may not be adequately transferred to the remaining crews.

PECO has high hopes that the Operator Training Simulator would provide some of the solutions to these concerns. As a result, PECo actively participated in the EPRI project (RP1915) for developing an Operator Training Simulator. In the next section, a brief discussion of the restoration features of the OTS will be given.

EPRI OPERATOR TRAINING SIMULATOR [2] AT PECo

System Configuration

The functional resemblance between the Energy Management System (EMS) and the Operator Training Simulator is given in Figure 1. The features and capabilities at the trainee consoles are replicas of the operator consoles in the EMS or SAMAC control room. For the EPRI OTS, the trainee consoles are linked to the SAMAC backup computer (Burroughs A10H) containing the Control Center Model (CCM).

To simulate the data from the SCADA RTUs and the PJM Interconnection to SAMAC, a separate computer (CYBER 962-11) containing the Power System Model (PSM) is added and is connected to the Instructor Position (IP). Finally, the Man-Machine Interface (MMI) and on-line display functions at the Instructor Position were provided by a third computer which is an Apollo DN3000 workstation. The overall EPRI OTS configuration at PECo is given in Figure 2.

Figure 1 Extension of an Existing EMS with an OTS.

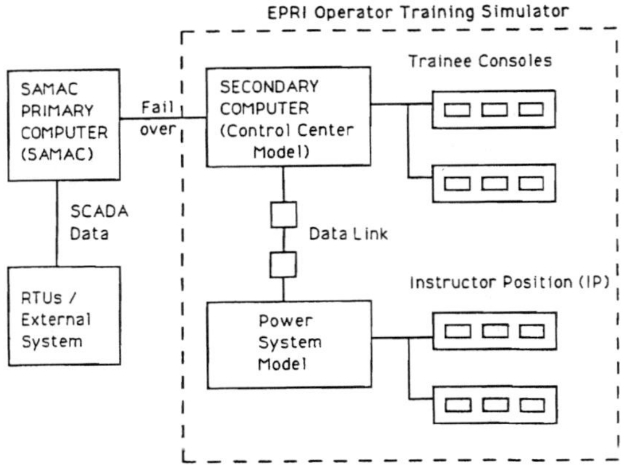

Figure 2. OTS Configuration at PECo.

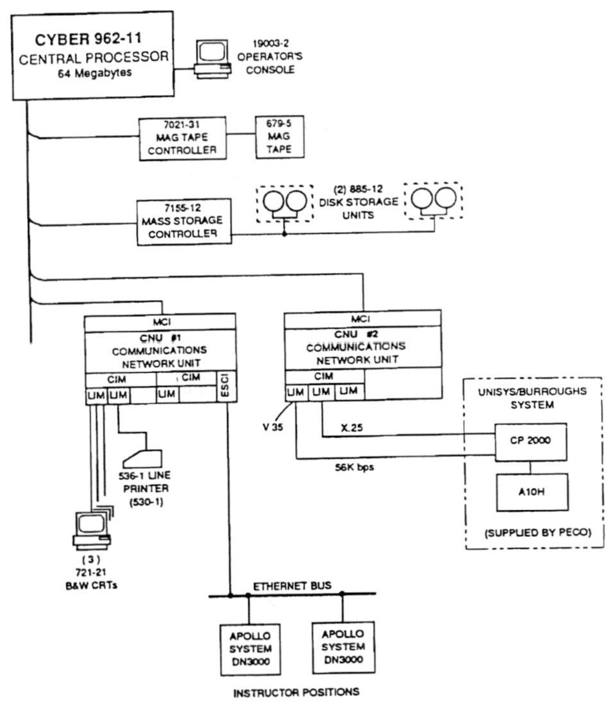

The OTS software was developed under EPRI sponsorship. The software is transportable and available to all EPRI members for assembling an Operator Training Simulator.

Control Center Model (CCM)

Typical functions of the CCM include:

- Data Acquisition
- Supervisory Control
- Man Machine Interface
- Alarm Processor
- Automatic Generation Control (AGC)
- Economic Dispatch (ED)
- Interchange Scheduling
- Network Applications

For restoration purposes, the AGC and ED functions will not be used until the PJM transmission system has been reenergized at some time during the system restoration.

Power System Model (PSM)

The PSM computer simulates the behavior of the PECo and PJM Interconnection Systems. Specific models include:

1. **Power Plant Models** - The power plant models employed in the OTS have three levels of complexity and details of these models are given in Reference [6]. They contain models of:

   0 Fossil units, including once-through subcritical, once-through supercritical, and drum boiler units.

   0 Nuclear units, including a boiling water reactor (BWR) model and a pressurized water reactor (PWR) model.

0 Combustion Turbines (CT).

0 Hydro units.

For restoration training purposes, the blackstart capability of CTs and Hydro units is represented by a constant power injection. The MW outputs are manually adjusted and the governor control is bypassed.

The nuclear units are not included since they will not be available for at least 36 hours. However, two outside sources are required to supply the nuclear generating stations. These two sources, in addition to power from their own emergency diesels, are mandated prior to their startups.

PECo does not have any once-through subcritical units. For the drum type units, they can be represented by a power injection equal to the ramping rate of the units. For a once-through supercritical unit, the available power is modelled as the sum of a step function equal to the minimum power output and the unit's ramping rate starting at the time when the minimum power output has been reached.

2. **Load Models** [7] - six different load models of varying complexity are available. They are:

   (1) Model O - constant MVA model,
   (2) Model OS - Model O with voltage and frequency sensitivities,
   (3) Model OM - Model O with load management capability in which fractions of load can be removed or reinstated,
   (4) Model OSM - Model OS with load management capability,
   (5) Model OSR - Model OS with voltage regulation capability, and
   (6) Model OSMR - Model OSM with feeder voltage regulation.

For restoration, Model OSM was found to be adequate.

3. **Protective Relay Models** [7] include:

   (1) Over/Under Voltage Relays,
   (2) Volts per Hertz Relays,
   (3) Inverse-time Overcurrent Relays,
   (4) Over/Under Frequency Relays,
   (5) Loss of Excitation and Overexcitation Relays,
   (6) Synchro-check Relays, and
   (7) Automatic Reclosure Relays.

4. **Network Models** [7] consist of network, transformer and generator models.

The interaction between these models is given in Figure 3. The simulation cycle is 5 seconds and the simulation algorithms are given in Reference [8].

Instructor Position

The features of the IP include: Instructor MMI, OTS initialization and scenario creation, simulation and scenario control, simulation of manual field operations, snapshots and snapshot review, base case save and restore, PSM and trainee monitoring, and PSM interface.

Figure 3. OTS Model Interconnection.

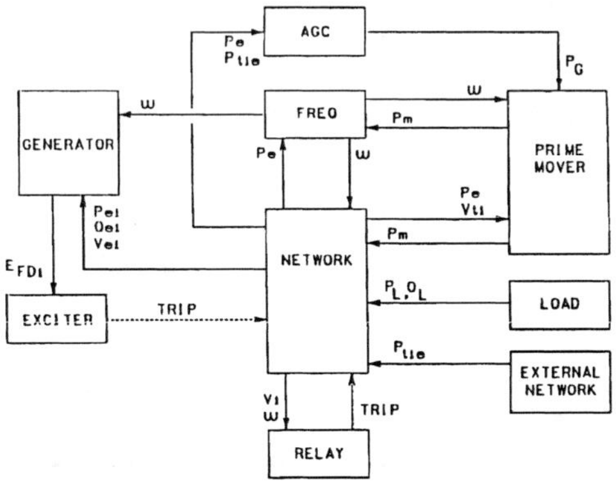

RESTORATION CASE STUDIES

The EPRI OTS was tested using the following restoration scenarios:

Case 1. Blackstart and Resynchronization of two Islands

The objectives of this case are to test the performance of the OTS on: (1) islanding strategies, (2) the energization of a "dead" island by an outside source, (3) blackstart of CTs, and (4) resynchronization of two islands.

The specific area of focus is the Delaware Generating Station as shown in Figure 4. This figure is taken from the CRT display at the Apollo workstation.

Specific steps taken in this exercise are:

1. (Tripping Line 130-02) - forming a Delaware island in which the steam units (Units #7 and #8) are supplying the local load. The result is shown in Figure 4. Note that the loads are not shown in the figure and no MW transfers are occurring through #1 and #2 transformers.

2. (Tripping Units #7 and #8) - blackout in the Delaware island.

3. (Opening Circuit Breaker (CB) #D1) - splitting up the Delaware island into left and right islands.

4. (Clearing all CBs on the dead bus in the right island)

5. (Closing Line 130-02 CB) - energizing the left island with an outside source.

6. (Start CTs: Units #11 and #12) - blackstart CTs.

7. (Closing all ring bus CBs in the right island) - generation and load balancing within an island.

8. (Close CB #D1) - Synchronizing two islands.

The final configuration is given in Figure 5. The exercise went smoothly and the results compared favorably with those from the off-line planning studies.

Figure 4. The Delaware Islanding Scenario.

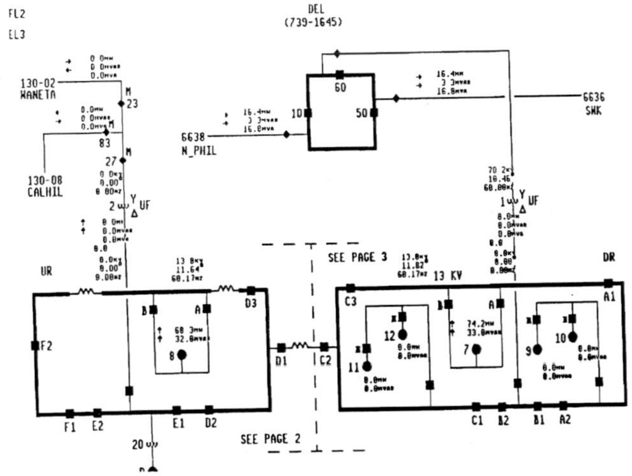

Figure 5. The restored Delaware Island.

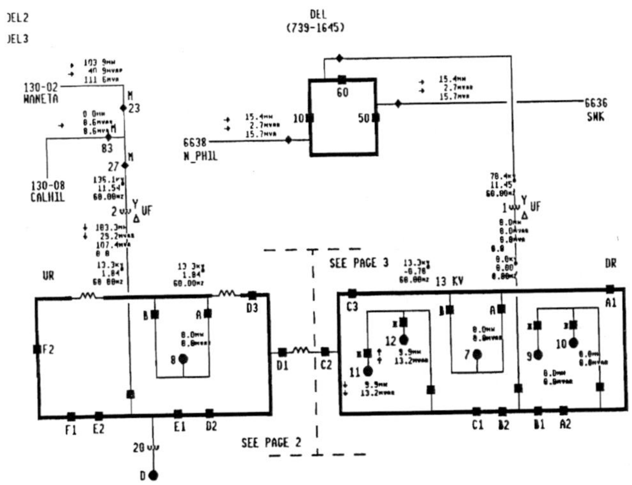

Case 2. <u>Blackstart of a Hydro Generating Station</u>

PECo relies heavily on generation from the Conowingo Hydro Generating Station for most of the system's blackstart capability. Since Conowingo Station is remote from the load center in Philadelphia and is also near Peach Bottom Nuclear Generating Station, it is important that the PECo procedure: (1) allows adequate cranking power to supply the fossil units in the city, (2) provides a second feed to Peach Bottom, and (3) at the same time, picks up as many critical customers as possible to stabilize the units.

Figure 6 is a display of the system diagram on the CRT of the IP console. The connection between Conowingo and Peach Bottom is via the 34KV line (Line 3305). The figure was recorded after the complete shutdown. The objective of this exercise is to test a portion of the PECo restoration procedure with reference to the Conowingo units. The specific procedure includes:

1. Trip Line 220-01 to Clay and Line 220-03 to Nottingham.
2. Blackstart Units #3, 4 and 6, at Conowingo.
3. Energize Line 220-01.
4. The following CBs are not shown in the display:
   a. Close CB 115 and #1 and #5 Transformer CBs at Colora.
   b. At Colora, close 3310 line CB and energize 3301 line.
   c. At Clay, close #100 and #120 Trans. CBs and Line 149-00 CB.
5. Close Line 220-03 CB to Nottingham.
6. The following CBs are not shown in the figure:
   a. At Colora, close #20 Trans. CB and 225 CB.
   b. At Bradford, close #230 Trans. CB, and 410 and 420 CBs.
   c. At Nottingham, close #6 Trans. CB, and 100 and 120 CBs, and energize Line 143-00.
   d. At Nottingham, close CB 90.
   e. At Bradford, close CB 250.
7. Stop.

Throughout the restoration process, the MW outputs from the Conowingo hydro units were raised to meet the MW demand. The voltages were also adjusted when the voltage limits had been violated. We felt that the OTS results are credible and we have been using this restoration scenario and the Delaware scenario for the EPRI OTS Workshop.

Figure 6. The Conowingo Restoration Scenario.

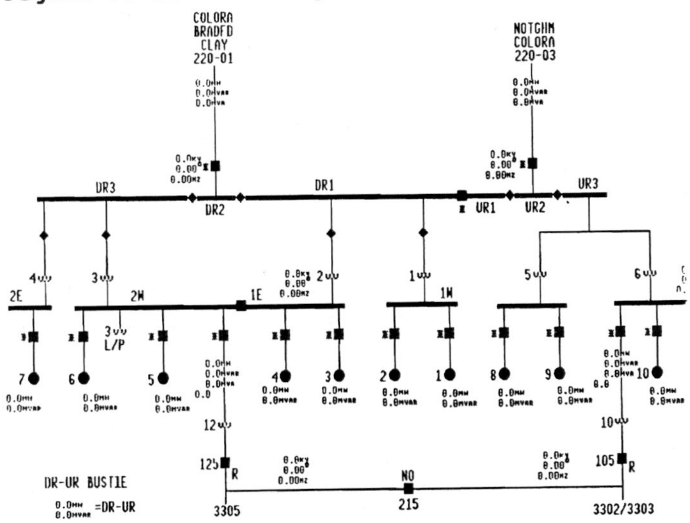

Case 3. <u>PECo Restoration Procedure at Emilie Substation</u>

Figure 7 shows an island within the PECo restoration procedure. The generation for the island consists of CTs at both Falls and Croydon Substations. The blackstart capability is provided by the Falls CTs. The Croydon CTs will start once the cranking power from the Falls CTs becomes available.

The objective of this exercise is to test the particular islanding procedure. Specific restoration steps after clearing all dead bus immediately following the blackout are:

1. Check and close Emilie CBs 140, 150 and 180.
2. Check and close Croydon CBs 235, 315, and 335 independently.
3. Check and open Emilie CBs 785 and 895.
4. Open Emilie CB 190 and close CB 120.
5. Close Eddington CB 125 independently.
6. Open Falls CB 215 and close #1 Transformer CB.

7. Start the CTs at Falls.
8. Close Emilie CBs 895, 785, 545 and 555.
9. Close #5 and #6 Transformer CBs to pick up some loads to stabilize the generating units.
10. Start the CTs at Croydon.
11. Close Eddington #4 and #6 Transformer CBs.
11. Restore the loads within the island and then wait for resynchronization opportunity at Richmond Generating Station (not shown) through Eddington Substation.

Once again, the OTS results agreed with those from the restoration planning studies. No limit violations were detected and the procedure appears to be feasible. The entire restoration process on the OTS takes less than 30 minutes.

Figure 7. The System Diagram for the Emilie Substation.

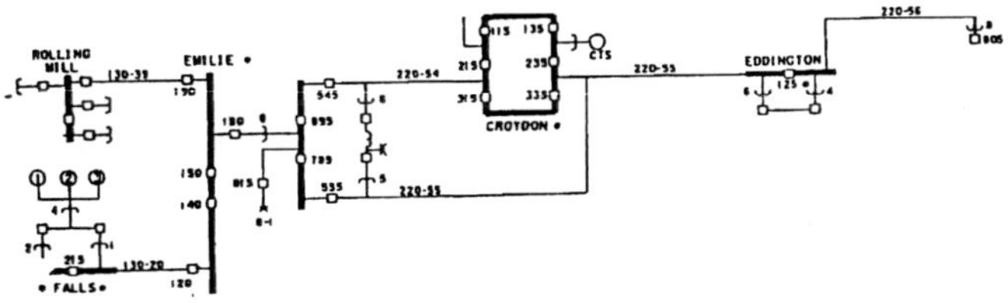

DISCUSSIONS

During these restoration exercises, several attributes in addition to those given in References [6-8] have been identified as follows:

1. Differing from the software used in the Dispatcher Simulator of Reference [9], the EPRI software is capable of simulating a complete shutdown and blackstart. In fact, the agenda of the EPRI OTS workshops call for: (1) a voltage collapse leading to a complete shutdown, and (2) blackstart and restoration.

2. Another attribute of the OTS software allows opening breakers on a dead bus.

3. In an OTS, all operators can be trained using the same training scenarios. As a result, all operators will have the same opportunities to learn how to handle a specific operational problem.

4. The OTS power flow program is capable of handling restoration conditions while almost all existing EMS power flow programs cannot handle these conditions. Unless the OTS software is on-line in the SAMAC computer, the operators may not have some of the information which is available during the OTS restoration training.

5. Since the OTS will generate an alarm if the water level at the pond falls below a preset level, the OTS would have identified the water level problem during the restoration drill on October 25, 1989.

At the same time, we have identified some shortcomings and they are:

1. The OTS simulation does not take into consideration the actual time factors involved in the restoration. Consequently, the actual restoration time could be many times those indicated in the OTS. The lack of timing factors could be very crucial in a restoration training program and need to be provided by the instructor.

2. For restoration purposes, external interactions and coordinations need to be included as a part of the training program. An example is that we have to develop a training program in which the instructor can play a "realistic" role of field personnel, substation operators, power plant operators, pool dispatchers and dispatchers from the neighboring utilities.

3. In energizing a circuit, the OTS load models do not include the effect of cold load pickup. Depending on the length of the outage, results from the OTS could be optimistic. However, this brings up an interesting dilemma since most EMS's, including PECo's SAMAC, do not monitor system loads.

4. If a specific restoration step cannot be executed, the OTS is presently not very "friendly" in providing alternative restoration actions and guidance. This capability is not required as a part of the original OTS project and is being developed in the new EPRI Power System Restoration project (RP3104-1).

At PECo, research activities are ongoing to develop solution approaches to overcoming some of these shortcomings. Independent from results of these activities, the OTS is still an excellent facility for checking a restoration plan and for training operators to get familiar with the execution of each step in the plan.

CONCLUSION

Because the social and economical consequences of a major blackout are devastating, utilities must take steps to prepare for such an unlikely event. From a system operational viewpoint, the availability of a well prepared restoration plan and an Operator Training Simulator for restoration training is a step in the positive direction.

PECo is currently evaluating all functional capabilities of the EPRI OTS in normal, emergency and restoration operations. For restoration purposes, the preliminary experience shows that there is room for improvement. Particularly in the situation when unforeseen events prevented the execution of predetermined restoration instructions, it is desirable to have a computer aid which resembles an expert operator to provide alternatives and guidance for continuing the restoration effort.

Currently, EPRI, PECo and the University of Washington are working together in the EPRI project (RP3104-1) to apply Expert System technology to system restoration [10]. The primary objective is to develop tools to assist operators to restore the system. Upon the completion of the Expert System prototype, the plan is to integrate the prototype into the OTS for restoration training. The progress and result of the research activities will be reported in the near future.

ACKNOWLEDGEMENT

The Operator Training Simulator project (RP1915-2) was sponsored by the Electric Power Research Institute. The EPRI managers for the project were Drs. C.J. Frank and D. Curtice. Contributors to the project include the EPRI Project Advisors, and many individuals from PECo, EMPROS, MACRO Corporation, Arizona State University, Power Math Associates, Power Computer Applications, Seville Training Systems and Hughes Training Systems.

REFERENCES

1. G.L. Wilson and P. Zarakas, "Anatomy of a Blackout", IEEE Spectrum, February 1978, pp.39-46.

2. J. Waight, et al., "Operator Training Simulator; Technical Report", a Draft of the Final Report to EPRI for Research Project RP1915-2, August 1990.

3. D. Boutacoff, "Simulating The Control Center", EPRI Journal, June 1990, pp. 26-35.

4. D. Scheurer, "Philadelphia Electric Company System Restoration Plan", presented at the Pennsylvania Electric Association Meeting, Pittsburgh, PA, Jan. 25, 1980.

5. R.J. Kafka, "Industry Practices in Power System Restoration", presented at the Power System Restoration Panel Session of the 1989 IEEE Summer Power Meeting at Long Beach, CA.

6. V. Kola, A. Bose, and P.M. Anderson, "Power Plant Models for Operator Training Simulators", IEEE Trans. on Power Systems, Vol. 4, No. 2, May 1989, pp. 559 - 565.

7. M. Prais, C. Johnson, A. Bose, and D. Curtice, "Operator Training Simulator: Component Models", IEEE Trans. on Power Systems, Vol. 4, No. 3, August 1989, pp. 1160 - 1166.

8. M. Prais, G. Zhang, A. Bose, and D. Curtice, "Operator Training Simulator: Algorithms and Test Results", IEEE Trans. on Power Systems, Vol. 4, No. 3, August 1989, pp. 1154 - 1159.

9. K.M. Yung, K.C. Lo, and J.W. Chang, "Operational Experience on the China Light and Power Company's System Operation Training Simulator", IEEE Trans. on Power Systems, Vol. 5, No. 2, May 1990, pp. 521 - 530.

10. R.F. Chu, C.C. Liu, K.L. Liou, and A.T. Holen, "An Expert System for Generation Capability Assessment During System Restoration" presented at the EPRI Workshop on Expert System Products for Power System Operations, September 19, 1990, Seattle, Washington.

AN ADVANCED TRANSPORTABLE OPERATOR TRAINING SIMULATOR

J.G. Waight, K. Nodehi, M. Rafian, H. Van Meeteren*
Empros Systems International

A. Bose
Arizona State University

R. Wasley
Macro Corporation

E. Stackfleth
Hughes Training Systems

E. Dobrowolski
Philadelphia Electric Company

ABSTRACT

Since the introduction of the first Dispatcher Training Simulator (DTS) at PICA in 1977, DTS systems have evolved in scope, complexity and performance. This paper outlines the design of an advanced, transportable Operator Training Simulator (OTS) which was developed as part of EPRI research project 1915-2. Advanced modeling technology and instructor capabilities are described along with training examples which demonstrate the capabilities of this OTS.

INTRODUCTION

During the past few decades, the complexity of the job of the power system control center operator has increased dramatically. Several factors have contributed to this change in complexity. The utilities of the United States have become increasingly interconnected, coalescing today into three major interconnections. Later environmental and economic factors combined to favor the construction of new generating stations at sites remote from existing load centers and with limited transmission access to these load centers. The increase in the cost of petroleum fuels and the availability of relatively cheaper hydroelectric power from Canada has led to transmission of electric power over long distances over a strained transmission network. During the late 1980's, large unforecasted increases in peak summer demand has further aggravated the situation. The nature of operating problems has changed in many instances from concern about thermal problems, to concern over Volt/VAR coordination and the potential for voltage collapse.

All of these factors have increased the complexity of the jobs of system control center operators. To meet these challenges, utilities have procured and installed computer based SCADA and EMS systems. Sophisticated tools such as state estimators, contingency analysis, and optimal power flows have been placed in the hands of system operators as operating aids in the new environment. These new tools have brought with them the need for more training of system operators in the use of these tools, especially during emergency operating conditions.

At the same time other economic factors and the deregulation of utilities have led to the downsizing of utility operations staff. In many cases early retirement options may lead to the loss of veteran operators with many years of operating experience. These positions are then filled with junior operators with less on-the-job experience.

All of these factors have accentuated the need for training system control center operators in all phases of system operations: normal operations, emergency operations, and in system restoration. Simulation based training, which has been widely accepted in other contexts, was first introduced as a tool for operator training with the announcement and demonstration of the Dispatcher Training Simulator (DTS) at PICA in 1977 [1]. This DTS had limited capabilities in terms of model size and in the fidelity of the simulation, the User Interface, and the control functions.

In the following years many vendors introduced DTS systems, mostly integrated with new energy management systems [2], [3], [4], [5]. Model sizes grew, and improvements were made in the fidelity of the training devices. Yet certain limitations persisted especially in the area of simulation of complex voltage phenomena, in instructor tools, and in capabilities for trainee evaluation. The Electric Power Research Institute undertook the development of an advanced Operator Training Simulator (OTS) with improvements over other simulators then on the market. Research Project 1915-1, [6] laid out the guidelines which could be used to develop a such a simulator.

This paper reports the research and development of the advanced transportable simulator.

*Mr. Van Meeteren is now with Energy Control Consultants.

REQUIREMENTS

The overall requirement was to develop an Operator Training Simulator suitable for simulation of normal, emergency, and restorative conditions of an electric power system and to have a useful simulation tool for operations, planning and after-the-fact analysis. The overall requirement was subdivided into five specific requirements.

The first requirement was to develop simulation capabilities needed for system operator training. This led to the development of requirements and specifications for models and algorithms suitable for simulation based training of operators in emergency conditions including complex voltage phenomena in real time. Fidelity requirements were stated in two ways. The simulator must behave as the real world does as seen by the operator. The man machine interface and the control functions used by the trainee must be the same as those used by the operator in the Energy Management System (EMS).

The second requirement was to develop the OTS on a practical, realizable platform. This meant that the OTS must run on commercially available, all digital computer hardware. Custom built hardware, such as analog or hybrid simulators, was not considered as a development option.

The third requirement was to develop improved instructor capabilities. These capabilities support the instructor in preparing the training sessions, executing the training sessions, and evaluating the performance of the trainee.

The fourth requirement was to develop a simulator which is transportable and applicable to a broad range of utilities. This objective was established to ensure that the developed OTS is of benefit to as many utilities as possible.

The fifth requirement was to demonstrate and evaluate the OTS in the operational environment of a host utility. Philadelphia Electric was selected by EPRI as the result of a formal process which included proposal solicitation, evaluation, and selection.

Reprinted from *IEEE Proceedings of the PICA Conference,* pp. 164–170, May 1991.

OTS CONFIGURATION

The OTS is comprised of three major hardware components: the Power System Modeling (PSM) computer, the Control Center Modeling (CCM) computer and the Instructor Position(s) (IP).

The PSM computer is used to simulate the Power System Modeling and Educational Subsystems. The Power System Modeling subsystem consists of the models for the electrical network, generating units, prime movers, loads, relays, local controls for the system, the AGC systems of external companies, and possible power pool controllers. The Educational Subsystem allows the instructor to set up events which occur during the simulation. The PSM has an instructor interface that allows personnel other than the system control center operator to control the electrical power system. Typically, such persons include substation operators, power plant operators, and system control operators of neighboring utilities and possible power pools. The instructor interface is provided through the Instructor Position(s). The PSM also has an interface to the CCM. This interface includes simulations of the SCADA RTU traffic and of inter-utility data links.

The CCM computer contains a model which replicates the user utility's EMS. Functions included in the user utility's EMS are modified to communicate data to and from the PSM instead of the RTUs and other utility or power pool computer systems. These functions typically include:

- Alarm Processor
- Data Acquisition
- Economic Dispatch
- Interchange Scheduling
- Load Frequency Control
- Man-Machine Interface
- Network Analysis Applications
- Scheduling Applications
- Supervisory Control

The Instructor Position is based on a workstation. This workstation has features to support instructor activities during simulation. These features include:

- Instructor MMI
- Education subsystem interface
- Logging

Figure 1 shows the generic configuration of the OTS.

In order to implement the OTS software, the Central Processing Unit (CPU) used for the PSM should be a computer with the following main features as a minimum:

- 10 MIPS
- 1 MegaFlop
- Floating point hardware
- 64 MB directly accessible memory
- A minimum of 4 data channels

CPU requirements for the CCM are identical to those of the EMS primary system. These requirements may be satisfied by using the secondary CPU of the EMS for the CCM.

The instructor console is based on an Apollo DN 3000 workstation or equivalent Hewlett Packard/Apollo workstation. A second workstation will be required if the user utility plans on having two instructor consoles. The workstation(s) are connected to the PSM by means of an ethernet (TCP/IP) connection.

The PSM handles the communication process with the CCM via data interface hardware. The communications protocol required is the CCITT X.25 protocol.

Data communication between the PSM and CCM uses two dedicated lines. Redundancy is supported. Typically, however, both lines are fully operational. For each line, two service classes (normal and high priority) are defined. The lines operate in full-duplex mode at a speed of 56 Kbs.

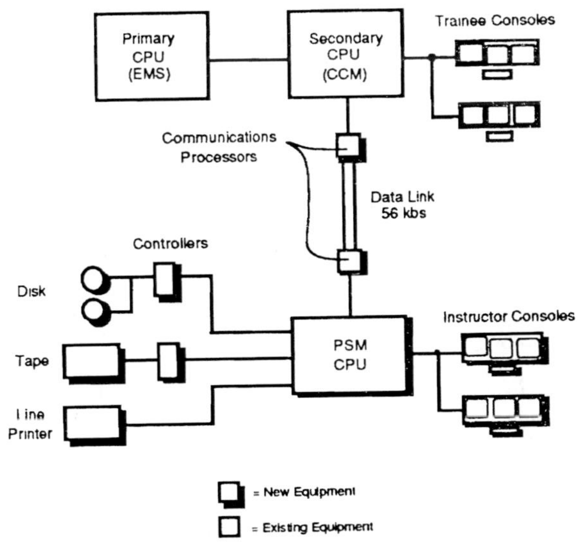

Figure 1. Generic Configuration of OTS

Software Implementation

One of the requirements was to develop an OTS that is transportable and can be made available to a broad range of utilities either as part of a new EMS or an an extension to an existing EMS. In order to fulfill these requirements a configuration was selected whereby the CCM could be implemented on redundant control center hardware and the PSM and IP could be implemented on additional hardware. A further requirement was that the PSM software could be ported from the selected hardware platform to other hardware with minimal efforts. At the time of the design, software standards for portable operating systems, user interfaces, and data base management systems were not widely accepted nor available.

In the OTS design the portability requirement was satisfied by the construction of an OTS Executive which includes the necessary subset of the capabilities available in proprietary EMS executives. The PSM software contains interfaces to the OTS Executive which in turn is interfaced to the proprietary operating system available on the selected hardware.

The PSM software is made up of a number of tasks which are executed concurrently, both on a periodic and demand basis. The OTS Executive was developed in ANSI Fortran to schedule tasks and to accomplish intertask communication. Intertask communication is handled using a messaging scheme, where one task can direct a message to any other task. Messages are placed in a mailbox in a queue which is processed in FIFO order. The flow of messages is unidirectional. Each task has a mailbox and can only read messages from its allocated mailbox, but it may send messages to any other task's mailbox.

When porting the PSM to a hardware platform with a different operating system it is only necessary to modify the interfaces of the OTS Executive to the new operating system.

The PSM has its own data management system which is independent from the one used on the EMS. Because of requirements for portability, the data management system is written in ANSI Fortran 77. The data management system is comprised of three main modules: a builder, a populator, and a mapper. The data base builder, mapper, and populator together perform the data base generation function.

The data base builder uses source data to size both the data base and program code and to create the data base directories used by the access function. The data base populator function uses source data to fill the data base areas needed by the PSM applications software. The source data is checked by the data management system so that errors can be found as early as possible during data base generation.

The data base mapping function builds maps which allow transfer of PSM simulated RTU values to the CCM SCADA data base and CCM supervisory control values to the PSM data base.

The interaction between the instructor and the OTS is supported by the IPMMI software and hardware. Control of the OTS, data entry into the data base, and presentation of data on displays are performed by the IPMMI. All data requests and updates performed by the IPMMI relate to data resident in the PSM database. The IPMMI also has the ability to initiate PSM tasks and to transmit program generated messages to the instructor.

Displays used by the IPMMI consists of static masks and linkages. Static masks are presented as created initially while data linkages allow the presentation of dynamic data on displays. Function key and poke point linkages are features of the IPMMI which are used for task executions, display selection and data entry. A windowing technique is used to view multiple displays simultaneously.

OTS CAPABILITIES

The capabilities of the OTS may be described in four categories: the Power System Modeling (PSM) subsystem, the Educational subsystem, the Scenario building Subsystem, and the Control Center Modeling (CCM) subsystem. The power system modeling subsystem consists of the models and algorithms which simulate the power system. The control center modeling subsystem is a replica of the EMS used by the trainee. The educational subsystem is comprised of the tools and capabilities which allow the instructor to use the other subsystems. The scenario building subsystem consists of tools which aid the instructor to construct training scenarios.

Power System Model

The power system model consists of the static and dynamic models which represent the slow dynamic behavior of the system. [7], [8].

Power Plant Models

The PSM includes energy source models for fossil, hydro, combustion turbines and nuclear units. Various boiler models are provided in tiers of complexity. The boiler models for fossil fuel come in three levels of complexity: a complex, an intermediate, and a simple model. There are only two tiers for the pressurized water reactor (PWR) and combustion turbines. Fossil fuel models represent once-through subcritical, once-through supercritical, and drum boiler. Boiler control is provided in a simplified form. Coordinated control as well as boiler following control are provided. Nuclear plant models include boiling water reactors and pressurized water reactors. A common steam turbine model is available for fossil fuel and nuclear plants. A hydro turbine model as well as a combustion turbine model is available. The PSM data base provides default data for the power plant models if some or no data are available.

Load Models

The load model calculates the network real and reactive loads from a specified load curve.

The power system is divided into several zones. A separate zone load curve may be defined for three day types (Weekday, Saturday, Sunday), and two seasons (Summer, and Winter). Distribution factors distribute load from zone level to load groups, and from load groups to feeders. Loads distributed to feeders are computed at nominal voltage and frequency. Feeder load may consist of a non-conforming portion, and a conforming portion. The non-conforming portion of load does not follow the load curve; however, it may have a time schedule to be turned on or off. A random noise portion can be added to the loads to represent random fluctuations in load. Voltage dependency of load can be represented by a second degree polynomial. Frequency sensitivity, load management, and voltage regulation by feeder tap changer and switchable shunts can be selected in various combinations. The effect of these features is included in the load model and therefore does not require the network model to extend into the individual feeders [7]. These features of the load model are important for the simulation of complex voltage phenomena such as voltage collapse.

Relay Models

Relays important for the simulation are modeled. Over/Under voltage relays trip generators due to high voltage and drop load at buses with low voltage. Volts/Hertz relays trip generators if a volt/Hertz ratio is violated. Over/under excitation relays allow units to violate their MVAR capability curves for a specified duration. Inverse time overcurrent relays trip branches if flow limits are violated for a given duration. Auto reclosure relays simulate reclosure attempts of branches due to faults. Over/under frequency relays trip generators due to high frequency, and shed load at low frequencies. The amount of load shed depends on the frequency. Synchro check relays check voltage magnitude and angle difference when two sides of the controlled circuit breaker are within the same island. Voltage magnitude and frequency are checked if the two sides of the controlled circuit breaker are in two different islands.

Network Model

Tap changing and phase shifting transformer actions are modeled outside power flow calculations because changes usually occur in a longer time frame than the OTS five second cycle time. Thus, separate modules compute the new tap settings and the power flow considers those values fixed. The network model of the transformer is a series impedance and unequal shunt legs. A piecewise linear capability curve is available for each generator. Thus, maximum and minimum reactive power output of the units are determined from the capability curves.

The network model of the AC line is a series impedance and unequal shunt legs. An HVDC multi-terminal model is also available. The DC system power flow is solved iteratively with the rest of the system. Each multi-terminal system must have one slack bus, but all of the other buses can have a power control or angle control converter. The current and angle limits are monitored and in case of violation, the control type is switched. There are also some soft angle limits which when violated, initiate transformer tap changes. The slack bus is automatically changed if limits are violated at that bus.

Solution Methods

The trapezoidal method of integration is used to solve the differential equations. A one second integration step was determined to be sufficiently accurate for the models represented in the OTS.

The topology processor determines the network connectivity and keeps track of changes to the network. It determines bus optimal order using Tinney scheme 2.

The power flow solution technique is the fast-decoupled power flow. Generator VAR limiting and remote voltage control are done within the power flow solution. The OTS uses a special topology processor which is integrated with the OTS power flow. Information about network topology changes is used to incrementally reorder and incrementally update the factors of the matrices used in the power flow solution. After the OTS has run for some time the effects of topology changes accumulate. Then matrices must be rebuilt and refactored to avoid buildup of computational inefficiency. This is done during a cycle when there are no topology changes and the operator is not sensitized to the response of the OTS.

Educational Subsystem

The Educational subsystem consists of tools which assist the instructor during pre-session, session, and post-session activities. A brief description of each session is as follows.

Pre-Session Activities

The main part of a pre-session activity is preparing base cases. Base cases or snapshots consist of sufficient network and model data such that when restored, the conditions at the time of save are recreated. An important source of base cases are snapshots of the system taken either manually, or by turning on the automatic time interval snapshot feature of the OTS. The instructor can prepare for a session activity by restoring a snapshot which is pertinent to goals that are intended to be achieved during a training session. Furthermore, the instructor can modify the restored snapshot to set the starting conditions. These modifications are for instance changing the date/time of simulation, or changes in load, generation, topology, etc.. Finally, the instructor can save the modified snapshot as a base case which can serve as a starting point for the session activity.

Another part of pre-session activities includes set up of scenarios. Scenarios are either designed and pre-tested by the instructor using the OTS Simulation capabilities to achieve certain operating conditions, or produced by scenario creating tools such as the Heuristic Scenario Builder or a transient stability program.

Session Activities

Session activities are comprised of the following steps

- Initialization of the simulator and restoration of the base case prepared as part of the pre-session activities.
- Execution of scenarios created for the session.
- Participation by the instructor in the training session.
- Setting criteria for performance measures.

Participation in the session activities includes execution of scenarios as required, modifying events as deemed necessary, and responding to trainee requests as power plant and substation operators for actions not supported by supervisory control. The instructor should also monitor the trainee's responses to the events and should set the system parameters intended to be monitored. For example, voltages at certain buses, frequency of a given island, etc. may be monitored in order to gain adequate information about the trainee's performance during the training session.

Post-Session Activities

Following the completion of a training session, the instructor can evaluate the performance of the trainee through the monitored parameters and snapshots which the instructor designated to be monitored throughout the simulation. Also the instructor can review comments he might have entered during the training session regarding the ability of the trainee to handle the emergency cases, or the efficiency of means advised by a trainee to bring the system back to normal.

Scenario Building Subsystem

This subsystem consists of the tools that aid the instructor in building training scenarios. These tools are: The EPRI ETMSP Transient Stability program, and the Heuristic Scenario Builder (HSB).

The Transient Stability (TS) program and its associated power flow are interfaced to the OTS such that they can be executed at any time during simulation without any manual data preparation. However, since the TS program cannot be executed in real time, the OTS must be paused during the execution of this program.

The output of the TS program can be directed to the PSM printer. Also, as a result of operation of the relays modeled in the TS program the post-transient network topology can be easily re-created for the OTS. This is done by translating these topology changes into circuit breaker trip events. These events are then transferred to the OTS event library upon request by instructor.

The HSB uses expert system technology to assist the instructor in building scenarios automatically. Given the condition of the system in a base case, the HSB searches for the combination of equipment outages in order to achieve a pre-specified goal. The goal must be specified by the instructor and may be a line overload, or a low voltage problem. The location of the goal state may be either specified by the instructor, or it may be selected by the HSB depending on the specified training objectives. These objectives include the degree of scenario difficulty and the trainee experience level.

OTS IMPLEMENTATION AT PHILADELPHIA ELECTRIC COMPANY

An important objective of this project was the demonstration of the developed software at a host utility. This meant that the power system models and algorithms and the educational subsystem had to be implemented in a real-time environment which included a real-time data base and a real-time man-machine interface. The PSM computer had to be connected to the secondary computer of an EMS and a CCM had to be developed starting with the host utility's EMS.

Philadelphia Electric Company was selected as the utility to host the demonstration of the OTS and to participate substantially in the development of the OTS.

PECO OTS Hardware Implementation

For the implementation of the OTS at Philadelphia Electric Company, a PSM computer was selected, the OTS data was collected, checked and populated into the OTS data base, and displays were created at the instructor's position. The PSM computer was connected to the secondary computer of PECO's EMS (System Automatic Monitoring and Control or SAMAC) in which a control center model (CCM) was implemented.

SAMAC was originally implemented by North American Rockwell, TRW and PECO. SAMAC is based on Unisys/ Burroughs A1OH processors and assorted peripherals and front-end devices.

The computer used for the PSM is a CYBER 962-11 with the following features:

- 64 MBytes memory
- 10 peripheral processors
- 12 data channels
- Operators console
- NOS/VE Operating System

The computer used for the CCM is a Unisys/Burroughs A10H which is also the secondary processor of the SAMAC system. The A10H is a dual processor with the CCM running in one CPU and PECO accounting and system maintenance running in the second CPU. The Unisys/Burroughs MCP Operating System was used.

Two instructor positions are supported at PECO. Each IP is based on an HP/Apollo DN3000 workstation. In addition, two CCM consoles are available for the instructors. These instructor CCM consoles can be initialized to either active or slave mode. In active mode, the instructor CCM console provides normal interactive capabilities. In slave mode, one of its CRTs displays exactly the same information as the master CRT of a specified trainee console.

Two communications processors were utilized to connect the PSM computer to the CCM computer. The PSM communications processor consisted of the following products:

- CDC Distributed Interface (Communications Network Unit)

- CDC CDCNET Communications Software running X.25 protocol (Telenet/Tymnet certified)

The CCM communications processor consisted of the following products:

- Unisys CP2000 Communications Processor
- Unisys Communications Management System (CMS)
- Unisys Network Architecture Network Services (BNA)
- Telenet/Tymnet CCITT X.25 Protocol

The OTS at PECO also includes two trainee consoles, a front-end system which drives the consoles, and a duplicate copy of the mimic boards used in SAMAC.

PECO OTS Software Implementation

To support the implementation of the OTS at PECO, data was provided for the PECO power system network, for the PJM interconnection, and for a simplified representation of the world external to PJM. Special models were developed to simulate the PJM power pool operations, and the operation and control of pumped storage and hydro units at Conowingo and Muddy Run.

The CCM has the same power system monitoring and control capabilities as SAMAC. The CCM was built from SAMAC and included customized software that allows data, normally transferred between SAMAC and its RTUs and PJM, to be transferred between the CCM and PSM. SAMAC functionality is retained including the same man-machine interface, steam generation control, economic dispatch, hydro plant control, system status processing and security analysis, supervisory and combustion turbine control, limit checking, CRT trending, report generation, logging, and alarm/event capabilities.

SAMAC functionality is not fully supported by the PSM, however, because implementation of some SAMAC capabilities is considered non-essential and would have incurred too much cost for too little training benefit. Thus, the PSM does not support: (1) on-demand reports between SAMAC and PJM other than the critical lines-out-of-service report, (2) SAMAC external device and communications circuit testing and (3) SAMAC scanning of pulse accumulator data (normally every hour). All other SAMAC functions are fully supported.

The CCM supports OTS instructional features. These features include initialization, simulation control, snapshot save and restore, base case save and restore, and snapshot review.

The initialization command originates from the Instructor Position MMI (IPMMI). Based on the parameters received with the command, the CCM initializes its data base and sets up the trainee consoles and the instructor CCM consoles accordingly for the training session. For PECO, either trainee console may be defined as a Power Director, Assistant Power Director, Shift Supervisor, North Load Dispatcher, Assistant North Load Dispatcher, South Load Dispatcher, or Assistant South Load Dispatcher position.

Simulation control supported by the CCM includes start, stop, pause and resume. These commands are normally initiated from the IPMMI. If enabled, they can also be initiated from the trainee consoles. The start command will be accepted by the CCM only after the completion of initialization to start a training session. The stop command stops simulation and all CCM activities. The pause command suspends simulation. After a pause command, the CCM is still active. It can be used to review SAMAC displays but accepts only stop, resume, and snapshot save commands. The resume command continues all simulation features after pause.

The snapshot save command originates from the IPMMI. It is used to take a snapshot of the CCM data during a training session. The snapshot saved in the CCM may be reviewed after initialization and before starting a training session via the snapshot restore command which allows the instructor and trainee to examine the state of the power system and the CCM at the time the snapshot was taken. The snapshot may also be saved permanently in the PSM, via a base case save command, as part of the initialization sequence. The base case restore and snapshot restore are activated during initialization to establish the starting point of a training session either from a permanently saved base case in the PSM (base case restore) or a previously saved snapshot in the CCM (snapshot restore).

The CCM provides broadcasting capability which allows the instructor to send text messages from the instructor CCM consoles to the trainee consoles.

PECO Testing

The CCM software and the PECO software modifications were developed on a Burroughs B6925 computer located at Empros using remote terminals located at Macro. A CCM/PSM interface and PSM simulator were developed to facilitate CCM testing. All unit testing and preliminary system integration were performed at PECO using an off-line Unisys A10H computer with limited Man-Machine Interface (MMI). The final PSM/CCM integration was performed at Empros on a CCM platform configured with a Burroughs B6925 computer, a subset of MMI equipment, and a CYBER 962-11. The Factory Acceptance Test (FAT) was conducted on the same CCM platform at Empros. The Site Acceptance Test (SAT) was performed at PECO on the final CCM configuration consisting of the Unisys/Burroughs A10H computer and all of the appropriate MMI equipment. After the completion of SAT, the CCM custom software and PECO software modifications were merged into PECO's SAMAC software using conditional compilation techniques so that PECO's SAMAC and CCM software will be maintained as a single source file and remain equally current as enhancements and changes occur to the SAMAC software and data base.

OTS RE-IMPLEMENTATION AT ANOTHER SITE

Re-implementation of the PSM at another site may require modification in some or all of the following aspects of the PSM.

a. Data Base

Format definitions for some data may require changes, or new data may need to be introduced. Limits for sanity checking of some data need to be re-examined.

The mapping function that establishes relations between data articles in the CCM SCADA tables and particular equipment in the PSM network model, both internal and external to the user utility may be of a different nature and may require modification or redesign. The SCADA messages exchanged over the CCM to PSM data link are subject to design modification. Extended mapping points that signify CCM points (analog and digital) which have no direct representation in the network model may vary from one user utility to another.

Although care has been taken to develop models for all commonly used utility equipment, any utility may have some special requirements, such as a particular type of pumped storage, power plant, or equipment which is not defined or is defined differently than the ones in the present OTS. These specials may require additional development.

b. Power System Models

Most of the items for re-implementation described previously affect the modeling software directly or indirectly, so the related modules should be adjusted accordingly.

Special features may require that new models be developed and new displays created. Unit control modes and the CCM's dispatch program may be different from one utility to another which may call for modification of the related PSM modules.

Power pool operations may have certain energy policies and constraints that impact the modeling software.

The CCM may be implemented on the reserve computer of a dual redundant EMS. Alternatively, it may be implemented on a standalone computer which is a replica of the one used in the EMS. When implemented on the secondary computer of an EMS, the failover capability of the EMS must be preserved. Some type of trainee console switchover capability may be required so that a trainee console may display simulated data and execute programs which run on the secondary computer.

The CCM development effort for PECO is somewhat simplified by the fact that the PECO EMS is a centralized system. The CCM development for another site must consider the implications of distributed processing and the availability of source code if it is a turnkey system. The design of the data link software may be transportable, but it is unlikely that all software will be. Basic differences between other EMS systems that will effect the data processing implementation are: external data links, data acquisition data types, and floating point formats. The level of CCM development effort will depend on the amount of necessary customization and the ease with which the custom software can be integrated.

Any part of the PSM which relates to the CCM directly or indirectly may be affected and may need to be re-examined.

EXAMPLE SCENARIOS

The capabilities of the OTS were demonstrated to a group of engineers, programmers, training program instructors, and operations personnel representing twenty-three EPRI member utilities. This seminar was sponsored by EPRI in June 1990 at PECO. The demonstrations were arranged to show the modeling capabilities of the OTS as well as the use of the OTS as an effective training tool. Experienced control center operators from Philadelphia Electric Company were actively involved in the preparation and conduct of the demonstrations. The following topics were the main theme of the seminar: Voltage Collapse, Blackstart/Restoration, Islanding, and Scenario Building. Also, an actual historical event on the PECO system was re-created to demonstrate a specific training session.

Voltage Collapse

Philadelphia Electric Company as part of the Pennsylvania, New Jersey, Maryland (PJM) interconnection covers a service area of 2475 square-miles with a population of 3,500,000 people. Vast amounts of power are imported from generation centers in the west of the interconnection to load centers in the east across high voltage transmission lines. According to a 1988 study conducted by PJM loss of a 500 KV line simultaneous with the loss of a large generator in the east could result in a serious voltage dip in the system.

This study was used to develop a voltage collapse scenario for the OTS. The system was set up with larger than normal power imports from the west. The given contingencies were programmed in the event scheduler to occur at a certain time after the start of simulation. The voltages at several key locations were monitored using the instructor displays. Voltage at a monitored bus dropped from 515 KV, before the event, to 485 KV after the event. This condition lasted for approximately one minute before the voltage dropped to 476 KV. During the one minute interval the generating units were trying to boost their controlling voltages by raising their MVAR output above maximum. Most of the over excitation relays modeled in the OTS have a trip time of one minute. This accounts for the fact that the voltages in the system were prevented from declining for approximately one minute because the units were supplying the needed reactive power. The capacitors which support and maintain the voltage, declined substantially in their MVAR output due to the low voltage profile in the system. The condition gradually deteriorated to lower values of voltage in many busses. This process lasted approximately one minute before some voltages dropped below 50% of their nominal value. At this point the system voltage collapsed and all generating units were tripped.

Another observation made in this scenario was that due to the initial impact of the original disturbance the voltages in the entire system were affected. The LTCs which are normally controlling their lower voltage sides, change taps in a direction to increase the voltage at the controlled busses. This further reduces the voltage at the high voltage busses. The LTC models of the OTS model the time delay associated with moving taps. Usually in the PECO system initial tap is moved within 90 seconds with subsequent taps within 15 seconds. This contributes to the fact that the voltage in the system collapsed approximately 2 minutes after the initial disturbance.

Islanding

A generating plant in the PECO system which consists of two 130 MW steam units, and four 20 MW combustion turbines was selected for this scenario. The generation of the plant was adjusted to supply the 130 MW of load fed by the plant. Breaker tripping which resulted in islanding the plant was performed by the event scheduler. The instructor displays reflected the existence of two islands with two different frequencies. The new island was next blacked out by tripping one of the steam units. The busses were cleared, and all loads were disconnected except for approximately 10 MW of load. Next, a combustion turbine was started to pickup the load. After the new island frequency stabilized, another combustion unit was brought on line to pick up an additional 10 MW of load. The newly formed island was next connected to the main island after the frequency had stabilized close to 60 Hz.

Blackstart/Restoration

A very large block of load was dropped in order to blackout the system. All units tripped and simulation continued with no generators on line. A section of the system was selected for blackstart. The selection was based on PECO's system restoration guidelines with some simplifications to accommodate for the limited time allocated to this demo. Five hydro turbine generators were gradually brought on line to pickup small amounts of load. Two parallel paths to a load center were established, and approximately 35 MW of load and generation was restored.

Scenario Building

This demonstration involved routine activities performed by an instructor. Actions such as changing taps, regulating voltage by generators, and dispatching generation were performed to obtain a training case.

Re-Creation of a Historical Event

On Friday, May 4, 1990 the Philadelphia Electric system experienced a fault at the Emilie substation. The fault occurred when a 230 KV breaker was closed to energize a capacitor bank. Subsequent investigations showed that the substation copper ground cable had been stolen. As a result, flashover in the capacitor bank developed extremely high voltage surges which destroyed the lightening arrestors of two 230 KV lines. The result was the loss of two 230 KV lines and complete loss of power to the Emilie 34 KV distribution substation. Inspection of the contingency analysis package at the control center showed that further loss of a 138 KV line would result in several lines violating their emergency ratings.

To re-create the system conditions on May 4th, the historical records were studied and adjustments to load and generation were made to make the conditions similar. All equipment that was out of service on May 4th, were taken out of service in the OTS. The new condition was saved as a base case. Using this case, the tripping of the lines and loss of load at the substation were programmed as scenario events. Two experienced operators were seated at the trainee positions, and two members of the PECO training staff were seated at the instructor positions. After the events occurred, the instructors, role playing different station operators and the supervisory control center operator, informed the trainees of the events at the substation. The interaction between instructor and trainees continued during the entire training exercise.

Inspection of the contingency report showed very close similarities to the report generated at the control center on May 4th. An instructor under direct order from one of the trainees started three combustion turbines. Although this did not eliminate the contingencies, the situation was slightly improved (this was also true on May 4th). At this point the OTS was taken one step further and the line in the contingency report was tripped (this event did not happen on May 4th). As the contingency report had predicted, two lines were loaded beyond their emergency ratings. The only option left to the trainee was dumping load at several substations (this was the plan on May 4th if ever the contingency had materialized). Soon after dumping load the overloads were relieved.

CONCLUSION

An advanced, transportable OTS has been developed and implemented at Philadelphia Electric Company. The requirements for simulation capabilities which allow training in normal, emergency, and restorative conditions have been met through advanced models and algorithms. The requirements for transportability have been met through the unique configuration. Software transportability has been achieved through the use of standards available at the time of design and the selected design techniques. The requirement for improvement of instructor capabilities has been met through the development of instructor tools. The demonstration of the OTS has been described.

ACKNOWLEDGEMENT

The work described in this paper has been funded by the Electric Power Research Institute as part of RP1915-2.

REFERENCES

[1] J.R. Latimer and R.D. Masiello, "Design of a Dispatcher Training Simulator System", IEEE PICA 1977, pp. 87-92.

[2] R. Podmore, J. Giri, M. Gorenberg, J.P. Britton, N. Peterson, "An Advanced Dispatcher Training Simulator", IEEE Transactions Power Apparatus and Systems, January 1982, pp. 17-25.

[3] F. Wu, R Mitten, K. Lui, and S. Vemuri, "System Operation Training Simulator in China Light & Power Energy Control System" IFAC Symposium on Power Systems & Power Plant Control, Beijing, China 1986, pp.158 -163

[4] D. Magee,et. al. "A Large Two Computer Dispatcher Training Simulator", IEEE Transactions on Power Apparatus and Systems, June, 1985, pp 1433-1438.

[5] E. Handschin, G.Schaffer, H.Van Meeteren, J.Waight, "Implementation and Integration of a Dispatcher Training Simulator", Paper No. 39-303, CIGRE, 1990.

[6] M.K. Enns, et al. Considerations in Designing and Using Power System Operator Training Simulators EPRI EL-3192, April 1984

[7] M.Prais, C. Johnson, A.Bose, and D. Curtice, "Operator Training Simulator Component Models", IEEE Transactions on Power Systems, August 1989 pp 1154- 1159.

[8] M. Prais, G. Zhang, Y. Chen A. Bose and D. Curtice, Opera-tor Training Simulator: Algorithms and Test Results", IEEE Transactions 'on Power Systems August 1989.

Restoration Simulator Prepares Operators for Major Blackouts

R.F. Chu[1], E.J. Dobrowolski[1], E.J. Barr[1], J. McGeehan[1], D. Scheurer[1], and K. Nodehi[2]

The operator training simulator is an excellent tool for checking a restoration plan and for training operators to be proficient in executing each step of the plan

Recent trends in utility generation planning have increased the likelihood of a major blackout and the complications of restoration efforts. Well-planned restoration procedures and trained operators familiar with using the procedures effectively are needed to minimize the social and economic impact of a prolonged blackout.

The restoration procedure should be up to date and well documented; the restoration instructions must be concise and easy to understand and execute. Restoration training should take place in a realistic environment in which operators can practice executing the restoration procedure.

Since it is impossible to wait for a blackout or to shut down the system to establish the necessary restoration training environment, alternative training methods are needed. Philadelphia Electric Company (PECO) decided to acquire an operator training system on which simulation-based restoration training could be conducted. As a result, PECO has actively participated in the EPRI research project (RP 1915) for developing an operator training simulator (OTS) and is its host utility. Presently, the OTS project is completed, the simulator is in operation, and two OTS workshops have been given.

This article describes the PECO system and operating environment, provides details of the simulator, and highlights some preliminary results of using the OTS for restoration training.

PECO System

PECO is a metropolitan utility company serving over 1.45 million customers in approximately 2,475 square miles of Philadelphia and its suburbs. PECO is a summer-peaking company with an all-time peak of 6,826 megawatts occurring on August 15, 1988. The total generating capacity (summer rating) is 7,759 megawatts consisting of nuclear (37.2 percent), mine-mouth-coal-fired (9.1 percent), coal-fired (8.8 percent), oil-fired (28.3 percent), and hydro and pumped-storage (16.6 percent) generating units.

PECO is a member of the Pennsylvania, New Jersey, and Maryland interconnection (PJM), which schedules units of its member companies on a daily basis and is responsible for pool automatic generation control (AGC) dispatch. PECO receives a steam megawatt regulation signal, a hydro megawatt regulation signal, and an economic megawatt loading signal from PJM every 14 seconds and uses these signals to assign online generation of the PECO units.

PECO's System Automatic Monitoring and Control (SAMAC) system or energy management system (EMS) consists of dual Unisys A10H computers and is located at the company's headquarters in downtown Philadelphia. Supervisory control capabilities are available at SAMAC and at the substations.

Restoration strategy

PECO has maintained a written restoration procedure for many years. This procedure was used to restore the system successfully without any significant problems following the 1967 blackout. Since then, the restoration procedure has been updated annually to incorporate system changes and checked using results from multiple runs of an offline ac interactive load-flow program.

The PECO blackout scenario is that a critical contingency occurs in the PJM transmission system during the summer peak loading period and the contingency leads

[1] Philadelphia Electric Company

[2] EMPROS Systems International

Reprinted from *IEEE Computer Applications in Power,* Vol. 4, No. 4, pp. 46–51, October 1991.

to a system voltage collapse. Consequently, a complete shutdown of the entire PJM power pool is assumed in the PECO procedure. With no outside generation available from PJM, the restoration instructions are to blackstart the combustion turbines and hydro and pumped-storage units to provide cranking power to the remaining generating stations and to reestablish the PECO network.

> ***Well-planned restoration procedures and trained operators familiar with using the procedures effectively are needed to minimize the social and economic impact of a prolonged blackout***

The basic restoration strategies employed by PECO are:

- **Divide the system into islands.** In the planning stage, each island is designed to have: (1) units with adequate blackstart capability (i.e., hydro units, combustion turbines (CTs), etc.) to provide cranking power to the remaining generating facilities within the island and to supply power to critical loads, and (2) sufficient generation to supply most of the load within the island.
- **Restore each island.** Immediately after the shutdown, the operators open almost all system circuit breakers and restart the blackstart units in each island. The next target in the restoration procedure is to establish a path between these units and fossil units, which require offsite cranking power. Along these paths, some critical loads are restored for stabilizing the blackstart units. After the fossil units have been restarted, other customers within the island can be reenergized.
- **Synchronize the islands.** After each island is reenergized, the last step in the restoration procedure is to resynchronize these islands.

Since nuclear generation will not be available for at least 36 hours, PECO can only carry about 60 percent of its peak summer day load and must get outside power purchases to restore the entire load. Consequently, the procedure *pauses* when a generation and load balance has been reached and *proceeds* when help from the neighboring PJM member utilities becomes available.

Restoration training

The last complete shutdown of the PECO system was in 1967. Some of the operators from that era are still with the company, and their experience provides valuable input to the restoration plan and training requirement. Since the system constantly undergoes changes, a continuous training effort has been imposed.

It is possible to stage a restoration drill in which a shutdown is assumed and the operators are to act through each step of the restoration plan. This is the approach taken by PECO as well as at other PJM member utilities.

Three or four times a year, PJM organizes a restoration drill in which each company individually simulates the following: complete shutdown with no external help, blackstart of the internal units, reestablishment of the individual company's system, restoration of customers, and reestablishment of the transmission system between companies. PJM assumes the pool dispatching position when two or more companies have been reconnected.

Throughout the drill, operators *act* through each restoration step as well as the communications to dispatchers from the PJM pool and other companies. The drill normally starts at 7 a.m. and ends after 7 p.m. on the same day. During the drill, each major restoration step is documented. On the following day, PECO holds a review meeting to discuss the drill results, to identify potential problems, and to make improvements on the existing procedure if needed.

The advantages of holding a post-drill meeting are evident. For example, during a post-drill meeting, the fact that Muddy Run pumped-storage hydro units would have run out of water after the ninth hour was identified. Assuming that these units are still available, as was the case during the drill, produces optimistic results on the restoration effort. As a result, another drill starting at the ninth hour of the same exercise was held. In this drill, a different shift of operators was involved and the new situation faced by the operators was a load shedding scheme to account for the loss of the pumped-storage hydro generation.

Needs and concerns

Despite the efforts in generating a well-prepared plan and in providing training, there are some concerns:

- The procedure has not been actually tested, and, during the drill, the system provides no feedback to the operators. The operators have to rely on their experience to envision the system behavior resulting from each step of the procedure.
- The existing procedure does not consider physical and operational constraints of the system, such as the availability of generating capacity, active and reactive balance, cold-load pickup, etc.
- The training of operators at these drills is not uniform, since a different shift of operators is involved

in different drills. The experience gained in one drill by one crew of operators may not be adequately transferred to the remaining crews.

PECO has high hopes that EPRI's operator training simulator will produce some of the solutions to these concerns and, therefore, actively participated in the project.

Operator Training Simulator

The OTS software, developed under EPRI sponsorship, is transportable and available to all EPRI members for assembling an operator training simulator.

The functional resemblance between the EMS and the operator training simulator is shown in Figure 1. The features and capabilities at the trainee consoles replicate those at the operator consoles in the EMS or SAMAC control room. For the EPRI OTS, the trainee consoles are linked to the SAMAC backup computer (Unisys A10H) containing the control center model (CCM).

To simulate the data from the SCADA RTUs and the PJM interconnection to SAMAC, a separate computer (CYBER 962-11) containing the power system model (PSM) is added and connected to the training instructor position (IP). Finally, the man-machine interface (MMI) and online display functions at the IP are provided by a third computer, an Apollo DN3000 workstation. The overall OTS configuration at PECO is shown in Figure 2.

Control center model

Typical functions of the CCM include data acquisition, supervisory control, MMI, alarm processor, automatic generation control (AGC), economic dispatch (ED), interchange scheduling, and network applications.

For restoration purposes, the AGC and ED functions are not used until the PJM transmission system is reenergized at some time during system restoration.

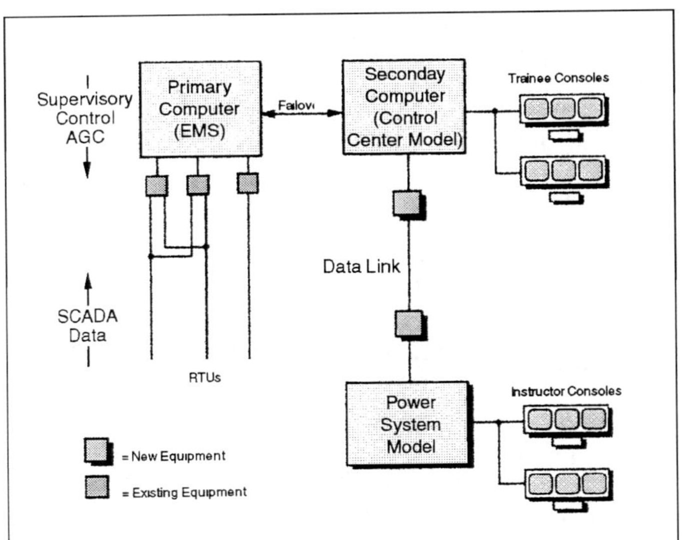

Figure 1. Extension of an existing EMS with an OTS

Power system model

The PSM computer simulates the behavior of the PECO and PJM systems. Specific models include:

- Power plant
- Load
- Protective relay
- Network.

Power plant models include fossil units (including once-through subcritical, once-through supercritical, and drum boiler units), nuclear units (including a boiling-water-reactor and pressurized-water-reactor models), combustion turbines (CTs), and hydro units. For restoration training purposes, the blackstart capability of CTs and hydro units is represented by a constant power injection. The megawatt outputs are manually adjusted, and the governor control is bypassed. The nuclear units are not included, since they will not be available for at least 36 hours. However, two outside sources with top priority are required to supply the nuclear generating stations. These two sources, in addition to power from their own emergency diesels, are mandated prior to their startups. PECO does not have any once-through subcritical units. For the drum type units, they can be represented by a power injection equal to the ramping rate of the units. For a once-through supercritical unit, the available power is mod-

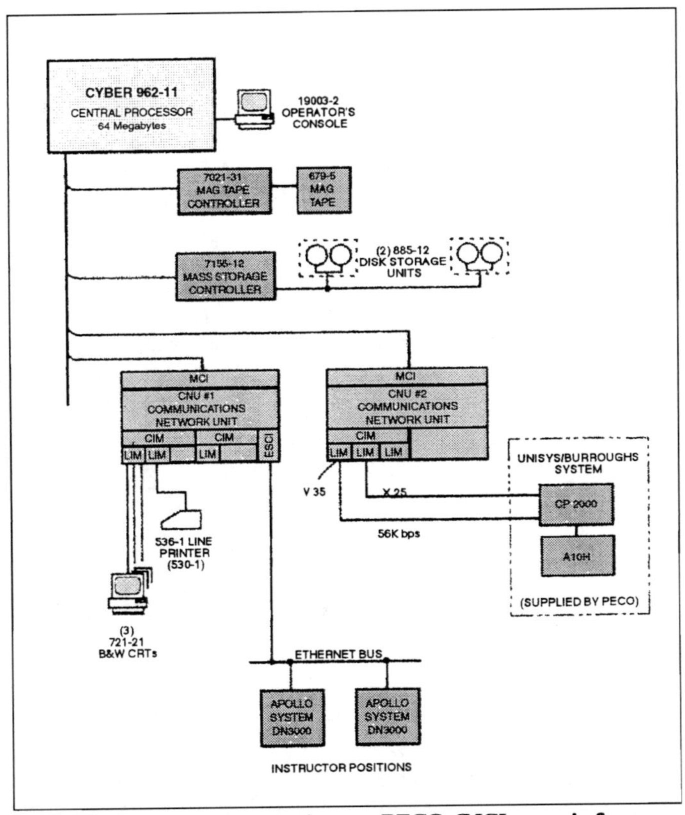

Figure 2. OTS configuration at PECO (MCI = mainframe channel interface, CIM = communication interface module, LIM = line interface module, and ESCI = Ethernet serial communication interface)

eled as the sum of a step function equal to the minimum power output and the unit's ramping rate, starting at the time when the minimum power output has been reached.

Load models of varying complexity are available: Model O (constant MVA), Model OS (Model O with voltage and frequency sensitivities), Model OM (Model O with load management capability in which fractions of load can be removed or reinstated), Model OSM (Model OS with load management capability), Model OSR (Model OS with voltage regulation capability), and Model OSMR (Model OSM with feeder voltage regulation). For restoration, Model OSM was found to be adequate.

Protective relay models include the following relays: over/under voltage, volts per Hertz, inverse-time overcurrent, over/under frequency, loss of excitation and overexcitation, synchro-check, and automatic reclosure.

Network models consist of network, transformer, and generator models.

Instructor position

The features of the IP include instructor MMI, OTS initialization and scenario creation, simulation and scenario control, simulation of manual field operations, snapshots and snapshot review, base-case save and restore, PSM and trainee monitoring, and PSM interface.

Restoration Case Studies

The OTS was tested using the following scenarios: blackstart and resynchronization of two islands and blackstart of a hydro generating station.

Blackstart and resynchronization of two islands

The objectives of this case were to test the performance of the OTS on islanding strategies, the energization of a *dead* island by an outside source, blackstart of CTs, and resynchronization of two islands.

The specific area of focus was the Delaware Generating Station, shown in Figure 3 (screen display at the SAMAC or trainee terminal).

Specific steps taken in this exercise were to:

1. Trip line 130-02, forming a Delaware island in which the steam units (7 and 8) are supplying the local load. (Note that the loads are not shown in this figure. No megawatt transfer through transformer 2 is shown, transformer 1 is assumed off due to maintenance.)
2. Trip units 7 and 8, causing a blackout in the Delaware island.
3. Open circuit breaker (CB) D1, which splits up the Delaware island into left and right islands.
4. Clear all CBs on the dead bus in the right island.
5. Close line 130-02 CB, energizing the left island with an outside source.
6. Blackstart CT units 11 and 12.
7. Close all ring bus CBs in the right island, balancing generation and load within an island.
8. Close CB D1, synchronizing the two islands.

The exercise went very smoothly, and the results compared favorably with those from the offline planning studies.

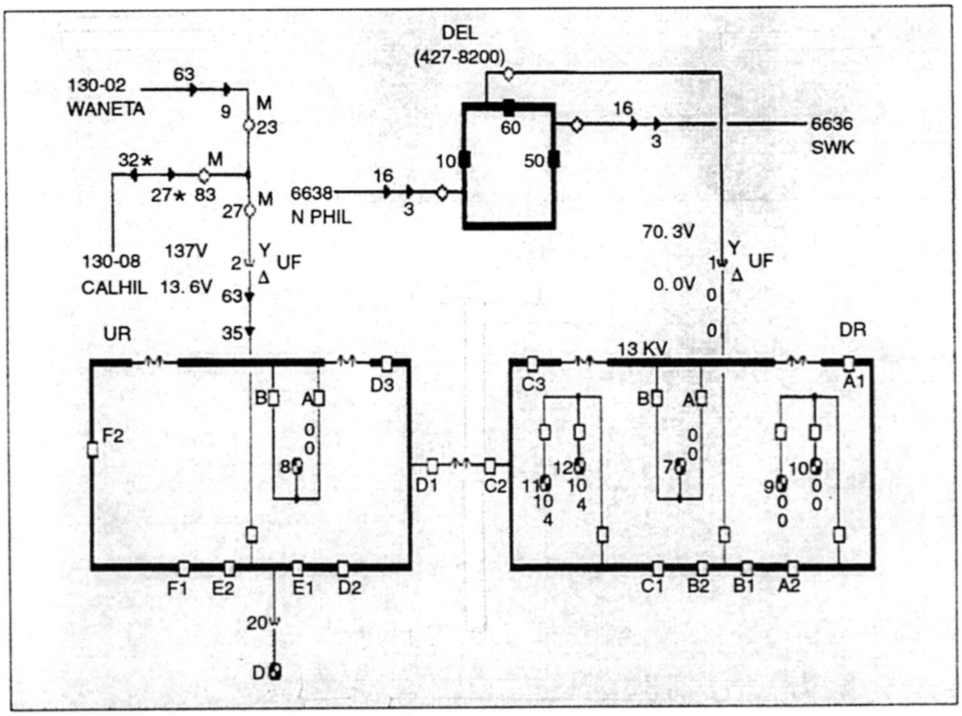

Figure 3. Delaware islanding scenario (DEL = Delaware Station, SWK = Southwark Station, CALHIL = Callowhill Substation, N PHIL = North Philadelphia Substation, UR = up-river bus, DR = down-river bus, UF = equipped with under-frequency relay, and M = motor operated disconnect. Megawatt directional flow is shown with black arrowheads and Mvar directional flow is shown with green arrowheads. All voltages are in kilovolts.)

Blackstart of a hydro generating station

PECO relies heavily on generation from the Conowingo Hydro Generating Station for most of the system's blackstart capability. Since Conowingo Station is remote from the load center in Philadelphia and is also near Peach Bottom Nuclear Generating Station, it is important that the PECO procedure allows adequate cranking power to supply the fossil units in the city, provides a second feed to Peach Bottom, and, at the same time, picks up as many critical customers as possible to stabilize the units.

Figure 4 shows the screen display (recorded after the complete shutdown) of the affected portion of the system diagram. The connection between Conowingo and Peach Bottom is via the 34-kilovolt line (line 3305).

The objective of this exercise was to test a portion of the PECO restoration procedure with reference to the Conowingo units. Specific steps taken in this exercise were to:

1. Trip line 220-01 to Clay, and line 220-03 to Nottingham.
2. Blackstart units 3, 4, and 6.
3. Energize line 220-01.
4. Close CB 115 and transformers 1 and 5 CBs at Colora (these CBs are not shown in this figure).
5. At Colora, close line 3310 CB and energize line 3301.
6. At Clay, close transformers 100 and 120 CBs and line 149-00 CB.
7. Close line 220-03 CB to Nottingham.
8. At Colora, close transformer 20 CB and line 225 CB.
9. At Bradford, close transformer 230 CB, and lines 410 and 420 CBs.
10. At Nottingham, close transformer 6 CB, and lines 100 and 120 CBs, and energize line 143-00.
11. At Nottingham, close CB 90.
12. At Bradford, close CB 250.
13. Stop.

Throughout the restoration process, the megawatt outputs from the Conowingo hydro units were raised to meet the megawatt demand. The voltages were also adjusted when the voltage limits had been violated. The OTS results were credible, and the Delaware and Conowingo scenarios are being used in the OTS Workshops.

Application Highlights

Several important benefits and shortcomings have been identified during the course of PECO's participation in this EPRI research project.

Benefits

The EPRI software is capable of simulating a complete shutdown and blackstart. In fact, the agenda of the OTS workshops calls for a voltage collapse leading to a complete shutdown, blackstart, and restoration.

Another attribute of the OTS software allows opening breakers on a dead bus.

The OTS allows all operators to be trained using the same training scenarios and to have the same opportunities to learn how to handle a specific operational problem.

The OTS power-flow program is capable of handling restoration conditions, while many EMS power-flow programs cannot handle these conditions.

The OTS generates an alarm if conditions fall below a preset level.

As a part of a new power system restoration project (RP 3104-1), PECO has recently concluded efforts with EMPROS to develop a cold-load pickup model in the OTS. Depending on the length of the outage, the effect of the cold-load pickup can be simulated. However, this brings up an interesting dilemma, since most EMS's, including PECO's SAMAC, do not have the cold-load information on each feeder.

Shortcomings

The OTS simulation does not take into consideration the actual time factors involved in the restoration. Consequently, the actual restoration time could be many times those indicated in the OTS. The lack of the timing factors could be

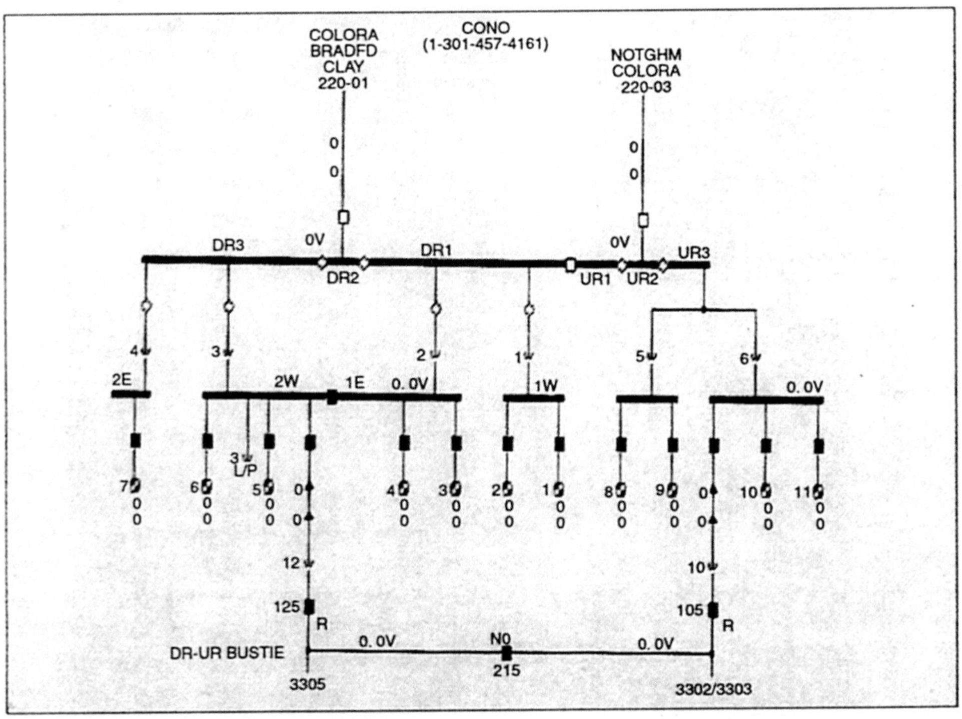

Figure 4. Conowingo restoration scenario (CONO = Conowingo Hydro Generating Station, BRADFD = Bradford Substation, NOTGHM = Nottingham Substation, L/P = station light and power, NO = normally open, R = automatic reclosure)

very crucial in a restoration training program and need to be provided by the instructor.

For restoration purposes, external interaction and coordination need to be included as a part of the training program. For example, a training program should be developed in which the instructor can play a *realistic* role of field personnel, substation operators, power plant operators, pool dispatchers, and dispatchers from the neighboring utilities.

If a specific restoration step cannot be executed, the OTS is not very *friendly* in providing alternative restoration actions and guidance. This capability is not required as a part of the original OTS project and is being developed in the new project (RP3104-1).

Ongoing Research

At PECO, research activities (e.g., the recent development of the cold-load pickup model) are ongoing to develop solution approaches to overcoming some of the shortcomings.

Currently, EPRI, PECO, and the University of Washington are working together in the EPRI project (RP 3104-1) to apply expert-system technology to system restoration. The primary objective is to develop tools to assist operators to restore the system. Upon the completion of the expert-system prototype, the plan is to integrate the prototype into the OTS for restoration training. The progress and result of the research activities will be reported in the near future.

Summary

Because the social and economic consequences of a major blackout are devastating, utilities must take steps to prepare for such an event. From a system operational viewpoint, the availability of a well-prepared restoration plan and an operator training simulator for restoration training is a step in the right direction.

PECO is currently evaluating all functional capabilities of the OTS in normal, emergency, and restoration operations. For restoration purposes, preliminary experience shows that there is room for improvement. Particularly in the situation when some unforeseen events prevented the execution of predetermined restoration instructions, it is desirable to have a computer aid that resembles an expert operator to provide alternatives and guidance for continuing the restoration effort.

Independent from results of these activities, the OTS is still an excellent facility for checking a restoration plan and for training operators to be familiar with the execution of each step in the plan.

Acknowledgment

Both the operator training simulator project (RP 1915-2) and power system restoration project (RP 3104-1) were sponsored by the Electric Power Research Institute. Drs. C.J. Frank and D. Curtice were the EPRI managers for the OTS project and G.W. Cauley is the EPRI manager for the restoration project.

For Further Reading

J. Waight, et al., "Operator Training Simulator: Technical Report," Final Report to EPRI for Research Project RP 1915-2, EL-7244, May 1991.

M. Prais, C. Johnson, A. Bose, and D. Curtice, "Operator Training Simulator: Component Models," *IEEE Transactions on Power Systems*, Volume 4, Number 3, August 1989, pages 1160-1166.

M. Prais, G. Zhang, A. Bose, and D. Curtice, "Operator Training Simulator: Algorithms and Test Results," *IEEE Transactions on Power*, Volume 4, Number 3, August 1989, pages 1154-1159.

BULK POWER SYSTEM RESTORATION TRAINING TECHNIQUES

A joint report by the Working Group on Operator Training - M. E. Cooper, Chairman; and the Power System Restoration Working Group - M. M. Adibi, Chairman.

Contributory authors are: M. M. Adibi, L. H. Fink, R. G. Wasley, R. T. Gonzales, A. B. Storey, F. J. Wood, R. J. Kafka, E. J. Dobrowolski, K. Nodehi, J. G. Waight, R. Hoffman.

ABSTRACT - This paper is presented to focus industry attention on various tools and techniques used to develop and implement effective bulk power system restoration training programs; the need to continue developing improved training tools is highlighted.

Six short note reports are presented. The first two reports focus on the preparation which should take place as a precursor to the development of a formal system restoration training program; this includes identification of the operating issues which should be addressed in a system restoration plan/training program and an analysis of how operators should recognize and respond to these conditions. Three reports describe the wide variety of training methods currently used in existing system restoration training programs; the methods described range from simple classroom lectures to the use of advanced power system simulators. The last short note provides a synopsis of the emergency preparedness training programs used in other industries.

INTRODUCTION

M. E. Cooper, Associate Member
Potomac Electric Power Company

With the increase in power system reliability, major disturbances occur less frequently, consequently operators receive little experience in restoration. The need to train power system operators to restore their electric system is well recognized. There are a variety of tools currently available which can be used to develop and deliver effective system restoration training programs. Each company or power pool must develop its own system restoration plan based on an analysis of how its electric system will behave under abnormal conditions; formal restoration plans can then be written and training programs can be developed to ensure that system operators can follow these plans.

There are many different methods of delivering system restoration training; each method has its place in a comprehensive system restoration training program. Simple methods such as self-study are most appropriate for learning 'facts' about system restoration; other methods such as computer-based instruction can be very useful in assessing understanding of basic concepts. Role plays and simple simulations are good tools for drill and practice of well-defined scenarios; more advanced classroom exercises and use of more sophisticated high fidelity simulators are extremely beneficial in training system operators to respond to situations that have never actually occurred.

While there are many useful training programs in existence, much remains to be done in the field of developing system restoration training. Many companies have outdated restoration plans that don't take into consideration the latest research; most companies have not conducted a job task analysis to assist in the development of a system operator training program. Few companies have dedicated the resources required to develop and utilize comprehensive training programs. The emergency preparedness training of airline pilots and nuclear plant operators is much more sophisticated than that offered to power system operators.

92 WM 147-9 PWRS A paper recommended and approved by the IEEE Power System Engineering Committee of the IEEE Power Engineering Society for presentation at the IEEE/PES 1992 Winter Meeting, New York, New York, January 26 - 30, 1992. Manuscript submitted September 3, 1991; made available for printing November 25, 1991.

The available generic power system simulators can provide procedural training in early stages of operator training; for exercising and preparing experienced operators to cope with system-specific and time-critical emergency situations, such as restoration, high-fidelity system-referenced simulators are needed.

ISSUES TO BE ADDRESSED IN POWER SYSTEM RESTORATION (PSR) TRAINING

M. M. Adibi, Senior Member
IRD Corporation

L. H. Fink, Life Fellow
ECC, Incorporated

PSR Problems

Major power disturbances are caused by a variety of initiating events, each impose different restoration problems and in each restoration case, power system operators play the most critical role. The impact of prolonged blackouts on the public, the economy, and power system equipment makes rapid, effective restoration very important.

During restoration, system operators are faced with a state of their system that is quite different from that to which they are accustomed in day-to-day operation, and for which the EMS application programs at their disposal were not designed and are not well adapted.

There are distinct differences between the normal state and the restorative state in the kind of models relevant to each, in the objectives pursued, and in the information available at the control centers to support the models. In the normal state the primary objective is to minimize the cost of producing power, subject to observing certain security constraints. EMS application programs that help attain this goal incorporate models that represent the simplified, primarily steady state behavior of power systems, and incorporate data which is principally obtained automatically from the system.

During the restoration process, by contrast, the objective is quite different--it is to minimize restoration time and the amount of unserved kilowatt-hours of energy, with security as a subsidiary objective, and without regard to production cost. At the same time many of the dynamic phenomena, neglected (or unimportant) during normal operation, play a critical role during restoration and need to be taken into account. For example, in restoring a power system, operators often need to consider information such as:

- the mix of available prime movers, including drum type, once-through subcritical, once-through super-critical, oil burning, coal burning, and nuclear;
- boiler-turbine start-up characteristics, including time intervals required for cranking power, paralleling, reaching minimum loads, soaking and reaching maximum load; and
- the prime movers' mode of operation and conditions just prior to the disturbance, such as base loaded, cycling, peaking, hot, cold, and very cold.

The present SCADA systems and EMS software do not provide all this information. As a result, additional data must be obtained, usually via voice communication between the operators in the EMS center and those in major facilities.

PSR Plans

To help power system operators in dealing with system restoration, most operating companies maintain restoration

Reprinted from *IEEE Transactions on Power Systems,* Vol. 8, No. 1, pp. 191–197, February 1993.

procedures based on certain operating philosophies and practices, and familiarity with characteristics of their power plant restart capabilities and their power system reintegration peculiarities. These plans are usually developed well ahead of time, on the basis of numerous simulations of the steady state and dynamic behavior of the power system under various operating conditions, and engineering-operating judgment reflecting many factors not readily modeled by those tools.

The resulting procedures are made available in the form of written manuals which prescribe steps or "rules" to be followed during restoration, along with appropriate check points along the way in order to verify that everything goes according to the plan.

PSR Issues

The issues of common concern which need top be addressed in PSR training include:

- Identification of the status of the collapsed system, components and equipment.
- Restart and supply of station service to plants, substations, cable pumping plants, compressed air, etc.
- Coordination of power plant start-up timings with load pick-ups to bring generators to their stable minimum levels and within the range of major analog controllers.
- Energizing large sections of transmission lines within the acceptable transient and sustained overvoltages.
- Picking up load in large increments without the risk of frequency decline.
- Reintegration of the skeleton of the bulk power supply with the requisite time-consuming switching operation.
- Deactivation of automatic load shedding and automatic switched capacitors during initial phases of restoration.
- Maintenance of steady state and transient stability as the system is being restored and when impedances are large.
- Reduction of standing phase angles when closing loops to firm up transmission paths.
- Probability of successful start-up of thermal units, in particular, combustion turbines.
- Stability of black start and peaking combustion turbines while they constitute a large portion of generation.

PSR Training

Available generic simulators can provide procedural training in the early stages of operator training. However, for exercising and preparing experienced operators to cope with system-specific and time-critical emergency situations such as PSR, high-fidelity system-referenced simulators are needed. It is important to note that during the restart and reintegration phases of PSR, a power system often consists of one or two islands, most of the automatic controls have tripped or are deactivated; and the system is primarily under manual control. During these early phases of PSR, wide voltage and frequency ranges are tolerated. Under these large perturbations of long duration, the models and simulations which have primarily been developed for small perturbations will not accurately represent the behavior of the power system and its components.

An on-line restoration guidance system, capable of guiding the operator in making decisions as to what steps to take and when to take them, would go a long way toward minimizing the duration of blackouts, and consequently their impact. The nature of the problem, in the sense of incomplete information and many ambiguous alternatives, suggests a need for an integrated tool, consisting of both expert systems and appropriate analytical tools. Such a tool would serve as an advisor to bring to the on-line environment many of the tools and capabilities planners/operators use in planning for restoration, while it would still be subject to operator approval for all major decisions.

This approach can be considered as a goal, but in the meantime, it is important to develop detailed PSR plans and keep them up-to-date for operator training.

JOB TASK ANALYSIS (JTA) FOR SYSTEM RESTORATION TRAINING PROGRAMS

R. G. Wasley, Member
MACRO Corporation

JTA is a process that identifies, analyses, and documents actual on-the-job tasks to define essential job activities including requisite knowledge and skills. JTA also provides a foundation for developing comprehensive training programs. For example, in competency-based training, where success is assured if and only if the trainee learns to perform a specific job, JTA provides the starting point for the overall systematic development of a training program. In this regard, JTA focuses on the desired outcome of training to define exactly what the trainee must accomplish in terms of specific tasks, performance criteria representing the standard or level of competency to be achieved, and the conditions under which the tasks must be performed. The results of JTA are then used to develop the training program's methodology, where the issue is how best to achieve the accomplishments or goals of the training program.

Relevance of JTA to System Restoration Training

Is JTA relevant to system restoration training? Restoring a power system from a partial or full blackout state to normal operation is a complex process. Nevertheless, it is a job. Someone has to do it!

Typically, the restoration process is guided by a set of operational procedures developed as part of a general system restoration plan. A plan that is based on postulated scenarios (after all, system restoration does not occur frequently), and a plan that can involve many implementation paths. The ideal plan should be flexible, covering as many initial system disturbance conditions and potential restoration problems as possible. The plan should also be kept up-to-date and implemented during regular restoration drills. These drills provide an opportunity to exercise and improve the plan and for utility personnel to gain valuable, albeit simulated, restoration experience.

Further, despite the complexity of restoration, the restoration process inevitably breaks down into a sequence of specific tasks and related activities. These tasks include monitoring system conditions, blackstarting units, establishing transmission paths between stations, reconnecting load, connecting electrical islands, and reconnecting tie lines. Task activities may be implemented by using SCADA facilities or by directing others such as plant and substation personnel.

Within this context, the answer to the topical question concerning the relevance of JTA to system restoration training is, 'yes'. System restoration training programs should result in system operators who have mastered the restoration skills of successful system operators by learning and performing the specific tasks used to restore their own power system. And JTA should be the starting point. But, remember, JTA does not determine how the goals of the training program are achieved; this aspect of system restoration training is considered in other sections of this paper.

Applying JTA to System Restoration Training

Applied to system restoration, JTA must first identify all known restoration tasks by producing a preliminary list of the task names. A task is a complete set of activities (steps or actions) that result in the accomplishment of a major goal. Tasks may be broken down into subtasks. For example, the task to supply cranking power to a generating unit from a blackstart unit may break down into: (1) select a path and (2) energize the selected path. Task (or subtask) activities, such as close breaker X, need not be documented at this time. They are analyzed later.

How is the list produced? A good starting point is the existing system restoration plan. Typically, the training specialist puts together a preliminary task list by reviewing this plan and by interviewing restoration experts, such as senior system operators and their supervisors. Task observation is much less amenable because system restoration is not routine. Nevertheless, in most cases, it would be helpful to observe the "simulated" restoration process during a drill. Other resources include consultants, other utilities, and technical literature.

The next step is to verify the preliminary task list to ensure completeness and inclusion of only those tasks that are really critical to system restoration. Verification techniques include surveys based on questionnaires submitted to operations staff and additional reviews by recognized restoration experts.

Once the restoration tasks have been identified and verified, they must be analyzed or broken down into the specific activities by which each task is performed or accomplished. This process should include analysis of the specific technical knowledge, implementation skills, and prerequisites that are needed to perform the restoration activities accurately. Relevant technical knowledge and skills are best mastered at the time they are needed by an activity, but may be sufficiently generic or applicable across several activities that they can be mastered separately.

Analysis can be based on having one or more restoration experts "walk through" each restoration task as it would actually be performed, describing each activity and identifying any technical knowledge that is essential to performing the activity correctly. Once documented, the task activities should be reviewed by operations staff and restoration experts until the training specialist is satisfied that they are complete, accurate, and up-to-date. The task and activity descriptions will require regular maintenance. Analysis should also define the overall structure of the system restoration program by grouping logically related tasks and by sequencing tasks to reflect their interdependencies (some tasks being prerequisites for others). Finally, analysis should add the necessary performance criteria and conditions that apply to each task.

Much of the technical knowledge related to system restoration involves actions, facts, and skills that are also relevant to normal operations. Therefore, a system restoration training program should build on "normal" training programs to avoid duplication and thereby foster greater efficiency and effectiveness. This means that the JTA for system restoration training should clearly distinguish between the normal and restoration aspects of task activities. For example, closing a breaker is not in itself a special system restoration activity, but checking the necessary conditions to be met before closing the breaker can be a special restoration activity.

Summary of JTA Benefits

JTA results in many important benefits. It provides the essential foundation for building a restoration training program. And, by documenting the job of restoration in terms of the actual restoration tasks accomplished, a common reference for easier understanding and communication is established.

This means that trainees, instructors, training specialists, restoration experts, and managers can immediately appreciate the significance of the system restoration program within their own areas of interest. Trainees know exactly what they will learn, instructors know exactly what they will teach, training specialists know exactly what must be accomplished, restoration experts know that what has been agreed upon incorporates correct restoration practices, and managers know exactly what the system restoration program entails for purposes of funding or allocating appropriate resources.

The benefits of JTA can also lead to greater trainee motivation, a more complete and consistent training program, greater instructional efficiency and effectiveness, less dependence on instructor capabilities, and improved ability to evaluate trainee performance.

BULK POWER SYSTEM RESTORATION TRAINING - CONVENTIONAL MEANS

A. B. Storey, Member
R. T. Gonzales, Member
New York Power Pool

F. J. Wood
Pennsylvania, New Jersey
Maryland Interconnection

Most utilities have broad restoration plans, which are based on fairly optimistic assumptions. It is presumed that the dispatcher can restore the power system from these plans - with appropriate adjustments, if necessary. Therefore, training must give the dispatcher a good understanding of the fundamentals, a thorough understanding of the policy, and a practical capability for handling the myriad of unpredictable events which can be expected to occur. The dispatcher must be equipped to handle conditions which are typically different from anything he or she has ever experienced before.

This short note paper addresses means for providing training for these unusual needs, utilizing conventional training methods.

Self-Study Methods

Portions of restoration training can be handled with written material or video tape presentations. Self-study methods should comprise a relatively small portion of the overall training effort. Although the dispatcher may learn some of the fundamentals through these methods, more practical means are needed for applying the principles learned.

Classroom Techniques

This form of training can range from one-on-one tutorial sessions to formal regional seminars. These programs should initially provide dispatchers with a practical understanding of the basic principles involved so that unexpected problems can be solved. The utility's restoration policy, and its goals, should be the core of these programs. The effectiveness of the training, therefore, will be closely tied to the effectiveness of the policy.

Coverage topics fall into two categories:

- Generic - This area includes necessary bulk power system restoration fundamentals, and must address major issues that the dispatcher needs to consider during a restoration. Training must be practical.
- Utility Specific - Training must address the utility's restoration plan and its priorities. Assessment methodologies, specific job responsibilities, communication techniques, and notification requirements must be covered.

Classroom style training should be conducted by a combination of system operation, engineering, generation plant, and field operation personnel. "What if" type situations must be discussed in detail. The trainers must be responsive to dispatcher feedback.

Informal Role Play - One method for enhancing classroom style sessions is to conduct role play scenarios where dispatchers "talk through" the stages of a system restoration. This training method effectively involves the participating dispatchers. Role playing uncovers problems and misunderstandings.

Drills

The role playing exercise should be expanded to include field, plant, and other utility personnel. The scope of the scenario can be extended to the Pool level as the Pennsylvania-New Jersey-Maryland (PJM) Interconnection, New York Power Pool (NYPP), and others have done. PJM conducts two drills per year, with each member utility participating. Every effort is made to simulate realism, and approximately 300 operating and support personnel are involved.

Drills provide an excellent testing ground for the restoration plan. If conducted realistically, they will uncover potential hurdles, and even problems with the existing plans. A key to good training and problem solving lies with the extent of the exercise. It should involve as many of the people and events, as would be involved in an actual bulk power system restoration. The exercises must be run frequently, and conditions must be varied so that dispatchers are able to train in the handling of unpredictable events.

Job Aids

The restoration process can be greatly enhanced with the inclusion of job aids. These may include reference items such as flow charts, color-coded system diagrams, or step-by-step operating instructions. Training must address the use of these aids, if they are to be used effectively on-the-job.

Evaluation

In order to measure the effectiveness of the training program, and individual progress under that program, evaluations are recommended:

- Fundamental concepts and generic material are best tested with objective written or oral tests.
- Drills and role playing are effective ways of measuring dispatchers' skills in handling cases involving no deviations from the plan. (Events requiring selective deviation from the plan can also be introduced during drills and role playing exercises, but care must be exercised in evaluating dispatcher responses, against various possible solutions.)

The restoration plan is also being "tested" throughout the training process. It is essential that management react quickly to all policy and technical discrepancies which are uncovered during training sessions.

Summary

The need for restoration training cannot be overstated. The training conducted must be practical and thorough. Dispatcher involvement in the development of the policy, will enhance not only the document, but also the dispatchers' understanding of it.

COMPUTER-BASED TRAINING TO SUPPORT RESTORATION TRAINING

R. J. Kafka, Senior Member
Potomac Electric Power Company

Potomac Electric Power Company (PEPCO), like many other operating companies, became increasingly aware of the problems related to system restoration in the late 1970s. A number of activities were taken with respect to modifications to the electric system to aid in system restoration, and studies were made to determine how the system could be restored following a substantial or complete shutdown. These activities were successful, but it was also necessary to transfer the knowledge to the system operators, and many of the training methods described in this short note paper were used. One of the methods explored was computer-based training, but very few relevant courses were found. Finally, PEPCO determined that it would undertake the development of a specific course to teach the key guidelines that had been uncovered during the planning studies.

Computer-based training is a self-learning method where the student interacts with a special computer program rather than a set of printed or recorded materials. Many of the computer-based methods are similar to the programmed learning systems developed in the 1950s and '60s, but with features that avoid the problems of those early programmed learning courses. In fact, computerization of this training method has made programmed learning practical while providing additional capabilities and features. Automatic program looping and control, superior display and simulation capabilities, and non-intrusive record keeping are key contributions of computer-based training.

Advantages

An obvious similarity between computer-based training and other programmed learning methods is that it is student-paced and scheduled. Given sufficient resources for the student population, students can determine when they will be engaged in training. Since training is on an individual basis, there is no inherent need to coordinate and schedule a class with an instructor and other students. This characteristic is very useful in situations where only a few people can be spared from a work organization for other activities, or when work schedules do not lend themselves to organized classes. Since system operators typically work on rotating shifts, the ability to allow a student to schedule a training session at odd hours or as opportunities arise is invaluable.

Properly designed computer aided training is truly interactive. The computer program reacts to student inputs or actions and responds quickly with appropriate feedback. Various correct and incorrect responses should be anticipated by the lesson designers, and feedback should reasonably supply additional information or prompt for a more appropriate student action or answer. When student input indicates that review is required, the program can automatically branch to a prior portion of the lesson or to a different description of the concept or skill to be learned. Depending upon the design of the lesson(s), sufficient demonstration of the student's mastery of a subject could also bypass lesson sections or drills. The interactive nature of the computer-based training is especially evident in simple simulations, discussed later.

In any well designed training program the learning objectives will have been determined before a single lesson is written. Computer-aided training programs virtually require this discipline due to the investment in equipment, computer systems and, especially, the technical matter experts, training personnel and computer professionals. The result is a structured approach to the mastery of the set of skills and educational goals of the training course. The goals will have been partitioned into a number of smaller goals and steps, and a logical sequence will have been developed so that each intermediate training objective will not depend upon one which appears later in the sequence. For greater student attention to and retention of each objective, the skill learned or knowledge gained will be referenced or used in subsequent, and ideally, immediately following lessons.

Because the advancement through each lesson and from lesson to lesson is under the control of some program shell or operating system, the mastery of an objective can be assured or enforced before the student is allowed to advance to the next lesson. The combination of a carefully laid out sequence of learning objectives and the requirement of mastering an objective before advancing to the next lesson section or next lesson is a significant advantage. Each student's progress can be determined by knowing where he is in the course, and the instructor or supervisor can be assured that all of the objectives will have been mastered at the end of the course.

The information regarding the current position of the student in the course is easily obtained from the computer if the program shell or operating system has been designed to collect this information. Even the most elementary program shell for computer-based training should have this capability. Furthermore, the student "path" to that point can be documented by maintaining a record of the number of attempts required to complete a lesson or by maintaining quiz or drill scores. This can all be done automatically so that it is not intrusive. Of course, students generally know when records are being maintained regarding their progress. There are differing philosophies concerning the effects of such knowledge on the students' acceptance of a self-paced training program. The advantages of collecting the data, even if student identities are concealed, are quite evident when evaluating the effectiveness and progress of the training.

Simulations

A significant advantage of computer-based training is the ability to include animation or simulations in the lessons. Animated simulations will depend upon the graphics display characteristics of the computer system, and computer-based training which uses simulation will begin to restrict the computer systems on which the training program can operate. Before the advent of quasi-standard personal computer systems, this restriction could be the dominating factor. For purposes of our discussion we are essentially addressing the capabilities of IBM compatible personal computers.

Interactive simulations are one of the best ways to illustrate dynamic processes, and system restoration is filled with dynamic processes including frequency response to load pick-up and line-end voltage variations. By being able to visualize the process that is desired to be controlled, the operator has a better understanding of cause and effect. This is reinforced if the simulations have been designed to include operator control actions. Learning is significantly enhanced by simulations which allow the student to experiment, discovering interactions, what works and what does not work. Simulations must generally be simplified or otherwise modified to reduce the program size and computation time. When it is not feasible to simplify the simulation dynamics, there are other techniques to reduce dead time and computational burden. One technique is to perform the mathematical simulations in advance and to store intermediate solutions in a file which can be accessed by the training program. The results can be quite convincing to the student so that it appears that the dynamics are occurring in approximate real-time. This technique also works well for step-wise simulations such as those involved with switching, load pick-up and network extension. Since all simulations are done in advance, careful planning is required, and student interactive control choices must be limited to those included in the simulation.

Disadvantages

As was stated earlier, there are some conditions which apply if computer-based training is to be attractive. The required equipment must be available to the student population. It must be available on some reasonable schedule or in reasonable quantities. Finally, the desired training program must be available.

In the discussion of simulations, the equipment was limited to IBM compatible personal computers, since that is expected to be the most widely available set of compatible equipment. There may be further requirements regarding display capabilities, execution speed, math co-processors or disk storage. If the required equipment is not already available to the students, there may be some additional expense or delay in obtaining it. This may also limit the number of students who can be accommodated. There may be schedule conflicts to resolve if the required quantity of equipment cannot be obtained. Whatever the required equipment is, whether it is IBM compatible PC or some other specific configuration, it must be available for training.

Of greater concern than the availability of the proper computer equipment is the availability of the proper training courses. There are many courses covering general or elementary subjects, and some of these are appropriate for operator training, but very little is available specifically for system restoration. The development of a specific course could be very expensive and time-consuming. The usual estimate of effort is one hundred hours of development time for each hour of course time, and the individuals involved include technical matter experts, training professionals and computer course development professionals. It would also be reasonable to expect that some type of program shell or course controller would need to be obtained.

An Example

As stated earlier, due to the lack of available material, PEPCO undertook the development of a computer-based system restoration course. The course was developed through a vendor on a proprietary training program shell based in mainframe computer equipment. (The course was later converted to IBM compatible PC equipment.) Training objectives were defined, and a team of training, computer and technical experts was formed to develop the course. While the development effort was considerable, the end result was a ten-lesson course which still meets PEPCO's requirements for the computer-based portion of the system restoration training. Because so many basic concepts must be considered for system restoration, the course has also been beneficial in introducing new graduate engineers to the concepts of electric systems.

The course covers many aspects concerning frequency-load balance, voltage control and system re-energization. It takes advantage of the graphics and interactive capabilities of the computer-based training system, and enjoys a high rate of operator use and acceptance. Coupled with the other training methods employed, the computer-based system restoration training is an integral part of PEPCO's operator training.

CAPABILITIES OF AN OPERATOR TRAINING SIMULATOR TO SUPPORT RESTORATION TRAINING

J. G. Waight, Member
K. Nodehi, Member
Empros Systems International

E. J. Dobrowolski, Member
Philadelphia Electric Co.

As described elsewhere in this paper, there exists the need for training system control center operators to restore the bulk power system following a system collapse. Since in most utilities, blackouts are an infrequent phenomena, there do not exist numerous opportunities for on-the-job training in system restoration. However the criticality of the situation during a system-wide blackout, requires system control center operators to be able to restore the bulk power system as quickly and as safely as possible. While there are many facets to appropriate training to ensure proficiency in restoration, an Operator Training Simulator (OTS) is a key part of a successful training program in this area. Briefly, an OTS permits the training of system operators while avoiding risk to human life, damage to power system equipment, or the cost of non-economic operation due to trainee errors during on-the-job training. Using an OTS, it is possible to expose trainees to a greater variety of critical situations, to allow practice and drills in solving operations problems, and to compress the equivalent of many years of on the job training into much shorter time periods. This short note paper describes the capabilities of a system control center OTS which are essential to supporting training in system restoration.

An OTS is made up of three major components: a Power System Model (PSM), a Control Center Model (CCM), and an Educational Subsystem (ES). The architecture and terminology have been previously described in reference [1].

Power System Models and Algorithms

The Power System Model provides the simulation of the power system. The specifications for this model are derived from its intended use, namely the training of the system operator. These models are required to simulate measured data which is viewed by the system operator in the control center by means of a real time SCADA or EMS system. This means that models are required which can simulate the dynamic behavior of the bulk power system as seen by the system operator through a system which scans data typically every 2, 10, and 30 seconds. The models should simulate data with accuracy similar to that of the data gathered by the instrumentation and telemetry of modern SCADA systems (typically 1% and 5%). The output required of these models are the electrical quantities such as MV, MVARS, MVA, amps, and digital statuses seen by the system operator in the control center. Fidelity requires that the models of the simulator behave as the real world would in response to the actions of system operators. A remarkable degree of fidelity is possible today, depending on the availability and accuracy of data for model parameters.

An OTS then, has models of the bulk power system network: transmission lines, transformers, phase shifters, shunt reactors, shunt capacitors, generating units, loads, etc. Dynamic models are required for generating units, prime movers, loads, relays, transformers, and external company AGC controllers and power pool controllers. It is important that the models and algorithms of the OTS give valid results over a wide range of voltage and frequency such as is likely to be experienced during system restoration.

Since system restoration training requires training in specific power system operating procedures at the user utility, the OTS must allow an exact representation of the user utility's system. The OTS must allow the representation of all relevant types of equipment with exactly the same connectivity as used in the utility.

Control Center Model

Procedure based training such as system restoration requires that the representation of the control system used by the trainee in the OTS be an exact replica of the on line system. This means that the displays, keyboard, strip charts, buttons, alarm bells, whistles, etc. should be the same in the OTS and the on line system. The OTS should allow the training of a team of operators simultaneously, as is required during system restoration. Multiple trainee consoles are thus required, each being able to present the view of the training exercise as would be seen at that console during actual system restoration. Together with the PSM, the CCM allows the trainee operator to execute the steps of the power system restoration plans and to see the results of his actions in real time exactly as if conducting an actual restoration. This is seen as a critical advantage of using the OTS for restoration training.

System Restoration is a communications intensive process. The system control center operator is in constant communication with numerous other utility operators outside the control center such as the operators of power plants, manned substations, field crews, and the operators of neighboring utilities. This means that the OTS should also have the ability for training the operator in such communications. A typical approach to providing this aspect of system restoration training, is to provide the instructor with the ability to play the role of those previously mentioned utility personnel outside the control room. The work load required of the instructor in these scenarios is very large and typically requires two or more instructors to conduct effective training.

Educational Subsystem Capabilities

The Educational Subsystem capabilities normally provided with an OTS are important for training operators to conduct system restoration. Time based pre-scheduled events are an important way to steer the system toward a collapsed or blacked out state. Other events or malfunctions of the power system equipment may occur during the restoration in several ways. Pre-scheduled events may occur during the course of the restoration according to the plans set up by the instructor. Additional events such as circuit breaker openings may be introduced into the scenario as a result of relay models in the power system simulation. A good example is over/under frequency relaying which trips generation/load if during the restoration scenario the actions of the operators result in system frequency going out of allowed ranges for certain time durations. There is also a requirement for the capability for the instructor to introduce events "on the fly" during the restoration scenario.

The Educational Subsystem should also have the capability to keep a record of all instructor and trainee actions, and to record key information about the state of the power system during the scenario. Such information is important in the post session debriefing and evaluation of the trainee.

Various scenarios may be established to conduct restoration training. References [1] and [2] provide several example scenarios of system restoration exercises which can be conducted with an OTS having the capabilities described above.

Summary and Conclusions

In summary, an OTS may be used as an important part of a training program for system operators in system restoration. The capabilities of an OTS which are particularly important for restoration have been summarized.

References

1. J. Waight, et al, "An Advanced Transportable Operator Training Simulator" IEEE PICA, Baltimore, 1991

2. E. Barr, et al, "The Uses of an Operator Training Simulator for Power System Restoration" IEEE PICA, Baltimore, 1991.

LESSONS FROM OTHER INDUSTRIES

R. Hoffman, Member
CAE Electronics, Ltd.

Preparing responsible people for emergency situations which they probably will never actually experience, yet with which they must be prepared to cope should they occur, is a concern in a large number of occupations. The following is a summary of information gleaned from a series of informal interviews with people involved in training airline pilots, nuclear plant operators, air traffic controllers, communication satellite controllers and airport security personnel.

Airline Pilots

The training of airline pilots is tightly regulated by governmental and international regulatory agencies. Not only are the training procedures defined but even the training devices and simulators themselves must be certified by the regulatory agencies.

A commercial airline pilot typically spends four sessions per year of four hours each in a flight simulator. Two of the sessions are considered as training, i.e., dry runs of the examination, and two are true examinations which he must pass in order to keep his license. The training curricula is designed such that during the course of a year he is exposed to a broad range of emergencies defined by the regulatory agency. For each defined emergency situation, the crew has memorized a sequence of initial checks and actions which they must execute immediately from memory. After going through the initial memorized procedure, the crew can refer to a manual for additional steps to perform.

As part of the checkout procedure at the airport for each flight, it is mandatory that the flight crew must review the steps of a set of principal emergency procedures (e.g., loss of an engine during runway acceleration before and after critical velocity is attained, rapid decompression above 10,000 feet altitude, rapid descent and emergency evacuation of the aircraft, etc.)

Nuclear Plant Operators

At a typical nuclear generating station after every six weeks of normal shift work, the operator spends one full week in a training program. During this week, about 25 to 28 hours are spent in a classroom situation and 12 to 14 hours on a simulator. Thus during the course of a year, the operator spends from 80 to 100 hours on the simulator.

There is an annual evaluation of operators for licensing purposes by the NRC but generally the trainee is unaware of when he is being reviewed since the training philosophy is to consider every simulator session as if it was a real plant situation.

There are two separate external organizations which are involved in defining and evaluating the training programs. The industry funded Institute of Nuclear Power Operations has identified some 700 specific tasks that need to be performed by the operator and has categorized them as being either knowledge or skill related. A set of training scenarios on the simulator has been defined such that during the course of a year, the trainee is required to exercise all of the defined skills at least once.

The April 1991 issue of Nuclear News has an interesting series of articles on some current issues in nuclear training.

Air Traffic Controllers

It is very difficult to become an Air Traffic Controller. For example, in 1989 approximately 25,000 individuals applied to enter the Canadian Government training program; 400 were accepted and only 115 successfully completed the program.

The formal training program at the National Training Center involves approximately 70 days of classroom training and 35 days of simulator-based training. Simulators are used primarily for basic procedure training while training for emergencies is primarily from printed materials. After completing the program at the National Training Center, the individual is assigned to a specific site where he undergoes additional training specific to the site. The major portion of this training is on-the-job. Approximately one year is required to become fully qualified.

Once the program has been completed, there is typically one week per year of refresher training consisting primarily of lectures and group discussion of topics of general interest.

Communication Satellite Operators

The chief responsibility of the satellite controller is to maintain the correct satellite position and orientation. Disturbances arise from the unpredictable solar activity which produces random radiation pressure and can cause electric charge build up and discharges which can put the satellite into unknown states. The worst situation is an uncontrolled satellite "spin-up". The operations manager of a company which has been operating communication satellites for 19 years described how they have developed very specific operating procedures for recovering from a spin-up; these are based on a very extensive analysis of possible causes.

New operators, typically ex-military technicians accustomed to following procedures to the letter, go through a six month training program consisting of private study (60%), classroom time (30%), and simulator-based training (10%). With each new satellite the associated training simulator tends to become more sophisticated and, therefore, its usefulness in training is enhanced. The formal training is followed by an additional six months of supervised on-the-job training before an operator is qualified to work entirely on his own.

Operators are encouraged to occasionally re-read the procedures manual on their own and once or twice a year they have ad-hoc group refresher training sessions.

Airport Security Personnel

Airport operations personnel must always be prepared to deal with emergencies ranging from bomb threats and fuel spills to major airplane crashes. The training supervisor at a major Canadian airport stated that the role of his department was to provide as much information as possible on the potential consequences of an emergency and to keep in constant touch with operations people just to remind them that emergencies could happen at any time. Information and training were stated as being the key components of emergency preparedness.

Underlying all the training programs is a very careful analysis of required procedures and defining the role of all the involved agencies. In addition to preparing and updating emergency procedures manuals, typical training methods include roundtable exercises where a group of people work through an emergency, discussing what each of the various parties would do. Every six months they organize a major simulated emergency to the extent acceptable by the public using the airport.

Training Pushed By Safety Issues

In each of these cases, the importance of careful analysis and development of detailed plans to deal with potential emergencies was highlighted. The emphasis on continued readiness training seemed to vary considerably, however, being the greatest for aircraft pilots and nuclear plant operators. In both of these applications, the underlying driver is public safety, as opposed to economics or efficiency. Because of the involvement of the regulatory and licensing agencies, the technical sophistication and degree of detail in the simulators far exceeds those used in other industries, including dispatch training simulators for power system operators

Experiences Using the Dispatcher Training Simulator as a Training Tool

Gary Miller Art Storey
New York Power Pool
3890 Carman Rd.
Schenectady, NY 12303

Subramanian V. Vadari Kathy Brewer
ESCA Corporation
11120 NE 33rd Pl.
Bellevue, WA 98004

Abstract

Power system operations are becoming increasingly complicated and critical. There is a need to provide more directed power system operation training than that provided by on-the-job experience. The experiences of New York Power Pool have shown the dispatcher training simulator to be an effective tool in the training of power system operators and dispatchers, particularly in areas such as voltage control, stability, and system restoration.

In this paper the authors share some thoughts and experiences gained in the area of operator training using a simulator. Detailed descriptions have been provided regarding the simulator capabilities, training objectives, and training environment. In addition, the authors have provided a case study which proved the value of the simulator as an operator training tool.

1 Introduction

Power system operations in the 1990's pose new challenges for dispatcher training. Re-establishing service after a large scale disturbance is more complicated because networks are increasingly interconnected and heavily loaded. Maintaining the relative stability of an increasingly tightly coupled power system has also become more complex. Operating errors now can have far-reaching effects from both system stability and economic perspectives. Effective training for dispatchers and operators has become particularly critical because of their crucial decision-making role in operating the system safely and economically [1].

System-specific training for power system operators and dispatchers requires a clearly defined and structured training program. The program should have specific, stated objectives. The strategies and procedures put into place must work towards achieving those objectives.

The training will be enhanced with a sophisticated training environment in which an operator is trained on a simulated system. This system should have the same look and feel as the real-time system, from both the user interface and system response perspectives.

The Dispatcher Training Simulator (DTS) is an effective tool to help meet these training challenges, particularly in areas such as voltage control, stability, and system restoration training.

92 SM 403-6 PWRS A paper recommended and approved by the IEEE Power System Engineering Committee of the IEEE Power Engineering Society for presentation at the IEEE/PES 1992 Summer Meeting, Seattle, WA, July 12-16, 1992. Manuscript submitted January 31, 1992; made available for printing April 24, 1992.

1.1 Objective of the Paper

The objective of this paper is to address the following:

1. Use of a simulator and common DTS configurations.
2. The configuration of the DTS at New York Power Pool (NYPP).
3. How the DTS has been successfully used at NYPP for the past three years.
4. Possible enhancements to a DTS to facilitate greater learning.

1.2 Organization of the Paper

The paper is organized into the following sections: Section 1 introduces the main topic of the paper (DTS and its importance as a training tool) and a list of the authors' objectives. The next section presents a brief overview of the DTS, its main subsystems and different possible configurations of the generic simulator. Section 3 presents a more specific version of the simulator, namely the NYPP DTS, its capabilities, and configuration. This section also presents a brief overview of the NYPP modeled system, followed by a description of the training strategies and procedures used by NYPP for conducting dispatcher training. Section 6 presents a test scenario which demonstrates the effectiveness of using the DTS for operator training. This section also analyses the scenario, presents the lessons learned, and proposes some improvements which would enhance the DTS' capabilities to simulate the NYPP system. Last, the conclusion section presents a summary of the findings presented in the paper.

2 DTS Overview

The DTS [2] simulates the electrical network, user interface, and power system behavior. It provides a realistic environment for dispatchers to practice operating tasks under normal, emergency, or restorative conditions.

Although the DTS is used primarily for dispatcher training, it may also be used for dispatcher evaluations, engineering studies, power system model evaluation, and offline testing of energy management system functions and operational procedures.

The DTS is comprised of three subsystems:

1. **Control Center Model** (CCM) - This is a replication of the energy control center environment. It is the portion of the DTS with which the trainee interacts, and includes SCADA, AGC, and security control applications.

Reprinted from *IEEE Transactions on Power Systems,* Vol. 8, No. 3, pp. 1126–1132, August 1993.

2. **Power System Model (PSM)** - This subsystem simulates the electrical network components and dynamic responses of power system equipment such as generating units, transmission lines, loads, transformers, relays, etc.

3. **Instructional Position (IP)** - This subsystem provides the software and displays for monitoring and controlling the training session.

2.1 DTS Configurations

The DTS can be configured in a number of different ways depending upon requirements. There are advantages to each of the different configurations. Figure 1 shows four possible configurations and a brief description follows:

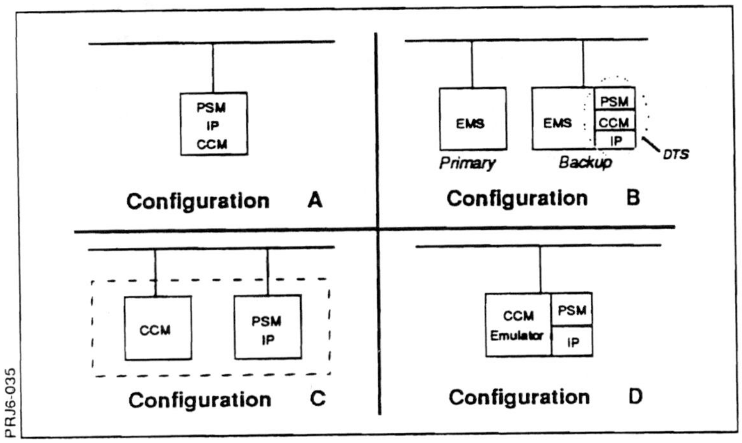

Figure 1 DTS Configuration

1. Configuration A - The entire DTS can be installed on a stand-alone computer and uses a specific power system model. This setup is particularly useful for presenting power system concepts and typically uses low-cost workstations.

2. Configuration B - The DTS can be installed on one of the backup Energy Management System (EMS) computers.

   This is a desirable configuration since the instructor and the trainees need only learn one user interface, and there is one set of databases to maintain. Also, a DTS configured in this fashion allows the capability of initializing with data from the real-time EMS. A disadvantage of this configuration is the increased loading on the backup computer.

3. Configuration C - The CCM and PSM/IP exist on different computers. The CCM is the same as the existing EMS model and communicates with the PSM/IP through a communications gateway. This configuration became field operational in the mid to late eighties. This configuration has also been reported in reference [5].

   Configuration C allows the operator to train with the same user interface used for day-to-day operations but requires the maintenance of two sets of databases (one for the CCM and the other for the PSM/IP). It also requires the instructor to learn two different user interfaces. In addition, this configuration does not allow for easy initialization with data from the real-time EMS.

4. Configuration D - This configuration is very similar to the Configuration C except for one difference. The PSM/IP and CCM are all installed on the same computer. A CCM emulator (also installed in the same computer) allows the trainee to practice with a familiar user interface.
   The same advantages and disadvantages as for configuration C apply here. However, configuration D reduces the utility's investment in hardware, for example, an expensive CCM computer. This may become a more widely accepted configuration as vendors move toward open architectures and standard interface protocols.

The NYPP DTS is using configuration C.

3 NYPP DTS

The dispatcher training needs at the NYPP differ somewhat from the needs of its member utilities and from the needs of a typical utility. The pool dispatchers do not direct any switching operations, nor do they communicate directly with generating plants. The operators accomplish their task of coordinating and controlling the operation of NYPP's Bulk Power System through their eight member systems. The NYPP dispatcher's effectiveness is primarily a function of his or her ability to recognize power system hazards, to choose effective corrective actions, and to communicate decisions to the member utilities who then implement the corrective actions.

Therefore, the training simulator must be capable of meeting this need. The dispatchers also train with the same user interface they use on shift. The primary difference between the real system and the DTS is that during the simulator training, the NYPP operator communicates decisions on corrective actions to the training instructor who acts in the role of the member utility and other power pool operators.

3.1 NYPP DTS Computer Configuration

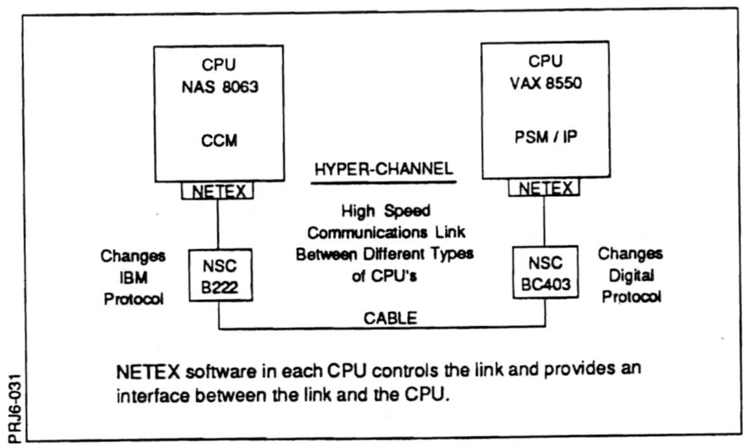

Figure 2 NYPP DTS Computer Configura

Figure 2 presents the NYPP simulator configuration and describes the use of the two computer systems:

1. The Digital Equipment Corporation VAX 8550 mini-computer provides the power system model (PSM) and the instructor position (IP).

2. The National Advanced Systems (NAS) 8063 Main Frame processor provides the Control Center Model (CCM). The NAS computer is IBM compatible.

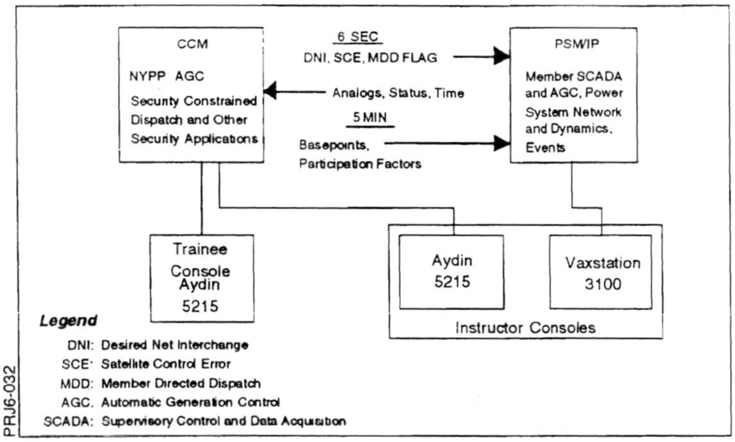

Figure 3 NYPP DTS System Overv

The tow machines are linked by a high-speed Network System Corporation (NSC) hyperchannel. The NAS and VAX systems communicate in much the same way as in the real-time operation between the Pool and the member systems (Figure 3).

This configuration is very similar to the configuration C described in section 2.1. The trainees work at consoles identical to those they use in the real-time environment.

The instructor interfaces primarily with the VAX machine, although he has NAS user interface capabilities. He acts in the role of member, or neighboring system dispatcher, and controls the training scenario progress by initiating "events" such as loss of generator or transmission lines. The events can be prescheduled or initiated at any time during an exercise.

3.2 NYPP Modeled System

The NYPP operating area includes all of New York State with small areas in New Jersey and Canada. It is made up of eight member utilities with a peak load of over 26,000 MW. The NYPP generation mix includes hydro , oil, gas, nuclear, and combustion turbines. The NYPP Control Center is responsible for the security of the bulk power system, which includes voltage levels up to 765kV.

Through the execution of the NYPP Security Constrained Dispatch (SCD) programs, basepoints and Satellite Control Errors (SCE) are provided to each to member system. This yields economical generation levels for the Pool without violating any security constraints.

The power system model is comprised of:

- Over 600 generating units
- a transmission network of over 25000 lines, 1700 buses, and 1300 breakers
- 600 transformers and 40 phase angle regulators
- 8 DC installations
- Over 1000 load buses
- System protection schemes including overcurrent, overvoltage, undervoltage, overfrequency, underfrequency, and synchro-check relaying.

3.3 NYPP DTS Capabilities

The NYPP DTS supports many operating scenarios, among them a system restoration scenario. Many of the capabilities listed below relate to NYPP's system restoration simulations.

- The real-time NYPP system has an online Security Constrained Dispatch (SCD) program which controls the pool's generation resources. The NYPP DTS has the capability to incorporate the output of the SCD program into the simulation.
- The DTS has a simulation clock which is maintained internally and can be set to any time or date either in the past or in the future. Various schedules including load, circuit breaker, and transaction schedules all use this clock to provide realism to the training. Since the clock is updated internally, it can also be used to speed up or slow down the simulation.
- The DTS can simulate energy management schemes specifically associated with NYPP's pooling arrangement and operations.
- The DTS provides realistic calculations of voltage which is essential to the meaningful simulation of the conditions involved in a system restoration.
- The DTS includes realistic reactive limit values ("D Curves"). This capability is essential since many of the generators involved in system restoration will be running against those limits.
- The capability of simulating multiple islands is necessary. The dispatcher must be able to operate the power system as one or a number of electrical islands. This involves control of frequency through proper load balancing, and the handling of the unusual voltage conditions often associated with extremely light load conditions, without help from outside sources.
- Frequency monitoring and control are important capabilities since frequency control is extremely precarious during a system restoration, especially where islanding is involved. The maintenance of system frequency within the narrow band width required for steam generators is essential to the success of a restoration effort.
- The underfrequency load shedding schemes are designed to help maintain system frequency, and they do so very effectively if properly located and set. Most dispatchers do not have much practical experience in handling island conditions. Therefore, it is essential that they experience the impact of this form of protection and take it into account in the restoration process.
- The impact of outage time on thermostatically regulated loads is most significant during a system restoration. The capability to simulate cold load pick-up phenomenon is required if the training is to be realistic. The forecasting of the amounts of load to be picked up "cold" is crucial to the maintenance of a load-to-generation balance. Adequate load modeling capability is needed. This includes the capability to model load blocks and load-voltage and load-frequency dependence.

- The synchro-check relay may prevent closure of breakers within a synchronized system when the voltage angle an/or magnitude exceeds the relay settings. These conditions often exist during system restoration where there are relatively long and heavily loaded transmission lines. Although it may be desirable to close breakers at some point in the system, the dispatcher may not be able to do so due to synchro-check blocking. It is essential that the dispatcher know where these conditions exist, and that decisions be made in advance as to when such relays should be bypassed. As such, it is necessary to simulate these protection relays in the DTS.

3.4 Using Transient Stability Analysis

The NYPP DTS also has an online transient stability analysis package [3] which allows the instructor to pause the DTS at any point in time and analyze the transient stability of the system for any particular component fault. The outages that result from the transient stability analysis can then be fed back into the DTS events scenario for the mid-to-steady state analysis.

NYPP is planning training sessions involving the use of the transient stability simulation. This training will prepare operators for another crucial aspect of system operation: the management of the system under transient conditions in the network. Because the topic is highly technical and the phenomena difficult to visualize, NYPP is combining classroom instruction with DTS dynamics and realism to present this concept to the dispatchers.

Training should address realistic conditions and be applicable to the power systems which the dispatcher operates. NYPP's training will, therefore, make use of its simulation capability and is expected to occur in four phases:

1. A practical explanation of what is involved in keeping the system synchronized and what happens when it becomes unstable. Although this is expected to be a classroom type session, it will be based on NYPP's system characteristics.
2. Simulator illustrations of the course material. This phase allows dispatchers to see the effects of low system voltage, heavy transfers, slow clearing times, and other system conditions on the stability of the NYPP system. These sessions are expected to be informal and will allow for considerable interaction.
3. A follow-up to the system restoration training sessions described earlier. Islanded system and limited tie scenarios will be used for illustrating stability principles and restoration considerations.
4. Incorporation into a scenario. The points will be driven home with a practical problem which must be solved using the knowledge gained from earlier sessions. This session will be more formally structured than the second phase of training.

During all four phases of this training, references will be given to show how these conditions apply to the NYPP system. Dispatcher will also be given opportunities to test specific self-developed scenarios.

4 Training

In utilizing the DTS most effectively for dispatcher training, the following considerations are important:

1. A formal training program with instructional support.
2. A systematic process for designing and developing simulator training sessions based on system operator performance requirements and training needs.
3. A lesson plan for DTS training sessions and a statement of the learning objectives.
4. An evaluation of whether the trainees met the stated objectives.

4.1 NYPP Training Strategies and Procedures

The training NYPP conducts is based on identified operator needs. Although much of NYPP's effort is directed to training against known hazards, the greatest challenges involve identifying the unexpected events which have often been the cause of power system disturbances.

NYPP's simulator training takes several different forms:

- Exercise/Critique - This involves the running of a structured scenario. Trainees are typically exposed to various emergency situations. Critiques are conducted during and following the exercise. The primary purpose of this kind of scenario is to condition the dispatchers to make rapid decisions and to implement corrective actions in a timely manner. The simulation may be paused at various times to discuss results.
- Lesson/Exercise - A structured course is developed that addresses a particular aspect of system operation. Simulator examples and exercises are run to provide a practical illustration of the material covered.
- Introduction of System Software Changes - When significant power system or dispatching software changes are being made, the simulator can be used to introduce them to the dispatchers prior to their actual implementation. Advanced Applications routines were incorporated into a recent set of training sessions. Dispatchers not only had a good opportunity to see the advanced applications but they also had input into their development. Dispatcher comments were significant and allowed the NYPP Computer Staff to receive meaningful suggestions which were incorporated into the finished product. This type of input reduces development time for new applications while increasing dispatchers' acceptance and understanding of the new applications.
- After-the-Fact Illustrations - These sessions involve modeling and utilizing actual power system events in an effort to illustrate the optimal handling of various situations. They provide lessons learned from past events and can expose all of the NYPP dispatchers to the experiences of the few that were on-shift when the event occurred. Developing such cases with two computers involves considerable time and effort. Nevertheless, it does have appreciable benefit for events which have special training value to all dispatchers.
- Dispatcher Constructed Scenarios - Scenarios and events based on inputs from the dispatchers are developed. In most

cases, the scenarios are handled on an informal basis, and the simulator is paused frequently to discuss the handling of the situation. An instructor is always present during these sessions and conducts all phases of this training.

- Informal Experimentation - Providing instructions for running the simulator and encouraging dispatcher interaction with the DTS establishes the conditions for the dispatchers to use the simulator as a dynamic means for studying the power system and the control programs involved. Such sessions can be conducted independently by the dispatchers.

4.2 NYPP's Restoration Experience

The topic of system restoration provides on of the best illustrations of how a simulator can be effectively used to train dispatchers [4]. Restoration exercises also test a simulator's capabilities under the most strenuous computational conditions. NYPP has conducted a series of restoration training sessions and the results have been excellent. Exercises have addressed not only the technical aspects of restoring the system, but also the methods of accomplishing this. Later phases of this training have emphasized the need for making decisions with limited amounts of information and for communicating those decisions effectively to the people who will implement them.

The simulation sessions address state-wide restoration policies, thereby reinforcing training in the techniques associated with these written plans. Situations are incorporated into the exercises which force the trainees to make modifications to the normally prescribed restoration plan. Key lines are made unavailable, and dispatchers are forced to develop alternative paths to achieve the desired end results.

During the simulated restoration exercises, dispatchers and instructors gain experience and have made discoveries in the following areas:

- The type of voltage control and the anticipated voltage conditions during system restoration.
- System frequency deviation in an islanded situation when small blocks of load are picked up with very little generation online.
- The potential of losing all restored generation and load with the loss of a single element during the restoration.
- Equipment limitations and capabilities during a restoration.
- Clarification of steps in the written restoration procedures.
- The effects of synchro-check and frequency relays during restoration.

The realism of the changing system conditions as the dispatchers implemented each step of the restoration procedure proved the DTS to be a very useful tool for testing and gaining experience on the NYPP restoration plan and control methods available.

4.3 Training Environment

Figure 4 shows both the existing layout and the new layout (which will become operational in a few months). As shown in the figure, the instructor's console is separated from the trainee consoles by a partition and one-way glass. There are two trainee consoles in the existing layout versus three in the new layout. (The NYPP control center has three operator consoles.) The instructor's console has an intercom system to communicate with the trainees and to listen to the trainees' conversations during a session. A telephone system similar to that existing in the control room allows the trainees to direct-dial the instructor's console (as if they were calling a member system or another pool). There is also a 'hot line' button on the trainee's telephone console which simulates calling all members at once for an emergency condition. Finally, all the telephone conversations are taped and can be used during the review of the trainee session. These taped conversations are also used in the classroom for improved communication training.

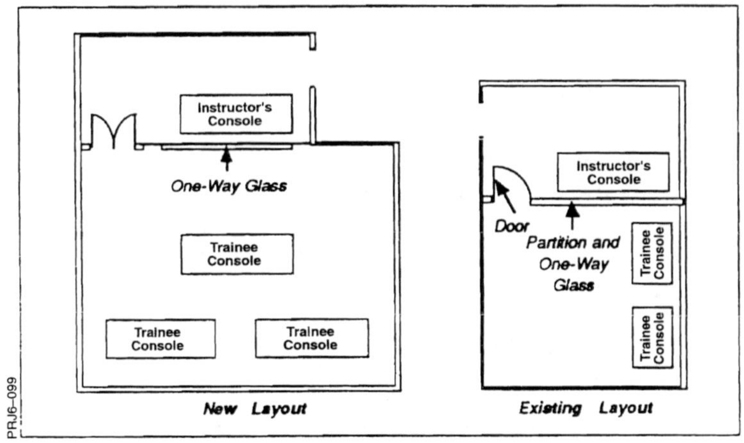

Figure 4 NYPP DTS Training Lay

Future enhancements will include a map board and charts. The additions will help to more clearly duplicate control room conditions.

5 Test Case Scenario

One training session in particular was very gratifying to the NYPP training department. A situation occurred in the power system for which the dispatchers had recently been trained using the DTS. They responded correctly reflecting the lessons they had learned in the simulation training.

5.1 Description of the Training Case

The training session was delivered in January and February of 1990. The simulation addressed 1990 summer peak conditions - the NYPP load was 25,850 MW.

The session began with one of three 345/115kV banks, all of which are critical in supplying load and voltage support to one of NYPP's member systems, out-of-service.

The company involved has generation that feeds directly into the 115kV transmission system and generation that feeds into the 345kV transmission system. The 345/115kV banks to supply company load. In addition to the 345/115kV banks, this system has 5-115kV ties to other areas (Figure 5).

Prior to the training session, the instructor had set up one of the 115kV tie lines to exceed its long-term emergency (LTE) rating if a specified problem occurred. The Senior Pool Dispatcher is required to notify a member's facility, even if it is not part of the

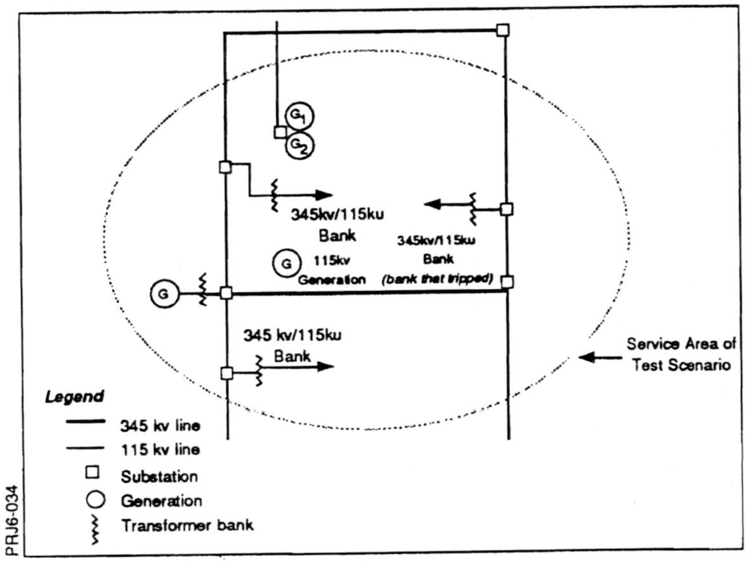

Figure 5 NYPP 115kv Service Area with 345kv Suppl (used in the test scenario)

bulk power system. The trainee found (by using a program that calculates generation shift factors) that the only way to correct this violation was by starting two gas turbines (GT1 and GT2) that are located in the area. They were started in the DTS.

Starting the gas turbines corrected the 115kV line contingency violation, but as the load was still increasing, the 115kV voltages in the eastern portion of the member's system continued to deteriorate. The member system dispatcher (simulator instructor) notified the Senior Pool Dispatcher of low voltage problems. The discussion involved how to correct the low 115kV voltage and what would happen to the member system voltage if a generator connected to the 115kV system was lost or one of the remaining 345/115kV banks tripped.

After experimenting with different scenarios, the trainee and instructor concluded that the member system should implement a 5% voltage reduction immediately to protect against the contingency discussed and possibly a voltage collapse. The voltage reduction was implemented in the DTS and the 115kV voltages returned to acceptable levels.

As an additional illustration, the instructor tripped one of the remaining 345/115kV banks to show what could have happened if the voltage reduction had not been implemented. When the remaining bank was tripped, the member system's 115kV voltage deteriorated to approximately 90kV which was indicative of a voltage collapse and portions of the member system being blacked out.

5.2 Description of the Actual Event

On July 18, 1990 at 22:24, the 345/115kV bank that was out-of-service in the training session caught fire and tripped. As the load increased the next morning, the member system's 115kV voltage deteriorated and the member vent into 5% voltage reduction, first in the eastern portion of his system and finally throughout the entire system. Also, the same 115kV tie line showed a post contingency overload in reality as it had in the training case. The member agreed to start both GTs in the area if the contingency occurred.

Since the NYPP dispatcher on duty had covered a very similar situation in simulator training in February, it gave him more insight in his decision-making process when the incident actually happened. He also knew what he could expect as the load increased the next morning. The training he received on the DTS proved to be a valuable asset for the dispatcher and for the NYPP operations.

5.3 Lessons Learned

Experience using the DTS for dispatcher training has provided the following valuable lessons:

- The closer the DTS replicates the power system dynamics and user interface, the more confidence the dispatchers have in its benefits. The credibility of the simulation contributes to the dispatchers' training and greater confidence in their own operational skills.
- Since the DTS can be used to create events on the system which rarely occur in real-time, it can also help identify weaknesses or errors in the energy management system software, modeling or operating procedures.
- Voltage metering point selection in the simulation must match real-world locations. NYPP discovered in simulation that significant system voltages could not be "read" because the metering device read a bus section other than the one being energized. In these cases, manual intervention would have to take place to change the metering point to the energized bus. Observations such as these can be valuable in improving actual restoration plans.
- Simulation training provides an environment to familiarize and get feedback from dispatchers regarding new displays and software programs before they are implemented in the real-time system.

  During restoration exercises, the dispatchers suggested several significant improvements to the existing CCM user interface. These suggestions may not have otherwise surfaced if the restoration exercise had not been conducted.
- Operating procedures and restoration procedures can be practiced using the DTS. This may point out inconsistencies in the procedures and facilitate improvements to them. Using the DTS for this purpose results in more questions and discussion than from reading an operating manual. It can also identify problems before they occur on shift.
- The DTS is valuable in preparing dispatchers for upcoming power system enhancements. New facilities can be modeled in the DTS for both study and operational training purposes in advance of their in-service dates. Therefore, dispatchers gain valuable experience before the new facility is placed into service.
- The DTS is a valuable study tool in assessing upcoming system outages. The impact of the outage and available corrective actions can be evaluated using the DTS prior to the facility being removed from service in real-time. Predicted DTS system conditions are very accurate when compared to real-time system results after the outage occurs.

5.4 Future DTS Improvements

Based on more enlightened experience, the authors believe that the following proposed improvements to the DTS would enhance the ability to create increasingly realistic training scenarios:

- Constant frequency control at the unit level. The ability of a unit to control frequency without AGC would typically be used in an islanded situation.
- Enhanced interface capability to other systems would provide the ability to interchange data easily between DTS databases and other loadflow type programs.
- Static VAR Compensator (SVC) model to be used during the solving of the DTS power flow and study power flow. A project is presently underway to implement this functionality.
- Additional exciter models for transient stability would provide increased simulation accuracy.
- More detailed DC model for use in the transient stability mode.

6 Conclusions

The New York Power Pool recognizes the need for exposing dispatchers to realistic operating experiences, and the fact that this cannot be effectively accomplished through on-the-job experiences alone. Classroom, tutorial, and self-study programs address the dispatchers' correct reactions to simulations which can occur on-shift. However, such programs are not a substitute for hands-on experience. Also, there are some situations, such as system restoration, where simulation can give dispatchers additional experience and realism beyond that obtained through written procedures and drills.

In many cases a utility does not really know what will happen on its system until it actually occurs. Many tools exist for modeling the power system. One of the most dynamic and realistic of them is the Dispatcher Training Simulator. Its value is best illustrated by creating simulated emergency conditions that provide dispatchers an opportunity to gain experience and confidence in the handling of these situations.

Although the primary purpose of a DTS facility is to train dispatchers and operators, the benefits derived from the DTS as a study tool in the development of programs and displays, and from the discovery of modeling and software errors in the real-time system are appreciable.

Acknowledgments

The authors would like to acknowledge the help of the following people without whom the submission of the paper would not have been possible: Jay Giri, David Sun, Bill Owen, Kendall Demaree, Lin Von-Ulm Paulson, and Mali Ellegala.

References

[1] James Smith, et. al., Development of Operator Training Curricula using the Instructional Systems Development Approach, **IEEE Transactions on Power Apparatus and Systems**, Dec 1985, pp 3446-3452.

[2] R. Podmore, et. al., An Advanced Dispatcher Training Simulator, **IEEE Transactions on Power Apparatus and Systems**, Jan 1982, pp 17-25.

[3] Subramanian Vadari, Kendall Demaree, Davis Hwang, Gary Miller, Robert Waldele, An Online Implementation of Transient Stability in a Dispatcher Training Simulator, **IEEE Transactions on Power Systems**, T-PWRS, Feb 1991, pp 135-144.

[4] Power System Restoration Working Group Report, New Approaches in Power System Restoration, Presented at IEEE PES Winter Meeting, Jan 1992.

[5] EPRI Report, EL-7244, Operator Training Simulator.

Verification of an Advanced Power System Restoration Support System Using an Operator Training Simulator

T. Kobayashi, D. Moridera
The Chubu Electric Power Company
Nagoya, Aichi, Japan

S. Fukui, K. Komai
Mitsubishi Electric Corporation
Kobe, Hyogo, Japan

ABSTRACT-This paper describes an advanced restoration support system for regional load dispatching and local supervisory control centers. A model-based diagnostic task shell is applied to diagnose network faults, and an inference algorithm not dependent on a specific power system has been developed exclusively for restoration plan determination. With this system, adjustments for network fault diagnosis, necessitated by power system facility expansion or alteration, can be made simply by adding or changing models. Restoration knowledge can be partially revised in case of restoration policy change. Connected on-line to an operator training simulator, this system's restoration support performance has been verified through daily restoration training. In this process, several problems of the system were detected and corrected. Consequently, it has become possible for the system to support restoration almost as well as a trainee, thereby confirming that connecting this system to an operator training simulator is an effective method for both verifying and improving restoration support system performance.

Key words ; restoration support, model base, knowledge base, operator training simulator, verification

1. Introduction

Several knowledge-based restoration support systems have been developed to assist power system operators in case of power system fault to speed up restoration operation. In these restoration support systems [1]-[3], they have tried to implement operators' heuristics regarding target power system restoration. Some of these systems are already in the field testing stage; however, they have several problems, such as:

They use knowledge that relies on a specific power system; therefore, it is difficult to adjust the knowledge in accordance with power system facility expansion or alteration, or operating policy change.

As for verification, these systems are tested in the field using on-line input from an actual power system. However, since low-frequency faults, such as bus or transformer breakdown, rarely occur within a given period of time, verification may not be possible for restoration of complex faults such as large-scale outages.

To solve these problems, the authors have developed an advanced restoration support system which does not depend on a specific power system. This system will be used at regional load dispatching and local supervisory control centers, and have been verified using an operator training simulator (OTS).

A model-based task shell is used to diagnose network faults; to determine restoration plans, a restoration inference algorithm has been developed which does not depend on a specific power system. Therefore, adjustments for network fault diagnosis, necessitated by power system facility expansion or alteration, can be made simply by adding or changing models. Restoration policy change can be easily accommodated by partial revision of restoration knowledge.

This system is connected on-line to an OTS, and its restoration support functions have been improved based on verification results of daily restoration training. The real-time power system simulation function of the OTS enables accurate reproduction of the system conditions during and after a fault. Moreover, the registration of fault cases allows generation of complex faults or those accompanying large-scale power failure anytime, thereby realizing highly efficient and reliable verification.

This paper shows how to realize restoration support functions not dependent on a specific power system, discusses verification using an OTS as well as how to select test cases and test restoration support functions, and demonstrates the effectiveness of a verification method using an OTS with verification results, and restoration support function improvement based on those results.

2. Restoration Support Method

2.1 Network Fault Diagnosis (NFD)

(1) Model-based Diagnostic Shell Application

Since the diagnostic task shell MELDASH [4] has a built-in general diagnostic algorithm which is not limited to one application area, diagnostic systems can be constructed simply by adding an application model of a diagnosis target. The built-in algorithm performs a generate-and-test strategy. The addition of a protective relay system model to MELDASH has made possible the NFD.

Thus, facility maintenance or protective relay mode change can be easily accommodated by adding or changing the relevant equipment or protective relay model. Moreover, when relay operation data at the time of a fault are summarized, the faulty sections can be diagnosed corresponding to the data level simply by changing the data summarization model of the data transmission system.

This paper was presented at the 1993 IEEE Power Industry Computer Application Conference held in Scottsdale, Arizona, May 4 – 7, 1993.

Reprinted from *IEEE Transactions on Power Systems,* Vol. 9, No. 2, pp. 707–713, May 1994.

(2) Diagnostic Algorithm and Protective Relay System Model

Fig.1 shows the general flow of the general diagnostic algorithm and the models incorporated into MELDASH. The diagnostic algorithm first generates hypotheses of faulty sections from action data provided by the protective relay at the time of fault occurrence. Then, while simulating the protective relay and circuit breaker actions, it compares them to action data at the time of a fault. The hypothesis is judged as a faulty section, whose simulation process corresponds to action data at the time of a fault. This simulation process is presented to an operator as reference data for determining which hypothesis is most likely to be correct.

The protective relay system model is comprised of power system components such as transmission lines and buses, protective relays, data transmission system, fault current circuits, and so on. The model contains various types of relays - not only fault-eliminating relays, but also transmission line re-closing equipment, trial cutting off relays, and substation protective relays - to enhance diagnosis accuracy.

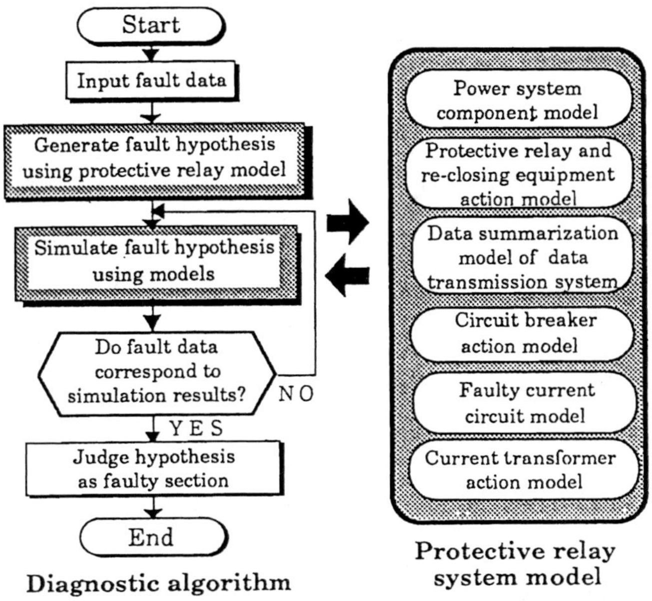

Fig. 1 Diagnostic algorithm and model structure

2.2 Restoration Plan Determination (RPD)

(1) Modeling of the Power System

Parts of systems which require restoration after a fault are modeled as overload, outage and separated blocks, as shown in Fig. 2. The overload block is a group of downstream side systems of overload equipment; the outage block is a group of systems which are in an outage state due to the fault; and the separated block is a group of systems which are separated from the upper systems due to the fault and have a balanced supply and demand relationship.

(2) Knowledge Base

The knowledge base consists of knowledge regarding restoration operation for the overload, outage, and

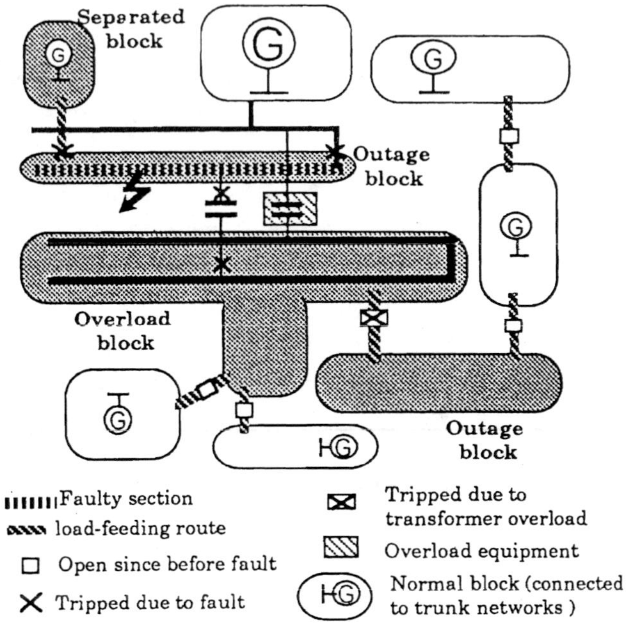

Fig. 2 Power system modeling

separated blocks and knowledge regarding restoration order of blocks. In some cases of actual restoration, restoration of two or more blocks is performed in parallel. It is pssible to perform such restoration with knwoledge of the restoration order for each block. Fig. 3 shows the knowledge structure.

Each block's restoration operation knowledge is essentially based on the knowledge base proposed in Reference Material [5]. Following are the summarized contents of each category of knowledge.

[Knowledge for outage block restoration operation]

- Knowledge for estimated faulty section isolation:
  Makes operation to isolate a section deemed faulty on the basis of NFD results
- Knowledge for load-feeding from the normal block:
  Selects a route for load-feeding from the normal block and makes load-feeding operation
- Advance operation knowledge for load-feeding:
  Makes operation to be implemented in the normal or outage block before load-feeding, so as to avoid equipment overload in the normal block during load-feeding
- Knowledge to restore remaining outage load:

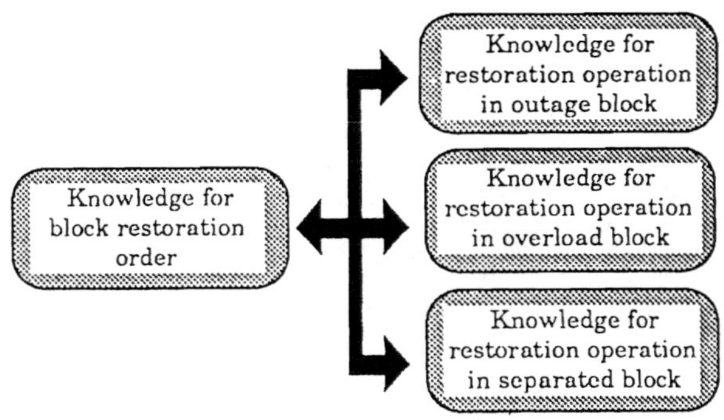

Fig. 3 Restoration knowledge base

Indicates methods of load-feeding from outside systems, such as power distribution systems
- Faulty section isolation knowledge for transmission line or substation faults:
  Makes operation of switching outaged distribution centers, substations, and users to the normal system by isolating the outage block and charging

[Knowledge for overload block restoration operation]
- Knowledge for power flow adjustment within the overload block:
  Makes operation to be implemented in the overload block in order to eliminate overload
- Knowledge for load-switching to the normal block:
  Makes switch operation by selecting load in the overload block to be switched to the adjacent normal block in order to eliminate overload
- Advance operation knowledge for load switch:
  Makes operation to be implemented in the normal block before switch operation in order to avoid equipment overload in the normal block during switching

[Knowledge for separated block restoration operation]
- Knowledge for power-off of the separated block:
  Makes parallel-off operation of hydroelectric power generators in order to eliminate the separated block

[Knowledge for block restoration order]
- Knowledge for deciding restoration sequence:
  Performs quantitative diagnosis (outage load elimination volume, overload elimination volume, number of operations, etc.) of conditions in each block after restoration operation and determines block to be given restoration priority

Since each category of knowledge is described as procedures for each of the modeled blocks, it does not depend on a specific power system. Moreover, since knowledge in one block is not related to that in another, the restoration operation knowledge necessitated by changes of power system structure or power system operating policy can be adjusted separately for each block.

(3) Inference Algorithm for Restoration

The inference algorithm for restoration can be explained as follows, in accordance with Fig. 4. First, restoration operations are made for each of the outage, overload, and separated blocks of post-fault systems by applying the knowledge for restoration operation in each block.

Next, deducing from hypothetical post-restoration conditions of each block as to when the restoration operation is implemented, a block to be actually restored is selected by applying the knowledge for deciding restoration sequence. Then, according to the simulation of restoration operation for the selected block, post-restoration conditions are revised. A restoration plan is determined by repeating the series of steps until there are no overload, outage, and separated blocks. Finally, a post-restoration system can be obtained. For the outage block which cannot be restored due to lack of a load-feeding route, a restoration plan is determined on the premises that restoration can be achieved through application of the knowledge to counter remaining outage load or faulty section isolation knowledge, which is discussed in the following section.

The restoration operation simulation process is retained as rationale for the results of RPD.

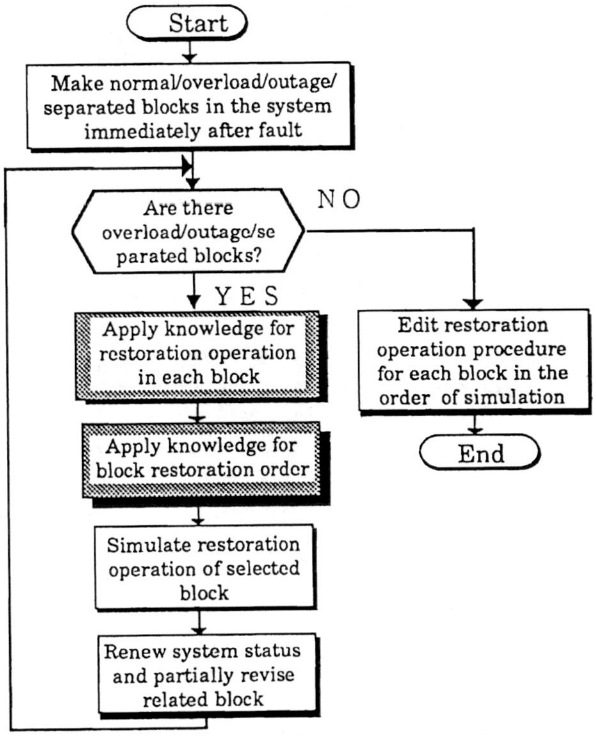

Fig.4 Restoration inference algorithm

(4) Faulty Section Isolation (FSI) Guidance

When unrestored loads remain within an outage block having a faulty section after restoration, FSI procedure is guided. Since subsequent operations are determined according to actual operation results, it is difficult to make FSI procedure off-line in advance. For this reason, each time the system status is changed by the FSI operation, the transmission line or substation FSI knowledge is applied to guide subsequent operations.

3. Verification Using OTS

3.1 System Configuration

Fig. 5 shows the configuration of a verification system connected to an OTS [6].

On the assumption that it is installed in a regional load dispatching and local supervisory control center, this system is connected to a simulated computer unit at a regional load dispatching and local supervisory control center, to verify the extent to which the system can achieve restoration support based on data from the current transmission network and data manually input by an operator.

3.2 Selection of Power System

Since this verification is carried out for one sub-transmission system and its results should be applicable to other systems, a power system which has

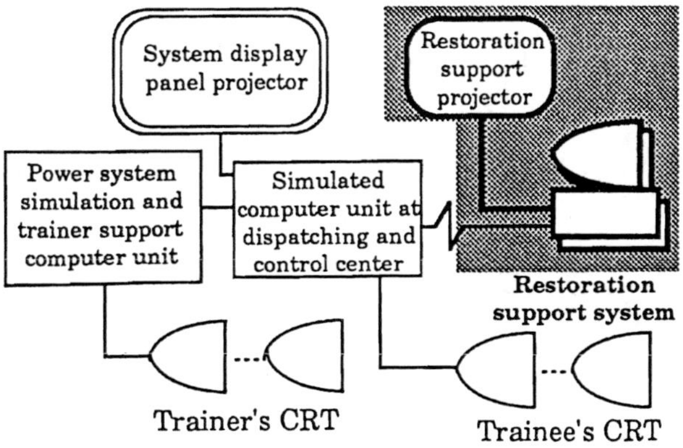

Fig. 5 Verification system configuration

standard sub-transmission system characteristics is selected.

Standard sub-transmission system characteristics include:

- Standard system structure (voltage class, etc.)
- Appropriate type of protective devices, including stabilizer
- An appropriate, not excessive, number of hydroelectric systems involved
- A large number of new devices (e.g. GIS)
- Presence of extra-high voltage loads
- Flexible system structure
- Standard operating principles

As a system which satisfies all these conditions, the operation simulator system at Okazaki Dispatching and Control Center (about 60% of the actual system in terms of scale is simulated) was selected out of six actual sub-transmission systems. The scale details of this system are shown in Table 1.

Table 1 Power system scale for verification

Type	No.
Thermal power generator (275 kv, 154 kv)	5
Hydroelectric power generator (154 kv, 77 kv)	3
Primary substation(275/77 kv, 154/77 kv)	13
Secondary substation(77/33 kv)	7
Intermediate substation (77 kv, no load)	14
Extra-high voltage load	61
Distribution substation	56
Transmission line route	108

3.3 Selection of Test Cases

To evaluate the results of system restoration support, it is necessary to know the correct answer in a given test case. In a test case of NFD, the correct answer can be easily prepared by setting the faulty section and corresponding data such as activated protective relays and tripped circuit breakers in consideration of reactive actions of the protective relays, circuit breakers, and re-closing equipments during a fault. On the other hand, in the case of RPD and FSI guidance, it is difficult to prepare correct answers in advance, since criteria are not established for evaluating appropriateness of restoration operation procedures.

Therefore, cases for this verification were selected as follows:

- Training cases in which a trainee had succeeded in restoration in past daily training were selected. (Success or failure judgment was made by the trainer.) The records of the trainee's hypothetical faulty section and the restoration principles employed in the training were used as the correct answers.
- In addition to cases of daily training, cases involving complicated operating sequences of protective relays were prepared exclusively for the verification of NFD, and were registered as training cases.

For the verification of NFD, RPD and FSI guidance, 15 cases of fault, based on the results of system operation simulator training, mainly on a bus fault at the primary substation, were prepared. Exclusively for the verification of NFD, 64 cases were prepared for each protecive relay and system operation mode of each type of device. Table 2 shows the breakdown of these cases.

Table 2 Breakdown of test cases

Type	No. of cases
For NFD	
Primary substation bus fault	33
Secondary substation bus fault	9
GIS distribution substation bus fault	9
primary substation transformer fault	4
Transmission line fault	21
Multiple faults	3
For RPD and FSI	
Primary substation bus fault	12
Secondary substation bus fault	3

3.4 Test of Restoration Support Functions

(1) NFD

A faulty section is automatically identified based on network data of the process from fault occurrence to transmission line re-closing equipment actuation. Network data include data on transmission line re-closing equipment actions, frequency relay actions and overload relay actions automatically generated by the power system simulation function of the OTS, in addition to actuated protective relay and tripped circuit breaker data registered in the training cases. Fig. 6 shows the flow of network data during verification.

The results of NFD are displayed in order of

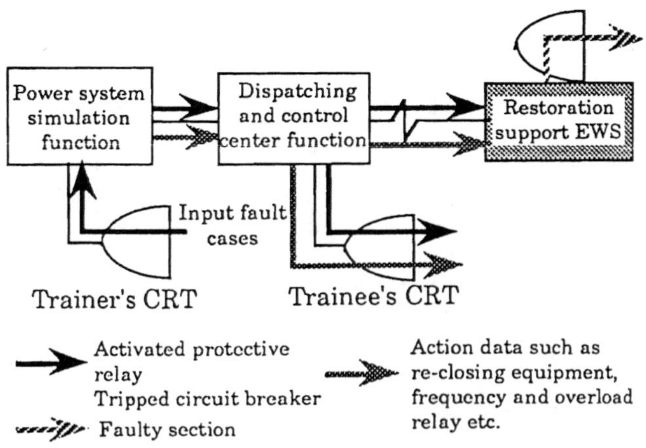

Fig. 6 Data flow in the test of NFD

accuracy, as judged from reactive actions of the devices during a fault, with sections divided by circuit breakers (CB), line separators (LS) and current transformers (CT) as the minimum sectional unit. The correct answer is included in the diagnosis results; when all simulation results correspond to network data at the time of a fault, the diagnosis is adjudged correct. If not, the model is modified by referring to simulation processes of protective relays and other devices that are used as judgment rationale, so that it corresponds to the actual action sequence. Fig. 7 shows an example NFD and simulation process.

NFD results

Faulty section	Fault type
■As/s AB-1L bus side LS-CB	Ground

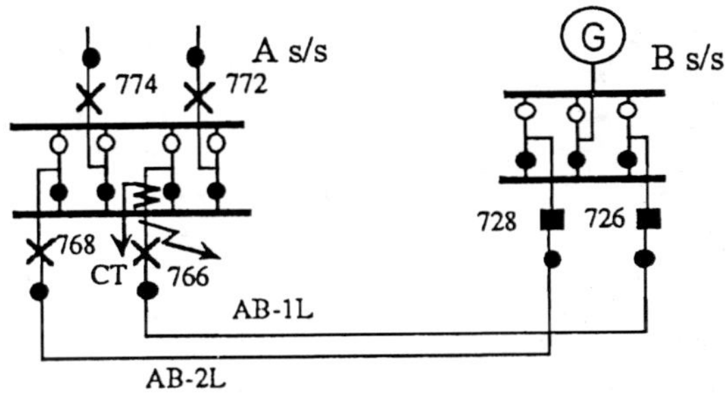

Magnified diagram of faulty section

NFD rationale

Case 1

Sequence : generated events

1: As/s AB-1L ground fault inside AB-1L supervisory CT
2: As/s AB-1L/ground transverse differential relay actuated
3: As/s 766 CB tripped
4: Bs/s AB-2L/ground transverse differential relay actuated
5: Bs/s 728 CB tripped
6: As/s ground substation protective relay tripped
7: As/s 768 CB tripped
7: As/s 772 CB tripped
7: As/s 774 CB tripped
8: Bs/s AB-2L/low-speed re-closing equipment actuated
9: Bs/s 728 CB re-closed

Fig. 7 NFD results and an example of simulation

(2) RPD

As shown in Fig. 8, restoration operation procedures and a post-restoration system are determined on an operator's demand. Post-restoration load values are deduced from load patterns prepared in advance and from transmission line and transformer power flow values periodically extracted from a dispatching and control center function.

When the restoration sequence of the overload, outage, and separated blocks and the restoration operations for each block, including load-feeding and load-switching, correspond to restoration operation procedures instructed by a trainee using an OTS, the sequence and operations are adjudged correct. If not, causes for discrepancies and countermeasures are examined by investigating the restoration operation from the point at which the discrepancies begin to appear, while referring to the restoration operation simulation process of the determination rationale. Then, the knowledge for determining the restoration sequence for

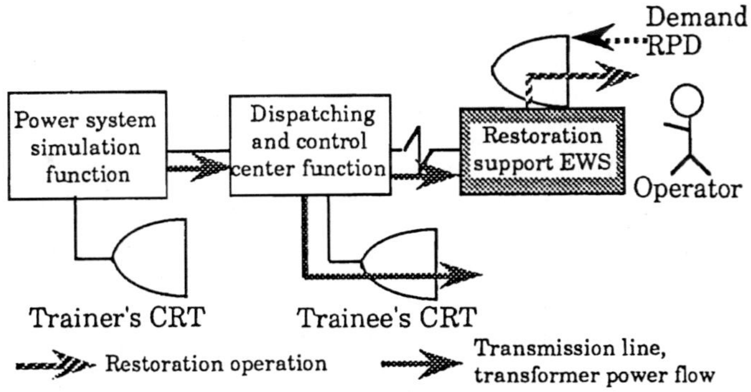

Fig. 8 Data flow in the test of RPD

each block and the restoration operation knowledge for each block are modified. Fig. 9 shows an example RPD and simulation process.

For an outage block where unrestored loads remain after restoration, FSI operation procedure is guided on an operator's demand. Different operation procedures follow FSI, depending on whether or not re-charging of a faulty section is executed for an operation at a given step. For this reason, when an operator demands guidance, the succeeding FSI procedure is automatically

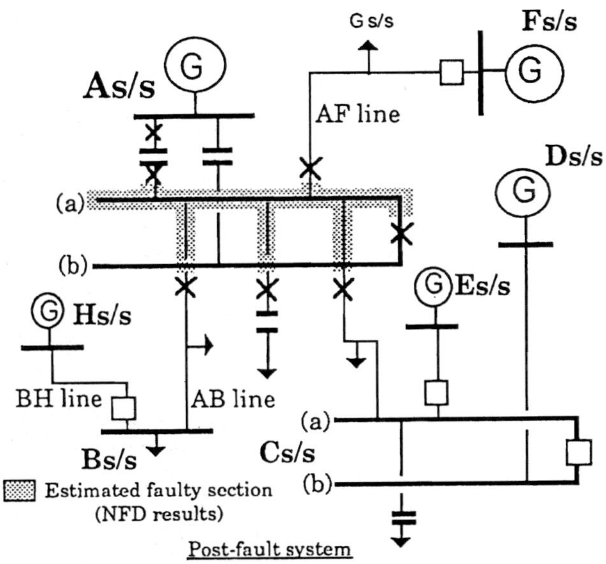

Restoration plan

No. : Restoration procedures

1: Isolate faulty section
2: Restore C s/s (a) bus system from (b) bus system
3: Restore G s/s from AF line
4: Isolate AB line from B s/s
5: Restore B s/s system from BH line
6: Restore A s/s (a) bus system by FSI

RPD rationale (summary)

- Outage block AF line; outage load volume 38.0 MW; connected to F/s; no overload
- Outage block B s/s; outage load volume 65.5 MW; connected to BH line; overload exists [8.5 MW]; advance operation: execute AB line isolation
- Outage block C s/s (a) bus; outage load volume 77.0 MW; connected to E s/s; overload exists [15.5 MW]; connected to D s/s; no overload

* From the above, restore outage block C s/s (a) bus from D s/s

Fig. 9 Example RPD and simulation process

determined each time the operator executes FSI operation on the trainee's CRT, as shown in Fig. 10. These processes are repeated until the unrestored loads are removed.

When restoration operation, including the selection of circuit breaker for trial charging, isolation of outage blocks, and load-switching, corresponds to the FSI procedures instructed by a trainee using an OTS, it is adjudged correct. If not, causes for discrepancies and countermeasures are examined by investigating the restoration operation from the point at which the discrepancies begin to appear, and the FSI knowledge which includes the restoration operation is modified.

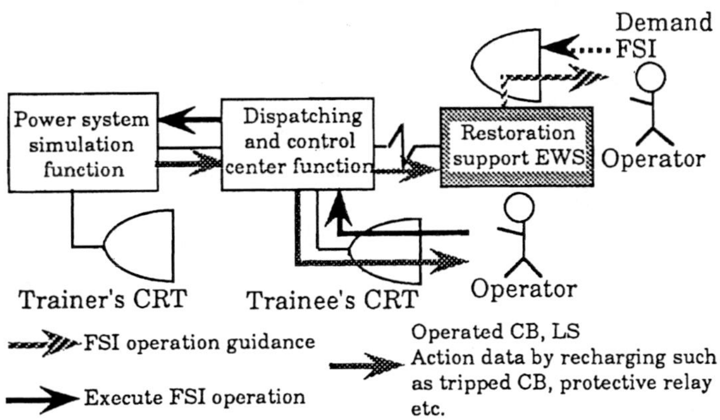

Fig. 10 Data flow in the verification of FSI guidance

4. Verification Results

4.1 Improvement of Restoration Support Functions

(1) NFD

In the following cases, initial models inputted into MELDASH differed from the actual action sequences of protective relay and other devices:

- In the case of a bus fault at the primary substation, the circuit breaker at the load side was tripped due to transformer overload relay; however, the circuit breaker model did not consider circuit breaker tripping by overload relay.
- In the case of a bus fault at a secondary substation, non-standard substation protecitive relay-caused circuit breaker tripping differed from the model. Only standard substation protective relays were considered in the initial model.
- In the case of a bus fault at a secondary substation, circuit breaker tripping sequence by trial cutting off relay action differed from the model.

Therefore, the models were modified as follows:

- To simulate circuit breaker tripping by action of overload relay and other fault expansion control relays, transition to the status in which circuit breaker tripping occurs due to causes other than protective relays was added to the circuit breaker action model.
- New models for non-standard substation protective and trial cutting off relay were added to the protective relay action model.

In addition, the following measures were devised to make it easier to determine whether incorrect identification has resulted from a model error or a protective relay or circuit breaker action sequence error in a simulator fault case.

- When all hypotheses failed, all hypothetically identified sections were defined as faulty and simulation processes which failed in test were preserved as the diagnosis rationale.

(2) RPD

This restoration operation differs from the restoration method employed by a trainee, mainly in the following points:

- Restoration priority is given to an outage block which has a large amount of restored loads. Therefore, an outage block with no loads tends to be left unrestored, and the load within its adjacent outage blocks is also left unrestored.
- Since the switch load from an overload block to its adjacent normal block is set as the load for one transmission line section, overload may not be eliminated.
- When support power is insufficient in all load-feeding routes of one outage block, a route with a large amount of support power is selected, regardless of advance operations such as load-dividing within an outage block or support power expansion in the normal block; therefore, an outage block may not be restored when overload during load-feeding cannot be corrected by advance operations, or when a load-feeding route with a large number of advance operations is selected.

Thus, the knowledge for block restoration order and restoration operation for each block was modified as follows:

[Knowledge for block restoration sequence]

- To the knowledge for determining restoration order, "an outage block with no loads should be restored before other outage blocks" was added.

[Knowledge for overload block restoration operation]

- The knowledge for selecting switch load within the knowledge for load-switching to the normal block was changed from "Switch load at time of system switch should correspond to one transmission line section from switch point exceeding overload volume" to "Switch load at time of system switch should correspond to up to two transmission line sections from switch point, depending on overload volume, and multiple switching is possible."

[Knowledge for outage block restoration operation]

- To the knowledge for selecting a load-feeding route within the knowledge for load-feeding from the normal block, "when support power is insufficient in all load-feeding routes of one outage block, first make advance operations by all load-feeding routes, then perform restoration using route which does not generate overload in normal block and requires only small number of advance operations" was added.

4.3 Observations

By improving restoration support functions based on verification results, correct identification of faulty sections and confirmation of the same restoration procedures as those of a trainee using an OTS were achieved in all the verification cases shown in Table 2.

Improvement of NFD was achieved only by revising the protective relay and circuit breaker action models. RPD was improved by partial revision of each knowledge framework within the knowledge base shown in Fig. 3.

Thus, it can be said that this system allows the improvement of functions, necessitated by power system facility expansion or alteration or operating policy change, simply by partial revision of the knowledge for RPD and protective relay system models.

As well, it is deemed a major achievement, in terms of verification using an OTS, that with the accurate real-time simulation of system conditions generated due to the passage of time after fault occurrence, overload relay actions overlooked in NFD were detected, and FSI guidance knowledge was verified under almost the same conditions as in actual operation.

NFD involves inferring a faulty section from the physical phenomena of the protective relay system and electrical cicuits. Since MELDASH simulates such phenomena in discrete state transition models, correctness of NFD depends on the accuracy of the models. This fact was confirmed through tests using an OTS.

On the other hand, RPD and FSI guidance basically involves combining conditions of system components; therefore, absolute evaluation criteria are not clear, and it is difficult to generate correct answers. In this verification, the results of restoration training which had been carried out using an OTS and adjudged by experienced operators and trainers as successful were employed as correct answers. For this, it is deemed that contents of support by this system have been made approximate to those of actual restoration.

5. Conclusion

This paper has introduced the advanced functions of a restoration support system for regional load dispatching and local supervisory control centers, and discussed a verification method using an OTS and verification results. Since this system has an inference algorithm which does not depend on a particular power system, its NFD model and knowledge for RPD can be revised partially during maintenance work on power system facilities or in the case of operating policy change.

As well, the improvement of restoration support functions based on verification results has made it possible for the system to support restoration almost as well as a trainee using an OTS, thereby confirming the effectiveness of this verification method. Currently, the authors have been making further research toward the goal of practical application, investigating the influence of this system on the computer system performance at a control center when installed there.

References

[1] K.Hotta et al.,"Implementation of a real-time expert system for a restoration guide in a dispatching center," IEEE Trans. Power Syst.,vol. PWRS-5, pp.1032-1038, Aug. 1990.

[2] S.Moriguchi et al.,"A large-scale SCADA system with real-time knowledge-based functions," in Proc. 1989 ESAPS, July 1989, pp21-27.

[3] H.Marathe et al.,"An on-line operational expert system with data validation capabilities," in Proc. 1989 PICA, May 1989, pp56-63.

[4] K.Komai et al.,"MELDASH: A dignostic application-specific expert systems shell for network fault diagnosis," International Journal of Electrical Power & Energy Systems, vol.14, no.2, April 1992.

[5] Y.Shimakura et al.,"A knowledge-based method of making restoration plan of bulk power system,"IEEE Trans. Power Syst., vol.7, no2, pp.914-920, May.1992.

[6] I.Takeyasu et al.,"Development of a real time training simulator for power system operators," in Proc. 1989 IFAC, Aug. 1989, pp439-444.

Evaluating A Restoration Tool Using Consolidated Edison's Training Simulator

Mehdi Rafian
Parvis Sigari

Siemens Energy & Automation, Inc.
Empros Power Systems Control
Brooklyn Park, MN 55428

Daniel Kirschen

UMIST
Manchester, United Kingdom

Steven Silverman
Ronald Hamlin
Richard Bednarik

Consolidated Edison Company of New York
New York, NY 10003

Keywords: power system restoration, knowledge-based system, operator training simulator.

Abstract - To facilitate the implementation of restoration procedures in the event of a blackout, Consolidated Edison is considering the implementation of a knowledge-based restoration tool. This tool would not only guide operators through the execution of predefined procedures, but would also have the ability to adapt and modify these procedures if needed. These requirements can be met if the tool is built on top of a model of the power system and if it decomposes the procedure into components with an explicit or easily identifiable purpose. This tool has been developed and integrated with ConEd's Operator Training Simulator so that it can be used to train the operators in the restoration procedures. After evaluation it may be enhanced to become an on-line tool for use during an actual restoration.

1. Introduction

In spite of stringent preventive security measures and heroic corrective actions, large scale system collapses do happen. Electric utilities must therefore be prepared to respond to such incidents in an orderly manner so that the electricity supply can be restored to all customers as quickly as possible. To achieve a sufficient level of preparedness, Consolidated Edison (ConEd) has devoted a significant amount of resources to the development of a restoration strategy and to the establishment of detailed restoration procedures. In addition, these procedures are updated each year to reflect the changes which have occurred in the configuration of the system. Finally, and this may be the most important, all affected personnel are regularly trained in the role they have to play during a restoration. In particular, ConEd's recently acquired training simulator is used to train system operators in the implementation of the detailed restoration procedures.

Since they are prepared in the serene atmosphere of the planning department, these restoration procedures are the product of a thorough analysis of the problem and of a manual "optimization" aimed at making them fast and easy to implement. Unfortunately, their optimality rests upon the assumption that it will be possible to implement them exactly as they were written. In practice, this may be impossible either because the configuration of the system has changed or because some devices may not be available. In addition, it is impossible to predict what the load level and the load distribution will be during the restoration. While it is possible to compensate somewhat for this uncertainty by specifying different transformer tap settings for different load levels, voltage deviations could seriously affect the speed of the restoration. Finally, the operator does waste valuable amounts of time transferring information from the paper documents describing the prepared procedures to the computer system through which they are implemented. Performed under stress, this manual operation also increases the risk of error.

96 WM 166-9 PWRS A paper recommended and approved by the IEEE Power System Engineering Committee of the IEEE Power Engineering Society for presentation at the 1996 IEEE/PES Winter Meeting, January 21-25, 1996, Baltimore, MD. Manuscript submitted July 21, 1995; made available for printing January 10, 1996.

Aware of these limitations, ConEd decided to explore whether a software tool such as the Restoration Assistant (RA) [7,8] could facilitate the implementation of the restoration procedures and palliate their limitations. The following goals were set for this project:

- To develop an effective tool for training operators in restoration procedures.
- To verify that the specific features of ConEd's restoration strategy can be integrated into RA.
- To check that RA is sufficiently reliable to warrant its upgrading to an on-line tool for use during an actual restoration.

This evaluation can be conclusive only if it relies on an extensive and realistic simulation of the conditions which would prevail during a restoration of the ConEd system. Since ConEd's operator training simulator (OTS) is designed to accurately model such conditions, it was decided to integrate the Restoration Assistant with the OTS and to carry out the tests in that environment.

The work described in this paper represents the convergence of two lines of research: the development of training simulators capable of realistically modeling restoration conditions [1-5] and the development of knowledge-based restoration tools [6-10]. It has also been facilitated by the work of the IEEE Power System Restoration Working Group (whose many papers have been collected in [11]) and by the IEEE System Restoration Training Task Force [12].

The remainder of this paper discusses the requirements that ConEd's restoration strategy places on a restoration tool and how these requirements were satisfied. It concludes with a brief description of the tests which are under way and an outline of future work.

2. Overview of ConEd's restoration strategy

ConEd's restoration strategy considers both the case where only the ConEd system has collapsed and the case where it went down together with its neighbors.

In the first case, restoration assistance is available from neighboring utilities under the form of cranking power and emergency power. Seven procedures have been defined to take advantage of this assistance. All these procedures are intended to be executed concurrently. Each procedure starts from a tie-line and re-energizes a portion of the system in several steps. Gas turbines are started and synchronized as soon as possible to assist in the load pickup. Larger steam units are also restarted and synchronized as soon as they are ready to come on-line. Finally, the areas re-energized by each procedure are reconnected together.

Reprinted from *IEEE Transactions on Power Systems,* Vol. 11, No. 3, pp. 1636–1642, August 1996.

In the second case, no assistance is available and the restoration must rely on black-startable generating units. Two concurrent procedures have been devised. The first one re-energizes the five boroughs of New York City while the other brings power back to the Northern 345kV and 138kV portion of the system. Each of these procedures begins with the start-up and synchronization of gas turbines. The energization is progressively expanded to other stations to make possible the start-up of other generating units. When enough capacity is available, load is progressively picked up and larger steam units are fired up. When both procedures have been completed, the system consists of two islands which are ready to be synchronized.

All nine of these procedures were designed to be able to restore as much of the load as possible in a relatively small number of steps. A large number of power flow studies was carried out to ascertain how much load should be picked up at each step to maintain the voltage within acceptable limits.

Each procedure consists of an average of about eight major steps. At each step, actions in one or more of the following categories must be performed:

- Breaker openings required to circumscribe the next step of the energization
- Breaker closings required to ensure that all the desired area will be energized
- Closing of the "hot breaker": this operation will energize either a portion of the system or some load
- Transformer tap adjustments: considering the large number of cables in the ConEd system, controlling the voltage during the restoration is a significant problem. Different tap settings are suggested for different to-be-restored load levels.
- Load sheddings: low voltage loads are shed in some procedures to limit the amount of load picked up as the system is re-energized.
- Generating units to be started up and synchronized.
- Load restorations: shed loads which can be restored when enough generation capacity becomes available.

It is important to note at this point that the procedures implicitly assume that some switching devices are in their default position.

3. Functional Requirements

From the onset of the project, it was decided that the restoration tool should not only assist operators in the implementation of the predefined procedures but also help them adapt or replace these procedures when they cannot be implemented as written. In particular, the tool should:

Check whether a procedure or parts of it is actually feasible. An operator should not waste time starting the execution of a procedure if it can be shown that this procedure (or parts of it) cannot be completed because a critical device is not operational.

Lead the operator through the procedure for which he or she is responsible. The operator should not have to manually enter the steps of the procedure. The tool should indicate what step should be executed next and whether this step is actually feasible.

Adapt a procedure to reflect the actual restoration conditions. For example, the tool should determine, based on the anticipated load pickup, how the transformer taps should be set to keep the voltage within the acceptable range.

Suggest a remedial action when the execution of a step in a procedure has caused an overload or a voltage limit violation.

Modify a procedure to make it feasible. This infeasibility may be due to a change in the configuration of the network or to the unavailability of a device. The tool should detect the problem and propose an alternative which deviates as little as possible from the predefined procedure.

Provide a simple mechanism for manually entering simple modifications to a procedure if the operator feels that such modifications are necessary.

Assist the operator in the development of a new restoration plan if the predefined procedure turns out to be totally inapplicable.

Help the operator visualize the status of the restoration and the consequences of the next step in the procedure.

4. Design Issues

4.1 Representing restoration procedures

After reading the functional requirements described in the previous section, the question which comes to mind is: how can restoration procedures be represented in a form which can be manipulated by a software tool?

The simplest method consists in organizing the procedure as a sequence of pieces of text, where each step of the procedure corresponds to one piece of text. If the procedure is organized in this fashion, a program can easily be written to present the procedure step-by-step to the operator. In this form, however, the restoration tool provides very little added value to the procedure because it does not "understand" it anymore than someone who has never studied music "understands" a partition. In particular, it cannot check whether the procedure is feasible because it does not know which devices must be available. It cannot even check whether a particular step has been completed because it doesn't know what devices are involved in that step and what was supposed to happen to them.

It is clear that the restoration tool needs to have access to a model of the problem domain, in this case a model of the portion of the power system which must be restored. As discussed in [13], an object-oriented approach is usually chosen to create this model. Each power system device is represented as an object of the appropriate class. Attributes of these objects specify the status of the corresponding devices or relations (such as connections) among devices. A procedure can then be represented as a series of steps were each step is represented by an abstract object of the class Step. Each object of this class has a set of attributes or slots corresponding to the various operations which can be performed as part of a step. These slots contain pointers to the objects representing the devices which should be the object of these actions. A textual description of the actions which must be performed as part of a step can easily be reconstructed from the definition of an object. For example, the textual description of the simplified step shown on Table 1 would read as shown on Table 2.

Table 1: Example of an object of the class Step

Name:	FER_1
Station:	FER
Open Switches:	8P8, 8P3, 8P6, 6P10
Closed Switches:	8P2, 8P4, 8P5
Hot Switch:	8P1

Table 2: Textual description of the restoration step given in Table 1.

Open breaker 8P8 in station FER
Open breaker 8P3 in station FER
Open breaker 8P6 in station FER
Open breaker 6P10 in station FER
Close breaker 8P2 in station FER
Close breaker 8P4 in station FER
Close breaker 8P5 in station FER
When all these operations have been performed and confirmed, do:
Close breaker 8P1 in station FER

Formulating restoration procedures on the basis of a model of the power system allows the Restoration Assistant to track the completion of a restoration step by comparing the actual position of all the switching devices listed in the step with their desired position. It also makes it possible for RA to check a priori whether the devices which must be activated are actually operational. This falls somewhat short of a full feasibility check of the restoration step because this description (like most written restoration procedures) is limited to the devices which are actually activated during the step and does not include a list of all the equipments which will be passively affected by the step, nor a list of the switching devices which are assumed to be in their default position.

This simple representation of the restoration steps has a major flaw: if a step cannot be executed as defined, it is impossible to develop an alternative unless the purpose of the original step is known. A method for representing and storing the goals of a step along with the actions to be taken during this step must therefore be devised.

Table 3 shows the two attributes which need to be added to the example of Table 1 to make the purpose of this step explicit. The attribute "Node origin" points to the electrical node from which a connection will be made and the attribute "Node destination" points to the node which is the target of this connection. Using the one-line diagram shown on Figure 1, the reader can check that the purpose of this step is indeed to make a connection from line 116 (which is assumed to be energized) to the breaker at the end of line 119.

Table 3: Expanded definition of the example of Table 1

Name:	FER_1
Station:	FER
Open Switches:	8P8, 8P3, 8P6, 6P10
Closed Switches:	8P2, 8P4, 8P5
Hot Switch:	8P1
Node origin:	FER_8
Node destination:	FER_9

If breaker 8P2 happens to be stuck in the open position, this step is clearly unfeasible as written. However, RA can implement a search procedure to try to find an alternative path between the same node origin and node destination. In this case, such an alternative exists and is described in Table 4.

Table 4: Definition of the step modified to take into account the fact that breaker 8P2 is blocked open.

Name:	FER_2
Station:	FER
Open Switches:	8P2, 8P7, 8P4, 8P10
Closed Switches:	8P8, 8P6, 8P5
Hot Switch:	8P1
Node origin:	FER_8
Node destination:	FER_9

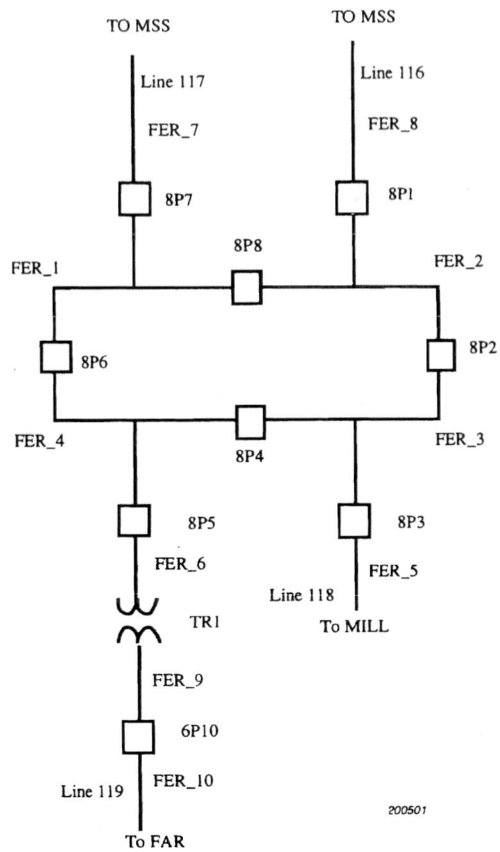

Figure 1. One-line diagram of the station used in the example

In some cases, it is impossible to fix a problem within the confines of one step. In our example, line 116 may be faulted or breaker 8P1 may be blocked in the open position. In this case, RA needs to go up one level in the hierarchical definition of the goals and define a new series of steps which achieve the same higher level goal. In this case, it may decide to energize station FER from station MSS using line 117 instead of 116. This will require the definition of a new step at station MSS and a new step at station FER. Once defined, the actions to be performed as part of these steps can be identified using the search procedure mentioned above. If this search is successful, these steps can replace the original ones in the procedure. If it is unsuccessful, stations cannot be reconnected in the predefined sequence. In this case, the operator can specify (by pointing and clicking) the stations which should be reconnected. Based on this information, RA defines a series of steps and instantiates them using the same basic search procedure.

Manually decomposing the ConEd's restoration procedures into steps and defining the purpose of each of these steps would have taken a large amount of time. It was therefore decided to develop a "reverse engineering" program capable of taking a quasi-textual description of a procedure and breaking it down into a series of steps with an identified purpose. A by-product of this reverse engineering program is a list of all the components passively affected by each restoration step. This information can then be used to determine with more certainty the feasibility of each step.

4.2 Correcting Voltage Problems

Since the ConEd system contains many kilometers of cables, overvoltage problems can be quite severe during the restoration. Some of these problems can be fixed by adjusting the transformer taps while others require the rapid connection of shunt reactors or additional loads. Once an overvoltage problem is detected or anticipated, RA can identify the most appropriate remedy, i.e. the de-energized reactor or load closest to the location of the overvoltage. The energization of this load or reactor can then be defined in the form of a step whose node origin is an already energized node and whose node destination is the node where this load or reactor is connected.

4.3 Visualization

Since the representation of the procedures is layered on top of a model of the power system, it is relatively easy to create a graphical user interface which helps the operator visualize the restoration process. This is achieved by associating with each object representing a power system device a graphical representation of this device. A change in the status of a device is then automatically reflected in the appearance (i.e. shape and color) of the associated graphical representation. By default, the user interface displays the current energization status of the network. However, RA can also highlight the portion of the network which will be energized during the next restoration step. The one-line diagrams can also be used by the operator to manually modify a restoration step or to mark a device as unavailable. Having marked a device as unavailable, the operator can ask RA to develop a procedure which does not rely on the use of this device.

Several displays have also been constructed to help the operator keep track of some of the global restoration parameters such as the amount of load restored, the available generation capacity, the frequency and the power transferred via tie-lines. Figures 2, 3, and 4 show examples of these displays. These displays are automatically updated every 5 seconds.

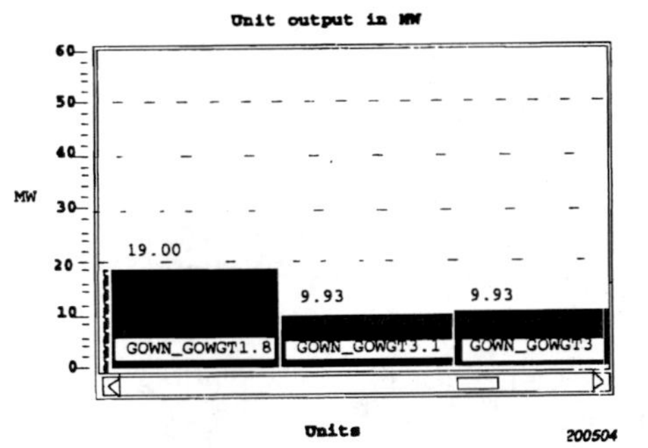

Figure 2. Bar chart showing the active power generated by each unit which has been restarted during the restoration procedure.

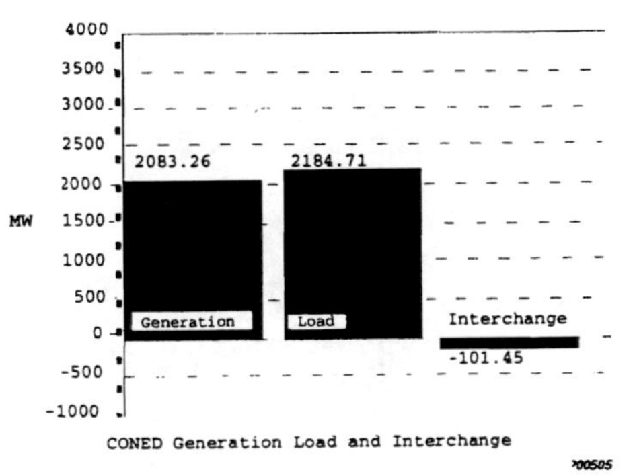

Figure 3. Bar chart showing the load/generation/interchange balance.

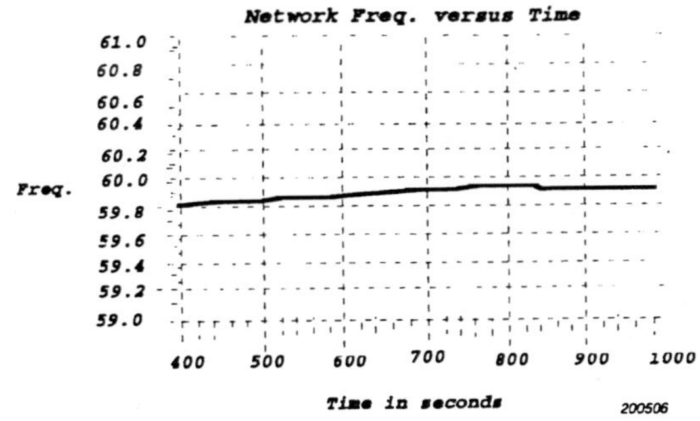

Figure 4. Line chart showing the evolution of the system frequency during the restoration.

5. Integration Issues

5.1 Power System Model

It has already been mentioned in a previous section that RA is built on top of a sophisticated model of the power system. As argued in [13] a significant portion of this model can be shared by several applications. It therefore makes sense to create an application environment which takes care of the model and into which applications such as the Restoration Assistant can be plugged. Figure 5 shows a schematic representation of this environment which is called ODE (Operator Decision Environment). ODE essentially consists of three parts: a server, a set of interfaces and the application program or programs. The server (OdeRT) is responsible for making an up-to-date model available to the applications. Every time an event (such as a switching operation) takes place in the power system, it updates the attributes of the affected objects and, if necessary, performs a topology analysis to keep track of the bus and island structure of the system and of the energization status of each device.

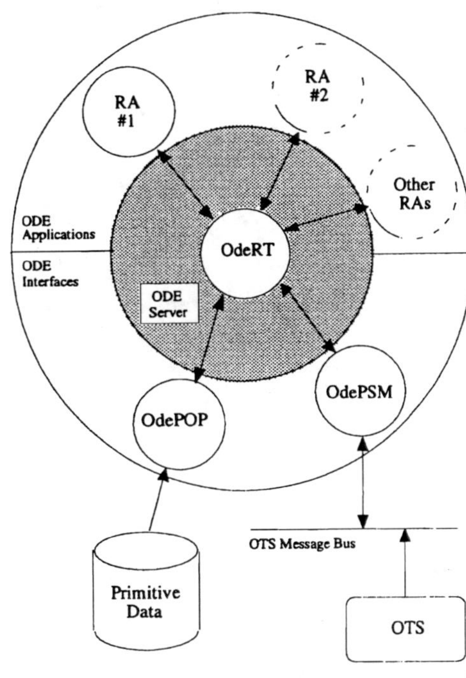

Figure 5. Schematic Representation of the ODE Environment

5.2 Modules and Interfaces

Standardized interfaces have been designed to handle all the interactions between ODE and the OTS with which it is integrated. The OdePSM interface handles all the run-time communications between ODE and its host environment. It connects to the OTS message bus and receives all the messages which might be of interest to ODE and its applications. It also handles the transfer of information in the other direction, i.e. the commands and requests that ODE applications send to the OTS. The OdePOP interface is used to create the power system model used by ODE from the primitive data description contained in the OTS database.

ODE applications such as the Restoration Assistant communicate with the OdeRT model server through a shared memory and remote procedure calls. This form of integration makes it possible for ODE to support several applications simultaneously. A more detailed description of the architecture of the system can be found in [13].

Figure 6 shows how the RA fits within the OTS configuration at ConEd. The OTS consists of three main subsystem:

- The Power System Model (PSM) which simulates the dynamic behavior of the system.
- The Control Center Model (CCM) which is a replica of the actual control center which the trainee are familiar with.
- The Instructor Position (IP) which is used by the instructor to manage the training session and to simulate the actions of operators outside the control center.

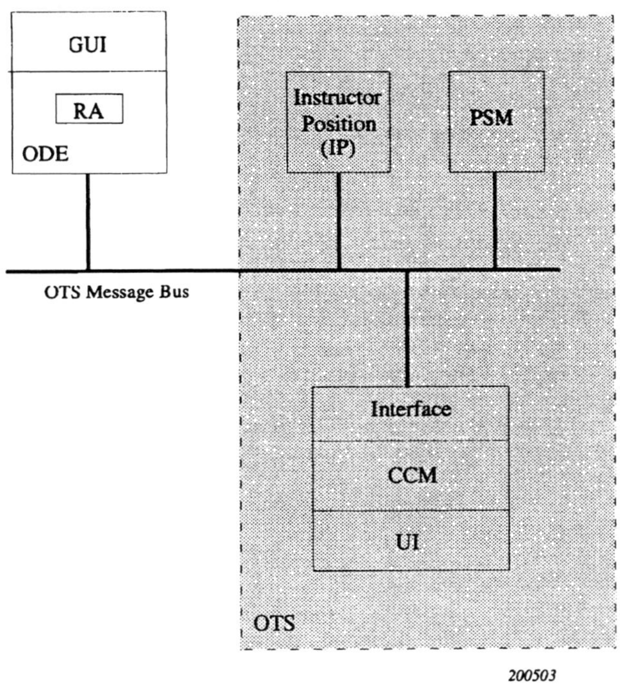

Figure 6. Integration of the Restoration Assistant with the OTS

At this point, the only subsystem with which the Restoration Assistant has direct interactions is the PSM. These interactions have been designed to look like the interactions that the RA would have with an actual power system. The RA/ODE model is initialized from the status of the PSM and is updated when qualitative changes take place in the PSM. If the user wishes to perform a control action through the RA (such as opening or closing a breaker), a message is sent to the PSM. The PSM simulates the execution of this command and sends a message back to the ODE informing it of the new status. It is only upon receipt of this message that ODE updates its model.

5.3 Initialization and Case Management

The OTS and the RA must be initialized from a previously saved case. For restoration training, these saved cases represent one of the following situations:

- The networks of ConEd and all its neighboring companies are completely blacked out.
- The network of ConEd is completely blacked out but the networks of some of its neighbouring companies are not.
- The ConEd network is partially blacked out, either as a result of a partial blackout or because the case reflects a partially restored from a complete blackout.

These cases can be prepared by the instructor from a snapshot of the system taken using the OTS educational subsystem functions. These functions include the Event Scheduler the Heuristic Scenario Builder or the transient stability tool. Snapshots taken during training sessions can also be used as saved cases ended in a blackout.

Since the OTS and the RA do not operate in the same environment, their initializations must be carefully coordinated. To facilitate the task of the user, this process has been made transparent through the development of an initialization protocol which is initiated and controlled from the OTS initialization display. When the user decides to activate the RA with the OTS, this protocol proceeds as follows:

- The OTS loads the dynamic data contained in the file corresponding to the selected case.
- The OTS triggers the execution of a script which starts the OdeRt part of the ODE environment if this environment is not yet running.
- OdeRt loads the static portion of its model of the power system and establishes the communication channel between ODE and the OTS via OdePSM process. OdeRt then starts the RA application.
- RA sends a request to the OTS for dynamic data.
- OTS creates a file containing the dynamic data necessary to initialize RA and informs RA of the availability of this data.
- RA loads this data and initializes itself.
- RA loads any additional dynamic data which may have been saved in conjunction with the saved case retrieved by the OTS.
- RA informs the OTS that it is ready.

Form this point on the OTS and RA are synchronized and remain synchronized because every event taking place in the OTS is relayed to ODE and RA. If OTS is stopped and restarted from a new saved case, RA is re-initialized.

5.4 Saving a Case in the RA

At any time during the restoration, the instructor can take a "snapshot" of the state of the power system. Snapshots can also be taken automatically and periodically at predefined intervals. Every time a snapshot is taken, the following actions are taken:

- The OTS saves the dynamic data necessary to described the state of the system into a saved case file.

- The OTS informs RA that it has created a new saved case and gives it the identification number of this case.
- RA saves the dynamic data which it needs to initialize its model and which is not saved by the OTS (e.g. number of hours since a generating unit has been started or shut down.)

6. Field Experience

The Restoration Assistant was installed in ConEd's training center and integrated with the OTS in October 1994. This integration has made it possible to check that RA meets the functional requirements described earlier in this paper. In particular, it was verified that the manner in which the RA guides the operator through a restoration procedure is not only correct but also effective and flexible. In other words, the operator is reminded of what needs to be done, understands what the consequences of the next step will be and has the ability to modify a step or series of steps if needed. Highlighting the part of the network which will be energized by the closing of a "hot switch" was found particularly useful. It was also verified that RA can find an alternate restoration path when the predefined path is impossible and when such an alternate path exists. RA's ability to perform rapidly and semi-autonomously many switching operations gives the operator more time to consider each move and analyze its consequences.

In addition to this evaluation of its usefulness as an on-line tool, RA has also been used for training. To date, all 27 ConEd operators have undergone RA-based training on all the restoration plans. It has been observed that RA's ability to graphically represent the consequences of a series of steps not only shows to the trainees what needs to be done but also helps them learn why it needs to be done in a certain way.

The flexibility of the RA user interface allows the trainer to customize the training exercise. For example, the trainer can easily enable or disable a switching device through the RA user interface. This feature can be used by the trainer to create a more difficult exercise where one or more switches which are used in the normal restoration procedure are disabled and where the student must develop an alternative procedure, possibly with the help of RA. Conversely, a certain number of devices which must be used during the restoration are not remotely controllable and are modeled as such in the Control Center Model of the OTS. When the trainee wants to operate such a device, he or she "calls" the trainer who is playing the roles of all the field and plant personnel. The trainer then operates the device through the Instructor Position workstation. While this procedure enhances the realism of the exercise, it does consume a significant portion of the time of the trainer. If the trainer wants to create a more free-wheeling exercise and spend more time evaluating the performance of the trainee, he or she can let the trainee perform all the operations through the RA interface.

Occasionally, a trainee will inadvertently perform a switching action which re-energizes the entire grid in one step. When the PSM portion of the OTS attempts to model the consequences of this action, the power flow diverges, or causes a severe voltage problem thereby indicating that the network has blacked-out again. When such an event is detected, the RA automatically rescinds the last switching operation and warns the trainer. This feature makes it unnecessary to restart the simulation and the restoration process and therefore saves a considerable amount of training time.

Finally, RA-based training has had an unanticipated benefit. Restoration procedures are updated only once a year while the model of the power system used by the OTS and RA is updated on an almost weekly basis to reflect the changes which are continuously being made in the network. Since RA tries to correlate the description of the procedure with the model of the system, areas where this procedure is out of date are automatically detected. Many problems of this type have been identified during RA-based training of ConEd personnel and the corresponding procedures have since been corrected.

7. Future Work

Based on the results of the tests which are carried out with the OTS, a list of the additional requirements which must be satisfied before the RA can be placed in the hands of the operators for use during an actual restoration will be developed. One of these features is certain to be the development of a multi-user version of RA where each user is responsible for the execution of one of the procedures. The ODE architecture makes the implementation of this feature fairly easy because multiple applications can already share the same model. Each user can therefore work independently on the execution of one procedure while being kept informed of the progress of the other operators.

The implementation of major restoration steps can result in a voltage or frequency change larger than anticipated because of the inevitable differences between the actual and postulated conditions. It is thus important to predict as accurately as possible these changes in voltage and frequency based on the information available during the actual restoration. The OTS is capable of predicting changes in the quasi-steady state voltage and frequency. This approach would have the advantage that the fidelity and robustness of the OTS have already been demonstrated. Such verifications would proceed as follows. Before major switching operations the PSM would be initialized using a real-time snapshot of the actual system. The consequences of this restoration step would then be simulated and the results of this simulation would be analyzed by a new module of the restoration assistant. If potential voltage or frequency problems are detected, the operator could adjust transformer taps, change the amount of load picked up by the re-energization, perform capacitor/reactor switching, limits the extent of energization, or finds an alternative route for energization. The RA would recommend the best remedy among the feasible possibilities to maintain an acceptable voltage/frequency level. While the validity of this verification would depend on the accuracy of the model used to predict the remedy, it could considerably reduce the risk of damaging equipment or suffering a setback in the restoration process.

An accurate on-line computation of the transient voltages which might occur after each restoration step would require considerably more computing power than can realistically be devoted to power system restoration at this point.

While the Restoration Assistant can probably be enhanced to support other important restoration activities such as the coordination with field crews, this difficult question requires a more detailed analysis.

8. Conclusion

An effective restoration tool should not only have the ability to guide the operator through predefined restoration procedures but also the capacity to adapt, modify or replace these procedures if necessary. Two conditions must be satisfied before a tool can achieve these goals:

- The tool must be built on top of a model of the power system which is updated in real-time to reflect the actual conditions of the system
- Procedures must be internally decomposed into steps whose purpose is either explicit or can easily be identified.

Such a tool has been built and integrated with ConEd's operator training simulator. The tool is presently used to train ConEd operators in restoration procedures. Validation and verification tests are being performed in this environment before a decision is made on making RA available to the operators for use during an actual restoration.

References

[1] R.F. Chu et al., "The Uses of an Operator Training Simulator for System Restoration," in Proc. of the 1991 IEEE Power industry Computer Applications Conference, Baltimore, Maryland, pp. 171-177.

[2] Operator Training Simulator," EPRI Report EL-7244, Volumes 1 and 2, May 1991.

[3] Miller et al., "Experience Using the Dispatcher Training Simulator as a Training Tool," IEEE Transactions on Power Systems, Vol. 8, No. 3, August 1993, pp. 1126-1132

[4] T. Kobayashi et al., "Verification of an Advanced Power System Restoration Support System Using an Operator Training Simulator," IEEE Trans. on Power Systems, Vol. 9, No. 2, May 1994, pp. 707-713.

[5] Rajagopal et al., "Workstation Based Advanced Operator Training Simulator for Consolidated Edison," IEEE Trans. on Power Systems, Vol. 9, No. 4, November 1994, pp. 1980-1986.

[6] CIGRE Task Force 38.06.04, "A Survey of Expert Systems for Power System Restoration," ELECTRA No. 150, October 1993.

[7] D.S. Kirschen and T.L. Volkmann, "Guiding a Power System Restoration with an Expert System," IEEE Trans. on Power Systems, Vol. 6, No. 2, May 1991, pp. 558-566.

[8] D.S. Kirschen and T.L. Volkmann, "Power System Restoration Tools," Paper 38-105 presented at the 1992 CIGRE session.

[9] C.C. Liu et al., "Generation Capability Dispatch for Bulk Power System Restoration: A Knowledge-Based Approach," IEEE Trans. on Power Systems, Vol. 8, No. 1, February 1993, pp. 316-325.

[10] M.M. Adibi, R.J. Kafka and D.P. Milanicz, "Expert System Requirements for Power System Restoration," IEEE Trans. on Power Systems, Vol. 9, No. 3, August 1993, pp. 1592-1600.

[11] IEEE Power Engineering Society, "Power System Restoration," IEEE Special Publication 93 TH0 605-6 PWR.

[12] M.E. Cooper, M.M. Adibi et al., "Bulk Power System Restoration Training Techniques," IEEE Trans. on Power System, Vol. 8, No. 1, February 1993, pp. 191-197.

[13] J. Bann et al, "Expert Systems Integration: Issues and Solutions," in Proc. of 1995 Power Industry Computer Applications Conference, pp. 458-463.

System Restoration Guidelines: How to Set-Up, Conduct, and Evaluate a Drill

A Report by the Operator Training Working Group
Joseph D. Willson, PJM Interconnection Association

ABSTRACT
Once an organization has developed a restoration plan it needs a method to evaluate that plan. The evaluation method described here is the use of a system restoration drill / exercise. Simulating restoration conditions will provide a measure of an organization's ability to respond to a real incident. Periodic drills and exercises help to maintain a high state of preparedness for those individuals involved. These drills also test and validate the organization's restoration plan before it is activated during an actual incident. This paper describes a set of guides for: How to Set Up a restoration Drill; How to Conduct a Drill; and then How to Evaluate a restoration drill.

INTRODUCTION
One of the three sessions conducted at the December 1994 IEEE Workshop on Operator Training concentrated on the development of a set of guidelines that could be used by utilities embarking on system restoration drills. The results of that workshop are presented in this paper.

The reliability of the electric power grid in North America is extremely high because system operators continually monitor the power flows and voltages across the system and maintain these values within established bounds. Thus electric power blackouts occur very infrequently. If a blackout does occur, every effort is made to rapidly restore the transmission system, restart generating units, and pick-up the customer load. The electric system operators must not only be trained to handle the routine daily functions but also be trained in system restoration and remain proficient in these restoration skills. One method to develop and hone these skills is through periodic practice sessions. However, blacking out the system to train operators hasn't been deemed acceptable. An appropriate alternative is to conduct system restoration drill.

96 WM 164-4 PWRS A paper recommended and approved by the IEEE Power System Engineering Committee of the IEEE Power Engineering Society for presentation at the 1996 IEEE/PES Winter Meeting, January 21-25, 1996, Baltimore, MD. Manuscript submitted June 26, 1995; made available for printing January 10, 1996.

The drill development process may begin with a single exercise but it needs to become an on-going commitment for progressive improvement. Exercises should not be one-shot deals; rather a part of an organization's on-going commitment to improving the overall restoration process. The existing restoration plan is used as the starting point for establishment of the drill and defining its content and objectives. a planning group should have a long term plan for guiding the focus of subsequent drills over time.

The drills successes and short comings are then channeled into modifying the overall dispatcher restoration training program. The training program, drills, and/or the restoration plan itself can be adjusted to better serve the requirements of the organization. Drills become an intricate part of the system's restoration training program. With each exercise more information is learned which contributes to continuing improvement of the overall plan. Communications will probably be the biggest obstacle to initially overcome. Clear concise understanding of what is going on must be maintained throughout the drill. So the first order of business is to set down and define what is to be done. Some examples are shown in the next section. An excellent reference document addressing the coordinated communications aspects of exercises has been prepared by the Federal Emergency Management Agency [1].

DEFINITIONS
Once you decide to proceed on developing a restoration drill as part of your overall restoration program, you need to ensure that everyone involved in the drills has the same understanding and training. To begin it is necessary to establish some basic terminology and definitions to be used prior to and during the various types of drills or exercises. It is also necessary to define roles and activities of the drill participants. Listed below are some sample definitions:

Type of Activities
Exercise: an activity designed to promote emergency preparedness; test or evaluate emergency operations plans, procedures or facilities; train personnel in emergency management duties; and demonstrate

Reprinted from *IEEE Transactions on Power Systems,* Vol. 11, No. 3, pp. 1619–1629, August 1996.

operational capability. Exercises consist of the performance of duties, tasks or operations very similar to the way they would be performed in a real emergency.

Orientation Seminar: focus is on training and familiarization with roles, procedures, and responsibilities. It is designed to elicit constructive discussion by the participants as they attempt to examine the plan and as they try to resolve any problems with the plan. During this session the group would resolve questions of coordination and assignment of responsibility.

Drills: characterized by activity to test, develop, or maintain skills in a single emergency response procedure. Focus is limited.

Tabletop Exercise: presents a simulated emergency situation which is solved by group discussions. The exercise is intended to evaluate plans and procedures and to resolve questions of coordination and assignments of responsibility.

Functional Exercises: intended to test or evaluate the capability of an individual function, or single complex activity with a function. Only applicable where the activity of function can be effectively evaluated in isolation from other emergency management activity.

Full-Scale Exercise: intended to evaluate the operational capability of emergency management systems in an interactive manner over a substantial period of time. Involves the testing of a major portion of the basic elements existing within emergency operations, plans and organizations in a stressful environment.

Progressive Full-Scale Exercise: intended to evaluate the operational capability of the system in an interactive manner over a substantial period of time and may involve introducing additional problems, fast forwarding (simulating time) operations, etc. It involves the testing of a major portion of the basic elements existing within operations plans.

Seminars: serves the dual purpose of familiarization and motivation. The orientation seminar involves bringing together those with a role or interest in a plan, problem or procedure. The variety of methods available to conduct the orientation include lecture, film, slides or other visuals, and panel discussion.

Simulation: tool to create the perception of a situation, event, or environment which will evoke responses similar to those that a real emergency would prompt. Simulation attempts to approximate reality by using symbols, maps, drawings, scripts, or, in more elaborate exercises,film, videotapes, or computer graphics. The key to good simulation is the perception of realism by the individual or group being exercised.

Scenarios: set of hypothesized chain of events. Provides the story and back drop of what has occurred. Provides the outline of what should occur during the course of the exercise.

Testing: specific activity (when associated with an exercise or drill) which is designed to assess actual readiness capability of procedures, personnel, facilities, or equipment against required capabilities described in emergency plans.

People

Identifying the participants and their roles is key to the exercise's success. Participants roles vary depending on the drill scenario.

Players: those individuals participating in the drill in their normal job functions and other entities who have been assigned active roles in the management or response to incidents. Who may play: system operators, production personnel, engineering staff, transmission and distribution personnel, public information staff, etc.

Drill Coordinator: individual who monitors the flow of the exercise. The drill coordinator's purpose is make sure that the exercise is conducted according to the objectives that were established. The drill coordinator would be responsible for developing the content of the debriefing sessions. The drill coordinator ensures that the level and pace of the activity are realistic given the limitations imposed by the scenario.

Evaluators: one or more individuals who have the responsibility for documenting and evaluating the actions of the players during the drill. They observe the actions and decisions of the exercise to later report what went well and what did not go well. They monitors the decisions made in the exercise and later reports on them. Evaluators do not prompt the players during the drill or exercise. As follow-up they may help identify objectives for future drills.

Observers: non-participants invited to view a drill/exercise. Observers may be allowed to view selected drill aspects or the entire drill's activities. Observers do not interfere with the actions of the players, drill coordinators, or evaluators.

Simulators: creates the simulated responses to the emergency by sending prescribed messages and/or spontaneous messages to the players. Simulators create an artificial reality through the delivery of their messages in order to evoke player reactions similar to those occurring in a real restoration situation. The role of the simulator is to portray the part of all "outside environment" to the participants. They can play the role of groups (i.e. plant control room operators) who are not participating in the drill.

Types of System Restoration Drills

Before beginning the development of the drill or exercise you must understand the relationship of certain elements to the entire system restoration process. The major categories being discussed are: orientation sessions, table top exercises, functional exercises, full-scale exercises, and progressive full-scale exercises.

Orientation sessions are considered a foundation necessary to lay the groundwork for future restoration exercise efforts. An exercise development program is based on the premise that you must crawl before you can walk, and walk before you can run. The surest route to failure is to take on an actual exercise before you are ready. As you progress from orientation sessions to full-scale exercises more preparation time, more personnel, and more planning is required. An exercise development program is a commitment to progressively more challenging exercises over a period of time.

HOW TO SET UP A DRILL

Before you launch an exercise, groundwork must be laid to make certain that the exercise will be successful. You need to review the current restoration plan and decide what aspects will be included in the exercise by conducting a needs assessment for the individuals and an assessment of your capability to conduct the exercise.

The next step is to define the scope of the drill and establish the drill's objectives, which are described later in this paper. Subsequently the detailed scenario is developed, a part of which includes the determination of the necessary participants. Establishment of the drill date may sound easy but it may prove to be one of the more difficult operations. The date will depend on the type of exercise, the skills of the participants, the training requirements, and the physical plant adjustments that must be made. Also, to be successful, drills should not conflict with other scheduled activities such as major management meetings.

Purpose of Drill / Exercise

There are several major kinds of outcomes of a drill or exercise: the first is individual training with people practicing their roles with the desire of getting better at them (a learning experience for the individuals). The second is to determine the involvement of others outside your immediate area that should and need to be involved in the restoration plan (eg. company Public relations personnel). Third, validate the plan Finally, expose problems.

Companies need to differentiate between drills for testing emergency procedures and drills for testing individual's proficiency to deal with a system restoration situation.

The fundamental purpose of an exercise program is to improve operational readiness. In support of that goal, exercises can also:

- -reveal planning weaknesses
- -reveal resource gaps
- -improve coordination
- -clarify roles and responsibilities
- -improve individual performance
- -develop proficiency and confidence
- -test plans and systems in 'live' situations
- -foster cooperation between departments,
- -help formulate readiness posture of co.
- -satisfy specific requirements of certain agencies or organizations
- -enhance utility-state/local emergency management relationship

Define the Drill Objectives

What are the objectives of the exercise? Before going too far it is essential that you clearly state and get a full understanding of what you're trying to accomplish with the drill. The drill may have a number of objectives but all objectives have to be measurable and all involved personnel must understand the objectives.

There needs to be agreement on objectives and these objectives must be in alignment from a pool, control area, and company/system basis! For each defined objective there should be an expected action or action plan to be followed. During the exercise the objectives will be continually evaluated considering these expected actions.

Identify who should be involved in drill planning

Participants required are from at least four major areas: senior management (even if the executive does not play in the exercise, her or his authority and support are essential), a team involved in exercise development (the exact size would depend on the extent of the drill), role of the emergency manager (must clearly be an active player in the exercise), and getting the departments and agencies involved as players (consider who would be required in a real emergency and include them).

A variety of tasks must be completed to successfully develop an exercise. These include:
1. Identifying Resources
2. Planning the exercise
3. Developing simulation materials
4.Preparing facilities, creating displays and materials, and considering exercise logistics
5. Identifying, selecting, and training the staff

Before undertaking the exercise development an overall drill coordinator needs to be established. This individual will organize and lead the exercise design team and assign tasks to its members and will take the lead in arranging for participation and in managing all administrative and logistical matters associated with the exercise. The role of the drill coordinator continues through the actual conduct of the drill and the development of the post-exercise evaluation report.

It is essential to have an experienced and capable individual to fill the important role of the drill coordinator. This person must be an individual who can devote a considerable amount of time over several months in pulling together all of the necessary elements for a successful exercise.

The exercise design team will assist the drill coordinator in developing the exercise content and procedures. Specifically, they will determine exercise objectives, tailor the scenario, and develop the sequence of events. Members of the design team should also assist in conducting pre-exercise training sessions. Ultimately, members of the team would be good candidates to function as either simulators or drill coordinators.

The participants should be those personnel (not alternates) who would function in that capacity in the event of an actual emergency. Displays, maps, status boards, etc. should be brought up to date and made operational for the exercise. Participant work spaces should be pre-designed and working supplies made available. No special or new forms should be introduced.

The exercise must be planned with the same thoroughness as any major organizational effort. There are many useful tools available to help plan the drill. Whichever one you choose it needs to depict the relationship of tasks versus time.

Designing the Drill Scenarios

The scenario narrative sets the scene for the simulated event and briefly describes what has happened up to that time in the exercise. The scenario can be constructed to be of a overview sequence of events nature or of a detailed sequence of events nature. However, it is critical that the narrative get the participants in the mood for the exercise. They need to be motivated to play the game.

To help in developing the narrative consider the following checklist:
1. What event is happening?
2. How did we find out?
3. was there advance warnings?
4. What time did it happen?
5. What was the sequence of events?
6. Where did the event occur?
7. How wide spread is the situation?
8. What damage occurred?
9. What is the impact to the public?
10. What response has occurred?
11. What does the future look like?
12. When will we return to normal?

The specific drill conditions should be based on the systems geographical location, past bad weather conditions, past experience, and/or case studies of other utilities. The scenario could include a total system shutdown, a partial system shutdown, isolated islands of load/generation, etc. If adverse weather is a concern use that as a starting point. What needs to factored into the scenario is realism. In developing starting position consider using engineering support to develop the case or to at least validate the problems that are being simulated.

Identify Types of Scenarios to Include in Plan

Operators deal day in and day out with actual response times for units and switching times for transmission work. During a drill use drill times that equal actual Time. Do not try and accelerate time. If it takes an hour for a person to inspect a substation before breakers can be closed then one hour needs to be included in the simulation. Attempting to accelerate time (1 hour drill time equals 4 hours real time) will probably only result in a failed drill. Accelerated time can only be effectively used if the drill is broken into pieces (such as a multi-day exercise or a table top exercise which has discrete unconnected learning objectives). Also don't ask operator to perform activities during the drill that he/she wouldn't perform in a real life emergency. The same tools, logs, and report forms should be used during the exercise.

Consideration of time needed by the system operator, plant personnel, and field personnel to analyze the situation must be considered to ensure realism. If the initiating events are within your system more personnel from other departments may be needed to play roles in the exercise. The scenario may also be constructed such that the system operator must have to use some decision making skills; the system operator will have to make alternate restoration selections because not all equipment is operational.

HOW TO CONDUCT A DRILL

Execution of the Drill Plan

The day of the drill is the culmination of all the planning efforts. The success of an exercise depends largely on the participants having a clear and consistent understanding of what is expected of them. Many exercises fail because the ground rules and/or simulation techniques to be used during the drill are inadequately explained. To ensure that all exercise objectives and procedures are understood, participants should be briefed before the start of the exercise. the briefing should include a statement and discussion of the exercise objectives, the simulated time period in which the exercise is to be conducted, a description of the environment, and an outline of the procedures and ground rules being used. This briefing should be held immediately before the start of the drill.

Players and Their Roles

An integrated drill will require different resource commitments depending on the type of exercise and the breath of the drill. Some drills will require minimal preparation time and will be simple to execute. Other will require much planning time, and thus months of lead time, and will entail integrated execution details. Participation raises four considerations: getting sponsorship from senior management; getting a team involved in exercise development; defining the roles of each participant; and getting the necessary departments and people involved as players.

Many complex components go into making an exercise successful, but without the support of the senior management implementation of the exercise will be a long hard battle competing with other projects and priorities that may interfere with established plans and will probably strip away necessary support people.

The exercise development team will most likely consist of two or three individuals responsible for developing the exercise but who use the skills and expertise of other departments/areas in planning the drill.

In developing the drill scenario, you need to determine which departments and which people need to be involved. Ideally, an effective exercise will need to include all of the organizations and departments and personnel that would respond in a real emergency operation, as either participants or simulated inputs.

Some suggestions for participants are as follows: Dispatchers, Plant Personnel, Network System Engineers, Power Plant Engineers, Control Center Engineers, Trainers, Management, Public Relations, Telecommunications, Observers, Simulators, Evaluators, Drill coordinators, Players, Coordinators, Clerical, Other Support.

Personnel placement in drill setting is also critical - make sure the observers are placed where they can reasonably accomplish their objectives; make sure adequate space and work space is provided to all personnel. Ensure open but controlled communications. Also, there needs to be a hierarchy of players in drill planning/execution.

An example of the staff required to perform a drill just at the control center of a utility might involve 16 people. During an actual system restoration that amount may grow to 36 or more. The question to ask is: Will the additional people understand their roles in real-time? If your answer is "I don't know" then perhaps they should be brought into the drill as participants.

Drill	Personnel Real Time	Titles
4	4	System Operators
4	6	Extra-System Operators
2	4	Engineers
2	6	Management
1	4	Clerical
2	6	Other Support
1	6	Computer / Telecommunication
16	**36**	**TOTALS**

Simulator Role

In order to create a real-life environment, simulators act as, and on the behalf of, the involved companies, systems, departments, etc. which would normally interact with the exercise participants. Simulators provide inputs to the exercise that would be made by these other organizations. Some of these inputs are scripted in advance while others will be contingent upon an event occurring during the exercise. Simulators must be prepared to reply to participants questions based on information provided them in the training materials and in the exercise briefing sessions. Simulators need to keep the action going in a drill. It may be necessary for the simulator to call a participant during the exercise to inquire about an event, a status, or for an update. In some cases it may be necessary to improvise a situation that would call attention to the lack of action in response to the event. Ensuring that key events are given full and timely attention will require ingenuity on the part of the simulator.

Dealing with overload of information is much easier to handle then to deal with gaps of information. In dealing with too much information, reassign its collection to another drill participant. You could request that available unit data be reported to one system operator while damaged equipment be directed to another individual. Divide the messages coming in into two categories: those Essential-to-Operations right now or Required-for-the-Future and those of an Informational nature. If gaps occur, the exercise participants must be given additional information by the simulators. During the exercise boredom must be avoided by creating optional information messages or requests.

Simulated Time

Functional exercise can depict events and situations that would actually occur over an extended time period. In order to include the several distinct phases of the emergency in an exercise, it would be necessary to stop the exercise and advance time by the required number of hours. These breaks in time should be kept to a minimum and should occur normal break points in the exercise. For example begin the first day of the drill at 7:00 am and continue until 3:00 pm at which time the drill participants could go home. The second day of the drill could continue as of 3:00 pm or could have the time advanced so that additional drill activity could be practiced. If simulated time is used, all activity that would have taken place must be scripted., and the participants briefed on what has occurred since the previous day. The drill coordinator is responsible for managing the simulated time transitions and preparing transition updates. All supporting material must also be updated to coincide with the simulated time.

Physical Plant

The physical facilities for the various types of exercise will vary greatly, but, in any exercise, the key question to ask is where do the players work? Where possible, exercise where you work. Use the same tools, phones, forms, etc. Remember, the drill is intended to simulate reality. The responses of the participants are planned to be as similar as possible to those that would occur in a real event. If this isn't possible replicate as much of the players facilities as possible. If your real-time restoration plan calls for an activity simulate it. Its better to find out in a drill that there aren't enough telephones then in a real situation.

Communication

Is the communications capability of drill site adequate? Does it replicate or come close to the capability the system operator uses every day? If the operator has the capability to simultaneously broadcast a message to a number of people in real life, has that capability been provided? Are additional phones or different phones required? Is the phone directory up to date if non dispatching circuits are needed? Are the charts and maps available to the operator? If not do you need a second set?

Evaluators, simulators, and the drill coordinator must work as a team for the drill to go smoothly and to achieve its objectives.

Date / Timing

Timing must be realistic throughout the drill. Where practical, participant and simulators should use the same actions that would be required if it where a real-life experience. The drill needs to portray to the participants the time delays that will be encountered. In some instances the scenario may even be designed to reflect possible additional delays such as would occur if the blackout occurred during a severe winter storm. Time to assess the situation and report on damage would greatly extend the time devoted to the operator's time to assess the system. The exercise needs to provide time for substation operators to travel to/between substations, time for generators to shut down and then restart, time to acquire people to aid in the restoration, and time perform all the necessary switching operations and for clearing all the alarms that I'm sure the operators will encounter.

HOW TO EVALUATE A DRILL

What to Evaluate

A realistic exercise provides the best opportunity to evaluate its restoration plan and its overall preparedness to restore the system. The extent and depth of the evaluation to be undertaken is determined by the drill coordinator and the drill participants. The reasons you need to evaluate a drill's performance are: (1) to identify needed improvements in the plan; (2) to identify needed improvements in the restoration process; (3) to identify needed training and staffing deficiencies; (4) to observe

whether the exercise has achieved its objectives; and (5) to identify needed operations equipment.
People improve just by practicing, but the system improves by identifying needed changes and then making them.

Evaluate what you set out to achieve in the drill. Objectives, clearly stated and precisely stated need to be evaluated because then and only then would it be possible to know when the objectives had been achieved. Therefore, you evaluate the attainment of objectives by observing the action of the participants and players. Being prepared to evaluate requires planning for the observation of the actions and decisions taken by players. One approach involves the following steps: Recall the specific objectives of the exercise, the detailed actions, and the actions or decisions they suggest; Identify the players expected to take the actions or decisions as those who should be observed; Locate evaluators in a position to observe players and actions; and Brief the evaluators on what actions or decisions are expected.

In order to effectively observe events in complex or less complex exercises, evaluators, simulators, and drill coordinators have to track response actions of participants. This is made easier by the designation of certain events, prior to the drill, as key events. These are events that most directly test the objectives of the exercise and which should be watched most carefully. Ideally, key events are developed as part of the overall effort to develop an exercise scenario and other exercise materials.To aid in the evaluation all observations to the key events should be recorded on a Key Event Response Form.

When to Evaluate

Conduct an immediate post exercise critique followed by a more detailed evaluation report. The advantages and, disadvantages are shown in Figure 1

	Immediate Critique	Follow Up Evaluation
WHEN	As soon as exercise is over	One to three weeks later
WHO	Participants only with drill coordinator as the discussion leader	Evaluators write report and the participants and coordinators review the report
FORMAT	Each player asked to describe his/her learning experience	Written report from Observers and other data is circulated to key players. Meeting held to review.
RESULT	Players satisfaction that management listens to them	Meeting to review report and agreement on recommendations for follow up improvements
ADVANTAGES	Players are already there, and their memories are fresh. High motivation to contribute	Time for thoughtful reflection and recommendations. More reasoned discussions. Written report gains status as action document.
DISADVANTAGES	Fatigue, can lead to a gripe session, concentrate on minor problems and not the big picture.	Problems fade with memories. Too much refinement and summarization. Difficult to reassemble the participants.

Figure 1

A problem log allows participants, drill coordinators, controllers, and simulators to note and document potential problems associated with responses as they occur during an exercise. These potential problems are then analyzed in a debriefing session to determine which ones are serious enough to require corrective action and to determine their source.

While the exercise is in progress, drill coordinators, simulators, observers, and participants should record on the problem log any observed action that may possibly create a problem. Entries in the problem log should be evaluated after the exercise to determine whether a problem really exists, the extent and source of the problem, and corrective action where applicable. Each event recorded must be identified by the time of the event, who observed the event, and where. Since drill must be evaluated against the pre-established objectives all individuals need to understand their roles in this evaluation process.

What Tools to Use in The Evaluation

Questionnaires are an excellent source for obtaining detailed insights from participants on the adequacy of the plan, level of training required, and functional information which might otherwise go unnoticed. Often personnel who hesitate to enter into group discussions will contribute significant information in response to a questionnaire. To be effective, participant questionnaires should be carefully prepared, pretested, and administered while participants are still in place at the conclusion of the exercise.

Immediately after a drill has been completed evaluation questionnaires can be distributed to all participants. The purpose is to obtain information from participants about their experiences during the exercise and their perceptions of the exercise. Evaluations can be anything from a narrative answer to a question to some form of numerical answer. A personnel critique sheet or evaluation questionnaire is an effective way to collect comments from a participants that might be willing to share in a public meeting. A sample critique sheet is shown in figure 2. The completion of this critique sheet could be facilitator led with the facilitator walking the participants through the process of the drill, the objectives, etc.

Critique Sheet

Name (optional) ____________________
Role (optional) ____________________

Please check the evaluation you feel best indicates the quality of performance. Please provide a reason for that rating and if possible provide your recommendation for improving performance.

EVALUATION	REASONS	RECOMMENDATION
Subject Exceeded Expectations ___ Met Expectations ___ Needs Improvement ___		
Subject Exceeded Expectations ___ Met Expectations ___ Needs Improvement ___		
Subject Exceeded Expectations ___ Met Expectations ___ Needs Improvement ___		

Figure 2

Combining a narrative answer with a numerical evaluation Can also yield additional information for evaluation report writers. Therefore in addition to specific action or decision evaluations you may want to include several open-ended questions about the performance in general. Players should be encouraged to make recommendations for problems raised in the exercise. A sample Questionnaire is shown in figure 3. At the conclusion of the drill after participants questionnaires have been collected, a formal drill evaluation debriefing should be conducted. During this session the drill coordinator should stimulate and encourage open communication and discussions. All attendees should be encouraged to provide their observations and comments. the drill coordinator should treat this sessions as another data source and should not try and direct the responses in any direction. the only direction would be to ensure that all objectives are at least brought up for discussion.The purpose of the debriefing is to examine the operational and management problems revealed during the course of the exercise.

The purpose of the immediate post exercise critique is to examine the actions and decisions of the players. You want players to provide comments on performance. However no matter how hard you try, players are going to want to critique the exercise itself. Avoid this situation but you'll probably fail.
Another useful tool is the Participant Evaluation Questionnaire. This report presents an analysis of needed improvements in the process based on an examination by all the players involved in the drill. A sample Questionnaire is shown on Figure 4. This tool needs to be framed by the exercise objectives. The final evaluation report is often used to summarize the entire exercise process. As such it is a very useful tool as a planning device for future drills.

Just as an exercise without recommendations makes the exercise incomplete, recommendations without follow up will keep the organization from getting the full benefit of the exercise.

Structured Evaluation Questionnaire

Please complete this form by placing an "X" in the appropriate box. Provide an explanation when you select boxes "sometimes", "seldom," or "never."

NAME: ____________________ TITLE: ____________________

DRILL POSITION: ____________________

1. Were the plans, procedures, and resources used during the exercise adequate?

[ ] Always [ ] Sometimes [ ] Seldom [ ] Never

Explanation / Comments ____________________

2. Did the participants have access to the information they needed in carrying out their responsibilities?

[ ] Always [ ] Sometimes [ ] Seldom [ ] Never

Explanation / Comments ____________________

3. Did the exercise address the objectives adequately?

[ ] Always [ ] Sometimes [ ] Seldom [ ] Never

Explanation / Comments ____________________

4. Were the voice and data communications systems adequate?

[ ] Always [ ] Sometimes [ ] Seldom [ ] Never

Explanation / Comments ______________________________

5. Was the exchange of information among the participants adequate (internal and external)?

[ ] Always [ ] Sometimes [ ] Seldom [ ] Never

Explanation / Comments ______________________________

Figure 3

Other Evaluation Tools to Use

When conducting the evaluation you need to consider a number of different elements, some of which are recorded automatically as well as the manually recorded information. Collect all the information. Some other things to consider in evaluating the effectiveness of the exercise use the following:

1) predefined set of performance indicators
2) filled out by participants
3) predefined set of restoration milestones
4) filled out by observers
5) Data collected and Feedback the information Learned
6) log sheets
7) Simulation (OTS/DTS)
8) tape recorders (phone/portable)
9) alarm history
10) log recorders

ADDITIONAL AREAS

Computer Programs may only be available on a limited basis during the restoration drill. The incorporation of a dispatcher training simulator (DTS) to help the system operator to safely restore the transmission system is an area receiving much attention [2], [3], [4]. Future DTS enhancements to help the operator in the suggestion of transmission restoration will be very beneficial. If a DTS is used additional DTS specific objectives should be developed and incorporated in the exercise.

CONCLUSION

Simulated emergency conditions can provide a measure of an organization's ability to respond to a real incident. However, as is often the case, there may be long periods of inactivity after initial training. Should an incident occur, personnel will find it difficult to respond appropriately. Periodic drills and exercises will help the operators maintain a high state of preparedness. Drills also are an excellent tool to test and validate your restoration plans and training program. A coordinated drill needs to include a variety of personnel from various disciplines in your organization. Drills require a commitment from senior management down through every participant in order to be successful. the result of thoughtful planning, serious training, and intense drilling will be a staff of individuals ready prepared to respond to the emergency situation.

References

[1] "Exercise Design Course, Guide to Emergency Management Exercises", Federal Emergency Management Agency, January 1989.

[2] M.M. Adibi, et al, "New Approaches in Power System Restoration", IEEE Transactions on Power Systems, Vol 5, No. 4, November 1992, pp 1428-1434

[3] J. Waight, et al, "An Advanced Transportable Operator Training Simulator", IEEE PICA, Baltimore MD, May 7-10, 1991

[4] E. Barr, et al, "The Uses of an Operator Training Simulator for Power System Restoration" IEEE PICA, Baltimore MD, May 7-10, 1991

PARTICIPANT EVALUATION QUESTIONNAIRE

Please take a few moments to fill out this form. Your opinions and suggestions will help us prepare better exercises in the future,

1. Please rate the overall exercise on the scale below:

   1 2 3 4 5 6 7 8 9 10
   Very Poor — Very Good

2. Compared to previous drills, this was:

   1 2 3 4 5 6 7 8 9 10
   Much Worse — Much Better

3. How well prepared for the drill were you:

   1 2 3 4 5 6 7 8 9 10
   Unprepared — Very Prepared

4. The drill scenario needs to be modified to include the following:

   ____

5. Please add any comments or criticisms on the following line

   ____

6. How do you think this process / procedure could be improved?

   ____

7. How would you improve the pre-drill training?

   ____

8. How could your participation in the drill have been more effective?

   ____

9. What tools or equipment should be provided to facilitate your participation in the drill? In real-life restoration?

   ____

Figure 4

Power System Restoration Training Questionnaire Results

A Report by the Power System Restoration Training Task Force
Joseph D. Willson, Chairman, PJM Interconnection Association

Abstract
The Working Group on Operator Training (WGOT) and the Power System Restoration Working Group (PSRWG) established a joint task force in 1992 to determine how to train System Operators in the area of system restoration. The initial assignment for this task force was to determine what is the present level of restoration training being given to system operators. To perform this task the new task force developed a survey and sent it to electric utilities across the NERC regions. The survey was designed to categorize existing programs and to surface additional training needs. The survey was also to serve as a reference for those utilities embarking on a power system restoration training program.

Introduction
The survey consisted of seven sections which included: (1) Profile of the responding systems, (2) the Energy Management Systems of those utilities, (3) information about their system restoration plans, (4) Data about their training programs, (5) How training evaluation is conducted, (6) The content of their restoration training program, and (7) Any information and experience they had in conducting restoration drills.

Four types of questions were utilized in the survey. They included: multiple choice responses; yes/no answers; fill-in-the-blank questions; and open ended questions. For the multiple choice and Yes/no responses the data reflects the frequency of the response; for the fill-in-the-blank and open-ended questions the answers are presented verbatim.

Of the 225 surveys distributed, 93 were returned representing a 41 percent return rate. The questionnaires were received from all NERC regions with the exception of Alaska. The results discussed in this paper therefore represents the restoration training programs currently being used in most NERC regions.

96 WM 165-1 PWRS A paper recommended and approved by the IEEE Power System Engineering Committee of the IEEE Power Engineering Society for presentation at the 1996 IEEE/PES Winter Meeting, January 21-25, 1996, Baltimore, MD. Manuscript submitted June 26, 1995; made available for printing January 10, 1996.

Company Profiles
Of the 93 surveys received, 22 percent came from the WSCC region, 77 percent came from the Eastern NERC Interconnected regions, and 1 percent from systems outside of North America. Half of the replies were from local utilities of which many (41 percent) are a part of a regional power pool. Approximately one-third reported a peak load of over 4,000 MW, one-third had peak loads in the 1,000 - 4,000 MW range, and the remaining one-third reported a peak load of less than 1,000 MW. Forty percent of the systems reported less than 200,000 customers, 23 percent had between 200,000 and one-half million customers, and 25 percent served 1 million or more customers.

Energy Management System (EMS) Capabilities
Three-quarters of the respondents' EMS systems do not differentiate between normal. emergency, and restorative states. Of the 25 percent that do differentiate states, only 9 percent use Expert systems to help define the states, the remaining differentiations are as a result of being defined as part of the alarm. Disturbance data collection and/or analysis capability is performed by two-thirds of the systems. Forty-four percent of the systems have a system operations trainer but only 31 percent have a dispatcher training simulator. Specific questions for this section are shown below:

System Restoration Plan
Most of the systems (84 percent) presently have a system restoration plan for their organization. The restoration plans were developed primarily by the system operators, system operations engineers, and system operations managers. Input in formulating the restoration plans came principally from experience (95 percent). On many systems computer simulations were used to help develop the plans (40 percent). Eighty-two percent of the plans are structured using detailed step-by-step procedures: most aspects are very detailed (32 percent) while the others (50 percent) contain detailed procedures for parts of the plan. The restoration plans are usually revised each year (24 percent) or as needed (59 percent). Software was only purchased by 1 percent of the systems to help in their restoration plan development.

Reprinted from *IEEE Transactions on Power Systems,* Vol. 11, No. 3, pp. 1630–1635, August 1996.

Other responses related to this topic are listed below:
1. Who developed your most current system restoration plan? (Check all that apply.) Note for this response and all similar questions where "all that apply" responses are detailed, the percent calculation is performed as follows: "number of checks divided by the number of responses to that question. For example, 66 of the 93 (71%) responses to this question indicated that system operators developed their current plan while only 4 out of 93 involved substation managers.

system operator/s	71 %
system operations engineer/s	62 %
system operations manager/s	65 %
power plant operator/s	8 %
power plant engineer/s	12 %
power plant manager/s	8 %
substation operator/s	6 %
substation engineer/s	14 %
substation manager/s	4 %
outside consultant/s	1 %
others	10 %

2. Does your plan include an overall restoration philosophy?
Yes 95 %
No 5 %

3. Is your restoration plan structured on the NERC Restoration Reference Document?
Yes 63 %
No 37 %

4. Have you ever had an opportunity to use your restoration plan in real-time operations?
Yes 29 %
No (If no, go to Question 9) 71 %

5. Did you make modifications to the plan after its real-time use?
Yes 67 %
No 33 %

6. How old is your system restoration plan?
Still in draft - has not been officially issued 11%
First issued 1-5 years ago 30%
First issued more than five years ago 59%

7. Does your plan include details which outline the actions required of personnel in other departments such as power plant personnel and substation personnel?
Yes 71 %
No 29 %

8. Does your plan include details which outline the actions required of other Systems / companies?
Yes 52 %
No 48 %

9. Does your plan include notifications and mobilization plans? Yes 76 %
No 24 %

Training Program
Restoration training is provided in three forms: initial training for entry level system operators (61 percent); refresher training for experienced operators (79 percent); and advanced training for experienced operators (26 percent). Ninety seven percent of the systems were able to either train all or at least some of their operators last year with the same training level of commitment being provided every year. Eighty-nine percent of the systems indicated that there has been an increased interest in system restoration training at their location. The two most important reasons were:(1) increased recognition of the need for this type of training(71 percent); and (2) culture changes(24 percent). Only 18 percent indicated an increase due to the purchase of a new dispatcher training simulator. Specific questions and responses were:

1. Do you have a system restoration training program for your system operators?
Yes 41 %
No (If no, do not complete the rest of the questionnaire - return it to IEEE as is.) 59 %

2. Do you train your system operators to use this plan?
Yes 97 %
No 3 %

3. Is your training generic, company-specific or Both?
Generic 11%
Company-specific 43%
A mixture of generic and company-specific 26%
Pool Coordinated 21%

4. Does your system restoration training program include specific training objectives?
Yes 76 %
No 24 %

5. Who participates in your system restoration training? (Check all that apply.)

System operators	100%
Operations engineers	40%
Operations managers	45%
Power plant operators	38%

Power plant engineers	10%
Power plant managers	15%
Substation operators	28%
Substation engineers	10%
Substation managers	10%
Neighboring utility personnel	25%
Communications - Press Release	20%
Others	8 %

6. Which of the following techniques are used in providing system restoration training?
(Check all that apply.)

Self-study manual/s	43%
Self-study computer based course/s	3 %
Self-study videotape based course/s	10%
Classroom lecture/s with engineers providing lecture/s	28%
Classroom lecture/s with system operators providing lecture/s	50%
Classroom exercise/s	53%
System specific simulator exercise/s	25%
Generic system simulator exercise/s	8%
Drill/s	33%

7. How often do you modify your restoration training program?

Training program never been revised	11%
Training program is revised each year	24%
Training program is revised as needed	66%

8.Who makes changes to the training program?
(Check all that apply.)

Trainer	66%
Engineer	34%
System operator	55%

9.Which of the following types of system restoration training do you provide to system operators?

Initial training program for entry level system operators
Yes = 61% No = 39 %

Refresher training program for experienced operators
Yes = 79% No = 21 %

Advanced training program for experienced operators
Yes = 26% No = 74 %

10.How many system operators received some training in system restoration in the past year?

All	52 %
Some, but not all	45 %
None	2 %

11.How many system operators will receive some training in system restoration every year?

All	52 %
Some, but not all	45 %
None	2 %

12.If there has been an increased interest in system restoration training at your company, has it been due to any of the following?
(Check all that apply.)

Purchased new Dispatcher Training Simulator (DTS) or Operator Training Simulator (OTS)	18%
Increased recognition of need for this type of training	71 %
Disturbance on own system	16%
Disturbance on another system	5 %
Other (please specify) ____________	8 %
Not applicable (no increased interest)	11%
Culture changes - increased use of Bulk Power System	24 %

13.What information, which is not readily available to you now, would be most useful to helping you develop or improve your power system restoration training activities?

A. A Dispatcher Training Simulator (6 comments)
B. Network Modeling
C. System details (capability of transmission system to black start large generating units
D. Drills
E. More field personnel control
F. Operators aids such as generic graphs of line loading when energizing open ended line voltages, equipment reactive requirements, rules of thumb to follow, etc
G. What other systems use in lieu of a DTS
H. Justification for money to be spent for training personnel
I. More information understandable by a system operator on frequency control; how it is done at a power plant.
J. Information exchange with other company/power pools.
K. Results from electric magnetic transients studies to let us know what the probability of success would be, if any, in re-energizing specific pieces of equipment under constrained system conditions.

Training Evaluation

Evaluation of the effectiveness of the various training programs was very different. More than sixty percent evaluate the effectiveness but only one-third used a formal process. This section elicited the most responses for the questions on" what would you change? " and "what is the most effective aspect of your program". Detailed responses to these and all other questions are detailed below:

1. Do you evaluate the effectiveness of your system restoration training?

Yes	62 %
No	38 %

2. Does your evaluation include formal assessment of individual system operators knowledge and skills?

Yes	37 %
No	63 %

3. If an operator does not perform as well as desired, is there a plan to provide individual tutoring?

Yes	62 %
No	38 %

4.Is the DTS used for evaluation?

Yes	29 %
No	71 %

5.How are the results of your evaluation/s used? (Check all that apply.)

To modify the training program	58%
To modify the system restoration plan	35%
To provide more training for individuals who did not score as well as expected	48%

6. If you could make one, or more, changes to your existing system restoration training, what would you change?

A. We plan to make use of a DTS during the next phase of our system restoration training
B. Use of a Simulator (purchase and use of a DTS)
C. More extensive: (this) requires management support for money and people
D. All suggestions on changes to the good have been acted upon
E. Make it system specific with our brand of SCADA
F. More refresher training
G. Make our DTS work : at present it doesn't support black start / system restoration drills due to power-flow crashing
H. Regular training sessions
I. Teach philosophy of restoration so that pre-determined goals can be reached without relying on pre-determined step by step restoration plans
J. Need a formal Training Program
K. Need more time
L. Perform more localized system collapses

7. What do you consider to be the most effective aspect of your existing training program?

A. More of it; more actual real-time exercises instead of just discussions
B. Be allowed to pay overtime for training
C. Actual testing start-up units
D. Cooperation
E. OTS
F. Availability of trainees: availability of Trainers
G. Expert classes
H. Acquaint operator with procedures
I. We get 2 weeks of classroom every 3 years which is very good. The DTS is also at that location (out of state) but the controls are so foreign that it isn't very functional. We do not have a DTS at our facility.
J. Breaking the class down into 4 small groups with a facilitator and going through the four phases of system restoration which we, the trainers, developed.
K. Drills (3 comments)
L. It exists and is not generic
M. The dispatchers get to practice restoring the system under a variety of conditions
N. Pull staff away from job requirements to perform quarterly or semi annual training
O. Training identified as a job requirement. Staff develop quarterly training goals on an individual basis.
P. Having a plan in place for black-start
Q. Having a general philosophy on how to control voltage, frequency, and in general, placing our system back together following a complete system shutdown
R. Only the overall basic understanding
S. Theory and exercises to simulate restoration activities
T. Build more restoration and islanding scenarios for DTS
U. All very effective . DTS and classroom undoubtedly best !
V. More extensive in system disturbance not just black-start W. The system operations department has a plan for restoration at several levels
X. Drills and classroom discussion enable us to improve or recommend improvements to the plan
Y. Drills or exercises conducted annually with ECC personnel and quarterly with power plant and substation personnel
Z. Classroom training
AA. Commitment of management to training
AB. The Power Pool coordinated drills
AC. Training is performed on a Pool wide basis
AD. Full time staff; a training center
AE. The training simulator provides the operators to perform under conditions which he rarely sees. Operating problems discovered at this level have

significant benefits if as a result the same problems are avoided in real time.

Training Content

System restoration training programs used by the utilities contain topics geared to (1) Dynamic performance of the system, (2) dynamic performance of equipment, (3) recognition of abnormal conditions, (4) remedial action, (5) recovery, and (6) multi-site sabotage awareness. Respondents were requested to check all topicsthat apply that are included in their System Restoration Training)

1.Dynamic Performance of System:

transient stability	55 %
oscillations	37 %
relay action	76 %
control-initiated swings	26 %
causes of disturbances	66 %
special protection systems	50 %
synchronization	89 %
power angle	61%
flow equations	32 %
MVAR flows	82 %

2.Dynamic Performance of Equipment:

governor response	68%
exciter response	47%
relays and breakers	76%
under-frequency relays	84%
under-voltage relays	61%
special protection systems	50%
metering	37%
automatic plant controls	47%
automatic generation control	63%
automatic voltage controls	61%
auto. generator and load tripping	55%
automatic system separation	39%
load shedding equipment	79%
communications equipment	71%
black-start equipment	79%

3.Recognition of Abnormal Conditions

loss of load	76%
breaker operations	79%
line faults	68%
generator trips	82%
frequency deviations	76%
interchange amounts	53%
voltage levels	76%
system separations	68%
communications with other plants and utilities	79%
parallel flows	37%
multi-site sabotage	11%
Equipment damage assessment	34%
Substation and switching station readiness (availability)	60%

4.Remedial Action

islanding	66%
load shedding	82%
shifting generation	55%
switching operations, including interconnection	79%
isolating system operation	61%
high and low frequency operation	55%
high and low voltage operation	63%
Phase Angle Regulator shifts	29%
load rejection on generators	47%

5.Recovery

generation start-up capabilities and pickup rates	76%
sectionalizing	79%
load pickup priorities and problems	82%
low voltage networks	39%
synchronizing within system and at interconnections	82%
under-frequency relays	76%

6.Multi-Site Sabotage Awareness

physical security of critical facilities	11%
multi-site sabotage techniques and strategies	3%
criteria for differentiating sabotage or vandalism from routine equip outages	11%
operating considerations unique to multi-site sabotage attacks	3%
operating the system for loss of the threatened facility	13%

Restoration Drills

Fifty-nine percent of the systems presently perform system restoration drills. Most of the drills are of a duration of between 4 to 8 hours. Most systems practice this form of training at least once a year. The new trend appears to be the use of analysis tools to help plan these restoration drills. The three kinds of tools mentioned the most were the dispatcher simulator (23 percent), off-line power flow programs (23 percent), and expert systems (9 percent). Evaluation of the success of a drill varied among the respondents. Specific questions and replies are:

1.Do you conduct Restoration Drills?

Yes	59%
No	41%

(For those systems that do conduct drills the following six questions apply)

2.What is a typical duration for a restoration drill?

less than four hours	32%
Four to eight hours	59%
Eight to sixteen hours	9%
Greater than sixteen hours	0

3.How often do you conduct restoration drills?

More frequently than once a year	46%
Once a year	34%
Less frequently than once a year	19%

4.How many employees take part in a restoration drill? Check all that apply.

Dispatchers / System Operators	95%
Plant Operators / Substation Personnel	50%
Technical support personnel	55%

5.Are any analytical tools used as part of the restoration drills? (check all that apply) Consider:

Operator Training Simulator	23%
Off-Line Power Flow Studies	23%
Off-Line Transient Stability Studies	9 %
Other ______	14% (9 % expert systems)

6.Do you interact with any external companies or organizations during a restoration drill?

Yes	71 %
No	29 %

7. How do you determine whether the objectives of the restoration drill were met?

A. Evaluation against objective and percent load restored vs time

B. By having system restoration exercises twice a year we hope that this provides the necessary training required.

C. Subjective

D. Operator performance / feedback

E. Critique Format (3 comments)

F. Time; Voltage Limits held; Cranking Unit Stability; Successful start up of a fossil fuel plant from a cranking start; simulated black lite and restoration plan

G. Completion of steps in correct order

H. Successful circumvention of prescribed circuits taken out of service

I. Review and evaluation by all personnel involved (2 comments)

J. Time it takes to get black-start units up and carrying test load

K. When all stations, transmission, and load is restored to the available capacity in relative amount of time

L. Observers critique operators during exercises or drills

M. Reviewed at power pool level but we never get much feed back

N. Debriefing meeting

O. Review of drill by instructors with participants after the fact

P. Completion of restoration drill with bulk power system restored

Q. By pre and post meetings to determine the objectives and if they were met and to determine what problems there were and possible solutions for them.

Conclusion

Power system restoration training is being included in many of the utilities basic training program. The use of simulators to help in this training process is growing and will probably serve a the basic building block for restoration training. System operators are looking forward to using these new simulators and there advanced capabilities. Evaluation tools need to be developed or enhanced in order to ensure that the training programs meet the desired training objectives. System operator training in power system restoration is a valuable module in a systems training curriculum.

Acknowledgement

The author wishes to express his thanks to the members of the Operator Training and Restoration Working Groups and to the Power System restoration Training Task Force for helping to prepare the questions contained in the survey.

PART V

Specific System Restoration

I. General Discussion

This part deals with the specific restoration plans of various organizations. Although each is different, they share many common characteristics. With respect to restoration planning, we see an evolution from plans based on the experience and expertise of system planning and operations personnel to heuristic trial and error simulations to computer-guided and optimized simulations. Many plans call for separating the system into subsystems during restoration. This breaks the large problem into smaller, more solvable ones. Many subsystems are expected to restore themselves to a certain level without guidance and coordination from the main control center, sometimes without internal communications.

II. Discussion of Reprints

5.01 Low Voltage Bulk Power System Restoration Simulation, 1981, p. 341

This paper proposes a method to energize substantial portions of the transmission system from one location without picking up any load. A simulation of energizing the Los Angeles Department of Water and Power (LA DWP) system from a set of underexcited hydro units indicated that such a process was possible. To determine whether the hydro units could actually be operated at the unloaded, underexcited level simulated, a number of tests were made by connecting the units to a set of large capacitors at a nearby DC conversion station. The tests showed that it was possible to energize the capacitors while maintaining an acceptable voltage and reactive power (var) absorbing level at the generators. Simulations determined that under the tested conditions there would be an acceptable voltage rise on the transmission system. This strategy was never implemented.

5.02 Summary of System Restoration Plan for the Pacific Northwest Power System, 1982, p. 347

This informal paper accompanies a presentation to the PSR Task Force at the 1982 Winter Power Meeting. The Task Force had invited anyone with a formal system restoration plan to present it to the Task Force as a means of educating the group and encouraging the development of formal plans. The paper describes the coordinated system restoration plan for the Northwest Power Pool. The intent is for each utility to restore as much of its own system with its own resources as possible and then to reintegrate the systems under Bonneville Power Administration (BPA) coordination. The presentation includes six formal steps to be taken, and it provides guidelines for load pickup.

5.03 Field Experiences in Reenergization of Electrical Networks from Thermal and Hydro Units, 1984, p. 354

This paper describes the procedures, guidelines, tests, and experience of the National Authority for Electrical Energy—Italy (ENEL) system. The philosophy for restoration of the ENEL system is described. Restoration of subsystems is to be autonomous, with no internal communication within the subsystem. All actions at a location are based on locally available information. Generators are designed to trip to house load when reaching the underfrequency limit. The characteristics of loads upon pickup following various durations of disconnection are described. Finally, the experience of an actual partial system blackout and its restoration are detailed.

5.04 System Restoration at Philadelphia Electric Company, 1984, p. 361

This paper is based on one of the earliest load flow simulations of a system restoration plan. It discusses what conditions and situations must be considered in simulating a system restoration plan. The paper also discusses what to consider when updating a previously developed and simulated plan. In particular, when company plans call for the installation of new units and the retirement of older units relied upon for system restoration, an entirely new strategy may be required. The paper concentrates on generation available in the first hour of restoration.

5.05 Service Restoration Following a Major Failure on the Hydro-Quebec Power System, 1987, p. 364

A power failure occurred in the Hydro-Quebec system in December 1982. Following that disturbance, Hydro-Quebec determined that different methods were required to reduce the risk of additional damage during restoration.

Since most of the load to be restored is supplied by the 735kV system, special consideration was given to switching, resonance, transients, and overvoltages. Transient and steady-state simulations were run to test alternative methods. A key conclusion was to use insertion resistors on circuit breaker energizing transformers. New guidelines were developed.

5.06 Policies for Restoration of a Power System, 1987, p. 374

This paper starts with a theoretical discussion of the possible "states" of a power system and a general philosophy of system restoration. This includes planning, preparedness (training and practice), and conditioning (necessary elements such as uninterruptible power supplies, communications, and blackstart capability). The philosophy is applied to the Central (control) Area of the Mexican Interconnected System. The changes made to the system following the January 1981 blackout of the Western and Central Area are generally described. The description of the restoration plan consists of definitions of tasks and responsibilities of various positions and the actions of each black-start subsystem.

5.07 Restoration in Sweden and Experience Gained from the Blackout of 1983, 1987, p. 381

This paper describes the power restoration philosophy that was developed in the 1950s and its evolution to the 1980s. The strategy has been to restart generation, reenergize transmission, and restore load with only local information. In the 1950s this was a manual operation but had been converted to automatic local (individual breaker) operation. The paper discusses how this system operated during the December 1983 blackout and describes changes to be made based on that experience.

5.08 Simulation and Field Tests of the Black Start of a Large Coal-Fired Generating Station Utilizing Small Hydro Generation, 1990, p. 388

This paper presents a primary example of a simulation followed by an actual test of the system that was simulated, with comparisons of simulated and tested state conditions. The simulation and test confirmed that two relatively small hydro stations could energize 230 kV and 115 kV lines connecting them to a large steam unit and restart the large motors needed to start the unit. The details of the special actions needed for the test are described, as are the procedure and system modifications developed as a result of the test. There is good agreement between simulated and actual response of the system.

5.09 Power System Restoration Incorporating Interactive Graphics and Optimization, 1991, p. 395

This paper describes a semiautomated and mathematically optimal system to develop a system restoration plan. It is one of the first applications of interactive graphics for development of the system restoration plan. It uses a graphic interface for the planner to select the scenario and view the results of the optimization and simulation. It is applied to a subsystem of the Hydro-Quebec system.

5.10 Simulations and Field Tests of a Reactor Coolant Pump Emergency Start-up by Means of Remote Gas Units, 1992, p. 402

This is another example of using small, remote units to restart large induction motors in a power plant. In this case, the purpose is not to restart the large nuclear unit, but to effect its safe shutdown during system blackout. The generators to be used are relatively small combustion turbines that must be paralleled to provide a sufficient safety margin. The simulation is detailed and dynamic. The paper explores the problems with paralleling two lightly loaded combustion turbine generators. Following development of the plan through simulation, the plan was tested on the system, and good agreement was found between the simulated and actual results.

5.11 Major Incidents on the French Electric System: Potentiality and Curative Measures Studies, 1993, p. 408

The investigators began with a review of incidents on large systems to determine a relatively small set of initiating mechanisms. They then applied these mechanisms to the French (EDF) system via transient and dynamic simulation to determine the susceptibility of the system. For threatening mechanisms, again through simulation, they tested various methods to halt or limit incidents that could lead to collapse of the system. The limitations of currently installed protection systems were discovered, and modifications were made to enhance the reliability of the system.

5.12 Black-start and Restoration of a Part of the Italian HV Network: Modeling and Simulation of a Field Test, 1996, p. 415

Starting with the existing restoration guidelines, the investigators determined what parts of the ENEL system would need to be modeled to enable simulation of restoration with a dynamic simulator. They developed their own dynamic models to permit accurate simulation of their equipment and system.

A restoration test was performed on part of the system. The test was then simulated with the models that had been developed, and the simulation was in close agreement with the test results. This close correspondence confirmed the validity of the model and enabled the development of restoration guidelines through simulation rather than tests on the real system.

5.13 Interactive Long-Term Simulation of Power System Restoration Planning, 1997, p. 424

The authors describe the short- and long-term dynamic model they have developed for the Hellenic system. The model has been developed specifically for system restoration. They then simulate the 1989 blackout of the system in order to test the validity of the model. While some problems were noted with the model such as nonconvergence, they occurred primarily during severe operating conditions that should be avoided. The investigators conclude that the model is sufficiently accurate and fast to be used as a training tool for restoration.

5.14 Analysis of the Power Blackout of June 8, 1995 in the Israel Electric Corporation, 1997, p. 432

The paper describes the blackout of the Israeli system and analyzes the initial causes and succeeding events. Recommendations are made to minimize the chance of future blackouts and to provide for the more rapid restoration of the system.

LOW VOLTAGE BULK POWER SYSTEM RESTORATION SIMULATION

Edward J. Simburger, Member, IEEE
Energy Systems Directorate
The Aerospace Corporation
P.O. Box 92957
Los Angeles, CA 90009

Francis J. Hubert
Operating Engineering Subsection
Los Angeles Department of
Water and Power
Los Angeles, CA 90054

Abstract - Time is the crucial element in restorative state control for electric power systems. One method for reenergizing large portions of a blacked out electric power system with a minimum of switching and thus time to restoration, is described. Results of a simulated system restart utilizing actual power system facilities are compared to power flow studies. The simulation tests were performed on the Los Angeles Department of Water and Power (LADWP) system in order to verify the concept of utilizing generation operating at extremely low voltage to energize as large a part as possible of the unloaded bulk power transmission system located within the service area.

INTRODUCTION

This paper describes a new method of reenergizing the bulk power system after a large scale system failure similar to the New York City blackouts. After the Northeast and New York City blackouts, many agencies became concerned about the general lack of procedures for energizing the bulk power supply system and restoration of service. Among these concerned was the LADWP.

The LADWP system serves the city of Los Angeles, and the bulk power system within the service area is comprised of a 230 kV and 138 kV transmission system with interconnections to the Southern California Edison Company at the Sylmar switching station and Velasco substation. Other interconnections exist as a part of the 500 kV western transmission system. The 500 kV system is not considered to be a viable part of a black start capability due to the excessive capacitive VARs. The major generating facilities, located within or close to the service area, include four oil- or gas-fired steam generating stations and the Castaic Power Plant, a large pumped storage hydroelectric facility.

Because the Castaic facility has black start capability and is close to the service area, basic black start policy at LADWP is to utilize this plant, if available, to energize the bulk power transmission system and to provide cranking power to the steam generation facilities. Due to the large capacity of the Castaic facility (1570 MVA, 1250 MW) and to the critical time constraints for providing cranking power to the steam units, it appeared time expedient to energize as large a part of the bulk power transmission system as possible at one time. This reduces the number of switching operations and, if successful, has the advantage of providing a large energized grid to which additional generators can be synchronized as they become available. This also reduces the need to synchronize a number of small islands.

The two primary concerns associated with reenergizing the bulk power transmission system are (1) excessive voltage and (2) associated excessive capacitive VARs developed from energizing the bulk power transmission system without a connected load. The voltage problem can be overcome by either (1) energizing the transmission system with a generation source operating at a voltage lower than normal or by (2) adding load to the power system which has a reactive component that tends to reduce the voltage at the point of application. The excess VARs must be absorbed by the generation available when energizing a transmission system with no load.

Power flow studies were run which indicated that four main units at Castaic could absorb the VARs of the entire bulk power system located within LADWP service area, with the exception of some of the 230 kV underground cables. These cables supply load transformers at substations which are radially connected to the overhead transmission system.

OPTIMUM SYSTEM STARTUP PROCEDURE

This procedure conceptually is designed to allow the reenergization of the bulk power transmission in the shortest possible time, assuming that no other power is available from neighboring systems. Therefore, the number of switching operations must be minimized. The minimal amount of switching required prior to energizing any lines is as follows:

1. All customer load must be disconnected from the bulk power transmission system.
2. All unnecessary high voltage underground cables should be disconnected from the bulk power transmission system.
3. Any faulted lines or equipment must be disconnected from the system to be energized.
4. All long transmission lines must be disconnected from the system.
5. All interconnections to other systems must be opened.

Once this is accomplished, the bulk power supply system can be energized, providing that generation with sufficient capacity is available to absorb the VARs of the remaining system.

81 SM 329-2 A paper recommended and approved by the IEEE Power System Engineering Committee of the IEEE Power Engineering Society for presentation at the IEEE PES Summer Meeting, Portland, Oregon, July 26-31, 1981. Manuscript submitted February 2, 1981; made available for printing April 7, 1981.

Reprinted from *IEEE Transactions on Power Apparatus and Systems,* Vol. 100, No. 11, pp. 4479–4484, November 1981.

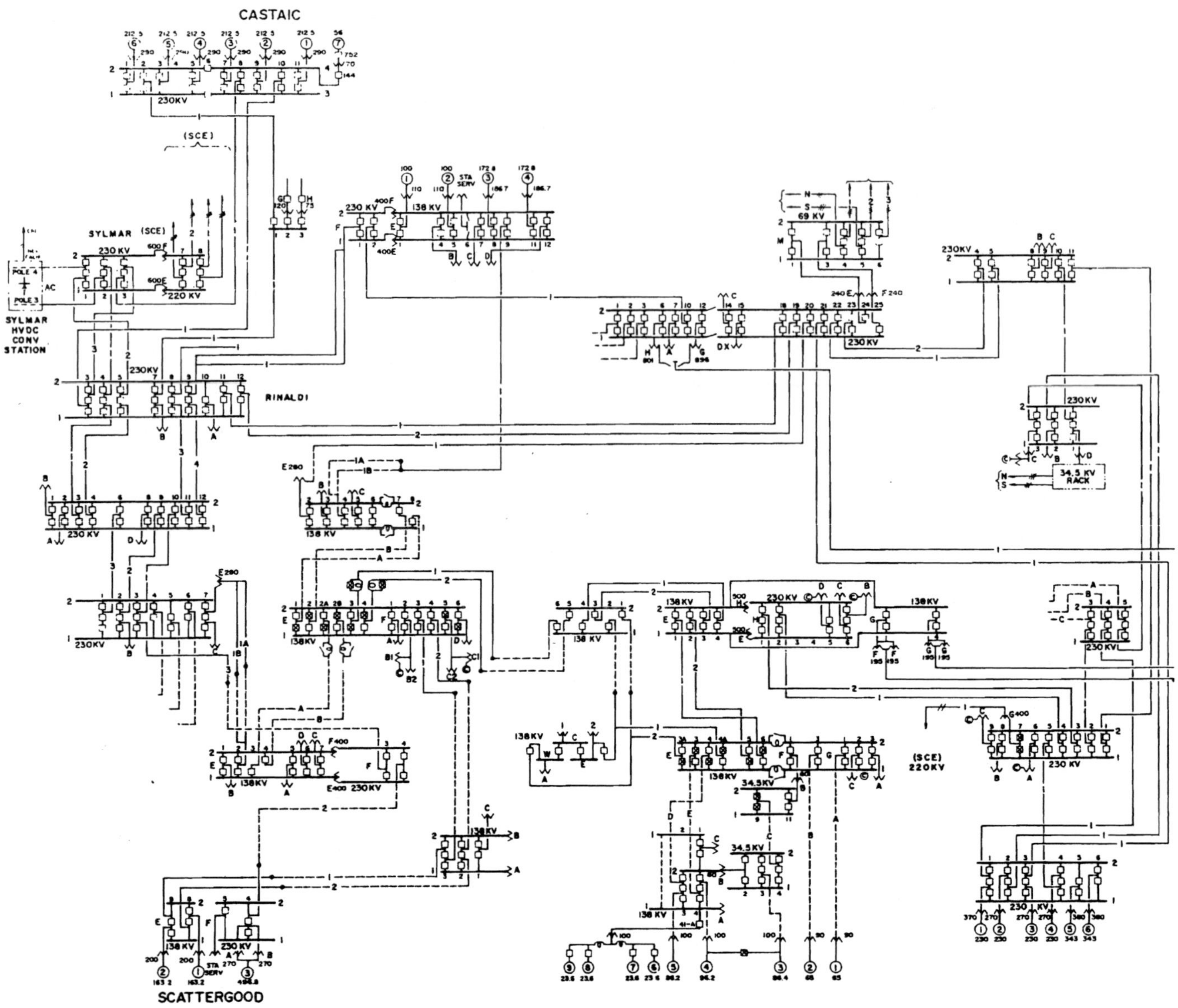

Figure 1. Portion of LADWP System to be Energized in One Step

Figure 1 above shows the balance of the LADWP system which would exist after the preceding steps are taken. This is the network which would be energized in one step in applying the low voltage startup procedure.

LOW VOLTAGE STARTUP PROCEDURE

The department's experience in energizing long transmission lines from isolated remote hydro facilities alerted the authors to a transient phenomenon which can be best described as a self-excitation of synchronous generators when supplying a purely capacitive load. This phenomenon expresses itself as a sudden increase in the terminal voltage of the machine when energizing a large capacitive load relative to the machine capacity.

In order to overcome the transient voltage problem, the low voltage startup procedure was proposed. This procedure has the following basic steps:

1. Synchronize the machines to be used to energize the system at normal frequency and voltage.
2. Block the governors to maintain constant frequency at no load.
3. Place the exciters on manual control.
4. Simultaneously reduce the voltage on all machines to the minimum obtainable without losing synchronism. (Zero VAR flow between machines indicates voltage balance.)
5. Energize bulk power system prepared as indicated in previous section.

Description of System to be Simulated

Power flow studies for the portion of the LADWP system shown in Figure 1 indicated that a maximum voltage on the 230 kV system of 1.05 per unit (PU) (242 kV) would occur at Scattergood Generating Station if the voltage on the high voltage bus at

Castaic Power Plant did not exceed 0.9 PU (207 kV). This rise in voltage as one gets further from the source is due to the essentially capacitive nature of the unloaded transmission system.

However, the power flow simulation accounts for only the steady-state values and does not predict the associated transient voltage rise when an unloaded synchronous machine is utilized to energize a purely capacitive load.

This information was calculated and empirically verified by actually energizing a capacitive load approximating the actual system to be energized. A capacitive load of 480 MVAR at normal voltage was energized, utilizing four Castaic main units (250 MVA each for a total of 1000 MVA). The transient voltage rise calculated was 134 kV. Thus in an actual black start situation on the LADWP system, an initial 230 kV bus voltage at Castaic of 40 kV should result in nearly normal voltage once the circuit breakers (CBs) are closed energizing the bulk power system shown in Figure 1.

UNIT CAPABILITY TEST

In order to (1) verify the concept of utilizing generation operating at low voltage to energize a large part of the bulk power transmission system and (2) gain some insight into the operational difficulties that may be encountered in implementing such a scheme, a series of tests were carried out on the LADWP system. The first test determined the practical operating range with respect to voltage for the units at Castaic Power Plant. Each unit is rated at 250 MVA with a nominal generator terminal voltage of 17.25 kV. The units are provided with static exciters. It was determined that the minimum excitation level resulted in a generator terminal voltage of 2 kV. At this voltage level, the frequency instrumentation and synchroscope gave unreliable readings. To overcome this it was necessary to synchronize the units at full voltage and then simultaneously reduce the voltage of all units to the desired level. The units remained stable for generator terminal voltages down to 3 kV, below which loss of synchronism and slipping of poles were encountered when more than three units were synchronized.

TEST DESCRIPTION AND POWER FLOW RESULTS

Initially, it was verified that the Castaic units could be operated in a mode where a number of units could be synchronized and remain synchronized at rated frequency with no load and at an extremely low voltage 0.16 PU (3 kV). Then a method to simulate the energization of the LADWP system was carried

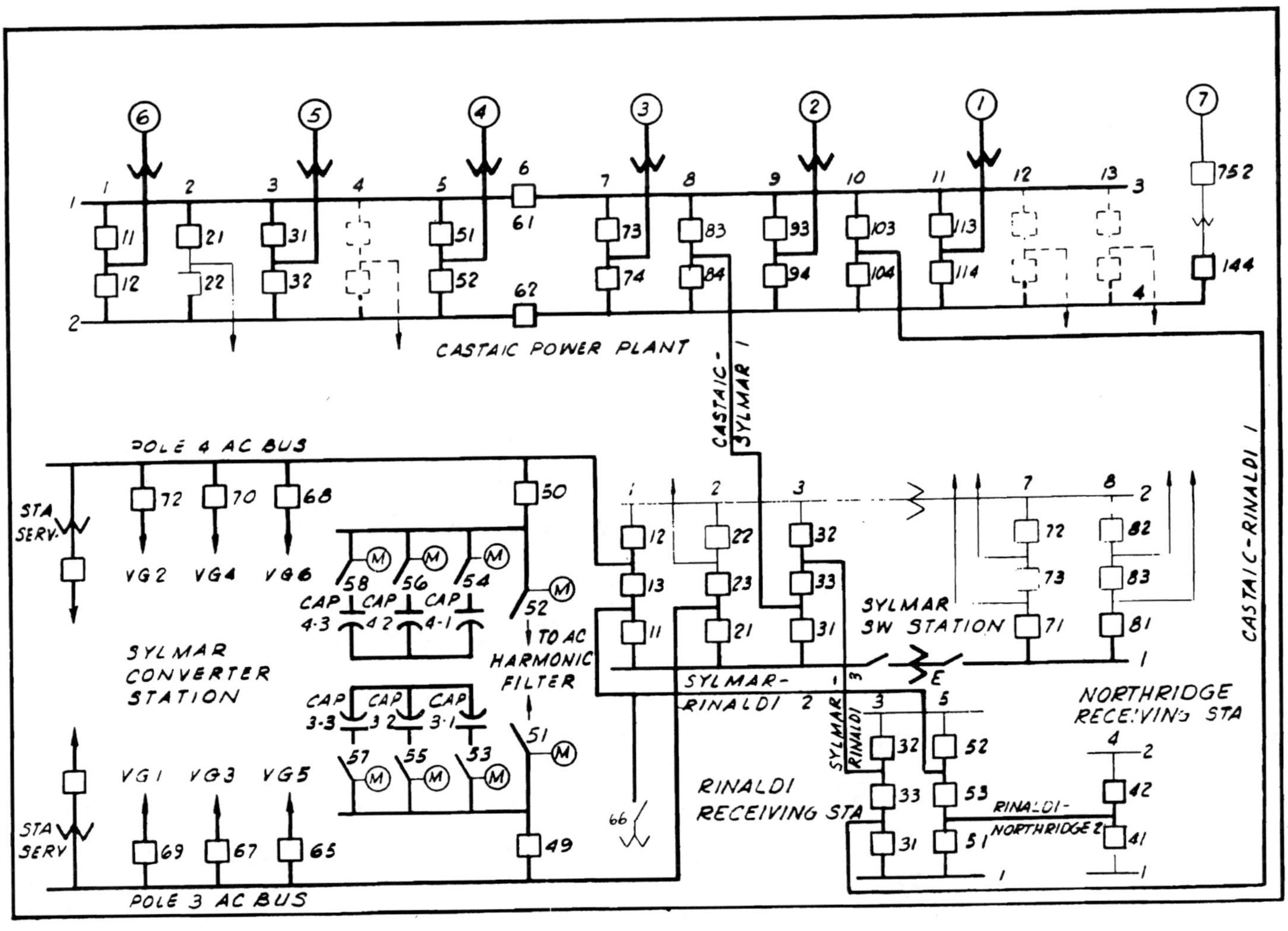

Figure 2. Single Line Diagram Simulation Test Circuit

out. In an actual black start situation, the unloaded bulk power transmission system presents essentially a capacitive load to the generators. Therefore, a reasonable simulation of a black start would be to energize a large capacitor utilizing the Castaic units as sources. The effective VAR load of the LADWP system determined in power flow studies is about 360 MVAR.

About 20 miles from the Castaic Power Plant is the Sylmar HVDC Convertor Station which has a total of 480 MVAR of power factor capacitors comprised of six capacitor groups of 80 MVAR each. Thus, the test to simulate a low voltage startup involved the electrical isolation of the Castaic units and the Sylmar capacitors from the system and necessary transmission between the two locations. Figure 2 shows the single line diagram of the test circuit.

Power flow studies were run for energizing five and six capacitor groups at Sylmar, utilizing four main units at Castaic. A summary of the results of the power flow studies is shown in Table 1.

Table 1. Summary of Power Flow Results

No. of Capacitor Groups	MVAR Sylmar	MVAR Castaic	MVAR At Gen. Term.	Voltage Sylmar	Voltage Castaic
5	-380	-348	- 87	224 kV	207 kV
6	-480	-420	-105	228 kV	207 kV

Transient Voltage Rise Calculations

The following procedure was used to determine the voltage rise which will occur when the capacitors are energized. First, let us look at the equivalent circuit for the open circuited generator [1]:

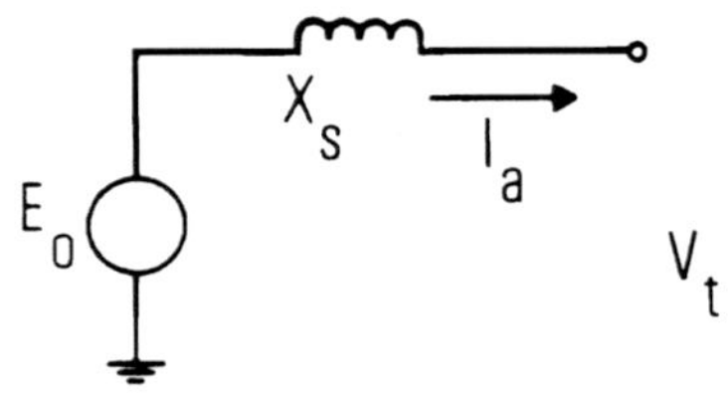

$E_o = V_t + j\ I_a\ X_s$

Since for an open circuit $I_a = o$

$E_o = V_t$

Therefore, for the Castaic machines, E_o is equal to the initial terminal voltage prior to energizing any capacitors.

E_o = 3kV = .1739 PU

When the capacitors are switched on, the following circuit approximates each generator and its load:

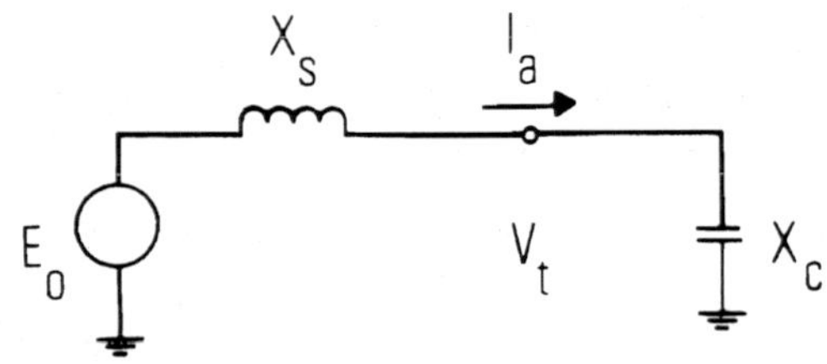

$V_t = E_o - j\ I_a\ X_s = j\ I_a\ X_c$

$E_o = j\ I_a\ X_s + j\ I_a\ X_c$

Solve for I_a:

$$I_a = \frac{E_o}{j(X_s + X_c)}$$

For the six capacitor test the following is true:

X_c = -1.779 PU per phase as seen by each generator

X_s = 1.338 PU

E_o = 3kV = 0.1739 PU

Thus, by knowing the per phase capacitive impedence seen by each generator, the generator synchronous reactance and the excitation of the generator, the voltage rise can be calculated. Table 2 provides the results of these calculations and a comparison with the values obtained during the five and six capacitor tests. Prior to the five and six tests, a preliminary test was run, whereby one, then two, then three and four capacitors were energized, utilizing three units synchronized at initial voltage of 2 kV. Figure 3 graphically illustrates the voltage rise phenomenon encountered when energizing a purely capacitive load.

The actual values obtained are in fairly close agreement to the calculated values. Any discrepancies between the calculated and average values for

Table 2. Calculated and Actual Generator Voltage Current and MVAR After Initial Energization of Capacitors

No. of Capacitor Groups	Calculated Values Voltage (kV)	Calculated Values Current (A)	Calculated Values Load (MVAR)	Actual Test Results Voltage (kV)	Actual Test Results Current (A)	Actual Test Results Load (Line MVAR/4) (MVAR)
5	9.1	1693	-26.7	7.5	1850	-22.5
6	12.1	3299	-68.0	12.3	3762	-82.5

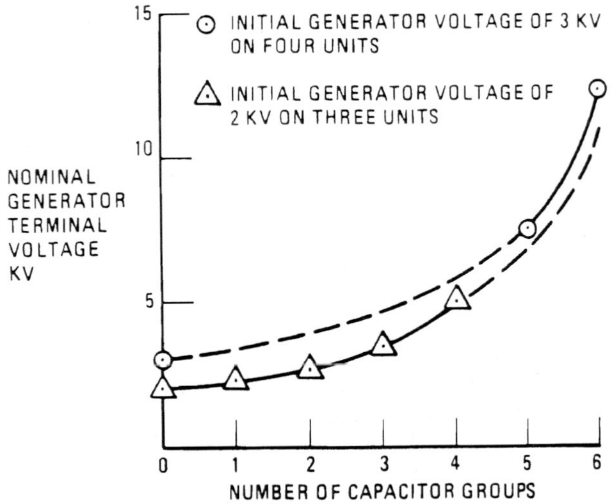

Figure 3. Plot of Voltage Rise Versus Number of Capacitor Groups Energized

the four machines can be accounted for by the fact that the model used is an approximation of the actual configuration and also due to the fact that there was some imbalance in the excitation levels between each machine, resulting in an unequal sharing of the VAR load.

TEST PROCEDURE AND RESULTS

The test system was electrically isolated from the system and all equipment deenergized. Once this was accomplished, the deenergized test system was connected, as indicated in Figure 2. A bus was selected at Castaic which was isolated from the test system and the rest of the power system. Four main units were synchronized to this bus at rated speed, no-load and full voltage. Then the units terminal voltages were simultaneously reduced to 3 kV. A series of runs were made whereby a first one, then two, and then three etc., power factor capacitors were energized by closing CBs 83 and 103 simultaneously at Castaic. This energized the entire test system. When the five capacitors at Sylmar were energized, the unit terminal voltage increased from 3 kV to 7.5 kV, and the bus voltage at Castaic increased from 40 kV to 105 kV. (No change was made to the excitation system.) The total MVAR loading was −90 MVAR at Castaic on the 230 kV bus. The voltage at Sylmar was 112 kV with a total MVAR load of −88 MVAR.

The voltage was then increased in 1 kV unit terminal voltage steps up to 15.5 kV. The bus voltage at Castaic was a maximum of 215 kV with maximum MVAR loading of −415 MVAR. At Sylmar, the maximum voltage was 232 kV with a maximum MVAR load of −423 MVAR.

Table 3 provides the generator test data and Table 4 provides the bus and line test data for the five capacitor test.

The test system was deenergized and reconnected with six capacitors. When the six capacitors at Sylmar were energized, the unit terminal voltage increased from 3 kV to 12.5 kV, and the bus voltage at Castaic increased from 40 kV to 176 kV. (No change was made to the excitation system.) The total MVAR loading was −330 MVAR at Castaic on the 230 kV bus. The voltage at Sylmar was 191 kV with a total MVAR load of −356 MVAR.

Table 3. Generator Data, Five Capacitor Test

Step	Generator Voltage (kV) U1	U2	U3	U4	Generator Current (A) U1	U2	U3	U4	Generator Field Current (A) U1	U2	U3	U4	Generator (MVAR) U1	U2	U3	U4
1	3	3	3	3	1200	500	200	800	200	60	100	45	+5	0	0	0
2	7.5	7.5	7.2	7.4	500	2200	2000	2700	200	60	110	45	-5	-30	-5	-10
10	15.0	15.5	15.0	15.5	3600	3700	4190	3820	225	230	230	250	-90	-90	-105	-97

Table 4. Line and Bus Data, Five Capacitor Test

Step	Castaic Bus Voltage (kV)	Castaic: Castair-Sylmar (A)	(MVAR)	Castaic-Rinaldi (A)	(MVAR)	Sylmar: Pole 3 Bus (kV)	(A)	(MVAR)	Pole 4 Bus (kV)	(A)	(MVAR)
1	40	0	0	0	0	0	0	0	0	0	0
2	105	100	-50	250	-40	110	356	-50	112	217	-38
10	215	650	-230	525	-180	232	721	-258	231	452	-165

Table 5. Generator Test Data, Six Capacitor Test

Step	Generator Voltage (kV) U1	U2	U3	U4	Generator Current (A) U1	U2	U3	U4	Generator Field Current (A) U1	U2	U3	U4	Generator (MVAR) U1	U2	U3	U4
1	3	3	3	3	--	--	--	--	150	100	130	130	0	0	0	0
2	12.3	12.5	12.0	12.5	3300	3900	3950	3900	150	100	130	130	-70	-75	-75	-75
10	14.3	14.6	14.2	14.7	4200	4350	4700	4500	150	150	150	165	-105	-100	-110	-108

Table 6. Bus and Line Test Data, Six Capacitor Test

		Castaic				Sylmar					
	Castaic Bus 1	Castaic-Sylmar Line		Castaic-Rinaldi Line		Pole 3 Bus			Pole 4 Bus		
Step	Voltage (kV)	Current	(MVAR)	Current	(MVAR)	(kV)	Current (A)	(MVAR)	(kV)	Current (A)	(MVAR)
1	40	0	0	0	0	0	0	0	0	0	0
2	105	100	-50	250	-40	110	356	-50	12	218	-38
10	215	650	-230	525	-185	232	721	-258	231	452	-165

The voltage was then increased in 1 kV unit terminal voltage steps up to 14.5 kV. The bus voltage at Castaic was a maximum of 207 kV with a maximum MVAR loading of -460 MVAR. At Sylmar, the maximum voltage was 225 kV with a maximum VAR loading of -482 MVAR.

Table 5 provides the generator test data, and Table 6 provides the bus and line test data for the six capacitor test.

CONCLUSIONS

The test results agree favorably with power flow studies and calculations conducted prior to the test. The power flows predicted that, with 207 kV at Castaic, the bus voltage at Sylmar would be an acceptable level of 228 kV. These values agree favorably to the actual values obtained for the six capacitor test. The calculated value of the generator terminal voltage after the transient step rise of 12.1 kV also agrees with the actual value of 12.3 kV obtained for the six capacitator test.

The test verified that a capacitive load similar in magnitude to that which would be expected when energizing the bulk power transmission system within the LADWP service area can be energized at one time, safely utilizing at least four Castaic main units.

This method for reenergizing bulk power transmission systems can be applied to systems where there is sufficient capacity in one plant to absorb the VARs of the network to be energized. This can result in a significant time savings by reducing the number of switching operations required in the restorative process.

ACKNOWLEDGMENTS

This work was performed at LADWP as a part of the formulation of a comprehensive set of operating procedures for the Department, to cover all aspects of power system operations.

REFERENCES

[1] Fitzgerald, A.E., C. Kingsley, Jr., A. Kusko, Electric Machinery, McGraw-Hill, New York, N.Y., 1971.

SUMMARY OF SYSTEM RESTORATION PLAN FOR THE PACIFIC NORTHWEST POWER SYSTEM

Paul F. Arnold, SM
Bonneville Power Administration

1. History and Characteristics of the BPA Grid

The Bonneville Power Administration's beginnings go back nearly 45 years to August 20, 1937 when President Franklin D. Roosevelt signed the Bonneville Project Act. The Bonneville Power Administration or BPA took its name from the first Federal hydroelectric dam in the Pacific Northwest region located on the Columbia River 38 miles upstream from Portland, Oregon.

Today, BPA serves 7.5 million people, transmitting power from 30 Federal dams, numerous Northwest utility dams and four large thermal power plants. The geographic service area covers over 300,000 square miles in Washington, Oregon, Idaho and Montana.

BPA is one of 19 major Northwest utilities comprising the Northwest Power Pool. BPA is by far the largest of the utilities serving one half the regions load. The BPA is the hub of the Northwest power grid, operating over 80% of the regions EHV transmission. A total of 13,000 circuit miles of transmission, including 3,154 circuit miles of 500 kV, join 14 Northwest generating

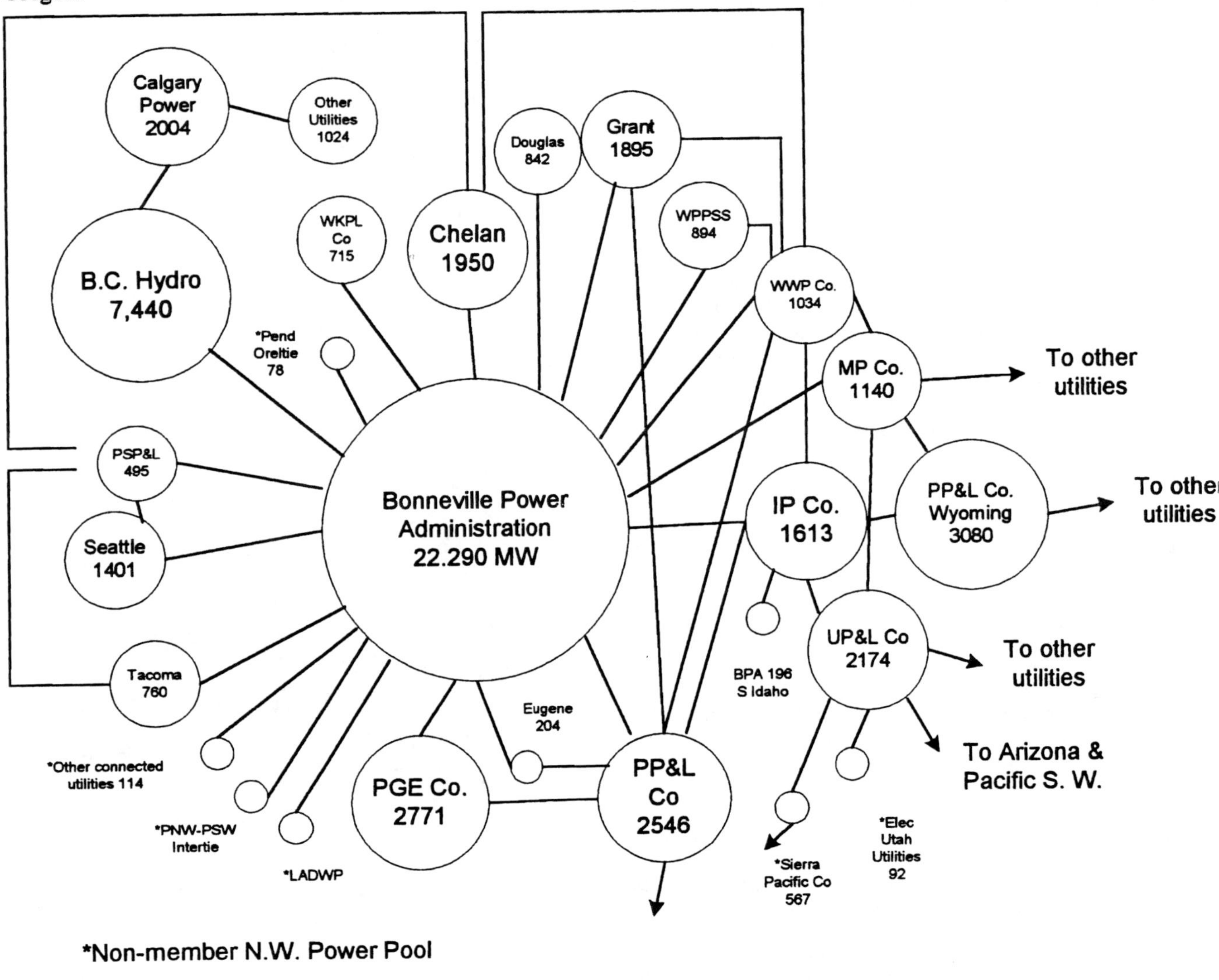

Figure 1. Northwest Power Pool Interconnection

IEEE 1982 Winter Power Meeting
Power System Restoration Task Force
January 1982, New York,

Reprinted from *1982 IEEE/PES Winter Power Meeting*, January 1982.

utilities at 150 points of interconnection. **Figure 1** shows the Northwest Power Pool interconnections.

In addition, regional interties connect the Northwest power grid with Canada, California, Montana and Idaho. The Pacific Northwest-Pacific Southwest interties consist of two 500 kV AC and one 800 kV DC lines connecting the Northwest to Pacific Southwest utilities. Construction of the Pacific Northwest-Pacific Southwest intertie was made possible by ratification of the Columbia River Treaty with Canada in 1964. The treaty provided for three large dams and reservoirs in British Columbia and one in Montana which effectively doubled the regions hydroelectric storage capacity. The BPA grid is connected to Canada by 500 kV and 230 kV transmission. Together, these ties provide seasonal diversity, emergency exchange of power during major generation or transmission outages, and sharing of transmission reserve. Since 1968, the Northwest has provided 150 billion kilowatt hours of surplus power to Southwest utilities, the equivalent of 250 million barrels of oil.

Other regional ties, consisting of 500 kV, 230 kV and 115 kV transmission connects West Coast utilities to owned thermal resources in Montana and Wyoming via the BPA gird.

BPA owns and operates a total of 371 substations. Supervisory control and dispatch of BPA's Main Grid and Subtransmission system is shared by two Control Centers. The Dittmer Control Center is responsible for 90 Main Grid and Western subgrid substations, all major interconnections and BPA controlled generation. The Eastern Control Center controls 35 Eastern subgrid substations. Approximately 73 substations are partially attended and only 7 substations are manned around-the-clock. Communication facilities consisting of 141 BPA microwave repeater stations provide a necessary and vital link for control of the power system during normal and emergency operation.

Because of BPA's dependence on communication networks, the microwave system is nearly as important as the transmission grid. BPA has developed a microwave restoration plan which reroutes critical circuits around trouble spots. The transmission grid restoration plan under development by Northwest utilities depends on a working communications system.

2. Blackout Scenario

Even with the Northwest's extensive transmission and hydro generation facilities it is possible to conceive of a region wide blackout. The most plausible scenario is developed from a typical operating pattern. During periods of light load, major Columbia River hydro generation is shut down to minimum levels. At the same time, large amounts of energy are imported from outside the region. These imports consist of energy exchange from Southwest utilities in return for peaking capacity as well as Eastern thermal generation scheduled to owner utilities on the West Coast. Heavy imports from Canada also occur in the form of displaced energy return from Southwest utilities and energy surplus to other Northwest utilities. During this operating pattern, a sudden loss of generation in the Northwest could cause a cascaded loss of transmission and islanding of the Northwest.

A stability control scheme is employed to prevent islanding of the Northwest. The scheme detects loss of transmission and generation and immediately trips up to 3000 MW of industrial load in the region. If this scheme fails, the system will island.

At this point underfrequency relaying will shed load in order to preserve thermal generation and interconnected transmission within the island. Boundaries of the island are shown in **Figure 2**. Failure of this scheme to coordinate with four major thermal plants in the region could result in a further breakup within the island and a possible blackout. An underfrequency load shedding and restoration program is coordinated among Northwest Power Pool utilities. However, recent studies show that miscoordination and subsequent system breakup is possible. Program refinements have been suggested, however, computer simulations still have recognized limitations.

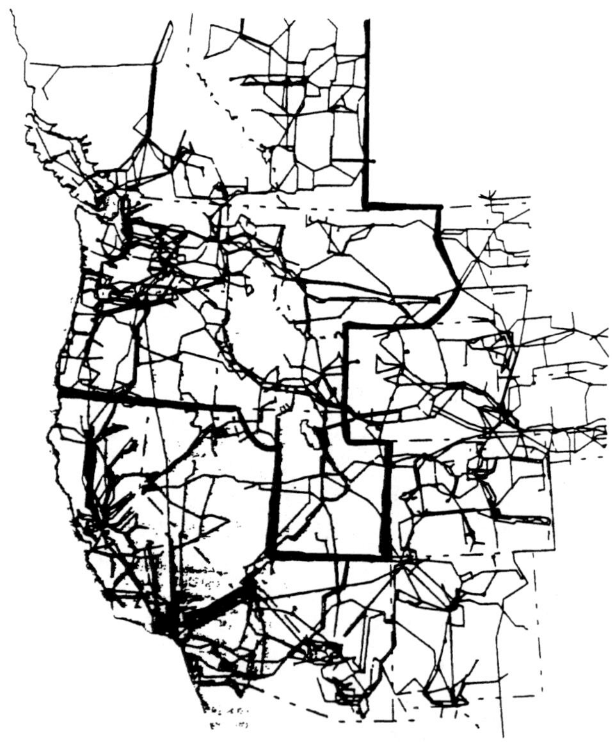

Figure 2. Western Systems Coordinating Council Map of Principal Transmission Lines

The NWPP formed a system restoration task force to develop a Coordinated Area Restoration plan. Ten utility representatives are charged with devising a coordinated plan to restore the entire Northwest power grid from a total blackout. The concept of the plan is diagramed in **Figure 3**.

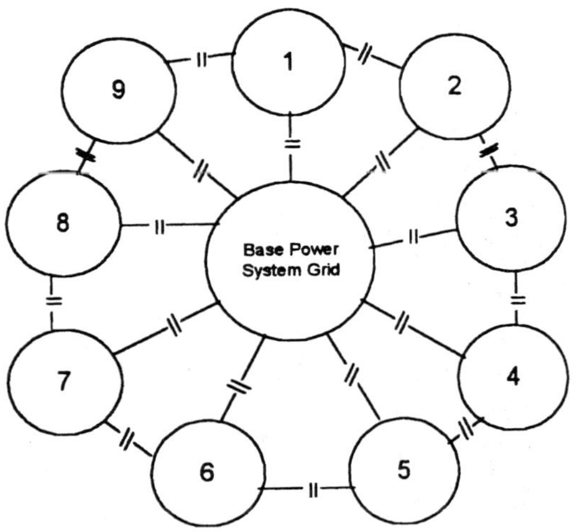

Note: Parallel signs (I I) indicate open interconnection points between Coordinated Areas and the Base System Power Grid. Also, interconnection points between Areas are drawn in only for illustration and do not necessarily reflect Area to Area interconnection points as they may finally be agreed to.

Figure 3. The Concept of Coordinated Area Restoratin

The philosophy behind the plan is this: Nine geographic areas have been identified, consisting of one or more member utilities. Each area will conduct restoration activities independently. BPA will act as the central coordinating utility once the systems are interconnected. But first, BPA's major role will be to establish a base power system grid.

In each of the coordinated areas, utilities will identify separation and interconnection points with each other and with BPA. A detailed plan for each area will outline procedures for restoring individual utility loads using their own generation resources. Once a base grid is established, all generation resources will be pooled and reallocated by BPA to restore blocks of Northwest utility load.

This restoration concept offers two advantages. First, it allows BPA time to restore its major transmission and generation facilities. This task requires independent action and undivided attention by BPA dispatchers and Federal generating plant operators. Second, pooling of resources and reallocation by BPA ensures fast, coordinated restoration of all Northwest loads, with a minimum of load sectionalizing activity.

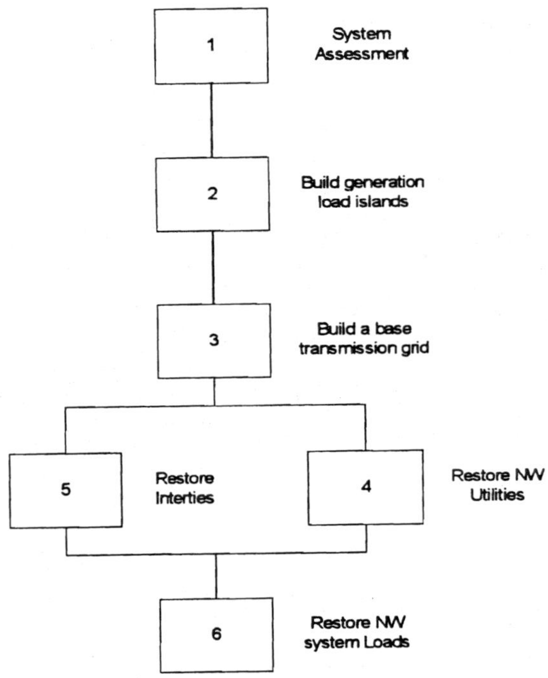

Figure 4. The Six-step Plan for Restoration of the BPA Grid

3.Restoration of BPA Base System Power Grid

Figure 4 outlines a six step plan for restoration of the BPA grid.

Step 1. System Assessment. This step involves identifying the system status following a system disturbance.

Step 2. The next step is to build generation-load islands. Critical generation-load paths will be restored independently. Dispatchers and SCADA controllers will team up in the two control centers to restore aluminum industry loads from the closest generation source. Aluminum plant loads will be restored first since they are time critical. These loads must be restored within 45 minutes or the molten aluminum will freeze. There are other advantages for restoring this load first. Aluminum loads are easily accessible from the 500 kV and major 230 kV grids. All substations serving these loads are under SCADA control. The loads can be picked up in small increments and provide an ideal base load when restoring remaining system loads.

All critical generation-load islands are listed in Table 1. This table specifies assigned dispatch authority to either the Dittmer Control Center or the Eastern Control Center. It also specifies who controls each station on the generation-load path. These are the people who must coordinate the restoration of these mini-islands.

Building generation-load islands requires several major tasks. They are:

1. Clearing of deenergized busses
2. Restarting generation
3. Line charging
4. Load pick-up.

Clearing of deenergized busses will be accomplished by means of a dead bus clearing scheme to be installed on over 50 busses on the BPA Main Grid. Locations for this scheme are indicated on a map, **Figure 5**. The scheme consists of a single SCADA control point which will trip all circuit breakers on a bus. Tripping is interlocked by undervoltage relays which must detect a dead bus. This scheme was necessary in order to clear generation-load paths in time to reach critical aluminum potline loads. Tripping circuit breakers one at a time from black start will be included in the NWPP would delay the overall restoration time by at least 2-3 hours.

The BPA system restoration plan identifies general restart procedures for hydro generating plants and also identifies which plants have black start capability. Detailed generation restart procedures Coordinated Restoration plan for each generating utility and Federal agency.

Table 1. Critical Generation-Load Islands

Island	Bus	Control
Dispatch authority assigned to DCC:		
1 G,M	Coulee 500	WPRS Operator
	Raver 500	SCADA 1
	Tacoma 230	SCADA 1
	Tacoma 13.8	SCADA 1/Kaiser
Dispatch authority assigned to ECC:		
2 G,M	Coulee 230	WPRS Operator
M	Bell 230	BPA Operator/SCADA1
	Bell 13.8	BPA Operator/SCADA1 /Kaiser
	Columbia 230	SCADA 2
	Columbia 115	SCADA 2
	Valhalla 115	SCADA 2
	Valhalla 13.8	SCADA 2/Alcoa

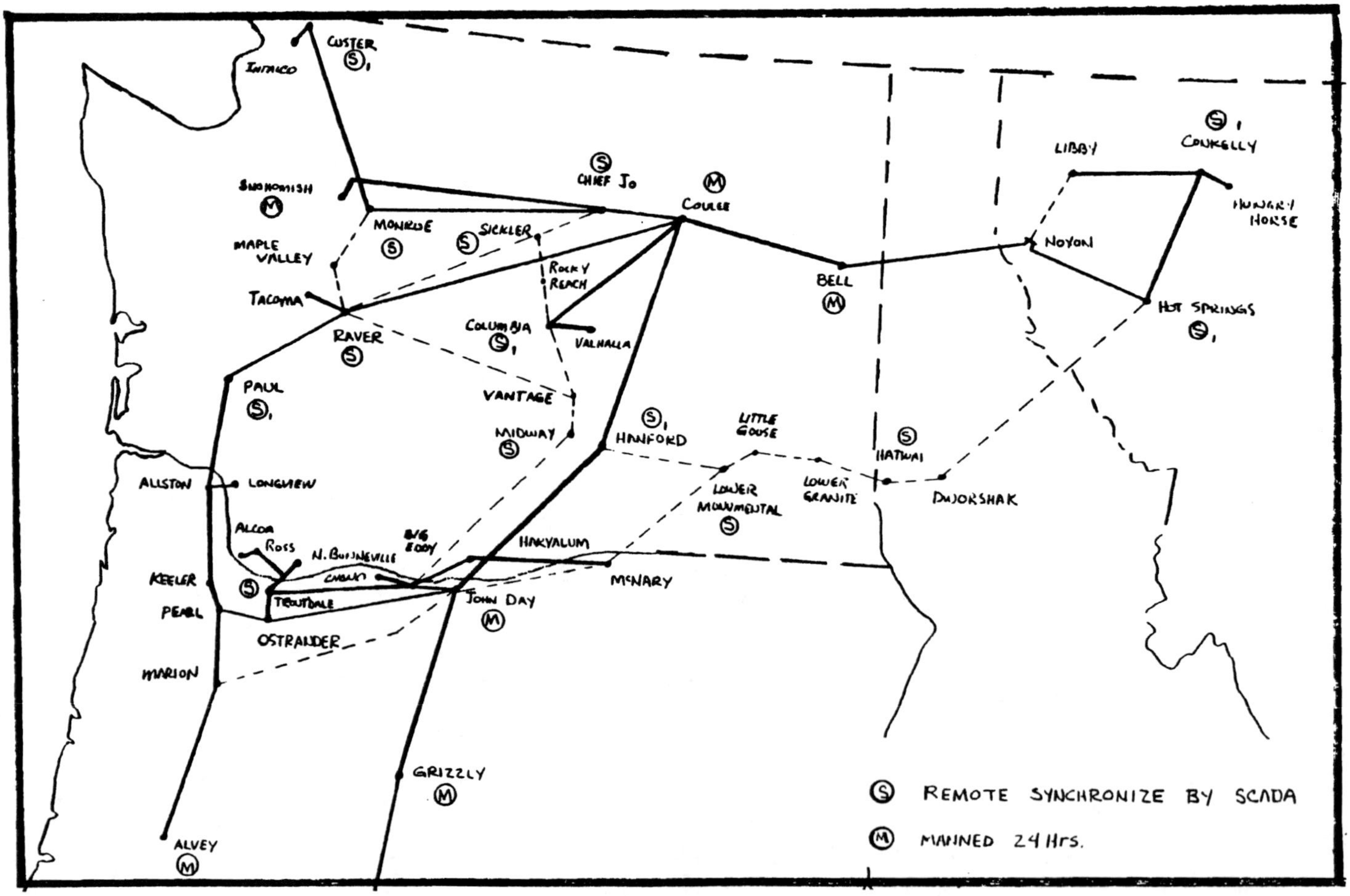

Figure 5. Dead Bus Clearing and Remote Synchronizing Location

Dispatch authority assigned to DCC:

3	G	Chief Joseph 500	SCADA 1
	G	Chief Joseph 230	SCADA 1
		Custer 500	SCADA 1
		Custer 230	SCADA 1
		Intalco 230	SCADA 1
		Intalco 13.8	SCADA 1/Intalco

Dispatch authority assigned to DCC:

4		Chief Joseph 230/345	SCADA 1
	M	Snohomish 345/230	BPA Operator
	M	Snohomish 115	BPA Operator

Dispatch authority assigned to DCC:

5	G,M	John Day 500	SCADA 1
		Ostrander 500	SCADA 1
		Troutdale 230	SCADA 1
		Troutdale 13.8	Scada 1/Reynolds

Dispatch authority assigned to DCC:

6		Troutdale 230	SCADA 1
		Ross 230	Scada 1
	G	North Bonneville 230	SCADA 1
		ALCOA 230/115	SCADA 1
		ALCOA 115	SCADA 1
		ALCOA 13.8	SCADA 1

Dispatch authority assigned to DCC:

7		Ostrander 500	SCADA 1
		Pearl 500	SCADA 1
		Keeler 500	SCADA 1
		Allson 500	SCADA 1
		Allson 230	SCADA 1
		Longview 230	SCADA 3
		Longview 115	SCADA 3
		Longview 13.8	SCADA 1/Reynolds

Dispatch authority assigned to ECC:

8	G	Big Eddy 230	SCADA 1
	G	Big Eddy 230	SCADA 1
		Chenoweth 115	SCADA 2
		Harvey 13.8	Martim Marietta- The Dallas
		Harvalum 230	SCADA 2
		Harvalum 13.8	Martin Marietta
	G	McNary 230	SCADA 1

Dispatch authority assigned to ECC:

9	G	Libby 230	SCADA 1
	G	Hungry Horse 230	WPRS Operator
		Conkelley 230	SCADA 1
		Conkelley 13.8	SCADA 1/Anaconda

Dispatch authority assigned to DCC:

10		Marion 500	SCADA 1
	M	Alvey 500/230	BPA Operator/SCADA 1
		Alvey 230	BPA Operator/SCADA 1
		Alvey 115	BPA Operator/SCADA 1

Dispatch authority assigned to DCC:

11	S	John Day	BPA Operator/SCADA 1
	S	Grizzly 500	BPA Operator/SCADA 1
	S	Malin 500	BPA Operator/SCADA 1

Note: G: Substation integrates generation. M: Substation manned 24 hours. Operator will clear bus on dispatcher' order. S: The AC intertie can restored manually by operators at John Day, Grizzly and Malin, under direction of the BPA dispatchwer. The Malin operator will synchronize the NW to the SW system. Speed control for the NW will most likely be via John Day generation. Grand Coulee generation may be used for speed control if an internally connected base transmission grid has been established.

Line charging and load pickup procedures are also being identified in the BPA plan. The procedure is outlined in **Figure 6**. Before a 500 kV transmission line is charged, all available generator units will be paralleled. The generator terminal voltage will be set to minimum levels with the voltage regulator in service. Generator speed will be set within a range of 60-61 Hz. Available shunt reactors will be connected to the line before it is charged.

Once the line is charged a system voltage check is made. Voltage adjustments can be made if the voltage is too low. The last step is to pickup load.

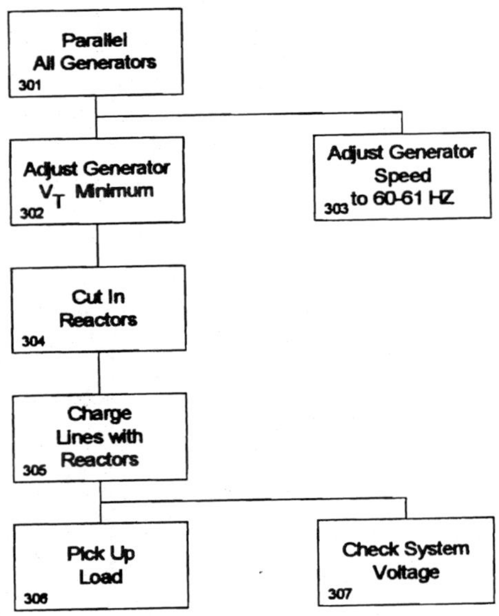

Figure 6. How to Build Generation-Load Islands

Table 2 lists minimum voltage levels for selected Federal hydro plants. It shows that in the BPA plan the highside transformer voltage, and terminal voltage limits imposed by the regulator and by the plant control computer.

Table 2. Generator Terminal Voltage Limits

Name	H-S	L-S	Reg.
The Dallas 230	209	12.0	11.7
John Day 500	481	12.4	11.7
McNary 230	206	12.4	11.7
Chief Joseph 230			12.5
Chief Joseph 500			12.5
Grand Coulee	210	12.5	12.5
Grand Coulee	500	14.25	14.25
Lower Monumental		12.4	12.4
Little Goose		12.4	12.4
Lower granite		12.4	12.4
Libby			
Hungry Horse			
Bonneville			

Note: H-S: No load high side voltage, L-S: No load low side voltage by computer, Reg.: By regulator.

Table 3 lists the Individual Plant Loading Guide to be used during load pickup. It shows that the BPA plan identifies the total plant capacity in MW and the size of load that can be picked up for a speed drop of 0.5 Hz. This value is expressed as a percent of rated plant capacity as well as in MW. For two or more plants connected in an island, the individual MW values can be added up. This method is valid for almost any combination of plants assuming adequate interconnecting transmission. The totals at the bottom line indicate the aggregate load pickup value for a large Northwest island with all generation at speed-no load.

Figure 7 shows how the allowable load increment increases as the system load increases. From speed-no load to an 80% base load, the load pickup value increases from about 3.5% to 14%. An additional criteria to the 0.5 Hz drop is that the momentary voltage drop does not exceed 20%.

3.Building a base transmission grid implies tieing the individual generation-load islands together. A major task here is synchronizing. Synchronizing control capability to SCADA is being installed at selected substations to connect generation-load islands without the assistance of a substation operator. These are called Remote Synchronizing schemes and utilize standard telemetry for speed, voltage and phase angle difference across an open circuit breaker. The dispatcher can use

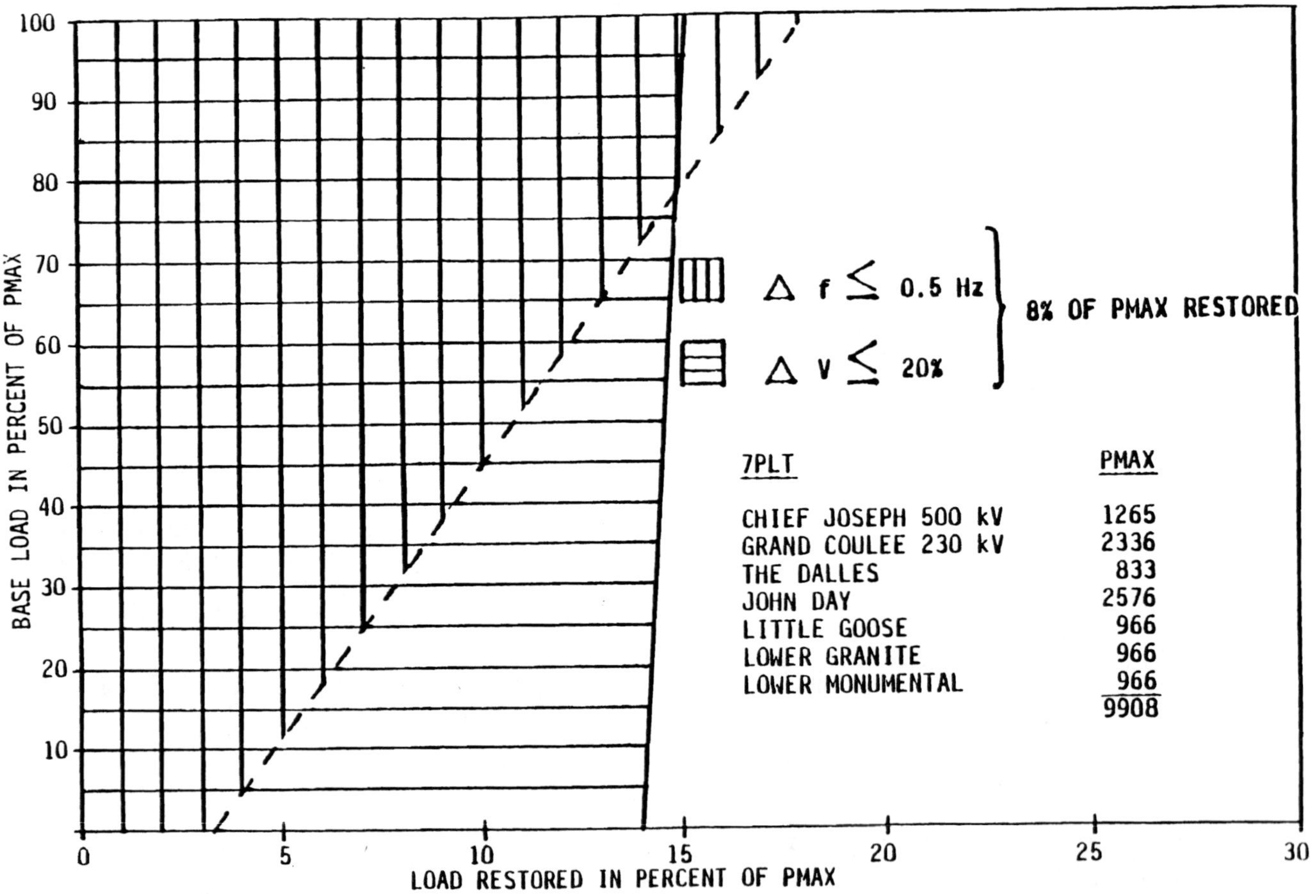

Figure 7. Normal Stress Seven Plant System

this information to coordinate system speed and voltage adjustments. When two island systems are adjusted within desired tolerance, an automatic synchronizer relay at the substation is enabled by SCADA. **Figure 6** indicates where Remote Synchronizing schemes are installed. Steps 4, 5, and 6 involve coordination with other utilities.

Table 3. Individual Plant loading Guide

Plant Name	Rated MW	Load pickup %	Load pickup MW
1 Bonnevile 230 kV	360	1.7	6
1 The Dallas 230 kV	1545	2.3	36
1 John Day 500 kV	2273	3.1	70
2 Priest Rapids 230 kv	830	2.97	25
2 Wanapum 230 kV	875	2.51	22
2 Rock Island	432	0.375	2
2 Rock Reach 230 kV	1277	2.87	36
2 Wells 230 kV	815	3.67	30
2 Chief Joseph 230 kV	1078	3.9	42
2 Chief Joseph 500 kV	800	2.3	18
2 Grand Coulee 230 kV	2125	6.3	133
2 Grand Coulee 500 kV	3900	5.3	206
3 Lower Mounumental	852	3.1	26
3 Little Goose	852	3.1	26
3 Lower Granite	852	3.1	26
Total	18866	3.73	704
4 McNary 230 kv	884		
4 Dworshak	421		
4 Hungry Horse	300		
4 Libby	422		

Note: Generation at speed-no load, Speed drop 0.5 Hz, 1: Tested, 2: Transient stability study, 3: Estimated from John Day units, 4: No information at this time.

4. Once a base transmission grid has been established, BPA will interconnect with other Northwest utilities. This is where the BPA plan melds into the NWPP coordinated plan.

5. Regional interties will be restored as they become available. However, initial efforts will be concentrated in building a base grid using Northwest hydro generation and aluminum industry load. The Pacific Northwest-Pacific Southwest intertie could be restored simultaneously with generation-load islands in the Northwest since the intertie stations are manned continuously. Standard procedures exist for restoring regional ties following a breakup of the WSCC region. Power schedules between interconnected utilities will not be resumed until adequate transmission capacity is restored and all regional ties closed.

6. With regional interties restored and internal ties made with Northwest utilities, load restoration will proceed rapidly. At that time load can be restored in very large blocks with a minimum of sectionalizing. However, a future step in the development of the NWPP plan may be to coordinate load sectionalizing activities with local utilities during load restoration to avoid blowing distribution transformer fuses on cold load pickup.

4. Conclusion

The components of the BPA plan are intended to give dispatchers and operators a perspective on goals and objectives, philosophies and procedures, and system capabilities and limitations. Actual system restoration tests were conducted, which provided on the job training for dispatchers and operators as well as providing data for comparison to off line computer simulations and development of procedures. Annual tests will be conducted to provide continued training and gather new data to refine operating procedures. On the other side of the coin, BPA will continue to strive towards system reliability in hope that a restoration plan of this magnitude and complexity will never have to be relied upon.

FIELD EXPERIENCES IN REENERGIZATION OF ELECTRICAL NETWORKS FROM THERMAL AND HYDRO UNITS

E. Mariani
member IEEE

F. Mastroianni

V. Romano

ENEL - National Authority for Electrical Energy
Naples - Italy

Abstract. The problem dealt with in the paper is the fast return to normal operation of a generation and transmission power system, after it has experienced a total or partial power failure. The aim is to illustrate the practices adopted in a hydrothermal power system and to present the outcomes and performances verified in some tests and field experiences. Considerations about the influence of the behaviour of the load and of the generating units involved in restoration are also given.

INTRODUCTION

The experience gained in the field of emergency control, as well as the developing theoretical approach both for preventive actions and restorative procedures, have allowed a reduction in frequency and amplitude of power failures.

The practices adopted are again evolving, but some aspects of them appear today to be accepted to quite a wide extent; an overview of the procedures and problems connected with the reenergization in ENEL network after black-outs may be of interest, even if not exhaustive, especially for stretched power systems, like the ENEL's one in its southern part (fig. 1). This part of system is characterized by medium-long distances (100 - 300 km) existing between hydro power stations, thermal power stations and major loads; the installed thermal power is by far more important than hydro power; thermal power stations have no black start capabilities, which are on the contrary present in most hydro power stations.

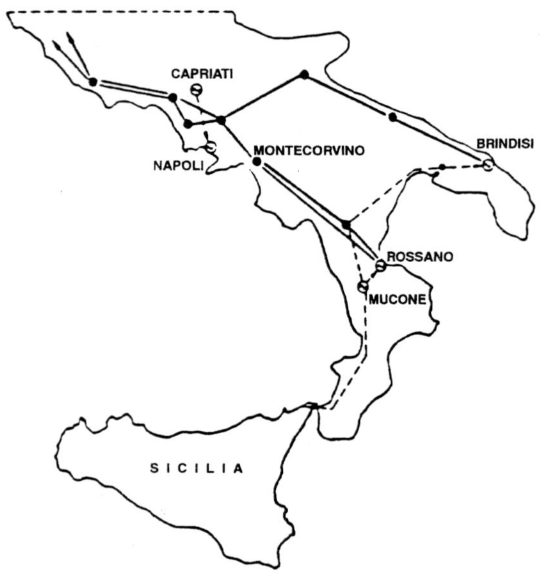

Fig. 1 - Southern part of ENEL national grid

84 WM 093-1 A paper recommended and approved by the IEEE Power System Engineering Committee of the IEEE Power Engineering Society for presentation at the IEEE/PES 1984 Winter Meeting, Dallas, Texas, January 29 - February 3, 1984. Manuscript submitted August 19, 1983; made available for printing November 29, 1983.

A fast network and load restoration may be achieved by assigning autonomous procedures to the staff of power stations and of outstations (or of the remote control centers controlling them), in order to reduce the burden of the dispatcher or making it unnecessary in the first phase of building-up of network.

The defence plan against power failures is conceived in such a way that either thermal and hydro power plants should eventually trip to houseload and keep alive, so being able to actively participate to network and load restoration. Should nevertheless a thermal power plant had come to a stop, due to unsuccessfull tripping to houseload, the aforesaid restoration procedures will guarantee a fast reenergization of its auxiliaries (from hydro power station or from tie lines), making possible the restart of thermal power plant itself.

The paper will present the criteria underlying the reenergization procedure, the problems posed by thermal and hydro units and the solutions adopted, the problems posed by the behaviour of the load when it is restored, the illustration of some tests and practical cases.

REENERGIZATION PROCEDURE

A preestablished partitioning of network is done in ''power restoration lines'' (PRL), each constituted by one or more generators (''early restoration generators'', ERG), circuits and loads; the configuration of each PRL is periodically (e.g. yearly) updated, as far as new elements of the electrical system are commissioned, or old ones are decommissioned.

Reprinted from *IEEE Transactions on Power Apparatus and Systems*, Vol. 103, No. 7, pp. 1707–1713, July 1984.

After a black-out each PRL, independently from one another, will reconstitute autonomously, that is with no exchange of information between the outstations belonging to it, or between the outstations and the dispatcher.

Assuming as an example the scheme in fig. 2, the following operations will be performed /1/, in the following order (the generator shown in power station 1 is assumed to be an ERG, that is preferably a hydro generator):

a) if lack of voltage lasts more than 3 min (this time interval is supposed to be sufficient to assume that the network is in black-out), the staff of each station will open all the breakers of the station itself, with the exception of those feeding the station auxiliaries.
   This operation may also be performed by an automatism. The aim is to avoid that the generator which will first energize the PRL will find to much active load to restore (excess of active power output), or to much line length to energize (excess of reactive power absorption).

b) The staff of station 1, after a further interval of 5 min (introduced to allow the manoeuvres of point a), will energize the circuit 1-2, closing the breakers A and B.

c) The staff of station 2, on the appearance of voltage on circuit from 1, will energize the circuit 2-3, closing the breakers C and D.

d) The staff of station 3, on the appearance of voltage on circuit from 2, will energize the circuit 3-4 closing E and G, and a preset load closing F.

e) The staff of station 4, on the appearance of voltage from 3, will energize a preset load closing the breakers H and I.

The load which is so restored in station 4 may well be constituted by the auxiliaries of a thermal power station.

The load being restored in station 3 may help a better regulation of voltage and speed of the ERG of station 1.

The time elapsed from the switching on of breaker A and that of breaker I will of course depend on the number of intermediate stations which, as far as possible, should be a minimum.

The synchronization of the various PRLs so reconstituted (fig. 3) will then be coordinated (even if not necessarily) by the dispatcher, as well as the gradual load restoration.

The speed achievable in load restoration depends on the state of generators; hydro units are of course, in this respect, in better conditions than the thermal ones; the last can nevertheless perform satisfactorily, especially if the interval between the run-back, occurred at the moment of tripping to houseload, and the beginning of load pick-up is short; this requirement stresses the importance of a fast reconstitution of network.

The scheme represented in fig. 2 shows a simple example; in practice the PRL may comprise also some meshed path.

An important thermal power station may have its auxiliaries reenergized by more than one PRL, to account for the possibility that a permanent fault make impossible the building up of a particular PRL, and also to reduce the delay in auxiliaries restoration.

PROBLEMS POSED BY THERMAL UNITS

As stated above, a thermal unit shall in general be available to network reenergization.

This depends /1/, /2/ on the ability to:

a) Keep alive on its own auxiliaries

b) Energize HV circuits, with the generator under AVR control, and terminal voltage set at the minimum value allowed by auxiliaries supply

c) Pick-up load in steps of little amplitude, with the turbine under governor control, and boiler regulator on.

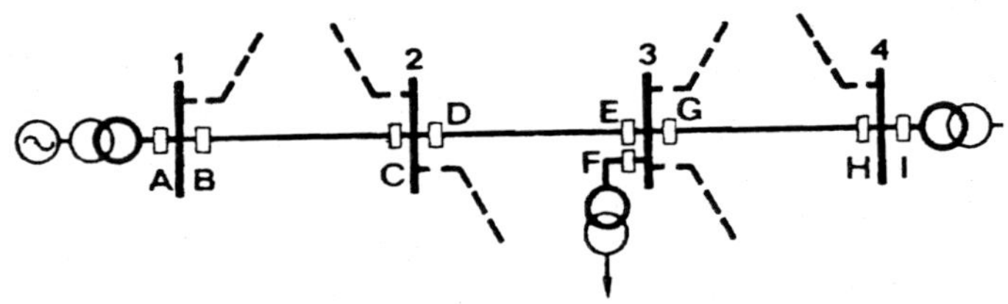

Fig.2 - Example of a power restoration line (PRL)

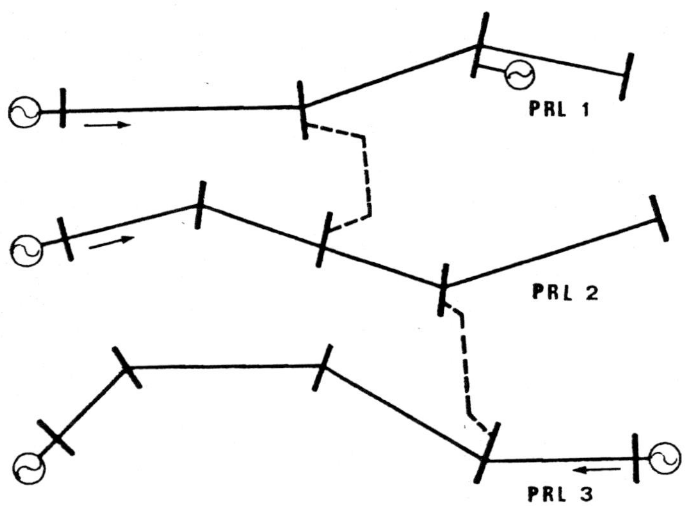

Fig.3 - Synchronization of various PRLs

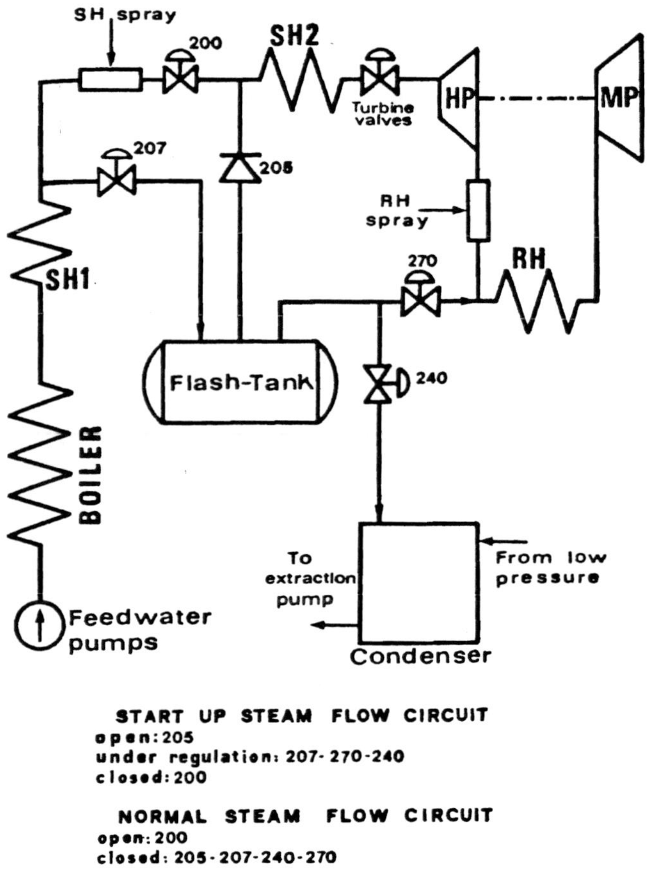

Fig. 4 - Normal and start-up simplified steam flow circuit of an once throug boiler

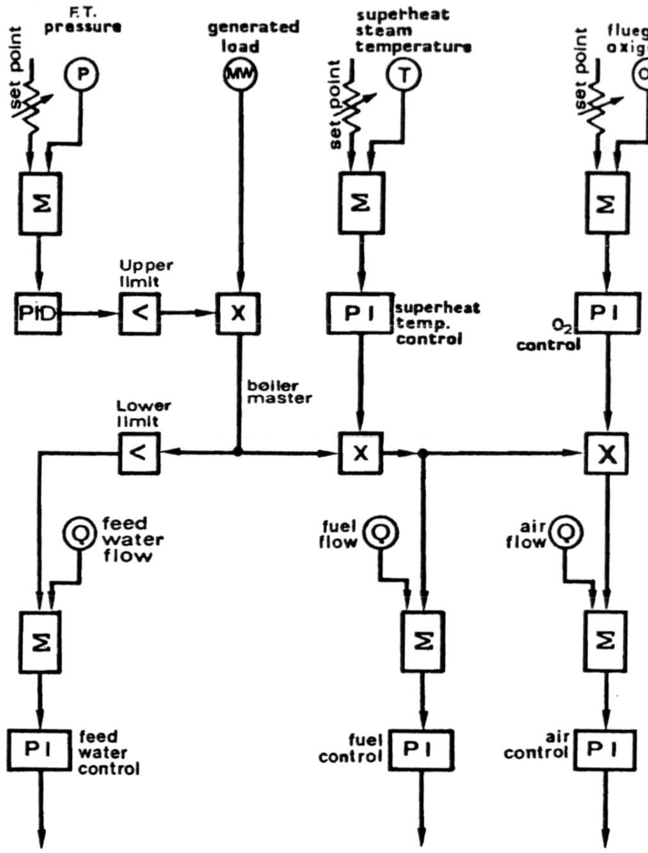

Fig. 5 - Simplified regulation scheme for an OTB on start-up circuit

As far as point a) is concerned, the main requirement is that the unit be not in a start-up phase; that is, valve 200 must be open (fig. 4).

The tripping to houseload takes place when frequency remains below 47.5 Hz for more than a preset time (typically 5 s) or below 46.5 Hz (instantaneously), or when the main breaker trips due to other perturbations (except those leading the unit to complete stop, e.g. due to internal failures).

For once through boilers (OTB) units the tripping to houseload takes place in the following steps. The fires are tripped and feedwater pump stopped; the maximum thermal energy must be stored, in order to allow feeding the unit's auxiliaries for 15 - 20 min, living on thermal storage; this is accomplished by delaying the closing of valve 200 and by keeping in closed position the valve 207 up to the moment in which the pressure on SH2 has decreased to an appropriate value (e.g. 35 kg/cmq); at this moment the turbine begins to be fed through 207 - flash tank - 205 - SH2, and 200 is closed. The feedwater pump is then started-up and fires reignited, in order to make the unit also able to pick-up load in appropriate steps (point c)). To this end, at the Rossano power station (320 MW units), the flash tank pressure is regulated at 60 kg/cmq, the steam excess being spilled through 240 to the condenser; a regulation scheme is adopted (fig. 5) of the boiler follows type, to control the frequency of the network which is being reenergized.

The differences in temperature, if any, between SH2 and RH are controlled by 270, which at 90 MW is fully open; load steps up to 5-10% may be picked-up in this phase.

Beyond 90 MW 270 must be closed, and this is performed only when the unit is synchronized with prevalent network.

As far as point b) is concerned, a 370 MVA generator can typically energize up to 200 km of 400 kV overhead circuit, with therminal voltage set at 0.9 V_n; the reactive power adsorbed in steady state is about 120 MVAr, a value which complies with the capability curve of the generator.

PROBLEMS POSED BY HYDRO UNITS

Hydro units are in general of limited power rating, with respect to thermal ones, and have a lower percentage requirement for auxiliary power; they are hence in general provided with black-start equipment (little hydro turbine, or diesel engine), even if the probability of successful tripping to houseload is very high.

A problem may be posed by reactive power, when energizing HV circuits; for this purpose some particular loads are restored early in the PRL reenergization, as illustrated before (fig. 2).

Some stability problem with high head plants are in general solved with an appropriate setting of accelerometer (or equivalent element) parameters.

The permanent frequency bias must in general be set at a low value, to make easier the frequency regulation and hence the synchronization of the various PRLs to each other.

The parameters (permanent bias, accelerometer constant etc.) will preferably be the same during normal operation and restoration phase, to avoid the need of switching them through intervention of the staff, when restoration conditions occur; they will be necessarily the same for remote-controlled units.

As an example, the following are the main characteristics and parameters of a hydro ERG (early restoration generator) and of its governor: Pelton type turbine rated 60 MW; head 600 m; penstock characteristic time T_w = 1.1 s; lowest resonance frequency of penstock 0.18 Hz; inertia constant of the unit H = 4.1 s; servomotor integration time T_x = 1.2 s; permanent bias b_p = 2%, accelerometer constant T_n = 2.1 s.

PROBLEMS POSED BY SOME LOADS

With the exception of purely resistive loads, all the other present when switched on a transient absorption of active and reactive power which may be many times the steady-state value. This transient is nevertheless in general of short duration (1 - 2 s), so that the frequency and the voltage of the ERG restoring them undergo little variations.

In some cases, especially when air conditioning an refrigerator apparatus are reenergized, the transient can on the contrary last longer time, as clearly illustrated in /3/, /4/. This might hinder a regular build-up of PRLs and a fast network and load restoration.

It must nevertheless be noted that the phenomenon of long duration (in the order of 5 s) of high power absorption (caused by failure of start-up of motors) presents itself only when the deenergization of these particular loads lasts less than 2 - 4 min, as shown in /3/ and /4/. On the other hand, from the moment of a complete network break-down to the beginning of load restoration a longer time interval normally elapses (in general at least 15 min, according to the reenergization procedure above described). Hence in case of restoration after black-out, refrigerator type loads do not cause serious problems related to initial (few seconds) power transients.

Two other important aspects, connected again to refrigerator and air conditioning type apparatus, must instead be considered.

The first one is the simultaneous start of their motors if the deenergization has lasted long time (more than, e.g., 15 min), due to diversity factor; this may cause, after a short initial transient of 1 - 2 s, a steady power demand of up to, e.g., 1.3 - 1.5 times the value requested before deenergization; the decay to previous value may take many minutes (see fig. 7). The load steps to be first restored must be evaluated keeping this in mind, in order to avoid excessive power request from ERGs, and possible problems in frequency regulation.

The second aspect is connected to the voltage drops, accompanying active power requests higher than normal, on MV feeders. High voltage drops, if lasting more than few seconds, may cause the stalling of motors driving compressor loads, and hence a further increase of power request from ERGs. The same considerations as before about load steps evaluation apply.

DESCRIPTION OF SOME FIELD EXPERIENCES

Load restoration from infinite network

A mix of urban and rural customers, fed by MV feeders, was disconnected and restored, with different deenergization intervals.

Fig. 6 shows two typical transients; in the first one (fig. 6a) the deenergization interval was long (5 min), and the transient shows a short peak (lasting less than 1 s) and a final value bigger than previous one (41 MW against 37); in the second (fig. 6b) the deenergization interval is very short (roughly 2 s), the initial peak decays in 5 - 6 s, and the final value is lower than previous one (36 MW against 41).

The outcomes may be interpreted as mentioned before: in case of long interruption refrigerator loads reconnect in higher number than normal average, due to temperature increase caused by the prolonged interruption itself; on the other hand the backpressures on refrigerator cycle had sufficient time to decrease, allowing a normal restart of driving motors and hence a very short initial transient. In case of short interruption, on the contrary, the driving motors fail to restart due to high backpressures, and they are gradually (in matter of seconds) switched off by their thermal protections; the initial transient lasts longer, and the load which remains connected after it is lower than the previous one.

In the first case the load then decreases to normal value in matter of minutes; in the second it increases similarly to normal, when thermal protections reset.

This fenomenon is illustrated in fig. 7, which shows another series of deenergization - energization of loads. The different duration of initial transients may not be distinguished due to time scale; it is on the contrary clearly illustra-

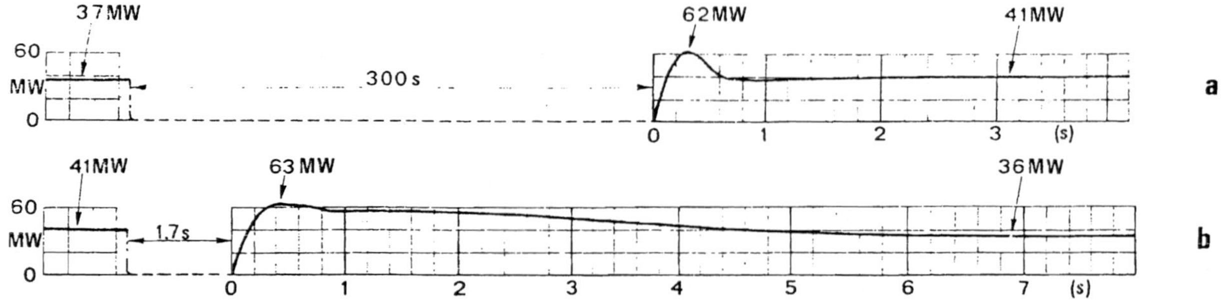

Fig.6 - a) Load restoration after long deenergization b) Load restoration after short deenergization

ted the behaviour of load after the initial transients, with lower values rising to normal after short interruptions, and higher values decreasing to normal after long interruptions.

Table I shows the main parameters registered in a third series of tests, which confirm the above considerations. Some tests took place in late morning of a work day, and other during the following night (the last are the same of fig. 7). The column headed "steady-state after" lists the values registered immediately after the initial transients; they are lower or higher than the values previous to interruption, respectively for short or long deenergization interval; the discriminating time interval appears to be around 4 min.

Building up of a PRL and load restoration from a thermal unit

Some tests have been performed in a thermal power station equipped whit four 320 MW units of OTB type (Rossano power station).

One of the units was islanded on its own auxiliaries by tripping the main breaker, and was then used to start up another unit of the same power station. The test was successfull. Both units remained on house load for more than 4 hours, without problems.

The same unit was then used to energize a 400 kV line of a lenght of 200 km about; the reactive power absorbed was about 150 MVAR, near the underexcitation limit of the alternator; the regulation of voltage was good.

At the receiving end of the line various loads were next restored in steps, through 400/150 and 150/20 kV transformers (Montecorvino station); it was ascertained that to avoid the risk of tripping on underfrequency, each step had not to exceed the value of 30 MW about; the average speed of load pick-up (with the unit on the start-up circuit) could reach a maximum of 10 MW/min.

In one of the first tests of load restoration, one load was switched from main (national grid) network and connected after a short delay (2 s about) to the PRL being fed by Rossano unit; the

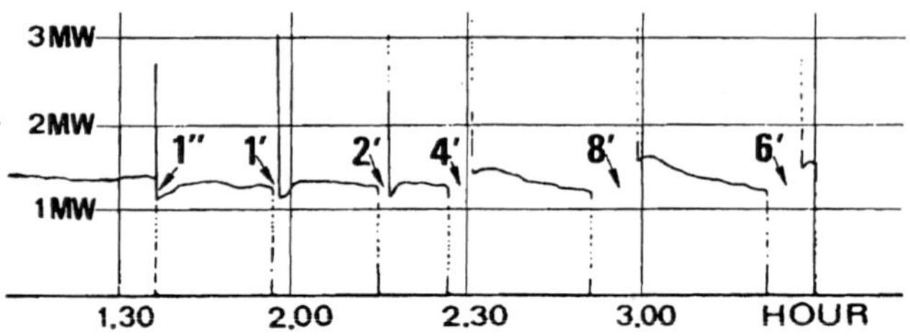

Fig.7 - Load restoration after various deenergization interval

Tab.I - Characteristic values at reenergization of a MV feeder

HOUR	WAITING TIME (s)	ACTIVE LOAD (MW)				REACTIVE LOAD (MVAR)			
		STEADY STATE		PEAK VALUE	AFTER 1 s	STEADY STATE		PEAK VALUE	AFTER 1 s
		BEFORE	AFTER			BEFORE	AFTER		
10.20	1	3.3	2.7	5.6	4.2	0.9	0.6	2.0	1.0
10.45	480		3.1	6.3	3.1		0.8	2.6	0.8
11.27	60	3.3	2.7	5.2	4.1	1.0	0.6	2.6	1.1
11.33	270	3.2	3.1	5.8	3.3	1.0	0.7	2.7	0.8
1.40	1	1.4	1.1	3.4	2.5	0.3	0.1	1.0	0.6
2.00	60	1.4	1.2	3.7	2.5	0.4	0.2	2.5	0.6
2.18	120	1.3	1.2	4.0	2.2	0.4	0.2	1.2	0.4
2.30	240	1.4	1.5	4.8	1.9	0.4	0.5	1,8	0.6
2.55	480	1.3	1.6	4.7	1.6	0.4	0.6	3.7	0.6
3.25	360	1.3	1.6	3.6	1.6	0.4	0.5	3.3	0.5

transient of power output of the unit is illustrated in fig. 8; it shows the phenomenon referred to in previous paragraphs (long initial transient due to failure to start of motor-driven compressors); the unit tripped on underfrequency after 9 s about.

The present feeling about the use of thermal units for reenergization is that they shall preferably pick-up load after their synchronization with other units in the same PRL, in order to reduce the probability of tripping on underfrequency; the

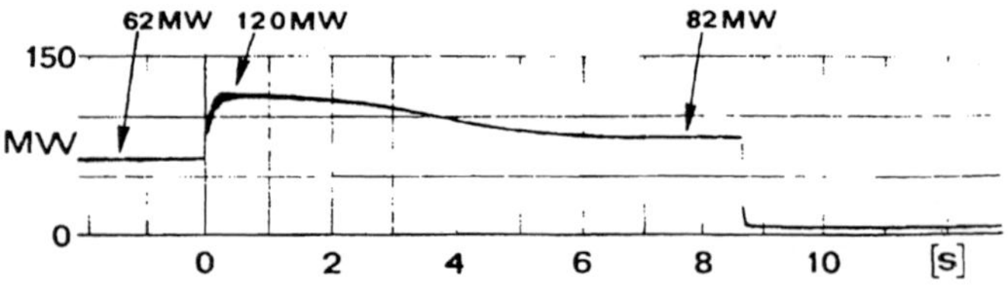

Fig.8 - Unit trip on underfrequency

other units will preferably be hydro.

In case of unavailability of hydro units, a resynchronization with the main grid is felt anyhow desirable, especially for OTB type units which are limited by the maximum power output relevant to start-up circuit.

Building up of a PRL and load restoration from thermal and hydro units jointly

Fig. 9 shows the test scheme, with two hydro units of 50 MW each (Mucone power plant) and one thermal unit of 320 MW (Rossano power plant). The two power plants are connected by means of a 150 kV circuit. The breaker positions indicated in the figure are those prevailing just before the beginning of the test. The 320 MW unit had an output of about 200 MW, flowing towards the main grid through B and A.

The two power plants have complementary characteristics: the thermal one is well suited to voltage and frequency regulation; the hydro one to fast load pick-up.

The staff of hydro plant was given the following instruction: "act on power output setting of governor in order to keep frequency equal to nominal"; and that of thermal plant: "pick-up load regularly, consistently with steam availability and watching at frequency". The aim of these instructions was to assign to thermal unit the task to pick-up load at a roughly constant speed, and to hydro units that of absorbing the variations of power request connected to the stepwise variation of load.

At the time zero the breaker A was switched off (LR in fig. 12); after 15 s about the thermal unit tripped automatically to houseload, switching off the breaker B.

At the instant 8 min (according to the reenergization procedure, as described above) the breaker C was closed and the 400 kV bus was energized.

At the instant 15 min the thermal unit was synchronized and breaker B switched on (fig. 12); in the sequence shown in fig. 12 the loads U_1 of 17 MW (fig. 10 and 12), U_2 of 30 MW (fig. 12) and U_3 of 36 MW (fig. 11 and 12) were restored by switching on the corresponding breakers.

The fig. 10 (restoration of load U_1) and 11 (restoration of load U_3) show the behaviour of the three generating units; the water pressure transient at the base of the penstock is also shown. The permanent bias of the hydro units was set at 2%. On the power output of hydro units (first 3 s about) the negative penstock effect and the positive effect of initial deceleration of rotating masses are superimposed, as well as an electromechanical decaying oscillation.

From fig. 12 it is apparent that between the instants 16 and 27 min about the thermal unit has picked-up some 67 MW, at an average speed of 6 MW/min.

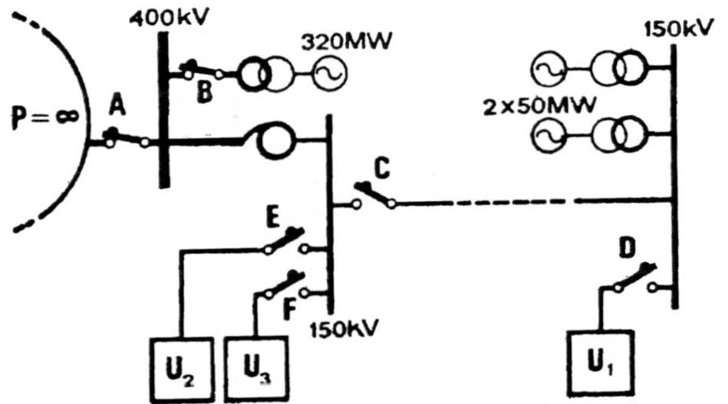

Fig.9 - Test scheme of a PRL with hydro and thermal units

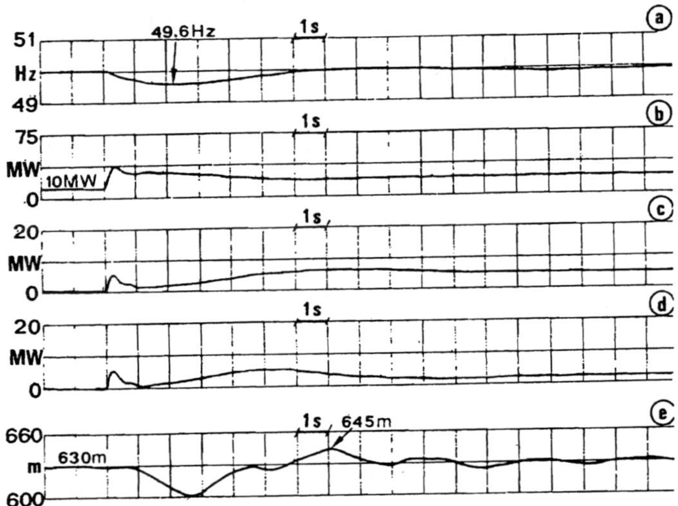

Fig.10 - Restoration of U_1 (fig. 9)
a- frequency
b- active output of thermal unit
c- active output of hydro unit 1
d- active output of hydro unit 2
e- pressure at the base of penstock

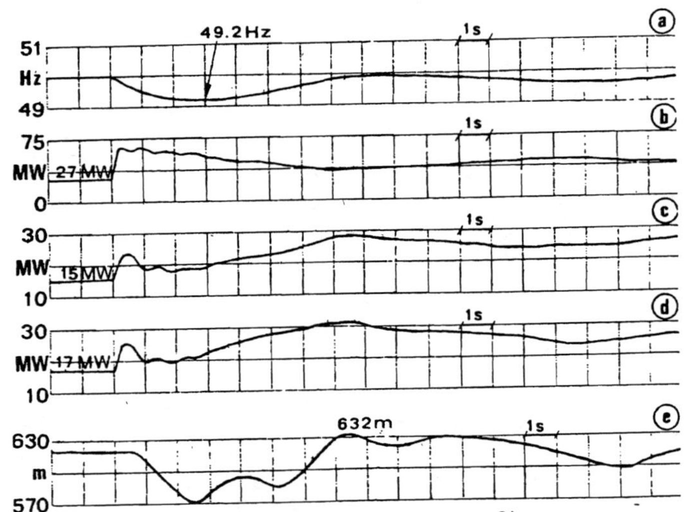

Fig.11 - Restoration of U_3 (fig. 9)
a- frequency
b- active output of thermal unit
c- active output of hydro unit 1
d- active output of hydro unit 2
e- pressure at the base of penstock

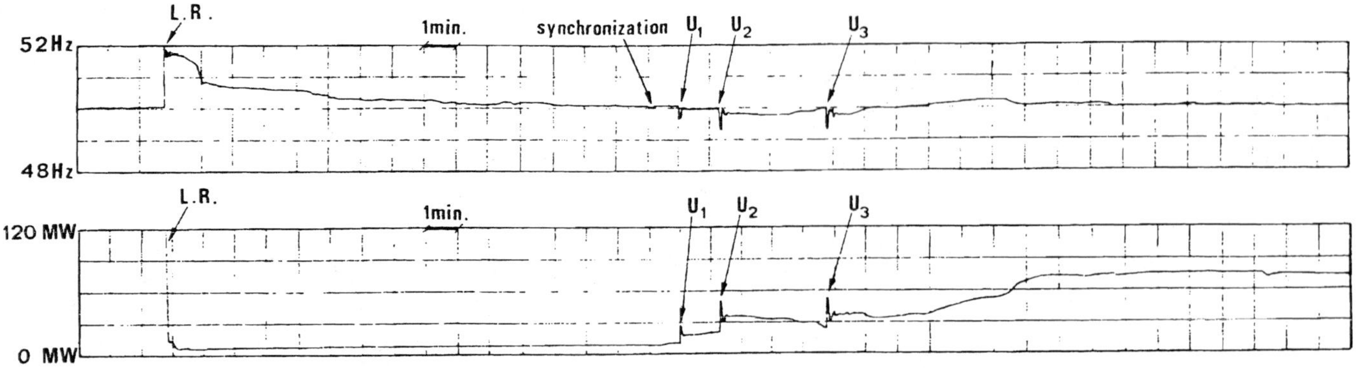

Fig. 12 - Joint load pick-up of hydro and thermal units: frequency and power output of thermal unit

Outcomes of one practical case

In August 1982 the network shown in fig. 1 experienced a complete black-out, after the tripping of the 400 kV circuit connecting it to the northern part of national grid.

The main PRLs involved in restoration were Capriati-Napoli and Mucone-Rossano-Brindisi.

The recovery time has been of about 30 min (complete load restoration 40 min); the resynchronization of the last two thermal units (one at Rossano and the other at Brindisi) took place 2 hours about after the initial event, due to internal troubles.

The overall performance may be considered satisfactory, even if this occurrence has demonstrated the opportunity to support the human operators with appropriate automatic apparatus (e.g. to implement the switching phase described in point a of reenergization procedure paragraph).

A careful maintenance and sufficiently frequent inspections and tests of auxiliaries apparatus of power plants and substations are of course mandatory for a reliable application of the restoration procedure.

C O N C L U S I O N S

After a network black-out a fast and reliable return to normal operation can be achieved if generator units, of hydro and thermal type, had a successfull trip to houseload, and the procedures of restoration are autonomous, that is they are implemented without being necessarily coordinated, at least in a first phase, by the network dispatcher.

To his end, after a preset time interval has elapsed from voltage disappearance, the elements of the network are disconnected through the switching off of practically all the breakers; an autonomous building-up of the network itself then begins, through the reconstitution of preset power restoration lines (PRL), composed of hydro and thermal power plants, HV circuits and some loads, and having the main purpose to insure the power supply of the auxiliaries of thermal plants.

Finally the various PRLs are synchronized to one another and with the main grid which remained alive, and load restoration is completed, as well as the re-start of generating units which had come to complete stop, if any.

A 400 - 220 - 150 kV network with a diameter of roughly 500 km, three main thermal power stations and four main hydro power stations, and a peak load of about 3,000 MW was completely restored in a practical case within 30 - 40 min.

Either hydro and thermal units (drum type and once through type boilers) can perform satisfactorily, if governor parameters, regulation schemes and steam flow circuits are set appropriately. The average load pick-up speed of a 320 MW thermal unit can reach 10 MW/min.

The correct intervention of the staffs wich are in charge of the operation of power plants and of substations may be made easier with the support of automatic apparatus, sensitive to voltage appearance and disappearance, for the switching operations.

R E F E R E N C E S

/1/ G. Fusco, D. Venturini, F. Mazzoldi, A. Possenti - ''Thermal units contribution to the electric power system restoration after a black-out'' - CIGRE Conference Sep 1982, paper 32-21

/2/ W.P. Gorzegno, P.V. Guido - ''Load rejection capability for large steam generators'' - PAS, March 1983

/3/ R.J. Frowd, R. Podmore, M. Waldron - ''Synthesis of dynamic load models for stability studies'' - PAS, January 1982

/4/ E. Mariani, F. Mastroianni, V. Romano - ''Transient behaviour of load in HV and MV networks restoration'' - L'Energia Elettrica, August 1982 (in italian).

SYSTEM RESTORATION AT PHILADELPHIA ELECTRIC COMPANY

David Scheurer, SM
Philadelphia Electric Co.

In 1979, Philadelphia Electric Company developed an updated System Restoration Procedure via a computerized interactive loadflow simulation. The objective of the simulation was to uncover weaknesses in the existing procedure and test its ability to restore the system following a total system shutdown. It was assumed that no sources of cranking power would be available from the outside world and that no internal generation remained on-line.

Prior to the simulation, the effects of a prolonged system shutdown were investigated, in order to determine the possible degradation to system facilities and support equipment at various time intervals following the loss of power. Analysis revealed concerns in the following areas:

1. **Supervisory Control/Data Acquisition**
   Backup power supplies for remote terminals (batteries) would become inoperable within 1-2 hours, resulting in loss of data acquisition and supervisory control capability.

2. **Circuit Breakers**
   During extreme cold weather conditions, loss of heaters in air or gas systems could render breakers inoperative within 1-4 hours due to gas liquification and loss of pressure. Upon the occurrence of this condition, these breakers may remain inoperable for several hours following power restoration.

3. **Telecommunications**
   Ringing current where supplied from A.C. dependent sources would be lost immediately. Dial systems may become overloaded due to large volume of traffic resulting in difficulty reaching key personnel.

Presented at
Workshop on Real-Time Monitoring and Control of Power Systems
McGill University
Montreal, Quebec
October 10-12, 1984

4. **Oil Static Cable System**
   Loss of oil pumps would result in loss of pressure and possible cable failure on energizing.

5. **Substation Manning**
   The ability to man substations in a timely manner may be severely diminished due to telecommunications and traffic patterns.

In order to address these concerns a new restoration strategy was required. Since complete insurance against failure of remote terminals and circuit breakers requires total system light and power restoration within one hour, the existing sequential restoration method was replaced with one consisting of a set of subsystems or islands. All subsystems are restored concurrently utilizing available blackstart capacity to energize transmission and restore light and power. A test procedure was developed and implemented for all dedicated telephone facilities and those found to be A.C. dependent were or will be modified.

In addition, an automatic notification system with speech synthesis was installed. This device, when initiated, sequentially calls key personnel, logs the call status, and delivers the appropriate message.

All oil-static cable nitrogen pumps and crossover valves are manned at the outset of a shutdown in order to insure that proper operating pressures are maintained. Correct pressures adjusted for elevation are listed in the procedure.

Finally, substation manning procedures have been developed to insure that available manpower is dispatched in the most timely and efficient manner possible.

After defining a course of action which considered the above concerns, the restoration simulation proceeded. Restoration of the various islands utilizes units with blackstart capability (CT and hydro) to energize transmission and supply cranking power to non blackstart units within each subsystem. The Philadelphia Electric's blackstart capability is as follows:

Reprinted from *8th IEEE Biennial Workshop on Real-Time Monitoring and Control of Power Systems,* October 1984.

BLACKSTART CAPABILITY

TYPE	UNITS	INSTALLED CAPACITY	
Diesel	6	18	MW
C.T.	42	863	
Hydro	19	1390	
Steam*	11	1730	
Total	78	4003	MW

* Started via C.T.

In order to maintain a suitable voltage profile during the restoration process, generators and synchronous condensers are operated with a leading power factor whenever possible. As voltage increases during transmission restoration, small blocks of critical load and station light and power are restored. Synchronization between subsystems takes place at generating stations where control of generation is available. After the first hour following the shutdown, approximately 1000-1200 MW of load is restored for station light and power and voltage control. The expected available generating capacity at the end of the first hour is shown below:

UNIT RECOVERY - 1ST HOUR

	Estimated	Maximum Capacity	
Hydro	512	1392	MW
CT/Diesel	615	948	
Steam	0	0	
Nuclear	0	0	
Total	1127	2340	MW

Assuming that all units recover as estimated, sufficient capacity is available for first hour load restoration. Although sufficient hydro capacity exists for first hour restoration, the Muddy Run pumped storage plant (880 MW) must be manned prior to initiating a blackstart. This capacity would therefore not be available until the second hour following shutdown. Also, due to their geographic location, both hydro plants are within the same subsystem and therefore a sequential restoration procedure would be required to fully utilize these units for first hour load restoration.

Blackstart tests of fossil and large C.T. plants from combustion turbines have been conducted where system conditions permit and have been successful. Testing of the islanding concept and subsystem synchronization however, presents a difficult problem. Difficulty exists in isolating designated area while maintaining system reliability, and any operating errors would be extremely costly. Much work is still needed in test development, validation and scheduling to insure that all equipment functions as expected.

Since the general restoration philosophy was developed in 1979, relatively few changes to the system have taken place and the present procedure (updated for the 1984 system) is felt to be adequate. The Philadelphia Electirc system of the 1990's however poses new uncertainties. The installation of Limerick Nuclear Station will result in the following equipment retirements by 1990.

RETIREMENTS BY 1990

		Summer Net Capacity	
C.T.			
	Richmond (West)	160	
	Richmond (North)	240	
Total C.T.		400	MW
Steam			
	Richmond 9	166	
	Southwark 1	168	
	Southwark 2	168	
	Delaware 7	126	
	Delaware 8	124	
	Schuykill 3	48	
Total Steam		800	MW

Other future retirements as yet unidentified may further reduce first hour and overall available capacity. The retirements shown reduce the first hour max capacity by 400 MW and the total maximum capacity by 1200 MW.

UNIT RECOVERY - 1ST HOUR

	1980		1990		
	Est.	Max.	Est.	Max.	
Hydro	512	1392	512	1392	
CT/Diesel	615	948	355	548	
Steam	0	0	0	0	
Nuclear	0	0	0	0	
Total	1127	2340	867	1940	MW

As shown above, the expected available capacity falls short of the required amount for first hour light and power restoration resulting in the possibility of equipment failures under adverse conditions. Additional concerns arise from these first retirements since the steam units represent all city generation and will result in subsystems having no fossil units. Without these units, several islands will have limited restoration capacity and little var control.

The estimated available capacity over the first 6 hours following shutdown is shown below.

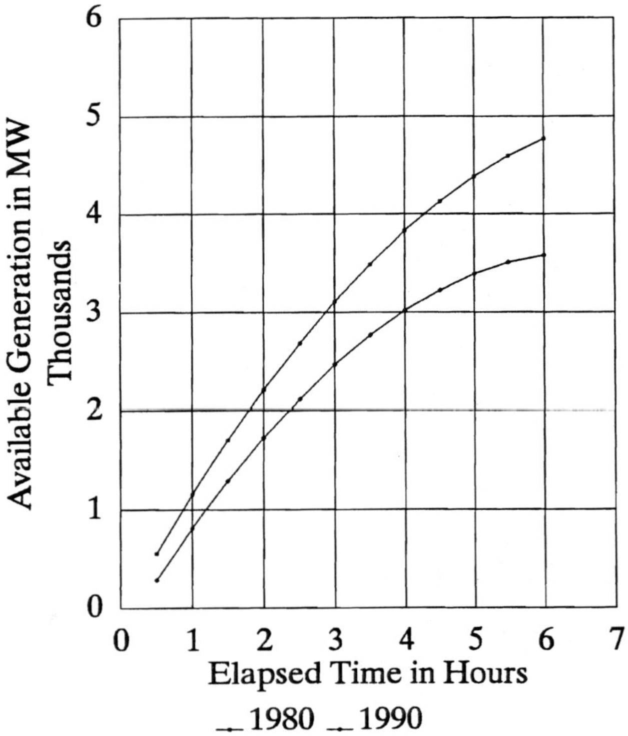

Estimating Available Generation Capacity

Conclusion

A new study and simulation of the Philadelphia Electric system restoration plan is presently being conducted to determine the required philosophy changes in light of the above retirements.

In conclusion, much work is still required to develop a suitable means of testing various phases of restoration procedures on a periodic basis without sacrificing system reliability. The validity and practicality of the procedure must also be evaluated as large economical generating units replace the smaller existing units, since these large units require longer startup times and their employment will result in the quicker starting units being retired or, at best cold.

It is hoped that investigation in these areas will result in expedient and practical future restoration plans.

SERVICE RESTORATION FOLLOWING A MAJOR FAILURE ON THE HYDRO-QUÉBEC POWER SYSTEM

Gaston Morin, member, IEEE
Hydro-Québec
Montreal, Canada

Abstract - Service restoration following a major power failure can occasion voltage problems on weak EHV systems from the point of view of both steady-state and temporary operation. This is due to the use of long EHV transmission links between the production sites and the load centres, which can cause large amounts of reactive power to be generated, and low short-circuit level, leading to low-frequency resonance, further excited by the presence of transformers. This paper describes an approach to service restoration which should speed up the process of restoring service to customers while at the same time limiting the risks of damage to equipment.

INTRODUCTION

A major power failure on the Hydro-Québec power system on December 14, 1982 emphasized the urgent need for closer investigation of the utility's service-restoration methods.

Continuous expansion of the 735-kV network over the last few years has seen several substantial revisions to the initial methods but the December 14 incident revealed the importance of providing alternative techniques and, in particular, the need to take account of the effect of temporary overvoltages on weak systems.

To the author's knowledge, the subject of service restoration using the EHV transmission system has not been dealt with very thoroughly in the literature so far. This paper will therefore outline the main problems experienced with this somewhat original procedure, from the choice of base systems to steady-state and temporary voltage control. A new service-restoration philosophy which should minimize the risks of damage to the equipment will be described. It will also present a short review of the greatest difficulties encountered with initial base system and examine the effects of transformer switching on a specific power system. The use of closing resistors as a good mean to reduce inrush currents and then, harmonic voltages and the effect of load pick-up as a good measure to reduce harmonic impedances and overvoltages will be pointed out.

Finally, planned future developments of these studies will be described.

86 WM 183-8 A paper recommended and approved by the IEEE Transmission and Distribution Committee of the IEEE Power Engineering Society for presentation at the IEEE/PES 1986 Winter Meeting, New York, New York, February 2 - 7, 1986. Manuscript submitted July 24, 1985; made available for printing December 16, 1985.

735-kV Service Restoration

Hydro-Québec's 735-kV transmission system comprises some 10 500 km (6 500 miles) of lines and 27 substations (figure 1). There are two basic production sites, namely the LG2, LG3 and LG4 power stations generating around 10 300 MW in the northwest of the province and the Manic-Outardes (5 400 MW) and Churchill Falls (adding some 5 000 MW) complexes to the east.

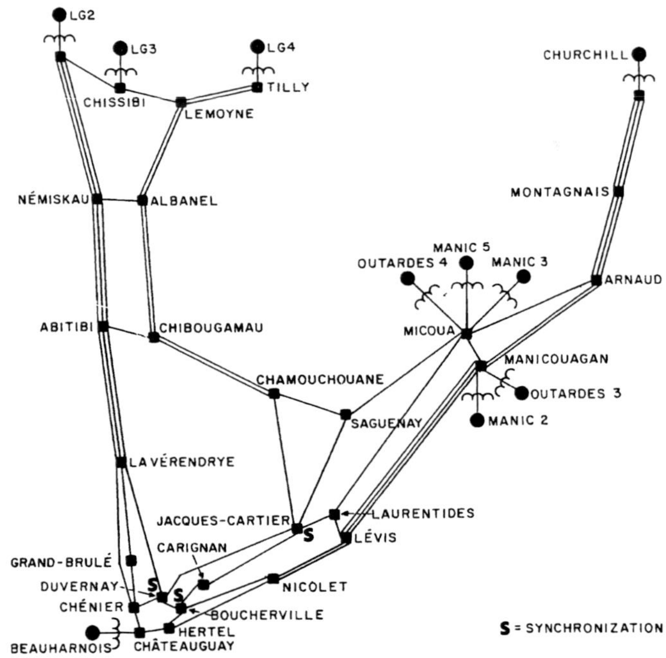

Figure 1: Hydro-Québec's 735-kV system.

To restore service to the Montreal and Québec City consumption centres after a power failure has occurred, Hydro-Québec's plan for rapidly strengthening the 735-kV skeleton and reducing the service-restoration time is to simultaneously re-energize five base systems (all 735 kV), namely:

1. Manicouagan
2. Micoua
3. Maisonneuve
4. James Bay, west corridor
5. James Bay, east corridor

So far, however, only three of these networks (1, 3 and 4) have been used to supply the load centres but recourse to the other two is planned in the near future.

The restoration of service on a transmission system depends on a number of criteria and practices related to the integration of the power produced, the load supply and the power-system equipment used. The

Reprinted from *IEEE Transactions on Power Delivery,* Vol. 2, No. 2, pp. 454–462, April 1987.

primary aim is to re-establish the production-load link as quickly as possible without sacrificing equipment by exposing it to the risk to damage. The procedures originally used for this purpose at Hydro-Québec were based on experience at lower voltage levels and with the 735-kV system under steady-state operating conditions but, as production sites gradually found more remote locations and the load increased, long 735-kV lines and high-capacity transformers made their appearance, creating new requirements.

The major problems with both the early load-restoration procedures and the more recent ones are related to the Ferranti effect and to temporary overvoltages caused by transformer switching. The most important problem is parallel resonance [2], which occurs whenever the system short-circuit level drops or if the lines are particularly long and capacitive. The low-frequency (120, 180,... Hz) resonances that appear can be excited when transformers saturate during a switching operation [4, 5, 6, 7] or when an excessive power-frequency (60-Hz) voltage occurs.

Ferroresonance creates an additional problem when several unloaded transformers are used on weak capacitive base systems such as the two James Bay corridors. Harmonic resonance can also occur under such conditions.

To reduce the risks of damage during service restoration, it is essential to control the switching on sequence of the equipment in every substation and generating station by avoiding the use of unloaded transformers, for example, or by adding closing resistors to the circuit breakers used to switch the transformers, etc. Such precautions do cause a certain delay in the restoration process but they are nevertheless fully justified from the long-term economic point of view.

There is no doubt that service restoration based on the 735-kV system involves a high degree of risk unless the application of appropriate corrective measures.

SERVICE RESTORATION PHILOSOPHY

A base system (figure 2) comprises three parts with very distinct characteristics:

- The source system, consisting of AC generators and step-up transformers;
- the transmission system, consisting of a source-load link;
- the load, including step-down transformers.

The source-system contribution comprises:

- voltage control;
- active- and reactive-power control.

Voltage control, in manual or automatic mode, is performed at the machine terminals so that it is essential to know the characteristics of the AC generator and the step-up transformer to be able to determine the limits imposed by this source system.

The transmission system usually comprises two types of line, namely noncompensated or compensated by means of static or dynamic components.

The load comprises different transformation levels and various scattered loads, thus offering a tool for controlling the voltage of the base system. The main parameters to be taken into consideration are:

- amount of load required for voltage control purposes;
- amount of load required for reducing the system impedance at resonance;
- minimum number of transformers to be energized;
- capacity (MVA) of the transformers to be switched;
- pre-switching voltage conditions;
- respect for supply of priority loads wherever possible.

Voltage control under steady-state and transient conditions

Since the base systems used for service restoration are constantly changing, it is difficult to assess the risks inherent in switching operations with a high degree of certainty. A study of steady-state, transient and temporary operating conditions reveals the main points that should be taken into account.

- The steady-state voltage should remain within the range of 0.9 to 1.05 p.u. once the switching operation has been performed.
- Switching overvoltages should be limited to 1.8 p.u. (e.g. for line switchings).
- Temporary overvoltages should not be more than 1.5 p.u. (8 cycles) so as not to exceed the settings of the fastest overvoltage protections or the electrical characteristics of the equipment.

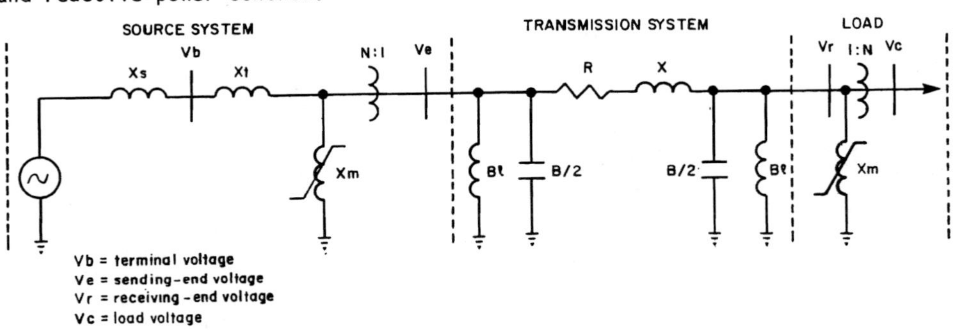

Figure 2: Simplified base system.

In order to keep the voltage within the specified ranges for the different operating conditions, the following procedure should be used:

- Set the voltage regulator of the machines to automatic mode.

- Check that the reactive-power absorption capacity of the generating units is not exceeded.

- Compensate the line at the receiving end before proceeding with energization if possible.

- Raise the source short-circuit level in order to limit voltage and frequency fluctuations.

- In the case of transformer switching, ensure there is a minimum load on the system so as to limit the impedance at resonance, or switch in a loaded transformer, or switch with a suitable closing resistor.

- Keep energized transformers loaded with a 10% nominal rated load to avoid ferroresonance.

In addition to voltage-control problems during load restoration on initial systems, account should also be taken of practical constraints. For instance:

- Communications should be reduced to a minimum.

- Auxiliary services should be restored as soon as power comes back at each installation, taking into account the transformer-switching criteria.

- The base-system synchronization points should be located at attended substations or substations with remote synchronization controls located at the regional control centre;

- Switching operations calling for the operator to leave the control room must be eliminated.

- Emergency instructions should be carried out according to prevailing voltage and frequency conditions.

CASE STUDIES

Some additional precisions should be given here, about non-linearities, before entering more deeply in the subject. First, the voltage profiles obtained from a load flow program (as in figure 3) neglect the non-linearities. This is also true for impedance curves coming out of any harmonic analysis program like HARMO or EMTP [8]. However, the non-linear characteristic of power transformers has been represented in switching studies in a similar way as in reference [8].

Manicouagan base system

Service restoration on this system has already been described elsewhere [1,2] and, since the procedure has not changed basically, only the steady-state operating voltage profile will be presented here as a reminder. This base system is supplied from Manic 2 and Outardes 3 generating stations (see figure 3).

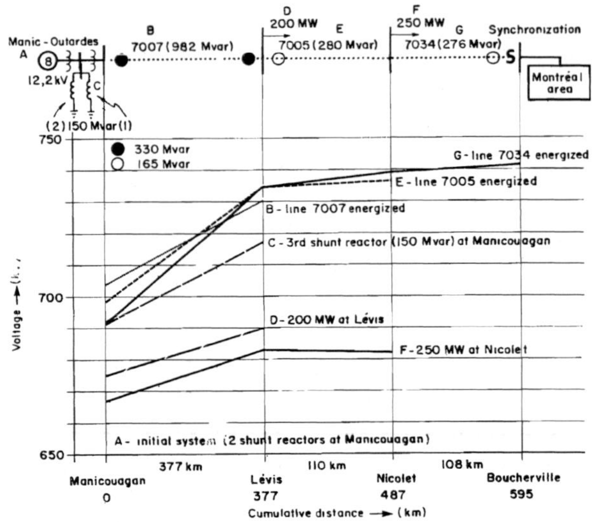

Figure 3: Voltage profiles for the Manicouagan base system.

Micoua base system

The Micoua base system is supplied from Manic 3, Manic 5 and Outardes 4 generating stations and is synchronized with the James Bay east corridor at Jacques-Cartier Substation. Figure 4 illustrates the voltage swings under steady-state conditions following various switching operations.

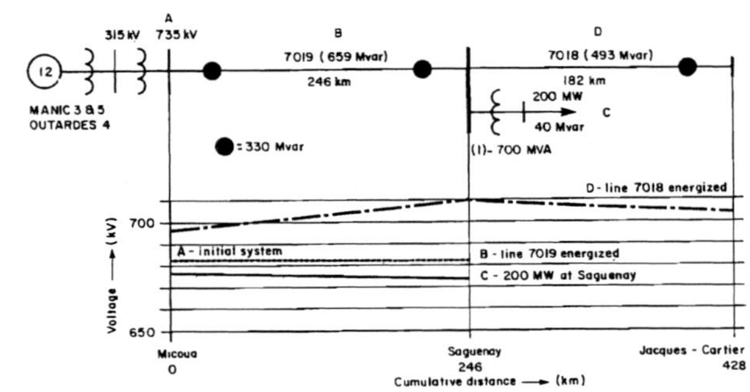

Figure 4: Voltage profiles for the Micoua base system.

Maisonneuve base system

The Maisonneuve base system consists of a 735-kV loop around the city of Montreal and is characterized by some 20 transformers rated between 600 and 1 650 MVA, most of which do not have dedicated circuit breakers, and by the generation of some 600 Mvar of reactive power.

Another feature of note is the winter peak load of over 9 000 MW, which has to be supplied from the 735-kV substations whereas the generation capacity is only 2 000 MW. In view of the fact that the other base systems cannot be synchronized with the Montreal loop for at least one hour after a failure has occurred, it is essential to restore as much of the load as possible from local generating stations.

The Maisonneuve system is critical, in fact, since the resonant frequency varies between 240 Hz and 120 Hz as a function of the source short-circuit level and the length of the 735-kV line energized. During the December 14 failure, overvoltages of up to 1.8

p.u. occurred, causing damage to surge arresters and a transformer.

Beauharnois generating station, from which energy is sent to the New York Power Authority via Châteauguay Substation, is repatriated within ten minutes after a failure starts. The voltage conditions at Châteauguay are then checked before connecting a load of up to 100 MW to the 120-kV bus and energizing the Châteauguay-Hertel line.

Each 735-kV substation comprises two busbars but only one is energized at a time in order to limit the number of transformers to be switched in because of the lack of transformer circuit breakers. Furthermore the lines to the south of Montreal (Hertel-Boucherville) are energized first, since the load is largely located in downtown Montreal and outskirts. These lines are not as long and therefore do not necessarily need compensation, at least at this stage.

Various steps in the restoration procedure will now be described, with reference to the system illustrated in figure 5.

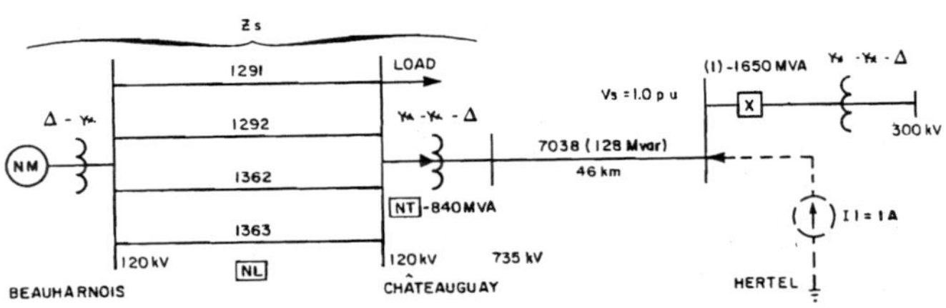

Figure 5: Beauharnois-Hertel system.

The main characteristics of the equipment are presented in appendix. When the above-mentioned overvoltages of 1.8 p.u. due to the switching of an unloaded 1 650-MVA transformer at Hertel Substation occurred, damaging a transformer and a surge arrester there, the load at Châteauguay Substation was zero. The breaker used for the operation (a receiving-end line breaker) was equipped with a 350-ohm/8 ms resistor. It is not sure how many AC generators were operating at the time.

The aim of the study described here was not to simulate the events that resulted in the failure of the equipment but to determine which load restoration method involves less risk of damage. The simulations were performed using the Electromagnetic Transient Program (EMTP) and IREQ's AC simulator. First the resonance conditions were evaluated for different system operating conditions by injecting a unitary positive-sequence current I1 on the Hertel and Châteauguay buses and by calculating the impedance of this system vs the frequency (figure 6) using a harmonic analysis program (HARMO) developed by IREQ.

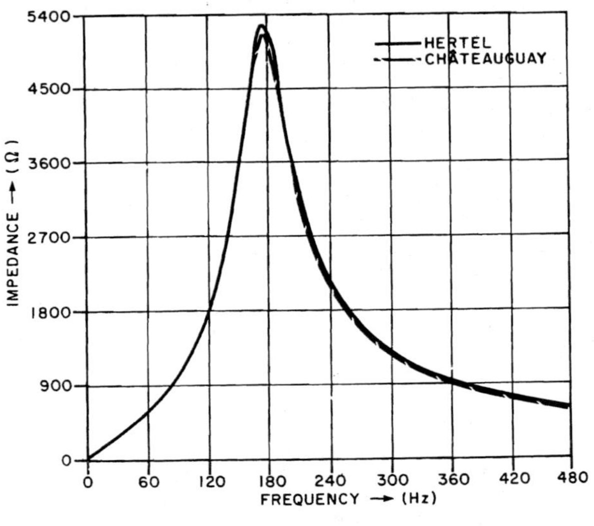

Figure 6: Z1(f) seen from Châteauguay and Hertel Substations.

To determine the effect of the source impedance Zs (i.e. influence of the number of lines, generators, and transformers comprising the source system) on the resonance conditions, a positive-sequence current I1 of 120 Hz, 180 Hz, 240 Hz and 300 Hz was injected for different values of Zs. The results are illustrated in figure 7 which shows Z1(f) seen at Hertel Substation for different source-admittance values (Ys = 1/Zs). The value Ys = 1.0 corresponds to two 120-kV lines (NL = 2) between Beauharnois and Châteauguay, 15 generators at Beauharnois (NM = 15), and one 125/765-kV transformer (NT = 1) at Châteauguay whereas Ys = 2.0 corresponds to NM = 30, NL = 4 and NT = 2. The most likely range of source admittance is 0.87 ≤ Ys ≤ 2.0, which means that the system can be tuned to 180 Hz or 240 Hz.

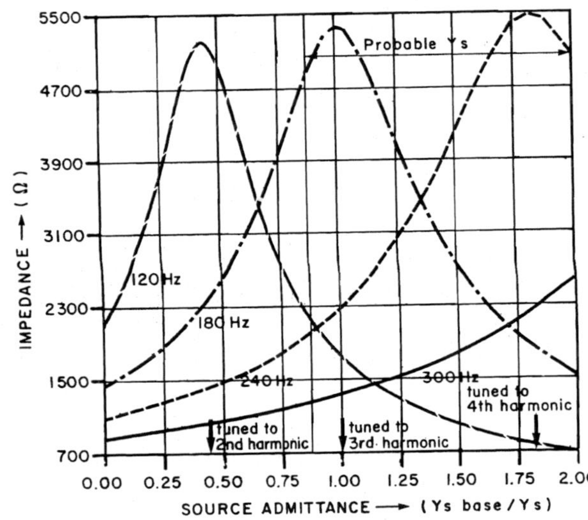

Figure 7: Z1(f) seen from Hertel Substation as a function of the source admittance.

As a result of the December 14 event, it was agreed to energize the loaded transformers in the Montreal loop and for this purpose, it was recommended that a load of 50 MW be connected on the secondary side of the transformer to be energized at Hertel Substation. The voltage at the switching bus was therefore reduced to 0.9 p.u. to minimize risk of damage. However, the simulations were done with a voltage of 1.0 p.u. on the switching bus in order to determine the overvoltages at a voltage equal to 1.05 p.u. of the transformer rated voltage.

Various alternatives were studied in the search for a more appropriate solution:

- switching the loaded transformers;
- picking up the load at Châteauguay substation;
- changing the resistance and the insertion time of the resistors installed on the circuit breakers involved in the transformer switching operation.

Simulations were then performed to assess the effect of the following parameters on overvoltages:

- value of the closing resistor;
- duration of resistor insertion time;
- amount of load on the system;
- source impedance.

Figure 8, which summarizes the effect of each of these parameters, shows that when a resistor of 350 ohm/8 ms is inserted in series with the transformer, the temporary overvoltages (12 cycles $\leq t \leq$ 30 cycles) are very dependent on the load at Châteauguay but far less so for closing resistors of 1 000 ohm/16 ms or more. The advantage of using higher resistance values is that they substantially reduce the temporary overvoltages but, on the other hand, on-load switching of the Hertel transformers, a measure which Hydro-Québec has used in its desire to limit the risks of damage, will therefore have to be abandoned because of the higher resistance values. There is no doubt that service restoration based on the 735-kV system involves a high degree of risk; for example, if just one series circuit breaker remains open, it can cause energization on no-load and deprive the system of the support expected.

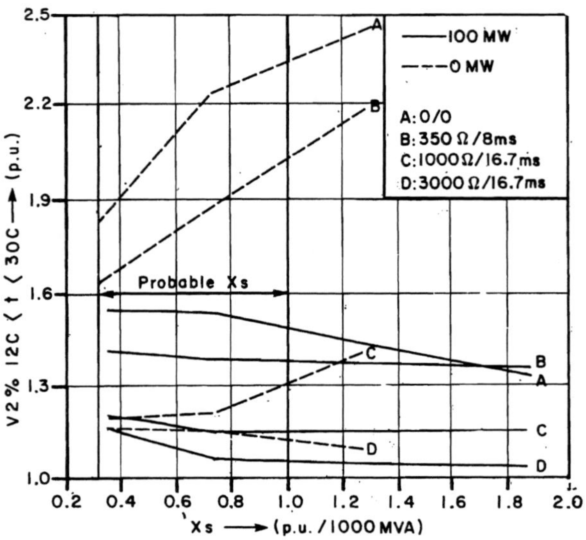

Figure 8: Influence of source impedance on the overvoltages as a function of the load at Châteauguay, the resistance value and the insertion time.

There are three solutions:

- to use a resistor of 1 000 ohm/16.7 ms or even a higher value;
- to pick up some of the load at Châteauguay before switching the transformer at Hertel; or
- to use a resistor combined with load.

The line circuit breaker used to energize the transformer at Hertel is equipped with a closing resistor of 350 ohm/8 ms but calculations show that the resistance could be increased to 2 000 ohm without exceeding 1.8 p.u. when this line is switched. As the line concerned is only 46 km (30 miles) long, a slight voltage drop occurs at the resistor terminals, resulting in a lower switching overvoltage. On the other hand, the guaranteed minimum insertion time does not exceed 14 ms, which is why a resistor of 2 000 ohm/14 ms was selected.

Extrapolating from the values illustrated in figure 8, the overvoltages obtained with this resistor should be between the results obtained with R = 3 000 ohm/16.7 ms (when the load = 0), namely approximately 1.4 p.u.

On the other hand, if R is kept at 350 ohm/8 ms and the load is raised from 0 to 100 MW at Châteauguay, a decrease in the temporary overvoltage from 2.0 p.u. to 1.4 p.u. is noted.

Thus the combination (load and resistor) solution manages to reduce the temporary overvoltages V2% to below 1.2 p.u. Figure 9 shows the effect of the load on the system impedance at resonance.

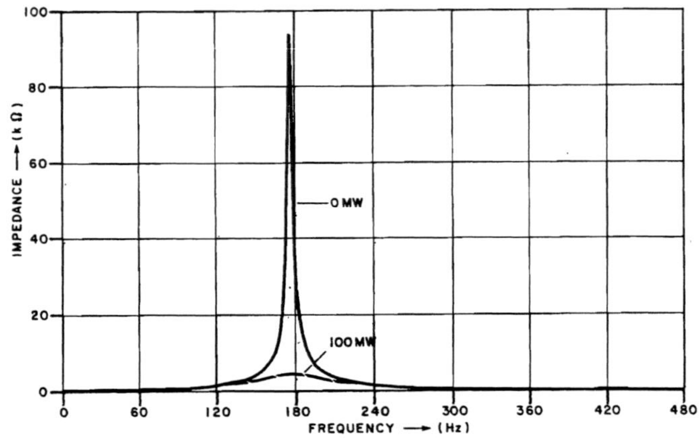

Figure 9: Influence of load on Z1(f).

Following the energization of a 1 650-MVA transformer at Hertel substation, an additional load of 150 MW is picked up at the same substation (total = 300 MW) before the Hertel - Boucherville line is energized, once the voltage conditions have been checked (figure 10). Three transformers with ratings varying between 600 and 1 100 MVA are then energized at Boucherville Substation.

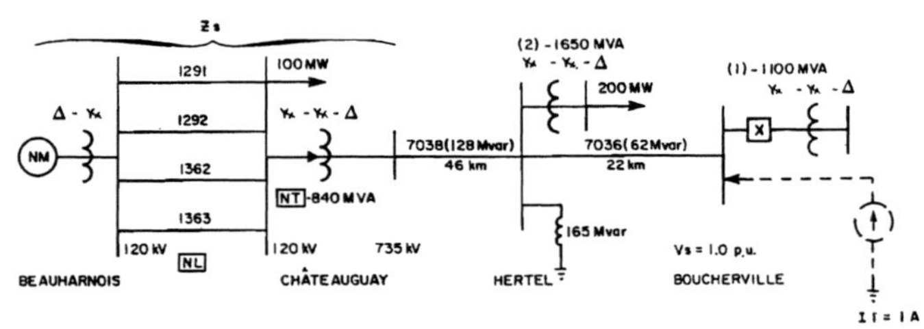

Figure 10: Beauharnois-Boucherville system.

The effect of Zs on the system impedance at different harmonic frequencies is given in figure 11, which provides a clear illustration of the strong reduction in maximum impedance compared with the previous example (100 MW, as seen in figure 7). The recorded overvoltages are below 1.2 p.u. in all cases.

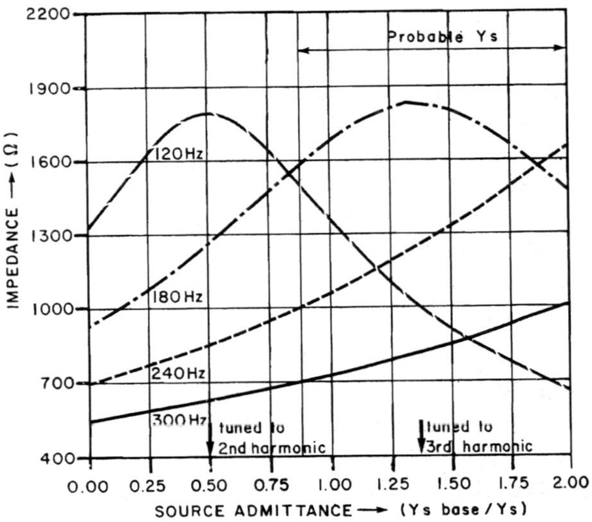

Figure 11: Z1(f) seen from Boucherville Substation as a function of the source admittance.

A minimum load of 130 MW is then picked up at Boucherville, the Boucherville-Duvernay line is energized and, then, an unloaded 1 650-MVA transformer (figure 12). A point of note in figure 13 is that the resonant frequency drops as the line length increases, although the source short-circuit level remains the same. Actually, tuning to 120 Hz and to 180 Hz is possible at this stage. The overvoltages remain lower than 1.2 p.u.

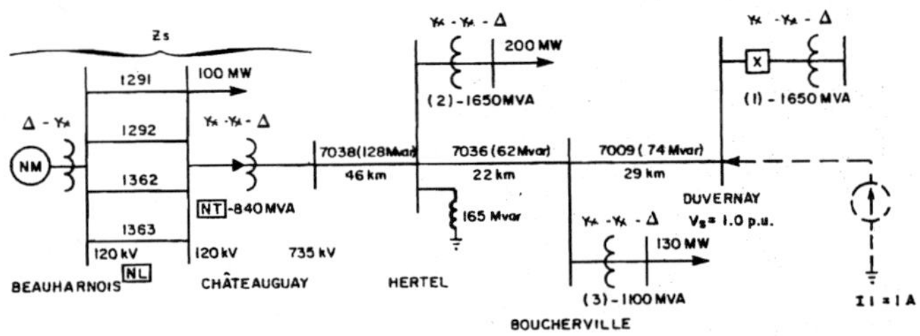

Figure 12: Beauharnois-Duvernay system with switching of an unloaded 1 650-MVA transformer at Duvernay.

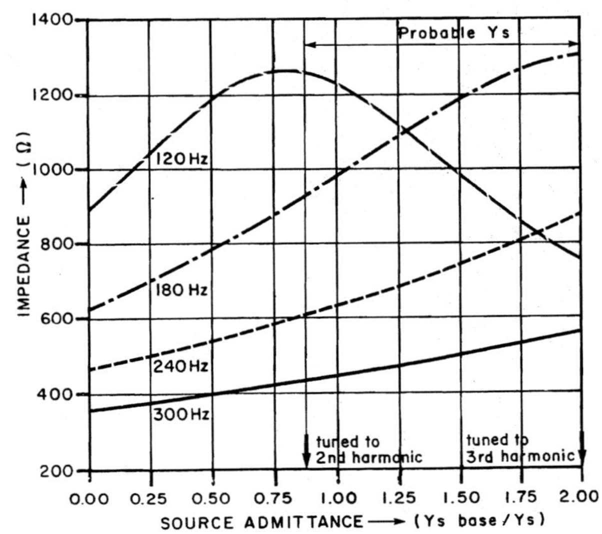

Figure 13: Z1(f) seen from Duvernay Substation as a function of the source admittance.

Subsequently, a 130-MW load is connected at Duvernay Substation and the Duvernay-Chênier line is energized followed by an unloaded 1 650-MVA transformer at Chênier (figure 14). The impedance curve in figure 15 shows 120-Hz tuning for Ys = 1.25. Again the overvoltages are below 1.2 p.u.

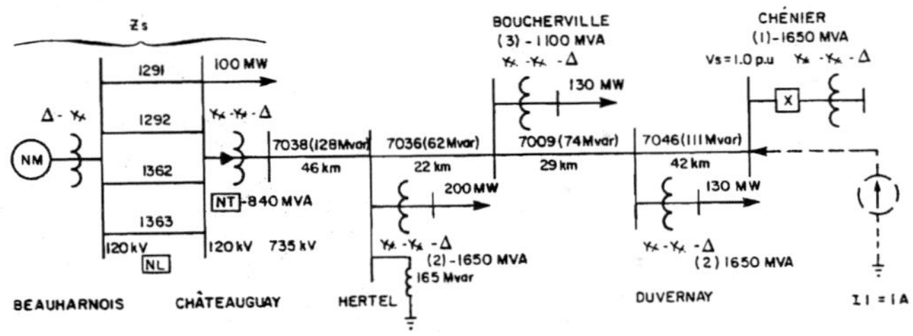

Figure 14: Beauharnois-Chênier system with switching of an unloaded 1 650-MVA transformer at Chênier.

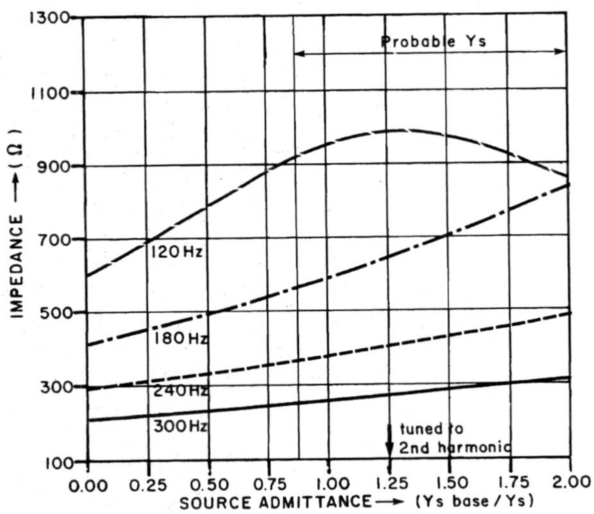

Figure 15: Z1(f) seen from Chênier Substation as a function of the source admittance.

Figure 16 shows the variation in the temporary overvoltages V2% in relation to the parameters of the insertion resistors and the loads on the different systems. The most critical switching operation seems to be the energization of a transformer at Hertel with no load on the system: the overvoltages recorded in this case are 2.25 p.u. when R = 350 ohm/8 ms and 1.48 p.u. with R = 1 000 ohm/16.7 ms.

The full lines in this figure represent the probable overvoltages when emergency measures are applied for different load conditions. The impact of the load on the Beauharnois-Hertel system is significant when R = 350 ohm/8 ms. When the load increases from 0 to 100 MW, the overvoltage drops from 2.25 p.u. to 1.45 p.u.

The broken lines represent variations in the emergency measures applied to the Beauharnois-Chênier base system. With a resistor of 350 ohm/8 ms, for example, ferroresonance can occur following the energization of a 1 650-MVA transformer at Chênier under low-load conditions (100 MW). This can be explained, at least in part, by the fact that this system is better tuned at 120 Hz so that it is essential to reduce its impedance to 120 Hz by picking up load. On the other hand, a 2 000 ohm/12 ms resistor has a similar effect on the overvoltages by reducing the saturation of the transformer being switched.

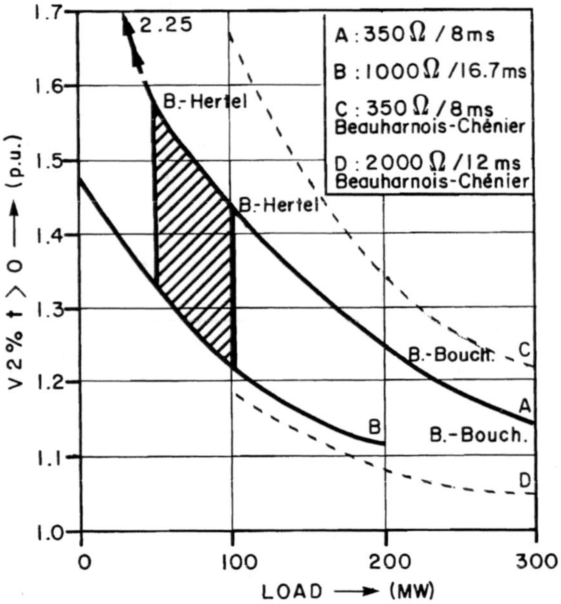

Figure 16: Effect of load on overvoltages for different resistance values and insertion times.

The conclusion drawn from figure 16 is that a load pick-up of 50 MW at Châteauguay Substation and the use of a 2 000 ohm/12 ms closing resistor reduce the overvoltages to below 1.35 p.u. following the switching of transformers in the Montreal loop.

Voltage conditions permitting, the Chénier-Châteauguay line can be energized to close the 735-kV Montreal loop but, since this line produces some 240 Mvar of reactive power, it will only be switched under heavy load conditions or following synchronization of this base system with another.

James Bay base system (west corridor)

The first LG2-Duvernay line is about 1 100 km (675 miles) long and generates some 2 800 Mvar of reactive power. At the present time it is 88% compensated with 165- and 330-Mvar shunt reactors. The rated power of each machine is 370 MVA and it takes a minimum of three such units at 12.5 kV to respect the maximum-voltage criterion of 1.05 p.u. Furthermore, the maximum reactive power absorbed is roughly 100 Mvar/machine wheras each machine can absorb almost double that with a voltage of 0.9 p.u. across its terminals.

The step-up transformers at LG2 are energized by gradually increasing the voltage, which means that the exciter breaker of the AC generator is open and the unit therefore starts up with only one transformer connected. When the rotation speed reaches 90% of the synchronous speed, the exciter breaker closes and the excitation voltage is gradually raised resulting in virtual elimination of the inrush current. This allows the transformer to be energized smoothly and avoids transformer switching with radial lines. Synchronization of the AC generators is possible at the high- and low-voltage levels; in fact, synchronization on the high-voltage side will eventually be remote-controlled.

Figure 17 shows the steady-state voltage swings recorded after different line-switching operations. An analysis of the temporary overvoltages that follow transformer-switching operations on this system will be published at a later date.

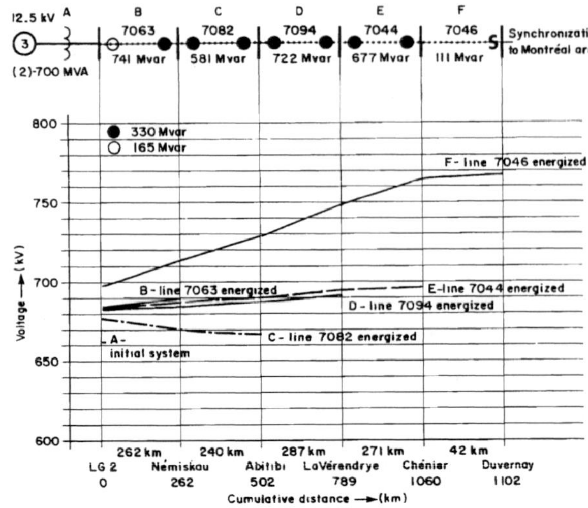

Figure 17: Voltage profiles for the James Bay base system (west corridor).

James Bay base system (east corridor)

Following the commissioning of the LG3 generation site in 1983 and the new line corridor linking it to Montreal, a study was undertaken to see how to use these new facilities for load pick-up. This line corridor has the drawback of having only 63% compensation between LG3 and Jacques-Cartier substation. The reactive-power generated by this first link is 2 600 Mvar and is compensated using shunt reactors for a total of 1 650 Mvar.

In order to make use of this corridor for service restoration, it was decided to install a 330-Mvar shunt reactor at Chibougamau Substation to boost the compensation to 75% to allow synchronization with the Micoua base system at Jacques-Cartier Substation. Figure 18 shows the voltage swings at different stages in the service-restoration procedure used on this system.

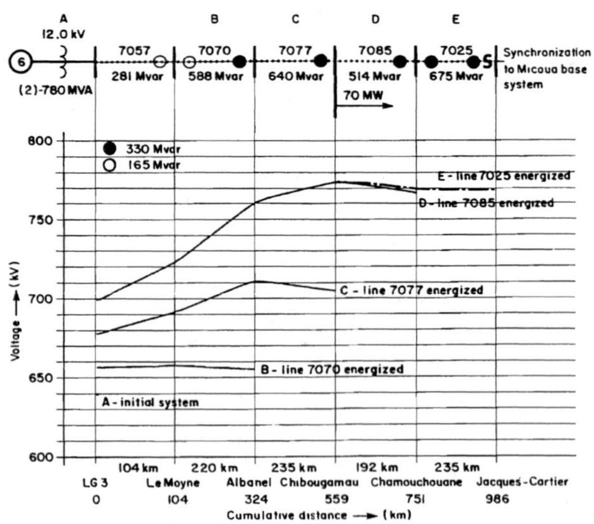

Figure 18: Voltage profiles for the James Bay base system (east corridor).

Future Development

An emergency-measures centre was set up at Hydro-Québec in 1983 to supply the technical assistance the system control centre staff may require during major incident or a provincial power failure. The atmosphere that prevails when emergency measures are required is usually tense because of the time constraints and mistakes can easily be made, causing even further delays in restoring service to the customers.

The centre is used to keep track of service restoration as it proceeds and to obtain information from the regions to enable the procedure to go on.

The last few years have witnessed a positive effort by Hydro-Québec to implement regional control centres throughout the province to collect the required data and send it to the control centre in Montreal which uses it for system management purposes. Thus all the information needed for automatic service restoration is constantly available and, furthermore, updated at regular intervals.

To provide the technical tools needed by the staff at the emergency-measures centre, it has been recommended that a real-time interactive load-flow program be implemented in order to verify the pre-switching voltage conditions. Furthermore, since transformer switching is critical on weak systems, Hydro-Québec is presently weighing up the possibility of implementing a system impedance vs frequency calculation program to determine resonance conditions for any combination of elements, in real time.

A number of criteria must be established for implementing a service-restoration procedure. These are:

- voltage swings;
- frequency variations;
- active- and reactive-power limits of generators;
- resonance conditions on the various systems.

Technicial analyses of these criteria were followed by the formulation of solutions designed to cover many system contingencies. It is therefore possible to preprogram the base-system restoration procedures and their alternatives using pre-established criteria.

Studies are now underway to determine the impact of automatic load pick-up on the time taken to restore service to customers throughout the province and, also on the decrease of damage to equipment. The economic viability of the proposed solution will then be evaluated.

CONCLUSION

The study described in this paper allowed a number of problems likely to occur in restoring service on 735-kV lines to be identified and solutions formulated. Closing resistors with a suitable resistance value and insertion time can be installed on the circuit breakers used to energize the transformers since they can reduce post-switching temporary overvoltages fairly considerably. The presence of load on base systems allows the system impedance to be reduced at resonance and, at the same time, the overvoltages and the risk of ferroresonance. The load/closing resistor combination appears to be the best solution. A new philosophy has been described which takes account of the main phenomena involved and should thus substantially curtail the risks of damage to equipment.

Problems that occur during load restoration make it essential to have efficient real-time tools, such as load-flow and impedance vs frequency calculation programs, tools which herald the possibility of eventually achieving entirely automatic load restoration.

APPENDIX

Beauharnois generator

X''_d = 0.30 p.u. 13.8 kV/50 MVA

Beauharnois step-up transformers

X_{HL} = 0.09 p.u. 13.2/127.4 kV/50 MVA

120-kV lines (60 Hz)

1291 and 1292: Z0 = (0.268 + 3.407 mH)/km
Z1 = (0.058 + 1.194 mH)/km

1362 and 1363: Z0 = (0.221 + 2.866 mH)/km
Z1 = (0.028 + 0.942 mH)/km

735-kV lines (60 Hz)

Z0 = (0.2618 + 3.005 mH)/km B0 = 3.354 s/km
Z1 = (0.0106 + 0.870 mH)/km B1 = 5.004 s/km

Châteauguay transformer

X_{HL} = 0.184 p.u. 765/125/12.5 kV/840 MVA
Xair core = 0.40 p.u. Knee voltage = 1.15 p.u.

Hertel and Duvernay transformers

X_{HL} = 0.123 p.u. 700/300/11.9 kV/1 650 MVA
Xac = 0.40 p.u. Knee voltage: 1.15 p.u.

Boucherville transformers

X_{HL} = 0.123 p.u. 700/300/11.9 kV/1 100 MVA
Xac = 0.40 p.u. Knee voltage = 1.15 p.u.

Chênier transformers

X_{HL} = 0.20 p.u. 700/300/11.9 kV/1 650 MVA
Xac = 0.40 p.u. Knee voltage = 1.15 p.u.

EHV Transformer overvoltage capability

700-kV and 765-kV transformers should sustain a 1.5 p.u. minimum overvoltage (peak value) for at least one second. This is considered by Hydro-Québec as conservative values.

Silicon Carbide arrester characteristics

Conventional arresters used on the 735-kV system have the following characteristics.

Nominal rating : 612-kV (R.M.S.)
Spark-over voltage range: 1.6-2.0 p.u. (peak value).
Maximum absorbed energy : 5kJ/kV

Closing resistors

Closing resistors used on circuit breakers are designed to sustain the short-circuit current capability of the breakers for two closing-opening sequences.

ACKNOWLEDGMENTS

The author thanks Marcel Harvey of Hydro-Québec, for his valuable collaboration in preparing this paper, and Jean Bélanger and Gilbert Sybille of IREQ for their contributions to the simulation work.

REFERENCES

[1] G. Morin, M. Perrachon, "Remise en service du réseau Hydro-Québec suite à une panne majeure". IEEE Conference, Montreal, October 1982.

[2] G. Morin, X.D. Do, "Reduction of temporary overvoltages in EHV weak systems". 26th IEEE Midwest Symposium on Circuits and Systems, Puebla, Mexico, August 1983.

[3] G. Morin, "Surtensions temporaires lors de de la remise en charge du poste Manicouagan", M.Sc.A. thesis, École Polytechnique de Montréal, June 1983.

[4] G. Sybille, M.M. Gavrilovic, J. Bélanger, V. Q.DO, "Transformer saturation effects on EHV system overvoltages", IEEE/PES 1984 Summer Meeting, July 15-20, 1984.

[5] D. Povh, W. Schultz, "Analysis of Overvoltages Caused by Transformer Magnetizing Inrush Current", IEEE PAS-97, July/august 1978, pp. 1355-1362.

[6] C.V. Thio, J.R. McNichol, W.M. McDermed, D. Povh, W. Schultz, "Switching Overvoltages on the Nelson River HVDC System - Studies, Experience and Field Tests", Proceedings of the IEEE Conference on Overvoltages and Compensation on Integrated AC-DC Systems, Winnipeg, Canada, July 1980, pp. 15-26.

[7] N.H. Huynh, "Surtensions de manoeuvre causées par la mise sous tension des transformateurs sur le réseau de la Baie James, - Hydro-Québec Report, October 1981.

[8] Electro Magnetic Transient Program, Rule Book, Bonneville Power Administration, Portland, March 1983.

Discussion

Adam Semlyen (University of Toronto, Toronto, ON, Canada): I would like to congratulate the author for his excellent paper. I found of special interest the information related to problems of harmonic resonance which may arise during the restoration process. The source of harmonics appears to be the nonlinear transformer magnetizing branch. However, synchronous machines may also produce harmonics under unbalanced conditions, since they act as harmonic converters between currents of different rotational sequences. For instance, a negative sequence 60-Hz loading will produce a positive sequence 180-Hz harmonic.

The author's opinion on the significance of the synchronous machine as a source of harmonics would be much appreciated. For the analysis, the transmission lines must be represented as untransposed or discretely transposed. The negative sequence component of the 60-Hz charging current is then the source of a third harmonic voltage produced by the generator and the latter may be strongly magnified by Ferranti effect or outright resonance. Using modal decomposition, one may find at some frequency that only one of the line modes, for instance the $\alpha(1, -2, 1)$ or the $\beta(1, 0, -1)$ mode, is in resonance so that the charging current has a strong negative sequence component which then will produce another higher harmonic current and voltage, and so on.

Does the author have data about the harmonic content of the voltages and currents and about their symmetry?

Manuscript received February 25, 1986.

B. R. Shperling (New York Power Authority, NY): The author analyzed service restoration procedures on a number of 765-kV lines which start at Chateauguay 765-kV substation, connected by four short 120-kV lines to Beauharnois hydro generating station. The study was partially initiated by a power failure in December 1982 during which a power transformer and a number of surge arresters failed.

In connection with these two tasks, the author's clarification of the following points will be greatly appreciated:

1) What was the sequence of switching operations before the equipment failures at Hertel substation? What was the voltage on the power transformer before its failure and why could the surge arrester not protect it?
2) Why was the 765-kV shunt reactor at Hertel substation not energized before switching in an unloaded 765-kV power transformer (Fig. 5)? This reactor should practically compensate the Chateauguay-Hertel 765-kV line and noticeably reduce switching and especially temporary overvoltages during the energization of Hertel transformer with no load on the system.
3) Temporary and switching overvoltages on the Chateauguay-Hertel-Chenier 765-kV transmission during restoration procedures depend not only on 765-kV lines but on short 120-kV lines as well. These lines between Chateauguay and Beauharnois are equivalent to quite long 765-kV lines, depending on a number of 120-kV lines in parallel. Therefore, upgrading of the 120-kV lines with adjacent substations to a higher nominal voltage might be an effective way for reducing system overvoltages. Were technical and economical aspects of this approach considered by the author?
4) What are the characteristics of gapless zinc oxide arresters on the Hydro-Quebec system and where are they installed in the southern part of the network?
5) Supplementing the paper with oscillograms illustrating strong influence of the closing resistors on the temporary overvoltages, as was mentioned in the conclusion, will be of great interest.

Manuscript received February 27, 1986.

Gaston Morin: The author thanks Mr. B. R. Shperling for his discussion. The discusser asked about the conditions prevailing before equipment failure. From the system shown in Fig. 5, only two 120-kV lines were in service over a maximum of four with an unknown number of generators. From this system, line 7038 was energized and then the 1650 MVA transformer. For an unknown reason the operator did not verify the voltage before switching in the transformer at around 800 kV, which is higher than the 1.05 pu limit. We had very high overvoltages lasting for many seconds, exceeding the maximum energy absorption capability of the SiC arresters.

Concerning the energization of the 765-kV shunt reactor at Hertel substation: the effect of a reactor is first to reduce Ferranti effect of the line and at the same time to lower the steady-state voltage, and secondly to shift the resonant frequency to a higher value without changing the losses too much, implying a tuning effect (for initial resonant frequency lower than a given harmonics) or a detuning effect (for initial resonant frequency equal to or higher than a given harmonics) with a comparable impedance value. In my opinion there is no general statement related to the effect of the reactor on temporary overvoltages connected to harmonic resonance. This shunt reactor has been recommended following that study to increase flexibility of the overall service restoration process and in some cases to decrease temporary overvoltages.

The effect of the 120-kV lines is to change the short-circuit power level at the switching bus. It can be seen from Fig. 7 by example that the source admittance variation can tune the system to a specific harmonic. I agree with the discusser by saying that these 120-kV lines are acting like 765-kV lines by looking only at the series inductance, because any long transmission line has the effect of reducing the short-circuit level at the receiving end. But the effect would be quite different by looking at the line charging and its effect on the resonance conditions. We did not see any reason to upgrade that 120-kV system because we came out with a relatively costless solution which is to pick up load at Chateauguay 120-kV and change some closing resistors, which was at least solving our problem adequately.

To the fourth question, there were no zinc oxide arresters on duty at that time. The effect of the closing resistors can be seen in the next figures. For a given network configuration, as shown in Fig. 5 of the paper, with a 100-MW load at Chateauguay 120-kV, we can see this effect from Fig. 1 (350 Ω/8 ms) and Fig. 2 (1000 Ω/16.7 ms). The closing resistor provides an initial damping of the inrush currents and at the same time of the harmonic voltages. This is not, however, a general statement because the effect of closing resistors on high impedance resonance is not that good. That is the reason why we recommended to pick up load also.

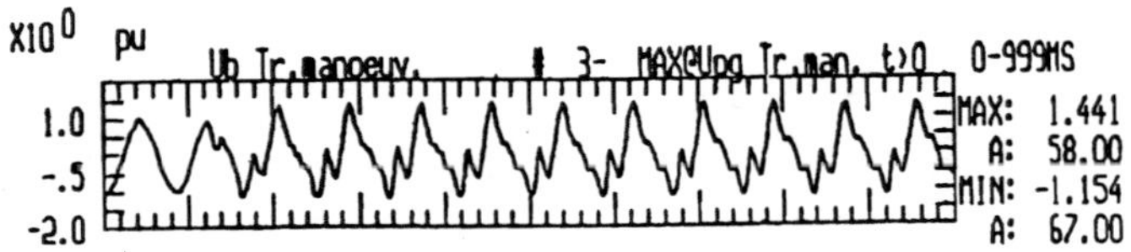

Fig. 1. Voltage on phase B at the switching bus (Hertel) with R = 350 Ω/8 ms.

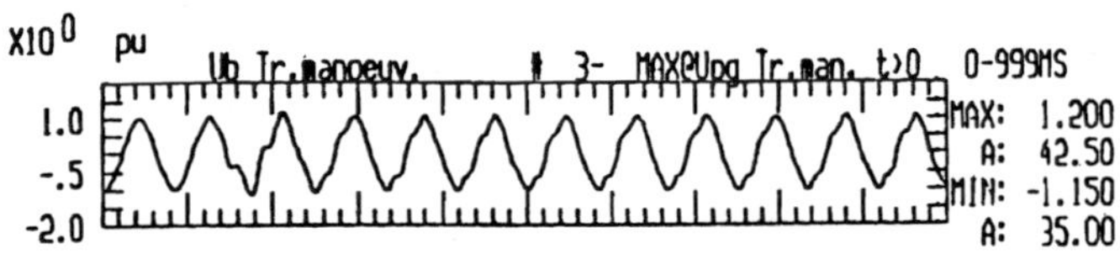

Fig. 2. Voltage on phase B at the switching bus (Hertel) with R = 1000 Ω/16.7 ms.

The comments of Dr. A. Semlyen about the unbalance effect of the line charging current is absolutely true on a theoretical basis. However, this effect has been neglected in our simulations and we did not represent the synchronous machine in detail. Use of this simplified model was based on the assumption that such effect is negligible for realistic transmission line lengths. In some other simulations, by using a detailed synchronous machine modeling on a system tuned near 180 Hz, we observed a very negligible effect for a 46-km untransposed 735-kV transmission line. In another case we observed for a 200-km 735-kV untransposed transmission line a ± 10 percent second harmonic component in the field current of the machine around the mean value, which is very significant. So, I think that it is possible to obtain a third harmonic amplified voltage on the system, with long transmission lines, tuned to that harmonic.

Manuscript received March 28, 1986.

POLICIES FOR RESTORATION OF A POWER SYSTEM

J. Gutiérrez, Member M. Staropolsky A. García

COMPAÑIA DE LUZ Y FUERZA DEL CENTRO, S. A.

(MEXICO, D. F.)

Abstract - Presentation of policies for effective restoration of a power system. Three plans implement these ideas, covering conditioning, preparedness and restoration. The problem is tackled with a triple approach, from the viewpoint of installations, personnel and the restoration plan.

SCOPE

The purpose of this paper is to establish general guidelines for action in effectively restoring an electric power system in an extreme state of collapse. Interest is centered on the whole problem rather than on detailed manoeuvers, since the latter vary for each power system, depending on the configuration, design, degree of reliability of the elements, operating practices, etc.

Explicitly, the final goal is restoration of all services in the shortest possible time. To achieve this, partial goals should be met consistent with the assembly of the bulk network starting with small zones having the capability for autonomous start-up.

STATING THE PROBLEM

Simultaneous action of three interactive systems working in coordination achieves the electric energy supply.
These systems are:

a) Personnel
b) Information and control system
c) Power system

In general, a collapse in the electric energy supply is due to the breakdown of the power system and it is understood that restoration will be achieved by personnel action on the power system through the information and control system.

From the power system standpoint, the problem is in the domain of Fink-Carlsen's ideas; it can be placed in any of the states shown in Figure 1 (Ref.1)

86 SM 327-1 A paper recommended and approved by the IEEE Power System Engineering Committee of the IEEE Power Engineering Society for presentation at the IEEE/PES 1986 Summer Meeting, Mexico City, Mexico, July 20 - 25, 1986. Manuscript submitted December 11, 1985; made available for printing June 27.

Printed in the U.S.A.

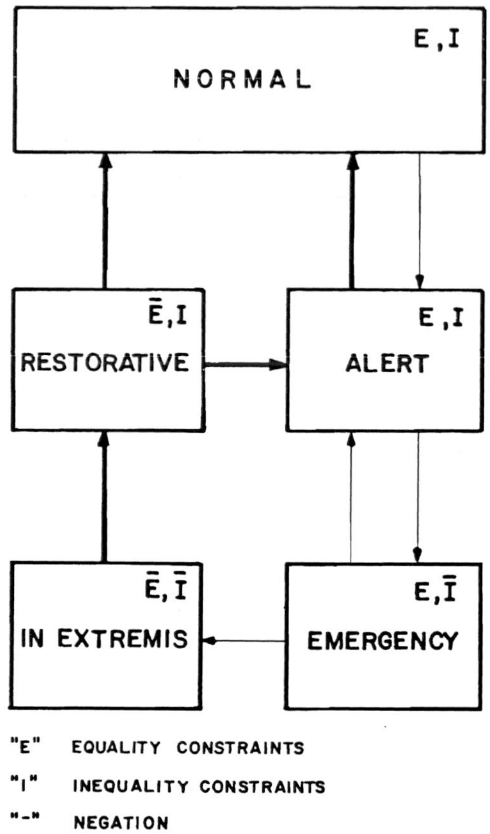

Fig. 1 System Operating States

This diagram shows exactly how, if a power system in a normal state operating in a secure mode loses its margins of generation and transmission reserve and for any reason the probabilities of a disturbance in any of its components increase, it goes into a state of alert.

In the state of alert the system operates insecurely, although the equality and inequality constraints continue to be satisfied.

From that moment on any disturbance moves the system from an alert state to an emergency state, as the inequality constraints are violated when it operates at a lower frequency with abnormal voltages or with some portion of the equipment overloaded. Obviously, secure conditions no longer prevail, although the system continues to be an assembled whole.

On rare occasions, the system degenerates, going into a new state - "in extremis" - where both inequality and equality constraints are violated;

Reprinted from *IEEE Transactions on Power Systems*, Vol. 2, No. 2, pp. 436–442, May 1987.

that is, the load is no longer satisfied since the system becomes disarticulated, as the generators have been tripped or the load shed, either from the low frequency protective effect or from the effect of tripping transmission circuits, thus directly affecting consumers, and eventually constituting total collapse. The purpose of this paper is to outline the policies or rules for action to bring the system out of this extreme situation, placing it in a restoration state where inequality constraints will be satisfied. In essence this signifies assembling of the network to operate as an articulated system with subsequent transfer to an alert or normal state, restoring a full load to the system.

In Figure 1 a dark line is used to show the change of state from "in extremis" to normal.

PRINCIPLES

Our general restoration philosophy is based on three basic operating principles: decentralized restoration, strength derived from unity, and preparedness, which is achieved through practices and tests. Each of these will be explained briefly.

Decay and paralysis of the system mean that it can no longer exist as a whole. Its elements must be rescued, to serve as the starting point from which decentralized restoration can start again. The system's elements can only be reestablished by dividing, making the network grow gradually with resultant formation of a large system that can bear the full load.

Strength of unity is the second principle applied, which attains its true dimension starting from full knowledge of the plan to be followed, which can only be achieved through a thorough, timely, continuous information program, accompanied by detailed training for station and control area operators.

Preparedness, being constantly alert, is the third ingredient required . Analytical simulation and actual, systematic testing of portions of the plan, done periodically to guarantee constant readiness, bring it closer to a condition of complete reliability.

It is our understanding that confidence in the plan as established can only be gained through practices.

Three plans - for restoration, preparedness and conditioning, - will be formulated in keeping with the principles established. Application of these plans on the time scale is shown in the next figure.

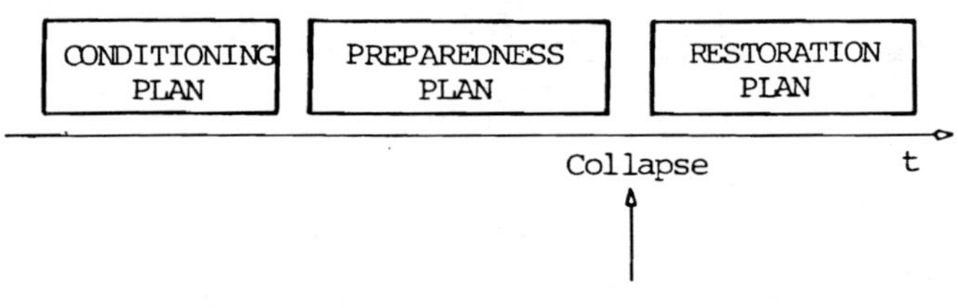

Fig 2 Sequence of the Conditioning, Preparedness and Restoration Plans.

The restoration plan is shown first because both the conditioning and preparedness plans are a function of it and take their meaning from it.

RESTORATION PLAN

As previously stated, the first concrete idea for restoration is that a system in a state of collapse as a whole is a nonviable entity since it cannot satisfy the inequality and equality constraints. Consequently, it must be fragmented in order to start again on the basis of its elements.

Various states, listed below for subsequent analysis, are established in this general idea.

a) Isolation of areas and zones
b) Black start-up and energizing zones
c) Feeding thermoelectric plant auxiliaries and zone synchronization
d) Reintegration of control areas into the bulk network.

Black start-up refers to one which occurs in a generating plant without any external support from the network.

a) Isolation of areas and zones

In collapse, since it cannot satisfy its constraints, a system must start automatically or manually at predetermined points so that the start-up attempts will not have to cope with an immense network which cannot be raised at a single point.

At that moment, both the fully assembled network and any intervention in it by various control centers are hindrances. Partitioning the network must be done at two levels: the control area and by zones.

Each zone will consist of the network portion satisfiable by a source with black start-up facilities.

It is desirable to accomplish zone isolation as soon as possible after the state of collapse occurs. However, this goal involves the risk that partitioning might occur at some point without prior verification of the general state of collapse. In any case, the two risks must be balanced out: a delay in partitioning against having some quick, undesirable openings.

In the Fink-Carlsen diagram this activity would take us from the "in extremis" state to "restoration", as shown in Figure 1.

b) Black Start-Up and Zone Energizing

The main objective at this stage is to get power at some point on the network, energizing a portion of it so as to satisfy part of the load in order to stabilize the generating plant's operation based on the largest possible number of synchronized machines.

The goal is a reliable, self-sufficient electric network that satisfies its inequality constraints in a reliable, secure manner, even though the equality constraints are not met; that is, even though there is an unconnected load. The idea is to assemble a strong network starting from secure zones. It would be senseless to integrate a bulk network immediately without prior stabilization and maturation of each of the zones as much as possible.

The greater the number of generating units that are connected, the more easily the maximum high voltage transmission lines would be assimilated. Under these conditions, load pick-up is an imperative for acheiving zone maturation and connecting machines.

c) Feeding Thermoelectric Plant Auxiliaries and Zone Synchronization

A mature zone can be used to feed the necessary portion of the network that energizes the thermoelectric plant's auxiliaries. Synchronization of the thermoelectric plant's machines will allow the zones already formed to pick up more load, up to the limit of their available capacity. From that moment on, zones may be synchronized as their growth and strengthening progress, until each control area can be assembled as a whole

d) Reintegration of the Control Areas into the Bulk Network

Reintegration of the control area, made up of already synchronized zones, into the bulk network, is the last step in the restoration plan. From that moment on, as a system, it can be expected to satisfy all consumer demand; being heterogeneously distributed, the generating resources do not permit all the areas simultaneously to satisfy their needs partially.

This action takes the system from the RESTORATION state to an ALERT state, satisfying equality constraints (service to all consumers) and the inequality constraints (the system and its parts working within their normal operating voltage and current ratings).

The system must still be rescued from the ALERT into a NORMAL state, incorporating a larger generating capacity and the circuits needed to reestablish the reserve margins required to operate in this new state.

Under the outlined restoration plan, there is special interest in duty assignment in order that each operating entity know what its authority is, its limits of action and the degree of autonomous operation it is given.

Quick and effective accomplishment of the autonomous work from the time the system's collapse is recognized is understood, which will take the generating plants from total shutdown to the point of supply to the power control center on the high voltage bus bars from the generating plant.

Working in this field, the operator must fully recognize his responsibility. He can only understand this in the context of thorough knowledge of his tasks, a result of continuous, adequate, timely training.

Similarly, we understand the tasks of both the area centers and the National Center to be supervision of the lower levels, supporting them rather than acting in their stead.

Throughout the proposal we observe the need for adequate information and control resources during any blackout to make the mentioned supervision possible.

PREPAREDNESS PLAN

The purpose of this plan is to make every member of the team operating the power system (regardless of his level) alert and constantly vigilant, always prepared for any contingency.This alertness attitude does not supplant the preventive control tasks intended to prevent blackouts through the actions that cause secure operation of the system. It simply considers that since collapse is possible, preparation to resolve it when it occurs is necessary, in addition to everything done to avoid it.

We identify three levels of training and one of information within this context. The objective at the first level is to define by analytical simulation the best network to be energized and to train the personnel at the National Center and the area control centers by means of these exercises. In this phase, the points for partitioning the network are established at the area level and the zones are delineated within each area.

The assembly sequence for each zone and generating plant is established at the second level. At this point the area control center personnel should interact with those from each power plant, and as for the previous level, we believe digital simulation practice obligatory for this exercise.

The third layer of training is at the substation operators level, since they are responsible for the partitioning task, when this is not automatic, and for reassembling the network.

We consider wide distribution of the plan indispensable to avoid people who are ignorant of the general reassembly procedure working like automatons. Each operator will only work enthusiastically at the personal level by understanding that individual actions form part of a precise, harmonious general plan.

The training of reference must be verified. The system demands viable plans, not theoretical, compatible with the actual state of the equipment, the workers' skills, and their time scale efficiency. Actual tests of black start-up and no-load energizing of long lines, as well as systematic shnchronization practices are best for this purpose. Experience shows that continuous exercises are necessary to retain skills. Unless put into practice, operating instructions will result forgotten, confused and sometimes ignored. Thus our final recommendation, that a plan of this nature be instrumented systematically over time at all operating levels.

CONDITIONING PLAN

This plan consists of the activities that permit availability to a power system of the indispensable minimum elements for efficient restoration. We stress the difficulties in the precise establishment of a plan usable by all systems considering the wide variety of configurations between one system and another, the diversity of operating practices and the different characteristics for using the reserve.

Under these conditions, we are only presenting the general guidelines that should be reviewed making possible formulation of detailed plans for each system based on its specific needs.

The first point is to provide control centers with both the equipment and models required for periodic, systematic simulation exercises. Most systems have met this need based on recognition of formal practice from the analytical standpoint by the operators, rather than by the analytical engineers, who have traditionally been in charge of these matters. The practice of the airplane industry, where the use of simulators has been a daily practice for several decades, has been emulated.

The second area of interest should focus on uninterruptible power supply systems and on emergency equipment, including transfer systems for the computer center and operators' room. These centers should be designed with at least one redundancy of the uninterruptible equipment and an emergency plant as a back-up for it.

The third aspect considers conditioning the generating plants for autonomous start up. Some plants, such as hydraulic plants, are now equipped with all these facilities. We believe that in thermoelectric plants, particularly those with gas turbines, the necessary investments to permit black start-up are worthwhile.

The next point is substations. Duplicate services are commonly used in stations and in some of the more important substations to provide additional support from the emergency plant. However, because they are not tested under a strict schedule and running loaded, they eventually fail when most needed. Therefore, we recommend that they also be subjected to systematic tests of start-up, transference and loading.

Finally, from the network standpoint, as was seen before, not only must the starting points be established, but a check must be made at those places that the corresponding synchronization equipment is physically available , to be installed where there is none, or the network's opening points should be relocated.

The network modelling should somehow include these limitations so the operator can study actual, feasible alternatives, getting away from ideal speculations that cannot be implemented in practice.

The main activities for each operating level are summarized below:

DIRECTIVE GROUP:
- General coordination of the conditioning, preparedness and restoration plans
- Issuance of the restoration plan
- Dispatchers Training
- Supervision of restoration by areas
- Supervision of the bulk network assembly

CONTROL AREA:
- Distribution of the restoration plan
- Operator training
- Duty assignment in the control center, generating plants and substations
- Test scheduling
- Supervision of black start-ups
- Area and zone synchronization

GENERATING PLANTS AND SUBSTATIONS:
- Testing
- Actual black start-up (autonomous)
- Disconnection and synchronization

APPLICATION PROJECT

Presentation of the restoration problem, its scope and the plan for restoring the Central Control Area (CCA) of the Mexican Interconnected System, with general indication of the division of labor in the control center, the sequence for network assembling, and the means available to the CCA for the black start-up of its plants.

The Central Control Area is one of the eight control areas the Mexican Electric System is divided into for operational purposes. This division is seen in Figure 3.

Six of the eight areas are connected, integrating the National Interconnected System. The Central Area is one of the most important due to its relative electric size, the number of tie lines to neighbouring areas, the strength of its network, and its high load density.

Table 1 gives the main characteristics of the Central Area from the standpoint of installations and demand.

Table 1

Characteristics of the Central Control Area

Installed capacity	5211 MVA
No. of substations	122
No. of lines	239
voltages	400/230/85 KV
length	6328 km
Maximum demand (1984)	4025 MW (12978 MW National)
Load factor	60.3% annually 68.8% monthly

Figure 4 shows the Central Area network in more detail.

As a result of the total collapse suffered in the Western and Central Areas and a large part of the Eastern Area at 6:58 a.m. on January 15, 1981, intensive work has been done to solve the problems arising from the blackout.

Analysis of that event evidenced the power system's limitations, both as to structure and as to operational procedures. All these problems are being overcome on a sustained basis, and our country's system and control centers are now more able to cope with the eventuality of a total collapse.

In fact, we now have static Var compensators available that avoid the need for daily assembly and disassembly of the power network, breakers have been installed instead of blades in the line reactors that previously could only be operated with a deenergized line, a better balanced, more effective low-frequency protection scheme has been elaborated, based on solid-state relays replacing the old electromechanical ones, and a real time information and control system has been purchased which will soon allow us to make on-line evaluations of the system's security. There are periodic practices of black start-ups in the turbogas and hydraulic plants, and a complete plan for system restoration has been designed, making it possible to restore it in an integral, planned way.

Much work remains to be done, including on-line emergency control and strengthening of the control center with effective supervision systems and security analysis.

In this paper we describe what the restoration plan consists of.

DECENTRALIZED TASKS

The Central Area's Operations and Control Center is staffed by personnel working 24 hours a day, 365 days a year. To distribute the work, five black start-up zones have been defined. Each of the 24 hour a day positions have three major zones assigned, to be covered independently, and the two secondary zones are to be supervised at a second level of priority.

We shall now outline the tasks assigned to each of the positions.

Systems Engineer

The system engineer is the Central Control Area's operational authority in real time, being the shift supervisor responsible for all operational decisions related to the tasks of coordination with the first level of the hierarchical structure, which is the National Energy Control Center. The first task assigned to this position is identification with the National Center of a general blackout condition. Next, this person is responsible for supervising the start-up and synchronization of the Infiernillo and Villita Hydroelectric plants, to send power to Mexico City.

Assistant Systems Engineer

He is the assistant to the first position, working with him in supervising the bulk network. He is assigned the mission of supervising the network assembly, starting from the J. Luque Thermoelectric Plant and the Lechería Turbogas Plant.

The purpose of operating the J. Luque - Lechería pair is to provide power via Victoria Substation to Tula Thermoelectric Plant, to supply its auxiliaries.

Subtransmission Engineer

In charge of the Central Area's Subtransmission network . Assigned the mission of coordinating and supervising sending power from the Necaxa Hydroelectric Plant to the Valle de Mexico Thermoelectric Plant to supply its auxiliaries.

Assistant Subtransmission Engineer

Two other sites where black start-up is possible are the Nonoalco 85 KV Turbogas Plant and the Valle de Mexico 85 KV Plant.

The general idea of restoration consists of independent, simultaneous start-ups of all the available sources that are reliable and significant for a black start-up. Following a successful start-up, there can be gradual loading so as to synchronize

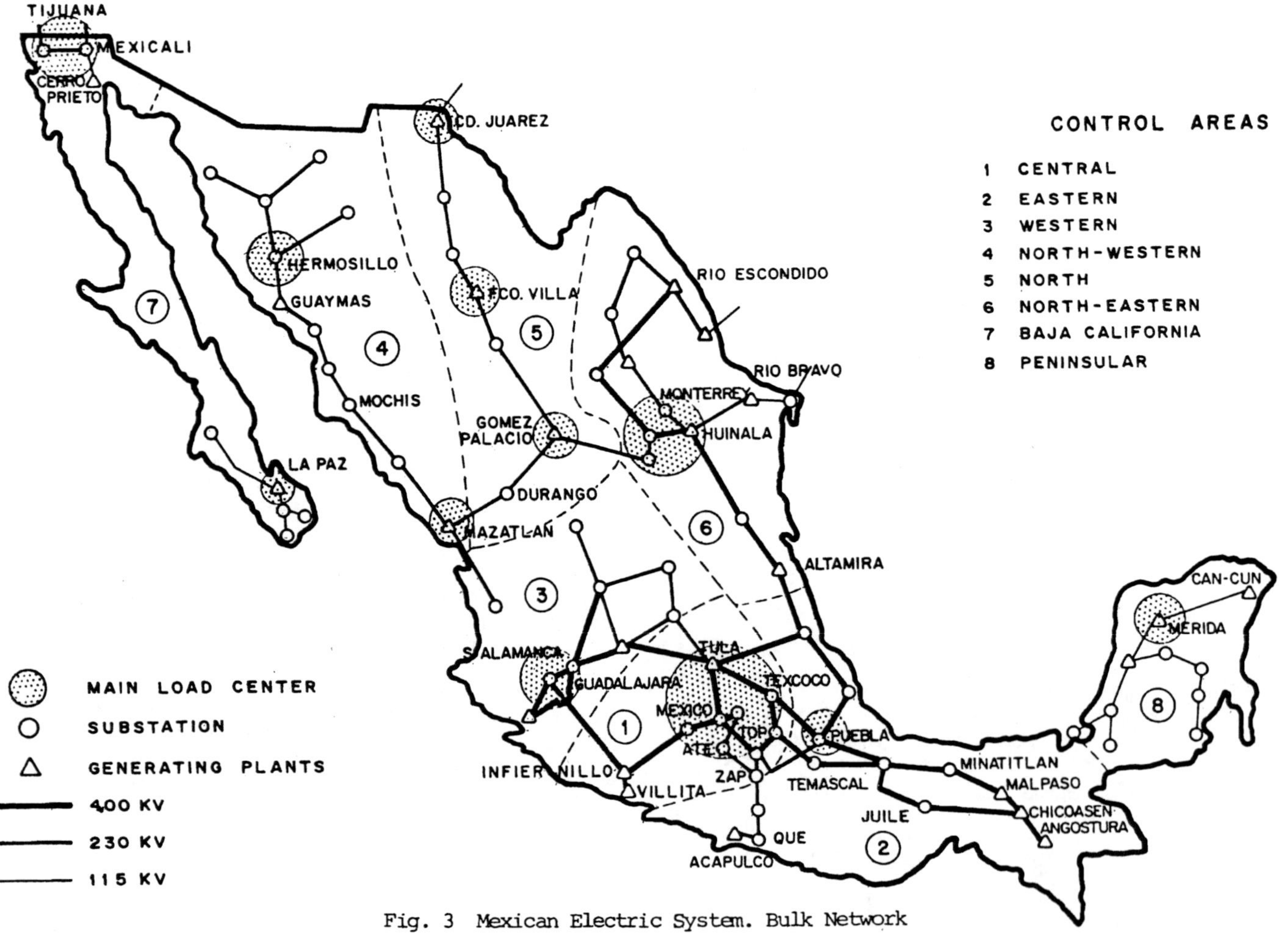

Fig. 3 Mexican Electric System. Bulk Network

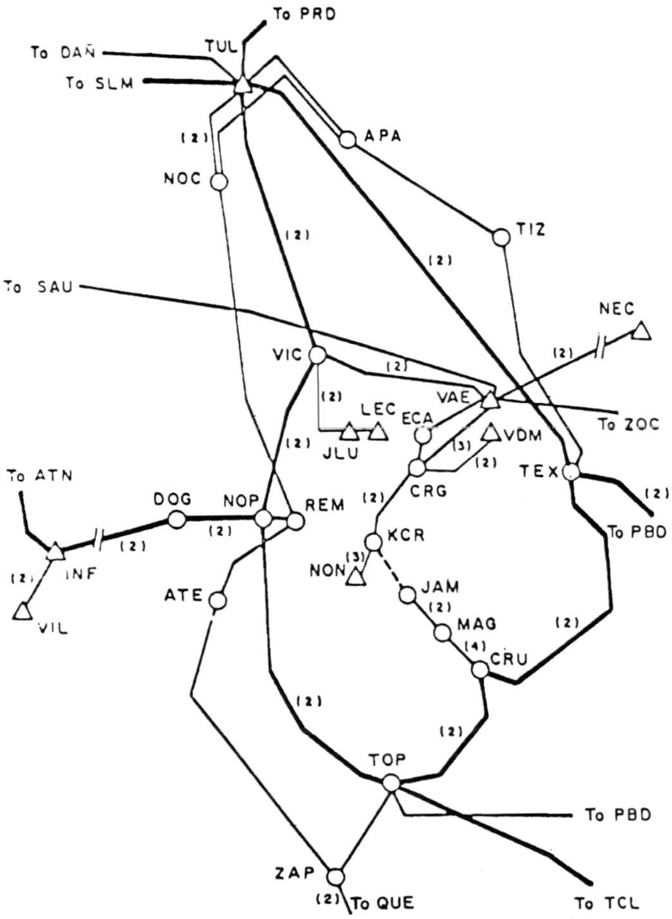

Fig. 4 Detail of Central Area Network

APA - APASCO
ATE - ATENCO
ATN - ACATLAN
CRG - CERRO GORDO
CRU - STA. CRUZ
DOG - DONATO GUERRA
DAÑ - DAÑU
ECA - ECATEPEC
INF - INFIERNILLO
JAM - JAMAICA
JLU - J. LUQUE
KCR - KM. CERO
LEC - LECHERIA
MAG - MAGDALENA
NEC - NECAXA
NOC - NOCHISTONGO
NON - NONOALCO
NOP - NOPALA
PBD - PUEBLA (DOS)
PIT - PITIRERA
PRD - POZA RICA (DOS)
QUE - QUEMADO
REM - REMEDIOS
SAU - EL SAUZ
SLM - SALAMANCA
TCL - TECALI
TEX - TEXCOCO
TIZ - TIZAYUCA
TOP - TOPILEJO
TUL - TULA
VAE - VALLE DE MEXICO 230 kV
VDM - VALLE DE MEXICO 85 kV
VIC - VICTORIA
VIL - VILLITA
ZAP - ZAPATA
ZOC - ZOCAC

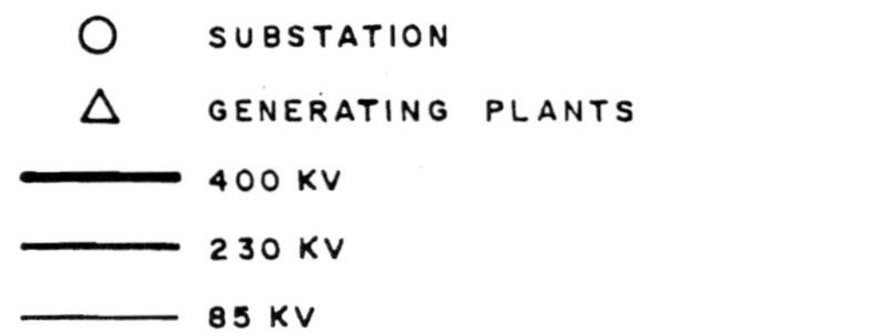

the largest possible number of units at the plant level in order to offer secure power to the large thermoelectric plants

From that moment on, synchronization of the groups that have started up independently can be considered, then proceeding to assemble the bulk network in coordination with the National Center.

Figure 5 shows the nearby zones of influence to the black start-up plants.

AUTONOMOUS TASKS

Each of the black start-up zones is supported by a power plant capable of performing a start-up without any external support from the network. In this case, precise local instructions are provided for performing a black start-up autonomously; that is, without a prior order from the control center in the case of a power failure at the substation high voltage bus bars.

In these cases, the scope of the autonomous tasks lies in the following concepts:

- Isolation of the power plants at the high voltage bus bars in the substations.
- Start-up of a generating unit
- Energizing the high voltage bus bars in the plants
- Communication with the control center to supply power to its bus bars.

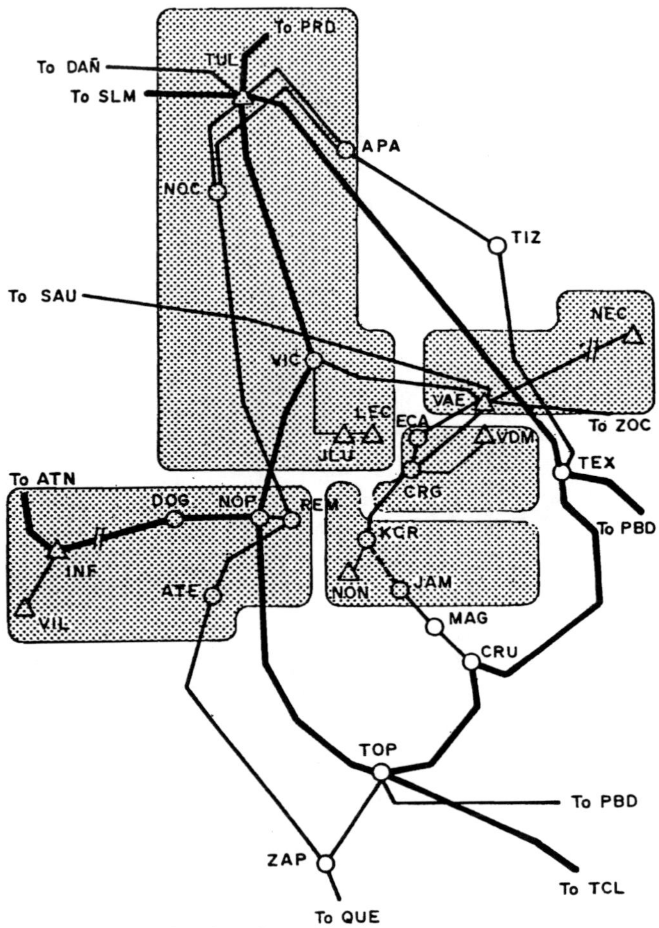

Fig. 5 The five black start-up zones

An annual black start-up program is available at the five points mentioned before to train the operators and ensure they will always be conscious of these procedures in the event of a collapse.

CONCLUSIONS

We have discussed the Central Control Area restoration plan in a general way, indicating the basis for network assembly and stressing the advantage of a large number of plants with black start-up capabilities.

We have stressed the operating practices followed routinely for autonomous start-up of plants. Nevertheless, we recognize the instrumentation and equipment limitation on providing remote supervision of plant start-up. With the implementation of the real time information and control system, these tasks can hopefully soon be handled from the control center.

REFERENCES

(1) L. Fink and K. Carlsen, ''Operating Under Stress and Strain'', IEEE Spectrum, March 1978.

RESTORATION IN SWEDEN AND EXPERIENCE GAINED FROM THE BLACKOUT OF 1983

Roger Kearsley
Swedish State Power Board
Operations & Sales
S-162 87 VÄLLINGBY
Sweden

Abstract. In Sweden the basic philosophy for system restoration after disturbances was worked out when hydro-power was still the main source of production. It gives a high degree of delegation to local personnel and reduces the need for intervention from the National Control Centre.

Since 1970 nuclear power has taken over an increasing share of the total production. Nevertheless the same basic restoration principles are still applied today.

In December 1983 the whole of southern Sweden was hit by a blackout. This provided a good test for the restoration procedures. On the whole restoration went well but some lessons were learned.

INTRODUCTION

The Swedish transmission system was originally designed to transmit hydro-power from the rivers in the north to the load centres in the south. This is still a major function. Today there are basically seven long parallel series-compensated 400 kV lines, plus several 220 kV lines between north and south. However during the 1970s and 1980s large nuclear power stations have been built near to the load centres in the south. Since the last two nuclear sets were commissioned in the summer of 1985, the hydro production in 1986 will probably be less than 50 % of the total for the first time.

The Swedish system is linked with those of Norway, Finland and Denmark (Zealand) to form the synchronous Nordic system. There are also HVDC links between Sweden and Denmark (Jutland), Norway and Jutland, and Finland and the USSR (fig 1).

The power flow depends on the subdivision of production between different types of power. This varies all the time owing to load variations, the water situation in the reservoirs, the availability of plant etc.

86 SM 328-9 A paper recommended and approved by the IEEE Power System Engineering Committee of the IEEE Power Engineering Society for presentation at the IEEE/PES 1986 Summer Meeting, Mexico City, Mexico, July 20 - 25, 1986. Manuscript submitted November 27, 1985; made available for printing April 23, 1986.

Printed in the U.S.A.

The basic philosophy is to operate the nuclear power stations at constant production and to use the more easily regulated hydro-power in the north to cope with the daily load fluctuations. The day-to-day operation of the system is supervised from the National Control Centre (NCC) in Stockholm.

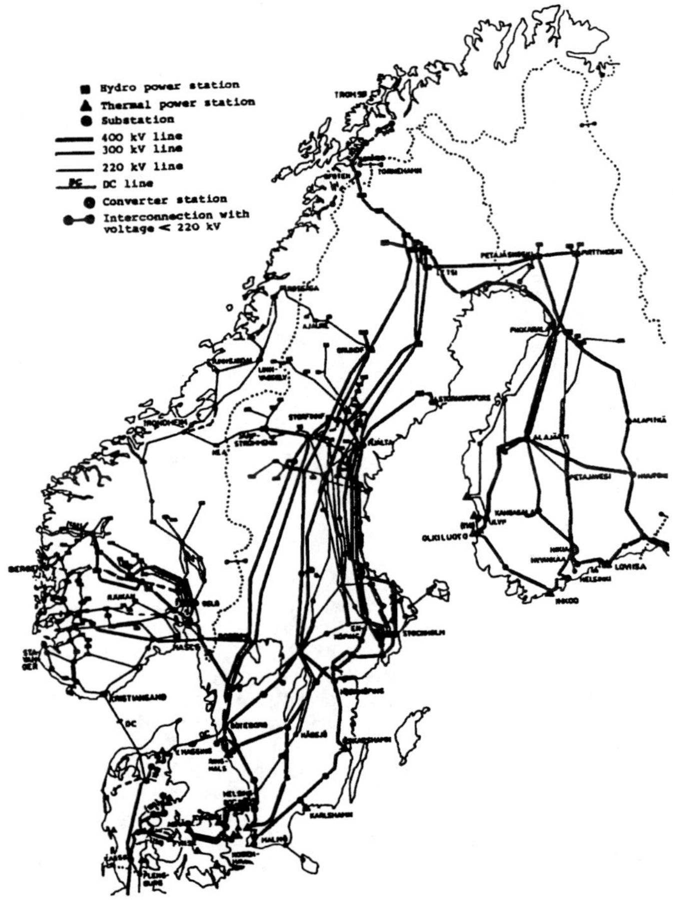

Fig. 1 The Nordic Power System

Restoration after disturbances is performed in a decentralized manner with a minimum of intervention from the NCC. The basis of restoration is an instruction system according to which switching is performed. Switching can be performed by a local supervisor or by programmed automatic equipment.

The restoration principles are the same irrespective of whether the disturbance is minor or severe. Basically the same principles have been applied since the 1950s with good

Reprinted from *IEEE Transactions on Power Systems,* Vol. 2, No. 2, pp. 422–428, May 1987.

results despite the changes which have taken place in the power system.

In December 1983 a major disturbance blacked out southern Sweden. The load centres were cut off from hydro production in the north and almost all nuclear production shut down. This provided a good test for the restoration procedures. On the whole experience was favourable but some new lessons were learned.

RESTORATION IN THE SWEDISH NETWORK.

Historical background

The restoration philosophy was worked out at a time when the hydro production in the north was dominant. The heavily-loaded system was sensitive to disturbances and blackouts were not uncommon. Restoration had often to be performed from scratch.

Supervision of operations was performed by personnel in each station. As telephone exchanges at the time were not automatic it was difficult for stations in different parts of the system to communicate quickly with each other and with the NCC. Also in the NCC there was very limited information on prevailing system conditions.

Consequently a system of standing instructions was worked out. Using this personnel on the spot in different parts of the system, judging from local criteria, could work together to restore the system after a breakdown. This has given the system a "self-healing" property which has proved very valuable, providing a favourable delegation of the work with a minimum of intervention from the NCC.

The situation today

Since this restoration strategy was introduced great advances have been made in communications and information systems. The public telephone network has been automized and power-lines have been provided with carrier-frequency telephony. Also advanced information systems have been introduced at the NCC and other dispatching centres. The character of the power system has changed too, notably by the introduction of nuclear power in the south. Many stations are now unmanned and more and more local supervision is being taken over by regional dispatching centres (RDC).

Despite these developments the same basic restoration philosophy is still applied. After major disturbances involving a split in the north-south transmission, and almost inevitably the shutdown of nuclear units in the south, the strategy is to re-energize from the north.

Basic Rules for restoration

When loss of voltage occurs all circuit-breakers shall be opened, so that the restoration can be controlled. This is done automatically by "zero-voltage tripping."

During a total blackout it shall be possible to perform "dead-network" start of a number of large hydro units in different regions in the northern network. These units are provided with auxiliary diesel sets for this purpose. The RDC which first succeeds in starting shall then energize lines towards the south and towards adjacent RDCs.

RDCs in the south shall wait for voltage on the incoming lines from the north. When voltage appears lines can be reconnected as long as sufficient active and reactive power is available. Priority shall be given to lines connected to nuclear-power stations.

Restoration instructions

The restoration instruction begins with a general introduction. In this definitions, principles, routines for reporting to NCC etc, are laid down. The main directive part of the instruction consists of a number of order sheets for each regional dispatching centre (RDC). These order sheets define measures to be taken during disturbances affecting one line only, as well as measures to be taken during disturbances affecting several lines. In addition there are special supplement sheets which describe the principles for restoration within the specific RDC area. As a reserve measure, there are also order sheets and supplementary sheets for each individual station, for use in case the remote control from the RDC is out of order and the station must be manned.

A simple example of an order sheet is shown in fig 2 and the corresponding supplement sheet is shown in fig 3. A sketch of the station referred to is shown in fig 4.

If restoration after disturbances is to function properly the instruction must be kept up-to-date. Extensions in the network and changes in layout must be followed by corresponding amendments to order sheets and supplement sheets. Temporary changes in operating configuration having a duration of at least two months also call for adjustments to the instructions.

The instruction system covers by necessity only the normal operating configuration in the network. However, in a large network it is seldom that the normal configuration applies. A certain amount of improvisation is therefore necessary. The supplement sheets define the general principles and thus serve as a guide to RDC personnel even in situations when the network configuration is not normal. In such situations direct orders or reminders from the NCC can also be needed.

A rapid and effective restoration depends on RDC personnel being well conversant with the instruction system. Recurring courses are therefore arranged, all such personnel attending a 3-5 day course every three years. The relevant part of the network is dealt with as regards developments, limitations, security etc. An important feature is a revision of the restoration instruction with the help of a disturbance simulation. In addition there is special training in restoration locally once a year.

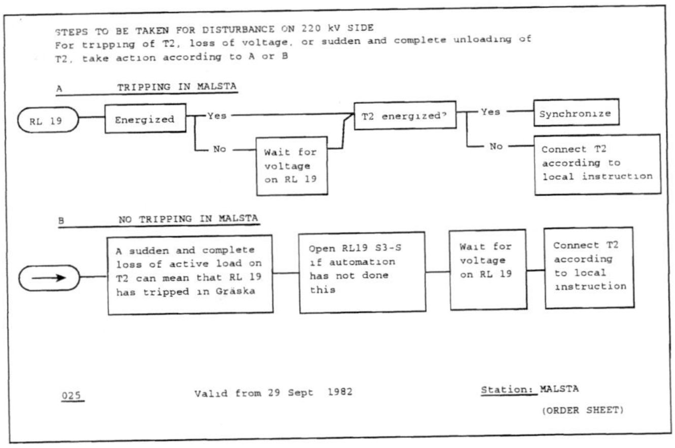

Fig. 2 Order sheet

SUPPLEMENT TO ORDER SHEET

PRINCIPLES

RL 19 shall be energized from Gräska

For disturbances operations are normally restored on the 220 kV side by automatic restoration equipment.

If a transformer fault on T2 is suspected in connection with the tripping of RL 19, no switching which involves energizing T2 shall be carried out before T2 has been inspected.

When parallelling, the following maximum limits shall apply:

Phase angle difference: 90°

Voltage difference on 220 kV system: 80 kV

Operating attendant for Malsta is in Dispatching Centre, Uppsala

REPORTING

Without delaying the steps according to this instruction, the NCC shall be informed after each 220 kV circuit-breaker tripping The NCC and Uppsala standby engineer shall be informed when normal operations have been restored.

Valid from 29 Sept. 1982 Station: MALSTA

(Supplementary sheet)

Fig. 3 Supplementary sheet

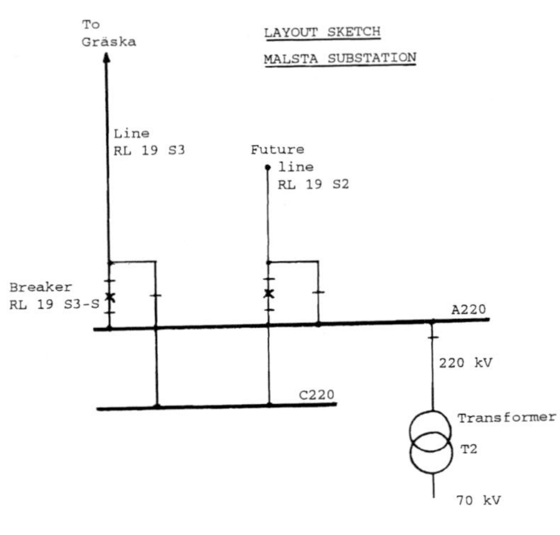

Fig. 4 Station layout

Automatic restoration equipment

Automatic restoration equipment has been in operation in some stations since the 1960s. Today such equipment is installed in most unmanned stations. The first automation was of the electro-mechanical type. Later designs were of the static type, and now all new equipment is computer-based.

The motives for automation are that it is fast and eliminates the "human factor". After a disturbance it is therefore preferable that the automation should perform the restoration. However, should this fail, remote control from the RDC is used. During major disturbances, the automatic restoration equipment can temporarily be blocked from the RDC. This may be necessary to avoid a too rapid reconnection of load to the system.

The basic principle is that automation shall work in exactly the same way as an operation attendant at an RDC. It is therefore programmed to strictly follow the restoration instruction. The automation is thus controlled wholly by local criteria.

The automation in a station has individual units each of which serves a circuit-breaker. Each unit is programmed separately. Start is triggered when a circuit-breaker is definitely tripped by its relay protection.

A number of security measures are built into the restoration automation. Thus the equipment blocks to prevent repeated reconnection on to a permanent fault, or closing a circuit-breaker which was open before the fault. There are also built-in checks that synchronizing, parallelling or energizing is performed within permitted limits.

When the restoration has been started, extreme voltages can arise. Shunt reactors and shunt capacitors can be automatically switched in and out to cope with this. For start and synchronizing of generators there is often special automation which can be activated by the automatic restoration equipment.

THE BLACKOUT OF 1983

Brief description

At about 13.00h on Tuesday December 27th 1983 the Swedish power supply system was subjected to a serious disturbance. A busbar fault in a vital substation near Stockholm caused cascade tripping of lines and cut off the load centres from hydro production and imports. Shutdown of nuclear production followed. The disturbance affected more people than any previous disturbance in Sweden. The entire southern half of the country including the major cities was blacked out (fig 5).

Before the fault

As it was the first working day after the Christmas break the load had increased more slowly during the morning than for a normal weekday. In the winter there is usually a dip in load after lunch, until it begins to grow dark at about 15.00 h. The load then rises again to reach its peak around 17.00h.

As the weather was also quite mild for the time of year ($0^{o}C$ in most of the country), the electric-heating load was not too high. The balance between production and consumption before the fault is shown below:

Production (MW)		Consumption (MW)	
Hydro	10 850	Basic load	17 060
Nuclear	5 800	Heating boilers	920
Fossil	300	Exports	320
Imports (Norway)	1 350	(Finland+Denmark)	
Total	18 300	Total	18 300

The transmission network was intact except for a 400 kV line to Finland and a 220 kV line near Stockholm. Transmission from north to south was 5 600 MW which is well below the stability limit of 6 100 MW. The voltage level in the 400 kV network was normal (400-405 kV) and the frequency was stable (50.01 Hz).

The breakdown

Just after the breakdown the transmission system was split into two parts and the lines to Norway and Denmark had tripped.

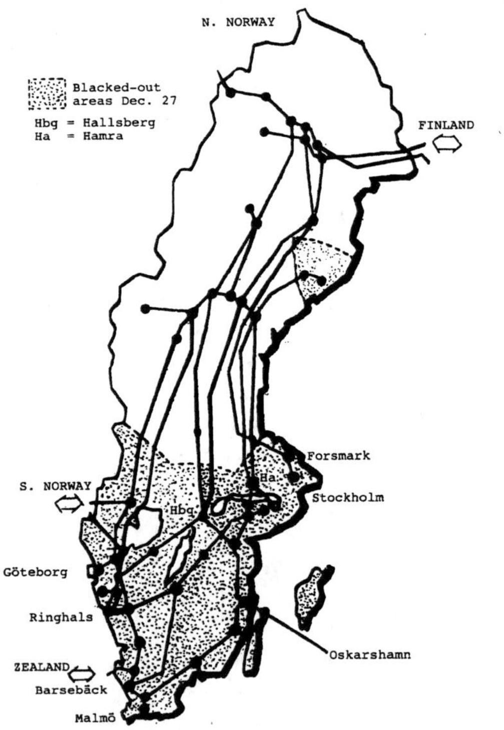

Fig. 5 Extent of the Blackout

The southern system was thus suddenly deprived of about 7 000 MW of production from the north and from Norway. Frequency and voltage fell rapidly. The frequency drop was so rapid that load shedding was unable to save the system. All nuclear units in operation in this part of the system were tripped. None of them succeeded in assuming house-turbine operation. Total breakdown was a fact. After a few seconds almost all circuit-breakers which had not already tripped were opened by zero-voltage protection.

The northern system, owing to the location of the split, included one nuclear power station. One block suffered a fast shutdown for reactor pressure and the only remaining unit in the other block continued to supply power to the north.

When the split occurred there was suddenly a production surplus of 6 000 MW in the north. This caused the frequency to rise to 54.0 Hz before overfrequency protection and turbine regulators could restore the balance. The interconnections with northern Finland and Norway also tripped and northern Sweden was isolated.

In one area of the north there was loss of supply to three major towns owing to a circuit-breaker failure and unselective clearance of fault currents. Also a few industries, mainly steel and cellulose, were affected. This was not due to loss of voltage, but because sensitive processes could not withstand the frequency and voltage transients which occurred.

The restoration

The network: Immediately after the disturbance the network in the north was largely intact and most hydro power stations were either still in operation or ready to start. This provided a basis for reconnection of 400 kV and 220 kV lines from north to south according to the restoration instructions. At least to begin with, load and production could be balanced by means of the easily-regulated hydro power.

Owing to the disturbance being so widespread it was sometimes necessary for NCC to intervene. An example of this was to prepare for reconnection to our Nordic neighbours. By 13.35h the 400 kV link with south Norway was reconnected and by 13.43h that with Zealand. By 13.48h the Swedish 400 and 220 kV networks were largely intact (fig 6).

The main job for the NCC during the recovery phase was otherwise to regulate the reconnection of load with regard to the network and production resources.

Thermal production: All nuclear production had been lost except for 310 MW which was at first connected to the northern system. Emergency power production in gas-turbine and oil-condense stations was immediately ordered to start up. Gas-turbines can be started up quickly (2-15 min) but for oil-condense stations the starting time is much longer (3-24 h). However as the oil-condense production increased the gas-turbines could be stopped until by 21.00h there were no longer any in operation.

The nuclear units were restarted as soon as safety demands allowed, but technical problems led to delays. The first unit to come back was synchronized on December 27th at 22.35h and the last on December 29th at 14.44h.

Imports: As soon as the network was intact and reconnected to the other Nordic systems maximum power imports were ordered from southern Norway and Denmark. 1200 MW was ob-

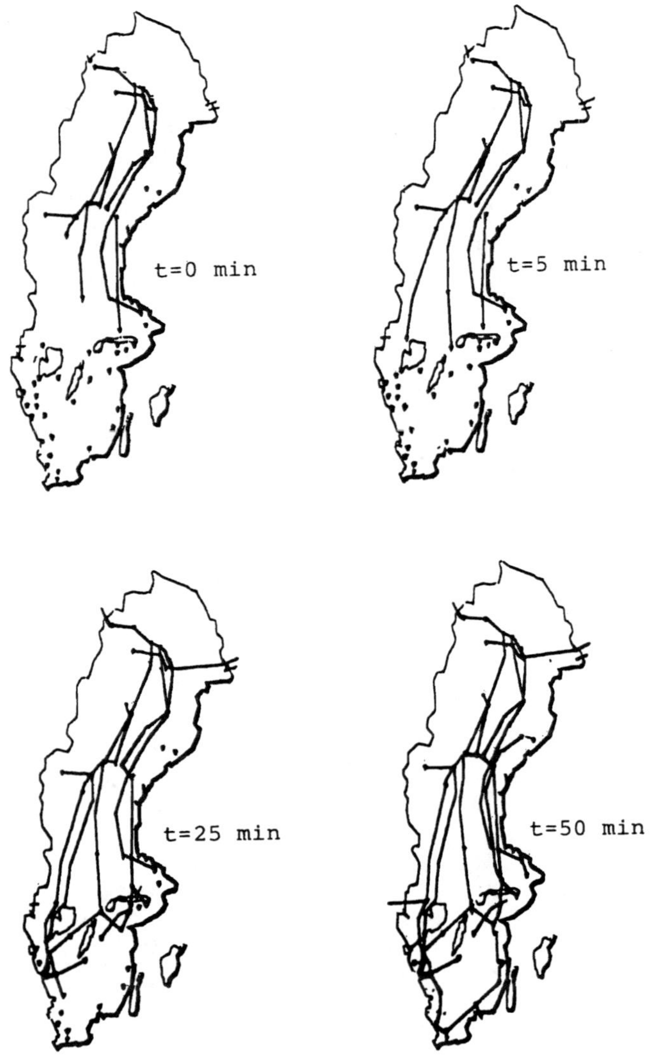

Fig. 6 Restoration of the Transmission Network

tained from Norway (hydro). Imports from Jutland and Zealand (thermal) began at 14.30h and soon reached the maximum limits of 280 MW and 700 MW respectively.

Load: The reconnection of load with regard to production and transmission reserves was supervised by the NCC in collaboration with the RDCs.

During the first hour no restrictions on successive load increases were necessary. Load was added both automatically and manually based on local voltage and frequency criteria.

When transmission from the north reached its limit further reconnection of load had to be matched by new production and imports in the south. No special priorities were applied. The aim was simply to obtain as even a distribution as possible. At distribution level a certain amount of priority was given to sensitive loads.

At 18.20h the NCC could announce that the situation allowed for unrestricted reconnection of load. Fig 7 shows the relation between production and consumption, disturbed and undisturbed, for the entire day.

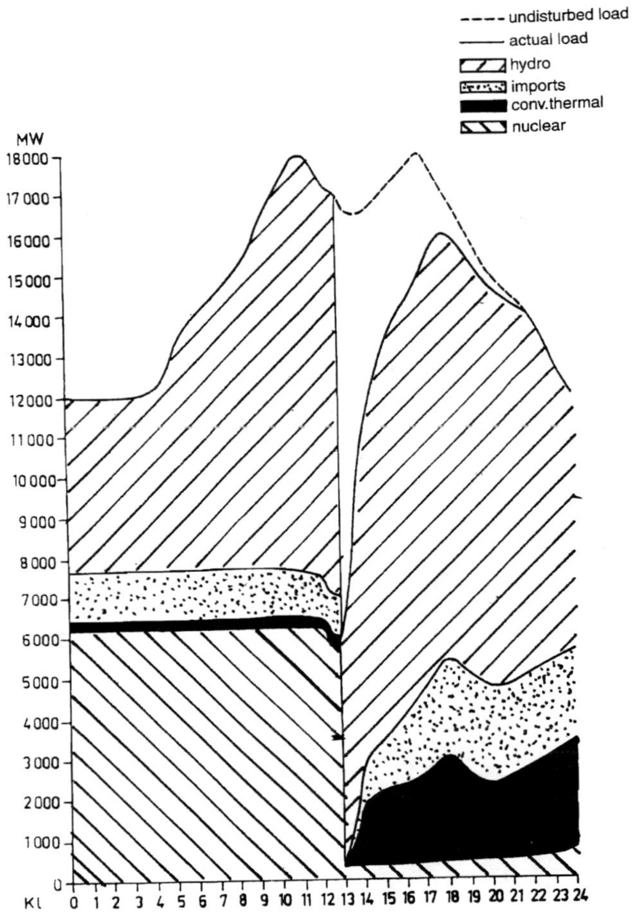

Fig. 7 Production & Consumption on December 27 1983

Preparations were made to reduce the load the following day by up to 25 % by selective disconnection in case production resources were inadequate. However, this was not necessary owing to the mild weather and the thermal power made available.

EXPERIENCE GAINED AND STEPS TAKEN

The progress in communications and information systems which has been made during the last decades has not diminished our need for the basic restoration principles. Decentralized automatic or semi-automatic restoration is still very valuable. Some new basic reasons for this are given below.

* Volume of information: During major disturbances in large meshed systems the sheer volume of information quickly available to personnel in dispatching centres can be a problem. The operator can "drown" in information and it is difficult for him to quickly survey the situation.

* Experience of personnel: Nowadays large-scale disturbances are relatively rare. Also people change jobs more often than they used to. These factors both contribute to personnel throughout the system not being so experienced as regards real-life disturbances. Therefore standing instructions and automatic

equipment, e.g. for restoration, can play a big part in limiting faults caused by personnel errors. In addition the need for regular training including disturbance simulations should be stressed.

---

* Information system at NCC: The number of incoming alarm signals to the computerized supervisory system at the NCC was greater than the system could swallow. The system was blocked and could not be used for the first half hour of the recovery.

Luckily an older system feeding a wall diagram was operative. Also data transport functioned despite the high data-load level and could serve the RDCs.

The information system at the NCC has been in operation since early 1977 and is heavily loaded even for normal operations. A new computer with greater capacity will be installed in 1987.

After the blackout 1983

After the blackout the rate of restoration was reasonably acceptable, being similar to that for other recent major disturbances at home and abroad (fig 8).

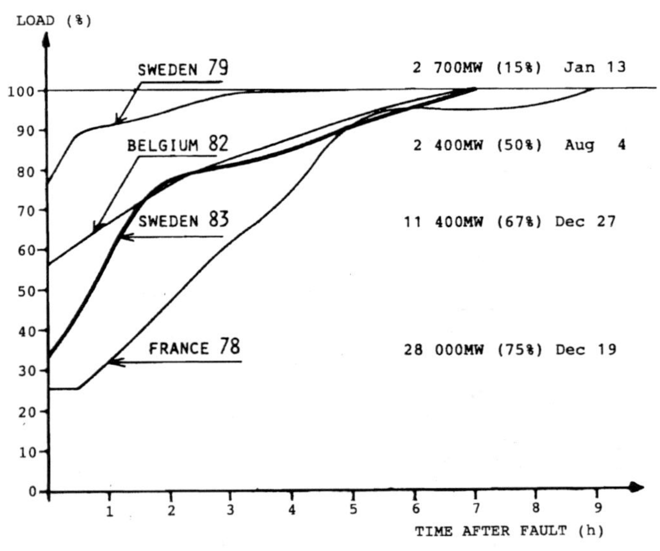

Fig. 8 Rate of Reconnection of Load for Recent Major Disturbances

The basic restoration principles giving the system its self-healing property proved to be invaluable. At the NCC there were problems with information and communication which would have slowed down the restoration process if this had been more centralized. Experience gained which should make it possible to improve preparedness for new major disturbances is outlined below.

* Telephone network: The public telephone lines to the NCC and to RDCs were intensively occupied by the media and private persons wanting to know what was happening. A separate line-carrier telephone system could however be used for important operative calls.

* External information: The non-operative communication with superiors, regions, the media and the public not only blocked telephone lines but also interfered with the important restoration work. The quality of the information to the public was also criticized.

New information routines have therefore been established. More permanent channels of communication with the media have been set up outside the NCC in order to relieve operational personnel of this additional task. The number of supervisory personnel at the NCC will also be temporarily increased during major disturbances for information purposes. One extra man will keep the media contact informed. Another, a network specialist, will inform the RDCs about the extent and consequences of the distubance.

---

* Reconnection of load: Although the experience of the restoration principles was mainly good, a weakness was revealed.

In the past there has been no problem in allowing load to be reconnected as the network is built up. The proportion of vulnerable nuclear-power has increased though, and in December 1983 5500 MW of nuclear production was lost. As it could not be restored quickly this led to problems. Load was reconnected quickly in the parts of the network to be restored first, thus using up all the available production resources. The remaining areas were then forced to wait until alternative production could be obtained before load could be reconnected.

Since 1983 more nuclear production units have been commissioned and nuclear production can now be as high as 9500 MW. The loss of this amount of nuclear production definitely would not allow for uninhibited reconnection of load.

Consequently a new instruction has been issued. This states that for "major disturbances" reconnection of load in all regions shall be limited to 50 % of the load before the disturbance. This should give a more balanced load distribution. For normal local disturbances there is no restriction on the reconnection of load. NCC personnel shall inform the RDCs when "major-disturbance" rules apply.

It is not yet clear how this limitation is to be achieved in practice. In some regions it is possible, via remote control, to block automatic frequency-controlled equipment for load reconnection after load shedding. In the same way network restoration equipment can be blocked if it causes reconnection of load. Other possibilities are remote disconnection of low-priority load, e.g. electric boilers.

* Nuclear power stations: The ability of nuclear power units to restart quickly after disturbances is limited. However it takes less time to restart after house-turbine operation than after emergency shut-down. It is therefore important to study how to modify equipment and routines in order to improve the probability for successful transfer to house-turbine operation. If this probability can be assessed for different contingencies (e.g. by testing) it will be easier to plan day-to-day standby production reserve.

CONCLUSIONS

Security of supply

The security of supply in Sweden is on a par with that in other industrialized countries. For low-voltage consumers the average annual interruption time for the whole country is about two hours. For urban consumers it is less than 30 minutes. The annual average interruption time for high-voltage consumers is less than 30 minutes. Only a fraction of this time is caused by faults in production units or in the 400 or 220 kV transmission networks, the larger part being due to faults in regional networks.

Faults occur in all systems, causing more or less serious disturbances. In the extreme case a fault can have far-reaching consequences and lead to a blackout. The risk for blackouts can never be eliminated, even in large strongly meshed systems. However, all available means for reducing this risk must be utilized, within reasonable economic limits. In this context the disturbance of December 1983 provided a lot of experience. As a direct result studies were started in several areas. These included sub-station and power-station design, protective systems, operations and maintenance, communications and external information.

Restoration

On the whole the restoration after the blackout went well, largely thanks to the prevailing restoration routines. However, experience gained indicated that improvements could be made in the following areas:

Dispatching centres:

- Operational personnel should be relieved of contacts with the media etc during stress situations such as disturbances.
- It is important that the information system can cope with the amount of data produced during a major disturbance.
- There should be routines for reinforcing certain functions during major disturbances.

Reconnection of load: After major disturbances involving loss of nuclear-power production, only a limited amount of load can be reconnected to start with. Without obtaining special permission from the NCC 50 % of the load before the disturbance may be reconnected.

Nuclear power stations: It is important to improve the probability for successful transfer to house-turbine operation.

REFERENCES

1 The Blackout in Sweden 27-12-1983, Vattenfall publication, February 1984.

2 S Larsson and R Kearsley, "The Blackout in the Swedish Power Supply System, December 1983".

5th CEPSI Conference, Manilla, Philippines, November 1984.

3 R Otterberg "Restoration after Disturbances in the Swedish Bulk-Power Network - Instruction System and Automatic Equipment".

CIGRE Study Committee 39 Colloquium, Toronto, Canada, September 1985 (Paper no EM85 04).

SIMULATION AND FIELD TESTS OF THE BLACK START OF A LARGE COAL-FIRED GENERATING STATION UTILIZING SMALL REMOTE HYDRO GENERATION

Randy R. Lindstrom, Member
Nebraska Public Power District
Columbus, Nebraska

Abstract

The ability to rapidly recover from a major, region-wide blackout is a significant concern of electric utility engineers and operators. Even though such blackouts are rare occurrences, it is prudent and necessary for an electric utility to prepare for such events and to develop plans for recovering from them. This paper describes a method which utilizes remote hydro generation to restart a major coal-fired generating station. Previous work on restoration techniques and the use of combustion turbines to blackstart a generating station has been documented [1,2]. This paper documents a comprehensive black start project. It includes considerations for a black start project, details of an analytical study, identification of critical factors, development and implementation of an actual black start test, comparison of field tests with the simulation results, and the required system modifications identified during the project.

Introduction

Nebraska Public Power District (NPPD) has undertaken a project to investigate a means by which the 1300 MW coal-fired Gerald Gentleman Station (GGS) could be brought on-line following a major region-wide blackout without assistance from neighboring utilities. Although large scale blackouts are uncommon, they have happened in the past, and it has taken several hours to restore electrical service [3,4]. In the event of a blackout, NPPD would like to minimize the amount of time required to restore electrical service.

Gerald Gentleman Station consists of two 650 MW coal-fired generating units. Alone, GGS can supply approximately one-third of the summer peak load requirements for the entire state of Nebraska. Also, the state's entire electrical grid could be re-energized with GGS, making it possible for other generating units to begin start-up procedures. Due to the fact that GGS is a base load resource and at least one unit would most likely be running prior to a region-wide outage, GGS is the most favorable and logical location to begin system restoration.

89 SM 670-1 PWRS A paper recommended and approved by the IEEE Power System Engineering Committee of the IEEE Power Engineering Society for presentation at the IEEE/PES 1989 Summer Meeting, Long Beach, California, July 9 - 14, 1989. Manuscript submitted January 13, 1987; made available for printing April 27, 1989.

Project Scope

The scope of the GGS Black Start Project was to determine if a small transmission system consisting of the North Platte and Kingsley Hydro generating stations and various 115 kV and 230 kV transmission facilities could be energized (Figure 1), and if this small system had the capability to start the largest auxiliary motors associated with a generating unit at GGS.

The North Platte Hydro Station is located approximately 23 miles east of GGS and consists of two 14.5 MVA Francis wheel turbine-generators. The Kingsley Hydro Station is located approximately 33 miles west of GGS and consists of a 52.6 MVA variable pitch Kaplan wheel turbine-generator.

Each GGS unit has six electric motors, 2000 HP and above, which would have to be started in order to bring each unit back on-line. These six motor loads and their sizes are:

7000/4000 HP Induced Draft Fan Motor
4000 HP Primary Air Fan Motor
3000 HP Forced Draft Fan Motor
2500 HP Reservoir Water Pump Motor
2500 HP Circulation Water Pump Motor
2000 HP Start-up Boiler Feed Pump Motor

The reasons for considering the start-up of only the six largest motors in the scope of the project were:

1. These six largest motors would present the biggest problems during start-up, since they draw high currents and require long starting times.

2. With these six motors operational, a GGS unit can be brought on-line and obtain a generation level of 300 MW.

3. Once these six large motors are up to speed, they would act like induction generators and assist with the starting of smaller auxiliary equipment motors.

4. Other motors in the 2000 HP size range associated with each GGS unit, such as soot blower air compressor motors, would

Reprinted from *IEEE Transactions on Power Systems,* Vol. 5, No. 1, pp. 162–168, February 1990.

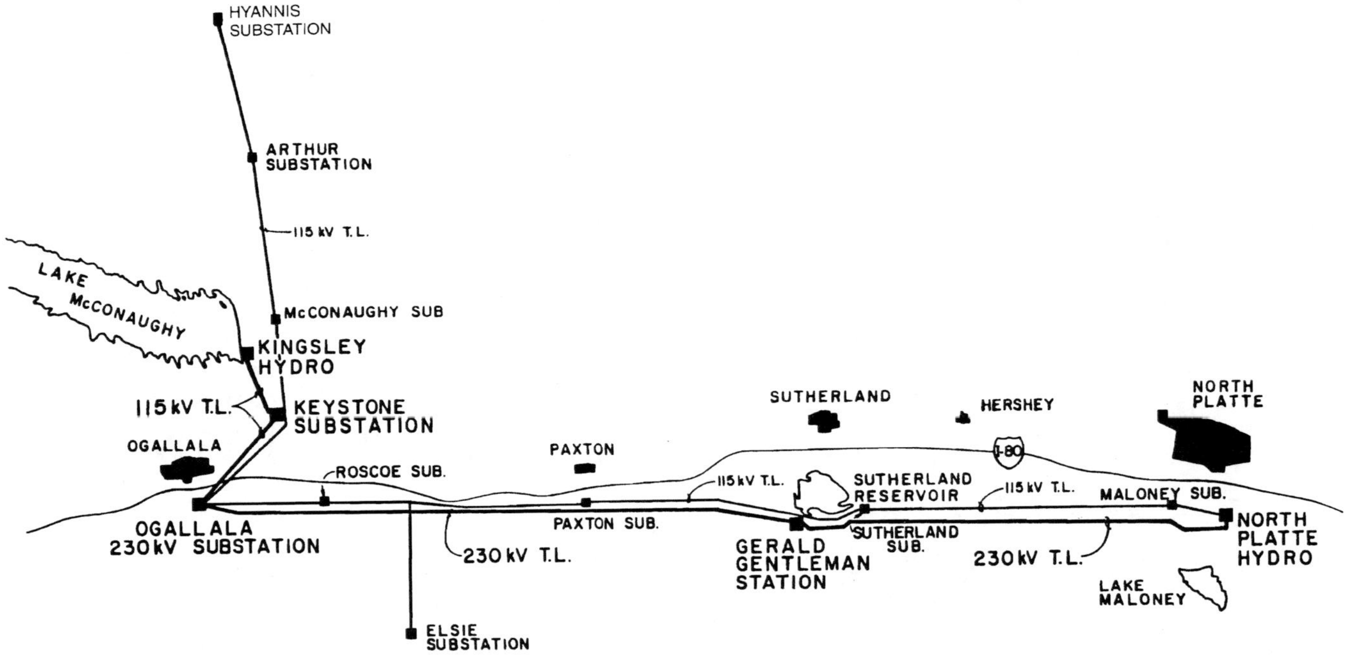

Figure 1: Kingsley-Gentleman-North Platte Area Diagram Depicting The Transmission Facilities Associated With The GGS Black Start Project

not be required for the initial start-up and would not have to be operated until a GGS unit was on-line and able to provide starting assistance.

Analytical Study

To gain further insight as to whether the black start plan would actually work, and if it did, its effect on the equipment involved, an analytical study was undertaken. This study involved detailed computer models of the motors, hydro generators, and the transmission system.

The governor and excitation system model data for the North Platte and Kingsley Hydros were developed in-house from field measurements and tests.

Manufacturers' motor data was gathered for each of the six largest motors required for start-up of GGS Unit #1. From this data, computer model data was developed and altered to match actual starting characteristics that were measured by GGS plant personnel.

There was concern about the time-overcurrent relaying associated with the 7000/4000 HP I-D fan motor since it had tripped in the past while starting the motor on high speed. This motor is a two speed motor which can be operated at either 7000 HP (10 pole operation) or at 4000 HP (12 pole operation). Rated speed for a 10 pole machine is 720 rpm, while rated speed for a 12 pole machine is 600 rpm. The starting current requirements for 12 pole operation are lower and it does not take as long to bring the same inertia up to a reduced speed. From the computer simulations (Figure 2), it was evident that starting the I-D fan motor on high speed would draw too much current and take too long, thus the timer would time-out and the relay would operate. It was decided that this motor had to be started on the slow speed setting in order to get it started during a black start condition.

After all of the models were combined, computer simulations of starting the six largest motors at GGS with the North Platte and Kingsley Hydros were performed (Figure 3). The motor starting order was determined by size, the largest (7000/4000 HP I-D fan motor) first, and then in descending order. The thought behind this procedure was the

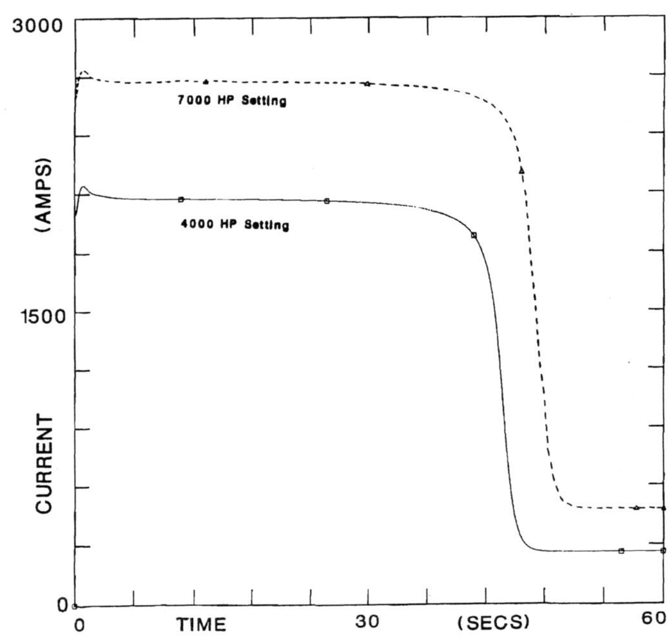

Figure 2: Starting Current Comparison For Starting The Induced Draft Fan Motor On Either Speed Setting

induction generator effect provided by the larger motors, as the other motors were being started. Also, starting a larger motor after a smaller one was on-line could possibly cause the smaller one to stall due to the large, sustained voltage dips associated with starting the larger motor (Figure 4).

Another concern that had to be addressed was relaying and protection. From the computer studies, system frequency dips associated with starting the large auxiliary motors were noticed (Figure 5). There is an underfrequency load shedding scheme at North Platte. To prevent dropping the North Platte load in the middle of the motor starting sequence, the underfrequency load shedding relays at North Platte had to be disabled prior to the test or prior to any restart sequence following a blackout condition.

Cold Load Pickup Capability

In order to obtain a stable operational level for the two hydros, they had to be isolated with some system load. In an actual blackout condition, the hydros would be picking up "cold" load and depending on the duration of the outage, this "cold" load could be as much as three times its original pre-outage value.

NPPD performed a "Cold Load Pickup Test" on the North Platte Hydro. The power output of the two units was reduced to 0 MW, then a small portion of North Platte area load was outaged and then picked up with the hydro units. There was a frequency dip of 1.4 Hz but it recovered back to 60 Hz within 7 seconds. The governor droop settings for the two units were changed to 0% so that they would be more responsive to frequency deviations.

With the success of the "Cold Load Pickup Test", no problems were anticipated with the concept of the North Platte and Kingsley Hydros coming on-line and picking up small portions of load in succession. This is especially true as more load is picked up, since with each additional amount, the wicket gates on the hydros will open further and provide more operating flexibility and stability.

An operating restriction associated with the Kingsley Hydro Station also had to be addressed. Operating Kingsley Hydro at a power output level of less than 15 MW results in cavitation in the turbine. This cavitation is a normal occurrence during the start-up and shut-down of this unit, and presents no problems as long as it does not persist for extended periods of time. Due to the projected length of the black start test, cavitation problems were not anticipated at Kingsley. Under an actual blackout condition, as much "cold" load as possible would be picked up at Kingsley to achieve the 15 MW minimum load level.

System Establishment

The final step in the analysis phase of the project was to evaluate the establishment of the small system which would be needed to start the motors at GGS. The 230 kV line from Ogallala to GGS to North Platte would have to be energized to provide station start-up power to GGS as well as to tie the two hydro stations together. Breaker reclosing restrictions imposed by sync-check relays at GGS would require the entire 50 mile 230 kV line to be energized from Ogallala in a single operation along with the 345/230 kV transformer, a 20 MVAR tertiary reactor for

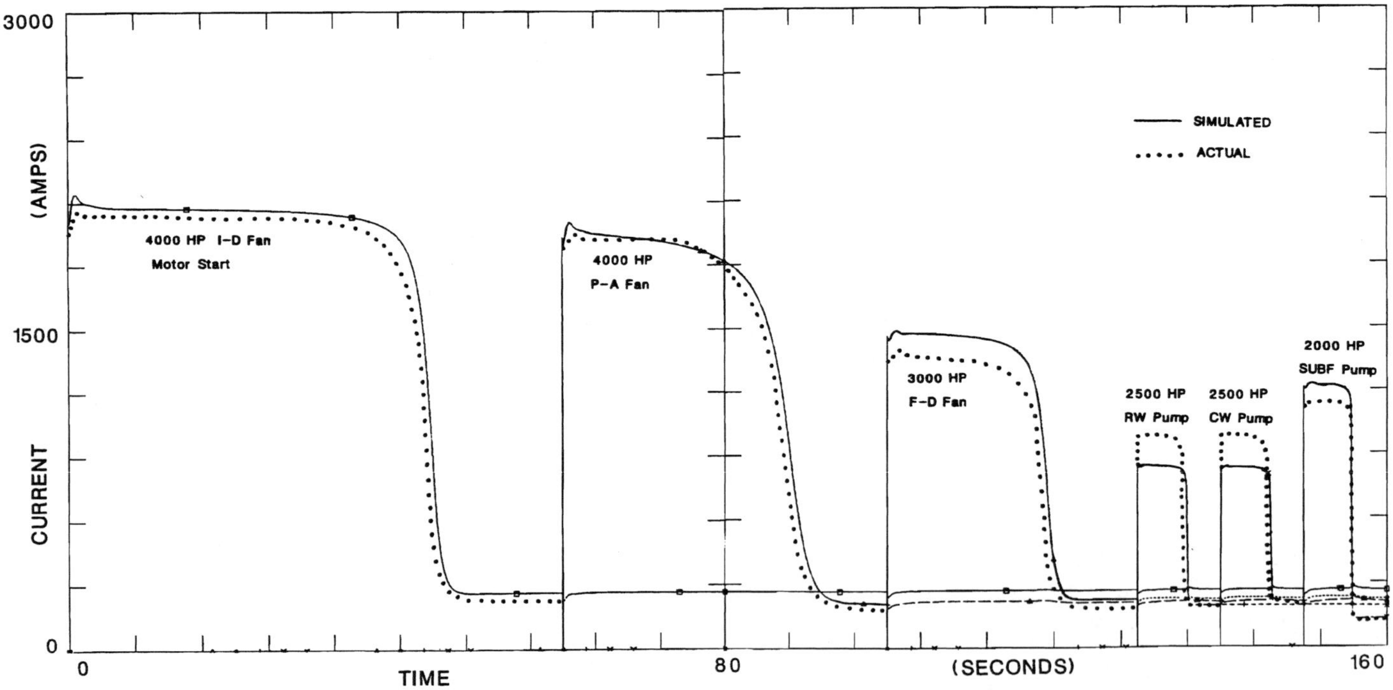

Figure 3: Motor Starting Current Comparison Plots Of The Computer Simulated Response And The Actual Measured Response While Starting The Auxiliary Motors With The Small Test System

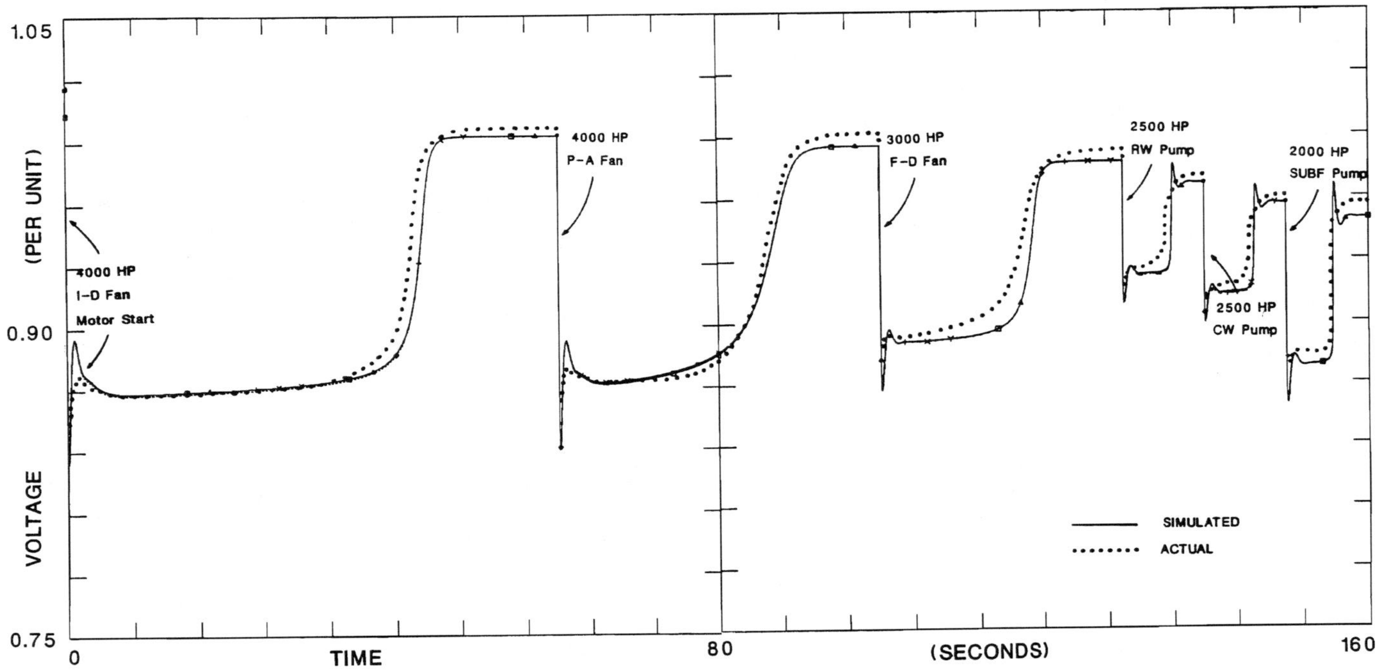

Figure 4: GGS 7.2 kV Bus Voltage Comparison Plots Of The Computer Simulated Response And The Actual Measured Response While Starting The Auxiliary Motors With The Small Test System

voltage control, and the Start-up Reserve Auxiliary Transformers at GGS.

The weak source condition at Ogallala coupled with the switching of this amount of 230 kV line could be a potential source of transient and dynamic overvoltages which might cause arrestor operation or damage equipment. Computer simulations showed a magnitude of 1.4 p.u. for these overvoltages and this condition only lasted for a couple of cycles, due to the existence of closing resistors in the 230 kV breakers at Ogallala.

Black Start Tests

Although the analytical studies indicated that the large auxiliary motors could be started with the small test system, NPPD could not be absolutely sure of the plan without an actual

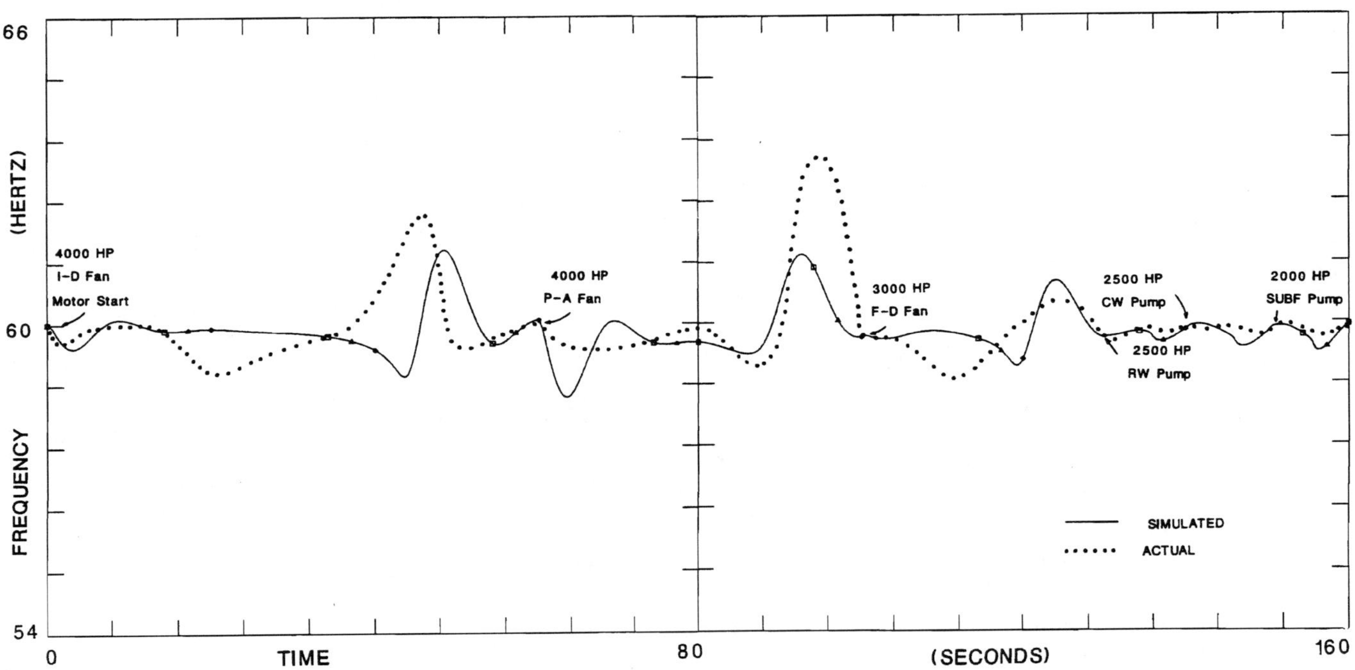

Figure 5: System Frequency Comparison Plots Of The Computer Simulated Response And The Actual Measured Response While Starting The Auxiliary Motors With The Small Test System

test. A black start test was performed on September 14, 1986 at 0200 hours. This date coincided with the annual maintenance outage of GGS Unit #1, so that there would be minimal impact on operations at Gerald Gentleman Station.

The first phase of the field test was to isolate the North Platte Hydro with some local area load and the Kingsley Hydro with some Ogallala area load (Figure 7). The next phase of the test was to tie the two isolated systems together by energizing the Ogallala-GGS-North Platte 230 kV line from Ogallala and then synchronizing the two systems together across a low side generator breaker at the North Platte Hydro. The North Platte Hydro is a manned station and was a logical place to coordinate any generation changes required to reduce any angular differences between the two isolated systems.

Overall, the GGS Black Start Test was a success, even though some minor problems were encountered. The problems were correctable and are as follows:

(1.) Upon isolation of the Ogallala area load with the Kingsley Hydro during the test system setup, frequency regulation was somewhat erratic and oscillated from 59-61 Hz.

(2.) The energization of the North Platte 230/115 kV transformer and the 115/13.2 kV generator step-up transformer from the 230 kV line, created some sustained dynamic overvoltages which lasted for six to seven seconds.

(3.) There was a problem on the first start of the I-D fan motor, in that an electro-magnetic relay contact hung up and didn't drop out properly, causing the motor to trip off-line after it had already reached rated speed. The motor did start on the second attempt.

(4.) Another problem surfaced at the Kingsley Hydro Station immediately after the P-A fan motor at GGS had started. The electronic governor at Kingsley blocked due to a large change in net head, causing the variable pitch blades of the turbine to lock in a stationary position. This caused the system frequency to oscillate between 55 and 65 Hz until the governor was manually reset.

Comparison of the field test data with the analytical study showed a very close correlation. Figure 6 depicts a portion of a visicorder recording of the motor current and the GGS 7.2 kV bus voltage as the Induced Draft fan motor is being started on it's slow speed setting (4000 HP) during the Black Start Test. Due to the overall length of the visicorder plots and strip charts they could not be included in this paper. The measured results are plotted along with the computer simulated responses in Figures 3,4, and 5. Comparison of Figure 6 with Figures 3 and 4 shows a very close correlation between the computer simulation and the "real world" response. Due to governor alignment problems and governor blocking problems experienced at Kingsley Hydro, the actual frequency response did not match the computer simulated response. The actual frequency plot in Figure 5 does not show all of the frequency excursions

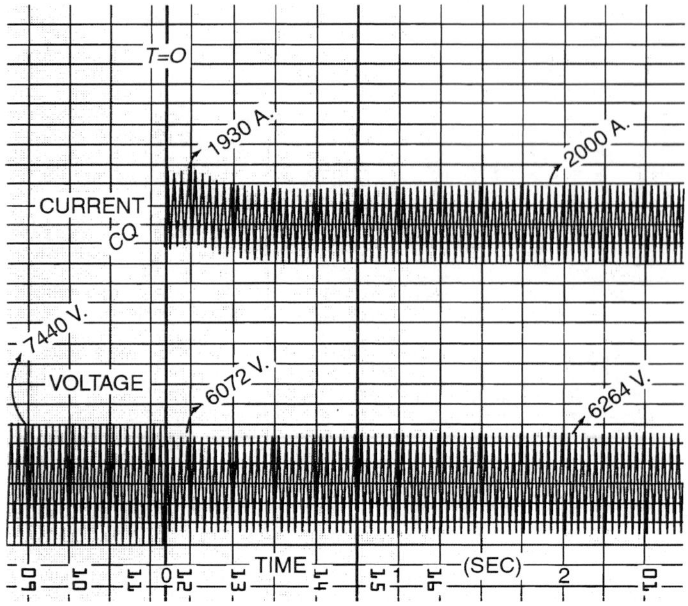

Figure 6: Actual Recording Of The Motor Current And The GGS 7.2 kV Bus Voltage As The Induced Draft Fan Motor Is Being Started During The Black Start Test.

experienced after the P-A fan motor was started since the time base for the computer simulation had to be followed. Modifications to the governor control circuitry at Kingsley have instilled confidence that the actual frequency response will correlate better with the simulated response (Figure 5).

System Modification and Procedure Development

The essentially successful completion of the Black Start Tests have provided NPPD with the confidence that GGS can be black started if the need ever arises. However, some system modifications were required and operating procedures had to be developed to make this a truly usable and viable option.

The first modification was to align the electronic governor on the Kingsley Hydro. This required making adjustments to the electronic governor control circuitry to optimize its operation while operating in the isolated system mode of governor control. This alignment was performed on October 25, 1986.

The first step of the governor alignment was an off-line test which required Kingsley Hydro to be running at rated speed but not connected to the electrical grid. The off-line test consisted of manually restricting the wicket gate limit and causing the frequency dip to 55 Hz (165 r/min). Then the wicket gate limit was released to allow the governor to respond to a simulated small net head change. The governor performance was monitored by examining the amount of frequency overshoot and swing damping.

The next step of the alignment procedure involved isolation of the Kingsley Hydro with a small portion of system load and starting a large induction motor at GGS. Governor response was monitored and changes were made to the off-line governor control gain

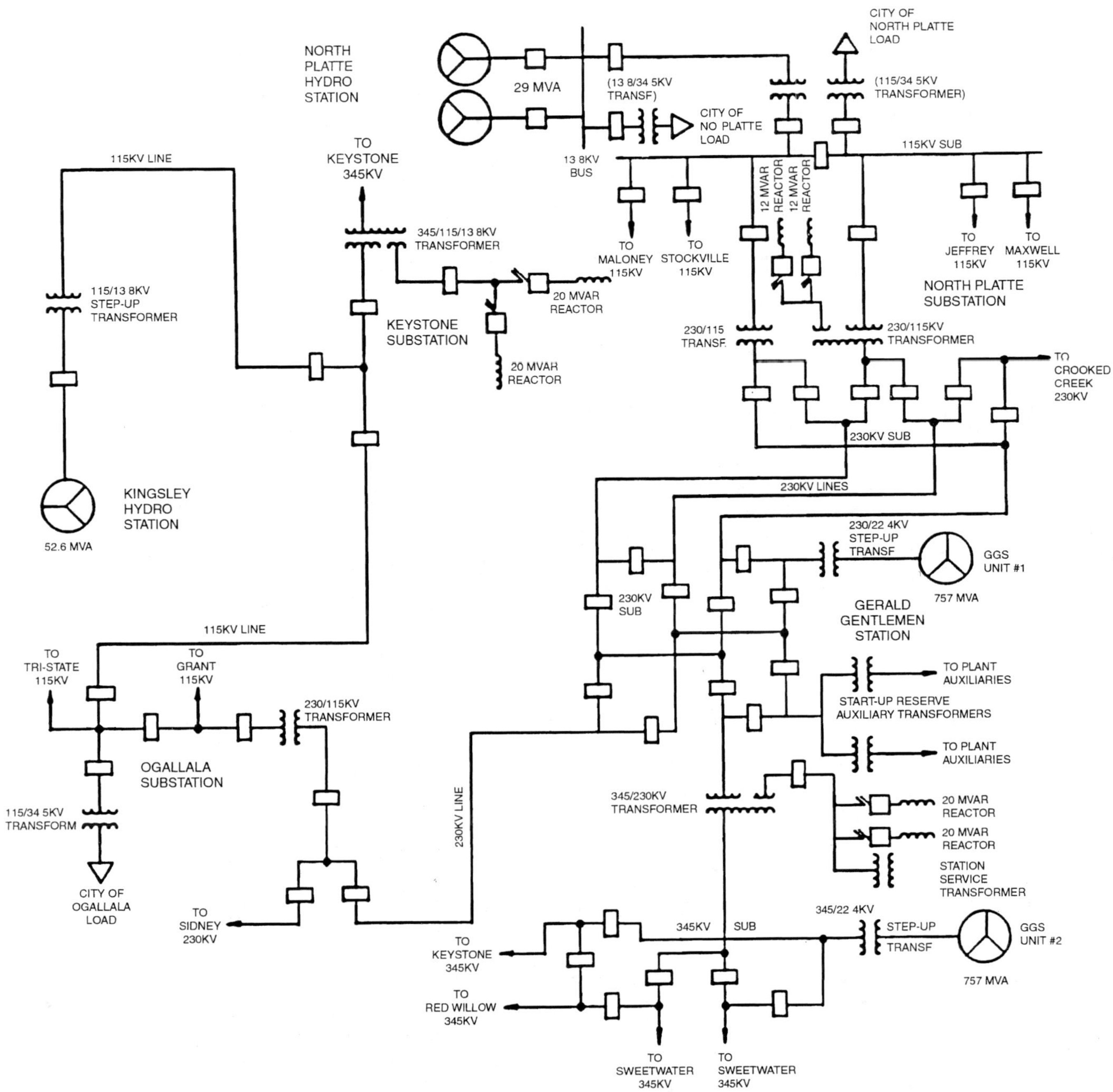

Figure 7: One-Line Diagram Of Facilities Associated With The GGS Black Start Project

settings. Several motor starts and setting changes were performed until acceptable governor response was achieved. The final step of the alignment procedure involved isolation of the Kingsley Hydro with a bigger portion of system load and starting a large induction motor at GGS. These tests gave an indication of governor response at different unit operating points. Adjustments made to the off-line governor control gain settings provided optimal governor response at the two levels tested.

For isolated system mode operation, the off-line governor control settings will be used. These gain settings are set so that the unit will not hunt as the system frequency deviates. The off-line mode of operation is an overcompensated mode for the governor. The off-line gain settings are normally quite lower than the on-line settings.

By optimizing the governor damping control circuit gains, Kingsley Hydro will provide good frequency regulation and load following response while operating in an isolated system mode and the frequency oscillations present during the test system setup phase should no longer exist. A switch was installed in the control room at Kingsley to permit switchover to isolated mode governor operation.

In order to eliminate the problem with the governor blocking and locking up the turbine blade position during large net-head changes, a new microprocessor was installed in the electronic governor control circuit that automatically resets the turbine blade lockup feature when the net head pressure differential alarm is cleared.

A procedural change was implemented to alleviate the problem of the sustained dynamic overvoltages which developed by energizing the North Platte 230/115 kV transformer and the generator step-up transformer together. Computer simulations conducted since the Black Start Test, indicated that the dynamic overvoltages did not develop when the North Platte 230/115 kV transformer and the generator step-up transformer were energized from the 230 kV system as completely separate steps.

The difficulty with starting the Induced Draft fan motor at GGS was due to relaying problems. Extensive investigation revealed that the trip was due to several circumstances. An auxiliary relay used to switch in the overcurrent relays used during motor starting on the slow speed winding, was incorrectly wired. Also, due to some calibration errors and a seal-in problem, the indicating control switch sealed in and didn't allow the relay to drop out when it should have. This tripped the motor off-line after it had already reached rated speed. Corrective action has been taken to eliminate all of the problems.

Another system modification involved the reclosing relay logic associated with the generator breaker at the Kingsley Hydro. During an actual black start condition, Kingsley Hydro will be required to energize a dead transmission system and pick up "cold load" in the Ogallala area. The reclosing logic for this breaker allowed for it to be closed only under a "hot" system, "hot" generator situation. If this relay package were to be replaced, it would be necessary for the reclosing provision for the breaker to be selectable, such that the breaker can be closed under "hot" generator, "dead" system conditions. Presently, this relay is jumpered to allow the breaker to be closed under a hot bus/dead line condition.

Modifications have been made to allow supervisory control of the sync-check reclosing relays on the 230 kV breakers at the Gentleman Substation. Previously, these breakers could not be closed under a "dead" bus/ "dead" line condition, which would exist under a blackout situation. Also, since the system could fail with these breakers open, this modification provides the dispatchers with the ability to properly configure the 230 kV breakers during a blackout situation.

It was also necessary to make sure that all critical voice communication channels would function under blackout conditions. The major change required in this area was to provide ring thru capability for the microwave phone at the North Platte Hydro which was independent of AC station service.

Conclusion

This paper has pointed out that there are numerous details to be considered before, during, and after, a black start project of this magnitude. It has been demonstrated with an actual test that the black start expectations for GGS are realistic and can be achieved. With some minor system modifications and procedural changes, any problems which might occur can be avoided. Results from the field test and the analytical study correlated very well. Hopefully, this paper will give insight to some considerations for future black start projects and will prompt other utilities to develop black start capabilities within their system.

Acknowledgment

The author wishes to acknowledge the technical staff members of the GGS Black Start Task Force which was chaired by J. H. Doudna. The task force included personnel from the following entities: Nebraska Public Power Dictrict's Planning Division, Power Operations Division, Protection and Control Division, Hastings Dispatch Center, Gerald Gentleman Station, North Platte Region, Western Region (Ogallala District),and the Central Nebraska Public Power & Irrigation District (Kingsley Hydro). The cooperation and support from these people made the GGS Black Start project a success and made this paper possible.

References

[1] R.D.Shultz, G.A.Mason, "Blackstart Utilization Of Remote Combustion Turbines, Analytical Analysis And Field Test". IEEE Transactions On Power Apparatus And Systems, Vol. PAS-103, pp. 2186-2191, August 1984.

[2] M.Adibi, P.Clelland, L.Fink, H.Happ, R.Kafka, J.Raine, D.Scheurer, F.Trefny, "Power System Restoration--A Task Force Report". IEEE Transactions On Power Systems, Vol. PWRS-2, pp. 271-277, May 1987.

[3] G.L.Wilson, P.Zarakas, "Anatomy Of A Blackout". IEEE Spectrum, pp. 38-46, February 1978.

[4] D.N.Ewart, "Whys And Wherefores Of Power System Blackouts". IEEE Spectrum, pp.36-41, April 1978.

POWER SYSTEM RESTORATION INCORPORATING INTERACTIVE GRAPHICS AND OPTIMIZATION

by

J.A. Huang, F.D. Galiana

Department of Electrical Engineering
McGill University
Montréal, Canada

G.T. Vuong

Hydro-Québec
Montréal, Canada

ABSTRACT - This paper presents a new power system restoration planning methodology which incorporates interactive graphics and an optimization algorithm. The interactive graphics provides a tool to define restoration scenarios, view the restoration process, and display the optimization and simulation results. By introducing the concept of optimal restoration scenario, the process of selecting the control variables can be formulated as a mathematical problem which can be solved by nonlinear programming. This approach removes most of the guesswork found in conventional restoration planning and provides a systematic and optimal solution in the sense of minimizing the variations from normal, and the number of control variable adjustments from stage to stage. The optimization reduces the duration of the restoration procedure and guarantees the security of all variables during the restoration process by respecting all steady-state operating constraints.

INTRODUCTION

With the increased dependence of the economy and society in general on a high level of power system reliability, not only is it important to avoid power failures, but it is essential to limit the scope and the duration of these failures. With this aim, utilities are becoming more interested in developing computer aided restoration methodologies, specially those that rely on interactive graphics to facilitate the creation, visualization and analysis of restoration plans.

Many electric companies have developed system restoration schemes to meet their particular needs. Most of these schemes contain general guidelines [1-3] which serve to define the required restoration plans, but there exist relatively few systematic procedures [4-6]. In recent years, expert systems have shown potential since heuristic knowledge plays an important role in power system restoration [7-12]. It seems, however, that no effort has been developed to represent and solve the restoration process as a mathematical problem. Also, the literature contains few implementations describing the use of interactive graphics to visualize and control the restoration process.

This paper presents a new power system restoration planning methodology which incorporates interactive graphics and an optimization algorithm. The interactive graphics provides a tool to define restoration scenarios, view the restoration procedures, and display the optimization and simulation results. This restoration scheme invokes, for the first time, an algorithm to optimize the settings of all controllable variables by respecting all steady-state operating constraints. These constraints are very hard to satisfy with conventional heuristic approaches, and the algorithm facilitates their satisfaction considerably. This approach therefore cuts back on trial-and-error procedures, thereby reducing the duration of the restoration process. The proposed strategy is tested on the Hydro-Québec system, but it is formalized in a general way so that it is applicable to other systems.

SYSTEM RESTORATION

During system restoration after a major black-out, the main problem that operators face is to re-energize the system as quickly as possible while maintaining voltages within acceptable limits, both for transient and steady-state conditions. Maintaining such limits is difficult because the networks during restoration are unloaded or very lightly loaded, so that the HV lines, which are usually under-compensated, generate an excess of vars, thereby increasing voltages to very high levels. This problem is particularly acute in the case of system such as the Hydro-Québec network.

Such networks are characterized by long transmission lines, remote hydraulic generating facilities and a quasi-radial main grid system. The main problems for such systems are the long overall restoration times due to sequential switchings and adjustments, and the various overvoltage conditions resulting from line charging, harmonics and switching surges. It is therefore essential that restoration be carefully planned before actions are taken.

In order to gain speed, Hydro-Québec's currently used strategy applies the following principles:

- The main grid network (spanning tree) is divided into sub-networks, or sub-systems, which may be energized independently and simultaneously.
- Energize sub-networks of main grid as quickly as possible, neglecting the load restoration aspect during this critical period.
- Minimize the number of re-adjustments and re-routings.

Reprinted from *1991 IEEE Power Industry Computer Application Conference*, pp. 216–222, May 1991.

The advantage of re-energizing the sub-networks independently and in parallel is to substitute a complex problem by several simpler sub-problems which are easier to solve and, at the same time, to speed up restoration processes.

OPTIMIZATION OF RESTORATION PROCESS

A sub-system is re-energized in several stages, each corresponding to closing a circuit breaker and energizing a transformer, a new line section, a shunt reactor or a load. Each closing of a breaker essentially creates a new network, a new set of voltages and a new active/reactive power balance. At each stage, it is essential that voltage and other limits be respected. To meet these constraints, the operator can control the generator bus voltage, transformer taps, switchable shunt reactors, and the amount of load restored. In the following, these quantities which help to control the network will be referred to as control variables.

Since, during restoration, the operator is under stress and speed is of the essence, it is preferable to keep the number of control variable adjustments at each stage as low as possible. In fact, an ideal solution is one where the control variables set at one stage are kept unchanged during all subsequent stages. In this way, the operator has, at the most, only one control variable to adjust per stage. For example, a sub-system from Churchill Falls to Manicouagan is energized in six stages (Figure 1). There are three

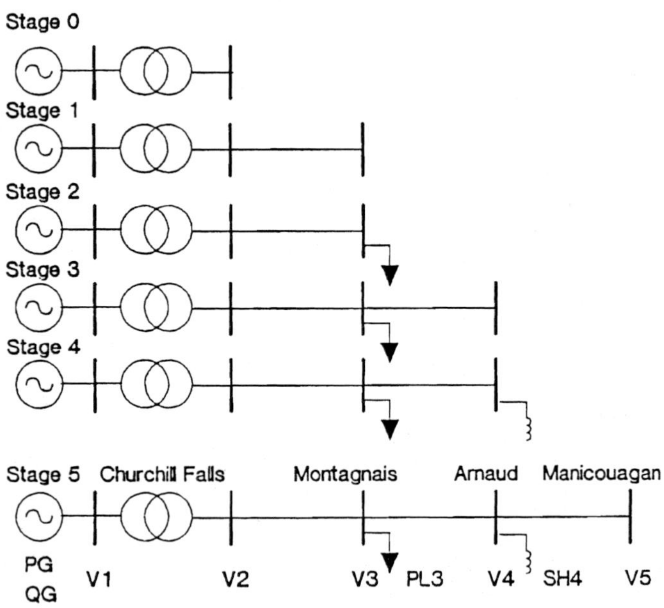

Figure 1 Restoration of a sub-system in stages

control variables: the generator voltage at Churchill Falls, the load at Montagnais and the shunt reactance at Arnaud. If the control variables were allowed to vary from stage to stage, as many as 12 adjustments would be required. So many adjustments are not acceptable in real-time since it would unnecessarily lengthen the restoration time and increase downtime costs. Finding a control generator voltage in stage 0, for example, which if kept fixed during all subsequent stages, guarantees that all constraints will be satisfied at all stages is not a trivial task. This step in restoration planning has usually been conducted using a trial-and-error approach with a power flow program. The multi-stage optimal restoration method proposed here was designed to simplify this task and to maximize operational efficiency [13].

Radial sub-systems are used below to illustrate the concept, but the proposed method is not limited to such systems. In fact, its derivation was not based on any specific assumption.

It should be mentioned that restoration planning also requires transient network analysis (TNA) studies to simulate transient behaviour, particularly after switching operations. The objective is to make sure that the restoration plan obtained from the above steady state methodology does not cause excessive switching overvoltages. The focus of this paper is however on the steady state aspects.

Power flow equations

At each stage k, the network equilibrium equations (power flow) can be written as:

$$F_k (X_k, U_k) = 0 \qquad (1)$$

Where X_k represents the bus voltage magnitudes and the phase angles in stage k, while U_k represents the controllable variables in stage k (e.g. generator voltages, loads and shunt inductances).

The dimension or the values of X_k and U_k vary with the stage. For each new line section switched on, X_k increases by two elements, corresponding to the new voltage magnitude and phase angle, and the value of all other elements changes since there is now a new network topology and a new set of power flow equations (1). If the new stage results from the switching of a load or a shunt reactor, the dimension of U_k changes with the addition of a new control variable. In this case, X_k changes in value but not in dimension, since the number of power flow equations remains the same.

The mathematical representation of the m-stage restoration process is therefore governed by m power flow equations in the form of equation (1).

Operating constraints

To guarantee that at each stage all voltage and reactive power limits are met, a number of inequality constraints are imposed. These take the form:

$$G_k (X_k, U_k) \geq 0 \qquad \forall k \qquad (2)$$

where (2) represents limits on the bus voltages, the real and reactive power for the generators, the inductance of the shunt reactors and the loads.

Minimum adjustment constraints

The ideal control strategy, which eliminates readjustments of control variables as mentioned above, can be expressed mathematically as follows:

$$U_{kj} = U_{ij} \qquad \forall k>i \qquad (3)$$

i.e. once a control variable U_{ij} at stage i is fixed, then it is kept unchanged for all subsequent stages k>i, j being the index of the bus with which the control variable is associated. This assumes that at most one control variable is introduced per stage.

While reflecting desirable characteristics for restoration plans, constraints of this form tend to reduce the solution space so that a feasible solution may not exist in some circumstances. Such failures only mean that the specified scenario is unrealistic and should be modified or replaced by a new one. With the previous trial-and-error approach, there was no clear indication whether a solution existed for a given scenario.

Objective function

The optimization criterion selected for measuring the performance of the restoration process during the various stages essentially expresses bus voltage deviations from the prescribed profile. Since voltage deviations are not of equal importance for different buses at different stages, weighting factors are introduced to strengthen the enforcement where it is most needed, such as at load buses and synchronization buses. It has been found that a quadratic function of the voltage deviations from the desired voltage profile gives satisfactory results. The objective function is therefore:

$$\text{Min.} \quad \sum_{i=1}^{M} \sum_{j=1}^{N_i} W_{ij} \, (V_{ij} - 1.0)^2 \qquad (4)$$

where:

N_i is the number of buses at stage i
M is the number of stages
W_{ij} is the weighting factor for bus j at stage i
V_{ij} is the voltage for bus j at stage i

GRAPHICS ENVIRONMENT

The interactive graphics facilitates the definition, visualization and analysis of restoration plans. This facility is essential for the generation of such plans because of the numerous heuristic and trial-and-error steps involved. For example, the sub-division of the main grid into sub-networks is greatly simplified through the graphics environment. Another area where graphics are a big help is in defining the sequence of switching operations (switching scenario) required to energize each sub-system. With the graphics environment we can not only define restoration plans, but we can visualize on one line diagrams of the grid, the energization process at its various stages, including the values of all variables. The final plan can also be thoroughly analyzed in many different ways with this facility.

Figure 2 illustrates the main components of the graphics environment developed for power system restoration [14]. These are, the graphics software and its data base, the man-machine interface (MMI), the external simulation programs and their data bases, and the translator between the graphics and external programs data bases. Communication among these modules is automatically handled by the general purpose graphics software EASE+ [15].

The graphics module allows users to create, modify and display network diagrams, to interface with external programs and data bases, as well as to display simulation results.

The data bases contain the necessary data for graphical representations (graphic symbols and associated numerical data) of the power systems and external simulation programs.

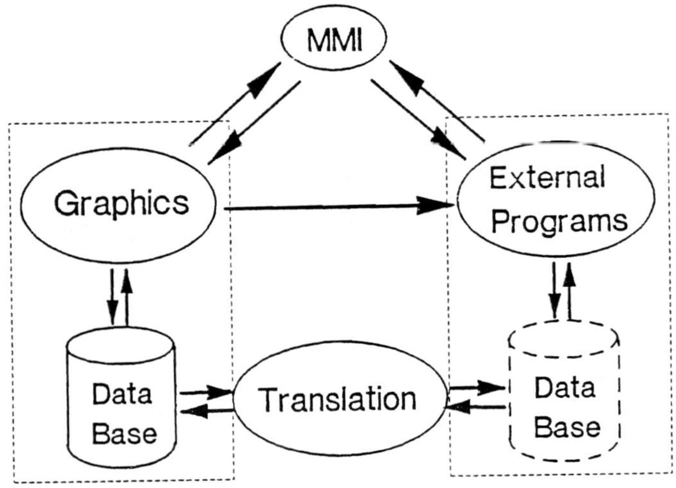

Figure 2 Graphics environment

External programs include load flow, optimal power flow, and path searching. The environment permits the addition of other simulation programs as well as of expert system modules.

The MMI allows the user to interface with the various components of the graphics environment through a mouse or keyboard, by selecting items from user-designed pop-up menus, or by pointing to and selecting specific components on the screen. Through the MMI, the user can also create icons to represent network components and interconnect them graphically. The MMI also permits the user to interactively enter on the screen the data characterizing each component, and to automatically store them in a data base.

The translation module is a program to convert data from the graphical format to and from the format of the external programs (the writing of this program is facilitated by a package available in the general purpose graphics software).

The control of the logic flow among the various components of the environment is done through the execution of compiled procedures programmed in the language of the general purpose graphics package. This package also contains numerous tools to build and modify the environment.

The main software requirements of the present environment are: (1) The general purpose graphics software EASE+, (2) Conventional power system simulation and optimization software written in Fortran.

The hardware requirements for the environment are relatively modest: A 286 or 386-based personal computer with 640k RAM, math co-processor (optional), VGA (or EGA) monitor, and a mouse (optional).

RESULTS

Manual generation of a switching scenario

A restoration switching scenario is defined by a sequence of components (e.g. a generator, a transformer, a line section, a shunt reactor, a load) which are connected to the sub-system being energized by the closing of its corresponding circuit breaker. The graphics environment allows the user to create a switching scenario interactively and easily on the graphic screen one element at a time. More specifically, in order to create a scenario interactively, the function "define scenario" is accessed (Figure 3). The first component to be energized must be a generator. This is accomplished by placing the cursor on a generator icon on the graphic screen and clicking the mouse. The color of the generator will turn to red which means the generator is energized. After energizing the generator, an adjacent sub-system can be energized by closing a series of circuit breakers. A circuit breaker is closed in the same way as energizing a generator, that is by selecting it with the mouse. Once a circuit breaker is closed, the color of the corresponding energized element becomes red. Therefore, after these operations the color of the energized sub-system turns red, and the restoration scenario is then defined. Figure 4 shows a defined scenario which is for energizing the sub-system from Churchill Falls to Manicouagan as described in Figure 1.

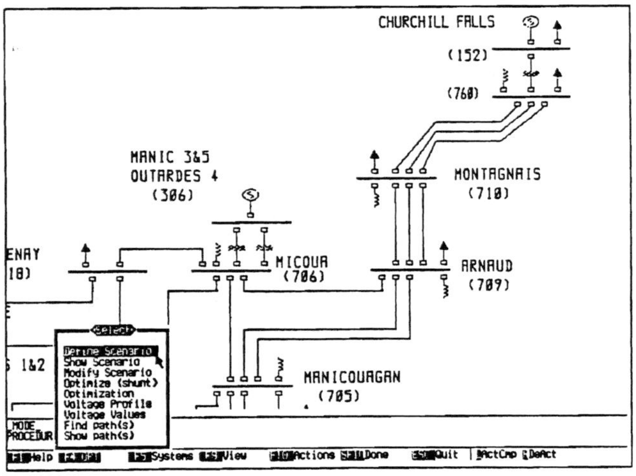

Figure 3 Creation of a restoration scenario

Optimum settings of controllable variables

Once a restoration scenario is defined, it is necessary to select the values of the controllable variables at all stages to guarantee satisfaction of the operational constraints. In order to speed up the restoration process, it is also important that the controllable variables once set, not be modified in later stages. This task is fulfilled by running the optimization program described earlier. This program can be called through the graphics environment by selecting a function key. The optimization program reads the defined scenario from the graphics data base through the translation module. It optimizes the settings of the controllable variables of the energized sub-system at each restoration stages. The program will tell user whether the defined scenario is feasible or not, and then sends these results back to the graphics data base.

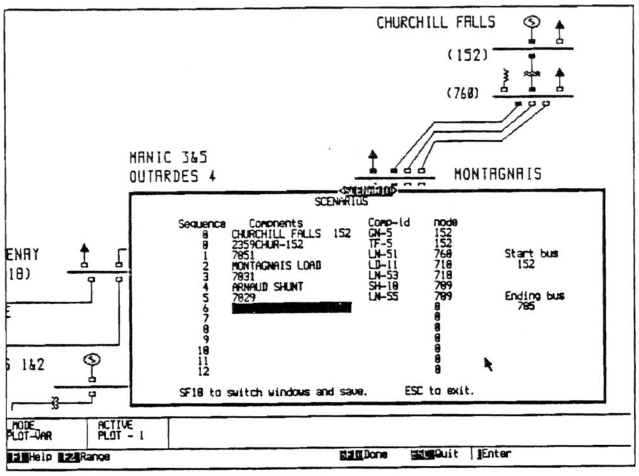

Figure 4 The defined restoration scenario sheet

Voltage profile display

After running the optimization program, the voltage profile of all restoration stages can be viewed on the graphics screen. The voltage profile shows how the voltages on buses of the sub-system perform at each restoration stage. Figure 5 is an example of the voltage profile of the scenario defined in Figure 4. The first column represents the generator voltage (Churchill Falls) of all restoration stages. The second column is the voltage at bus 2. The third column displays the voltage values of all restoration stages at bus 3 (Montagnais), and so on. There are several bars in a column each of which has a different color and represents the voltage value of a restoration stage. In this case, there are six restoration stages, and therefore there should be six bars in column one (the generator voltage) for each restoration stage. The height of the bars, which represents the voltage value, in the first column is identical which means that the generator voltage remains unchanged for all restoration stages. In addition, other displays exist to show the optimum values of shunt inductors (Figure 6) and loads added to the network at each stage of the scenario.

Modification of restoration scenarios

A defined restoration scenario sometimes may not be

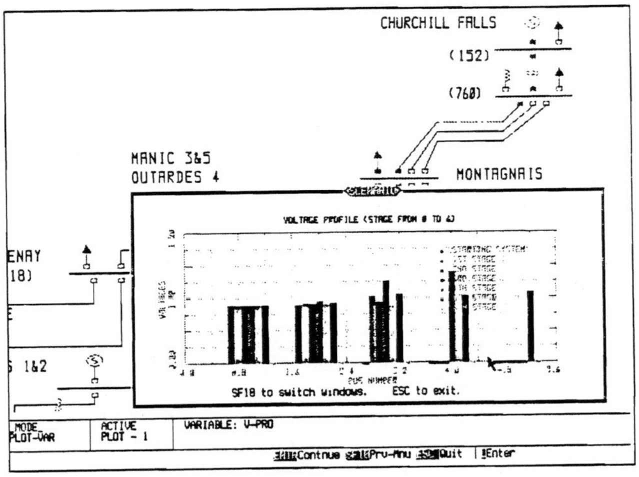

Figure 5 Display of the voltage profile

feasible if there are insufficient reactive power sources at each restoration stage, or if the voltage variation from stage to stage is too high at a bus. To avoid such situations, the user may want to modify the scenario based on his experience and knowledge. This environment provides a function to modify scenarios very easily.

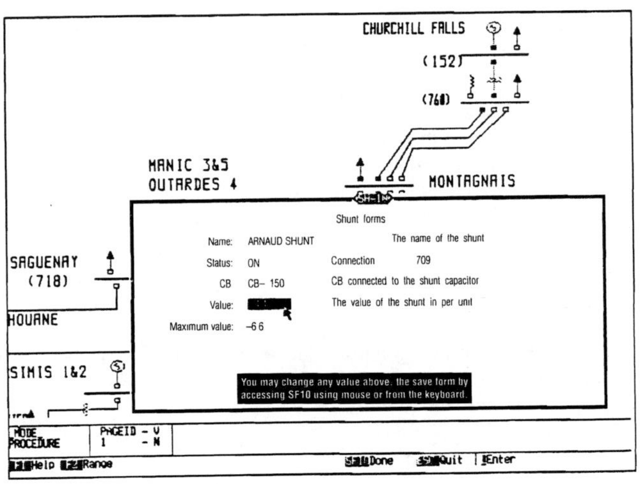

Figure 6 Optimum value of shunt element

For example, the restoration scenario defined in Figure 4 shows that the voltage variation at bus Arnaud (column 4 in Figure 5) is too high from stage 3 to stage 4. This is due to the fact that stage 3 adds a line section at Montagnais, thus raising the voltage, while stage 4 energizes a shunt reactor at Arnaud which decreases the voltage. In order to reduce the voltage variation, the

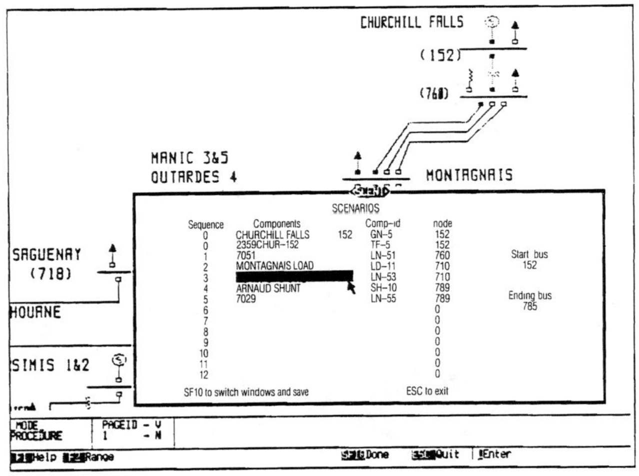

Figure 7 Modification of scenario

user may change the order of energization, that is, before energizing the line section, a shunt reactor is added at the end of the line section. This can be done by accessing the function "Modify

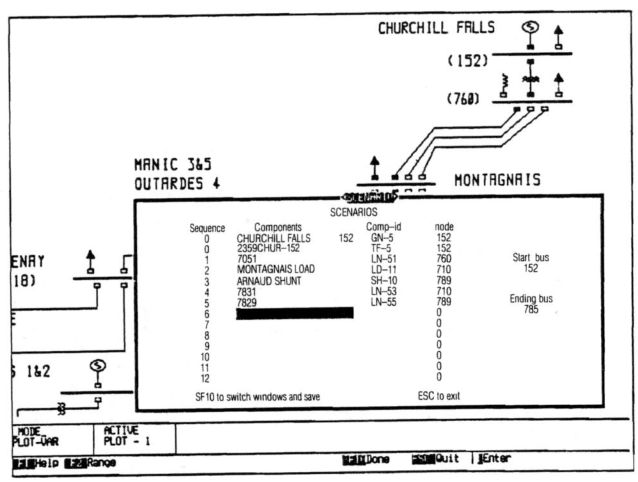

Figure 8 Modified scenario

scenario" shown in Figure 3. The scenario sheet pops up. The user then places the cursor on the component at the stage and erases it

(Figure 7). The scenario sheet is updated automatically, erasing the components in subsequent stages. The status of the components and corresponding circuit breakers is also updated automatically on the graphics screen. The user can then define the rest of the stages of the scenario in the same way as described before. Figure 8 and Figure 9 show the modified scenario and its voltage profile which is much flatter than that in Figure 5.

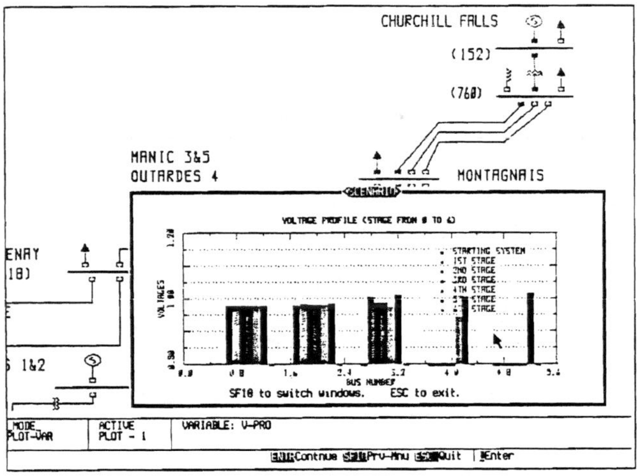

Figure 9 Voltage profile of modified scenario

Finding a set of available paths

Path searching is another external program in the environment. The search uses the algorithm of breadth first to find a set of paths with available equipment. For example, if a circuit breaker is out of service, then the line connected to the circuit breaker is not available, and a path cannot go through that line. This work is done by the function "Find path(s)". A starting point and an ending bus are selected with the mouse. The path searching program reads the data from the graphics data base together with the availabilities of all equipment in the system and writes back the 5 shortest paths in the sense of minimum number of circuit breakers. These can be viewed on the graphics screen one at a time starting from the shortest path. With the help of these available paths, the user can then define switching scenarios without worrying about the equipment availability. This path finding facility also contains a number of heuristic rules, such as choosing among possible parallel paths.

Load flow to help synchronization

Energized sub-systems have to be synchronized to the main grid system. Each sub-system is synchronized with the grid at its synchronization point where the voltage has to be equal to the grid voltage. There are two ways to satisfy this condition: 1) Increase the weighting factor on the synchronization bus in equation (4). This weighting factor can decrease the deviation of the voltage from the nominal value [13], however in order to satisfy the operation constraints, the deviation of the voltage sometimes cannot be decreased even if the weighting factor is increased; 2) The alternative is to run a load flow on the sub-system which is the final stage of the restoration scenario. If the voltage on the synchronization bus is not equal to the grid voltage, the user can manually adjust the generator voltage in the sub-system, and then run a load flow which can bring the voltage closer to the value of synchronization.

CONCLUSIONS

This paper has described a new power system restoration planning methodology which incorporates interactive graphics, an optimization algorithm, path searching procedures, heuristic rules, and interactive load flow simulations.

The graphics environment is particularly powerful and useful. It greatly facilitates the creation of restoration plans visually on the screen, and the dynamic display and analysis of simulation results. For example, through the graphics environment, we can easily sub-divide the grid to be restored into sub-networks. We can request the execution of external programs such as finding available paths from generation to load for energization purposes. We can define scenarios of switching sequences, and execute the scenario optimization. Finally, we can also modify and test a restoration plan interactively with a load flow simulator.

By introducing the concept of optimal restoration scenario, the process of selecting the control variables can be formulated as a mathematical problem which can be solved by nonlinear programming. This approach removes most of the guesswork found in conventional restoration planning and provides a systematic and optimal solution in the sense of minimizing the variations from normal, and the number of control variable adjustments from stage to stage. The optimization reduces the duration of the restoration procedure and guarantees the security of all variables during the restoration process by respecting all steady-state operating constraints.

Future extensions of this application will be linked to expert systems modules which will incorporate heuristic rules for the sub-division of the grid into sub-networks, and for the automatic creation of switching scenarios. Rules exist based on years of experience which can be exploited to further improve the efficiency of the restoration process. Heuristic rules which could be programmed in a knowledge based system also exist to provide advice to the operator under emergency restoration conditions.

ACKNOWLEDGEMENTS: This research was supported by the Natural Sciences and Engineering Research Council of Canada, and by the Fonds pour la Formation de Chercheurs et d'Aide à la Recherche, Québec.

REFERENCES

[1] R.J. Kafka, M.M Adibi, et al., "**ROLE OF INTERACTIVE AND CONTROL COMPUTERS IN THE DEVELOPMENT OF A SYSTEM RESTORATION PLAN**", IEEE Transactions on Power Apparatus and Systems, Vol. PAS-101, No. 1, January 1982, pp. 43-52

[2] M.M. Adibi, et al., "**POWER SYSTEM RESTORATION - A TASK FORCE REPORT**", IEEE Transactions on Power Systems, Vol. PWRS-2, No. 2, May 1987, pp. 271-277

[3] M.M. Adibi, et al., "**POWER SYSTEM RESTORATION - THE SECOND TASK FORCE REPORT**", IEEE Transactions on Power Systems, Vol. PWRS-2, No. 4, November 1987, pp. 927-933

[4] J.A. Huang, F.D. Galiana, G.T. Vuong, "**SYSTEM RESTORATION IN POWER SYSTEM**", Proceedings of IASTED Power High Tech'89, Valencia, Spain, 1989, pp. 17-25

[5] E.J. Simburger, F.J. Hubert, "**LOW VOLTAGE BULK POWER SYSTEM RESTORATION SIMULATION**", IEEE Transactions on Power Apparatus and Systems, Vol. PAS-100, No. 11, November 1981, pp. 4479-4484

[6] H. Kodama, H. Suzuki, "**INTERACTIVE RESTORATION CONTROL OF ELECTRIC POWER SYSTEMS**", Report 514-04, **CIGRE-IFAC Symposium**, Control Applications for Power System Security, Florence, September 26-28, 1983

[7] F.F.Wu, A. Monticelli, "**ANALYTICAL TOOLS FOR POWER SYSTEM RESTORATION CONCEPTUAL DESIGN**", IEEE Transactions on Power Systems, Vol. PWRS-3, No. 1, February 1987, pp. 10-16.

[8] T. Sagaguchi, K. Matsumoto, "**DEVELOPMENT OF A KNOWLEDGE-BASED SYSTEM FOR POWER SYSTEM RESTORATION**", IEEE Transactions on Power Apparatus and Systems, Vol. PAS-102, No. 2, February 1983, pp. 320-329

[9] K. Hisayoshi, K. Masahiko, "**AN EXPERT SYSTEM ASSISTING FAULT ANALYSIS AND RESTORATION OF A POWER SYSTEM**", Workshop on Expert Systems, Milan, Italy, January 15-16, 1987.

[10] N. Kakimoto et al., "**AN APPLICATION OF AI TO RESTORATION OF BULK POWER SYSTEMS**", Symposium on Expert Systems Application to Power Systems, Stockholm-Helsinki, August 22-26, 1988.

[11] D.S. Kirschen, T.L. Volkmann, "**RESTORING A POWER SYSTEM WITH AN EXPERT SYSTEM**", IEEE/PES 1990 Summer Power Meeting, Paper 90SM 341-8 PWRS.

[12] S.M. Wang et al. "**A DECISION-SUPPORT EXPERT SYSTEM FOR BULK POWER SYSTEM RESTORATION**", Proceedings of the Tenth Power Systems Computation Conference, Graz, Austria, August 19 - 24, 1990, pp. 966-971.

[13] J.A. Huang, F.D. Galiana, G.T. Vuong, "**AN OPTIMIZATION ALGORITHM FOR POWER SYSTEM RESTORATION**", Proceedings of the Canadian Electrical Association, Fall 1990 Meeting, St. John's, Newfoundland, September 9-12, 1990

[14] J.A. Huang, F.D. Galiana, "**AN INTEGRATED PERSONAL COMPUTER GRAPHICS ENVIRONMENT FOR POWER SYSTEM ANALYSIS AND DESIGN**", IEEE/PES 1991 Power Winter Meeting, Paper 91WM 075-2-PWRS.

[15] "**EASE+ REFERENCE MANUAL**", Version 3.0, Expert-EASE Systems,Inc., Belmont, California, December 1989.

APPENDIX

System data used in Figure 1:

Line	R	X	B
1-2	0.00003	0.00226	-0.0340
2-3	0.00060	0.01411	3.8180
3-4	0.00058	0.01351	3.5550
4-5	0.00045	0.01042	2.2020

shunt inductance: -4.4 p.u. at bus 4.

Results shown in Figure 4:

	Stage 0	Stage 1	Stage 2	Stage 3	Stage 4	Stage 5
PG	0.00000	0.00264	2.00352	2.02120	2.00639	2.00756
QG	0.03205	-3.6486	-2.7635	-6.6401	-2.0087	-4.1846
V1	**0.97088**	**0.97088**	**0.97088**	**0.97088**	**0.97088**	**0.97088**
V2	0.97085	0.97941	0.97730	0.98633	0.97555	0.98061
V3		1.00653	0.99096	1.05613	0.97827	1.01485
PL3			*-2.0000*	*-2.0000*	*-2.0000*	*-2.0000*
V4				1.08211	0.94480	1.00929
V5						1.02100

Note: V1 (bold) and PL3 (underline) are remained the same for all restoration stages.

SIMULATIONS AND FIELD TESTS OF A REACTOR COOLANT PUMP EMERGENCY START-UP BY MEANS OF REMOTE GAS UNITS

Pavel Omahen
Elektroinštitut "M. Vidmar"

Ferdinand Gubina
Faculty for Electrical Engineering and Computer Science

Ljubljana, Slovenia, Yugoslavia

Abstract

The problem of the reactor coolant pump start-up in case of emergency by means of remote gas power plant units was analyzed. A simulation model was developed which enabled a detailed simulation of the transient process occurring at the start-up. The start-up of the RCP motor set was simulated in case of available one and two gas units. The field tests were performed and the measured variable values complied well with the simulation results.

Two gas units have been determined as a safe start-up scheme of the RCP motor set considering for safety reasons accepted busbars and motor protection settings. A derived model for deep rotor bars was experimentally confirmed as effective means for the RCP motor set start-up transient simulation. Start-up procedures has been designed and adopted to the safety procedures of the Nuclear Power Plant Krško.

1. INTRODUCTION

Safety of a nuclear power plant depends on a reliable electric power supply to the safety related components and drives in case of emergencies. According to safety regulations, the 632 MW Krško Nuclear Plant (NPP) has to meet the requirements to assure the power supply from an internal emergency generation source and two external sources /1/. The external sources have to be mutually independent. This requirement could be met by two different electrically remote voltage levels, which are not subject to the same consequences of a disturbance occurring at one of them.

After a loss of the external power sources, the nuclear power plant has to resort to the reactor shut-down. The safety related house load for a safe reactor shut-down consists of two diesel generator sets (DG) and a DC battery source. However, the capacity of these internal sources does not suffice for driving the reactor coolant pumps (RCP) and the reactor has to cool off by a natural water circulation process.

In order to increase the reactor shut-down safety in case of a power system black-out, two possibilities were analyzed: a rapid restoration of the 110 kV power network to which the NPP house load is connected, and a start-up of another power plant to ensure the NPP safety related house load supply. Since the restoration time can not be controlled in all cases, the second alternative had to be adopted.

In case of the power system of Slovenia, the 400 kV and 110 kV networks can not be treated as completely independent sources since they may be affected by the same black-out simultaneously. Therefore, another independent source of electric power was needed to meet the safety requirements. The closest Brestanica Gas Power Plant (GPP) could be utilized as a second independent source after adjusting it to be able to exhibit a "black start" and continue a stable isolated operation with a specific load pattern.

92 WM 077-8 EC A paper recommended and approved by the IEEE Nuclear Power Engineering Committee of the Power Engineering Society for presentation at the IEEE/PES 1992 Winter Meeting, New York, New York, January 26 - 30, 1992. Manuscript submitted August 30, 1991; made available for printing November 20, 1991

The load involved in a safe shut-down of the NPP consists predominantly of the asynchronous motor drives. A start-up of induction motors with their nonlinear load characteristic by means of a gas unit had to be evaluated carefully. For a started generating gas unit, the most difficult case represents the duration of the RCP induction motor start-up along with adjusting to the active and reactive pump load variations. After a black- out and unsuccessful transition of the NPP unit to the house load, the diesel generators can not supply power to the RCP motor and they halt before they succeed to be connected to a gas plant. Therefore, a start-up of a RCP induction motor rated 5.2 MW by means of one or more of 32 MVA gas units had to be simulated and field tested. The question posed was how many units are needed to ensure a reliable start-up of a RCP, meeting the stringent protection restrictions imposed on the voltage and currents of the RCP and on 6.3 kV busbars of the safety related house load.

A detailed simulation was performed giving directions for field tests which were conducted in turn. On that basis, the necessary requirements and regulations for the RCP start-up procedure by means of a remote gas power plant were expected to be developed.

2. MODELING THE REACTOR COOLANT PUMP START-UP

2.1. The power system outlay

The location of the Slovene power system is specific, connecting a longitudinal structure of the cascaded Greek and Yugoslav systems to the UCPTE interconnection. Disturbances and transients in any of those systems reflect in oscillations of power and overloads on the interconnection tie-lines resulting in an occasional partial or complete black-out.

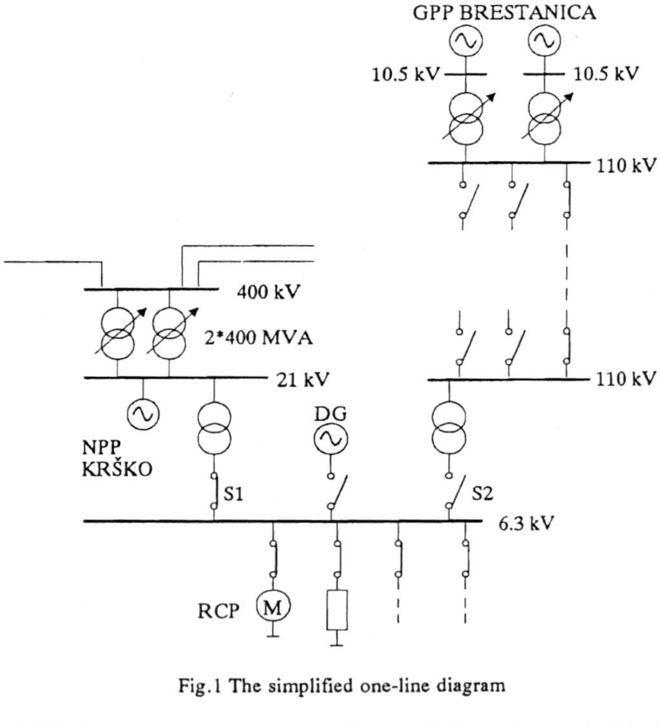

Fig.1 The simplified one-line diagram

The island operation of the gas units starting-up a RCP can be organized in the 110 kV network by a sequence of switching maneuvers resulting in the scheme given in Fig.1. The power network including the house load busbars and corresponding loads are shown in a simplified one line diagram, Fig 1.

Reprinted from *IEEE Transactions on Energy Conversion,* Vol. 7, No. 4, pp. 691–697, December 1992.

2.2. Reactor coolant pump

2.2.1. RCP mechanical torque characteristics

The reactor coolant pump nominal active power of 5.2 MW roughly corresponds to the necessary start-up power drawn from a synchronized gas unit. The characteristic of the RCP is given in Fig. 2. The mechanical torque of the centrifugal pump resembles a quadratic dependence on slip and is given by the following approximation

$$M_m(s) = a + b\,s + c\,s^2 \qquad (1)$$

The parameters a,b, and c were calculated from the measured values at the nominal temperature. They are given in Appendix B.

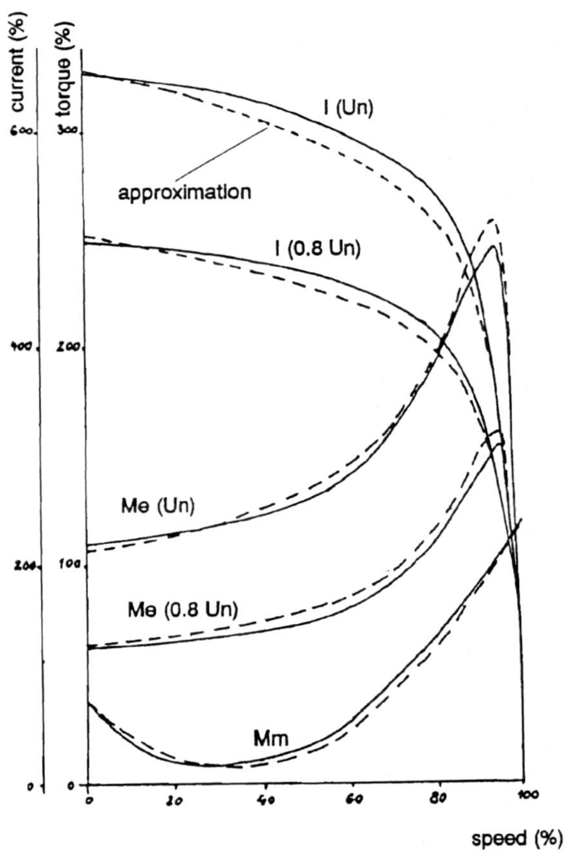

Fig. 2 Reactor coolant pump characteristics

2.2.2. Basic model of the induction motor

The reclosure of an induction motor to the voltage source exhibits highly nonlinear characteristics. Its start-up is followed by an electromagnetic transient in the stator and rotor windings and an electromechanic transient during acceleration of the RCP. The start-up produces a change in the induction motor terminal voltage and asymmetry in the stator currents and fluxes. Close to the nominal frequency, the stator direct current (magnetic flux) component produces together with the rotor currents and fluxes a pulsating moment and an average moment proportional to Ohmic losses of the motor. Since the start-up takes place at a range of speed, their influence on the motor transients is interesting. Brereton /2/ and Humpage /3/ found that influence of the stator transients on the final result is negligible. This result is important in the sense of the induction motor transient model simplification by neglecting the stator transients and taking into account the rotor transients.

The electromagnetic dynamic model of the induction motor can be adopted in the form /3/

$$pE' = -j\omega\, s\, E' - (E' - j(X-X')\, I_1)/T_o' \qquad (2)$$

where E' is a complex voltage proportional to the magnetic flux, X' is computed from the equivalent given in Appendix A, and s represents the slip.

The electromechanic part of the model is based on the dynamic equilibrium of moments

$$ds/dt = (M_m - M_e)/\,(2H_m) \qquad (3)$$

where H_m combines the inertia time constants of the entire motor set. The induction motor is connected to a flywheel which enables maintaining the pump operation and a continued forced coolant flow in case of a short energy supply outage.

2.2.3. Motor windings type influence

The induction motor model given in the preceding paragraph can be utilized in case of a single cage rotor which exhibits an even distribution of current in the rotor bar cross section within the entire speed range. The model can not be applied in case of a double cage or deep rotor bar induction machines which are prone to nonlinear distribution of current in the rotor windings at a high slip due to the skin effect. A high difference between the rotor and synchronous speeds increases the current density at the conductor surface and, in turn, enlarges the rotor resistance and decreases the rotor reactance. In addition, a high motor start-up current brings the iron to saturation and causes a further reduction of the motor reactance. However, a lower reactance and higher resistance of a double cage or deep rotor bar induction motor result in an increased moment which improves its performance in the observed case.

Modeling these phenomena was investigated by Klingshirn /4/, Smith /5/, and Lund /6/ who introduced additional parallel R-L branches to the rotor model. That model is very useful if there are corresponding data available. Experiences with the distributed model of rotor windings led to approximations in modeling the rotor resistance and reactance. That approach is useful in case of lack of data and in case of an available moment-slip and current-slip characteristic. The field tests showed that, in case of a deep rotor bar induction motor, the current-slip and moment- slip characteristics can be approximated satisfactorily by a basic equivalent circuit given in Appendix A and with a variable resistance and reactance following the relation

$$R_2(s) = R_2(0)\,(1 + K_r\, s)$$
$$X_2(s) = X_2(0)/(1 + K_x \sqrt{s}) \qquad (4)$$

where $R_2(0)$ and $X_2(0)$ represent their corresponding values at the zero slip (s=0). The constants K_r and K_x are obtained from the available motor characteristics experimentally. Utilizing the parameters given in Appendix B, less than 5 % deviation of the motor model characteristics (moment - slip and current - slip) from the measured curves can be obtained for the given range of voltages in case of applying the relations of equation (4).

2.2.4. RCP motor set protection limitations

The entire pump-flywheel-motor set time constant amounts to 11 s for 100 % of the nominal voltage and is increased to 24 s at 80 % of the nominal voltage. A further decrease of voltage results in a considerable increase of the motor set time constant and may activate the delayed overcurrent protection.

During the start-up, the maximal active power in the peak of the moment curve reaches approximately 2.5 times the motor starting power depending on the starting voltage value. The start-up current amounts to 650 % of the nominal current in case of the nominal voltage.

A considerable voltage drop on the transformer and transmission line in case of an isolated RCP start-up with a synchronized gas unit represent a severe problem. Large voltage drops at the motor terminals can excite the motor and house load busbar undervoltage protection. The corresponding characteristics had to be taken into account within the model.

2.3. Gas unit modeling

The gas units have been equipped accordingly to be able to "black start" and take over the load within less than 12 minutes. Their regulating capabilities can meet the requirements of isolated operation due to their high flexibility. A difficult problem that a gas unit encounters is a large voltage dip due to high start-up induction motor currents in case of isolated operation.

2.3.1. Generator model

Three 32 MVA gas units are available in the Brestanica Gas Power Plant for a remote RCP start-up over a direct 110 kV voltage line connected to the Krško NPP. The performance of the cold start-up depends on the generating unit parameters and its governor and voltage regulator characteristics. For simulation, a well known generator model of 6-th order with damping windings was utilized. The synchronous generator saturation has also been taken into account.

The dynamic balance of the generator rotating masses and load is given by the equation

$$T_a \, d^2\delta / dt^2 + D \, d\delta / dt = \omega_o (P_t - P_e) \qquad (5)$$

where T_a represents the unit inertia time constant, D the unit damping which is low compared to the damper winding contribution, ω_o is the nominal frequency, P_t and P_e are mechanic and electric powers.

The gas unit parameters are given in Appendix B. The units are equipped with the AC excitation system based on rotating diodes (Rotaduct). The voltage regulator is of a thyristor type and complies with the IEEE model 1 type. Its data are shown in Appendix B.

2.3.2. Turbine governor

The start-up of a RCP induction motor represents a relatively difficult task for a gas unit regarding frequency and voltage control. The task of meeting the requirements has to be carried out by the governor and voltage controller. The gas unit governor is very fast and could be adjusted for two operating conditions:

- Integral governing characteristic for isolated operation
- Proportional governing characteristic for parallel operation in case of more gas units.

The governor is based on an operational amplifier and enables a simple change of its characteristic according to the above requirements. Its main blocks are given in Fig.3. The governor parameters are given in Appendix B.

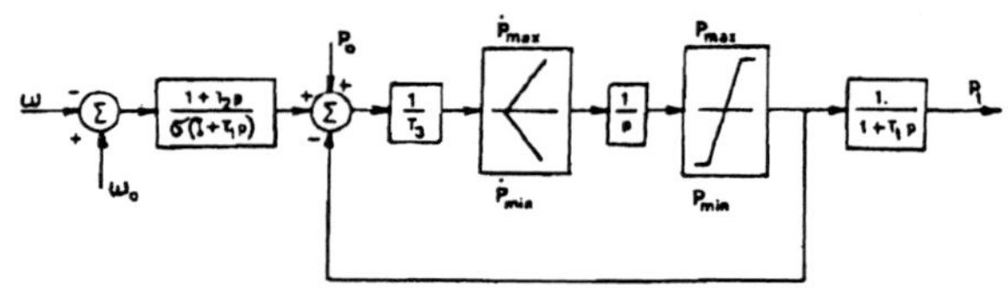

Fig. 3 A gas unit turbine governor model

3. SIMULATION AND FIELD TEST RESULTS

3.1. Simulation package

A special dynamic simulation package DINSIS was developed for simulation of various dynamic phenomena in the power system. It can handle 100 generators, Ohmic, inductive and asynchronous loads connected to the network. The generator can be represented by models of various order. The lines are represented with a π equivalent. The loads can be represented as a nonlinear frequency and voltage dependent.

Various events can be handled by the package ranging from different element switching, start-up of induction motors, isolated operation of power system segments, short circuits at various locations, switching of an induction motor to buses of different voltage amplitudes and phases etc. Furthermore, various protection schemes are integrated in the simulation package. By means of the package, an asynchronous operation of the power system parts could be analyzed.

The implicit trapezoidal procedure was adopted for solving differential equations. For the network calculations, the sparse matrix techniques with an optimal ordering algorithm were utilized. The package was successfully tested for a wide range of applications including the dynamic behavior and control analysis of a cascaded hydro power plants equipped with the bulb units.

3.2. Simulation results

The following questions had to be answered by simulation: is one gas unit capable of starting the RCP motor set within the protection limitations? The results of simulation were further used for field tests.

3.2.1. RCP start-up with one generator unit

The gas unit governor was set to the integrator scheme producing a zero speed droop. The generator voltage had to be set within voltage limits to a higher than the nominal value in order to reduce the RCP motor set voltage dependent time constant and to avoid the protection activation.

The RCP motor set was connected to its 6.3 kV busbars and through the 110 kV transmission line to the synchronized generator running at 50 Hz. The simulation results for one gas unit starting a RCP set are given in Fig. 4. The generator voltage was set to 1.08 U_n producing 6.8 kV at the RCP busbars. Immediately after the RCP start-up, the busbar voltage dropped to 4.4 kV and the induction motor current reached 4.8 p.u., The generator load at that moment was 2.5 MW and 28 Mvar. After 0.25 s, the RCP busbars voltage reached the lowest value of 4.2 kV. Activation of the voltage regulator improved the voltage profile and at that moment the induction motor reached its peak current value of approximately 5.4 p.u. After the induction motor passed its moment characteristic break point the voltage regulator restored the house load busbar voltage to 6.68 kV.

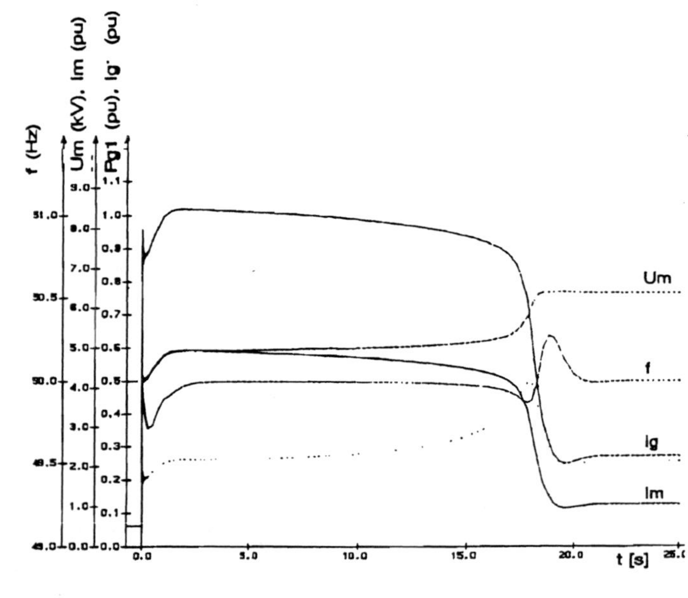

Fig. 4 RCP motor start-up with one gas unit

Simulation showed that a single gas unit could eventually start a RCP motor set within the protection set limitations according to the stringent protection requirements for nuclear components operation. However, the current and voltage values were close to the preset protection limits. A solution with two gas units is preferable for safety reasons.

Starting the second RCP motor set with a gas unit already loaded with 5.6 MW and 2.4 Mvar was simulated. However, the voltage at the 6.3 kV busbars dropped to 4 kV and activated the undervoltage protection. The start of the second RCP motor set with one gas unit is therefore not possible.

3.2.2. RCP motor set start-up with two generator units on line

Parallel operation of two gas units is operationally more demanding. A careful synchronization procedure has to be conducted in order to avoid operation in a motor regime of one generator activating its protection. Due to the 4 % governor speed droop, a frequency deviation has to be taken into account. Since both generating units are connected to the transmission line via a separate transformer, a half of the voltage drop is expected in case of a RCP start-up making the arrangement in this respect more favorable.

For a RCP motor start-up, the generator initial voltage was set to 108 % U_{ng} resulting in 105% U_{nb} at the RCP busbars. In order to avoid additional problems of the frequency decrease due to the governor speed droop, the frequency of the units is preset to 50.5 Hz before the RCP motor start-up.

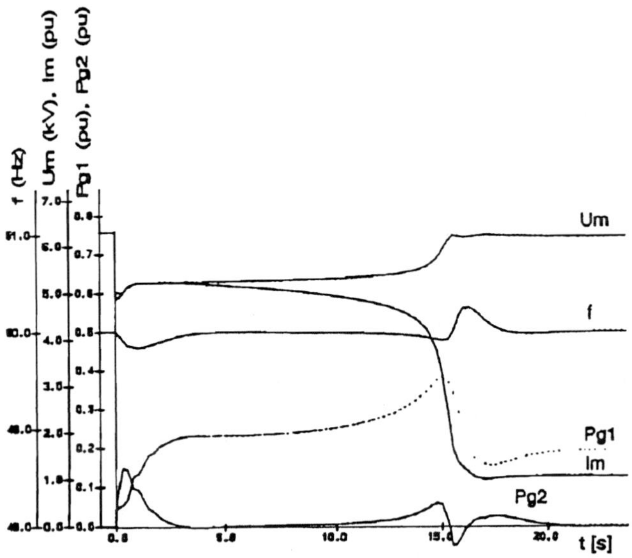

Fig. 5. RCP motor start-up with two gas units

Voltage and current time dependence after a RCP motor start-up is given in Fig. 5. The motor terminal voltage dropped to 80% of the nominal value. One second later, the voltage regulator raised the voltage to 87% of nominal value. Moreover, the generator currents reach approximately 60 % of their nominal value. The increased initial busbar voltage resulted in less than 14 s of the start-up time. The frequency deviation was low (0.4 Hz) due to fast gas turbine governors. The simulation proved that the RCP motor start-up could be easily conducted in an emergency situation.

Start-up of the second RCP motor set after the transient settled down at the generator load of 5.6 MW and 2.4 Mvar was simulated at 6.7 kV of the initial RCP busbars voltage. The minimal voltage was 4.9 kV and its average value during the start-up was 5.6 kV. At the end of the transient process, the voltage reached 6.6 kV. The transient duration somewhat exceeded 13 s. The frequency dropped from the initial 50.3 Hz to 50.1 Hz an none of the motor protections could be excited.

3.3. Results of the full scale field test

The most difficult task at a regular start-up of a RCP motor set is maintaining two gas units after synchronization of the second one in stable operation at no load condition or at a small load. A gas unit could easily drift into a motor operation after which the corresponding protection would disconnect it.

The field tests were conducted with two generating units utilizing the simulation results. Prior to the second generating unit synchronization, the first gas unit was loaded in order to prevent the motor operation of the gas unit and to ensure a stable operation of both units. A part of the Krško NPP house load of approximately 2 MW was connected to the first unit after it reached the nominal frequency.

The initial generator voltage was set to the same values as at simulation. A comparison of the measured and calculated current and voltage time dependence is shown in Fig. 6a and 6.b. During the first second, the registered current and power time function is not adequate. It is due to the fact, that the measuring transmitters do not respond to rapid changes and transmit an average value.

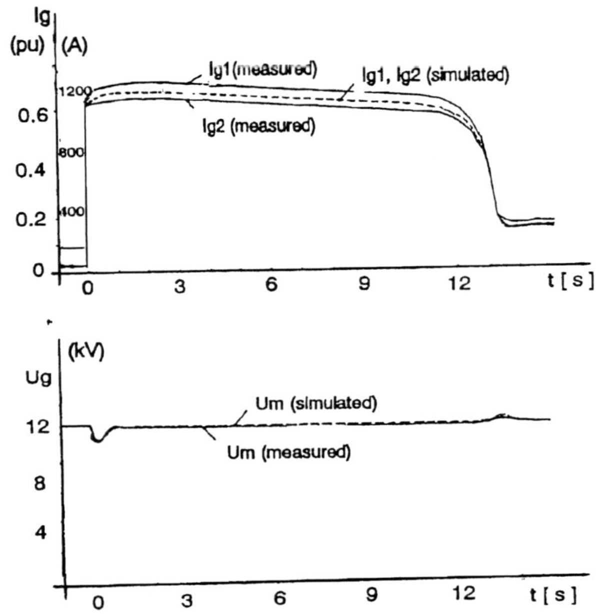

Fig. 6 Measured and calculated generator variables

4. DISCUSSION OF THE RESULTS

The measured curves coincide very well with the simulated ones. The currents of both generators differ but their average value is almost the same as the calculated one. Coincidence of both curves along the entire transient process confirms the initial supposition based on the simulation results that the stator transients of the induction motor are not necessary to be taken into account at simulation of this type of transients.

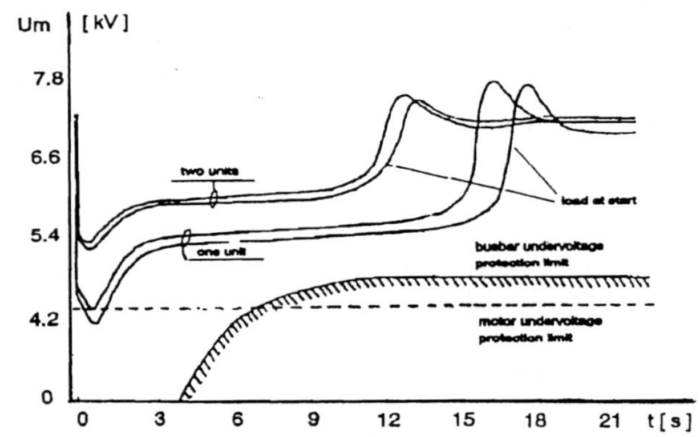

Fig. 7. Voltage trajectories and protection limits

Minimal deviations between the measured and calculated variables in case of two generators and a RCP motor start-up confirms also the conclusions obtained by simulation that the case with one gas unit might present a less reliable way of starting the RCP. Its simulated variables being close at the verge of protection settings as shown in Fig. 7 indicate that the arrangement with one gas unit is not suitable for a safe start-up of the RCP motor set which is not the case for two gas units. In the figure, both the undervoltage protection of the induction machine as well as the busbar protection are shown as a function of time.

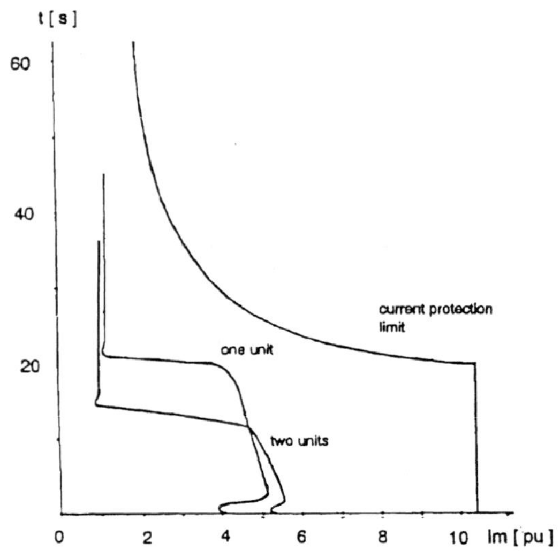

Fig.8 RCP motor current versus time

The start-up currents of the motor sets are given in Fig. 8. They show an available margin before the current protection could be activated in case of one gas unit. However, the voltage drop has to be taken into account which in turn increases the start-up time and diminishes that margin.

5. CONCLUSION

Simulation results and subsequent field tests confirmed validity of the applied models for RCP start-up simulation. It was shown that the complex electromagnetic process in the deep bar rotor winding can be represented by the empirically obtained equations (4). For the resistance and reactance of this type of winding, only the motor moment and current characteristics are requested.

The simulation model has been successfully applied in determination of the emergency start-up procedure of the reactor coolant pump by means of remote gas units. Two gas units are recommended for that purpose to ensure a reliable pump start-up in case of an emergency.

REFERENCES

1. Krško Nuclear Power Plant, Final Safety Analysis Report, part II, chapter 7&8.
2. D.S. Bereton, D.G. Lewis, C.C. Young, "Representation of Induction Motor Loads During Power System Stability Studies", *AIEE*, August, 1957.
3. W.D. Humpage, K.E. Durani, V.F. Carvalho, Dynamic-response Analysis of Interconnected Synchronous-asynchronous Machine Group", *Proceedings IEE*, Dec. 1969.
4. E.A. Klingshirn, H.E. Jordan, "Simulation of Polyphase Induction Machines with Deep Rotor Bars", *IEEE PAS*, July/Aug. 1970.
5. J.R. Smith, G.J. Rogers, G.W. Buckley, "Application of Induction Motor Simulation Models to Power Station Auxiliary Pump Drives", *IEEE PAS*, Sept./Oct. 1979.
6. P. Lund, " Induction Machine Models for Practical Use in Simulation of Transients in Power Station Auxiliaries", *Electric Power System Research*, No. 11, 1986.
7. J. Arrillaga, C.P. Arnold, B.J. Harker, Computer Modeling of Electrical Power Systems, J. Wiley and sons, New York, 1983.
8. P.C. Krause, Analysis of Electric Machinery, Mc Graw-Hill, New York, 1986

APPENDIX A: Induction motor model

The rotor transient and the electromagnetic transient of a single cage induction motor are based on the stationary electric equivalent circuit given in Fig A1. The resistance depends on slip $s = (\omega_o - \omega)/\omega_o$ in p.u. where w_o is the synchronous and w current speeds respectively.

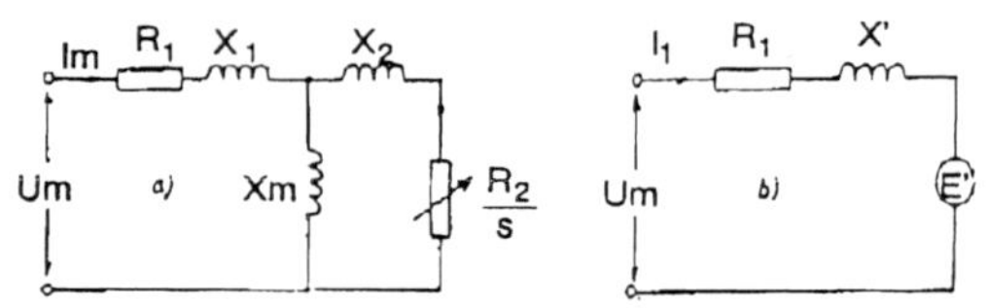

Fig. A1 Stationary and transient equivalent circuits of an induction motor

The rotor transient can be represented by an adequate phasor of the internal voltage behind its transient reactance as seen from Fig. A1b. The transient reactance is approximately defined by

$$X' = X_1 + (X_2 X_m)/(X_2 + X_m) \quad /p.u./ \qquad (A.1)$$

$$X = X_1 + X_m \quad /p.u./ \qquad (A.2)$$

Where X is the induction machine reactance at synchronous speed (s=0). The rotor time constant is defined as

$$T_o' = (X_2 + X_m)/R_2 \quad /s/ \qquad A.3)$$

The induction motor current follows the expression

$$I_1 = (U - E')/(R_1 + jX') \quad /p.u./ \qquad (A.4)$$

APPENDIX B: The power system and elements data

Tab. A.1. Gas unit generator, RCP motor and transformer data

	Generator	Motor	Units
S_n	32	6	Mva
P_n	25.6	5.22	MW
U_n	10.5	6.0	kV
X	2.22		p.u.
X'	0.113		p.u.
R_s	0.002		p.u.
T_{do}'	9.6		s
T_{do}''	0.025		s
H	2.28	5.55	s
u_k	11.51	6.5	%

Model parameters of RCP and induction motor:

R_1 = 0.018 p.u., X_1 = 0.005 p.u., X_m = 5.0 p.u., $R_2(s=0)$ = 0.013 p.u., $X_2(s=0)$ = 0.157 p.u., K_r = 1.0 p.u., K_x = 0.4 p.u., a = 1.1, b = -3.2, c = 2.35

Tab. A.2. Governor and voltage regulator parameters

Governor (PI type)	Governor (Isol. oper.)	Voltage reg. (IEEE Type 1)
$I = 1$	0	$K_a =$ 400 p.u.
$T_1 =$ 1.4 s	0.7 s	$T_a =$ 0.02 s
$T_2 =$ 0.7 s	0.7 s	$U_{r\,max} =$ 8.1 p.u.
$T_3 =$ 0.02 s	0.02 s	$U_{r\,min} =$ -8.1 p.u.
$T_t =$ 0.1 s	0.1 s	$K_f =$ 0.02 p.u.
$\sigma =$ 0.04	0.04	$T_f =$ 1 s
$P_{min} = 0$	0	$K_e =$ 1.0
$P_{max} = 0.8$	0.8	$T_e =$ 0.2 s
$S_{e\,max} =$ 0.66		
$S_{e\,min} =$ 0.46		

Discussion

N. Trehan (U.S. Nuclear Regulatory Commission, Washington, D.C.): Since this paper is to be published in the IEEE Transaction, it will be prudent for the authors to use the terminology and symbols recognized by the nuclear industry in the USA.

(1) Page 1 Title, "Simulations and Field Tests of a Reactor Coolant Pump *Emergency* start-up by means of Remote Gas Units,"—The term emergency is associated with the starting of the emergency core cooling systems (ECCS) on an engineered safety feature (ESF) actuation signal. The ECCS protects the fuel cladding following a loss of coolant accident (LOCA) by providing a timely, continuous and adequate supply of borated water to the Reactor Coolant System and, ultimately, the reactor core. In a PWR, ECCS provides high head (safety injection pumps and centrifugal charging pumps) injection, low head (residual heat removal pumps) injection and accumulator injection immediately following an accident. Since the gas turbines (GPP) will provide a second source of power in case of a unit trip, the title should be modified as "---Reactor Coolant Pump Startup by means of Remote Gas Units on a Unit Trip."

(2) Page 1, Title—Instead of starting a reactor coolant pump (RCP), a fast or slow transfer scheme to the second source of power from GPP could be utilized. This would avoid the RCP starting transient.

(3) Page 1, Introduction—Safety related loads are only used following an accident like a pipe break accident. Therefore, the sentence, "The safety related load for a safe reactor shutdown," should be modified to "The loads used for shutdown of the reactor on a unit trip."

(4) Page 1, Figure 1—The diesel generator is shown as connected to a non-Class 1E (non-safety) bus. Are these diesel generators different from the emergency diesel generators (EDGs) which are used exclusively to provide borated water following a LOCA? If they are the same, then the EDGs should not be challenged to start the RCPs which are balance of plant (non-safety) loads. EDGs should be used excluisvely to shut the reactor down in case of an accident or loss of offsite power like a pipe break.

(5) Page 1, Figure 1—It depicts ordinary switches. I believe these are circuit breakers and as such should be represented by the symbols or

I believe the above comments will help you to publish a better paper.

Manuscript received June 15, 1992.

Pavel Omahen, Ferdinand Gubina: The authors wish to thank Dr. N. Trehan for his valuable comments.

We agree that the term "emergency" could eventually be associated with starting of the emergency core cooling system. In our paper, however, it has been used to describe the network emergency status (both external sources simultaneous blacked-out) and the necessity for a start-up of the reactor coolant pump by means of a reliable source of power. As it is well known, a start-up of the reactor coolant pump is not an engineered safety feature. It is therefore clear that the title can be misleading only in the case of considering terms in a narrow sense of nuclear safety.

Dr. Trehan supposed that the gas power plant has been already in operation, which is not the case in the emergency network conditions as it is clearly stated in the paper. The gas power plant is used as an independent source of power in the form of a fast starting cold reserve because both primary external sources could black-out simultaneously under special conditions mentioned in the paper.

We agree that the arrangement of the diesel generator sets (only one is shown due to lack of space) in the one-line diagram is oversimplified from the pure nuclear safety point of view. The diesel generators are in fact connected to a class 1E bus. In case of a complete black-out of both external sources, which is treated in the paper, these diesel generators automatically start supplying the class 1E bus. After the remote gas units start, however, all safety and non-safety loads can be supplied from this reliable source. In the paper, Fig. 1 is simplified in the electrical sense for the situation described. The same applies for all the elements (busses, circuit breakers, loads, etc.). It has not been the primary intention of the paper to show the exact one-line diagram in the nuclear safety sense but to communicate our experience related to the dynamic problems of the reactor coolant pump start-up by means of gas units serving as a source of reliable power supply in the network emergency conditions.

Manuscript received June 15, 1992.

MAJOR INCIDENTS ON THE FRENCH ELECTRIC SYSTEM:

POTENTIALITY AND CURATIVE MEASURES STUDIES

by

C. COUNAN M. TROTIGNON
Electricité de France

E. CORRADI G. BORTONI
Cesi

M. STUBBE J. DEUSE
Tractebel

Abstract

The steady care for electric systems security improvement will more and more incite the grid designer to take the major failure risks into account. Electricité de France has therefore adopted a methodology based upon the study of possible degradation mechanisms, then upon the verification of their plausibility by simulating the system dynamic behaviour. Only this approach can fit with the complexity of great meshed systems. It enabled the performances of a new protection device against losses of synchronism to be defined; this "defence plan" will be operational in 1995 on a part of the FRENCH system.

KEY WORDS : transient stability, system collapse, protection.

I. INTRODUCTION

The electric systems are planned, built and operated in such a way that the customers should not be affected by the consequences of the most frequent hazards. That is especially true for the electrical incidents occurring on transmission lines, generating sets or substations. A number of rules stem from these considerations, like the well-known "n-1 rule". Expressed in a simple way, this rule stipulates that the electric system remains operational even after the loss of an element such as the tripping of a 400-kV line.

However, the satisfactory operation of an electric system stems from complex balances that cannot be guaranteed in all adverse conditions. Thus, despite precautions taken to cope with the usual events, extremely precarious situations can be met ; these situations result from critical operating conditions unfortunately joined with combinations of incidents occurring at near intervals. These combinations may be worsened by failures of protection or regulating devices. The probability of occurrence of such situations is very low but they may considerably reduce the system operation safety margins and even bring it near to a collapse. All the conditions then exist to give rise to a serious incident causing a large-scale disabling of the generation and transmission system, and leading to an interruption of power supply to many customers. Moreover the greater the failure is, the longer the time to restore a normal supply is.

92 SM 432-5 PWRS A paper recommended and approved by the IEEE Power System Engineering Committee of the IEEE Power Engineering Society for presentation at the IEEE/PES 1992 Summer Meeting, Seattle, WA, July 12-16, 1992. Manuscript submitted February 3, 1992; made available for printing May 13, 1992.

At a time of increasing dependence on electric energy, the economic and social consequences of such incidents are considerable. It should be noted that the impact of large-scale incidents might increase with the existing conditions for the development of tie-ins between companies. This trend is certainly a factor likely to favour the system security following a limited incident but, on the other hand, propitious to the spreading of large-scale failures. Though it is not likely to occur, a collapse on international scale might become increasingly realistic in future.

The FRENCH electric system set in the UCPTE network (West European interconnected system) does not escape from the rule.The probability of a large-scale incident is very low on this system without it being possible to evaluate it with the aid of reliable probabilistic methods. One can only estimate that an incident might occur every ten years. This is corroboted by the operation feedback during the past years; the two last great incidents in France occurred on 19/12/78 [1] and 12/01/87 [2].

Electricité de France decided to re-examine the breaking and spreading potentialities of such incidents within its own system on horizon 2000, and to deduce the contingent evolutions of its existing so-called "defence plan" against great collapses.This study was conducted by EDF with the collaboration of European partners, CESI (Centro Elettrotecnico Sperimentale Italiano S.p.A.) and TRACTEBEL (BELGIAN Company).

II. A BASIC APPROACH : DETERMINATION OF SYSTEM COLLAPSE MECHANISMS

As said earlier, in-depth incidents occurring on large-size systems are the outcome of always complex combinations of adverse factors; as there are a lot of factors involved, their combinative number is still greater; to try to list these combinations for the purpose of deducing the most appropriate measures would be a considerable work doomed to failure. The only possible approach is different; it consists in identifying the potential rupture lines of the system, the prevailing electrotechnical phenomena involved, and their sequencings.

Associated phenomena and sequencings will be later referred to as the system collapse mechanisms. Thus an incident will no more be characterized by its primary events, but by its mechanisms; in the same way the new defence devices will be designed such as they oppose these mechanisms.

The mechanisms have first been searched from the worldwide collection of large-scale incidents. Their likelihood has secondly been checked in the context of

Reprinted from *IEEE Transactions on Power Systems,* Vol. 8, No. 3, pp. 879–886, August 1993.

the FRENCH system. Finally, for each possible mechanism, curative actions have been proposed and tested. On the two last study steps many simulations of the system behaviour have been carried out on the basis of incident scenarios representative of the mechanism under study.The simulations had to comply with strict specifications :

- recording of a large-size system,
- deep, far-reaching transients,
- display of mechanism sequencings corresponding to phases of possibly different durations,
- easy operation and test of new palliative measures.

III. PRELIMINARY STUDY OF DEEP INCIDENTS

About forty complex incidents, adequately documented, have been analyzed to take down their mechanisms. They occurred on the UCPTE, NORDEL (North European interconnected system), Great-British, North American or Japanese electric systems, mainly during the last decade. The mechanisms are confirmed to be:

- frequency drops,
- loss of dynamic stability (transient or long-term),
- loss of steady state stability,
- loss of voltage stability,
- transmission structure trippings especially owing to a cascade load transfer.

It also evidences some thirty detrimental factors. Some of these factors refer to the network states; the others concern the transitions which make the network quickly step from one state to another.
Lastly, a dozen of typical events prone to initiate the degradation process can be isolated.
A review of the 26 most significant retained incidents proves that between 2 and 13 (5 on the average) of this fundamental factors or events participated in that collapses.

III.1 Detrimental factors and events

In order to alleviate this report, only the occurrence percentages of the factors and events met in 20% at least of the considered incidents will be presented.

a) Factors concerning the pre-incident state

- Incidents often occur in bordering areas (in the electrical meaning) of large networks : 31% (46% if the particular case of longitudinal networks is included);
- existence of an intermediate busbar on a high power transfer axis : 31%, and, especially large-scale transmission of energy on parallel circuits: 27% ;
- system weakening resulting from circuits out of commission: 31%.

b) Events initiating the incident

- Standard faults like line or busbar short-circuits : 42%,
- Tripping due to an overprotection of generators or to a disturbance of auxiliary services : 31%,

c) Factors promoting the spreading of incidents

- tripping due to cascade overload transfers : 35%,
- poor setting of protections including an overprotection : 23% ,
- existence or creation of areas showing an important energy-related unbalance : 38%,
- improper or ill-adapted load sheddings : 27%,
- Voltage instability due to a shortage of reactive power reserves : 20%.

III.2 Degradation mechanisms

The degradation mechanisms listed in III.1 do not develop at the same rate. This characteristic is important. First from the analysis standpoint, as the same incident may involve several mechanisms; the simulation of their sequencing must be possible. On the other hand, if curative measures are proposed, the rate of the phenomenon plays a decisive role in the choice of technologies adapted to their implementaton.

a) Fast system degradation: loss of synchronism

The fastest degradation processes within an electric system are the frequency drop and the loss of transient stability. With regard to frequency collapse, if the original energy-related unbalance is high, a frequency incompatible with the generator operation can be reached within a few seconds. The loss of synchronism is very fast, too; it may occur within a second after the original disturbance.

The high rate of such phenomena prevents any human action; curative action can be taken only using automatic devices. In France, these devices are known as "defence plan". The existing FRENCH defence plan already comprises a number of actions against the two mechanisms mentioned above:

- a load shedding plan on a frequency criterion, is aimed at fighting frequency collapses; it sheds nearly 60% of the load in four equal steps between 49 and 47.5 Hz;
- local relays called DRS (FRENCH acronym of "area islanding protections in case of loss of synchronism") should detect any loss of synchronism and isolate immediately the ill network portion from the healthy one.

Frequency collapses have not been detailed in so far as frequency is equivalent within the whole tied-in system, provided no loss of synchronism exists. Planning for changes in the existing curative actions would involve a redefinition of the process for the whole UCPTE, which was not the purpose of the study.

Losses of synchronism, however, have been particularly studied. They tend indeed to spread from an epicentrum which is premonitory of remedies whose application to the only FRENCH network means an improvement for all the interconnected system. Such a behaviour also enables, from the simulation standpoint, the FRENCH system and the nearby networks to be differently sharpened so as to be consistent with the quality of data supplied. The adequacy of DRS devices, as far as curative actions are concerned, has been checked.

b) Slow degradation of the networks: voltage instabilities

The other mechanisms listed may be considered to be slower. The collapse rate of a network where cascade overloads occur depends on the overload protection time lags. In the case of voltage collapse, inadmissible values may be reached within 5 to 20 minutes, and even less under adverse circumstances.

Curative actions with a human component are then possible; in France, these "safeguarding" actions consist in a fast rise of generators voltage and active power set points, in blocking network transformer tap changers or, lastly, in a load rejection (supply voltage drop and even load shedding). The rapidity of tap changer blocking has recently been improved by use of automatic devices.

Safeguarding actions might however not suffice if the above mechanisms developed more quickly. Such a possibility has been investigated, especially in the field of fast voltage collapses.

It must be underlined that static or dynamic instabilities can be superposed to slow phenomena.

IV. SIMULATION METHODOLOGY

The simulation work consisted in working out and in studying a number of scenarios representative of the possible mechanisms on the FRENCH system. In order that the scenario remains plausible, one tried to let the system develop freely on the basis of a weakened initial state and of a combination of primary events; the deviations noted in the progress of the incident then only result from the responses of system components and from actions taken by its relays. Thus to make this approach possible the simulation constraints are:

- to record all the automatic devices and relays likely to modify the progress of the incident.
- to report all the mechanisms likely to occur whatever their dynamic ranges and development rates.

Thus the automatic devices are modelled with the details and the accuracy needed. One can mention for instance the generator rotor current limitation whose performances may be important not only from the viewpoint of transient and static stability but also during voltage collapse phenomena. One can also mention some relays scattered over the network:

- generation unit islanding relay acting on voltage, current or frequency criteria,
- line or transformer overload protection,
- automatic transformer tap changer,
- area islanding protections in case of loss of synchronism (DRS).

Two modelling levels have been adopted. The first one has been designed to analyze for 5 to 10 seconds the network high-rate behaviour; it corresponds to the simulation field of conventional transient stability programs; the second one, more comprehensive, makes it possible to extend the preceding field to the medium and long term analysis.

In accordance with these modelling levels, two simulation tools have been used, EDF's present transient stability program STABTRANS [3] on the one hand, and EUROSTAG [3] [4] on the other hand. EUROSTAG is a new power system simulation package. It simulates simultaneously both electromechanical transients and long term dynamics, thanks to an automatically variable stepsize. The risks of fast-developing incidents, whose evaluation had started before EUROSTAG was available, have been reviewed, especially with the aid of the STABTRANS software, which have been long available within EDF.

EUROSTAG, has been used for the simulation of system behaviour in the short-, medium- and long-term. For instance, gradual voltage collapse mechanisms likely to result in a loss of static or transient stability or to lead to protection or automatic devices operation have been considered.

The nature of the phenomena has implied the representation of not only the FRENCH electric system (whole 400kV and major 225kV structures) but also its environment within the UCPTE; that leads to a network comprising from 1100 to 1600 nodes, over 2000 links and about a hundred generation units.

V. LOSSES OF SYNCHRONISM

V.1 Disturbance propagation due to a loss of synchronism

Several scenarios have been worked out to test the resistance of the FRENCH system to losses of synchronism and to evaluate the need of dedicated countermeasures. The grid has been weakened by the absence of important links. High power flows have been considered owing to the unavailability of generating sets. Moreover a long three-phase short-circuit brought the system to be unstable. The simulations have shown that in such severe conditions a loss of synchronism can occur. At once local, affecting a few units, this loss of synchronism creates large disturbances which may involve the nearby generating sets. The system becomes more and more disturbed, causing the possible tripping of some generators; these trippings increase the system weakness which may rapidly lead to a wide collapse. Without dedicated countermeasures the loss of synchronism can spread and affect a large part of the system and bring it in unbearable operation conditions. An example of such bad conditions induced by a 600ms three-phase fault occurring on R substation is given in figure 1 A.

V.2 Short-term curative measures

As seen earlier, failing any appropriate measures, the loss of synchronism can spread over most of the network, thus causing its collapse. On EDF's network, special devices have been in use since the seventies to limit the expansion of such incidents: the DRSs . These local devices, set at the ends of predefined EHV lines, aim at tripping them automatically when voltage beats, characteristic of a loss of synchronism, occur on these lines. The DRS detect these beats by monitoring the variations in amplitude of the local voltage. The DRSs are scattered over the network according to a preset configuration. Their arrangement is based on the principle according to which the electric system can be structured into areas. The generators of an area have an homogeneous dynamic behaviour in case of disturbances. Then, a major loss of synchronism affects not only a generation site but all the sites within one or a set of areas. To prevent the disturbance from spreading, these areas must be islanded. The network fragmentation must match the borderlines of the areas involved. Thus DRS are located at the ends of all the lines which cross an area border.

The DRS plan has acted in various occasions since its first commissioning during regional and national scale incidents. It has proved its advantages but also its limits ; these limits confirmed by simulation will be depicted in paragraph V.3.

During the last decade, the 400kV network meshing has been reinforced, so that it implies the updating of areas topography and DRS adjustment. This updating has been reviewed on the base of simulations of losses of synchronism caused by three-phase short-circuits exceeding the critical clearing time. An accurate modelling of DRS has been introduced into the EUROSTAG software used for this study. The short-circuits have been applied to most of the 400kV substations of the formerly weakened EDF's network in turn. They must be regarded only as a mean of causing losses of synchronism which should evidence the generic fragmentation diagram of the system. In a first modelling, non-activated (ie. being unable to open the line) DRS relays have been connected in any node of the network. At the end of each simulation, the number of voltage beats (which more or less corresponds to the number of pole slippings) detected by each DRS and their detection time have been listed. A list of the generation sites where a loss of stability occurred has also been worked out. The DRS areas have been determined through a review of all these informations. This analysis has also given the setting of each DRS ie. the number of beats causing the line tripping. The reliability and the selectivity of the DRS protections are based on these choices.

Validation tests have then been carried out to check the robustness and selectivity of the solution. All the simulations have been repeated with active DRSs, namely controlling the line tripping, for various short-circuit durations.

For example, figure 1.B shows the behaviour of the network in the same condition than figure 1.A, but when the system is equipped with DRSs. The disturbance is now prevented from spreading throughout the network owing to the DRS partitioning. The occurrence of untimely line openings should however be noted.

V.3 Inadequacy of the renovated DRS plan

An analysis of the DRS plan operation using checking scenarios has moderate results. On 29 significant cases, 21 progress satisfactorily, 17 of them (59%) without undesired line tripping, 1 (3%) with an islanding of sites where no loss of synchronism occurred, and 3 (10%) with undesired line tripping. The other eight cases (28%) are a failure. The success rate therefore exceeds 70%. Failures mostly result from the limits of detection of a loss of synchronism by DRSs. DRSs, in fact combine the functions of synchronism loss detection and of system fragmentation. If they are adequately located to best fulfil the second function, this can prove inadequate to the first one. Furthermore, the electrical centres move according to the operating conditions, which makes the detection on the area boundaries uncertain. Lastly, the selectivity of the nearby DRSs, based upon their location and their own threshold of detected voltage beats, cannot be guaranteed.
The simulation have proven an ultimate limit to the principle of DRSs: under given circumstances, the islanding of out-of-synchronism regions even if entirely successful does not suffice to prevent the incident from spreading ; if the region where the loss of synchronism occurs initially plays an important part in the balances required by the electric system (owing to its tieing-in role or to its participation in the generation), its disconnection within an already disturbed context might be hazardous for the other sections of the system. The remotest portions of the system which were originally safe might give rise to an induced incident resulting in another loss of synchronism. A very fast load shedding in these regions would be helpful to circumvent these secondary incidents; This load shedding cannot be worked out by the existing underfrequency relays; in such cases the variations in frequency are not sufficient within the period of time requiring an action.

V.4 The future defence plan against losses of synchronism

On the basis of the above analyses, a new defence plan, better adapted to the

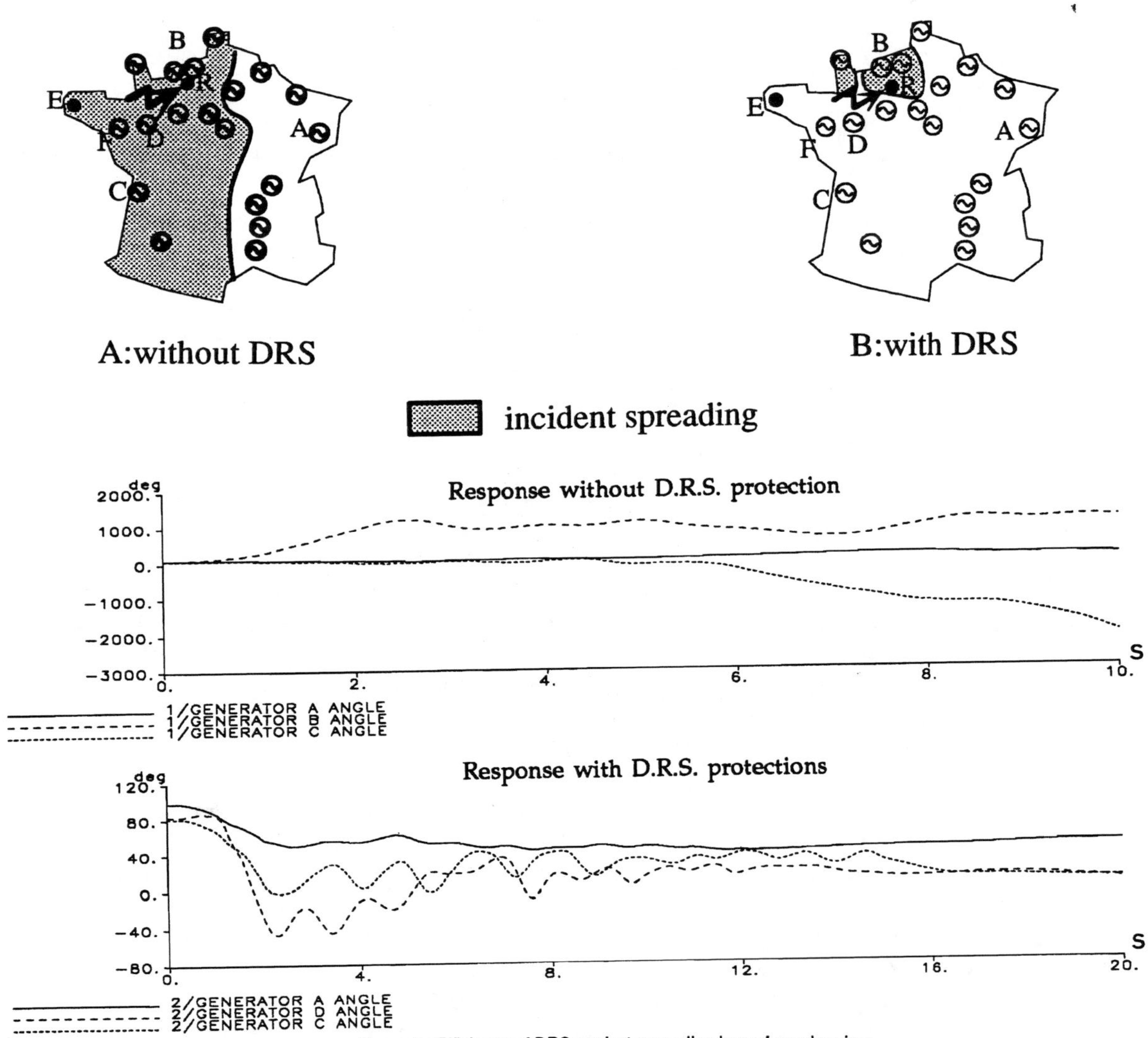

Figure 1 : Efficiency of DRS against spreading loss of synchronism

operating conditions of the FRENCH system beyond the year 2000, must fulfil the following functions:

- to ensure a reliable, simultaneous fragmentation of all out-of-synchronism areas,
- to promote the restoration of system balances under steady-state, and also transient, conditions immediately following the fragmentation of out-of-synchronism areas so as to avoid any induced incident.

The first function is an improvement of the existing defence plan; it is still based on the organisation of the network into "islandable" areas. This improvement will be obtained through a loss of synchronism detection based on a criterion much closer to the feared phenomenon than the voltage beats: a comparison between the voltage phases measured within the elementary areas of the electric system. The architecture selected which is best suited to this novel detection principle is briefly represented in Figure 2 ; it uses network phase measurements [5] that are pooled within a Central computer. This computer decides the necessary fragmentation and delivers the appropriate orders for line tripping.

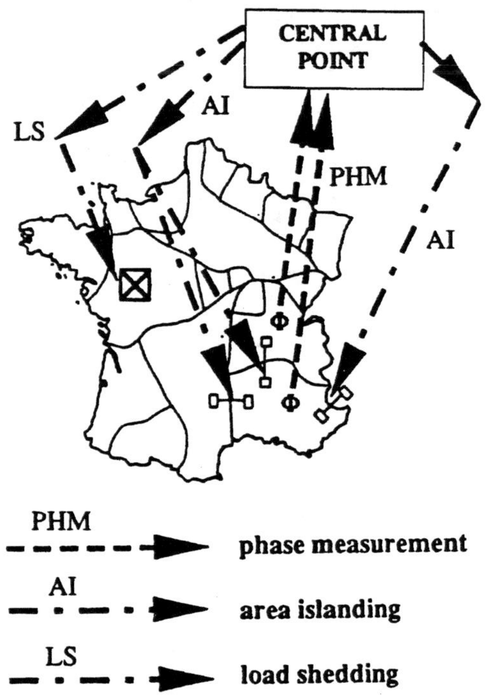

Figure 2 : Principles of the new system of defence against losses of synchronism

The second function is more unusual; it will aim at distributing the load sheddings over the safe portion of the system which are necessary to speed up the balance restoration after a fragmentation; the required load shedding algorithm (volume and distribution) implies the knowledge of the regional generation/demand unbalance state just before the incident; such knowledge supposes a global view of the system which can be achieved within the computer already responsible for the fragmentation.

This new defence plan has been modelled on the STABTRANS software. Simulations have then been carried out in order to test its effectiveness and to sharpen its characteristics and required performances. Thus, the response time of the device, from the physical occurrence of the loss of synchronism (characterized by an inter-area phase difference above 180°) up to fragmentation and load shedding actions, must be as short as possible to be fully efficient. In view of the technological possibilities, 1.3 second seems a judicious compromise.

Simulations also enabled the load shedding operation law to be defined; this law is based upon an inter-area pre-disturbance phase deviation criterion . Such criterion is representative of the load flow constrainsts applied to the system and of the possible topological weakenings. In the load shedding areas, about 30 % of load can be shedded.

Figure 3.A shows a good application of the coordinated plan. Figure 3.B shows the unsatisfactory coordination between the partitioning and load-shedding actions, leading to an unacceptable fragmentation of the network.

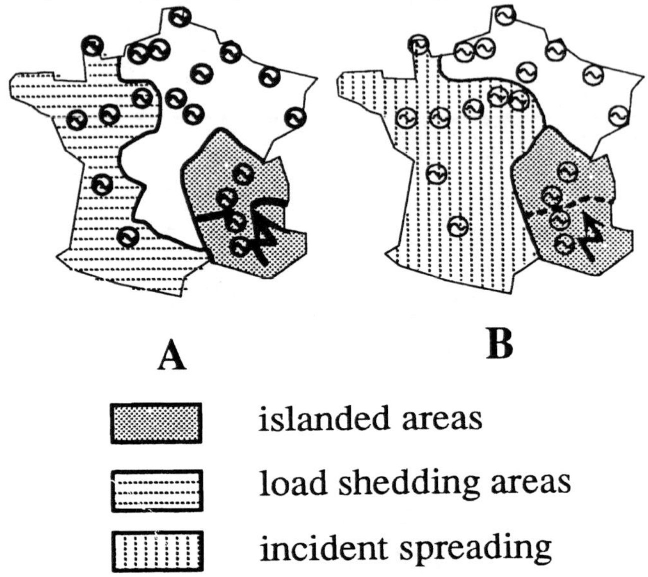

Figure 3 : Efficiency of new defence plan against spreading loss of synchronism

VI. OTHER TYPES OF INCIDENTS

Among the network degradation mechanisms mentioned in III.2b), voltage instability seems to be the most frequent cause of deep incidents in the large-size interconnected systems. Thus, a voltage collapse due to local reactive power shortages, can lead to the loss of close generation units, aggravating the process.

Simulations made in this study required the greater modelling effort because of the various phenomena that might occur during the gradual voltage drop. Beyond the different automatic devices partly mentioned in paragraph 3, a particular load model has been adopted; it incorporated an equivalent admittance behind a transformer with a continually variable transformer ratio automatically tuned. Furthermore all the loads were distributed among areas, the loads within an area being capable of varying in the same ratio as a pilot load. A hand-operated remote load-shedding device had also been added.

VI.1 Significant case selected

In order to study the phenomena associated with the voltage collapse, the western part of the FRENCH system has been considered. This network is weakened by the opening of interconnection links, the unavailability of active and reactive power generating sets. In addition, a rise in the demand at load peak (at time t=0) of about 15% within an hour is simulated so that the system is operated under unsafe conditions. Lastly, the manual or automatic safeguards actions described in section III.2.b) have been deliberatly inactivated. In such extreme conditions, the unscheduled loss of a generator in the **F** power plant (see map in figure 1) at t=4600s unavoidably causes a voltage drop and the successive islanding of several units. Within about two minutes, the voltage at the end of the 400kV network dropps to less than half its initial value.

VI.2 Fast voltage collapse: which curative actions?

The previous scenario results from an extremly severe combination of constraints, leading to a rather fast voltage collapse; only automatic actions could withstand such an incident; the functional features of this device have been explored.

In extreme conditions, near to voltage collapse, the load shedding is the ultimate solution. In order to be efficient and secure this load shedding must answer the triple question: where, when and which load amount has to be shed ? to get this answer a global view of the system is necessary. Thus the location of the remote load sheddings is determined by analyzing the trend of the marginal costs of the reactive power at each consumer node. This analysis is made during the simulation taking account of the actual situation of the generation units (rotor current limitation or normal AVR operation). An optimal power flow program is used to compute the marginal costs. Loads are assumed to be independent of voltage and have the value defined for the simulation time considered. The marginal costs undergo a discontinuity just as the voltage instability occurs (corresponding to the criterion dq/dv=0). The transition is very quick - within a few seconds, and can even be instantaneous in the case of topology modification (circuit tripping). After marginal costs inversion the load shedding becomes unavoidable. Load shedding must be performed in the substations where the reactive power marginal costs are the highest just before their inversion. Its amount is determined using a simplified modelling of each concerned area.

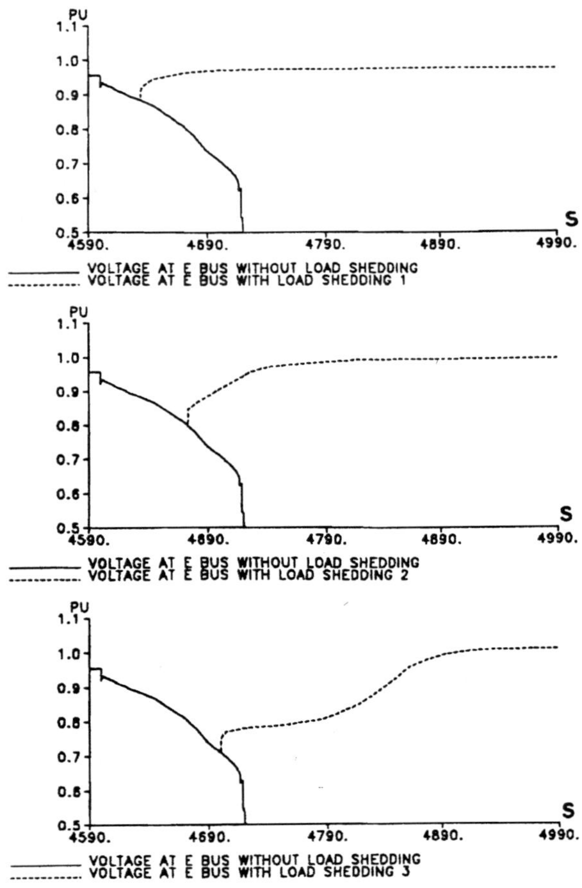

Figure 4 : Fast voltage collapse ; load shedding efficiency

The analyses have shown that the later the load shedding happens, the greater its quantity must be to prevent the system from collapsing. Simultaneously, the more the load shedding is delayed, the more the system is weakened and the less it is capable of withstanding a large-scale load shedding without risking a loss of synchronism. Beyond a given period of time, these two constraints become incompatible and the recovery of the system is unfeasible. It is therefore mandatory to carry out the load shedding with a minimum delay to reduce the quantity of load rejected and to make the operation more likely to succeed. The example proposed in figure 4 shows the voltage variations in substation E (map in figure 1) without load shedding and with one of the three following load shedding options :

- 15% of the load just as voltage instability is detected, (A)
- 20% with a delay time of 39s, (B)
- 30% with a delay time of 60s. (C)

The simulations have shown that even in the presented case of fast voltage instability, a load shedding can theoretically save the system from the collapse. However this action needs on line computing of the system state from informations which are not all available nowadays. This difficulty and the very small probability of this collapse (the existing safeguards actions have been considered out of service) has lead EDF not to consider such an action for the moment.

VII. CONCLUSION

The large-scale incidents will be increasingly taken into account in the safeguarding and protection policy followed for the large electric systems owing to the development of such systems and to the ever increasing safety or quality targets they are assigned to. This policy is however costly, due to the already high security reached level. However the electric systems will never become invulnerable and it is necessary to clearly define the need for improvement of their fitness. The upstream approach for expressing the need entails an identification of the drawbacks of the electric system. These drawbacks are not obvious as the operation feedback in relation with the major incidents occurring on a large system is low. Therefore a complementary analytical approach is necessary. Electricité de France has tried to make such an approach and proposed, and tested, a methodology which it has applied to its own particular system.

On the basis of a worlwide bibliographic review of the most recent large incidents, the main degradation mechanisms of a system have been investigated; the plausibility of such mechanisms for the FRENCH system was still to be established. It implied a considerable effort of dynamic behaviour simulation of a large system exposed to severe constraints corresponding to electrotechnical phenomena of different natures and dynamics.

This methodology enabled the functional performances of a novel device for preventing losses of synchronism from spreading to be proposed and defined.The studies presented here have been completed by experiments on the feasibility of this new "defence plan" (local voltage phase measurement, centralized computing, communication links). All the studies make it possible to decide to put in place this device in the Alps and south-east areas of France. It will be in operation in 1995.

Out of this concrete result, the approach made could be applied to other electric systems, especially to those whose closely meshed structure eliminated any event- and dedicated arrangement-related approach.

However, this methodology needs the processing of a great number of data hard to obtain, especially in the context of international networks.

REFERENCES

[1] "The incident of 19 December 1978"
SEE meeting - 20/12/79
Rev. Gen. Electr. Tome 89 n°4 - Avril 1980

[2] HARMAND Y, TROTIGNON M, LESIGNE J.F., TESSERON J.M., LEMAITRE C, BOURGIN F.
Analyse of a voltage collapse incident and proposal of new philosophy of parades based on different time scales
CIGRE 1990, Report 38/39-02

[3] M. STUBBE, A. BIHAN, J.C. BAADER, J. DEUSE
STAG a new unified software program for the study of the dynamic behaviour of electrical power systems.
IEEE Transactions and Power system - Vol 4 n° 1 - Feb 89.

[4] P. BORNARD, B. MEYER, J.P. ANTOINE, M. STUBBE
EUROSTAG a major step in power system simulation.
IERE meeting Rio - Mai 1991.

[5] DENYS Ph., COUNAN C., HOSSENLOPP L., HOLWECK C.
Measurement of voltage phase for the FRENCH future defence plan against losses of synchronism
IEEE Summer Meeting San Diego 31/7/91
91 SM 353-3 PWD

Discussion

BOZA AVRAMOVIC, (ECC, Inc., Fairfax, VA): This paper provides a comprehensive discussion of a new control scheme intended to contain a spread of serious disturbances.

Most schemes in use for this purpose are based on local measurements and the local (uncoordinated) controls. In contrast, the scheme of the paper is based on direct angle measurements, on fast communications, and on a central, computer based, coordination of controls. Simulation results presented in the paper show superior performance of the centralized scheme in comparison with the conventional, decentralized scheme.

Our concerns are with the possible impact of degradations or failures in the very tight communications/control loop. Would the authors please comment on the possible risks and remedies in this area.

C. COUNAN, M. TROTIGNON, E. CORRADI, G. BORTONI, M. STUBBE and J. DEUSE : The authors wish to thank Mr. Avramovic for his interest in the paper.

What are the risks that might induce a malfunction of the new scheme ?

On one hand, what would happen if the scheme doesn't operate the D day (the day it is needed) ? As simulations have shown, the loss of synchronism may spread to other parts of the power system. Probably after some seconds, the out-of-step generators, and may-be others, would be isolated or tripped, for instance by undervoltage relays, letting the rest of the system with a very important lack of power. So perhaps it would result in a large-scale voltage collapse, or heavy overloads, or frequency drop generalized to the whole UCPTE system, or a combination of the above.

On the other hand, what would happen if the scheme operates when it should not operate ? There would be transmission line trippings or, for the worst case, an area-islanding, with possible load-shedding in some part of the system.

So, how do we deal with reliability and availability aspects of such a scheme ? We have been doing specific studies, with a method derived from the one used in the nuclear domain. The chosen architecture of the system is an actual result of these studies taking into account precise requirements on this subject. We use redundancy, when necessary, to achieve the required performances. These dependability studies will be updated regularly during the building stage, with the manufacturers. We forecast also systematic on-line tests, open-loop tests of course, i.e. without area-islanding or load-shedding actions.

Manuscript received September 23, 1992.

Black-start and Restoration of a Part of the Italian HV Network: Modelling and Simulation of a Field Test

B. Delfino, G.B. Denegri, M. Invernizzi, A. Morini
Electrical Engineering Dept.
University of Genoa, ITALY

E. Cima Bonini
ENEL S.p.a., DPT
Regional Control Center
Turin, ITALY

R. Marconato
ENEL S.p.a.
Planning and Strategies
Department, Rome, ITALY

P. Scarpellini
ENEL S.p.a., DSR
Automatica Research Center,
Cologno Monzese, ITALY

***Abstract*-ENEL (Italian Electricity Board) planned a set of operator guidelines for the restoration of parts of the national grid starting from selected hydro power plants and progressively reaching thermal units. In order to improve system restoration reliability, ENEL periodically executes field tests taking advantage of the maintenance outages of the thermal groups.**

The paper concerns with modelling and simulation of an experimental test performed on the Italian HV network near the French border. The study makes use of a stability code provided with user defined model facilities, enabling an accurate description of loads and regulation equipments during the restoration process.

Keywords: Power System Restoration, Computer Modelling and Simulation.

I. INTRODUCTION

The chance to rapidly recover the electric power service after a diffuse blackout appears as a significant aspect both for planners and control center operators, according to the need for a more and more reliable and secure supply.

Even if service interruption is a rare event, it is essential for an electric utility to prepare itself to face such emergencies, developing strategies to minimize the impact and to restore energy to the users as quickly as possible.

In Italy, ENEL (Italian Electricity Board) formerly produced operator guidelines to restore parts of the national electrical grid, beginning from hydro power plants, able to self restarting with no external voltage provided.

The definition of a restoration plan for an electric network involves decisional problems of combinatorial nature, usually including several aspects, often contradictory.

Service restoration must be performed taking account of several system constraints, namely of dynamic nature (power system stability, frequency control, synchronization problems), linked to load reconnection, thermal generators start-up and loading, and sequences of network meshing as well.

All these aspects cause the restoration problem to be so complex that it can not be studied completely through an analytical approach. Nevertheless, by means of simulation programs it is possible to systematically analyze the most important features of power system restoration, especially concerning the behavior of control systems of hydro and thermal power units.

The importance of digital codes, both to define restoration plans and to train prospective control center operators, has been reckoned in international field by IEEE [1, 2] and CIGRE [3], which devoted working groups to the problem of modelling and simulating the different time frames in which service recovery can be organized.

The introduction of flexible power system simulators and programs, which allow the user to define own models, have been stimulating more deep investigations in order to foresee the system behavior, operating in critical conditions, and the optimal tuning of the involved controls.

A trend of large electric companies is to use the simulation instruments to test the performance of field experiments which, however, for evident reasons, could not regard large network areas.

ENEL, and especially the Northern-Western Department, exploiting programmed outages of thermal units for maintenance, reproduces area blackout conditions with the aim to validate, or at least to update, the relevant restoration plan.

The availability of fairly recent field test results encouraged the present investigation, which preliminary concerns modelling of components and restoration procedures, and successively compares simulation results with the experimental data

The possibility of simulating a complete power system recovery indeed allows to identify unsatisfactory voltage and frequency transient behaviors, and to propose alternatives both in terms of regulations and reconnection sequences as well.

All the work done at this stage is preliminary to the formulation of guidelines for an operator decision and training support system, to be developed using Artificial Intelligence techniques [4].

II. THE ITALIAN RESTORATION PLAN

Power system restoration requires to activate an overall integrated and co-ordinated procedure which involves personnel staff at power plant, transmission and distribution level. This co-ordination is achieved through the teletransmission system whose reliability and availability are of primary importance for the restoration being successful. This control organization is realized in Italy by means of three hierarchic levels, as follows:

- The National Control Center (NCC), which is responsible for system security, frequency control, load dispatch and the supervision of the transmission level (220-380 kV);
- The Regional Control Centers (RCCs), which control the subtransmission networks (132-150 kV),
- The Area Control Centers (ACCs) which are responsible for local switching on the transmission network and the hydro power plants.

Relying on appropriate communication facilities between these centers, ENEL has conceived a plan of system restart which considers an integrated procedure to be implemented in a orderly and progressive manner. Prepared on the basis of off-line computer studies, the plan considers a wide range of operative conditions although it cannot account for the events that have originated the

95 SM 549-6 PWRS A paper recommended and approved by the IEEE Power System Engineering Committee of the IEEE Power Engineering Society for presentation at the 1995 IEEE/PES Summer Meeting, July 23-27, 1995, Portland, OR. Manuscript submitted January 3, 1995; made available for printing May 5, 1995.

Reprinted from *IEEE Transactions on Power Systems,* Vol. 11, No. 3, pp. 1371–1379, August 1996.

blackout. All system components involved are assumed available; the unavailability of some elements will cause part of the plan to fail, delaying the complete recovery of the users, but without invalidating the overall efficiency.

In order to enable parallel restoration, the Italian HV network has been partitioned in several and independent parts, each containing hydro and thermal generation, connecting lines, transformers and loads. Because of the particular generation mix in Italy, the plan relies on black-start of selected hydro power plants, provided with special speed governor and named "early restoration plants".

The plan is split into five steps. The first is relevant to the detection of the blackout status at the different control levels and to the switching sequences to be performed for network preparation by the ACCs and power plant under surveillance. It is of prime importance the role played by the RCCs and/or the NCC in recognizing the blackout and its boundaries, in starting the plan, by disconnecting loads and most of the lines, or stopping it if necessary. At the end of this step, "early restoration plants" must be started up. The objective of the second phase is to quickly constitute a HV path between black-start units and thermal plants in order to recover them into service as soon as possible, depending on the success of the actions performed during system degradation. At this stage, the RCCs communicate with ACCs in order to get ready the successive steps for load re-energization. Indeed, cranking power (called "ballast loads") must be supplied along the path not only to maintain the voltage profile inside an acceptable range but also to prepare easily a load base for the thermal unit which will be next resynchronized. The amount of load to be fed corresponds to the minimum load above which the thermal unit can operate (normally, thirty percent of the rating) in boiler-turbine co-ordinated regulation. Moreover, the RCC will re-energize, through the path, other hydro plants, not essential for the success of the plan, but useful for improving voltage and frequency regulation. The third phase is entirely co-ordinated by the RCCs. The plant operators, upon approval of their RCC, resynchronize the thermal units to the path and start the load pick-up. Generator loading up to its minimum load under manual control is of particular concern owing to the critical operation of the units in islanded conditions.

The fourth phase is managed by the RCC supported by the NCC if necessary. As the thermal units have reached their lower capability limit, connection between independent paths is performed. At the end of this phase, the network will have sufficient power and stability to withstand the transients related to further load reconnection and recovery of other generating units. Finally, the last phase, supervised by the NCC, regards the complete interconnection of the network.

Among these phases the most critical are the initial ones, in which the system is growing up and is virtually much more vulnerable, owing to frequency and voltage stability problems. At this time a wrong maneuver can jeopardize the effectiveness of the plan. Therefore, in order to refine the strategy, it becomes very important to simulate these tasks with proper modelling of system components and controls.

III. Modelling for Restoration Simulation

The simulation of a restoration procedure requires a specific modelling of the incoming network together with its special regulation equipments.

Due to the topological algorithms used by the digital programs, it is much more difficult to reconnect step by step an electric system than to account for the opposite process of separation, common side effect of protection interventions.

In addition, representation of reconnected loads plays a significant role in the safe evolution of the restoration plan, with particular reference to the island frequency behavior; in fact, there is a close relationship between the duration of the outage and the power demand profile.

Finally, as previously stated, during the system recovery unconventional control channels are involved, the representation of which is not usually present into standard digital codes. Facilities provided by user defined models must be available to account for such peculiar devices.

A. Network Modelling

The main trouble in power system modelling is the difficulty of including, in the transient stage, buses not previously present in the power flow run.

Consequently, it is quite useful to perform the load flow analysis on a power system with lines and transformers out of service, with so many reference buses as the number of involved generators, which are imagined as successfully restarted, with rated voltage magnitudes at their terminals.

The simulation of the restoration plan is performed by means of a transient stability program: electrical paths are energized, beginning from buses of hydro generators provided with black-start facilities, by progressively reclosing branches and loads.

A special difficulty is how to model the parallel connection of energized islands; with regard to this point, a synchronization device has been implemented, that compares the voltage magnitude and phase angle at the parallel switch terminals and performs the connection as soon the mismatch falls within a pre-defined tolerance.

B. Load Modelling

If field experiments are examined to detect the load demand when recovering from a blackout condition, overshoots reached in a very short transient and following decays to values similar to the pre-emergency ones can be noticed. This last phenomenon shows time constants of some seconds, depending on the time interval during which the user was not supplied [5].

Air conditioners and refrigerators are primarily responsible for these different possibilities: they roughly behave as asynchronous motors with load torque dependent on the disconnection time. If the interruption is long (more than 5 minutes), the pressure resets and new motor starting is possible with usual initial peak of power, rapidly decreasing to a new steady-state value, higher than the initial one, due to the increased number of users which need to recover a correct working temperature.

On the contrary, if the disconnection is shorter, most loads of small size fail to restart, because of the significant backpressure causing motor overload and consequent cascading trips of internal thermal protections. A longer and heavier transient of active power demand is therefore detectable, and the new steady-state condition presents a smaller value than the pre-contingency one, since a certain amount of loads are safely recovered only after some minutes, when the load torque allows a correct restarting.

These qualitative considerations lead to the conclusion that field experiments of reenergization of electric networks often show worse underfrequency behaviors than those detectable in actual blackouts: in order to reduce the disturbs to the users, they are disconnected for a short time during tests, producing however a more critical demand profile.

Anyway, while simulating the restoration on part of the Italian network, since the composition of the single cold loads was unknown, experimental data were used to model equivalent load demands with linear (active power) and exponential (reactive power) fitting.

Load real and reactive powers are therefore described as shown in Fig. 1, that is:

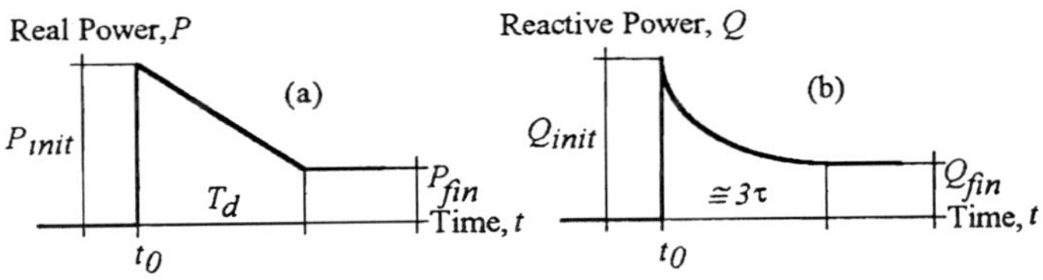

Fig. 1. Real (a) and reactive (b) power demand of a cold load pick-up.

$$P(t) = Q(t) = 0, \text{ when } t < t_0,$$

$$P(t) = P_{init} - \frac{P_{init} - P_{fin}}{T_d} \cdot (t - t_0), \text{ when } t_0 \le t < t_0 + T_d,$$

$$P(t) = P_{fin}, \text{ when } t \quad t_0 + T_d;$$

$$Q(t) = Q_{init} - (Q_{init} - Q_{fin}) \cdot (1 - e^{-\frac{t - t_0}{\tau}}), \text{ when } t \quad t_0$$

where

t_0 = instant of insertion of the cold load;

T_d = time interval of the real power linear decay;

τ = time constant of the reactive exponential decay;

P_{init}, Q_{init} = peak values at cold load pick-up;

P_{fin}, Q_{fin} = final steady-state values of the cold load.

C. Modelling of Frequency Regulators for Early Restoration Generators

ENEL hydro power plants with black-start facilities are equipped with a special control system, named AURET. It is an auxiliary device used for frequency regulation, particularly effective during islanded operation; in fact, while energizing a restoration path, AURET works to limit steady-state underfrequencies due to the cascading insertions of cold loads. A reduced amount of the island frequency deviation is required in order to perform a correct parallel operation with the thermal unit at the end of the restoration path.

Such a control equipment acts as a local secondary frequency regulator, characterized by a very low value of permanent droop (usually 0,5%) and long time constants (about 10 s and 200 s).

The simulation model of frequency regulation on early restoration generators is reported in Fig. 2.

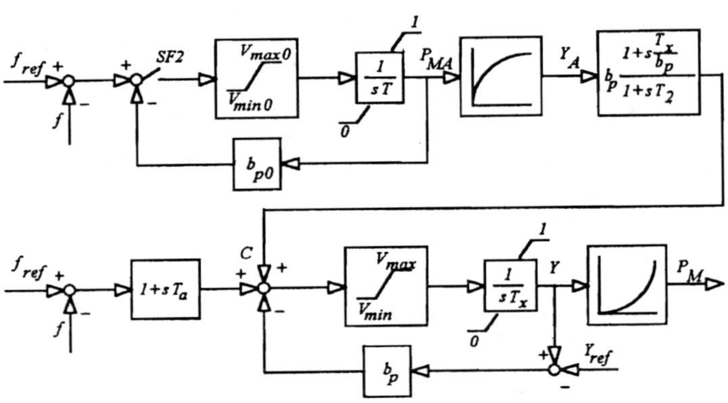

Fig. 2. Frequency regulators on early restoration hydro generators.

In the lower part of the block diagram a conventional speed governor is presented: the frequency error operates on a proportional-derivative block, that drives the regulation servo, whose output position Y is related to the mechanical power through a single non-linear block describing together gate actuator, water supply system and hydro turbine.

The upper part of the block diagram shows the AURET channel, which is firstly activated as soon as the frequency falls out of the range 49.5 - 50.5 Hz. Next, this loop is either switched off (see switch SF2 in Fig. 2), when the frequency enters the band ± 0.075 Hz and does not leave it for at least 10 seconds, or immediately reactivated, when the frequency overtakes the specified dead band.

A regulation servo is present, similar in the form to the one of the speed governor but with a different value of the permanent droop; two blocks perform both the compensation of the non linearity gate position - mechanical power and the dynamics of the primary regulation servo. This causes the AURET output C to be, apart from the time constant T_2, a power set point for the hydro unit, similar to the contribution coming from the automatic generation control.

IV. Field Experiences and Simulation Results

After having introduced the simulation models, let us examine a restoration field test performed by ENEL in the Northern-Western Department. A tenth of days before the experience, a large emergency condition was detected in the same district, so that it was found useful to verify the correct behavior of the network in recovering state, particularly for thermal power plants after the load rejection.

A. Field Test Scenario

The layout of the transmission and subtransmission network in Northern Western Italy, involved in the restoration test, is depicted in Fig. 3.

In the test, unit 3 of the thermal power station at Vado Ligure received voltage from its restoration path, including hydro power plants with black-start facilities (Entracque), conventional hydro power plants (Casteldelfino, Brossasco, Andonno) connected to the subtransmission level, and a group of substations of about 150 MW, in order to provide sufficient load to properly starting the unit itself.

The purposes of the experimental procedure were basically the following ones:

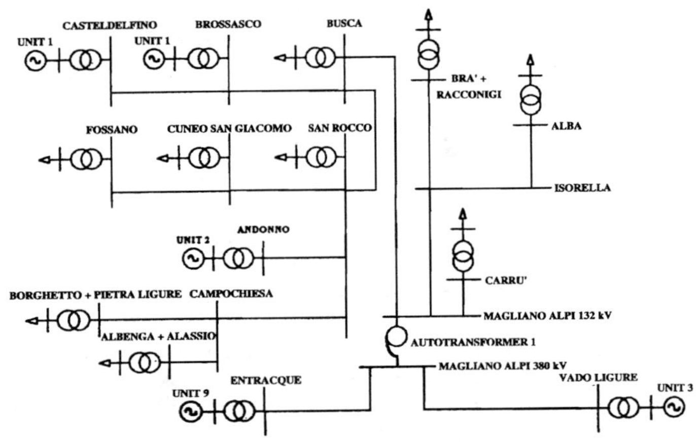

Fig. 3. One line diagram of the system under the restoration test. Thin lines represent subtransmission level (132 kV), while thick ones draw the transmission system (380 kV).

- to verify the black-start and restoration capability of the hydro units, that is the starting logic and time, as well as their effect on frequency and voltage during cold load pick up, in islanded condition with a maximum load of about 150 MW.
- to confirm the expected time necessary to send voltage to the thermal unit in Vado Ligure.
- to check the load rejection operation of the thermal unit and its stable working on its auxiliaries.
- to examine the behavior of the combined hydro-thermal power system during parallel connection and generation ramping of the thermal unit.

B. Description of the Operative Sequence

The creation of a restoration island for the field experience was made taking advantage of the double busbar structure of the majority of ENEL power stations; one bus was used to hold the interconnection with the national grid, whereas the second one was devoted to the separated system.

This configuration allowed not to affect operations of units not involved in the restoration path, as well as to supply loads before their commutation from the national grid to the restoration island. In this way, disturbs to the users were dramatically reduced, although the real blackout conditions were not completely matched.

The initial condition for service recovery was reached through a sequence of breaker openings, and a complete stop of the involved units, apart from the thermal plant, which successfully performed the load rejection, thus feeding its own auxiliaries only.

The first intervention done was the black-start of unit 9 at the hydro power plant of Entracque, with its own auxiliaries supplied by local batteries. This generator is an early restoration one, that is provided with the AURET equipment, and its size is sufficient to recover load for the later ramping of the steam turbine generator.

After the positive conclusion of the start-up, energization of unit transformer, 380 kV busbars, line connecting the power plant to the station of Magliano Alpi, and local autotransformer 380/132 kV/kV was performed.

An EHV no load branch was therefore supplied, behaving as a capacitive load; sustained and transient overvoltage phenomena were detected, but of limited amount, due both to the short length of the line itself and the reduced reference setting on the unit AVR (about 16.6 kV instead of rated 17 kV).

The energization of the line Magliano Alpi - Carru' included the assumption of the first cold load (rated 12 MW), placed at Carru'. After this occurrence the hydro unit placed at Entracque experienced the first significant frequency transient, with the intervention of the AURET channel.

Cold load pick-ups caused significant underfrequency operations of the island, lasting in the average about 15 seconds.

The test continued with progressive reinsertion of other pre-set loads (Alba 24 MW, Bra' + Racconigi 24 MW, Busca 6 MW and San Rocco 11 MW), with similar dynamic behavior, until the other hydro units were reached by the restoration path.

Parallel operation of such smaller generators, not provided with black-start facilities, was performed feeding their auxiliaries from the island, in order to bring the unit to the rated speed. Thanks to the reduced permanent droop of the AURET channel, which is able to maintain the island frequency not less than 49.8 Hz, safe parallel connection of the units is assured.

The new steady-state condition of the integrated subsystem was characterized by an intermediate area frequency and a new distribution of the generated active powers, according to the ratings and the permanent droops of the committed machines.

However, the overall generation, after the insertion of all the hydro units, was not properly shared, due to the reduced permanent droop of unit 9 at Entracque.

In order to prevent saturation of the active generation of Entracque when further loads would be picked up, the power setting of the generator of Andonno was locally increased.

No troubles were detected in such procedures for what concerns the electromechanical oscillations; in fact, measurements generally showed sufficient damping.

At this point new load was inserted (Cuneo San Giacomo 18 MW, Fossano 14 MW, Borghetto + Pietra Ligure 22 MW, Albenga + Alassio 20 MW) bringing the total demand of the separated network to about 150 MW, to be transferred to the thermal unit after its connection to the system under test.

In the next step, voltage was recovered to the unit of Vado Ligure, through a final switching on of the 380 kV line Magliano Alpi - Vado Ligure.

As previously stated, since this generator had succeeded in the load rejection operation, its resynchronization with the separated system was correctly performed. The reconfiguration of the overall island took about 35 minutes.

Finally, thermal generator ramping was actuated, with a slow rate of about 3 MW/min, until most of the system load was acquired (about 130 MW).

In this phase a deep involvement of the AURET channel was detected, which forced the frequency to values around 49,9 Hz.

The restoration test was therefore successfully ended, with following interconnection of the separated system with the national grid.

C. Simulation Results

The restoration test just discussed has been analysed by computer simulation based on the modelling described in Section III. Appendix 1 summarizes the parameter values of generator speed governors, included the AURET regulator. From the whole operative sequence three particular transients are here examined in detail, as they collect significant aspects of the restoration process.

1) Insertion of the First Cold Load: After having reached the station of Magliano Alpi and reenergized the local autotransformer, subtransmission lines are switched on and cold loads are progressively picked up.

The first load to be connected is placed at the station of Carru'.

Active and reactive power demands are modelled according to the relevant equations presented in the previous chapter, with parameter values derived from the best fitting of experimental results:

$$P_{init} = 24 \text{ MW},\ P_{fin} = 12 \text{ MW},\ T_d = 10 \text{ s}$$

$$Q_{init} = 30 \text{ MVAR},\ Q_{fin} = 1 \text{ MVAR},\ \tau = 0.4 \text{ s}$$

The comparison between experimental and simulation diagrams is shown in Figs. 4 and 5 with reference to voltage magnitude and frequency respectively, evaluated at the generation bus of Entracque.

Voltage curves present an oscillating behavior, with considerable damping and good correspondence between the minimum and maximum transient value.

Besides, the cold load pick up causes a frequency transient conditioned both by the load active power variation and the regulation effects.

In terms of comparison, steady-state frequency deviation results nearly equal, whereas the transient behavior denotes higher damping in the simulation than in the field test.

To better understand the role played by the involved regulation loops, a linearized model of the system during this operation has been analytically treated, too. Insight of the whole available outputs suggests the present considerations:

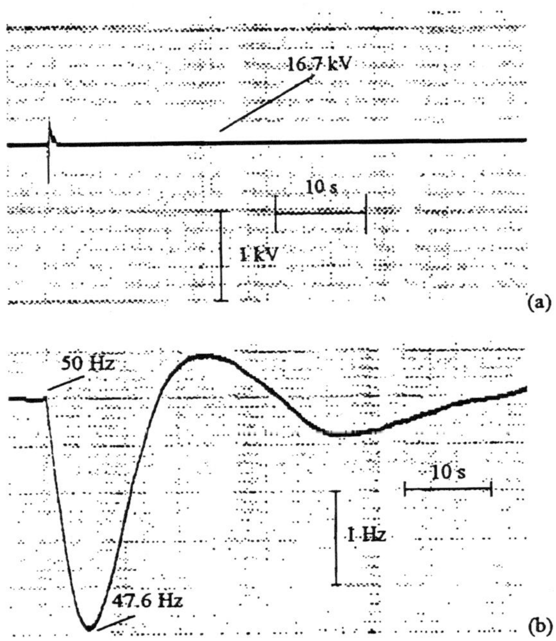

Fig. 4. Field test measurements at Entracque busbars while picking up the first cold load (a = voltage magnitude, b = frequency). Scales are made comparable with the ones of Fig. 5.

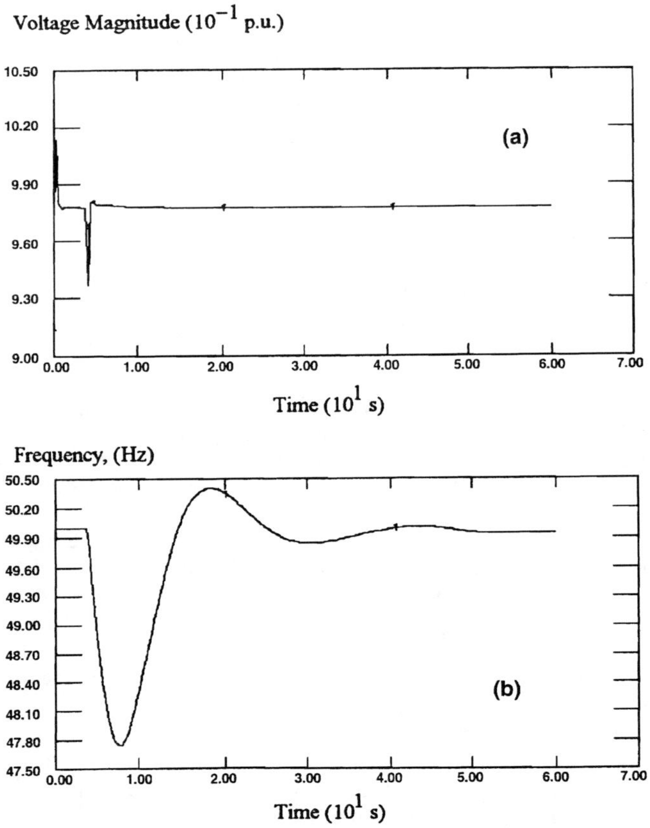

Fig. 5. Simulation plots evaluated at Entracque busbars while picking up the first cold load (a = voltage magnitude, b = frequency).

- in the first part of the frequency transient the contribution of the primary speed governor, much quicker than the AURET channel, is higher. The underfrequency peak depends on the active power demand but is strongly affected by the saturation in opening speed of the regulation actuator. The non-linearity drives the frequency to a lower value and delays the whole transient with respect to the ideal linear response.
- frequency overshoot is caused by the complex poles of the resulting loop and the presence of a zero in the primary frequency regulator.

2) Generation Ramping on a Hydro Unit: The restoration procedure goes on with the insertion of further lines and loads, thus stimulating transients similar to the ones above described, until a load amount of about 90 MW is reached.

In the field experience the hydro groups of Brossasco, Casteldelfino and Andonno have been connected to the growing system, as far as voltage is brought at their buses.

After the insertion of such units, the Entracque generator, which was previously producing about 88 MW, reduces its active power to 83 MW, with negligible effects on voltage and reactive power profiles; the new steady-state frequency of the extended island settles down to 49,87 Hz.

At this stage of the strategy, if additional cold loads were inserted in order to provide sufficient load (Cuneo San Giacomo, Fossano, Borghetto + Pietra Ligure and Albenga + Alassio) to reconnect the thermal unit , the frequency of the separated system would evolve as shown in Fig. 6.

Insight of the graph reveals an increasing frequency fall, particularly noticeable after the final pick up.

In fact, after the last insertion, the total load of the island would be around 162 MW. Such an amount would be compatible with the whole rotating power, but the main impact would be on the Entracque unit, due to its reduced permanent droop, thus leading to saturation the AURET channel with consequent significant rise of the frequency deviation.

To prevent therefore the system from too large frequency excursion, as well as to exploit in a more convenient way the available resources of the small hydro units, local adjustments on the gate opening settings can be made, in order to increase the mechanical power, regardless of the contribution determined by the speed governor.

Experimental results indicate, for example, that after the parallel operation of the Andonno unit, this one is driven to a power generation of about 20 MW.

The simulation of this action is made through a modification of the valve opening reference setting, resulting in the mechanical power profile indicated in Fig. 7.

In connection, the generated active power and the system frequency, evaluated as usual at the Entracque busbars, are reported in Fig. 8.

While the real power decreases to compensate the loading of Andonno unit, the damped oscillations of the frequency are associated to the dynamic behavior of the speed regulation loop in response to a decreasing load power, with a new steady-state frequency much closer to its rated value.

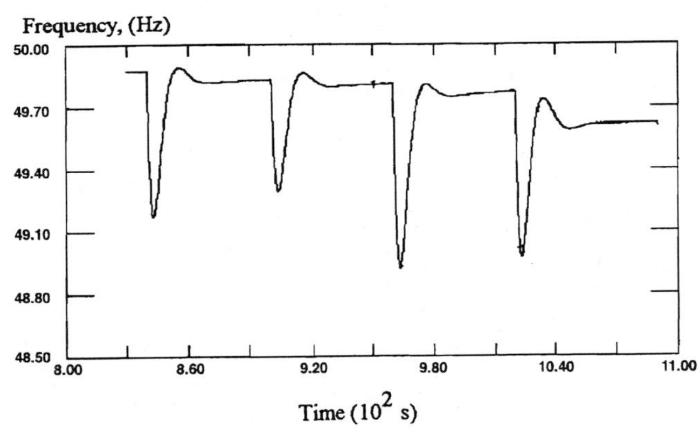

Fig. 6. Frequency decay due to the insertion of subsequent cold loads.

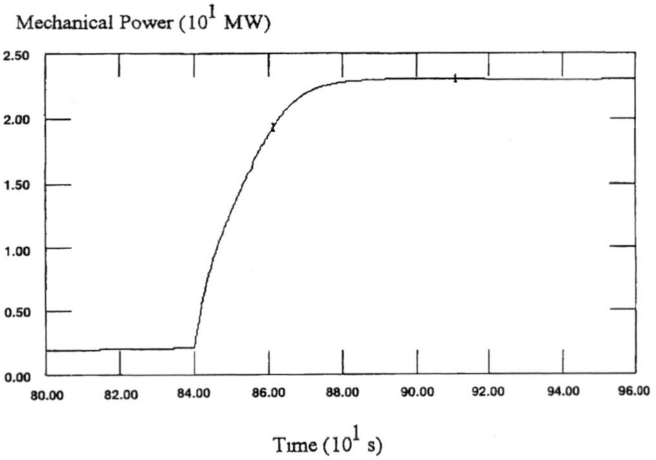

Fig. 7. Mechanical power of the Andonno unit caused by the action on its valve opening reference setting.

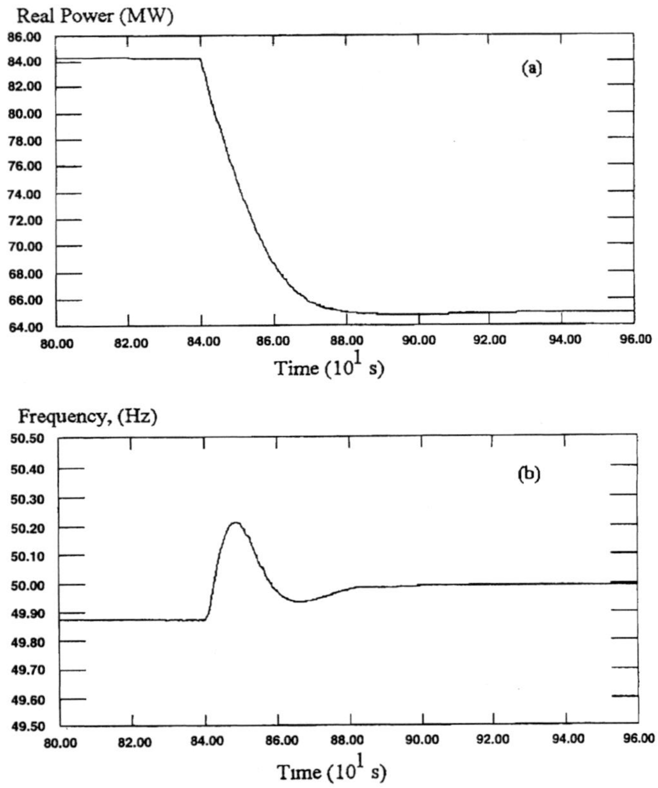

Fig. 8. Effects on the Entracque unit caused by the power ramping of Andonno (a = electrical power, b = busbar frequency).

Simulation results allow to detect the final values of frequency and active power sharing among the four operating units:

$$f_{fin} = 49.98 \text{ Hz},$$

$$P_{Entracque} = 67.30 \text{ MW}, \; P_{Brossasco} = 0\ 19 \text{ MW},$$

$$P_{Casteldelfino} = 0.37 \text{ MW}, \; P_{Andonno} = 20.32 \text{ MW}.$$

The reported values can be confirmed on the basis of analytical considerations. After the Andonno generation ramping, the frequency enters the range 49.925 Hz - 50.075 Hz and remains inside it for at least 10 seconds; consequently, the AURET channel is switched off From this time on, the AURET output remains constant, acting on the Entracque speed governor as a power reference setting

In other words, the real power balance in the final steady-state condition must satisfy the following equation:

$$P_{Load} = \sum_{i=1}^{i=N} E_i (f_n - f'_{fin}) + P_{M,Andonno} + P_{M,AURET},$$

where:

N = number of involved hydro units,

P_{Load} = real power demand of the separated network;

f_n = rated frequency;

f'_{fin} = calculated steady-state system frequency,

E_i = regulating energies of the units, considering for Entracque only the contribution of the primary speed governor;

$P_{M,Andonno}$ = steady-state contribution to the Andonno mechanical power due to its ramping,

$P_{M,AURET}$ = steady-state contribution to the Entracque mechanical power, frozen at the instant when the AURET channel is switched off (66.33 MW).

From the balance relationship, the steady-state value for the network frequency can be obtained

$$f'_{fin} = 49\ 98 \text{ Hz}$$

whereas the generators produce:

$$P'_{Entracque} = 67.32 \text{ MW}, \; P'_{Brossasco} = 0.19 \text{ MW},$$

$$P'_{Casteldelfino} = 0.17 \text{ MW}, \; P'_{Andonno} = 20.32 \text{ MW}.$$

These calculated values (mechanical powers) show quite good accuracy with the simulation outputs (electrical powers); it may be stressed that generation losses are negligible. The only significant difference regards the unit of Casteldelfino, where a complete steady state condition is not yet reached.

3) Generation Ramping on the Thermal Unit: After having reconnected the last cold loads, the islanded system layout is completed with the energization of the 380 kV line Magliano Alpi - Vado Ligure, which allows the connection of the thermal unit of Vado Ligure.

This plant is supposed to have successfully performed the load-rejection; it is therefore operating isolated, on its auxiliaries, which require about 10 MW.

After the transient due to the synchronization of the thermal power plant, the frequency is around 49 8 Hz and the unit is ready for the following power ramping.

Accordingly, the mechanical power of the generator is firstly increased, operating on the steam flow admitted in turbine, from the initial 10 MW to about 24 MW. In a second phase the machine increases its power following a ramp of the reference setting of the pressure regulation loop, until the lower capability limit is reached.

The experimental curve of the power rise is represented in Fig. 9.

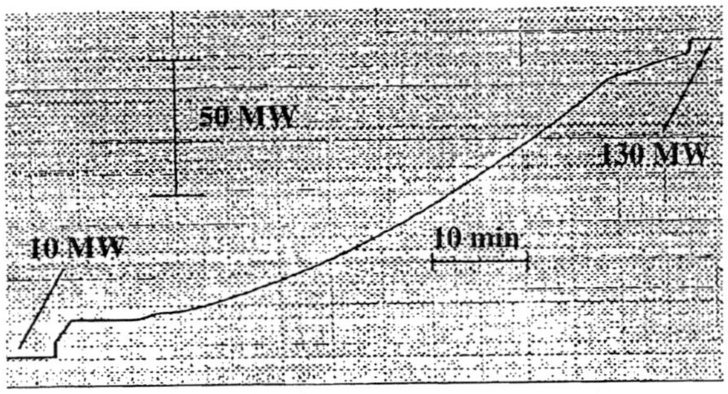

Fig. 9. Electrical power of Vado Ligure thermal unit during the generation ramping. Scales are made comparable with the ones of Fig. 10.

For simulation purposes, the ramping of the thermal unit is modelled as an injected active power into the network, obtained from piecewise linear interpolation of the experimental curve.

The corresponding electrical power generated at Entracque is presented in Fig. 10, which for completeness includes the power oscillations during the resynchronization phase.

As shown in Fig. 10, the generated active power decreases to compensate the thermal unit ramping.

The associated frequency transient appears in Fig. 11.

A progressive increase of the machine frequency can be noticed, towards a final steady-state value of 50.035 Hz.

Such overfrequency is caused by the disconnection of the AURET channel, which leaves a too high power reference on the Entracque unit of about 17.8 MW.

Since the total production of Entracque is about 16 MW, a negative contribution from the speed governor must be produced (1.8 MW approximately), which leads to a frequency, greater than 50 Hz.

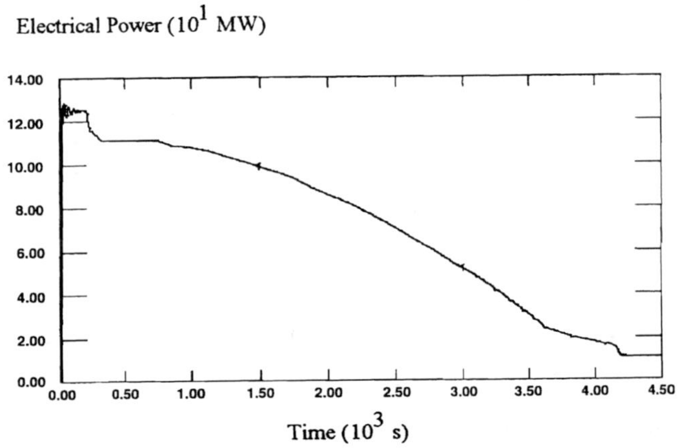

Fig. 10. Electrical power supply at Entracque during the thermal unit ramping.

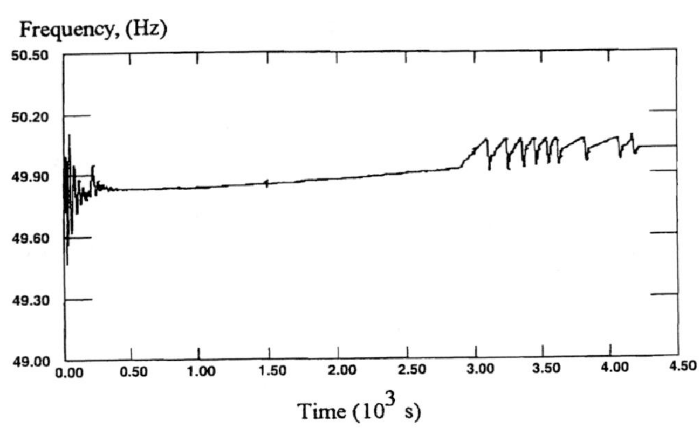

Fig. 11. Frequency transient evaluated at the Entracque busbar.

In the end, it is also important to comment the presence of frequency perturbations, typical of a bang-bang control, during the final part of the thermal unit ramping up.

In fact, along this time frame the AURET regulator is alternatively inserted and excluded, depending on the state of the switch SF2.

To clarify the reasons for such oscillations it is useful to analyze the frequency behavior in detail, as shown by the zoomed transient reported in Fig. 12, where the AURET channel begins its on-off commutations.

During the first part of the thermal unit ramping, AURET is working and contributes to the decrease of the mechanical power at Entracque, even if the frequency error is positive.

However, the accelerating power of this last unit is still positive, so that the system frequency tends to increase remaining nevertheless out of the AURET insensibility range.

When the frequency enters this range and remains inside it for 10 seconds, the AURET channel is switched off, leaving its contribution nearly constant at the value relevant to the instant of disconnection.

The lack of the decreasing contribution of the AURET channel causes an increment of accelerating power, which results in an increased rate of change of the system frequency.

As soon as the frequency reaches the upper bound of the AURET range, the channel is immediately reactivated so that a significant deceleration can be detected, being maximum the negative frequency error input to the loop.

The speed swings inside the range and the AURET is disconnected again, with the delay of 10 seconds; this behavior lasts until the unit ramping is finished, causing the saw-teeth frequency profile shown in the Figures.

V. Conclusions

The presented activity has been developed in order to simulate the behavior of an electric power system experiencing a restoration plan following a blackout condition.

In the past, ENEL worked out relevant strategies by validating and improving them on the basis of field tests, performed over the actual restoration paths.

The possibility of using a computer simulator allows to rationalise future field tests, to perform a parameter sensitivity analysis and to assist operator training.

The availability of field test data allows to base the simulation studies over actual behaviors of critical inputs, particularly concerning the cold load models and the thermal unit power ramping.

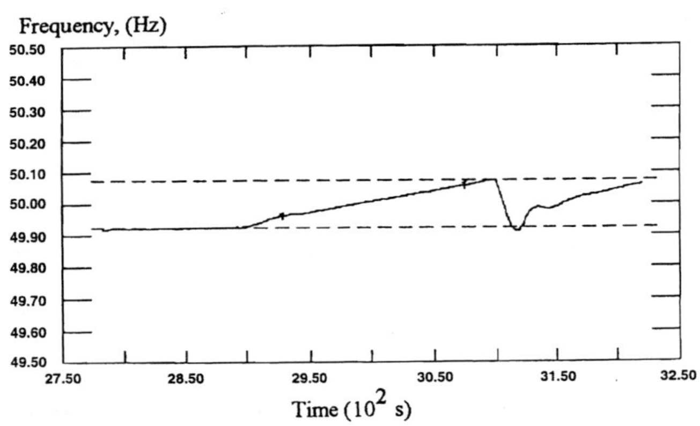

Fig. 12. System frequency at the Entracque busbar during the on-off switching of the AURET.

The overall results obtained appear quite satisfactory; indeed, comparison between frequency transients at the first cold load pick-up confirms the accuracy of the adopted modelling, in particular for the regulation devices.

Improvement in the system modelling regards cold load pick-up and thermal unit loading, to better simulate the examined restoration steps and to extend the analysis to the following stages, where there is a lack of experimental data.

In addition, further aspects to be considered are the compatibility of the progressive network meshing and loading with the presence of load shedding relays, activated by excessive underfrequency transients, and the unavailability of equipments during the reenergization process, which should be counteracted by a strategy modification.

This last in particular involves a decisional procedure, which could be optimally helped by an expert system approach; in this concern a knowledge based module is under study, with the prospective goal of integration in an energy management system.

VI. References

[1] IEEE Power System Restoration Working Group, "Special Consideration in Power System Restoration - The Second Working Group Report", *IEEE Trans. on Power Systems*, vol. PWRS-9, no. 1, February 1994, pp. 15-21.

[2] M. M. Adibi, J. N. Borkoski, and R. J. Kafka, "Analytical Tool Requirements for Power System Restoration", *IEEE Trans. on Power Systems*, vol. PWRS-9, no. 3, August 1994, pp. 1582-1591.

[3] CIGRE' Task Force 38.02.02, "Modelling and Simulation of Black-start and Restoration of Electric Power Systems", *Electra*, no. 147, April 1993, pp. 20-42.

[4] B. Delfino, M. Invernizzi, and A. Morini, "Knowledge-Based Restoration Guidelines", *IEEE Computer Applications in Power*, vol. CAP-5, no. 3, July 1992, pp. 54-59.

[5] E. Mariani, F. Mastroianni, and V. Romano, "Field Experiences in Reenergization of Electrical Networks from Thermal and Hydro Units", *IEEE Trans. on Power Apparatus and Systems*, vol. PAS-103, no. 4, July 1984, pp. 1707-1713.

VII. Appendix 1

Each hydro unit is provided with a speed governor. Moreover, the Entracque unit, which is an early restoration generator, is equipped with the AURET channel. Table I presents the regulator parameter settings.

TABLE I
PARAMETERS OF THE GENERATOR FREQUENCY REGULATORS

Generator	Regulator type	Regulator parameters					
		T (s)	T_1 (s)	T_2 (s)	b_{p0} (p.u./p.u.)	V_{max0} (p.u./s)	V_{min0} (p.u./s)
Entracque	AURET	1	27.5	10	0.005	0.01	-0.01
		T_x (s)	T_a (s)	b_p (p.u./p.u.)	V_{max} (p.u./s)	V_{min} (p.u./s)	
Entracque	speed governor	1.17	4	0.06	0.04	-0.25	
Brossasco	speed governor	0.53	2.5	0.05	0.04	-0.25	
Casteldelfino	speed governor	3	10	0.04	0.04	-0.25	
Andonno	speed governor	0.78	3	0.05	0.04	-0.25	

Discussion

N.I.VOROPAI (Siberian Energy Institute, Irkutsk, Russia):There is the important task in the problem of rapid restoration of electric power grid after blackout concerning the creation of restoration plan that are used in training the dispatching personal as well as in validating required software for the modelling and simulation of restoration process. Authors presented very interesting and detail example of such restoration plan dealing with experimental test performed on the Italian HV network near the French border.

I should like to make the following comments to this paper:

1. Authors are discussing three hierachical levels of control organization in Italy and the main principles of control action coordination at these levels. The example of network for the restoration test includes the HV lines relating to different control levels. It would be intresting to see the separation of restoration decision between the different levels for the presented test.

2. There is a very short information in a paper about the network modelling especially concerning the long-term restoration process simulation by the load-flow and transient stability analysis. It would be interesting to know about restoration process simulation in more detail.

Manuscript received August 14, 1995.

B. Delfino, G.B. Denegri, M. Invernizzi, A. Morini (Electrical Engineering Department, University of Genoa, Italy); **E. Cima Bonini, R. Marconato, P. Scarpellini** (Automatica Research Center, ENEL S.p.a., Cologno Monzese, Italy): The authors wish to thank Dr. Voropai for raising interesting points of discussion with regard to restoration field experiences and simulation.

Concerning the early stages of the Italian plan, it must be pointed out that during restoration tests and in real black-outs, ACCs execute sequences of switching operations firstly to condition the network for subsequent reenergization and afterwards to build the restoration paths until thermal units are reached. During these time frames, RCC and NCC only supervise the development of the process, without intervention if everything goes on successfully. RCC and NCC act therefore as a sort of feedback block of an imaginary control system for network restoration.

In other words, until the end of the second step of the plan, ACCs follow written guidelines, so that they can proceed in recovering part of the system even in the extreme case of communication failure. RCCs and NCC stop or delay the process if some components are unavailable for restoration, or if alternative paths are detectable to more quickly reach large thermal units.

In the presented field test, all network elements were successfully reenergized by the involved ACCs, which performed switching on transmission and subtransmission (to provide appropriate sizes of cranking power) networks, so that no conflicts in supervision were detectable.

Turning now to the second comment, the proposed simulation covered only a part of the restoration strategy, according to what is actually performed during field tests.

The use of traditional codes for dynamic analysis poses additional problems owing to the progressively changing network topology. The algebraic equation system that describes the electrical network must be continuously updated following the actions (circuit breaker closures) performed on the system.

To manage this situation, the initial conditions are represented with units supposed either after successful black-starts (hydro) or load rejections (thermal), all interconnected by branches declared open in the load flow study, just to maintain the whole network connectivity.

Dynamic simulation is performed by progressively closing branches and reenergizing loads. Such simulation covers the whole restoration process, without intermediate analyses, and it is not a sequence of transients beginning from steady-state calculations for each recovered system configuration.

The synchro-check devices described in the paper enable island resynchronization and are typically located in branches modelled without shunt admittances, e.g. by generator step-up transformer.

In the paper it is also stated that a more detailed model of thermal unit prime movers and related controls is desirable, particularly for the simulation of generator loading up to its minimum load. As a general comment, to overtake most of the difficulties in representing the network under restoration, advanced network simulators for on-line operation [1] reveal more efficient and versatile than the ones built using "conventional" transient stability programs, even if provided with user defined model features.

[1] B. Delfino, G.B. Denegri, M. Invernizzi, A. Morini, "Enhancing Flexibility of Restoration Guidelines Through an Integrated Simulator KB Approach", *Proceedings of the IEEE Stockholm Power Tech*, June 18-22, 1995, paper SPT IC 04-03-0209, pp. 129-134.

Manuscript received October 19, 1995.

Interactive Long-Term Simulation For Power System Restoration Planning

N.A. Fountas — Student Member
N.D. Hatziargyriou — Senior Member
C. Orfanogiannis
A. Tasoulis

Electric Energy Systems Laboratory
Department of Electrical and Computer Engineering
National Technical University of Athens

National Dispatching Center
Public Power Corporation

Athens, Greece

Abstract - **The problem of restoring a power system following a complete blackout is complex and multi-faceted. Many control actions have to be performed on time, while constraints such as power balance and system stability have to be carefully respected. In this paper, the application of long-term dynamic analysis in studying frequency and voltage responses due to load and generation mismatches in isolated systems or during extension of the existing system in the restoration phase is presented. Simulation results covering the main steps of the Hellenic power system restoration process following a recent total blackout are presented and discussed.**

Keywords: Power System Restoration, Restoration Planning, Fast Simulation, Dispatcher Training.

I. INTRODUCTION

The problem of Power System Restoration (PSR) following a complete or partial blackout has been studied for many years with special emphasis on developing effective restoration plans. A considerable number of technical papers concerning PSR practices as well as experiences from real incidents has been published, [1-19]. Some of the primary problems the system operator faces during restoration are :

- *reactive power imbalance*: it is crucial that system voltages remain within allowable levels as energising long distance transmission lines causes large MVAr line charging, which can cause severe reactive power imbalance resulting in, for example, generator self-excitation and runaway voltage rise, [2, 5, 11, 14].
- *active power imbalance:* loads must be picked up in small increments to avoid large frequency deviation and hence recurrence of the system collapse, [2, 8-9, 16].
- *malfunction of protection and control systems*, such as underfrequency and switched capacitor relays, and proper coordination of the number and location of synchronising facilities, [2, 8, 12, 16].

Since blackouts seldom occur, the personnel in control centers and generating plants has little experience in handling such extreme disturbances. It is therefore necessary for an electric utility to set-up orderly restoration procedures for such events and to focus on training and education of engineers and dispatch operators. The increasing need for competent long-term dynamics (LTD) analytical tools to serve as supporting means to power system planners and dispatchers for the above purposes has been well recognised, [2, 7-8, 10, 13, 15-16]. To this end, a dynamic analysis program is specially adapted and used for the simulation of restoration procedures as followed by the Hellenic Public Power Corporation, PPC.

The paper is organised as follows: The first part briefly describes the modelling capabilities of the program used. A comprehensive overview of the applied fast simulation method is also provided. The second part comprises a review of the Hellenic power system and reports on PPC restoration philosophy. Special care is devoted to a recent major shutdown. Simulation results are presented and discussed. It is shown that the developed software can assist in operator training and for restoration planning purposes.

II. POWER SYSTEM DYNAMIC ANALYTICAL STUDIES

LTD analytical studies address system performance under large-scale upsets which invoke the actions of slow processes, protective systems and controls not present in conventional transient stability programs, [22-23]. A typical range of long-term time period can exceed several tens of minutes.

PSR usually entails sectionalisation of the system into several islands in order to develop a rapid response in performing restoration. Stability is a question of whether or not each island will reach an acceptable operating equilibrium with minimal disruption to services.

A. Modelling capabilities. The program offers two simulation modules. In the Full Integration (FI) module,

96 WM 175-0 PWRS A paper recommended and approved by the IEEE Power System Engineering Committee of the IEEE Power Engineering Society for presentation at the 1996 IEEE/PES Winter Meeting, January 21-25, 1996, Baltimore, MD. Manuscript submitted August 1, 1995; made available for printing November 16, 1995.

Reprinted from *IEEE Transactions on Power Systems,* Vol. 12, No. 1, pp. 61–68, February 1997.

short term dynamics are simulated. This module utilises a 5th order generator model regarding the transient and subtransient influences and uses an implicit integration method for the algebraic-differential equations describing the dynamic response of the system. In the second Quasi-Stationary (QS) module only longer-term dynamics are simulated. The option exists to automatically switch from quasi-stationary mode to full transient simulation mode whenever there is a need for more detailed examination of the phenomena involved.

For LTD studies an adequate representation of the prime mover and energy supply systems should be incorporated. Appropriate models for the wide range of protection and control systems that are invoked when the system is in the emergency state should also be included. The program incorporates the following component models, [21-25],:

For *thermal power plants,* steam turbines are represented by a 3 stage single reheat turbine.

Hydroelectric plants are simulated using a second order model that includes the water hammer effect, as well as the transient and steady state droops.

ULTC models (with timers, delays and deadbands), and *generator field current limiters* (overexcitation limiters) are provided.

Basic *protective schemes* are utilised such as: undervoltage load-shedding relay, undervoltage generator protection, turbine underfrequency protection and underfrequency load-shedding relay.

All controller models include important nonlinearities, such as excitation voltage ceilings, governor gate velocity limitations, minimum and maximum permissible power etc. Furthermore, low-order models for a drum type or once-through boiler with an integral boiler-turbine control system are currently under development.

B. Brief description of the Quasi-Stationary algorithm. For the long-term simulation required for restoration studies, a quasi steady-state approach, that is the approximation of the system trajectory through successive transient equilibrium points has been used, [23-24]. Following a disturbance or a change in the control and protection devices of the system, the transient dynamics are not simulated, but only the corresponding equilibrium point after the transients have died out is computed. In consequence, the various long-term components and the protective devices that might lead to new operating points (e.g. overexcitation limiters) are scanned to determine the time of the next transition. The differential equations are replaced by the corresponding equilibrium equations. All generator buses are modelled as PQ nodes for the dynamic load flow. The steady-state active power response of the generators in the new stability point is derived using the frequency response characteristics after the primary speed control action, while the active losses are redistributed iteratively. The reactive output of the machine after the primary voltage regulation is evaluated from the steady-state power-angle relationship. To ensure convergence, the consideration of all generators as PQ nodes is taken into account in the construction of the fast decoupled load flow Jacobian. The system (island) frequency calculation employs the uniform frequency assumption used in long-term dynamics studies. This is accomplished each time there is a dynamic load flow run and it offers an accurate feeling for the frequency evolution during the restoration process.

C. Interactive graphical package. The power system simulation program is integrated into the interactive graphical analysis package of GINA, [21]. GINA has been used for the last 10 years in performing analytical studies and for educational purposes. Its current version runs on PC platfroms and it provides facilities for power flow, stability and recently LTD studies. For PSR studies, the package is equipped with a data base of past incidents experienced in the Hellenic Power System and the restoration procedures followed which can be retrieved and studied. In addition, new island boundaries can be established, a time sequence of events can be set up and the capability even exists to interactively change the scenario as the simulation progresses. The behaviour of the system is monitored by the evolution of voltages, frequencies and load in a number of selected locations. It should be noted that the interactive simulation program allows simultaneous solution of all electrical islands formed.

III. SIMULATION STUDIES

A. Review of the Hellenic Electrical system. A large scale model of the Hellenic power system corresponding to the actual summer peak of June 1994 (5470 MW/2665 MVAr) has been developed in the Electric Energy Systems Laboratory of NTUA, Fig.1. The power system model consists of 235 buses, 334 branches, 51 transformers and 38 equivalent generators. The total power is provided by 8 thermal plants and 13 hydro plants. A strong 400 kV transmission network of 1400 Km superposed and interconnected with the older 150 kV transmission lines of 6600 Km is in operation to provide the necessary power flows. The system is interconnected in the northern part of the country with Albania (400 and 150 kV lines), Yugoslavia (400 and 150 kV lines) and Bulgaria (400 kV line). There are four 400 kV main links between generating plants and the major load centers: i) Ptolemaida-Athens, ii) Ptolemaida-Thessaloniki, iii) Acheloos-Athens, iv) Lavrio-Athens. The link between 150 kV and 400 kV lines is achieved by three winding 400/150/30 kV autotransformers. System loads are modelled as mixed type constant power/constant admittance.

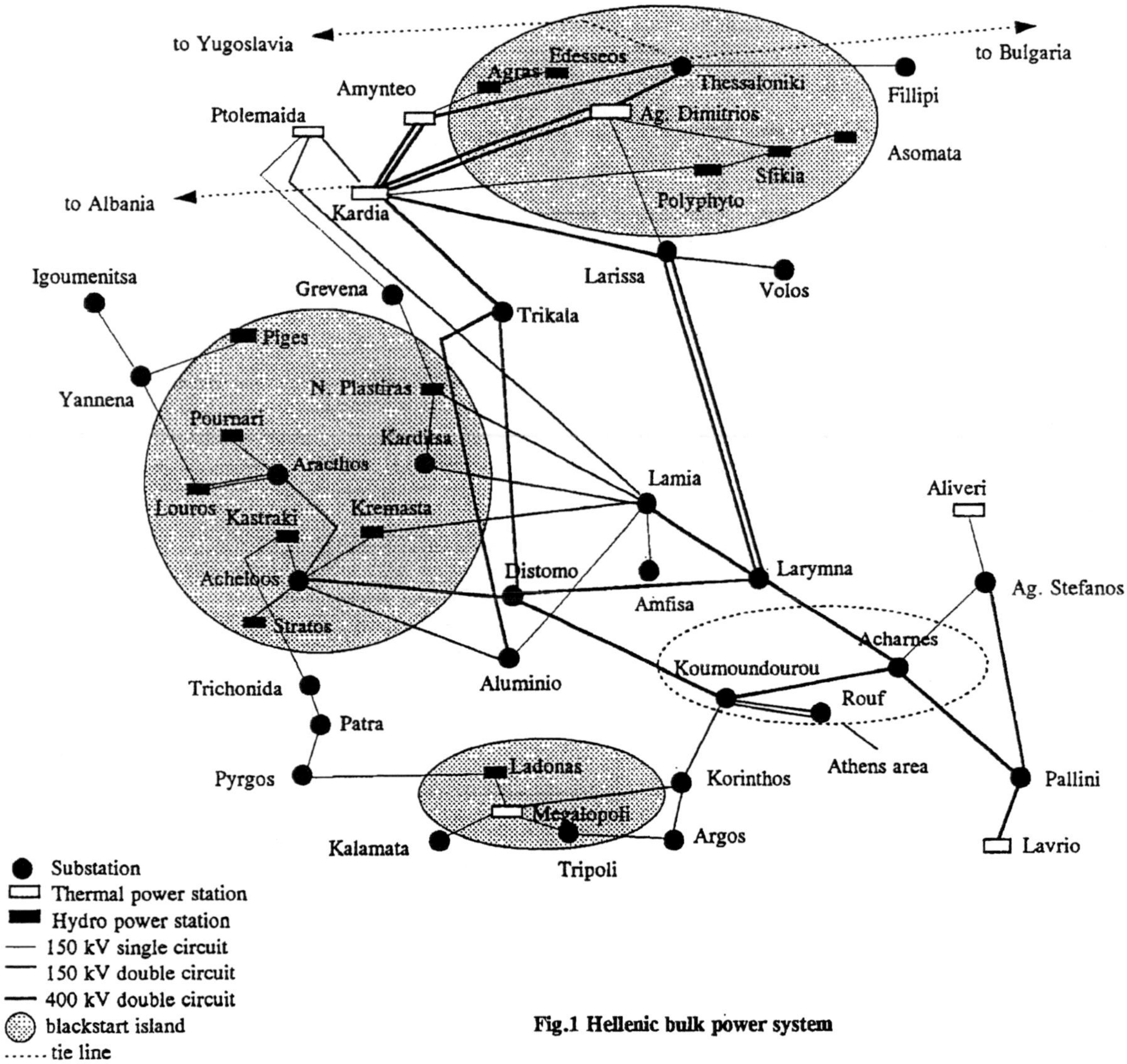

Fig.1 Hellenic bulk power system

For system voltage regulation about 700 MVAr shunt reactors on the 30 kV autotransformers windings along with 1030 MVAr shunt capacitor banks on 15/20 kV step-down substations are installed.

B. PPC restoration strategy. PPC utilises an "all-open" switching strategy following a major system failure. The adapted "build-up" strategy entails the main guidelines described in [15, 18], namely, [20]:

- assessment of power system status; faulted area isolation if it is known or search for its identification.
- starting as soon as possible of all hydro stations - insure power supply of all thermal plant auxiliaries from hydro power stations. Supply of station service for the hot restart of drum type boilers before elapse of the maximum critical time.
- sectionalisation of the de-energised system into several island sub-systems, so that each sub-system includes at least one black-start hydro unit and/or one drum type boiler; each island should have the ability to match the load and generation with frequency limitations.
- interconnection of generating stations within each island after units obtain minimum generation but prior to restoration of full load within the island.
- independent gradual load and generation build-up inside each subsystem. Higher priority is assigned for immediate service to the critical load center of Aluminio representing an aluminium process industry, Fig.1, to capital Athens area and to other big load centers.
- resynchronisation of islands.
- completion of system rebuilt by picking-up the remaining unserved load and closing of ties with the interconnection systems.

C. Special problems encountered during restoration. In the course of system reenergisation, power plant and control center operators have reported the following major concerns, [20]:

- *Failure of thermal stations auxiliaries start-up* due to low voltage supplied from small hydro stations.
- *Fuzzy boundaries of formed islands.* The formulation of islands is basically random. It is not certain that the all-open switching strategy before restoration has been successfully implemented, either due to human error or equipment malfunction. Therefore it is necessary for operators to have a clear outlook of the boundaries of each island (e.g. long distance transmission lines energised and loop closures).
- *Violation of island stability limits.* Careful consideration should be exercised to continual monitoring and maintenance of subsystem load-generation and reactive balance.
- *Poor coordination* of the consumer load reconnection rate and of the island plants load pickup capability. The resupply of the consumers must be completed in such a way, so that each time instant the available island generation along with the in-service generators ramp rate will be explicitly known. Special attention should be given to the load value at the time of reconnection due to increased coincident factor (cold load pick up).
- *Excessive differences at the magnitude and the phase angle of bus voltages when closing loops.* During system reintegration these differences could approach or even exceed the stability limits characterising loop reclosure (e.g. 20% and 30°).
- *Excessive differences at voltage/frequency magnitude during islands reconnection.* The same occurs during synchronisation of independent randomly formed islands with generation-load mismatch. Often,. the inability to regulate frequency at two adjacent islands and voltages at the bus where synchronisation is attempted, causes delays and efforts should be undertaken to improve the prevailing conditions or to search for new synchronisation locations.

D. The Blackout of 1989. The Hellenic power system experienced a total blackout in 1989, [20]. At 14.17 h a busbar fault occurred in the vital substation of Kardia. In order to isolate the fault, overcurrent protection was activated. 0.84 s later, strong oscillations were recorded. This caused cascade tripping of lines and cut off many load centers. In fact, 1.58 s after the fault 1429 MW system generation was lost, while the interconnection with Yugoslavia had tripped. The total net production loss for fault isolation in an elapsed time interval of 2.5 s, was 1822 MW out of the available 3592 MW production (51%). 8 thermal generating stations were tripped due to undervoltage, underfrequency, overloading and overcurrent machine protection relays. This resulted in frequency drop and consequently more units were shutdown. The frequency drop was so rapid that load shedding was unable to save the system from collapsing. In order to restore service, the station personnel followed the previously described parallel restoration guidelines. The formulated islands - in the West, North and South of the power system respectively - are illustrated in Fig. 1. The first hydro station started at about 14.25 h. After the gradual parallelism of islands, at 16.50 h started the reconnection of the 150 kV network with the rest of the 400 kV system. Restoration of customer load was completed in 3 hours after fault occurrence.

E. Simulation results. A set of cascading operating snapshots of the Hellenic power system during the restoration phase following the 1989 blackout has been replicated utilising the program previously described. In the scenario of Table 1., a selective sequence of restoration actions has been established. The simulation time covers the first hour after fault occurrence. The chosen time instants of Table 1. correspond to actual PPC time recordings of the restoration procedure. In Fig. 2. frequency evolution in the three islands is given. The voltage profile of three major generating stations belonging to different islands is rendered in Fig. 3. System active generation as well as active and reactive restoration load are illustrated in Fig. 4-6.

action number	time instant (s)	restoration action
1.	100	Kremasta-Acheloos line energised
2.	500	Aluminio load pickedup
3.	800	Polyphyto-Sfikia line energised
4.	1000	Ag.Dimitrios-Sfikia line energised
5.	1200	Amynteo load pickedup
6.	1500	Kalamata load pickedup
7.	1800	Lamia load pickedup
8.	2000	Tripoli-Argos line energised
9.	2100	Amfisa-Lamia line energised
10	2700	Ptolemaida load pickedup
11.	3000	N.Plastiras-Lamia line energised
12.	3300	Karditsa load pickedup
13.	3600	Amfisa load pickedup
14.	4000	West - South islands synchronised at Korinthos

Table 1. Sequence of Restoration Actions

Discussion. Because of the long-term nature of the simulated phenomena, PSR simulation has high CPU requirements. The QS module proved to be robust exhibiting high computational efficiency. It required about 10 iterations on average for each restoration action in order to reach system's new equilibrium point. The 70-minute simulation depicted in Table 1. needed less than 4 minutes CPU-time on an IBM PC 486.

In different occasions the algorithm failed to converge to new equilibrium point. Such inadequacy is attributed to

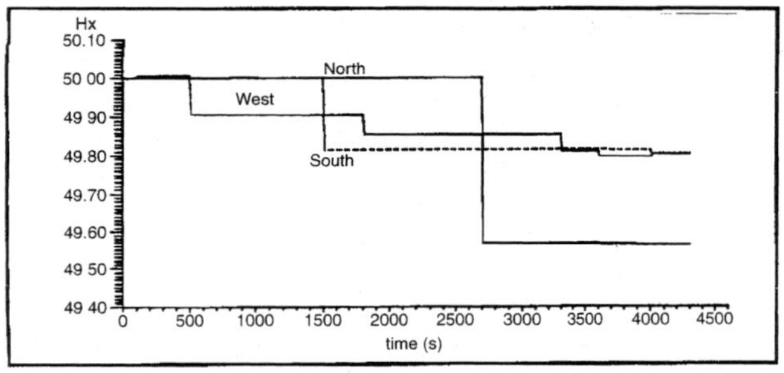

Fig. 2. Island frequencies trajectory

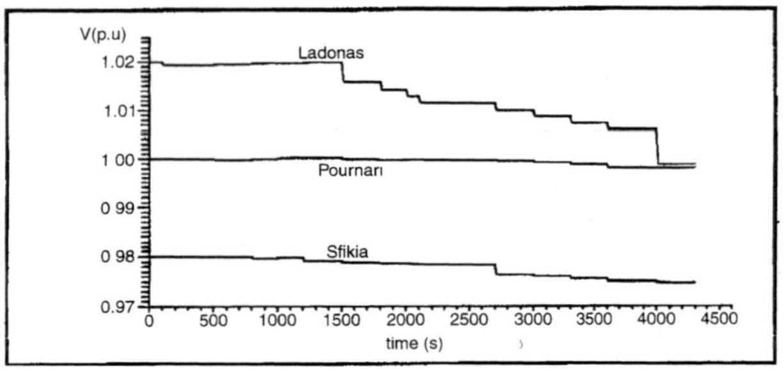

Fig. 3. Voltage profile at three island generating stations

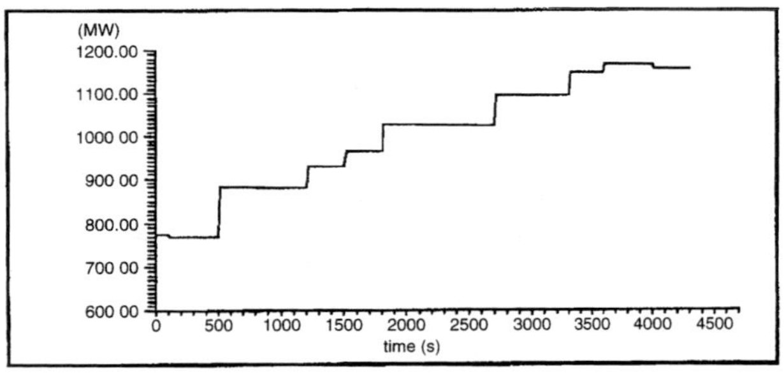

Fig. 4. System active generation

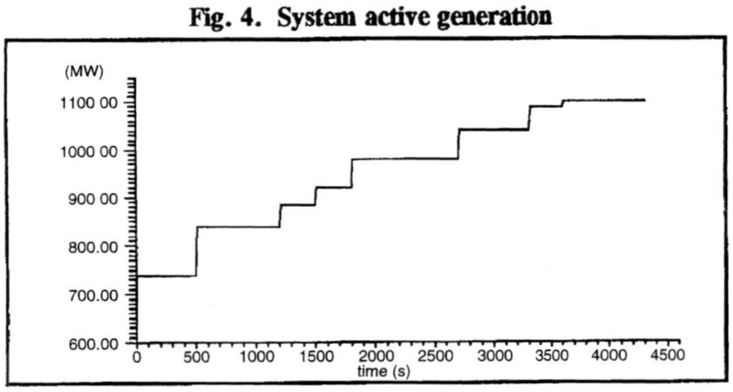

Fig. 5. System active restoration load

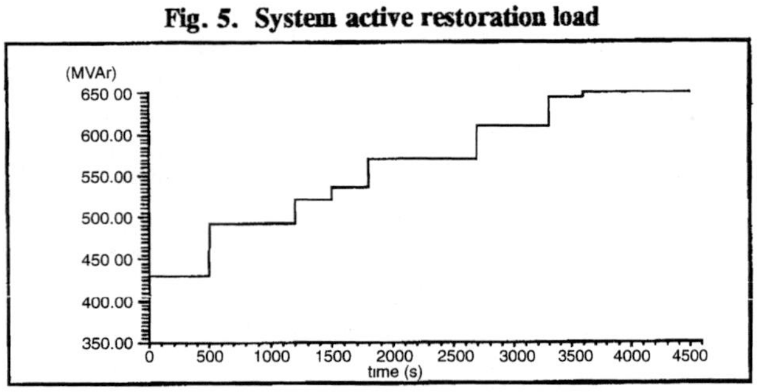

Fig. 6. System reactive restoration load

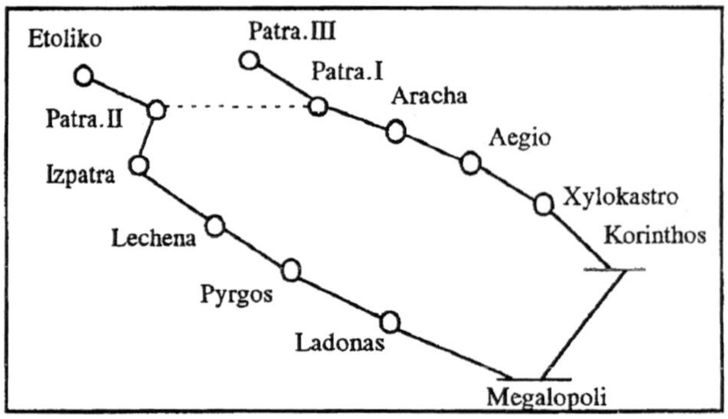

Fig. 7. South island loop

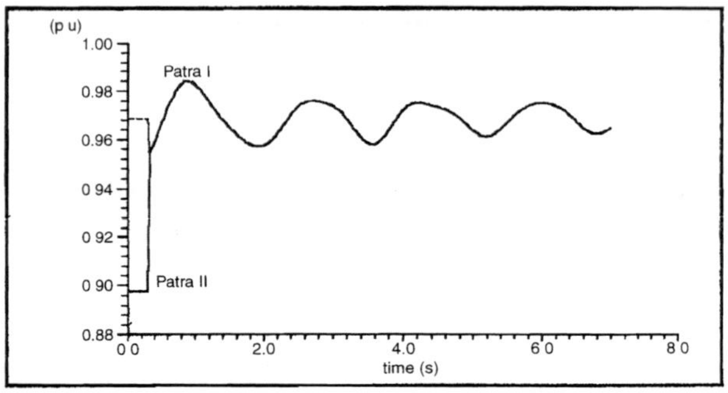

Fig. 8. Loop closure transient

operating constraint violation; this conclusion has been verified in the FI module of GINA that utilises a detailed integration. Some selected problems that can be identified in the simulation of the restoration after the blackout of March 1989, and have been reported in Section C. are:

1. *Excessive load pick up rate.* In trying to restore large load blocks the algorithm fails. This is attributed to poor coordination between load pickup and the rate of response of prime movers which causes unacceptable frequency excursions. To avoid the risk of tripping on underfrequency, frequency drop should be kept at maximum of about 0.5 Hz from its previous value, as discussed in [8] and ascertained by the results of Fig. 2.

2. *Long transmission line energisation.* In certain occasions the operators failed to energise long-distance transmission links due to excessive MVAr line charging and absence of generator reactive absorbing capability. By using the program, the amount of the reactive absorbing capability of the generators can be easily monitored. This information can be used to effectively enable the energisation of circuits having a large line charging value.

3. *Unsuccessful loop closure.* Restoration process was often interrupted or delayed by excessive standing phase angles across a breaker which connects two adjacent lines. In the example of Fig. 7, in trying to close the depicted South island loop, there was an algorithm failure, the reason being an excessive standing phase angle of 50 degrees between Patra.II-Patra.I. By switching to the FI module

corresponding to this operating point it was found that it was infeasible to reach a new equilibriun point with this level of loading. By altering line loading and increasing the generation profile of the nearby stations, the energisation was successfully implemented, Fig. 8.

IV. Conclusions and Perspectives of Future Use

In this paper, the application of a Long-Term-Dynamics simulation tool based on a fast Quasi-Stationary solution algorithm for Power System Restoration studies is presented. The program is specially adapted to simulate the restoration procedures followed by the Hellenic Public Power Corporation. A real total blackout incident in the Hellenic power system has been explored. It is shown that integrated LTD programs can provide an effective means by which the operator can be trained to deal with restoration situations in a secure, controllable and off-line simulation environment and they can be used for effective restoration planning. It is expected that the combination of the analytical tool described, supported by artificial intelligence techniques and augmented by the experience gained from past blackouts, could serve as a valuable aid to system operators and dispatchers.

V. Acknowledgement

This work is a part of the project 'Computer Programs for Power System Restoration Following a Major Disturbance', financed by the Power Public Corporation (PPC) and the General Secretariat of Research and Technology of Greece.

VI. References

[1] L.H. Fink, K. Carlsen, "Operating under stress and strain", IEEE Spectrum , March 1978, pp. 48-53.

[2] R.J. Kafka *et* al. "System Restoration plan development for a metropolitan electric system", IEEE Trans. on PAS, Vol. PAS-100, No.8, Aug. 1981, pp. 3703-3713.

[3] IEEE Committee Report, "System restoration - Deploying the plan", IEEE Trans v. PAS-101 Nov 1982, pp.4263-4271.

[4] E. Mariani et al, "Field experiences in reenergization of electrical networks from thermal and hydro units", IEEE Trans. on PAS, Vol. PAS-103, No. 7, July 1984, pp. 1707-1713.

[5] G. Morin, "Service restoration following a major failure on the Hydro-Quebec power system", IEEE Trans. on PD, Vol. PWRD-2, No.2, April 1987, pp. 454-463.

[6] R. Kearsley, "Restoration in Sweden and experience gained from the blackout of 1983", IEEE Trans. on PS, Vol. PWRS-2, No.2, May 1987, pp. 422-428.

[7] R.J. Kafka *et* al. "Role of interactive and control computers in the development of a system restoration plan", IEEE Trans. on PAS, Vol. PAS-101, No.1, Jan.1982, pp. 43-52.

[8] IEEE Committee Report, "Power System restoration - A Task Force report", SRTF, M. M. Adibi, chmn, IEEE Trans. v. PWRS-2, May 1987, pp. 271-277.

[9] IEEE Committee Report, "Power System restoration - The Second Task Force Report", SRTF. M. M. Adibi, chmn, IEEE Trans. v. PWRS-2, n.4, Nov. 1987, pp. 927-933.

[10] IEEE Committee Report, "New Approaches in Power System Restoration", SRWG, M. M. Adibi, chmn, IEEE Trans. v. PWRS-7, No.4, Nov. 1992, pp .1428-1434.

[11] IEEE Committee Report, "Overvoltage Control During Restoration", SRWG, M. M. Adibi, chmn, IEEE Trans. Vol. PWRS-7, No.4, Nov 1992, pp.1464-1470.

[12] IEEE Committee Report, "Special Considerations in Power System Restoration", SRWG, M. M. Adibi, chmn, paper 93 WM 202-2-PWRS, IEEE/PES 1993 WPM, Colombus, OH, Jan 31 - Feb 5, 1993.

[13] IEEE Committee Report, "Bulk Power System Restoration Training Techniques", WGOT, M. E. Cooper, chmn and SRWG, M. M. Adibi, chmn, IEEE Trans. Vol. PWRS, Feb 1993, pp.191-197.

[14] IEEE Committee Report, "Special Considerations in Power System Restoration, The Second Working Group Report", SRWG, M. M. Adibi, chmn, paper 93 WM 202-2 PWRS, IEEE/PES 1993 WPM, Columbus, OH, Jan 31 - Feb 5, 1993.

[15] M. M. Adibi, L. H. Fink, "Power System Restoration Planning", IEEE Trans. Vol. PWRS-9, No.1, Feb 1994, pp. 22-28.

[16] M. M. Adibi, J. N. Borkowski, and R. J. Kafka, "Analytical Tool Requirements for Power System Restoration", paper 94 WM 220-0-PWRS, IEEE/PES 1994 WPM, New York, NY, Jan 30 - Feb 3, 1994.

[17] M. M. Adibi, R. J. Kafka, and D. P. Milanicz, "Expert System Requirements for Power System Restoration", paper 94 WM 223-8-PWRS, IEEE/PES 1994 WPM, New York, NY, Jan 30-Feb 3, 1994.

[18] J. J. Ancona, "A Framework For Power System Restoration Following a Major Power Failure", paper 95 WM 143-8 PWRS, IEEE/PES 1995 WPM, New York, NY, Jan 29-Feb 2, 1995.

[19] CIGRE SC 38, WG 02, TF 02, "Modelling and Simulation of Black Start and Restoration of Electric Power Systems", Electra No. 147, April 1993, pp. 21- 42.

[20] "Report on Hellenic Power System Blackout", 1989, PPC internal correspondence.

[21] C.D. Vournas, N.D. Hatziargyriou, B.C.Papadias, "Interactive Power System Simulation program - Application to the Hellenic system", paper 38-206, CIGRE 1990 Session, 26 Aug-1 Sep, Paris.

[22] M. Stubbe (Convenor), CIGRE TF 38-02-08, "Long-Term Dynamics Phase II", Final Report, March 1995.

[23] T. Van Cutsem, Y, Jacquemart, J.-N. Marquetr, P. Pruvot, "A comprehensive analysis of mid-term voltage stability", paper 94 SM 511-6 PWRS, IEEE/PES 1994 SM, San Francisco, CA, July 24-28, 1994.

[24] N. A. Fountas, N. D. Hatziargyriou, "Fast Simulation of Power System Mid-Term Dynamics", IFAC LSS'95 proceedings, pp. 527-532, London, UK 11-13 July, 1995.

[25] P. Kundur, *Power System Stability and Control*, McGraw-Hill, 1994.

Discussion

M. M. Adibi (IRD Corporation, Bethesda. The authors are to be commended for an excellent paper on the use of a dynamic simulator in developing restoration plan for the Hellenic Bulk Power System. The paper describes the recent major power system disturbance of the Hellenic PPC and then develops a restoration plan which addresses many steady-state and dynamic constraints posed by the restoration process. This paper is an important addition to the power system restoration literature. The following points need further clarification:

1. Would variable step integration avoid the switching between the FI and QS modules ? If the fast dynamic behaviors (transients) are not simulated in the QS modules, how are the corresponding equilibrium points determined ?

2. What steps are taken to identify the reasons for non-convergences of power flows in an island? How is the non-convergence described to the operators during training and when using the simulator ? Noting that the island boundaries are "fuzzy" and not predictable, unless controlled islanding is used.

3. Recognizing that in the initial phase of restoration, due to load-generation mismatch, the tendency for generator tripping is high and in view of the differences in dynamic responses of the prime movers to sudden increase in loads, and in view of the different underfrequency load shedding schedules, how is the response reserve of an island determined and how is it maintained ? Have authors developed response reserve indices or guides for the use by operators ?

4. In providing remote cranking power to thermal units by the hydro electric stations, have the authors timed the cranking process to ascertain its timely completion within the maximum critical time allowed for the hot restart of the drum-type boiler-turbine-generators ?

5. Noting the significant differences between the designed values and operational limits of reactive capabilities of synchronous machines, what steps have been taken to determine the adequacies of: (a) over-excitation when starting large auxiliary motors (over 6,000horse power) in the thermal plants, and (b) under-excitation when energizing the 400kV lines between the hydros and thermal plants ?

6. Have the authors established a procedure for reducing the standing phase angle in the South Island Loop so that it can be implemented without the time consuming iterative process of altering line loadings and generation ?

The authors' comments to the above questions will be greatly appreciated.

Manuscript received January 31, 1996.

N.A. Fountas, N.D. Hatziargyriou, C. Orphanogiannis, A. Tasoulis:

We wish to thank Mr. Adibi, a leading expert in the field of restoration, for his interest in our paper and his comments. We will respond to each one separately:

1. We agree that a variable step integration algorithm would avoid switching between the Full Integration (FI) and Quasi Stationary (QS) modules. However, the approach proposed in this paper provides trajectories of the system variables saving a large amount of computational time, since only algebraic equations after each discrete action are solved instead of the time consuming integration of differential equations. The basic features of the algorithm are shown in the following flow chart, [24].

2. The operator becomes aware of the load flow non-convergence in the QS module by appropriate warning messages. The non-convergence of the power flows can be investigated by switching back to the Full Integration module. In this module, the evolution of the system profile is illustrated in detail and the prevailing conditions can be searched in order to find out the reasons for non-convergence. It should be stated that in our simulation runs we have established scenarios with pre-defined islands, i.e. controlled islanding.

3. It is certainly true that the response reserve during island reconnection is crucial. In the dynamic simulator developed, the difference between required and available real power at each instant during restoration can be displayed taking into account the permanent droop characteristic of the prime mover of each machine. Load shedding due to underfrequency relay action has not been taken into account. This project has focused, however, on the use of a training tool to develop a feeling of the power system operators for the evolution of load-generation mismatch during gradual system reconnection. More advanced indices and guidelines would be certainly needed in order to exploit the simulator as an on-line restoration aid.

4. We realise that the completion time for the cranking process of the drum-type boiler-turbine-generators is of extreme importance in the course of system restoration. In our simulator it is possible to estimate this time, however, it has not been possible so far to compare it with actual time recordings or critical time intervals.

5. a) During start-up of the large auxiliary motors in the thermal plants, as it was ascertained by PPC operators, no problems were noticed regarding auxiliary malfunction. This was also verified in our simulation runs. b) According to PPC restoration policy, the 400kV lines are energised at the latest stages of the restoration process, when all big thermal stations are back in service and therefore enough absorbing reactive capability is available.

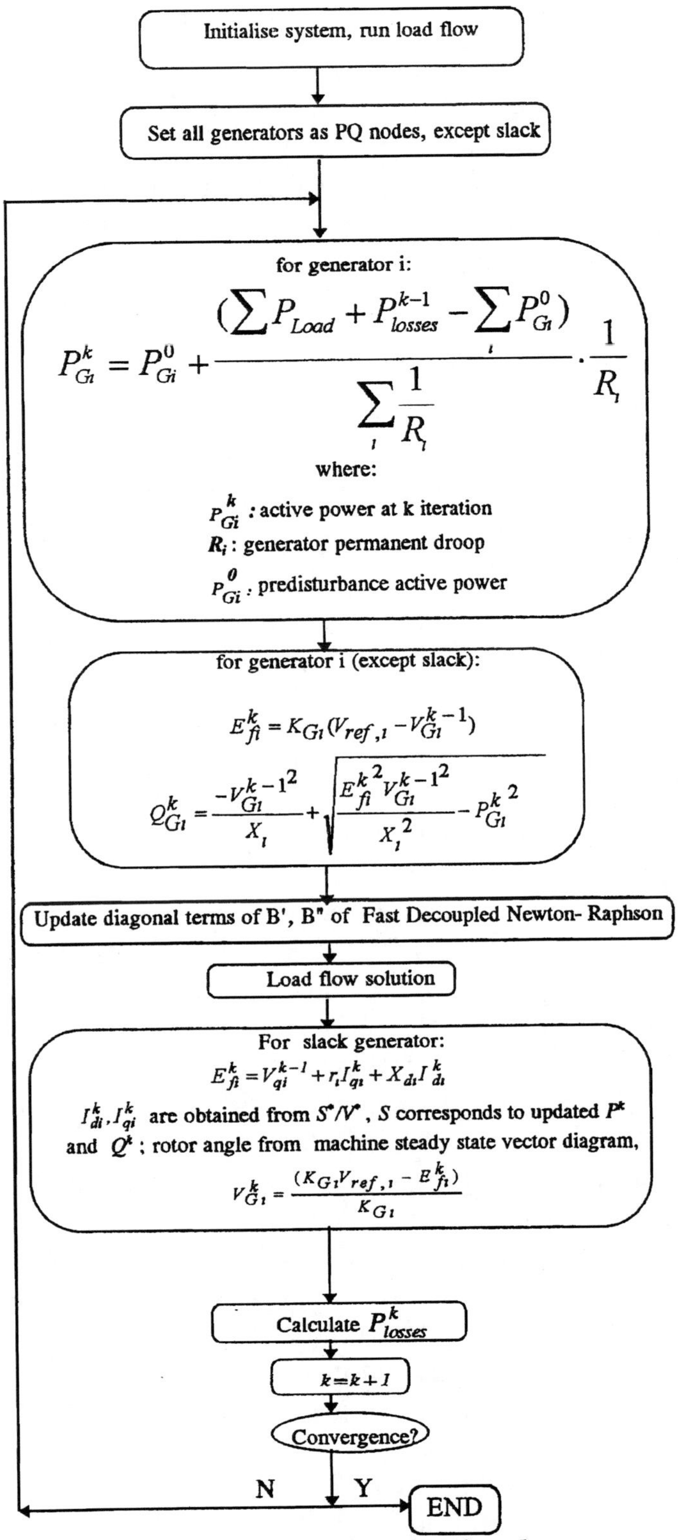

Fast simulation algorithm flow chart

6. Standing phase angle problems during system re-integration are faced using an error and trial approach. We consider, however, this point together with points 3 and 4 of Mr. Adibi's discussion very interesting and worth implementing in order to enhance the flexibility and usefulness of the developed dynamic simulator.

Manuscript received April 2, 1996.

ANALYSIS OF THE POWER BLACKOUT OF JUNE 8, 1995 IN THE ISRAEL ELECTRIC CORPORATION

Y. HAIN   I. SCHWEITZER

The Israel Electric Corporation
P.O.Box 10, Haifa, Israel

ABSTRACT

This paper is a technical survey of the power failure which affected the Israel electrical network on June 8, 1995.
As a result of the failure about 70% of the consumers were disconnected from the energy supply.
In the first part of the paper, the general system characteristics and the condition of the network before the incident are summarized.
In the second part, the events during and after the failure are described.
The third part summarizes the lessons learned from the power blackout and outlines future strategies to be adopted by IECo.

1. INTRODUCTION

1.1 General System Characteristics

The Israel Electric Corporation is a national company responsible for generation, transmission and distribution of energy throughout the country.
The Electrical network is isolated from the neighboring countries and consists of:

187 km of 400 kV circuits
3325 km of 161 kV circuits
352 km of 110 kV circuits

Installed generation capacity is 6388MW; peak load during last winter 1994/95 was 5538MW.
The commonly used N-1 design criteria is applied for 161 kV network design. Due to high demand growth during the last five years (about 7.6% yearly), and the prognosis for the continuation of such demand growth for the next several years, IECo plans to rapidly increase the installed generation capacity as well as transmission and distribution capability. According to the current expansion plan, the Main system figures in the year 2000 will be:

installed capacity - 9534MW
763 km of 400 kV circuits
4210 km of 161 kV circuits

1.2 System Condition Prior to the Power Failure on June 8, 1995

On June 8th, the predicted peak demand was 4500MW, the weather was hot - 35^{0}C. There was some deficit of power production in the southern part of the country due to four generation units being out of service in the Eshkol power station.
An additional problem was a low voltage profile in the central and southern part of the system because of a reactive power sources deficit.
At 13:00 the actual power produced by the entire system was 4482MW. Power flowed from the northern part of the country to the southern part through four 161KV circuits connected to the largest northern switching station, Caesarea (two additional circuits were under reconstruction). The power flow was 701MVA (640 + j286). The North-South interconnection through 400kV circuits was not yet completed.
At 13:15 two generators (40MW each) tripped in the central part of the country because of a bus bar short circuit. The above failure caused an increase of power flow from north to south. At 13:30 the power produced in the entire system was 4534MW, while the North-South power flow was 806MVA (730 + j342) (Fig. 1).
The failure that finally caused the power blackout occurred at 13:36.

PE-724-PWRS-0-12-1997 A paper recommended and approved by the IEEE Power System Engineering Committee of the IEEE Power Engineering Society for publication in the IEEE Transactions on Power Systems Manuscript submitted December 5, 1995, made available for printing December 16, 1996

Reprinted from *IEEE Transactions on Power Systems,* Vol. 12, No. 4, pp. 1752–1758, November 1997.

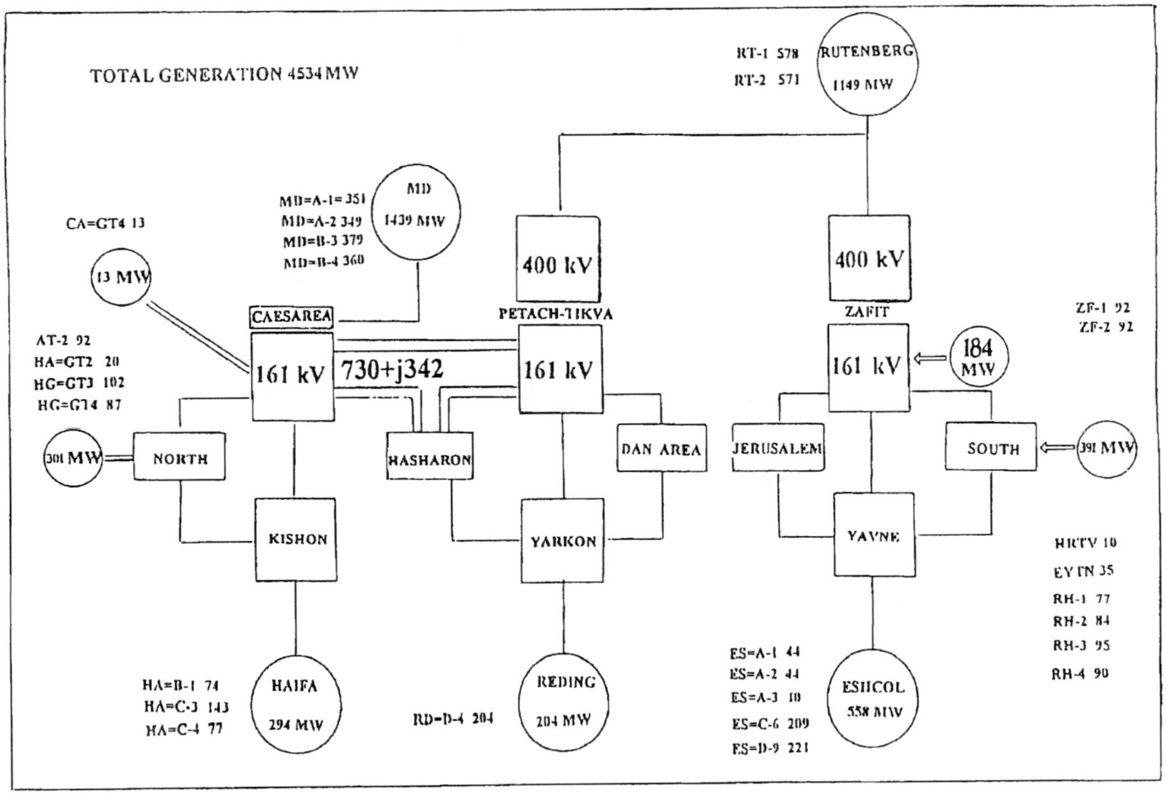

Fig. 1: system condition prior to the failure on 08/06/95, at 13:30

2. EVENTS DURING AND AFTER THE FAILURE

2.1 Description of the Events that Caused the Blackout of the Southern Part of the Country

At 13:36:36 one of the 161 kV circuits connecting northern and southern parts was tripped due to a two phase fault (tripping time - 100 ms).
Thirteen seconds later, at 13:36:49, another circuit of the same double circuit line was tripped because of a two phase fault (tripping time - 140ms).
Both circuits were tripped by the first zone of their distance protection devices. Both short circuits were caused by a fire under the double circuit line.
After the trip of the above circuits, the North-South interconnection was provided by two 161kV lines carrying 706MVA from north to south. In summer conditions these circuits are capable of transferring 330MVA each for half an hour.
Due to summer peak hours the demand continued growing, and at 13:50 the North-South power flow increased up to 902MVA (818 + j381).

At 13:50:29 the Ramat-Hovav gas turbine unit in the Southern part of the country was tripped by over-excitation protection devices, increasing the North-South power flow by an additional 93MW and 120MVAR.
After the Ramat-Hovav unit trip, the voltage drop and the increase of the current led to conditions in which the fourth (starting) zone of the distance protection on one of the North-South interconnecting lines operated (13:50:33) (Fig. 2).

The last North-South interconnecting line was tripped by the distance relay because of power oscillations.At 13:50:37 the system was split into two separate subsystems: a northern "island" and a southern "island" (Fig. 3).

After the split, the southern island was left with a deficit of more than 900MW. This deficit caused a frequency decrease and load shedding actuation. Unfortunately, 300ms after system separation, two large southern units of 550MW each tripped (within an interval of 200 ms).

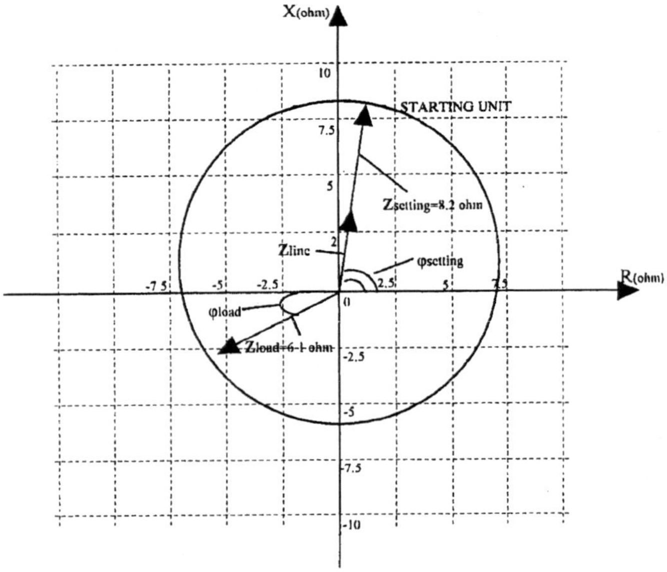

Fig. 2: Operating characteristics of distance protection installed in Petach Takva on 161kV line Caesarea-Petach Tikva

The tripping of these two units, located in the Rutenberg power station, was caused by unnecessary operation of the stator coolant protection. After the trip, both units still supplied their auxiliaries. This additional loss of 1100MW in the south led to total collapse of the southern part of the country and trip of all power units (Fig. 4). The population of more than 3 million people including the main cities of Jerusalem and Tel-Aviv remained without electricity.

In the northern part, three power units tripped after the separation of the system. Load shedding operated properly and led to stabilization of the northern part while the demand was about 1377MW.

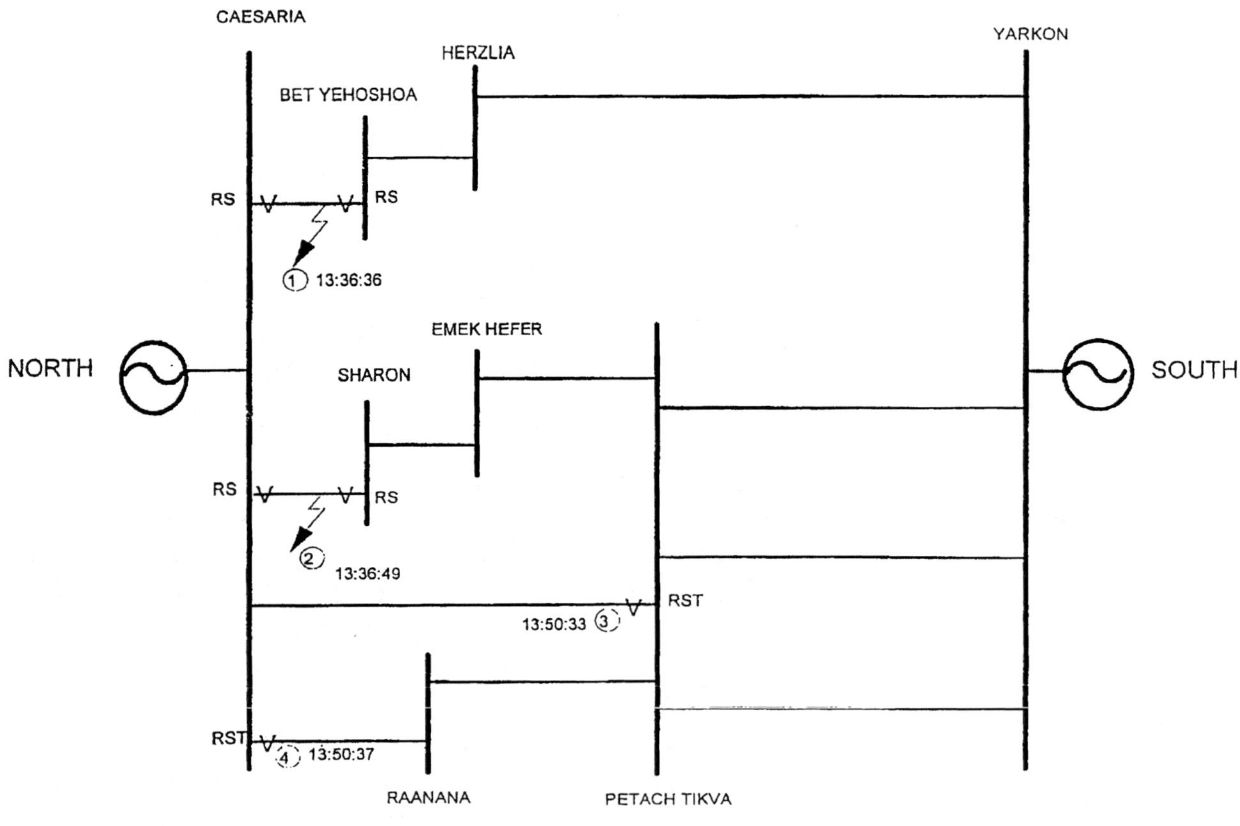

Fig. 3: North-South interconnection diagram

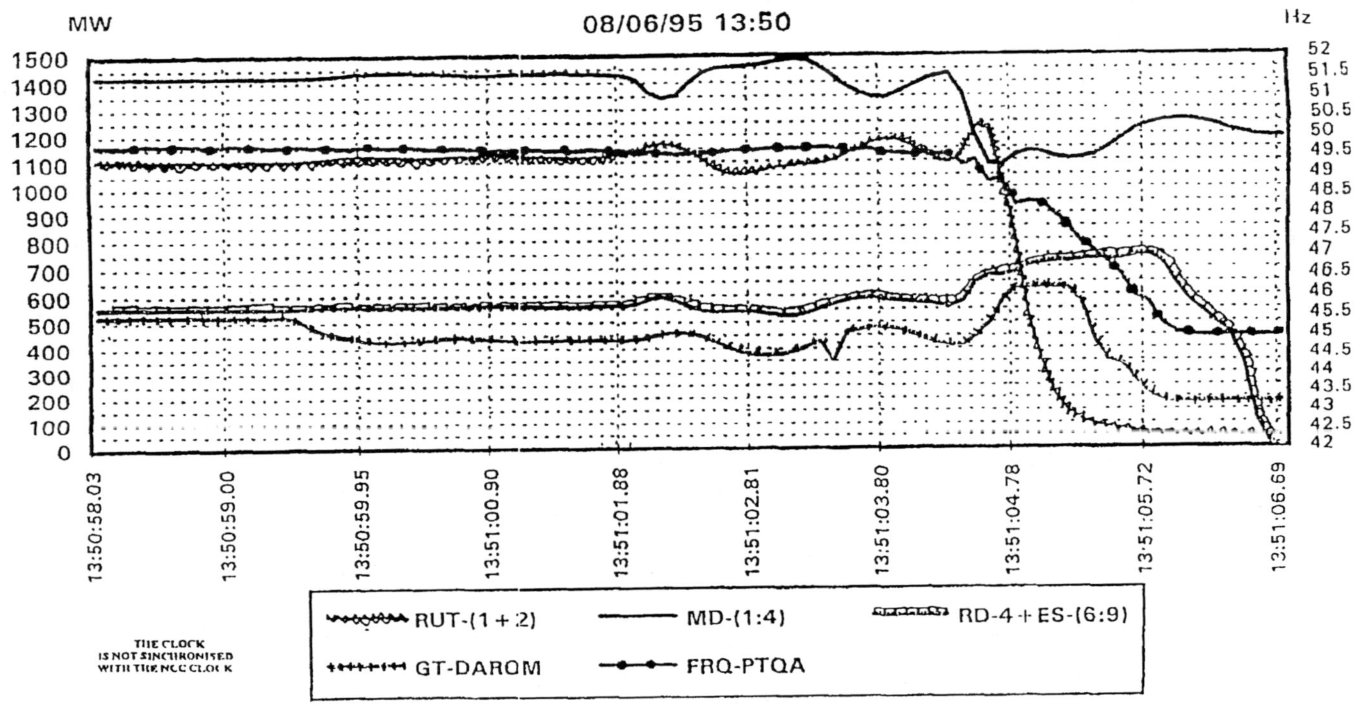

Fig. 4: Steam and gas turbine units load diagram during the failure

2.2 Restoration of the System

The restoration of the system started from actuation of 50MW Gas turbines. During the first hour after the blackout, voltage was supplied to the main switching stations of the south. At 15:39 and 15:45 two units of power station Rutenberg were synchronized. At 17:00, about three hours after the failure, 70% of the consumers were reconnected to the network. At 18:00, 90% of the consumers had power restored.
800MWh of energy were not supplied due to the failure, equivalent to 160 minutes of non-supplied energy.

3. LESSONS LEARNED FROM THE BLACKOUT, AND FUTURE STRATEGIES

3.1 Organizational Response

After the blackout, two investigation committees were appointed: one internal to the IECo and another external committee appointed by the minister of energy.

Generally, both committees arrived at the same conclusions. In addition several subcommittees of IECo issued recommendations on the following subjects:
- organization of management and operational tools to cope with severe failures in the network.
- reactive power and voltage control in the system.
- improvements in power stations' response to severe failures in the system.
- restoration of the system, including resynchronization at different points in the system.
- load shedding development on voltage profile, tap-changer behavior during voltage collapse.
- coordination of protection and automatics operation in the system, including power stations.

3.2 Technical Analysis of the Failure

Being comparatively small isolated system, the IECo paid a lot of attention to load frequency control. Short distances between the production and load centers never caused problems of reactive power control and voltage profile. However, high demand of reactive power during the last 5–6 years has led to low voltages on the 161kV bus-bars in the central and southern parts of the country. The growth of the reactive power demand can be explained by the large increase in the number of air conditioning devices (motors) during these recent years. The rate of power capacitor installation dropped significantly behind the rate of reactive power demand growth.

The above severe failure that affected the Israeli system occurred because of the coincidence of the following circumstances and failures:

- deficit of power generation in the southern part of the country due to urgent maintenance works in Eshkol power station.
- low voltage profile in the system due to reactive power sources deficit.
- fire under the double circuit line which caused the trip of two circuits.
- trip of Ramat-Hovav Gas Turbine by the over-excitation protection devices. The maximum excitation limit (MEL) unit was not installed in the gas turbines. The presence of the MEL unit would have prevented the protection operation and the consequent generator trip. The above G.T. trip resulted in additional power flow from north to south and shortened the time available for operator intervention in the national control center.
- unnecessary operation of the stator coolant protection in the Rutenberg power station after system separation. The stator coolant protection logic was initially incorrectly designed by the manufacturer of the generator and was changed by local staff after the failure.

The trip of the Rutenberg power station prevented the load shedding automatics from stabilizing the southern system.

It is necessary to mention that during the failure process, which lasted for fourteen minutes, the power flow from north to south increased and the voltage profile became lower.

After the Ramat-Hovav unit trip, the impedance vector of the load entered the operating zone of the starting unit of the distance protection and caused the tripping of the third circuit which interconnected northern and southern areas. Under these conditions there was no chance for the only remaining circuit between north and south to survive.

The last interconnecting circuit was tripped by the distance protection due to power oscillations.. This distance protection device is an old design without a power swing block element. In any event, even if there had been such a blocking element, under the conditions of network described in 2.1 the last interconnecting circuit would have been tripped by the distance relay due to high currents and low voltage - similar to the trip of the third interconnecting circuit. Generally speaking, the distance protection of all interconnected circuits operated satisfactorily.

In summary, a slow voltage collapse process led to system separation.

3.3 Summary of Recommendations of Future Strategy

The following recommendations were issued by the Committees in IECo which dealt with the above incident:

- reconsider the indices, criteria and countermeasures for voltage collapse.
- speed up the development of the 400kV network.
- optimize the reactive power and voltage profile control using software (preferably on-line real-time calculations).
- increase the amount and rate of installation of power capacitors at the medium voltage level (substations and distributions lines).
- install MEL and VAR control systems on all gas turbine units.
- provide the possibility of synchronizing the subsystems at different points of the system.
- implement undervoltage load shedding.
- produce a new, updated protection coordination study of the network with special emphasis on the severe failures regimes.

SUMMARY

This paper describes the power failure which affected the Israel electrical network on June 8, 1995.

As a result of the failure, about 70% of the consumers were disconnected from the energy supply.

Before the failure occurred, there was already a low voltage profile in the central and southern parts of the system because of a high reactive power demand and a reactive power sources deficit.

The failure started from the trip of two 161kV circuits connecting the northern and southern parts of the system. Both circuits were tripped because of a fire under the double circuit line. After the trip of the above circuits, the North-South interconnection was provided solely by two 161kV lines. Due to the summer peak hour, demand continued growing, and during the next 14 minutes the North-South power flow increased by 28%. The additional trip of the Ramat-Hovav gas turbine in the southern part of the system (tripped by over-excitation protection) led to conditions in which the distance protection device of one of the North-South interconnecting lines operated. The last North-South interconnecting line was tripped by a distance relay due to power oscillations. From that moment, the system was split into two separate subsystems - northern and southern "islands".

After the split, the southern island was left with a deficit of more than 900MW.

About 300ms after system separation, two southern units of 550MW each were tripped by the unnecessary operation of the stator coolant protection device. This additional loss of 1100MW in the south led to the total collapse of the southern part of the country and the trip of all power units.

A geographical area inhabited by more than 3 million people, including the main cities of Jerusalem and Tel-Aviv, remained without electricity.

The following recommendations were issued after the fault investigation:

- Reconsider the indices, criteria and countermeasures for voltage collapse.
- Speed up the development of the 400kV network.
- Optimize the reactive power and voltage profile control unit software.
- Install MEL and VAR control systems on all gas turbine units.
- Implement system automation against severe failure regimes.

PART VI

Knowledge-Based Systems

I. General Discussion

POWER system restoration is one of the important applications for knowledge-based systems. The pioneering paper [6.01] authored by Sakaguchi and Matsumoto generated much interest in the field. Knowledge-based systems as applied to restoration covers bulk power systems, subtransmission systems, and distribution systems. Because of the complexity and scope of the subject, there are numerous subproblems, covered by the following papers, including identification of the fault location, handling of uncertain/inconsistent information, restoration strategies, power system representation and modeling, generation of unit characteristics, transmission path finding, determination of the switching sequence, optimal field crew dispatch, tie line assistance, integration with EMS and analytical tools, and integration with OTS.

II. Discussion of Reprints

6.01 Development of a Knowledge-Based System for Power System Restoration, 1983, p. 443

This paper is a pioneering effort to develop a knowledge-based system (KBS) for power system restoration. The "context" tree concept for reasoning method and the top-down procedure are used to build the restoration actions. The paper divides restoration patterns into whether or not a line and a bus have been energized before the fault event. Five contexts and the related parameters are developed with 16 rules. The paper also discusses the possibility of the "guidance" level approach to system restoration. Many of the subsequent research and development efforts in the KBS area have been inspired by this visionary paper.

6.02 Development of a Guidance Method for Power System Restoration, 1989, p. 450

This paper proposes a guidance method for restoration and reports on the implementation of a prototype system. Since unforeseen disturbances or events are always possible during restoration, the reported method focuses on automatic modification of the current restoration plans to cope with the unforeseen events as they take place. Detailed processing flow of the guidance method and the dispatchers' heuristics approach are reported. The paper states the importance of accuracy of the input data in restoration plans and emphasizes the use of a state estimator to identify bad data. The authors report that dispatchers have demonstrated and evaluated the developed prototype.

6.03 The Development of Power System Restoration Method for a Bulk Power System by Applying Knowledge Engineering Techniques, 1989, p. 458

The expert system developed in this paper focuses on fast and safe bulk power system restoration. The paper proposes to identify a relationship between system components and the restoration policy. To make the expert system efficient, the paper utilizes a "blackboard" software structure, arguing that with a blackboard model, it is not necessary to check/search all pieces of equipment to develop the restoration actions. The developed system covers all required restoration criteria, and it is implemented using an object-oriented programming technique. The authors report that their prototype system has been tested and implemented.

6.04 Implementation of a Real-Time Expert System for a Restoration Guide in a Dispatching Center, 1990, p. 466

An on-line expert system is proposed for operator restoration guidance and for the off-line validation of the acquired restoration knowledge. Traditionally, most restoration actions are performed manually by dispatchers. Hence, the computer-based expert system reported in this paper is a new concept. In their study, the authors use the "context" concept and reasoning procedures to build and verify restoration guidance. Apparently, the proposed system has been tested and implemented. The developed expert system then has been integrated with an existing EMS.

6.05 Guiding a Power System Restoration with an Expert System, 1991, p. 473

Most utilities maintain written restoration procedures. To avoid the disadvantages of bulky and inflexible written restoration procedures, this paper proposes an expert system that has two complementary modules: the strategic planning and the tactical planning. The strategic module is selected by the dispatcher, and it includes detailed restoration actions in the specific tactical reasoning module. In tactical reasoning, the expert system prompts the dispatcher to choose the restoration path and presents switching sequences based on the selected path.

6.06 Knowledge-Based Systems as Operational Aids in Power System Restoration, 1992, p. 481

This survey paper discusses the background and capabilities of various expert systems proposed. The survey covers many restoration papers. The authors split the proposed restoration systems into three groups: conventional restoration approaches, algorithmic methods, and knowledge-based approaches. The knowledge-based approaches are also divided into three subgroups: operational aids methodology, algorithmic methods, and knowledge-based approaches. Verification and evaluation of the developed restoration systems are also discussed. The survey covers both the on-line and off-line methods.

6.07 Operational Experience and Maintenance of an On-Line Expert System for Customer Restoration and Fault Testing, 1992, p. 489

This paper reports the test results of a service restoration and fault testing system. The system is intended to serve as a restoration guide for subtransmission systems. For maintenance of a rule-based system, the current rules may need to be modified or new rules may have to be added. A method for maintaining an existing rule base system is reported. To avoid redundancy and conflict in the rule base, the rule-checking algorithm is proposed. When new rules are added or existing rules are modified, tests are performed to validate and verify the accuracy and performance of the expert system.

6.08 A Knowledge-Based Method for Making Restoration Plan of Bulk Power System, 1992, p. 497

This paper proposes two modules for power system restoration: Load-Feeding Knowledge (LK) and Augmentation Knowledge (AK). The LK module includes the subknowledge block "Load-Division" which divides and groups outage areas in terms of the availability of power supply and equipment. Using the LK block, the proposed system builds a restoration plan to minimize the number of outage areas. If more power is required, then the AK block supports the restoration plan by using equipment and generators outside the outage areas.

6.09 Generation Capability Dispatch for Bulk Power System Restoration: A Knowledge-Based Approach, 1993, p. 504

This paper introduces a comprehensive knowledge-based expert system for restoration, with a special focus on the maximization of generation capability during bulk power system restoration. To maximize the generation capability, a number of knowledge modules are developed, such as real power (MW) and reactive power (MVAR) managements. Heuristic rules for a generator startup sequence are developed, and the search algorithm is used to find an optimal path between the interrupted load and the available generating units. To minimize the unserved load (MWH), several important system parameters and constraints are considered. A prototype system has been tested using the actual system data.

6.10 Expert System Requirements for Power System Restoration, 1994, p. 514

This paper describes the heuristics, requirements, and constraints needed for power system restoration. It defines three periods of restoration: establishment of initial sources of power, reintegration of a power network into a bulk power supply, and minimization of the unserved load. This paper emphasizes the efficient assignment of human resources during restoration, which is rarely dealt with in the literature. The role of expert systems in each period is illustrated.

6.11 Tie Line Utilization During Power System Restoration, 1995, p. 523

The specific issues concerning tie line utilization during power system restoration are described. In some cases, it is impossible to serve all the unserved loads without importing power from tie lines. To build a restoration strategy, this paper proposes a mathematical model for maximization of generation capability involving tie lines. Both import and export of power using tie lines are considered. The developed tie line utilization model is integrated with a restoration expert system developed by the authors. Using test cases, the authors show that generation capability can be optimized and the number of outage areas can be reduced.

6.12 Power System Restoration by Joint Usage of Expert System and Mathematical Programming Approach, 1995, p. 531

Based on transmission paths available for system restoration, the paper proposes a method to find a shortest-time-lowest-cost transmission path for the target system. The study proposes an "operation cost" weight for every breaker and switch on the transmission paths. The total computation time to find a target system and the corresponding restoration actions has been improved over the previous method.

6.13 A Petri Net Algorithm for Scheduling of Generic Restoration Actions, 1997, p. 538

This paper proposes the generic restoration actions (GRAs) model. Specific restoration strategies can be formed by combining GRAs. GRAs include: start_black_start_unit, find_path, energize_line, synchronize X with Y, connect_tie_line, crank_unit, and energize_busbar. GRAs can be applied to various components of a power system. To actually apply the concept to system restoration problems, this paper uses the Petri Net model to represent the GRAs. In the proposed Petri Net technique, the required time for each restoration action is assumed to be available so that the proposed algorithm can be used to determine the most efficient restoration sequence. To this end, the Best First search algorithm is used. However, when the outage area is large, the restoration actions/plans represented by the Petri Net may be complicated.

6.14 Hierarchical Time-Extended Petri Nets As a Genetic Tool for Power System Restoration, 1997, p. 546

This paper proposes Petri Nets with time variables in order to incorporate the actual times for restoration actions. Because of the difficulty in maintaining knowledge-based systems, model-based methods are proposed. The Petri Net model proposed in the paper is able to handle unforeseen changes in the power system. When the size of a power system is large, it is difficult to represent restoration plans. Therefore, the authors propose a multiple inputs and multiple outputs (MIMO) concept. The proposed Petri Net model is expected to help develop guidance expert systems. The proposed system has been tested with the actual power system data.

6.15 Decision Aid Function for Restoration of Transmission Power Systems: Conceptual Design and Real Time Considerations, 1998, p. 553

This paper reports a system that supports dispatchers during restoration. It focuses on the guidance method. A number of practical and useful methods are proposed for development of an on-line restoration system. The developed system is organized with two blocks. In the Reasoning Kernel block, the system validates the dispatcher's control actions through interface with the dynamic simulator and the knowledge base module. Some real-time issues are addressed for the other block, the Real Time Update Kernel block. The paper discusses the frequency with which the power system status has to be updated using real-time information and the relationships between the cycle for updating data and the cycle for finding a restoration plan.

The system has been tested on an Operator Training Simulator.

DEVELOPMENT OF A KNOWLEDGE BASED SYSTEM FOR POWER SYSTEM RESTORATION

T. Sakaguchi, Nonmember K. Matsumoto, Nonmember
Mitsubishi Electric Corporation
Hyogo, JAPAN

Abstract - A knowledge based system is developed for automatic control of a power system in the restorative state. System operator's heuristics seem to play a central role during the course of system restoration. So efforts are made to represent those heuristics by a set of knowledge the goal of which is to restore the power supply without overloading any transmission line. While the overload check is a numerical computation, reasoning in the knowledge based system is a symbolic computation. A method to link these two different types of computation is proposed. The knowledge based system is expected to serve managing a large scale power system in the restorative state.

INTRODUCTION

It is well known that an electric power system has three operating states : preventive, emergency and restorative state. The power system is normally in the preventive state at most of the time. A large disturbance may bring the system to the emergency state. If the disturbance is large enough, the system goes into the restorative state. Blackout is the extreme of this state. The most likely look is a brownout, which means a partial power outage. Once this happens, system dispatchers try to re-energize the area where the power is down. The control to re-energize is referred to system restoration. It is needless to say that the modern society requires both rapid and safe system restoration. As system restoration at the present time is performed mostly through manual operation, developing an automatic system for restorative control is a challenging subject to power system engineers.

Why has the automatic system in restorative control been so slow? Unlike the load flow and stability problems, it is not easy to formulate the restorative control problem within the analytical framework. The solution method must have a capability of treating some combinatorial nature of the problem. It must also be able to take care of not only physical laws but also artificial laws in the power system. The artificial laws mean the operating procedures which are defined arbitrarily among system operation personnels.

This paper presents a "knowledge based system" approach to the restorative control problem. The idea of the knowledge based system (simply KB system) was developed by the research groups of artificial intelligence (or machine intelligence)[1], one of the branches in computer science. A KB system called MYCIN [2] is widely used for medical consultation. As shown in Fig. 1, two elements comprise a KB system: knowledge base and inference engine. Ignore the dashed lines, system designer and user for the time being. Every component of the knowledge base describes a piece of knowledge on numerical and symbolic computations to solve a problem. In other words, the knowledge base is another expression of a computer program. An inference engine interpretes the knowledge base, reasoning values of several specified parameters. The prime purpose of this paper is to show that the KB system is an effective means to write the physical and artificial laws of the power system which are needed to solve the restorative control problem. The KB system is developed on DEC System 20/20. As symbolic computation is dominant to the reasoning mechanism, the KB system is written by a list processing language known as Interlisp [3]. Elementary knowledge is assumed on Interlisp.

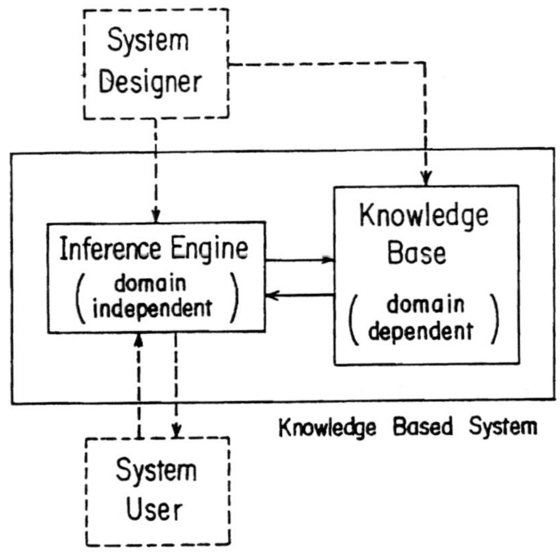

Fig. 1. A structure of the knowledge based system.

KNOWLEDGE BASE AND REASONING

Knowledge Base

All of the knowledges in solving the power system restoration problem are represented by either a property list or a production rule. The property list has a form of (object attribute value). For instance, if the voltage of BUS-A is 275,400 volt, this knowledge is stored in the knowledge base by a triple:

$$(\text{BUS-A VOLTAGE } 275{,}400) \quad (1)$$

Therefore the triple of (BUS-A NAME JOHN-DAY-A) means that the name of BUS-A is JOHN-DAY-A. Online data and system constants are usually represented in this way. It is noted that the value portion of the triple for online data will be updated when system changes its operating point.

82 SM 351-5 A paper recommended and approved by the IEEE Power System Engineering Committee of the IEEE Power Engineering Society for presentation at the IEEE PES 1982 Summer Meeting, San Francisco, California, July 18-23, 1982. Manuscript submitted January 28, 1982; made available for printing April 14, 1982.

Reprinted from *IEEE Transactions on Power Apparatus and Systems,* Vol. 102, No. 2, pp. 320–326, February 1983.

Another form of knowledge is a production rule, which has a form of (if-clause then-clause). The if-clause is called premise part and the then-clause is called action part of the knowledge. For instance, a production rule:

```
premise: (AND (NOTSAME CNTXT COMPCHARGE ISCHARGED)
              (NOTSAME CNTXT COMPFAULT ISFAULTED))
action : (CONCLUDE CNTXT RESTREQUIRE YES)
                                                    (2)
```

expresses the knowledge on judging if the component of the power system is required to restore. That is,

> if the charging-status of the component is not charged and the fault-status of the component is not faulted, then it is concluded that restoration-requirement-status of the component is yes (= the component must be restored).

CNTXT acts as a variable, and an object is substituted when a rule is interpreted. Suppose that BUS-A is substituted to CNTXT of the rule. Then, the rule means that

> if the charging-status of BUS-A is not charged and the fault-status of BUS-A is not faulted, then it is concluded that the restoration-requirement-status of BUS-A is yes (= BUS-A must be restored).

The inference engine examines the first predicate of the rule. If BUS-A is not charged, the engine goes to check the second predicate. If BUS-A is not faulted, then the engine concludes a triple of (BUS-A RESTREQUIRE YES) in the knowledge base.

The objects involved in a problem are structured by a tree called context tree. Fig. 2 shows a concept of the context tree. Buses, transmission lines, circuit breakers and protective relays will be contexts in the system restoration problem. Every context has one or many parameters such as name, voltage, real power flow and so on. A context at the top of the tree is called a root context, and parameters of the root context are called goal parameters. In Fig. 2, CONTEXT1 is the root context and the goal parameter is PARM1.

Reasoning

The task of the inference engine is to trace the goal parameters. The following is a reasoning procedure. First, the engine looks into the knowledge base to see if the value of PARM1 of CONTEXT1 has been determined. If it has already been traced, then the engine terminates reasoning. If not, the engine picks up a set of production rule whose action part concludes PARM1 of CONTEXT1. One of the rules picked up may have a form of

```
premise: (AND (SAME CNTXT PARM2 value2))
action : (CONCLUDE CNTXT PARM1 value1))          (3)
```

At this time, the engine tries to evaluate the premise part of the rule. If PARM2 of CONTEXT2 has been determined and the predicate SAME is true (PARM2 of CONTEXT2 = value2), then the engine sets the triple (CONTEXT1 PARM1 value1) in the knowledge base. If PARM2 of CONTEXT2 does not equal value2, then the engine stops evaluating the rule and goes to the next rule. It may happen that PARM2 of CONTEXT2 has not been traced yet. When this happens, the engine sets a subgoal at tracing PARM2 of CONTEXT2 and starts picking up the rules which will help the engine determine the value of PARM2. Once all parameters of the contexts on the bottom of the tree are traced, the goal parameters of the root context will be determined. In Fig. 2, PARM3 of CONTEXT3 will determine PARM2 of CONTEXT2, and PARM1 of CONTEXT1 will be traced by either PARM2 of CONTEXT2 or PARM4 of CONTEXT4.

The reasoning process explained above is known as a top-down procedure. There is another reasoning method called a bottom-up procedure. It is recognized that the top-down is generally much faster than the bottom-up. Therefore, the top-down procedure is used throughout this paper.

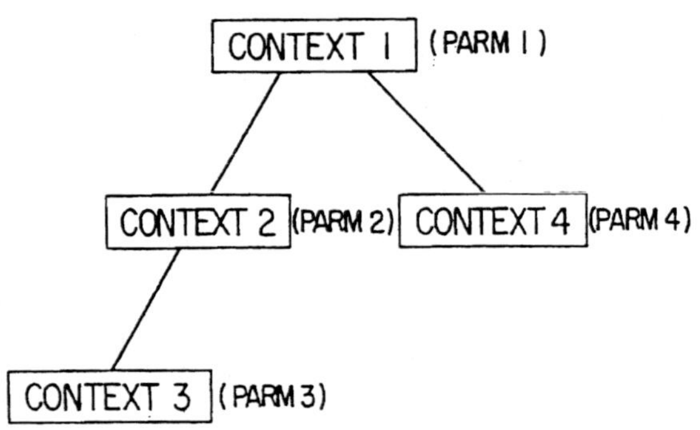

Fig. 2. A concept of the context tree.

KNOWLEDGE BASE FOR SYSTEM RESTORATION

Outline of Idea for System Restoration

A standard strategy is to restore all the elements which had been energized before the brownout. One way to do this is that the bus shall be reconnected to the transmission line through which it was fed before. However, the line may be faulted permanently and unable to feed the bus again. So the other way is to depend on a stand-by transmission line to the bus. Summarizing these, there are two restoration patterns:

- pattern-1: to energize a bus with a line which had fed the bus before the fault
- pattern-2: to energize a bus with a stand-by line which had not fed the bus before the fault.

If a bus to be restored meets either pattern, it is concluded that the bus can be restored.

The next thing to be considered is to check if an overload will happen after the restoration is performed. This is examined by the load flow computation. Although there are other check items essential to the system restoration like overvoltage and synchronization, they are all ignored in developing a complete knowledge based system in this paper. Those check items will be included easily in the KB system. The KB system of this paper is designed to take into account the restoration pattern and overload check.

Context Tree of Knowledge Base

Fig. 3 shows the context tree of the KB system created to determine the restoration procedure of the

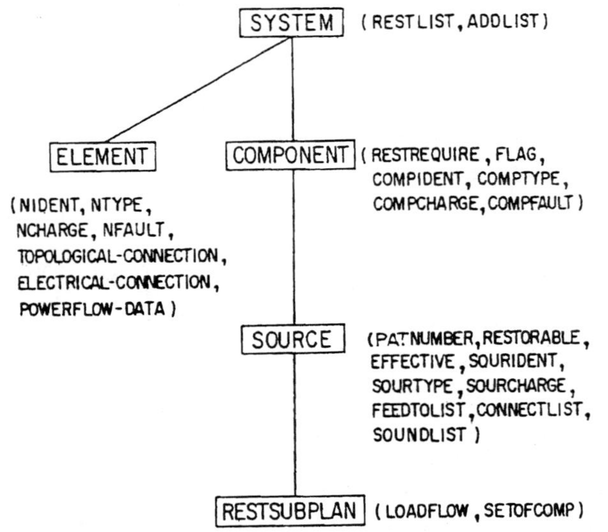

Fig. 3. Context tree of the knowledge based system for power system restoration.

power system. SYSTEM is the root context and it has two goal parameters: RESTLIST and ADDLIST. RESTLIST of SYSTEM gives an area which can be re-energized without overloading any part of the transmission network. Once it has been traced, the system operators know a restoration plan for the partial blackout area. The knowledge base tells that four types of context are needed to infer the goal parameters of SYSTEM. ELEMENT is a context for node/branch of the power system. NIDENT is a parameter showing the identity of ELEMENT. For instance, if the identity of ELEMENT1 is JOHN-DAY-A, then NIDENT of ELEMENT1 = JOHN-DAY-A. POWERFLOW-DATA is a parameter for load flow computation, and it covers impedance, real/reactive power injection, specified voltage and loading limit.

COMPONENT is a context for an element of the power system which is picked up as an object to restore. In our scheme, bus becomes a COMPONENT. RESTREQUIRE tells if the component is required to restore. RESTREQUIRE of COMPONENT = YES means that the component must be restored. There may be many components as a candidate for restoration. FLAG is used to record a mark showing that the component has already been searched. Other parameters of COMPONENT are identity, type, charging-status, fault-status, voltage, and connecting topology.

SOURCE is a context from which a component is re-energized. In our KB system, a restoration procedure is permissible if it is one of the categorized patterns. So RESTORABLE of SOURCE = YES is that the source can energize the component. There are two production rules to decide the restoration pattern number, and the number is set to PATNUMBER of SOURCE. There may be a case that overloading would result in the network when the restoration is performed through the source, even though the source/component pattern is permissible. EFFECTIVE of SOURCE checks whether an overload will result or not. The load flow computation is essential to determine the value of this parameter. SOUNDLIST of SOURCE refers an area which is topologically connected to the source and stays energized after the fault.

RESTSUBPLAN is a context for a portion of power system to be restored. SETOFCOMP of RESTSUBPLAN specifies a plan of restoration. It lists up all the elements of the power system that are re-energized. Once SETOFCOMP of RESTSUBPLAN and SOUNDLIST of SOURCE have been determined, all data are ready for the load flow computation. The load flow program is run and the overload is checked. If the overload happens, LOADFLOW of RESTSUBPLAN is set to OVER-LOADED. Otherwise it equals NO-OVERLOAD.

Production Rules in Knowledge Base

Sixteen rules are included in the knowledge base. Fig. 4 lists all the rules. A rule has premise, action and subject. The subject part means a context type to which the rule is applicable. If the premise part is true, the action part of the rule is performed. Otherwise it is turned down. The following are more natural English style expressions of the rule.

RULE1 : to conclude RESTREQUIRE of COMPONENT
If the component is not charged and it is not faulted, then it is concluded that the component must be restored.

RULE2 ; to conclude RESTORABLE of SOURCE
If the source is charged and its restoration-pattern-number is either 1 or 2, then it is concluded that the source is restorable.

RULE3 : to conclude COMPTYPE, COMPCHARGE, COMPFAULT of COMPONENT
If there exists an element in the network such that its identity equals the identity of the component, then set the type, charging-status, and fault-status of the element to those of the component.
(Note that this rule looks into the online data bank to transfer the values of parameters of an element to the component.)

RULE4 : to conclude SOURTYPE, SOURCHARGE, FEEDTOLIST, CONNECTLIST OF SOURCE
If there exists an element in the network such that its identity equals the identity of the source, then set the type and charging-status of the element, list of other elements fed by the element before the fault, and list of other elements topologically connected to the element, to those of the component.
(Note that this rule also transfers the values of parameters of an element to the source.)

RULE5 : to conclude PATNUMBER of SOURCE
If the identity of the component is included in the list of elements that were fed by the source before the fault, then it is concluded that the restoration-pattern-number of the source is 1.

RULE6 : to conclude PATNUMBER of SOURCE
If the identity of the component is included in the list of elements which are topologically connected to the source and it had not been fed by the source before the fault, then it is concluded that the restoration-pattern-number of the source is 2.

RULE7 : to conclude EFFECTIVE of SOURCE
If the source is restorable and the restoration-subplan does not result in overloading, then it is concluded that the source is effective for the restoration-subplan.

RULE8 : to conclude RESTORABLE of SOURCE
If the source is not charged or it is not permissible to re-energize, then it is concluded that the source is not restorable.

RULE9 : to conclude EFFECTIVE of SOURCE
If the source is not restorable, then it is concluded that the source is not effective for restoration.

RULE10: to conclude FLAG of COMPONENT
If the component is not required to restore,

then it is concluded that the search flag is set to the component.

RULE11: to conclude RESTREQUIRE of COMPONENT
If the component is charged or it is faulted, then the component is not required to restore.

RULE12: to conclude FLAG of COMPONENT
If the source is effective for restoration, then it is concluded that the search flag is set to the component.

RULE13: to conclude LOADFLOW of RESTSUBPLAN
If the list of elements connected to the source in the sound portion of network is known and the list of elements to be restored with the source is known, then it is concluded that whether overload will result or not is determined by the result of load flow computation.

RULE14: to conclude RESTLIST of SYSTEM
If all components have been searched, then it is concluded that the final list of elements to be restored equals another list called ADDLIST.
(Note that restoration subplans so far have been recorded in ADDLIST.)

RULE15: to conclude FLAG of COMPONENT
If the component is required to restore and the source is not restorable, then it is concluded that the search flag is set to the component.

RULE16: to conclude PATNUMBER of SOURCE
If the restoration-pattern-number of the source is not known yet, then it is concluded that the number is NONE.

A comment is made here how to link symbolic computation of reasoning to numerical one of load flow computation. The link is performed in RULE13. If the premise of the rule is true, then the inference engine executes its action part one argument of which is a Lisp function powerflow [arg1; arg2; arg3]. The function transfers the data for load flow computation and initiates the load flow program. It receives the overload check as well as the output of load flow in the power system problems.

Example of Determination of Restoration Procedure

The KB system is applied to determine the restoration procedure of a network shown in Fig. 5 (a). Suppose that the area enclosed by the dashed line in Fig. 5 (b) has lost power because of a permanent fault on BUS1. The problem is how to restore the power supply to the area without overloading any element. The KB system determines the procedure in the following way.

The KB system starts tracing the value of RESTLIST of SYSTEM1 shown in Fig. 5 (b). Remember that RESTLIST is intended to be a list of elements to be restored safely. RULE14 is picked up to trace the parameter, but the knowledge base at this time is not rich enough to do it. So the KB system creates COMPONENT which is needed to trace RESTLIST. Let COMPONENT1 be BUS2. It has several subgoal parameters like RESTREQUIRE and FLAG. The goal of reasoning here is switched to determine all parameters of COMPONENT1. The KB system chooses RULE1 to trace RESTREQUIRE of COMPONENT1. As the knowledge of online data has been gathered, the inference engine can evaluate the premise part of RULE1. BUS2 is neither charged nor faulted, and RESTREQUIRE is traced positively (= BUS2 must be restored). Next parameter is FLAG. RULE10, RULE12 and RULE15 are selected to trace it. First, RULE10 is rejected because its premise is inconsistent with the knowledge base supporting restoration of BUS2. The KB system then creates SOURCE upon evaluation of RULE12. Here the goal of reasoning is switched again to the parameters of SOURCE. Let SOURCE1 be LINE2. PATNUMBER of SOURCE1 is traced as 2 by RULE6, because the online data and network topology at this time satisfy the premise part of the rule. Now it has been known that BUS2 can be restored with LINE2. But how effective is it? RULE7 and RULE9 answer this question.

In order to evaluate RULE7, the KB system creates RESTSUBPLAN which defines an area to be restored down from BUS2. Let RESTSUBPLAN1 be (LINE3 LINE4 LINE5 BUS3 LINE8) of Fig. 5 (c). Overload is checked by the powerflow function in RULE13. As the powerflow on LINE2 exceeds its limit (see Table 1), LOADFLOW of RESTSUBPLAN1 is traced as OVERLOADED. The KB system creates another subplan. Let RESTSUBPLAN2 be (LINE3 LINE5) of Fig. 5 (d). As no overload has resulted, it is concluded that LOADFLOW of RESTSUBPLAN2 equals NO-OVERLOAD. Suppose that there is no other subplan. As all parameters of RESTSUBPLAN1, RESTSUBPLAN2 and SOURCE1 have been traced, FLAG of COMPONENT1 is set to SEARCHED by RULE12. The KB system has concluded that the area of (BUS2 LINE3 LINE5) can be restored through LINE2 without overloading any line.

There is still an area which is not restored yet. The KB creates another COMPONENT to see if the restoration is possible. Let COMPONENT2 be BUS3. Similar reasoning is performed, and it is concluded that the area of (BUS3 LINE8) can be restored safely. Now that all candidates for COMPONENT have been searched, the KB system concludes RESTLIST of SYSTEM1 to be (BUS2 LINE3 LINE5 BUS3 LINE8) by RULE14. The final restoration plan of Fig. 5 (e) is created in this way.

Sixteen rules are sufficient to create a restoration plan for the system of Fig. 5 (a), because the system is not very large and permissible patterns are simple. A large number of rules will be needed for more complex systems. However, it will be dealt easily within the framework of a KB system.

name of component	loading limit	restoration subplan 1	restoration subplan 2	final restoration plan
LINE 2	150.0	197.8	115.2	115.2
LINE 3	100.0	47.0	47.0	47.0
LINE 4	100.0	82.2	0.0	0.0
LINE 5	100.0	68.0	68.0	68.0
LINE 7	150.0	0.0	0.0	58.1
LINE 8	100.0	58.0	0.0	58.0

All values in MVA

Table 1. Line loading limits and expected line loading in the restoration subplans.

```
RULE1 PREMISE (AND (NOTSAME CNTXT COMPCHARGE ISCHARGED)
                   (NOTSAME CNTXT COMPFAULT ISFAULTED))

      ACTION (DO-ALL (CONCLUDE CNTXT RESTREQUIRE YES)
                     (PRINTCONCLUSIONS CNTXT RESTREQUIRE
                              REQUIRES-RESTORATION))

      SUBJECT (COMPONENT)

RULE2 PREMISE (AND (SAME CNTXT SOURCHARGE ISCHARGED)
                   (SAME CNTXT PATNUMBER (ONEOF 1 2)))

      ACTION (DO-ALL (CONCLUDE CNTXT RESTORABLE YES)
                     (PRINTCONCLUSIONS CNTXT RESTORABLE
                              RESTORABLE))

      SUBJECT (SOURCE)

RULE3 PREMISE (THEREXISTS (GETALL ELEMENT)
                          (AND (SAME FREEVAR NIDENT
                                (VAL1 CNTXT COMPIDENT)))
                          FREEVAR FOUNDVAR)

      ACTION (DO-ALL (TRANSDIFPARM CNTXT COMPTYPE
                                   FOUNDVAR NTYPE)
                     (TRANSDIFPARM CNTXT COMPCHARGE
                                   FOUNDVAR NCHARGE)
                     (TRANSDIFPARM CNTXT COMPFAULT
                                   FOUNDVAR NFAULT)

      SUBJECT (COMPONENT)

RULE4 PREMISE (THEREXISTS (GETALL ELEMENT)
                          (AND (SAME FREEVAR NIDENT
                                (VAL1 CNTXT SOURIDENT)))
                        . FREEVAR FOUNDVAR)

      ACTION (DO-ALL (TRANSDIFPARM CNTXT SOURTYPE
                                   FOUNDVAR NTYPE)
                     (TRANSDIFPARM CNTXT SOURCHARGE
                                   FOUNDVAR NCHARGE)
                     (CONCLUDE CNTXT FEEDTOLIST
                      (GETIDENT (CADR (VAL1 FOUNDVAR
                             ELECTRICAL-CONNECTION))))
                     (CONCLUDE CNTXT CONNECTLIST
                      (APPEND (CAR (VAL1 FOUNDVAR
                              TOPOLOGICAL-CONNECTION))
                             (CADR (VAL1 FOUNDVAR
                              TOPOLOGICAL-CONNECTION))
                             (CADDR (VAL1 FOUNDVAR
                           TOPOLOGICAL-CONNECTION)))))

      SUBJECT (SOURCE)

RULE5 PREMISE (AND (SAME CNTXT COMPIDENT
                         (VAL1 CNTXT FEEDTOLIST)))

      ACTION (CONCLUDE CNTXT PATNUMBER 1)

      SUBJECT (SOURCE)
RULE6 PREMISE (AND (SAME CNTXT COMPIDENT
                         (VAL1 CNTXT CONNECTLIST))
                   (NOTSAME CNTXT COMPIDENT
                            (VAL1 CNTXT FEEDTOLIST)))

      ACTION (CONCLUDE CNTXT PATNUMBER 2)

      SUBJECT (SOURCE)

RULE7 PREMISE (AND (SAME CNTXT RESTORABLE YES)
                   (SAME CNTXT LOADFLOW NO-OVERLOAD))

      ACTION (DO-ALL
             (CONCLUDE CNTXT EFFECTIVE YES)
             (CONCLUDE CNTXT ADDLIST
              (APPEND (GETPROP
                       (SUBVAL CNTXT (QUOTE ADDLIST))
                       (QUOTE ADDLIST))
                      (VAL1 CNTXT SETOFCOMP)))
            (PRINTCONCLUSIONS CNTXT SETOFCOMP
                   RESTORED-SYSTEM))

      SUBJECT (RESTSUBPLAN)

RULE8 PREMISE (OR (SAME CNTXT SOURCHARGE NOTCHARGED)
                  (SAME CNTXT PATNUMBER NONE))

      ACTION (DO-ALL (CONCLUDE CNTXT RESTORABLE NO)
                     (PRINTCONCLUSIONS CNTXT RESTORABLE
                              RESTORABLE))

      SUBJECT (SOURCE)

RULE9 PREMISE (AND (SAME CNTXT RESTORABLE NO))

      ACTION (CONCLUDE CNTXT EFFECTIVE NO)

      SUBJECT (SOURCE)

RULE10 PREMISE (AND (SAME CNTXT RESTREQUIRE NO))

       ACTION (DO-ALL (CONCLUDE CNTXT FLAG SEARCHED)
                      (CONCLUDE CNTXT ADDLIST
                       (LOOKUPVALUEOF (SUBVAL CNTXT
                                      (QUOTE ADDLIST))
                                      (QUOTE ADDLIST))))

       SUBJECT (COMPONENT)

RULE11 PREMISE (OR (SAME CNTXT COMPCHARGE ISCHARGED)
                   (SAME CNTXT COMPFAULT ISFAULTED))

       ACTION (DO-ALL (CONCLUDE CNTXT RESTREQUIRE NO)
                      (PRINTCONCLUSIONS CNTXT RESTREQUIRE
                              REQUIRES-RESTORATION))

       SUBJECT (COMPONENT)

RULE12 PREMISE (AND (SAME CNTXT EFFECTIVE YES))

       ACTION (CONCLUDE CNTXT FLAG SEARCHED)

       SUBJECT (SOURCE)

RULE13 PREMISE (AND (KNOWN CNTXT SOUNDLIST)
                    (KNOWN CNTXT SETOFCOMP))

       ACTION (CONCLUDE CNTXT LOADFLOW
               (POWERFLOW CNTXT (VAL1 CNTXT
                                  SOUNDLIST)
                                (VAL1 CNTXT SETOFCOMP)))

       SUBJECT (RESTSUBPLAN)

RULE14 PREMISE (FORALL (GETALL COMPONENT)
                       (AND (SAME FREEVAR FLAG SEARCHED))
                       FREEVAR)

       ACTION (DO-ALL (CONCLUDE CNTXT RESTLIST
                                (VAL1 CNTXT ADDLIST))
                      (PRINTCONCLUSIONS CNTXT RESTLIST
                              TOTAL-RESTRED-SYSTEM))

       SUBJECT (COMPONENT)

RULE15 PREMISE (AND (SAME CNTXT RESTREQUIRE YES)
                    (SAME CNTXT RESTORABLE NO))

       ACTION (DO-ALL (CONCLUDE CNTXT FLAG SEARCHED)
                      (CONCLUDE CNTXT ADDLIST
                       (LOOKUPVALUEOF
                        (SUBVAL CNTXT (QUOTE ADDLIST))
                        (QUOTE ADDLIST))))

       SUBJECT (SOURCE)

RULE16 PREMISE (AND (NOTKNOWN CNTXT PATNUMBER))

       ACTION (DO-ALL (CONCLUDE CNTXT PATNUMBER NONE)
                      (PRINTCONCLUSIONS CNTXT PATNUMBER
                              APPLICABLE-CONDITION))

       SUBJECT (SOURCE)
```

Fig. 4. Production rules of the knowledge base for power system restoration.

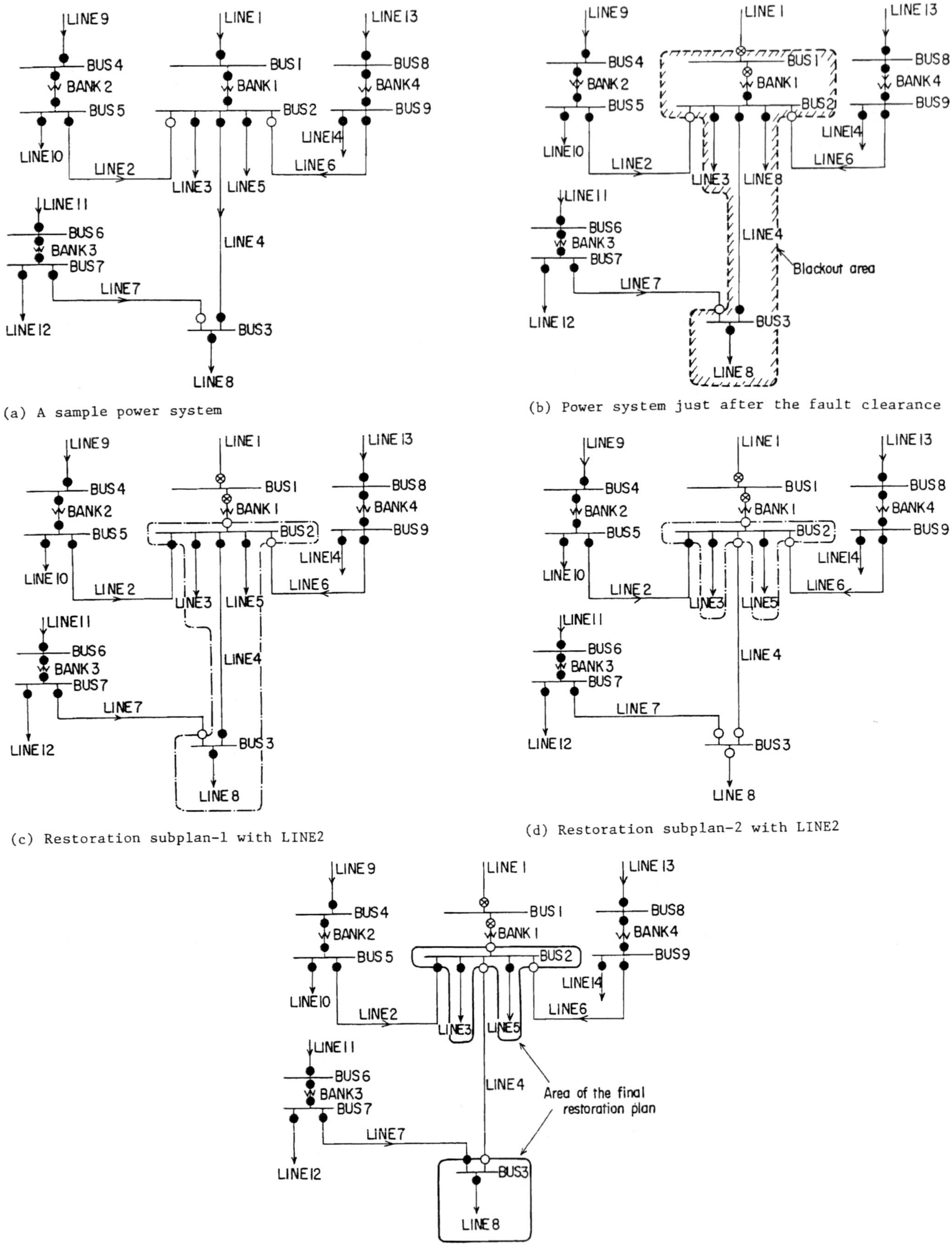

(a) A sample power system

(b) Power system just after the fault clearance

(c) Restoration subplan-1 with LINE2

(d) Restoration subplan-2 with LINE2

(e) Final restoration plan

Fig. 5. Sample power system and reasoning process for determination of the restoration procedure.

● CB in
○ CB off
⊗ CB tripped

DISCUSSION

Advantage of KB System Approach

The KB system can program heuristics. Unlike the preventive and emergency control, it is difficult to formulate the restorative control problem within analytical framework. The reason for this is that the connecting topology changes during the course of restoration and the problem must be solved through searching a solution tree. Heuristic programming is a nice technique to solve a problem having such kind of nature. As the cognitive science indicates, heuristics are essential for a human being to solve a problem. The critical role of system operators in the restorative control is a positive evidence to this. As programming by a knowledge base can express the problem solving process of human being, particularly his heuristics, much better than programming by an algorithmic language like Fortran, it is an important strength over conventional approaches.

Modularity of a knowledge. The most primitive form of a knowledge is a production rule. Each rule can be thought as a module for a piece of knowledge. In Fig. 1, the dashed lines from system designer show a means to create the knowledge base and also to reinforce the inference engine. A rule is added to, deleted from , and modified in the knowledge base very easily because of its modularity. Therefore a large knowledge base will be created in a short time period.

A chain of reasoning is transparent to the user. Unlike the Fortran-written system, the KB system can tell how it comes to the conclusion. In other words, it can explain its reasoning process. The explanation capability is usually embedded in the inference engine, not in the knowledge base. Looking over the explanations given by the engine, the user (= system dispatcher) can confirm or may correct the conclusion of computer systems. This is very important to apply the KB system to other power system problem, such as a computerized diagnosis of nuclear power plant.

Universality. It is easy to understand that the knowledge base is problem domain dependent. It should be noted that the inference engine is domain independent. So a different KB system can be developed by replacing the knowledge base only. This kind of universality is a nice feature of the KB system approach.

Disadvantage of KB System Approach

Techniques have not been developed yet to check the consistency of knowledge base as a whole. The modular construction may create an inconsistent knowledge base, which means a conflict among the rules. At this time, the conflict resolution must be made by the system designer, using the explanation capability of the inference engine. It is an open problem of the KB system approach.

Symbolic computation is slow. Computing speed is essential to any online control system. Unfortunately, the Lisp program runs slowly on the medium size computer. This weakness will not be a bottleneck for long, because the Lisp machine coming up in the marketplace is expected to run the Lisp program much faster than the computers today.

CONCLUSION

Although the computer systems are common in the power system control in the preventive and emergency state, they have been slow in penetrating into the restorative control. A reason for this seems that the problem solving ability of a humen being is far more superior to the computer systems today, and very few efforts were made to program such a human ability. This paper has presented a heuristic programming approach with a KB system to serve managing a large scale and complex power system. The KB system has just started penetrating into the power system as a new engineering technique. It is hoped that it will be used widely in the management of the power system.

REFERENCES

[1] E. A. Feigenbaum, "Knowledge Engineering : The Applied Side of Artificial Intelligence", Stanford Heuristic Programming Project Memo HPP-80-14, July, 1980.

[2] E. H. Shortliffe, "Computer-Based Medical Consultation: MYCIN", Artificial Intelligence Series 2, Elsevier, 1976.

[3] "Interlisp Reference Manual", XEROX Palo Alto Research Center, October, 1978.

DEVELOPMENT OF A GUIDANCE METHOD FOR POWER SYSTEM RESTORATION

Y. Kojima S. Warashina
Tokyo Electric Power Company
Choufu, Tokyo, Japan

S. Nakamura K. Matsumoto
Member, IEEE
Mitsubishi Electric Corporation
Amagasaki, Hyogo, Japan

Abstract - This paper proposes a guidance method for the restorative control problem. It gives operators an appropriate restoration plan and restorative procedures by using the knowledge base of expertise and heuristics on power system operations. An approach that combines the knowledge engineering technology with conventional programming has been adopted. This architecture gives various kinds of flexibility. It is possible to modify the restoration plan automatically if unexpected occurences such as new faults happen and to accept operator's changes to the proposed plan. A prototype has been developed to verify these functions.

Keywords : Power system restoration, Decision making, Knowledge base, Guidance, Problem-solving

INTRODUCTION

Power systems are becoming more complicated, and greater reliability is required these days. Under these circumstances, it is very important to restore the power system quickly and safely when faults happen. This situation has imposed control systems to provide support to system dispatchers in the decision making process. To fulfill this requirement, many approaches to supporting or automating restoration have been studied, and the practical use stage of the restorative control in the subtransmission system has been developed in recent years [1]. In general, restoration in a subtransmission system can be formulated as a large scale combinatorial problem. There are few developments in large scale system theory applicable to the problem which deals with this combinatorial nature of the problem. The knowledge based systems in [2,3,4] aimed at encoding the heuristics for solving the restoration problem. Several patterns of restorative control have been extracted from operational experiences and manuals.

On the other hand, the restoration problem gets much more complex in a bulk power system than in a subtransmission system. Bulk power systems are different from subtransmission systems at many points. They have power sources and loop circuits, but they are not connected directly to customer loads. Therefore, operators must take the following into account for the restoration of bulk power systems. Generator outputs should be determined to keep the power balance between generation and load. To restore the components of a bulk power system does not mean to energize the lost loads. They have to send a message to restore loads to the lower control centers. There are so many uncertain factors such as possibility of generators re-start and expected loads that it is difficult to make a fixed restoration plan based on a state estimation of power system a few hours hence. Operators should consider voltages and stability of the system too.

Among studies which deal with bulk power system restorations, some propose guidelines to make restoration plans before a fault should happen[5,6]. Another presents a method to establish a target system for future time points[7]. This paper aims at proposing a restoration guidance method which can generate on-line restoration plans and restorative procedures for bulk power system when a blackout occurs. The proposed method has the following characteristics:

(1) This method allows operators to actively participate in problem solving.

(2) It adopts an approach to make restoration plans and operational sequences adaptable to changes in the situation.

(3) It gives real-time responses which are fit for practical use.

An approach that combines knowledge engineering technology with conventional analytical programming has been adopted to realize these functions. A prototype has been developed to verify the effectiveness of this approach. It has been confirmed that it can give operators appropriate restoration plans and restorative procedures.

BASIC CONCEPT OF GUIDANCE METHOD FOR RESTORATION

To develop an effective restoration guidance method, it seems the best approach to analyze and model operators' decision making processes of restoration. The following are the results of this analysis.

(1) Operators always have an image of ideal system conditions (target system) using all equipment expected to be available as effectively as possible.

(2) If an outage should occur, they try to make a restoration plan based on the forecast of load demand and available equipment.

(3) After making the restoration plan, that is, target system, they find out the procedure of switchings and loadings to reach the target system making a comparison between the real system and the target system.

(4) They modify the restoration plan if they feel it needs change due to unexpected occurences. And they begin operations to reach the new target system.

89 WM 162-9 PWRS A paper recommended and approved by the IEEE Power System Engineering Committee of the IEEE Power Engineering Society for presentation at the IEEE/PES 1989 Winter Meeting, New York, New York, January 29 - February 3, 1989. Manuscript submitted January 28, 1988; made available for printing December 1, 1988.

Reprinted from *IEEE Transactions on Power Systems,* Vol. 4, No. 3, pp. 1219–1227, August 1989.

This analysis has produced the basic concept of the guidance method shown in Fig. 1. The "construction of restoration plan" in the figure is utilized when a reconstruction of the target system is required due to a situational change such as a new fault. And the guidance method reconstructs a new target system having continuity between the old and new target systems, referring to the old target system and the forecasted load demands and available equipment based on the current state of the real system. The "restoration action (guidance)" in Fig.1 is utilized periodically or when a situational change occurs, and it suggests proper restorative operations for reaching the target system in its current state, comparing the target system with the real system. The basic concept shown in the figure is very suitable for realizing the required objects of the restoration guidance method.

OBJECTS OF THE GUIDANCE METHOD

The main object of the guidance method is to assist operators in problem analyses and solutions, especially:

(1) Basic functions

a) to continually inform about the condition of the system and progress of the restoration
b) to construct a restoration plan and show it
c) to suggest proper restorative operations to reach the target system in its current state
d) to detect needs for amendments of the restoration plan according to changes in the situation

(2) Interactive functions

a) to allow operators' modifications of the proposed restoration plan
b) to allow operators' commanding restorative actions different from the proposed ones
c) to verify operators' commanding actions by means of simulation before they are put into action

Restoration processes are separated into four phases in this guidance method. To attain the above-mentioned objects, the necessary functions are realized in each phase.

(a) Recognition phase

The first tasks following a blackout are to recognize the current state of the system and to determine the faulted components which should not be used in restorative operations by using the information of the active relays and tripped circuit breakers.

(b) Premise phase

The guidance method summarizes the fault state in this phase. A diagram of the system shows which portions are deenergized and which loads are lost. Operators can recognize the state of the system just after fault clearance. And they designate premises to make restoration plans. For example, operators estimate the time for the start-up and restart of thermal plants. If some plants are able to restart and resume service within several hours, they are regarded as available equipment.

(c) Target phase

The computer establishes a target system a few hours later using these premises (that is, judgement and designation of which elements are available) and shows it to operators.

(d) Operation phase

The appropriate action sequences to reach the target are generated and presented to operators in this phase.

Among these four phases, target phase (c) and operation phase (d) are the major parts of the guidance method and so full details of them are given hereafter.

PROCESSING FLOW

The processing flow of the guidance method is shown in Fig. 2. Throughout all phases, operators can recognize the progress of restoration by means of a

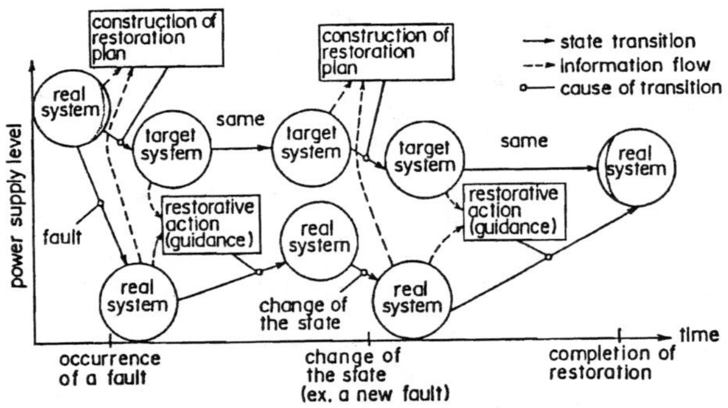

Fig. 1 Basic concept of guidance method for restoration

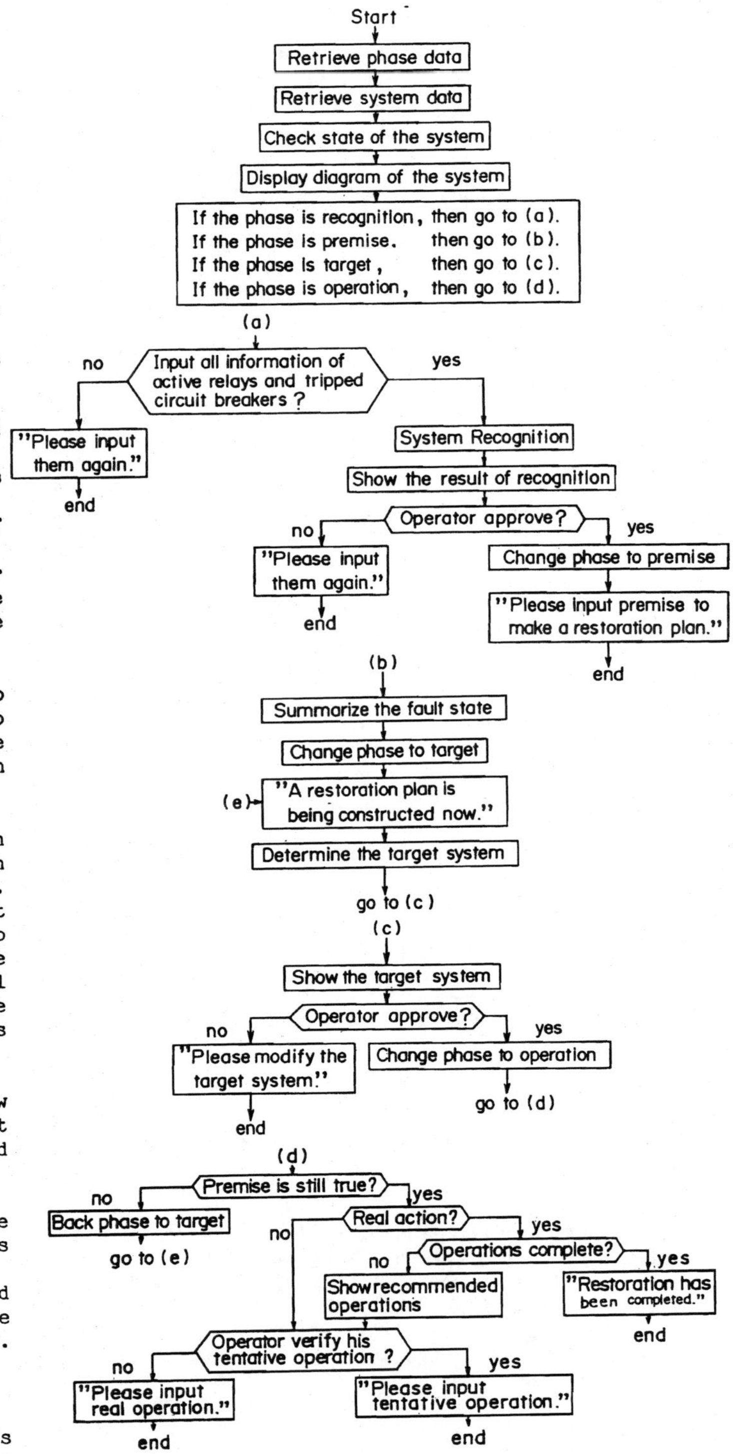

Fig. 2 Processing flow of the guidance method

diagram of the system which shows power flows, bus voltages, outage loads, restored loads, states of switches, deenergized portions and so on. Operators are allowed to change the data which express the states of a target system in the target phase. In this way they can modify the restoration plan which is created by the computer. In the operation phase, the computer checks whether the premises are still true at first. If the premises are not true, that is, unforseen events happen, the phase is returned to the target and the restoration plan is remade. The computer does not prepare fixed operation sequences in advance, but provides feasible actions to reach the target at the moment for operators' demands each time. As a result, it allows operators' commanding restorative actions which are different from the proposed ones. In addition, it can estimate states of the system after performing tentative actions by means of simulation. Then they can verify their commanding actions before they are put into action and continue restoring the system safely.

DEVELOPMENT OF TARGET SYSTEM

The approach to develop a target system is described in this section. If there are n switches in the system and all switches are considered, 2^n combinations must be exhaustively dealt with to determine a final configuration. There are few analytical algorithms which can solve such large scale combinatorial problems in a short time. Therefore, operators' heuristics play an important role in this problem.

Analyzing operators' measures against many hypothetical faults, five steps are clarified through which restoration plans are developed. The following actions are tried to minimize the amount of outage loads at first.

(1) Keeping demand supply balance, operators adjust all generator outputs without overloadings.

(2) They change the power sources for substations. There are two kinds of power sources: neighboring energized buses and stand-by lines.

(3) They separate buses to receive power from multiple power sources.

(4) They transfer loads in the subtransmission network.

Taking into account only active power, they determine the system configuration minimizing the amount of outage loads.

(5) As the next step, they choose the appropriate number of reactive power suppliers to prevent abnormal voltages.

Based on the above analysis of operators' ways of thinking, the process to develop a target system consists of two stages. One is a stage to determine a final configuration taking only active power as an index. The other is a stage to choose the appropriate number of reactive power suppliers to prevent abnormal voltages. As a result, the active power problems and reactive power problems are separated from each other. Therefore, it is possible to realize fast processing speed by using high speed algorithms such as the DC load flow method. In addition, to determine the number of reactive power suppliers to be used to prevent abnormal voltages, we use a fast load calculation algorithm which does not need to modify Jacobian for changing amount of reactive power.

Because the final configurations determination is a major part among these two stages, details of this stage are given in the following. The developed algorithm is shown in Fig. 3. The algorithm has some useful characteristics:

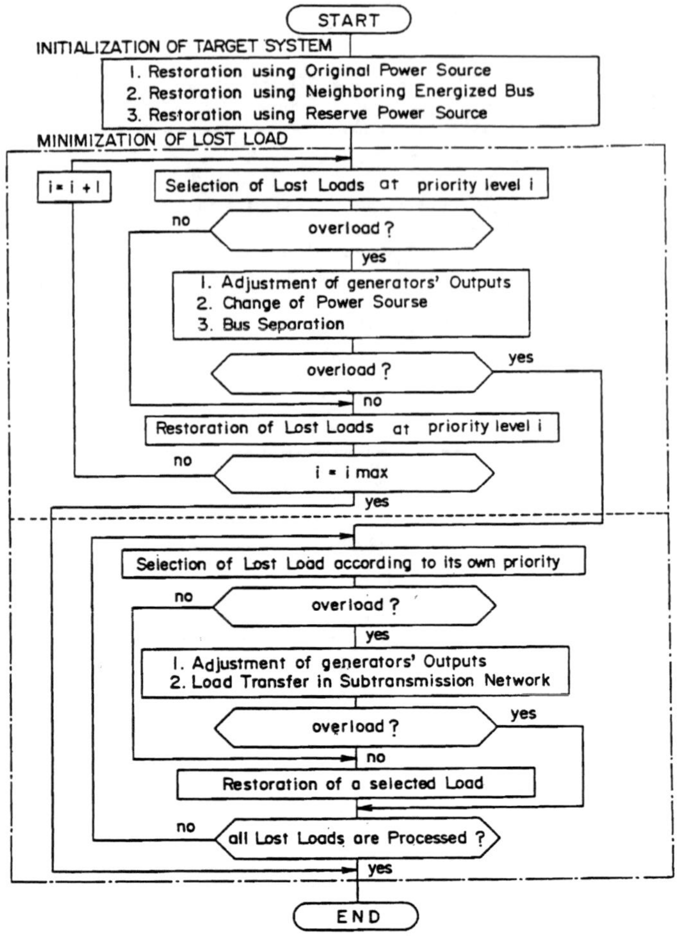

Fig. 3 Flowchart of determination of final configuration

(a) Bulk power systems are not connected directly to customer loads, so that the computer makes an initial condition of the target system and then tries to minimize the lost loads.

(b) To achieve high speed processing of restoring loads, the computer restores a set of the same priority level loads collectively at first. If overloads occur at some level, then an individual load is selected to be restored according to its own priority.

(c) A DC load flow method is used to check or resolve overloads.

(d) The stability constraints are transcribed into load flow constraints based on the off-line case studies. A stability check is also done in the same way as overload check.

OPERATIONAL GUIDANCE

This section presents an approach to create operational guidances. When the prototype makes a restoration plan, it determines a final configuration of restored system at first. This mechanism makes setting intermediate passing points easily and avoiding redundant operations at the restorative control possibly. Comparing the configuration of the real system with that of the target system, the computer finds switches which should be operated. If there are m switches to be operated, m! combinations must be checked to determine the order of them in a blind search. Then operators' heuristics are important in this case, too. The operational guidance method which is based on the analysis of operators' expertise is given below.

The switching sequence is determined without violating any operating constraints. Operating constraints are listed here;

- -interlocking condition between switches
- -no charging of faulted equipment
- -no charging of equipment deenergized for maintenance
- -no overloading of any lines, banks, or generators
- -prohibition of new power interruption
- -prohibition of new system splitting
- -prohibition of occurrence of abnormal voltage
- -phase angle constraint of synchronized-closing

Operational guidance consists of switching operational guidance and load-dispatching instruction guidance.

SWITCHING OPERATIONAL GUIDANCE

The switching sequence is determined based on the following rules.

(1) If there are switches to be opened in the de-energized area, then open them.

(2) If there are switches (bus-tie CB and LS) to be closed in the deenergized area, then close them.

(3) If there are switches to be closed for jointly using lines, buses or transformers in the energized area, then close them.

(4) If there are switches to be closed for making a closed loop circuit in the energized area, then close them.

(5) If there are switches to be opened for breaking a closed loop circuit in the energized area, then open them.

(6) If there are switches to be closed for combining two islands in the energized area, then close them.

(7) If there are switches to be closed to energize equipment, then close them.

(8) If there are switches to be closed to let generators participate in the network, then close them.

Operations which are safe and able to strengthen the system are selected prior to others in these rules. These rules are applied in this order. A set of operations are shown which can be operated at present. They contain at most one operation which changes power flow conditions. Therefore, operators can put these operations into practice at the same time and in any order without overloadings. In advance the computer checks whether abnormal voltages due to high harmonics generate. This is checked by using the ratio of capacitances to ground, demand power and total capacity of generators on-line.

LOAD-DISPATCHING INSTRUCTION GUIDANCE

A list of loads to be restored are shown collectively without causing overloads. These loads are selected by taking into account of generators' ramp rates and the priority of loads. And the total amount of loads per instruction is limited to 500MW to prevent an excessive frequency fall.

METHOD TO COMBINE KNOWLEDGE ENGINEERING TECHNOLOGY WITH CONVENTIONAL PROGRAMMING

While the human heuristics are crucial to the problem solving in restoration, we have other sources of knowledge about power systems as well. They include how the power system is structured, and how each component functions. One of the most important subjects in the knowledge-based approach to the power system problem is how this type of knowledge is encoded in the rules.

In a power system network, power flow is one of the most important factors for finding the solution. However, it seems difficult to encode the knowledge necessary to estimate the power flow in the rule form. Therefore, analytical tools such as the standard power flow program have been linked to the knowledge base. This method links analytical tools to the inference engine, calling the Fortran written programs and receiving numerical data such as the power flow.

The restoration of a power system is a matter of great urgency. A software architecture has been selected in which it is possible to realize the necessary processing speed, considering the size of the actual power system and required software functions. As a result, one architecture, as shown in Fig.4, has been obtained. The prototype implemented in OPS83-based meta-knowledge and Fortran-based analytical programs is designed to be used interactively.

The prototype was constructed on an engineering workstation. The hardware architecture is shown in Fig. 5. The knowledge base part was written in OPS83 [8] which had enough processing speed to control the power system and can be easily linked to other languages. Analytical tools and the user interface part were written in Fortran.

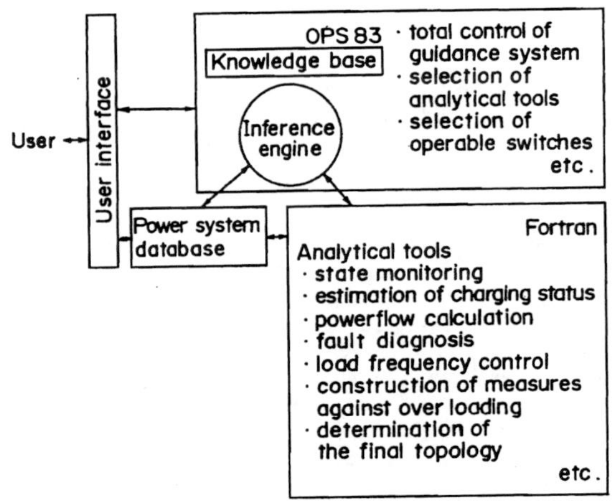

Fig. 4 Software architecture of the prototype

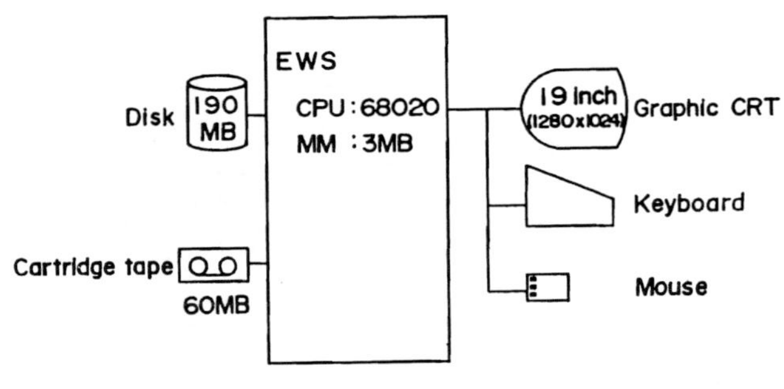

Fig. 5 Hardware architecture of the prototype

Outer routines can be called subroutines if their interfaces are contained in OPS83. This function enables us to combine knowledge engineering technology with conventional programming. For example, one of the rules is shown in Fig. 6. This rule corresponds to a part of processing (b) shown in Fig. 2. The roles of this rule are given in the following. At first, it renews the phase to target after setting the premises. Then it presents the message that a restoration plan is being developed. At last, it calls the Fortran program rest_plan which uses the load flow program many times to make a target system. Analytical tools are called by rules in this way and the results are returned to the knowledge base.

The power system database consists of several uniform styles of data files for each phase and for temporary use. System conditions and restoration plans are stored in these data files. This structure enables operators to move from one phase to another and to modify the data.

This archtecture of the prototype allows operators to actively participate in many points of processing as shown in fig. 2. And combination of procedural programs with rules has brought about high speed processings. Because not all processes are dealt with by rules, about 200 rules are enough to implement the prototype.

```
rule premise_to_target
{
 &p (phase name=premise);
 &u (user  response=yes);-- Setting of premises
                          -- is over.
-->
     modify &p (name=target);
     write() |A restoration plan is
               being constructed now.|, '\n';
     call rest_plan();    -- Fortran subroutine
                          -- to make plan
};
```

Fig. 6 An example of a rule combining knowledge engineering technology with conventional programming

EXPERIMENTAL RESULTS

An example of restorative control guidance is shown in this section. A real size power system model (36 generators, 48 substations) is taken as a model system in this example. The faults are assumed to have occurred on lines A-B 1L and 2L in Fig. 7. The developed prototype has constructed a restoration plan shown in Fig. 8. Figs 7 and 8 are hard-copies of the CRT display of the prototype. Fig. 9 is an example of the guidance which the prototype has produced.

In the target system shown in Fig.8, buses at B S/S and C S/S are separated to receive power from two different power sources. One is used to restore its own loads and the other is transported to the neighboring substations. The bus-tie has been closed to restore the bus and loads at F S/S. These restorative actions can restore all 2050MW outage loads in this example. The effect of 'bus separation' and 'load transfer in the lower power system' is shown in Table 1. These restorative means are introduced to improve the quality of restoration plans to the operator's level. In this way, the prototype gives an appropriate plan which is similar to one operators expect.

All operations listed in Fig. 9 are first steps from the present configuration shown in Fig. 7 to the target configuration. It is possible to put them into operation in the current state. (1) is separation of the buses of B S/S. (2)-(4) are opening of circuit breakers which are located on the F S/S side of line E-F. All these four actions are opening operations in the deenergized area. (5) is closing of the circuit breaker on the D S/S side of line G-D as the closing operation to make a closed loop circuit. (6)-(9) are closing operations to energize the outage lines of lines D-B and C-B. These are appropriate operations which can be put into practice without regard to order.

Only an example of the guidance to a partial outage is shown in this section. It has been confirmed for the prototype to give appropriate guidances in case of blackouts.

DEMONSTRATIONS TO OPERATORS

The prototype was demonstrated repeatedly to a total of about 100 persons concerned with dispatching in order to evaluate the performance through off-line computer simulations. In all tested scenarios which were prepared and designed extempore according to dispatchers' requests, the prototype gave appropriate responses.

Dispatchers' comments are summarized here: The following are necessary and good functions for an actual system.

-to allow operators' modifications of restoration plan

-to allow operators' actions different from proposed ones

-to show operations which can be operated at the same time and in any order

-to verify operators' commanding actions by means of simulation before they are put into action.

The time occupied on making a target system depends largely upon the difference of conditions between the target system and a pre-fault system. It takes about 20 seconds if some cases require bus separations or load transfers. Otherwise the computer produces it in about 10 seconds. And a processing time to create operational guidances depends upon the number of operations which are shown at the same time. It is usually about 10 - 20 seconds. Therefore, they can estimate the time to restore the system in a short time, and its information is useful for responding to customers. At present, each one of four control centers is alloted a portion of the power system to be controlled in the bulk power system of Tokyo Electric Power Company. On the other hand, the prototype is able to deal with the system as a whole. In some cases, it has constructed good restoration plans which the operators did not anticipate.

Because shortage generation is usual when a wide-

Table 1. The effect of restorative means

	Restorative means to use				Residual outage loads (MW)	Number of swithching operations
	Generation adjustment	Change of power source	Bus separation	Load transfer		
case 1	○	○	○	○	0	29
case 2	○	○	—	○	210	60
case 3	○	○	—	—	1200	12

○: used
—: unused

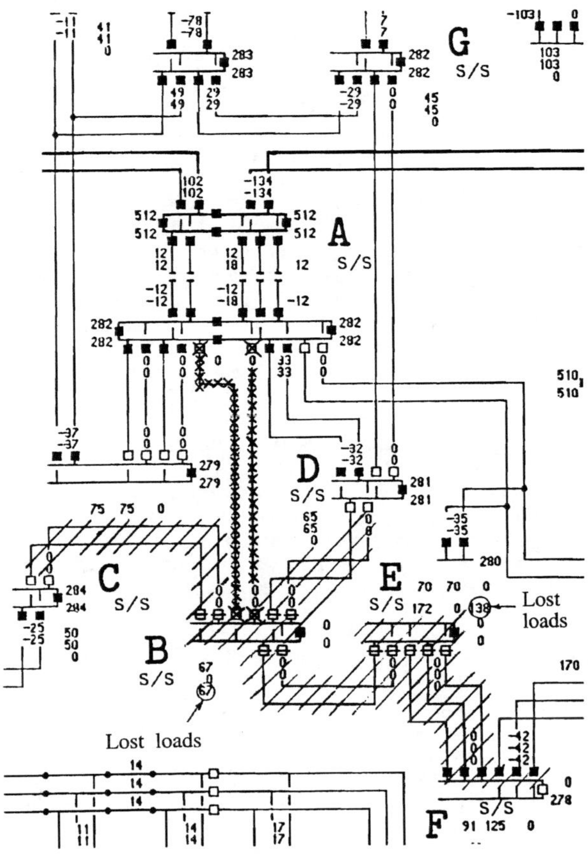

×××× : Faulty equipment

//// : Deenergized equipment

■ : Closed breaker

□ : Open breaker

⊠ : Tripped breaker

⊟ : Breaker opened by the staff of the outage station in accordance with the manual

Fig. 7 Current status of the model system

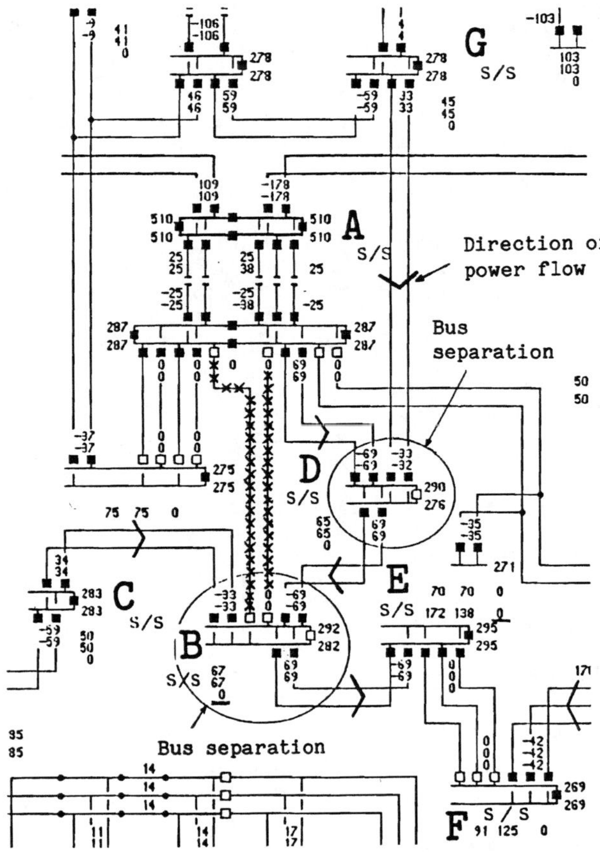

Fig. 8 Target system

(1) Changing topology
275KV bus-tie of B S/S can be opened.
All LSs in B S/S must be the same state as those of the target system.
(2) Changing topology
CB on F S/S side of line E-F 1L can be opened.
(3) Changing topology
CB on F S/S side of line E-F 2L can be opened.
(4) Changing topology
CB on F S/S side of line E-F 3L can be opened.
(5) Loop in
CB on D S/S side of line G-D 1L can be closed.
LS near bus B should be closed and the opposite LS opened.
Difference of phase angles = 0.295 degrees
(6) Charging
CB on D S/S side of line B-D 1L can be closed.
(7) Charging
CB on D S/S side of line B-D 2L can be closed.
LS near bus A should be closed and the opposite LS opened.
(8) Charging
CB on C S/S side of line B-C 1L can be closed.
(9) Charging
CB on C S/S side of line B-C 2L can be closed.

Fig. 9 An example of restorative switching guidance

spread outage occurs, sequence or timing of switching and amount or timing of load-dispatching become important factors to restore the system safely. There were multiple operators' views of these things so that the prototype could not get a paticular authorized estimation. But it has been estimated well enough for practical use in case of partial outages.

CONCLUSIONS

This paper proposed a restoration method for bulk power systems and a framework fcr the restoration guidance system. It has been developed with the following intentions. As it is impossible to forecast all the events for hours during the restoration of the bulk power system, the guidance system should be able to deal with every unforeseen event. It also should accept many kinds of operators' directions during all phases of restorative control. And it should perform well in real-time. What approach can satisfy all of these specifications? Our answer is to combine the knowledge engineering technique and conventional analytical tools. This is done in a way that the analytical tools are linked to the knowledge base, calling the Fortran written programs and receiving the results.

All on-line input data in this paper are assumed to be correct. As a result, the prototype might create a restoration plan which does not correspond to real conditions if some on-line data are incorrect. To avoid this trouble, it remains as a future subject to develop a method of bad data detection and switching status identification by state estimation of on-line data.

The prototype based upon this model has shown very satisfactory results and provided prospects of practical applications to the actual system restoration. We are going to enhance the man-machine interface and test the prototype in the field to create a practical restoration guidance system.

REFERENCES

[1] M. Kato et al., " Automatic Restorative Operation Method for Subtransmission System," National Convention of the Institute of Electrical Engineers of Japan, paper No.1578, April 1987 (in Japanese).

[2] T. Sakaguchi and K. Matsumoto, "Development of a Knowledge Based System for Power System Restoration," IEEE Transactions on Power Apparatus and Systems, Vol.PAS-102, No. 2, pp.320-329, February 1983.

[3] C. C. Liu, S. J. Lee, and S. S. Venkata,"An Expert System Operational Aid for Restoration and Loss Reduction of Distribution Systems," Proceedings of IEEE 1987 PICA Conference, pp.79-85, May 1987.

[4] F. F. Wu and A. Monticelli, "Analytical Tools for Power System Restoration - Conceptual Design," IEEE Transactions on Power Systems, Vol.3, No. 1, pp.10-16, February 1988.

[5] R. J. Kafka, D. R. Penders et al., "System Restoration Plan Development for a Metropolitan Electric System," IEEE Transactions on Power Apparatus and Systems, Vol.PAS-100, No.8, pp.3703-3713, August 1981.

[6] R. J. Kafka, D. R. Penders et al., "Role of Interactive and Control Computer in the Development of a System Restoration Plan," IEEE Transactions on Power Apparatus and Systems, Vol.PAS-101, No.1, pp.43-52, January 1982.

[7] J. Zaborszky et al., "Restoration of the Large Electric Power System Using a Computer Generated Sequence of Target System," IFAC Symposium on Power System and Power Plant Control, Beijing, pp.301-306, August 1986.

[8] C. L. Forgy, "The OPS83 User's Manual," Department of Computer Science, Carnegie Mellon University, March 1985.

Discussion

Adibi, M. M., IRD Corp., Bethesda, Maryland. The authors should be commended for developing an efficient "prototype guidance system" which combines operator's heuristic problem solving capability with the analytical tools, such as power flow program, to provide real-time plans and procedures for restoration of bulk power systems following partial to complete collapse.

Apparently, the guidance system accomodates operator's interventions during restoration procedures and also can automatically update the plans & procedures to allow for occurances of unexpected events. An interesting feature of the authors approach is the verification of operator's commands by simulation prior to its implementation.

The paper covers the significant problem of integrating "knowledge based experties", analytical tools and operator interactions. However, it is not clear how important problems such as coordination of the restart of thermal power plants (fossil & nuclear) with reintegration of the bulk power supply, the switching transients of the 500kV and 275kV lines, the prime mover's frequency response to the sudden cold load pickup, or the actions of the various major & minor control & protective systems are addressed.

In its third report [1], the Power System Restoration - Task Force, has outlined the top twelve restoration issues which are based on the expriences and practices of the power industry in USA. It would be of interest to learn how the authors view these problems. For instance, under the "Recognition Phase", authors place importance on identification of the faulted components, while the PSR-TF report states that in the initial phase of restoration, it is important to assess the state of the power system but not to spend the critical time in fault identification which most probably has been transient in nature and has cleared itself.

The authors are again complemented for a timely and significant paper in a field which is in its infancy and needs more formal investigations.

[1] Adibi, M. M. (Chairman), "System Operation Challenges: Issues and Problems in Power System Restoration", IEEE Transaction on Power Systems, Vol.PWRS-3, No.1, February 1988, PP 123-124.

Manuscript received February 14, 1989.

B.F. WOLLENBERG and D.S. KIRSCHEN, Control Data Corporation, Plymouth MN: The authors have done an excellent job of designing an automated scheme for a power system restoration guidance system. The fact that the operators play an integral role in the restoration process and can override results of the expert system seems to be quite important. From what we have determined, this is absolutely necessary for operator acceptance.

Many utilities have done extensive restoration plan development and would like the expert system to try the preestablished plans as base options and then alter them during restoration to fit the situation. Can the author's scheme accommodate preestablished restoration plans, or does their system always construct a plan from scratch each time? Using preestablished plans could cut down on the search space for a restoration plan. As the authors indicate, with n breakers there are 2^n combinations to determine a final configuration, however, with preestablished plans there would be far fewer combinations to search over.

We would be interested to learn how "hard" the operating constraints are in the restoration. Would the restoration plan adopted by the program violate one or more constraint if that were the only way to achieve a restoration?

We would be interested to learn how far the author's have carried their scheme. Is the restoration advisor interconnected to a working SCADA system? Is it in use now in a utility operations office?

Manuscript received February 24, 1989.

Y. KOJIMA, S. WARASHINA, S. NAKAMURA and K. MATSUMOTO: The authors would like to thank the discussers for their comments and questions. Mr. Adibi is interested in the details of dealing with power system characteristics relevant to restoration. In response to his queries, the authors would like to add a few words. Our system is a prototype to verify the guidance method and a rough power system model is provided. In our system an outline of course of restoration will be determined in a short time without checking each operation in details. These checks would be done in the "Operation Phase". Further research on the topics is presently being conducted for practical use and it is hoped that the ability of the system will be upgraded in the near future.

With respect to the second question, the authors would like to point out that fault identification forms a part of assessing the state of the power system. As a result, it determines equipment which should not be used in restorative operations. If a fault is permanent, connection to the faulty component may cause a secondary accident. And in the paper, "component" means a section which is separated by circuit breakers, such as line, bus, transformer and so on. The diagnosis method does not take so much time, usually less than one second. In addition to fault diagnosis the system estimates switching status, system configuration, charging status and the amount of outage loads in the "Recognition Phase."

Drs. Wollenberg and Kirschen raise three points. We will respond to each one individually. Our system always constructs a plan from scratch each time. In our opinion our approach is better than the approach suggested by discussers. The reasons are given in the following. As a matter of fact, using preestablished plans could cut down on the search space. But this approach needs quite a few preestablished plans to respond to all the faults. It may be almost impossible to prepare plans without omission for all faults in the large system such as the bulk power system of Tokyo Electric Power Company. And the preestablished plans must be modified if the system configurations or operational conditions are changed.

Our system creates "Target System" which satisfies all constraints. In other words there are no violations of constraints after all operations are put into action. If operator cannot help violating any constraints during each operation, it will be dealt with on operator's own judgment.

On the question of practical use of the system, we are investigating what functions must be upgraded or added for on-line use and the system has not been connected to the field yet.

Manuscript received April 5, 1989.

THE DEVELOPMENT OF POWER SYSTEM RESTORATION METHOD FOR A BULK POWER SYSTEM BY APPLYING KNOWLEDGE ENGINEERING TECHNIQUES

by

Yukio Kojima   Shigeru Warashina

Tokyo Electric Power Co., Inc.
(Japan)

Masakazu Kato   Hadime Watanabe
Member, IEEE

Toshiba Corp.
(Japan)

Abstract In this paper, a prototype expert system for bulk power system restoration is discussed. Restoration of a bulk power system involves operation of various equipment that is complexly inter-related. For this reason, object-oriented programming techniques employing a frame representation of knowledge and heuristics has been adopted to achieve efficient processing. The prototype system has successfully satisfied the requirements of speed, stability and flexibility for a complex bulk power system restoration by applying knowledge engineering techniques.

Keyword System restoration, Knowledge engineering, Frame representation, Production rule, Blackboard model, Energizing phase, Load supplying phase.

INTRODUCTION

A stable supply of power is required from power systems as the importance of electrical energy increases. Particularly in the case of high voltage systems (500kV/275kV bulk power systems), this requirement is high. Accordingly, a variety of technology for equipment and operation is being developed to improve the level of supply reliability. Excellent results have been achieved.

There are, however, cases when power system faults cannot be avoided. In these cases, it is necessary to quickly end the interruption of service. In present bulk power systems, restoration is performed by preparing various restoration methods for predicted faults. The operators perform the actual restoration based on these prepared procedures. However, when a fault too complex to predict occurs, it is necessary to quickly determine the required procedures based on operation rules and past experience while processing a large amount of information. It is thus desirable to develop a system to support the operator in making judgments and performing processing during emergencies. A power system restoration method for all bulk power system faults ranging from limited outages to outages over a wide area has been developed on a model power system.

System restoration procedures can be formulated as follows:

Let x_s represent a system condition immediately after a fault
x_f a system condition already restored and
u_i $(i=0, 1, 2, \cdots)$ each restoration operation.

The restoration process $\{x^{(i)}\}_{i=0,1,\cdots}$ can then be expressed as

$$x^{(0)}=x_s$$
$$x^{(1)}=u_1 \cdot x^{(0)}$$
$$\vdots$$
$$x^{(n)}=x_f=u_{n-1} \cdot x^{(n-1)}$$

89 WM 163-7 PWRS A paper recommended and approved by the IEEE Power System Engineering Committee of the IEEE Power Engineering Society for presentation at the IEEE/PES 1989 Winter Meeting, New York, New York, January 29 - February 3, 1989. Manuscript submitted January 26, 1988; made available for printing December 1, 1988.

The restoration procedures are $\{u_i\}_{i=0,1,2,\cdots}$.

As mentioned above, the following expression is commonly used :

$$\sum_i t(u_i) \rightarrow \min$$
$$\text{s.t.}\ \ g(x_f)=0$$
$$\text{operation rules.}$$

where $t(u_i)$: time required for restoration operation u_i
$g(x)$: not to supply demand under system condition x

Next, the following must be considered :

- $t(u_i)+t(u_j) = \max(t(u_i), t(u_j))$ when operation u_i and u_j are executable simultaneously. Even though the number of restoration operations is larger, the time required to restore may be less if some operations can be executed simultaneously.
- Operation u_k is not always executable under system condition $x^{(i)}$. Operation u_k may not be executable because of the presence of overvoltage and so on. This indicates that a restoration procedure $\{u_i\}_{i=0,1,\cdots}$ may not be executable if operation u_j and u_{j+1} change each other. Operation timing and order are very critical.

The authors have developed a power system restoration method based on the following strategy since the above problem is too difficult to solve directly.

- Optimization based on practical operation rules has been performed (e.g., finding the "optimum" restoration route to minimize operating time) to attain faster restoration procedures. Furthermore, all operations that can be executed simultaneously are. (Speedy restoration)
- Importance has been placed on the possibility of safely executing system operations. Safety is ensured by preventing system operations from causing overvoltage, overload or instability of the system. Attention is also paid to operation timing and order. (Safe restoration)

An additional consideration is made. Restoration procedures can be modified flexibly in accordance with changes in the conditions of restoration and unsure elements.

Restoration methods for lower voltage systems have been developed using simulation, mathematical programming method and so on [1] [2].

Since the generation of restoration procedures is considered a type of a combinatorial problem, restriction of a possibly huge searching space is required. Constraints relative to practical operation must also be considered. For these reasons, knowledge engineering techniques have been applied to power system restoration [3] . Such a system is now undergoing field testing [4] .

A bulk power system restoration has been studied [5] [6] including procedures for speedily energizing a de-energized network [7] .

Reprinted from *IEEE Transactions on Power Systems,* Vol. 4, No. 3, pp. 1228–1235, August 1989

In [5], [6], a detailed and successful methodology is described considering various conditions such as active, reactive power balance and so on. But interaction with the operators is required to generate restoration procedures.

In [7], reenergization procedure of a whole system is based on energizing a preestablished partitioning of system with one or more generators. Although it is effective, it can not be applied to the bulk power system of Tokyo Electric Power Company, as will discussed later, because of its characteristics of the system configuration.

This paper proposes a method by which speedy, safe and flexible restoration procedures are generated automatically without interaction with the operators for a large scale bulk power system.

Restoration of a bulk power system involves operation of various equipment that is complexly inter-related in comparison to a lower voltage system. For example, examination of many operations is required simultaneously in case of single restoration operation. A single restoration operation may also influence many other operations. To alleviate this problem, object-oriented programming techniques employing a frame representation of knowledge and heuristics have been adopted in order to realize this prototype system more efficiently.

A group of frames is controlled by production rules which were introduced in this prototype system, in order to generate detailed restoration procedures taking into consideration operation timing.

In addition, introduction of a blackboard model for division of knowledge in the knowledge base allows efficient processing. This is necessary because the knowledge base grows large when system components such as buses, switches, transformers and so on are represented as frames.

APPLICATION OF KNOWLEDGE REPRESENTATION AND PROCESSING IN THE RESTORATION SYSTEM

Relationship between System Components

In restoration, several components are complexly interrelated and change during the course of the restoration process. This relationship is illustrated in Fig. 1.

During restoration, power must be quickly supplied from energized systems to loads that are not being supplied. In the interest of maintaining supply capability, the restoration of power plants is essential. When these loads and plants are restored, the restoration route configured by the transmission lines and substations must be energized. Overvoltage is a danger during this time. During supplying of loads, overloading is also a danger. Both of these must be checked for. In order to prevent overvoltage, there are use of VAR equipment and a load restoration method. In order to prevent overload, there are measures such as system changeover, bus separation, adjustment of generating power and so on.

In the restoration process, a number of individual restoration operations can be performed simultaneously, but energizing of the restoration route and supplying of loads dependent on the generating capacity of the power plants must be checked.

System conditions such as system capacity change according to supplying of loads, synchronizing power plants with the system, and interconnection of isolated subsystems.

The restoration policy and individual restoration operations are determined based on the relationship between these system components.

Modeling of the Restoration System – Frame Representation of System Components

Since the knowledge is organized in a hierarchy, the frame representation is good for the restoration system modeling. The class frame representations [8] for the elements that configure the system, substations, transmission lines, power plants, loads, are made. Under these, the various substations, transmission lines, plants and loads are placed as instance frames. In addition, a hierarchical structure taking into consideration the system is employed as the super frame for these frames. This is described in Fig. 2.

In the frame, not only such data as connection information, equipment/facility capacity, but also knowledge concerning restoration are embedded in slots in the frame.

Knowledge Processing

Object-oriented Programming

As is indicated in Fig. 1, a large number of the system components are complexly interrelated. Object-oriented programming is effective in flexibly processing these.

Each of the frames for substations, transmission lines, plants and loads are considered as single objects. Messages are sent to related objects and the receiving object forwards the message to other objects as required. If required, the receiving object sends a message to the following object based on its own information. For example, in the case of a substation object, it uses its energized/de-energized status information and faulted equipment information to prepare the message. The required information is thus obtained and processing can be performed.

Production Rules

The energized state of the system, system capacity, and size of supplied loads all vary with time. In the case of restoration that requires a long time, such as bulk power system restoration, the order of operations and establishment of operation timing are extremely important in determining the restoration operation procedures.

In the present system, rules that evaluate the transition of time are created and the change of objects over time is controlled as indicated in Fig. 3.

Following are examples of actual production rules:

- Operation B may be performed if operation A is completed.
- Operation C may be performed if the system capability increases from the current level.
- Plant E may be synchronized with the system if D minutes pass.

Blackboard Model

Thus far, a substation has been represented as a single system component. However, a substation is actually a single system that comprises buses, transformers, VAR equipment and switches.

During restoration, substations must perform such operations as use of VAR equipment, bus separation and changeover of systems. When a substation has no faulty equipment and no operation constraints are violated, the above operations can be performed freely. However, if constraints are placed on the substation due to faulty equipment or operation rules, it is necessary to consider the possibility of operation at the device level. As with system changeover, if operation of the substation will affect other substations, consideration at the substation device level is also required.

The blackboard model presented in Fig. 4 has been adopted. ks1 is a knowledge base concerning restoration that expresses a substation as a single system component, the form that has been so far discussed. ks2 through ksm are knowledge bases used to investigate the possibility of operations at the device level of each substation.

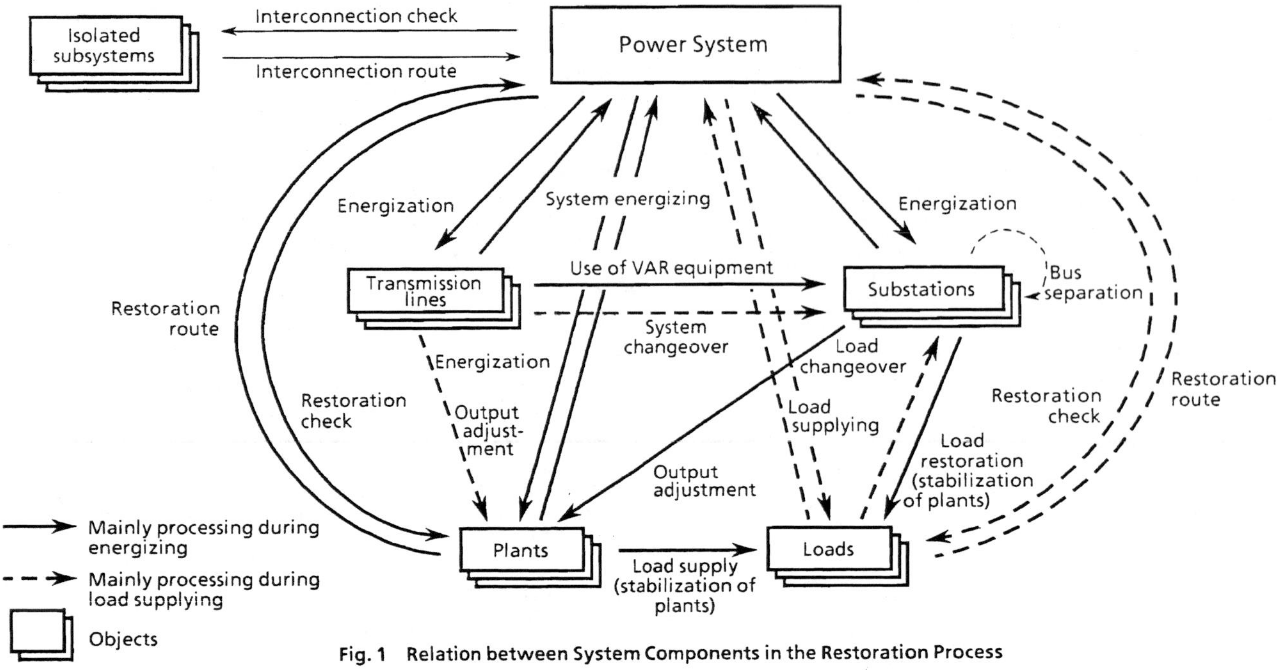

Fig. 1 Relation between System Components in the Restoration Process

ks1 performs the inference and processing for the generation of restoration procedures. However, when it becomes necessary to operate a substation that has faulty equipment or operation constraints, communication with the knowledge bases concerned is carried out to check the corresponding restoration operations.

In this manner, an operation check is only performed for the substations that require it. Since this check is performed at the substation device level it is not necessary to search massive amounts of data. This allows dependable, efficient processing.

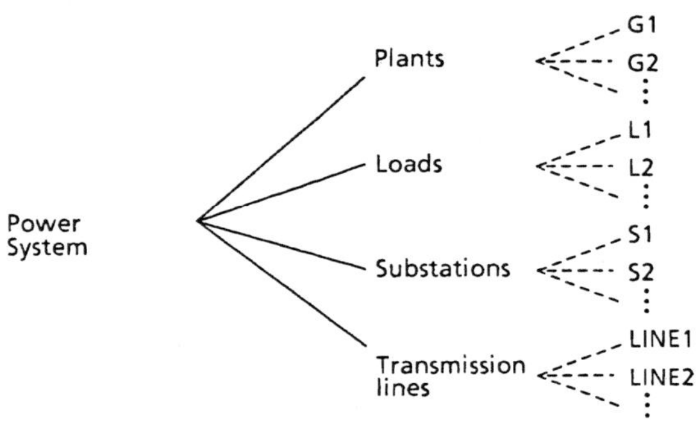

Fig. 2 Frame Representation of System Components

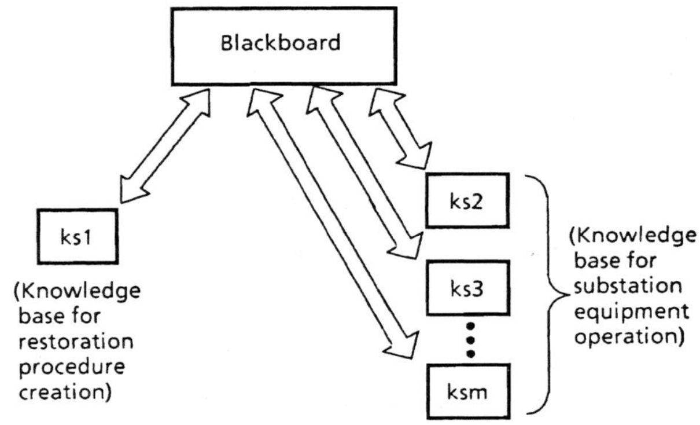

Fig. 4 Blackboard Model of Knowledge Representation

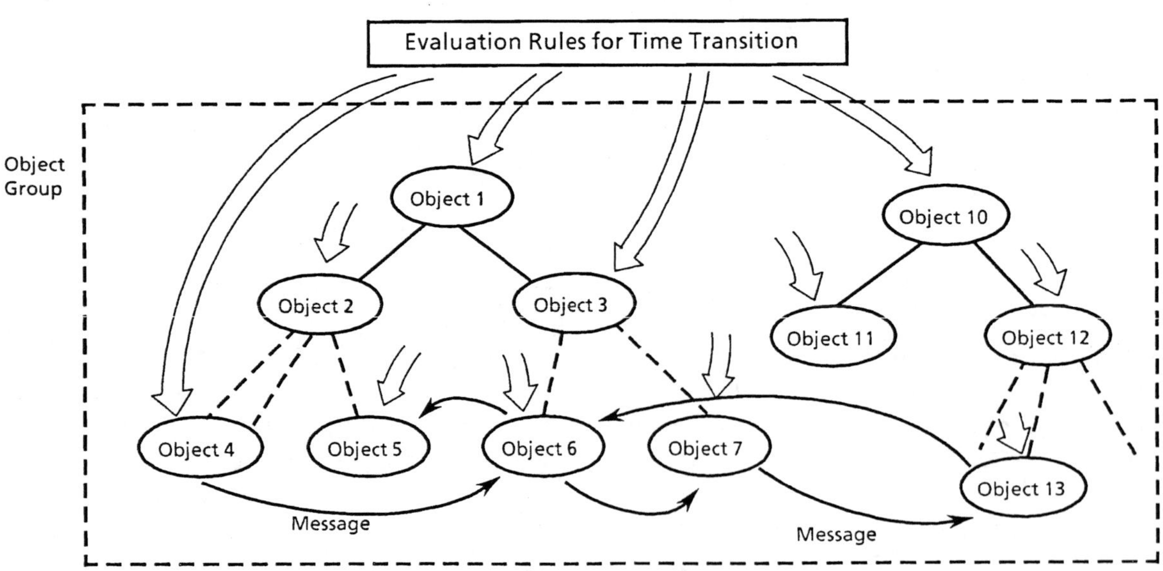

Fig. 3 Method of Inference

FLOW OF BULK POWER SYSTEM RESTORATION

The bulk power system of TEPCO (Tokyo Electric Power Co., Inc.) is characterized by large thermal power and nuclear power capacity to the east, and large hydro power capacity to the north and west as indicated in Fig. 5. This power is first transmitted by a 500 kV loop system and then to areas of large demand such as Tokyo and Yokohama.

One of the neighboring power systems, Tohoku Electric Power Co., is a 50Hz system. It is directly linked to TEPCO's power system. The other neighboring system belongs to Chubu Electric Power Co. Since this system is a 60Hz system, it is linked to TEPCO's system via a frequency converter station.

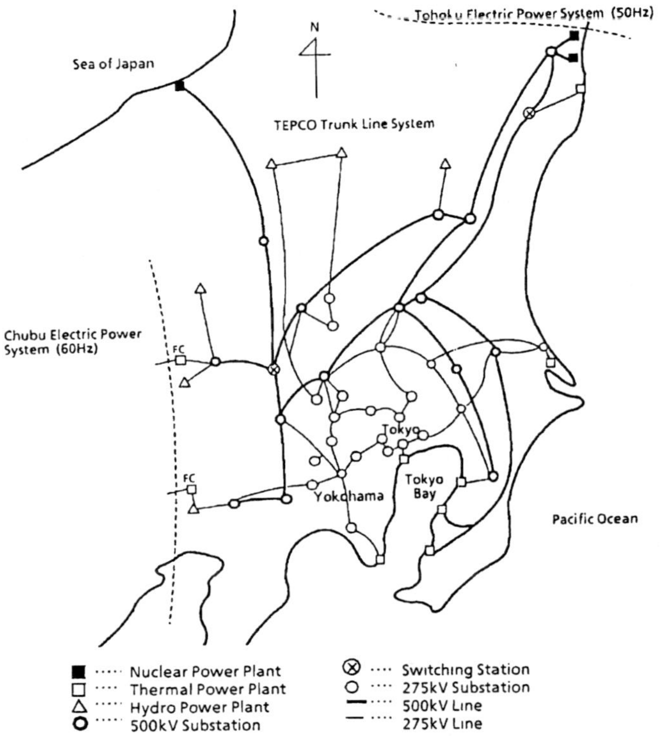

Fig. 5 Bulk Power System of TEPCO

In restoring the bulk power system, first the 500kV, 275 kV transmission lines and 500kV substations are energized in sequence and power for restoration is supplied to the shutdown large-capacity thermal power and hydro power plants in order to increase supply capability. In the case of outage over a wide area, power for restoration is received from a neighboring system or a hydro power plant for trial power transmission. This hydro power plant is a plant which can perform leading phase operation with no active power. Supply power is gradually supplied to loads in this manner as supply capability increases.

The restoration process for bulk power systems can be divided into two phases based on the restoration condition of the power system – energizing phase and load supplying phase. After determining the restoration policy for each phase, the individual operations conforming to the policy for each phase are repeated.

In the energizing phase the major goal of the restoration policy is to quickly supply power for restoration. For example, the order of restoration of shutdown power plants is determined. In the load supplying phase, a system configuration that does not cause overloads is determined. Dependent on system conditions, it is possible to perform several restoration operations simultaneously. The order and timing of execution are important.

The energizing phase is defined as the phase in which the system state that will permit starting of restoration is confirmed, and supplying of loads by quickly restoring stopped substations and shutdown power plants is started. Supplying of loads necessary for stabilizing the system is also included in this phase. Since the load supplied is small in the energizing phase, it is more important to consider voltage and reactive power rather than active power.

The load supplying phase is defined as the phase following the occurrence of a fault in which loads are supplied in a manner that overloads do not occur whether or not there is a preceding energizing phase. In the load supplying phase, primary consideration is given to active power. For this purpose, the supply-demand balance and overload of system equipment is mainly checked.

When system conditions change during restoration process, such as restoration of faulted equipment, restoration policy and procedures are modified immediately.

The basic configuration of the restoration algorithm is shown in Fig. 6.

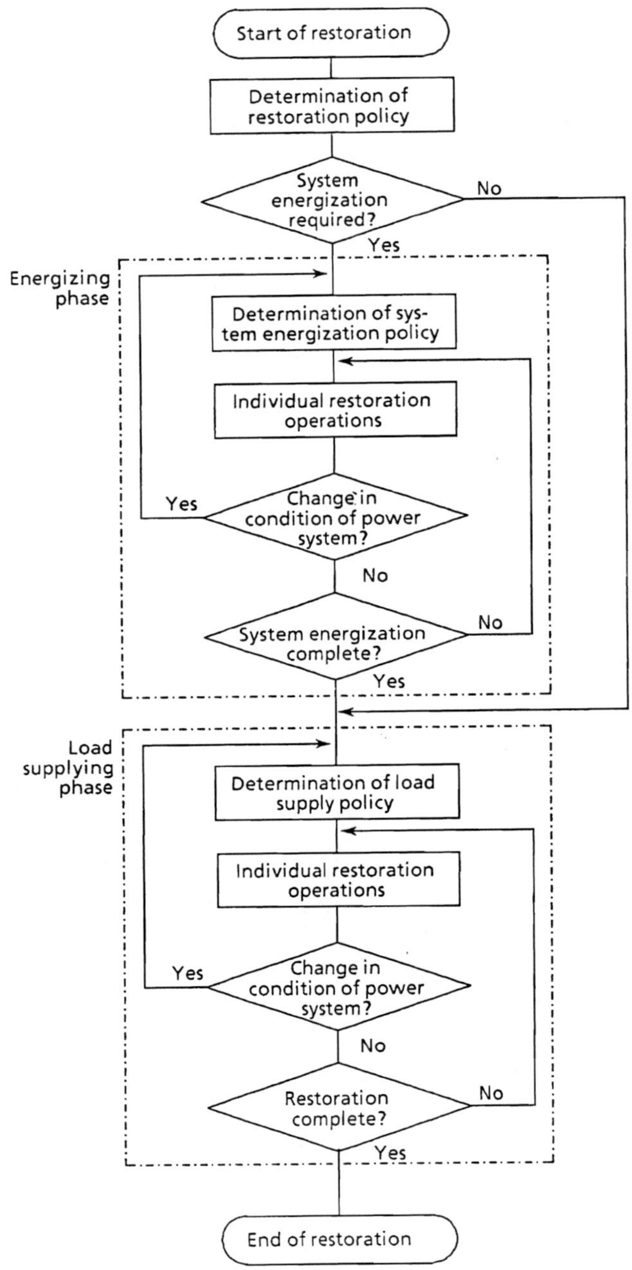

Fig. 6 Configuration of Fault Restoration Algorithm

PROCESSING FLOW IN THE PROTOTYPE SYSTEM

– ENERGIZING PHASE –

Processing in the energizing phase takes place as follows :

1) Verification of system condition after fault

Checking of energizing is performed here. Checking for isolated systems is also performed here.

2) Determination of restoration policy

If an energized system is not present, startup of the plant for trial transmission or reception of power from a neighboring system is investigated.

An interconnection route between isolated subsystems or the restoration route in the direction of the shutdown power plants is selected. Evaluation of the possible restoration routes is performed based on the criterion of minimizing operating time. In addition, energizing from the low-voltage side to the high-voltage side of a transformer must be avoided.

The route for the plants and substations that performs quick restoration (recovery of power) is selected based on the actual operation rules.

Next, the restoration priority for shutdown plants is determined.

For shutdown plant i and j,
if $t_i \cdot W_j \leqq t_j \cdot W_i$,
then the restoration priority of plant i is higher than that of plant j,
where t_i is a time required for plant i to synchronize with the system,
W_i is a capacity of plant i.

In doing so, the maximum supply capability is expected and speedy restoration is thus possible.

3) Operation execution check

Based on the energizing capability of the restored plants, an investigation is performed to determine the number of plants and substations that can be simultaneously restored from the charge capacity (considering the employed VAR equipment) along the restoration route. The restoration scheduling for each plant and substation to be restored can be calculated along the restoration route. In calculating the time, the operation time required for energizing a single transmission line is estimated. Determining the restoration timing in this manner allows the system energizing state to be grasped from multiple isolated subsystems with the same timing when generating operation procedures in 4). During energizing, the voltage of the receiving-end substation is checked. If overvoltage is predicted by a simulation, the countermeasure is to use the VAR equipment located in a substation along the restoration route. If the energizing capability of the restored generators is small, there are cases when energizing of some plants cannot be performed.

Safe and speedy restoration is possible by performing operation execution checks and simultaneous operation to the fullest extent.

4) Generating operation procedures

The restoration time scheduled in 3) is used as the operation timing and the restoration process for the entire system is simulated. The operation procedures are generated in accordance with operation timing.

Since several operations cannot be performed simultaneously at the same substation, the operations are carried out in order of priority.

If a plant is to be restored, that is to receive power for restoration, the operation to synchronize with the system is performed. After a fixed time it is synchronized with the system. At this point charging capability increases and the system condition thus changes. This means that some number of plants and substations can now be restored or energized. Restated, the process of synchronizing generators allows the system capacity to be expanded and, in turn, for additional stopped substations to be restored. Not only generators and substations, but also loads that can be supplied to stabilize the system or strengthening of the system by forming a loop configuration are also investigated. Operation procedures for these operations are generated as well.

– LOAD SUPPLYING PHASE –

Processing Flow in Load Supplying Phase

In the present system, a simulated restoration of the load to be restored is performed and the system configuration is changed so as to reduce the overload. Occurrence of an overload in the loop system is avoided. That is to say, since it is difficult to relieve an overload in a loop system by system operation, supply margin from substations configuring a loop system by adjustment of generating power is requested. It is planned to relieve the overload by system operation of the radial system downstream. If an overload occurs in the loop system, it is assumed that the overload has occurred in the substations configuring the loop system.

Processing in the load supplying phase takes place as follows:

1) Determination of restoration policy

First, the supply margin of each substation in the loop system is calculated by following formulation.

$$\sum_{i \in J} (\triangle P_L)_i \rightarrow \max$$
$$\text{s.t. } \underline{P}_G \leqq P_G \leqq \bar{P}_G$$
$$P_B = Y_B \cdot A \cdot Y_N^{-1} \cdot P_N$$
$$P_N = \begin{bmatrix} P_G \\ P_L + \triangle P_L \end{bmatrix}$$
$$|P_B| \leqq \bar{P}_B$$

where $\triangle P_L$: supply margin vector at each plant
$(\triangle P_L)_i$: i-th element of $\triangle P_L$
P_G : plants' output vector
$\bar{P}_G, \underline{P}_G$: plants' maximun, minimum output vector
P_B : branch power flow vector
Y_B : branch admittance matrix
A : incidence matrix
Y_N : nodal admittance matrix
P_L : supplied load vector
P_N : nodal injection power vector
$\bar{P}_B$: branch capacity vector
J : a set of substations from which power can be directly supplied to outage area

In other words, if a load that exceeds this supply margin is restored, no matter how much generating output is adjusted an overload will occur in the loop system.

Next, the supply margin of the substations in the radial system downcurrent is calculated from the capacity of the transmission lines and transformers based on the previous supply margin.

A restoration simulation of supplying loads from the same substation as before the fault is performed.

2) Method of relieving overload

If an overload occurs, its clearance or relief is investigated. Methods of clearance include system changeover, bus separation and load changeover (load is supplied at designated substation different from before fault). The algorithm for this is described later.

When system changeover or bus separation is to be carried out, a loop switching method is used in which the system is temporarily configured as a loop and the system is then switched off. At this time, it is sometime not possible to perform switching due to overload.

When the overload cannot be relieved no matter what method is employed, all the loads cannot be supplied as planned. This is dealt with by forgoing to supply some loads downcurrent from the overload. Concerning loads for which supply has been delayed, the restoration of additional facilities is awaited and the restoration method is reinvestigated.

3) Generation of the operation procedures

If the operations indicated by the inference process in 1) through 2) are performed an overload will always occur and in the last stage the overload will be cleared. Therefore, this type of operating procedure cannot actually be performed. It is thus necessary to modify the operating procedures so that they can actually be performed.

The actual operation procedures are generally the reverse of the above inference process.

Measures for Relieving Overload

Three methods can be considered for relieving an overload: system changeover, bus separation and load changeover. The general flow for relieving an overload is as follows :

(i) To investigate substations downstream of an overload

(ii) To investigate substations to be changed over from each substation given in (i)

(iii) To check whether to decrease the total overload value (in case of bus separation, to investigate load combinations, too)

Concerning adjustment of generating power, this is considered during calculation of supply margin and thus need not be considered here.

In relieving an overload it is problematic to determine the combination of the above three methods. In this prototype system, the order of investigation of combinations is from largest to smallest contribution to relief; namely, system changeover, bus separation and load changeover. The order of producing combinations can, however, easily be changed.

In this prototype system, in investigating the above methods of clearance, the clearance measure is considered possible for investigation if the total overload value decreases even if overload occurs at other equipment.

PROTOTYPE SYSTEM

The prototype system was constructed based on a nearly actual-size bulk power system with 36 power plants, 68 transmission lines and 48 substations.

Although some power plants are simplified, this system involves all transmission lines and substations in TEPCO's bulk power system and the exact bus-bar system at 500kV substations.

The scale of the prototype system is 350 frames, 11 production rules, 20 demons and 500 methods.

Example of results

An example of restoration results by this prototype system is presented.

In this example, the order of overload relief method, that is to say, system changeover, bus separation and load changeover, is changed for three fault cases. Final overload value and the number of operations are indicated is Table 1.

Fig.7 is an example of a character terminal screen and Fig. 8 is an example of a graphics terminal screen.

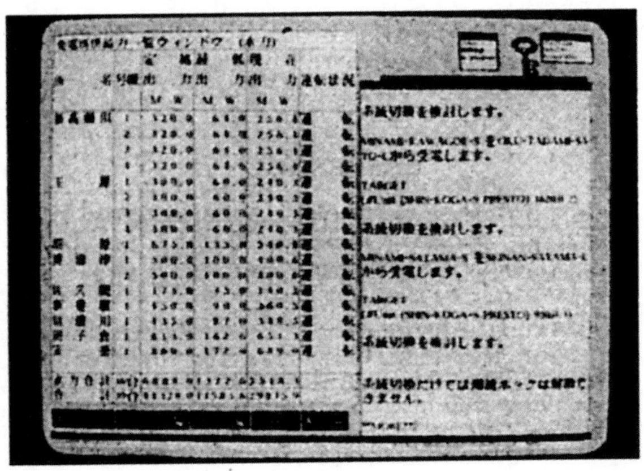

Fig. 7 Photo of Character Terminal Screen

Table 1

	CASE A					CASE B					CASE C				
*	1	2	3	4	5	1	2	3	4	5	1	2	3	4	5
number of operations	SC 1 BS 2 LC 3	BS 1 SC 1 LC 3	SC 1 LC 3	SC 1 BS 2	LC 5 SC 1 BS 1	SC 1 BS 2 LC 4	BS 1 SC 1 LC 5	SC 1 LC 9	SC 1 BS 2	LC 11 SC 1 BS 1	SC 1 BS 1 LC 0	BS 1 SC 0 LC 0	SC 1 LC 4	SC 1 BS 1	LC 5 SC 1 BS 1
final overload value (MW)	0	346.4	346.4	325.0	2.4	0	0	183.0	926.0	0	0	0	0	0	0

* Overload relieving strategy

	FIRST	NEXT	LAST
1 :	SC –	BS –	LC
2 :	BS –	SC –	LC
3 :	SC –	LC	
4 :	SC –	BS	
5 :	LC –	SC –	BS

SC : system changeover
BS : bus separation
LC : load changeover

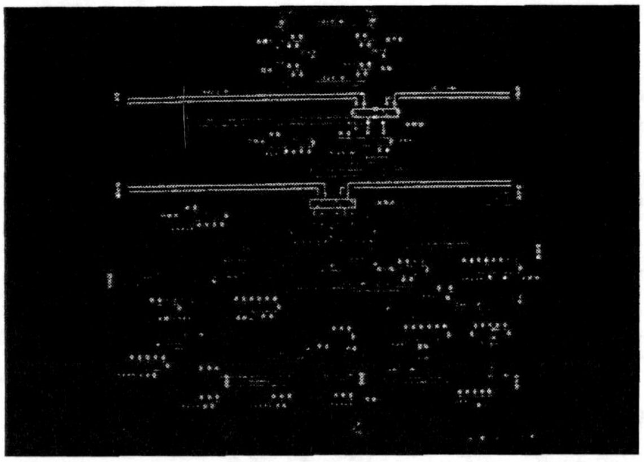

Fig. 8 Photo of Graphics Terminal Screen

CONCLUSION

The present prototype system has confirmed that applying knowledge engineering techniques to power system restoration is effective. That is to say, employing knowledge engineering techniques has provided a complex restoration system for bulk power systems that is speedy, safe and flexible.

Validity of restoration method is now being verified for various fault cases, using this system.

In the future, the authors intend to investigate the linking of such a system with an ongoing EMS (Energy Management System). Topics include increasing processing efficiency and more fully utilizing knowledge. As the system approaches commissioning it is hoped that the level of completion can be increased.

REFERENCE

[1] H. Suzuki, T. Haba " Development of Automatic Operation Logic for Power System Emergency Condition " IEE of Japan Vol. 93-B No.8 Aug., 1973.

[2] H. Suzuki, H. Kodama et.al " Power System Switching Logic for Restorative Operation on Secondary System " IEE of Japan Vol. 97-B No.3 March, 1977.

[3] T. Sakaguchi, K,Matsumoto " Development of A Knowledge Based System for Power System Restoration " IEEE Trans. on, PAS, Vol. PAS-102 No.2 Feb., 1983

[4] S. Moriguchi, H.Sakaguchi et.al " An Expert System for Power System Fault Analysis and Restoration " CIGRE SC 39 Tokyo Meeting, Oct., 1987

[5] R. J. Kafka, D. R. Penders, et.al " System Restoration Plan Development for a Metropolitan Electric System " IEEE Trans, on PAS, Vol. VAS-100, No.8 Aug. 1981

[6] R. J. Kafka, D. R. Penders, et.al " Role of Interactive and Control Computers in the Development of a System Restoration Plan " IEEE Trans. on PAS, Vol. PAS-101, No.1. Jan., 1982

[7] E. Mariani, F. Mastroianni et.al " Field Experiences in Reenergization of Electrical Networks from Thermal and Hydro Units " IEEE Trans. on PAS, Vol. PAS-103, No.7, July 1984.

[8] A. Barr, E. A. Feigenbaum ed., " The Handbook of Artificial Intelligence " William Kaufman, Inc. 1981

Discussion

Adibi, M. M., IRD Corp., Bethesda, Maryland. The authors have presented an interesting approach to restoration of bulk power supply which apparently is fast, safe, flexible and automatic without the need for operator's interaction and intervention. Their approach is certainly a departure from the current thinking of the Power System Restoration-Task Force [1, 2 & 3], which allows for operator's heuristic problem solving capabilities augmented by verification through high-fidelity system-referenced simulation. Granted, that the latter approach is inherently not as fast, but it does take advantage of operator's intuition and judgement, based on his experience which can not readily be interpreted into an expert system.

Evidently, the TEPCo's eastern thermal and possibily the nuclear plants do not have load rejection capabilities and for restart need external power to be provided by the northwestern hydros. Under such a condition, the restart of thermal plants, energizing of the 500kV lines and the load pickup to the minimum generation of thermal plants, would be performed simultaneously and not in separate phases as suggested in the paper.

Although restoration procedures are developed to meet the specific requirements of an individual power system, yet many power systems have certain characteristics in common and, to that extent, certain general approaches apply. The authors comments are invited on the three reports by the PSR-TF which have been prepared with the above commonality consideration.

[1] Adibi, M. M. (Chairman), "Power System Restoration - A Task Force Report", IEEE Transactionon on Power Systems, Vol.PWRS-2, No.2, May 1987, PP 271-277.

[2] Adibi, M. M. (Chairman), "Power System Restoration - The Second Task Force Report", IEEE Transaction on Power Systems, Vol.PWRS-2, No.4, November 1987, PP 927-933.

[3] Adibi, M. M. (Chairman), "System Operation Challenges: Issues and Problems in Power System Restoration", IEEE Transaction on Power Systems, Vol.PWRS-3, No.1, February 1988, PP 123-124.

Manuscript received February 14, 1989.

Y. Kojima, S. Warashina, M. Kato and H. Watanabe : The authors would like to thank Mr. Adibi for his comments and interest in this paper.

Task Force Report [1] outlines the following common concerns regarding power system restoration:

(1) Immediate resupply of station service
(2) Time consuming switching operations
(3) Start-up timing of thermal units
(4) Voltage rise problems in energizing unloaded transmission lines
(5) Frequency response of prime movers to a sudden load pickup
(6) cold load inrush, power factors and coincident demand factors

Task Force Report [2] shows restoration problems relative to these concerns.

The restoration method or strategy presented in this paper directly reflects projections of (1), (2), (5). The reactive power balance presented in the Task Force Report [1] [2] is checked in detail.

Now, it is very difficult to predict precisely cold load pickup because of insufficient data. The response of prime movers is very complicated and it may be impossible to simulate it in detail within the allowable time limits. Excess frequency decline at restoring loads must absolutely be avoided, as presented in the Task Force Reports [1], [2], [3].

The system frequency will be maintained within allowable limits during load restoration even though the predicted error of cold load pickup is relatively large, when all power plants are synchronized with the system. Items (4) and (6) are not directly reflected in the restoration method presented in this paper because they are very difficult to determine. Therefore we divided the restoration process into two phases: energizing phase and load supplying phase. In the former phase, synchronization of shutdown plants is given higher priority. Although this restoration strategy may prolong the operation, it is thought to be moderate disadvantage, because there is little risk of an excessive frequency drop and consequent recurrence of system outage.

Task Force Report [1] refers to detailed characteristics of steam units and combustion turbines. In this paper, hydro-power plants are given priority at energizing bulk power system because they can absorb charging currents in the transmission system better than hydro-thermal plants explained in Task Force Report [1], and thermal units (steam units) are used to supply loads rapidly.

There are many items in common between the method presented here and the one presented in the Task Force Reports. These items must be taken into consideration when planning any restoration method.

However, from the stand point of processing efficiency and practical operations, we think it insufficient to take only these items into consideration, when planning restoration for a large scale bulk power system. So, we have introduced heuristics developed from operators' experience, and we have aimed for efficient and practical restoration planning.

Manuscript received April 3, 1989.

IMPLEMENTATION OF A REAL-TIME EXPERT SYSTEM FOR A RESTORATION GUIDE IN A DISPATCHING CENTER

K. Hotta, H. Nomura, H. Takemoto

Kansai Electric Power Co., Inc.
3-3-22, Nakanoshima, Kita-ku
Osaka, Japan

K. Suzuki, S. Nakamura, S. Fukui

Mitsubishi Electric Corporation
1-1-2, Wadasaki-cho, Hyogo-ku
Kobe, Japan

Abstract - This paper describes a real-time expert system for a restoration guide which will be practically used in a dispatching center to evaluate the performance of the knowledge base system.

The system configuration is decentralized by workstations and one of them is attached to SCADA computer in order to get on- and off-line information on the current state of power system.

A restoration guide of this system has two operation modes. One is on-line guide mode, which gives operator an appropriate restoration plan and restorative procedures by using the knowledge base of expertise and heuristics on power system restoration.

The other is off-line simulation mode, which is useful for knowledge engineer to verify the validity of the acquired knowledge.

Implementation issues of this system such as system function, knowledge acquisition, knowledge representation, interface of SCADA system and man-machine interface are discussed. Some execution results from the on-line guide mode are provided to demonstrate capability and efficiency of this system.

Keywords; Expert system, real-time, restoration guide, dispatching center, implementation, knowledge acquisition, knowledge representation, verification

1. INTRODUCTION

In recent years, with an increase in the number of power systems under the control of a power dispatch center, the behavior of faults has been complicated and the fault-affected area has broadened as a result. For these reasons, it is increasingly necessary that dispatching center operators, in the event of a fault in a power system, quickly and accurately identify the faulty section and determine the nature of the fault, based on information about the fault, and then to issue appropriate restoration instructions.

Currently, operators manually make decisions on restoration methods, relying on their own technical knowledge and experience of system operation and protective relays. It is very difficult using the current method to identify the faulty section in the event of multiple faults, blind faults or faults accompanied by malfunction of protective relays or breakers. In addition, the calculation of system's marginal supply power or overload check, is currently carried out manually and making restoration plan is solely based on human experience. Thus, using the current manual restoration method, it is difficult to shorten the blackout time.

This paper was sponsored by the IEEE Power Engineering Society for presentation at the IEEE Power Industry Computer Application Conference, Seattle, Washington, May 1 - 5, 1989. Manuscript was published in the 1989 PICA Conference Record.

Hence, vigorous efforts have been made to use expert systems as an effective approach to automate power system restoration with computers.[1,2,3,4]

Currently, however, practically no such systems are in operation at dispatching centers for large-scale power systems.[5,6]

Thus, we have developed a restoration guidance system which harnesses expert system technology to assist operators in their quick and accurate understanding of faults and their remedies.

The system, scheduled to be test-operated in 1989 in the Kobe Dispatching Center, is presently undergoing various tests.

2. OVERVIEW OF THE SYSTEM

The present system, connected on-line to a SCADA computer system, assists operators in their fault handling in real time. Its major features are as follows:

(1) With the use of real on-line data, it is possible to execute restoration guidance within one minute after a fault occurs.

(2) With a knowledge base created from the system operation knowledge of veteran operators, it is possible to quickly and correctly execute restoration guidance most appropriate to current power system state.

(3) With restoration guidance executed on distributed computers (engineering work stations), it is possible to minimize effects on performance of SCADA systems.

2.1 System Functions

This system includes two operational modes, i.e., on-line operation and off-line operation. The on-line operation mode permits faulty equipment identification, restoration planning and restorative operation guidance in the event of a fault in a power system. The off-line operation mode allows operators to learn remedies, such as restoration planning and restorative operation, using current on-line data by setting an imaginary fault in a power system.

(1) Security monitoring function

a) Detection of changing status and occuring fault: Analysis is made as to the contents of on-line information from Super cyclic Digital data Transmission (SDT) and off-line information provided by operators.

b) Outage equipment identification: The section where power supply fails by a fault can be identified based on information from tripped breakers and voltage levels.

c) Overload and outage load identification: Overload level of equipment and amount of outage load caused by a fault are identified.

(2) Restoration guidance functions

a) Faulty equipment identification: Based on information as to protective relays (main and backup), equipments may be identified as possibly subject to a fault, out of a group of outage equipments which has been caused by a fault.

Reprinted from *IEEE Transactions on Power Systems,* Vol. 5, No. 3, pp. 1032–1038, August 1990.

b) Restoration planning:

- Overload removal planning: If overload occurs in a sound system, a plan is made indicating post-restoration system state in which overload is removed by means of generator output adjustment, system switching and so on.
- Outage load charge planning: If outage load occurs, a plan is made indicating post-restoration system state in which the overload is removed and amount of outage load minimized by means of a trial charge, system switching and so on.

c) Restorative operation guidance: When the restored system resulting from restoration planning is compared with the current system, the restorative operations required currently to reach the restored system are elaborated.

(3) Man-machine interface function

a) Graphics and message indication: With regard to power system states before and after a fault, and to current and post-restoration states, the image of the entire system or that of an arbitrary part of it is indicated in a graphics (Fig. 1). Faulty equipment identification, restoration planning, restorative operation guidance results and restoration guidance flows are indicated by messages.

b) Confirmation and correction of restoration guidance results: During restorative operation, operators confirm and correct the restoration guidance results on a CRT.

c) Recording of restoration guidance results: Restoration guidance results are output to a laser printer automatically.

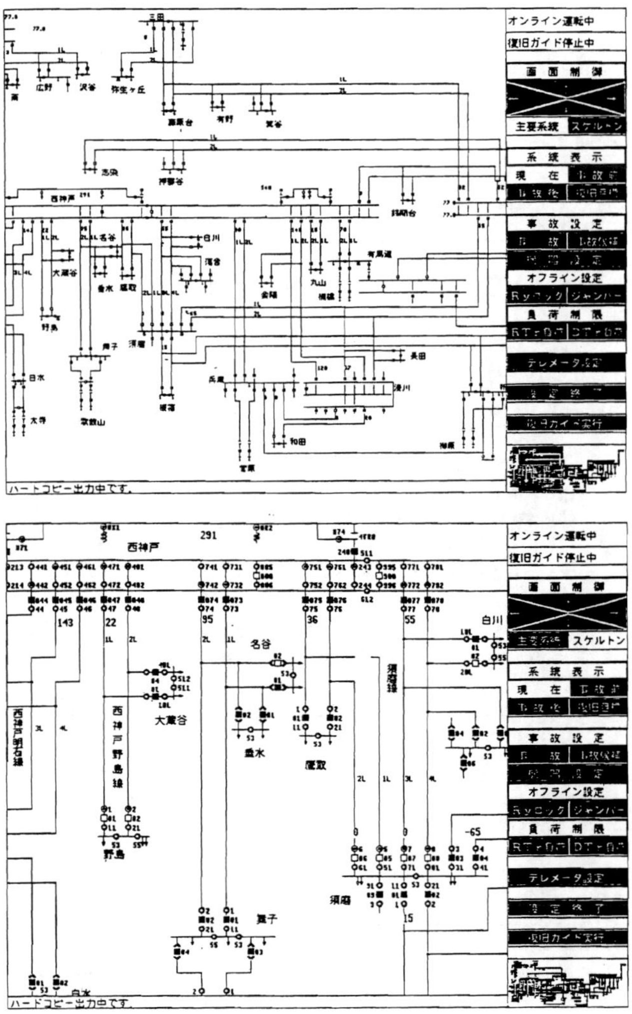

Fig. 1 An Example of a Graphics Indication

2.2 Hardware Configuration

The hardware configuration is as shown in Fig. 2. The functions of each CPU are as listed in Table 1.

The SCADA CPUs and the communication CPU are linked via GPIB, thereby permitting high-speed data transmission on line.

The three CPUs (processing speed 4MIPS respectively) are connected to a ring-shaped computer network which permits distributed processing for security monitoring, restoration guidance and man-machine interface. Due to this configuration, time required for CRT display (2 to 3 seconds) is kept constant even in the event of blackout over a wide area, at which a great amount of on-line information is transmitted. Also, the processing capacity of the system at the occurrence of accidents in rapid succession can be improved.

Table 1 CPU Functions

CPU Name	Functions
Communication CPU	Security monitoring
Restoration guidance CPU	Restoration guidance Man-machine I/F (graphics indication)
Guidance indication CPU	Man-machine I/F (message indication, record output)

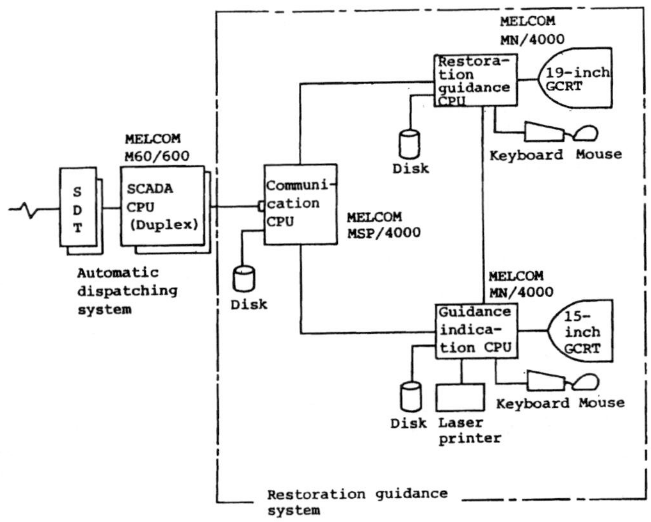

Fig. 2 Hardware Configuration

2.3 Software Configuration

Fig. 3 indicates the system software configuration. The restoration guidance, coded in OPS83, a production system language capable of high-speed reasoning and in C, a procedural language strong in numerical analysis, consists of some 500 rules and about 200 functions.

Data base is divided into two main classes, namely the equipment data base and the calculation data base. The equipment data base stores information on names, types and shapes of power system components required for man-machine I/F, which is not changed via on-line information from SCADA. The calculation data base stores information on connection between components, voltage level and amount of load for the execution of various types of numerical analysis (load flow calculation etc.) for restoration guidance.

Information on telemeter readings and the switching condition of the circuit breaker and the line switch, which constitute the variable portion of the calculation data base, is constantly renewed via security monitoring, using on-line information from SCADA.

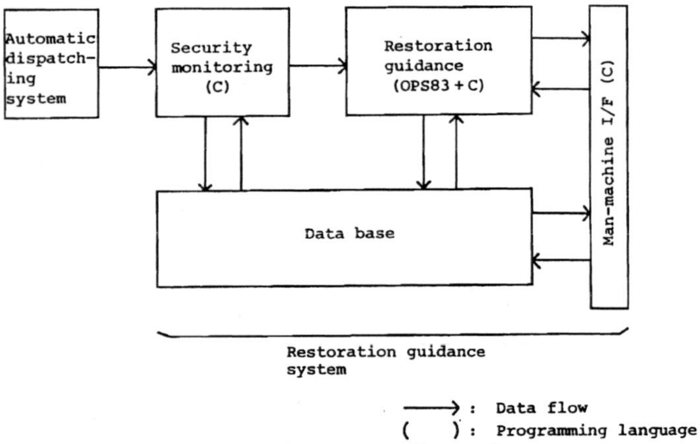

Fig. 3 Software Configuration

3. THE IMPLEMENTATION OF THE KNOWLEDGE BASE SYSTEM

In order to create a knowledge base for the restoration guidance function, it is necessary to acquire power system operation knowledge for fault restoration, such as about faulty equipment identification and restoration planning, and to determine the methods for structuring and representing the acquired knowledge. This section describes restoration guidance knowledge acquired from operators and the knowledge base system developed to implement the knowledge on the computer.

3.1 Acquiring Restoration Guidance Knowledge

Knowledge was gathered in cooperation with operators by desk-simulating ca. 100 typical faults extracted from those which are prepared for a fault prevention during summer by operators and from the records of past serious faults. The following includes the basic knowledge thus obtained with regard to faulty equipment identification, overload removal planning, outage load charge planning and restorative operation guidance.

(1) Faulty equipment identification:

a) Faulty equipment is identified based on the result of outage equipment identification (i.e., identification of the outage equipment group) and the information of activated protective relays, assuming that the faulty equipment is, without exception, located within the outage section.

b) The identification is accomplished on the assumption that all pieces of the information of the activated protective relays are correct.

c) If main protective relays (MRy) are activated, the equipment which the relays protect is identified as the faulty equipment.

d) If backup protective relays (BRy) are activated without activation of the main protective relays, such equipment, if any, as located within an area covered by the backup protective relays and not protected by the main protective relay, is deemed as the faulty candidate.

(2) Overload removal planning

a) If parallel system use (in lines and transformers) removes overload, such parallel use is employed.

b) If the adjustment of generator output removes overload, generator output is adjusted.

c) If loop formation removes overload, a loop is formed.

d) If system switching removes overload, system switching is performed.

e) If any of these methods from a) to d) is not effective, overload removal is attempted by combining those which are likely to be effective.

f) If overload is not removed at step e) above, overload is removed by load control.

(3) Outage load charge planning

a) If restoration without occurring overload is possible with the source which had fed current outage area before the fault, that source is used.

b) If a) above is not the case, an attempt is made to use adjacent source which had not fed current outage area before the fault. If this source feeds outage area without occurring overload, restoration is accomplished with this source.

c) If a) and b) above are not the case, power is restored with the largest of the stand-by power sources, accompanied by load control.

d) Steps a) to c) above are performed from the higher voltage downward.

(4) Restorative operation guidance

a) When the current system state are compared with post-restoration state indicated by overload removal planning and outage removal planning, a currently executable restorative operation is determined.

b) Restorative operation procedures are determined in accordance with a priority fixed in advance depending on the purpose of the operation and with the reliability indicator of the power system, such as occurrence of overload. (Fig. 4)

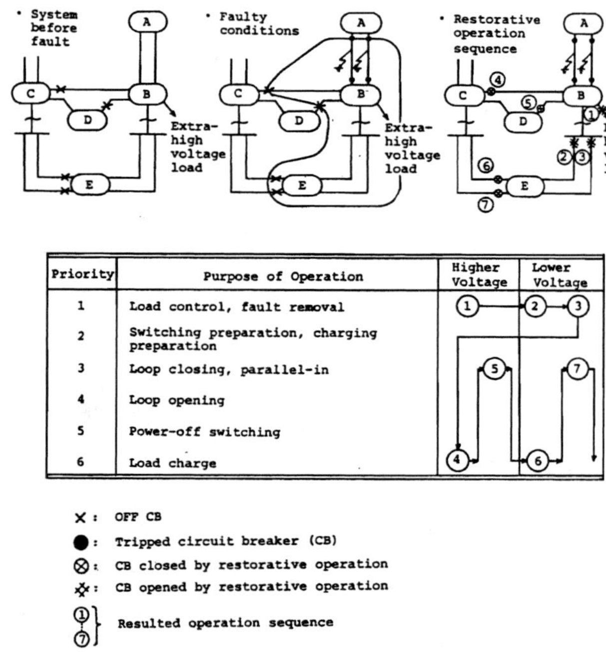

Fig. 4 An Example of the Determination of Switching Device Operation Procedures

3.2 The Architecture of the Knowledge Base System

The knowledge base system in the restoration guidance system consists of four inference engines and a section to process the numerical analysis of power system as shown in Fig. 5.

The restoration guidance control engine, by starting and stopping reasoning on the other three engines, controls the progress of restoration guidance during restorative operation.

If a new fault occurs during restorative operation for example, the restorative operation guidance engine will be stopped and the faulty equipment identification engine will be started. Also, if it is discovered that the circuit breaker cannot be switched, the restoration planning engine will be started. Through these measures, a new plan for restoration is framed, assuring uninterrupted execution of restoration guidance.

Since these three engines are of different types, namely the diagnostic type (faulty equipment identification), the planning type (restoration planning), and the control type (restorative operation guidance), the improvement of reasoning speed and the maintainability of the knowledge bases as a whole is attempted by dividing into engines with individual knowledge base. These knowledge bases are given a common basic structure and common reasoning procedures, thereby allowing them to be consistent and to accept additions and corrections uniformly.

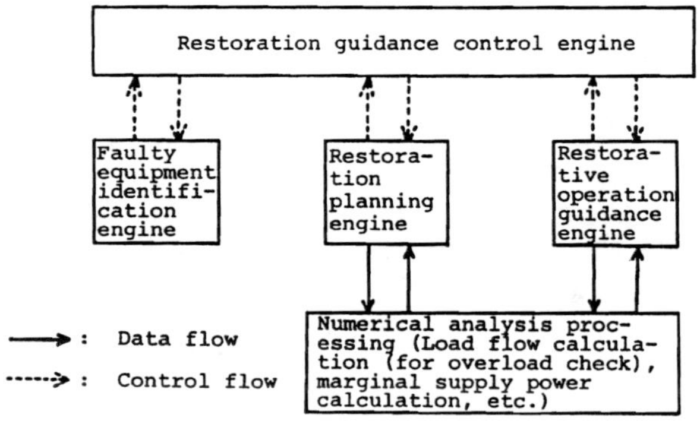

Fig. 5 The Architecture of the Knowledge Base System

3.3 The Basic Structure of the Knowledge Bases

Each procedure in restoration guidance consists of groups of steps and at each step the gathering of online information and reasonings are concentratively accomplished to meet the purpose of the step. It may occur that reasoning at upper steps requires the conclusions and data obtained at lower steps. If this is the case, processing is at once brought to the lower steps and then transferred back to upper steps.

In the present knowledge bases the reasoning processes are structured using the concept "context".[1] Context types vary and each type corresponds to the characteristics and the concept of each step in the restoration guidance.

Since an upper-lower relationship is generated among contexts as the restoration guidance operation proceeds, the contexts are expressed in a tree structure. The contexts in Fig. 6 are expressed by an AND graph. To process context A, for example, it is necessary that the processings of contexts B and D have been completed.

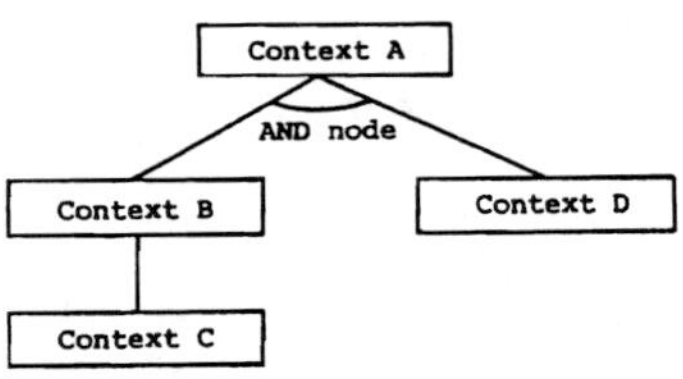

Fig. 6 An Example of a Context Tree

The completion of a context requires the identification of various pieces of information as to the context. These pieces of information are called parameters. An example of a context tree and parameters are as shown in Fig. 7. The subjects of restorative operation guidance are expressed in three contexts, i.e., generation power adjustment, load power control and operation of switching device such as circuit breakers and line switches. The processing of the context "generation power adjustment", for example, is completed when all parameters, i.e., the generator name and the adjusted amount of output, have been determined.

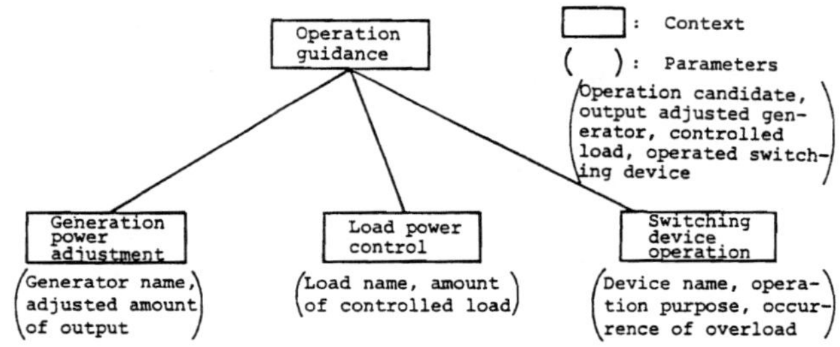

Fig. 7 The Context Tree and Parameters for Restorative Operation Guidance

3.4 Reasoning Procedures

The context trees mentioned above give the contexts involved a static order to determine the parameters for each context.

Reasoning is carried out as the context parameters are determined. In this process a practical context tree as shown in Fig. 8 is dynamically generated from the static context tree as shown in Fig. 7.

The example in Fig. 8 indicates that obtaining operation candidate as a parameter of operation guidance-1, four lower contexts for operation candidates (four in this case), i.e., generation power adjustment, load power control and two switching device operations are created. Then, those restorative operations which need guidance are determined in consideration of system reliability and operational priority among the lower contexts; subsequently the result of determination is transferred to parameters of operation guidance-1, namely output adjusted generator, controlled load and operated switching device.

Since all context parameters are determined at this point, reasoning for one restorative operation guidance ends.

3.5 Knowledge Representation in OPS83

In order to implement the knowledge base structure and the reasoning procedures in OPS83, the representation forms of WM (Working Memory) and PM (Production Memory) below are used in this system.[7]

With these forms used, WMEs (Working Memory Elements) and the rules can be represented in only three types respectively, thereby permitting simple maintenance of the knowledge base.

(1) The construction of the WM

The WM is divided chiefly into the following three types of WME.

a) Context WME: This expresses context parameter settings.

```
<General form>
 (Context   par_ind:  parent context     ]Common to
  name                index              ]all
            sel_ind:  self context       ]contexts
                      index
            parameter name 1
            parameter name 2        = ||(not yet determined)
                 :
                 :              :   ≠ ||(already determined)
            parameter name N   );
```

b) Parameter WME: This stores detailed information about parameters.

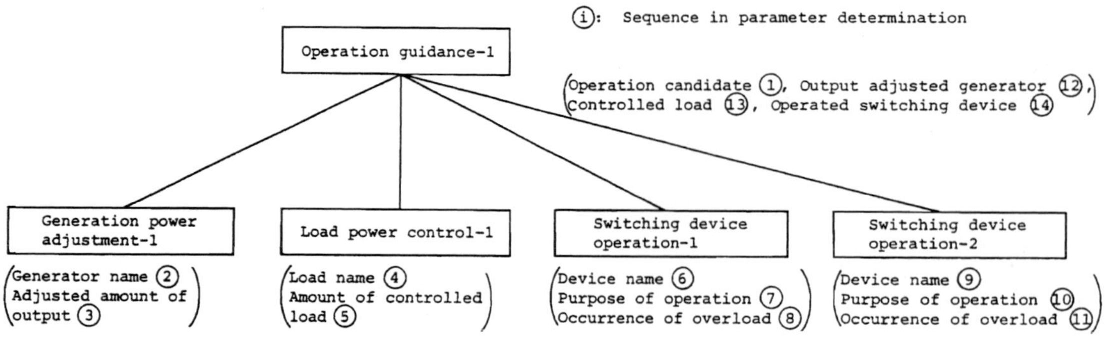

Fig. 8 A Context Tree Generated in a Reasoning

```
<General form>
 (Parameter    con_ind:  index of context      ┐ Common to
  name                   to which the pa-      │ all
                         rameter to be de-     ┘ parameters
                         termined belongs
               attribute 1:  value 1
               attribute 2:  value 2
                   ⋮             ⋮
               attribute N:  value N        );
```

c) Flow control WME: This controls the sequence to determine parameters among contexts.

```
<General form>
 (goal         con_typ:  type of context to which the
                         parameter to be determined
                         belongs
               con_ind:  index of above context
               tra_par:  name of parameter to be deter-
                         mined              );
```

(2) The construction of the PM

The PM is divided chiefly into the following three types of rule.

a) Rule to determine parameters: This determines context parameters.

```
<General form>
 rule    rule name
 {
 &1 (goal  con_typ = context name;          ┐The name of
           tra_par = parameter name  );     ┘parameter to be
                                             determined and
                                             its context
 &2 (context name sel_ind = &1.con_ind;
                  parameter name = || );
                     ⋮
 -->
     modify &2 (parameter name = value of parameter);
                     ⋮
 };
```

Fig. 9 shows an example of the rule to determine the load_name of load_power_control in the context tree in Fig. 7.

```
rule   tracing_load_name_of_load_power_control
{
 &1 (goal  con_typ = load_power_control;
           tra_par = load_name            );
 &2 (load_power_control    sel_ind = &1.con_ind;
                           load_name = ||);
 &3 (operation_guidance    sel_ind = &2.par_ind);
 &4 (operation_candidate   con_ind = &3.sel_ind;
                           object_type = load);
 -->
     modify &2 (load_name = &4.object_name);
};
```

Fig. 9 An Example of the Rule to Determine Parameters

b) Rule to change parameters to be determined: This changes the sequence to determine parameters within a context.

```
<General form>
 rule   rule name
 {
 &1 (goal con_typ = context name;
          tra_par = parameter name);
    (context name   sel_ind = &1.con_ind;
                    parameter name<> ||);
 -->
     modify &1 (tra_par = the name of parameter to be
                          next determined);
 };
```

c) Rule to create a context anew: This creates anew WMEs for a child context and flow control so as to execute depth first search to determine parameters.

```
<General form>
 rule    rule name
 {
 &1 (goal con_typ = parent context name;
          tra_par = parameter name);
 &2 (context name   sel_ind = &1.con_ind;
                    parameter name = ||);
                    ⋮
                                       ┌ Sets a new
 -->                                   ↓ context index.
 make (child context name sel_ind = genatom ( );
                          par_ind = &2.sel_ind;
                             ⋮                 );
                    ⋮
                                       ┌ The newest
                                       ↓ context index
 make (goal           con_ind = &GENATOM;
                      con_typ = type of child context;
                      tra_par = parameter name of
                                child context);
 };
```

4. AN EXAMPLE OF RESTORATION GUIDANCE EXECUTION IN A SIMULATED FAULT

4.1 Simulated Fault

The power system, having the 77 kV/33 kV network under the control of the Kobe Dispatching Center, includes some 120 power stations/substations, major extra high-voltage loads and about 80 power transmission routes.

As shown in Fig. 10, it was assumed that a fault occurred to a line (AB1L or AB2L) in the transmission from As/s to Bs/s, causing blackout of Bs/s and Cs/s Substations.

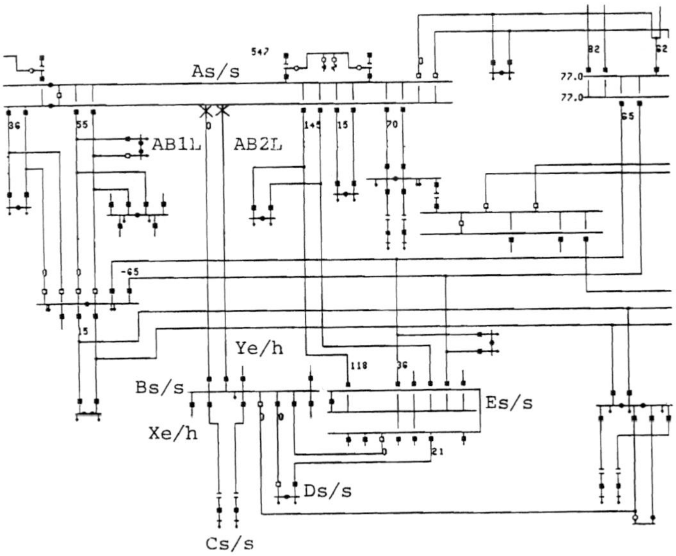

X : Tripped CB
s/s: Substation
e/h: Extra-high voltage load

Fig. 10 Simulated Fault

4.2 An Example of Restoration Guidance Execution

Fig. 11 indicates the power system state immediately after the fault when information about the fault is generated by a simulator and input into the present system. Fig. 12 shows the relevant messages.

Figs. 13 to 14 show the contents of operation guidance during the restoration operation.

Fig. 15 shows the contents of operation guidance after CB03 at Es/s is closed for load charging and the power system state after restoration.

The processing time from faulty equipment identification to restoration planning completion was approximately 50 seconds. One restorative operation guidance session lasted approximately 5 seconds.

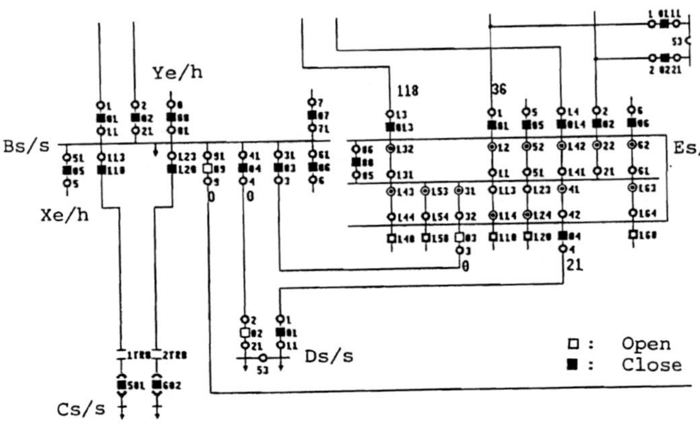

Fig. 11 Power System State Immediately after the Fault

** Identifying Faulty Equipment ... **

AB1L phase fault occurred
Bs/s CB01 failed to open

AB2L phase fault occurred
Bs/s CB02 failed to open

[Fault information]

8/20	13:10	As/s	CB083	MRy, BRy
8/20	13:10	As/s	CB084	MRy, BRy
8/20	13:10	Bs/s		MRy

[Amount of Outage Load]

Xe/h	38 MW
Ye/h	15 MW
Bs/s	16 MW
Cs/s	21 MW

** Making Restoration Plan ... **

(1) System switching

Bs/s
As/s -> Es/s 16 MW switching

(2) Load control

Xe/h 38 MW
Ye/h 15 MW

Fig. 12 Messages Immediately after the Fault

[Restorative Operation Guidance]

Fault removal	Bs/s	CB01	OFF
Fault removal	Bs/s	CB02	OFF
Load control	Bs/s	CB05	OFF
Load control	Bs/s	CB06	OFF
Load control	Bs/s	CB07	OFF
Load control	Bs/s	CB08	OFF

Fig. 13 Operation Guidance for Removing Fault, Controlling Load

[Restorative Operation Guidance]

Load charge	Es/s	CB03	IN	Confirmation of current value
Load charge	Ds/s	CB02	IN	Confirmation of current value

Fig. 14 Operation Guidance for Charging Load

[Restorative Operation Guidance]

Loop closing	Ds/s	CB02	IN	phase angle difference -2.0°

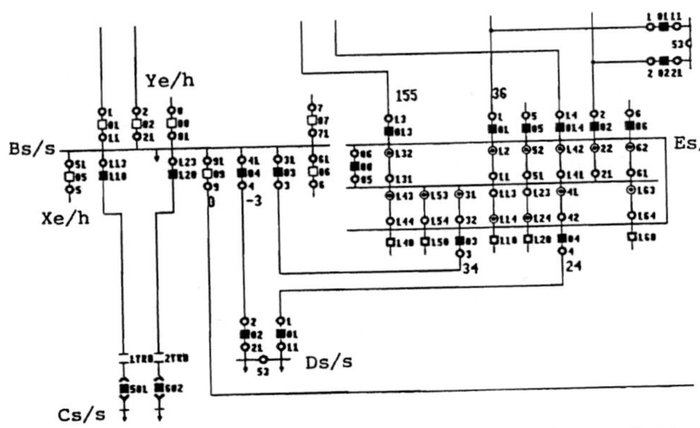

Fig. 15 Operation Guidance for Loop Closing and the Power System State after Restoration

5. CONCLUSIONS

This paper has introduced a real-time expert system now undergoing considerable fault case for restoration guidance. The progress so far indicates a hopeful future for a quick and accurate fault restoration support system. The system will continue to be thoroughly tested in the field until it can be introduced into practical service.

Areas of future study include:

(1) Improvement of the man-machine interface system through field tests

(2) Design of a practicable system which integrates SCADA and the entire restoration guidance system (ex. centralized or decentralized)

(3) Development of program tools for maintaining and handling of acquired knowledge bases

REFERENCES

[1] T. Sakaguchi et al., "Development of a Knowledge Based System for Power System Restoration," IEEE Transactions on Power Apparatus and Systems, Vol. PAS-102, No. 2, pp. 320-329, February 1983.

[2] F. F. Wu et al., "Analytical Tools for Power System Restoration - Conceptual Design," IEEE PES 1986 Winter Meeting, No. 86 WM 105-1, February 1986.

[3] C. C. Liu et al., "An Expert System Operational Aid for Restoration and Loss Reduction of Distribution Systems," Proceedings of IEEE 1987 PICA Conference, pp. 79-85, May 1987.

[4] K. Tomsovic et al., "An Expert System as a Dispatchers' Aid for the Isolation of Line Section Faults," IEEE Transactions on Power Delivery, Vol. PWRD-2, No. 3, pp. 736-743, July 1987.

[5] N. Koike et al., "A Real-time Expert System for Power System Fault Analysis," IASTED International Conference on High Technology in the Power Industry, pp. 376-380, 1986.

[6] C. C. Liu et al., "Development of an Expert System as Power System Operational Aid," IEEE International Workshop on Artificial Intelligence for Industrial Applications, pp. 70-75, May 1988.

[7] C. Forgy, "OPS83 Users Manual," Computer Science Department, Carnegie-Mellon University, 1983.

GUIDING A POWER SYSTEM RESTORATION WITH AN EXPERT SYSTEM

Daniel S. Kirschen
Empros Systems International
Minneapolis, Minnesota

Terry L. Volkmann
Northern States Power Company
Minneapolis, Minnesota

Abstract - Our civilization has become so dependent on a steady supply of electric power that a blackout bears an enormous economic cost. Minimizing the amount of time needed to restore a power system helps reduce this cost and the public's resentment over the interruption of service. An expert system can help the operator achieve this goal by providing actualized restoration plans, dependable recommendations and a clear picture of the situation.

A hierarchical structure has been adopted for the prototype expert system described in this paper. This structure separates the selection of the next objective (strategic reasoning) from the development of a plan showing how this objective can be achieved (tactical reasoning). This separation enables the system to quickly adapt its recommendations when a problem arises during the restoration.

Planning connection paths through substations of widely different configurations is greatly facilitated by the adoption of a hierarchical description of the topology of these substations. This knowledge representation scheme is the symbolic equivalent of the one-line diagrams used by the operators.

Keywords: power system restoration, power system operations, expert systems, knowledge representation.

Introduction

Although electric utilities constantly strive to maintain the stability and the integrity of their power system, absolute security can never be guaranteed. A catastrophic sequence of events that defeats the preventive measures and overcomes the corrective actions will occasionally unfold and cause a collapse of the entire power system. If such a blackout does occur, electrical service must be restored as soon as possible to minimize the economic loss, the inconvenience to the public and the direct cost to the utility. At the same time, this restoration must proceed carefully to avoid endangering lives, damaging equipment or jeopardizing the progress already accomplished in the re-energization.

While performing the numerous operations required to rebuild the power system following a blackout, the operator must be attentive to many unusual voltage, line flow, stability and resource constraints. It is important to note that most operators have probably never had to work with the system when it is so far removed from its normal state. References [1] to [5] present very interesting descriptions and discussions of the problems which commonly arise during restorations.

90 SM 341-8 PWRS A paper recommended and approved by the IEEE Power System Engineering Committee of the IEEE Power Engineering Society for presentation at the IEEE/PES 1990 Summer Meeting, Minneapolis, Minnesota, July 15-19, 1990. Manuscript submitted January 31, 1990; made available for printing June 21, 1990.

Many utilities have concluded that improvising a restoration plan under the stress of a blackout situation could lead to unacceptable delays. After a careful study of their individual situation, these utilities usually develop one or more written restoration procedures reflecting the most effective ways of restoring the power system [6-12]. Although it is commonly agreed that carefully prepared procedures considerably simplify the restoration, these procedures have serious limitations and the development of more powerful methodologies has been suggested [2].

The main disadvantages of a written restoration procedure derive from its static nature. It is most likely that the status of the power system following the blackout will differ from the postulated conditions used to develop the plans. Developing separate procedures for different initial conditions alleviates but does not solve the problem. Difficulties almost always arise during a restoration and it is clearly impossible to prescribe alternatives for every conceivable problem. At some point, the operator in charge of the restoration is thus likely to be forced to stray without guidance from the prepared plan. Such detours can be large or small but their preparation always takes time and their implementation increases the probability of errors.

Another disadvantage of written restoration procedures is the amount of time required to reflect modifications in the power system. It is interesting to note that a review of 48 major disturbances lists outdated procedures as the second leading cause of restoration problems [2].

The format of conventional restoration procedures also becomes a hindrance if the procedure cannot be followed as written. Textual information is intended to be read sequentially and does not lend itself to other modes of consultation. Following a written decision tree forces the reader to jump back and forth in the text, querying a written document is absurd, and switching to another level of detail is impossible.

Finally, this textual format and the paper medium commonly used to record these procedures force the operator to enter the restoration instructions into the control computer. Transferring information manually is obviously a slow and error-prone process.

Sakaguchi and Matsumoto [13] were the first to suggest that an expert system could be used for planning switching operations. Since then this concept has been extended to the restoration of distribution systems [14] and generalized to the creation of optimal switching sequences in substations [15]. Other authors [16, 17] have proposed frameworks for the development of expert systems applicable to the restoration of high voltage networks. Reference [18] shows how an expert system has been used to assist in the automation of the restoration of a small power system.

The restoration of a power system is often described as a two-phase process. During the first phase, a skeleton network linking the generating stations and the main substations is re-energized. This energization is then expanded to the rest of the network and customers are reconnected as power becomes available.

This paper describes the Restoration Assistant, a prototype expert system designed to reduce the amount of time required to complete

Reprinted from *IEEE Transactions on Power Systems,* Vol. 6, No. 2, pp. 558–566, May 1991.

the first phase of the restoration of a large power system. This reduction is achieved by:

- presenting restoration plans that reflect the actual state of the power system to the operator;
- relieving the operator from the more mundane tasks of the restoration to let him or her concentrate on more complex activities and on tasks which require human interactions;
- providing a clear picture of the status of the restoration to facilitate the decision-making process.

This expert system was developed on the basis of the newly-defined restoration procedures of Northern States Power Company and may reflect some of this utility's restoration philosophies. In particular, these procedures do not call for the simultaneous creation of several restoration "islands" as is practiced by other utilities. Consequently, the problems related to the existence of separate islands have not yet been studied in detail. However, every effort has been made to keep the architecture flexible and the rules generic to ensure the applicability of this system to a wide range of situations.

Description of the System

In artificial intelligence terminology, the restoration of a power system is classified as a planning problem. These problems have traditionally been solved by determining a goal state and searching heuristically for a sequence of actions leading from the current state to this goal state [19, 20]. This sequence of actions constitutes the plan. If a problem surfaces during the execution of a plan, which is very likely in the case of a restoration, it must be discarded and the time-consuming planning process must be repeated. This approach is thus not practical in situations where the environment is unpredictable and uncertain.

However, the restoration can be decomposed into a set of goals or objectives. Examples of objectives which must be achieved during the first phase of the restoration include the supply of cranking power to each generating station and the creation of backup links to each of these stations. It is then preferable to plan only as far as the completion of the next objective. Using this approach, the amount of planning which must be repeated is much smaller if a plan fails at execution. Furthermore, if the expert system detects that, due to the current conditions, no feasible plan can be found for a given objective, this objective can be temporarily replaced by another.

This decomposition also creates a clear and useful separation between strategic and tactical matters. At the strategic level, the expert system suggests to the operator which objective appears to be the most urgent and broadly outlines the different ways of reaching it. At the tactical level, the expert system prepares a detailed plan showing how a given objective can be achieved following a given approach. In the remainder of this paper, strategy thus refers to the decisions which influence the global outcome of the restoration while tactic pertains to the decisions which affect only the success of a particular objective.

Once a plan has been reviewed by the operator, the expert system can take care of issuing the commands necessary to implement it. If a problem develops during the execution of these commands, the expert system will notify the operator and suggest an alternative.

The expert system consists thus of three modules (strategic planning, tactical planning and execution control) and a graphical human interface. A detailed description of each of these modules is presented in the following sections.

Strategic Planning

The identification of an internal or external source of power is the first step in restoring a power system. Once this source has been activated, through a blackstart or an agreement with a neighboring utility, it is used to supply the auxiliaries of other power plants and make possible their restart. Backup paths between the energized stations must also be established to reduce the probability of a catastrophic set-back.

For the purpose of strategic planning, the expert system decomposes the creation of this skeleton network into a set of objectives. Each of these objectives corresponds to the creation of an energization or backup connection between two stations. While the list of objectives is given, the order in which they are achieved is not imposed. Except for the first two or three steps following the activation of the initial source of power, there are usually several objectives that can be reached directly and often several ways of reaching them.

The expert system keeps track of which stations have been energized and which backup paths have been established. The strategic level module also relies on a predefined set of paths which have been determined through engineering studies as being suitable corridors for the restoration. A path is either a single line or a series of lines and minor substations linking two major substations. Once a substation has been energized, the paths that emanate from it become feasible. If such a feasible path leads to a de-energized substation, the energization of this substation becomes a reachable objective. If this objective was already reachable, the newly feasible path is recorded as an alternative way of reaching this objective. Similarly, if a feasible path links two energized substations, it provides a way of creating a backup path between these two stations. A path is not considered to be feasible if it comprises a faulted line, a substation whose communication system has failed, or a substation that has been disabled by a major incident (e.g. a fire.)

The predefined paths are not a detailed description of how a connection can be established between two substations. They are defined solely in terms of lines and stations and their sole purpose is to avoid having to consider too many details when searching for a possible approach to an objective.

The feasible paths and the corresponding objectives are displayed on a high level diagram of the power network. The operator can thus decide which objective will be undertaken next based on a clear display of the options. If a problem arises in the execution of an objective, the operator can temporarily abandon it, return to the high-level diagram and select a substitute objective.

If the restoration is managed from a central location, all the reachable objectives cannot be worked on at the same time and the most urgent one must be selected. The relative urgency of the various objectives depends in part on their intrinsic importance which can be assessed off-line. However, this urgency is also affected by time-dependent factors which can change drastically during the restoration. For example, the priority assigned to the reenergization of a particular station may be high until the expiration of the deadline for a hot restart. On the other hand, the priority given to the supply of auxiliary power to a nuclear plant would increase dramatically if the emergency generators of this plant were to fail. Upon request from the operator, the strategic planning module will reassess the time dependent factors pertaining to each reachable objective and combine them with their static importance to establish a list of priorities. While all the reachable objectives are shown to the operator, the most urgent objective is brought to his or her attention through a special highlighting. The knowledge encoded in the expert system currently is capable of only a rudimentary priority assignment. More knowledge must be acquired to refine these recommendations.

The final decision as to which objective to pursue and which path to use rests with the operator and is communicated to the expert system by pointing and clicking on the graphical representation of the objective. The strategic planning module records the selected objective and the chosen path and passes this information to the tactical planning module which is thus assigned the task of establishing a detailed list of actions which must be carried out to achieve this objective. Possible actions include opening and closing breakers, checking the value of a measurement or calling another human operator to request the execution of a particular operation.

It must be emphasized that at the strategic level, the expert system reasons only about abstract concepts (such as objectives and tasks) and high level objects (such as stations and paths).

Figure 1 illustrates the type of information provided by the strategic planning module. In this example, several objectives can be reached from the substations already energized. Energizing the station BDS is recommended by the expert system as the most urgent objective. However, the operator could instead choose to energize the stations BLL, HBR, ASK or WHT or to create a backup connection between the substations RRK and IVH via the substation CGR.

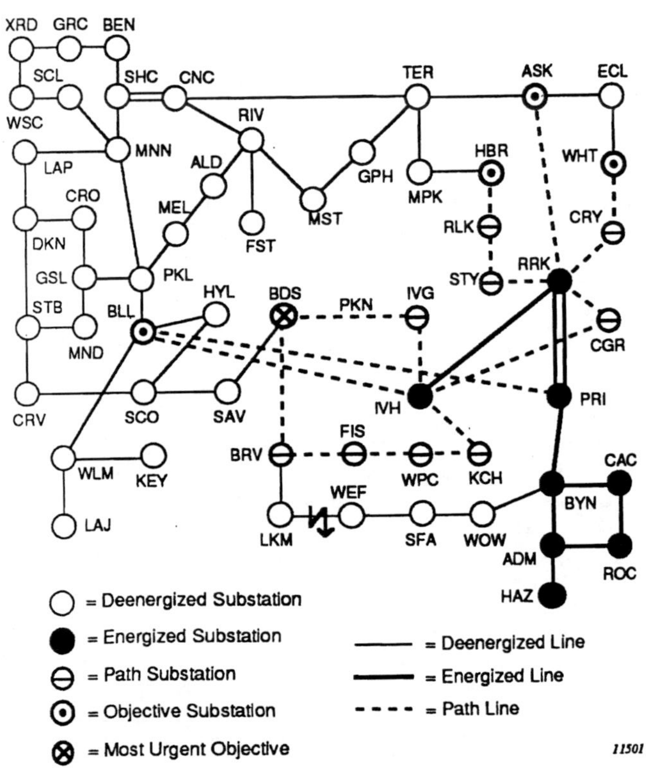

Figure 1. An Example of Strategic Information and Recommendations Provided by the Restoration Assistant.

The expert system suggests two paths for the energization of BDS. The first goes through IVG and PKN while the second goes through KCH, WPC, FIS and BRV. Once the energization of BDS is achieved, the path which was not chosen will be recommended for a backup connection. The path between BYN and BDS is not suggested by the expert system because a fault on the line between LKM and WEF makes it infeasible.

Figure 1 is a black and white approximation of one of the dynamic color displays that have been designed for the expert system. On the actual displays, the user can switch to another display, trigger a planning task, or start the execution of a plan simply by pointing and clicking at the appropriate location.

Tactical Planning

The tactical planning module is responsible for translating into a list of actions the tasks defined by the strategic planning module. A task usually prescribes the connection of two stations via a path. A path is either a single line or a sequence of lines and intermediate substations connecting the two stations. Each task is decomposed into subtasks, one for each substation affected by the connection. Each subtask thus prescribes the creation of a connection between two points within a single substation. These two points are:

- the incoming and outgoing lines for an intermediate substation;
- the outgoing line and the energized bus closest to this line for an energized station;
- the incoming line and the auxiliaries bus for a deenergized station.

Tactical planning consists therefore primarily of determining the switches that must be opened or closed to connect two points of a substation.

The knowledge base used for tactical planning is complete if it covers all substation configurations and any pair of points in any substation. Furthermore, the presence of malfunctioning breakers or faulted busses may not prevent it from producing a feasible connection plan if such a plan exists.

A tactical plan is correct if it is equivalent to what an experienced operator would do under the same circumstances. In general, this implies avoiding the energization of unrelated lines and minimizing the number of switching operations. However, this minimization is not carried out with a short term perspective. Some extra switching operations are often performed before they are actually needed to facilitate future connections and reduce the number of operations needed in the long run.

In addition to the traditional criteria of completeness and correctness, the tactical planning module must also be efficient. It is indeed likely that the expert system would be rejected by the operators if it took more than a few seconds to construct a plan.

The development of an expert system which satisfies these requirements has been greatly facilitated by the design of a new knowledge representation scheme. The topology of a substation is commonly defined for algorithmic applications by specifying the components to which each device is directly connected. For expert systems, this approach is not practical because it does not provide a global view of the substation. Planning a connection path based on this information requires a search analogous to finding one's way in a city by relying strictly on the street signs at the end of each block. Such a search is feasible [15] but very different from the mental process used by an operator to plan a connection path. It is easier and more intuitive to rely on a "symbolic map" of the substation. A hierarchical knowledge representation scheme has been designed to provide this map to the tactical planning module.

The main features of this scheme will be explained using the simplified diagram of the RRK substation shown on Figure 2. This station is divided into two levels separated by transformers (RRK-345 and RRK-115). Each level consists of busses (e.g. 115-TOP and 115-BOT for RRK-115, 345-TOP and 345-BOT for RRK-345) and branches (A, B,C, E and F for RRK-345, E,F, G, H, I, J and K for RRK-115). A branch is defined as one, two, or three switches connected in series. Branches belong to different classes depending on how many switches they contain and whether or not they include a transformer. For example, C is a three-switch branch while E is a two-switch-transformer branch.

If both ends of a branch are in the same level, the branch is an internal branch of this level (e.g. A , B and C for RRK-345, G, H, I, J and K for RRK-345). If the ends are at different levels (e.g. E, F) it is an external branch of each level. A branch is also external if one of its ends is connected to a line.

Several attributes are defined for each branch. For example, the definition of branch J is:

from-end:	115-TOP	(a bus)
to-end:	115-BOT	(a bus)
first-switch:	J1	(a breaker)
first-midpoint:	E	(a branch)
second-switch:	J2	(a breaker)
second-midpoint:	RRK-CGR	(a line)
third-switch:	J3	(a breaker)

Similarly, the definition of branch E is

from-end:	345-TOP	(a bus)
to-end:	J	(a branch)
first-switch:	E1	(a breaker)
first-midpoint:	TR-1	(a transformer)
second-switch:	E2	(a disconnect)

The line RRK-CGR has the following definition:

from-station:	RRK	(a station)
to-station:	CGR	(a station)
from-end:	J	(a branch of RRK)
to-end:	Z	(a branch of CGR)

Some operations which must be repeated often when determining a path through a substation become very simple using this scheme. For example, the definition of line RRK-CGR indicates that this line has a terminal at station RRK and that this terminal is on branch J. The definition of branch J shows that this terminal is separated from bus 115-BOT by only one breaker. 115-BOT is thus the "closest bus" to line RRK-CGR at station RRK and the energization of this line will usually require the energization of that bus.

This scheme is applicable not only to breaker-and-a-half type substations but also to ring-bus and other types of substations. Since a unified representation of all substations is now available, a single method can be used to plan a connection path between two points in any substation. This method proceeds as follows:

- If the two points are in different levels, the problem is first decomposed into the selection of an external branch linking the two levels and the planning of a path in each level.
- The location of the endpoints of each path are then pinpointed using the branches where they are located and the closest busses.
- Using this information, the type of connection is determined. For example, if the two endpoints are closest to the same bus, a connection will be established along this bus. Otherwise, a more complex connection type is suggested. The rule base is organized in such a way that the most desirable types of connection are considered first. If a path would involve a faulted bus or a breaker that is blocked open, it is rejected and another path is determined.
- Finally, a list of all the switching operations required to create this path is constructed on the basis of the selected connection type and the status of the substation.

Example 1:

Suppose that it has been decided to energize the station CGR from the station IVH and through the substation RRK. A plan showing how to connect the points labeled IVH and CGR on Figure 3 must be developed. These points being in different levels, the expert system begins by identifying an external branch linking these two levels. Branch F is chosen because it links directly the busses closest to the endpoints. The problem is thus decomposed into three subproblems: closing all the switches of branch F, connecting branch F to the line to CGR and connecting branch F to the line to IVH. The first subproblem is trivial. For the second, it is determined that the connection should be made on the bus 115-BOT while for the third it should be made on the bus 345-BOT. Figure 3 summarizes the operations required to create this path.

Example 2:

Suppose that a connection between the same two endpoints is desired but that it is known that breaker F1 is blocked open. The expert system determines that branch F cannot be used and thus chooses branch E to link the two levels. The line to IVH must therefore be connected to the bus 345-TOP. This path is inferior to the one constructed in the first example because it requires the energization of the line to PRI. Figure 4 summarizes the operations required in substation RRK to create this path. The system also determines which breakers must be opened in station PRI.

Figures 3 and 4 are only black-and-white approximations of dynamic color displays. On the actual displays, the operator can modify a plan simply by pointing and clicking on the desired switches. If the planned connection involves a neighboring station, clicking on the name of that station brings up a similar diagram which highlights the operations required in that station.

It must be emphasized that the knowledge used to develop the tactical plans is purely generic. It reflects only standard operating procedures and does not include any information specific to a particular substation. Therefore, the knowledge base does not need to be updated each time the configuration of a substation is modified.

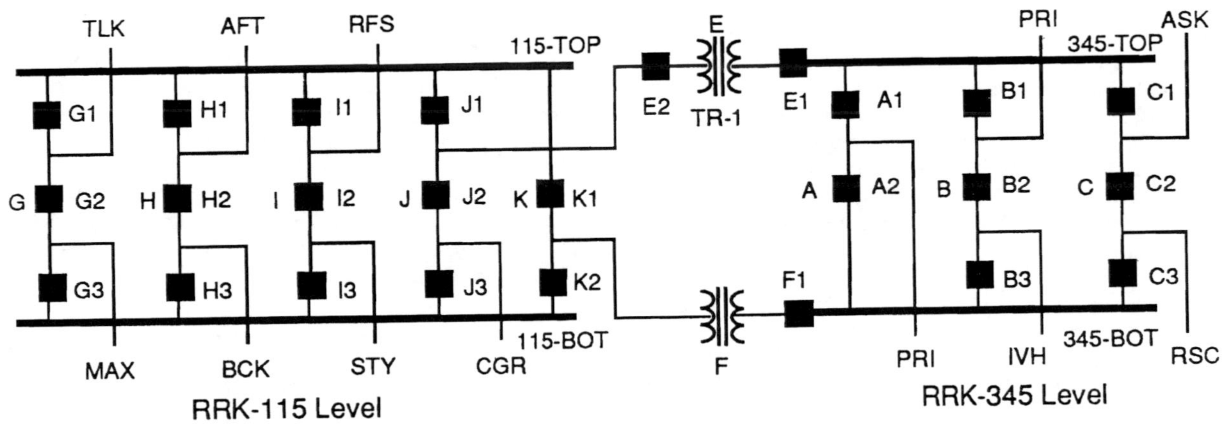

Figure 2. Description of the Substation RRK using the Knowledge Representation Scheme of the Restoration Assistant.

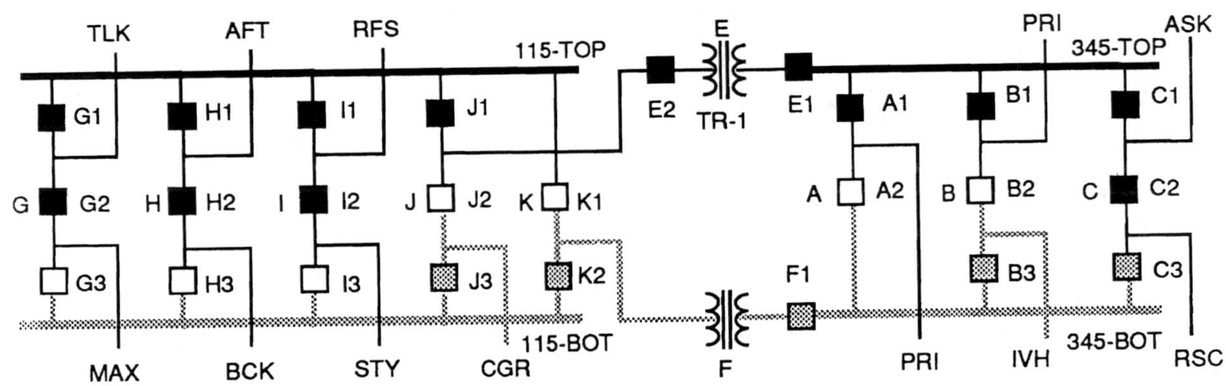

Figure 3. Tactical Plan for the Condition of Example 1.

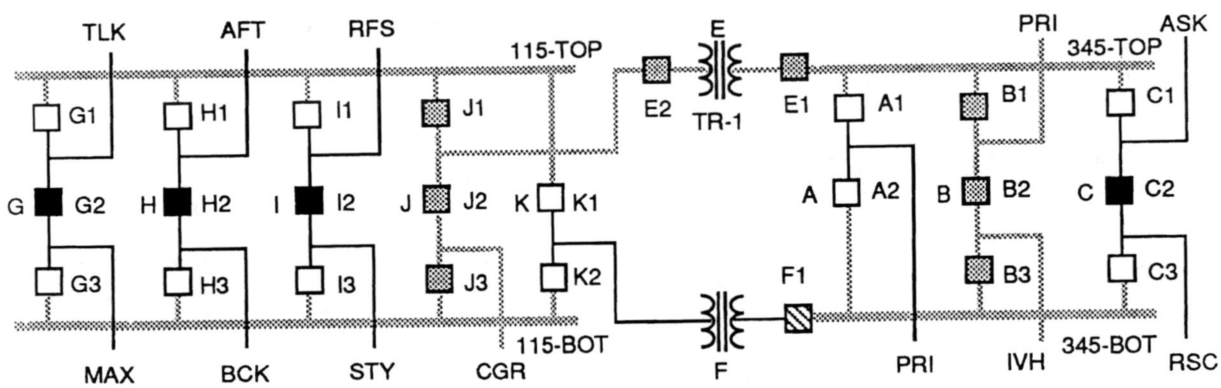

Figure 4. Tactical Plan for the Condition of Example 2.

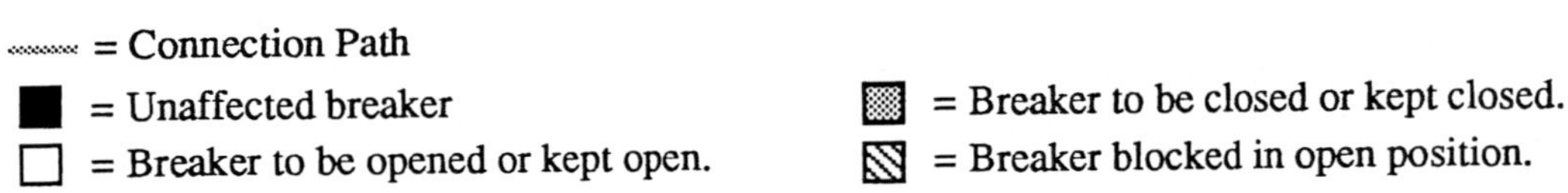

11502

Execution Control

Once a plan has been reviewed and approved, the operator has the option of letting the expert system take care of its execution. Performing switching operations manually is indeed a slow process. The one-line diagram of each substation must be displayed, the cursor must be positioned on the appropriate switch, the switching command must be issued and confirmed. A substantial amount of time can clearly be saved by automating these manual operations.

If the responsibility for implementing the switching commands is delegated to the expert system, the amount of time required to issue an extra command is very small. The expert system can thus be used to execute commands that might be neglected when the switching is performed manually. For example, issuing an "open" command to a breaker which is already open resets the automatic reclosure mechanism which might have been activated by a relay during the system collapse. If this mechanism is not reset, the breaker may close automatically when one of its sides is energized depending on the last breaker operation which took place before the blackout. This closure could reconnect a significant amount of load and cause a major set-back in the restoration.

In order to properly schedule the execution of the switching operations, the expert system monitors the completion of the commands which have already been issued. If the event processor of the SCADA system indicates that a command did not produce the expected result, the expert system brings this fact to the attention of the operator and attempts to develop an alternate plan. If the failure is due to a switch blocked in the closed position, a small modification to the plan can usually be found. On the other hand, if the failure is due to a switch blocked in the open position, the entire plan may have to be modified.

Implementation and Tests

The prototype has been implemented as a rule-based expert system on a Symbolics 3675. Its knowledge base was developed using the PROTEUS expert system shell [21] and currently contains approximately 350 rules. A significant portion of the Northern States Power system has been modelled to validate the expert system.

Each module was first tested separately. In particular, the tactical planning rules were checked on different substation configurations. A large number of paths were generated within each substation and verified by comparing them to standard operating practices. If the expert system was informed of the unavailability of some devices, it generated alternate paths which were carefully examined and found acceptable.

The various modules of the expert system were then integrated to simulate the restoration of the NSP system. If no equipment trouble was introduced, the expert system produced a plan identical or equivalent to the written procedures.

Planning the energization of a station along a path which crosses three large substations typically takes about five seconds. A particularly complex connection or the presence of known equipment problems may require a slightly longer time. Planning a restoration path on-line is therefore not only theoretically feasible, but can also be done in a sufficiently short time to be practical and acceptable to an operator.

Conclusion

Reducing the amount of time required to restore a power system following a blackout is the ultimate goal of the expert system described in this paper. Each of the following features of this expert system contribute to this reduction:

- The plans generated by the expert system reflect the current configuration of the substations, the true status of the power system, and the known equipment problems.
- Some operations which are usually performed manually can be delegated to the expert system. The expert system not only performs these tasks faster than the operator, but the time saved can be used by the operator for more complex duties or for functions requiring human interactions.
- Since the plans generated by the expert system are already in an "electronic" format, the amount of information which must be manually entered is minimal and the probability of error is considerably reduced.
- The expert system reasons and interacts with the operator at both the system level and the equipment level. This feature allows the operator to separate minor difficulties from major decisions.

Experiments with the prototype system have shown that it is capable of preparing and displaying complete and correct tactical plans much faster than an experienced engineer could point on a substation diagram to the breakers that should be operated.

The Restoration Assistant is intended to supplement, not replace, the design of a restoration strategy for each utility. A significant amount of knowledge is acquired from the results of these studies and implicitly incorporated in the expert system under the form of restoration objectives, paths, and priorities. Inserting this knowledge explicitly in the expert system would further increase its power and flexibility.

References

[1] M. M. Adibi et al. "Power System Restoration - A Task Force Report," IEEE Transactions on Power Systems, Vol. PWRS-2, No. 2, pp. 271-277, May 1987.

[2] M. M. Adibi et al. "Power System Restoration - The Second Task Force Report," IEEE Transactions on Power Systems, Vol. PWRS-2, No. 4, pp. 927-933, November 1987.

[3] M.M. Adibi, D. Scheurer, "System Operations Challenges: Issues and Problems in Power System Restoration," IEEE Transactions on Power Systems, Vol. PWRS-3, No. 1, pp. 123-124, February 1988.

[4] W.A. Johnson et al., "System Restoration - Deploying the Plan," IEEE Transactions on Power Apparatus and Systems, Vol. PAS-101, No.11, pp. 4263-4271, November 1982.

[5] U.G. Knight, "Aids for the Emergency Control of Power Systems," CIGRE Electra, No. 67, pp. 101-134, December 1979.

[6] R.J. Kafka, D.R. Penders, S.H. Bouchey, M.M. Adibi, "System Restoration Plan Development for a Metropolitan Electric System," IEEE Transactions on Power Apparatus and Systems, Vol. PAS-100, No.8, pp. 3703-3713, August 1981.

[7] R.J. Kafka, D.R. Penders, S.H. Bouchey, M.M. Adibi, "Role of Interactive and Control Computers in the Development of a System Restoration Plan," IEEE Transactions on Power Apparatus and Systems, Vol. PAS-101, No.1, pp. 43-52, January 1982.

[8] D. Scheurer, "System Restoration at Philadelphia Electric Company", paper presented at the IEEE PES Workshop on Real-Time Monitoring and Control of Power Systems, Montreal, Canada, October 10-12, 1984.

[9] S. Peach, "Hydro-Québec System Restoration Synthesis," paper presented at the IEEE PES Workshop on Real-Time Monitoring and Control of Power Systems, Montreal, Canada, October 10-12, 1984.

[10] P.F. Arnold, "Summary of System Restoration Plan for the Pacific Northwest Power System," paper presented to the Power System Restoration Task Force at the 1982 IEEE Winter Power Meeting.

[11] E.J. Simburger, F.J. Hubert, "Low Voltage Bulk Power System Restoration Simulation," IEEE Transactions on Power Apparatus and Systems, Vol. PAS-100, No.11, pp. 4479-4484, November 1981.

[12] E. Mariani, F. Mastroianni, V. Romano, "Field experience in Reenergization of Electrical Networks from Thermal and Hydro Units," IEEE Transactions on Power Apparatus and Systems, Vol. PAS-103, No.7, pp. 1707-1713, July 1984.

[13] T. Sakaguchi, K. Matsumoto, "Development of a Knowledge Based System for Power System Restoration," IEEE Transactions on Power Apparatus and Systems, Vol. PAS-102, No.2, pp. 320-329, February 1983.

[14] C.C. Liu, S.J. Lee, S.S. Venkata, "An Expert System Operational Aid for Restoration and Loss Reduction of Distribution Systems, " IEEE Transactions on Power Systems, Vol. 3, No. 2, pp. 619-626, May 1988.

[15] Z.Z. Zhang, G.S. Hope, O.P. Malik, "A Knowledge-based Approach to Optimize Switching in Substations," IEEE Transactions on Power Delivery, Vol. 5, No. 1, January 1990.

[16] T. Kojima et al. "Restoration Guidance System for Trunk Line System," IEEE Transactions on Power Systems, Vol. 4, No. 3, pp. 1228-1235, August 1989.

[17] I. Takeyasu et al. "An Expert System for Fault Analysis and Restoration of Trunk Line Power Systems," in Proceedings of the First Symposium on Expert Systems Applications to Power Systems, Stockholm-Helsinki, August 22-26, 1988, pp. 8-24 to 8-31.

[18] L.R. Blessing, C.K. Bush, S.J. Yak, "Automated Power System Restoration Incorporating Expert System Techniques," in Proceedings of the Second Symposium on Expert Systems Applications to Power Systems, Seattle, July 17-20, 1989, pp. 133-139.

[19] R.E. Fikes, N.J. Nilsson, "STRIPS: A New Approach to the Application of Theorem Proving to Problem Solving," Artificial Intelligence, Vol. 2, No. 3/4, pp. 189-208, 1971.

[20] E. D. Sacerdoti, "Planning in a Hierarchy of Abstraction Spaces," Artificial Intelligence, Vol. 5, No. 2, pp.115-135, 1974.

[21] C. Petrie, D. Russinoff, D. Steiner, "PROTEUS: A Default Reasoning Perspective", in Proc. 5th Generation Conference, National Institute of Software, October 1986.

Discussion

M. M. ADIBI Application of expert systems for assessing the status of distribution systems and determining the appropriate switching operations following major disturbances has been explored in [13,14 & 15] and the authors paper is a significant contribution in this area. The restart and reintegration of bulk power supply however, requires a combination of expert systems and analytical tools so that during the "strategy" and "tactical" planning, the following concerns would be properly addressed [3,16 & 17]:

1. energizing large sections of transmission lines within the acceptable transient and sustained over voltages without the risk of generator self excitation,
2. load pick-up in large increments without the risk of frequency decline and recurrence of outage,
3. maintenance of steady state and transient stabilities as system is being restored and when impedances are large,
4. reduction of standing angles when closing loops to firm up transmission paths, and
5. coordination of thermal plant start-up timings with load pick-ups to bring generators to their stable minimum levels and within the range of major analog controllers.

The existing analytical tools are large, comprehensive and inefficient for on-line strategy and tactical planning. Does the authors approach consider the above and other power system constraints and what existing or new analytical tools are used in planning the "skeleton" network?

Under Strategic Planning it is stated that the "strategic level modules" rely on a predefined set of paths which have been determined through engineering studies. This contradicts the statement under Introduction that the "status of the power system following the blackout will differ from the postulated conditions used to develop the plans". Could the authors clarify these two apparently contradictory statements?

In order to avoid losing critical times such as hot restart of thermal plants and to take advantage of several sources of black start capabilities within the power system, generally two or more subsystems are restored in parallel. Could the authors clarify Northern State Power's philosophy in adopting sequential restoration?

While restoration procedure has a definite beginning, its completion is not so clear cut. Therefore, minimizing the restoration duration as the goal is not as meaningful as maximizing the restoration of service mega-watt-hour as shown in [2]. The curves are reproduced here for convenience. It can be seen that both scenarios or strategies reach the 1600 Mw load level 10 hours after the start of restoration.

However, during the 10 hours, Scenario I serves 17.4 GWH or 42% more than 12.3 GWH for Scenario II. Would the authors comment on the effect of change in goal on their strategy planning?

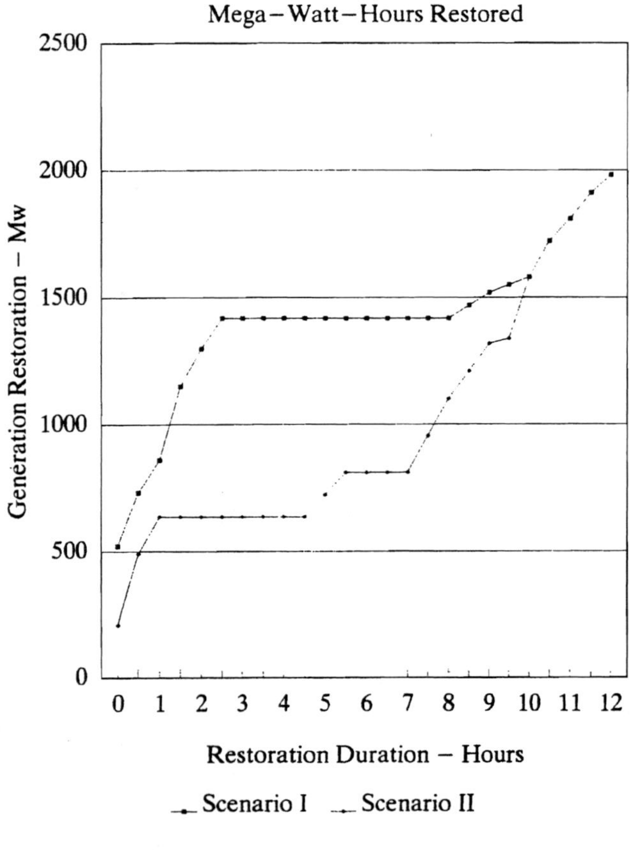

Manuscript received July 27, 1990.

D.S. Kirschen, T.L. Volkmann: The interest that Mr. Adibi shows for our work and his encouraging comments are very gratifying. His remarks highlight some of the areas which we will have to investigate in more details as we continue the development of our expert system. In particular, it is clear that the use of some analytical tools would increase the confidence of the operator in the recommendations of the Restoration Assistant. At this time, we believe that streamlined versions of the analytical tools will be used to check the plans developed by the expert system. Considering the uncertainty about the value of some parameters and the questionable validity of the traditional models during a restoration, simplified computations might be sufficient.

The contradiction between the design of the Restoration Assistant and the introductory statement concerning the uncertainty as to the state of the power system after a blackout is only apparent. If this state was totally predictable, prepared plans would be effective and there would be little need for an expert system. If it was totally unpredictable, the expert system would have to have complete flexibility to adapt to any situation and would thus be extremely complex. In practice, the uncertainty is such that the expert system must have some flexibility but does not have to invent a new restoration strategy for each case. It has thus been designed to develop reconnection plans based on a choice between paths which have been shown to be viable alternatives through engineering studies.

Although a parallel restoration of two or more subsystems could theoretically shorten the restoration, the implementation of such an approach is difficult when the control of the power system is centralized as is the case for Northern States Power. The operator in charge of the restoration would have to split his or her attention between two areas and it is doubtful that the end result would be beneficial.

Mr. Adibi is entirely correct to point out that restoration strategies should be compared on the basis of the amount of energy delivered to the customers. It is clear that the choice of restoration objectives should be guided by this criteria.

Manuscript received September 27, 1990.

Knowledge-Based Systems as Operational Aids in Power System Restoration

KEINOSUKE MATSUMOTO, MEMBER, IEEE, TOSHIAKI SAKAGUCHI, MEMBER, IEEE, RICHARD J. KAFKA, SENIOR MEMBER, IEEE, AND MIKE M. ADIBI, LIFE SENIOR MEMBER, IEEE

Invited Paper

Today's society depends heavily upon a high level of power system reliability. Accordingly, it is very important not only to avoid power failures, but to restore power systems quickly and safely if these failures happen. Since power systems and their operation are becoming more complex, and there is a shortage of experienced personnel, utilities are becoming increasingly interested in developing computer aided restoration systems. To fulfill this requirement, many approaches to supporting or automating restoration have been developed in recent years. Among them, some approaches have adopted knowledge-based technology, which is likely to become one of the main technologies for power system restoration. This paper begins by analyzing the background of power system restoration and discusses how this process might benefit from the use of knowledge-based systems. After describing the state of the art of knowledge-based systems for power system restoration, efforts are made to review knowledge-based techniques to handle system restoration problems. Then the experiences of system development are reported and evaluation of the systems are addressed. Finally, some needed technological innovations are listed that are required before knowledge-based systems can extensively be applied to power system restoration.

I. Introduction

Today's power systems provide a highly reliable supply of electric power. However, there is still the possibility of a system wide outage. In order to reduce the possibility, extent, and duration of major failures in power systems, preventive, corrective, and restorative actions are required [1], [2]. Restoration of a power system after a system blackout is a difficult task. It is complex, delicate, and time-consuming, yet losses to customers and to the utility mount rapidly until service is restored. Current control system software has been developed primarily to facilitate monitoring and control of the power system under normal conditions. The system restoration task is still performed manually, according to guidelines prepared in advance. This means that operators, in conditions with which they are least familiar, and during which they are under the greatest stress, are most on their own, and depending on previously prepared plans. Comprehensive plans are complex and convoluted as to be very difficult to interpret, especially for operators under stress. As system restoration at the present time is performed mostly through manual operation, developing an automatic system for restorative control is a challenging subject to power system engineers. The purpose of power system restoration is to restore customers interrupted during a disturbance as quickly as possible.

Why has the development of an automatic system for restorative control been so slow? Unlike the load flow and stability problems, it is not easy to formulate the restorative control problem within the analytical framework. The solution method must be capable of treating some combinatorial nature of the problem. It must also be able to take care of not only physical laws but also artificial laws in the power system. The artificial laws mean the operating procedures which are defined arbitrarily among system operators. System operator's heuristics seem to play a central role during the course of system restoration.

If one tries to formalize system restoration from the mathematical point of view, it could be described as a mixed integer, multiobjective, multistage, nonlinear optimization problem subject to numerous constraints. However, the limitations of present algorithms require an operator to judge the accuracy and feasibility of the assessments and solutions. In case of emergency, it is difficult for the operators to get correct data and to draw an appropriate conclusion if there is no computer aid. This situation has imposed control systems to provide support to system operators in the decision making process. The knowledge base technology can be considered suitable for these restoration problems [3], [4]. Knowledge-based systems are expected to help manage a large scale power system in the restorative state. The technology could be regarded as a goal not only for system restoration, but for eventual automatic operation of

Manuscript received February 5, 1991; revised November 15, 1991.

K. Matsumoto and T. Sakaguchi are with the Mitsubishi Electric Corporation, Amagasaki, Hyogo 661, Japan.

R. J. Kafka is with the Potomac Electric Power Company, Bethesda, MD 20817.

M. M. Adibi is with the Industrial Research and Development Corporation, Bethesda, MD 20817.

IEEE Log Number 9107270.

Reprinted from *Proceedings of the IEEE,* Vol. 80, No. 5, pp. 689–697, May 1992.

the power systems [5]–[9]. These knowledge-based systems are able to:

1) use natural reasoning with knowledge that experienced operators use;
2) diagnose, plan, advise and explain their line of reasoning;
3) provide high performance in a relatively specialized field such as power system restoration;
4) expand, continually absorbing new knowledge as it becomes known;
5) outperform human operators under stress, and do so more reliably, especially in those situations which occur infrequently.

To develop good knowledge-based systems for power system restoration, one needs to understand the kind of problems that power system restoration (PSR) poses, the degree of success in addressing such problems, and the need for further endeavor. In this paper power system restoration is separated into three phases: State recognition, restoration plan development, and restorative operations.

In the recognition phase, the status of the collapsed power system is determined, and by using information furnished by the protective system, i.e., relays and circuit breakers, the faulted pieces of equipment which can not be used in restoration are identified. In the plan development phase, an ideal system configuration (i.e., target system [10]) is established using all equipment expected to be available as effectively as possible. The appropriate action sequences to reach the target system are generated in the operation phase.

The subsequent section briefly describes the state of the art of applications of knowledge-based systems to power system restoration. This is followed by the review of knowledge-based techniques to deal with the specific system restoration problems. Then the experience in system development is reported and evaluation of the systems is addressed. Finally, the future areas of research are discussed.

II. Survey of Restoration Systems

There are conventional approaches which base on general restoration guidelines and algorithmic methods besides knowledge-based approaches. Some papers described policies and plans of system restoration. For example, one strategy is to divide the system into several subsystems, each containing at least one black-start capability unit to restart power plants [1], [2], [11], [12]. Another strategy is to receive the initial power for the restoration process from one of the adjacent power systems [13]. In addition to them, a different strategy was proposed by Simburger *et al.* [14]. The idea of this paper is to reenergize the power system at extremely low voltages. Several systematic algorithms for restoration [10], [15], [16] are based on these guidelines.

Sakaguchi and Matsumoto [17], [18] proposed the first application of knowledge-based system technology to power system restoration. These papers discuss the use of a knowledge-based system to capture heuristic knowledge from operators and operations manuals that direct the steps taken in restoring power after a large system failure. Since then, a large number of papers on knowledge-based system applications to power system engineering have dealt with power system operation including restoration.

The restoration problems in bulk, subtransmission and distribution systems share some common issues. One needs to develop a restoration plan which is subject to certain constraints and to find transmission paths and switching sequence to pick up loads. However, each subsystem also has its own specific issues in some aspects according to each system's characteristics. For example, the following items are especially important factors at the bulk power system level [11], [12], [19], [20]:

1) restart and supply of station service to plants, substations, cable pumping plants, compressed air, etc.;
2) coordination of power plants start-up timings with load pickups to bring generators to their stable minimum levels and within the range of major analog controllers;
3) energizing large sections of transmission lines within the acceptable transient and sustained over voltages;
4) picking-up load in large increments without the risk of frequency decline;
5) reintegration of the skeleton of the bulk power supply with the requisite time-consuming switching operation;
6) deactivation of automatic load shedding and automatic switched capacitors during initial phases of restoration;
7) reduction of standing angle when closing loops to firm up transmission paths;
8) probability of successful start-up of thermal units, in particular, combustion turbines;
9) stability of black-start and peaking combustion turbines during the time when they constitute a large portion generation.

On the other hand, these issues are not as critical in distribution systems. Instead, load rerouting, overload relief, etc., become the main problems to be solved due to distribution systems characteristics.

Some knowledge-based systems for system restoration are classified according to voltage levels and their proposed main subjects as shown in Table 1. Comments on some of them are given in the following.

Kojima *et al.* [21] proposes a guidance method for the restorative control problem. It provides operators with an appropriate restoration plan and restorative procedures. The approach combines the knowledge engineering technology with conventional programming. This architecture provides several kinds of flexibility, e.g., it is possible to modify the restoration plan automatically if unexpected events occur and to accept operator's changes to the proposed plan.

Dijk *et al.* [27] describes system restoration as a decision problem and developed a general decision framework for solving network restoration.

Liu *et al.* [30], [33], [41] deals with diagnosis of transmission lines with automatic switching protection. The papers

Table 1 Knowledge-Based Systems for Power Systems Restoration

Main Subject / Voltage Level	Guidance Method	Restoration Strategy	Object-oriented Programming
Bulk Power System	Kojima [21] Wang [25] Ito [26] Kirschen [29]	Kakimoto [23] Takeyasu [24] Dijk [27]	Kojima [22] Darnault [28]
Subtransmission System	Tomsovic [30] Liu [31] Hotta [32]	Sakaguchi [17] Marathe [33] Moriguchi [34]	Keronen [35]
Distribution System	Liu [36] Wada [37]	Liu [38] Monticelli [39]	Matsumoto [40]

also suggest test actions to localize faulted sections aimed at restoring the largest number of substations served from the line.

Hotta *et al.* [32] reports a restoration system for a local dispatching center (77–33 kV system). In this system, security monitoring (faults, overload and outage) and restoration guidance are supported using on-line information. The system has been developed as a decentralized system using three workstations with OPS83 (a rule-based system development tool) and C language. It is connected to the Supervisory Control and Data Acquisition (SCADA) system of the dispatching center and the results have been evaluated.

Ito and Hata [26] develop and implement an expert system which can serve as a substation operator's guide at a 500 kV substation. The system performs diagnosis based on relay and circuit breaker information and suggests restoration procedures. The diagnostic rules are tabulated for on-line use to improve the performance.

At the distribution system level, the system developed by Wada *et al.* [37] aims at the optimization of load rerouting calculation, service restoration and overload relief respectively. The system has been tested extensively and, at present, it is installed in the master station of the distribution automation system at a local sales office.

Huang *et al.* [42] provides a review of power system restoration. This paper reviews the conventional restoration approaches and algorithmic methods, as well as the knowledge-based approaches.

III. Knowledge-Based Techniques for Restoration

Various knowledge-based techniques are necessary to handle power system restoration. These techniques can be divided into three groups: Operational aids methodology, restoration level methodology, and system development methodology. Operational aids methodology is related to the guidance or decision-support for operators. Restoration level methodology includes the essential and proper technology for power system restoration plans and procedures. System development methodology provides us with the way to build knowledge-based systems for system restoration.

A. Operational Aids Methodology

As it is impossible to predict all the events which may occur during the restoration of the power system, restoration systems should be able to cope with unexpected events. They also should be able to accept many kinds of operators' directions during all phases of restorative control. They should perform well in real time. To fulfill these requirements, guidance systems or decision-support systems are needed. The interactive and explanatory functions of knowledge-based systems could play important roles in the operational aids.

A typical guidance method [21], which allows operators to actively participate in problem solving, adopts an approach to make restoration plans and operational sequences adaptable to changes in the situation. Operators modify the restoration plan if they feel it needs changes due to unexpected occurrences. Then they begin operations to reach the new target system. This analysis produced the basic concept of the guidance method shown in Fig. 1.

The main objective of the guidance method is to assist operators in problem analyses and solutions to:

1) continually inform operators of the condition of the system and progress of the restoration;
2) construct a restoration plan and show it;
3) suggest proper restoration operations to reach the target system in the current state;
4) detect the need for amendments to the restoration plan according to changes in the situation;
5) allow operators' modifications of the proposed restoration plan;
6) allow operators' command of restorative actions different from the proposed ones;
7) verify operators' command of actions by means of simulation before they are put into action.

User interface is important to make good use of the guidance method. Since power systems have been growing and becoming more complex, the burden of the operator to recognize system status from much information is also increasing. This so-called "human recognition barrier" can be alleviated by using a good user interface. The reasoning processes of knowledge-based systems can be made clear to operators since a good user interface is easy to design. The interactive and explanatory functions of the knowledge-based systems are indispensable to the good user interface. One typical way is to use a graphic interface. Some applications are using on-line diagrams for status reporting and data retrieval.

Although the graphic interface is effective for the recognition problem, it takes much effort to develop. Sometimes the development of the user interface requires more effort than that of the knowledge base. An efficient manner of development of the graphic interface is also required.

One way to improve this issue is to separate the human interface from the knowledge base, both in software and hardware configurations. The separation of software improves modularity. Therefore maintainability can be improved. The separated user interface can be reused for other

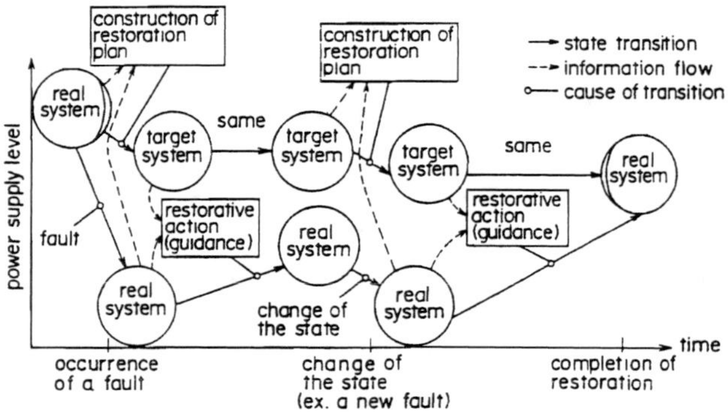

Fig. 1. Basic concept of guidance method for restoration.

applications. The separation of hardware devices improves response time and extensibility.

B. Restoration Level Methodology

State recognition, plan development, and restorative operation are the main tasks of power system restoration as mentioned above. Concerning state recognition, other papers in this issue describe alarm processing and fault diagnosis and therefore these topics are not addressed here in detail.

1) State Recognition: The first tasks following an outage are to determine the status of the collapsed power system, and to identify the failed equipment which can not be used in restoration by using information furnished by the protective system. Most diagnostic systems fall under one of two categories: symptom-based diagnosis [43] or model-based diagnosis [44].

2) Restoration Plan: Review of the literature shows that the strategies to restore power systems are as follows: Priority is given to the supply of all thermal power plants. After satisfying this initial priority, the restoration process is to be continued to recover the power network. At this stage, four basic actions can be performed:

- synchronizing generators;
- picking-up loads;
- reinforcing networks;
- interconnecting islands.

Various methods have been used to reenergize lines and substations based on shortest path algorithms [22]–[24] or prepared plans [28]. One approach which is somewhat different from these is based on the review of operator's restorative actions which consists of five steps:

1) keeping load and generation in balance while adjusting generator outputs to avoid line and equipment overloads;
2) making use of interconnection assistance when available and feasible;
3) supplying loads from multiple power sources;
4) transferring loads in the subtransmission network;
5) preventing system overvoltage by balancing the reactive power supply.

Based on the above observation of operators approach and philosophy, the process to develop a target system could consist of two stages. One is to determine a final configuration taking only active power as an index. The other is a stage to choose the appropriate number of reactive power suppliers to prevent abnormal voltages. As a result, the active power problems and reactive power problems are separated from each other. Therefore, it is possible to realize fast processing speed by using high speed algorithms such as the direct current load flow method [45].

Local load shedding is another methodology to relieve the overloaded equipment and to prevent the power system from further degradation. Such a methodology should be able to respond within the short-time rating of equipment and to minimize the amount of load to be curtailed in specific locations. Shah and Shahidehpour [46] propose a heuristic approach to the local load shedding method.

3) Restorative Operation: Knowledge-based technology can be used to derive a sequence of switching actions to restore and strengthen the power system. The switching sequence is determined based on the following rules [21]:

1) If there are switches to be opened in the deenergized area, then open them.
2) If there are switches (bus-tie CB and LS) to be closed in the deenergized area, then close them.
3) If there are switches to be closed for jointly using lines, buses or transformers in the energized area, then close them.
4) If there are switches to be closed for making a closed loop circuit in the energized area, then close them.
5) If there are switches to be opened for breaking a closed loop circuit in the energized area, then open them.
6) If there are switches to be closed for combining two islands in the energized area, then open them.
7) If there are switches to be closed to energize equipment, then close them.
8) If there are switches to be closed to let generators participate in the network, then close them.

Operations which are able to maintain security and to strengthen the system are selected prior to others in these rules. For example, the operations in the deenergized area do not affect other elements. Therefore, the operations are given the highest priority. These rules are applied in this order. A set of operations are shown which can be performed at present. They contain at most one operation which changes power flow conditions, that is, comes under rules 3)–8) but does not cause overloads. Therefore, operators can put these operations into practice at the same time and in any order without overloading. In advance the computer checks whether abnormal voltages will be generated by harmonics or transients. This is checked by using the ratio of capacitances to ground, demand power and total capacity of generators on-line.

A list of loads to be restored are shown collectively without causing overloads. These loads are selected by taking into account generators' ramp rates and priority of loads. The total amount of load per instruction is limited to a certain quantity to prevent an excessive frequency decline.

C. System Development Methodology

1) Knowledge Acquisition: The restoration needs various kinds of knowledge such as faulty equipment identification, overload removal, load pickup, restoration planning, and so on. This knowledge will be acquired by interview of experienced operators and by review of the operations manuals. Since operations manuals for operators have been documented already, this knowledge is relatively available compared with other applications such as power system planning. Typical test cases are extracted from the case list made by operators considering the past record. However, these methods are not sufficient to construct a practical knowledge base. For example, in the stage of field test, it was found that the lack or delay of on-line data degraded the performance of the system. Knowledge about this matter has not been acquired thoroughly before putting the system in the field. The feedback from the verification of the system is indispensable to construct practical knowledge bases.

2) Integration with Existing Environment: Integration with the existing environment primarily consists of two issues: one is the integration with existing equipment such as SCADA systems for on-line use, the other is the integration with existing programs. For the first problem, the appended approach and decentralized configuration are preferred because of its efficiency and flexibility. For the second problem, argument passing and data file are commonly used for off-line use. Shared memory and inter process communication might be required for on-line applications.

Since the restoration systems require the on-line information, the connection to existing equipment such as the SCADA system is crucial. Figure 2 shows three configurations proposed for real time applications [33], [34], [47], [48]: Appended approach, embedded approach, and back end approach. Many restoration systems have adopted the appended approach because it is not necessary to modify the existing systems and has much flexibility and extensibility. The appended knowledge-based systems are decentralized and therefore the efficiency is higher.

An example of hardware configuration [32] is as shown in Fig. 3. The function of each central processing unit (CPU) is listed in Table 2. The three CPU's (processing speed 4 MIPS, respectively) are connected to a ring-shaped computer network which permits distributed processing for security monitoring, restoration guidance and user interface. Due to this configuration, the time required for display is kept constant (2 to 3 seconds) even in the event of blackout over a wide area, during which a great amount of on-line information is transmitted. Also, the processing capacity of the system upon the occurrence of accidents in rapid succession can be improved.

Development of system restoration plans entails many careful simulations using analytical tools [49] such as; power flow, short circuit, electromagnetic transient, transient stability, and dynamic programs; and extensive en-

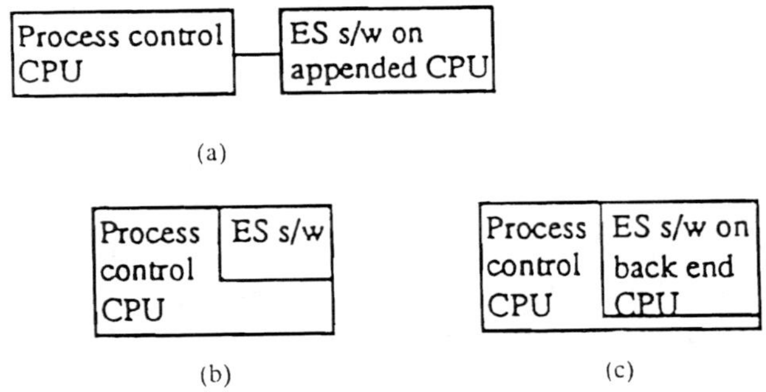

Fig. 2. Configurations for real time application. (a) Appended approach, (b) embedded approach, and (c) back end approach.

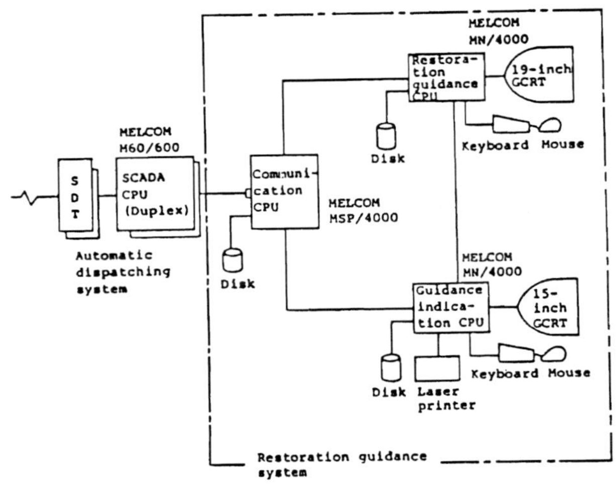

Fig. 3. Hardware configuration.

Table 2 CPU Functions

CPU name	Functions
Communication CPU	Security monitoring
Restoration guidance CPU	Restoration guidance User interface (graphics indication)
Guidance indication CPU	User interface (message indication, record output)

gineering judgment reflecting the many factors not readily modeled by existing tools. During the last decade knowledge-based system technology emerged as a tool by which researchers are trying to shrink this planning process into the confines of the on-line environment. The data communication between the knowledge base and Fortran analytical programs is one of the key issues to construct good support systems. Such systems would be capable of updating the restoration plan and procedure to allow occurrences of unexpected events and verification of operators' commands by on-line simulation prior to its implementation. Some utilities are making use of sophisticated simulation facilities [50] to analyze a number of network restoration plans. For the simulators to be an effective training tool and prepare the operators to cope with restoration problems, it must have the capability of modeling phenomena in the restoration processes [51], [52]. Unit dynamic constraints are considered in [53].

The efficiency, maintainability, readability and portability of the support systems highly depend on data communication system. The following three methods have been evaluated [9]: Interprocess communication, conventional file method and argument passing method. An interprocess communication tool called "Mailbox" was used to exchange information between these two processes as shown in Fig. 4. In the conventional file method, knowledge base calls analytical tools as shown in Fig. 5. From the previously mentioned experiences the authors currently feel that the argument passing method is most suitable for the engineering workstations because of its efficiency.

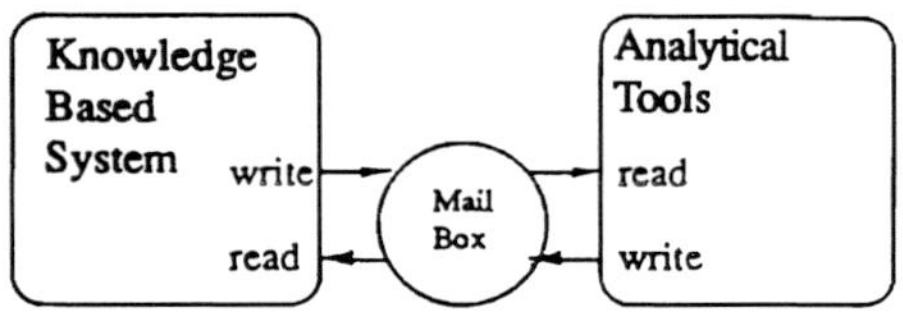

Fig. 4. Communication using interprocess tool.

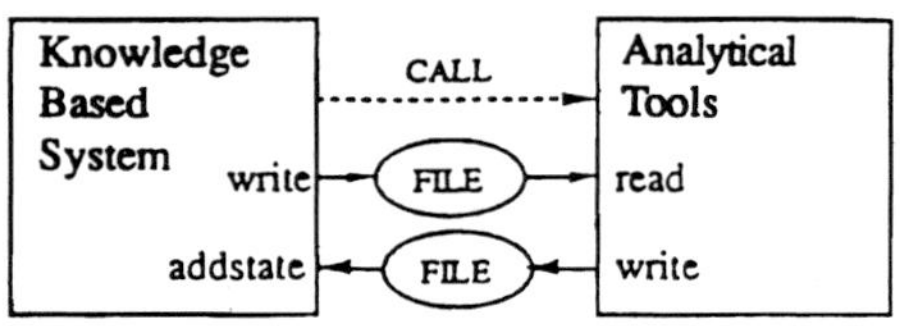

Fig. 5. Communication using files.

IV. Field Experience and Evaluation

Some restoration systems are able to deal with much more data at the same time than human operators. As a result, they can construct good restoration plans which operators could not anticipate in some cases. Sequence of switching and amount or timing of load-dispatching become important factors for safe system restoration because insufficient generation is the general case during a widespread outage. There were multiple operators' views on these issues so that a single, most applicable plan is difficult to develop. Restoration systems should produce multiple alternatives so that operators can judge which is the best by comparison.

The verification and the evaluation of the knowledge base may be one of the most important issues to develop practical systems. In the development of the restoration system [32], first the knowledge base has been verified using typical test cases specified by operators in an off-line situation.

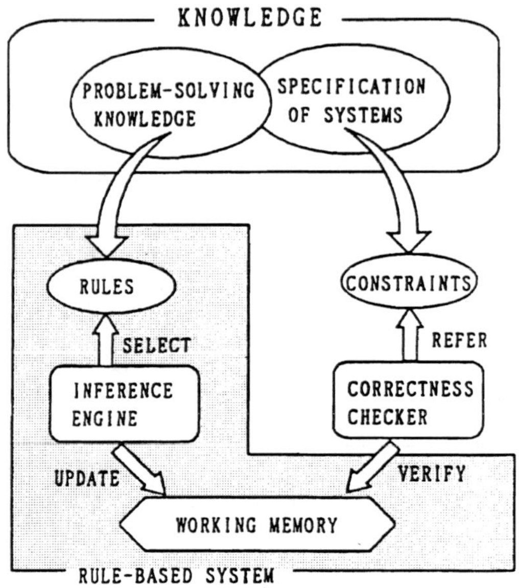

Fig. 6. Framework of dynamic verification method.

Then, the simulated fault data have been injected to the connection point of the SCADA system to check the total support system. Finally the system has been connected to the SCADA system and evaluated using actual on-line data.

Some problems of the knowledge base were identified by these methods. For instance, it was mentioned that the lack or delay of the on-line data degraded the performance. Even if a part of input data is absent, the knowledge-based systems can cope with the difficulties by sacrificing the precision of solutions. The above evaluation method appears to be the best approach to develop the feasible systems at this time. However, a more logical verification of the knowledge base would also help the verification process.

Some approaches for relation checking of rules [54], [55] have been reported for logical verification of knowledge bases. In addition, a dynamic verification method [56] is also proposed. In this method, a "Correctness Checker" shown in Fig. 6 traces the reasoning process by watching the system working memory (WM), which stores the reasoning results. System specifications are transformed into constraints to WM and they are adopted as evaluation criteria of inference. Examples of constraints are as follows:

1) There are not two WM elements which show open and closed status of one switch at the same time.
2) A line switch operation with making or breaking the load current is prohibited.
3) Final configuration of power system must be attained at the last step.

These constraints can be easily understood by examining an inference process as shown in Fig. 7.

V. Conclusion

Past studies of the application of knowledge base methodology to power system restoration have focused on selected

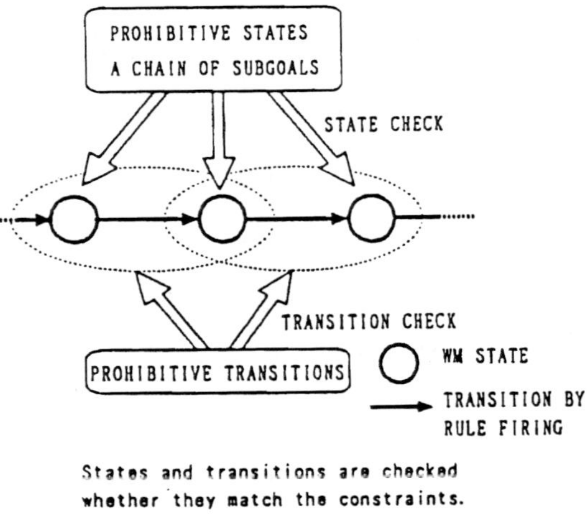

Fig. 7. Inference process and constraints.

elements of the problem. Based on extensive experience in the development of system restoration plans, and recognizing the maturing capabilities of knowledge base technology, it is believed that judicious integration of such methodology with network simulator, analytical tools, guidance method, communication facilities, data base, graphical user interface, etc. will enable many elements of restoration planning to be carried out in an on-line environment. The reasons why these functions are necessary to the restoration system are described in the following.

With the increase in power system reliability, major disturbances occur less frequently, and therefore operators receive little experience in restoration. The authors have identified the need for high-fidelity simulators for exercising and preparing experienced operators to cope with system-specific and time-critical restoration problems. The simulators must be highly interactive and provide responses to the operator commands similar to those of an actual power system.

The need is to find ways to integrate the capability of knowledge-based systems with the analytical tools required by power system restoration. Such a system would be capable of updating the restoration plan and procedure to allow occurrences of unexpected events and verification of operators' commands by on-line simulator prior to its implementation. A sophisticated guidance method must be established and on-line, real time inference facility is required to develop these features. In addition, communication facilities are needed to gather detail data from various places. These will provide powerful on-line assistance to operators of the type they most need, at the time they most need it. The result will be a significant reduction in the time required for major restorations, and dramatic reduction in the losses that are incurred.

We should also explore task-specific or domain-specific shells [57] for power system restoration. These methods seem to be an approach to removing the bottleneck of knowledge acquisition, verification and maintenance, because there is no need to extract knowledge to solve problems from experts. Such knowledge is already abstracted and built in shells. Already prepared modules are regarded as bug free and they do not have to be verified and maintained at all. The domain shell has also a standard data base called domain model which provides knowledge presentation schemes familiar to operators of power systems.

References

[1] R. J. Kafka, M. M. Adibi *et al.*, "System restoration plan development for a metropolitan electric system," *IEEE Trans. Power App. Syst.*, vol. PAS-100, pp. 3703–3711, Aug. 1981.

[2] R. J. Kafka, M. M. Adibi *et al.*, "Role of interactive and control computers in the development of a system restoration plan," *IEEE Trans. Power App. Syst.*, vol. PAS-101, pp. 17–25, Jan. 1982.

[3] R. P. Schulte, "Artificial intelligence solutions to power system operating problems," *IEEE Trans. Power Syst.*, vol. PWRS-2, pp. 920–926, Nov. 1987.

[4] K. Komai, K. Matsumoto, and T. Sakaguchi, "Analysis and evaluation of expert's knowledge for power system restoration by mathematical programming method," in *Proc. IEEE Int. Symp. Circuits and Systems*, June 1988, pp. 1895–1898.

[5] B. F. Wollenberg and T. Sakaguchi, "Artificial intelligence in power system Operations," *Proc. IEEE*, vol. 75, pp. 1678–1685, Dec. 1987.

[6] T. Sakaguchi *et al.*, "Prospects of expert systems in power system operation," *Int. J. Electrical Power and Energy Systems*, vol. 10, pp. 71–82, Apr. 1988.

[7] Z. Z. Zhang, G. S. Hope, and O. P. Malik, "Expert system in electric power systems—A bibliographical survey," *IEEE Trans. Power Syst.*, vol. PWRS-4, pp. 1355–1362, Nov. 1989.

[8] C. C. Liu and T. S. Dillon, "State of the art," in *Expert System Applications in Power Systems*, T. S. Dillon and M. A. Laughton, Eds. Hertfordshire, UK: Prentice Hall, 1990, ch. 11, pp. 383–408.

[9] Y. Kono, H. Suzuki, and K. Matsumoto, "Expert system applications to power systems operation," in *Proc. Fifth IEEE Int. Symp. Intelligent Control*, Sept. 1990, pp. 1139–1144.

[10] J. Zaborszky *et al.*, "Restoration of the large electric power system using a computer generated sequence of target systems," in *Proc. IFAC Symp. Power Systems and Power Plant Control*, Beijing, China, Aug. 1986, pp. 301–306.

[11] M. M. Adibi *et al.*, "Power system restoration—A task force report," *IEEE Trans. Power Syst.*, vol. PWRS-2, pp. 271–277, May 1987.

[12] J. Gutierrez *et al.*, "Policies for restoration of a power system," *IEEE Trans. Power Syst.*, vol. PWRS-2, pp. 436–442, May 1987.

[13] A. Johnson *et al.*, "System restoration—Deploying the plan," *IEEE Trans. Power App. Syst.*, vol. PAS-101, pp. 4263–4271, Nov. 1982.

[14] E. J. Simburger and F. J. Hubert, "Low voltage bulk power system restoration simulation," *IEEE Trans. Power App. Syst.*, vol. PAS-100, pp. 4479–4484, Nov. 1981.

[15] H. Kodama and H. Suzuki, "Interactive restoration control of electric power systems," in *CIGRE/IFAC Symp. 39–83* paper 514–04, Sept. 1983.

[16] G. Morin, "Service restoration following a major failure on the Hydro-Quebec power system," *IEEE Trans. Power Delivery*, vol. PWRD-2, pp. 454–463, Apr. 1987.

[17] T. Sakaguchi and K. Matsumoto, "Development of a knowledge-based system for power system restoration," *IEEE Trans. Power App. Syst.*, vol. PAS-102, pp. 320–329, Feb. 1983.

[18] K. Matsumoto *et al.*, "Heuristic management of complex and large scale power systems in restorative states," in *Proc. CIGRE/IFAC Symp. 39–83* paper 514–01, Sept. 1983.

[19] M. M. Adibi *et al.*, "Power system restoration—The second task force report," *IEEE Trans. Power Syst.*, vol. PWRS-2, pp. 271–277, May 1987.

[20] M. M. Adibi *et al.*, "System operation challenges," *IEEE Trans. Power Syst.*, vol. PWRS-3, pp. 118–126, Feb. 1988.

[21] Y. Kojima, S. Warashina *et al.*, "Development of a guidance method for power system restoration," *IEEE Trans. Power Syst.*, vol. PWRS-4, pp. 1219–1227, Aug. 1989.
[22] Y. Kojima, S. Warashina *et al.*, "The development of power system restoration method for a bulk power system by applying knowledge engineering techniques," *IEEE Trans. Power Syst.*, vol. PWRS-4, pp. 1228–1235, Aug. 1989.
[23] N. Kakimoto *et al.*, "An application of artificial intelligence to restoration of bulk power systems," in *Proc. 1988 ESAPS*, Aug. 1988, pp. 8.17–8.23.
[24] I. Takeyasu, T. Fukuta *et al.*, "An expert system for fault analysis and restoration of trunk line power system," in *Proc. 1988 ESAPS*, Aug. 1988, pp. 8.24–8.31.
[25] S. M. Wang *et al.*, "An decision-support system for bulk power system restoration," in *Proc. Tenth PSCC*, Aug.1990, pp. 966–971.
[26] S. Ito and I. Hata *et al.*, "Application of expert system to 500 kV substation operation guide system," in *Proc. 1988 ESAPS*, Aug. 1988, pp. 6.14–6.21.
[27] H. E. Dijk, van der Weegen, and G. P. T. Roelofs, "An expert system for power system restoration—Restoration strategies," in *Proc. 1988 ESAPS*, Aug. 1988, pp. 8.9–8.16.
[28] P. H. Darnault *et al.*, "An expert system as a guide for information synthesis and decision making in the restoration of power system after a blackout," in *Proc. 1989 ESAPS*, July 1989, pp. 399–404.
[29] D. S. Kirschen and T. L. Volkmann, "Guiding a power system restoration with an expert system," presented at the IEEE/PES Summer Meeting, 90 SM 341–8 PWRS, July 1990.
[30] K. Tomsovic, C. C. Liu *et al.*, "An expert system dispatchers' aids for the isolation of line section faults," *IEEE Trans. Power Delivery*, vol. PWRD-2, pp. 736–743, July 1987.
[31] C. C. Liu, K. Tomsovic, and S. Zhang, "Efficiency of expert system as on-line operating aids," in *Proc. Ninth PSCC*, Sept. 1987, pp. 695–701.
[32] K. Hotta *et al.*, "Implementation of a real-time expert system for a restoration guide in a dispatching center," *IEEE Trans. Power Syst.*, vol. PWRS-5, pp. 1032–1038, Aug. 1990.
[33] H. Marathe, C. C. Liu *et al.*, "An on-line operational expert system with data validation capabilities," in *Proc. 1989 PICA*, May 1989, pp. 56–63.
[34] S. Moriguchi *et al.*, "A large-scale SCADA system with real-time knowledge-based functions," in *Proc. 1989 ESAPS*, July 1989, pp. 21–27.
[35] J. J. Keronen, "An expert system prototype for event diagnosis and real-time operation planning in power system control," *IEEE Trans. Power Syst.*, vol. PWRS-4, pp. 544–550, May 1989.
[36] C. C. Liu *et al.*, "An expert system operational aids for restoration and loss reduction of distribution systems," *IEEE Trans. Power Syst.*, vol. PWRS-3, pp. 619–626, May 1988.
[37] M. Wada *et al.*, "An expert system applied to the distribution automation system," in *Proc. 1989 ESAPS*, July 1989, pp. 50–56.
[38] C. C. Liu, S. Lee, and K. Vu, "Loss minimization of distribution feeders: optimality and algorithms," *IEEE Trans. Power Delivery*, vol. PWRD-4, pp. 1281–1289, Apr. 1989.
[39] A. Monticelli, A. L. Morelato, and M. Rodriguez, "Expert system and heuristic search approaches to distribution system restoration," in *Proc. 1988 ESAPS*, Aug. 1988, pp. 12.1–12.7.
[40] Y. Matsumoto *et al.*, "An expert system for restoration of distribution lines," in *Proc. 1989 ESAPS*, July 1989, pp. 57–61.
[41] T. K. Ma *et al.*, "Operational experience and maintenance of an on-line expert system for customer restoration and fault testing," in *Proc. 1991 PICA*, May 1991, pp. 350–357.
[42] J. Huang *et al.*, "System restoration in power systems," in *Proc. IASTED Int. Symp. High Technology in the Power Industry*, May 1989, pp. 9–17.
[43] E. H. Shortliffe, Computer-based medical consultations: MYCIN. NY: American Elsevier, 1976.
[44] R. Davis, "Diagnostic reasoning based on structure and behavior," *Artificial Intelligence*, vol. 24, pp. 347–410, July 1984.
[45] B. Stott, "Review of load-flow calculation methods," *Proc. IEEE*, vol. 62, pp. 916–929, July 1974.
[46] S. Shah and S. M. Shahidehpour, "A heuristic approach to load shedding scheme," *IEEE Trans. Power Syst.*, vol. PWRS-4, pp. 1421–1429, Nov. 1989.
[47] M. J. Damborg and M. Chen, "An example of integrating an expert system into a control center using a dispatcher training simulator," in *Proc. 1988 ESAPS*, Aug. 1988, pp. 3.21–3.26.
[48] M. Chen, M. J. Damborg, and T. M. Athay, "Full integration of an expert system into an energy management system using a dispatcher training simulator," in *Proc. 1989 ESAPS*, July 1989, pp. 474–479.
[49] F. Wu and A. Monticelli, "Analytical tools for power system restoration—Conceptual design," *IEEE Trans. Power Syst.*, vol. PWRS-3, pp. 10–16, Feb. 1988.
[50] I. Susumago *et al.*, "Development of real-time simulation method for power system restoration after major disturbances," presented at the Int. Conf. Large High Voltage Electric Systems, Paris, France, Aug. 1986.
[51] K. Dickers and D. Rumpel, "Modeling for grid restoration studies," *Electric Power & Energy Systems*, vol. 9, pp. 45–55, Jan. 1987.
[52] G. Krost, D. Rumpel and E. N'Guessan, "Design of a network restoration expert system with training facilities," in *Proc. 1989 ESAPS*, July 1989, pp. 392–398.
[53] K. Kaneko, Y. Ueki, and Y. Fukuyama, "An expert system for automatic restoration of power systems considering dynamic constraints to thermal plants," *Trans. of the IEE of Japan*, vol. 110-B, pp. 286–293, Apr. 1990.
[54] H. Marathe, C. C. Liu, and T. K. Ma, "A relation-checking algorithm for maintenance of rule-based systems," *Electrical Power & Energy Systems*, vol. 12, pp. 235–241. Oct. 1990.
[55] H. Sasaki *et al.*, "A novel scheme for validation and verification of rule bases in expert systems," in *Proc. 1989 ESAPS*, July 1989, pp. 416–422.
[56] K. Matsumoto and T. Sakaguchi, "An approach to the dynamic verification of knowledge-based systems," in *Proc. 1989 ESAPS*, July 1989, pp. 423–427.
[57] K. Matsumoto, K. Komai, and T. Sakaguchi, "DASH: A diagnostic application-specific expert systems shell," in *Proc. Tenth Int. Workshop on Expert Systems and their Applications*, Avignon, France, May 1990, vol. 1, pp. 197–211.

Operational Experience and Maintenance of an On-Line Expert System for Customer Restoration and Fault Testing

Ta-Kang Ma
Chen-Ching Liu
Men-Shen Tsai

Department of Electrical Engineering
University of Washington
Seattle, WA

Ruth Rogers
Stephen L. Muchlinski
Jim Dodge

Puget Sound Power and Light Company
Bellevue, WA

Abstract

CRAFT (Customer Restoration And Fault Testing) is a rule-based expert system capable of locating the faulted section of a transmission line based on the status of breakers and automatic switches. CRAFT has been operating on-line in the Puget Sound Power and Light control center since July 1988. In this paper, the on-line operational experience is reported.

Many of the developed expert systems will be maintained or modified by utility engineers. This study proposes a methodology for maintaining rule-based expert systems. The following capabilities are incorporated: (1) Detecting the relations of conflict, redundancy, etc., to maintain consistency of the rule base, (2) Tracing the reasoning process to understand the problem solving approach, (3) Reducing the size of test cases after changes are made to the rules. The proposed method has been implemented for CRAFT.

1.0 Introduction

CRAFT (Customer Restoration And Fault Testing) is a rule-based expert system (ES) capable of locating the faulted section of a multi-tapped transmission line based on the status-change alarms from the energy management system (EMS). The development of CRAFT started in 1985 under the sponsorship of Electric Power Research Institute (EPRI) and Puget Sound Power and Light (PSPL) Company. The rule-based expert system has been operating on-line in the PSPL control center since July 1988. PSPL power dispatchers believe that CRAFT can reduce manual tasks in an emergency operating state, reduce the outage time of customers, serve as a double check, and be used for training. This paper provides the operational experience of CRAFT and the PSPL evaluation. The prototype of CRAFT was reported in [1]. References [2,3] discuss the on-line implementation and knowledge base enhancement.

Papers presented at the Seventeenth PICA Conference at the Hyatt Regency Baltimore Hotel, Baltimore, Maryland, May 7 - 10, 1991
Sponsored by the IEEE Power Engineering Society

After successful implementation of CRAFT, the focus of our study is directed toward the maintenance of the rule-based expert system. Currently, the on-line version of CRAFT has about 500 rules; a new rule module with about 250 rules will soon be added. All 69 PSPL transmission lines with automatic switching are included in the data base. Software maintenance is a routine task in the control center environment. Software engineers are trained to maintain the data base and they also support application software. The study issue in this paper is the maintenance of rule-based systems, which is new in the utility operational environment. As the power network structure and/or device settings are modified, the accuracy of the ES needs to be maintained through simulation of the ES for the new scenarios. If rules are modified or added as a result of correction/modification of the rule base, testing also needs to be performed to maintain the accuracy and consistency of the rule base.

The application of expert systems to power systems has been an area of great interest for the last ten years [4]. The ES application areas include alarm processing, diagnosis, restoration, design, etc. [5,6,7]. At present, a number of expert systems have reached the stage of practical use, e.g., [8,9]; most of them are rule-based. However, the maintenance issue of rule-based expert systems is not fully resolved. Maintaining a large rule-based system can be difficult since (1) the ES development tool, such as OPS83, is new to the utility software engineer, (2) persons in charge of maintenance may not fully understand details of the rule base. Changing rules in the ES may result in inaccuracy, conflict, or redundancy in the rule base.

Higher level ES development tools, e.g., KEE and G2, provide some facilities to perform syntax checking and/or to indicate violations of the allowable values of variables. Some of them also provide good MMI to allow the user to obtain a global view of the knowledge base. Algorithms for checking relations such as conflict, redundancy, etc., have been developed [10,11,12]. However, they are not standard facilities in the ES development tools yet. The need for ES maintenance/evaluation tools is well recognized in the power system area; the relation checking and verification issues have been investigated in [13,14,15].

This paper proposes a generic maintenance methodology for rule-based system which is applied to the CRAFT system. The proposed methodology is designed to assist utility software engineers in maintaining the ES following data/rule changes. The maintenance procedure incorporates testing, relation checking and explanation facilities developed by the UW. The maintenance software has been implemented with the CRAFT system at the PSPL control center. Several demonstrations and training sessions have been given to the PSPL EMS personnel.

Reprinted from *IEEE Transactions on Power Systems,* Vol. 7, No. 2, pp. 835–842, May 1992.

In the remaining of this paper, section 2 provides the operational experience of CRAFT. The maintenance issues and procedures are discussed in section 3.

2.0 Operational Experience of CRAFT

The on-line implementation of CRAFT was completed in July 1988. Since CRAFT is one of the first on-line ESs in the power system control center environment, it is felt that an analysis of its performance and operational experience would be useful for utilities who are involved in ES projects. Therefore, the UW has been collecting and analyzing PSPL outage reports, dispatchers' comments, and CRAFT responses to EMS alarms. The results are described in this section.

The frequency of faults in the PSPL system during July 1988- July 1990 is shown in Table 1. The total number of faults related to automatic switching is 74. From the table, more faults occurred in January and February; they were mainly caused by snow/wind storms. In an emergency condition with numerous alarms, CRAFT provides assistance to dispatchers.

Table 1 Frequency of Faults in the PSPL System

Month		Number of Faults
September	1988	1
January	1989	10
February	1989	7
March	1989	5
April	1989	2
June	1989	2
January	1990	22
February	1990	15
March	1990	1
June	1990	4
July	1990	5
Total No. of Faults		74

For each fault scenario, dispatchers documented the outage conditions and their decisions in restoring the customers. The outage conditions and dispatcher actions are compared to CRAFT suggestions. Out of the 74 cases, 7 are outside the CRAFT design capabilities. Five cases involve stuck breakers, while CRAFT does not handle malfunctioning breakers/automatic switches. Two other cases are related to single phase faults which result in low fault currents. CRAFT reported erroneous results since circuit breakers reclosed successfully.

The remaining 67 cases are within the CRAFT design capability. For evaluation purposes, the scenarios are divided into the following categories:

- Type 1 : Suggestions made by CRAFT are identical to the actions taken by PSPL dispatchers.
- Type 2 : Suggestions made by CRAFT are appropriate for the fault, but the suggestions are different from the actions taken by the dispatchers.
- Type 3 : CRAFT failed to detect the fault or gave wrong suggestions.

The performance of CRAFT is shown in Table 2. Out of the 67 cases, 55 belong to Type 1, i.e., CRAFT responses are identical to dispatchers' actions. The 9 cases in Type 2 are related to test actions. The difference here is that dispatchers may start restoration of customers before the exact faulted section is located, while CRAFT may suggest test actions first to detect the fault. CRAFT is being modified to allow dispatchers to override the CRAFT suggestions.

Table 2 Classification of Faults

Type of Fault	1	2	3	Total No. of Faults
No. of Cases	55	9	3	67

CRAFT failed in 3 cases. All 3 are due to maintenance switching, which results in EMS alarms. Opening or closing a supervised switch by the field personnel leads to alarms for ``Unauthorized'' actions. These alarms are interpreted by CRAFT as those triggered by faults. Hence, CRAFT reported false fault locations. Improvements are being made to allow CRAFT to better understand maintenance switching.

2.1 PSPL Evaluation

Almost all dispatchers received training from the UW project team on how to use CRAFT and the nature of ES tools. Many dispatchers look forward to having CRAFT as an operational aid, even though their confidence in CRAFT has to be established as time goes on.

In its present implementation, CRAFT is used by power dispatchers as an after-the-fact tool for the analysis of disturbances. CRAFT resides on a VAX computer appended to the SCADA computer through a datalink. In the event of a fault on a transmission line, the power dispatchers are too involved in restoring the system to access CRAFT on a different computer system, which incorporates different man-machine and display standards. However, it must be said that in at least one instance, CRAFT diagnosed a substation outage which was not apparent to the power dispatchers. In this case, service was restored in a much more timely manner than would have otherwise happened.

CRAFT has been very useful for training dispatch personnel and support engineers. CRAFT is able to simulate faults on the transmission system and demonstrate the behavior of automatic switches for sectionalizing line segments. Hence it is considered to be a valuable tool for demonstrating the methods PSPL uses for maintaining the integrity of the electrical system.

In 1988, PSPL started to upgrade the EMS. An approach was selected which allowed PSPL to keep the existing system running while adding new computers, MMI, and communication front end processors. One of the requirements is to integrate the User Interface into a dispatcher oriented information support tool for monitoring and controlling. PSPL plans on accomplishing this task by using workstations to unify the presentation of information. In this way, many applications can run on a variety of computer systems transparent to the dispatcher and be viewed through one common interface.

As part of the continuing work with the UW, migration from the current MMI and computer hardware is taking place. CRAFT has been demonstrated in a new real-time database environment with new enhanced graphics which more nearly simulate the real-time displays the power dispatchers currently use. The current version of CRAFT displays information through the same workstations that will be used by dispatchers.

With a different MMI, power dispatchers will be able to call up displays immediately for CRAFT's diagnosis of faults. At present, CRAFT uses interrupts from device changes of the state to begin the diagnostic process. As the database and computer migration continues, CRAFT will have access to additional power system data from real-time power applications to enhance the capability.

Design and control engineers have shown an interest in using CRAFT as a design tool for protection schemes on new and re-built transmission lines. They feel the abilities of CRAFT for demonstrating faults in a real-time scenario are far better than design tools they presently have.

2.2 CRAFT User Requirements

Since ES tools are new to power utilities, the user requirements are not always understood. The CRAFT user requirements are listed below:

1. VAX computer with 6 MB RAM
2. OPS83 software for VAX computer
3. OPS83 training, 3 days class & exercises
4. Data base requirements
   UWRIM (Univ. Wash. Relational Information Management System), 1 day training
   Other database facilities can be used.
5. Implementation of interface between SCADA alarms & CRAFT,
   2 person-months of a senior software engineer (SCADA side interface software)
6. Testing of faults on lines,
   effort depends the number of lines and number of line sections
7. Training users/support staff, 3 hours/person
8. Addition of utility specific rules, maintenance of knowledge base, 6 months from introduction to competency

3.0 Maintenance of an ES

The operation and planning of a power system involve complex tasks. Much domain knowledge is required to make an ES practical. As the rule base of an existing ES grows, the ES maintenance also becomes more difficult. Our intention is to help maintain consistency and accuracy of the rule and data bases when modifications have to be made. Basically, an ES contains five major parts: rule base, data base, inference engine, display set, and interface programs. The display set can be considered as a part of the data base. Based on our experience in constructing ESs, a generic maintenance procedure for rule-based ES is proposed, which is applied to the maintenance of CRAFT.

Section 3.1 describes the proposed ES testing procedure following data base modifications. In the testing process, if the ES fails to solve any of the new scenarios, then the knowledge base of the ES needs to be modified. Section 3.2 provides a procedure for rule base modification.

Maintaining consistency and accuracy of a rule base is a primary objective in maintenance. When a rule is modified or a new rule is added, the consistency and accuracy of the knowledge base must be maintained. In this study, a Rule Checking Algorithm (RCA) has been developed and implemented as a tool to help maintain consistency of the rule base. Possible relations between rules are defined in section 3.3. The RCA and its implementation are discussed in sections 3.4 and 3.5 respectively. A realistic example from CRAFT is given in section 3.6.

For data base modifications, the testing needs to be performed only for the modified portions of the data. However, if a change is made to the rule base, all fault scenarios that may be affected by the change must be tested.

To facilitate the testing and maintenance, an explanation module has been developed, which provides a summary of the reasoning using high level English rules. The developed explanation facility is presented in section 3.7. Several rules proposed to help reduce the number of test cases are described in section 3.8.

3.1 ES Testing Following Data Base Modifications

A generic data base modification and testing procedure is shown in Fig. 1. After changes are made to the data base, the ES needs to be simulated for the new scenarios. In the testing process, data inconsistency may be found. Moreover, the rule base of the ES may not be able to handle all the new scenarios. In this case, the rule base modification and testing procedure (section 3.2) needs to be followed.

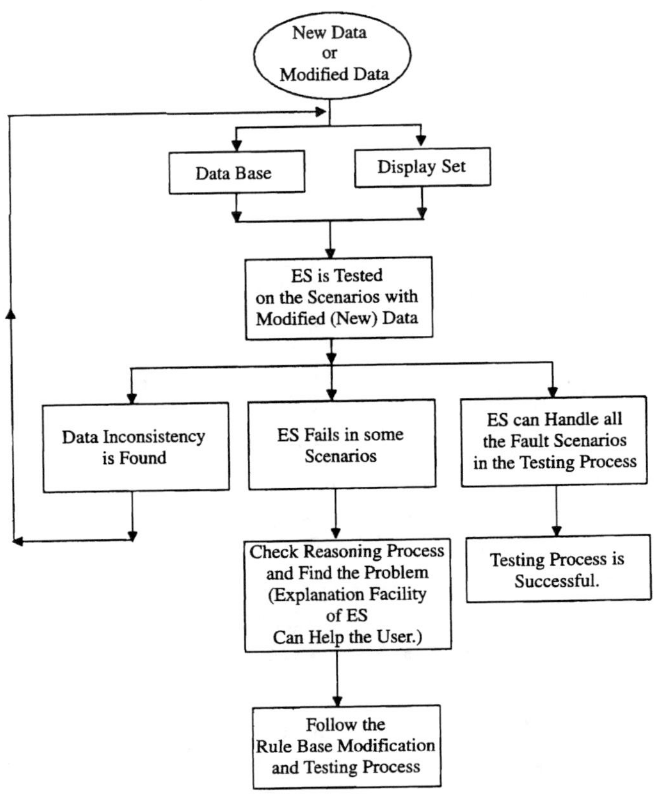

Fig. 1 Data Modification and ES Testing Procedure

For example, PSPL engineers may make the following three types of changes to the power system:

(1) Change the device information on a line: A manual switch on a line may be changed to an automatic one and reclosure time settings of a circuit breaker may be adjusted.
(2) Modify an existing line configuration: Engineers may add new devices or buses to a line.
(3) Model a new line to power network.

Once any of the above situations occurs, the CRAFT data base will be modified.

3.2 Rule Base Modifications and Testing Process

Fig. 2 shows the proposed knowledge base modification and testing procedure for rule-based systems. In Fig. 2, the 5 stages are: rule base modification, rule consistency checking, program linkage, scenario testing, and result analysis. For the rule base modification stage, a program has been implemented to direct the user to select a task from Fig. 3 and then display the common rule conditions used in the selected task with example rules. Following the rule writing stage, the rule consistency checking algorithm will be executed. As mentioned previously, the purpose here is to identify possible conflict or redundancy, etc., among rules. The third stage is to compile and then link ES modules for execution. Stage 4 deals with the simulation of ES for various fault scenarios. Stage 5 shows that the modification/testing process is iterative in nature.

The processing of a rule-based ES can be described in terms of the tasks performed during the solution process. For

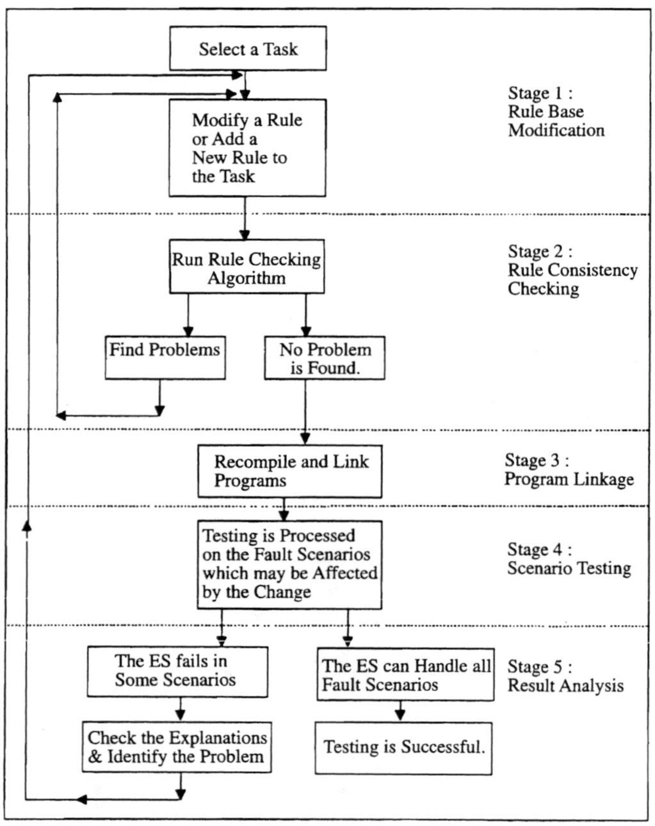

Fig. 2 Rule Base Modification and Testing

example, the CRAFT rules are grouped into tasks, each representing a number of rules performing a specific step in reasoning. In Fig. 3, each block in the dotted box refers to a specific task. When a new rule is to be added, it should be added to the appropriate task. It is recommended that only one task be modified at a time.

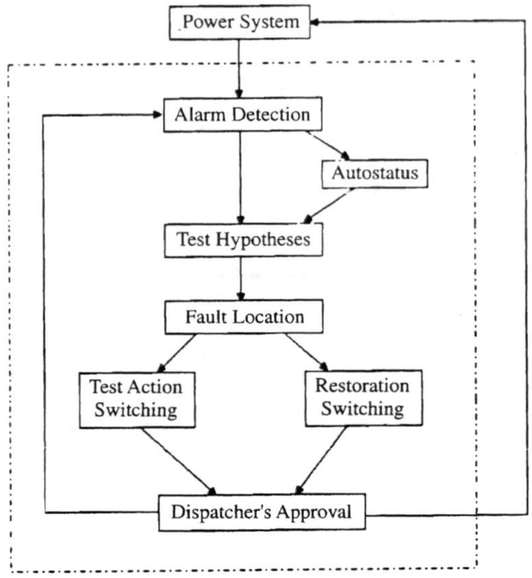

Fig. 3 CRAFT Problem Solving Approach

Specifically, when any of the following situations takes place, the rule base of CRAFT needs to be modified.

(1) CRAFT fails to detect the fault location.
(2) CRAFT gives wrong suggestions for the line testing or restoration action.
(3) CRAFT capability needs to be extended.

3.3 Relations Between Rules

This section defines six relations that may exist between a pair of rules or a rule and a chain of rules. The relations are cause-effect, mutual exclusion, conflict, subsumption, redundancy and implication. The mathematical definition of each relation can be found in [12].

Cause-Effect
A rule R2 is said to be the effect of rule R1 if rule R2 can be satisfied every time the action of rule R1 is taken.

Mutual Exclusion
Two rules are said to be mutually exclusive if the two rules can not be satisfied at the same time.

Redundancy
A rule R2 is redundant with respect to rule R1 if the condition and action parts of the two rules are equivalent.

Conflict
A rule R2 is in conflict with a rule R1 if the condition parts of R1 and R2 are equivalent but the actions of R2 are different from the actions of R1.

Note that differences in actions in the ``conflict'' relation may not imply any actual conflict in reality. The proposed algorithm simply identifies the possibility of a conflicting situation.

Subsumption
A rule R2 is subsumed with respect to a rule R1 if the actions of the rule R2 and R1 are equivalent but the R2 conditions are more restrictive than the R1 conditions.

Implication
A rule R2 is implied by a rule R1 if the condition parts of the two rules are equivalent but the actions of R2 are a subset of actions of R1.

3.4 Rule Checking Algorithm (RCA)

Each rule is modeled by two sets, one representing the condition part and the other the set of actions. The sets are defined by the variables (or attributes) used in the rule formation. For example, the condition set is the collection of all data combinations which satisfy the conditions of the rule. The rule relations rely on the set relations between the condition and actions parts of different rules. Table 4 shows that set relations in condition and action parts can be combined to conclude relations among rules. The cause-effect relation requires the actions of a rule to cause the condition of another rule to be satisfied.

When a relation holds only for some scenarios, it is called a "partial" relation [12]. (In this case, the set equivalence in Table 4 is not required.) Since the condition set represents ALL possible values of the attributes, the proposed RCA can detect partial relations and identify ALL scenarios under which the relations hold. This is important since the need to check the relation for each specific scenario is eliminated. Another feature of the RCA is that it is able to detect the relation between a rule and a rule chain. A mechanism developed for successive derivation of rule/rule relations allows RCA to detect the rule/rule chain relation. Finally, this algorithm is designed in such a way that the same inference algorithm used by OPS83 is used to derive the rule relations.

Table 4 Deriving Relations among Rules

IF the relation between Condition sets is	AND the relation between Action sets is	THEN the Relation Between Rules is
Equivalent	Equivalent	Redundant
Equivalent	Superset or Subset	Implication
Equivalent	Different	Conflict
Superset or Subset	Equivalent	Subsumption
Disjoint	———	Mutual Exclusion

3.5 Implementation of the RCA

The efficiency of RCA depends on an effective method for comparison between a new rule and the existing rule base, which may contain a large number of rules. Based on our experience with OPS83, a RETE representation of the rule base is very efficient for comparison of the new rule with the existing rule base. Such a network shares checking of common conditions between rules. Thus, RETE simultaneously compares the new rule with all rules in the rule base. RCA is implemented on a VAX 3100 using the C language, which is chosen for its high processing efficiency and powerful symbol manipulation capability.

RCA performs four tasks: 1) Form the RETE network for rules in the rule base, 2) Find the relations between condition parts of new rule and rules in the rule base, 3) Select a rule in the rule base which is a candidate for a relation with respect to the new rule, and 4) Compare the action part of the new rule with the action part of the selected rule to conclude the rule relation.

3.6 An Example of Inconsistent Rules

This following example illustrates a conflict relation among rules which may be introduced during modification/ expansion of the rule base.

Example
When a fault occurs on a line, PSPL dispatchers may have different opinions on the line test actions, depending on the actual circumstances. Currently, CRAFT suggests the test actions which, if successful, would restore as many substations as possible. A PSPL line is given in Fig. 4. When a fault occurs anywhere between devices 431 and 170, circuit breakers 498 and 170 trip with switches 432 and 431 opening on dead line. At 10 seconds after the fault, circuit breaker 170 will reclose and trip again. On the other end of the line, breaker 498 will reclose successfully and, 3 seconds later, switch 432 will close on hot line. Since the faulted section can not be identified immediately and three substations (Sequoia, Lea Hill, and "M" Street) are deenergized, test actions should be considered. CRAFT suggests the actions of "open switch 1001" and "close circuit breaker 170" based on 2 rules, setup15 and setup19, shown in Fig. 5. If the fault is between 431 and 1001, then all the substations would be reenergized. However, on the line in Fig. 4, the test actions taken by dispatchers may be "open switch 796" and

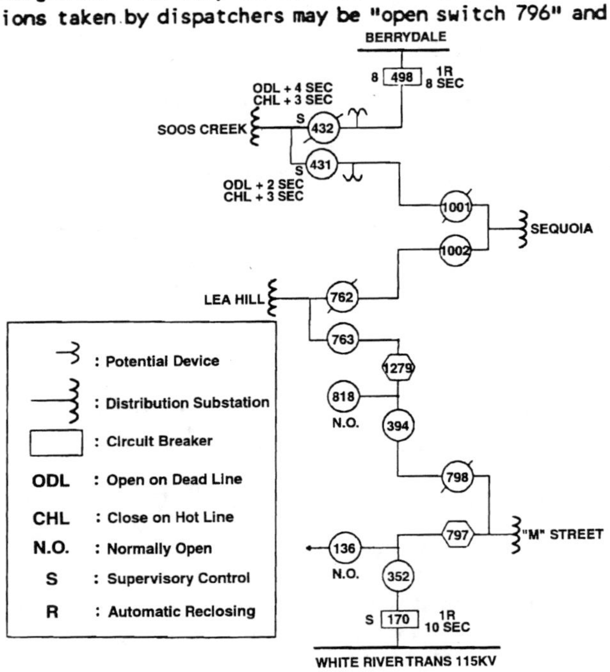

Fig. 4 A Sample PSPL Line

"close circuit breaker 170." This allows them to test whether the faulted section is between 431 and 796. Rule new_setup1, shown in Fig. 6, is a new specific rule to incorporate dispatchers judgement for this line, which is in conflict with the rule chain setup15 and setup19.

RCA detects the conflict relation between rule new_setup1 and the chain of rules setup15 and setup19. Note that this conflict relation is partial, i.e., it holds only for line 107. However, based on dispatchers' opinion, the rule new_setup1 needs to be added to the rule base. This provides an alternative test action for on-line operation of CRAFT.

```
rule setup15
-- IF   task is to set up testing
-- AND  the substation to be restored is upline from
--      the source
-- AND  the closest switch to this sub is closed
-- THEN schedule an event for opening of the isolation
--      switch
{
&task   (task solving=y; id=test_hypothesis;
         status=setup; );
&restore(restore subject = &task.subject;);
&sub    (substation subject=&task.subject;
         id=&restore.sub; to<=&restore.source;);
&device (device subject=&task.subject; line=&sub.line;
         status=closed; location= &sub.from; );
    ~   (event subject=&task.subject; id=&device.id;);
-->
make  (event subject=&task.subject; id=&device.id;
       action=open;);
};
```

```
rule setup19
-- IF   task is to set up testing
-- AND  the source is downline
-- AND  there is an open device on the down side of
--      the line
-- AND  this device has not been scheduled for a test
--      closing
-- THEN schedule an event for closing of this device
{
&task   (task solving=y; id=test_hypothesis;
         status=setup; );
&restore(restore subject = &task.subject;);
&sub    (substation subject=&task.subject;
         id=&restore.sub; to<=&restore.source;);
&device  (device subject=&task.subject; status=open;
    location > &sub.from; location<=&restore.source;);
  ~     (event subject=&task.subject; id=&device.id;);
-->
make (event subject=&task.subject; id=&device.id;
      action=close;);
};
```

Fig. 5 CRAFT Rules Setup15 and Setup19

3.7 Design of the Explanation Facility

The explanation facility of an ES is intended to provide a summary of the key arguments in the reasoning. The proposed method for an explanation facility is discussed here. For each task of an ES, a number of "key" rules are selected. Each key rule represents a block of rules which

```
rule new_setup1
-- IF   task is to set up testing
-- AND  the testing is on line 107
-- AND  the substation to be restored is upline from
--        the source
-- AND  the fault is between 431 and 170
-- THEN schedule an event for opening switch 796
-- AND  schedule an event for closing of device 170.
{
&task  (task solving=y; id=test_hypothesis;
        status=setup; );
&restore (restore subject = &task.subject; line=107;);
&sub     (substation subject=&task.subject;
          id=&restore.sub; to<=&restore.source;);
&device1 (device subject=&task.subject;
 line=&sub.line;status=closed; location= &sub.from; );
&device2 (device subject=&task.subject; status=open;
    location > &sub.from; location<=&restore.source;);
&device3 (device subject=&task.subject; line=&sub.line;
status=closed; id = 796;);
~ (event subject=&task.subject; id=&device1.id;);
~ (event subject=&task.subject; id=&device2.id;);
-->
make (event subject=&task.subject; id=&device3.id;
          action=open;);
make (event subject=&task.subject; id=&device2.id;
          action=close;);
};
```

Fig. 6 Rule New_Setup1

implement an English rule acquired from the dispatchers. A number of explanation rules are written for the key rules in each task and stored in an explanation module. As the user asks "why," the OPS83 rule firing sequence is examined. The key rules are selected from the sequence. Each key rule provides sentences in English which will be filled with the data from the scenario being studied.

The following example describes the explanation facility of CRAFT. A PSPL line is shown in Fig. 7. Suppose a fault occurs between devices 970 and 613. CRAFT provides explanations for the fault scenario when the user asks "why" (Fig. 8). Rule locate_fault11 is one of the key rules in the "fault location" task of the CRAFT rule base. Its corresponding explanation rule is rule exp_locate_fault11. Both rules locate_fault11 and exp_locate_fault11 are given in Fig. 9.

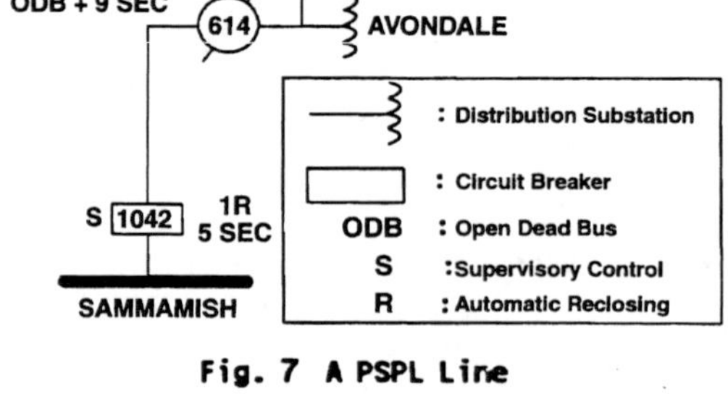

Fig. 7 A PSPL Line

CRAFT uses rule locate_fault11 in the reasoning process to identify the fault. The first two lines "Fault lies between open switches furthest apart." and "Fault lies between 970 - 1042." in the explanation of Fig. 8 come from rule exp_locate_fault11.

Explanation facility of CRAFT has been tested on all the 69 PSPL lines. A total of 101 rules were written for the explanation of CRAFT decisions.

```
Fault appears to lie between 970 - 613
*************** Proposed Solution *******************
Open 613 locally.
Close 1042 by supervisory.
Close 614 locally after automatic function is deacti-
vated.

why
Fault lies between open switches furthest apart.
Fault lies between 970 - 1042.

Switch 614 was open when
first reclosure of breaker 970 was unsuccessful.
Fault must lie between 970 - 614.

Bus fault on bus Avondale is not considered.
Fault lies to the left of 613.
```

Fig. 8 Sample Output of CRAFT Explanation

3.8 Reducing the Number of Test Cases

In the reasoning of an ES, each fault location or restorative action is identified by a set of rules. As discussed in section 3.2, all the rules in the rule base can be grouped into a number of tasks, say, T1, T2,.., Tn. Let T be the set of {T1, T2, .., Tn}. For a specific fault scenario S, only a portion of the tasks in set T are needed to solve the problem.

Let Si be the set of tasks which are needed to solve the scenario i. If a rule X in task Tj is modified, then ES testing is needed only for the fault scenarios which may be affected by the change. In other words, given a fault scenario i, if task Tj is not a component of Si, then this fault scenario i does not have to be tested after a change is made to rule X.

For instance, in CRAFT rule base, a specific task "Test Action Switching," which contains quite a few rules, is used to suggest test actions when a fault can not be identified immediately. Therefore, the fault scenarios which do not require any test actions do not need to be tested again after rule changes are made to the task "Test Action Switching."

The following three rules are used by CRAFT to reduce the size of the test cases.

Rule 1
IF
a rule is added to the AUTOSTATUS task,
(AUTOSTATUS is a task which derives the status of un-supervised automatic switches following a fault.)
THEN
the lines without unsupervised switches do not need to be tested. (Several fault scenarios are possible on each of these lines.)

Rule 2
IF
a rule is added for suggesting switching operations in the testing process,
THEN
the fault scenarios which do not involve any test actions do not need to be tested.

Rule 3
IF
a rule is added for identification of a fault on a three terminal line and the rule can not be applied to two terminal lines,
THEN
the scenarios which occur on two terminal lines do not need to be tested.

```
rule locate_fault11
-- IF   task is to locate a single fault
-- AND  device 1 is the furthest open device up the
--      line
-- AND  device 2 is the furthest open device down the
--      line
-- AND  the possible line faults have not been
--      considered
-- THEN obtain as a list of possible faults the line
--      sections between these 2 open devices
{ &task   (task solving=y; id=test_hypothesis;
           status=locate_fault;);
  &hypo   (hypothesis subject=&task.subject;
           id=single_fault; status=test; );
  &fault  (line_fault subject=&task.subject;
           status=radial; start=0; end=0; );
  &device1 (device subject=&task.subject;
            status=open; );
  &device2 (device subject=&task.subject;
           location>&device1.location; status=open; );
      [int_real(&device2.location-&device1.location)]
  --> modify &fault  (start=&device1.location;
                     end=&device2.location; );
     if (&explain)
     { make (explain subject=&task.subject; rule=lf11;
           no= &expno; val1=&device1.id;
           val2=&device2.id; );
     &expno = &expno + 1;  };
};
```

```
rule exp_locate_fault11
-- If   rule locate_fault11 is used in the reasoning
--      process to locate the fault,
-- THEN output the following two lines when user
--      enters "why."
--
-- "Fault lies between open switches furthest
--   apart."
-- "Fault lies between "val1" - "val2"."
{ &task (task id=explain_craft_actions; );
  &exp  (explain subject=&task.subject; rule=lf11;
        status<>|used|; );
    ~  (explain subject=&task.subject; status<>|used|;
        no < &exp.no; );
  --> write()| Fault lies between open switches|,
             | furthest apart.|,'\n';
      write()| Fault lies between |, &exp.val1,| - |,
             &exp.val2,|.|, `\n';
      modify &exp (status=used; );
 };
```

Fig. 9 Rule locate_fault11 and Its Explanation Rule

4.0 Conclusion

The real time operational experience of an ES, CRAFT, is reported based on actual fault scenarios collected during the last two years. The on-line experience and PSPL dispatchers' comments helped to enhance the capability of CRAFT. CRAFT has achieved a high degree of accuracy; only 3 failures out of the 67 cases analyzed. The computational performance of CRAFT is also greatly improved with the new VAX 6000 computer. On the average, CRAFT responds to a command in 2-3 seconds. Dispatchers identified the MMI as the weakness of CRAFT. The improvement of CRAFT MMI is an on-going task at PSPL. Our goal is to make CRAFT accessible to dispatchers from their usual console and MMI.

In this study, a new ES maintenance and testing procedure was developed to help utility software engineers maintain the ES. The generic maintenance procedure has been implemented for the CRAFT system. The proposed procedure has received initial acceptance by the PSPL EMS personnel. However, since the ES maintenance is a new task in the power system environment, further improvement of the procedures is expected.

Acknowledgement

The CRAFT implementation is sponsored by EPRI (RP 2944-1) and PSPL. The development of maintenance methods is sponsored by PSPL and National Science Foundation. The authors would like to acknowledge the contributions from Dr. H. Marathe, who was an ESCA Fellowship recipient. The support and assistance from Mr. Mike Hunter, Director of Operations, PSPL, and power dispatchers are greatly appreciated.

5.0 References

[1] K. Tomsovic, C. Liu, et al., ``An Expert System as a Dispatchers' Aid for the Isolation of Line Section Faults,'' IEEE Trans. on Power Delivery, July 1987, pp. 736-743.

[2] H. Marathe, C. Liu, et al., ``An On-Line Operational Expert System with Data Validation Capabilities,'' Proc. PICA Conference, 1989, pp. 56-63.

[3] C. Liu and M. Damborg, CRAFT: On-Line Expert System from Customer Restoration and Fault Testing, EPRI Report EL-6880, Volumes 1-5, March 1990.

[4] Y. Pao and S. Oh, ``A Rule-Based Approach to Electric Power Systems Security Assessment,'' Proceedings of IEEE Conference on Pattern Recognition and Image Processing, 1981, pp. 340-342.

[5] B. Wollenberg and T. Sakaguchi, ``Artificial Intelligence in Power System Operations,'' Proceedings of the IEEE, 1987, pp. 1678-1685.

[6] C. Liu, ``Knowledge-Based Systems in Power Systems : Applications and Development Methods,'' Trans. IEE of Japan, April 1990, pp. 241-250.

[7] C. Liu and T. Dillon, ``State of the Art,'' Expert System Applications in Power Systems, Prentice Hall, UK, 1990, Eds T. Dillon and M. Laughton, Chapter 11.

[8] S. Moriguchi, et al., ``A Large-Scale SCADA System with Real-Time Knowledge-Based Functions,'' Proceedings of Second Symposium on Expert Systems Application to Power Systems, July 1989, pp. 21-27.

[9] R. Bijoch, S. Harris, and T. L. Volkmann, ``Intelligent Alarm Processor at Northern States Power,'' Proceedings of Second Symposium on Expert Systems Application to Power Systems, July 1989, pp. 79-83.

[10] T. Nguyen, et al., ``Knowledge Base Verification,'' AI Magazine, Summer 1987, pp. 69-75.

[11] M. Suwa, A. Scott, and E. Shortliffe, ``An Approach to Verifying Completeness and Consistency in a Rule-Based Expert System,'' AI Magazine, Fall 1982, pp. 16-21.

[12] H. Marathe, T. Ma, and C. Liu, ``An Algorithm for Identification of Relations Among Rules,'' Proc. IEEE International Workshop on Tools for Artificial Intelligence, 1989, pp. 360-367.

[13] H. Marathe, C. Liu, and T. Ma, ``A Relational Representation of Complex Knowledge with Applications to Power System Problems,'' Proceedings of First Symposium on Expert Systems Application to Power Systems, 1988, pp. 2.4-2.9.

[14] H. Sasaki, K. Kawahara, et al., ``A Novel Scheme for Validation and Verification of Rule Bases in Expert Systems,'' Proceedings of Second Symposium on Expert Systems Application to Power Systems, July 1989, pp. 416-422.

[15] K. Matsumoto and T. Sakaguchi, ``An Approach to the Dynamic Verification of Knowledge-Based Systems,'' Proceedings of Second Symposium on Expert Systems Application to Power Systems, July 1989, pp. 423-427.

A KNOWLEDGE-BASED METHOD FOR MAKING RESTORATION PLAN OF BULK POWER SYSTEM

K. Shimakura, J. Inagaki, Y. Matsunoki

Hokuriku Electric Power Co., Inc.
15-1, Ushijima-cho, Toyama
Toyama, Japan

M. Ito, S. Fukui, S. Hori

Mitsubishi Electric Corporation
1-1-2, Wadasaki-cho, Hyogo-ku
Kobe, Japan

ABSTRACT - A method is proposed for use in event of power system outages which uses general-purpose restoration knowledge not dependent on pre-outage system states in order to generate post-restoration target systems in which post-outage systems are taken as initial states.

Conventionally post-outage system states are formed to emulate as closely as possible pre-outage system states, with system operations performed only within blackout systems. Therefore, depending on the amount of pre-outage load, some outage loads may be experienced in the restored system.

Proposed here is a method by which system operations in both blackout systems and sound systems are combined according to the amount of load in the pre-outage systems, so that post-restoration system states with minimal outage loads from post-outage systems will be generated.

A prototype system incorporating actual power systems and utilizing this method was built and tested under simulated conditions. The effectiveness of the proposed system is discussed on the basis of the test results.

Keywords: power system, knowledge base, restoration knowledge, general-purpose rule.

1. INTRODUCTION

This paper describes a method for making post-restoration system states to be set as target (hereafter referred to as the "restoration target system") in power system faults.

In the conventional method, no system operations are made on sound systems; moreover, a restoration target system is generated according to the system state before the outage. For this reason, there are cases where outage loads continue to exist in the restoration target system, depending on the loads in the power system before the outage. [1], [2], [3], [4], & [5]

The method proposed here has the features listed below; depending on the size of the loads in the power system, post-outage system states are considered, and restoration target systems minimizing the amount of outage load are generated.

(1) There is no dependence on the system state prior to the outage.
(2) Segmentation of the outage load is carried out for a controlled load power transmission.
(3) The amount of support power is increased through the operation of sound systems.
(4) Knowledge used to generate post-restoration systems is described with general-purpose rules not based on particular systems.

Papers presented at the Seventeenth PICA Conference at the Hyatt Regency Baltimore Hotel, Baltimore, Maryland, May 7 - 10, 1991
Sponsored by the IEEE Power Engineering Society

A prototype system adopting the proposed method and designed for actual systems was constructed with engineering workstations. The effectiveness of the proposed method is demonstrated by test results.

2. PROPOSED METHOD

2.1 Modeling of Power System

The model for the power system used in this method is shown in Fig. 1, and keywords used are defined below:

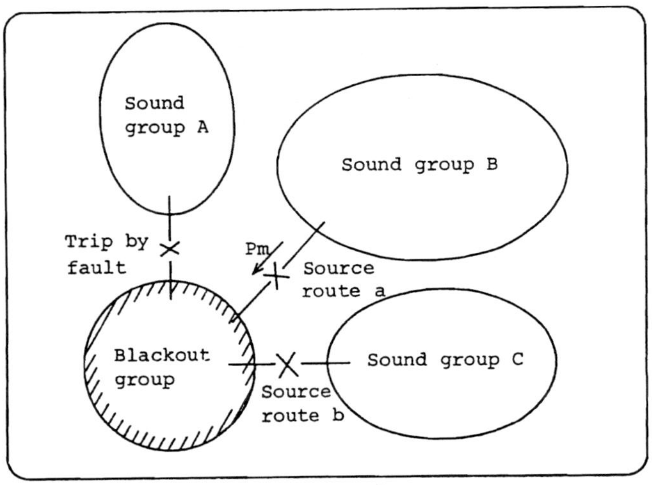

Fig. 1 System Model

- Charging group: A group of charging equipment electrically connected by means of circuit breakers and line switches.
- Blackout group: A group of blackout equipment electrically connected by means of circuit breakers and line switches.
- Sound group: A charging group which includes no overloaded pieces of equipment.
- Route: The generic name including all power transmission lines, transformers, and bus-ties.
- Source route: This refers to the route connecting the sound group with the blackout group. Power transmission lines are considered in a unit of one to two lines, and transformers are considered in a unit including parallel transformers.
- Support power: If no overloading occurs within sound group B when source route a is fed with its maximum allowable power flow value Pm, the afore-mentioned maximum power flow value Pm is adopted as the support power value for source route a. If an overload should occur within sound group B, the power flow value of source route a when the power flow rate in the route causing the overload is held at maximum power flow value is adopted as the support power value for source route a. The route in sound group B with the overload is then called the power flow neck route.

Reprinted from *IEEE Transactions on Power Systems,* Vol. 7, No. 2, pp. 914–920, May 1992.

• Restoration route: This refers to a source route where support power is equal to or greater than loads in the blackout group, with the following restrictions:
  ① No outage equipment is charged.
  ② No overload can occur in the system by connecting the source route.
  ③ No charging is made from the lower voltage side to the higher voltage side.

2.2 Constitution of Knowledge Base

Knowledge base constitution is explained with reference to Fig. 2.

The knowledge base consists of load-feeding knowledge, support power augmentation knowledge, and flow control knowledge, which carries out the reasoning control for those sets of knowledge, and these are described with production rules.

Load-feeding knowledge consists of blackout restoration knowledge and load-division knowledge.

Load-feeding knowledge is implemented in carrying out power transmission while attempting segmentation of the blackout load, while support power augmentation knowledge is the knowledge used for augmenting the amount of power that can be transmitted via the restoration route.

The simulation system takes the post-outage system as the initial state and in the course of its reasoning changes the system via the knowledge-base interface to eventually arrive at the restoration target system.

The loads in the system after restoration is equal to the loads immediately before the outage. Power generation volume is set to the value immediately before the outage for tripped hydraulic power generators, and "0" for tripped thermal power generators.

The present method aims to generate a system with a minimum of post-restoration outage loads through repeated application of load-feeding knowledge and support power augmentation knowledge on the system initialized to the post-outage state. The knowledge for flow control contains specific sets of knowledge.

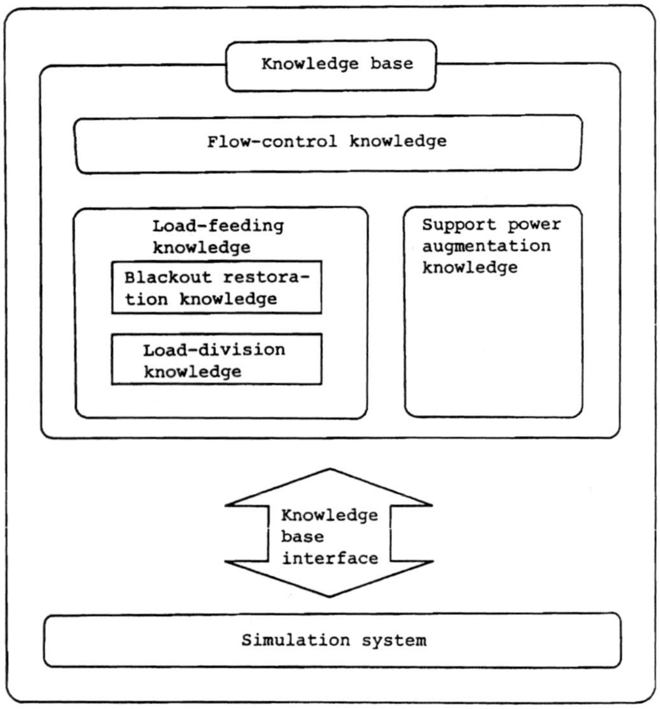

Fig. 2 Knowledge Base Configuration

Flow-control knowledge is explained with reference to Fig. 3.

(a) Load-feeding knowledge is applied on the simulation system taken as the post-outage system state to restore outage loads.
(b) When outage loads cannot be restored with load-feeding knowledge alone, support power augmentation knowledge is applied.
(c) After sets of knowledge applied in step (b) are registered, the simulation system is restored to the post-outage system state.
(d) Support power augmentation knowledge registered in step (c) is applied on the simulation system, after which load-feeding knowledge is reapplied.

The simulation system is taken as the restoration target system when all blackout groups have been restored, or when load-feeding knowledge and support-power augmentation knowledge are no longer applicable.

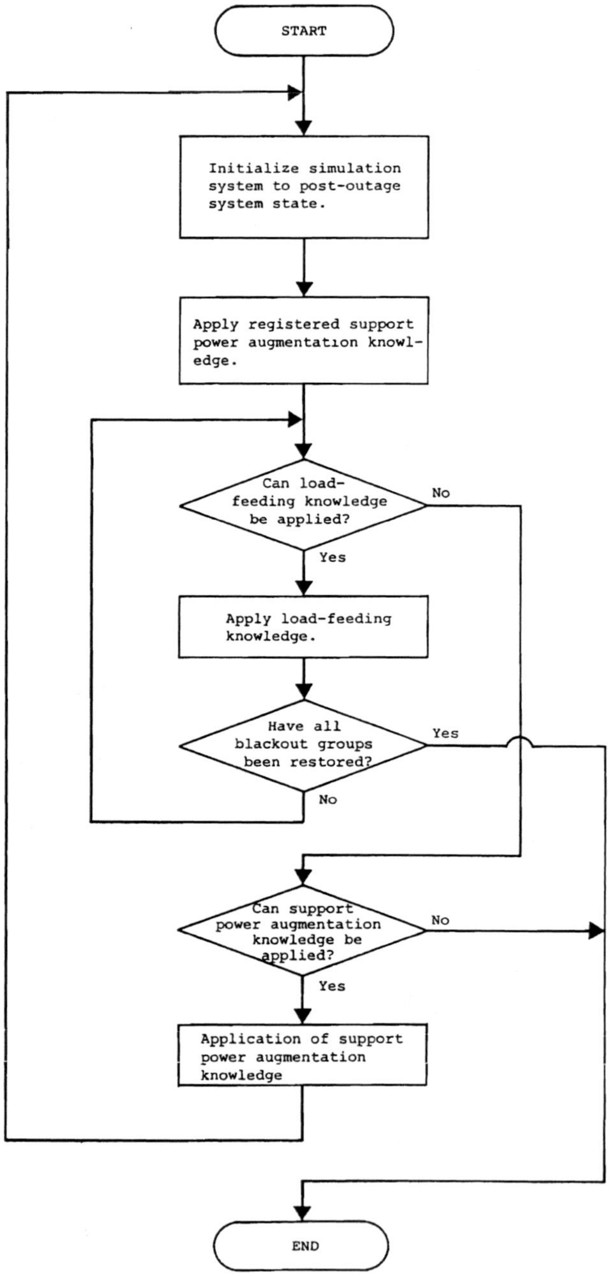

Fig. 3 Flow-control Knowledge

2.3 Knowledge Used for the Generation of Restoration Systems

Both load-feeding knowledge and support power augmentation knowledge are generally applicable, being independent of any particular system, as described below.

2.3.1 Load-feeding knowledge

(1) Blackout restoration knowledge

Blackout restoration knowledge is used for selecting restoration routes for and feeding power to blackout groups.

(a) If a restoration route exists, power is fed to the loads in the blackout group.

(b) If two or more restoration routes for the blackout group exist, a route is selected and loads are fed according to the following list of priorities.

1: Priority is given to restoration routes of higher voltages.

2: Priority is given to restoration routes which can restore blackout groups not restorable by other restoration routes.

3: Priority is given to restoration routes that have greater capacity for restoring outage loads.

4: Priority is given to restoration routes with greater support power supply rates.

(c) If no restoration routes exist, load-division knowledge is applied.

(d) If all blackout groups are restored, application of load-feeding knowledge is terminated.

(2) Load-division knowledge

Load-division knowledge is used to divide the loads so that they do not exceed support power. Specifically, a substation in the blackout group is singled out and considered, and load division is carried out by the severance of all routes connected to the bus in that substation.

(a) If substations exist which have not yet been considered, loads are divided according to the following criteria.

1: Priority is given to the substation of higher voltage.

2: Priority is given to the substation with the greater amount of support power available via adjacent restoration routes.

3: Priority is given to the substation with greater loads connected to the bus.

(b) If load-division knowledge has already been applied to one of the substations, blackout restoration knowledge is applied again.

(c) If all substations have been considered, application of load-feeding knowledge is terminated.

(3) Examples of application of load-feeding knowledge

Here is an example of application of load-feeding knowledge to systems after outage.

Step 1: Since no restoration route existed for the blackout group Z0 (Z-zero) in the post-outage system (as shown in Fig. 4-(1)), load division knowledge is applied. [blackout restoration knowledge (c)]

Step 2: Of the substations A S/S and B S/S which have not yet been singled out for consideration, A S/S, whose support power volume available from adjacent restoration routes is greater than that of B S/S, is focused on, and all routes connected with the bus in the substation are severed. [load-division knowledge (a)-2]

Step 3: Load-feeding knowledge is applied to the system on which load-division knowledge has already been applied (Fig. 4-(2)). [load-division knowledge (b)]

Step 4: Restoration route to the blackout group Z1 is performed via route a, whose amount of restored loads is greater than that of route b. [blackout restoration knowledge (b)-3]

Step 5: Power is fed to the loads in the blackout group Z2 via route b, whose support power rate is the greater of restoration routes b and d. [blackout restoration knowledge (b)-4]

Step 6: Power is fed to the loads in the blackout group Z6 via route g, whose amount of restored loads is greater than that of restoration route c, d, e, or f. [blackout restoration knowledge (b)-3]

Step 7: Power is fed to the loads in the blackout group Z5 via route f, whose amount of restored loads is greater than that of restoration route c, d, or e. [blackout restoration knowledge (b)-3]

Step 8: Power is fed to the loads in blackout group Z3 via route d, whose support power supply rate is greater than that of restoration route d. [blackout restoration knowledge (b)-4]

Step 9: Power is fed to the loads in blackout group Z4 via route e, which makes restoration of the blackout group feasible. [blackout restoration knowledge (a)]

Step 10: As all the blackout groups have been restored, application of load-feeding knowledge is terminated. [blackout restoration knowledge knowledge (d)]

Application of knowledge is carried out in the sequence described above, and a system in which AS/S has undergone bus-splitting is generated as the restoration target system. (Fig. 4-(3))

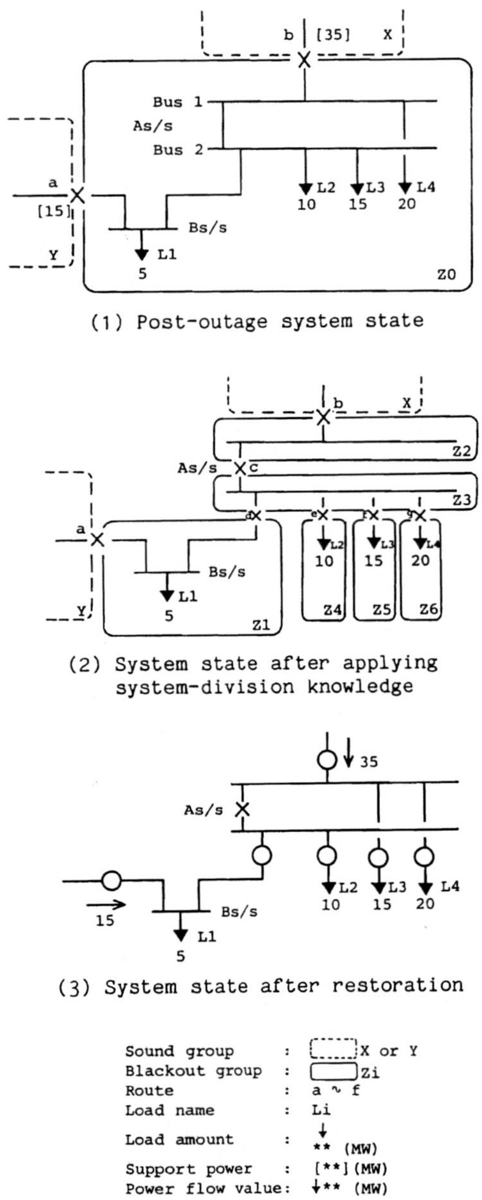

Fig. 4 Example of Application of Load-feeding Knowledge

2.3.2 Support power augmentation knowledge

Augmentation of support power takes place in cases where support power is insufficient for restoring outage loads. The knowledge includes the following:

(1) Equipment at recess utilization knowledge

Equipment arranged in parallel with the route presenting a power flow neck to the source route is added to increase the amount of support power.

(a) If a unit of parallel equipment at recess can be utilized in combination within the sound groups, the unit is utilized to increase the amount of support power.

(b) If two or more units of equipment at recess are available for use, the one offering a greater increment in support power when used in combination is given priority.

An example of the application of this knowledge is shown in Fig. 5-(1).

In the system shown in the figure on the left, route b, the power flow neck to source route a, constitutes an A S/S transformer, and a second transformer installed in parallel is in recess. By combined usage of the unit in recess, maximum power flow for route b, the power flow neck, is increased from 100 MW to 200 MW, with an increase of support power via source route a from 50 MW to 150 MW. (Refer to figure on right.)

(2) Generator output adjustment knowledge

Output of the generator locating at load side of the route presenting a power flow neck with respect to the restoration route is stepped up to increase support power.

(a) If there is any generator adjustable for its output at load side of the route constituting the power flow neck within the sound group, power output is increased for that generator.

(b) If there are two or more generators whose outputs are adjustable, output is increased for the one promising a greater increment in support power.

An instance where this knowledge is applied is shown in Fig. 5-(2).

In the system shown on the left, the generator C P/S locates at the load side of route b constituting the power flow neck to source route a. If the output of this generator is increased from 80 MW to 100 MW, the power flow in route b constituting the power flow neck is decreased from 70 MW to 50 MW, causing support power along source route a to be increased from 30 MW to 50 MW. (Refer to figure on the right.)

(3) Overload allowance knowledge

This knowledge allows overload power transmission along the route constituting a power flow neck to the source route, thus increasing amount of support power.

(a) If overload power transmission is possible along the route constituting the power flow neck within the sound group, such overload power transmission is performed.

(b) If there are two or more routes presenting power flow necks which can accommodate overload power transmission, the one affording the greatest increment in support power when overload power transmission is allowed is selected.

An instance where this knowledge is applied is shown in Fig. 5-(3).

In the system shown on the left, route b presents a power flow neck to source route a. Allowing maximum power flow for the route to increase from 100 MW to 120 MW entails an increase in support power for the source route from 50 MW to 70 MW. (See figure on the right.)

(4) Sound group load-switching knowledge

With this knowledge, support power is increased by switching the load at load side of the route presenting the power flow neck to the sound group to an adjacent sound group.

(a) If it is possible to switch the load locating at load side of the route presenting a power flow neck within the sound group to an adjacent sound system, a load-switching is performed for that load.

(b) If there are two or more cases where load-switching is possible, priority is given to the one offering a greater increment in support power.

An instance where this knowledge is applied is shown in Fig. 5-(4).

In the system shown in the figure on the left, there is substation C S/S contacting with adjacent sound group Y at the load side of route b presenting the power flow neck to source route a. If the load of this C S/S is switched over to the sound group Y, the power flow through route b presenting the power flow neck is reduced from 100 MW to 50 MW, causing an increase in support power along source route a from 50 MW to 100 MW. (See figure below.)

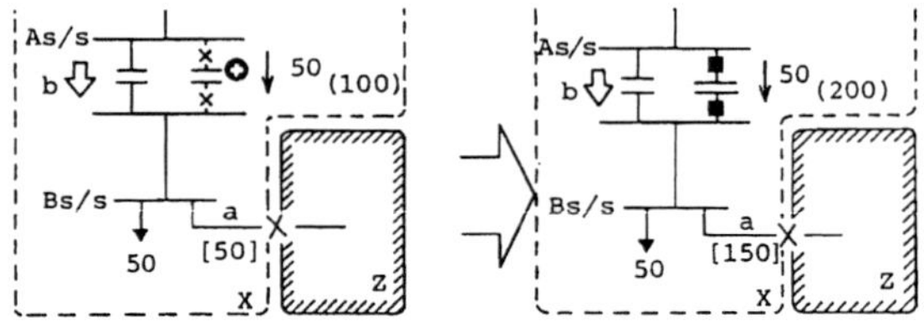

(1) Combined use of equipment in recess

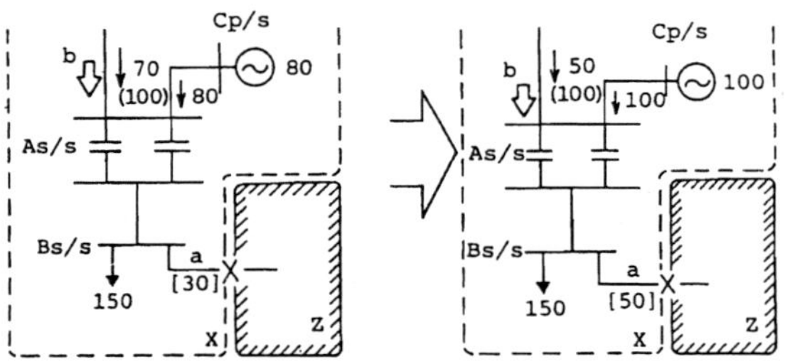

(2) Adjustment of generator output

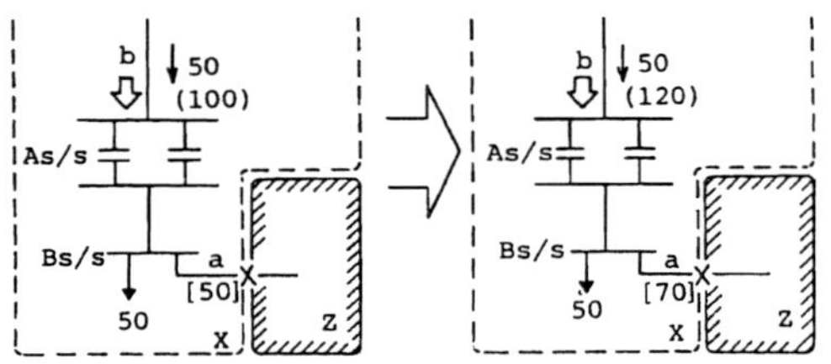

(3) Overload allowance

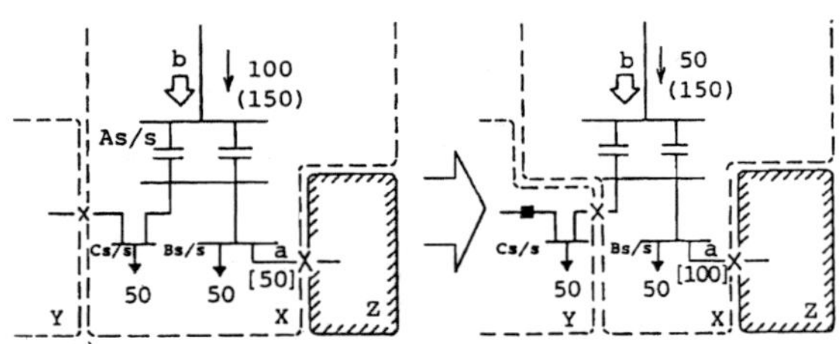

(4) Load-switching of sound group system

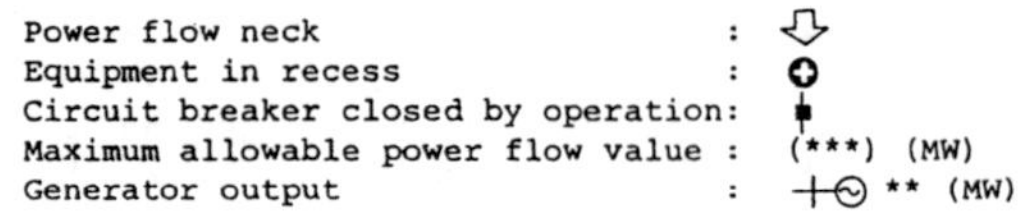

Fig. 5 Example of Application of Support Power Augmentation Knowledge

3. TESTING THE PROTOTYPE SYSTEM

3.1 Test Subject

The power system selected for prototype is a real-scale model system including thermal and hydraulic power generators of 275 kV to 77 kV classes, with five units of thermal power generators and 10 units of hydraulic power generators and substations at 40 locations. The power system has a loop configuration of 275 kV parts, and parts of 154 kV or less form a radial system configuration. X S/S and Y S/S in the system diagram are other sound groups adjoining the system. (Fig. 6)

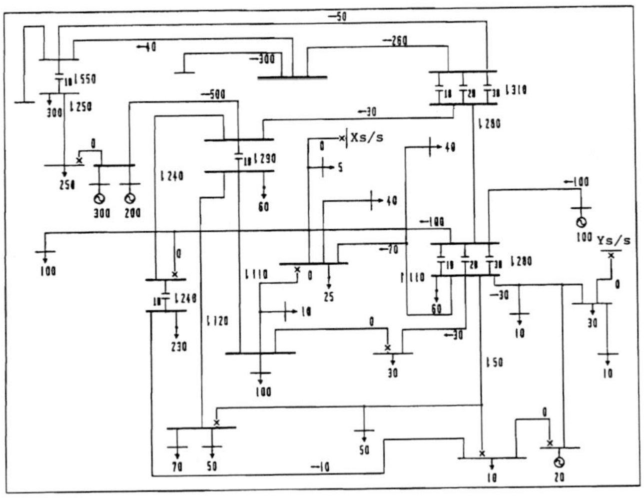

Fig. 6 Pre-outage System State

3.2 Selection of Test Cases

Faults occurring in the system can be classified into three major categories: line fault, bus fault, and transformer fault. They can also be classified by voltage ratings for equipment used (i.e., 275 kV, 154 kV, or 77 kV). In selecting fault cases, types of faults and equipment voltage ratings were taken into consideration, with 9 (3×3) cases of outage equipment selected from among cases likely to occur in actual systems and cases of accidents experienced in the past. Furthermore, system states were classified according to total loads in the pre-outage system into three stages: light load, medium load, and heavy load. For each of these loading conditions, outage equipment were tested for the nine cases mentioned above, with a total of 27 (3×9) cases.

3.3 Examples of Testing and Observations

3.3.1 Examples of testing

In carrying out actual testing on cases of transformer fault (in 154 kV rating), results of application of the present method on two cases of system state, one with medium load and the other with heavy load, were evaluated.

(1) Execution result under medium load

With the post-outage system under medium load (Fig. 7), a condition was assumed in which a transformer fault at A S/S caused total blackout at stations B S/S and C S/S. There is only one blackout group under this condition with outage loads of 200 MW. Routes a, b, and c are available as restoration routes for the blackout group. Support power to route a is 160 MW with route g constituting the power flow neck. Power flow necks for routes b and c are routes f and e, respectively, rendering 180 MW and 40 MW, respectively, in support power.

Below is the procedure with which restoration knowledge was applied to the system.

Step 1: Since no restoration routes are available for the blackout group, load-division knowledge is applied. [blackout restoration knowledge (c)]

Step 2: Of the power stations A S/S, B S/S, and C S/S that have not been considered yet, all routes connected with C S/S, which with its adjacent restoration routes can provide a greater amount of support power, are separated. [load-division knowledge (a)-2]

Step 3: Load-division knowledge is applied to the system from which all routes connected with C S/S are severed. [load-division knowledge (a)]

Step 4: Power is fed through route b to load L1 in A S/S and loads L4 and L5 in B S/S. [blackout restoration knowledge (b)-3]

Step 5: Power is fed via route a in the following sequence: bus in C S/S, load L2, and finally load L3. [blackout restoration knowledge (b)-3]

Step 6: Since all the blackout groups have been restored, application of knowledge is terminated. [blackout restoration knowledge (d)]

By the application of these sets of knowledge, the post-restoration system (Fig. 8) is obtained.

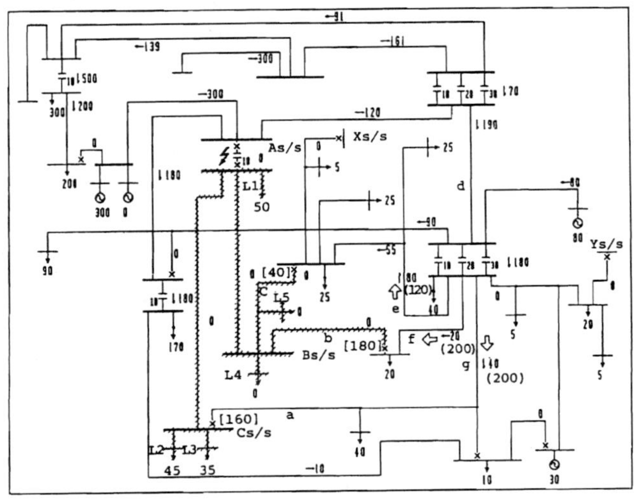

Fig. 7 Post-outage System State under Medium Load

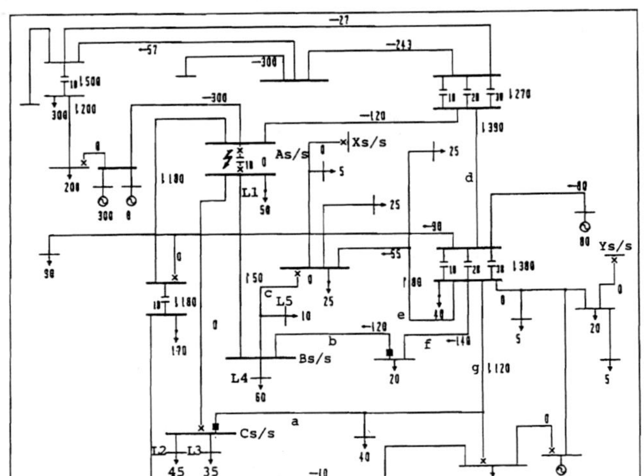

Fig. 8 Restoration Target System under Medium Load

(2) Execution result under heavy load

The same fault as in the medium-loaded state is assumed for the post-outage system state (Fig. 9) which is subject to heavy load.

Under the conditions, the number of blackout groups is 1, with outage loads amounting to 290 MW. In this case routes a, b, and c are available as source routes for the blackout group. Route d becomes the power flow neck for routes a and b, and the amount of support power comes to 120 MW. Route e becomes the power flow neck for route c and its support power amounts to 20 MW.

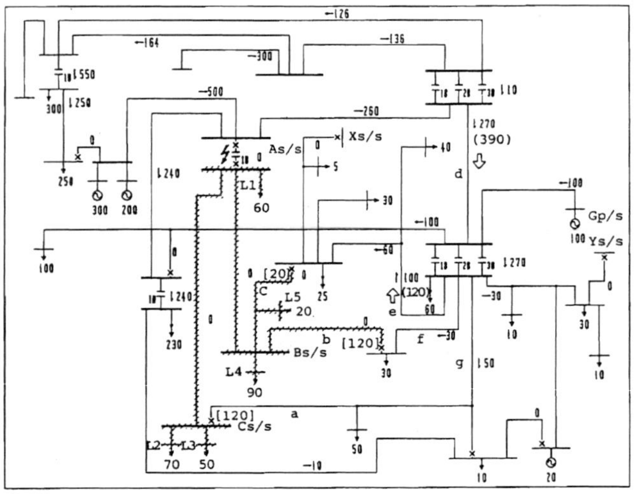

Fig. 9 Post-outage System State under Heavy Load

Step 1: Since no restoration routes are available for the blackout group, load-division knowledge is applied. [blackout restoration knowledge (c)]

Step 2: Of the power stations A S/S, B S/S, and C S/S that have not yet been considered, all routes connected with C S/S, whose loads connected with the bus is greatest, are separated. [load-division knowledge (a)-3]

Step 3: Blackout restoration knowledge is applied to the system from which all routes connected with C S/S were separated. [load-division knowledge (b)]

Step 4: Power is fed via route a in the following sequence: bus in C S/S, load L2, and finally load L3. [blackout restoration knowledge (b)-3]

Step 5: By feeding power to loads L2 (70 MW) and L3 (50 MW) in C S/S, route d becomes the power flow neck for all source routes in the simulation system, and the amount of support power becomes 0 MW. Accordingly, outage loads cannot be restored solely by the application of load-feeding knowledge, so support power augmentation knowledge is applied. [flow-control knowledge (b)]

Step 6: At this point, power generator G P/S which is adjustable for its output at load side of route R4 constituting the power flow neck becomes available. [generator output adjustment knowledge (a)]

Step 7: The adjusted generator output obtained in 6 (generator output increase from 100 MW to 130 MW) is registered as preprocessed data, and the simulation system is restored to the post-outage system state. [flow-control knowledge (c)]

Step 8: The support power augmentation knowledge regis-tered in step 6 is applied to the simulation system, and subsequently load-feeding knowledge is reapplied. [flow-control knowledge (d)]

Step 9: In the simulation system to which support power has been augmented, support power increases by 30 MW, providing 150 MW via routes R2 and R3. It should be noted, however, that the mere application of load-feeding knowledge results in repetition of steps 1 through 5 above, since outage loads can no longer be restored with load-feeding knowledge alone. Therefore, support power augmentation knowledge is reapplied.

Step 10: At this stage, overload power transmission is allowed for routes d and e constituting the power flow necks. [overload allowing knowledge (a)]

Step 11: Overload allowance for the power flow necks obtained in step 10, i.e., power flow limit: 390 MW to 430 MW for route d; 120 MW to 135 MW for route e, is added as pre-processed data, and knowledge that has been applied to the simulation system so far is discarded to restore the simulation system to the post-outage system state. [flow-control knowledge (c)]

Step 12: Support power augmentation knowledge registered in steps 7 and 11 is applied to the simulation system and load-feeding knowledge is subsequently reapplied.

Step 13: In the simulation system to which support power have been augmented, the amount of support power available via routes a and b becomes 190 MW, and that available via route c becomes 35 MW. Even if load-feeding knowledge is applied, however, the result is the same process as in steps 1 through 5 above, since load-feeding knowledge alone is insufficient for restoring outage loads, and therefore support power augmentation knowledge is applied again.

Step 14: At this point, load-switching is performed by switching the loads in the sound groups locating at load side of the routes constituting the power flow necks to adjacent sound systems. Such system switching as described here includes switching to the sound group containing X S/S and to the sound group containing Y S/S. The load-switching to the side containing X S/S allows for a load-switching of 60 MW, while that to the side including Y S/S allows for a load switching of 40 MW. This means that the switching to the side including X S/S can provide a greater increment in support power. [sound group load switching knowledge (a)]

Step 15: The load switching obtained in step 14 above (i.e., switching the 60 MW load to the sound group containing X S/S) is added as pre-processed data, and the simulation system is restored to the post-outage system state. [flow-control knowledge (c)]

Step 16: Support power augmentation knowledge registered in steps 7, 11, and 15 is applied to the simulation system, followed by application of load-feeding knowledge. [flow-control knowledge (d)]

Step 17: In the simulation system for which support power has been augmented, the amount of support power available via routes a and b becomes 250 MW. The application of load-feeding knowledge alone is unable to restore the state of outage, and only results in the same process as steps 1 to 5. Support power augmentation knowledge is therefore applied again.

Step 18: At this point, load-switching is performed. Changeover to the sound group including Y S/S is available for switching of the system in question, enabling switching of the 40 MW load. [sound group load-switching knowledge (a)]

Step 19: The load-switching obtained in step 18 (the switching of the 40 MW load to the sound group including the Y S/S) is added as pre-processed data, and the simulation system is restored to the post-outage system state. [flow-control knowledge (c)]

Step 20: Support power augmentation knowledge registered in steps 7, 11, 15, and 18 is applied to the simulation system, followed by application of load-feeding knowledge. [flow-control knowledge (d)]

Step 21: The amount of support power available via routes a and b becomes 290 MW in the simulation system to which support power was augmented. Of substations not considered yet within the blackout group (A S/S, B S/S, and C S/S), all routes connected with C S/S, whose loads connected with the bus is the greatest, are separated. [load-division knowledge (a)-3]

Step 22: Blackout restoration knowledge is applied to the system for which all routes connected with C S/S have been severed. [load-division knowledge (b)]

Step 23: Power is fed via route b to load L1 in substation A S/S as well as to loads L4 and L5 in station B S/S. [blackout restoration knowledge (b)-3]

Step 24: Power is fed via route a in the following sequence: bus in C S/S, load L2, load L3. [blackout restoration knowledge (b)-3]

Step 25: Since all blackout groups have been restored, application of knowledge is terminated. [blackout restoration knowledge (d)]

A post-restoration system (Fig. 10) is obtained by application of the above-mentioned sets of knowledge.

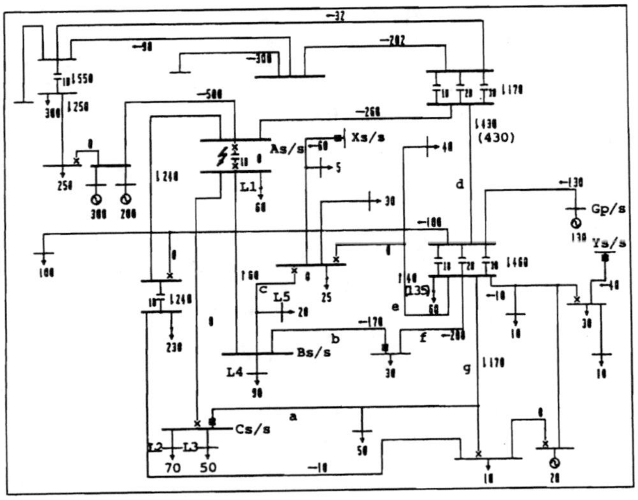

Fig. 10 Restoration Target System under Heavy Load

3.3.2 Observations

When the proposed method is applied to the system state under a medium load, restoration is accomplished solely with operations performed within blackout groups and no operations are performed within sound groups.

When the system is applied to the system state under heavy load, augmentation of support power used for the operations in sound groups is achieved not only with operations within blackout groups but also with application of support power augmentation knowledge.

In test cases involving heavy loads, not all of outage loads can be restored with methods that do not rely on augmentation of support power. Therefore, with the proposed method, augmentation of support power was attempted by applying generator output adjustment knowledge, overload allowance knowledge, and sound group load switching knowledge included in the support power augmentation knowledge in order to generate restoration target systems without leaving outage loads.

As a result of testing, the present method exhibited a high degree of practicality even in cases of support power deficiency as experienced with outages occurring under heavy load, not to mention cases with ample amount of support power for outage loads as experienced with outages under light to medium loads.

For other test cases as well, good agreement with actual post-restoration system states has been confirmed by veteran operators.

In terms of processing time, testing results of 6 seconds for the system state under medium load and 50 seconds for the system state under heavy load were obtained. In other cases, processing times obtained have been three seconds at the shortest and 100 seconds at the longest, which is considered to be quite practicable. Note that processing times obtained represent actual records taken at engineering workstations capable of 7 MIPS, and the implemented knowledge base consists of about 200 rules in OPS83. [6]

4. CONCLUSIONS

This paper proposes an approach for augmenting support power that better restores outage loads by not only performing load-feeding to blackout systems but also performing operations in sound systems. A prototype system was prepared and tested to confirm the effectiveness of this proposal.

Areas of future study include a review on the realization of more practical restoration target systems by incorporating knowledge about the start-up of thermal power generators and on the voltages and stability at the time of restorative operations.

REFERENCES

[1] T. Sakaguchi et al., "Development of a Knowledge-based System for Power System Restoration," IEEE Trans. on PAS, Vol. PAS 102, No. 2, pp 320-329, Feb. 1983

[2] Y. Kojima et al., "Development of a Guidance Method for Power System Restoration," IEEE Trans. on PS, Vol. PWRS 4, No. 3, Aug. 1989, pp 1219-1227

[3] K. Hotta et al., "Implementation of a Real-time Expert System for a Restoration Guide in a Dispatching Center," PICA '89, May 1989, Seattle

[4] S. Moriguchi et al., "A Large-Scale SCADA System with Real-time Knowledge-based Functions," Second Symposium on Expert Systems Application to Power Systems, July 17-20, 1989

[5] J. Zaborasky et al., "Restoration of the Large Electric Power System Using a Computer Generated Sequence of Target System," IFAC Symposium on Power System and Power Plant Control, Beijing, pp 301-306, August 1986.

[6] C. L. Forgy, "The OPS83 User's Manual," Department of Computer Science, Carnegie Mellon University, Maroh 1985

Generation Capability Dispatch for Bulk Power System Restoration: A Knowledge-Based Approach

Chen-Ching Liu and Kan-Lee Liou
Department of Electrical Engineering
University of Washington

Ron F. Chu
Philadelphia Electric Company
Philadelphia, PA

Arne T. Holen
Electric Power Systems
Norwegian Institute of Technology
Trondheim, Norway

Abstract - This paper presents our results of the development of a knowledge-based system (KBS) for power system restoration. An extensive list of critical constrains and parameters is identified. Based on these constraints and parameters, a goal-driven strategy for system restoration is developed. Part of the strategy is a procedure to determine the generator starting times, which helps maximize the MWH load served over a restoration period. The KBS is implemented in the KES software. Simulations of the KBS are performed with data from the Philadelphia Electric system.

Keywords - Power System Operation, Knowledge-Based System, System Restoration.

1. Introduction

Restoration of a power system following a blackout is a complex, stressful and time-consuming task [1-4]. Power system dispatchers coordinate with power plant operators to start the generating units with black-start capability. Transmission paths need to be identified so that the power can be used to crank other non-black-start generating units and to pick up the necessary loads to stabilize the power system. Switching actions should follow guidelines established by off-line studies to minimize switching transients, instability problems, damage to equipment, etc. Much of the complexity of system restoration is introduced by the large number of generation, transmission and distribution, and load parameters/constraints. Some constraints are functions of time. For example, a generating unit can become unavailable as the boiler-turbine set cools down following a blackout.

At present, there is no commonly-accepted computer method for the solution of the overall restoration problem. The application of knowledge-based techniques to system restoration has been an area of great interest. The pioneering work in [5] deals with identification of alternative paths to pick up the de-energized loads. Their system was further developed to provide intelligent "guidance" to system dispatchers during restoration [6]. The approach proposed in [7] consists of an energizing and a load supplying phases. Reference [8] reports a conceptual framework which integrates analysis programs and knowledge-based systems around a blackboard structure. The method proposed in [9] separates system restoration into a strategic reasoning step which determines the objective and leaves construction of the switching path to the tactical reasoning. An automated knowledge-based system for restoration was developed in [10].

92 WM 108-1 PWRS A paper recommended and approved by the IEEE Power System Engineering Committee of the IEEE Power Engineering Society for presentation at the IEEE/PES 1992 Winter Meeting, New York, New York, January 26 - 30, 1992. Manuscript submitted August 29, 1991; made available for printing November 20, 1991.

References [11-12] report knowledge-based approaches to restoration of the subtransmission and distribution systems. None of the above-mentioned approaches consider the dynamic (time-varying) constraints related to power plants. A recent paper (in Japanese) reports an approach to handling dynamic constraints related to thermal plants [13]. In the paper, five types of generating units are considered; the first type includes the units operating in the "hot" state, ready to be synchronized with the power system. The second and third types are determined by the vacuum state of the condenser; if vacuum is maintained as in Type 2 units, then starting can be achieved quickly. Otherwise, a longer delay is expected, which is the case for Type 3 units. Types 4 and 5 are hydro units with and without self-starting capability, respectively. Needless to say, Type 1 units are preferred during restoration. The consideration of dynamic constraints in the approach is very significant. However, the approach does not systematically deal with the worst case - a total system shutdown, which requires complex coordination to allocate the cranking power from the black-start units to other units. Furthermore, it considers only a limited number of system constraints; for example, the critical minimum interval constraints of super-critical-once-through units are not incorporated.

The purpose of this research is to develop a knowledge-based system (KBS) for generation capability dispatch during restoration of a bulk power system. This paper reports on the results of a survey of critical constraints/parameters during restoration, handling of constraints/parameters, sequencing of generator start-ups, transmission path-finding, software design, and simulations. The proposed system considers a total system shutdown and hence it is necessary to handle complex sequencing of generator start-ups. Restoration of a partial shutdown can be considered a subproblem of the total shutdown and can be handled by the proposed method as well. Specifically, the contributions are:

1. Identification of an extensive list of critical constraints and parameters.
2. Development of a goal-driven restoration strategy incorporating generation, transmission and load models and constraints.
3. Development of a systematic method to help maximize MWH load served by generating units, considering critical, including dynamic, constraints and parameters.

This project is sponsored by Electric Power Research Institute (EPRI). The prototyping of the proposed KBS has been completed and tested with data from Philadelphia Electric Company. The integration with other power system application software (e.g. power flow, dynamics) is being pursued.

The organization of this paper is as follows. Section 2 reports the critical parameters/constraints identified in this study. The restoration strategy and the sequencing of generator start-ups are discussed in Section 3. The software design and implementation is presented in Section 4. Section 5 provides the simulation results based on the Philadelphia Electric (PECo) system.

Reprinted from *IEEE Transactions on Power Systems,* Vol. 8, No. 1, pp. 316–325, February 1993.

2. Problem Formulation

Generating units have different physical characteristics and requirements. A proper sequence of unit start-ups can avoid violation of constraints while increasing the system MW output. Providing cranking power to a non-black-start unit does not serve load directly but it generates more MW as soon as this unit ramps up. To illustrate the handling of generation constraints, an example with four generating units is given [14]. For this example, fictitious data are used for power and time. The data should be considered normalized quantities with respect to some base values of power (MW) or time (minute).

G1: Fossil, start-up power requirement: 1, critical maximal interval: 5.

G2: Super-critical-once-through (SCOT), start-up power requirement: 1, critical minimum interval: 5.

G3: Fossil, start-up power requirement: 2 , critical maximal interval: 4.

G4: Black-start unit.

The MW capability curves are shown in Figure 1. In Figure 2, the system MW capability is the total system MW capability minus the start-up requirements. The term "capability" is used here since the load is not considered and, therefore, the actual generation may be lower. The black-start unit cranks G3 first since the starting of G3 will be delayed much longer if the critical maximal interval, 4 units, expires. In contrast, G2 can not be cranked until after the critical minimal interval. It can be verified that the unit start-up sequence shown in Figure 2 results in the maximal MWH load served.

System restoration requires coordination among generating units, transmission and distribution switching actions, and load pick-ups. This section provides the specific restoration actions and system parameters/constraints incorporated in the proposed methodology.

Objective: To maximize the MWH load served over a restoration period

Restoration Actions:

Start up generating units with black-start capability
Provide cranking power to non-black-start generating units
Switch breakers/switches
Pick up/Shed loads

System Parameters/Constraints

The following list of parameters and constraints is compiled from utility interviews and literature review conducted for this study. The reference(s) following each item indicate(s) the sources which specify the need to consider the item. The symbol A or S means that a specific constraint is suggested by M. M. Adibi, IRD Corp., or J. States, retired PECo dispatching supervisor, respectively.

1. Power Plant
   - Max MW output (normal or emergency) [6, 7, 13, 15-16, A, S]
   - Rated power factor (from generator capability curves available for different H_2 pressure conditions) [15, A, S]
   - Reactive power over- and under- excitation limits (from generator capability curves) [6, 7, 15, A, S]
   - Start-up and houseload requirement (MW & MVAR) [7, 15-16, A, S]
   - Start-up times [7, 15-16, A, S]
   - Ramping rates [13, 15-16, A, S] (Ramping from paralleling to releasing) (Ramping from releasing to max output)
   - Frequency response of prime mover to sudden load pick-up [6, 15, A]
   - Critical minimum & maximum intervals for different types of boilers [13, 16, A]
2. System Considerations
   - Reactive power balance [6, 7, A, S]
   - Load and generation balance [6, 7, 13, 15, A, S]
   - Under-frequency load shedding (Relay Modeling) [A, S]
   - Stability limits during restoration [13, A, S]
   - High and low sustained voltage limits [13, A, S]
   - MW and MVAR reserve requirement during restoration [A]
   - Phase difference in a loop configuration [7]
3. Lines & Substations
   - Switching operations [6, A, S]
   - Line thermal limits (overload of system equipment) [6, 7, A, S]
   - Line stability limits [13, A]
   - Voltage transients when energizing lines [13, A]
4. Load Behavior
   - Cold load pick-up (load models) [A]
   - Recovering load models which store previously evaluated dynamic data for different load types [15]

The parameters/constraints internal to power plants identified in our survey can be found in [14].

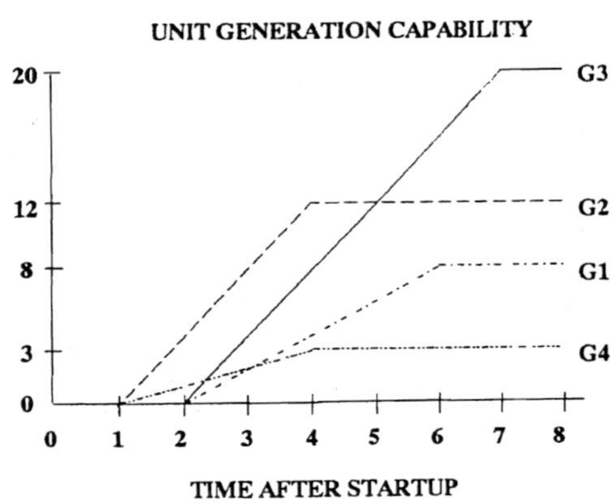

Figure 1. Generation capability curves of 4 units

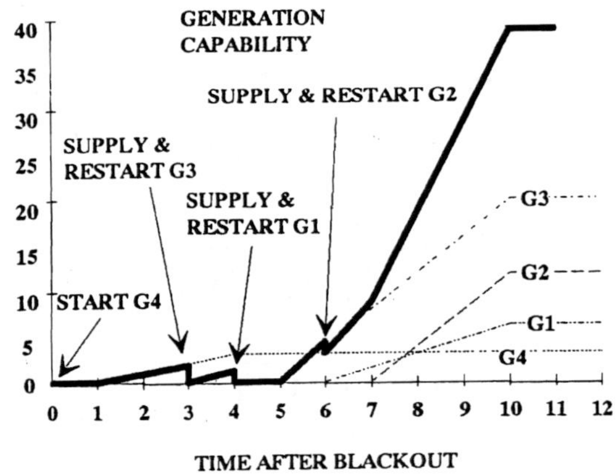

Figure 2. System generation capability of 4-unit example

Flexibility and Optimality

During restoration unexpected situations can occur, e.g., faults, failures, stuck breakers, shortage of man-power. During our knowledge acquisition process, dispatchers from PECo indicated that a knowledge-based system would be valuable if it can generate alternative actions as unexpected situations arise. For example, if network help is unavailable, finding an alternative path is a major issue. In other words, a restoration KBS should be able to adapt to the new environment as the system condition changes. Flexibility and optimality can be conflicting criteria in restoration planning. Optimal switching actions result in higher system MW output; however, it is time-consuming to search for the true optimal sequence. Flexibility is judged to be a higher priority criterion for the proposed methodology.

3. Knowledge-Based System Methodology

This section describes the proposed restoration methodology, sequencing of generating units, and path finding.

3.1 Restoration Strategy

In our approach, restoration planning is viewed as a goal-driven process depicted in Figure 3. The highest level goal, "System Restoration," is to maximize the MWH load served while maintaining a secure operating condition. In a practical environment, this module corresponds to the Power Director (system restoration planner) who coordinates the process. To achieve the goal, "System Restoration" acquires information from "MW Management," "MVAR Management," "Path Management," "Load Management," and "Stability Inspection," etc. Each module represents the rules and procedures to handle the specific task delegated by the "System Restoration" module. Each module also represents a lower level goal, or subgoal. For example, to achieve the subgoal, "MW Management," it is necessary to obtain the MW capability curves of black-start and non-black-start units. To prepare these curves, the status of the units, i.e., mechanical, fuel, and man-power, has to be checked. The communication among the modules is interactive. The goal-driven strategy leads naturally to an implementation with the inference mechanism provided by KB system development tools.

The detailed functions served by the main modules are described in the following.

MW Management

After the blackout, black-start units are started as soon as possible. Units with critical maximum intervals are ranked higher in priority to receive cranking power.

Some units have preassigned black-start units to supply cranking power. If all such preassigned units are available and the transmission paths are also clear, the assignment problem is trivial. The problem becomes difficult when some units become unavailable or equipment failures occur.

Some non-black-start units have the critical maximum or critical minimum interval constraints. As mentioned in Section 2, if a unit with a critical maximum interval is not started within the interval, there may be a delay of hours or days before the unit becomes available again. A unit with a critical minimum interval can not be started until the interval ends. A simplified start-up curve of a unit requires only the time between supply of the cranking power and the minimal generation. An elaborate model includes time instants for (1) supply of cranking power, (2) paralleling, (3) releasing and (4) maximal output. Both models are incorporated.

The power plant operators in charge of non-black-start units are asked to respond to the system restoration module's question: "How far away are you ?" The answer indicates the time required for the plant to finish preparations to receive cranking power. The time will be entered manually to the specific constraints for the units. This is the situation at PECo.

Once a unit is started, its generation capability is reported to the system restoration module. External sources from tie lines are handled by "MW Management."

MVAR Management

The MVAR management module assesses the MVAR capabilities of generating units and the MVAR which would be produced if a transmission path is energized.

This requires calculations to determine the MVAR amount. The MVAR capability of a generator can be found from the generator capability curve. Combustion turbines (CTs) have limited MVAR absorbing capability; hence, it is infeasible to start a CT if the MVAR produced by the transmission path can not be absorbed.

Path Management

This module identifies a path and examines the status, i.e., faults, line capacity, and transients due to switching. The shortest path, i.e., the path that requires the minimal number of switching actions, between the supplying and receiving units is desirable unless it is not feasible.

This module suggests regulated operations after a blackout without command from the system restoration module, e.g., closing the sectionalizing and tie breakers and opening the ring breakers in a substation.

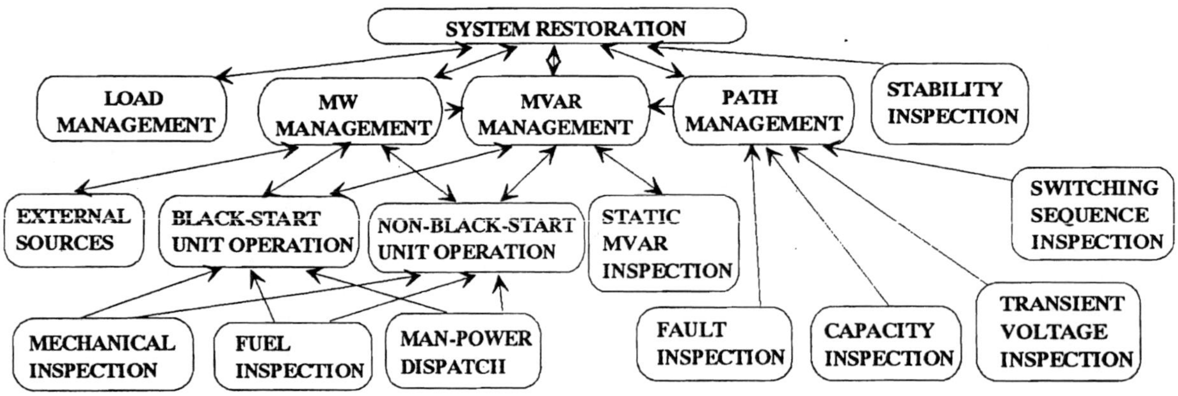

Figure 3. Restoration strategy

Load Management

During restoration, loads are restored based on the priority and system security considerations.

Loads can be restored to absorb MVAR produced by transmission lines in order to maintain the system voltages. This module also includes models for the load behavior such as cold load pick-up.

Switching Sequence Inspection, etc.

The switching actions are coordinated to minimize problems such as overvoltage, voltage transients.

The energization from the low voltage side to high voltage side of a transformer is prohibited. Line and transformer capacities should not be exceeded during restoration. The insulation of oil-pressured cables can be ensured only when the oil pressures are higher than the critical values.

3.2 Start-Up Sequence & Temporal Reasoning

The start-up sequence of generating units is determined according to Figure 4. The proposed search method uses heuristics to identify a feasible sequence. If it is desirable to obtain the optimal sequence, search can continue by finding the next promising sequence. The process can iterate until the optimal sequence is found. Even though the proposed search is able to obtain the optimal sequence, the process may be very time-consuming. The primary goal here is to find a good, feasible solution by heuristics.

Finding a feasible start-up sequence

The following heuristic procedure is proposed:

1. *The black-start units are started first. The units with critical maximum intervals are higher in priority to receive cranking power.*
2. *If the cranking power is insufficient, the critical maximum interval unit with the most urgent time constraint is cranked first.*
3. *If two or more units are equally urgent, the one with lower MW rating is cranked first.*

A unit with lower MW rating can reach its minimum generation level sooner and the generation can be used to crank other units. In contrast, cranking a large but slow-ramping unit may not pay off since the time constraints of some critical maximum interval units may be violated.

4. *If all units with critical maximum intervals are started, then the unit which can generate the highest MWH within the restoration period is given a higher priority.*
5. *If any critical maximum interval unit is not started, say, due to insufficient cranking power, starting of a different unit can be considered. However, if cranking a unit x prohibits the start of a critical maximum interval unit y sometime before the interval expires, then unit x will not be started.*

The temporal relation between starting of a unit x and a critical maximum interval unit y is interesting [17]. The temporal logic requires search to determine if starting a unit now will prohibit the start-up of a critical maximum interval unit later. The temporal reasoning is performed by the search shown in Figure 5. The search starts with a possible sequence of units beginning with x; this sequence does not indicate the exact starting time instants. The starting times are determined by checking the system capability and the start-up requirements. If the unit y can not be started before the critical maximum interval ends then "starting x implies inability to start y."

The search process for temporal reasoning is straightforward at this time, but an efficient search algorithm can be used in the future to implement the temporal reasoning command.

Finding the optimal start-up sequence

The above heuristic procedure does not guarantee optimality of the start-up sequence. The optimality is measured by the total MWH that can be served during the restoration period, say 4-8 hours. The total MWH is the area under the system MW capability curve over the period. After the feasible sequence is obtained based on the above procedure, if optimality is desired, search can continue.

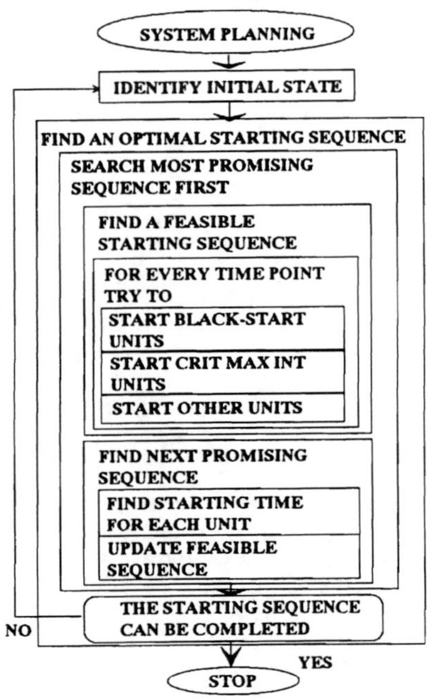

Figure 4. Start-up sequence determination

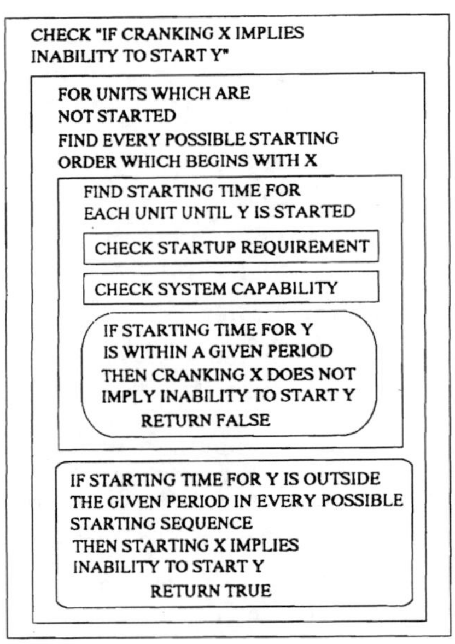

Figure 5. Temporal reasoning

3.3 Restoration Procedures

To deliver cranking power to non-black-start units, it is necessary to build transmission paths between black-start and other non-black-start units. System restoration starts with sectionalization of the system into islands and the restoration of individual island. Suppose the cranking power allocation is determined by the procedure in Section 3.2. Let BU be a black-start unit and MU be a main unit to be cranked.

1. *Close sectionalizing breakers and tie breakers. Open ring breakers. Open the line breakers connected to the stations.*
2. *Set tap changers to predetermined settings.*
3. *Perform A* search to build a path from BU to MU.*

A* search is a heuristic algorithm for the identification of the shortest path between given nodes in a network. Under some conditions, the optimality of the path is guaranteed [18]. Normally, Path Management suggests the default path. If, along the path, there is a fault on a bus, then find a sub-path of which the starting point is the bus before the fault location and the ending point is the bus after the fault. A* search is used for the optimal path identification. Now suppose such a sub-path cannot be found. Then, if the starting point is the beginning of this path and the ending point is the destination, then the path does not exist. Otherwise, move the starting bus backward and/or move the ending bus forward and keep searching. Similarly, if there is a line fault, find another line or sub-path.

To find a sub-path, Path Management not only asks MVAR Management to check MVAR requirements, and Load Management to find loads which can satisfy MVAR requirements, but also asks Fault Inspection, Capacity Inspection, Transient Voltage Inspection, Switching Sequence Inspection, etc. to check if there is any problem along the path.

Inside MU station, prepare the breakers for this path. If a fault or failure is encountered, repeat step 3.

Build the path section by section. A section is composed of two neighboring stations and the path between them. There is no other station between two neighboring stations. Restore the first section of the path. Start BU. Pick up/drop sufficient real and/or reactive loads along this section to stabilize frequency and voltages. If the whole system has been restored, stop. Else, restore the next section, breakers, lines, and buses.

4. *If the MU station is ready and it receives cranking power, start MU. If it is not ready, repeat Step 3 for another unit MU.*
5. *Restore other substations in this island.*
6. *Synchronize the islands.*

Generating units with local cranking power can not be started unless there is enough load to stabilize the unit.

4. Software Implementation

This section describes the proposed models for the critical parameters/constraints given in Section 2 and implementation of the knowledge-based system.

In the prototype developed with the Knowledge Engineering System (KES) software, the power system components are organized into "classes," which is a term for object-oriented data structure. Each class is a group of members which share the same "attributes," or characteristics. Attributes of a class can be inherited from a parent class in the hierarchy. For instance, Figure 6 shows a high level class, "Generating units." Non-black-start units include those with critical maximum intervals and others with critical minimum intervals. Attributes in the class "Units with critical minimum intervals" include those attributes inherited from its parent classes, "Non-black-start units" and "Generating units."

Values and thresholds related to constraints and parameters are entered into the "slots" of attributes. The information can be retrieved during the reasoning process of the KBS. In the following, the structure of some classes is discussed.

(**Class** Generating Units)
Attributes
    _max MW generation
    _min MW generation
    _max MVAR generation
    _max MVAR absorption
    ... (total 43 attributes)

The class "Non_Black_Start Units" inherits all attributes of the parent class, "Generating Units." In class "Non_Black_Start Units," the attribute, "_start_up MW requirement," represents the start-up and houseload requirements of a generating unit without black-start capability. A child class of "Non_Black_Start Units" is "Units with Critical_Max_Int." The value in a slot of the attribute "_critical maximum interval" is the time period within which the specific unit should be started. This value can change with time, depending on the power plant condition. The values of the attributes are used to evaluate the start-up priority.

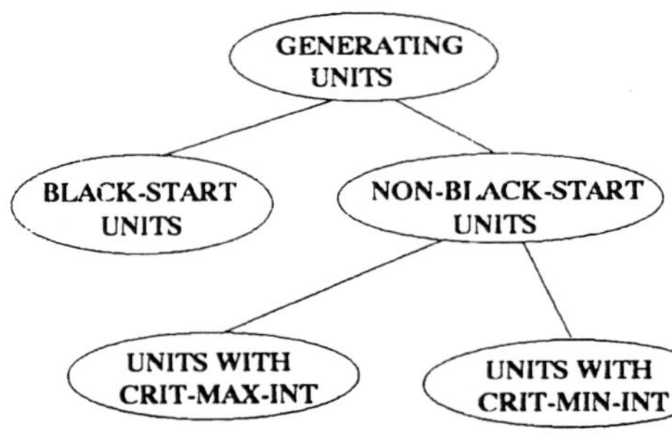

Figure 6. Class inheritance

This KBS is implemented in rules, demons and a section called "Actions," which is the main procedure in the KES terminology. The implementation of a module, "System Restoration" in Figure 3 is discussed. For illustration, part of a demon in this module related to temporal reasoning is shown below:

Start Other Units Unbx:

if Cranking Unbx Implies Inability to Start Any Crit Max Int Unit And Not to Crank Unbx now This Crit Max Int Unit can be Started *Sometime*
then
    reassert Unotnowother = Unbx.
    \consider Unbx later
else
    reassert Uaddother = Unbx.
    \add Unbx to starting plan
endif.

This demon uses the value of critical maximum interval in the class, "Units with Critical_Max_Int," to perform temporal reasoning. All constraints in Section 2 are modeled in the KBS using classes and attributes as shown above. Values such as MVAR generated by a transmission path are calculated with a demon. Some values such as the critical maximum interval are assumed to be available from power plant operators. Other values such as the limit on the voltage transient due to switching are assumed to be available from off-line simulations.

Remarks:

1) The frequency response of prime movers to sudden load pick-up can be considered by two methods. The first one uses an equation which fits the frequency response obtained from off-line simulations. The other method simply stores the minimal frequencies for various load pick-up values in tables [15].

2) The load MW and MVAR as functions of time after cold load pick-up can be studied with two similar techniques. The KBS is designed to incorporate both. The first option is to use equations which fit the cold load variation following an outage. The second option uses tables to store recovering load models which are previously evaluated dynamic data for different load types [15].

5. Simulation Results

Three cases are used to demonstrate the cranking power assignment and cranking path building capability of this knowledge-based system. In all cases, it is assumed that load is available during the restoration.

5.1 Schuylkill-Southwark

This case is the restoration of PECo's Schuylkill-Southwark island, which is shown in Figure 8. There are 2 CTs and 2 fossil units at Schuylkill. Schuylkill-10 and Schuylkill-11 are black-start CTs. Schuylkill-1 and Schuylkill-3 are non-black-start steam units. The ramping rates, maximum generation and start-up requirements of the 4 units are shown in Table 1.

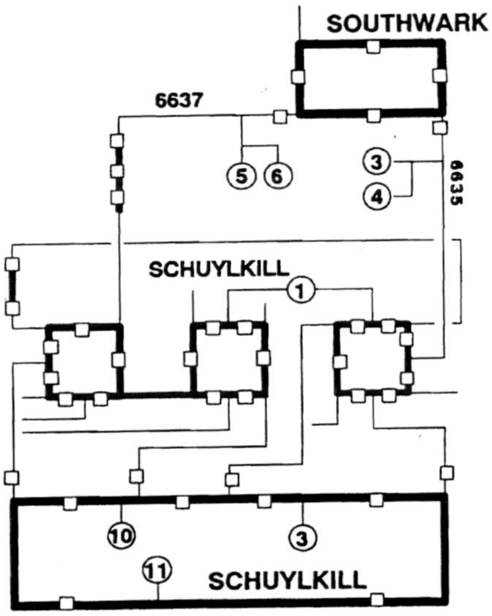

Figure 7. Schuylkill-Southwark island

Unit	Type	MW Cap. (MW)	Ramp. rate (MW/hr)	Start-up req. (MW)
Schuylkill_1	Steam	135	135	15
Schuylkill_3	Steam	135	135	15
Schuylkill_10	CT	15	45	0
Schuylkill_11	CT	15	45	0
Southwark_3-4	CT	30	90	0
Southwark_5-6	CT	30	90	0

Table 1. Units in Schuylkill-Southwark island

There are 2 CTs, Southwark_3-4 (aggregated) and Southwark_5-6, at Southwark; both are black-start units. Their ramping rates and maximum generation are also in Table 1.

Two scenarios have been studied. The first one assumes both CTs at Schuylkill are available; therefore, cranking power is available to the steam units. In this case, the KBS suggests starting of CTs, Schuylkill-10 and Schuylkill-11, first and then starting the steam units, Schuylkill-1 and Schuylkill-3, as soon as cranking power is delivered.

As shown in Figure 8, 20 minutes after the CTs are on, sufficient power is available to crank Schuylkill-1 and Schuylkill-3. The KBS suggests switching operations for the cranking paths. If there is no further delay in switching, the generation MW capability of this island will reach the maximum 80 minutes after the blackout. Figure 8 also shows the MW capability curves of the individual units.

The second scenario assumes that Schuylkill-10 is not available and the power generated by Schuylkill-11 can supply cranking power to Schuylkill-3 only. In this case, the KBS tries to find another cranking source for Schuylkill-1. The only available units at this time (starting of restoration) are the black-start CTs at Southwark. These units, however, have low MVAR absorbing capability. In order to transmit power from Southwark to Schuylkill-1, the power must go through either line 6635 or line 6637. These lines are underground cables which produce reactive power. If the MVAR cannot be absorbed, there may be voltage problems. Therefore, the KBS finds that it is infeasible to provide power from Southwark to Schuylkill-1.

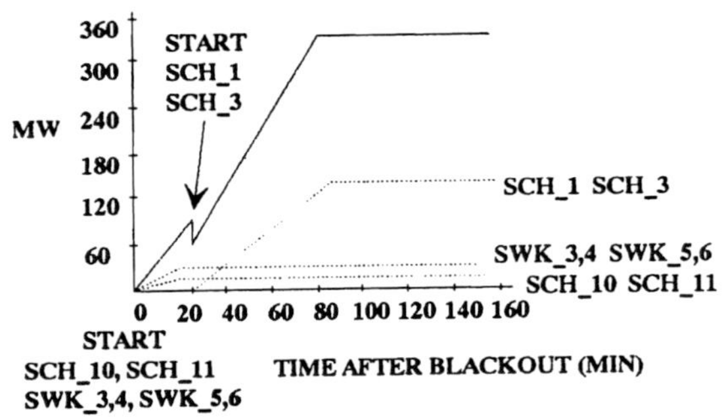

Figure 8. Generation capability for Schuylkill-Southwark (scenario 1)

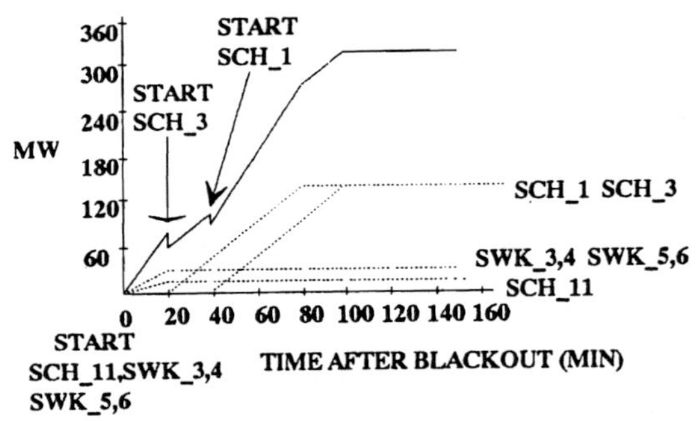

Figure 9. Generation capability for Schuylkill-Southwark (scenario 2)

However, after Schuylkill-3 is on, it can supply cranking power to Schuylkill-1. The KBS suggests supply of cranking power from Schuylkill-3 to Schuylkill-1. Referring to Figure 9, 20 minutes after Schuylkill-11 is on, it can supply cranking power to Schuylkill-3. Another 20 minutes later, Schuylkill-1 can start with the cranking power from Schuylkill-3. The switching operations for building cranking paths are also suggested by the KBS. The system generation capability reaches the maximum 100 minutes after the blackout, as shown in Figure 9.

In both scenarios, after all units at Schuylkill and Southwark are started, the KBS attempts to connect the two stations. Due to the limited MVAR absorbing capability of the CTs at Southwark, the KBS suggests these CTs to be shut down. The CTs can be restarted after the units at Schuylkill absorb the MVAR generated by lines 6635 and 6637.

5.2 Falls-Croydon

This case is the restoration of PECo's Falls-Croydon island. The system is shown in Figure 10. There are 2 CTs, but only the CT at Falls is a black-start unit. (The 3 CTs at Falls are aggregated into one.) The goal is to supply cranking power from Falls to Croydon. The default path is obtained from PECo's restoration manual. While building this path, stuck breakers or faults are found on lines or buses. Therefore, the KBS has to come up with an alternative path.

The default path from Falls to Croydon is along line 130-20, Emilie, and line 220-54. If this is feasible, the KBS will suggest breaker operations which include closing breakers 120, 140, 150, 180, 895, 545, 315, 335, and 235.

Whenever a fault or failure is encountered, the KBS will find an alternative path by searching the neighboring buses and lines. Two faulted scenarios are considered, each with a different fault location. The first scenario is a fault on the bus between breakers 895 and 545. After finding a fault on this bus, the KBS tries to find an alternative path from the bus between breakers 785 and 895 to Croydon. The KBS finds the alternative path through line 220-55 and suggests closing of breakers 785, 555 and 235.

The second faulted condition is a fault on the bus between breakers 140 and 150. Searching the neighboring buses and lines, the KBS cannot find any alternative path. Hence, the KBS concludes that the power from Falls cannot be delivered to Croydon through the paths inside this island.

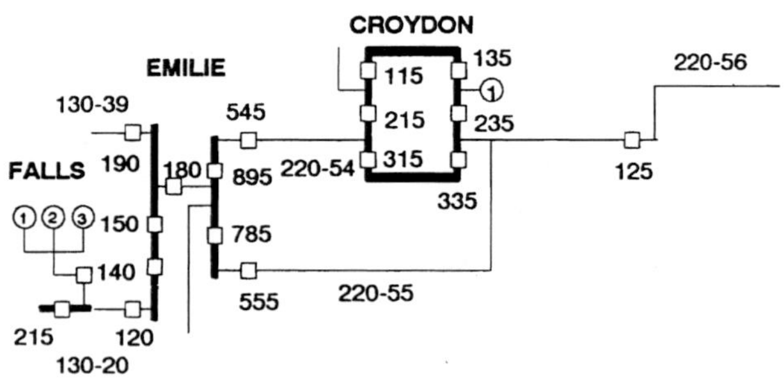

Figure 10. Falls-Croydon island

5.3 PECo System

The last case is to apply the KBS to the entire PECo system. For simplicity, units at the same station with similar characteristics are aggregated into one. The characteristics of a unit are obtained from the records of an actual PECo restoration drill. After aggregation, 15 generating units are considered in this case. Their characteristics are listed in Table 2. They include 9 black-start and 6 non-black-start units. Among the 6 non-black-start units, 2 have critical maximum intervals and 2 other units have critical minimum intervals.

Unit	Type	MW cap. (MW)	Ramp. rate (MW/hr)	Crank to paral. (hr)	Crit. max. int. (hr)	Crit. min. int. (hr)	Start-up req. (MW)
Chester_4-6	CT	39	120	0	N/A	N/A	0
Conowingo_1-11	Hydro	385	384	0	N/A	N/A	0
Cromby_1-2	Steam	345	148	1:40	N/A	N/A	8
Croydon_1	CT	302	120	0:30	N/A	5:00	6
Delaware_7-8	Steam	252	108	2:40	3:20	N/A	5
Delaware_9-12	CT	54	162	0	N/A	N/A	0
Eddystone_1-4	Steam	600	157	1:40	N/A	3:20	12
Eddystone_10-40	CT	56	168	0	N/A	N/A	0
Falls_1-3	CT	45	135	0	N/A	N/A	0
Moser_1	CT	30	90	0	N/A	N/A	0
Muddy Run_1-8	Hydro	660	246	0:30	N/A	N/A	13.2
Richmond_3-4	CT	96	288	0	N/A	N/A	0
Schuylkill_1	Steam	135	135	2:00	2:30	N/A	2.7
Schuylkill_10-11	CT	28	84	0	N/A	N/A	0
Southwark_3-6	CT	52	156	0	N/A	N/A	0

Table 2. Units of PECo system

UNIT	CRANKING UNIT	START TIME
Chester_4-6	(Black-start)	0:00
Conowingo_1-11		0:00
Delaware_9-12		0:00
Eddystone_10-40		0:00
Falls_1-3		0:00
Moser_1		0:00
Richmond_3-4		0:00
Schuylkill_10-11		0:00
Southwark_3-6		0:00
Cromby_1-2	Moser_1	0:10
Delaware_7-8	Delaware_9-12	0:10
Muddy Run_1-8	Conowingo_1-11	0:10
Schuylkill_1	Schuylkill_10-11	0:10
Eddystone_1-4	Eddystone_10-40	3:30
Croydon_1	Falls_1-3	5:10

Table 3. Cranking units and start times of PECo units

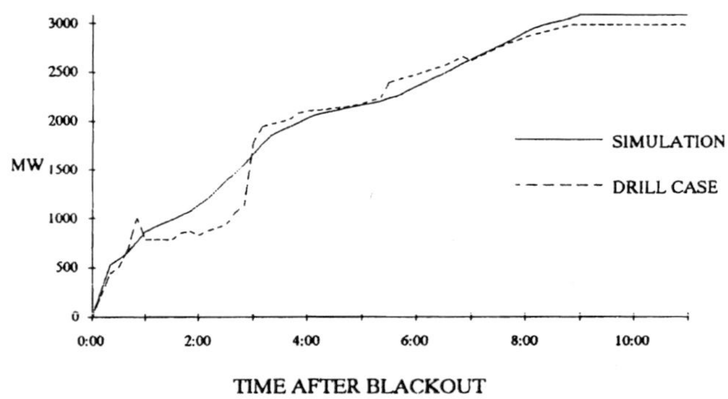

Figure 11. PECo system generation capability

After a blackout, the first suggestion from the KBS is to start all available black-start units which include 8 CTs and 1 hydro unit. Then the KBS assigns cranking power to the non-black-start units which include 1 jet-engine type CT, 1 pump-storage hydro unit, and 4 steam units. This assignment considers unit constraints and generation capabilities. Some predetermined transmission paths are assumed to be available in this case. The cranking unit and the start time for each unit are shown in Table 3. The system generation capability curve resulting from the KBS recommendations is given in Figure 11, together with data from the drill. The results from the KBS are very close to those from the drill in PECo, which is based on the restoration manual and the power director's decisions during the drill. In the actual drill, a hydro unit was shut down, and later restarted, because a regulation was violated. This explains why the MW curve of the drill case is lower between 1:00 and 3:00. The KBS needs to be modified to include specific regulations.

6. Conclusion

This study results in a knowledge-based system which can serve as a system dispatchers' aid during restoration. The contributions of this research are the survey results of critical restoration constraints/parameters, generator start-up sequencing, and the knowledge-based system restoration strategy.

Some issues remain to be addressed. The values of some constraints, such as stability and voltage transient limits, are left to off-line simulation. This can be obtained more conveniently if the KBS is integrated with power system simulation programs. The other issue is the integration of the proposed KBS in the operator training environment [19]. These subjects are to be investigated in the future.

Acknowledgement

This research is sponsored by Electric Power Research Institute, Palo Alto, CA, under contract no. RP3104-1. We appreciate the comments from the reviewers and Mr. Gerry Cauley, EPRI project manager.

References

[1] M. M. Adibi, P. Clelland, et al, "Power System Restoration - a Task Force Report," IEEE Transactions on Power Systems, Vol. 2, No. 2, 1987, pp. 271-277.

[2] M. M. Adibi, D. Barrie, et al. "System Operations Challenges," IEEE Transactions on Power Systems, Vol. 3, No. 1, 1988, pp. 118-126.

[3] M. M. Adibi, J. N. Borkoski and R. J. Kafka, "Power System Restoration - the Second Task Force Report," IEEE Transactions on Power Systems, Vol. 2, No. 4, 1987, pp. 927- 933.

[4] Working Group 02, of Study Committee 38, Task Force No. 2, "Modelling and Simulation of Black Start and Restoration of an Electric Power System: Results of a Questionnaire," Electra, CIGRE, No. 131, July, 1990, pp. 156-169.

[5] T. Sakaguchi and K. Matsumoto, "Development of a Knowledge Based System for Power System Restoration," IEEE Transactions on Power Apparatus and Systems, Vol. 102, No. 2, 1983, pp 320-329.

[6] Y. Kojima, S. Warashina, et al., "Development of a Guidance Method for Power System Restoration," IEEE Transactions on Power Systems, Vol. 4, No. 3, 1989 pp. 1219-1227.

[7] Y. Kojima, S. Warashina, et al., "The Development of Power System Restoration Method for a Bulk Power System by Applying Knowledge Engineering Techniques," IEEE Transactions on Power Systems, Vol. 4, No. 3, 1989 pp. 1228-1235.

[8] F. F. Wu and A. Monticelli, "Analytical Tools for Power System Restoration - Conceptual Design," IEEE Transactions on Power Systems, Vol. 3, No. 1, 1988, pp. 10-16.

[9] D. S. Kirschen and T. L. Volkmann, "Guiding a Power System Restoration with an Expert System," IEEE 90 SM 341-8 PWRS.

[10] L. R. Blessing, C. K. Bush, and S. J. Yak, "Automated Power System Restoration Incorporating Expert System Techniques," Proc. Second Symposium on Expert Systems Application to Power Systems, Seattle, July, 1989, pp. 133-139.

[11] H. Y. Marathe, C. C. Liu, et al., "An On-line Operational Expert System with Data Validation Capabilities," Proc. 1989 PICA, Seattle, pp. 56-63.

[12] C. C. Liu, S. J. Lee, and S. S. Venkata, "An Expert System Operational Aid for Restoration and Loss Reduction of Distribution Systems," IEEE Transactions on Power Systems, Vol. 3, 1988, pp. 619-626.

[13] Kazuhiro Kaneko, Yoshiteru Ueki, and Yoshikazu Fukuyama, "An Expert System for Automatic Restoration of Power Systems Considering Dynamic Constraints to Thermal Plants," Trans. IEE Japan, Vol. 11-B, No. 4, 1990, pp. 286-293. (in Japanese)

[14] R. F. Chu and C. C. Liu, et al., "An Expert System for Generation Capability Assessment During System Restoration," Presented at EPRI Seminar, Expert System Products for Power System Operations, September 1990, Bellevue, Washington.

[15] K. Dickers and D. Rumpel, "Modelling for Grid Restoration Studies," Electric Power and Energy Systems, 1987, pp. 45-55.

[16] D. Rumpel, "Performance of Power Plants in Cycling and Disturbed Grid Conditions," CIGRE 1981, 81 PP 01.

[17] Fred Kroger, *Temporal Logic of Programs*, Springer-Verlag, Berlin, 1987.

[18] S. L. Tanimoto, *The Elements of Artificial Intelligence: An Introduction Using LISP*, Computer Science Press, 1987.

[19] R. F. Chu, E. J. Dobrowolski, et al., "The Use of an Operator Training Simulator for System Restoration," Proceedings of the IEEE Power Industry Computer Application Conference, Baltimore, Maryland, May 7-10, 1991, pp. 171-177.

Discussion

M. M. ADIBI, IRD Corporation, Bethesda, Maryland, **J. N. BORKOSKI**, Baltimore Gas & Electric Company, Baltimore, Maryland, and **R. J. KAFKA**, Potomac Electric Power Company, Washington, D.C.: In response to Dr. Chen-Ching Liu's request we are pleased to provide general comments and discuss the results of the paper. The discussers are particularly interested in this paper, since the Generation Dispatch Scenarios (GDS) approach on which it is primarily based was first developed at PEPCo and BG&E [A & B]. The primary objectives at PEPCo were to reach a certain level of real power generation prior to reaching the minimum critical time of super-critical once-through boilers. At BG&E the objectives included reaching a certain level of reactive power absorbing capabilities within a specified time, while maximizing the MWH of load served (see Discussion to [9]). These objectives were addressed by combining the expertise of the power plant operators, power system operators, system analysts, with the extensive use of steady state and dynamic simulations of the power system as well as power plants [2] and [C].

Some of the analytical tools such as frequency response of prime movers (hydro, steam and CT), switching surges of high voltage transmission lines, and cold load dynamics had to be developed during the GDS project. The GDS approach for achieving the above objective was time consuming, requiring the integration of power plants and power system (thermal, mechanical and electrical) dynamics; this integration is not important nor generally used under normal or emergency operating conditions, where, for instance, power plant thermal transients are not explicitly taken into account by AGC algorithms.

During these and other studies undertaken by the discussers at their individual companies and as members of the Power System Restoration Working Group (PSRWG), it was recognized that several of methods or techniques for power system restoration were good candidates to be addressed by Expert System applications. The involvement of the first discusser in the planning stages of EPRI RP 3104-1 was valuable in nominating candidate applications and their required knowledge bases for the research project. In the course of two workshops and several meetings, the knowledge gained at PEPCo and BG&E was transferred to the authors. The authors are commended for formalizing some of these features in their Knowledge-Based System (KBS).

The following points do not appear to have been addressed or clearly stated in the authors' otherwise excellent paper:

c1. It is recognized that no two start-ups of the same boiler-turbine-generators describe the same time-Mw output curve. Therefore, it is important that the interaction between generation and loads be coordinated by an on-line feedback from the plants. These dynamic phenomena (plant, system, loads), as stated above, are neglected (or unimportant) during normal operation but play critical roles during restoration.

c2. Power systems have different characteristics, with each having a specific set of critical first objectives following a partial or a complete collapse. As such, it appears to be more realistic and practical to develop generic building blocks and procedures so that each operating utility could tailor its restoration program to meet its own specific needs rather than use a generic restoration program.

c3. The available generic models such as operator training simulators are adequate to simulate power systems under normal and emergency operating conditions. However, they do need to be complemented with long- term dynamic models in order to simulate power systems under large perturbations of long duration during which wide voltage and frequency ranges are tolerated.

Manuscript received February 21, 1992.

The authors response to the following specific questions or comments are very much appreciated.

q1. Is it sufficient to consider restoration of a partial shutdown to be a sub-problem of the total shutdown as the authors suggest?. How can this statement be reconciled with the industry's prevailing view that restoration following a partial collapse when two or more systems are interconnected, necessitating a top-down restoration, is quite different from the complete collapse which requires bottom-up restoration of only one system?

q2. In maximizing the MWH of load served by generators it is stated that the critical, including dynamic, constraints and parameters have been considered. It should be noted that constraints of the system on network have not been included in this current implementation. Constraints of the system could be more significant than individual unit constraints.

q3. The optimality of the start-up sequence is not well defined in the paper. This definition depends on the specific objective(s) in a given power system, as described under c2 above for PEPCo and BG&E systems. In an all hydro system the restoration of service to aluminum industry loads within a specified period has been the objective [D].

q4. To what degree can the comparison of system generation capability curve obtained from the authors' KBS with those from the PECo's drill, shown in Figure 11, verify the validity of either curves, since both are based on the same assumptions regarding power plant and power system prevailing dynamic characteristics, which assumptions appear not to have been simulated or considered in either case? For instance, under Simulation Results, Case 1, Scenario 1: the assumption that with the simultaneous start--ups of Cts #10 & #11, and simultaneous start-up of Steam Units #3 and #1, the maximum generation of 270 MW can be reached in 80 minutes, appears unreasonable. First, the start-up requirement of 15 MW or 11% for this type of steam unit is very high, one half of this value is more reasonable; second, the start-up of the two steam units simultaneously requires significant additional manpower; more likely, the two units in one plant will be started up by the same crew, sequentially; and third, a 15 MW CT can be started and reach full generation in 5 to 10 minutes, provided that load is available. Allowing for these corrections and considering other factors in steam unit start-ups, will show no differences between the two scenarios in Case 1. However, for restoration purposes, having two Cts at the same plant does increase the probability of successful start-up.

References:

[A] Kafka, R.J., et al, "System Restoration Plan Development for a Metropolitan Electric System," IEEE Trans. PAS pp. 3703-13, Aug. 1981.

[B] Borkoski, J. N., et al, "Power System Restoration - The Second Task Force Report," IEEE Trans. PAS pp. 927 - 933, 1987.

[C] Adibi, M. M. and R. J. Kafka, "Power System Restoration Issues," IEEE Computer Applications in Power, Vol.4, No.2, Apr. 1991.

[D] Arnold, P. F., "BPA System Restoration Plan," Power System Restoration Task Force, IEEE PES '82 WM, in Dallas.

Closure

K. L. LIOU, C. C. LIU, R. F. CHU: The authors appreciate the discussion by Messrs. M. M. Adibi, J. N. Borkoski, and R. J. Kafka. Following a blackout, the generic restoration procedure calls for the restart of generators, re-energization of transmission paths and restoration of loads. In this paper, our focus is on the generation capability dispatch during system restoration. Some issues related to path energization and load restoration will be addressed in greater detail in our future work.

At this time, the ability of the knowledge-base system (KBS) is limited; it is able to provide operators with feasible generator start-up sequences after a blackout. To produce such sequences, the KBS has to consider the constraints of the power system. In this stage, the numerical algorithms for power system simulation have not been integrated with the KBS. Therefore, the best projection that our KBS can make is based on pre-stored system data and power plant characteristics.

We are aware that the characteristics of generating units may change due to various factors, such as thermal stress, man-power, etc., especially after a blackout. Therefore, the proposed modules "Black-Start Units" and "Non-Black-Start Units," shown in Figure 3, contain essential parameters and constraints related to generating units. In these two modules, the generation capability may be reduced if the problem of mechanical failures or man-power shortage arises. In this case, the two modules will be notified by submodules such as "Mechanical Inspection", "Man-Power Dispatch." The proposed KBS considers power system responses and is able to modify its solution, if necessary.

Our responses to the discussers' questions are:

q1. We believe the modules in our restoration strategy are able to handle partial blackouts to a great extent. As a matter of fact, earlier expert systems developed by other researchers are aimed at partial blackouts, e.g., a transformer fault causing an outage at several substations. This involves search for generation resources and transmission paths, both were included in our strategy. We recognize that there will be special requirements for a specific power system; they need to be added to the existing KBS.

Presently, we are enhancing the module of "External Sources." The enhancement will enable the KBS to deal with tie lines more effectively. As a result, we recognized that we will also need to enhance the "System Restoration" module of the existing KBS to make the strategy more flexible. The enhanced KBS should be able to determine the priority of actions depending on the outage conditions. However, we believe that the basic building blocks for the generic strategy are complete.

q2. The network constraints are important when a cranking path is to be energized or a load is to be picked up. These constraints are considered within the module "Path Management." As shown in Figure 3, the submodules, "Fault Inspection", "Capacity Inspection", etc., provide "Path Management" with the necessary information. "Path Management" also needs help from "MVAR Management" to prevent voltage problems while energizing paths. All constraints listed in Section 2 are included in the proposed KBS, except under frequency load shedding and phase difference. Again, the KBS will not be able to perform numerical simulation to obtain values for these constraints until the algorithms are integrated.

q3. Our goal is to maximize the MWH generation capability over a restoration period. The KBS is designed to achieve this goal using heuristic rules and temporal logic search techniques. We recognize the tradeoffs between the computer simulation time and optimality of the resulting sequence. Other system-specific objectives for restoration such as the restoration of service to selected loads mentioned by the discussers can be accommodated. This requires modification of the restoration strategy utilized by the KBS.

q4. In the PECo drill case, the operators followed instructions of the restoration manual and adapt them to the drill environment. The restoration manual is the result of extensive off-line studies and is expected to provide "optimal" restoration performance. The KBS was initialized with the PECo system configuration and generating unit characteristics. Although the KBS start-up sequence is different from that of the drill case, the generation capability compares favorably with that of the drill case. Additionally, such comparison adds to our confidence in the developed KBS.

In the example of section 5.1, the start-up requirements of generating units are fictitious values. These values are made large enough to test the capability of this KBS in finding alternative cranking sources. The man-power issue is considered in our KBS. However, in this example, we assume that sufficient man-power is available. A similar assumption is made for a sufficient amount of fuel storage. Therefore, if necessary, two CTs can be started at the same time. In the example of section 5.3, the reader will find more realistic values of start-up requirements for different types of generating units.

We would like to comment on the points c1-c3 raised by the discussers:

c1: In its ultimate development stage, the proposed KBS would serve as an on-line operators' aid. When the on-line plant and loading data becomes available, the user can re-run the KBS to update the restoration sequence.

c2: Refer to our response to q1.

c3: We agree that the integration of the KBS with long term dynamic simulation programs would be valuable.

Finally, this project benefited from the contributions of many individuals. In addition to M. Adibi (IRD) and J. States (retired from PECo), we would like to thank C. Frank and G. Cauley (EPRI) and the project advisors for their valuable guidance and practical insights which help define the restoration problem, constraints, and needs. The advisors are C. K. Bush (Atlantic Electric), W. A. Fuller (Texas Utilities), H. S. Grewal (Los Angeles Department of Water and Power), S. Muchlinski (Puget Sound Power and Light), T. L. Volkmann (Northern States Power), and R. F. Yates (PSI Energy).

Manuscript received April 10, 1992.

EXPERT SYSTEM REQUIREMENTS FOR POWER SYSTEM RESTORATION

M. M. Adibi, LM
IRD Corp.
Bethesda, MD

R. J. Kafka, SM
Potomac Elec Power Co
Washington, D.C.

D. P. Milanicz, SM
Baltimore Gas & Elec Co
Baltimore, MD

Abstract: This paper is one of series presented on behalf of the System Operation Subcommittee with the intent of focusing industry attention on power system restoration. Expert systems are being considered for restoring bulk power supplies. In general, there are three restoration periods following a major power disturbance: establishment of initial sources of power, re-integration of a skeleton of the bulk power supply, and minimization of the unserved loads. Expert systems together with analytical tools have the potential of addressing the restoration procedures over these three periods. This paper describes the expert system requirements from the point of view of the practicing power engineers with emphasis placed on the initial power sources and requirements. The paper draws on the previous reports by the Power System Restoration Working Group.

INTRODUCTION

Restoration procedure following a major power system disturbance spans three time periods: (i) interconnections of the individual "initial sources of power" with the "critical requirements for power" within the first 30 to 60 minutes, (ii) integration of generation and transmission to achieve the target system within 3 to 4 hours, and (iii) minimization of the unserved customer loads within 8 to 12 hours. These rather optimistic time periods respectively correspond to sectionalization and execution of restoration subsystems [1], establishment and achievement of the post-restoration target system, and restoration of loads. Power System Restoration Expert system (ES) in conjunction with the analytical tools, should play a significant role in the planning, training and operation of a restoration procedure over these three periods [2,3,4]. In this paper, the three periods of restoration are described and the roles of ES in each period are covered. Emphasis is placed on the initial power sources and requirements period which is most critical. The paper draws on previous reports by the Power System Restoration Working Group.

THE FIRST RESTORATION PERIOD

In the first restoration period the "initial sources of power" are simultaneously, independently and individually connected to the "critical requirements for power," to form a number of restoration subsystems.

The Initial Sources of Power

The initial sources of power can be provided within 5 to 15 minutes by quick starting combustion turbines, run-of-the river hydro units, and low head pumped storage hydros. The probability of successful start up of combustion turbines (CT) based on industry's experience is about one third to one half - a partial motivation to have two or more CTs in one location. Their restarts, if the first or second attempt is unsuccessful, require considerable wait period. The probability of starting hydro units and pumped-storage hydros are much higher and their restarts are much shorter, however, there are fewer plants of these types in the industry and their use depends on the availability of water.

The initial sources of power may be immediately available from (i) steam units equipped with full load rejection capabilities which stay in operation and supply their own house load, (ii) low frequency isolation schemes which isolate steam units with the "matching" local load by frequency sensitive relays at low frequency, Hz, and (iii) intentional islands which are initiated by distance relays, separating several generators with "matching" load during system disturbance. Full and partial load rejections are more prevalent in Europe, where the probability of successful load rejection at EdF is reported to be between 20 to 80%, and in Germany at about 50% [5]! Low frequency isolation schemes are primarily used in thermal power systems. Their probability of successful operation depends on the extent of mismatch between generation and the local load at the time of disturbance. It is enhanced by the partial load rejection capabilities of steam units and the availability of local non-conforming (fixed) loads. The probability of successful intentional islanding depends on the distance relaying schemes used, and type and severity of the disturbance. Although load rejection capability, low frequency isolation, and intentional islanding are installed in some power systems, they are not viewed by the industry as the most reliable initial sources of power. However, ES must be capable of utilizing all of these sources when available.

An important initial source of power which may be immediately available is interconnection assistance through a tie-line with adjacent power systems. The availability of this initial source depends on the extent of power system disturbance and any bilateral agreement between the two interconnected companies. As long as the assistance is limited to supplying the "critical requirements for power," there should be no technical impediment to such assistance. However, most utilities in planning for restoration are reluctant to rely on tie-lines as the <u>initial</u> sources of power, because it is very likely that a <u>total</u> company blackout would effect the neighboring utilities.

Briefly, the initial sources of power in order of priority include: (i) the "primary sources" consisting of combustion turbines, hydro units, and pumped-storage hydros, (ii) the "secondary sources" consisting of steam units with full load rejection capabilities, low frequency isolation schemes, and intentional islanding, and (iii) tie-lines with adjacent power systems which when available, shall also be considered a "primary" initial source of power. ES should be able to utilize all the primary and secondary "initial sources of power" including the available tie-lines to adjacent systems.

94 WM 223-8 PWRS A paper recommended and approved by the IEEE Power System Engineering Committee of the IEEE Power Engineering Society for presentation at the IEEE/PES 1994 Winter Meeting, New York, New York, January 30 - February 3, 1994. Manuscript submitted July 16, 1993; made available for printing November 29, 1993.

Reprinted from *IEEE Transactions on Power Systems,* Vol. 9, No. 3, pp. 1592–1600, August 1994.

The Critical Requirements for Power

The critical requirements for power include drum-type units, pumping plants for pipe-type cables, industrial loads, transmission stations and distribution substations. The maximum elapsed time for the hot restart of drum-type units is between 30 to 60 minutes, beyond which the units need to be rolled for a cold startup, 2 to 3 hours later [6]. The pumping plants maintain oil pressure in pipe-type cables. Following an interruption and in the absence of the source of heat which under normal operation is provided by the cable losses, oil in the steel pipe containing the cable cools down and loses pressure. The pumping plants need to be re-supplied and re-started within 30 to 45 minutes to maintain the required 12 to 14 atmospheres (about 200 psi) pressure and to prevent generation of vacuum and the release of the dissolved gases which would result in very costly damage to the cable insulation and cable pot heads if re-energized before adequate re-pressurizing, and perhaps even after [7]. The critical industrial loads such as electrolytic production of aluminum require re-supply of power to the pot-lines within 30 to 45 minutes to prevent freezing of the molten aluminum in the pots and the release of their carbon anodes by force. The transmission and distribution stations need stored energy in the forms of batteries and compressed air and gas (SF_6); and low voltage AC for heating and preventing SF_6 from liquefying; cooling fans for transformers; control and operation of relays, breakers, and disconnect switches. Many stations rely on the power system to supply and replenish the depleting energy storage. A substation may not necessarily be on the path between an "initial source of power" and the "critical requirement for power" which it supplies, but it may be needed to pick up real and particularly reactive loads for stabilizing the generation. Therefore, some of the transmission and distribution stations which have no or depleting energy storage and which are required in the first period of restoration need to be energized (depending on their capacity and depletion rates) within 30 to 60 minutes.

Briefly, ES should be able to arrange for power supplies to the "Critical Requirements for Power," in the following order of priority: (i) drum-type units and pipe-type cables, (ii) transmission and distribution stations, and (iii) industrial loads.

Initial Sources - Critical Requirements Interconnection

During restoration all the "critical requirements for power" which have priorities 1 and 2 need to be supplied. This requirement dictates that the primary "initial sources of power" should be started at about the same time and the "critical requirements for power" be supplied simultaneously. In addition, independent and parallel actions are required to immediately tap the secondary "initial sources of power" and the available tie-lines.

In selecting a critical requirement for supply by an initial source of power, ES will play a significant role and will face many restoration constraints. The ES actions and the number of constraints posed for providing cranking power to a drum-type boiler-turbine-generator unit by a remote black-start CT, encompass most of the activities and constraints faced by the other types of initial source/critical requirement combinations. Therefore, ES requirements in restoration of a typical CT-BTG subsystem will be described [8].

CT-BTG Restoration Coordination

The steps involved in a typical CT-BTG restoration coordination include: decomposing the CT-BTG restoration into individual activities with appropriate level of details, specifying the predecessor successor relationships that exist among the activities, estimating the time that elapses between the start and finish of each activity (including time for manual operations), scheduling activities to achieve the CT-BTG restoration goal of 30 to 60 minutes, monitoring progress of activities (particularly in training and operation modes), and adapting the CT-BTG restoration plan to a new environment when an unforeseen event occurs.

The CT-BTG restoration activities include:

- Placing the CT under manual control mode of, operation. During normal operations CTs are under automatic mode and used in parallel with larger units for peaking purposes, they cannot operate in the automatic mode when they are the only source of power [9].

- Deactivating dead-bus interlocks and sealed relays (in case of previous automatic tripping) at both ends of the path, i.e., at the CT bus and the BTG bus.

- Adjusting the tap positions on the step-up transformer at the CT end and the standby (step-down) transformer at the BTG end. Generally these transformers are not equipped with tap changers under load (TCUL) and should be preset [8].

- Selecting the CT-BTG path, preferably a low voltage single circuit of a double circuit line to minimize charging currents while improving chances of a successful operation.

- Selecting loads and their related paths which are needed for stabilization of the CT.

- Opening all the circuit breakers which are needed in the CT-BTG restoration to disarm all the automatic reclosure relays, including those breakers which connect the path to other part of the network [7].

- Over-riding voltage sensitive switched capacitors on radial feeders which are to be used for load pick ups.

- Deactivating load shedding under-frequency relays to prevent their undesired and inadvertent operation.

- Starting the CT, closing the breaker on the dead bus, and energizing the step-up transformer.

- Energizing the CT-BTG path one section at the time, picking up loads to stabilize the CT and to prevent it from under-excitation.

- Energizing the BTG standby transformer to supply power to the auxiliary equipment.

- Avoiding generation of harmonic resonance by connecting the standby transformer (also referred to as the station service transformer) after energization of the path [10].

- Picking up BTG auxiliary equipment in a descending power requirement order, i.e., boiler feed pump, induced draft fan, forced draft fan, condensate pump, circulating water pump, and auxiliary oil pump [8,9].

- Synchronizing the BTG unit to the high voltage bus using the unit transformer, or if the high voltage bus is sectionalized, placing the BTG on the dead bus section.

• Picking up radial or small loads to reach the BTG's minimum generation (about 30% of full load) capability, at which point the BTG can be placed under automatic control, i.e., combustion controller, temperature controller, feedwater controller, etc. The BTG is released to the power system operator when the plant operator has visibility of the subsystem formation by e.g. the subsystem frequency.

• Coordinating the rate of load pick up so that the time-load trajectory is enveloped by the minimum and maximum ramp rates of the BTG, thus avoiding flame instability, low temperature operation and loss of vacuum in the BTG.

CT-BTG Restoration Constraints

A large number of primary initial sources of power including CTs are unattended, and under normal operation can be started remotely from the power system control center. CT operation during restoration requires attendance for manual adjustments of various automatic controllers. Operation of CTs and BTGs are responsibilities of power plant operators. Power system operators basically perform three types of actions during a typical CT-BTG restoration: opening and closing breakers and disconnect switches, changing transformer tap positions, and if automated and possible, remotely adjusting set points on relays and local controllers in the switching stations and substations. For the performance of these tasks, ES should consider the physical constraints which are posed by electromagnetic, mechanical and thermodynamic behavior of the CT-BTG components in response to restoration actions. Some of these constraints can be determined from the characteristics of the various power plant and power system components and archived for use during restoration, others are functions of the prevailing system configuration and have to be determined by the use of analytical tools as described in [2].

(a) Component Constraints

ES should be able to use the parameters/constraints of power system components in establishing subsystems such as the CT-BTG. The following power system characteristics should be considered by ES in restoring a typical subsystem (CT-BTG) [1]:

Boiler-Turbine-Generator

• BTG auxiliary requirements in terms of voltage, frequency, active and reactive powers,

• Boiler-turbine's minimum generation and maximum/minimum ramping rates,

• Generator's reactive absorbing capabilities for different hydrogen pressures,

• Hot re-start time from the start-up to parallelling, and minimum and maximum time from paralleling to the minimum load.

Combustion Turbine

• Location with respect to the BTG (local or remote), and if remote, the voltage levels at the BTG,

• The number of CTs in the plant and whether attended,

• Reactive power absorbing capability of the CT generator as compared to the charging currents of path(s) to the BTG,

• Modes of voltage and frequency controls, and

• Relay settings including the minimum excitation relays, cross compensation relays for parallel operation of CTs, etc.

CT-BGT Path

• Sectionalization capability of the CT bus for isolating the CT and the desired path,

• The step-up transformer rating, tap ranges and reactive power losses,

• The voltage levels and charging currents of the CT-BTG paths,

• The number, size and power factor of radial loads on the CT-BTG path, and

• Thermal limits of interconnecting lines and cables.

Transmission/Distribution Stations

• Types of stored energy in terms of batteries and compressed air or gas for operation of breakers, disconnect switches and transformers,

• Sizes of stored energy in terms of number of operations and elapsed time after the disturbance,

• Deactivation capabilities for under-frequency relays and switched capacitors on the radial loads, and

• Availability of shunt reactors for reactive power balance.

(b) System Constraints

ES should be able to use analytical tools to determine the physical constraints posed in restoring subsystems, i.e., the initial sources of power and the requirements for power (such as CT-BTG) and store these constraints for use during training and operation modes. ES should also be able to use analytical tools during restoration to determine the constraints in case of unforeseen events. In planning for restoration of a typical subsystem (e.g., CT-BTG), the following analytical tools should be used to evaluate certain constraints prior to closing circuit breakers, changing transformer tap positions and adjusting set points:

• Restoration power flow program [11] should be used when energizing a line section to check the line charging current against the reactive absorbing capability of the CT and the sustained overvoltages against the voltage limits.

• Restoration transient voltage program [10] should be used when energizing a line section to check the transient voltages caused by switching surges against lightning arrestor ratings and transformer insulation levels.

• Restoration cold load pick up program [9] should be used when picking up loads to check the cold load magnitude and time variance behavior.

• Restoration dynamic simulator program should be used when picking up load to check the frequency response of CT and/or BTG to the sudden increase in load.

• The restoration power flow program should be used to provide checks against thermal limits of lines, and the reactive power balance between load and generation.

• Restoration short circuit program should be used to adjust protective relays for detection and isolation of possible faults under the minimum prevailing generation provided by CT and/or BTG, providing sufficient fault current exists.

• Restoration transient stability program should be used to check that the starting frequency and voltage of the induction motor driving the boiler-feed-pump (BFP) do not fall below the allowable limits causing locked rotor.

• The restoration transient stability program should be used to pre-select the CT's step-up transformer and BTG's standby transformer tap positions and adjust the CT's frequency prior to the BFP start up.

• The restoration dynamic simulator and restoration transient stability programs should be used to check the steady state and transient stability of the CT-BTG configuration before synchronizing these two prime movers.

ES Role in the Planning Mode

In the pre-planning phase of restoration, the planning team should be able to postulate a number of realistic pre-disturbance power system operations and conditions, based on their experience and familiarity with their own specific power system, and validate such operating conditions using analytical tools [2]. The post-disturbance power system status is not entirely unrelated to the pre-disturbance condition. The planning team, using experience with previous actual major power system disturbances, should be able to use analytical tools and simulate a selected number of post-disturbance status for each pre-disturbance condition [1,12].

For given pre- and post-disturbance conditions, ES should be able to pair the initial sources of power and the critical requirements for power into many different subsystems, then rank and store the subsystems for use in the training and operation modes. The criteria to be used by ES in ranking a subsystem should include:

• The time estimated to complete the execution of the subsystem as a fraction of the specified time needed by the "critical requirement for power," the smaller the fraction the higher the rank of the subsystem.

• The probability of successful execution of the subsystem activities such as start-ups of the CT, BTG and closing of the many breakers.

• The availability of a restoration team consisting of control center, power plant and field operators to implement the subsystem.

The industry's practice is to assign a team for restoration of each subsystem and also to limit the number of restoration teams to facilitate communication between the teams [1], i.e., the number of subsystems to be selected for simultaneous execution should be equal to the number of restoration teams.

In developing the subsystems, ES should be able to use a set of generic "restoration actions" [13] and to consider the characteristics and constraints of the power system of interest.

In the training mode, the pre-disturbance condition/post-disturbance status of the power system is selected as a scenario by the instructor. On some occasions the instructor may need to assume the role of power system restoration planner by using ES and analytical tools to verify scenarios or to develop new scenarios, or he may need to use ES and analytical tools to develop a subsystem which has not been developed in the planning mode. However, in general both the instructor and the trainee should be able to rely on and use the pre- and post-disturbance conditions planned, the subsystems developed and archived by ES, and the tabulated analytical tool results [14].

In the operation mode. both the pre-disturbance condition and post-disturbance status are the prevailing condition and status of the power system. ES should be able to select and present the subsystems which meet the current power system status. In the absence of a needed subsystem and provided time allows, the operator can use the generic restoration actions and with the help of ES and tabulated analytical tool result develop the needed subsystem.

Briefly, ES should be able to:

• accept the current power system pre- and post-disturbance conditions and provide the operator with a set of applicable subsystems, and

• facilitate the use of the generic restoration actions and the tabulated analytical tool results to develop an unforeseen subsystem.

THE SECOND RESTORATION PERIOD

The second period begins when the generating units, such as drum-type BTGs, within the energized subsystems have attained their minimum generation and have been released by the plant operators to the restoration teams for further loading. The second period ends with the hot restart and synchronization of the large generating units such as supercritical once-through boiler-turbines and the cold restart of the drum-type unit (after rolling period). Depending on the minimum time allowed for the hot and cold restarts of these thermal units and maximum time allowed for the completion of subsystems, the second period may take 3 to 4 hours following an outage [6.7]. The return of nuclear units before entering their cooling down cycle may not be possible in this 3 to 4 hours time frame.t

The primary objective in the second period is the interconnection (synchronization) of subsystems to achieve the post-restoration target system. In the process of achieving this objective, all the subsystems should be expanded by energizing additional lines, picking up loads, and linking all the online and the necessary offline generating plants, transmission stations and distribution substations. At the completion of these activities, the subsystems would be synchronized and a skeleton of the bulk power supply established.

The primary function of ES in the second period is to establish paths between all the online and the necessary offline stations. In performing this function, ES should be able to provide secure paths between the above stations by energizing lines, determining switching sequences, timing each operation, checking the various constraints, and if possible ranking the alternative paths.

ES requirements in the second period are somewhat similar to those of the first period with the following significant differences:

• there will be many more situations for which the prepared and tabulated analytical tool results are not applicable,

• the relatively larger and loosely interconnected networks require more analysis of voltage and frequency transients, and

• the relatively longer duration of the second period, should allow the use of analytical tools for the on-line real-time analysis.

ES should be able to expand the energized subsystems and integrate the subsystems to achieve the target system using both the prepared and tabulated analytical tool results and the on-line real-time analysis. In addition ES should be able to establish secure paths between all the online and the necessary offline stations, determining the switching sequences, and timing the individual operation.

ES Requirements in the Second Period

ES in conjunction with analytical tools should be able to carefully consider the probable effects of each action on the rest of the system and on the subsequent restoration measures [1]. Transmission line loadings, reactive power balance, over- and under-voltage conditions, generator loadings, transformer tap selections, shunt capacitor and shunt reactor switching, determination of phase angles between load centers and generators, and adjustments of relay settings for minimum source operation should all be evaluated using the restoration power flow, short circuit and cold load pickup programs [2].

The rate at which generators should be loaded, the mismatch between load pick-up and generator response, the change in system frequency and their effects on under-frequency relays, transient stability of the power system, and the switching transients should all be evaluated using restoration transient stability, dynamic simulator and voltage transient programs [2].

An important ES requirement in the second period is the simultaneous restoration of several subsystems. ES should be able to allocate resources, coordinate activities, and revise procedures to ensure an orderly progress in the time intervals between the release of the BTGs by the plant operators to the restoration team and the hot re-start of the large thermal units. The simultaneous restoration of several subsystems appreciably reduces the duration of an outage but it also demands efficient use of manpower-machine resources and effective communication/coordination between the teams/subsystems. Critical path methods (CPM) have been extensively used to address similar requirements in other projects having similar objectives. ES should be able to use CPM or equivalent scheduling capabilities to achieve the target system within the estimated 3 to 4 hours, by allocation of resources, coordination of activities, and revision of plans, as the restoration progresses in the second period.

ES Procedure in the Second Period

Many of the activities and constraints to be considered by ES in expanding and interconnecting the subsystems to achieve the target system are similar to the first restoration period. For example, in both periods, loads are primarily picked up to stabilize the in-service thermal units and to reduce the charging currents. In the second period when all the plants and stations within a subsystem have been interconnected and other subsystems are not ready for synchronization, the slack time is used to pick-up additional loads. The three sets of activities which somewhat differ from the first period will be briefly described in the following subsections.

Picking up Loads

Loads should be restored in small increments to avoid excessive under-frequency deviation. This can be done by restricting the load pick up to radial loads and preferably to those with low power factors. Switched capacitors and under-frequency relays should be either deactivated or avoided. In a subsystem, the rate of load pick up should be controlled so that its load trajectory does not exceed the maximum ramp rate of the unit(s) within that subsystem. This calls for careful time estimation of breaker operations in energizing lines and feeders. In the first period, load trajectory is checked against both the maximum and minimum generator ramp rates.

Restoration of the large low voltage AC network loads should be avoided and postponed to the third restoration period when large units are in operation and can provide the required frequency response. ES should be able to use the analytical tool results to maintain load-generation balance and reactive power balance, avoiding instability, sustained overvoltages, and frequency excursions [1].

Energizing Transmission Lines

In interconnecting plants, stations and substations, a very few low and high voltage transmission lines and cables should be energized to minimize the charging currents and to avoid sustained over-voltages. If at all possible, energization of the high voltage, extra high voltage, double circuit lines, and underground cables should be delayed until the third period when large generators with ample reactive absorbing capabilities would be in operation [10,18].

ES should be able to prepare a list of actions required to energize a path between two stations. The list should determine the number of circuit breaker and disconnect switch operations, transformer tap selections, set point adjustments, and relay setting changes. The list should also include alternative path(s) which may be selected in case of unsuccessful operation of the first path.

ES should be able to use tabulated analytical tool results or on-line analysis to determine and consider the various constraints.

Synchronizing Subsystems

Subsystems should preferably be synchronized at generating plants where synchronizing facilities are available and the subsystems' frequencies and voltages can readily be regulated. This may require sectionalization of the main generator bus and availability of two connecting lines, one for each subsystem to be synchronized. Synchronization at switching stations is time consuming and during restoration should be avoided even if they are equipped with synchro-check relays. Prior to synchronization, the combined subsystems' transient stability, voltage transient and adequacy of response reserve should be checked.

In the training and actual restoration, in addition to using the prepared and tabulated analytical tools results on equipment and power system constraints, ES should be able to use analytical tools to determine the system constraints for the unforeseen situations such as a stuck breaker, unsuccessful energization of a path, synchronization of unplanned subsystems, or load pickup larger than planned. Also, ES should revise and/or adapt schedules developed during planning, or develop new ad hoc schedules!

THE THIRD RESTORATION PERIOD

The integration of the skeleton of the bulk power supply should be completed at the end of the second period, thus achieving the post-restoration target system. This should include: synchronization of all the subsystems, interconnection of all the online and offline power plants with the energized transmission stations and substations, and energization of all the stations needing stored energy. The third restoration period begins with the availability of the large thermal units.

The primary functions of ES in this period are: (i) scheduling the large thermal units including the super-critical once-through boiler-turbines, nuclear units, and the cold restart of the drum-type units, (ii) energizing the high voltage and extra high voltage lines to firm up the transmission system, (iii) Picking up large low voltage AC network loads, and (iv) activating tie-lines with the adjacent power systems. The complete restoration of load in a primarily thermal power system may take 8 to 12 hours.

The functional requirements of ES in this period differ from those of the first two periods in the following areas:

- there will be no need for several restoration teams and their simultaneous restoration activities as was the case in parallel operation of several subsystems in the previous two periods,

- there will be no prime movers with minimum critical times, all the large thermal units can be scheduled for hot restarts or cold restart any time after the elapse of their maximum time limits,

- the load pick up trajectory will be well within the maximum ramp rate of the large units, relaxing the requirements for a tight load-generation coordination, and

- the larger number and greater inertia of the prime movers in-service allow larger load pick up increments.

Scheduling Generating Units

ES should be able to use the real and reactive generators' capabilities to schedule their start ups and operations optimally [6,7]. The availability of real and reactive generation is based on the type of prime movers, fuel used, modes of operation, their post-disturbance conditions, and the start-up timings. There are two or more sets of timings for normal and emergency operation, the latter should be used for restoration purposes. The characteristics of prime movers which should be considered in the third period include [15]:

- Types; recirculating (drum-type), super- and sub-critical circulating (once-through), and boiling- and pressurized-water nuclear units.

- Fuel; coal, gas and oil fired.

- Operations; base load, cycling, and peaking.

- Conditions; a relative combination and range of boiler to turbine versus hot and cold

- Timings; time to provide cranking power, interval between paralleling successive units in a plant, time from start to paralleling, time from paralleling to minimum load, soaking time required, and time from release to maximum load.

ES should be able to use the above information to:

- schedule the start up and synchronization of the various prime movers to maximize the availability of the active and reactive (absorbing) generation,

- coordinate load and generation between synchronization of units and their release to the power system operator, and

- maintain adequate response reserve to withstand loss of the largest in service generator loading.

Firming up Transmission System

As the large thermal units are brought on line, ample reactive absorbing capability becomes available. ES should be able to firm up the transmission system by scheduling the energization of high voltage and extra high voltage overhead lines, closing of the second circuit of double circuit lines, and energization of pipe-type cables. In performing these functions, the following constraints should be considered [10,16]:

- the sustained overvoltages caused by the lightly loaded lines,

- the transient voltages caused by the switching surges,

- the harmonic resonance overvoltages caused by transformer saturation caused by low tap position and high voltage operation, and

- the excessive standing phase angle when closing a loop.

During the third period, the restored power system is more prone to occurrences of unforeseen events, however time does allow for the online analysis of the situations. ES should be able to use the restoration power flow, short circuit, voltage transient and electro-magnetic transient programs to determine the needed constraints.

Picking up Network Loads

Low voltage AC network loads are much larger than radial loads, have near unity power factor, are picked up by group circuit breakers, and therefore are more difficult to pick-up during the first two restoration periods, when few generators with small inertia are in service and the reactive absorbing capability is low. Prior to picking up the network loads, ES in conjunction with analytical tools should be able to [17]:

- estimate the size of the network load with the cold load pickup program using the elapsed time from the interruption, ambient temperature, etc,

- evaluate the frequency response of the generators in-service to the sudden increase in load, and

- determine the steady state and transient stabilities of the system.

Activating Tie Lines

There are no hard rules or criteria for activating the tie-lines during restoration. The timing and the extent of interconnection assistance is based on agreements between the companies. In general, the present practice is to activate the extra high voltage (EHV) interconnections in the system being restored when:

- units with large reactive power absorbing capabilities are in service,

- the automatic generation control (AGC) is in operation and is set to flat-tie control, and

- the two combined systems are stable.

The last requirement entails availability of the dynamic equivalents of adjacent systems.

CONCLUSION

The review of restoration activities over the three periods shows that their durations become longer in going from the first to the third period. It extends over 30 to 60 minutes in the first, 3 to 4 hours in the second, and 8 to 12 hours in the third period. The possibility of occurrence of unforeseen events and the probability of unsuccessful operations also become greater in going from the first to the third period. For these reasons the expert system and analytical tools become increasingly more important during the second and third restoration periods.

REFERENCES:

[1] R.J.Kafka, et al, "System Restoration Plan Development for a Metropolitan Electric System," IEEE Trans.v. PAS-100 n.8, 1981. pp.3703-13.

[2] M.M.Adibi "Power System Restoration Analytical Tools," Proceedings of Pennsylvania Electric Association, System Operation Committee, January 21-22, 1993

[3] K. Matsomuto, et al, " Knowledge-Based System as Operational Aids in Power System Restoration," IEEE Proceedings of May 1992.

[4] IEEE Committee Report, New Approaches in Power System Restoration, SRWG, M.M.Adibi, chmn. T-PWRS Nov 92 1428-1434[5]

[5] E.Mariani et al, "Field Experience in Reenergization of Electrical Networks from Thermal and Hydro Units," IEEE Trans. v.PAS-103 n.7, 1984, pp.1707-13.

[6] IEEE Committee Report, Power System Restoration - The Second Task Force Report, SRTF, M.M. Adibi, chmn. T-PWRS Nov 87 927-933

[7] IEEE Committee Report, Special Considerations in Power System Restoration, SRWG, M.M. Adibi, chmn, T-PWRS Nov 92 1419 - 1427

[8] R.D.Shultz and G.A. Mason, "Blackstart Utilization of Remote CTs, Analytical Analysis and Field Test" IEEE Trans.-PAS 84, 2186 - 2191

[9] IEEE Committee Report, Special Considerations in Power System Restoration - The Second Working Group Report, SRWG, M.M. Adibi, chmn, paper 93 WPM 202-2PWRS, presented at IEEE 1993 WPM, Columbus, OH, Jan 31 - Feb 5, 1993

[10] IEEE Committee Report, "Overvoltage Control During Restoration," SRWG, M.M. Adibi, chmn, T-PWRS Nov 92 1464 - 1470

[11] R. J. Kafka, et al "Role of Interactive and Control Computer in the Development of a System Restoration Plan," IEEE Trans. PAS, Vol PAS-101, Jan 31 - Feb, 1993

[12] M.M. Adibi and L. H. Fink "Power System Restoration Planning," paper 93 WM 204-8-PWRS, presented at IEEE 1993 WPM, Columbus , OH, Jan 31 - Feb 5, 1993

[13] C. C. Liu et al "Generation Capability Dispatch for Bulk Power System Restoration: A Knowledge-Based Approach," T-PWRS Feb 93 316 - 325

[14] IEEE Committee Report, Bulk Power system Restoration Training Techniques, WGOT, M.E. Cooper, chmn and SRWG, M.M. Adibi, chmn, T-PWRS Feb 93191 - 197

[15] F.P. De Mello and J.C. Westcott "Steam Plant Startup and Control in System Restoration," paper 93 WM 201-4-PWRS, presented at IEEE 1993 WPM, Columbus, OH, Jan 31 - Feb 5, 1993

[16] S. Wunderlich, et al, "An Approach to Standing Angle Reduction," IEEE Trans. 93 WM 205-5 PWRS.

[17] IEEE Committee Report, System Operation Challenges, COPWG, E.K. Nielson, chmn, Feb 87, IEEE Trans, v. PWRS-3 n.1, 1987 pp.118 - 26

[18] M.M. Adibi and D.P. Milanicz, "Reactive Power Limitation of Synchronous Machines," paper 93 WM 203-0-PWRS, presented at IEEE 1993 WPM, Columbus, OH, Jan 31 - Feb 5, 1993

Discussion

K. Matsumoto (Mitsubichi Electric Corporation, Amagasaki, Japan): Authors are commended for their efforts in abstracting expert system requirements for power system restoration. These requirements are based on concrete restorative activities that have not been classified yet and have not been known to this discusser. Expert systems have various useful characteristics: reasoning, guidance, decision support, explanation, heuristic, and so on. Would the authors please respond to the following questions:

- In the opinion of the authors, which expert system function(s) are applicable in the restorative procedure?
- What is the authors evaluation of the state of the art of expert systems as applied to restoration?
- Assuming that the current restoration expert systems are in their development phases, what are the authors recommendation for the future work?

Manuscript received February 10, 1994.

W. R. Lachs (University of NSW, Kensington, Australia): An important restoration issue has been the retirement of city power stations on many power systems. Not only has this eliminated a back-up supply to essential services but has added considerably to the time for restoration of supplies to the city. When there is a complete loss of supply within a city it becomes more difficult and takes longer because of the need to re-energize the transmission network and convey electricity from the now more distance stations. The introduction of battery energy storage plant (BESP) within city networks (which may become feasible with the future proliferation of electric vehicles) could provide a useful alternative to the retired city power stations. Not only could BESP provide a back-up to essential services but portions of the city's electricity network could remain energized. This would considerably speed up the restoration of supplies to city consumers. With the energy storage capacity and fast responses of their power electronics, BESP could absorb the full impact of load blocks as they would be reconnected, as well as tightly controlling voltage levels. This would shield generators from sudden load changes so lessening the risk of the loss of boiler ignition and unit trip [A].

Reference

[A] Lachs, W. R. and Sutanto, D., "Battery Storage Plant Within Large Load Centers," IEEE Trans. Power Systems, Vol. 7, No. 2, May 1992, pp. 762–769.

Manuscript received February 10, 1994.

J. J. ANCONA (Niagara Mohawk Power Corp., Syracuse, NY). The author's are to be congratulated for presenting a coherent insight into the application of an expert system (ES) to power system restoration. Compared to operators manually using analytical tools (AT) to help in restoration, an on-line ES integrated with AT holds the promise for offering operators a better, faster path (with more potential options investigated) to restoration following a major power failure.

One of the most important aspects of this paper is that it has been written from a power system operator's point of view. Rather than explaining what an ES can do or how it would work, the authors describe what a restoration ES needs to be able to do for successful restoration to occur. In this context, they help bridge the gap between the operator user and the programmer.

It has created a standard with which to judge the usability of a proposed restoration ES. Also, it should serve as an aid for utilities progressively moving through the following hierarchy of restoration preparation (from relatively simple to sophisticated):

1) Rely solely on existing emergency procedures and operators' experience.
2) Prepare general restoration guidelines and objectives, and a data base of helpful restoration information for a specific power system to supplement existing emergency procedures and further aid operators.
3) Prepare more specific restoration guidelines, plans, training sessions, and practice drills by postulating major power failure scenarios and using off-line AT to address restoration issues.
4) Integrate AT with SCADA/EMS to provide a set of on-line restoration tools to further help the power system operator.
5) Develop an ES which integrates on-line AT, SCADA/EMS, operator experience, and restoration guidelines and objectives to produce a more effective and efficient restoration system.

In the process of explaining the requirements for restoration ES, the authors provide an excellent discussion by defining: (1) three subsequent phases of restoration (i.e.: initial power sources are restarted to pick-up critical loads and form a sub-system; subsystems are interconnected to achieve a larger, more stable target system; and finally, larger generators are brought on-line and all load is restored); (2) a detailed sequence of steps that need to take place in each phase; (3) and issues that need to be addressed for both manual and ES restoration to be effective. Some important points that can be gathered from the paper include:

1) Compared to normal operation, power system interactions during restoration are more complex, and the opportunity for problems increases. Therefore, using integrated AT, ES must be able to reconcile a large number of conflicting issues, and provide balance to an already unstable system.
2) For any restoration ES to be truly effective, operators must become familiar with both its capabilities and its deficiencies.
3) If an ES will be heavily dependent upon data from SCADA/EMS, operators will obviously need to fall back upon a less sophisticated restoration method if the SCADA/EMS also malfunctions during a major power failure.
4) The use of and need for an on-line ES increases as restoration progresses through the three phases defined in the paper. At first, information may be incomplete and both time and options may be limited; consequently general guidelines and intuitive operator reaction may be the only choice. By the third phase of restoration, more alternatives may exist, and more time and information will available to allow an ES to more actively participate in the restoration process.
5) Certain refinements can be made to the existing power system to help improve the restoration process with or without ES (eg.: install emergency generators at critical substations rather than being forced to include them on the critical path in Restoration Phase I).
6) Critical public loads need to be considered in the restoration process. I would question whether critical industrial loads should necessarily receive a high priority for restoration.

Some would argue that low voltage AC network loads are more critical due to public safety concerns (even though they may require a stronger more stable system to adequately pick-up). I would appreciate the authors commenting on whether industrial loads should be picked-up at the expense of prolonging the restoration process and delaying the rebuild of a large stable system so AC networks can be restored. Perhaps the ES could be used to determine which public loads are most critical for restoration priority.

In conclusion, the authors are again to be commended for helping advance the development of a restoration ES by describing its required attributes from the perspective of the power system operator, and for providing an excellent outline of the sequence of steps and potential problems involved in the restoration process.

Manuscript received February 24, 1994.

M. M. Adibi, R. J. Kafka and D. P. Milanicz: The authors thank the discussants for providing the System Restoration Working Group with the opportunity to clarify issues which their questions and comments raise.

Mr. Ancona questions restoration priority of critical industrial loads vis-a-vis low voltage AC network loads. In the first two restoration periods, preferably radial loads which are small and have lagging power factor are picked up to stablize generation and to assist with reactive power balance. In comparison, the low voltage AC network loads are large (30 to 250 MW) with unity or leading power factors. They are usually picked up in the third restoration period when their sizes are within the online generators' frequency response and ample reactive absorbing capability is available. The industrial loads and in particular those with thermal inertia loads such as aluminum pots are controllable and can be picked up in smaller increments. In some power systems they are used as control means to assist with the restoration process. However, if feasible, and if it meets the criterion for the restoration process, there is no reason why the networks could not be restored in the early stages.

Dr. Lachs comments on replacing the retired city power stations with battery energy storage plant (BESP), certainly should be considered when they become feasible on the scales required to support city loads.

Dr. Matsumoto raises three question to which Dr. D. S. Kirschen of Empros Power Systems Control, Minneapolis, MN and Dr. C. C. Liu of University of Washington, Seattle, WA, as members of SRWG responds:

Dr. Kirschen: While restoring the power system following a major incident, the operator must make a large number of decisions and perform a wide variety of tasks. As the authors point out in the paper, knowledge-based systems can facilitate the decision-making process and expedite the accomplishment of many of these tasks. Dr. Matsumoto asks to which of these activities knowledge-based systems will be applicable. I believe that knowledge-based systems, assisted by analytical tools, will ultimately assist the operator in most of these activities. Instead of thinking in terms of individual expert systems designed to solve specific restoration sub-problems, it is probably more effective to design a decision-support environment which would bring together all the information and the tools that the operator may need during the restoration. Early versions of such an environment might provide only a limited number of reasoning tools dealing with relatively simple issues such as determining connection paths and checking operator-designed plans. More complex tools dealing with the strategic aspects of the restoration could be added later, when the simpler tools have gained acceptance among the users.

At this point, a few restoration expert systems have moved from the prototype stage to the development stage and are being tested at utilities in conjunction with training simulators.

The most important activity that developers can perform to prepare for future work is to observe how their system behaves during realistic restoration drills. This observation stage is crucial not only to detect possible reasoning flaws, but also to verify that the system can interact with the user *on the user's terms*.
Chen-Ching Liu, University of Washington

Dr. C. C. Liu: (a) System restoration requires close coordination among generation, transmission and distribution personnel/facilities. The coordination decision-making process involves information, experience, off-line studies, guidelines, and resource availability. General guidelines are usually available at utilities. However, power outages can occur in different ways, which may not be covered in the guidelines. Expert systems can help develop the restoration strategy on-line for the specific outage scenarios based on the general guidelines. An expert system should be able to decide when to run the power flow or stability simulation programs and prepare the scenarios for the analysis. The expert system should also determine what data generated by the analytical tools needs to be checked.

(b) Several expert systems dealing with different aspects of system restoration are included in a CIGRE survey [A]. These expert systems can handle partial shutdowns, switching substation, subtransmission restoration, distribution restoration, etc. Probably none of the existing expert systems can serve all the functions described.

(c) Expert systems integrated with analytical tools can provide opportunities for operators and planners to use the tools in training or planning. Full integration of restoration expert systems is still uncommon in the on-line operational environment of bulk power systems. Many practical issues need to be addressed before restoration expert systems can become a standard application in the utilities.

A fundamental issue is that no restoration expert system can serve all or even most of the desirable functions today. To achieve a mature and portable software tool, the partially solved problems need to be solved more completely. Also, system specific techniques or tools should be generalized so that tools can be used for many utilities. The generic tool should also accommodate system specific natures.

A generic software ready for customization is needed if expert systems are to be widely used in utilities in the future. This can be achieved if some "generic restoration actions" and a flexible software structure are provided. The user can use the generic restoration actions to customize the strategy to their own needs.

[A] C. C. Liu et al., "Practical Use of Expert Systems in Planning and Operation of Power Systems," Electra, Feb. 1993, pp.30-67. (CIGRE TF 38-06-03 Report)

Manuscript received April 19, 1994.

Tie Line Utilization During Power System Restoration

Kan-Lee Liou and Chen-Ching Liu
Department of Electrical Engineering
University of Washington

Ron F. Chu
Philadelphia Electric Company
Philadelphia, PA

Abstract - This paper identifies some of the critical roles of tie lines during power system restoration. Based on these roles, a systematic approach is developed to evaluate the impact of the availability of tie line assistance on restoration. Using this approach, a tie line expert system module has also been developed to supplement the previously developed EPRI restoration expert system (RES). The enhanced RES can now include available tie line power in recommending a restoration strategy to maximize the system generation capability by modifying the generator startup sequence. Simulation results confirm that outside assistance will improve the overall generation capability and provide an extra dimension to the flexibility of a company's restoration strategy.

Keywords: Restoration, Tie Line, Knowledge-Based System, Generation Capability Dispatch

1. Introduction

System restoration problems and their solutions have been the focus of many research activities in recent years. General concerns and strategies related to system restoration have been discussed extensively [1, 2, 3, 4]. At the same time, artificial intelligence techniques have been acknowledged in various studies [5, 6, 7] to be promising solutions for the system restoration problems. Sponsored by the Electric Power Research Institute (EPRI), we have been developing a knowledge-based system (KBS) for producing a generation capability dispatch strategy during system restoration.

With the exception that in one early IEEE working group report [8] the role of "other utilities" during restoration was mentioned, most studies conducted so far are confined within one utility. As a result, the benefits of interconnections with neighboring utilities were not capitalized for planning the restoration strategy or procedure for a particular utility. Note that, inter-system assistance can be indispensable during restoration especially for those with insufficient generation capability. Such tie line assistance is beneficial at any given time after the blackout and provides an extra dimension in planning the restoration procedure.

The tie line issue was identified as a subproblem of system restoration in a recent report [9] about the practical use of KBS in power systems. In general, existing KBS [10, 11, 12] has not addressed or solved the tie line utilization problem. As an enhancement to our recent developed EPRI Restoration Expert System (RES) [13], the tie line issues are addressed, the problem is formulated and the solution is developed and implemented. The research activities and results are presented in this paper.

In Section 2, the roles of tie lines during system restoration were identified during interviews with participating utilities and power pool. Based on these roles, a systematic approach is developed to evaluate the impact of tie lines on the restoration process. Using this evaluation approach, a generic tie line utilization strategy is developed to search for a set of generating units whose startup times and sequence can be modified leading to (1) a maximum increase in system generation capability when tie line assistance is available and (2) a minimum decrease in capability if tie line assistance has to be provided. The proposed approach and the tie line utilization strategy are given in Section 3 and 4, respectively.

The tie line utilization strategy has been implemented as a new module of the EPRI RES. Simulation results are presented to illustrate the enhancement to the RES due to this new module. The implementation and simulation results are given in Section 5 and 6, respectively. Note that the simulations are performed using data supplied by Philadelphia Electric Company (PECo). It should also be pointed out that the planning of the startup sequence is the main focus of this work. Within the existing RES, checking of voltage, stability, var support, limit violation, etc., are done by other modules. Using the RES as a planning, training or on-line restoration aids, these modules will perform cohesively to produce a useful restoration guidance to system operators.

2. Role of Tie Lines During Restoration

After a blackout, a utility with limited generation capability has to rely on its neighboring utilities to supply power to restore customer loads. For a utility with insufficient black-start capability, tie line assistance is crucial to restart and develop its generation capability. Even for those with sufficient generation capability, tie line assistance may be used to expedite the restoration progress or to guard against any unexpected contingencies in which crucial unit fails to start.

As a part of EPRI restoration project, we conducted an interview with operators from the PJM Interconnection, Philadelphia Electric and Puget Sound Power and Light. The objective is to identify the critical roles of tie lines during restoration. Basically, these roles are identified to be:

1. to crank non-black-start units,
2. to respond to requests for power from neighboring utilities,
3. to serve customer loads.

The details of each role and how existing RES modules can be used are described in this section.

2.1 Using Tie Lines to Crank Non-Black-Start Units

The overall restoration strategy of the EPRI RES is given in Figure 1 [13]. In the situation where the "Restoration Planning" module plans to crank a non-black-start unit using the available tie line power, several conditions needed to be checked.

94 WM 221-2 PWRS A paper recommended and approved by the IEEE Power System Engineering Committee of the IEEE Power Engineering Society for presentation at the IEEE/PES 1994 Winter Meeting, New York, New York, January 30 - February 3, 1994. Manuscript submitted August 2, 1993; made available for printing December 6, 1993.

Reprinted from *IEEE Transactions on Power Systems,* Vol. 10, No. 1, pp. 192–199, February 1995.

- Check if there is a path between the tie line and the non-black-start unit.

The "Restoration Planning" module will request the "Path Management" module to check on path's availability. "Path Management" will use its submodules to verify the availability of the cranking paths between the tie line and the next-to-go non-black-start unit.

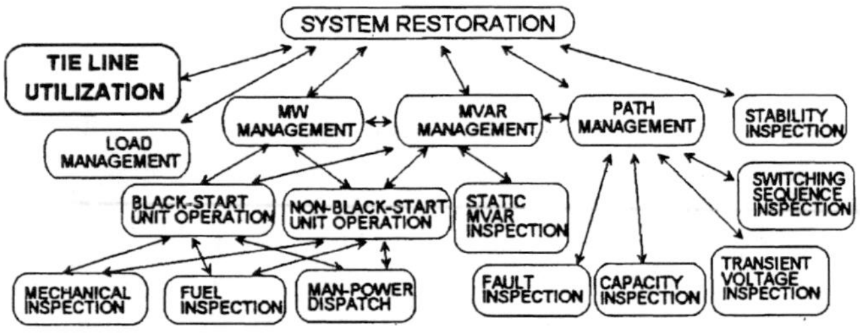

Figure 1. Restoration Strategy

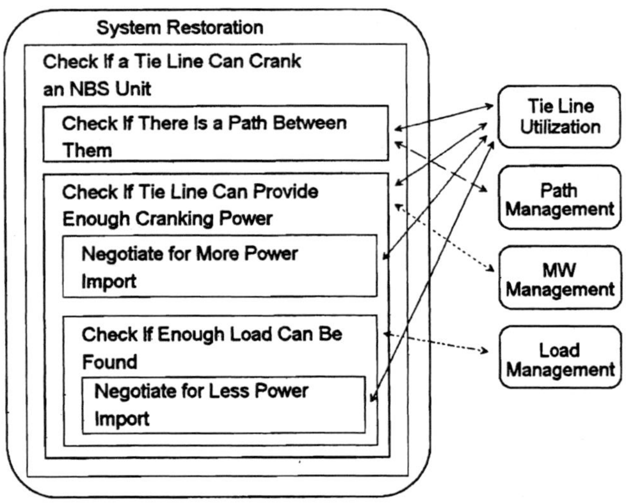

Figure 2. Using Tie Lines to Crank Non-Black-Start Units

- Check if the tie line can provide enough cranking power

Obviously, the tie line has to satisfy the startup requirement of the non-black-start unit. Otherwise, more power import will be needed. Due to the fact that the available power import is, in general, in fixed blocks, tie line utilization strategy must consider that (1) if the fixed imported power exceeds the required cranking power, enough loads along the cranking path need to be found to absorb extra power for stabilizing the system and (2) if such load cannot be found, a reduction of imported power is necessary. In either case, both the "Tie Line Utilization" and the "Load Management" modules will be invoked. Finally the interactions between these modules as well as the logics governing the use of tie line for providing cranking power are shown in Figure 2.

2.2 Responding to Tie Line Request from Neighboring Utilities

If the blackout affects more than one utility, some utility may have to respond to tie line requests from its neighbor(s) during restoration. In this case, as shown in Figure 3, the following conditions need to be considered:

- Check if system can provide enough power

To identify the amount of power available for assisting a neighboring utility, the utility has to evaluate its own generation capability with the information provided by the "MW Management" module.

- Check transmission capacity

If the amount of the tie line power is large, the capacity limits of the tie line and transmission lines may be violated. "Path Management" has to make sure that no such constraints will be violated.

- Decide priority of tie line

A utility may decide to provide power through the tie line(s) even though its own system is not completely restored. The user has to assign the priorities among tie line exports, cranking non-black-start units, and serving loads. The priorities may also be suggested by the expert system, which is able to evaluate the percentage of system load restored or the number of units restarted.

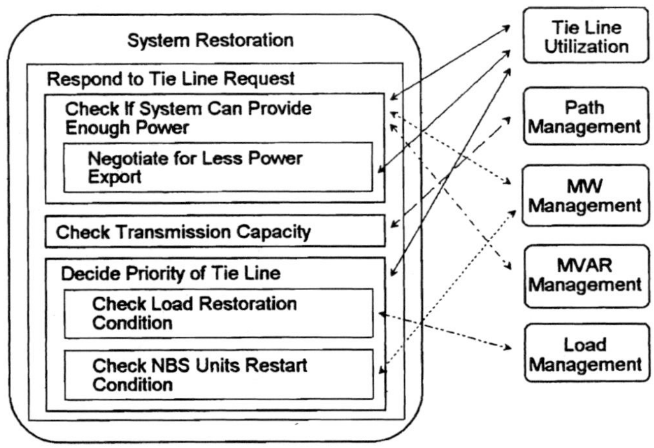

Figure 3. Responding to Tie Line Request from Neighboring Utilities

2.3 Using Tie Lines to Serve Loads

For those utilities with limited generation facilities, restoring all loads cannot be done without tie line imports. Some utilities may request tie line imports in order to speed their restoration processes. When the "Restoration Planning" module plans to serve loads with a tie line, the following conditions need to be checked as shown in Figure 4:

- Check if there is a path between the tie line and the loads

The "Restoration Planning" module will request the "Path Management" module to check this condition. "Path Management" will be assisted by its submodules to verify the availability of the cranking paths between the tie line and the loads.

- Check if the tie line can provide enough power

The tie line has to be able to provide enough power to serve the loads. If the tie line cannot provide enough power, a request for more power import may be necessary. As indicated in our interview with system operators, a tie line import power usually is provided in a fixed MW block of power. If the imported power is more than the power required by the loads, more loads need to be found to absorb the extra power. If enough load cannot be found, less power import is needed.

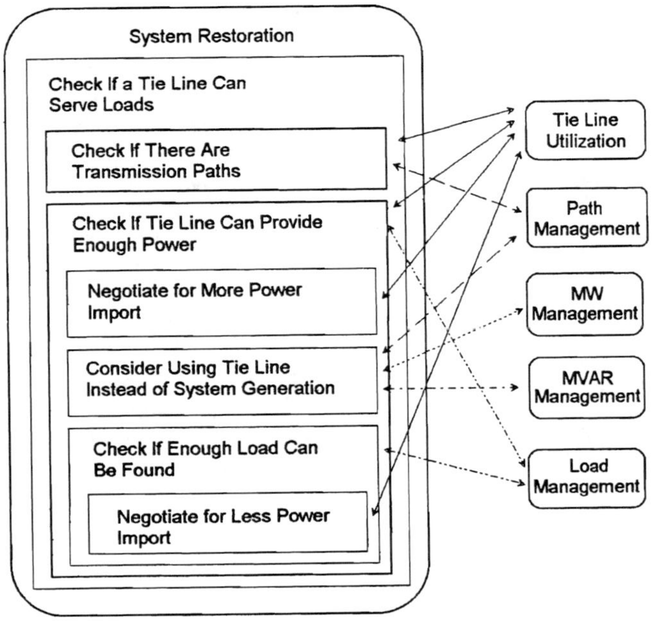

Figure 4. Using Tie Lines to Serve Loads

3. Proposed Tie Line Utilization Approach

The problem of how to utilize tie lines to maximize system generation capability during restoration can be described as follows.

Generation Capability Involving Tie Lines

The goal is to maximize

$$G = \sum_{i=1}^{n} \int_{t_i}^{t_\infty} g_i\left(t - t_i\right)dt$$

where

t_i is the start time for the generating unit i whose generating capability is $g_i(t-t_i)$

t_∞ is the planning ending time

Given an optimal generation restart sequence $(t_{1o}, t_{2o}, \ldots, t_{no})$, t_∞, g_i, $i=1,\ldots,n$, cranking paths used in the sequence, and a tie line import TL_i and export TL_e at t_{TL}, it is desirable to use TL_i and supply TL_e most efficiently to increase G.

Note that both power import (TL_i) and export (TL_e) are modelled. Both terms are used in cases where import tie line power is needed to enhance the generation capabilities and, at the same time, tie line export is necessary to assist another neighboring utility for its recovery.

In G, t_∞ is a positive constant and g_i is always positive and cannot be changed by TL_i or TL_e. Therefore, G will increase if a smaller t_i is found, i.e., to start unit i earlier than t_{io}, and decrease if a larger t_i is necessary, i.e., to start unit i later than t_{io}.

Definitions

1. TL_i is used to provide TL_e and crank a set of non-black-start units $\{g_{k1}|k1 \in S1\}$.
2. Another set of generating units $\{g_{k2}|k2 \in S2\}$ are used to provide TL_e - TL_i when $TL_e > TL_i$.
3. The cranking paths of a third set of non-black-start units $\{g_{k3}|k3 \in S3\}$ are completely used for cranking with TL_i and supplying TL_e. $S3$ is a set of units whose cranking paths are used for tie line utilization.

Conditions

If it is desirable to find a combination of such units that can increase G, then the following conditions must be satisfied:

1. There are transmission paths for TL_i to crank $\{g_{k1}|k1 \in S1\}$.
2. TL_i can provide sufficient cranking power for $\{g_{k1}|k1 \in S1\}$.
3. The original start times of $\{g_{k1}|k1 \in S1\}$ and $\{g_{k3}|k3 \in S3\}$ are later than t_{TL}.
4. There are transmission paths for $\{g_{k2}|k2 \in S2\}$ to supply TL_e.
5. $\{g_{k2}|k2 \in S2\}$ can provide sufficient power for TL_e.
6. The start times of $\{g_{k2}|k2 \in S2\}$ are earlier than t_{TL}.

Optimal Utilization of Tie Lines

Suppose that the above conditions 1-6 hold. The system generation capability (MWh) will have the maximum increase if the startup sequence is modified by changing the startup times of a combination of units that can satisfy the following condition:

The MWh increase provided by $\{g_{k1}|k1 \in S1\}$ minus the MWh decrease caused by $\{g_{k2}|k2 \in S2\}$ and $\{g_{k3}|k3 \in S3\}$ is higher than or equal to that of any other combination that satisfies conditions 1-6.

The MWh increase provided by $\{g_{k1}|k1 \in S1\}$ includes two parts:

(1) the summation of the MWh increase from the earlier startup of g_{k1}

(2) the summation of the MWh increase "induced" by g_{k1}

The MWh increase "induced" by a unit g_{k1} is calculated recursively as the summation of the following items:

(1) the summation of the MWh increase from the earlier startup of a unit $g_{k1'}$ that is cranked by g_{k1}

(2) the summation of the MWh increase from the earlier startup of a unit $g_{k1''}$ that is cranked with the power that was used to crank g_{k1}

(3) the summation of the MWh increase "induced" by $g_{kl'}$ and $g_{kl''}$

The MWh decrease caused by $\{g_{k2} | k2 \in S2\}$ ($\{g_{k3} | k3 \in S3\}$) includes two parts:

(1) the summation of the MWh decrease from the delayed startup (including loss) of $g_{k2}(g_{k3})$

(2) the summation of the MWh decrease "induced" by $g_{k2}(g_{k3})$

The MWh decrease "induced" by a unit $g_{k2}(g_{k3})$ is defined recursively as the summation of the following items:

(1) the summation of the MWh decrease from the delayed startup of every unit $g_{k2'}(g_{k3'})$ that is cranked by $g_{k2}(g_{k3})$

(2) the summation of the MWh decrease "induced" by $g_{k2'}(g_{k3'})$

Approximations

To reduce the search space, the following assumptions are adopted:

- Assume the MWh increase "induced" by a unit is small and can be neglected. Therefore the MWh increase from g_{kl} is $\int_{t_{TL}}^{t_\infty} g_{kl}(t - t_{TL})dt - \int_{t_{klo}}^{t_\infty} g_{kl}(t - t_{klo})dt$ and can be calculated with given t_{TL}, t_{klo}, and g_{kl}.

In other words, the MWh increase calculated here comes from earlier unit startup with the tie line power. Further increase that may be achieved with the resulting MWh increase is ignored. This does not lead to any inaccuracy in the PECo case.

- Assume that a unit in the startup sequence can be started only by an assigned black-start or non-black-start unit. Therefore the MWh decrease "induced" by a unit g_{k3} can be calculated with its relationship in the original startup sequence, i.e., $\Delta'_{k3} = \sum_{k3' \in S3'} \left(\int_{t_{k3'o}}^{t_\infty} g_{k3'}(t - t_{k3'o})dt \right)$, where $\{g_{k3'} | k3' \in S3'\}$ are units cranked by g_{k3}.

- Assume a minimal number of generating units are used to provide TL_e and the units with the "shortest"(fastest) paths will be considered first. Therefore the number of possible $\{g_{k2} | k2 \in S2\}$ will be smaller.

- Assume the "shortest"(fastest) paths are used in determining whether a transmission path goes through cranking paths and assume only the cranking paths for the units with start times later than t_{TL} are considered. Therefore the number of possible $\{g_{k3} | k3 \in S3\}$ will be smaller.

4. Tie Line Utilization Strategy

Based on the approach in Section 3, a generic tie line utilization strategy has been developed, which is shown in Figure 5. This strategy considers both tie line import and export in response to the restoration strategy or wheeling requirement. When tie line import is available and can be used to crank non-black-start units, this KBS will look for a set of non-black-start units and suggest changes in their start times that will contribute the maximum increase to the system generation capability. Note that the system configuration and startup requirements of non-black-start units need to be considered. At this time, the breadth-first method is used for the search. The amount of generation capability increase is calculated with the generation startup characteristics of each non-black-start unit. If such units can be found, a modified restoration plan will be suggested. Similarly, if tie line export is requested, this KBS will identify a set of generating units that can be used to supply the tie line. The identified units will result in the minimum decrease of system generation capability. Prior to recommending the tie line restoration strategy, both the generation capability of each unit and status of transmission paths from generating units to the tie line will be checked.

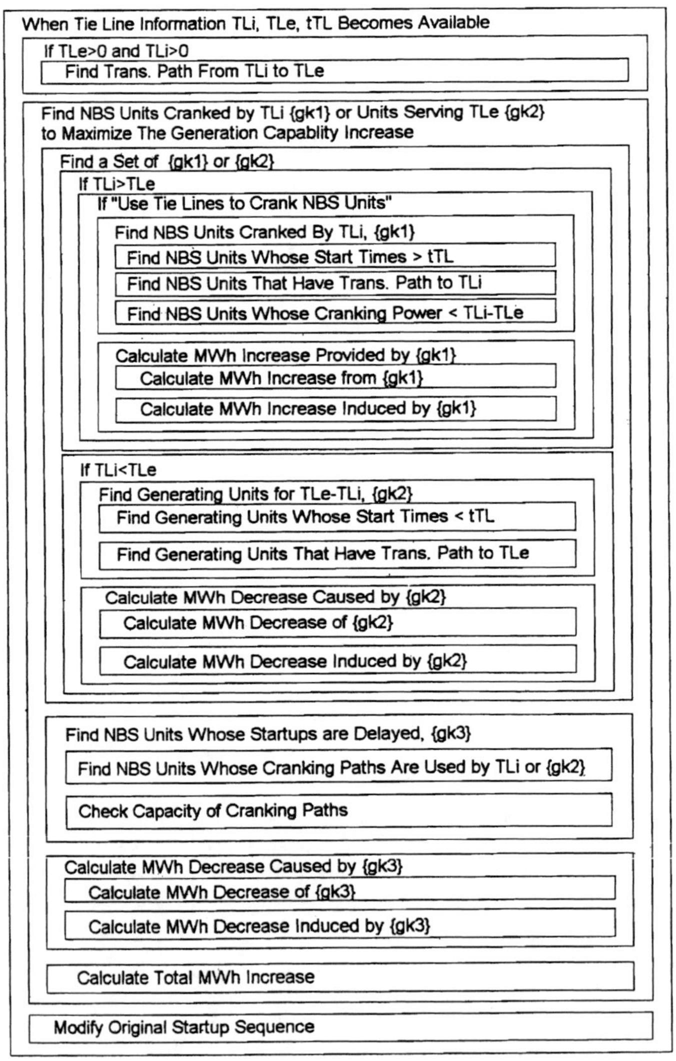

Figure 5. Tie Line Utilization Strategy

5. Implementation

The tie line utilization strategy with the proposed approximations described in Section 4 has been implemented as a new module of the existing EPRI RES. The RES design is based on a goal-driven strategy which is illustrated in Figure 1. It is developed using a commercially available knowledge-based system shell called KES, which is a C-based software tool. Object-oriented data structures are used to represent power system elements. The overall objective of the RES restoration strategy is to maximize the system generation capability during restoration. As shown in Figure 1, the top module, "System Restoration", which was used to coordinate decisions from other modules for maximizing the generation capability, has been modified to incorporate the tie lines. The interactions between the "Tie Line Utilization" module and other modules are shown in Figure 6. In this enhanced RES, if tie line assistance is available or the new "Tie Line Utilization" module is activated, the "System Restoration" module will be asked to provide the original startup sequence. If the system generation capability can be improved, the "Tie Line Utilization" module will return a modified startup sequence to the "System Restoration" module. The "Tie Line Utilization" module will also request the "MW Management" module to calculate the MWh increases or decreases of specified generating units. The "MW Management" module has access to the characteristics of specified generating units. The "Path Management" module provides information on the availability of transmission paths.

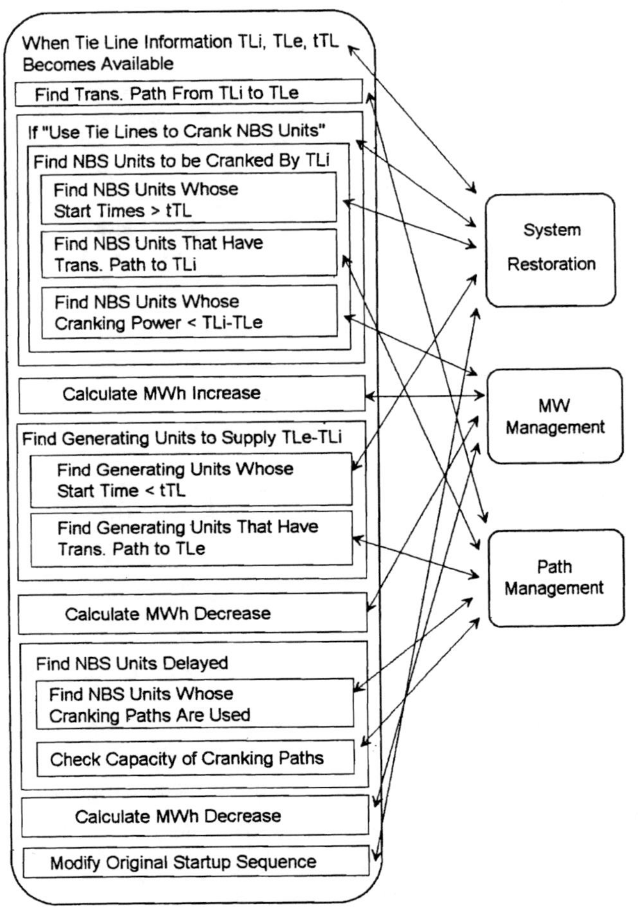

Figure 6. Relations between "Tie Line Utilization" Module and Other Modules

6. Simulation Results

Four cases are selected to illustrate the cranking power assignment and cranking path search capability of the enhanced RES that considers tie lines. These cases also demonstrate the use of the enhanced RES as a planning tool for studying various scenarios involving tie lines during system restoration. One of the test cases considers different import/export levels at a specific tie line location. Another case is about the appropriate levels for various tie line locations. Two additional cases are used to test the strategy when subjected to contingencies that involve one combustion turbine and one hydro unit, respectively. In all four cases, Philadelphia Electric Company system and data were employed.

6.1 Different Capacities for the Tie Line at Peach Bottom

In this case, assume that the tie line at Peach Bottom can provide different MW levels at 10 minutes after a blackout. RES can utilize the available tie lines, modify the generator startup sequences, and obtain improved system generation capabilities. For generation imports at Peach Bottom at a fixed MW blocks of 10 MW, 50 MW, or 100 MW, the resultant MW capability curves are shown in Figure 7.

As expected, the results indicate that a higher tie line import level results in a higher system generation capability. If the tie line assistance is mainly for providing cranking power to non-black-start units, the improvement in system generation capabilities will be limited by the MW cranking power requirements needed. Any extra MW will then be used to supply loads which are identified by the Load Management module.

In this case, the non-black-start units to be cranked by this tie are Cromby, Muddy Run Hydro, and Eddystone generating units. Since Eddystone units have critical minimum interval time constraints [13] and are not ready for restart. The total cranking power requirement at the Peach Bottom tie is 29.2 MW which is for units at Cromby and Muddy Run. Hence, 29.2 MW is the minimum MW import for supplying cranking power and additional power will be used to restore customer loads.

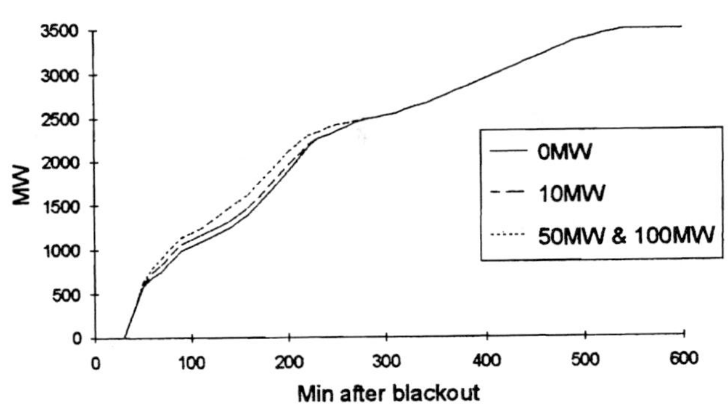

Figure 7. Generation Capabilities with Different Tie Line Import Levels at Peach Bottom

6.2 Different Tie Line Locations for the Same Import Level

In this case, two tie line locations were evaluated: Peach Bottom and Cox's Corner. Assuming that 50 MW can be made available at either location 10 minutes after a blackout. RES can utilize this information and project the resulting system generation capabilities. The resulting system generation capabilities for both locations are shown in Figure 8.

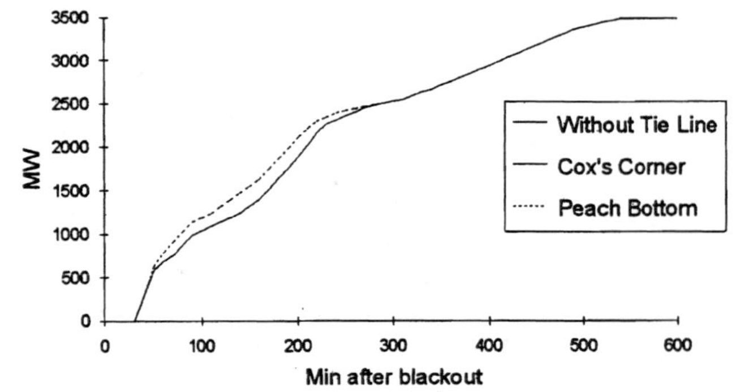

Figure 8. Generation Capabilities with 50 MW Available at Two Tie Line Locations

The simulation results show that an improvement of the system generation capability can be achieved with the tie line import from Peach Bottom. Here the tie line power is used to restart Cromby and Muddy Run at an earlier time. However, the tie line import at Cox's Corner does not increase the system generation capability because the Croydon unit to be cranked by this tie line is not ready due to a critical minimum interval constraint. For restoration planning purposes, the 50 MW tie line assistance at Peach Bottom is more preferable if a choice has to be made. However, if assistance at both locations are available, this RES results indicate that the 50 MW at Cox's Corner is for restoring customer loads. As such, operators need to prepare the system to restore loads at this time. Finally, these simulation results show how RES can be used to evaluate the priority of MW import at each tie line connection and to plan the next restoration step if the tie lie assistance in one or all connections becomes available.

6.3 Contingency I: The Combustion Turbine at Falls is not Available

The combustion turbine at Falls is the default cranking source for the non-black-start unit at Croydon. In this case, the combustion turbine at Falls is assumed to be unavailable. As shown in Figure 9, no alternative, except the tie line at Cox's Corner, is available for cranking the non-black-start unit at Croydon.

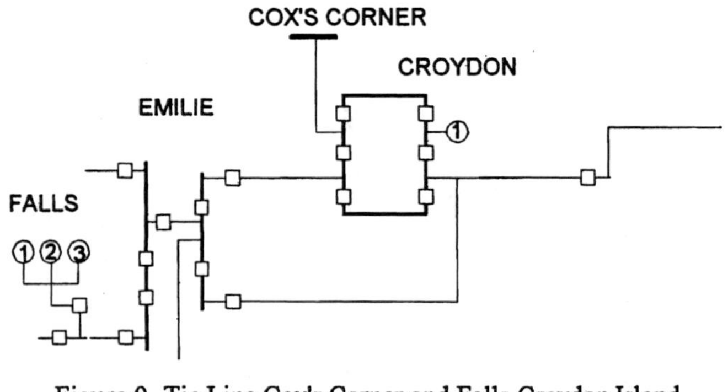

Figure 9. Tie Line Cox's Corner and Falls-Croydon Island

If the tie line at Cox's Corner is not available, then the Croydon unit can not obtain its startup power and therefore it can not be started. The system generation capability is decreased by the generation capabilities of both units at Falls and Croydon as shown in Figure 10.

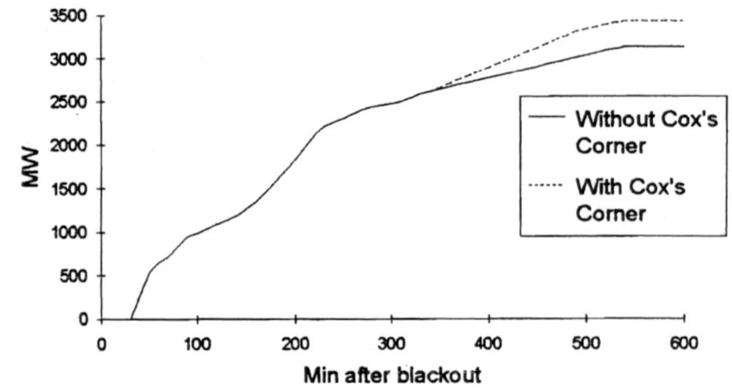

Figure 10. System Generation Capabilities for the Cox's Corner Case

Now suppose that Cox's Corner can provide 100 MW at 310 minutes after the blackout. (Note: the unit at Croydon has a critical minimum interval of 300 minutes.) Then RES will automatically modify the generator startup sequence and utilize the tie line to crank the non-black-start unit at Croydon. Therefore, the unit at Croydon can be started with the startup power from Cox's Corner and the system generation capability is increased by the tie line import and the Croydon unit while subtracting the MW capability of the combustion turbine at Falls. Once again, the RES demonstrate its ability in planning the restoration strategy at a time when tie line import becomes available for cranking non-black-start units.

6.4 Contingency II: Conowingo Hydro Units are not Available

In PECo's restoration procedure, the hydro units at Conowingo can provide cranking power to the non-black-start units at Muddy Run. In this case, it is assumed that the units at Conowingo are not available. Without tie line assistance, RES needs to identify alternative cranking power source which is from the Cromby station as shown in Figure 11. However, the Muddy Run units cannot be restarted until the Cromby units ramp up. The system generation capability shows such a delay in Figure 12.

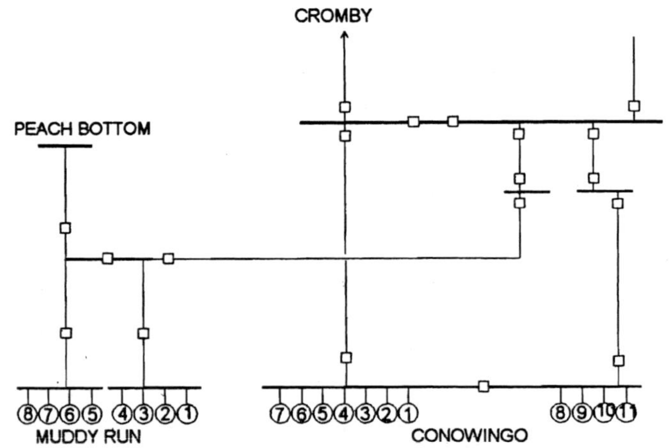

Figure 11. Tie Line Peach Bottom and Muddy Run Station

Suppose that Peach Bottom tie can provide 50 MW at 30 minutes after the blackout. RES can modify the generator startup sequence and have the non-black-start units at Muddy Run started earlier with cranking power from Peach Bottom tie line. Therefore, the system generation capability can be improved as shown in Figure 12. In this example, RES, when employed in a restoration planning environment, can provide information to planners to modify a company's procedure by negotiating tie line assistance if contingencies occur.

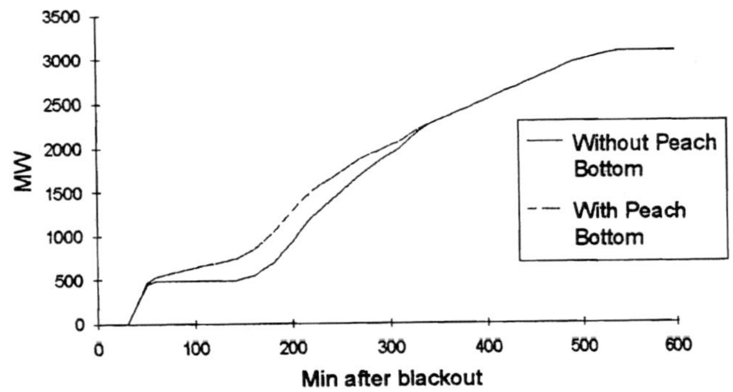

Figure 12. System Generation Capability (Peach Bottom)

7. Conclusion

In an earlier paper, we presented a knowledge-based system [13] for assisting dispatchers to restore their power systems under the assumption that no outside assistance is available. Recognizing its value and importance, the previously developed EPRI restoration expert system (RES) has been enhanced to include available tie line power in formulating the restoration strategy. Simulation results show that the availability of tie line assistance, as expected, will increase the system generation capability and the number of customers being restored. The RES is a useful restoration planning tool for planning and negotiating tie line assistance to and from neighboring utilities.

The contribution of this research are (1) the identification the critical roles of tie lines during restoration, (2) the development of tie line module and (3) the implementation of the tie line module within the EPRI RES.

Presently, we are applying the enhanced RES to identify the locations and capacities of tie line imports for maximizing the benefits derived from the availability of tie line power. The issue of wheeling power during restoration and the integration of the enhanced RES within the EPRI Operator Training Simulator are also being addressed.

Acknowledgment

This research is sponsored by Electric Power Research Institute under research project RP3104-1. We thank the operators from the PJM Interconnection, Philadelphia Electric and Puget Sound Power and Light for their assistance in identifying the critical roles of tie lines during restoration. We also appreciate the comments and suggestions provided by the project advisors and Mr. Gerry Cauley, EPRI project manager, during the course of this work.

References

[1] M. M. Adibi, P. Clelland, et al, "Power System Restoration - a Task Force Report," IEEE Trans. on Power Systems, Vol. 2, No. 2, 1987, pp. 271-277.

[2] M. M. Adibi, J. N. Borkoski and R. J. Kafka, "Power System Restoration - the Second Task Force Report," IEEE Trans. on Power Systems, Vol. 2, No. 4, 1987, pp. 927- 933.

[3] M. M. Adibi, L. H. Fink, et al., "Special Considerations in Power System Restoration," IEEE Trans. on Power Systems, Vol. 7, No. 4, Nov. 1992, pp. 1419-1427.

[4] R. Nadira, T. E. Dy Liacco, and K. A. Loparo, "A Hierarchical Interactive Approach to Electric Power System Restoration," IEEE Trans. on Power Systems, Vol. 7, No. 3, Aug. 1992, pp. 1123-1131.

[5] M. M. Adibi, L. H. Fink, et al., "New Approaches in Power System Restoration," IEEE Trans. on Power Systems, Vol. 7, No. 4, Nov. 1992, pp. 1428-1434.

[6] K. Matsumoto, T. Sakaguchi, R. J. Kafka, and M. M. Adibi, "Knowledge-Based Systems as Operational Aids in Power System Restoration," Proc. of IEEE, Vol. 80, No. 5, May 1992, pp. 689-697.

[7] K. Shimakura, J. Inagaki, Y. Matsunoki, et al., "A Knowledge-Based Method for Making Restoration Plan of Bulk Power System," IEEE Trans. on Power System, Vol. 7, No. 2, May 1992, pp. 914-920.

[8] W. A. Johnson, F. S. Prabhakara, et al., "System Restoration - Deploying the Plan," IEEE Trans. on Power Systems, Vol. 101, No. 11, Nov. 1982, pp. 4263-4271.

[9] C. C. Liu, T. K. Ma, K. L. Liou, and M. S. Tsai, "Practical Use of Expert Systems in Power Systems," Presented at Fourth Symposium on Expert Systems Application to Power Systems, Jan. 1993.

[10] Y. Kojima, S. Warashina, et al., "Development of a Guidance Method for Power System Restoration," IEEE Trans. on Power Systems, Vol. 4, No. 3, 1989 pp. 1219-1227.

[11] Y. Kojima, S. Warashina, M. Kato, and H. Watanabe, "The Development of Power System Restoration Method for a Bulk Power System by Applying Knowledge Engineering Techniques," IEEE Trans. on Power Systems, Vol. 4, No. 3, 1989 pp. 1228-1235.

[12] D. S. Kirschen and T. L. Volkmann, "Guiding a Power System Restoration with an Expert System," IEEE Trans. on Power Systems, Vol. 6, No. 2, 1991, pp. 558-566.

[13] C. C. Liu, K. L. Liou, R. F. Chu, and A. T. Holen, "Generation Capability Dispatch for Bulk Power System Restoration: A Knowledge Based Approach," IEEE Trans. on Power Systems, Vol. 8, No. 1, Feb. 1993, pp. 316-325.

DISCUSSION

M. M. Adibi, IRD Corporation, Bethesda, Maryland: Authors are commended for including tie lines as sources of power in their restoration expert system. Based on interviews with an interconnection, a primarily thermal power system and a heavily import dependent power system, they have divided the roles of tie lines to cranking non-black-start units, assistance to the adjacent systems, and serving blocks of loads. The authors also observe that interconnection assistance has not been considered in the previous restoration procedures. The following comments pertain to the above roles of tie lines and their absence in the restoration procedures. These comments are intended to supplement the paper content.

Practically all the previous restoration procedures have been of the "bottom-up" type and in fact this discussor knows of no publication on "top-down" type of restoration in which tie lines do play an important role. In the initial phase of bottom-up restoration procedure the primary sources of power, which include combustion turbines, hydro units and pumped-storage units, supply the critical requirements for power such as drum-type units, pipe-type cables and transmission/ distribution stations. Although, tie-lines with the adjacent power system when available are also considered a primary initial source, most restorations are prepared for area-wide outages and therefore do not rely on tie line sources.

Regarding the third role for the heavily import dependent systems, serving large blocks of loads, should technically be the responsibility of the exporter which is in control of the prime movers.

As to the major point of this discussion, the second role, the issue of interconnection assistance of where and when to close tie-lines has been with us since the 1965 Northeast Blackout and does not have a generic solution. There are concerns with respect to transient and dynamic stabilities of both systems after they are interconnected, sustained and transient over-voltages which closing of the tie line breaker may produce, the negative sequence voltages which the imbalanced EHV lines generate, etc. One of the present criteria is to activate the EHV interconnections in the system being restored when units with large reactive power absorbing capabilities are placed in service and when with the aid of AGC the tie line flow can be controlled.

In order to learn more about the present practices of the industry a survey was conducted in which 2 interconnections and 8 utilities participated. Below are the survey text, results and respondents' comments.

SURVEY: During restoration, a utility may interact with its neighboring utilities or pool. Such interaction may be essential for a utility that does not have sufficient cranking power to start critical units or serve critical load. Please indicate which of the following are of particular concern on your system and describe how they are considered during restoration (4/10 means 4 out of 10 responded positively):

1. Availability and details of tie line information. 4/10
2. How decisions are made as to when and where tie lines will be reclosed. 4/10
3. Coordination and negotiation issues between you and the neighboring utility if a tie is to be closed. 7/10
4. Stability or reliability issues involved in tie lines for each utility system in the interconnection. 3/10
5. Preparations before receiving (sending) power through a tie line. 6/10
6. Precautions or preparations in case a tie line trips. 5/10

Please provide any additional comments on the above that would be helpful, and/or describe any other such concerns that pertain to your system:

- The power coordination center coordinates major transmission tie activation between power systems.
- All tie lines are opened before the start of restoration process.
- Ties are closed when both systems are in stable condition.
- When ties are closed to improve system stability, the flow across ties are kept close to zero.
- Tie flows are maintained at a pre-determined value.
- Reactive power supply coordination between the two systems is most important.
- Before closing a tie, the recipient is advised of the amount and length of load pickup and its voltage level.

Manuscript received February 10, 1994.

Ron F. Chu (PECO Energy Co.) and Chen-Ching Liu (University of Washington): The authors appreciate Mr. Adibi's comments on the use of tie lines during system restoration. We would like respond to his comments by showing how the EPRI Restoration Expert System (RES) can help to address some of the issues resulting from the survey mentioned by Mr. Adibi.

- **Availability and details of tie line information:** RES assumes that the information is available through communications and exchange of the tie line information.
- **How decisions are made as to when and where tie lines will be reclosed:** RES, with the tie line module reported in this paper, is able to determine how to utilize the imported power or provide the power to be exported. To do so, RES needs to know the tie line locations, rates of imports and exports and the preparation time for receiving the import.
- **Coordination and negotiation issues:** RES helps to evaluate the level, timing and location of imports/exports that can be accommodated by a system.
- **Stability or reliability issues involved in tie lines for each utility in an interconnection:** RES does not consider the assignment of tie line scheduling and frequency control but RES can be customized if the specific policy is known. Operating reserve during restoration can be handled by RES; however, the level of reserves is an unresolved problem for many utilities.
- **Preparations before receiving (sending) power through a tie line:** RES is able to identify the loads to be served (or the units to provide generations) and their corresponding transmission paths.
- **Precautions or preparations in case a tie line trips:** Load shedding and/or rotating blackout policy when a contingency occurs during restoration can be simulated by RES.

We would like to thank Mr. Adibi for his discussion of our paper.

Manuscript received April 19, 1994.

Power System Restoration by Joint Usage of Expert System and Mathematical Programming Approach

T. Nagata, member, IEEE
Dept. of Electrical Engineering
Matsue College of Technology
Matsue, 690, Japan

H. Sasaki, member, IEEE
Dept. of Electrical Engineering
Hiroshima University
Higashi-Hiroshima, 724, Japan

R. Yokoyama, member, IEEE
Dept. of Electrical Engineering
Tokyo Metropolitan University
Tokyo, 159, Japan

ABSTRACT - This paper aims to develop a suitable restoration method by a joint usage of an Expert system (ES) and a Mathematical Programming (MP) approach. Under this devise, the optimal target system for storation is formulated as an MP problem and solved efficiently by the decomposition-coordination strategy; that is a whole problem is decomposed into subproblems according to the generic knowledge of experts. The introduction of a new concept "restorative operation cost" that reflects the various restoration strategies has decreased the number of rules in the knowledge-base to one third that of typical existing ES's. The validity and effectiveness of the combined method have been demonstrated by applying it to a practical size restoration problem.

Keywords: Power system restoration, target configuration, Expert system, Mathematical programming, Mixed integer programming

I INTRODUCTION

Power systems are becoming more complicated and higher reliability is required in recent years; under these circumstances, it is mandatory to restore the power system quickly and safely after a fault. The problem of restoration after service interruption is a complex decision-making and control problem for power system operators. Power system restoration consists of two sequential steps with switching operations excepted: The first step is to determine an optimal configuration as a restoration target and the second to make up a sequence of switching operations (restoration procedure) to bring a faulted system into the thus obtained target system while maintaining a certain level of security. This paper is concerned with the first step due to its primal importance and inherent complexity. Although the second issue is very important from the point of view of power system operations, it requires far less computation time since the first step takes into consideration supply capabilities during the restoration operations.

95 WM 142-0 PWRS A paper recommended and approved by the IEEE Power System Engineering Committee of the IEEE Power Engineering Society for presentation at the 1995 IEEE/PES Winter Meeting, January 29, to February 2, 1995, New York, NY. Manuscript submitted July 25, 1994; made available for printing January 18, 1995.

Conventionally, the determination of a target system for restoration and associated restorative control schemes are carried out by experts in the power system control center; however, computerized automatization has gradually but steadily been promoted to improve power system reliability. In general, the restoration problem may be described as a multi-objective, multi-stage, combinatorial, nonlinear, and constrained optimization problem [5]. The complexity of such a problem precludes the development of a generic method of determining restorative controls. To resolve this problem, various approaches have been investigated for the objective of the support and automatization of restoration Conventionally, the determination of a target system for restoration and associated restorative control schemes are carried out by experts in the power system control center; however, computerized automatization has gradually but steadily been promoted to improve power system reliability. In general, the restoration problem may be described as a multi-objective, multi-stage, combinatorial, nonlinear, and constrained optimization problem [5]. The complexity of such a problem precludes the development of a generic method of determining restorative controls. To resolve this problem, various approaches have been investigated for the objective of the support and automatization of restoration schemes [1-9]. The methods proposed so far for automatization can be roughly classified into three types: methods through the application of plural restoration logics, knowledge based systems, and mathematical programming approaches.

Applications of combined restoration logics belong to a class of procedure-oriented algorithms, in which the expert's knowledge and experiences are translated into pure procedures [6]. However, it has been pointed out that this method has great difficulties in software maintenance because of its scale and complexity.

The method based on knowledge bases is well known as an expert system (ES). ES is basically composed of two essential components: a knowledge base and an inference engine. The separation of knowledge-base (data) from inference engine (procedure) offers a profound progress in software maintenance compared with conventional software programs. So far, ES approaches can be regarded as successful approaches to solve the restoration problem to some extent. In fact, a few restoration expert systems based on the knowledge-based technique are actually operating in some control centers [7]. However, this method has no guarantee to find an optimal target configuration; moreover the maintenance of a large-scale ES has turned out to require heavy cost, thus greatly offsetting the advantage of introducing expert systems.

Reprinted from *IEEE Transactions on Power Systems,* Vol. 10, No. 3, pp. 1473–1479, August 1995.

The capabilities of mathematical programming to the restoration problem have also been studied extensively [4,9]. Generally, the problem of determining the target configuration can be formulated as a mixed integer programming problem. Whether or not each branch is included in a target system is expressed by a 0-1 variable, whereas other constraints such as the demand and supply balance condition are formulated in terms of continuous variables. Although an MP based method can obtain an optimal solution under specified operational constraints, solution time often exceeds practically allocated computation time for a large scale problem due to combinatorial explosion. In fact, the optimal target configuration for restoration could be solved by using Mixed Integer Programming (MIP) in our previous literature [9]. However, it is necessary to curtail drastically its computation time when to be implemented on a practical system. This fact must be a good reason why there is no practical MP based restoration systems.

The objective of this paper is to overcome the above mentioned difficulties. In our new algorithm, the whole problem is decomposed into a set of subproblems by using experts' knowledge, and then each subproblem is efficiently solved by the branch-and-bound method. Needless to say, the decomposition process is handled by an expert system with far less size of knowledge-base compared with the existing ES only systems. As a result of this approach, the computation time has reduced to 30 % of that of the existing methods. Hence, it is clear that the combined ES-MP approach must be suitable for exploiting their individual advantages. We believe that the joint usage of ES and MP methods for power system restoration such as proposed here can not only be of great assistance to operators, but reduce maintenance éfforts after delivery.

Another noteworthy feature of this research is the introduction of a new concept "restorative operation cost" to be allocated to each switching device. This facilitates a natural formulation of power system restoration as an MP problem. By adjusting the costs, it is possible to obtain a target configuration similar to what would be created by human experts. The suitable allocation of restorative operation costs needs good engineering sense and nullifies a certain portion of the knowledge to be included in the knowledge-base of the ES, thus significantly reducing the number of production rules. This in turn greatly improves the maintainability of the system.

The proposed method has been applied to a practical restoration problem of 82 nodes, 177 branches and 28 loads. It has made clear from simulation results that this method can obtain the optimal restoration target systems in all cases. Moreover, the number of rules in the knowledge-base has decreased to one third that of typical existing ES's.

II BASIC CONCEPTS FOR RESTORATION PROBLEM

When electric power supply is interrupted due to a fault, it is imperative to promptly restore the power system to an optimal target configuration. In general, power system restoration consists of two consecutive phases of tasks: the determination of a target configuration and the construction of a feasible operation sequence leading to the target system. This paper has focused on the first phase of power system restoration.

The problem of determining a target configuration for restoration can be formulated as a problem of simultaneously minimizing the restorative operation cost and outage load capacity under the restrained conditions shown below.

(i) limited capacity of available power source
(ii) power balance between supply and demand
(iii) branch (line and transformer) capacity constraints
(iv) radial configuration ("must be kept" operational constraints)

Table 1 Restorative operation cost

Device Type	Cost	Priority
DS-for-line-feeder	close=0.0 open=1.0	4 (lower)
CB-for-line-feeder	close=0.0 open=0.6	3
Bus-tie	close=0.0 open=0.5	2
DS-for-load	close=0.0 open=0.3	1 (higher)

DS : disconnecting switch CB : Current breaker

To minimize discrepancies between the switching patterns of experienced operators and those by the proposed method, we introduce the concept "restorative operation cost" which must be assigned to every switching device (i.e., bus-tie-breaker, feeder-circuit-breaker, disconnecting-switch-for-changing-bus, or disconnecting-switch-on-line). The assignment of an appropriate cost to each switching device makes it possible to obtain a target configuration similar to the one which would be created by human experts [6,9]. Table 1 shows an example of cost assignment which is decided from the restorative strategies of human experts. It is obvious that the smaller the cost of switching device, the higher its operation priority.

III FORMULATION OF THE PROBLEM

A. Target configuration

The problem of determining the target configuration is formulated as an MIP problem. In the following formulation, integer decision variables are used to denote branches in the restoration path and which load should be restored. By using two types of decision variables, we can describe a restoration problem in a natural way.

(1) List of nomenclature

- N : number of nodes
- M : number of directed branches
- N_s : number of source nodes
- N_ℓ : number of loads
- NF_k : initial node of branch k
- NT_k : terminal node of branch k
- F_i : a set of branches whose initial node is i
- T_i : a set of branches whose terminal node is i
- X_k : $X_k=1$: branch k constitutes the restoration path
  $X_k=0$: otherwise

Y_p : Y_p=1: load p is to be restored, Y_p=0: otherwise

P_k : active power flow of branch k

C_k : restorative operation cost of branch k (C_k=0 : if branch k is closed in the post-fault network)

U_k : flow capacity(limit) of branch k

L_p : capacity of load p

G_q : available power at source node q

z : objective function

(2) Objective function

The objective of restorative control is to restore all customer services as quickly as possible, which means to minimize the restoration time and un-served loads. Generally, it is difficult to directly specify the minimization of restoration time as the objective function. For such reason, in conventional ES's this goal is met by adopting a large number of rules associated with switching operations derived from experienced operators. In the proposed method, we transform these rules into "restorative operation costs" to qualitatively express the rules as shown in Table 1. Hence, the shortest restoration time may be realized by minimizing the restorative operation cost. The objective function can be expressed as

$$z = \min\left\{\alpha \sum_{k=1}^{M} C_k \cdot X_k + \beta \sum_{p=1}^{N_\ell} L_p\left(1-Y_p\right)\right\} \quad (1)$$

where, α and β are positive constants.

In this equation, the first term is to minimize the switching costs, while the second term to minimize the capacity of un-served energy. If all outage loads are restored, the value of the second term will become zero. By deleting the constant term, eq. (1) becomes

$$z = \min\left\{\alpha \sum_{k=1}^{M} C_k \cdot X_k - \beta \sum_{p=1}^{N_\ell} L_p \cdot Y_p\right\} \quad (2)$$

(3) Constraints

(a) Limited capacity of available power source

$$\sum_{k \in F_q} P_k \le G_q \qquad (q=1,2,\cdots,N_s) \quad (3)$$

(b) Power balance between supply and demand

$$\sum_{k \in T_i} P_k - \sum_{k \in F_i} P_k = L_i \cdot Y_i \qquad (i=1,2,\cdots,N) \quad (4)$$

(c) Branch capacity constraints

$$P_k - U_k \cdot X_k \le 0 \qquad (k=1,2,\cdots,M) \quad (5)$$

(d) Radial configuration

$$\sum_{k \in T_i} X_k \le 1 \qquad (k=1,2,\cdots,M) \quad (6)$$

B. Illustrative Example

For ease of understanding the above-mentioned scheme, we will explain the operating principle on a simple network in Fig. 1, where the hatched area has lost power supply due to a fault. Assume that BUS-2 and three loads (Loads 1, 2, and 3) are elements to be restored. This network can be represented by the node-branch expression in Fig. 2, consisting of 6 nodes and 9 branches. In Fig. 2, node #1 represents the source node; nodes #2 and #3 correspond to BUS1 and BUS2, respectively. "LineA" and "LineB" are the designation of two transmission lines connected to the source, each of which consists of parallel lines 1L and 2L. These parallel lines are lumped to one equivalent line to form branch (1) and (2) in Fig. 2. Note that, in the faulted area, a connecting point among switching devices must be defined as a node for the purpose of restoration switching, which is different from that in load flow models. That is, nodes #4, #5, and #6 correspond to mark "▲" in the blackout area in Fig. 1. The restorative operation cost allocated to each branch and the pre-fault load values are also shown in the figure.

Eqs. (2) to (6) corresponding to the network are solved by the branch-and-bound method. Let X_1-X_9 be 0-1 variables to express the adoption or rejection of the branches as the restoration route, and Y_1-Y_3 0-1 variables representing whether or not loads L_1-L_3 should be picked up. Fig. 3 shows the branching tree of the branch-and-bound method. It has turned out that the optimal solution has been obtained after the problem is decomposed into eight subproblems. Here, z = - 9.5 is the optimal value of the objective function and the optimal solution is given by X=(1 0 1 1 1 0 0 0 0), Y=(1 1 0). That is, the target paths are $X_1=X_3=X_4=X_5=1$, while the loads to be picked up are $Y_1=Y_2=1$.

Fig. 4 shows the optimal target system for restoration. According to our expectation, the bus-tie restorative strategy (which has higher priority than others in this case) is adopted (X_3=1), and the lowest load L_3 is disconnected (Y_3=0).

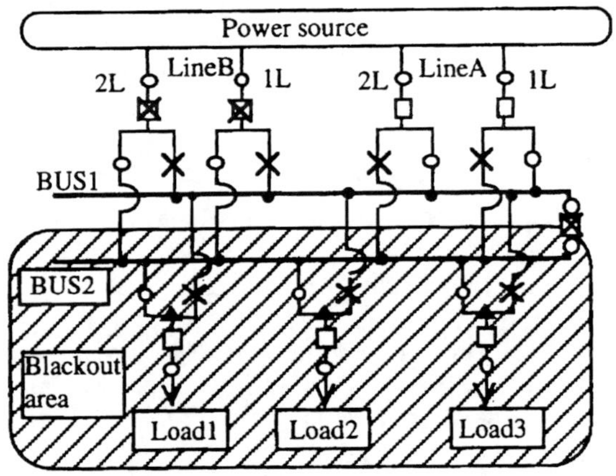

Fig. 1 Simple example network

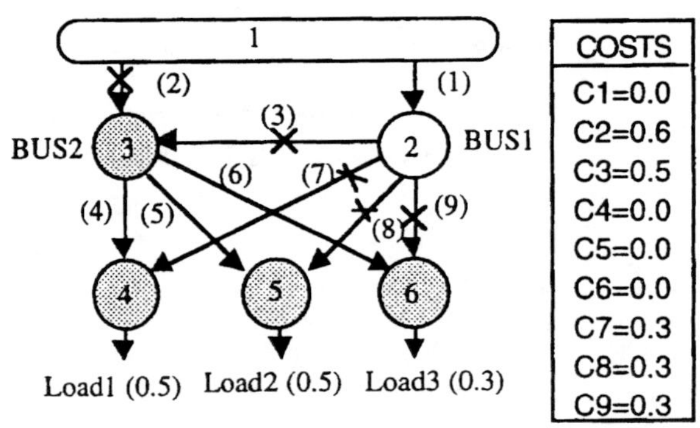

Fig. 2 Node-branch expression of the example system

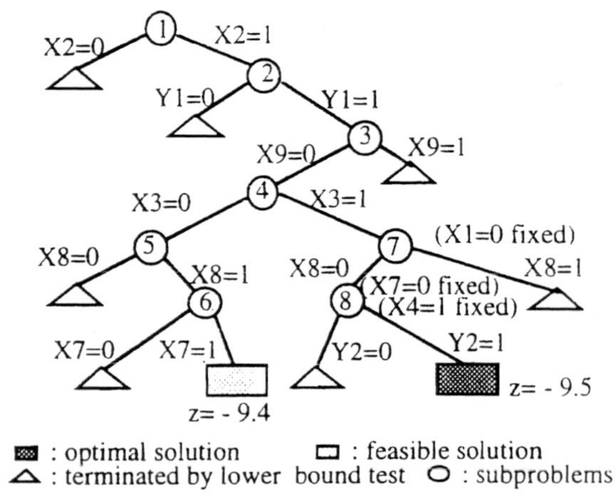

Fig. 3 The branching tree

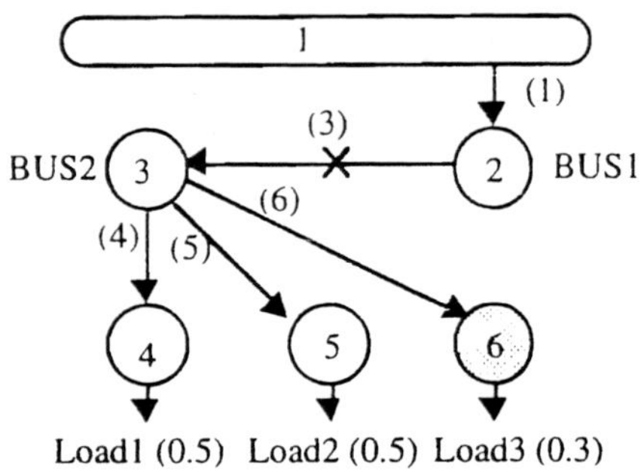

Fig. 4 A target configuration system for restoration

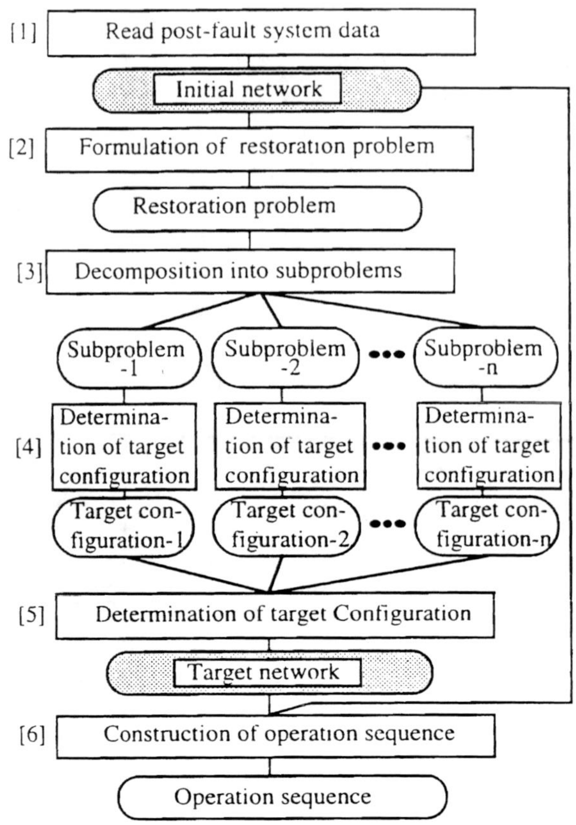

Fig. 5 The processing flow

IV SOLUTION METHOD

A General Flow

The solution process shown in Fig. 5 may be explained as:

[Step 1] The post-fault network connections and the pre-fault load values are read in. The operation costs of branches are also input from the system database.

[Step 2] Problem formulation as the MIP problem.

[Step 3] By using the specific features of restoration, the problem is decomposed into a set of subproblems (see Section IV B).

[Step 4] Each subproblem is solved by the branch-and-bound method to determine partial target configuration.

[Step 5] The final target configuration is decided by aggregating the obtained partial target configurations.

[Step 6] The operation sequence is constructed by comparing the post-fault network with the target network.

B Problem Decomposition by means of ES

It is difficult to solve a problem formulated as eqs. (2) to (6) as it is. As a general rule, a class of algorithms of obtaining the exact optimal through enumeration such as the branch-and-bound method requires an amount of calculation in the order of exponential function of the number of 0 - 1 variables, and therefore exact methods cannot solve a large scale problem within permissible computation time. This difficulty has been circumvented through the reduction of the problem scale by utilizing the characteristics inherent to fault restoration. In our method, the whole problem is decomposed, according to experienced operators' strategies, into a series of subproblems so that each subproblem can efficiently be solved by the branch-and-bound method.

In actual restorative operations on a regional power system, the following strategies are used:

(S-1) The sound (energized) area cannot be a target of switching operation during restoration.

(S-2) The substation cannot be energized from the power sources of different power systems irrespective of the status of the bus-tie-breaker.

These simple strategies are useful to divide a whole problem into subproblems. In this particular expert system, they are transformed into the following rules:

(D-1) **If** there exist nodes in the sound (energized) area, **then** nodes with the same voltage should be aggregated to a single node.

(D-2) **If** there exist nodes related to primary buses in a substation in the blackout area and the associated bus-tie branch is open,
then these nodes should be reduced into one node with the bus-tie artificially closed.

(D-3) **If** there exists a closed branch (tie-line) that connects the primarily buses of different substations in the blackout area. **and** the both buses have one or more candidate branches for restoration,
then the branch should be artificially opened.

The rule (D-1), related to the strategy (S-1), greatly decreases the number of 0-1 variables. Next, the rule (D-2) is derived from the strategy (S-2), in which closing a bus-tie breaker is equivalent to decrease one 0-1 variable. Finally, the rule (D-3) is combined with the strategy (S-2). Normally, a radial network configuration is adopted in a regional power system. Therefore, un-served area caused by a fault cannot be spread out. However, if a tie-line connecting primary buses is closed for some reason, the blackout area would be enlarged over the network. In such a case, the "vertical splitting" strategy is commonly used in actual restorative operations. That is, a human expert splits one voltage level of power system into several islands. Consequently, by opening the tie-line, a large un-served area could be separated into several small areas. Then, the restoration problem corresponding to each small un-served island can be solved in acceptable time by the branch-and-bound method.

This idea will be illustrated by taking the power system shown in Fig. 6, which can be considered as a practical size system as far as restoration problems are concerned. The system consists of 82 nodes, 177 directed branches, and 28 loads, and the voltage levels are 154kV and 66kV. In the figure, "Ass" to "Mss" stand for substations and the arrows indicate loads. We assume that quadruple circuit 1L-4L of line D-H are tripped by faults simultaneously, which has caused a wide outage area depicted in halftone.

Fig. 7 shows the node-branch expression of the model network. Here, two new nodes whose voltage levels are 66kV and 154kV are generated by applying the decomposition rule (D-1). Then, double circuit lines are reduced to a single branch in the same manner as in load flow models. L_1 to L_{11} are loads that are aimed to restore. By applying the decomposition rule (D-2), primary buses in "Hss" are reduced to one node with the bus-tie artificially closed. Then, according to the decomposition rule (D-3), line G-H (being reduced into a single branch) is opened artificially. Therefore, the network could be decomposed into the two sub-networks as shown in Fig. 8. The target configuration of the each sub-network is solved by the branch-and-bound method. Here, the number of 0-1 variables in Fig. 7 is 42, whereas the numbers of 0-1 variables in Fig. 8(a) and (b) are respectively 24 and 16. If the decomposition rules would not have been applied, the number of 0-1 variables is 147, resulting in unrealistic computation time (more than ten minutes).

V RESULTS AND DISCUSSION

In order to demonstrate the effectiveness of the proposed method, it has been applied to the model network of practical size shown in Fig. 6. This restoration problem is extremely difficult to solve due to the presense of a numerous number

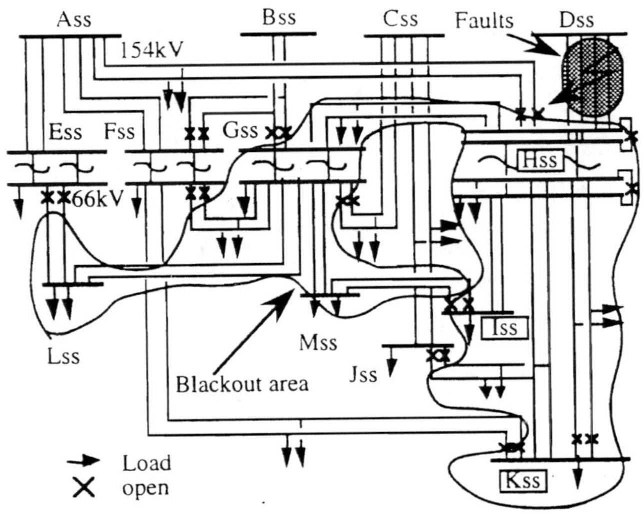

Fig. 6 Model network

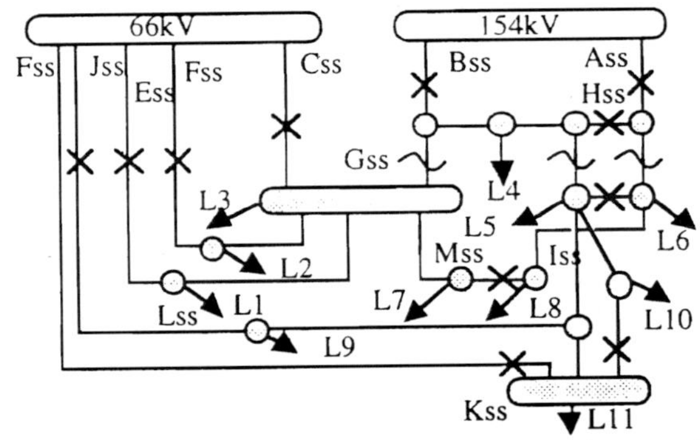

Fig. 7 Node-branch expression of the model network

of possible restoration schemes. A large number of simulations have been carried out on this network with changing network conditions. Here, two typical examples will be shown due to page limit: One is a case where all loads are reconnected. The other is a case where part of loads are left outage due to overloading. These test cases are the same as in our previous work [9], which does not use the decomposition-coordination technique.

A *CASE-I: - Restoration of all loads*

Fig. 9 shows the obtained target configuration for CASE-I, where the bold lines signify the restoration routes. It requires six closing operations (two 154kV lines, one bus-tie, and three 66kV lines are used to restore the network), and two opening operations (one 154kV tie-line and one 66kV line).

B *CASE-II : - Partial restoration of loads*

Next, let us consider the specific case; we assume that line B-G is on maintenance work in CASE-I. The target configuration of CASE-II is shown in Fig. 10. Although one load (L1) in Lss remains disconnected unfortunately, this is the optimal solution under this condition. The secondary bus in Gss is restored from 66kV line instead of the primarily bus. It is natural to lead to this goal since B-G line has the insufficient margin of power flow capacity. We call it "secondary bus around" strategy.

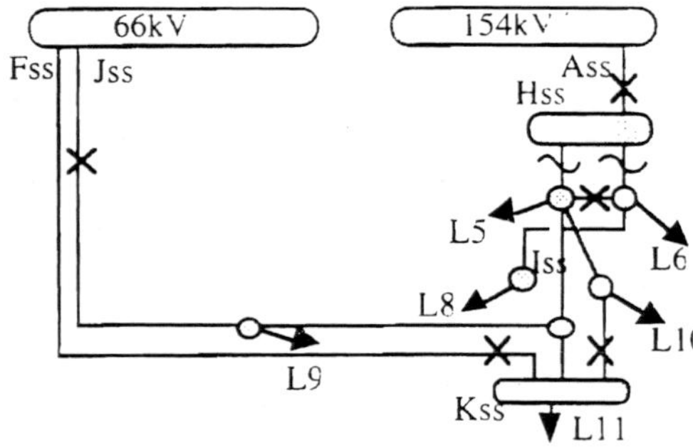

(a) Partial network-1

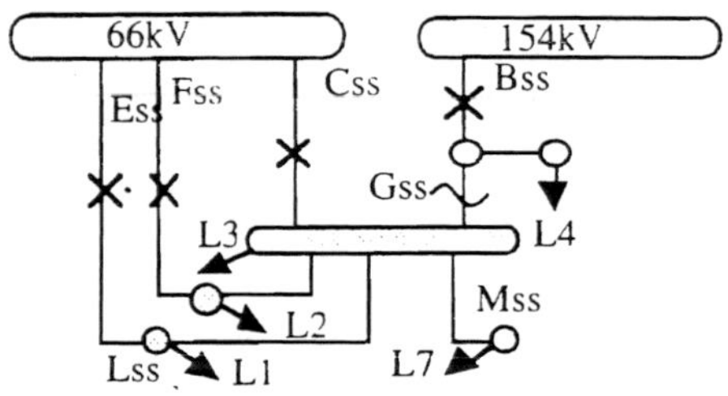

(b) Partial network-2

Fig. 8 Partial network

C Computation time

The computer used for simulation of the examples is a 68030 based EWS (Engineering WorkStation) of 3 MIPS. The computer program is written in Prolog and FORTRAN. Table 2 shows the computation time of above the cases. In the table, the straightforward method is the MIP approach without the decomposition-coordination technique. The previous method means our former approach [9], which is adopted only the decomposition rule (D-1). These values include the time for constructing operation sequence. As shown in this table, the solution has been solved about 15 [s] even for the difficult case where the load shedding occurs being 30% of the previous method. In the fault restoration of regional power supply system, the standard requirement for computer processing time from the fault detection to the completion of the operation sequence should general be within 60[s]. Therefore, it is good enough to apply the proposed method to the fault restoration system.

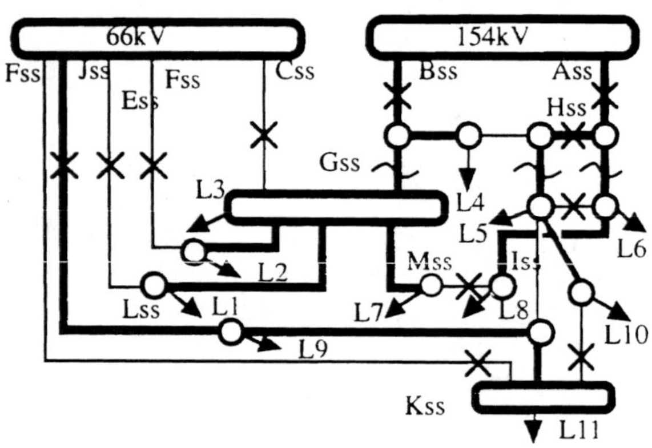

Fig. 9 Case-I : Restoration of all loads

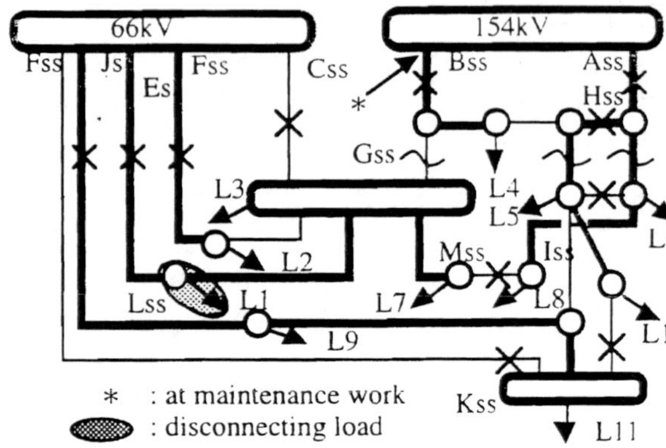

Fig. 10 Case-II : Partial restoration of loads

Table 2 Comparison of computation time (s)

CASE	Straightforward method	Previous method [9]	Proposed method
I	70	38	11
II	824	68	13

VI CONCLUSION

In this paper, we have presented the novel combined approach to power system restoration in which Mathematical Programming methodology is integrated into an Expert System. In this method, the optimal target configuration for restoration is determined efficiently by the decomposition-coordination strategy. As a result of this approach, the computation time has reduced to 30% of our earlier approach. Moreover, through the introduction of the new concept "restorative operation cost", a power system restoration can be expressed in a natural form as an MP problem, and the number of rules in the knowledge-base has decreased to one third that of typical existing ES's. The validity and the effectiveness of the proposed method have been demonstrated by simulating on a practical size restoration problem.

In this research, a method of adequately allocating a cost to each device has not been studied. It may change from system to system and also depends on restoration strategies adopted. It might seem the cost allocation can be subjective, nevertheless the great merit of introducing the cost lies in that once costs are assigned, whatever they may be, the proposed method gives rise to an optimal solution under the assumed cost values. Nowadays, an EWS of about 100 MIPS is not rare which can obtain an optimal solution for systems as large as that tested in this research within 1 second or 2. Hence, reasonable costs can be obtained by off-line simulations with quite affordable computational burden.

The next steps of this work would be to link the system with an expert system for fault diagnostic and to improve theman-machine interface. Future work should be directed to extending the method to handle bulk power restoration problems which involve the control of generators.

VII REFERENCES

[1] R.J. Kafuka, *et al.*, "Role of interactive computers in the development of a system restoration plan", IEEE Trans. on Power Apparatus & Systems, Vol. PAS-101, No. 1, pp. 43- 52, 1983.

[2] T. Sakaguchi and K. Matsumoto, "Development of a knowledge based system for power system restoration", IEEE Trans. on Power Apparatus & Systems, Vol. PAS-102, No. 2, pp. 320-329, 1983.

[3] A.Adibi, *et al.*, "Power system restoration - A task force report", IEEE Trans. on Power Systems, Vol. PWRS-2, No. 2, pp. 271-277, 1987.

[4] K. Komai, and T. Sakaguchi, "Analysis and evaluation of human knowledge for power system restoration by mathematical programming method", Trans., IEE Japan, Vol. 107-B, No. 6, pp. 269-275, 1987.

[5] F.F.WU, and A. Monticelli, "Analytical tools for power system restoration - Conceptual design", IEEE Trans. on Power Systems, Vol. 3, No. 1, pp.10-16, 1988.

[6] T. Kasuya, H. Oda, H. Inoue, and S. Iizuka, "Development of automatic restorative operation applicable to power system local area dispatching", Trans. IEE Japan, Vol. 109-B,No.4, pp. 177-184, 1989.

[7] Y. Arita, *et al.*, "Development of an operation supporting expert system for power system networks in an integrated control center", Trans. IEE Japan, Vol. 110-B, No. 6, pp. 504-510, 1990

[8] M. M. Adbi, *et al.*, "Special considerations in power system restoration", IEEE Trans. on Power Systems, Vol. 7, No. 4, pp. 1419-1424, 1992

[9] T. Nagata, H. Sasaki, and M. Kitagawa, "A method of determining target configuration for power system restoration by means of mixed integer programming approach", Trans. IEE Japan,Vol. 114-B, No. 2, pp. 179-185, 1994 (to appear in Electrical Engineering in Japan).

A Petri Net Algorithm for Scheduling of Generic Restoration Actions

Jaw-Shyang Wu*, Member **Chen-Ching Liu**†, Fellow **Kan-Lee Liou**†, Member **Ron F. Chu**‡, Senior Member

* National Kaohsiung Institute of Technology, Taiwan

† University of Washington, Seattle, WA 98195

‡ PECO Energy Company, Philadelphia, PA 19101

***Abstract*- This paper is concerned with scheduling of the restoration actions. Generic restoration actions (GRAs) have been identified in our previous research based on a variety of restoration strategies used by utilities. This paper proposed a scheduling method using the Petri Net (PN) techniques. The PN model incorporates the conditions that have to be met before an action can be taken. The time required to perform an action is also modeled in the net. A token passing and a backward search processes are used to identify the sequence of restoration actions and their timing. The advantages of the proposed methods are (1) the time required to restore a subsystem can be estimated, (2) a systematical method is obtained for identification of the sequence of actions. The PN method has been tested with simulation cases.**

Keywords: System restoration, Petri Net, scheduling, generic restoration actions.

1. INTRODUCTION

Power system restoration following a complete shutdown is a complex problem involving system operators, field crews and other utility personnel. Numerous components, actions, and constraints have to be dealt with in the process. A thorough overview of the critical issues can be found in a recent publication of IEEE Power System Restoration Working Group [1].

System restoration may involve condition assessment, restarting generating units, cranking non-black-start units, establishing the transmission skeleton, picking up load, and synchronization of islands. The specific strategy depends on the nature of a specific utility and the resources available to the utility. Generally speaking, restoration strategies can be categorized into top-down strategy, bottom-up strategy and various combinations of the top-down and bottom-up strategies [2]. In the simplest form, a top-down strategy starts the restoration by building the transmission skeleton. On the other hand, a bottom-up approach would provide cranking power to restore the generations first.

It has been widely recognized that knowledge-based systems (KBSs) are promising tools for power system restoration [3-7]. In general, the feasibility of KBSs for system restoration has been established. Knowledge-based restoration methods have been developed for transmission, subtransmission and distribution systems. A potential opportunity that was identified in [2] is the development of a *generic* software tool for system restoration. On this issue, our previous research resulted in a collection of Generic Restoration Actions (GRAs), i.e., actions that are common for various restoration strategies [2].

In our previous work, however, GRAs were only conceptual. In this paper, GRAs are modeled by a scheduling method based on Petri Nets (PNs). The PN models the states of power system components with 'Place' nodes and the restoration tasks to be performed by 'Transition' nodes. Once a PN model is established , the sequence of actions can be generated systematically. Furthermore, the timing of restoration actions can be estimated based on the time for each task.

PNs are powerful modeling tools; they have been successfully applied to many areas, e.g., fault section estimation [8], rule-based system evaluation [9], and power system protection [10]. In this paper, PNs are used to explicitly model the *states* and *transitions* of a power system during the restoration process. The proposed PN method using GRA concepts can be applied to various restoration strategies. Moreover, if the time to complete each action can be estimated, the time to complete a goal during restoration can also be estimated.

2. GRAs FOR POWER SYSTEM RESTORATION

Generic restoration actions (GRAs) are actions that form the basic tasks to develop different restoration strategies for power systems [2]. A set of GRAs was identified in [2]. Since the GRA concept is the foundation of the Petri Net method in this paper, these actions are summarized in this section:

96 WM 177-6 PWRS A paper recommended and approved by the IEEE Power System Engineering Committee of the IEEE Power Engineering Society for presentation at the 1996 IEEE/PES Winter Meeting, January 21-25, 1996, Baltimore, MD. Manuscript submitted August 1, 1995; made available for printing January 10, 1996.

Reprinted from *IEEE Transactions on Power Systems,* Vol. 12, No. 1, pp. 69–76, February 1997.

start_black_start_unit(X): to start a black-start unit (BSU) X if the requirements/constraints are met.

connect_tie_line(X): to connect a tie line X if the conditions on the tie line connection are met. *crank_non_black_start_unit(X,Y).* to crank a non-black-start-unit (NBSU) X from busbar Y if the requirements/constraints are met.

energize_line(X,Y): to energize a line X from a busbar or line Y if the constraints are met.

energize_busbar(X,Y): to energize a busbar X from a generator, tie line, or busbar Y if the constraints are met.

pick_up_load(X,Y): to pick up a load X from busbar Y if the requirements/constraints are met.

synchronize(X,Y): to synchronize two busbars, lines, or generating units X and Y if the constraints are satisfied.

In this paper, a restoration building block (RBB) [2] is an initial step to restore generation or transmission. For example, an RBB may represent a subsystem to be established in an initial step of restoration where some non-black-start generating units (NBSU) are cranked by black-start units. It is noted that various GRAs, including *start_black_start_unit(X)*, *energize_line(X,Y)*, and *crank_non_black_start_unit(X,Y)*, may have to be taken to establish an RBB. Furthermore, a GRA may require specific actions such as opening/closing a breaker or switch, adjusting transformer taps, placing tap changers under automatic or manual operations. The detailed action planning for each GRA will be necessary.

This paper is focused on the development of the PN-based scheduling method; the constraint-checking tasks can be handled by the KBS software developed in our previous EPRI-sponsored research [4].

3. PETRI NET MODEL FOR GRA SCHEDULING

A Petri Net has two sets of nodes, i.e., Place (P-) nodes and Transition (T-) nodes, and one set of directed arcs. P-nodes represent states and T-nodes stand for the actions or tasks that are performed to change the system states. Directed arcs connect P-nodes and T-nodes. The direction of an arc indicates a transition of the system state. There is a token at each P-node to indicate the current state by a specific value which is called the 'color.' If all input P-nodes of a T-node are colored, the T-node is 'activated' and the token color will propagate to all output nodes. A final goal is achieved when the color reaches the goal node.

A power system component can be in more than one state during restoration. For example, a busbar can be in an energized or a de-energized state. As a result, two P-nodes are used to designate the busbar states. Each GRA is denoted by a T-node. The incoming arcs of the T-node are represented by a T-node. Inputs from the P-nodes represent the conditions that must be met before the action can be performed. The outgoing arcs of the T-node point to the new

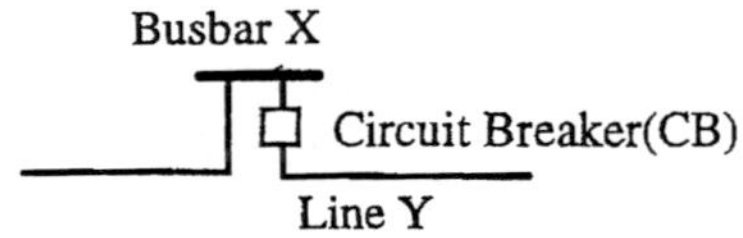

Figure 1 Energizing Line Y from Busbar X

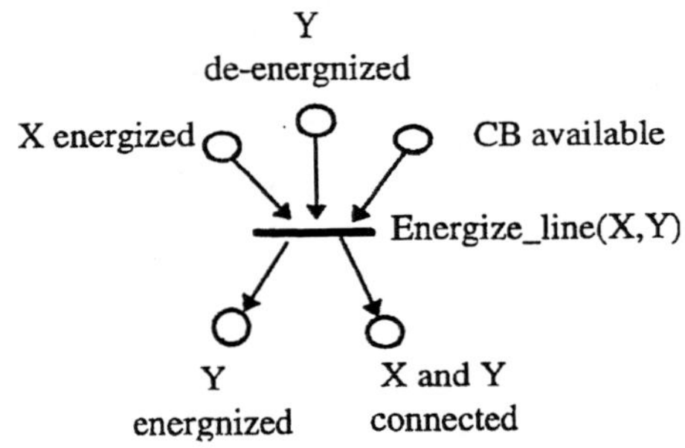

Figure 2 Petri Net Model

states. In Figure 1, suppose Line Y is de-energized and Busbar X is energized. As shown in Figure 2, the GRA 'energize_line(Y,X)' is represented by a T-node and the incoming arcs are from 3 P-nodes, i.e., (Line Y is de-energized), (Busbar X is energized), and (Breaker between X and Y is available). Following the transition, the outgoing arcs point to two nodes: (Line Y is energized) and (X and Y are connected).

Place Nodes

For every generator unit(G) or tie line(T), 3 nodes are used to represent its status: G/T is available, G/T is not connected, G/T is connected. For every busbar or line X, 2 nodes are created to represent its status: X is de-energized, X is energized. Two nodes are created for a load Z: Z is not served, Z is served. For every pair of neighboring busbar/line, 2 nodes represent their connection status: X1 & X2 are not connected, X1 & X2 are connected. As a goal state, one node is created to indicate that the transmission ring has been built, i.e. Ring is built. One node is created for each RBB to denote its completion.

Transition Nodes

T-nodes represent the restoration actions to change the states of a power system during restoration. The following T-nodes are created. For each black start generator unit B or tie line T, a node is created: connect_BSU(B) or connect_tie_line(T). A node corresponds to each non-black start generator unit N: Crank_NBSU(N). One node is created for each neighboring busbar/line NX of X: energize_busbar(X,NX). For each busbar B connected to tie line T, a node is created: energize_busbar(B,T). For each load Z, 1 node is created: pick_up_load(Z). For each pair of neighboring busbar/line, busbars, or lines N1 & N2, synchronization is represented by synchronize(N1,N2).

Connection Arcs

The connection arcs between T-nodes and P-nodes are based on their causal or sequential relation. Table 1 lists the input/output P-nodes for each type of T-node.

Numbers of Nodes & Arcs

Suppose that the numbers of system components are denoted by N_{TL}: number of tie lines, N_G: number of black-start unit, N_W:number of non-black start units, N_B:number of busbars, N_L: number of lines, N_Z: number of loads, Nb: number of neighboring busbar(s) and line(s) of a busbar b, and N_l: number of neighboring busbar(s) and line(s) of a line. The total number of P-nodes, T-nodes, and connection arcs for the GRAs can be determined for the entire PN. The number of pairs of neighboring busbar/line, busbars, and lines are given by:

$$N_{PR} = \frac{\sum_{N_B} Nb + \sum_{N_L} Nl}{2}$$

The number of P-nodes is:

$$3N_G + 3N_W + 3N_{TL} + 2N_l + 2N_Z + \sum_{N_B} Nb + \sum_{N_L} N_l$$

The number of T-nodes is

$$2N_G + 2N_W + 2N_{TL} + Nz + N_{PR} + \sum_{N_B} Nb + \sum_{N_L} N_l$$

The number of connection arcs is

$$6N_G + 6N_W + 6N_{TL} + 3Nz + 5\sum_{N_B} Nb + 5\sum_{N_L} N_l$$

As the size of a power system grows, the number of busbars and lines increase. The number of neighboring busbars and lines for each busbar or line should remain small. Hence, the last two summation terms grow linearly with the number of busbars and lines. In other words, sparsity is preserved in the PN models.

The power system in Figure 3 has 3 buses, 1 tie line, and 3 loads. Based on the synthesis process in the previous section, a PN can be constructed with 34 P-nodes, 24 T-nodes, and 109 connection arcs for the goal of "Building a Ring." Figure 4 shows part of the PN for this system.

Table 1. Connection Arcs of GRAs and RBBs

T-Node	Input P-node	Output P-node
Connect T	T is available. T is not connected.	T is connected.
Connect BSU	BSU is available. BSU is not connected.	BSU is connected.
Crank NBSU from B	NBSU is de-energized. B is energized.	NBSU is cranked.
Synchronize NBSU with B	NBSU is available. B is energized.	NBSU and B are synchronized and connected.
Energize B from NB	B is de-energized. NB is energized.	B is energized. B and NB are connected.
Energize B from T	B is de-energized. T is connected.	B is energized.
Energize B from BSU	B is de-energized. BSU is connected.	B is energized.
Energize L from NL	L is de-energized. NL is energized.	L is energized. L & NL are connected.
Pick up Z	Z is not served B is energized	Z is served
Synchronize N1 & N2	N1 is energized. N2 is energized. N1 & N2 are not connected.	N1 & N2 are synchronized and connected.
Build RBB R	BSU in R is connected. NBSU in R is connected. All busbars in R are connected. All lines in R are connected.	RBB R is completed.

4. ALGORITHM OF GRA SCHEDULING

In the proposed PN model, the states of power system components associated with the restoration process are specified by the 'token colors' of the corresponding P-nodes.

PN Model Construction

The proposed algorithm automatically constructs the PN model using (1) bus data, (2) line data, (3) generator data, (4) load data, (5) time required for each restoration action. Nodes are created based on the principles described in the subsections labeled *Place Nodes* and *Transition nodes* in Sec. 3. Arcs are generated based on Table 1.

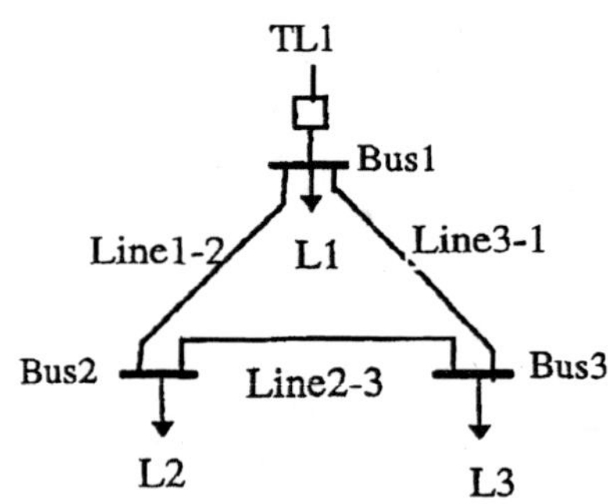

Figure 3 3-bus system

Initial Conditions

Under the initial condition, all originating P-nodes are black. The other P-nodes are white.

Token Passing

If all P-nodes entering a T-node have black tokens, then the T-node is activated. In other words, the GRA corresponding to the T-node is included in the next list of candidate actions. In Figure 4, in the first token passing step, only T1 is activated. The second step of token passing activates node T2. In the third step, 3 nodes are activated, T5, T14, and T15. If only one T-node is activated, the token will be passed to the P-node(s) following the T-node. Once a token is passed, the color of the originating nodes may have to be changed. If P15 is colored black, P2 has to be changed back to white but P1 remains black. When two or more T-nodes are activated, only one token will be passed at a time. The token will be passed to the node selected by the best-first search step. The solution is reached after the goal T-node is activated and the goal P-node receives the token.

Time Label

A time label is assigned to each P-node. Each T-node is given a time value that corresponds to the time required for

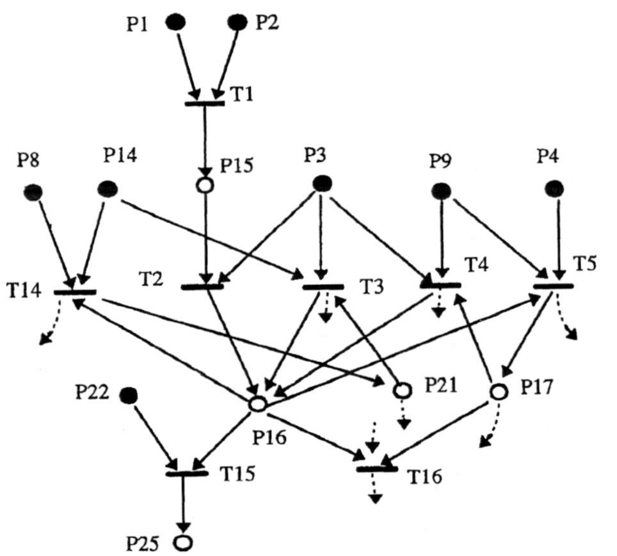

P1 TL1 is available
P2 TL1 is not connected
P3 Bus1 is not energized
P4 Line 1-2 is not energized
P8 Line 1-3 is not energized
P15 TL1 is connected
P16 Bus1 is energized
P17 Line 1-2 is energized
P21 Line 1-3 is energized
P22 L1 is not served
P25 L1 is served
P26 L2 is served
P9 Bus 1 and Line 1-2 are not connected
P14 Bus 1 and Line 1-3 are not connected
T1 Connect TL1
T2 Energize Bus1 from TL1
T3 Energize Bus1 from Line 1-3
T4 Energize Bus1 from Line 1-2
T5 Energize Line 1-2 from Bus1
T14 Energize Line 1-3 from Bus1
T15 Pick up L1
T16 Synchronize Bus1 and Line 1-2

Figure 4 PN Model of the 3-Bus system

completion of the specific action represented by that T-node. The T-node also has a time label that represents the cumulative time that the action will be complete. The time label of the P-node is cumulative as the token is passed. For instance, the initial time of the P-nodes is t=0 min. in Figure 4. If the time required to take the action represented by T1 is 25 min. The time label at P15 will be 25 min. Two P-nodes enter T2. The time label at P3 is 0 min. The algorithm will select the largest time label since all input conditions to a T-node must be met before the T-node is activated. Now if the time to complete action T2 is 5 min., then the time label at T2 will be 25+5=30 min.

In practice, it may be difficult to specify the exact time required to complete a task represented by a T-node. In this case, a range of time may be more realistic, e.g., 30-45 minutes. (The time to complete the action can be viewed as a random variable with a uniform distribution over the specified range.) When a range of time is specified for a T-node, the cumulative time label also becomes a range. Suppose a P-node has an exact time label, t. The minimum of the range for the T-node will be t+min time to complete the action, and the maximum of the range is t+max time to complete the action. If a P-node entering the T-node also has a time range, the range for the T-node will then be [min+min, max+max]. Clearly, the time to achieve the goal will also be specified by a time range.

Best-First Search

If two or more T-nodes are activated in a token passing step, the best T-node should be selected. In this research, a best-first search algorithm is developed. The heuristic criterion for the best solution is the one that requires the shortest restoration time. In every step, all activated T-nodes are checked and the one that can be completed within the shortest time from the time of starting of restoration is selected.

The evaluation function developed for this algorithm is:

$$Teva = Tlst + Tcnd$$

where Teva is the time needed from the starting of restoration to completion of the T-node action. Tcnd is the time to complete the candidate action and Tlst is the time from starting of restoration to completion of the predecessor of the node that is being evaluated. In Figure 4, after the second action T2, nodes T5, T14 and T15 are activated. Their time label is the same, i.e., 30 min. from P16. The Tlst value for all 3 T-nodes is 30 min. The Tcnd value for T5 and T14 is 5 min. but the value for T15 is 10 min. The algorithm will chosen T5 or T14 but not T15 for the token passing.

Based on the best-first algorithm, the solution is a timed sequence of restoration actions. The heuristic search does not guarantee global optimality but it is likely to achieve an efficient sequence of actions.

To perform the best-first search for cases where time ranges are used for P-nodes and/or T-nodes, one can select

Tlst to be the maximum of the range corresponding to the P-node, e.g., P16, and Tcnd to be the minimum of the high limits of the ranges for all T-nodes being evaluated, i.e., T5, T14, and T15 for P16.

Sequence of Restoration Actions and Their Timing

At any instant, many tokens may exist and several T-nodes may be activated. Therefore, tokens can be passed in several paths. Each path contains P- or T-nodes with their respective time labels. After the goal node receives a token, all paths have to be traced backward from the goal (bottom) node one by one. On each path the actions to be taken are specified by the T-nodes on the path. The time to start the action can be determined with the cumulative time labels of the T-nodes. Since two or more T-nodes may have the same time label, these actions can be implemented in parallel.

The proposed PN model construction and token passing algorithms are implemented in the object-oriented language C++, which provides higher expandability and portability. Nodes are implemented as objects. P-nodes and T-nodes form the P-node class and T-node class, respectively. Arcs are modeled by attributes of the P-node and T-node classes using the pointer structure. Based on our investigation, the data encapsulation capability of C++ makes the programming easier than that in C.

5. SIMULATION RESULTS

Two examples are used to demonstrate the sequence of actions that should be taken to restore transmission or generation after a blackout. Assume that all breakers are opened after the disturbance. In both cases, the (fictitious) time to complete an action for all GRAs is given in Table 2. The restoration process starts at time 0.

Case 1: 3-Bus System

The 3-bus system in Figure 3 is simulated by the top-down restoration strategy. The tie line which is the only power source will be connected first.

Table 3 shows the sequence of GRAs to take from starting of the restoration process. Several actions can be taken in parallel: 2 actions at t=30 min., 2 actions at t=35 min., and 3 actions at t=65 min.

Case 2. Schuylkill-Southwark Subsystem at PECO Energy

This case is the restoration of PECo's Schuylkill-Southwark island, which is shown in Figure 5 [10]. There are 2 CTs and 2 fossil units at Schuylkill. Schuylkill-10 (CT10) and Schuylkill-11 (CT11) are black-start CTs. Schuylkill-1 (ST1) and Schuylkill-3 (ST3) are non-black-start steam units. In Southwark there are 2 aggregated CTs, Southwark_3-4 (CT3-4) and Southwark_5-6 (CT5-6). Both of the CTs are black-start units.

Table 2 Time to Complete GRAs

Generic Restoration Action (GRA)	Time(min.)
Restart BSU	15
Energize Busbar from BSU/busbar/line	5
Connect Tie Line	25
Crank a NBSU from a Busbar	15
Synchronize between Busbars/Lines	20
Pick up Load	10

Table 3 Sequence of Actions for 3-Bus System

Time (min.)	Actions
0	Connect tie line TL1.
25	Energize bus 1 from tie line TL1.
30	Energize line 1-2 and line 3-1 from bus 1.
35	Energize bus 3 from line 3-1 and bus 2 from line 1-2
40	Energize line 2-3 from bus 3
45	Synchronize bus 2 with line 2-3
65*	Pick up load L1 at bus 1, L2 at bus 2, L3 at bus 3

* The priority of this specific top-down strategy is to establish the transmission skeleton before load pick-ups.

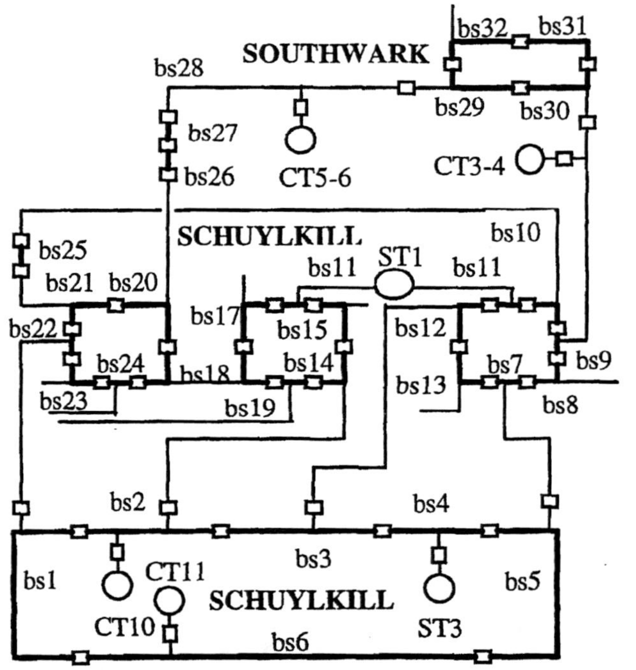

Figure 5 PECO's Schuylkill-Southwark Subsystem

Suppose the busbars bs6, bs18 and bs29 are not available due to maintenance and/or faults. The RBBs to provide cranking power for NBSUs are: RBB1 (CT10 cranking ST3 via bs2, bs3, and bs4), RBB2 (CT3-4 cranking ST1 via bs9, bs10, and bs11). The restoration sequence for this scenario is shown in Table 4.

Table 4 Sequence of Actions for Schuylkill-Southwark

Time (min.)	Actions
RBB1:	
0	Connect BSU CT10
15	Energize bs2 from CT10
20	Energize bs3 from bs2
25	Energize bs4 from bs3
30	Crank NBSU ST3 from bs4
45	Synchronize ST3 with bs4
RBB2:	
0	Connect BSU CT3-4
15	Energize bs9 from CT3-4
20	Energize bs10 from bs9
25	Energize bs11 from bs10
30	Crank NBSU ST1 from bs11
45	Synchronize ST1 with bs11
65	Energize bs1 from bs2, bs14 from bs2, bs12 from bs3 bs5 from bs4, bs30 from bs9, bs8 from bs9, bs17 from bs11, bs15 from bs11, bs25 from bs10
70	Energize bs22 from bs1, bs19 from bs14, bs13 from bs12, bs7 from bs5, bs31 from bs30, bs21 from bs25, and synchronize bs14 with bs15
75	Energize bs32 from bs31, bs20 from bs21, bs23 from bs22, and connect bs7 with bs13, bs8 with bs7, and bs22 with bs21
80	Energize bs24 from bs23, bs26 from bs20,
85	Energize bs27 from bs26
90	Energize bs28 from bs27

6. CONCLUSIONS

In this paper, a scheduling methodology for the sequence of restoration actions is proposed. Our method is based on the Petri Net model and the GRA concept. The PN method is used to determine the sequence of actions and the estimated time for each action from the starting time. The proposed method also identifies restoration actions that can be taken in parallel. Practically, restoration is likely to be interrupted by unexpected problems. In this case, the solution can be updated with the time attributes of the associated T-nodes.

To apply the proposed method, a bulk power system with numerous components should be partitioned; otherwise the PN model may be very complex. The proposed PN model works well for limited tasks such as scheduling the actions for an RBB. However, system restoration is a complex process involving several stages. Further analysis is needed to develop a practical scheduling method for various stages of system restoration. Manpower scheduling is critical during restoration. The proposed method provides a tool to identify critical tasks that are bottlenecks of a restoration strategy and tasks that can be performed in parallel. The PN software can be integrated with the KBS developed in our previous EPRI-sponsored work.

Acknowledgment: The authors would like to thank the reviewers for their helpful suggestions.

7. REFERENCES

[1] IEEE PES Power System Restoration Working Group (M. M. Adibi. Chair), "Power System Restoration," 93 THO 605-6 PWR.

[2] L.H. Fink, K.L. Liou and C.C. Liu, "From Generic Restoration Actions to Specific Restoration Strategies," *IEEE Trans. on Power Systems*, May 1995, pp. 745-752.

[3] T. Sakaguchi and K. Matsumoto., "Development of a Knowledge Based System for Power System Restoration," *IEEE Trans. on Power Apparatus and Systems*, Feb. 1988, pp. 320-329.

[4] C.C. Liu, K.L. Liou, R.F. Chu, and Arne T. Holen, "Generation Capability Dispatch for Bulk Power System Restoration: A Knowledge-Based Approach," *IEEE Trans. on Power Systems*, February 1993, pp. 316-325.

[5] M.M. Adibi, R. J. Kafka and D.P. Milanicz, "Expert System Requirements for Power System Restoration," *IEEE Trans. on Power Systems*, August 1994, pp. 1592-1600.

[6] Y. Kojima, S. Warashina, S. Nakamura, and K. Matsumoto, "Development of A Guidance Method for Power System Restoration," *IEEE Trans. on Power Systems*, August 1989, pp. 1219-1227.

[7] D.S. Kirschen, T.L. Volkmann, "Guiding a Power System Restoration with an Expert System," *IEEE Trans. on Power Systems*, May 1991, pp. 558-566.

[8] C.L. Yang, A. Yokoyama, et al., "Fault Section Estimation of Power System Using Color Time Petri Nets, " *Proc. Expert Systems Application to Power Systems*, Australia, Jan. 1993, pp. 321-326.

[9] Tomsovic, K., "Evaluation Rule-Based Systems," *Proc. ESAPS*, 1988, pp. 2.1-2.3

[10] L. Jenkins and H.P. Khincha, "Deterministic and Stochastic Petri Net Models of Protection Schemes," *IEEE Trans. on Power Delivery*, Jan. 1992, pp. 84-90.

Discussion

M. M. Adibi (IRD Corporation, Bethesda. The generic restoration actions (GRAs), in Section 2 of the paper are defined as e.g., "...cranking a non-black-start unit .... ." Later in the same section, restoration building blocks (RBBs) are defined as e.g., cranking of non-black-start units by local or remote black-start units. These definitions do not simplify the restoration process but add to its complexity for the following reasons:

1. The local or remote cranking of non-black-start units such as the drum-type boiler-turbine-generators are major undertakings consisting of a number of restoration levels, each level consisting of many simultaneous and sequential tasks, and each task consisting of many predecessor/successor, parallel/serial actions. No two such operations are alike or similar enough to be defined as a restoration building block. These unique operations typically comprise of over thirty actions, they are usually planned, analyzed, time estimated, simulated, tested and then included in the overall restoration plan [a].

2. The restoration actions typically include: opening or closing of breakers and disconnects, raising or lowering of voltage regulators, raising or lowering of phase-shifters, placing transformer tap changers under automatic or manual control, adjusting automatic set points for excitation, frequency, etc. These simple actions which are basic to the restoration process, may be termed "generic actions." The cranking of a "non-black-start" unit is too comprehensive an operation to be termed "generic."

Would the authors clarify the advantages of the Petri Nets over the accepted methods used in the industry for estimating restoration such as Program Evaluation and Review Technique (PERT) or Graphic Evaluation and Review Technique (GERT) ? In these techniques, duration of various restoration levels and tasks are estimated with some degrees of certainty. Such estimations are very essential in developing the overall restoration plan.

Could the authors breakdown the Petri Net in a more detailed and basic activities, assigning optimistic, most likely and pessimistic times to each action for determining the feasibility of the overall restoration plan? For example, in a recent operation [b], plans were to hot restart a steam electric unit using remote combustion turbine for cranking. Using GERT, the expected duration for the completion of the required 29 parallel and serial actions was determined to be 201 minutes with the standard deviation of 12 minutes. This estimation which was verified by test, indicated that the operation for hot restart of the steam unit within the required maximum critical time of 45-60 minutes was not feasible and that such an operation should only be considered for the cold restart of the unit after the minimum time of 3-4 hours is elapsed. This finding affected the overall restoration plan.

References:

[a] Adibi, M.M., D.P. Milanicz and T.L. Volkmann, "Remote Cranking of Steam Electric Stations," paper 96 WM162-4-PWRS, presented at IEEE 1996 WPM, Baltimore, MD, Jan 21 - 26, 1996.

[b] Milanicz, D.P. "Westport CT - Brandon Shores SES Remote Cranking Planning and Testing," presented at PSR-WG, IEEE 1995 SPM, Portland, OR, July 23-27, 1995

Manuscript received January 31, 1996.

R. F. Chu (PECO Energy), **Chen-Ching Liu** (University of Washington) and **Jaw-Shyang Wu** (National Kaohsiung Institute of Technology, Taiwan): The authors would like to thank the discussor for the helpful comments. Our response is as follows:

1. In this paper, our focus is on the concept of GRAs and its scheduling algorithm using Petri Nets; the restoration building blocks (RBBs) are only used for illustration of the proposed GRA scheduling algorithm.

2. At PECO Energy Company, there are restoration procedures for individual generating plants and a system restoration procedure for the power system. From the system operators' viewpoint, the system restoration procedure is more concerned about providing the necessary station power and the cranking power for restarting and the availability of the plant for restarting other plants and possible load pickups. Hence, a complex plant restart procedure such as craking or restarting a non-black-start unit can be simplified and modeled as a Generic Restoraion Action (GRA) in the *system* restoration procedure. Using the proposed approach, the total time for restarting the plant can be included in the time of the GRA and the overall system restoration time can be estimated.

We also point out that a GRA, such as "cranking_non_black_start_unit (X, Y)," can be customized by using different units, represented by X, and different busbars, represented by Y. Different times may be assigned to various customized versions of the same GRA.

3. We would like to compare the Petri Nets over the accepted method such as the Program Evaluation and Review Technique (PERT) as follows. PERT was developed for scheduling and evaluation of activities in a project. Relations exist which preclude the start of certain activities until others are completed. The computation of a critical path provides information on the shortest possible time to complete the overall project. As a result, PERT is an activity scheduling tool. On the other hand, the development of Petri Nets has a broader context. As a graphical tool, Petri Nets can be used as a visual communication aid similar to any flow charts, block diagrams and networks. Yet, as a mathematical modeling tool, it is possible to set up state equations governing the behavior of the system. Petri Nets are general modeling tools. For example, the

placenode can be interpreted as preconditions, input data, or resources needed. The transition can be used to represent an event, a computation step, a task or job, a step in logic, or a processor. Extensive analysis has been performed for Petri Nets. Examples are "reachability" from a token state to another token state and the net reduction techniques. Also, stochastic Petri Nets are found to be related to Markov Chains; therefore, it is possible to study the probabilistic nature of these Petri Nets. An extensive survey of the Petri Nets theory and applications is given in [A]. We also point out that the Petri Net model described in our paper is constructed *automatically* using the bus, line, generator, and load data and the time required for each restoration action. Even though we believe the Petri Nets theory is more general and more analytical tools for Petri Nets are available, we must add that both PERT and Petri Nets can be used as basic scheduling tools.

4. It is possible to break down the Petri Nets into a more detailed, basic activities and assign different time labels to the place and transition nodes to determine the overall restoration time for various scenarios. To achieve this, one would have to define the "basic" activities in a more detailed level. The actions mentioned by the discussor, e.g., changing tap settings, opening/closing a breaker, are suitable examples.

[A] T. Murata, "Petri Nets: Properties, Analysis and Applications," *Proceedings of the IEEE*, April 1989, pp. 541-580.

Manuscript received July 3, 1996.

Hierarchical Time-Extended Petri Nets as a Generic Tool for Power System Restoration

N.A. Fountas
Student member

N.D. Hatziargyriou
Senior member

K.P. Valavanis*
Senior member

Electric Energy Systems Laboratory
Department of Electrical and Computer Engineering
National Technical University of Athens, Greece

*Robotics and Automation Laboratory, The Center for Advanced Computer Studies
The University of Southwestern Louisiana, Lafayette, USA

Abstract - **Power System Restoration following a system collapse, is a very complex task. A wide variety of control actions has to be performed on time, while constraints such as power balance and system stability have to be carefully respected. In this paper, Hierarchical Time-Extended Petri Nets are exploited, in order to support the restoration process. The proposed framework has been applied to the blackstart of the Hellenic power system. Crucial issues pertaining to the restoration procedures are identified and the corresponding Petri Nets models are presented.**

Keywords: Power System Restoration, Hierarchical Time-Extended Petri Nets, Restoration Planning.

I. Introduction

Restoring service following a system collapse is a difficult decision and control problem for both system operators and operation planners, [1-2]. The features of the Power System Restoration (PSR) problem, i.e its combinatorial nature, the use of knowledge from a broad range of sources and of different types along with the number of criteria to satisfy, make it a very difficult decision making process. In the last years, knowledge-based techniques, such as expert systems, have been used, since heuristic knowledge and empirical judgement play an important role in power system restoration, [3-8]. These systems are based on rules or empirical associations obtained from experts in the field of restoration. The disadvantages of these (shallow reasoning) *rule-based* systems are: i. maintenance of large knowledge bases is difficult , ii. inability to deal with unexpected cases not covered by their heuristic rules, [9]. In recognition of these issues, *model-based* (deep-reasoning) methods are proposed as an alternative solution. These approaches: i. Are more maintainable than heuristic systems, ii. the knowledge acquisition is more straightforward, since the system structure and behaviour is known, iii. are more robust than shallow reasoning systems. This paper aims at proposing an innovative approach to provide support to system planners and operators when a blackout occurs and enable a qualitative analysis of the feasibility of the restoration procedures followed. The approach employs an extension of the *model-based* Petri Nets theory - Hierarchical Time-Extended Petri Nets, H-EPNs - [12-15], a reasoning tool that has gained increasing acceptance over the last decade in modelling of discrete event systems, mainly in the areas of computers (software and hardware), communication protocols and factory automation. Petri Nets exhibit a certain clearness and cleanness, permitting thus a simple and natural representation of many systems. By analysing a H-EPN model, useful information is obtained that can reveal bottlenecks, deadlocks and other problems that may exist within the underlying system and procedures, and suggest changes in them.

In Power Systems, Petri Nets have been applied to the areas of *computer-based automatic control design in substations*, [16-18], *sequence and process control*, [19-21], *power system protection*, [22-23], *fault section estimation*, [24] and *reliability evaluation*, [25]. In [26], ordinary PNs have been used for scheduling of *generic restoration actions*. However, this class of PNs has limiting modelling capabilities and several aspects of the PSR process cannot be properly included. The objective of this paper is to investigate and model the PSR problem using H-EPNs that have a superior performance in terms of modelling and decision power. The presentation of the proposed framework is conceptual. The rest of the paper is organised as follows: Section II. introduces the basic concepts relative to Petri Nets. In Section III. power system restoration procedures are demonstrated in the frame of the proposed hierarchical modelling. The restoration of the Hellenic system following a recent total blackout, [27], is investigated with H-EPN models. Discussion on the advantages and perspectives of H-EPNs follows.

II. Basic Notions of Hierarchical Time-Extended Petri Nets Framework

Petri nets (PNs) development goes back to 1960's. Since that time, they have gained increasing popularity and they have proved to be a valuable graphical and mathematical

96 SM 514-0 PWRS A paper recommended and approved by the IEEE Power System Engineering Committee of the IEEE Power Engineering Society for presentation at the 1996 IEEE/PES Summer Meeting, July 28 - August 1, 1996, in Denver, Colorado. Manuscript submitted December 28, 1995; made available for printing June 27, 1996.

Reprinted from *IEEE Transactions on Power Systems,* Vol. 12, No. 2, pp. 837–843, May 1997.

modelling tool applicable to many areas. As a *graphical* tool, PNs can be used as a visual-communication aid similar to flow charts, block diagrams, and networks with the additional ability to simulate *the dynamic and concurrent activities* of systems utilising tokens. As a *mathematical* tool, it is possible to set up state equations, algebraic equations, and other mathematical models governing the behaviour of systems. For a comprehensive, formal introduction to Petri Nets the reader is referred to [10-11].

A PN is a particular kind of directed graph, together with an initial state calle the initial marking, M_o. The underlying graph of a PN is a directed, weighted, bipartite graph consisting of two kinds of nodes, called *places* and *transitions*, where arcs are either from a place to a transition or from a transition to a place. In graphical representation, places are drawn as circles, ○ and transitions as bars, | . A *marking* (*state*) assigns to each place a nonnegative integer. If a marking assigns to place p a nonnegative integer k, we say that p is marked with k tokens. Pictorially, k black dots (tokens) are placed in p. By representing events with transitions and conditions with places, a system may be *mapped* into a PN model. The presence of a token in a place is interpreted as holding the truth of the condition associated with the place. The only execution rule in a PN is the rule for *transition enabling and firing*. A transition t is said to be enabled if each input place p of t is marked with at least w(p,t) tokens, where w(p,t) is the weight of the arc from p to t. An enabled transition may or may not fire. A firing of an enabled transition t removes w(p,t) tokens from all its input places p, and adds w(t,p) tokens to each of its output places, where w(t,p) is the weight of the arc from t to p. The movement of tokens through the PN graph represents the flow of information or control in the system.

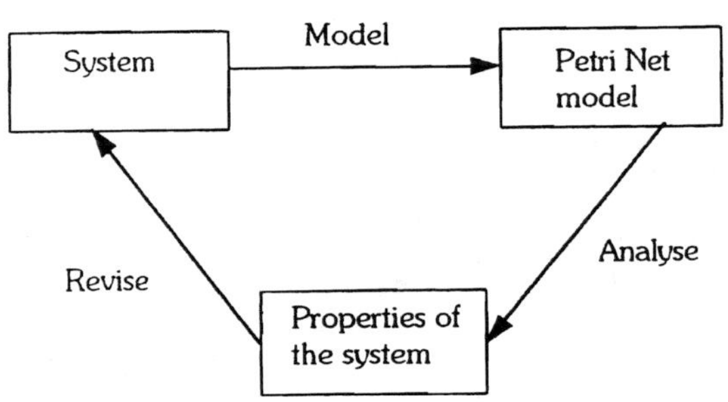

Fig. 1. H-EPN approach

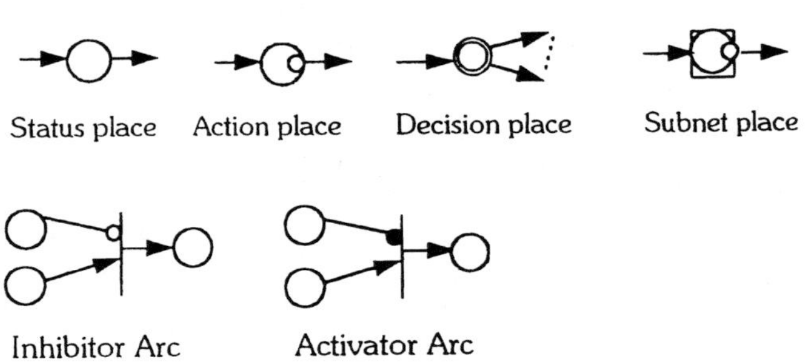

Fig. 2. Places and arcs in a H-EPN

As shown in Fig. 1, the studied system is first modelled as a Petri net, and then the model is analysed. This analysis can hopefully lead to important insights into the behaviour of the modelled system. This information can then be used to evaluate the system and suggest improvements or changes.

Classical PN based models have certain limitations in system modelling, [10-12], and therefore, their modelling power is limited, when they are used to model real-life complex systems with many components. To overcome this limitation, an enhanced class of PNs called Hierarchical Time-Extended Petri Nets (H-EPNs), has been proposed and presented by Valavanis, [12-15]. H-EPNs incorporate the following additional features: a) delays in processing associated with transitions/places, b) satisfaction of the need for multilevel representation leading to a better understanding of the various components of the system, c) assignment of priorities, d) utilisation of multiple types of tokens. In summary, H-EPNs account for all classes of conditions that may arise in the system through the introduced concept of *multiple* types of places, *multiple* classes of tokens and *multiple* arcs. In H-EPNs four different types of places are defined to model all classes of conditions that may arise in a system, Fig. 2: *a.* Status Place: it indicates the status of a resource or a part of the system. *b.* Action Place: this place indicates a process being performed by the system. *c.* Decision place: this place uses information external to the PN to resolve a conflict. A decision place has several outputs, but avoids conflicts by forcing its tokens along a particular path. *d.* Subnet place: This place is used as a modelling convenience. A H-EPN subnet is a representation of the operations of a subsystem that minimally interacts with other such subsystems. A subnet at the lowest level will perform an elementary operation of the system. In order to have a consistent control logic and increase the modelling power of PNs, activator and inhibitor arcs are also employed. An inhibitor (activator) arc is an arc from a place to a transition, such that the transition may fire only if that place is empty (full), thus enabling the resolution of conflicts that may arise during the system modelling. These arcs facilitate assigning priorities to certain transitions over other transitions.

Time Incorporation. Timed Petri nets comprise a particular type of PNs that associate delays with each transition t_j by introducing two times, $\tau_{1,j}$ and $\tau_{2,j}$, [10-11, 14-15]. A constant value for these times can be considered, i.e. a normalised one corresponding to the actual time settings. In a general case, however, these values can be variable. A transition t_j can fire only if it has been enabled for at least time $\tau_{1,j}$ and it must fire before $\tau_{2,j}$ (if it is enabled).

Petri Net Analysis. By analysing a H-EPN model, bottlenecks, deadlocks and other problems that may exist within the underlying system and procedures can be revealed, and suitable changes can be suggested. Reachability is a fundamental basis for studying the dynamic properties of the system under consideration. The firing of an enabled transition will change the token distribution (marking) in a PN according to the transition rule, [10]. A sequence of firings will result in a sequence of markings. A marking M_n is said to be reachable from a marking M_0 if there is any sequence of transition firings which will change the marking from M_0 to M_n. A set of reachable markings consists of a reachability set, which can be represented by a reachability graph. Each node is associated with a state and arcs(arrows), labelled by t_i , indicating transitions. A marking M_1 can be transformed into the marking M_2, by a directed arc, if there exists a transition t_i whose firing changes the initial marking M_1 to the marking M_2.

III. Power System Restoration Process

The problem of Power System Restoration following a complete or partial blackout has been studied for many years with special emphasis on developing effective restoration plans. Since blackouts seldom occur, the personnel in control centers and generating plants has little experience in handling such extreme disturbances. It is therefore necessary for an electric utility to set-up orderly restoration procedures for such events and to focus on training and education of engineers and dispatch operators.

A. H-EPN Methodology

System Restoration itself is a hierarchical problem. It embodies an inherent 'time hierarchy' associated with the determination and execution of the restorative control actions in the various stages of system re-establishment, [1,5]. The *hierarchical modelling approach* used by H-EPNs leads to restoration plans which are better structured and easier to modify, while a better decision-making process can be achieved. This modelling is readily related to the three-level knowledge-based procedure of "strategic", "tactical" and "operational" restoration planning described in [3-5]. A direct analogy can be observed between H-EPN key aspects and PSR process: *system resources* may be represented by electrical islands formed during system rebuilt, *independent operations* are demonstrated as switching actions for island reconfiguration and the *ultimate goal* is the full re-supply of all loads with a minimisation of the required time. For the formulation of the restoration strategies, the hierarchical modelling utilising H-EPNs follows three phases: 1. *top-down decomposition of the system into a set of component nets, 2. component net modelling, and 3. bottom-up synthesis of the system H-EPN model.*

The top-down decomposition is accomplished by a functional breakdown of the overall restoration process that eventually leads to a set of well defined operations common in diverse restoration strategies. In fact, in a typical restoration scenario, the following operations (i.e, generic restoration actions, [6]), most often take place:

- start blackstart unit
- energise busbar
- energise line
- pick up load
- crank non-blackstart unit
- synchronise two subsystems
- find path
- connect tie line

In the component net modelling, the previous primitive restoration actions constitute a sequence of *subnets* utilising action places at the lowest level of the H-EPN modelling. An entire net may be replaced by a single place or transition for modelling at a more abstract level (abstraction), or places and transitions may be replaced by subnets to provide more detailed modelling (refinement) facilitating thus a better understanding of the restoration process.

The restoration scenario is, therefore, *decomposed* into a set of subnets with parallel activities. In each subnet, *constraints violations* are checked by means of decision places with the form of an empty/full - no/yes -, status condition. In the final stage, after the H-EPN subnet for each generic restoration action has been validated, the resulting overall H-EPN power system model can be constructed by using bottom-up synthesis techniques, [12-14].

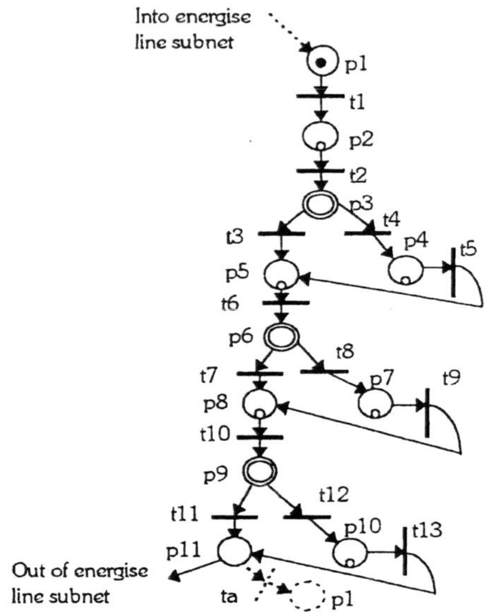

Fig. 3. **H-EPN model for the "Energise line" subnet**

Indicatively, the suggested H-EPN model for the energise line subnet is demonstrated in Fig. 3. A comprehensive description of the various places and transitions for the studied cases is given in Table 1 of the Appendix. The token in place p1 indicates entry into the subnet. The problems of transient voltage, sustained overvoltage and charging current that may appear during energisation of long, lightly loaded transmission lines are checked within the H-EPN model with the aid of the decision places, p3, p6 and p9. If an operating constraint is violated, the model will suggest a modification to the plan in the form of a "corrective action", depending on the actual status of the system. The flow of information continues through the flow of tokens after each of these problems has been successfully checked. By using the reachability graph, it can be shown that the net is *live*, *safe* and *reversible*. The existence of these important H-EPNs properties within the frame of the PSR problem assures that the proposed sequence of restoration actions is feasible. The H-EPN models for the other primitive restoration actions are similar to that of Fig. 3, preserving the same basic properties. The *dotted* transitions and arcs represent the modifications carried out on the individual nets, like looping back, by passing the tokens back to the input places, in order to validate the net properties.

Considering the sequence of transition firings and all markings reachable from the initial marking, the reachability graph of the PN model of Fig. 3 is drawn in Fig. 4. The H-EPN structural properties of *safeness*, *liveness* and *reversibility* of the constructed model, are validated by examining this *reachability graph*, [10-11]. The verification of these important H-EPNs properties within the frame of the PSR problem assures that the proposed sequence of restoration tasks is feasible and deadlock-free, [12-15].

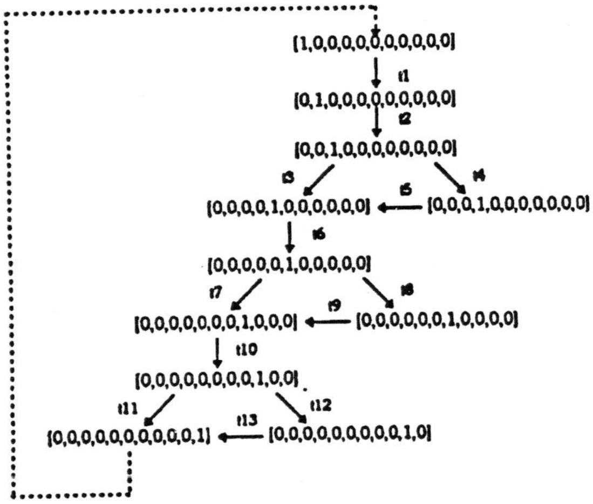

Fig. 4. Reachability graph of the H-EPN model of Fig. 3.

B. Application to the Hellenic Power System Blackstart

The Hellenic power system experienced a total blackout in 1989. A busbar fault that occurred in a vital substation in Northern Greece triggered a series of protection schemes which caused cascade tripping of lines and generators and finally led to a total shutdown. In order to restore service, the station personnel of the Public Power Corporation (PPC) followed the restorative procedures of the "build-up" strategy, [27]:

- assessment of power system status; isolation of faulted area or search for its identification.
- starting, as soon as possible, of all hydro stations - power supply of all thermal plant auxiliaries from hydro power stations. Supply of station service for the hot restart of drum type boilers before elapse of the maximum critical time.
- sectionalisation of the de-energised system into several island sub-systems, so that each sub-system includes at least one black-start hydro unit and/or one drum type boiler; each island should have the ability to match the load and generation within frequency limitations.
- interconnection of generating stations within each island after units obtain minimum generation, but prior to restoration of full load within the island.
- independent gradual load and generation build-up inside each subsystem. Higher priority is assigned to a critical industrial load center representing an aluminium process industry, to capital Athens area and to other big load centers.
- resynchronisation of islands.
- completion of system rebuilt by picking-up the remaining unserved load and closing of ties with the interconnection systems.

The proposed H-EPN framework of section A. has been applied to investigate the blackstart procedures following the blackout of 1989. An illustrative example is given in Fig. 5, where a restoration island formed in the Southern Greece is depicted. It comprises 7 buses, one blackstart hydro unit, Ladonas, and two non-blackstart thermal units, Megalop1 and Megalop2. There are also 3 local loads to be picked up, L1, L2 and L3. The H-EPN model for this system configuration is shown in Fig. 6. It incorporates the hierarchical modelling methodology through the introduction of action and subnet places.

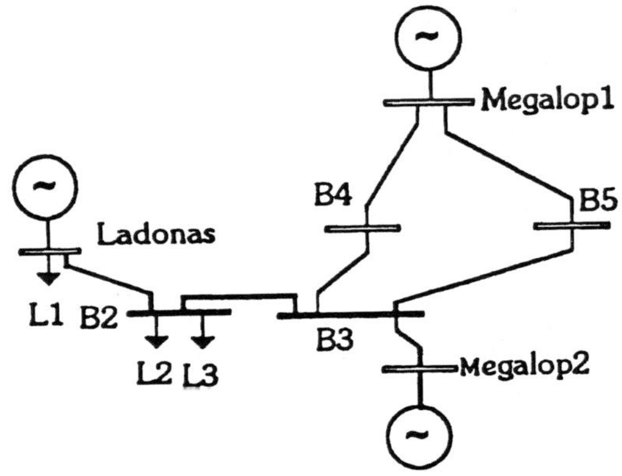

Fig. 5. South island subsystem

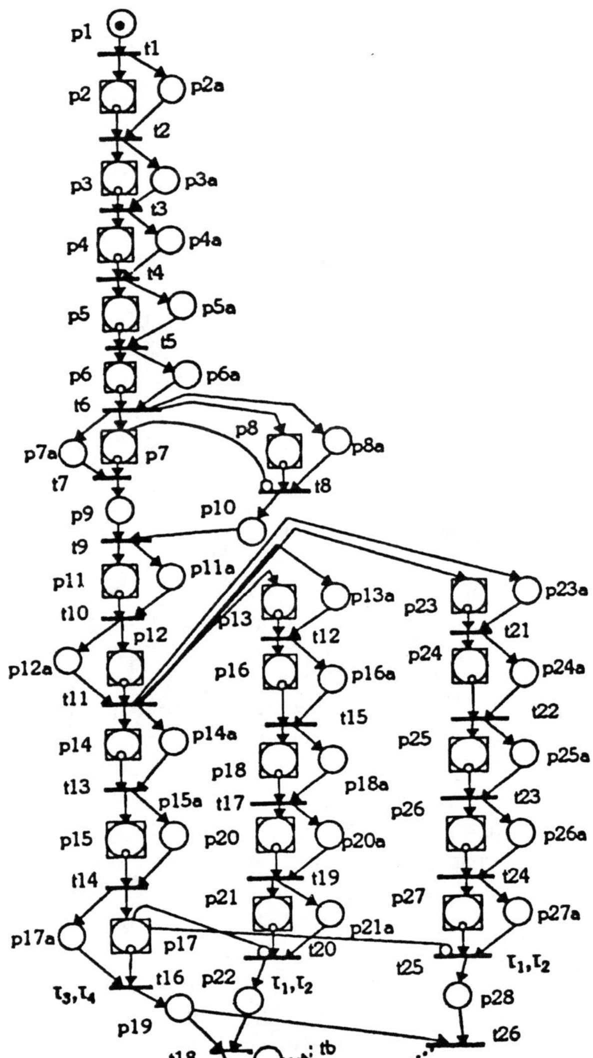

Fig. 6. H-EPN model for island reconnection

A *synthesis* of the primitive restoration actions (subnets) has been performed. After blackout occurrence, the suggestion from the H-EPN model, is to start the available black-start hydro unit, Ladonas, place p2. The places and the corresponding transitions of the model illustrate the restoration actions of unit blackstart, bus energisation, line energisation, load pickup and unit crank that take place during island reintegration. This is achieved through a cascade connection of the pre-defined subnets. For example, when t6 fires, load L2 is picked up, Appendix. This sequence of actions goes on along the available transmission paths until power supply is cranked to the thermal stations. The firing of t16 corresponds to the start up of the non-blackstart unit Megalop2, while t20 or t25 denote the start-up of unit Megalop1, as seen in the Appendix. The model also suggests that two alternative paths are available for the start up of the thermal unit Megalop1. The first energises bus B5 and the second bus B4. What transmission path should be followed, depends on the prevailing system conditions, while the final judgement belongs to the operator. It has been assumed that during island re-integration all switching actions are feasible, that is all operating constraints imposed are met.The list of the operating constraints checked is by no means complete. Ongoing work is under way to take into account other significant system constraints, as well, such as generator reactive over- and under- excitation or overload limit of system equipment utilising weighted arcs and threshold values. The following features of the PSR problem have been incorporated in the H-EPN models:

Priorities and time constraints. An issue of primary concern for power operators is the setting of *priorities* for optimum generator startup sequence and load restoration (auxiliaries, certain critical loads). This is addressed by incorporating *inhibitor arcs* in the H-EPN model of Fig. 6. For example, transition t7 has priority over transition t8, indicating that load L_2 representing a critical industrial load, has been given priority over load L_3. Moreover, unit Megalop2 has priority over unit Megalop1 because it has a smaller critical maximum interval compared to Megalop1, [2, 7]. Thus, transition t16 has priority over transition t20/ t25.

A related aspect is the recognition and co-ordination of critical time intervals. For example, the *maximum* elapsed time for the hot restart of drum-type units is usually between 30 to 60 minutes, beyond which the units need to be rolled for a cold startup, 2 to 3 hours later, [1,2,6]. Another constraint is the *critical minimum* interval, after which a super critical, once through unit can safely be restarted. By using the time logic described in Section II, if two transitions t_j and t_k are present and $\tau_{2,j} < \tau_{1,k}$ is set , then transition t_j has priority over transition t_k since t_j must fire (if it is enabled) before t_k will be allowed to fire. These time delays, τ_1 and τ_2 , are referred to as the critical minimum and maximum generator intervals, respectively. As it is shown in the model of Fig. 6, the time delays, τ_1, τ_2 and τ_3, τ_4 have been assigned to the transitions t16 and t20 or t25, corresponding to the black start of the two thermal units.

MIMO Nets. One disadvantage related to the applicability of Petri Nets in system modelling, is the fact that even for a moderately sized problem, the complexity of the model makes it difficult to handle. To overcome this problem, H-EPN Multiple Input Multiple Output (*MIMO*) subnets, [14], have been used to represent the overall system restoration process. The Single Input Single Output (*SISO*) subnets that are independently defined, are combined through *common places, transitions and arcs* to define a MIMO subnet. The common places and transitions represent the common generic restoration actions that take place within all electrical islands during system rebuilt. In such a case, every point of entry is associated with a corresponding point of exit. Such a MIMO subnet is then transformed to a corresponding SISO subnet utilising a powerful transformation formula (a detailed mathematical formulation can be found in [15]) that preserves the same basic properties for easy integration (as a SISO subnet) into the top-down decomposition of the restoration process. Finally, the blackstart of the Power System involves the synthesis of these bigger MIMO subnets. This approach has greatly facilitated the overall system modelling.

C. Comments on the H-EPN Approach

Visualisation and Properties. H-EPNs represent an efficient, interactive, graphical modelling and analysis tool for developing restoration plans providing *a)* easy visualisation of the current status of the power system during restoration, *b)* simulation of the *dynamic behaviour* of the power system by the flow of tokens exploring the analysis techniques of reachability graph, [10-11].

Utilising graphical facilities, the H-EPN model can show the energised part of the system, guiding thus the operator in his next steps, in order to find viable restoration paths. The verification of the H-EPNs structural properties, such as liveness and safeness, assures the stable performance of the power system following system collapse. This is fundamental for the successful completion of the PSR actions, in order to achieve the overall goal of load reconnection, as soon as possible.

H-EPNs have definite *advantages* over conventional representation tools, like flow charts. Namely, PNs can capture the *dynamic* behaviour of the system through the flow of tokens, while a flow chart is simply a static visualisation of the system profile. In addition, net 's structural properties can be proved taking advantage of the mathematical background of PN theory, properties essential for the deadlock free operation of the modelled system. Further, H-EPNs can more accurately replicate the behaviour of the power system encoding a broad range of aspects critical to the PSR process, such as priorities, operational limits violation, time dependency and multi-level hierarchical representation.

Genericity. Of primary importance is the fact that the proposed H-EPN methodology provides *generic* restoration guidelines rather than system-dependent. These general guidelines can be easily customised to each utility's restoration strategy and be applied on specific facts. In that sense, the proposed framework can serve as a *generic tool* applicable to several restoration scenarios and policies. In addition, H-EPN models exhibit high maintainability, in case of power system reconstruction and modification.

Restoration Planning. H-EPN based platform can be used as a planning restoration tool for studying various scenarios. It can be implemented as a decision support system for the power system operator in order to: **i)** help him choose among various ways of initiating the restoration **ii)** propose several alternative courses of action for a given system condition, **iii)** assist him in selecting intermediate restoration objectives, **iv)**

alert him when dangerous situations arise, v) explain how the system arrived at these alternatives with schematic visualisation, vi) evaluate the feasibility of alternative restoration strategies based on analytical tabulated data from off-line studies in conjunction with H-EPN structural properties, vii) display the applicable - pre-defined restoration procedures.

Implementation and Perspectives. At the present stage, the developed H-EPN model can be used for PSR planning, in an *off-line, stand-alone mode*. The encoded reasoning in the H-EPN model is based on logical interrelations represented by multiple classes of places and transitions. The issue of *path finding* is addressed with pre-defined paths assumed to be available during system rebuilt. These switching sequences are based on established practices.

The models developed for subsystems of rather moderate size can be analysed by inspection. For bigger systems, however, computer simulation is necessary. Various commercial software packages have been developed with the aim of proving the properties of a H-EPN and simulating the performance of the obtained model, [12-15]. These packages are equipped with user-friendly editing facilities and are generally constructed of two parts: model edit and actual analysis. The model can be edited either in textual or graphic form providing a full description of the places and the transitions. Analysis facilities are capable of simulating thousands of markings and transition firings per second.

On-going work concerns the development of an integrated software environment which will make use of the above software and at the same time will be able to interchange data and results from system simulation programs, [27]. In this way a more accurate checking of the feasibility of the restoration actions will be possible. At its ultimate stage, H-EPN modelling could be viewed as an interactive assistant to the operator, being part of a sophisticated decision-support environment. It is expected that the H-EPN platform will be able to accept data on-line from the SCADA-EMS database and have the flexibility to adapt to new situations, when arise and modify its suggestions under the prevailing conditions and operator's opinion. Dynamic path searching heuristic algorithms, such as A* search method, [6,7], will be used. It is believed that analytical tools combined with the H-EPN modelling will profitably complement each other for guiding the operator in the problem-solving process during emergency situations in an on-line environment.

IV. CONCLUSIONS

Restoration of a power system following a major power failure is a complex and stressful task with many diverse facets. This paper has concentrated on the investigation of the applicability of Hierarchical Time-Extended Petri Nets to model and analyse the PSR process. At the present stage of development, this endeavour has revealed a promising potential for providing a qualitative analysis of the PSR problem. As a future objective, an integrated environment of the proposed H-EPN framework in conjunction with powerful analytical tools and the SCADA/EMS database would help the power operators to cope with the PSR problem in a more efficient and better co-ordinated way.

V. ACKNOWLEDGEMENT

The financial support of the Power Public Corporation (PPC) and the General Secretariat of Research and Technology of Greece is greatfuly acknowledged.

VI. REFERENCES

[1] M. M. Adibi, L. H. Fink, "Power System Restoration Planning", IEEE Trans. on Power Systems, Vol. PWRS-9, No. 1, Feb. 1994, pp. 22-28.

[2] M. M. Adibi, J. N. Borkoski, R. J. Kafka, "Analytical tool requirements for power system restoration", paper 94 WM 222-0-PWRS, IEEE/PES 1994 WM, NY, Jan. 30 - Feb. 3, 1994.

[3] K. Matsumoto, T. Sakaguchi, R. J. Kafka and M. M. Adibi, "Knowledge-Based Systems as Operational Aids in Power System Restoration", Proceedings of the IEEE, Vol. 80, No. 5, May 1992, pp. 689-697.

[4] "A survey of expert systems for power system restoration", CIGRE TF 38.06.04, Electra No. 150, October 1993, pp. 87-105.

[5] M. M. Adibi, R. J. Kafka, D. P. Milanicz, "Expert System requirements for Power System Restoration", paper 94 WM 223-8-PWRS, IEEE/PES 1994 WM, NY, Jan. 30 - Feb. 3, 1994.

[6] L.H. Fink, Kan-Lee Liou, Chen-Ching Liu, "From Generic Restoration Actions to Specific Restoration Strategies", paper 94 SM 539-7 PWRS, IEEE/PES 1994 SM, San Francisco, CA, July 24 - 28, 1994.

[7] Chen-Ching Liu and Kan-Lee Liou, R.F. Chu, A.T. Holen: "Generation Capability Dispatch for Bulk Power System Restoration: A Knowledge-Based Approach", IEEE Trans. on Power Systems, Vol. 8, No.1, Feb. 1993., pp. 316-325.

[8] Kan-Lee Liou and Chen-Ching Liu, R.F. Chu, "Tie Line Utilisation During Power System Restoration", paper 94 WM 221-2 PWRS, IEEE/PES 1994 WM, NY, Jan. 30 - Feb. 3, 1994.

[9] X. Wang, T. Dillon, "Combining Deep and Shallow Reasoning in Second Generation Expert Systems for Power Systems", 10th PSCC, August 1990, Graz, Austria, pp. 959-965.

[10] J.L. Peterson, *Petri Net Theory and the Modelling of Systems*, Prentice-Hall Inc., Englewood Cliffs, NJ 07632, 1981.

[11] Tadao Mourata, "Petri Nets: Properties, Analysis and Applications", Proceedings of the IEEE, Vol. 77, No.4, April 1989, pp. 541-580.

[12] K.P. Valavanis, "On the hierarchical modelling, analysis and simulation of flexible manufacturing systems", IEEE Trans. on Systems, Man, and Cybernetics, Vol. 20, No. 1, Jan./Feb. 1990, pp. 94-110.

[13] S. Ramaswamy and K.P. Valavanis, "Modelling, Analysis and Simulation of Failures in a Materials Handling System With Extended Petri Nets", IEEE Trans. on Systems, Man, and Cybernetics, Vol. 24, No. 9, Sept. 1994, pp. 1358-1373.

[14] S. Ramaswamy, K.P. Valavanis, "Hierarchical Time-Extended Petri Nets (H-EPN) Based Error Identification and Recovery for Hierarchical System", IEEE Trans. on Systems, Man, and Cybernetics- Part B: Cybernetics, Vol. 26, No. 1., Feb. 1996, pp. 164-175.

[15] S. Ramaswamy, "Hierarchical time-extended Petri Nets for integrated control and diagnostics of multilevel systems", Ph.D. Thesis, *The Center for Advanced Computer Studies*, The University of Southwestern Louisiana, Lafayette, Spring, 1994.

[16] J. Clade, J. Miroux, J.M. Tesseron, L. Tourres, C. Corroyer, J. Kowali, "Recent Developments in Substation Control, Internal Communications and Automatic Devices as Used in the French Distribution Network", Paper No. 34-01, CIGRE International Conference on L.H.V.E.S., September 1978.

[17] K.P. Brand, J. Kopainsky, "Systematic Design of Automation, Protection, and Control in Substations", IEEE Trans. on PAS, Vol. PAS-103, No. 9, September 1984.

[18] J.L. Pinto de Sa and J. Damasio, "Coordination of Automatic Control Functions in Transmission Substations Using Petri Nets", IEEE Trans. on Power Delivery, Vol. 7, No. 1, pp.262-268, January 1992.

[19] K.P. Brand, J. Kopainsky, "Principles and Engineering of Process Control with Petri Nets", IEEE Trans. on Automatic Control., Vol. 33, No. 2, pp. 138-149, February 1988.

[20] J.L. Pinto de Sa and S. Paiva, "Design and Verification of Concurrent Switching Sequences with Petri Nets", IEEE Trans. on Power Delivery, Vol. 5, No. 4, pp.1766-1772, October 1990.

[21] J.L. Pinto de Sa, J.P. Sucena Paiva, "A Multitasking Software Architecture to Implement Concurrent Switching Sequences Designed with Petri Nets", IEEE Trans. on Power Delivery, Vol. 6, No. 3, July 1991, pp. 1058-1064.

[22] L. Jenkins and H.P. Khincha, "Deterministic and Stochastic Petri Net Models of Protection Schemes", IEEE Trans. on Power Delivery, Vol. 7 No. 1, Jan. 1992, pp. 84-90.

[23] F. Wang, J. Tang, "Modeling of a transmission line protection relaying scheme using petri nets", *paper 96 WM 022-4 PWRD*, IEEE/PES 1996 WM, Baltimore , MA, Jan. 21-25, 1996.

[24] C.L. Yang, A.Yokoyama. Y. Sekine, "Fault Section Estimation of Power System Using Colored and Timed Petri Nets", Proceedings of Expert Systems Applications to Power Systems, Australia, Jan. 1993, pp.321-326.

[25] H. Salehfar and R. Rodick, "A new and Fast Electric Power Generating System Reliability Evaluation Model Using Petri Nets", International Journal of Reliability, Quality and Safety Engineering, World Scientific Publishing Company, Vol 1, No .4., pp.459-473, 1994.

[26] Jaw-Shyang Wu, C-C Liu, K-L Liou, R.F. Chu, "A Petri Net Algorithm for Scheduling of Generic Restoration Actions", paper 96 WM 177-6 PWRS, IEEE/PES 1996 WM, Baltimore, MA, Jan. 21-25, 1996.

[27] N.A. Fountas, N.D. Hatziargyriou, C. Orfanogiannis, A. Tasoulis, "Interactive Long-Term Simulation For Power System Restoration Planning", paper 96 WM 175-0 PWRS, IEEE/PES 1996 WM, Jan. 21-25, 1996, Baltimore, MD.

VII. APPENDIX

All the tables and transitions that constitute the developed H-EPN models are fully described in a sequential manner.

Energise line subnet

p1: energise line subnet
t1: "Into" subnet for energise line
p2: energise line
t2: line energised
p3: check for transient voltage problem
t3: no transient voltage problem
p4: do corrective action
t4: there is transient voltage problem
p5: check for sustained overvoltage problem
t5: corrective action completed
p6: sustained overvoltage problem?
t6: sustained overvoltage problem checked
p7: do corrective action
t7: no sustained overvoltage problem
p8: check for charging current problem
t8: there is sustained overvoltage problem
p9: charging current problem?
t9: corrective action completed
p10: do corrective action
t10: charging current problem checked
p11: out of subnet energise line
t11: no charging current problem
t12: there is charging current problem
t13: correction action completed

Island Reenergisation

p1: into island integration
t1: into island integration
p2: blackstart unit Ladonas
p2a: "into" subnet blackstart unit
t2: unit Ladonas started
p3: energise bus Ladonas
p3a: "into" subnet energise bus
t3: bus Ladonas energised
p4: pickup load L1
p4a: "into" subnet load pickup
t4: load L1 picked up
p5: energise line Ladonas
p5a: "into" subnet energise line
t5: line B1-B2 energised
p6: energise bus B2
p6a: "into" subnet energise bus
t6: bus B2 energised
p7: pickup load L2
p7a: "into" subnet load pick up
t7: load L2 pickedup
p8: pick load L3
p8a: "into" subnet load pickup
t8: load L3 pickedup
p9: load L2 picked up
t9: load pickup completed
p10: load L3 picked up
t10: line B2-B3 energised
p11: energise line B2-B3
p11a: "into" subnet energise line
t11: bus B3 energised
p12: energise bus B3
p12a: "into" subnet energise bus
p13: energise line B3-B5
p13a: "into" subnet energise line
t12: line B3-B5 energised
p14: energise line B3-Megalop2
p14a: "into" subnet energise line
t13: line B3-Megalop2 energised
p15: energise bus Megalop2
p15a: "into" subnet energise bus
t14: bus Megalop2 energised
p16: energise bus B5
p16a: "into" subnet energise bus
t15: bus B5 energised
p17: crank unit Megalop2
p17a: "into" subnet crank unit
t16: unit Megalop2 cranked
p18: energise line B5-Megalop1
p18a: "into" subnet crank unit
t17: line B5-Megalop1 energised
p19: Megalop2 unit cranked
t18: island integration completed
p20: energise bus Megalop1
t19: bus Megalop1 energised
t18: island integration completed
p21: crank unit Megalop1
p21a: "into" subnet crank unit
t20: unit Megalop1 cranked
p22: unit Megalop1 cranked
p23: energise line B3-B4
p23a: "into" subnet energise line
t21: line B3-B4 energised
p24: energise bus B4
p24a: "into" subnet energise bus
t22: bus B4 energised
p25: energise line B4-Megalop1
p25a: "into" subnet energise line
t23: line B4-Megalop1 energised
p26: crank unit Megalop1
p26a: "into" subnet crank unit
t24: unit Megalop1 energised
p27: crank unit Megalop1
p27a: "into" crank unit subnet
t25: unit Megalop1 cranked
p28: unit Megalop1 cranked
t26: island integration completed
p29: island integration completed
ta, tb: temporary transitions to validate net properties

Table 1. Description of H-EPN places and transitions

DECISION AID FUNCTION FOR RESTORATION OF TRANSMISSION POWER SYSTEMS: Conceptual Design and Real Time Considerations

Tatjana Kostić, Rachid Cherkaoui, Alain Germond
Laboratoire de Réseaux Électriques
École Polytechnique Fédérale de Lausanne
Switzerland
e_mail: tatjana.kostic@lre.de.epfl.ch

Patrick Pruvot
Direction des Études et Recherches
Électricité de France, Clamart
France
e_mail: patrick.pruvot@der.edfgdf.fr

Abstract **- This paper deals with the conceptual design of a Decision Aid Function FOr Restoration (DAFFOR) after major black-outs, and in particular with real time considerations. Some important characteristics of the proposed DAFFOR are: (a) capability of taking into account the dynamic evolution of the network state, due to either a *control action* or an *unforeseen event*; (b) interactive mode, which permits users to request a validation of their own suggestion; (c) coupling with an internal dynamic simulator which is solicited to validate any suggestion. At present, the DAFFOR is coupled with the French National Utility Operator Training Simulator (EDF OTS). After intensive testing within the OTS environment, the operational version of the DAFFOR will be developed.**

Keywords: decision support, power system restoration, real time operation, interactive mode, dynamic simulator

1. INTRODUCTION

Restoration after a major disturbance is one problem operators in control centres are faced with. Their task in this case is to lead the restoration process by applying control actions to the power system, one by one, until full service is restored. This should be reached as soon as possible, and without any violation of the operating limits. Electric power utilities usually have restoration plans with guidelines [1], which should make operation during the restoration easier for the operators. However, different unforeseen events may occur during the restoration process, which can disable the applicability of the defined guidelines, forcing the operator to act according to his/her own experience and knowledge about the power system he/she operates.

The complexity of the restoration process itself, along with the stressful conditions in which it is carried out, makes a decision support tool to assist operators in this task very useful [2]. Since the very first publication on the subject [3], numerous works have been reported [4,5]. Some of these decision support systems have been developed up to the on-line operation level, and are integrated in the OTS [6] or the Control Centre environment [7].

PE-111-PWRS-16-09-1997 A paper recommended and approved by the IEEE Power System Engineering Committee of the IEEE Power Engineering Society for publication in the IEEE Transactions on Power Systems. Manuscript submitted May 27, 1997; made available for printing September 30, 1997.

1.1 Objective

The objective of this work is to propose and develop a decision support tool, intended for use by the operators of a control centre during transmission power system restoration after a major disturbance. The decision support tool should satisfy the following requirements:

1) **Use in the real environment.** This implies: (a) easy coupling with a power system simulator, for validation purposes, (b) easy implementation in a control centre, for real-time purposes, and (c) suitability for the actual working habits of the operators.
2) **Real-time operation.** This implies the capability to take into account the evolution of the power system, including the unforeseen events that may occur during the restoration process.
3) **Guidance function and interactivity.** These imply the capability to: (a) generate one control action at a time which, having been validated and proven not to cause any violation of the operating limits if executed on the power system, is suggested to or by the operator, (b) explain any result it generates, and (c) cope with the initial situation that can be a total or partial black-out.

1.2 Proposed Architecture

According to the above objectives, the **Decision Aid Function FOr Restoration (DAFFOR)** has been proposed and developed. The architecture shown in Figure 1 satisfies the first requirement, concerning the DAFFOR's use in a real environment, while its constituate elements, Real Time Update kernel and Reasoning kernel, satisfy requirements 2) and 3), respectively.

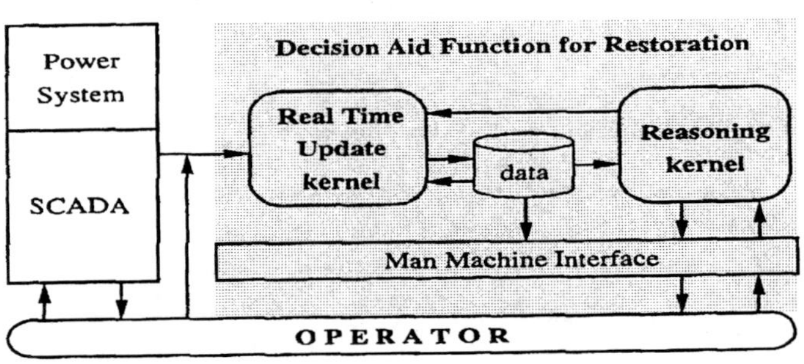

Figure 1. Overall structure of the DAFFOR integrated in a control centre

The **Real Time Update (RTUpd) kernel** is DAFFOR's data I/O interface. It steadily receives data coming from different sources, performs comparisons among them and generates a coherent power system state description at a given moment. This state description is the only real-time information made available to the Reasoning kernel.

Reprinted from *IEEE Transactions on Power Systems,* Vol. 13, No. 3, pp. 923–929, August 1998.

The **Reasoning kernel** guides the restoration process in real-time by suggesting one control action at a time, according to data generated by the RTUpd kernel and the restoration strategy. It can also assess a control action proposed by the operator. A combined knowledge-based and procedural approach was chosen for these purposes.

The present paper focuses on the real-time considerations in DAFFOR, which cover the **Real Time Update** kernel and its **real-time interaction** with the Reasoning kernel.

1.3 Overview of the Paper

To better understand the overall structure and the operation of DAFFOR, the Reasoning kernel will only be briefly discussed in Section 2. **Section 3** deals with the items of the main concern in this paper: the operation of the Real Time Update kernel and its interaction with the Reasoning kernel. Sections 4 and 5 present the implementation issues and the validation architecture, respectively. Final remarks and conclusions follow in Section 6.

2. REASONING KERNEL

As announced in Section 1.1, DAFFOR should produce as output **one control action at a time**. This requirement depicts the concern to (a) provide the operator with a suggestiön which fits his/her own manner of operating the power system, and (b) make the DAFFOR capable of following the dynamic evolution of the power system, step by step. The control action to be suggested to the operator in a given time is searched for in the Reasoning kernel.

The Reasoning kernel is the interactive advisory part of DAFFOR. It can operate in either ***research*** or ***validation*** mode. In *research* mode, its function is to suggest a control action to the operator, and in *validation* mode it is to assess a control action suggested by the operator. The Reasoning kernel consists of three modules, as shown in Fig.2.

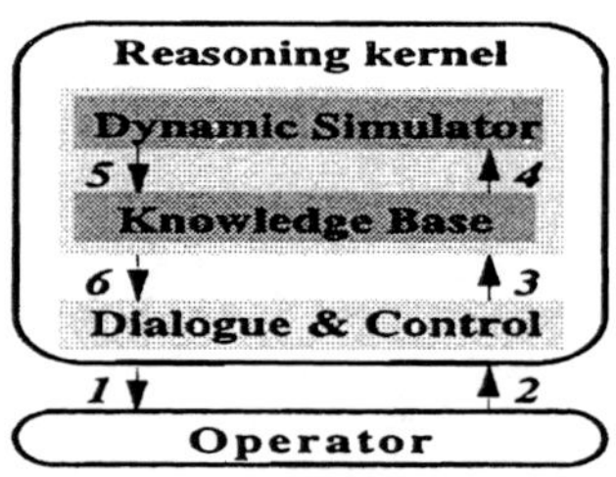

Figure 2. Reasoning kernel, the interactive advisory part of DAFFOR

The **Dialogue&Control (D&C) module** represents DAFFOR's user interface. It offers to the user (step 1) the possibility to chose the operation mode (*research* or *validation*), to *accept* or *reject* DAFFOR's suggestion, to ask for the *explanation* and to *end* the session. Depending on the user's choice (step 2), the Dialogue&Control module activates the corresponding task (step 3).

The **Knowledge-Based module** contains rules describing the restoration strategy and the manner in which it is applied to the power system. According to the French National Utility (EDF) strategy, the restoration process is divided into two stages. In the first, the so-called skeleton of the power system is built by synchronising the available power plants and energizing the main lines connecting them. In the second, most of the load is picked up, and transmission capacity is simultaneously reinforced by energizing additional circuits.

At present, the Knowledge-Based module contains about 120 rules, which guide DAFFOR's process of search for a control action. The considered actions are the following:

- for lines and transformers: change of operating status
- for production units: change of power and voltage set-point, change of operating status of the unit and of its primary frequency/power (f/P) control
- for loads: pick up or shed a consumer load by shedding level (*1-5*) or by absolute value (*x MW*), supply unit auxiliaries, (dis)connect a shunt compensation device

Each control action, either found during the search in the Knowledge-Based module or suggested by the operator, is tested with the Dynamic Simulator (step 4).

The **Dynamic Simulator** [8] generates the response of the power system (step 5) to a control action (step 4) in terms of the evolution of frequency, voltages, production level, power flow, losses etc., as well as the statistics about the violation of operating limits (overload, over-voltage, under-voltage, frequency excursions). It also contains a number of tests which filter the control actions that can be executed at a given moment from the set of all possible control actions. These tests are either deterministic (e.g., if a line is in service, the action that would try to put it in service again is disabled) or heuristic (e.g., the maximum quantity of a load that can be picked up in a given moment is defined with respect to the current spining reserve).

2.1 Search Process

Given the heuristic nature of the rules that the Reasoning kernel contains, and the huge number of possible search paths, particular attention must be paid to the reliability of a single control action that is to be suggested to the operator.

Let us denote a_r as the currently searched control action. It should represent the first element of a sequence of actions that, if performed one after another, would lead to the complete service restoration of the power system. The search process in the Reasoning kernel could (a) continue until the whole sequence is found, or (b) stop as soon as the action a_r is found. The first approach could be very time consuming and, in addition, the obtained sequence could become useless if an unforeseen event occurs. The problem with the second approach is that there would be no guarantee that the action a_r is on a good search path.

Given these disadvantages, a compromise solution for the search process has been adopted: the search continues until a **sub-sequence of length N_{act}, starting with the action a_r**, is found. The sub-sequence satisfies the following condition:

There will be no violation of any operating limits if the control actions a_i, $i=(r,r+N_{act}-1)$, are performed on the power system, one after another.

To illustrate the adopted search control approach, refer again to Figure 2. Consider that operators request the search for action a_r (step 2) and that the sub-sequence length is set to $N_{act}=3$. The D&C module will consecutively launch the KB module three times (steps 3 and 6) in order to obtain three control actions, i.e., the sub-sequence $[\boldsymbol{a}_r \mid a_{r+1}\ a_{r+2}]$. The KB module is expected to search for a single action according to the actual power system state and the strategy implemented in its rules. Each action that is likely to belong to the subsequence is simulated (step 4) in order to determine if the conditions on operating limits are satisfied. If the sub-sequence $[\boldsymbol{a}_r \mid a_{r+1}\ a_{r+2}]$ is found, its first action a_r will be suggested to the operators (step 6) and the other actions may be used as a departure point in the later search process.

If an action is suggested by the operator (*validation* mode), it will be considered as the head of the sub-sequence, i.e. the action a_r. In this case, the search process continues for the actions a_i, where $i=(r+1,r+N_{act}-1)$.

Note that the requirements on the solution feasibility and the speed of finding the solution are contradictory, but both depend on the sub-sequence length N_{act}. The solution feasibility increases, while the speed of finding the solution decreases proportionally with N_{act}. Therefore, the sub-sequence length N_{act} can be considered as a parameter and should be set appropriately after a number of tests.

2.2 Reasoning Cycle

The Reasoning kernel works in terms of reasoning cycles. A reasoning cycle r is the time interval during which the Reasoning kernel performs recursive search, guided by different heuristic criteria, in order to find the sub-sequence of N_{act} control actions, starting with the action a_r (as described in Section 2.1). During the reasoning cycle, the progress of the search is shown to the operator, as well as the evolution of some power system variables (frequency, voltage, spinning reserve).

If the search is successful, the action a_r is suggested to the operator. Otherwise, in *research* mode, the information about DAFFOR's failure is shown, or in *validation* mode, the operator's action is declared invalid. At this point, the operator may ask for an explanation.

If DAFFOR has failed to suggest an action, the operator has only the possibility to continue the current reasoning cycle r in *validation* mode (i.e., the operator should suggest an action a_r). In all other cases, the operator can decide either to terminate the current reasoning cycle or to ask for a new action a_r in *research* or *validation* mode. A reasoning cycle r is terminated as soon as the operator accepts the action a_r. For DAFFOR this means that the action a_r will be performed on the power system. Then the next reasoning cycle can start.

To sum up, **the final result of a reasoning cycle r is a control action a_r that the Reasoning kernel generated and suggested to the operator, and which the operator accepted to perform on the power system**.

The Reasoning kernel, as presented so far, does not take into account real-time data (its operation is based only on simulated data, generated by the Dynamic Simulator) and can be considered as an off-line version of DAFFOR. However, its structure, operation and output were conceived having in mind the final objective: decision assistance in real time.

3. REAL TIME OPERATION

In order for the Reasoning kernel to operate on-line, it needs a coherent data set describing the actual power system state. Furthermore, the Reasoning kernel should be warned when real-time data changes in order to be able to adapt its own operation to the evolution of the power system variables. The **Real Time Update (RTUpd) kernel** enables the Reasoning kernel to follow the real-time evolution of the power system state. The results of the RTUpd kernel represent the only data available to the Reasoning kernel. The RTUpd kernel generates a data set representing the power system at a given moment, and a warning for the Reasoning kernel to adapt the operation if any unforeseen events occur.

3.1 Update Cycle

The RTUpd kernel works in terms of update cycles. An update cycle u is the time interval of constant duration T between two consecutive runs of the RTUpd kernel (Fig.3). The time the RTUpd kernel needs to get real-time data and process it for the current update cycle u is denoted by e_u. During the interval $(T-e_u)$, the RTUpd kernel is in stand-by. Since the next real-time data processing is handled during the interval e_{u+1}, the Reasoning kernel considers that data does not change during the interval $\lambda(u)=T+[e_{u+1}-e_u]$. Note that the duration e_u may differ from one update cycle to another.

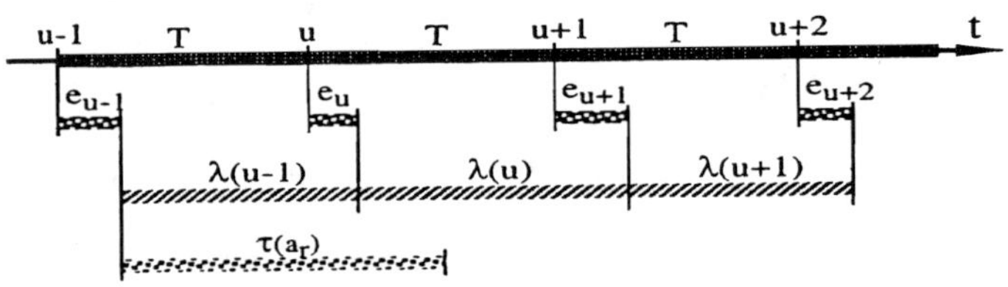

Figure 3. Illustration for update and reasoning cycles on the real-time scale

The duration of update cycles has been adopted as constant T to ease the implementation. It must be longer than or at least equal to the interval within which the RTUpd kernel processes data e_u. A variable duration of update cycles could be considered as well (e.g., interval e_u), but it could have the following consequence: if the data set generated by the RTUpd kernel changes from one update cycle to another, the Reasoning kernel might be forced to adapt its own operation too often if the update cycle is short (i.e., it would not have enough time to find a solution).

3.2 Reasoning in Real Time

Now the reasoning cycle r, introduced in Section 2.2, can be drawn on the real-time scale too (Fig.3). It can be defined as the time interval $\tau(a_r)$ whose duration is variable and depends on three factors: (1) the time the Reasoning kernel needs to find and suggest the control action a_r, (2) the moment when the operator accepts the action a_r, and (3) the result generated by the RTUpd kernel in the current update cycle. The first and the second factors were discussed in Section 2.2, the third will be addressed in Section 3.4.

As the operation of the Reasoning kernel depends on a data set, generated by the RTUpd kernel, the reasoning cycle r can start as soon as the RTUpd kernel terminates real-time data processing, i.e. in general case, after the interval e_u.

Note that the duration $\tau(a_r)$ of the reasoning cycle r may endure several update cycles (e.g., if the update cycle length T is small, if the search process is slow).

3.2.1 Expected Power System Response

From the point of view of the operator in a control centre, all control actions require time to be performed (e.g., when the operator requests a manoeuvre, such as by telephone, a certain time is necessary for a manoeuvre to be physically executed by another person). This time depends on the type of the action and will be denoted by $\tau_{operator}(a_r)$. Furthermore, each executed manoeuvre results in some transient events in the power system during a certain time, denoted by $\tau_{ees}(a_r)$. After that, the power system response to the action a_r should be within the time interval $\tau_{response}(a_r)=\tau_{operator}(a_r)+\tau_{ees}(a_r)$.

This time interval of the power system response is taken into account in the Reasoning kernel. The time of the operator's manoeuvre $\tau_{operator}(a_r)$ is associated as a parameter to every possible control action, and the time for transient events $\tau_{ees}(a_r)$ is computed by the Dynamic Simulator. Note that in reality, the exact moment of the power system response cannot be identified, so the choice of parameters has been such that **$\tau_{response}(a_r)$ represents an upper time limit** (i.e., the most pessimistic case).

Suppose now that the operator has accepted the suggested control action a_r, which terminated the reasoning cycle r at moment $t_{rc\_end}(a_r)$ within the update cycle u, as shown on Figure 4.a. The action a_r has been found in the Reasoning kernel taking into account the maximal response interval $\tau_{response}(a_r)$. In other words, the response interval $\tau_{response}(a_r)$ has been simulated during the reasoning cycle r, and therefore is represented on a separate simulation time axis t' (Fig.4.b).

As the operator has accepted the suggested action a_r at the moment $t_{rc\_end}(a_r)$, the **simulated** maximal response interval $\tau_{response}(a_r)$ for the action a_r may be used to estimate the **real** time by which the steady state after the action a_r would be reached. Therefore, the power system steady state after the action a_r can be expected to be $t_{exp\_ss}(a_r)=t_{rc\_end}(a_r)+\tau_{response}(a_r)$. This is represented as dashed links between Figures 4.a and 4.b.

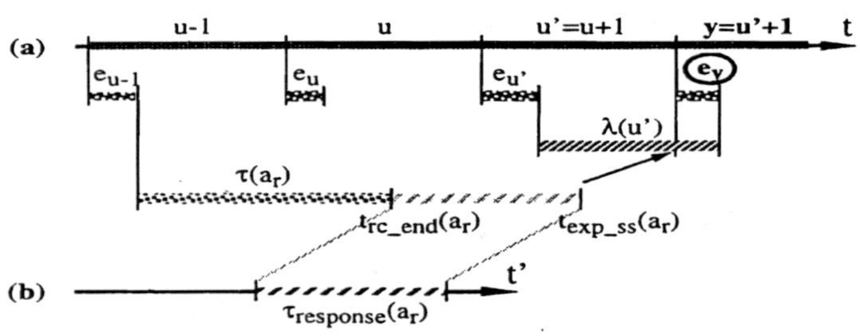

Figure 4. Determination of the moment when the power system steady state after the control action a_r can be expected. The axis t represents real time, and the axis t' corresponds to the simulated time

Let us denote u' as the update cycle within which falls the expected moment $t_{exp\_ss}(a_r)$ of the new steady state. For the given example $u'=u+1$, but in general case u' may be any update cycle that starts after the current one or the current update cycle itself ($u'\geq u$). Since data does not change during the interval $\lambda(u')$, the expected steady state can be identified at latest in the next update cycle, $y = u'+1$, within the interval e_y.

The update cycle y will be referred to as the expected update cycle, and the corresponding simulated steady state after the action a_r will be denoted *Sim(y,a_r)* and referred to as the expected data set. The idea is to use the simulated power system response to the action a_r in the RTUpd cycle for comparison with the real power system response, in order to identify the unforeseen events.

It can be concluded that the expected data set $Sim(y,a_r)$ can only be generated if the reasoning cycle r is terminated (i.e., the operator has informed DAFFOR that the suggested action a_r will be executed on the power system). The expected update cycle y may therefore have one of the following values:

- if the reasoning cycle r terminated, the expected data set $Sim(y,a_r)$ has been created and the expected update cycle y is computed as explained above
- if the reasoning cycle r is still running, there is nothing yet to expect and the update cycle y is set to 0

Both expected data set and update cycle are results of the terminated reasoning cycle, and represent one of the sources of information for the RTUpd kernel.

3.3 Real Time Update (RTUpd) kernel

The task of the RTUpd kernel is to generate a coherent power system state ***Real(u)***, which will be referred to as the **coherent data set**, in the current update cycle u. Since the Reasoning kernel operation is completely based on this data, it is very important that the coherent data set *Real(u)* be as close as possible to real-time data. Therefore, in addition to the generation of the coherent data set ***Real(u)***, the RTUpd kernel sets the value of a flag, denoted by ***close***, which controls the operation of the Reasoning kernel (this will be discussed in Section 3.4).

In order to generate the coherent data set *Real(u)* and set the value of the flag *close* for the current update cycle u, the RTUpd kernel has three inputs, as shown on Figure 5:

- power system real time data (switch status, tele-measurements) and information entered by the operator (step 8),
- its own output generated in the previous update cycle, i.e. the last coherent data set *Real(u-1)* (step 10'), and
- Reasoning kernel's output, described in Section 3.2.1, generated in the last terminated reasoning cycle, i.e. the expected update cycle y and data set $Sim(y,a_r)$ (step 7').

Step numbers marked by an apostrophe denote RTUpd kernel feedback inputs from DAFFOR, as opposed to those from external world (i.e., power system real time data). The above three inputs are processed in the RTUpd kernel sequentially, in two passes. In the first pass, the RTUpd kernel creates the coherent data set *Real(u)* and determines whether there have been some changes in the real-time data since the previous update cycle. The diagnosis from the first pass is then used together with the expected data set $Sim(y,a_r)$ in the second pass in order to identify an event (if any) and consequently to set the flag *close*.

3.3.1 RTUpd kernel: Pass 1

The RTUpd kernel first gets real-time data from the power system and the information provided by the operator (step 8). The information provided by the operator actually comes from the devices external to SCADA (telephone, radio, synoptic map, printer) and may be threefold: (a) switch status not available in SCADA, (b) corrections of invalid switch status, and (c) unavailability of a power system component (transformer, switch, etc.). This information can be introduced by the operator at any time, but will only be processed at the same time as information coming from the power system. If some real-time data is missing, the corresponding items from *Real(u-1)* will be assumed (step 10'). The real-time data set obtained in this way is the coherent data set *Real(u)* (steps 11, 9).

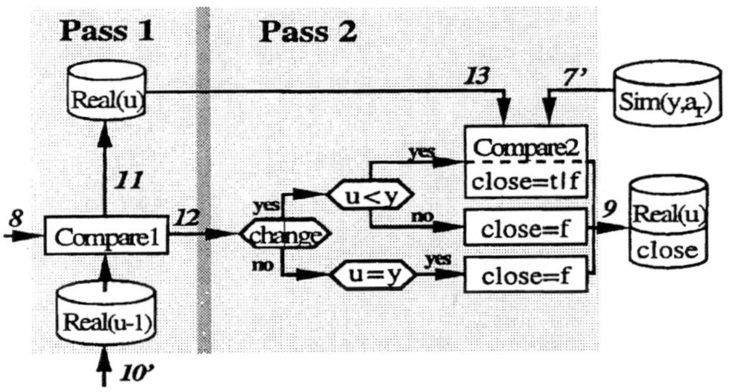

Figure 5. RTUpd kernel inputs, data processing and output. The flag *close* may be set to *True* (t) or *False* (f); its default value is *Ignore*

The coherent data set *Real(u)* is compared to *Real(u-1)* (function *Compare1*) in order to determine whether the power system state has changed from the previous update cycle or not (step 12) and will be replaced with *Real(u)* only if the data has changed.

3.3.2 RTUpd kernel: Pass 2

As discussed in Section 3.2.1, if there is an ongoing reasoning cycle *r*, the expected update cycle *y* is set to *0*. For the RTUpd kernel it means that the operator has not executed on the power system an action that is **known** to DAFFOR. Otherwise the reasoning cycle *r* has terminated, and this indicates to the RTUpd kernel that the operator has accepted **DAFFOR's suggestion** and executed the action a_r on the power system. If this is the case, the expected data set $Sim(y,a_r)$ has been generated and the RTUpd kernel will try to identify the new steady state due to the action a_r in all the following update cycles *u* that start earlier than, or simultaneously with the expected update cycle *y* (i.e., while $u \leq y$).

The following list explains the conditions under which the RTUpd kernel changes the default value (*Ignore*) of the flag *close* (step 9). Refer to Figure 5.

a) If data **has not changed** since the previous update cycle and the current update cycle is the expected one ($u=y$), then one of the following is true:
- the operator has executed the action suggested by DAFFOR, but the response from the power system did not arrive before the expected update cycle *y* (e.g., the switch position has not changed after the manoeuvre), or
- although the action suggested by DAFFOR had been accepted (in order to be executed on the power system), the operator did not execute it at all, for whatever reason.

Since some changes were expected in any case, but were not identified, the flag *close* is set to *False*

b) If **data has changed** since the previous update cycle and the current **reasoning cycle has not terminated** ($y=0, u>y$), then the operator has not executed an action known to DAFFOR, but there have been some changes for some other reason (e.g., the switch has been operated without a manoeuvre requirement, i.e., by a protection device). In this case, the flag *close* is set to *False*

c) If **data has changed** since the previous update cycle and there is a **terminated reasoning cycle** ($y \neq 0, u \leq y$), then the operator has executed the action a_r suggested by DAFFOR, and the power system response to this action can be already expected. Therefore, the generated coherent data set *Real(u)* will be compared to the expected data set $Sim(y,a_r)$ in the function *Compare2* (steps 13 and 7'), and the result may be twofold:

c.1) if the generated coherent data set differs significantly from the expected data set, the flag *close* is set to *False*

c.2) if these two data sets are found close enough, the flag *close* is set to *True*

Note that in Pass 2 the RTUpd kernel identifies the unforeseen events (all above-mentioned cases except **c.2**). From its standpoint, the unforeseen events are any events which change the power system data **except** the action validated by DAFFOR, and accepted **and** executed by the operator.

At present, the functions *Compare1* and *Compare2* take into account only the topology changes, computed on the basis of the switching status. As the operating status of the power system elements have discrete nature (open/closed), it is easy to compare two power system states topologically. The terms "close enough" and "significantly different", used in the points **c.1)** and **c.2)** above, in this case mean simply the equality and non-equality of the states, respectively. However, this is not sufficient for the creation of a reliable enough coherent data set. The tele-measurements should also be considered in comparison functions, which is much more complicated to manage. It would be necessary to determine the threshold values for the continious power system variables (powers, frequency, etc.) which could overcome the simplicity of DAFFOR's models compared to the real power system. In this case, two states are "close enough" if all the system variables are inside the determined threshold values, and "different" otherwise. The problem of determination of the threshold values is very complex and demands further research.

Another factor that strongly influences the reliability of the coherent data set generated by the RTUpd kernel is the power system's real-time data itself. The data on the power system is gathered by the SCADA function and treated inside the state estimator. During the restoration process, or at least in its first stage, the information available to the state estimator is likely to be incomplete (e.g., inobservability), which might put in question the validity of its output: real-time data as it is seen from the standpoint of the RTUpd kernel. Therefore a special, restoration-dedicated state estimator should be available.

3.4 DAFFOR's Real-time Operation

Up to now, DAFFOR's kernels have been discussed more or less separately. Figure 6 shows them placed together, for the purpose of describing the overall operation of DAFFOR. Figure 3 can be checked for a synchronisation between the update and reasoning cycles.

Suppose DAFFOR has entered the update cycle u. The RTUpd kernel would then get the power system real-time data and information entered by the operator (step 8), the coherent data set it had generated in the previous update cycle (step 10') and the expected data set, if there is one (step 7'). It generates the coherent data set *Real(u)* and sets the flag *close* to one of the values *[True,False,Ignore]* (step 9). If the current coherent data set has changed since the previous update cycle, it is saved to be used in the next update cycle (step 10'). All these operations are performed within the interval e_u. During the interval $(T-e_u)$, the RTUpd kernel is in standby. It resumes by executing the same operations in the next update cycle $(u+1)$.

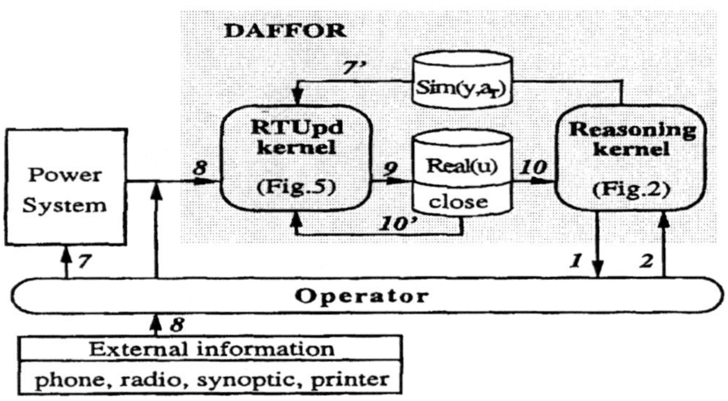

Figure 6. The real-time control flow in the DAFFOR

Now suppose that there is an ongoing reasoning cycle r during which the Reasoning kernel searches for the control action to suggest, or assesses the action suggested by the operator. During its operation, the Reasoning kernel will check the value of the flag *close*, set by the RTUpd kernel as explained in Section 3.3.2. **The value of the flag *close* determines the behaviour of the Reasoning kernel** as follows:

- if *close = False*, the operation of the Reasoning kernel must be adapted to the new situation in the power system, i.e. the **current** reasoning cycle r is **restarted** with new data in order to search for another control action a_r (cases **a**, **b** and **c.1**)
- if *close = True*, the **new** reasoning cycle can be **started** (case **c.2**)
- if *close = Ignore*, the Reasoning kernel continues its operation

During a reasoning cycle, the operator interacts with DAFFOR (steps 1 and 2). When the operator accepts the action suggested by DAFFOR, the expected data set is generated (as explained in Section 3.2.1) which will be used by the RTUpd kernel (step 7').

Finally, it is important to note that the operator is the only one allowed to execute any action on the power system (step 7). The only direct link between the power system and DAFFOR (step 8) is unidirectional.

4. IMPLEMENTATION

As was shown in Sections 2 and 3, each of DAFFOR's kernels has its own cycle and its own data set. The RTUpd kernel is executed in a synchronous manner, while the operation of the Reasoning kernel strongly depends on the RTUpd kernel's diagnosis. Furthermore, each of the data sets are initialised by their own kernel, but they must be available to other kernels in read mode. Consequently, DAFFOR is implemented as a programme with two processes which share data, as shown in Figure 7.

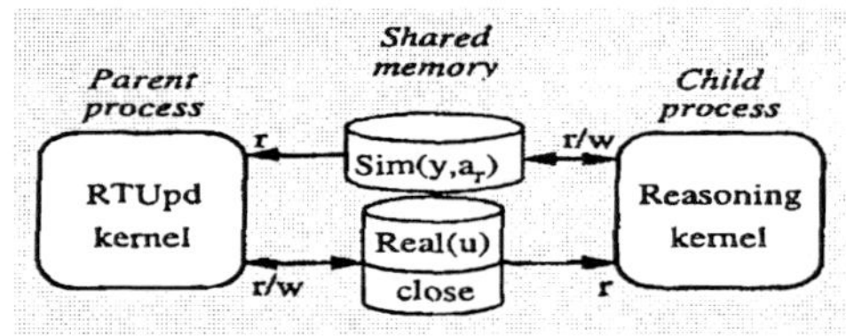

Figure 7. The DAFFOR's kernels are two processes that share data. Each data set can be read (r) by both kernels, but can be written (w) only by its own kernel

The **RTUpd kernel** is the parent process which initialises the data set *Real(u)* and the flag *close*. It is implemented entirely in the C programming language. It controls its child process execution by means of the flag *close*.

The **Reasoning kernel** is the child process which initializes the data set $Sim(y,a_r)$. Its Dialogue&Control and Knowledge-Based modules are implemented in an NEXPERT *OBJECT*™ shell, while all the functions that are called from them (including the Dynamic Simulator) and written in C (ANSI standard). The Reasoning kernel is controlled by its parent process via the flag *close*.

Data sets *Real(u)* and $Sim(y,a_r)$ are C structures and are implemented in the shared memory segment, as well as the flag *close*. All of these may be read by both kernels.

DAFFOR has been developed and is running on an HP workstation, under the UNIX operating system.

5. VALIDATION

As was mentioned in the introduction, DAFFOR should run in a transmission power system control centre. Before the installation in a real environment, it is coupled for validation purposes to the French National Utility's Operator Training Simulator (EDF OTS) [9]. The EDF OTS plays the role of the power system, with data from a regional control centre. The validation architecture is shown in Figure 8.

The real-time data supplied by EDF OTS corresponds to data that is available by SCADA, i.e., tele-measurements and switching status. In addition, the operator can introduce some unforeseen events either in the scenario file, or online via the user interface of EDF OTS. In this manner, DAFFOR can be tested for unforeseen events handling as well. The interprocess communication is through UNIX sockets.

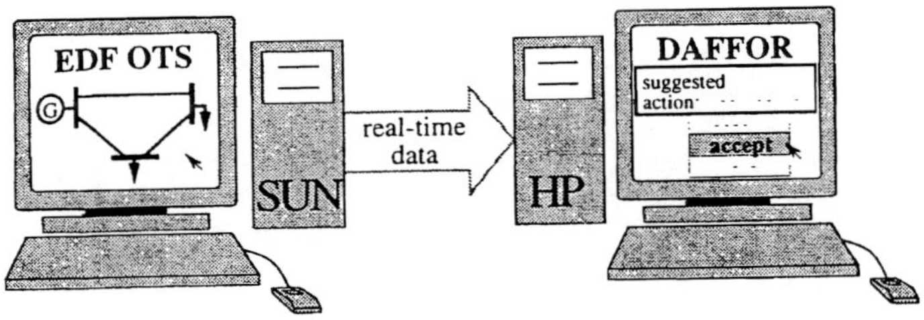

Figure 8. Validation configuration

It is important to note that DAFFOR and EDF OTS are two computer applications which run on different machines. The relevant information from EDF OTS is made accessible to DAFFOR in read mode only, which confirms the principle that only the operator can perform an action on the power system.

Comparing the **validation** architecture (Figure 8) and the **real environment** architecture (Figure 1), it can be concluded that moving from the former to the latter would be quite easy, requiring changes only in the real-time data access functions of DAFFOR's Real Time Update kernel.

6. CONCLUSION

In this paper, a Decision Aid Function FOr Restoration (DAFFOR) of transmission power systems in real time has been proposed. DAFFOR consists of two kernels. The Reasoning kernel (Section 2) has the task of assisting the operator during the restoration process, taking into account the real-time dynamic behaviour of the power system. This behaviour is analyzed by the Real Time Update (RTUpd) kernel in order to generate a coherent data set, representing the power system state in a given moment (Section 3.3). One of the sources of information for the RTUpd kernel that enables it to identify the unforeseen events is the simulated expected power system state, resulting from the execution of an action proposed by the Reasoning kernel (Section 3.2). Thanks to the coherent data set, the Reasoning kernel can adapt its own operation to the evolution of the power system variables (Section 3.4).

Taking into account the present state of DAFFOR, there are three main directions for future work:

1) More complex comparison functions in the RTUpd kernel should be developed, in order to obtain more reliable representation of the real-time data for the Reasoning kernel (Section 3.3.2).
2) Since DAFFOR is supposed to run during an abnormal power system state (restoration), a dedicated state estimator needs to be available (Section 3.3.2).
3) After intensive testing within an OTS environment (Section 5), the operational version of DAFFOR is to be developed.

ACKNOWLEDGEMENTS

This work has been supported by the Swiss National Foundation for Energy Research (NEFF) and Électricité de France (EDF).

REFERENCES

[1] J.J.Ancona, "A Framework for Power System Restoration Following a Major Power Failure", *IEEE Transactions on Power Systems,* vol.10, pp. 1480-1485, 1995
[2] M.M.Adibi, R.J.Kafka, D.P.Milanicz, "Expert System Requirements for Power System Restoration", *IEEE Transactions on Power Systems,* vol.9, pp. 1592-1600, 1994
[3] T.Sakaguchi, K.Matsumoto, "Development of a Knowledge Based System for Power System Restoration", *IEEE Transactions on Power Apparatus and Systems,* vol.PAS-102, n.2, pp. 320-329, 1983
[4] K.Matsumoto, T.Sakaguchi, R.J.Kafka, M.M.Adibi, "Knowledge-Based Systems as Operational Aids in Power System Restoration", *Proceedings of the IEEE,* vol.80, n.5, pp. 689-697, May 1992
[5] "A Survey of Expert Systems for Power System Restoration", Task Force 38.06.04 Report, TF Convenor B.Heilbronn, *Électra,* n.150, pp. 87-105, October 1993
[6] S.König, L.Mohr, G.Krost, D.Rumpel, U.Spanel, "An Expert System in Training Power System Restoration at the Stadtwerke Duisburg AG", *Proc. 1994 ISAP,* vol.2, pp.631-638, September 1994
[7] K.Kuroda, Y.Takegoshi, K.Yamashita, M.Ito, S.Fukui, "Verification of a knowledge-based restoration guidance system in a local dispatching centre", *Electrical Power & Energy Systems,* vol.15, n.3, pp. 185-192, 1993
[8] T.Kostić, "Dynamic Simulator as a Part of an Expert System for Power System Restoration", *M.Sc. Thesis,* Faculty of Electrotechnics, Belgrade University, Belgrade, 1994
[9] C.Jeanbart, Y.Logeay, M.Musart, "EDF's Dispatcher Training Simulator", *Cigré Conference,* August 28 - September 3, 1988, paper 39-12

PART VII

Distribution System Restoration

I. General Discussion

THIS part covers restoration techniques used in both distribution and subtransmission systems. A distribution area is typically serviced by several substations to which one or more primary feeders are radially connected. To improve reliability, adjacent feeders are connected through normally open tie switches. Distribution system restoration includes models for cold load pickup, transformer thermal behavior, and methods for determining optimal energization order. The distribution restoration involves a combination of analytical, heuristic, combinatorial, and knowledge-based systems. The emphasis on any one of these techniques has been a matter of choice. For example, in some papers optimization techniques have been selected which include analytical models (e.g., power flow) and to which special practical rules have been added. In other papers, a knowledge-based system has been selected to which heuristical and numerical algorithms have been attached.

II. Discussion of Reprints

7.01 Adaptation of the Simplex Algorithm to Modeling of Cold Load Pickup of a Large Secondary Network Distribution System, 1983, p. 565

This paper proposes a linear programming-based algorithm to determine the load composition of secondary distribution networks during cold load pickup. The data required by the model is a survey of connected loads of the same type used for load growth analysis. Upper and lower limits for all types of loads are based on engineering judgment. The model yields the composition of the load under worst-case conditions. The method has been applied to a commercial load to obtain a realistic worst-case model. To validate the mathematical model, a field model based on information gathered during on-site visits to representative enterprises has also been developed.

7.02 Statistical Synthesis of Physically Based Load Models with Applications to Cold Load Pickup, 1984, p. 569

An equivalent load model has been obtained in three steps: (1) modeling of elementary component loads, (2) classification into homogeneous groups, and (3) aggregation of the loads in each group. The functional model developed generates the operating state (e.g., the room temperature in the case of a space heater) of the component by solving a differential equation model, which represents the dynamics of the energy flow into and out of the component. Depending on the state yielded by the model, a control mechanism is activated (on/off). The fraction of loads in a group that are on at any particular time is then obtained by the solution of a statistical model based on the Fokker-Planck equation. A numerical example given shows how the fraction of space heaters that are on varies before, during, and after the outage.

7.03 Effects of Cold Load Pickup at the Distribution Substation Transformer, 1985, p. 577

The paper discusses a simulation model of the thermal behavior of thermostatically controlled loads. The simulation involves the period of time from when restoration begins until load diversity is regained. Transformer loading, top oil and hot spot temperatures, and insulation loss of life can be computed by the simulator for a variety of operating conditions. The degree of penetration of electric space heat is a major concern. Simulation results have been compared with data from actual outages with good agreement. The paper also analyzes the possible effects of cold load pickup on transformer protective equipment.

7.04 An Expert System Operational Aid for Restoration and Loss Reduction of Distribution Systems, 1987, p. 584

This paper presents a rule-based expert system for distribution system restoration involving switching actions for fault isolation, load transfer between adjacent feeders, and restoration of service to unfaulted loads that are temporarily out-of-service. The restoration plan is established in stages. First, groups of load sections are formed and tentatively assigned to adjacent feeders. Power flow simulation is then used to detect possible overloads and voltage limit violations. If violations are detected, load transfer is attempted considering one load section at a time. Given the radial network constraints, power flow is solved only for the relevant parts of the network. As a byproduct, the proposed method can also minimize the losses.

7.05 A New Algorithm for Service Restoration in Distribution Systems, 1989, p. 591

The authors consider that (1) optimal switching sequences obtained via combinatorial methods may be unfamiliar to distribution system operators, which makes their practical implementation more difficult, and (2) the load levels are never accurately known during restoration time, which allows for a certain degree of approximation in searching for the best switching sequence. Sensitivity analysis is then proposed as a means of selecting effective load transfers, that is, transfers that reduce constraint violations. The objective is to restore as much load as possible. Three types of constraints are considered: line/transformer capacity, voltage drop, and radial network constraint. It is further assumed that (a) the maximum voltage drop occurs at the end points of the feeders and (b) the maximum loadings occur at the roots of the feeders serving a group of load sections.

7.06 Heuristic Search Approach to Distribution System Restoration, 1989, p. 599

The paper considers that the load at each feeder zone, the primary feeder loading limits, the post-fault network configuration, and the feeder section impedances are known. Problem variables are the primary feeder loadings and the load voltage magnitudes. The decision variables are the status of switching devices. The problem is to distribute the loads among the feeders such that the load imbalance is minimized. The problem is solved by a branch-and-bound approach based on a binary tree search. It is used in a depth-first search in which an initial solution is obtained that closely follows the practical procedures normally used by system operators. Improved alternative solutions are then searched for. A list of all solutions found by the algorithm is presented to operators.

7.07 Distribution Transformer Overloading Capability under Cold-Load Pickup Conditions, 1990, p. 606

A guide for cold load pickup is proposed. Simulation has been carried out based on a typical cold load profile (load current vs. time) to determine how the maximum hot spot temperature and the insulation loss of life vary with the initial load of the system and for different outage durations. The resulting graphs are used for a given maximum hot spot temperature, the outage duration, and the external temperature. From these graphs, one can determine the corresponding initial load level and for a given value of loss of life. The basic transformer data have been obtained in a test program, which involved metering the thermal behavior of 18 transformers under common load pickup conditions.

7.08 MARS: An Aid for Network Restoration after a Local Disturbance, 1992, p. 614

A discrete event simulator is used to determine which parts of the network are energized after a postulated restoration action. Components such as switching devices are represented by logical transfer functions, which are activated by events. The simulator visits a list containing all events. The event with the highest priority is activated, and the corresponding network topology changes are performed. When insufficient information is available to determine the state of the system, an alarm is generated, and the subsystem planning will yield a tentative restoration plan. In this case the operator proceeds by trial-and-error until the actual network configuration is correctly identified. Then the discrete event simulator activates a power flow simulator that checks the limits and operating conditions of a postulated configuration. If a limit violation is detected, an alarm is sent to the planning subsystem. The planning module is a knowledge-based system. All relevant data are kept in a blackboard. (*Note*: MARS is an acronym for a French Expert System used in restoration of distribution systems.)

7.09 An Analytical Approach for Step-by-Step Restoration of Distribution Systems Following Extended Outages, 1994, p. 620

Due to the loss of diversity after a long outage, transformer loading limits can be reached when only a fraction of the total load is restored. This is typical of substation transformers that operate near rated capacity under normal conditions. Step-by-step restoration is then necessary. The procedure benefits from the fact that the served loads will regain diversity with time after restoration and the total load will decrease gradually. This will allow service to the remaining loads. During cold load pickup, the loads are then represented by a delayed exponential function. Overloading of transformers for a short duration is also considered in the model. The loading limits of transformers are determined by a thermodynamic model that takes into consideration the variation of critical temperatures with time. Under certain simplifying assumptions, an explicit analytical expression has been derived for the restoration time. Both the total time of restoration and the customer interruption time are affected by the order in which the sections are restored. The objective is to minimize either one, taking into consideration the dynamics of both the load diversity and the loading limits of distribution substation transformers.

7.10 Knowledge-Based Behavior Interface: Its Application to Power Network Restoration Support System, 1996, p. 627

An expert system has been described to assist system operators during restoration. The authors first recognize that the behavior of skilled operators cannot be reduced to a set of rules. Hence, the behavior of human experts has been analyzed and decomposed into skills, rules, and knowledge. The proposed expert system is used to restore any part of the network, from subtransmission systems up to the 500 kV trunk lines. Their restoration plan preparation consists of: (1) identification of the post-fault network configuration, (2) search for a restoration path, (3) determination of a hypothetical plan, (4) elimination of overloads in the hypothetical plan, (5) revision of the restoration path (which may become necessary as a result of the actions used to eliminate predicted overloads), and (6) preparation of the restoration procedures, which provides the steps necessary to take the system from its current state to the target configuration.

7.11 An Object-Oriented Approach for Implementing Power System Restoration Package, 1996, p. 634

The restoration package for the Taiwan Power Company, an island system, is developed. Object-oriented programming is used to facilitate both software and database maintenance. The package assists system operators in: (1) identification of the status of the network after a blackout, (2) restart and supply station service to plants, (3) removal of all sources of reactive power and deactivation of automatic load shedding, (4) coordination of the timings between power generation startup and load pickup, (5) hierarchical energization of the transmission network and energization of the 161 kV system before the 345 kV system, (6) picking-up load in large blocks provided that the low frequency constraint is met, (7) reintegration of the main power network using a preestablished switching sequence (determined off-line), and (8) reduction of standing phase angles to allow safe switching operations (closed loops).

7.12 Distribution System Design Optimization for Cold Load Pickup, 1996, p. 641

This paper formulates the design of a distribution system as an optimization problem in which both cold load pickup and transformer loss of life are taken into account. The decision variables considered are the substation transformer size(s) and the number of sectionalizing switches. The objective is to minimize the annual cost function representing cost of transformers, cost of switches, cost of energy interruption, and cost of transformer loss of life. The resulting combinatorial problem has been solved by enumeration. The proposed formulation is extended to include voltage drop and over current constraints. Further research is being pursued to develop more efficient optimization techniques to include the specific characteristic of this problem.

7.13 Distribution Feeder Reconfiguration for Service Restoration and Load Balancing, 1997, p. 647

The primary feeders are reconfigured to balance the load and to minimize the losses. The proposed algorithms are based on a mix of optimization techniques, heuristic rules, and fuzzy logic. The fuzzy component makes the approach less sensitive to errors (increased robustness). The objective is to perform restoration in the shortest period of time with the minimum number of switching operations. Remote controlled switches are given high priorities. Other switches are considered only when the remote controlled switches are not available. If the adjacent feeders cannot pick up the loads as required, secondary feeders are used. A relay coordination program is used to verify whether the protection requirements are met for a proposed restoration sequence.

7.14 Application of Expert System to Power System Restoration in Sub-control Center, 1997, p. 653

A rule-based expert system restoration is presented to be conducted by the regional control center. The outages considered are caused by local faults. The knowledge base is organized hierarchically, involving objects such as cells (e.g., a bus, a line, etc.) and subsystems (groups of cells). Cells are connected by switching devices (breakers and disconnect switches). A restoration plan is obtained, carrying out a number of simulations involving alternative switching sequences. Four types of rules are considered: energization rules, overload management rules, voltage correction rules, and flow-control rules. The flow-control tree considers that a restoration plan is satisfactory when (1) the entire network is energized, (2) there is no overload, and (3) there are no voltage violations. Overload management is carried out by switching operations or load shedding. Voltage correction is performed via shunt capacitor switching. The numerical (power flow) simulations are embedded in the expert system. To keep the numerical effort manageable, sensitivity analysis is used to predict the effect of postulated switching operation on system state.

ADAPTATION OF THE SIMPLEX ALGORITHM TO MODELING OF COLD LOAD PICKUP OF A LARGE SECONDARY NETWORK DISTRIBUTION SYSTEM

Melih H. Yolcu — Zivan Zabar, S. Member — Leo Birenbaum, S. Member

Polytechnic Institute of New York
Brooklyn, N.Y.

Seymour A. Granek, Member

Con Edison Co.
New York, N.Y.

Abstract - This paper describes a mathematical approach to the determination of the electrical load composition of a secondary network during cold-load re-energization. It is intended to answer two questions: What fraction of the normal steady-state demand must be restored at the time of re-energization in the most pessimistic case? What is the composition of this worse case restoration load?

The answers were obtained by making use of the Simplex Algorithm, a mathematical, but physically based, optimization procedure. The input data consisted of a customer survey of connected-load; maximum demand for this customer sample; experience-based power requirement limits as a function of the time-of-day.

For comparison with the mathematically-based results, on-site visits were made to representative commercial enterprises. The results were used to construct, independently, a second cold-load restoration model. The two models were surprisingly close.

INTRODUCTION

The re-energization of a large electric secondary network distribution system poses a severe problem for a utility [1-6]. The entire load area has to be re-energized at the same time since the secondary distribution network cannot be sectionalized. The duration and magnitude of the inrush current must be known to properly set network feeder relays to withstand the starting period. Also, the load tap changers on substation transformers must be set prior to energization to provide safe margins for both starting and running voltages. It is also necessary to determine the maximum load that the primary network should supply so that the secondary network can be re-energized.

The initial phase of power restoration depends on three major factors: the actual connected load, individual equipment start-up characteristics and the loss of load diversity. All three factors enter into the construction of a model that satisfactorily accounts for the starting phase.

The first and crucial step in the formulation of such a cold-load pickup model is to identify the electrical equipment which is supplied by the secondary distribution network. This information may then be used to make a realistic estimate of the actual load composition at the time of power restoration. Since there is no practical way for a utility to control equipment usage by the customer, it must be prepared for the worst start-up condition. Therefore the load model should be based on this pessimistic start-up condition. To construct the model it was found possible to use a mathematical linear optimization method called the Simplex Algorithm [7]. As input, the method requires a survey of connected load such as used for load growth analysis. Some engineering experience is needed to select upper and lower limits for various categories of connected load. The output of the program is a description of the worst-case on-line steady-state load composition. This breakdown is then used to determine system re-energization requirements.

In order to compare this prediction of the (steady-state) load that must be re-energized with the actual breakdown of electrical loading that might be connected at the time of power restoration, special on-site visits were made to many representative enterprises and locations. From these visits, a realistic field model was constructed. The steady-state cold load pickup model forms the basis for calculating the duration and magnitude of the inrush current by a method described elsewhere [3]. The results are presented below.

83 WM 180-7 A paper recommended and approved by the IEEE Transmission and Distribution Committee of the IEEE Power Engineering Society for presentation at the IEEE/PES 1983 Winter Meeting, New York, New York, January 30-February 4, 1983. Manuscript submitted August 27, 1982; made available for printing November 19, 1982.

GENERAL DESCRIPTION OF THE METHOD

There exist a great variety of individual types of loads that add up to the known steady-state overall system load. In the work described here, all of the electrical equipment which was connected to the line prior to the outage was conservatively assumed to be still on-line at the time of power re-energization, excluding manual-restart loads. These loads would normally be restarted later, manually, in a random fashion, following the initial power restoration.

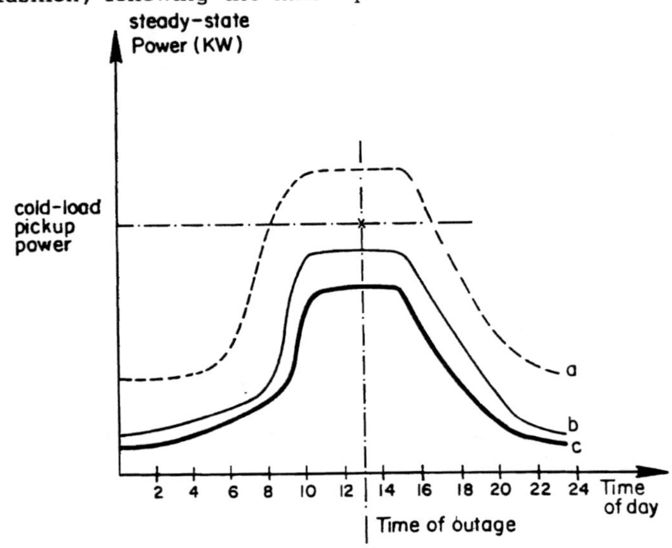

Fig. 1: General load cycle demand-curve. a) Max. connected load, b) Total demand, c) Pre-outage load.

Reprinted from *IEEE Transactions on Power Apparatus and Systems,* Vol. 102, No. 7, pp. 2064–2068, July 1983.

The graphical presentation in Figure 1 explains succinctly the methodology adopted. The total demand curve (solid line) of a secondary network is given for an average hot summer day [8]. The demand corresponding to the manual restart equipment is subtracted from it to obtain a modified steady-state pre-outage curve (bold solid line). If it is conservatively assumed that all diversity is lost, and that every piece of equipment constituting the total load has to be re-energized, then, it is possible to draw a curve showing the maximum possible steady-state restoration load (dashed line) as a function of the time of day. The applied method chooses a realistic worst-case steady-state load composed of the various types of equipment on-line at the time of the outage. Then it converts this load into a system start-up description. This steady-state cold-load is located between the lower and upper curves.

APPLICATION OF THE SIMPLEX ALGORITHM

A variable x_i representing the power demand in kW and a re-energization coefficient c_i are assigned to each of n types of connected loads (on-line equipment). The objective of the algorithm is to search for a set of x_i that maximizes the total demand, subject to constraint equations for each variable and for their sum. The demand equation is

$$Z = \Sigma_n c_i x_i \qquad (1)$$

where:

Z - Worst-case restoration load. All diversity is assumed to be lost [kW].

x_i - Pre-outage power demand of a load type, including diversity [kW]

c_i - "re-energization coefficient" (mathematically termed the objective coefficient).

The re-energization coefficient is defined as:

$$c_i = \frac{1}{d_i} \qquad (2)$$

where:

$$d_i = (\text{diversity-factor}) * (\text{use-factor}) =$$

$$= \frac{\text{coincident demand}}{\text{actual connected load}} * \frac{\text{actual connected load}}{\text{maximum connected load}}$$

Experience permits a reasonable estimate to be made of the value of each d_i. For example: for the noon hours of an average summer day, where the system has peak demand, window type air conditioners may have about 0.7 diversity factor, and a 0.9 use-factor.

Each x_i variable is constrained by practical considerations to a limited range of values. For example, one can assume that not all of the window type air conditioners are in operation (actually connected to the line) at the same time; perhaps at most only 90%. On the other hand, not all of them are off at the same time, either; perhaps the minimum number in use is 50%. These practical considerations provide a set of constraint equations:

$$\begin{gathered} x_{1min} \le x_1 \le x_{1max} \\ x_{2min} \le x_2 \le x_{2max} \\ \cdot \quad \cdot \quad \cdot \\ x_i = x_{ieq} \\ \cdot \quad \cdot \quad \cdot \\ \cdot \quad \cdot \quad \cdot \\ x_{nmin} \le x_n \le x_{nmax} \end{gathered} \qquad (3)$$

The values x_{ieq} correspond to those loads whose values are known precisely.

The steady-state power of the equipment which was connected to the line prior to the outage, and is assumed to be re-energized, cannot be larger than the maximum demand P_{max}. Also, the manual-restart equipment is excluded during cold load pick up response. Therefore:

$$\Sigma_n x_i \le P_{max} - P_{man} \qquad (4)$$

where P_{man} is the steady-state power required to operate the manual restart equipment. For example, for a commercial load, P_{man} includes equipment such as central air conditioners, elevators, escalators, etc.

The formulation of the linear optimization method is now complete. Mathematically, the problem to be solved may now be phrased more precisely: Maximize the objective function Z defined by eq. 1, subject to the constraints on the variables x_i, given by eqs. 3 and 4. The solution yields the worst-case steady-state power demand (Z) and the individual pre-outage powers (x_i) of each load type.

It is useful to define a quantity P_i for each load type. This is the steady-state power (no diversity) needed at the time of re-energization, corresponding to the worst case-condition:

$$P_i = c_i x_i \qquad (5)$$

The formulation of the cold-load pickup model is now complete. The load composition is represented by the values P_i, in kW, the individual equipment powers at the time of system startup.

COMMERCIAL LOAD IN THE NYC AREA

The above method was applied to the commercial portion of the Consolidated Edison load in the New York City (NYC) area. A 500 customer survey that had been carried out in 1980-81 was available [8]. This connected-load survey was conducted by the utility for an electric load characteristic study. The survey results are summarized in Table 1 (Enterprise Identification) and in Table 2 (Equipment Identification).

Table 1
Enterprise Identification
(% of total demand of commercial load)

No.	Name	Percent
1.	Offices	39.3%
2.	Department Stores	21.6
3.	Schools	8.7
4.	Banks	7.7
5.	Small Stores	5.2
6.	Hospitals	4.6
7.	Supermarkets	4.6
8.	Restaurants	3.5
9.	Hotels	3.4
10.	Others	1.4
		100.0%

Table 2
Equipment Identification
(% of total connected commercial load)

No.	Type of Load	Percent
1.	Fluorescent lighting	35.6%
2.	Incandescent lighting	4.5
3.	Central air conditioners	35.6
4.	Window air conditioners	1.7
5.	Refrigerators	2.8
6.	Fans	0.3
7.	Pumps	2.2
8.	Cooking equipment	0.6
9.	Water heaters	0.1
10.	Elevators & Escalators	6.4
11.	Office equipment	0.7
12.	Computers	2.4
13.	X-ray machines	0.1
14.	Television sets	0.1
15.	Miscellaneous	6.9
		100.0%

The data of this table (excluding manual-restart loads such as central air-conditioners, elevators, escalators, main computers and X-ray machines) together with practical constraints and re-energization coefficients (see Table 3) were used as input data for the optimization program. The results obtained are listed in Table 4 in the right-hand column.

Table 3*
Input data for sampled problem

Var.	Conn. load [kW]	x_{imin} (% of individual conn. load)	x_{imax} (% of individual conn. load)	x_{ieq} (% of individual conn. load)	c_i
x_1	95,221	70	95		1.05
x_2	12,120	70	95		1.25
x_3	4,534	50	90		1.59
x_4	7,470			100	1.42
x_5	698	50	90		2.09
x_6	1,612	20	70		1.23
x_7	5,919	20	70		1.25
x_8	373	20	70		1.25
x_9	1,900	20	70		1.39
x_{10}	3,255	20	70		1.11
x_{11}	82	20	70		1.11
x_{12}	18,431	20	70		1.00

151,615 kW - Restoration connected load
(Corresponding demand is 96,413 kW).

*Notes: Manual restart component not included.
Variables x_i correspond to P_i in Table 4.

In addition, a special field survey was made to determine which equipment might be connected after an outage. The survey consisted of visits to 1 or 2 of each of the enterprises listed in Table 1. At each site, a thorough technical interview of a responsible individual, generally the engineer on location, was followed by a tour of the plant. Although the field survey was limited in its scope, the consistency of the findings at each location indicated that the results were well-based. Therefore, this survey was taken to provide a realistic picture of the restoration situation.

The results are given in Table 4 (left-hand column), where the steady-state power requirements of all the equipment being re-energized are listed. The total demand was used as the reference for calculation of percentages since it is the practical factor of interest from a system viewpoint.

Table 4
On-line requirement at restoration
(% of total demand of commercial load*)

Quan.	Type of load	Field survey results	Simplex method results
P_1	Fluorescent lighting	42.6%	44.2%
P_2	Incandescent lighting	9.5	8.1
P_3	Window type A/C	3.0	4.0
P_4	Refrigerators	3.2	6.7
P_5	Fans	2.0	0.8
P_6	Cooking equipment	0.4	0.2
P_7	Pumps	0.4	0.9
P_8	Water heater	0.0	0.0
P_9	Office equipment	1.3	1.1
P_{10}	Computer terminals	0.3	0.4
P_{11}	Television sets	0.2	0.0
P_{12}	Miscellaneous	0.2	2.3
	Total:	63.1%	68.7%

* In the work described here, the ratio of total demand to total connected load was about 0.6.

It is obvious that except for the miscellaneous load the results were quite close. This suggests that the Simplex Algorithm can be a useful mathematical tool to help in the evaluation of the electrical loading during restoration.

RECIPE FOR CONSTRUCTION OF COLD LOAD MODEL

The specific steps that need to be carried out by a utility which desires to construct a cold-load pickup model according to the method outlined in the paper are listed below.

1. From the utility's Class Demand Study, identify the enterprises and the equipment involved (see Tables 1 and 2). Also, define the maximum demand P_{max} and the manual-restart power P_{man} of the network to construct constraint equation 4.

2. Select, for the appropriate time-of-day, the power requirement limits (see Table 3) to construct constraint equation 3.

3. Select and calculate, based on field experience, the re-energization coefficients (eq. 2) to construct the objective function Z of equation 1.

4. Run the linear program to obtain the pre-outage power demand x_i of each load type.

5. Calculate, preferably as part of the Program, the set of worst-case steady-state power quantities P_i (eq. 5) which now define the cold-load that needs to be re-energized.

SUMMARY AND CONCLUSION

A mathematical model of the load composition of a secondary distribution network during cold load pickup was developed. The analysis is based on the Simplex Algorithm and gives the details of the worst-case power requirements at startup. Two factors contribute to the practicality of the model: a) the existence of standard connected load surveys which utilities routinely conduct for electric load characteristic studies; and b) the availability of reasonable estimates, based on some degree of experience, of normal diversity and actual connected load at the time of the outage.

This method was applied to a commercial load in NYC to obtain a worst-case power-restoration model. Then, for comparison, on-site visits were made to many representative enterprises and locations to construct an independent realistic field model. The mathematical and the actual models, were found to be in surprisingly close agreement. This suggests that the Simplex Algorithm can be a useful mathematical tool to help in the evaluation of the load composition during a cold startup of a secondary distribution network.

ACKNOWLEDGMENTS

This work was supported by the Consolidated Edison Company of NYC under contract No. M850-0-0132-00. The authors wish to express their appreciation to Anita Merani for her contributions to the linear program; to John Pinto of Con Edison for his helpful suggestions and to Enrico Levi of the Polytechnic for his careful review of the paper.

REFERENCES

[1]. R.L. Wilde, "The Effect of Cold Load Pickup on Distribution Transformers," IEEE PES Winter Meeting, Atlanta, Georgia, Feb. 1981, No. 81 WM 137-9.

[2]. S. Ihara, F.C. Schweppe, "Physically Based Modeling of Cold Load Pickup," IEEE PES Winter Meeting, Atlanta, Georgia, Feb. 1981, No. 81 WM 084-3.

[3]. Z. Zabar, J.F. Pinto, "Startup Response of a Secondary Distribution Network, Part I: Dynamic Model of Appliance Start up" and "Part II: Load Analysis, Lab Tests and Startup Simulation," IEEE Trans. on PAS, Vol. 100, No. 6, June 1981, pp. 3143-3150.

[4]. E.J. Simburger, F.J. Hubert, "Low Voltage Bulk Power System Restoration Simulation," IEEE Trans. on PAS, Vol. 100, No. 11, Nov. 1981, pp. 4479-4484.

[5]. J.E. McDonald, A.M. Bruning, W.R. Mahieu, "Cold Load Pickup," IEEE Trans. on PAS, Vol. 98, No. 7, July 1979, pp. 1384-1386.

[6]. M.S. Chen, "Determining Load Characteristics for Transient Performances. Volume 3: Procedure for Modeling Power System Loads," EPRI Report No. EL-849, May 1979.

[7]. D. Whittle, "Optimization Under Constraints," John Wiley and Sons Ltd., 1972.

[8]. Consolidated Edison Company of New York, "Class Demand Study 1980-81."

Discussion

Daniel J. Ward, (General Electric Co., Schenectady, NY): The authors present an interesting technique for dealing with a complex subject. The analysis was aided by the fact that almost 45% of the original load could be neglected, i.e., the manual restart component. It appears that an adequate representation of the total lighting load for their system would yield a result beyond which further refinements aren't necessary.

It would be helpful to know what length of outage was assumed for the values of Table 3. In addition, how do the values of the re-energization coefficients change over time following re-energization?

Manuscript received February 17, 1983.

M. H. Yolcu, Z. Zabar, L. Birenbaum and **S. A. Granek:** We would like to express sincere thanks to Mr. Ward for his interest and for his valuable comments on the paper.

For a commercial load, the on-line equipment that must be re-energized does indeed consist principally of lighting, as Mr. Ward points out. Lighting represents about 52% of the pre-outage demand, and about 63% of the pre-outage power has to be restored initially. In contrast, motor-type loads (window-air-conditioners, fans, pumps, typewriters) represent only about 10% of the pre-outage demand.

The current inrush picture is quite different. The motor-type load current inrush may be 7 times its steady state value. This component of the total inrush actually exceeds that of the lighting load. For this reason, the motor component of commercial load should not be neglected in a start-up model.

In addition, during power restoration, commercial load represents only part of the total that must be supplied by the secondary network distribution system. Residential load [3] has a completely different characteristic in which most of the restoration load is of the motor type. Only if a network model consists of a combination of commercial, residential and industrial loads can it be a realistic representation of a system.

The outage time assumed for Table 3 was greater than 15 minutes in order to study the most severe case, where all system diversity has been lost.

We wish to address Mr. Ward's question of how the re-energization coefficients change over time. Suppose, following power restoration after a prolonged outage, there occurs a second outage of short duration. The reenergization coefficients will not be the same for the second restoration. These coefficients are defined as the product of the reciprocals of the diversity and use factors. With partial restoration of diversity, the values of these coefficients may be expected to decrease.

Manuscript received March 29, 1983.

STATISTICAL SYNTHESIS OF PHYSICALLY BASED LOAD MODELS WITH APPLICATIONS TO COLD LOAD PICKUP

Chee-Yee Chong, Member, IEEE
Advanced Information & Decision Systems
Mountain View, California
U.S.A.

Roland P. Malhami, Member, IEEE
University of Quebec at Chicoutomi
Chicoutomi, Quebec
CANADA

Abstract. Physically based load models are useful in the planning, operation and control of power systems, especially when it is necessary to predict the behavior of the system load due to changes in the system. This paper presents a methodology for predicting the load at a point in the system starting from elementary component load models. The methodology consists of three key steps: modeling of elementary component loads; classification into homogeneous (similar) groups; and aggregation of the load models in each homogeneous group using statistical techniques. The methodology is illustrated with the cold load pickup problem, where a mathematical model for the evolution of the fraction of "on" space heaters is developed. The resulting model consists of a system of coupled ordinary and partial differential equations for each homogeneous group of component loads. Numerical solutions of the equations are used to predict the fraction of "on" loads. The methodology is also applicable to direct load management.

INTRODUCTION

The modeling of power system loads is a very important problem since loads are ultimately the driving force behind the entire system. A traditional approach is to use historical data for the system demand to arrive at empirical models [1], [2]. While this approach has been adequate for many applications, the fact that it does not use dynamic models of the elementary component loads makes it unsuitable for some applications where it is necessary to predict the behavior of the load due to sudden changes in the system. Such applications include cold load pickup and direct load management. In cold load pickup, following an outage on a cold day, a large fraction of the installed space heaters will demand power from the system. Similarly, in direct load management, the payback phenomenon causes the load to increase abruptly after a power outage.

In such applications a bottom-up synthesis of the system load from the elementary component loads is a more attractive way of modeling the system load. End-use models [3] fall into this category. Another approach, [4] - [10], which is physically based, models the elementary components in the system and aggregates these demands to obtain the demand at a point in the system. One synthesis method, proposed for the evaluation of load management programs, is simulation based [4]. Others, such as [5], [6] and [7], use more analytical techniques to study the cold load pickup and load management problems. However, the inherent stochastic aspect of the individual customer loads, which contributes to the diversity factor, has not been treated adequately.

84 WM 068-3 A paper recommended and approved by the IEEE Power System Engineering Committee of the IEEE Power Engineering Society for presentation at the IEEE/PES 1984 Winter Meeting, Dallas, Texas, January 29 - February 3, 1984. Manuscript submitted February 4, 1983; made available for printing November 11, 1983.

In this paper we describe a physically based load modeling methodology which emphasizes the uncertain or stochastic nature of the problem and its application to the cold load pickup problem. The methodology starts with the modeling of the elementary component loads. This is followed by the classification of the elementary loads into similar (homogeneous) groups which can be considered together. Statistical techniques are then used to aggregate the load models within a group to obtain an equivalent model for the entire group. When the methodology is applied to the cold load pickup problem, the time evolution of the fraction of space heaters which are on can be modeled by a system of coupled ordinary and partial differential equations. The partial differential equations describe the dynamics of the population of loads which may be on or off. The ordinary differential equation gives the total fraction of the loads which are on. These equations are solved numerically to illustrate the effect of a power outage on the load dynamics.

Organization

The rest of the paper is as follows. We first review the overall load modeling methodology. The functional model for space heating of the house, which generates the demand at the component level, is then given. The next section describes the aggregation of the demands in a homogeneous group, followed by ways for treating nonhomogeneous groups. We then present some numerical results for the cold load pickup problem. Finally the data requirements for using this approach are discussed. An appendix contains the derivations of the main aggregation results.

METHODOLOGY FOR SYNTHESIS OF LOAD DEMAND MODELS

In this section we present the basic steps in our load modeling methodology. Our objective is to develop models for predicting the short term system load due to a large number of component loads. The overall methodology consists of the following three basic steps: modeling of elementary component loads, classification into homogeneous groups and aggregation of the loads in each group. These steps will be discussed separately in the following.

Modeling of Elementary Component Loads

The first step consists of developing models for the individual elementary component loads which constitute the overall system load. Many components have some kind of energy storage so that a demand for the energy of the component may not result in an instantaneous demand for electrical power. Rather, a control mechanism such as a thermostat determines the operating state of the component as to whether it should be on or off. Thus in general a component load model can be

Reprinted from *IEEE Transactions on Power Apparatus and Systems,* Vol. 103, No. 7, pp. 1621–1628, July 1984.

decomposed into two parts: a functional model and an electrical device model. The functional model generates the functional (operating) state of the component based on the demand for energy, the environmental conditions and the state of energy storage in the component. It corresponds to the energy storage portion of the load (plus the part of the environment that affects energy storage) and the control mechanism. The electrical device model has as its external inputs the voltage and frequency of the system and generates the actual electrical load as the output. The interaction of the two models is shown in Figure 1.

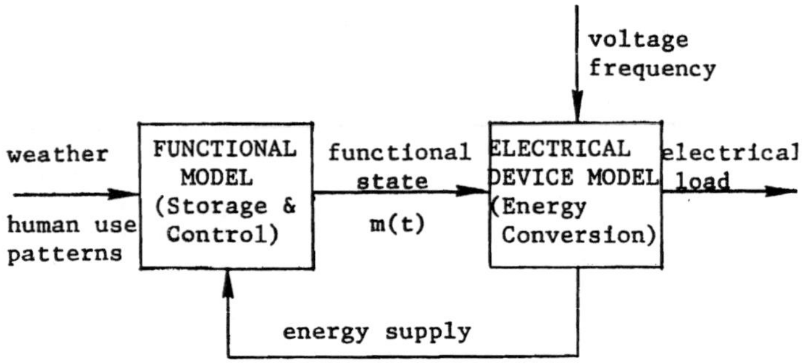

Figure 1. Decomposition of Elementary Component Load

Classification of Component Loads

Our goal is to represent the behavior of a group of loads by an equivalent model. Obviously components with widely different characteristics cannot be grouped together. Thus the next step in the load modeling methodology is to classify the component loads into groups with similar characteristics. In terms of the decomposition discussed above, this means similarity in the functional and electrical device models. Each of these groups is called a homogeneous group. An aggregate load model will then be found for each group.

Aggregation

The last step in the methodology is to find an equivalent model for each homogeneous group of loads. Let $P_i(t)$ be the (real) power demand of the i-th component in a homogeneous group and N be the number of loads in the group. Then $P_i(t)$ can be written as

$$P_i(t) = m_i(t)\ P(V,f) \tag{1}$$

where $m_i(t)$ is the functional state of the i-th load at time t and P(V,f) is the electrical demand for the load as a function of the voltage V and frequency f. $m_i(t)$ is 1 when the load is on and 0 when it is off. Note that P(V,f) can be assumed to be the same for all the loads in the group. Equation (1) is actually a simplified model since in some situations the electrical part may affect the functional state. For many applications, however, the electrical dynamics is so much faster than the functional dynamics that the simplification is valid.

The aggregate load for the group is then given by

$$P(t) = \sum_{i=1}^{N} P_i(t) = \sum_{i=1}^{N} m_i(t)\ P(V,f)$$

$$= [\frac{1}{N} \sum_{i=1}^{N} m_i(t)]\ NP(V,f)$$

$$= \bar{m}(t)\ NP(V,f) \tag{2}$$

where $\bar{m}(t)$ is the fraction of loads in the group which are on and N P(V,f) is the total installed power demand from the loads in the group.

In many applications, such as in cold load pickup or direct load management, we are particularly interested in forecasting $\bar{m}(t)$, the fraction of loads which are on at any time rather than how the loads respond to changes in the system voltage or frequency. This is the emphasis for the rest of the paper, where we present a model for the dynamic evolution of $\bar{m}(t)$.

FUNCTIONAL MODEL FOR SPACE HEATING

In this section we present a functional model for the space heating of a house. This model predicts the on-off behavior of the thermostat and serves as the basis for finding the total fraction of space heaters which are on in the cold load pickup problem. Air conditioning loads can be handled similarly.

Many possible models can be used to describe the heating and cooling of houses [11], [12]. These models vary in complexity depending on their intended use. For example, detailed simulation models [12] are more useful for the design of heating or cooling systems or for studying the specific energy demands of a house. Since our objective in modeling the demand process is for use in aggregation, a detailed model is not needed. Instead, the following simplified model is chosen. The energy storage portion of the load (which describes the heat loss and gain in the house) is given by a linear model and a nonlinear relay is used to represent the control mechanism which accounts for the switching between states of the thermostat. The linear part has a continuous state corresponding to the stored energy and is fairly traditional but the nonlinear part has a discrete state (either on or off) corresponding to the functional state. The result is a hybrid state system which is quite typical of functional models for components with energy storage. Specifically, the following models are used.

Energy Storage Portion

We use the temperature x(t) to represent the state of energy storage in the house. x(t) varies according to the following linear differential equation

$$C\frac{dx(t)}{dt} = -a'(x(t) - x_a(t)) + R'm(t)b(t) + w'(t) \tag{3}$$

where:

- C: the average thermal capacity of the house
- a′: the average heat loss rate through the floors, walls, ceiling, etc
- x(t): the average temperature of the house
- $x_a(t)$: the ambient temperature
- R′: the rate of heat supplied by the heater
- m(t): the functional state - 1(on) or 0(off)
- b(t): the state of the power supply (1 or 0)
- w′(t): a noise process

b(t) represents any possible interruption in the power supply due to power outage or direct load management. w′(t) represents modeling errors and all other processes of heat loss or heat gain which have not been accounted for explicitly. We assume w′(t) to be a white (wide band) random noise with zero mean.

Division of (3) by C yields

$$\frac{dx(t)}{dt} = -a\ (x(t) - x_a(t)) + Rm(t)b(t) + w(t) \tag{4}$$

where the definitions of a and R are obvious. w(t) is now assumed to have a noise variance parameter of σ^2.

when the heater is in the "on" state ($m(t) = 1$)

$$\frac{dx(t)}{dt} = r_1(x,t) + w(t) \tag{5}$$

where $r_1(x,t)$ is the mean rate of temperature gain.

When the heater is in the "off" state ($m(t) = 0$),

$$\frac{dx(t)}{dt} = r_0(x,t) + w(t) \tag{6}$$

where $-r_0(x,t)$ is the mean rate of temperature loss. The rates $r_1(x,t)$ and $r_0(x,t)$ include effects of weather, the thermal properties of the house, and the rating of the heater, etc.

Control Mechanism

The control mechanism changes the functional state $m(t)$ of the component load depending on the temperature $x(t)$ and the thermostat settings. The thermostat is assumed to have a deadband with an upper limit of x_+ and a lower limit of x_-. $m(t)$ switches from on to off, i.e., from 1 to 0 when the temperature $x(t)$ reaches x_+. It switches from off to on, i.e., from 1 to 0 when $x(t)$ reaches x_-.

Our model is similar in spirit to the linear models used in [5] and [7]. We have, however, introduced a noise process to take care of anything which have not been modeled explicitly. Another difference is in the way we treat the nonlinear part of the functional model. The switching phenomenon will be used directly in the aggregation process.

AGGREGATION OF A HOMOGENEOUS GROUP

In this section we present a mathematical technique for aggregating the functional states of a large number of loads in the same homogeneous group. Specifically, we would like to compute $\bar{m}(t)$, the fraction of loads which are on at any particular time. Equations for the computation of $\bar{m}(t)$ will be developed. The aggregation is based on the following assumptions.

Assumptions for Aggregation

Common External Inputs. The loads are geographically close together so that any variations in the common inputs such as weather can be ignored.

Elemental Independence. Conditional on the weather and other common inputs such as a system outage, the devices behave independently. More precisely, conditional on the common input processes such as the ambient temperature $x_a(t)$ and the power supply state $b(t)$, the demand processes $m_i(t)$ are statistically independent. The causes the natural diversity of the load under normal conditions.

Model For $\bar{m}(t)$, the Fraction of "ON" Loads

With these assumptions, we can obtain an aggregate model for the functional state $\bar{m}(t)$ when a large number of loads are present. The quantity we are trying to compute is

$$\bar{m}(t) = \frac{1}{N}\sum_{i=1}^{N} m_i(t) \tag{7}$$

where $m_i(t)$ is the functional state of the i-th device. When N is large, by Kolmogorov's law of large numbers [13], $\bar{m}(t)$ can be approximated by

$$\bar{m}(t) = E[m_i(t)|\xi] \tag{8}$$

where $E[\,.\,|\xi]$ denotes the conditional mean given the common inputs and parameter vector ξ. Heuristically, (8) states that the average of a large number of independent and identically distributed random variables is the statistical mean of a random variable. Since $m_i(t)$ can take on only the values of either 1 or 0,

$$\bar{m}(t) = \Pr[m_i(t) = 1|\xi] \tag{9}$$

i.e., $\bar{m}(t)$, the fraction of "on" loads, is the same as the probability that an individual load is on. The following results, which are derived in the Appendix, can then be obtained.

Equations for $\bar{m}(t)$. $\bar{m}(t)$ evolves according to the following ordinary differential equation.

$$\frac{d\bar{m}}{dt} = \frac{\sigma^2}{2}\left[\frac{\partial}{\partial\lambda} f_{1b}(x_+,t) + \frac{\partial}{\partial\lambda} f_{ob}(x_-,t)\right] \tag{10}$$

where $f_{1a}(\lambda,t)$ and $f_{1b}(\lambda,t)$ are the solutions of the following partial differential (Fokker-Planck [13],[14]) equation in the regions a and b in Figure 2:

$$\frac{\partial}{\partial\lambda} f_1(\lambda,t) = \frac{\partial}{\partial\lambda}\left[(a(\lambda - x_a(t)) - b(t)R)\, f_1(\lambda,t)\right] + \frac{\sigma^2}{2}\frac{\partial^2}{\partial\lambda^2} f_1(\lambda,t) \tag{11}$$

and $f_{ob}(\lambda,t)$ and $f_{oc}(\lambda,t)$ satisfy the following equation in the regions b and c in Figure 2:

$$\frac{\partial}{\partial\lambda} f_o(\lambda,t) = \frac{\partial}{\partial\lambda}\left[a(\lambda - x_a(t))\, f_o(\lambda,t)\right] + \frac{\sigma^2}{2}\frac{\partial^2}{\partial\lambda^2} f_o(\lambda,t) \tag{12}$$

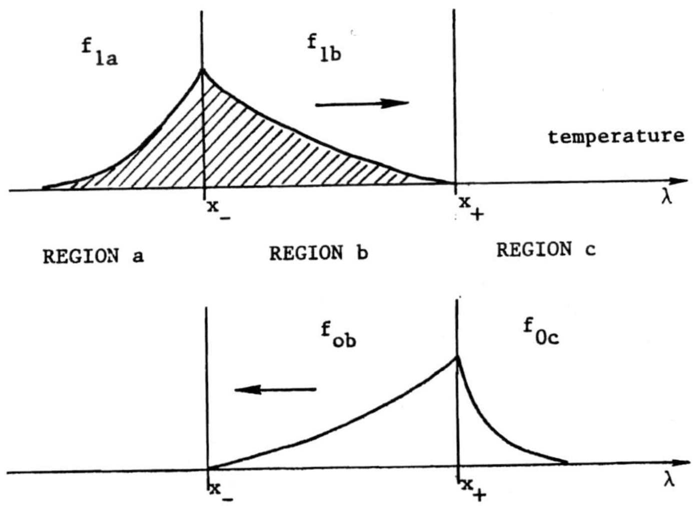

Figure 2. Propagation of Probability Densities

The boundary conditions to be satisfied for all t are:

$$f_{1a}(-\infty,t) = f_{oc}(\infty,t) = 0 \tag{13}$$

$$f_{1b}(x_+,t) = f_{ob}(x_-,t) = 0 \tag{14}$$

$$f_{1a}(x_-,t) = f_{1b}(x_-,t) \tag{15}$$

$$f_{ob}(x_+,t) = f_{oc}(x_+,t) \tag{16}$$

$$-\frac{\partial}{\partial\lambda} f_{1a}(x_-,t) + \frac{\partial}{\partial\lambda} f_{1b}(x_-,t) + \frac{\partial}{\partial\lambda} f_{ob}(x_-,t) = 0 \quad (17)$$

$$\frac{\partial}{\partial\lambda} f_{oc}(x_+,t) - \frac{\partial}{\partial\lambda} f_{ob}(x_+,t) - \frac{\partial}{\partial\lambda} f_{1b}(x_+,t) = 0 \quad (18)$$

Interpretations of the Equations

The above equations can be interpreted in the following way. $f_1(\lambda,t)$ and $f_o(\lambda,t)$ are actually probability densities for the "hybrid" states $\{x(t), m(t)\}$ of the functional model. Intuitively,

$$f_1(\lambda,t)d\lambda = \Pr\{\lambda < x(t) \leq \lambda + d\lambda,\ m(t) = 1\} \quad (19)$$

is the fraction of loads which are on and with temperature between λ and $\lambda + d\lambda$ and

$$f_o(\lambda,t)d\lambda = \Pr\{\lambda < x(t) \leq \lambda + d\lambda,\ m(t) = 0\} \quad (20)$$

is the fraction of loads which are off and with temperature between λ and $\lambda + d\lambda$. Equations (11) and (12) describe how these fractions evolve with time. The evolution depends on the heat(temperature) gain and loss rates as well as the noise variance. Equation (10) for $\bar{m}(t)$ can be written as

$$\frac{d\bar{m}}{dt} = r_{on}(t) - r_{off}(t) \quad (21)$$

where

$$r_{on}(t) = \frac{\sigma^2}{2} \frac{\partial}{\partial\lambda} f_{ob}(x_-,t) \quad (22)$$

is the rate at which the density $f_o(\lambda,t)$ reaches x_-, i.e., the rate at which loads are being turned on, and

$$r_{off}(t) = -\frac{\sigma^2}{2} \frac{\partial}{\partial\lambda} f_{1b}(x_+,t) \quad (23)$$

is the rate at which the density $f_1(\lambda,t)$ reaches x_+, i.e., the rate at which loads are being turned off.

The boundary conditions fall into three classes: Equation (13) gives the conditions at infinity which have to be satisfied for the total probability to be finite. Equations (14), (15) and (16) are continuity conditions. Equations (17) and (18) are conservation equations. They couple the two partial differential equations together and state that the loads leaving the "on" state move into the "off" state and vice versa.

A traveling wave model similar to the above has been proposed in [5] to treat the cold load pickup problem. Our model is, however, more general and can be shown to reduce to that model when proper approximations are made.

These coupled equations can be solved numerically to model the behavior of a large number of loads in the same homogeneous group. Some numerical results will be given later.

AGGREGATION OF NONHOMOGENEOUS GROUPS

In a general case, the loads within a group are not identical. Thus there may be a spread in the values of the parameters such as thermostat settings, building insulation values, noise variances, etc.

The following are two possible ways of handling this problem depending on the magnitude of the parameter variations.

Small Parameter variations

Let $\underline{\xi} = \{\xi_1,..,\xi_p\}$ be the vector of parameters. Assume some statistics of the parameters are known. The aggregation equation of (8) should now be modified to read as

$$\bar{m}(t) = E[E[m_i(t)|\underline{\xi}]] \quad (24)$$

i.e. after averaging over the noise and the initial conditions given the parameters, we have to average over the parameters. Denote the functional state for the homogeneous group with parameters $\underline{\xi}$ as $\bar{m}(t,\underline{\xi})$, i.e.,

$$\bar{m}(t,\underline{\xi}) = E[m_i(t)|\underline{\xi}] \quad (25)$$

Then for parameters narrowly distributed about ξ_0, $\bar{m}(t,\underline{\xi})$ can be approximated to second order by

$$\bar{m}(t,\underline{\xi}) = \bar{m}(t,\underline{\xi}_o) + \sum_{i=1}^{p} \frac{\partial \bar{m}}{\partial\xi_i}(\xi_i - \xi_{io}) + \sum_{i=1}^{p}\sum_{j=1}^{p} \frac{1}{2}\frac{\partial^2\bar{m}}{\partial\xi_i\partial\xi_j}(\xi_i - \xi_{io})(\xi_j - \xi_{jo}) \quad (26)$$

Let the parameter vector have means and covariances given by

$$E[\underline{\xi}] = \underline{\xi}_o \quad (27)$$

and

$$E[(\xi_i - \xi_{io})(\xi_j - \xi_{jo})] = \sigma_{ij}^2 \quad (28)$$

Then $\bar{m}(t)$ for the group can be approximated by

$$\bar{m}(t) = \bar{m}(t,\underline{\xi}_o) + \sum_{i=1}^{p}\sum_{j=1}^{p} \frac{1}{2}\frac{\partial^2}{\partial\xi_i\partial\xi_j}\bar{m}(t,\underline{\xi})\bigg|_{\underline{\xi}=\underline{\xi}_o}\sigma_{ij}^2 \quad (29)$$

In general, if the covariances or the derivatives involved are small, the group can be approximated by a homogeneous group. If this assumption is not true, it becomes necessary to introduce the correction terms given by the second derivatives and the variances of the parameters. The derivatives can be computed numerically if needed.

Large Parameter Variations

If the parameter spread is too large, it may be necessary to split the group into several smaller homogeneous or nonhomogeneous groups in which the parameter spreads are smaller. Then the overall functional state can be computed as as a weighted sum of the individual functional states for each group.

NUMERICAL RESULTS

In this section we present some numerical results of applying the developed model to the cold load pickup problem. Three cases are considered: a homogeneous group, a nonhomogeneous group and a general group. The results display the evolution of $\bar{m}(t)$, the fraction of space heaters which are on before, during and after the power interruption. The loads are allowed to reach a steady state and then the outage is introduced by setting $b(t) = 0$ in (11). The partial differential equations are solved by discretizing with respect to both space and time.

Homogeneous Group (Similar Loads)

The parameters for the base case are:

$$\Delta = x_+ - x_- = 2^{\circ}F$$
$$x_-/\Delta = 35$$
$$x_a(t)/\Delta = 15$$
$$a/\Delta = .00976\ (^{\circ}F\ \min)^{-1}$$
$$\bar{R} = R/\Delta = .4\ (\min)^{-1}$$
$$\bar{\sigma} = \sigma/\Delta = .3\ (\min)^{-1/2}$$

With these parameters, the average "on" and "off" durations of the heater are roughly in the neighborhood of 5 mins.

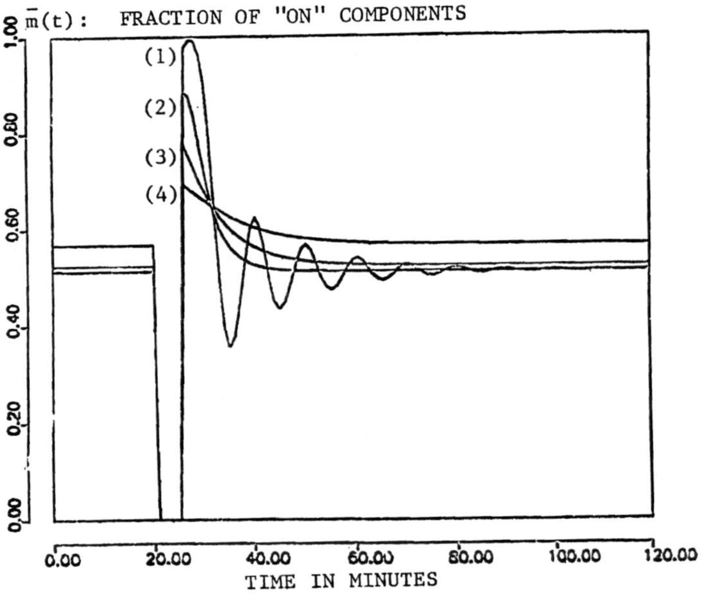

Figure 3. Load Dynamics (5 min. outage) as a Function of Normalized Noise Variance Parameter $\bar{\sigma}$. Cases: $\bar{\sigma}$=.1 for (1), $\bar{\sigma}$=.3 for (2), $\bar{\sigma}$=.5 for (3) and $\bar{\sigma}$=1.0 for (4).

Figure 3 shows the effects of changing the noise variance (which represents the degree of independence from one load to another) for a given outage duration of 5 mins starting at 20 mins. We notice that the noise variance has important effects on the dynamic response. As it increases, there is a corresponding decrease in the post-outage dynamic fluctuations and the system reaches its post-outage steady state faster. The noise acts as a damping factor to the dynamics. An increase in the noise tends to increase the diversity which has been reduced due to the outage. The dependence of the response on the noise indicates that it should be included for a better model of the aggregate load dynamics. Results which ignore this may be too pessimistic.

Figure 4 shows the effect of changing the heating rate $\bar{R}$. As the heating rate increases, the steady state value of the fraction of "on" loads decreases. This is to be expected since the loads can stay on for a shorter length of time.

Figure 5 shows the effect of changing the length of the power outage. As the duration of the outage increases, the fraction of loads which comes on after the interruption also increases, and so does the length of time needed to return to the steady state.

Nonhomogeneous Group (Small Parameter Variations)

In this case we use the results for two sets of neighboring parameters to compute the derivatives needed in (29). Equation (29) is then used to generate the post-outage dynamics in the nonhomogeneous group for various levels of parameter variance. Figure 6 considers only one variable parameter, the thermostat setting x_-. The parameter values are the same as before except that $\bar{\sigma}$ = .2 $(\min)^{-1/2}$.

The post-outage dynamic fluctuations decrease as the parameter variance i creases. This is again due to an increase in the diversity of the group.

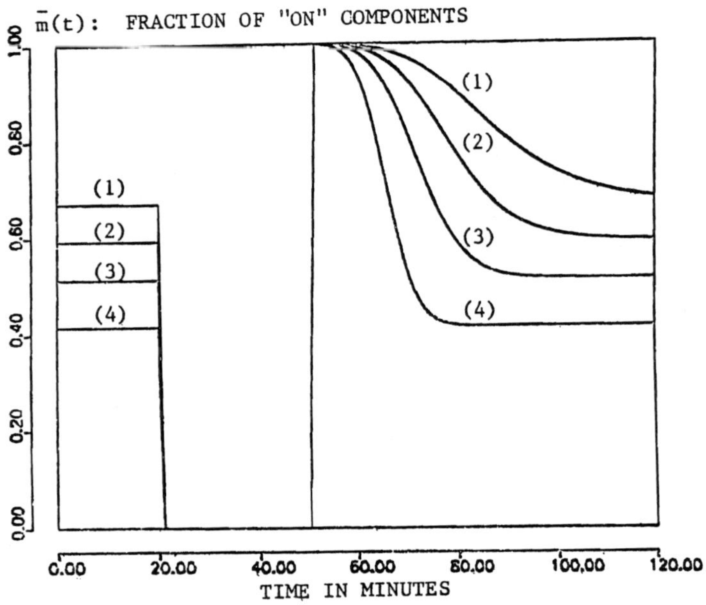

Figure 4. Load Dynamics (30 min. outage) as a Function of Heating Rate Parameter $\bar{R}$. Cases: $\bar{R}$=.3 for (1), .344 for (2), .4 for (3), and .5 for (4).

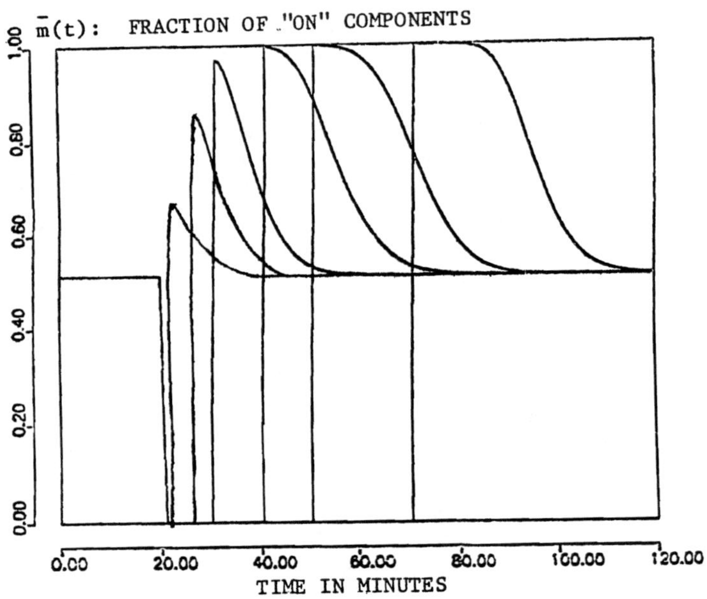

Figure 5. Load Dynamics as a Function of Outage Durations: 1, 5, 10, 20, 30, and 50 minutes.

General Group (Large Parameter Variations)

The general group is assumed to consist of sixteen homogeneous groups. The parameters for each homogeneous group are assumed to be the same as the base case parameters, except that $R/\Delta = r_i$ and $\sigma/\Delta = \sigma_j$ are now dependent on the group ij.

The values for r_i and σ_j are $r_1 = .35$, $r_2 = .4$, $r_3 = .5$, $r_4 = .6$, $\sigma_1 = .1$, $\sigma_2 = .3$, $\sigma_3 = .5$, and $\sigma_4 = .8$.

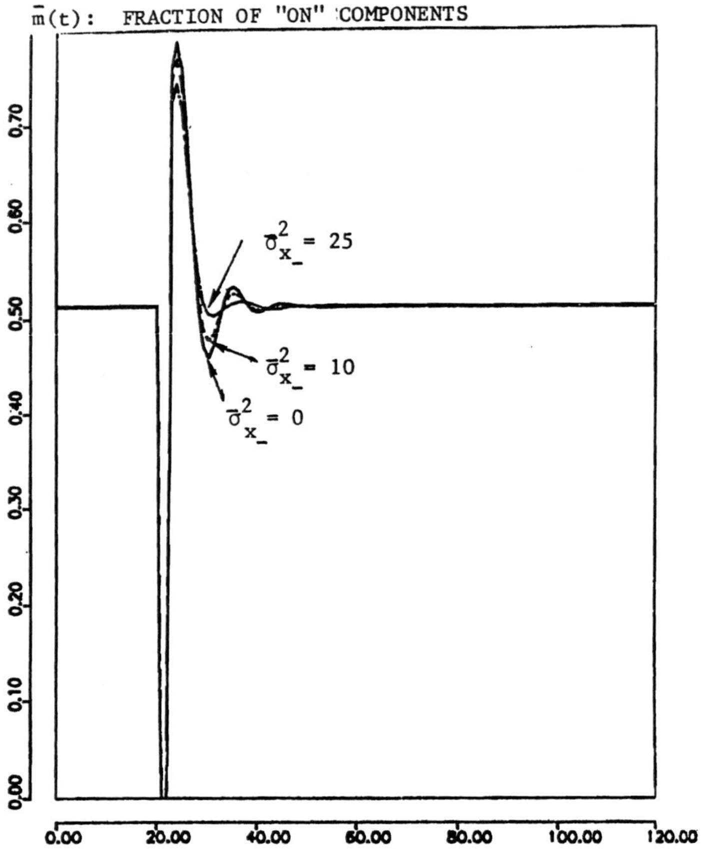

Figure 6. Load Dynamics (2 min. outage) as a Function of Parameter (Thermostat Setting) Spread. $\sigma_{x_-}^2$= normalized set-point variance.

The relative weight of a group with the parameter (r_i, σ_j) in the general group is given by w_{ij}. Let $\underline{W}$ be the matrix whose elements are the weights w_{ij}. Three sets of w_{ij}'s are chosen for the simulations, corresponding to uniform, triangular and skewed triangular distributions of the homogeneous groups within the general group.

Uniform distribution

$$\underline{W}_1 = \frac{1}{16}\begin{bmatrix}1&1&1&1\\1&1&1&1\\1&1&1&1\\1&1&1&1\end{bmatrix}$$

Triangular distribution

$$\underline{W}_2 = \frac{1}{64}\begin{bmatrix}1&3&3&1\\3&9&9&3\\3&9&9&3\\1&3&3&1\end{bmatrix}$$

Skewed triangular distribution

$$\underline{W}_3 = \frac{1}{64}\begin{bmatrix}9&3&3&1\\9&3&3&1\\9&3&3&1\\9&3&3&1\end{bmatrix}$$

The actual load dynamics is computed by means of

$$\bar{m}(t) = \sum_{i=1}^{4}\sum_{j=1}^{4} w_{ij}\bar{m}_{ij}(t) \tag{30}$$

where $\bar{m}_{ij}(t)$ is the aggregate functional state of the homogeneous group with parameters r_i and σ_j. Figure 7 shows the results of the simulation for a 5 minute outage. It is seen that as long as the distribution of the parameters is symmetric, the dynamics of the group is not that sensitive to the actual distribution of the parameters. On the other hand, the skewed triangular distribution shows some difference in the dynamics.

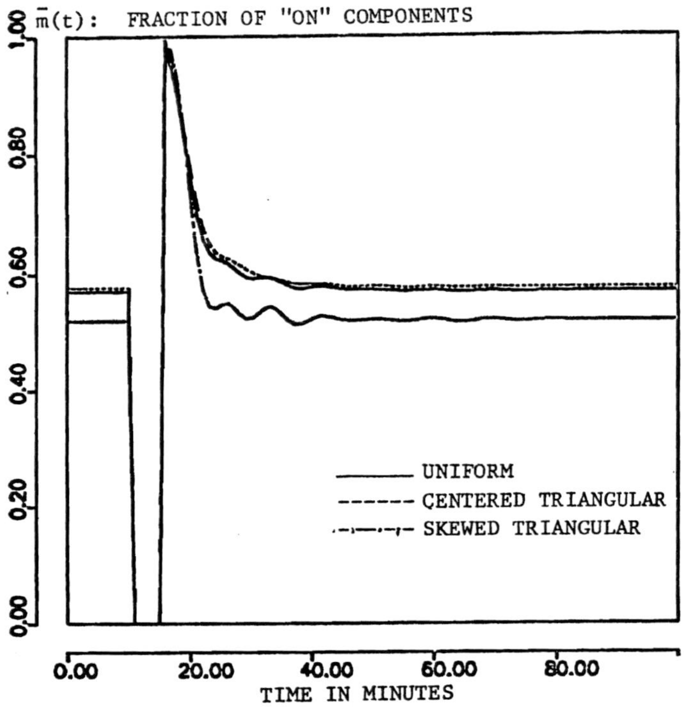

Figure 7. Load Dynamics (5 min. outage) of a General Group as a Function of Parameter Distribution.

DATA REQUIREMENTS

The individual functional model used for space heating is very similar to the first order models used in [5] and [7]. As such, the data requirements are also quite similar. Parameter data can be obtained in several ways, either from the basic specifications of the house, or an identification approach can be used to estimate the model parameters by conducting experiments on a representative house in the group. We would make some comments on the second approach.

During normal operations, r_1 in (5) and r_0 in (6) can be approximated by constants. The average "on" duration of the component is then given by

$$d_1 = (x_+ - x_-)/r_1 \tag{31}$$

and the average "off" duration is given by

$$d_0 = (x_+ - x_-)/(-r_0) \tag{32}$$

These quantities can be measured experimentally for a particular ambient temperature. If needed, the quantities a and R can also be computed. It can be shown that for a homogeneous group, the steady-state $\bar{m}(t)$ is

$$\bar{m}(t) = d_1/(d_1 + d_0) \quad (33)$$

This can be used to check the parameter values.

The "on" and "off" durations are actually the first passage times of x(t) from x_+ to x_- and from x_- to x_+. It can be shown [14] that the variances of the first passage times depend on the variance of the noise. Thus, by measuring the variances of the "on" and "off" durations, we can estimate σ, the variance of w(t). Details can be found in [15] or [16]

CONCLUSIONS

We have presented a methodology for predicting the behavior of a large number of loads starting from their individual models. This modeling approach is useful for applications such as the study of cold load pickup and direct load management when sudden changes in the system occur. The methodology provides models in which the parameters of the individual components appear explicitly. The effects of varying these parameters on the aggregate dynamics can then be studied. This model has been demonstrated on the cold load pickup problem. Data requirements for the use of this approach are similar to those proposed elsewhere in the literature and can be obtained from measurements on the individual loads.

The main limitation of this approach is that the model is given in terms of partial differential equations. The partial differential equations can, however, be solved numerically, such as by finite dimensional approximations. In addition, these models have not been validated in field tests.

We feel, however, that this approach provides a useful alternative for load modeling and can serve as a starting point for future research on practical issues such as data requirements and validation through field tests.

ACKNOWLEDGEMENTS

This research was supported by the U. S. Department of Energy, Division of Electrical Energy Systems, under Contract No. DE-AS01-77ET29116. The authors are grateful to Mr. L. Fink and Dr. T. Trygar for comments during the earlier part of the project, and to Drs. F. Wu, M.L. Chan and S. Ihara for many useful comments in preparing this paper.

REFERENCES

1. F.D. Galiana, et al., "Identification of Stochastic Electrical Load Models from Physical Data," IEEE Trans. Automat. Contr. vol. AC-19, pp. 887-893, Dec. 1974.

2. M.S. Sachdev, et al., "Representative Bibliography on Load Forecasting," IEEE Trans. Power Appar. Syst. vol. PAS-96, pp. 697-700, Mar./Apr. 1977.

3. C.W. Gellings and R.W. Taylor, "Electric Load Curve Synthesis - A Computer Simulation of an Electric Utility Load Shape," IEEE Trans. Power Appar. Syst. vol. PAS-100, pp. 60-65, Jan. 1981.

4. M.L. Chan, et al., "Simulation-based Load Synthesis Methodology for Evaluating Load-Management Programs," IEEE Trans. Power Appar. Syst. vol. PAS-100, pp. 1771-1778, April 1981.

5. S. Ihara and F.C. Schweppe, "Physically Based Modeling of Cold Load Pickup," IEEE Trans. Power Appar. Syst. vol. PAS-100, pp. 4142-4150, Sept. 1981.

6. W.W. Lang, et al., "An Analytical Method for Quantifying the Electrical Space Heating Component of a Cold Load Pickup," IEEE Trans. Power Appar. Syst. vol. PAS-101, pp. 924-932, April 1982.

7. T.M. Calloway and C.W. Brice, III, "Physically-based Model of Demand with Applications to Load Mangement Assessment and Load Forecasting," IEEE Trans. Power Appar. Syst. vol. PAS-101, pp. 4625-4631, Dec. 1982.

8. H.T.T. Nguyen and J.D. Birdwell, "A Physically-based Low-Order Model for Aggregate Air Conditioner Loads," Proceedings of 1982 American Control Conference, Arlington, Virginia.

9. Georgia Institute of Technology, "Development of an Electrical Load Demand and Response Model Based on a Rational Synthesis from Elementary Devices," Report to U.S. Department of Energy, No. DOE/ET-5109-1, 1979.

10. C.Y. Chong and A.S. Debs, "Statistical Synthesis of Power System Functional Load Models," Proceedings of IEEE Conference on Decision and Control, Fort Lauderdale, Florida, 1979.

11. J.B. Woodard, Jr., "Electric Load Modeling," Ph.D. Thesis, Department of Electrical Engineering, MIT, September 1974.

12. University of Wisconsin, "TRNSYS: A Transient Simulation Program," Engineering Experiment Status Report 38-9, Feb. 1978.

13. A. Papoulis, Probability, Random Variables, and Stochastic Processes, McGraw-Hill, New York, 1965.

14. D.R. Cox and H.D. Miller, The Theory of Stochastic Processes, Wiley, New York, 1965.

15. R.P. Malhami, "A Statistical Approach for Modeling a Class of Power System Loads," Ph.D. Thesis, School of Electrical Engineering, Georgia Institute of Technology, Feb. 1983.

16. Georgia Institute of Technology, "Statistical Approach to Physically Based Load Modeling," Final Report to U.S. Dept. of Energy, Division of Electrical Energy Systems, Contract No. DE-AS01-77ET29116.

17. R. Malhame and C.Y. Chong, "Stochastic Hybrid State Systems for Electric Load Modeling," Proceedings of 22nd IEEE Conference on Decision and Control, San Antonio, Texas, 1983.

APPENDIX

We present a brief derivation of (10) to (18). A more rigorous proof can be found in [15] to [17].

Derivation of (11) and (12)

The discrete state m(t) can only switch when x(t) reaches the limits x_+ or x_-. Therefore, in the interior of the regions a, b or c, the probability density of x, $f_i(\lambda,t)$ depends only on (4) and not on the thermostat. From the standard results in random processes [13], [14], $f_i(\lambda,t)$, i = 0,1 are given by the Fokker-Planck equations (11) and (12).

For convenience, we write (11) and (12) as

$$\frac{\partial}{\partial t} f_i(\lambda,t) = \frac{\partial}{\partial \lambda} g_i(\lambda,t) \tag{34}$$

for i = 1, 2. Integrating (34) over the interval $[\lambda,\lambda+\delta]$, we obtain

$$\frac{\partial}{\partial t} \Pr[\lambda < x \leq \lambda + \delta,\ m = i] = g_i(\lambda + \delta,t) - g_i(\lambda,t) \tag{35}$$

where $g_i(\lambda,t)$ can be interpreted as the rate at which the probability of m=i(1 or 0) crosses point λ from left to right. The difference between the rates at $\lambda + \delta$ and λ gives the net increase of the probability in the interval (Figure 8).

Derivation of (13)-(18)

Equation (13) has to be satisfied so that the probability is finite. Equations (14) to (16) are continuity conditions. Thus only (17) and (18) need to be derived. Consider the boundary x_+ (Figure 9). f_{ib} cannot cross the boundary. Thus

$$\frac{\partial}{\partial t} \Pr[x_+ - \delta < x \leq x_+,\ m = 1] = -g_{1b}(x_+ - \delta,t) \tag{36}$$

Similarly,

$$\frac{\partial}{\partial t} \Pr[x_+ - \delta < x \leq x_+,\ m = 0] = -g_{ob}(x_+ - \delta,t) \tag{37}$$

and

$$\frac{\partial}{\partial t} \Pr[x_+ < x \leq x_+ + \delta,\ m = 0] = g_{oc}(x_+ + \delta,t) \tag{38}$$

where g_{1b}, g_{0b} and g_{0c} are g_i defined in the various regions.

By the conservation of probability injections at x_+, the left-hand sides of (36) to (38) have to sum up to zero. Thus,

$$-g_{1b}(x_+ - \delta,t) - g_{ob}(x_+ - \delta,t) + g_{oc}(x_+ + \delta,t) = 0 \quad (39$$

When $\delta = 0$, we obtain (18) by using the continuity conditions.

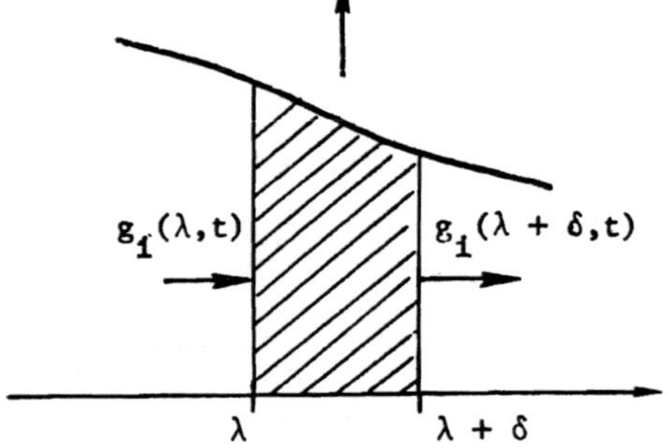

Figure 8. Probability Injection over an Interval

Derivation of (10)

$$\bar{m}(t) = P[m(t) = 1]$$

$$= \int_{-\infty}^{x_+} f_1(\lambda,t)\, d\lambda \tag{40}$$

Thus,

$$\frac{d\bar{m}(t)}{dt} = \int_{-\infty}^{x_+} \frac{\partial}{\partial t} f_1(\lambda,t)d\lambda$$

$$= \int_{x_-}^{x_+} \frac{\partial}{\partial t} f_{1b}(\lambda,t)d\lambda + \int_{-\infty}^{x_-} \frac{\partial}{\partial t} f_{1a}(\lambda,t)d\lambda$$

$$= g_{1b}(x_+,t) - g_{1b}(x_-,t) + g_{1a}(x_-,t)$$

$$= \frac{\sigma^2}{2} \frac{\partial}{\partial \lambda} f_{1b}(x_+,t) - \frac{\sigma^2}{2} \frac{\partial}{\partial \lambda} f_{ob}(x_-,t)$$

$$+ \frac{\sigma^2}{2} \frac{\partial}{\partial \lambda} f_{1a}(x_-,t) \tag{41}$$

Using (17) we obtain (10).

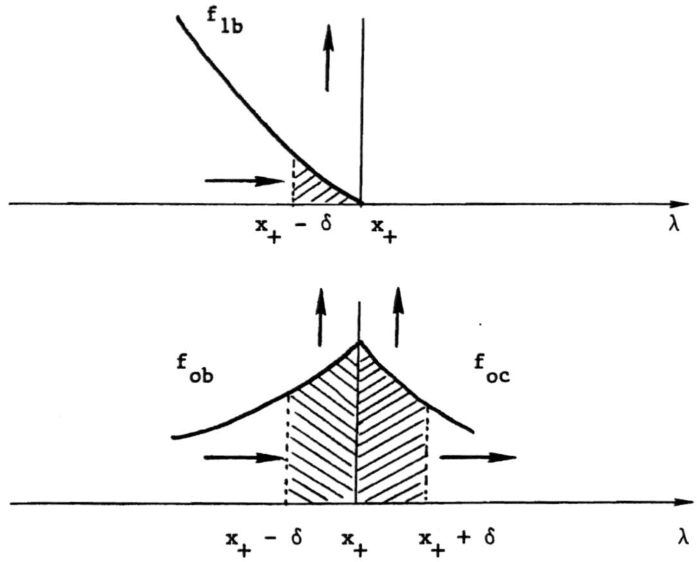

Figure 9. Conservation of Probability Injections at the Boundary x_+

EFFECTS OF COLD LOAD PICKUP AT THE DISTRIBUTION SUBSTATION TRANSFORMER

R. L. Wilde, Member, IEEE

RLW Engineering Ltd.,
Regina, Saskatchewan, Canada

ABSTRACT

The cold load pickup characteristics of electric space heating loads and its effect on the distribution substation transformer and its protective equipment is presented. A computer simulation of cold load pickup characteristics as seen at the substation level is used to determine the impact of saturation of electric space heat load on the various thermal and loading limitations of transformers.

INTRODUCTION

Uncertainty in the price and availability of fossil fuels, coupled with government incentives to convert from oil-derived heating systems to non-oil-derived heating has resulted in a dramatic increase in the use of electricity for space heating. In Canada, from 1971 to 1981, electric space heating increased by over 500%. Over the next ten years there will be even greater increases in this type of heating. The impact of this activity on the power system is very noticeable. While in new subdivisions the distribution system will be designed in anticipation of high saturation of electric space heating loads, conversions from non-electric space heating can increase the peak demand on existing feeders and transformers by a factor of five or more. Cold load pickup from this type of load will place even greater strains on existing distribution plant. Since cold load pickup (CLPU) is hopefully a rare occurrence, the saturation of electric space heating can increase to the extent that serious problems may be encountered during CLPU before problems are encountered during steady state operation.

Obviously with these heating system conversions taking place we require an advance knowledge of the capability of the distribution system to supply the load, both in the steady state and during CLPU. In fact, even in new subdivisions where the distribution system may be designed for high saturation of electric heating, our unfamiliarity with the CLPU characteristics of this load prevents optimizing the system for electric heating loads.

The goal of this project was to analyze the CLPU characteristics of residential customer loads and to analyze the effects of CLPU on: (a) the distribution substation transformer, (b) its protective equipment, and (c) on protective equipment along the distribution feeder. Of particular interest in this study was the effect of various levels of penetration of electric space heating load on this equipment. i.e. At what penetrations or saturation of electric space heating will CLPU problems be observed. In addition, we are also interested in the effect of heat pump loads. For example, during CLPU how does a heat pump load compare to conventional electric resistance heating? Another consequence of a high saturation of heat pump load would be a high saturation of air conditioning. This could subsequently result in summer CLPU problems.

84 T&D 377-8 A paper recommended and approved by the IEEE Transmission and Distribution Committee of the IEEE Power Engineering Society for presentation at the IEEE/PES 1984 Transmissiona dn Distribution Conference, Kansas City, Missouri, April 29 - May 4, 1984. Manuscript submitted October 31, 1983; made available for printing March 2, 1984.

COLD LOAD PICKUP SIMULATION

Cold load pickup refers to the loading imposed on a portion of the electric system following a service interruption during which some loss of diversity has occurred. CLPU consists of two distinct components of the restoration load. The first is the transient portion which is a result of transformer and motor magnetization current and lasts a very short time, typically a few cycles. The second portion, the enduring component, results primarily from a loss of diversity among thermostatically controlled loads. This portion of CLPU will be a function of the number, size (electrical capacity) and thermal behaviour of the connected loads. With electric space heating load the time constant of the load will be that of the heated space. Consequently, following a protracted outage, the length of time when all diversity remains lost can be quite long.

The computer model of CLPU which was developed in this work is a physically based, aggregate model of the appliances connected on the distribution system. One portion of these loads consists of appliances whose use is discretional and normally follows some daily pattern (such as lighting, washing). These loads will not have any additional flyback upon restoration. In most cases an appliance which was on when the outage occurred will still be on when power is restored; and vice versa.

Another category of loads is the thermostatically controlled group. This group of appliances includes furnace or heating plant, electric water heater, refrigerators, freezers, heat pump, air conditioner and attic fan. These appliances cycle on and off automatically in response to thermostatic control. Diversity among these appliances is achieved by the diversity of the steady state duty cycle of the appliances.

During an outage the temperature of the controlled space will decay towards the local ambient temperature and gradually diversity will be lost as more and more thermostats demand power. The length of time for this to occur will be dependent on the temperature differential between the controlled space and the outside ambient as well as other factors. The maximum length of time for loss of diversity for any controller will be equivalent to the normal "off time". Typically then the maximum length of time during an outage until all diversity will be lost will be around 20 minutes. Of course, the longer the

Reprinted from *IEEE Transactions on Power Apparatus and Systems,* Vol. 104, No. 3, pp. 704–710, March 1985.

outage lasts, the longer the load will remain undiversified after restoration.

The computer simulation models the thermal behaviour of thermostatically controlled appliances and aggregates the electrical demand from these and the non-thermostatic appliances. The calculation is made for the time period following restoration until diversity is achieved. The loading on the transformer; before, during and after the outage is used to predict transformer hot spot and top oil temperatures as well as to calculate the loss of insulation life attributable to the CLPU occurrence.

Figure 1 shows the general response of a typical thermostatically controlled appliance. The top line shows the response of a controller which was at its upper limit when the outage occurred. The bottom line represents a controller at its lower limit (i.e. just demanding power). From this diagram we can see there will be a time, t_1, after restoration when all appliances demand power. This will persist until such time as the warmest thermostat is satisfied and switches off. Gradually after this time the load would reduce, as more thermostats switch off, until diversity is restored. It is possible that some oscillation of demand would take place due to an, initially, somewhat synchronous cycling of the thermostats; however this is not a significant phenomenon in relation to the effects on transformers or protective equipment. We should also note that; all the rates shown here:

- temperature rise or drop during outage
- times until diversity is lost, remains completely lost and until it is restored

are all dependent on many external factors such as appliance insulation level, heat capacity, capacity of heating or cooling plant, temperature differential and controller hysteresis. This will also help to diversify this oscillation tendency.

Appliance thermal characteristics used in the model were obtained by laboratory and field measurement of the characteristics of individual appliances. The house thermal characteristics were obtained from data on the hourly heating requirement from a computer monitored house heating system during the 1982 winter heating season in Saskatchewan.

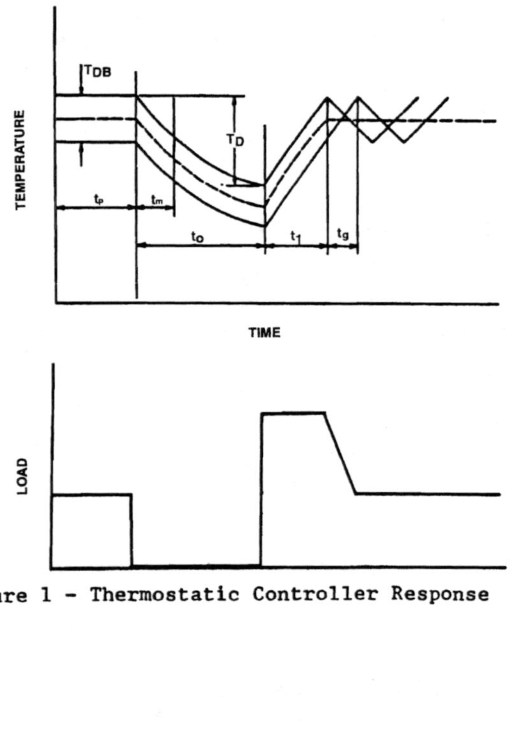

Figure 1 - Thermostatic Controller Response

Transformer thermal behaviour is modelled according to the transient thermal response as described in ANSI/IEEE Standard C57.92-1981. The loss of insulation life computation is also based on data presented in the guide.

Model Verification

In order to verify the computer model it is necessary to have restoration data from several actual outages. Restoration data adequate for this purpose is extremely rare. Since the CLPU flyback component will typically last less than one-half hour the instrumentation must have adequate time resolution to accurately reproduce the restoration waveform. Generally, recording demand instrumentation presents only a 15 minute average demand which is not even synchronized to the restoration time. Consequently the data from these instruments are inadequate for verifying a CLPU model. By the time a 15 minute average is taken most of the restoration event will be over.

A few utilities now have SCADA systems which are capable of high time resolution recording of substation loading. The Ottawa Hydro SCADA system captured the restoration loading for two outages which occurred during the winter of 1982.

Figure 2 shows the restoration loading following the two outages. The first outage occurred on Jan. 11, 1982 at an ambient temperature of $0^{o}C$ and lasted 90 minutes. The second outage; on March 14, 1982, lasted 53 minutes at an ambient temperature of $1^{o}C$. Both outages occurred on the same circuit.

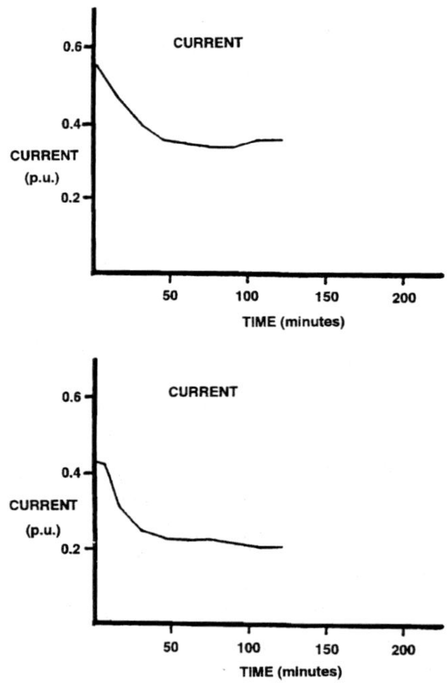

Figure 2 - Outage Restoration Loading; Ottawa Hydro

Figure 3 shows the outage restoration loading predicted by the computer model with the actual loading superimposed on the graph. The agreement of the model to the actual restoration is very good in these cases. The saturation of electric heating is fairly low on this feeder; at 5%, however both the flyback magnitude and duration are accurately predicted.

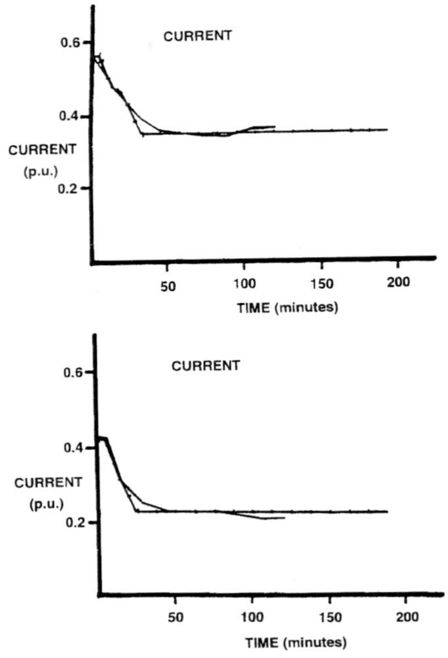

Figure 3 - Restoration Loading Predicted by Computer CLPU Model

Simulation Runs

The CLPU simulation was subsequently used to predict the response for several hundred outages in order to determine the sensitivity of CLPU restoration to the various influencing parameters and to quantify the effects of outages on the substation transformer and protective equipment. The parameters of most interest were the saturation of electric space heating equipment and its maximum demand rating (capacity).

Initially, some basic assumptions were made to arrive at a base case situation. The base case situation established approximately a worst case outage situation on a feeder serving a typical present day loading situation. This involved establishing the following parameters:

- time of day of outage. A preliminary investigation showed the worst case restoration time to be around 11 a.m.
- outage duration. A check of data available on outages at the substation showed 2 hours duration to be typical.
- ambient temperature. Chosen at -40°C; assumed worst case.
- hot water heating. According to a Statistics Canada Appliance Survey[4] 44% of non-electric heat customers have electric water heating. This relationship was used throughout the study.
- number of customers, heating system capacity, transformer capacity and other parameters. These are the main parameters of interest and the base case values represented approximate midrange values.
- House size and insulation levels. These parameters were set at levels to be expected in residences which would convert to electric heating, i.e. R10, R20, R28, $125m^2$(1320 ft^2).

Initial simulation runs were made to establish the approximate sensitivity of CLPU to these parameters. For example Figure 4 shows that from the base case position, (-40°C) an increase in ambient temperature results in a monotonic decrease in the effect on the transformer. Therefore, -40°C is, in fact, worst case (assuming it is the lowest temperature encountered in the geographic area of interest).

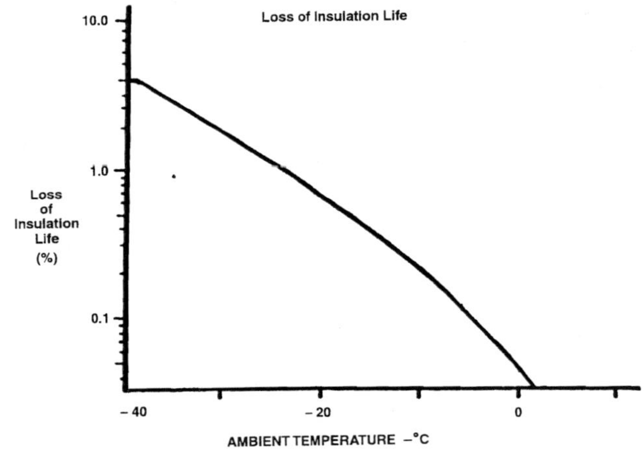

Figure 4 - Effect of Ambient Temperature on Loss of Insulation Life during Restoration

Likewise changing the time of day (coupled with typical diurnal temperature variation) resulted in a worst case situation occurring with restoration at 11 a.m.

From this base case situation the main parameters were then varied to quantify their effect. The next few graphs display these results.

Figure 5 shows the loss of insulation life (LIL) that a transformer would sustain as a function of electric space heat penetration. The capacity or rating of the electric heating system is fixed at 20 kW for this graph. The lines on this graph represent (from left to right) substation capacity of 2.5 kVA per customer, 5, 7.5 and 10 kVA per customer. It is noteworthy that the scale on the vertical axis is logarithmic.

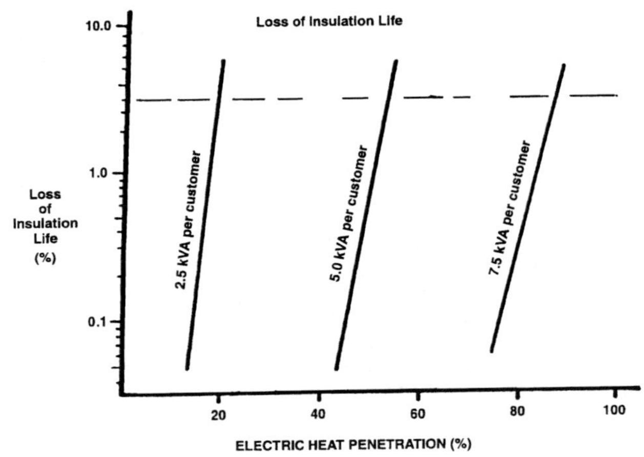

Figure 5 - Loss of Insulation Life vs Electric Space Heat Penetration

Also, if we note that the slope of the lines is very steep we can see that the effect of increasing electric heat penetration is very abrupt and very severe. For example on the 5 kVA line (1000 customers on a 5 MVA unit) with electric heat saturation (EH) = 44%, an outage (at -40°C, 120 minutes, etc.) will result in 0.05% LIL. At EH = 54% the LIL is 5%. At EH = 57% the LIL is 30% for a single occurrence. Transformer load immediately prior to the outage in these cases would be approximately 120%-140% of nameplate (at -40°C). Clearly, when the critical level of electric heat penetration is reached it is an abrupt and severe limit.

The ANSI/IEEE LIL limit of 4% for a single emergency situation is shown on the graph. Due to the steep slope of these curves this limit must be considered an absolute maximum.

Figure 6 shows the maximum hot spot temperature (HST) achieved during restoration displayed in a format similar to the last figure. Once again the increase in hot spot temperature is very steep with respect to increases in electric space heat penetration.

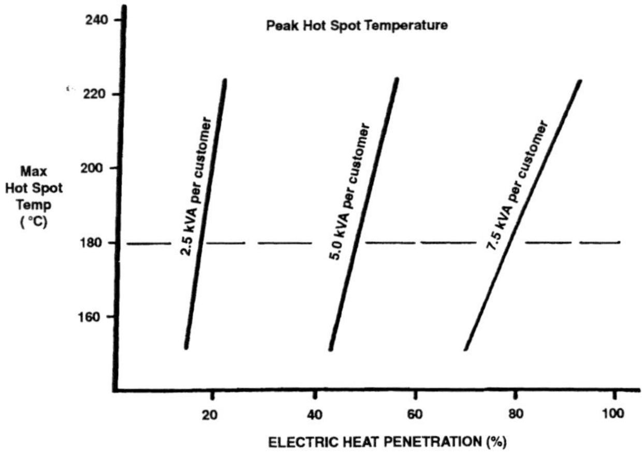

Figure 6 - CLPU Hot Spot Temperature vs Electric Space Heat Penetration

The ANSI/IEEE recommended limitation for maximum hot spot temperature is 180°C. On this graph, at 5 kVA of installed substation transformer capacity per customer, this limit would be reached at an EH penetration of 47%. Pre-outage load in this case would be approximately 130% of nameplate.

Similarly figure 7 depicts the maximum current during restoration expressed as per unit of nameplate rating. Once again the guide limitation of 200 % of nameplate is shown.

The 200% of nameplate limitation is reached, for the same conditions as the previous two figures, at EH penetration of 29% and a pre-outage load of about 100% of nameplate.

All the transformer limitations depicted in the last three charts have been consolidated on Figure 8. This figure shows the critical electric space heat penetration for each limiting criterion as a function of installed substation capacity. Once again it should be noted that this graph refers to the case where the average electric furnace rating is 20 kW and the outage situation is as discussed previously. The effects on the transformer in this case also refer to a 65°C rise unit. A 55°C rise unit sustains damage at a different rate and also has different loss characteristics. A similar chart for 55°C units will be presented shortly.

We can note on this figure that at, say 5 kVA per customer, the 200% of nameplate limitation is reached at EH = 29%; the 180°C limit is reached at EH = 47% and the 4% loss of insulation life limit is reached at EH = 53%. This situation is true for a two-hour outage. For longer outages, since all diversity is lost after only 20 minutes or so, the flyback magnitude does not change. i.e. for a 10 hour outage the 200% limit will remain at 29%.

However, since the flyback will last longer following a longer outage, the transformer temperature will rise higher during restoration and the loss of insulation life will also increase. This will lower the critical EH penetration for each of these limits. In effect, these two limit lines would be lower on figure 8 and would approach the 200% line. These limits will not coincide with the 200% line though, until the outage duration exceeds 10 hours.

Consequently the 200% of nameplate limitation can be expected to be a safe maximum electric heat penetration level for any anticipated outage situation.

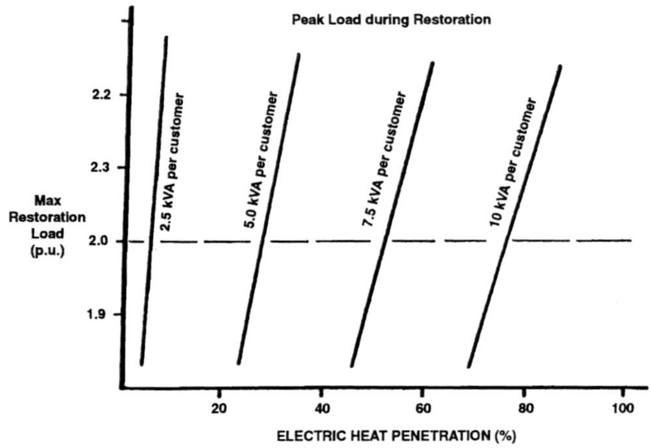

Figure 7 - Peak Restoration Loading vs Electric Space Heat Penetration

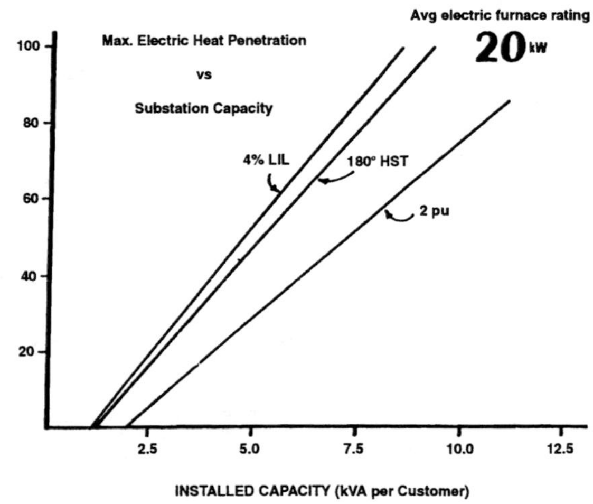

Figure 8 - Critical Electric Space Heat Penetration vs Installed Substation Transformer Capacity

55^oC Rise Transformers

These limitations cannot be applied directly to 55^oC rise transformers. 55^oC rise transformers entail a different insulation system which ages at a different rate than 65^oC rise transformers at a given temperature. Loss characteristics are also different for the two types of transformers.

The ANSI/IEEE recommended limits are also slightly lower for 55^oC rise transformers and are summarized as follows:

- 4% loss of insulation life for single occurrence (same as for 65^oC rise)
- 150^oC maximum hot spot temperature
- 200% of nameplate rating (maximum loading) (same as the 65^oC rise)

Running through the simulation runs for transformers with these characteristics we obtain limitations as shown on figure 9. It can be seen on this graph that the limitations for the 55^oC rise transformer are very similar to those for the 65^oC. In fact the ruling limitation, that of twice nameplate for maximum load, will be identical; since this limitation is load dependent and not transformer dependent.

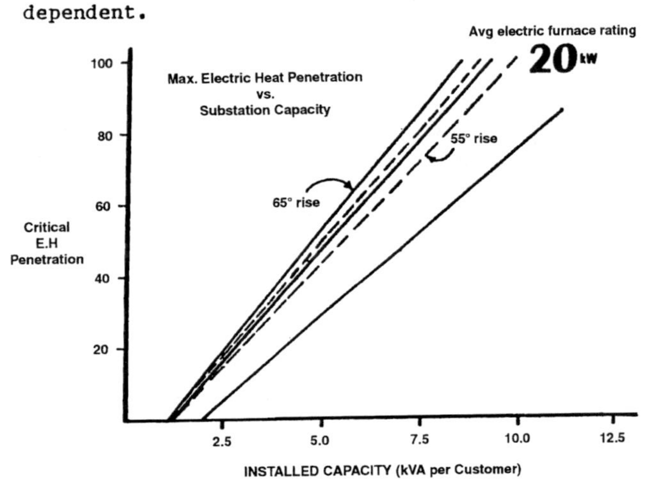

Figure 9 - Comparison of 55^oC Rise Limits To 65^oc Rise Limits

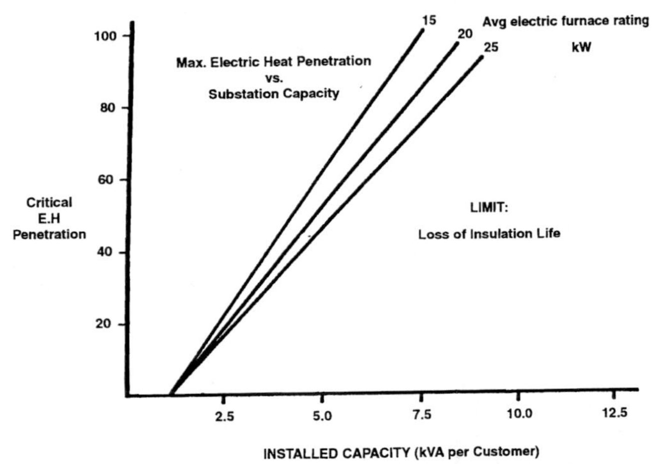

Figure 10 - Relative Effects of Electric Furnace Rating on Loss of Life Limit

15 kW and 25 kW Furnace ratings

Up to this point the discussion has been confined to an average furnace rating of 20 kW. In practice average furnace capacity can range from a low of around 15 kW in the case of baseboard heating up to 25 kW in the case of some large central units.

The next few figures compare the effects of restoration with average furnace capacities of 15 kW to 25 kW.

- Figure 10 shows LIL as a function of electric space heat penetration for 15, 20 and 25 kW. The 20 kW line has been seen previously. We can see here that, as expected, higher furnace capacities result in limiting levels of LIL being reached at lower electric heat penetration and vice versa.
- Figure 11 shows the variation of the maximum hot spot temperature limitation for 15kW, 20kW and 25 kW furnace ratings.
- Figure 12 shows the limitation of 200% loading during restoration for each case.

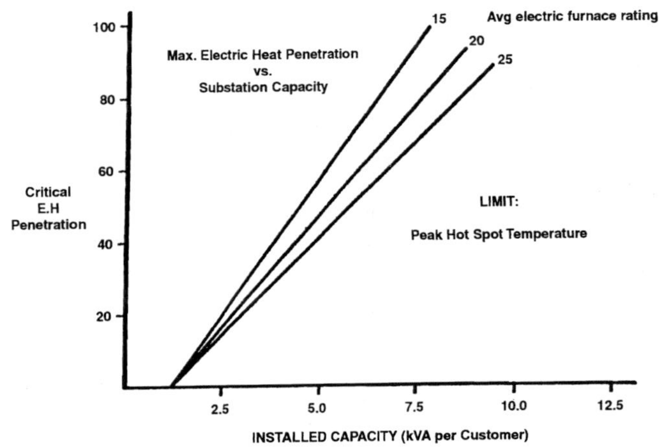

Figure 11 - Relative Effects of Electric Furnace Rating on Hot Spot Temperature Limit

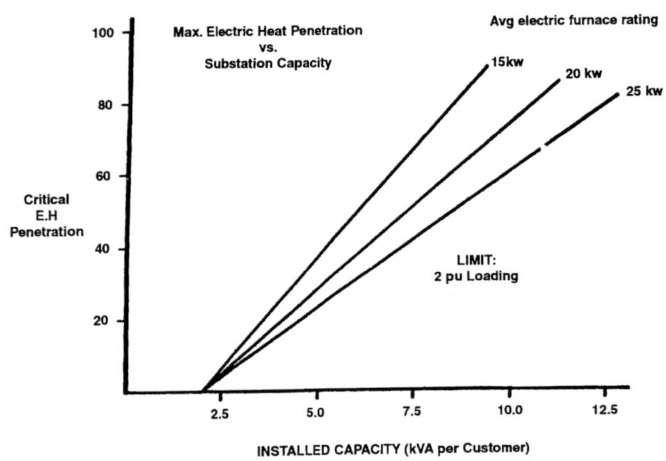

Figure 12 - Relative Effects of Electric Furnace Rating on 2 p.u. Loading Limit

In all cases of furnace demand, the curves have the same relationship seen previously; i.e. The 2 p.u. limitation is the lowest or ruling limit, followed by the 180 degree hot spot temperature and finally by the 4% LIL limitation.

Forced Cooling Limitations

Up to this point the discussion has also been confined to naturally cooled transformers. We are also interested in how CLPU will affect transformers operating in the forced cooling mode.

With forced cooling operative, transformer ratings are substantially increased. At these higher ratings losses are significantly higher than at the self-cooled rating due to copper loss being proportional to the square of the load. With these higher losses the internal operating temperatures will be higher and the hot spot temperature (HST) to top oil temperature differential will be higher. Likewise, similar percentage overloads will result in much higher internal temperatures and loss of insulation life (LIL) compared to natural cooled.

In this study the transformer characteristics used were those suggested in ANSI/IEEE Guide C57.92. The results for forced air, forced oil, and natural cooling are compared for the situation where the average furnace demand is 20 kW. Figure 8 seen previously showed the limitations (LIL, HST and 2 p.u.) for the naturally cooled situation.

In the forced cooling case the 2 p.u. limit will not change since it is load dependent and not a function of temperature ratings. The other limitations though will be affected. The higher losses will result in hotter operating temperatures and higher loss of life. Consequently the 4% LIL and 180°C HST limits will be reached at lower levels of electric heat penetration.

Figure 13 shows the limits for the forced air cooling situation (at less than 133% of natural cooled rating).

Notable effects and limits on this figure (relative to the natural cooled case) are:

- no change in the 2 p.u. curve
- both 180°C HST and 4% LIL curves have moved closer to the 2 p.u. curve.

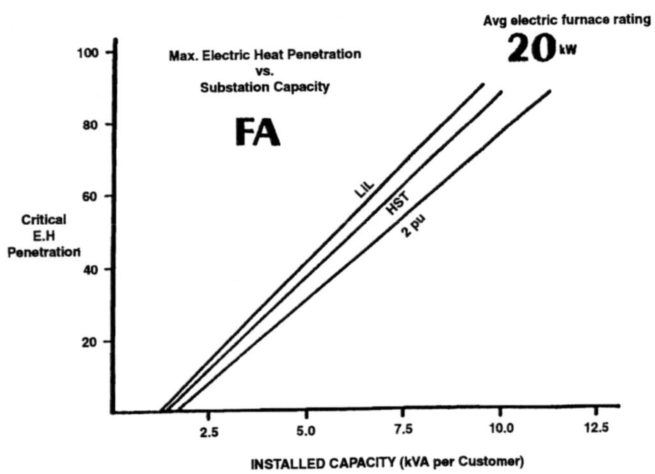

Figure 13 - Critical Electric Heat Penetration for Units Operating at Forced Air Ratings

This is depicted, as previously, for 2 hour outages. For longer outages the curves showing the thermal limits will fall lower and for outages of 8 hours duration would coincide with the 2 p.u. curve. For longer outages obviously the thermal limitations would have to be considered.

Figure 14 shows the same information for the forced oil cooling and forced air at greater than 133% of natural cooled rating. In this case all curves are essentially coincident and for outages longer than 2 hours duration hot spot temperature will be the limiting factor.

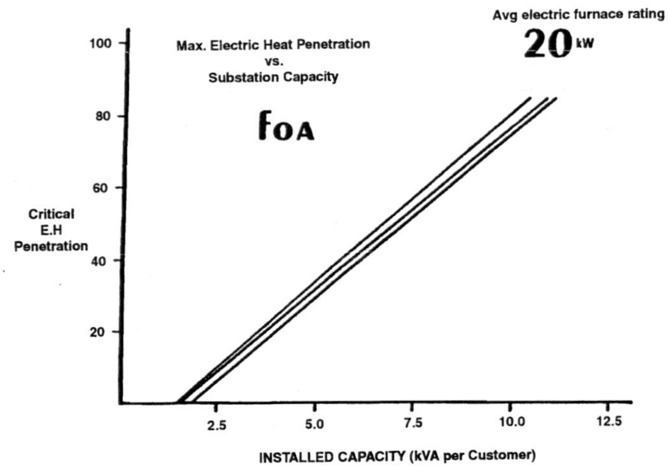

Figure 14 - Critical Electric Heat Penetration for Units Operating at Forced Air (above 133% of natural cooled) and Forced Oil Rating

The Effect of Transient Restoration Loading

The transient component of CLPU restoration current consists primarily of very short duration (less than 1/2 cycle) magnetizing currents for motors, transformers, ballasts, etc., followed by a period of about 0.3 seconds during which motor starting currents are supplied. After this time the transient component can be considered complete as the current abruptly reduces approximately to the level of the enduring component.

The "transient component" then is made up of motor starting currents and other electromagnetic transient currents. The effect of increasing the penetration of electric space heating then will be, for all intents and purposes, practically nil. In converting from non-electric to electric heating no new fan loads will be added without the elimination of water and oil pumping loads.

Even the additional resistance elements from the electric furnaces are usually staged on over a period of several minutes following restoration. Consequently the additional electric space heating load will have no effect on this "transient" component of restoration loading.

The "enduring component" of electric space heat load, though, can have an impact on the protective equipment (fusing or relay). The problem arises because the enduring component increases, in the first several minutes due to staging of heating elements, up to possibly twice nameplate at the substation transformer. The fuse characteristic, on the other hand, decreases during the first 300 seconds to a level of, normally, twice its nominal rating. Generally this level can be tolerated indefinitely.

If the electric heat penetration was such that the restoration loading reached 2 p.u. and the nominal fuse rating at the transformer was equal to the transformer nameplate current, we could expect fuse operation about 5 minutes after restoration. However the transformer will not normally be fused at 1 p.u., so this will not normally be a problem, at the substation. Even if a fuse did operate, it would do so at the level we want to limit the overload to anyway.

In essence then, unless feeder voltage drop results in problem motor starting, the transient component of CLPU will not pose any problems and, at the substation, the enduring component should not pose any problem with fusing. If problems are encountered it is indicative of either excessive feeder voltage drop during restoration or of equipment overload.

On the feeder, on the other hand, there may be laterals or portions of the feeder where the local electric heat penetration is significantly higher than at the substation. On these laterals the flyback could be much greater than the overall average flyback which the substation supplies. Since fusing levels along the feeder are usually based only on coordination of protective equipment, generalizations cannot be made here.

However, if we consider the capacity of the feeder or lateral, at the fuse nominal rating (kVA/customer) we can determine the maximum electric heat penetration.

Figure 15 shows the threshold level of electric heat penetration at which fuse problems could be encountered during CLPU. As indicated, this graph shows the level of electric heat penetration at which fuse melting may occur.

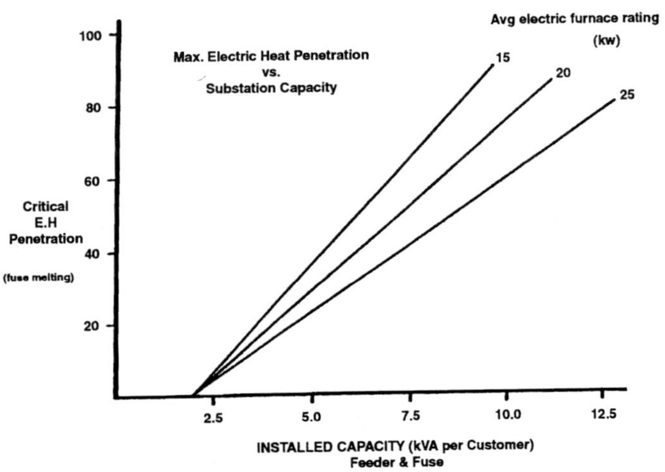

Figure 15 - Critical Electric Heat Penetration for Fuse Melting on Feeders

CONCLUSIONS

As the electric space heating component increases on the distribution system, we may begin to see CLPU problems at loading levels where this problem did not previously exist. Several of the graphs shown previously depicted the critical levels of electric space heat penetration as a function of the installed capacity. These limits are shown in Figure 12 for a range of installed electric furnace capacity from 15 kW to 25 kW. These limits are based on the ANSI/IEEE loading guide for power transformers.

At the substation no adverse effects should be seen on transformer protective equipment; due to CLPU unless the transformer is in fact being unduly overloaded during the CLPU restoration or a severe voltage drop problem exists. Excessive voltage drop could result in motor starting problems which could then interfere with proper operation of protection equipment.

Portions of a feeder or a lateral may be exposed to a very high local penetration of electric heating which may result in fuse operation during CLPU condition.

ACKNOWLEDGEMENTS

The author wishes to acknowledge the support of the Canadian Electrical Association (CEA) which funded the research reported in this paper. The advice of the CEA project monitors, Mr. V. Steciuk of Manitoba Hydro and Mr. R. Fulton of B. C. Hydro and Power Authority is also appreciated.

REFERENCES

1. "Guide for Loading Oil-Immersed Distribution and Power Transformers", ANSI/IEEE C57.92 (1981).
2. "Cold Load Pickup Problems Relating to Distribution Transformers", Wilde, R. L., CEA Final Report RP77-46, October, 1979.
3. "Effects of Cold Load Pickup at the Distribution Substation Transformer", Wilde, R. L., CEA Final Report 131 D 247, June, 1983.
4. Statistics Canada Appliance Survey, April, 1982.

AN EXPERT SYSTEM OPERATIONAL AID FOR RESTORATION AND LOSS REDUCTION OF DISTRIBUTION SYSTEMS

Chen-Ching Liu, Member Seung Jae Lee, Student Member S. S. Venkata, Senior Member

Department of Electrical Engineering
University of Washington
Seattle, WA 98195

ABSTRACT

This paper presents a new application of expert system techniques to the restoration of distribution systems. Primary feeders are typically radial in structure. To increase system reliability, neighboring feeders are connected through a normally open tie switch. When load zones on a feeder are interrupted due to a fault, system operators need to identify neighboring feeders and try to restore customers through the tie switches. To restore maximal number of zones, several steps are followed: group restoration, zone restoration and, if necessary, load transfer. Based on the methodology, an expert system is developed which can be utilized as an on-line aid to system operators in a distribution SCADA environment. The proposed expert system is implemented in Prolog. Implementation issues on knowledge representation, portability of the system, and computational efficiency are discussed. Several examples are used to illustrate capabilities of the proposed system.

1. INTRODUCTION

In recent years, expert system techniques have been successfully applied to power system problems. An area of particular interest is the development of expert systems as power system operator's aids in control center environments. The scope of this area covers alarm processing, system diagnosis, control, and restoration after a major system outage. The literature is briefly reviewed here. Alarm processing deals with three main issues: reduction of unnecessary alarms, interpretation of alarm messages, and presentation of concise diagnostic results to dispatchers[1-4]. An "intelligent" alarm processor to perform these tasks would greatly enhance a system's capability to deal with emergency conditions. The development of such a processor is an active area of research. A workshop in the 1986 CIGRE conference was devoted to the application of expert systems in alarm handling.

Once an abnormal system condition is detected, dispatchers need to determine appropriate controls to eliminate any violation of operating constraints. The expert system proposed in [5] incorporates empirical rules adopted by dispatchers in resolving abnormal voltage profiles. In general, the restoration of an interconnected power system is a complicated problem. Many factors need to be considered, e.g., load requirements, frequency variations, system voltages, unbalanced conditions, protection, stability, etc.[6,7,8]. We have developed an expert system to assist dispatchers in locating faults and restoring customers in power lines equipped with automatic protective switches[9].

The restoration of a distribution system is, relatively speaking, a more manageable problem. On the one hand, primary feeders are typically radial in structure. Also, in urban areas, feeders are designed such that customers can be restored through neighboring feeders connected through normally open tie switches. However, if a large-scale outage occurs, say, due to a storm, more customers may be interrupted and hence system operators would have to find a number of adjacent feeders and try to group customers so that they can be picked up by the feeders. In doing so, the operator needs to ensure feasibility of the restoration plan, i.e., no capacity or operating constraints should be violated.

In this paper, we propose an expert system for the distribution system restoration. We have constructed a knowledge base of approximately 180 rules which can be used by operators in planning a restoration scheme. The incorporated knowledge is acquired from the literature and discussions with distribution engineers. The proposed method is a generalization of the procedures and priority scheme reported in [10]. In particular, our method allows restoration of feeder zones in groups. The acquired knowledge is implemented in Prolog, which is gaining popularity among languages available for expert system development due to its flexibility and friendliness. To evaluate the computational performance, two slightly different versions were produced; one for the mainframe and the other for personal computers. For a typical scenario, the expert system can formulate a feasible restoration plan within a few seconds. The computational speed should be satisfactory for the future application in actual control centers.

This paper also proposes some new rules related to a problem encountered in normal operation: finding a feeder configuration with reduced line losses[11]. The proposed rules are simple but require the evaluation of distribution load flow. However the load flow needs only be carried out for a pair of feeders, instead of the entire system. This is due to the observation that only at most two primary feeders are affected when the location of an open tie switch is moved.

The organization of this paper is as follows. Section 2 describes the operation envi-

Reprinted from *1987 IEEE Power Industry Computer Application,* pp. 79–85, May 1987.

ronment of the proposed expert system. In Sec. 3, the developed knowledge base for system restoration is summarized. Our results on the selection of system configuration with reduced losses are presented in Sec.4. Section 5 deals with issues on implementation of the expert system. Finally, several examples are provided based on a subsystem of Puget Sound Power and Light Company, Washington.

2. PROPOSED EXPERT SYSTEM APPROACH

In a dispatching center, system operators monitor the system through the distribution SCADA and the real-time database. An expert system operational aid can be used to provide suggestions to operators so that customers can be restored quickly. This is particularly meaningful in emergency conditions when a large number of customers are interrupted.

In general, the expert system approach is beneficial for problems that involve non-trivial logic reasoning and human knowledge. For such problems, the expert system technique provides a natural way to represent the knowledge and implement the inference procedure. This paper deals with the control of primary feeder configuration. The problem possesses the following features which suggests an expert system approach:

1) The restoration problem deals with grouping customers and searching among neighboring feeders. The problem involves more logic reasoning and only minor numerical computation.
2) The grouping requires operator's judgement on system capacities and specific system or customer information. It is easy to incorporate the information in the rule base.
3) The software is to serve as an operator's aid; the expert system can help provide a user-friendly interface.
4)It is difficult to formulate the configuration restoration as a mathematical programming problem because of its combinatorial nature.

The function of the proposed expert system can be described by the block diagram in Fig.1. This study deals with the system restoration and therefore it is assumed that the fault location is already known. The expert system component can be developed to perform as a fault diagnostician(FD)[9].

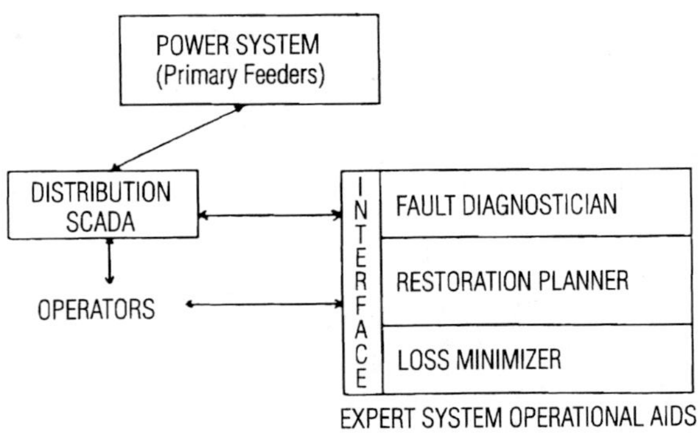

Fig. 1 Functional description of expert system operational aid

Two basic functional blocks dealt with in this paper are Restoration Planner(RP) and Loss Minimizer(LM). During an emergency condition when power outages occur, the former is capable of formulating a restoration plan to restore service to as many customers as possible. The capabilities include grouping zones in the primary feeders, checking capacity and operating constraints to ensure feasible restoration and planning load transfer to allow more customers to be restored. On the other hand, Loss Minimizer deals with the normal system condition when it is desirable to reduce the line losses while maintaining a reliable system configuration. The task of this block is to find the locations of open switches on the primary feeders so that the loads are distributed in such a way that the total loss is reduced. In Fig.1, Interface contains rules for user interface and those used to determine which function(FD,RP or LM) should be performed for a given system condition.

The assumptions made in this study are given here. First, it is assumed that the data and status of primary feeders, zones, switches and loading conditions are available through the distribution SCADA system or personnel communication. Secondly, only permanent line faults are considered since temporary faults can be eliminated by breaker reclosures and therefore no outage condition will sustain. Finally, the scope of this program is to deal with the restoration problem and therefore it is assumed that the fault location is already identified.

3. RESTORATION PLANNER(RP)

3.1 General Strategy

The function of RP is to come up with switching actions necessary for restoring customers affected in an outage. Based on the assumptions that the fault location is known and that the system data and switch status are available, RP is capable of formulating a set of breaker and/or switch actions so that the maximal number of outage zones in the feeders can be restored. To do this, the general strategy is hierarchical; one first identifies the whole outage area and divides it into groups of zones based on the adjacency of feeders. In other words, it is intended to restore each group from a neighboring feeder. However, the capacity and operating constraints will be checked to ensure the restoration is feasible.

Now if any feeder is unable to pick up an assigned group, the zones in this group will be attempted one by one. This process will be called zone restoration. After this step, if some zone still can not be restored, then a load transfer would be necessary. By transferring enough load from a neighboring feeder to a farther one, the neighbor will be able to pick up a zone. The transfer will be attempted for every feeder adjacent to the neighboring feeder. In this sense, the program attempts to restore the maximal number of outage zones.

It is remarked that in this study a configuration is considered feasible if the substation transformer MVA and line current capacities are not exceeded and voltage limits are satisfied at the remote ends of feeders.

In the evaluation of a restoration plan, voltages at the remote ends are calculated by approximate distribution load flow commonly used by engineers. Finally, primary feeders are usually designed to have a capacity margin enough to back up its neighboring feeders in case of emergencies. Therefore the proposed grouping idea would lead to a simpler and faster solution.

3.2 The Knowledge Base

The knowledge base of RP contains several segments: Group Restoration, Zone Restoration, Load Transfer, Switching Actions Determination and Inference Control. The first four segments are described in the following:

1) Group Restoration

It is assumed that the faulted section of a feeder is isolated. Due to the radial structure, one side of the zone can be restored by reclosing the breaker. The other side of the zone farther away from the breaker would have to be picked up by neighboring feeders. For ease of discussion, The open switch at the far end will be called the root switch. Note it is possible to have more than one root switches. The configuration of the interrupted feeder section with neighboring feeders can be viewed as a tree rooted at the root switch. Each tree branch terminates at an open tie switch adjacent to a neighboring feeder.

To find the groups with the corresponding back-up feeders, the following general rules are developed:

1. If a path from a root switch to the termination of a branch contains the minimal number of zones among all such paths, then identify the zones in this path as a group and include the switches at the branching points to other feeders into the list of root switches.

2. If a feeder is assigned a group of zones and the operating constraints are satisfied, then restore the group from this feeder and delete the corresponding root switch from the list.

3. If rule 2 above fails, then delete the corresponding root switch from the list and put these zones in this group into the list of unrestored zones.

The detailed work of the above general rules is performed by a set of rules related to the search process, calculating the approximate remote-end voltages, checking operating constraints, etc.

The system in Fig.2 will be used to illustrate the function of group restoration. For the indicated fault location, the root switch in the beginning of the search will be S2. Two groups will be identified here; the first one from S2 to S3, and the second from S4 to S5. They will be assigned to Feeder F2 and Feeder F3 respectively, if the operating constraints are satisfied.

2) Zone Restoration

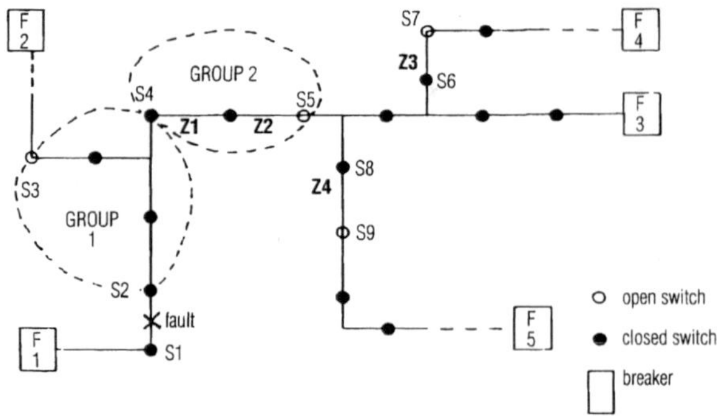

Fig. 2 Example system for service restoration

Suppose that a group of zones can not be restored from the results of applying the Group Restoration rules. Rules in this segment can be utilized to find a plan to restore as many zones as possible. Again, the method here will be summarized by a set of general rules.

1. If a zone has an identified candidate feeder and the feeder can pick up the zone without violating any operating constraints, then assign the zone to this feeder and delete this zone from the list of unrestored zones.

2. If a zone has more than one adjacent feeders, then the feeder with highest operating margin will be selected as the candidate feeder.

3. If there is any zone in the list of unrestored zones, then attempt to restore the next zone in the list.

As an example, in Fig.2, suppose group 2(S4-S5) can not be picked up by F3. Since F3 is the only adjacent feeder to Z2, the restoration of Z2 through F3 will be attempted. If it turns out successful, then Zone Z1 will be considered next. Note Z1 has two adjacent feeders, F3 and F2. By the rules in this segment, F2 is the only candidate since the margin of F3 is insufficient for this zone.

3) Load Transfer

After performing Zone Restoration, some zones may still be unserved since neighboring feeders have their loading limits and operating constraints need to be satisfied. In this case, some load of a neighboring feeder need be transferred so that the restoration can continue. If a particular zone can not find enough transferrable loads, then the zone will be declared as unrestorrable and the dispatcher will be informed. The general rules for Load Transfer are shown below :

1. If a zone is unrestored and it is intended to restore the zone from a feeder, referred to as the main feeder, which does not have enough margin, then choose a back-up(adjacent) feeder which is capable of supplying the load in the neighboring zone which is being served by the main feeder.

2. If a back-up feeder is able to pick up a zone to be transferred then assign the zone to this feeder and attempt to restore the unserved zone from the main feeder.

3. If the restoration in Rule 2 fails, then continue the transfer with another back-up feeder.

4. If the zone can not be restored by transferring load to its adjacent feeders then declare this zone unrestorrable.

Consider again the example in Fig.2. Suppose Zone 1 can not be picked up by F2. Based on results of the previous restoration steps, there is no other adjacent feeder which has enough margin to pick up Z1. So, transfer of end zones of F3 will be attempted. There are two possible transferrable loads, Z3 and Z4. First Z3 is transferred to F4 and F3 is checked to see if it is now able to pick up Z1. Suppose not, then Z4 will be transferred to F5, and the process will continue.

4) Determining Switching Sequence

The switching sequence will be determined based on the following rules.

1. If the fault is not isolated, then open the switch(es) next to the fault location.

2. If the fault is isolated and a breaker is separated from the fault by an open switch then close the breaker.

3. If Load Transfer was used in restoration, then determine the switching actions for the transfer.

4. Open switches before closing other switches.

For the system in Fig.2, the sequence of switching actions are given below.

1. Open S1 and S2 for fault isolation
2. Close the breaker on F1
3. Close S7 and open S6 to transfer Z3
   Close S9 and open S8 to transfer Z4
4. Open S4 and close S3 and S5

4. LOSS MINIMIZER

In a normal condition, it is desirable to operate a system with minimal power losses. A system configuration is determined by the status of all breakers and switches. Therefore, the issue here is to find a particular combination of open and close switches to achieve the operational goal. Practically, however, even if such a configuration is found, several factors may affect the decision of whether to adopt the configuration; for instance, reliability consideration, past experience, etc. In practice, system operators tend to make minor adjustments rather than drastic changes. In this section, a set of rules is proposed which can be used to find locations of open switches resulting in lower total system losses. We will present the proposed rules below.

We will define a feeder pair in a radial system first. A feeder pair is a pair of feeders joined by an open switch. In Fig.3, a feeder pair is shown as a subsystem in a distribution system with a radial structure. Based on this concept, the following general rules are discovered.

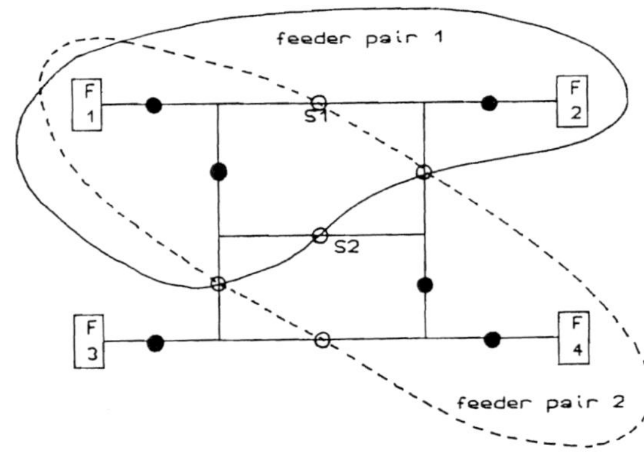

Fig.3 Feeder pairs

General Rule 1 :
If the losses in a feeder pair are reduced then the total system losses are also reduced by the same amount.

The above rule is a direct result of the observed disjoint property of a feeder pair with the rest of the system.

General Rule 2 :
If the feasibility of a feeder pair is maintained in a move, then the system also remains feasible.

General Rule 3
If the location of an open switch in a feeder pair can be moved to the neighboring switch and the move results in a reduction of feeder pair losses, then move the switch position.

If the move process is repeated for different switches based on rules 1 and 2, the system losses can be reduced step-by-step. To apply the above rules, the following observation is useful. For a given open switch position, there are at most two candidate close positions for the move. They are the closest switches on the two feeders in the selected pair. Note it is impossible to move into a branch of these feeders; if this is the case, one would encounter a situation where some zones are disconnected from sources. Furthermore the selected feeder pair will be directly connected and the radial structure will be destroyed. To evaluate each reduced-loss move, it is necessary to calculate the line losses that would result from a move. In this study, we use an approximate distribution load flow step to obtain the result.

The proposed rules are simple but quite useful for evaluating options of minor configuration changes. One possible scenario is that the loading in a zone increases by an amount significant enough to cause a concern on loss increase. System operators may decide to reconfigure the system by making minor changes in open switch locations.

The proposed scheme, however, does not guarantee global minimality of line losses. For this purpose, one would have to resort to an algorithmic method.

5. IMPLEMENTATION

The implementation of an expert system involves a number of issues; for example, computer language, knowledge representation, validation of accuracy and performance evaluation. In this study, we chose the Prolog language. The software package is widely available for personal computers at a very low cost. The language selection is appropriate because in planning a restoration one would set up a goal, e.g., group restoration, and then try to satisfy this goal. For group restoration, one would have to satisfy the subgoals of availability of neighboring feeders, satisfaction of operating constraints. If it happens that a subgoal is not satisfied,e.g., an operating constraint is violated, then the inference procedure will backtrack to search for another candidate. The above backtracking scheme is embedded in the Prolog inference engine{16}.

In Prolog, a goal to be satisfied and the corresponding subgoals can be represented by a rule. The following rule associated with the group restoration is given.

```
group_restoration :-
    group(Zones_in_group,Neighbor,Tie_sw),
    feasibility_check(Zones_in_group,
                      Neighbor,Tie_sw),
    delete_restored_zones(Unrestored_zones,
                          Zones_in_group),
    add_restored_zones(Restored_zones,
                       Zones_in_group).
```

Portability of an expert system is an important issue. The developed rules for restoration are general in the sense that it implements the restoration methodology rather than the specific structure or procedures of a particular system. Our only assumption on the power system structure is that the system possesses a group of feeders joined by normally open tie switches. It is noted that the proposed method is valid for systems with automatic switches. This is because the switches are designed to operate following the faults and therefore, once the system configuration(on,off status of switches) settles down, the restoration process can be initiated.

Besides rules, we use frames to represent the knowledge about a power system. Conceptually a frame is similar to a table summarizing relevant attributes pertaining to an object. An example is given here. Suppose a particular zone in a system can be described effectively with a table, such as the following:

frame zone

slot	value
name	Z32
ampacity	600
kW load	920
kvar load	280
line R	0.145
line X	0.106

The actual Prolog representation of this frame would be

```
zone(name(Z32),amp(600),kw(920),kvar(280),
                      r(0.145),x(0.106))
```

or simply zone(Z32,600,920,280,0.145,0.106).

Two slightly different versions of the Prolog programs were developed for RP; one for a medium-size mainframe IBM-4341 and the other for an IBM-PC. Between the two versions, the differences are some symbols, availability of If-then-else structures and garbage collection facilities in the mainframe. Also the mainframe version is interpretive while the PC one is compiled. Normally the compiled version is expected to be faster, however, in this case the difference between a mainframe and a PC prevents a meaningful comparison. Following the illustrative examples, the computation time of various scenarios will be presented for both Prolog versions. In general, the developed Prolog system appears to be efficient enough for this application; a typical program execution would take 2-5 seconds on IBM4341 mainframe and 3-9 seconds on IBM-AT.

The subprogram LM is implemented in Pascal due to the computational requirements of the approximate load flow and line losses. Both the Prolog and Pascal programs can be executed on the PC.

6. EXAMPLES AND RESULTS

In order to demonstrate the usefulness of the proposed method, a real system provided by Puget Sound Power and Light Co., Washington, was used. The example system is an underground system with a primary voltage of 12.5 kV and 3 substations with a total of 6 primary feeders. There are 43 zones and 43 switches as shown in Fig.4 and 6 open tie switches maintain a radial structure of the system. This section provides three test cases for service restoration and one for loss reduction. The test results are summarized in the following:

EXAMPLE 1. System Restoration

Case 1. Group Restoration

The fault is assumed to have occurred on Z27 and the feeder breaker has tripped. Once the fault is isolated, the breaker will be closed to restore the zones Z25 and Z26. The group restoration rules are used to identify the groups with the feeder which is intended to pick up the group. Three groups were successfully restored.

```
Results :
  Switching actions for fault on Z27 =>
          * fault isolation :
                 open U2117a, U553a
          * close breaker F1
          * open   U2122b
                   U2118b
          * close  UG1506a
                   U2121c
                   X1251
```

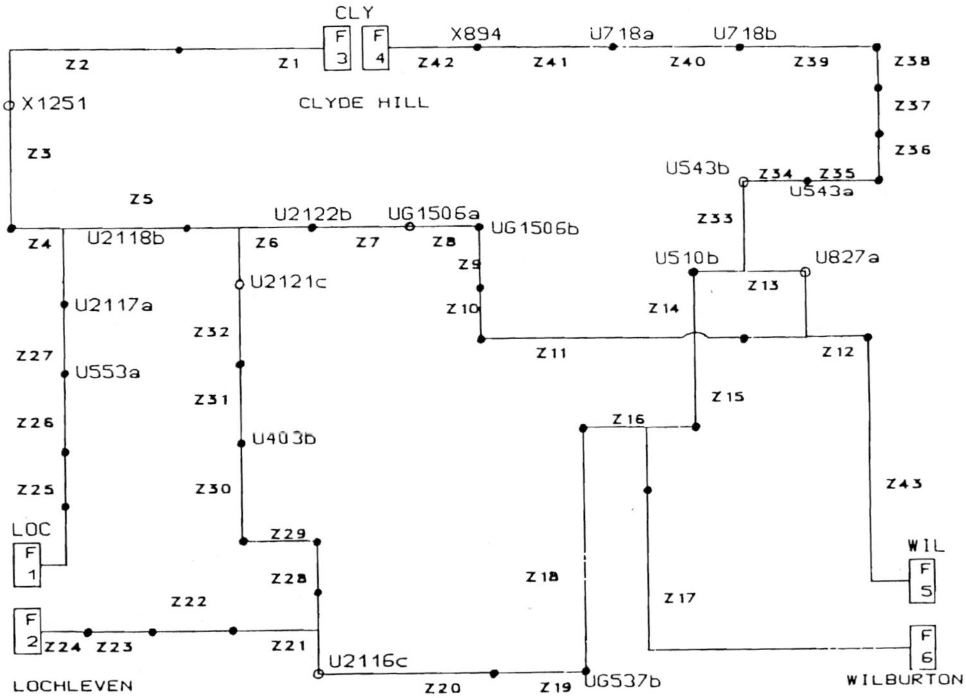

Fig. 4 A sample distribution system

Case 2. Zone Restoration and Load Transfer

The fault is assumed to be located in Z41. After isolating the fault and closing the breaker for restoration of Z42, group restoration rules are invoked , but the attempt fails for lack of adequate margin at 'WILBURTON' transformer. As a result, the Zone Restoration rules will be used and F6 feeder is assigned all interrupted zones except Z40. At this step, the capacity of WILBURTON does not allow further loading. By transferring the loads in Z20 and Z19 from F6 to F2, restoration of Z40 was achieved through feeder F6.

Results :
Switching actions for fault on Z41 =>
* fault isolation :
open X894, U718a
* close breaker F4
* load transfer
close U2116c
open UG537b
* close U543b

Case 3 Minimal Outage

Suppose the fault lies in Z42, In this case, after Zone Restoration, two zones Z41 and Z40 remain unrestored and the utilization of Load Transfer rules leads to the restoration of Z40. To achieve this, Z19 and Z20 are transferred from F6 to F2. The program would attempt to transfer Z18 in order to further pick up Z41. This was unsuccessful because F6 is unable to pick up Z40, even if zone Z18 can be transferred to F2. Furthermore all transferrable loads(those zones in the branches of F6) have been exhausted.

Results :
Switching actions for fault on Z42 =>
* fault isolation :
open X894
* leave breaker F4 open !
* load transfer
close U2116c
open UG537b
* open U718a
close U543b
** unrestored zones
Z41

Table I shows the comparison of processing time for two programs.

Table I Processing time measurements

	IBM 4341	IBM PC-AT
Case 1	1.68* (1.73)+	3.03 (3.56)
Case 2	4.51 (4.56)	6.82 (7.45)
Case 3	4.62 (4.69)	7.34 (7.99)

unit sec.
* output messages deleted
+ output messages included

EXAMPLE 2. Loss Reduction

The same system in Fig.4 is used for demonstration of the proposed rules for configuration adjustment. The scenario here is fictitious; two feeders are assumed to be loaded at 80 % of the capacity and the others at 50 %. Based on the proposed rules and the approximate load flow, five open switch positions are moved, resulting in 12 % reduction in total system losses(from 40.6 kW to 35.4 kW per phase). The results are shown in Table II. For lightly loaded cases(35 - 50 %), the loss reduction was about 6.5 %.

Table II Results of loss reduction

Number of moves	From SW	To SW	From Loss (kW)	To Loss (kW)
1	u2121c	u403b	8.86	8.36
2	u827a	u510b	13.96	13.91
3	u543b	u543a	19.33	16.51
4	u2116c	ug537b	13.22	12.68
5	ug1506a	ug1506b	14.28	13.11

7. CONCLUSION

A new application of expert system techniques to distribution system restoration is presented in this paper. We have constructed a knowledge base which contains rules that implement a solution approach that system operators can use in order to restore as many load zones as possible. The developed rules are general or portable in the sense that they do not depend on any particular system. The accuracy of the knowledge and reasoning process was validated through computer simulations. The computational speed is also satisfactory since it was able to come up with a plan in a few seconds, which is adequate for this application.

The actual implementation in a distribution SCADA environment has not been attempted. However, in an on-going project sponsored by EPRI, the University of Washington is investigating the hardware and software issues involved in embedding expert systems in control centers. To incorporate the rule-based programs, one option is to install them in the existing SCADA computer. This, however, requires an adequate computational power and memory storage. Another approach is to introduce an add-on machine to support the expert systems. In this case, interface issues between machines need to be considered and computational speed may suffer to some extent. For distribution systems, personal computers are attractive options since the area covered by a distribution SCADA is smaller, and personal computers are low-cost and easy-to-use.

ACKNOWLEDGEMENT

The authors would like to thank Mr. Bill Kotsogean, Puget Sound Power and Light Co., for his assistance and helpful discussions.

REFERENCES

[1] B.F.Wollenberg, "Feasibility Study for an Energy Management System Intelligent Alarm Processor", IEEE Trans. on Power Systems, Vol.PWRS-1,No.2, pp.241-247, May 1986.

[2] K.Komi, T.Sakaguchi and S.Takeda, "Power System Fault Diagnosis Expert System Enchanced by the General Problem Solving Method", Proc. IASTED Internatioal Conference, High Technology in the Power Industry, Montana, Aug. 1986.

[3] S.N.Talukdar, E.Cardozo and T.Perry, "The Operator's Assistant- An Intelligent, Expandable Program for Power System Trouble Analysis", IEEE Trans. on Power Systems, Vol.PWRS-1, No.3, pp.182-187, Aug. 1986.

[4] C.Fukui and J.Kawakami, "An Expert System for Fault Section Estimation Using Information from Protective Relays and Circuit Breakers", 86 WM112-7, IEEE PES Winter meeting, New York, Feb. 1986.

[5] C.C.Liu and K.Tomsovic, "An Expert System Assisting Decision-Making of Reactive Power/Voltage Control", IEEE Trans. on Power Systems, Vol.PWRS-1, No.3, pp.195-201, Aug. 1986.

[6] T.Sakaguchi and K.Matsumoto, "Development of a Knowledge Based System for Power System Restoration", IEEE Trans. on PAS, Vol.PAS-102, No.2, pp.320-329, Feb. 1983.

[7] Second Draft of International Questionnaire on Factors involved in Modelling and Simulation of Black Start and Restoration of an Electric Power System, presentation by Y.Ichida, WG38-02, CIGRE, Paris, 1986.

[8] F.Wu and A.Monticelli, "Analytical Tools for Power System Restoration - Conceptual Design", 86 WM105-1, IEEE PES Winter Meeting, New York, Feb. 1986.

[9] K. Tomsovic, C. C. Liu, P. Ackerman and S. Pope," An Expert System As a Dispatchers' Aid for the Isolation of Line Section Faults", 1986 IEEE PES Transmission and Distribution Conference, Anaheim, CA.

[10] S.Kato, T.Naito, H.Kohno and et. al, "Computer-Based Distribution Automation", Proc. IEEE 1985 PICA Conference, pp.374-380.

[11] K.Aoki, T. Ichimori and M. Kanezashi, "Normal State Optimal Load Allocation in Distribution Systems", 86 WM151-5, IEEE PES Winter Meeting, New York, Feb. 1986.

[12] F.Hein, "Expert System using Pattern Recognition by Real Time Signals", Proc. CIGRE Conference, Paris, 1986.

[13] R.J.Kafka, D.R.Penders,S.H.Bouchey and M.M.Adibi, "System Restoration Plan Development for a Metropolitan Electric System", IEEE Trans. on PAS, Vol. PAS-100,No.8, pp.3703-3713, Aug.,1981.

[14] C.H.Castro, J.B.Bunch and T.M.Topka, "Generalized Algorithms for Distribution Feeder Deployment and Sectionalizing", IEEE Trans. on PAS, Vol.PAS-99, No.2, pp.549-557, Mar./Apr. 1980.

[15] K.L.Clark and F.G.McCabe, "PROLOG : A Language for Implementing Expert Systems", Machine Intelligence, Vol.10, pp.455-475.

[16] W.F.Clocksin and C.S.Mellish, Programming in Prolog, 2nd ed. New York, Springer-Verlag,1984.

A NEW ALGORITHM FOR SERVICE RESTORATION IN DISTRIBUTION SYSTEMS

K.Aoki, Member; K.Nara, Member; M.Itoh, Student; T.Satoh, Member
Faculty of Engineering
Hiroshima University
Higashi-Hiroshima, Japan

H.Kuwabara
Member
Faculty of Engineering
Kinki University
Kure, Japan

Abstract - This paper proposes a new algorithm to quickly restore the de-energized loads in a distribution system by using the sectionalizing switches. The computational burden and the solution accuracy of the algorithm is improved by using the concept of the dual effective gradient method. Test results on practical system are given to demonstrate that the algorithm can be used in actual, large scale urban distribution system applications.

Key Words - Distribution System, Distribution Automation, Automatic Reconfiguration, Restorative Operation, Combinatorial Optimization

1. INTRODUCTION

Urban distribution systems in Japan are usually composed of many radial type feeders. Each feeder is divided into load sections with sectionalizing switches and has connections to other feeders via several open cut(tie) switches. Fig.1(a) shows such a small distribution system. The system in the figure is radial; the feeders f_1, f_2, f_3 and f_4 are separated by each other through cut(tie) switches, $C_1,\dots,C_5$ and the loop in feeder f_1 is kept open by loop switch L_1.

The main purpose of sectionalizing switches is the isolation or restoration of loads during a fault or a maintenance. For example, when a fault occurs on a bus, on a transformer, or on one of the load sections of the feeder, all the load sections which are surrounded by the cut switches, including fault point, are isolated from the system by the following procedure.

(i) a protective relay detects the fault and "trips" the Circuit Breaker.
(ii) a fault section is detected by reclosing CB and sectionalizing switches one by one (when fault section is energized, protective relay detects the fault again).
(iii) the load sections between the bus and the fault section are energized again.

In Fig.1(b) such a case is illustrated: the fault between the sectionalizing switches A and B causes these switches to open to isolate the fault and leaves the sections surrounded by switches C_1, C_2, C_4, and B de-energized. Note that de-energized loads form an isolated island.

Service restoration involves energizing the loads in the isolated island by transferring these loads to adjacent feeders, called "support feeders", by using the cut switches. For example, in Fig.1(b), support feeders are feeder f_3 with cut switches C_1 and C_2 and feeder f_2 with cut switch C_4.

89 WM 085-2 PWRD A paper recommended and approved by the IEEE Transmission and Distribution Committee of the IEEE Power Engineering Society for presentation at the IEEE/PES 1989 Winter Meeting, New York, New York, January 29 - February 3, 1989. Manuscript submitted August 25, 1988; made available for printing November 28, 1988.

Service restoration after a fault, usually referred to as emergency service restoration, has become more important especially as the distribution systems are becoming automated to improve the service quality. Use of remotely controlled sectionalizing switches in such environments makes it possible to perform emergency service restoration from a control center in much shorter time. It is clear that the emergency reconfiguration should be completed quickly to shorten the black-out. However, the reconfiguration problem is too complicated and too time consuming to solve. Because the reconfiguration problem is basically a large scale combinatorial optimization problem, reasons being that the urban distribution system is usually large in scale and has numerous sectionalizing switches to be operated. Therefore, it is very important to avoid solving the combinatorial problem in the reconfiguration calculation of a large scale distribution system. In this paper, an advanced systematic and non-combinatorial algorithm to restore a fault state large scale urban distribution system is proposed.

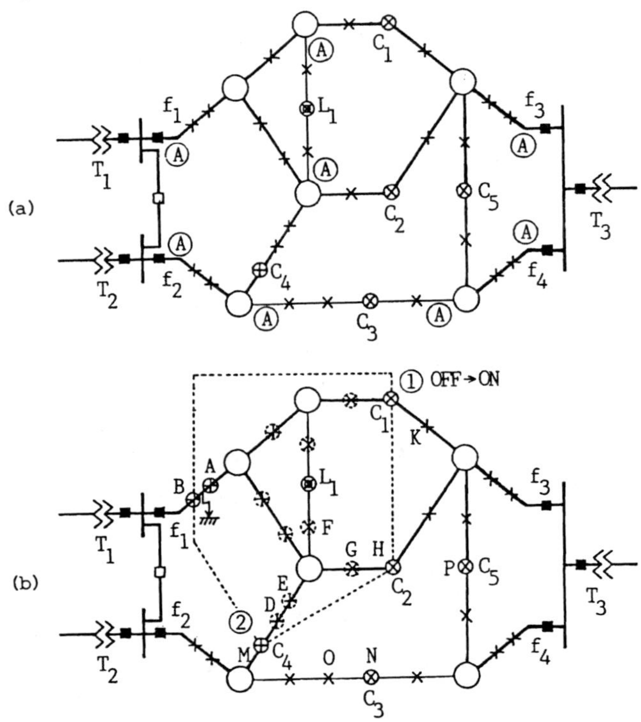

— : distribution line with large diameter
— : distribution line with small diameter
○ : branching point
× : sectionalizing switch "ON"
⊗ : sectionalizing switch "OFF" (Cut Switch)
⊗ : sectionalizing switch "OFF" (Loop Switch)
◌ : sectionalizing switch of de-energized area (actually "OFF" in physical system, whereas "ON" is assumed in the algorithm.)
□ : circuit breaker

Fig.1 Example of radial distribution system

Reprinted from *IEEE Transactions on Power Delivery,* Vol. 4, No. 3, pp. 1832–1839, July 1989.

Few studies about this problem appeared in the literature before 1980[1]. As the distribution automation gained more acceptance in the eighties, more studies have been reported.

The method proposed in references [2],[3] can be applied only for small distribution systems. Even though the method in reference [4] can be applied to further large scale systems, its algorithm is still consumes too much computation time to solve a problem of large scale distribution system. The methods proposed in references [2]-[4] need to assume that each section load is accurately known. However, it is impossible to predict the precise information of each section load. Therefore, the predicted section load is not always reliable, and the accurate solution of the combinatorial problem is not always correct. Even if each section load magnitude is predicted accurately, it is difficult to predict the load-voltage characteristic and the accurate load distribution pattern along the line section. Considering the inaccuracy of a predicted section load, sub-optimal solution of this problem is practical enough in real operation.

From the view point above, to solve these combinatorial problems in power distribution area, the authors have developed several practical non-combinatorial approximate solution algorithms[5]-[9]. In reference [5], an approximate solution algorithm is proposed for loss minimum load allocation. In references [8],[9], a method is proposed for load balancing of substation transformers. In reference [6], the authors have proposed a basic algorithm for emergency service restoration. This algorithm is generalized in reference [7] by including operating constraints in the problem. Although these algorithms are computationally efficient, they have some shortcomings: for example, the algorithm sometimes makes unnecessary switching operations and thus the solution indicates switching sequences which may not be familiar to the operators.

The proposed algorithm in this paper is a refinement of the previous methods and overcomes their shortcomings. This new algorithm tries to transfer the de-energized loads to other feeders in stages. First, all the de-energized loads are transferred to adjacent feeders, called main support feeder. Then, if necessary, "multi-stage support" is implemented: that is, in first stage, loads are transferred from the main support feeder to first stage support feeders and in second stage, from first stage support feeders to other feeders. Load curtailment is also considered when it becomes impossible to transfer all de-energized loads to other feeders without causing constraint violations in the system. Finally, some of the curtailed loads are tried to be restored by considering some new load transfer possibilities. Various measures are used to determine which loads to transfer or curtail and which feeders to use. The measures used are calculated by effective gradient method[10],[11].

It should be noticed that, after all the load transfer sequence is determined by utilizing this algorithm, the actual switching operations are performed, taking care not to make violations. When the actual switching operations are performed, only one loop can be constructed simultaneously in the middle of a load transfer procedure for the purpose of the protection.

The distinctive features of the algorithm proposed are as follows.

(i) Only the load transfers which are effective for reducing the constraint violations are performed, and hence the unnecessary switching operations are avoided.

(ii) Therefore, the computed result is more familiar to the operators than that by other methods.

(iii) Supporting priority of the support feeder can be included in the algorithm.

(iv) Extremely fast solution is available if the de-energized load is small enough to restore the load without creating violations.

Formulation of the service restoration problem is given in the next chapter. In Chapter 3, the proposed solution algorithm is presented. Test results are given in Chapter 4. Chapter 5 contains the conclusions.

2. PROBLEM FORMULATION

The main objective of service restoration problem is to restore as much load as possible (or minimize the loads in the isolated islands) by transferring de-energized loads in the isolated areas to other feeders that can take extra load. However, when transferring loads one has to check that the operating constraints are respected. Since the feeders in practice are usually operated radially, radiality of the system topology is an important constraint in the problem. There are also constraints that must be considered for each feeder: voltage drop limits and the line/transformer capacity limits.

To determine the power flow in the distribution system, we assume that the loads on the feeder sections are uniformly distributed and can be represented as constant current flows. The constraints for feeder i then can be represented as follows

line/transformer capacity constraints: $I_{ik} \leq \bar{I}_{ik}$ (1)

voltage drop constraint: $\Delta V_i \leq \Delta \bar{V}_i$ (2)

where ΔV_i and I_{ik} are the maximum voltage drop on the feeder i and current through the k-th supervising point of the line/transformer i respectively, and a bar on a quantity indicates an upper limit of that quantity.

By using radiality of the feeders, the constraints considered above for the feeders are simplified. It will be assumed that the maximum voltage drop will occur at one of the end points(tip) of the feeders as no capacitors ordinarily are facilitated in urban distribution system. Also the line capacity violations are assumed to be observed only at the roots of the branches where the diameters of the wires change. Therefore, it is enough to check the voltage drops and the current capacities at the points specified above. For example, for the system in Fig.1(a), voltage drop checks are needed at the both sides of the switches $C_1,\ldots,C_5$ and L_1, and current capacity needs to be supervised at only the points on the feeders marked by "Ⓐ". Detailed general formulation of the problem is shown in the Appendix 1.

3. SOLUTION ALGORITHM

The flowchart of the algorithm showing the basic steps is given in Fig.2. As the figure indicates, there are four basic steps in the algorithm. In the first step, de-energized loads in an isolated area are connected to an adjacent feeder, called the main support feeder. In the second step, called the multi-stage support, the main support feeders that have constraint violations, due to extra loads transferred, are declared as violation feeders and the extra loads from these feeders are tried to be transferred to other feeders to eliminate the constraint violations.

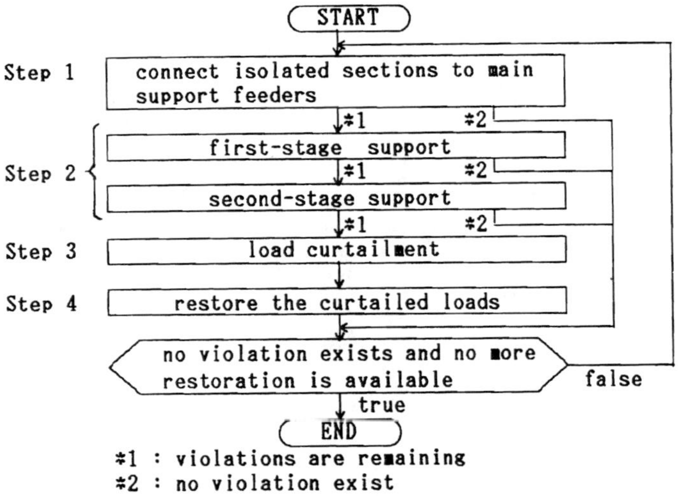

Fig.2 Flowchart of the overall solution algorithm

This step can be separated into two stages: "first stage support" and "second stage support". If the violations still exist at the end of second step, some loads are curtailed in the third step. Finally, in the fourth step, some of the curtailed loads are tried to be restored by considering some new load transfer possibilities.

To illustrate these basic steps of the algorithm, we will use the small system in Fig.1(b). Let us suppose that a fault occurs on the feeder section (A,B). Protection scheme will isolate this fault by opening the sectionalizing switches A and B and will create the isolated area enclosed in dotted lines in the figure. The algorithm will restore the loads in the enclosed area as follows: In the first step, the switch C_1 and feeder f_3 is chosen to connect the isolated area to main support feeder f_3. If no constraint violation occurs, the procedure ends here. This is one of the superior points of this method to the other ones. However, let us assume that this load transfer will cause voltage drop violation at the point H of the section (G,H). In step 2, loads will be tried to be transferred from f_3 to other feeders. Load transfer is tried in two stages: in the first stage load transfer from f_3 to one of the adjacent feeders, feeder f_2 is tried. Suppose only section load (M,D) can be transferred without causing any constraint violation in f_2 and such a transfer does not eliminate constraint violation in f_3. Then a second stage support is tried by first transferring section load (D,E) from f_3 to f_2 by tolerating a constraint violation, current capacity violation in f_2. Then violation in f_2 is eliminated by transferring section load (O,N) to another support feeder, feeder f_4. Assuming that the voltage problem in f_3 is still remaining, load curtailment is done in step 3 by de-energizing load in section (G,H). In the last step, curtailed load is restored again after changing the open position of loop switch L_1 to the point F.

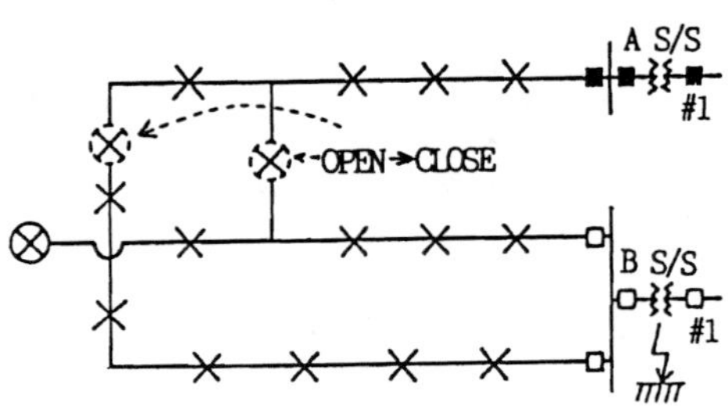

Fig.3Connecting the isolated islands to support feeder

The convergence property and computational efficiency of this algorithm depends very much on how one selects the switches to transfer the loads during the solution. For this purpose, various measures are developed and used in this algorithm. These measures are based on so called "dual effective gradient method" which is given for the general case in Appendix 2 [10],[11].

The detailed explanation of the steps of the proposed algorithm is given below.

Step 1 : Connecting the isolated sections to main support feeders

Isolated sections of a distribution system, created by protection scheme to isolate a fault, may in general be more than one isolated island. Therefore, at the beginning of the algorithm these isolated islands are identified together with the cut switches that can connect the islands to adjacent feeders.

To give an example, consider the single transformer outage at the substation B in Fig.3. In this case, the feeders supplied by the faulted transformer are de-energized by opening the CBs on both sides of the transformer and hence these two feeders become isolated islands and the cut switches on these islands can be used to connect the feeders to the adjacent feeder.

Connecting an isolated island to an adjacent support feeder via one cut switch assures that the radiality of the network is maintained. However, the power flow in the system and the constraints depend on which switch is selected for connection and some constraints may be violated after the closure of such a switch. To be able to compare the use of different cut switches available, we use a vector, called violation vector, **R** which shows how much the constraints are violated in a feeder. Each component of the vector measures how much a constraint is violated. For example, if k-th de-energized island is energized by closing j-th cut switch, when two constraints, current capacity and voltage limit, supposed to be violated, the components of violation vector are

$$R_{kj1} = \Delta V_k - \Delta \bar{V}_k, \quad R_{kj2} = I_k - \bar{I}_k \tag{3}$$

A schematic illustration of **R** is given in Fig.4. When a constraint is not violated, the corresponding component of **R** will be negative as shown in the figure. But since we want to measure only the violations, the component of **R** is taken as zero whenever it is negative. Thus the magnitude of **R** is zero if only if all the constraints are satisfied.

The violation vector is used in selecting the cut switch as follows: given an isolated island, say island k, and a set of cut switches that can be used to connect the island to adjacent feeders, the cut switch and the associated adjacent feeder is chosen as the one that minimizes the magnitude of the corresponding violation vector, i.e., selection is done by calculating the minimum $|\mathbf{R}_{kj}|$ where $|\mathbf{R}_{kj}|$ means the norm of the violation vector of k-th de-energized island after the isolated island is connected to the feeder via cut switch j.

If the supporting priority of feeders is to be considered, $|\mathbf{R}_{kj}|$ must be multiplied by coefficients α_j to reflect the priority order of the support feeders.

Step 2: Multi-stage support

If constraint violations occur after the load restoration procedure in step 1, the main support feeders with violations are declared as violation feeders. In this step, violations in violation feeders are tried to be eliminated by transferring some of the

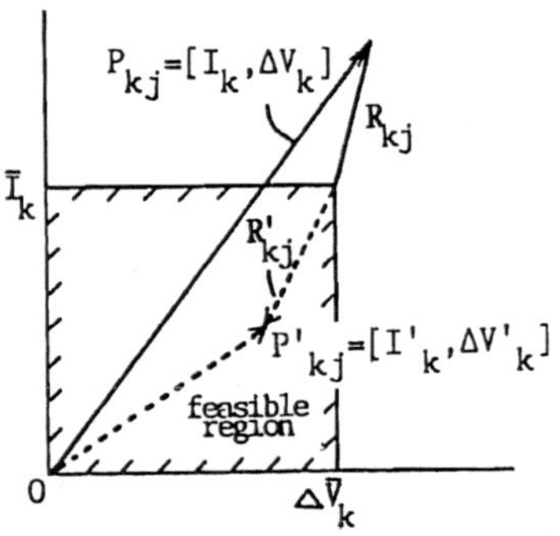

Fig.4 Schematic illustration of R_{kj}

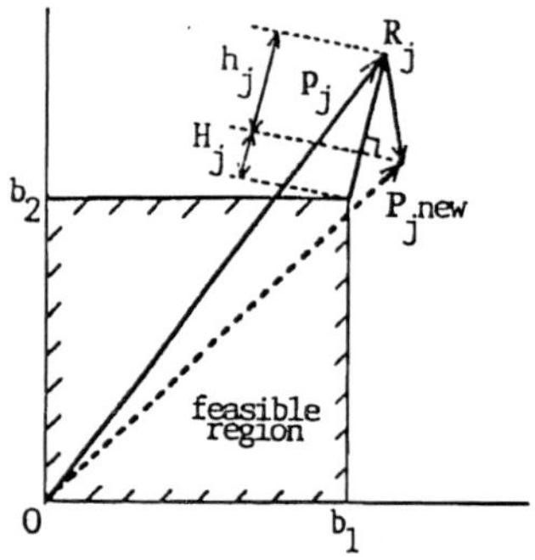

Fig.5 Effective length of violation withdrawal

loads on these feeders to other adjacent feeders. The load transfer is carried out in stages as follows.

first stage support

Flowchart of the algorithm for first stage support is given in Fig.6. Given a violation feeder, cut switches that can be used to transfer loads from the violation feeder to the adjacent feeders are identified first and then a cut switch and the associated feeder, called first stage support feeder, are chosen to transfer a section load. The cut switch which is most effective in eliminating the constraint violation is selected for load transfer. That is, the most effective load transfer is selected by solving the following optimization problem.

$$\text{Max} \sum_{j\in I_b} [h_j / \{\alpha_j(H_j+\beta)\}]y_j \qquad (4)$$

$$\text{Sub. to} \quad \sum_j A_{ij}y_j \leq b_i \quad (i=1,2,......) \qquad (5)$$

$$y_j \in \{0,1\} \qquad (6)$$

Where, the variable y_j is a 0-1 variable; it indicates the cut switch status by 1(closed) or 0(open). The notation I_b denotes a set of cut switches between sound feeders and all the de-energized islands. The quantities h_j and H_j are the effective length of violation withdrawal and the effective length of remaining violations respectively. Fig.5 illustrates these values. H_j is taken as zero if all the violations are eliminated after the load transfer and that is why the constant β is introduced in the above equation: to avoid zero divide. The parameter α_j represents the priority of support. Eq.(5), which shows eq.(App.2)-eq.(App.4) in Appendix 1 all together, denotes the constraints for the current capacities and the voltage drops of the support feeder corresponding to the load transfer by the switching operation of cut switch j. This optimization problem can be solved as follows by the effective gradient method.

The section loads are transferred from violation feeder to the first stage support feeders one by one, in the descending order of the following measure under voltage and current capacity constraints.

$$h_j / \alpha_j(H_j+\beta) \qquad (7)$$

This procedure is continued until all violations in the violation feeder are eliminated or until no more load can be transferred without creating new violations on the first stage support feeders.

second stage support

If the first stage support ends with a failure in eliminating all violations in some violation feeders, a second stage support is provided to those feeders.

Flowchart of the algorithm for second stage support is given in Fig.7. Second stage support to a violation feeder is done in two steps: first a load transfer from the violation feeder to a first stage support feeder is forced by tolerating some violations in the first stage support feeder, then violations in the first stage support feeder are eliminated by using second stage support feeders which are connected to the first stage support feeder through cut switches.

To determine which one of the first stage support feeders to be forced to take load, the following value is calculated for each cut switch considered,

$$h_j / \alpha_j(a_j+\beta) \qquad (8)$$

where, a_j is the magnitude of candidate section load that can be transferred to the first stage support feeder through cut switch j. Then the cut switch with the largest value of eq.(8) is selected first as the

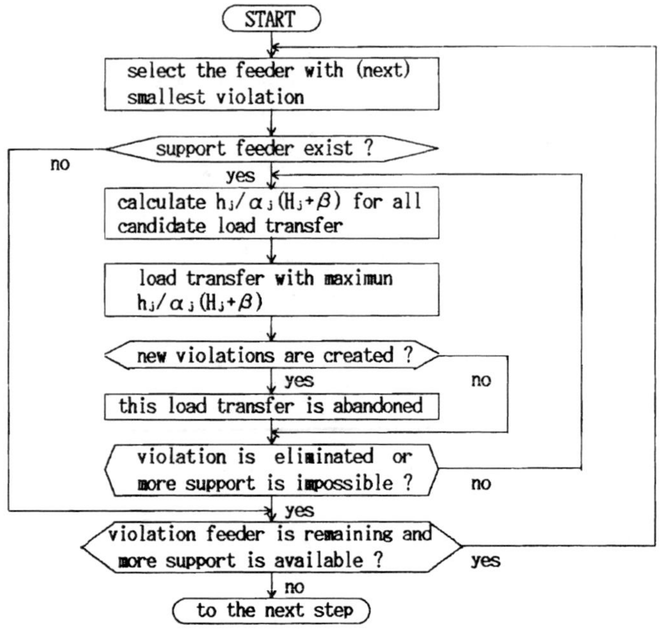

Fig.6 Flowchart of first-stage support

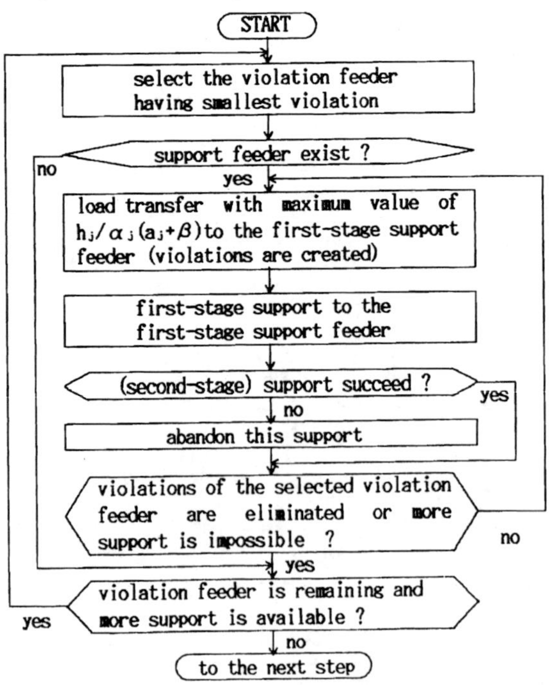

Fig.7 Flowchart of second-stage support

supporting point. This procedure corresponds to solving the optimization problem just same as eq.(4)-eq.(6), except the coefficients of the objective function, by the effective gradient method. From the definition of eq.(8), the load having small magnitude and causing large violation withdrawal is selected for the transfer.

After the load transfer, the support to the first stage support feeder is provided by treating it as a violation feeder and applying the first stage support procedure.

Second stage support is performed until all existing violations in the violation feeder are eliminated or it becomes impossible to transfer a load to any of its second stage support feeders without causing violations on them.

As indicated in the flowchart, when there are several violation feeders in the system, the feeder with smallest violation is supported first.

higher stage support

The first and second stage support procedures can be applied successively to provide higher stage support to the violation feeders. However, the distribution systems are usually designed so that all de-energized loads in an island can be restored without creating serious constraint violations. Therefore, in the implementation of the algorithm, higher order stage supports are not considered.

Step 3: Load curtailment

After the second stage support, if some constraint violations still remain in the system, load curtailment is performed.

Flowchart of the load curtailment procedure is given in the upper half of Fig.8. As indicated in the flowchart, loads in the end sections of the violation feeders are curtailed one by one until the violations are eliminated. The loads to be curtailed are selected by using the following measure,

$$a_j(H_j + \beta) \quad (9)$$

At each step the load with smallest value of eq.(9) is selected for curtailment. This means that the load having small magnitude, a_j and causing small remaining violation, H_j is selected.

Step 4: Restoration of curtailed loads

Load transfer among violation feeders has never been tried in above steps. Therefore, there are some possibilities in restoring the curtailed section loads by load transfer among former violation feeders. This is done in two steps: first one of the curtailed loads is energized by tolerating constraint violations, then these violations are tried to be eliminated by using the cut switches among former violation feeders and associated support feeders together. Flowchart of this procedure is given in the lower half of Fig.8.

As shown in the flowchart, the curtailed load to be energized is selected as the one which maximizes the following measure,

$$a_j/(|R_j| + \beta) \quad (10)$$

This means that the restored load should be as large as possible and the created violations should be as small as possible. After the load is connected to the feeder selected, this feeder is treated as a violation feeder and a first stage support is tried. If the violations cannot be eliminated by this support, the restoration of this section load is abandoned.

This step ends when all candidates are tried and no more restoration is expected.

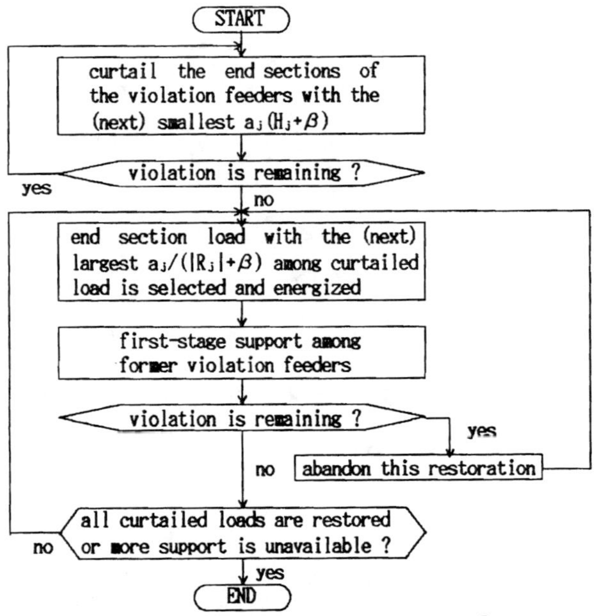

Fig.8 Flowchart of load curtailment and restoration of curtailed loads

4. NUMERICAL RESULTS

In this chapter, test results for two systems are given. First the test results of a small test system will be introduced to illustrate the algorithm. Then the test results of a large scale urban distribution system will be given to demonstrate the properties of the algorithm. The computer used for the test runs is a HITAC M-680/H (Gibson Mix: 32nsec). The amount of FORTRAN77 code of this algorithm is 4200 steps. In the algorithm, estimated maximum section load magnitude in the period of several hours (4 hours for example) after the fault, is used. The constant parameters are set to $\alpha_j=1$ for all j and $\beta = 0.0001$.

Small Test System

One line diagram of the small test system is given in Fig.9 together with the sectionalizing switches. Section load data is given in Table 1, but the section impedances are left unmentioned due to lack of space.

For the test run, it is assumed that a fault on transformer at substation A occurred. As a result, feeders f_5 and f_6 are left de-energized. In Fig.9, the

Table.1 Section load for the example system

Sec. No.	Load [A]	Sec. No.	Load [A]	Sec. No.	Load [A]	Sec. No.	Load [A]	Sec. No.	Load [A]
1	25	17	29	33	3	49	23	65	25
2	51	18	4	34	10	50	32	66	41
3	36	19	3	35	18	51	48	67	67
4	35	20	1	36	10	52	18	68	21
5	43	21	52	37	3	53	15	69	42
6	71	22	27	38	10	54	13	70	20
7	34	23	11	39	10	55	8	71	31
8	23	24	2	40	12	56	25	72	8
9	37	25	8	41	20	57	14	73	15
10	51	26	2	42	2	58	22	74	29
11	48	27	1	43	4	59	30	75	4
12	30	28	6	44	5	60	22	76	12
13	40	29	21	45	8	61	13	77	59
14	21	30	14	46	10	62	16	78	2
15	2	31	12	47	20	63	9		
16	23	32	1	48	25	64	14		

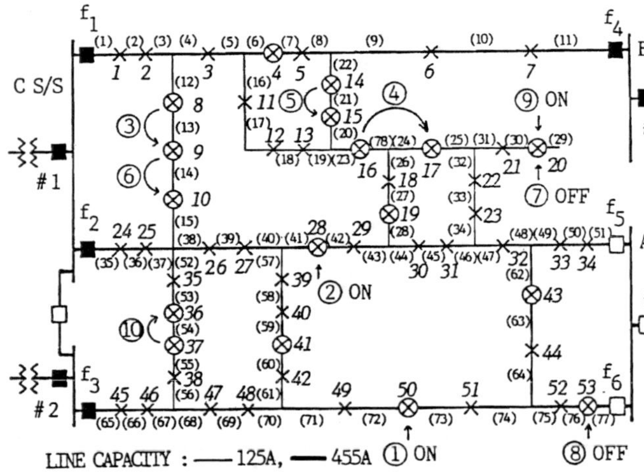

Fig.9 Small scale example system

simulated switching sequence is indicated by curved arrows and numbered according to the sequence they are implemented in the algorithm. The de-energized feeders f_6 & f_5 are restored by closing the cut switches #50 & #28 tolerating constraint violations in feeders f_3 & f_2. Hence, this causes the current capacity violations at their feeding points. Since the feeder f_3 does not have the support feeders, support is performed from the feeder f_1 to f_2 only. As the first stage support, the load transfer is performed by changing the position of the open cut switch from the point #8 to #9. If more load is transferred from the feeder f_2 to the feeder f_1, new violation would be created. Therefore, the first stage support is terminated and the second stage support is performed. That is, after making violations in the feeder f_1 by changing the position of the open cut switch from the position #16 to #17, the violation is eliminated by the first stage support (changing the position of the open cut switch from the position #14 to #15). After the second stage support, a first stage support is tried for the violation feeder f_2 again and as a result one more section load is transferred to f_1 by changing the position of the open cut switch from the position #9 to #10. After this load transfer, no more load transfer from the feeder f_2 to the other feeders is possible, so the load curtailment is performed by opening the sectionalizing switches #20 on the feeder $f_2(f_5)$ and #53 on the feeder $f_3(f_6)$, and all the violations are eliminated. The curtailed load on feeder $f_2(f_5)$ is restored by closing switch #20 tolerating violations in f_2. The violation is eliminated again by changing the cut switch position between feeders f_2 and f_3: from the position #37 to the position #36. Here, the solution is found for the small test system.

Large Test System

The performance of the algorithm is also tested by using a 6.6 KV real, large scale urban distribution system data. One line diagram of the system is given in Fig.10. The system has 5 substations, 11 substation transformers, 87 feeders, and 1188 load sections. The

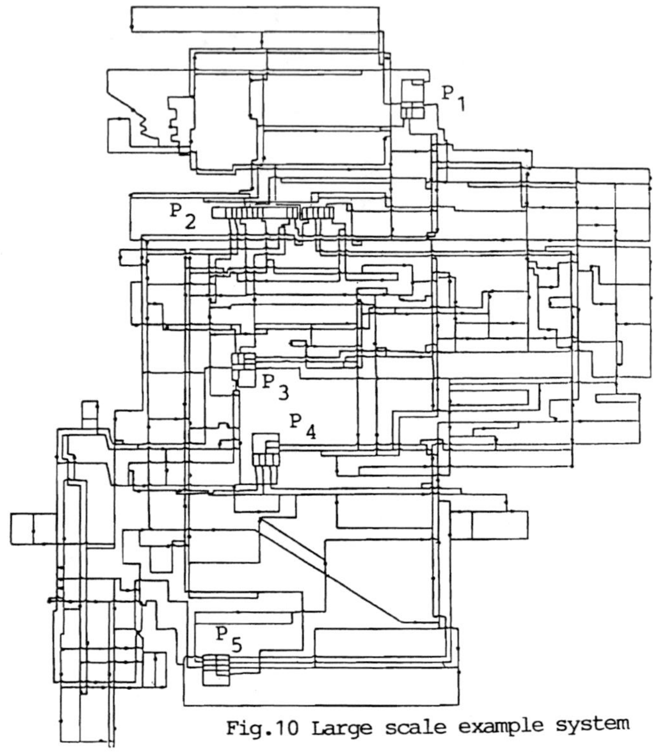

Fig.10 Large scale example system

Table 2. Transformer capacity and total load

Substation	P_1		P_2				P_3		P_4		P_5
Transformer	T_1	T_2	T_3	T_4	T_5	T_6	T_7	T_8	T_9	T_{10}	T_{11}
Capacity[A]	900	900	2000	2000	900	600	1500	2310	900	1100	1200
Load[A]	772	633	1539	1697	726	327	1258	2023	674	924	1050

Table 3. Comparison of the computing performance

	Result of this algorithm				Result of reference [7]			
Outage transformer No.	T_1	T_4	T_7	T_{10}	T_1	T_4	T_7	T_{10}
Total de-energized load[A]	772	1697	1258	924	772	1697	1258	924
Total supported load[A]	884	1708	1262	927	837	1793	1326	1045
Total switching operations	7	24	13	10	16	42	32	29
Computing time[sec]	0.055	0.284	0.326	0.149	0.408	0.919	0.494	0.490

Table 4. Comparison of the load transfer result for the fault of transformer T_1

Result of this algorithm						Result of reference [7]					
T_f	T_t	Switching operation	Transferred load[A]	Outage load[A]		T_f	T_t	Switching operation	Transferred load[A]	Outage load[A]
				772						772
1	4	2	173	599		1	3	8	388	384
1	5	1	86	513		1	4	3	96	288
1	3	3	358	155		3	5	3	(124)	
1	2	1*	267	0		1	3	1	86	202
						1	2	1*	267	0

(T_f: Supported Transformer, T_t: Support Transformer, * : Bus-tie CB ON)

position of open sectionalizing switches is indicated by "•" in the figure, closed switches are not marked in the figure for clarity. Transformer capacity and load data is given in Table 2.

Test results for different type of faults are listed in Table 3 and Table 4 together with the results obtained by using the old algorithm in reference [7]. In Table 3, test result for four different transformer faults are listed. By comparing the results of the two methods, the followings are seen. For each fault case, both the total number of switching operations and the computing time of this method are smaller than that of reference [7]. Also, less load is transferred in this method than the other except the T_1 fault case in which the section with large load (178 [A]) is selected and transferred to eliminate the over current of the transformer T_2 at the last support. The comparison of the more detailed load transfer results for the fault of transformer T_1 is shown in Table 4.

In Table 3 and Table 4, the superior points of this algorithm are demonstrated and verified. In the proposed method, a number of switching operations are reduced. The reason may be due to the fact that only the loads that are effective for reducing the constraint violations are selected and transferred. Computation time is also improved by the characteristics of this algorithm. That is, the minimum number of load transfers are selected and performed without calculating the combinatorial process to eliminate the constraint violations.

5. CONCLUSIONS

A new service restoration algorithm based on the effective gradient method for power distribution system is proposed. The algorithm is practical in the sense that it satisfies the requirements for actual on-line implementation of such algorithms; in particular the speed of computation and avoidance of unnecessary switching operations. An illustrative and a real system test results are given to confirm these conclusions.

ACKNOWLEDGMENT

The authors would like to express their acknowledgment to the Chugoku Electric Power Co.(Japan) for providing the valuable data used in the numerical calculations, and to Prof. F.F.Wu, Univ. of California, Berkeley for correcting English sentences.

REFERENCES

[1] Task Group on State of the Art Distribution System Design Working Group on Distribution System Design, Distribution Subcommittee : "Biography on Distribution Automation, 1967-1982" IEEE Trans.PAS Vol.PAS-103, No.6, June (1984)

[2] S.Kato, T.Naito, H.Kohno, H.Kanawa and T.Shoji : "Computer Based Distribution Automation" IEEE PICA San Francisco, May (1985)

[3] C.A.Castero Jr., and A.L.M. Franca : "Automatic Power Distribution Reconfiguration Algorithm Including Operating Constraints" IFAC Electric Energy Systems, Rio de Janeiro (1985)

[4] D.W.Ross, M.Carson, A.I.Cohen: "Development of Advanced Methods for Planning Electric Energy Distribution Systems", U.S. DOE Report ET-73-C-03-1845 Feb. (1980)

[5] K.Aoki, T.Ichimori and M.Kanezashi : "Normal State Optimal Load Allocation in Distribution Systems" IEEE Trans. PWRD, Vol.PWRD-2, No.1(1987)

[6] K.Aoki, H.Kuwabara, T.Satoh and M.Kanezashi : "Outage State Optimal Load Allocation by Automatic Sectionalizing Switches Operation in Distribution Systems", IEEE Trans. PWRD, Vol.PWRD-2, No.2, Oct. (1987)

[7] K.Aoki, T.Satoh, M.Itoh, H.Kuwabara and M.Kanezashi :"Voltage Drop Constrained Restoration of Supply by Switch Operation in Distribution Systems" IEEE PES Summer Meeting 87-SM 544-0 (1987)

[8] K.Aoki, H.Kuwabara, T.Satoh and M.Kanezashi : "An Efficient Algorithm for Load Balancing of Transformer and Feeders by Switch Operation in Large Scale Distribution Systems" IEEE PES Summer Meeting 87-SM 543-2 (1987)

[9] K.Aoki, H.Kuwabara, T.Satoh and M.Kanezashi : "Load Smoothing of Transformers and Feeders by Switch Operation in Distribution Systems" Trans. IEE Japan, Vol.107-B, No.9 Sep. (1987)

[10] S.Senju and Y.Toyoda : "An Approach to Linear Programming with 0-1 Variables", Management Science, Vol.15, No.4 (1968)

[11] Y.Toyoda : " A Simplified Algorithm for Obtaining Approximate Solution to Zero-One Programming Problems", Management Science Vol.21, No.12 (1975)

APPENDIX 1 General Formulation of the Problem

The emergency reconfiguration problem is mathematically formulated as follows.

$$\text{Maximize} \sum_{i} \sum_{j \in J_{i0}} a_{ij} x_{ij} \qquad \text{(App.1)}$$

Subject to

(line capacity constraint)

$$\sum_{j \in J_{ik}} a_{ij} x_{ij} \leqq b_{ik} \qquad \text{(App.2)}$$

(transformer capacity constraint)

$$\sum_{j \in J_t} a_{ij} x_{ij} \leqq b_t \qquad \text{(App.3)}$$

(voltage drop constraint)

$$\sum_{l \in T_e} \{(\sum_{q \in J_{il}} s_{iq} x_{ij}) z_{il}\} \leqq v_{ie} \qquad \text{(App.4)}$$

Where,

x_{ij}: such 0-1 variable as 0 if section j at feeder i is de-energized and 1 if it is energized. x_{ij} cannot arbitrarily be set to 1. $x_{ij}=1$ only if the adjacent section is energized and the sectionalizing switch connected to it is closed. x_{ij} is a function of the sectionalizing switch status.

a_{ij}: load magnitude of section j at feeder i

b_{ik}: line capacity at k-th point of feeder i

b_t : transformer capacity of transformer t

z_{il}: impedance of section l at feeder i

s_{ik}: $s_{ik} = a_{ik}$ (if $k \neq l$), $s_{ik} = a_{ik}/2$ (if k=l) as uniformly distributed load is assumed

v_{ie}: voltage drop limit at the e-th end of feeder i

J_{i0}: index set of de-energized load sections on feeder i

J_{ik}: index set of load sections connected to the leaf side of point k of feeder i

J_{il}: index set of load sections which exist at the leaf side of section l (included) of the feeder i

J_t : index set of load sections ij connected to transformer t

T_e : index set of load sections which exist at the trunk of the tree between bus and section e

APPENDIX 2 Dual Effective Gradient Method[10],[11]

The integer programming problem formulated as eq.(App.1)-eq.(App.4) is re-written by generalized form as eq.(App.5)-eq.(App.7). This kind of problem is referred to as the "multi-dimensional knapsack problem".

$$\text{Maximize} \quad z=c_1x_1+c_2x_2+\dots\dots+c_nx_n \qquad \text{(App.5)}$$

$$\text{Sub. to} \quad \sum_j d_{ij}x_j \leqq b_i \ (i=1,2,\dots,m) \qquad \text{(App.6)}$$

$$x_j \in \{0,1\} \quad (j=1,2,\dots,n) \qquad \text{(App.7)}$$

Where, x_j represents the variable, and c_i and d_{ij} represent the generalized coefficients of the objective function and the constraint equations respectively. In the case of m=1 or in the case of "knapsack problem", this problem is approximately solved by two folds.

(1) Initializing $x_j=0$ $(j=1,2,\dots,n)$, and x_j is set to 1 in order of decreasing (c_j/d_j) while eq.(App.6) can be satisfied.
(2) Initializing $x_j=1$ $(j=1,\dots,n)$, and x_j is set to 0 in order of increasing (c_j/d_j) until eq.(App.6) is satisfied.

The method (1) is referred to as "primal method" whereas the method (2) is referred to as "dual method".

The dual effective gradient method is an extension of the method (2) above. The basic conception of the solution algorithm of the dual effective gradient method is as follows.

$x_j=1$ $(j=1,2,\dots,n)$ can be an initial solution of the problem. If eq.(App.6) is satisfied, the problem is solved. However, ordinarily, eq.(App.6) is not satisfied. In such a case, several x_j must be set to zero. To select j, the increasing order of $(c_j/|\mathbf{d}_j|)$ may be used corresponding to the method (2), where, $|\mathbf{d}_j|$ means a norm of the vector $\mathbf{d}_j=[d_{1j},d_{2j},\dots,d_{mj}]^t$. In this case, as $\mathbf{d}_j$ has its direction, it is important to evaluate the magnitude of infeasibility reduction. The easiest method to evaluate it, is to introduce vector $\mathbf{r}$ which gives a distance from the point $\mathbf{p}=\sum_{j\in N}\mathbf{d}_j$, $(N=[j|x_j=1])$ to the feasible region as shown in Fig.App.1. The projection h_j, i.e. the projection of $\mathbf{d}_j$ to $\mathbf{r}$, can be used instead of $|\mathbf{d}_j|$. The projection h_j, the effective length of violation withdrawal, is referred to as "effective gradient" and is calculated by the following equation.

$$h_j=(-\mathbf{d}_j)^t(-\mathbf{r}/|\mathbf{r}|)=(\mathbf{d}_j{}^t\mathbf{r})/|\mathbf{r}| \qquad \text{(App.8)}$$

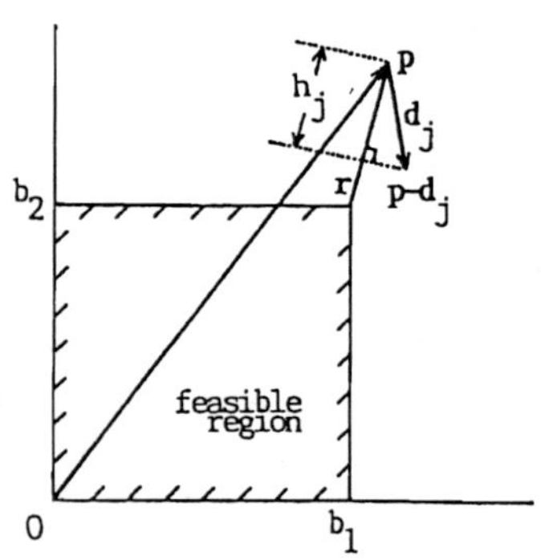

Fig.App.1 Illustration of effective gradient

$$\text{Where,} \quad r_i=\begin{cases} p_i-b_i & : p_i-b_i>0 \\ 0 & : p_i-b_i\leqq 0 \end{cases}$$

Therefore, the approximate solution of eq.(App.5)-eq.(App.7) is obtained by the following method, i.e., after setting $x_j=1$ $(\forall j)$, set $x_j=0$ in increasing order of (c_j/h_j) until eq.(App.6) is satisfied.

HEURISTIC SEARCH APPROACH TO DISTRIBUTION SYSTEM RESTORATION

A.L. Morelato and A. Monticelli
UNICAMP-Campinas-Brazil

Abstract

Service restoration, system reconfiguration, and other related problems are formulated and solved by heuristic search, which is a search strategy (e.g. depth-first search) armed with practical rules (say, based on operator experience) to guide the search. The proposed framework allows designing algorithms to solve a wide range of distribution restoration problems. System operator procedures can be accommodated as part of the search process. Test results are presented. An illustrative example is given in the Appendix.

Keywords: heuristic search, combinatoric optimization, expert systems, control center functions, distribution system restoration.

INTRODUCTION

It has been recognized that the introduction of automation techniques to electric utility distribution systems will allow better investment utilization as well as improved service quality. More especifically, automation may reduce the time for fault isolation and service restoration to unfaulted areas. In addition, load balancing procedures may offer opportunities for deferring investments; by transferring load from one feeder or substation to another one may delay the introduction of new equipment [1,2].

The conventional approach to power system computer applications relies on the use of numerical algorithms coded in procedural languages such as Fortran and C. More recent developments rely on symbolic programming languages, such as Lisp and Prolog, which allows for qualitative reasoning, in addition to plain numerical calculations [3]. We say that we have a procedure when all the steps required to solve a problem are specified. This is in contrast with AI languages, such as Prolog, in which the user specifies what a solution should be, but does not give an explicit procedure to reach the desired solution (in this case, we say that the inference mechanism is imbedded into the language, and, by that, it is not programmed by the user). The great attraction of the AI methodologies is that they bring us the opportunity to represent and use expert knowledge in the same machines (computers) that traditionally have been applied to solve numerical problems. By expert knowledge, we mean qualitative knowledge generally represented by practical rules and inference mechanisms normally used by engineers, including the "cookbookish" procedures adopted by some experienced operators. Simply stated, we could say that while the conventional approach is suited for quantitative (numerical) applications, the AI approach is more oriented to qualitative (symbolic) reasoning – we should say that this distinction is not strict in the sense that, for example, numerical computations could also be performed using a symbolic language like Prolog, but that would not be the best use for the language.

89 WM 111-6 PWRD A paper recommended and approved by the IEEE Transmission and Distribution Committee of the IEEE Power Engineering Society for presentation at the IEEE/PES 1989 Winter Meeting, New York, New York, January 29 - February 3, 1989. Manuscript submitted February 1, 1988; made available for printing January 4, 1989.

Heuristic programming techniques, such as the ones used in this paper, fall somewhere between the conventional and the AI approaches, in the sense that they can be coded in a procedural language (say, Fortran or C), but they also allow for the use of some practical knowledge. Heuristic programming techniques are used to solve combinatorial problems [6], that is, problems in which finding a solution normally involves the analysis of a large number of alternatives (the number of alternatives, or possible combinations, may increase exponentially). Notice that this is not to be blamed on the technique itself. Instead, it is a characteristic of the problems we are trying to solve; for example, how many ways do we have to carry out a series of switching operations, say, to relief a line overload? Also, very importantly, which is the best way? And that is the reason why practical knowledge is so important: it allows us to cut down the number of alternatives to be evaluated to a manageable number (this elimination of unpromissing alternatives is also referred to as pruning, for this can be seen as the pruning of a decision tree). Finally, it should be emphasized that the heuristic approach of this paper is not intended to be a replacement for AI technology, but an evolutionary effort in that same direction – basically, we are trying to do is to extend the ability of our conventional programs to perform qualitative reasoning, in addition to the existing numerical capabilities.

This paper proposes a general framework for developing algorithms that may be used to carry out functions such as service restoration and network reconfiguration. Having a common framework to deal with a variety of functions performed by a distribution system control center is important for it reduces the complexity of the software. They may be used in remote control of feeder sectionalizing and in other planning and operating tasks. The family of algorithms is quite flexible in the sense that they may include a wide range of strategies: at one end, we may have practical operating procedures, and, at the other end, optimization procedures. Even though the proposed approach uses some techniques that are also used in building Expert System [3,4], it is not an Expert System. The basic difference is that we are dealing with well defined procedures (algorithms).

To be practically implementable, a network reconfiguration method has to take into account a wide range of constraints: some of them are qualitative ones; others are not precisely defined. In addition, practical operating procedures have to be considered. Though the ability to represent knowledge is not as general as in an Expert System, the proposed approach can

Reprinted from *IEEE Transactions on Power Delivery,* Vol. 4, No. 4, pp. 2235–2241, October 1989.

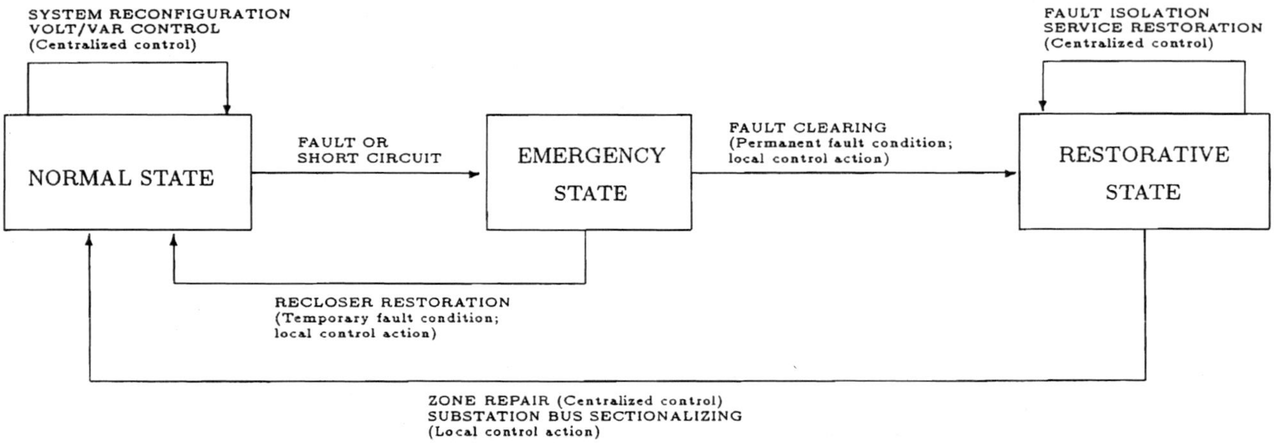

Fig. 1 Operating states of a distribution system.

handle the systems constraints while allowing the inclusion of important human procedures to guide the search for problem solutions.

BASIC CONCEPTS

The functions carried out through a distribution system control center depend on the state the system is in. For example, if the load is being supplied and all the operating constraints are being satisfied, then the objective may be, say, minimize system losses. But, if some load has been disconnected due a major fault, then the objective will be to restore service as soon as possible. Such apparently distinct functions share a common feature, however: both require solving a combinatorial problem whose solution is a set of switching operations. Nowadays, in most existing systems, those tasks are performed by experienced operators. Not withstanding this fact, increasing system complexity in conjunction with a broadening scope of distribution system operation are motivating the introduction of the new technologies generally referred to as distribution automation. This section is aimed at presenting the basic system operating states and the associated state transitions. These concepts will be used in the following sections in formulating and solving the problems regarding the various functions that can be performed by a distribution system control center. Functions such as network reconfiguration (say, for loss minimization) and customer service restoration, will be treated as voluntary state transitions, as opposed to involuntary state transitions (say, caused by faults).

System Constraints

Distribution systems are designed to operate under two sets of constraints: load constraints and operating constraints. That is, it is required that the system meet customer load demand, while operating within acceptable electric limits, such as primary feeder current limits and load voltage limits. Unpredictables happen however, and so the system may be led to a state where part of the constraints are not satisfied. Operating limits, such as feeder maximum current, can be further classified into normal condition limits and emergency limits. Violations in normal condition feeder current limits can be tolerated for some time, whereas emergency limits usually are avoided by proper overcurrent protection schemes.

Operating States

Based on the definitions of system constraints, one can introduce the concept of system state, as illustrated in Figure 1. The system is said to be in a *normal state* if all load is supplied and the emergency limits are not violated. The system is said to be in an *emergency state* if the emergency operating limits are violated (say, due to a fault or short circuit). The distribution system is said to be in a *restorative state* if there is service interruption (Part of the primary feeder system is disconnected in order to isolate a faulted section).

State Transitions

Figure 1 illustrates some of the possible transitions among the previously defined system states. Part of the state transitions are involuntary in the sense that they are not commanded by centralized control. This category includes events such as faults and load variations, as well as local control actions (say, protection based control actions). This paper emphasizes the development of strategies to carry out voluntary state transitions that can be implemented through distribution control centers.

(a) Normal→Normal: When the system is in the normal state, though emergency limits are satisfied, normal operating limits may be violated. If so, the situation can be corrected through system reconfiguration. In this case, the objective is to find a feasible operating point in the sense that normal operating limits are also satisfied. (In addition to the emergency limits which are broader). Even when all the operating limits are satisfied, one may try to improve the system operating condition by, say, minimizing losses or balancing the feeder loads resulting from system reconfiguration. In both cases, we will have normal-normal state transitions.

(b) Normal→Emergency: An emergency state occurs when a fault or short circuit exists on a primary feeder. High magnitude fault current on a feeder should be quickly cleared to avoid equipment damage.

(c) Emergency→Normal: When the fault condition is temporary, reclosing takes the distribution system back to the normal state.

(d) Emergency→Restorative: A permanent fault will activate a fault clearing action that takes the system to the restorative state (load has been disconnected, i.e., load constraints are violated).

(e) Restorative→Normal: The distribution system can be taken back to the normal state either by zone repair or substation bus sectionalizing, depending on fault location. In the first case, after repair, the faulted zone is reenergized. Substation bus sectionalizing is carried out when the fault is located in the substation bus.

(f) Restorative→Restorative: After clearing the fault it is necessary to identify and isolate the faulted zone. After that, partial service restoration may be carried out by breaker/recloser operation. Service restoration to other dark feeder zones can be carried out using tie switches. While the faulted zone is not repaired, however, the system will remain in the restorative state.

Part of the state transitions previously described are called ***involuntary transitions***: for instance, transition caused by faults or load variations, and transitions that follow local control actions (fault clearing). Other transitions are called ***voluntary*** in the sense that they are carried out through the distribution control center (automatically or not). The automation of distribution control centers will rely on development of algorithms for analyzing and synthesizing centralized control strategies. In the next section, we propose a generalized framework to develop such algorithms.

OUTLINE OF THE APPROACH

Voluntary transitions between operating states, carried out through a distribution control center, require the determination of appropriate control strategies. Thus, transition problems such as service restoration, network reconfiguration, volt/var optimization, feeder load balancing, loss minimization, or a combination of two or more these individual problems, involve the determination of control strategies that are implemented either automatically or with the help of a system operator. Though different problems require different solution procedures (algorithms), most of them share the same common base, i.e., we are dealing with discrete decision problems. As a result, we propose a generalized framework, based on heuristic programming, that can handle a wide range of voluntary state transition problems. The solution strategy involves the search on a binary decision tree. Specific knowledge about the problem (heuristics) is used to guide the search and to prune the tree. Pruning is important to avoid the uncontrolled growth of the number of alternatives to be investigated (combinatorial explosion). The advantage of the proposed approach is, we believe, the capability to incorporate practical operating procedures as part of the overall solution process. As a matter of fact, it is possible to design algorithms that generalize practical procedures utilized by system operators. In their general form, the algorithms are able to find nearly optimal solutions; in this case, operator solutions can be used to initialize the general procedure.

PROBLEM FORMULATION

To illustrate how the proposed framework can be utilized to design algorithms that generate control strategies, let us consider the system as being in a restorative state. The problem consists in determining a control strategy (comprising a series of switching operations) which will restore service to loads that are not in the feeder section affected by the fault. (This corresponds to a transition inside the restorative state. To move back to normal state, it will be further necessary to repair and reconnect the faulted section.) We will consider a hybrid restoration/reconfiguration problem in which the objective is two-fold: to restore service to customers and, at the same time, balance the feeder loadings. (It would be simpler to consider the plain restoration problem by itself, but the inclusion of the load balance as an additional goal will serve to illustrate the capabilities of the proposed methodology.)

Data: Load (kVA) at each feeder zone; Primary feeder loading limits; System configuration after fault isolation; Feeder section impedances.

System variables: Primary feeder loadings (kVA) at substation breakers; Load voltage magnitudes.

Decision variables: Switch/breaker status.

System constraints: Load demand supply; Feeder loading limits; Voltage magnitude limits; Radial network.

Problem statement: Distribute the load among the feeders in such a way that the load unbalance can be expressed by the Load Balancing Index (LBI) given by

$$LBI = \frac{1}{n}\left(\sum_{i=1}^{n}(\overline{y} - y_i)^2\right)^{1/2}$$

where n is the number of primary feeders, y_i is the normalized loading on feeder i (actual loading divided by the loading limit) and $\overline{y}$ is the average of the normalized loadings y_i .

Solution: The status of system switches/breakers (all the decision variables are declared either **closed** or **open**).

Feasible solution: A solution leading to a operating point that satisfies the system constraints.

Optimal solution: A feasible solution that minimizes load unbalance.

Local optimal solution: Practical rules based on specific knowledge about the problem can be used to limit the growth of the decision tree. Such practice may lead to a local optimal solution (usually a nearly optimal solution).

Comment: It may happen that there is no feasible solution to the problem. If so, we can change the objective: for example, we can try to minimize the number of unserved feeder zones. A different strategy would consist of trying to find a feasible solution first, and only then try to go on with, say, minimizing load unbalance. In any case, the important thing to keep in mind is that the basic approach remains the same.

HEURISTIC SEARCH

The decision problem we face involves deciding on switch/breaker status (open or closed). We propose to solve this problem by heuristic search on a binary decision tree [6]. By heuristic search we mean a general search method armed with domain-specific knowledge to guide the search. The search methods allow us to traverse the space of possible system states, whereas domain- specific knowledge is essential in limiting the size of the decision tree [7,8].

Binary Decision Variables: The decision whether the *k-th* switch is open/closed is denoted by the binary decision variable:

$$x_k = \begin{cases} 1 & \text{if the k-th switch is closed.} \\ 0 & \text{if the k-th switch is open.} \end{cases}$$

Decision Vector: There are 2^m possible combinations of switch positions in the system, each one corresponding to a decision vector:

$$\underline{x} = (x_1, x_2, ..., x_m)$$

where x_k is equal to 0 or 1.

Decision Tree: The decision process is best illustrated by a binary tree as shown in Figure 2. At the beginning of the process (root node), all the decision variables are undeclared, i.e., we have the original problem with m unknowns. As we move down on the tree, the decision variables are declared either 0 or 1: at level i of the tree we have m-i undeclared variables left. At the bottom of the tree (level m) all the decision variables are declared: each one of the 2^m nodes (leaves of the tree) is associated with a possible solution (not necessarily a feasible one).

Though the tree that must be searched could, in principle, be built explicitly, this is neither practical nor necessary. An implicit representation of the tree, given by the rules used to generate new nodes from the existing ones, is adopted instead. Thus, as the search process is carried out, only part of the tree is constructed explicitly and most of the tree never needs to be traversed. Domain-specific knowledge is used to avoid unnecessary search.

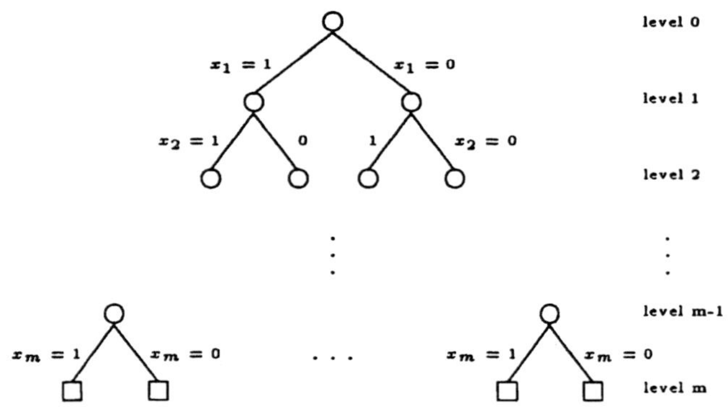

Fig. 2 Decision tree.

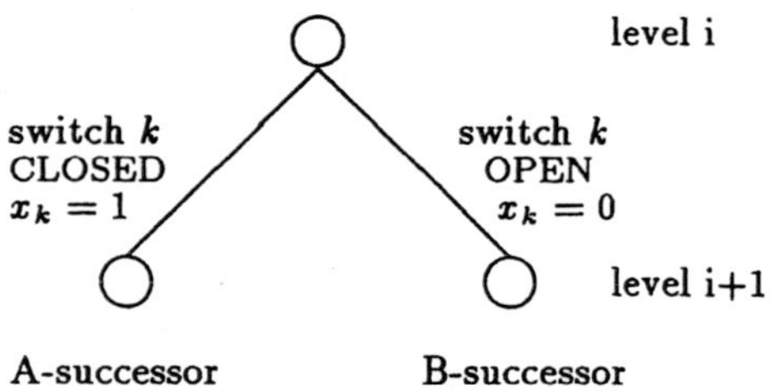

Fig. 3 Successors of a decision tree.

At each node of each tree level, we can choose any undeclared variable to be declared next. When a variable is declared,two new nodes (called successors) are created. The process of tree expansion (called branching) is illustrated in Figure 3. The choice of the next variable to be declared will depend on:

- Previous decisions taken on a particular path;
- Practical rules, based on operator experience, used to guide the search.

Thus, it is possible to simulate what an experienced operator would do under the same circunstances. In addition to guiding the search, practical rules can be used to prune the tree, which may be important in limiting the growth of the number of alternatives to be considered.

Search Strategy

The effectiveness of the method depends on the strategy used to control the search. Strategies such as breadth-first, depth-first and best-first can be adopted. In the present implementation of the method, we are using a depth-first search, which allows the direct simulation of operator procedures as part of the search process. First, we go down on the tree until a solution is reached; then the solution is tested for feasibility; after that we backtrack trying to find other solutions. Used without heuristics, the depth-first search is simply an exhaustive search on the decision tree. Even though exhaustive search is not suited for practical implementation, it is very useful in testing the effect of heuristics on the optimality of the solutions. As for practical implementations, the appropriate use of heuristics (the ones whose effect on the solution process is known) may be crucial regarding the efficiency of the method.

THE ALGORITHM

In this section, we describe an implementation of the proposed approach. Initially we present a basic algorithm, which uses a depth-first search strategy. For simplicity, neither heuristics nor pruning rules are considered at this stage. In the second part of the section, we comment on how domain-specific knowledge can be used to guide the search and to prune the decision tree. (By domain specific knowledge, it is meant the knowledge an expert has about his specific professional domain – in our application, this knowledge includes practical rules and procedures normally applied by system operators to carry out switching operations aimed at restoring service or improving system operating conditions. This kind of knowledge, also referred to as heuristic knowledge (or "non-mathematical" knowledge), plays an essential role in solving combinatorial problems by avoiding combinatorial explosions, and keeping the problems within manageable sizes.)

A few new concepts have to be introduced before presenting the basic algorithm:

- **Subproblem** is a problem derived from the original one, in which some of the decision variables have been declared. A subproblem is a simpler problem than the original one because it has fewer variables.
- **CCSP**, the Current Candidate SubProblem, is the subproblem which is being dealt with by the algorithm. The other problems either have been discarded or are in a waiting list.
- **LIST** is a list of subproblems waiting to be chosen as CCSP's.
- **INCUMBENT** is the best feasible solution obtained up to the present stage of the search.
- **OBJECTIVE** is the best value of the objective function up to the present stage of the search.

Basic Algorithm

1. Initialization

   1.1 Initialize LIST with the original problem.

   1.2 Initialize OBJECTIVE as infinity.

   1.3 Initialize INCUMBENT as none.

2. Stop test

   2.1 If LIST is empty, then stop.

   2.2 Else choose a subproblem from LIST to be the CCSP.

3. Solution test

   3.1 If all the decision variables of CCSP have already been declared, go to Step 4.

   3.2 Else, go to Step 5.

4. Feasibility test

   4.1 If the decision variables of CCSP lead to a feasible solution, then

   (a) If the feasible solution is better than INCUMBENT, update INCUMBENT and OBJECTIVE.

   (b) Discard CCSP.

   (c) Go to Step 2.

   4.2 Else, discard CCSP and go to Step 2.

5. Branching

   5.1 Determine the successors of CCSP, discard CCSP, and

   (a) Take the A-successor as the new CCSP.

   (b) Put the B-successor on LIST.

   5.2 Go to Step 3.

(Please refer to the Appendix for a detailed illustrative example showing how the basic algorithm can be used to solve a restoration/reconfiguration problem).

Use of Knowledge about the Problem

Now we give some examples on how domain-specific knowledge can be used to improve the efficacy of the search. For example, at Step 5 (Branching), the successors of CCSP will depend on the decision variables chosen to be declared. The choice of the decision variable is made according the following rules:

(a) Choose to declare the decision variable that corresponds to a closed switch in the initial configuration (after fault isolation); take the A-successor (closed) as the new CCSP and put the B-successor (open) on LIST.

(b) Next, choose to declare the decision variable corresponding to the switch that connects the unserved zone with the biggest load to the feeder with biggest loading margin (loading limit minus actual loading); take the A-successor as the new CCSP and put the B-successor on LIST.

(c) Next, choose to declare the variable corresponding to remainders switches; take the B-successor as the new CCSP and put the A-successor on LIST.

In addition to the rules used to select the branching variable (Step 5), we can also use rules at other steps of the algorithm. For example: at Step 1 (Initialization), we can consider part of the decision variables as being dormant; only switches that are more likely to be effective are left to be declared. At Step 2 (Stop test), we can use knowledge about the system to choose the CCSP most likely to lead to an adequate solution. At Step 4 (Feasibility test), when a new INCUMBENT is found, we can test whether the feeder with maximum loading (percentage of the limit) presents a satisfactory loading margin. If so, we can find it worthwhile to stop the search, taking the present solution as the final one.

TEST RESULTS

The proposed algorithm has been tested on the distribution network depicted in Figure 4. The total demand of each zone is listed in Table 1. The loading capacity of each feeder is 10 MVA. The following fault condition has been considered for testing: A permanent fault occurred in Zone 6; Breaker 10 cleared the fault; Switches 6,11, and 15 were opened to isolate the faulted zone; As a consequence, zones 7,8,9,10,13,14,15 and 20 became unserviced.

The problem solved was to restore service to the unserviced zones minimizing the loading unbalancing on distribution feeders. We have used the algorithm described in the previous section considering rules at Step 5 only (Rules for selecting the branching switch). Voltage limits have been ignored.

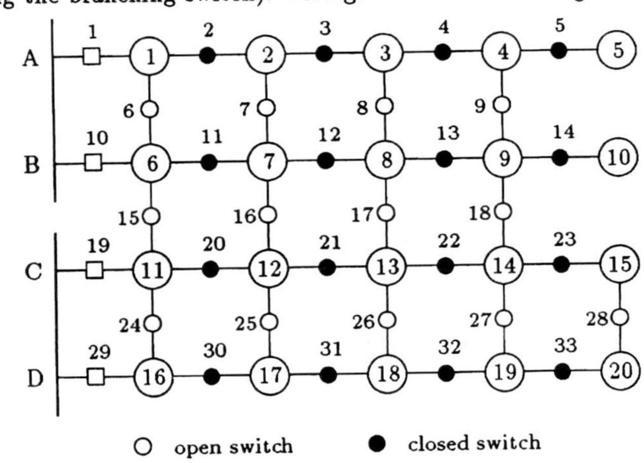

Fig. 4 The distribution network.

ZONE NUMBER	ZONE LOAD (kVA)	ZONE NUMBER	ZONE LOAD (kVA)
1	1600	11	3000
2	700	12	3500
3	1800	13	700
4	500	14	1000
5	1900	15	600
6	500	16	1500
7	1500	17	2000
8	1000	18	1800
9	500	19	1500
10	800	20	500

Table 1: Zone Loads

The results are summarized in Table 2. In the final configuration, feeders A, C, and D are loaded at 88%, 87%, and

89% of their loading capacity, respectively.

SW	STATUS	SW	STATUS	SW	STATUS
1	Closed	12	Open	23	Closed
2	Closed	13	Closed	24	Open
3	Closed	14	Closed	25	Open
4	Closed	15	Open	26	Open
5	Closed	16	Closed	27	Closed
6	Open	17	Open	28	Open
7	Open	18	Open	29	Closed
8	Closed	19	Closed	30	Closed
9	Open	20	Closed	31	Closed
10	Open	21	Closed	32	Closed
11	Open	22	Open	33	Closed

Table 2: Final Configuration

Notice that in this case the solution produced by the method is also the optimum solution of the problem. But, in general, this is not necessarily so. As mentioned before, a characteristic of combinatorial problems is that the number of alternatives increases exponentially with the number of decision variables (A good analogy is the nuclear chain reaction). Heuristic knowledge is used to cope with the increase in the number of alternatives, in the very same way experts do. Practical knowledge about the problem can be used, for instance, to prune the decision tree. (In the nuclear chain reaction analogy, this would be similar to using moderators to "prune" part of the triggering particles, in order to have a controlled reaction.) In the applications we have in mind, it may well happen that an optimum solution is involuntarily eliminated by pruning, which would lead to a suboptimum solution. In practice, however, normally we are interested in solutions that satisfy a certain constraints, rather than the optimum solution. Incidentally, this is what Expert Systems are all about, that is, methods that produce suboptimal solutions by mimicking the behavior of experts in specific knowledge domains.

CONCLUSIONS

A heuristic search approach to distribution system restoration has been proposed. The method uses a knowledge guided search strategy to solve problems such as service restoration and system reconfiguration.

The method is based on operator procedures and usually may generate the same solutions as operators would do. However, one can also investigate alternatives that normally would not be considered by system operators, which may be very helpful under certain critical operating conditions. Moreover, the proposed framework makes it possible to investigate the effect of practical rules on the optimality of the final solution, and so it may be a handy tool in designing new algorithms.

The preliminary results obtained reveal that the proposed technique is very promising. So, we feel encouraged to further test and improve the method. Undergoing research involves the extention of present capabilities to improve the ability to represent practical knowledge, including knowledge that is not suited for algorithmic representation. This, we believe, will bridge the gap between our heuristic search approach and Expert System techniques.

REFERENCES

1. IEEE Committee Report, "The Distribution System of the Year 2000", IEEE Trans. on Power Apparatus and Systems, vol. PAS-101, no. 8, pp. 2485- 2490, August 1982.
2. C.H.Castro, J.B.Bunch, T.M.Topka, "Generalized Algorithms for Distribution Feeder Deployment and Sectionalizing", IEEE Trans. on PAS, vol. PAS-99, no. 2, pp. 549-557, March/April 1980.
3. C.C.Lin, S.J.Lee, S.S.Venkata, "An Expert System Operational Aid for Restoration and Loss Reduction of Distribution Systems", 1987 PICA Conference, Montreal, 1987.
4. C.A.Podbury and T.S.Dillon, "An Intelligent Knowledge Based System for Maintenance Scheduling in a Power System", Proceedings of the Ninth Power Systems Computation Conference, pp. 708-714, Portugal, 1987.
5. A.Monticelli, F.F.Wu and M.Yen, "Multiple Bad Data Identification for State Estimation by Combinatorial Optimization", IEEE Trans. on PAS, vol. PWRD-1, pp. 361-368, July 1986.
6. T.C.Hu, Combinatorial Algorithms, Addison-Wesley, Reading, Mass., 1982.
7. E.Rich, Artificial Intelligence, McGraw-Hill, New York, 1983.
8. P.H.Winston, Artificial Intelligence, Addison-Wesley, Reading, Mass., 1984.

APPENDIX: AN ILLUSTRATIVE EXAMPLE

In this appendix, we will use the small example of Fig. A.1 to illustrates how the basic algorithm works: for simplicity no heuristics is considered, and so the method will simply carry out an exhaustive search. (In such a case, it is guaranteed that the solutions eliminated by pruning are suboptimum solutions, and by that optimality is preserved.)

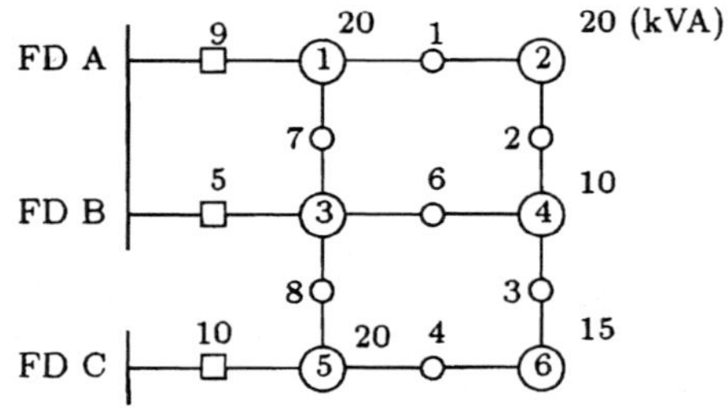

Fig. A.1 The illustrative example data.

Let us assume the following sequence of events. Phase 1: the system is in a normal state; breakers 5,9, and 10 are closed; switches 1,4,and 6 are closed, and switches 2,3,7,and 8 are open; the state vector is given by x = (1,0,0,1,1,1,0,0,1,1)) Phase 2: the system is in an emergency state; fault has occurred in zone Z3, and as a consequence breaker 5 has open; switches 6,7,and 8 have been open to isolate the faulted zone. Phase 3: the system is in the restorative state; in addition to the faulted zone Z3, zone Z4 is unserviced; service restoration will be carried out in two stages; in the first stage, service will be restored to zone Z4; in second stage, zone Z3 will be repaired.

The problem we will solve corresponds to the first stage of Phase 3, that is to say, we will determine a switching strategy to restore service to zone Z4. We will consider the objective of restoring service and, at the same time, minimizing the Load Balancing Index. Needless to say, the desired solution should also satisfy the operating constraints, i.e. (1) the final configuration should be radial; (2) unfaulted zones should be serviced; and (3) primary feeders loading limits should be satisfied (In this example, limits for feeders A,B, and C are assumed to be 50, 30, and 45 KVA, respectively.). Moreover, to further reduce the size of the search space, we will consider breakers 9 and 10 will be kept closed (thus, variables x9 and x10 will be made equal to 1). This will leave us with four variables to decide upon (x1,x2,x3, and x4), because variables x5, x6, x7, and x8 will be kept equal to 0, during the search process. Thus the initial decision vector will be x = (x1,x2,x3,x4,0,0,0,0,1,1).

Now, let us move to the solution procedure. Figure A.2 illustrates the branching mechanism: initially, we have a problem with four variables to be declared; the two successors yield by branching have only three variables left to be declared (These is the main characteristic of the branching process, the complexity of the successor problems is reduced at the expense of increasing the number of alternatives to be analyzed).*Subproblem 1* is discarded;*Subproblem 2*, which becames the new CCSP (Current Candidate SubProblem), is obtaining by making x1 = 1 (this corresponds to a A-successor, which is always obtained by closing the respective switch); *Subproblem 3*, which will be temporarily kept in a waiting list, is obtained by declaring x1 = 0 (this is the B-successor).

Figure A.3 illustrates the depth-first strategy. The search stops descending when all of the decision variables have been declared (either 0 or 1). In this case, we have a infeasible solution (*Subproblem 8* corresponds to a nonradial network) and a feasible solution (*Subproblem 9*) which becomes the incumbent solution (the best solution found until now). Now we backtrack to *Subproblem 7*, which was the last one to be added to the waiting list. And so on. Eventually, we will reach the situation given by Figure A.4.

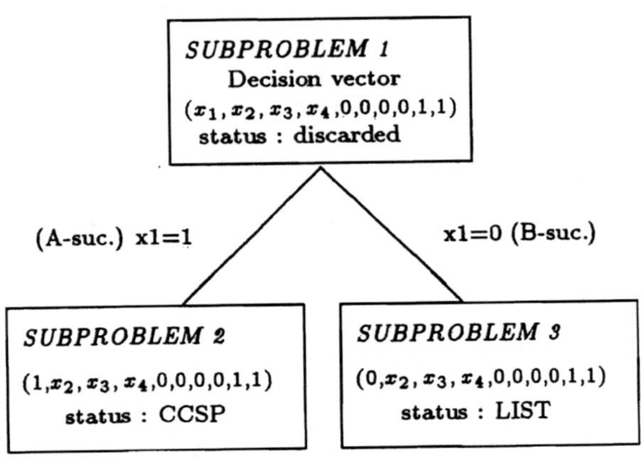

Fig. A.2 Decision tree after the first branching.

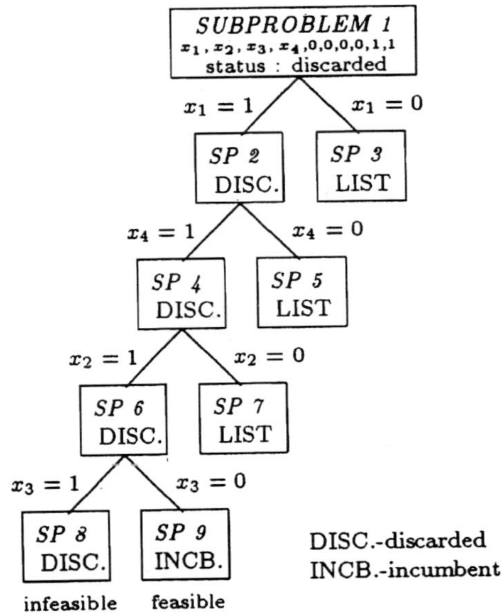

Fig. A.3 Decision tree after the first feasible solution is obtained.

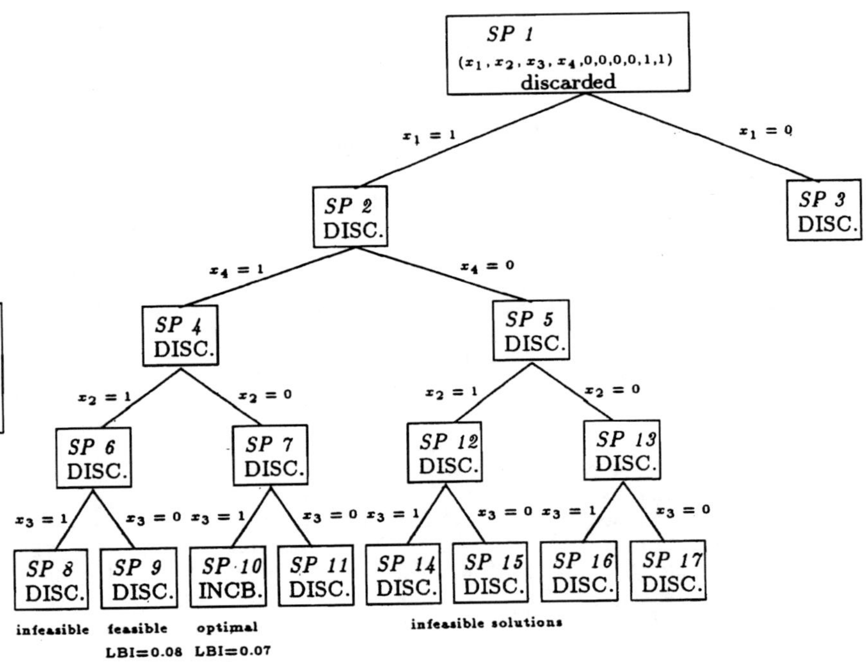

Fig. A.4 Decision tree at an intermediate stage of the search.

DISTRIBUTION TRANSFORMER OVERLOADING CAPABILITY UNDER COLD-LOAD PICKUP CONDITIONS

J. Aubin, Member R. Bergeron R. Morin
Institut de recherche d'Hydro-Québec (IREQ)
Varennes, Québec, Canada

Abstract - Greater use of electric heating in residential areas in North America has led to the appearance of a new type of thermal stress resulting from service restoration following a power outage. This loading condition, known as cold-load pickup, was simulated on a sample of 18 distribution transformers by applying overloads up to three times the rated current at ambient temperatures as low as -40°C. The data gathered was used to develop a thermal model representing the transient behavior of distribution transformers. The results were then generalized to produce a loading guide for evaluation of maximum hot-spot temperatures and loss of life under given cold-load pickup conditions.

Keywords - distribution transformer, cold-load pickup, electric heating, loading guide.

INTRODUCTION

Greater use of electric heating in North America has led to the appearance of a new type of stress on power distribution networks. In residential neighborhoods characterized by single-family dwellings, electric heating systems are generally oversized, so that load is diversified even at the coldest temperatures. For example, a recent Hydro-Québec study of 57 single-family houses built after 1975 in the Montreal area showed that at a temperature of -20°C the average heating load is 8 kW per house, whereas installed capacity averaged 18 kW.

When a power outage occurs as the result of a breakdown or a planned shutdown for maintenance purposes, this diversity is rapidly lost. When the power supply is restored, demand on the network will temporarily equal the entire connected load, gradually dropping to stabilize at the level of the original load. Cold-load pickup (CLPU) is therefore characterized by an overload whose magnitude is determined by the total connected load and whose duration is dependent on the ambient temperature and the length of the outage. This operating condition has been the object of a number of studies [1-6] which, to a certain degree, describe the nature of the overload that occurs under such conditions. Considering that most public utilities permit a winter load level that is higher than the rated power of their distribution equipment, CLPU can result in serious thermal stresses.

The first aim of the project described here was to establish a representative sample of distribution transformers and submit them to a series of severe CLPU tests in order to evaluate their behavior under realistic conditions. On the basis of the experimental results, a thermal model was then to be developed to allow generalization of the results to cover all CLPU conditions.

89 TD 403-7 PWRD A paper recommended and approved by the IEEE Transformers Committee of the IEEE Power Engineering Society for presentation at the IEEE/PES 1989 Transmission and Distribution Conference, New Orleans, Louisiana, April 2 - 7, 1989. Manuscript submitted October 7, 1988; made available for printing January 19, 1989.

TEST PROGRAM

The transformers were selected in terms of best representing the wide range of such equipment commonly manufactured in Canada while at the same time limiting the size of the sample. The sample shown in Table 1 consisted of 13 pole-type and five pad-mounted transformers roughly representing five Canadian manufacturers. All units but two were built after 1984, the two exceptions dating from 1971 and 1974. Power ratings of 50, 100 and 167 kVA were chosen as typical of the current population of transformers serving residential districts. The rated voltage of all the transformers was 14,400/120-240 V with a basic insulation level of 125 kV. All were equipped with ONAN cooling systems and only the two older units (100-1 and 167-3) had cooling radiators. The rated temperature rise of the windings was 65°C in all cases. The impedance ranged from 1.2 to 2.5%.

Table 1
THERMAL CHARACTERISTICS OF TRANSFORMERS UNDER RATED CONDITIONS

Transformer kVA-No.	Year of manufacture	Weihgt kg	Oil volume L	No-load losses W	Load losses W	Average winding rise °C	Hot-spot temperature rise °C
Pole-type							
50-1	1984	304	75	132	286	61.5	68.2
50-2	1984	304	75	138	286	63.3	69.8
50-3	1984	305	75	160	270	61.4	65.7
50-4	1984	305	58	166	236	55.9	61.8
50-5	1985	292	91	159	366	64.4	69.7
100-1	1974	427	95	276	732	79.6	89.8
100-2	1985	529	127	268	514	57.4	64.1
100-3	1985	560	150	274	581	57.9	65.0
100-4	1985	555	145	277	499	48.3	54.1
100-7	1985	485	87	280	469	61.2	70.7
167-1	1985	655	145	408	1610	49.6	55.1
167-2	1985	676	170	340	1800	54.8	63.5
167-3	1971	590	118	341	2500	68.1	80.8
Pad-mounted							
50-6	1986	400	172	163	263	39.6	45.4
100-5	1985	1300	154	280	528	27.7	36.2
100-6	1985	570	215	274	527	52.1	60.1
167-4	1986	1395	695	362	1700	28.4	35.6
167-5	1985	899	391	404	2290	53.0	62.9

Position of the Thermocouples

Each transformer was removed from its tank and equipped with 18 constantan-copper thermocouples. The thermocouples to be used inside the winding were first soldered to a 25-mm-square copper sheet, wrapped with Teflon tape, then positioned as shown in Fig. 1, all on the same side of the core, along the axis of the winding. To place each thermocouple in the windings, a space was made by inserting a wooden wedge, allowing the copper sheet with its attached thermocouple to be inserted. The HV winding was equipped with three thermocouples at the top and three at the bottom. The inner and outer LV windings (LV_1 and LV_2) were each equipped with only one thermocouple.

The temperature of the oil at the top of the cooling ducts was also monitored: two cooling ducts of the HV winding and those of the inner LV winding were equipped with thermocouples for this purpose. Seven

Reprinted from *IEEE Transactions on Power Delivery,* Vol. 5, No. 4, pp. 1883–1891, November 1990.

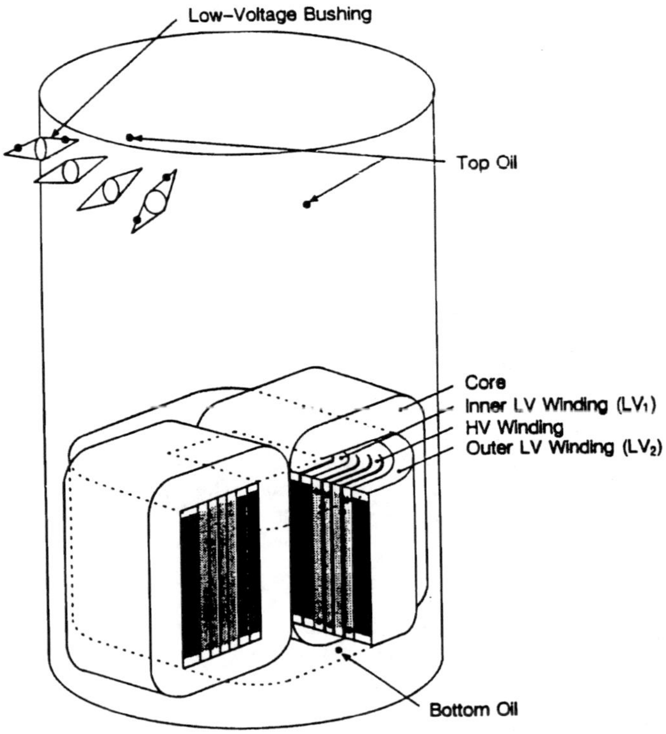

Figure 1 : Location of Thermocouples Inside Each Transformer

other thermocouples were installed outside the windings: the bottom-oil temperature was measured about 2 cm below the windings, the top-oil temperature by two thermocouples placed about 5 cm below the surface. All transformers had four LV bushings, the two end ones being equipped with thermocouples, both inside and outside the tank.

Test procedure

The no-load and load losses were first measured for each transformer; the results are shown in Table 1. Groups of four or five units were then submitted to the temperature rise tests simultaneously. For this test the HV winding was shorted out and connected to ground then the secondary windings of the various transformers were connected in series and powered from a variable, programmable voltage source. The test current for all the transformers in a given group was therefore the same.

A first series of tests was carried out in accordance with ANSI standard C57.12.90 [10] at an ambient air temperature of 30°C. After an initial preheating period at twice the rated current, the test current was adjusted to produce power losses in the transformers equal to total losses under rated conditions. This current was maintained until temperatures stabilized, at which point it was reduced to rated levels for a 1-h period. A data logging system recorded the temperature registered by each thermocouple at 5-min intervals.

To simulate CLPU at low temperatures, the transformers were installed in a cold chamber 5 m by 7 m whose walls were covered in stainless steel, unlike the painted concrete walls of the chamber where the standard temperature rise tests were carried out. The resulting difference in the radiation characteristics combined with the slight air movement in the cold room, made it difficult to later compare the thermal characteristics obtained in these different environments.

The test program simulating the CLPU stresses is shown in Fig. 2. First, the prescribed ambient temperature was maintained for approximately 30 h to allow the transformers to stabilize at this temperature. The rated current was then applied for a 6-h period in order to observe transient behavior with transformer oil at a very low temperature. Three CLPU conditions were then applied in succession to the transformers. The tests were carried out at constant currents, leading to a variation in losses. This loading cycle was repeated with three ambient temperatures: -40°C, -20°C and 0°C. To limit the risk of fire, an alarm signal sounded when the top-oil temperature rose above 140°C or the HV winding temperature rose above 200°C.

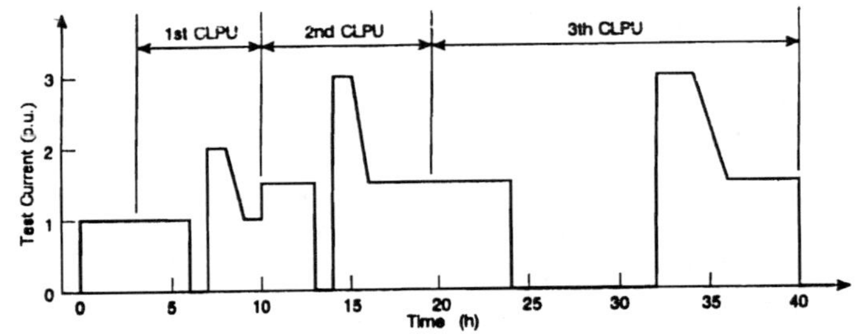

Figure 2 : CLPU Simulation Program

TEST RESULTS

Standard Temperature Rise Test

Temperature readings during the test indicated that the HV winding is consistently hotter than the LV winding. Attention was therefore focused on the HV winding. The nominal temperature rise of the windings was not determined by measuring the resistance, as is usually the case, because this method is impractical when simultaneously testing a number of transformers. The nominal temperature rise of the windings and the hot-spot temperature rise were determined through temperature measurements by thermocouples.

From the data supplied by six thermocouples located in the HV winding, two reference temperatures may be defined: θ_g, the average of all six thermocouples, and θ_f, the average of the three thermocouples at the top of the winding.

An analysis of the thermal flow inside the winding was performed in order to evaluate the temperature gradient between the hottest spot under the core and the section of the winding that is accessible to thermocouples. The results indicate that it is conservative to assume that, under rated current, the hot-spot temperature is 3°C higher than θ_f and the average winding temperature 1.5°C higher than θ_g. These corrections are taken into account in the rated temperature rise reported in Table 1.

Results of CLPU tests

When a load is suddently applied to a thoroughly cooled transformer at an ambient temperature of -40°C, there is reason to fear that the viscosity of the oil could temporarily lead to an excessive temperature rise in the winding. During the test program, the initial load was set at the rated level. An initial delay was noted between the heating of the top oil as well as an increase in the difference between the winding and oil temperatures, but no dangerous temperature level was reached.

The test program's severest overload condition occurred during the final cycle of the CLPU simulation. After the transformer had virtually stabilized at a load of 1.5 p.u., an 8-h outage was simulated, followed by a 2-h overload at 3.0 p.u. and a gradual return to the initial load. Transformer 100-1 was not subjected

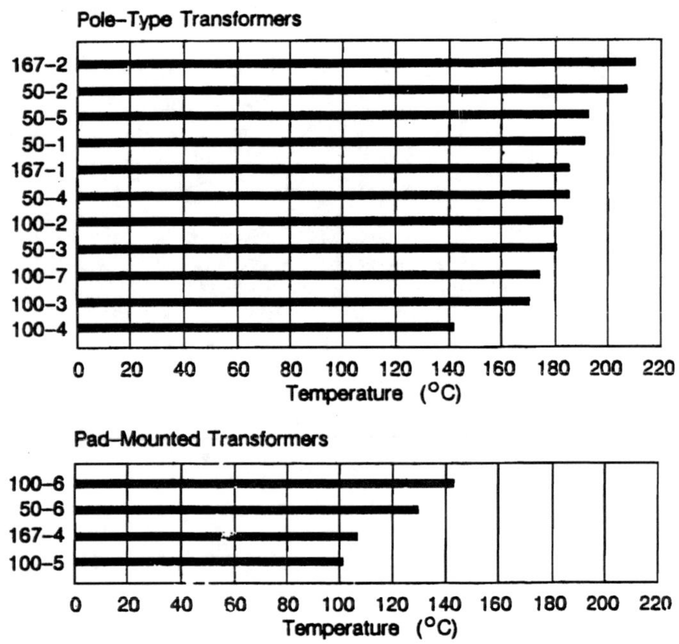

Figure 3 : Maximum Temperature Measured in HV Winding During 2-h Overload at 3.0 p.u.

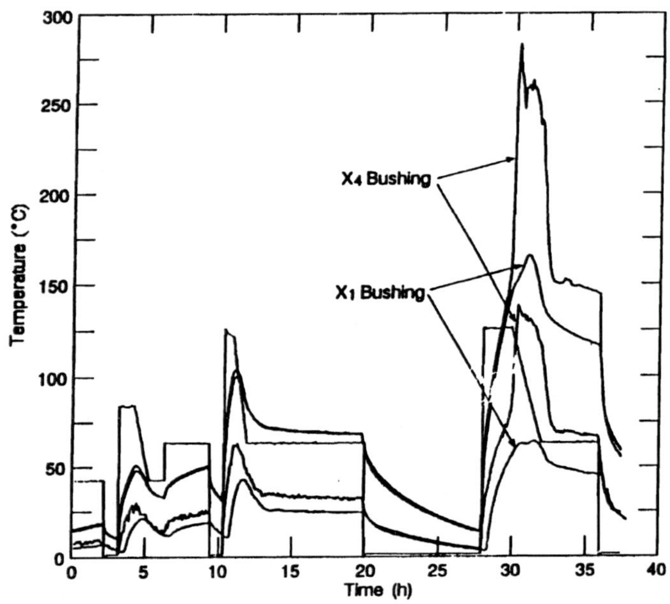

Figure 4 : Thermal Runaway of LV Bushings

to this test because extrapolation from previous tests indicated prospective temperatures of approximately 200°C for the top oil and 260°C for the HV winding. Transformers 167-3 and 167-4 were also not tested, owing to a failure of the power supply system.

The maximum temperatures measured in the HV winding during the tests at an ambient temperature of 0°C are shown in Fig. 3. It should be noted that at this load level the hot-spot temperature is estimated to be about 27°C higher than the recorded temperature θ_f since the nominal difference of 3°C increases with the square of the load factor.

During the tests at 3.0 p.u., five cases of thermal instability were noted in the LV bushings equipped with thermocouples. Figure 4 illustrates this phenomenon on transformer 100-7 during the 0°C test. Normally the temperature of the LV bushings is slightly higher than that of the top oil. With a tightening torque between 33 and 37 N•m, certain LV bushings are incapable of carrying three times the rated current for 2 h. In some cases, temperatures over 300°C were measured at the end of the bushing inside the tank, and those parts of the bushing made of synthetic material had permanent damage to them. Nonetheless, since most of the bushings underwent this stress with no apparent damage, minor modifications could conceivably eliminate this limitation on CLPU capability.

ANALYSIS OF THE THERMAL MODEL

Generalization of the test results calls for a mathematical model that allows the transformer temperature to be calculated for any loading profile.

The model used is similar to the one found in the ANSI and IEC loading guides [7, 8]. The temperature difference between the winding hot spot and the ambient temperature is calculated using the top-oil temperature as an intermediary value. The hot-spot temperature is therefore equal to the sum of the ambient temperature, the top-oil temperature rise, and the temperature difference between the hot spot and the top oil:

$$\theta_h = \theta_a + (\theta_o - \theta_a) + (\theta_h - \theta_o) \qquad (1)$$

Under continuous loading, it is assumed that temperature differences will vary with load as follows:

$$\theta_o - \theta_a = (\theta_o - \theta_a)_r \left[\frac{K^2R + 1}{R + 1}\right]^n \qquad (2)$$

$$\theta_h - \theta_o = (\theta_h - \theta_o)_r \, K^{2m} \, . \qquad (3)$$

Under dynamic loading it is assumed in a first approximation that these two temperature differences will react in an exponential manner when step loading is applied.

Oil Temperature Rise as a Function of Load

To determine the nominal temperature rise and the value of the exponent n, equilibrium is normally established for three different loading conditions and these points are plotted on a log-log graph to establish the characteristic temperature rise curve. In the test program, values for the temperature rise under rated load were available only for the test at rated losses and an ambient temperature of 30°C. In the CLPU test cycle, the current levels were not maintained for a sufficiently long period to permit the oil temperature to stabilize. Nonetheless, assuming that the temperature rise has exponential characteristics, it is possible to extrapolate from the observed temperatures in order to evaluate the final temperature that would be reached if a given load level were maintained for a sufficiently long period.

The top-oil temperature was evaluated in this way for three segments of the test program: the 6-h initial loading at 1.0 p.u., the 8-h loading at 1.5 p.u. and the 2-h loading at 3.0 p.u. The values of $(\theta_o - \theta_a)_r$ and exponent n derived from this exercise are shown in Table 3 for the three transformers used to validate the thermal model under transient conditions.

Oil Time Constant

It is found that when a given transformer is subjected to a step variation of load, the top-oil temperature rise does not behave exactly as an exponential function and the oil time constant can therefore not be treated as a fixed value. If the heating or cooling curve is approximated by a single exponential function,

Table 2
OIL TIME CONSTANT CALCULATED FROM EXPERIMENTAL RESULTS

Previous current (p.u.) Test current (p.u.) Duration (h)	0 1.2 2	0 3.0 2			1.0 0 24	1.5 0 8			
Ambient temperature (°C)	30	0	-20	-40	30	0	-20	-40	
Pole-type									
50-1	5.2	2.2	-	1.8	5.4	4.8	-	5.7	5.3
50-2	4.9	1.8	-	1.7	5.6	4.8	-	5.4	5.3
50-3	5.2	2.3	1.9	1.8	5.6	4.4	5.3	6.1	5.3
50-4	5.3	2.3	2.5	1.7	5.7	4.3	5.6	6.2	5.5
50-5	5.8	3.3	3.1	5.5	6.5	5.0	5.6	5.3	5.3
100-1	4.3	-	-	-	5.2	5.2	5.4	5.7	5.4
100-2	4.0	2.6	-	2.3	7.5	5.3	-	6.7	6.5
100-3	3.9	2.3	-	2.1	8.1	5.4	-	7.2	6.9
100-4	4.7	1.8	-	1.4	8.2	4.9	-	6.1	6.4
100-7	5.9	3.0	2.5	2.1	7.8	4.8	6.5	7.5	6.6
167-1	5.7	2.0	1.8	1.8	6.9	4.4	5.2	5.8	5.6
167-2	5.7	1.9	2.7	2.7	6.9	4.1	4.9	6.8	5.7
Pad-mounted									
50-6	5.2	3.6	1.2	4.2	6.1	5.9	5.5	4.5	5.5
100-5	5.9	7.8	-	-	11.0	8.7	9.2	12.0	10.2
100-6	4.2	2.1	-	1.5	7.5	5.6	-	6.8	6.6
167-4	-	1.1	2.8	1.2	10.7	10.5	9.2	-	10.1

the time constant is seen to vary with both the magnitude of the applied load and the ambient temperature, as seen in Table 2. It should be noted that, since the tests were carried out at constant current, power losses in the transformer gradually increased as the resistance of the windings increased. The reduction in oil viscosity compensates only partially for this increase and, as a result, perfectly exponential temperatures cannot be expected.

During the cooling-down period, however, the resistance of the windings does not intervene. In some instances, temperatures were recorded for up to 40 h and the apparent time constant for a specific segment of the curve can be seen to increase by some 50% in a 20-h period following the moment of shut-down. The overall time constant during cooling down therefore increases significantly with the duration of the observation.

A better approximation of the top-oil temperature behavior is obtained if the calculation takes account of a variable heat transfer coefficient between the tank surface and the cooling air. For natural convection it is accepted that this heat transfer coefficient h varies with respect to the temperature difference $\Delta\theta$ between the tank surface and the cooling air as follows:

$$h = \alpha\,(\Delta\theta)^{0.25} \tag{4}$$

where α is the convective coefficient factor.

Introduction of this variable into the energy pattern of the transformer tank yields an expression of the time constant that is dependent on the initial oil temperature rise $\Delta\theta_{oi}$ and ultimate rise $\Delta\theta_{ou}$ that would occur if the step load were maintained indefinitely. This expression also is related to a reference time constant τ_{or} and to the rated temperature rise $\Delta\theta_{or}$ as follows:

$$\tau_o = \tau_{or}\left|\frac{\dfrac{\Delta\theta_{ou}}{\Delta\theta_{or}} - \dfrac{\Delta\theta_{oi}}{\Delta\theta_{or}}}{\left(\dfrac{\Delta\theta_{ou}}{\Delta\theta_{or}}\right)^{1.25} - \left(\dfrac{\Delta\theta_{oi}}{\Delta\theta_{or}}\right)^{1.25}}\right| \tag{5}$$

The foregoing equation is similar to that suggested in the ANSI loading guide [7] if n = 0.8. However, it was found that this time constant τ_o should not remain fixed throughout the heating (or cooling down) period but should be readjusted for each time increment in the calculation process to make $\Delta\theta_{oi}$ equal to the actual top-oil temperature rise at that moment. This method shows good correlation with experimental results, as seen in Fig. 8.

Winding Temperature Rise with Respect to Oil

The thermal model defined above makes use of the temperature difference between the hot spot and the transformer top oil. Direct measurement of the hot-spot temperature was impossible but temperatures measured by thermocouples in the windings were available and the variation in these temperatures as a function of load can be examined.

The analysis is limited to temperatures observed in the upper part of the HV winding, which is where the highest temperatures were measured. Figure 5 shows a typical record of the difference between the average temperature measured by the three thermocouples in the upper part of the HV winding (θ_f) and the top-oil temperature (θ_o) for three transformers subjected to a step load of 3.0 p.u. at an ambient temperature of -20°C. It is evident that this temperature difference tends to stabilize after a high initial value, which is largely due to the reference used to represent the oil temperature. Throughout the test series, it was observed that the top-oil temperature showed a time delay of ≈ 0.2 h before reacting to a sudden overload. In the mathematical model used, this delay is not taken into account. Rather, it is assumed that the oil begins to increase in temperature immediately, an assumption which is perfectly reasonable as far as the oil in the cooling ducts is concerned. When the initial delay in the top-oil temperature rise is considered, it largely compensates for the high initial value for the temperature difference between the winding and the top oil.

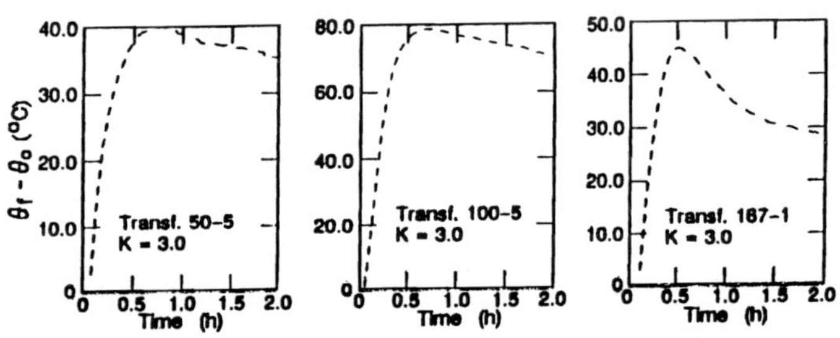

Figure 5 : Temperature Difference Between HV Winding and Oil During Load Application

In the thermal model adopted for this work, the temperature difference between the hot spot and the top oil is assumed proportional to the load level K to the power of m. This exponent is determined by plotting temperature rises as a function of load on a log-log graph. Figure 6 shows the results of this calculation for three transformers. Normally, for a given ambient temperature, the values corresponding to the three load levels should be found on the same curve with a slope of 2m. In fact, the values corresponding to rated current do not fall along this curve, especially at very low ambient temperatures, which implies some difficulty in establishing oil circulation under these conditions. Given the particular importance for the mathematical model to be reliable at high current levels, underevaluation of the winding temperature rise at low loads and low temperatures is acceptable.

With loadings of 1.5 and 3.0 p.u., the temperature difference in pole-type transformers does not appear to be directly influenced by the ambient temperature but

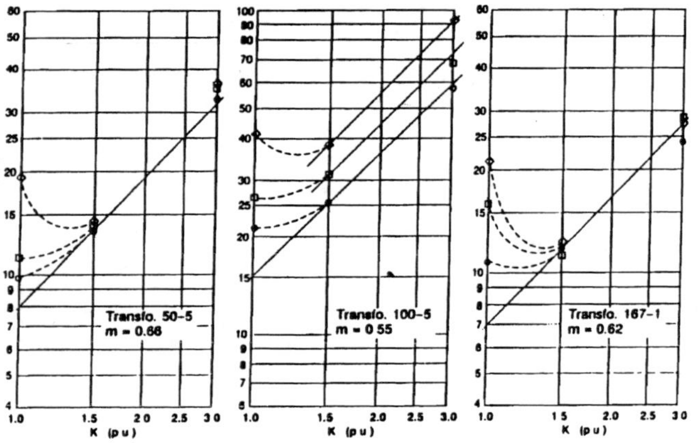

Figure 6 : Temperature Difference Between HV Winding and Oil vs. Load

with lower loads the difference increases significantly when the temperature drops to -40°C. Figure 7 shows the relationship between the temperature difference and top-oil temperature for various loads (K). For pad-mounted transformers (100-5), the temperature difference between the HV winding and the top oil is generally greater and appears to vary consistently with the top-oil temperature, even at 3.0 p.u.

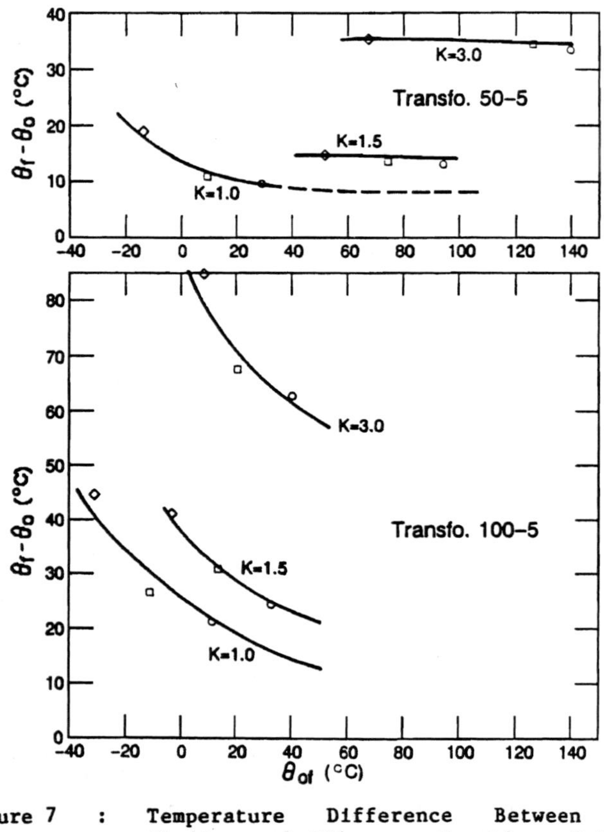

Figure 7 : Temperature Difference Between HV Winding and Oil as a Function of Load and Oil Temperature

Winding Time Constant

The winding time constant τ_w used in the thermal model cannot be calculated on the basis of experimental data, since the temperature difference between the winding and the oil cannot be adequately represented by an exponential function but can only be established approximately, respecting the steepness of the initial

Table 3
SUMMARY OF THERMAL CHARACTERISTICS
OF THE THREE REFERENCE TRANSFORMERS

Transformer / θ_a (°C)		50-5			100-5			167-1	
		0	-20	-40	0	-20	-40	0	-20	-40
R			1000			1000			1000	
$(\theta_o - \theta_a)_r$	°C	46	48	49	15	16	18	52	54	56
$(\theta_f - \theta_o)_r$	°C		8		15	19	24		7	
n			0.76			0.9			0.72	
m			0.66			0.55			0.62	
τ_o	h		4.5			7.5			5.1	
τ_w	h		0.1			0.2			0.1	

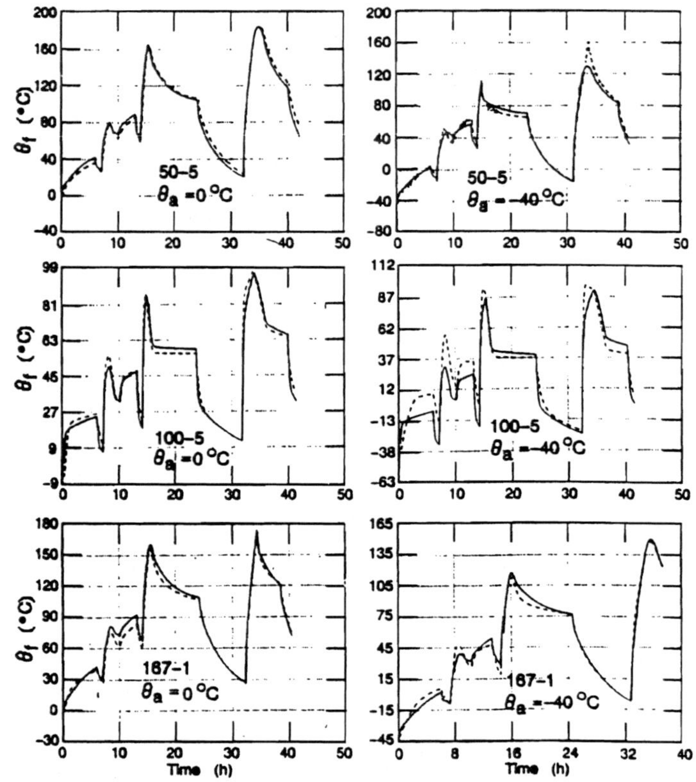

Figure 8 : Agreement Between Modeling Results (solid line) and Experimental Data (dotted line)

rise. In every case, a time constant of the order of 0.1 or 0.2 h gives satisfactory results for large overloads. With low loads (K = 1.0 p.u.), this approximation will give a pessimistic evaluation of temperature rise.

Validation of the Thermal Model

The hot-spot temperatures were evaluated according to the thermal model and the results of these calculations compared with the experimental observations of the CLPU simulation. The thermal characteristics of the transformers tested were derived from the test results and some of these are shown in Table 3.

Some thermal characteristics vary with the ambient temperature and is important to take these variations into account in the mathematical model, since these temperature rises vary with a power of the load and lead to considerable differences in the calculated value. The value R = 1000 was set arbitrarily to eliminate the correction factor $(K_2R + 1)/(R + 1)$, given that during the temperature rise tests the transformers were short-circuited and the iron losses consequently negligible. The maximum oil temperature rise values here correspond to the experimental values at rated current, rather than at rated losses.

In Fig. 8 experimental results are compared with calculated values for the temperature of the upper part of the HV winding. In view of the simplifications made in the thermal model, agreement was expected to be poor for the lowest loads, especially at -40°C ambient. The thermal model is primarily intended to correctly predict the temperature of the winding under the most severe overload conditions and, in this application, agreement is satisfactory.

It is therefore confirmed that the hot-spot temperature under CLPU loading can be correctly predicted with the usual thermal model as long as a variable time constant is introduced and a small correction made in the top-oil temperature rise to take acount of the ambient temperature.

COLD-LOAD PICKUP GUIDE

Typical Transformer Characteristics

In order to apply the thermal model to different CLPU conditions, it is necessary to define the thermal characteristics of a typical transformer and the characteristics chosen must be conservative in order to cover the majority of transformers currently manufactured. In view of the characteristics of the sample and the standard characteristics in the industry, the values shown in Table 4 were selected.

Table 4
THERMAL CHARACTERISTICS CHOSEN AS A BASIS FOR THE LOADING GUIDE

Ratio of load to no-load losses	R :	2.7
Nominal top-oil temperature rise	$(\theta_o - \theta_a)_r$:	$55 + (\theta_a \times 0.1)$
Nominal temperature difference between the hot spot and the top oil	$(\theta_h - \theta_o)_r$:	20
Top-oil exponent	n :	0.8
Winding exponent	m :	0.8
Oil time constant	τ_{or} :	4.5
Winding time constant	τ_w :	0.1

The nominal hot-spot temperature rise is 55 + 20 = 75°C, which corresponds to ANSI C57.91, Table 3; 85% of the tested transformers have a hot-spot temperature rise below or equal to this limit. However, pad-mounted transformers appear to have a lower temperature rise, and in this case a value of 65°C could be more appropriate. A correction factor of 0.1 θ_a is applied to increase the nominal top-oil temperature rise by 2°C for each 20°C drop in ambient temperature. This naturally applies only to temperatures below 0°C. As for the oil-time constant used in the thermal model, 4.5 h appears to be an appropriate value, on the conservative side.

The loss of life is calculated using the life-expectancy curve applicable to a 65°C insulation system, as described in the ANSI loading guide [7]. The additional loss of life is computed for the 24-h period, including the outage and recovery time.

Typical Cold-Load Profile

The profile of the overload that follows a power outage can be represented as shown in Fig. 9, where the following variables must be considered:

t_1 = duration of the outage
t_2 = duration of the overload
t_3 = duration of the return to normal conditions
K_1 = magnitude of the initial load
K_2 = magnitude of the overload

Since the aim was to produce a loading guide for predicting the maximum hot-spot temperature and loss of life as a function of the ambient temperature and the above variables, the latter must be reduced in number by identifying interdependency relations.

Several authors have attempted to identify the overload profiles that follow power outages [1-6], but on the whole very little information is available concerning loading in residential areas with predominantly electric heating in cold climates. The best is found in [1], where a large number of cases are simulated and a study of the thermal characteristics of residences provided the following values for a 1-h outage:

θ_a (°C)	t_1 (h)	t_2 (h)	t_3 (h)	K_2/K_1
0	1	0.17	0.17	3.7
-20	1	0.34	0.17	2.74
-40	1	0.52	0.33	2.1

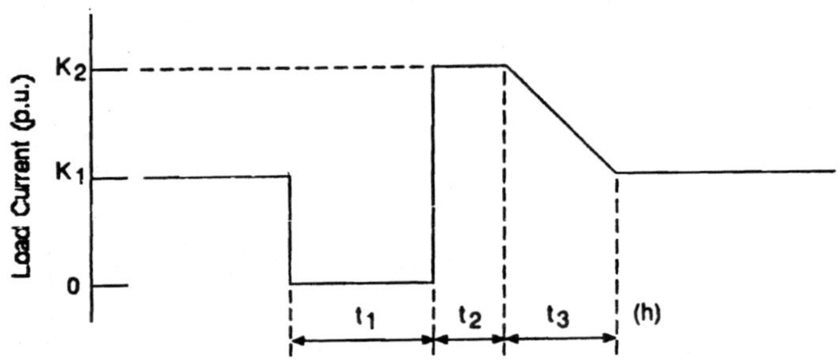

Figure 9 : Characteristic CLPU Variables

While it is somewhat risky to use these same proportions to characterize outages lasting as long as 8 h, they do allow the number of variables to be reduced:

$t_2 = (20°C - \theta_a) \times 0.0083\ t_1$
$t_3 = t_2$
$K_2 = 3.7\ K_1$ for $\theta_a = 0°C$
$K_2 = 2.74\ K_1$ for $\theta_a = -20°C$
$K_2 = 2.1\ K_1$ for $\theta_a = -40°C$

With this definition of the cold-load profile, the maximum hot-spot temperature and loss of life can be calculated for any set of initial load, outage duration and ambient temperature. The results can be presented graphically as shown in Fig. 10 for the case of an ambient temperature of 0°C.

To use these graphs, one must first determine the hot-spot temperature or loss of life that can be considered acceptable. For example, if the hot-spot temperature is to be limited to 200°C, with a 1-h outage at an ambient temperature of 0°C, the initial load must be limited to 1.0 p.u. If the possibility of an 8-h outage is to be considered, while maintaining the same hot-spot limit of 200°C, the initial load should be limited to 0.76 p.u. In this case the additional loss of life caused by the CLPU would be on the order of 0.5%. By deciding on a reasonable length of outage, the distribution system planner can determine acceptable load levels for various ambient temperatures.

If the CLPU profile differs significantly from that used here, the user would be wise to redo the calculation using the same thermal model in order to determine more realistic initial load levels.

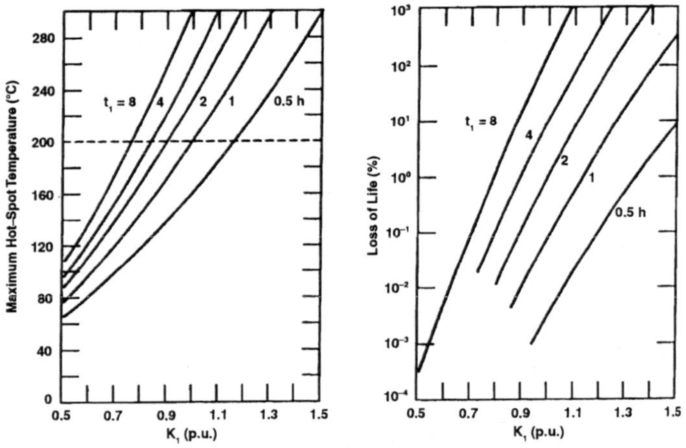

Figure 10 : Maximum Hot-Spot Temperature and Loss of Life for Typical CLPU Conditions

CONCLUSION

CLPU tests were carried out on 18 fully instrumented distribution transformers at overloads up to three times the rated current, following an 8-h outage. The transformers were equipped with thermocouples in the accessible portion of the windings so that temperatures could be measured directly during a test cycle simulating CLPU conditions up to three times the rated current for over 2 h. Under these conditions, the maximum temperatures observed in the winding varied from 102 to 210°C. Overload capability varies considerably from one transformer to another, depending on their rated temperature rise and the oil-time constant. The pad-mounted transformers tested here appear to have a greater overload capability than pole-type transformers.

A number of cases of thermal runaway were noted in the LV bushings. Temperatures of 300°C were observed, and permanent damage to some bushings resulted.

Between 0°C and -40°C, the nominal oil temperature rise should be increased by 2°C for each 20°C drop in ambient temperature in order to take into account the increased viscosity of the oil and the decrease in the heat transfer coefficient.

In all the transformers tested, the HV winding reached a higher temperature than the LV winding.

While the ANSI loading guide predicts a 15°C difference between the average temperature of the winding and the winding hot-spot temperature when the transformer is loaded at rated current, measurements would indicate that a value of 10°C would be more appropriate.

When a high-magnitude overload is applied to a cooled transformer, its dynamic behavior cannot be simulated correctly by assuming a fixed value for the oil time constant. The latter should be adjusted continuously, depending on the oil temperature at that time and the final temperature the oil will reach if the same overload level is maintained indefinitely. The equation proposed to this effect in the ANSI loading guide permits a good correlation between calculated temperatures and experimental observations.

The test results were generalized to produce guidelines for evaluating thermal stresses in the case of CLPU. The graph presented allows the maximum hot-spot temperature and loss of life to be evaluated directly as a function of the initial load and the outage duration. Thus, with a hot-spot temperature limit set at 200°C and a 60-min outage, the initial load should not be higher than 1.0 p.u. at an ambient temperature of 0°C and 1.5 p.u. at -40°C.

Oversizing of residential heating systems would appear to be the determining factor in evaluating thermal stresses during CLPU following a power outage. For specific applications, the hot-spot temperature and loss-of-life predictions could be improved if a better evaluation of the magnitude and duration of the CLPU were available. In all cases involving loads comprising electric heating in residential areas, the total connected heating load should be taken into account to an even greater degree than winter peak loading when determining the rated power of the transformer to be installed.

REFERENCES

[1] R.L. Wilde, "Cold-Load Pickup Problems Relating to Distribution Transformers," CEA Report 77-46, 1979.

[2] R.L. Wilde, "Effects of the Cold-Load Pickup at the Distribution Substation Transformer," CEA Report 131-D-247, 1983.

[3] C. Kropp, "Impact of Conversion to Electric Heat on Municipal Utility Systems," 1982 AMEU Summer Conference for Management, Ottawa, June 1982.

[4] J.N. Brooks, "Electric Heating and its Impact on Residential Distribution Systems," 1982 AMEU Summer Conference for Management, Ottawa, June 1982.

[5] R.L. Wilde, "State of the Art Study on Cold-Load Pickup on the Distribution System," CEA Report 176-D-366, 1983.

[6] G. Au Yeug, D. Walker, "Transformer Testing Under Cold-Load Pickup Conditions," CEA Report 77-46 Phase II, 1983.

[7] ANSI Standard C57.91-1981, "Guide for Loading Mineral-Oil Immersed Overhead and Pad-mount Distribution Transformers," 1981.

[8] IEC Publication 354, "Loading Guide for Oil-Immersed Transformers," 1972.

[9] P.G. Buchan, M.A. Green, "Thermal Behavior of Hot Spots in Distribution Transformers," CEA Report 103-D-202, 1984.

[10] ANSI Standard C57.12.90-1980, "Test Code for Liquid-Immersed Distribution, Power and Regulating Transformers and Guide for Short-Circuit Testing of Distribution and Power Transformers," 1980.

[11] C.F. Kropp, "Cold-Load Pickup with Particular Reference to Residential Electric Heating," CEA Report SD176/SD177.

ACKNOWLEDGMENTS

The authors would like to thank Mohamed Chaaban and Quoc-Anh Ton-That for their contribution in thermodynamics, Gilles Thivierge for planning and conducting the cold chamber tests, and Nancy Lessard for developing the many computer programs necessitated by the project.

This project was financed jointly by the Canadian Electrical Association and Hydro-Québec, whose interest and support allowed the work to be carried out.

LIST OF SYMBOLS

θ	:	temperature
$\Delta\theta$	:	temperature rise ($\theta - \theta_a$)
θ_a	:	ambient-air temperature
θ_b	:	bottom-oil temperature
θ_f	:	average of three thermocouples at top of HV winding
θ_g	:	average of six thermocouples in HV winding
θ_h	:	winding hot-spot temperature

θ_o : top-oil temperature
θ_w : average temperature in HV winding
τ_o : oil time constant
τ_w : winding time constant
K : load factor (p.u.)
R : ratio of load losses to no-load losses
n : exponent of oil temperature rise vs. total losses
m : exponent of winding temperature rise vs. load

Second suffix

r : at rated losses
i : initial value
u : ultimate value
f : at the end of a test

Discussion

L. W. Pierce, (General Electric Company, Rome, Ga.): The authors have conducted a comprehensive investigation of loading of distribution transformers under low ambient temperatures. The authors' data and discussion could be very useful for developing future loading guides. I have the following questions and comments for the authors.

A previous study by Aubin and Langhame (Discussion Ref. 1) described low ambient temperature tests with Voltesso 35 oil. What oil was used in the transformers tested by the authors?

It was reported by the authors that the high voltage winding was hotter than the low voltage winding on all the units. Short circuit type heat runs without losses in the core were performed. The total loss was input into the windings so that the tested oil rise should be correct. As shown in Table 1 the core loss for the pole type and small pad mount units is a significant portion of the total loss. Figure 1 shows that the inner low voltage winding may contact the core and the outer low voltage winding could also be in contact with the core. The core loss could thus have a significant effect on the temperatures in the low voltage winding unless adequate space was allowed for circulation of the oil between the core and winding. Did the analysis of the thermal flow inside the windings include the effect of core loss on the temperature of the windings? Was a heat flow analysis performed with and without core loss and what were the results?

It appears that there is a need to investigate heat run test procedures to determine if there is any difference in thermal test results between the loading back (bucking) and short circuit type heat run test methods for these type of transformers. This is needed because transformer designs and loss ratios have changed since the short circuit test method was adopted by the transformer test standards.

The authors report a temperature difference of 3°C between their thermocouple at the accessible location and the hot spot at rated conditions and state that this difference increases to 27°C during the overloads. The 27°C is significant. J. Aubin stated in the presentation of this paper that one unit was instrumented with imbedded thermocouples during manufacture. This data was used to develop the thermal model. This was not stated in the paper. Would the authors elaborate on this since it helps substantiate and increases the credibility of their test results?

My last question pertains to the m and n exponents given in Table 3. The Loading Guide equations and exponents are being reviewed by the IEEE PES Working Group on Guides for Loading. Temperature correction of losses is being considered in an alternate set of equations. The m exponent of .55 to .66 differs from the Loading Guide reccommended value and the 0.8 given in Table 4. The n exponent in Table 3 varies from .72 to .90 and .8 is used in Table 4. If the authors would review their data and provide an answer to the following questions in their closure it could be beneficial to the IEEE Working Group. If the losses were corrected for temperature in equations 2 and 3 and m and n exponents calculated from the test data, what would the values be? Would there be more consistancy between units? Would the values approach the values given in the Loading Guide?

Reference

J. Aubin and T. Langhame, "Effect of Oil Viscosity on Transformer Loading Capability at Low Ambient Temperatures," IEEE paper 89 WM 080-3 PWRD presented at IEEE Power Engineering Society Winter Meeting, New York, Jan. 28–Feb. 3, 1989.

J. Aubin, R. Bergeron, R. Morin: The authors are grateful to Mr. Pierce for the interest shown in our work. In answer to his queries, the oil used was Voltesso 35, as this is required in Hydro-Québec's specification for distribution transformers.

The analysis of heat flow in the winding was designed to simulate the heat-run test condition and the core losses as such were therefore not simulated. It is true that a heat run test performed with the loading-back method would somewhat increase the temperature rise of the low-voltage winding, especially the inner one, as the oil circulation is virtually, nonexistent between this winding and the core. Investigation of the magnitude of this effect was beyond the scope of this project, but futher investigation is definitely warranted to confirm the validity of the normal heat-run test procedure on transformers of contemporary design.

In any event, the impact of core losses is significantly reduced when the transformer is under severe overload conditions, as is the case in the cold-load pickup situation.

In order to gain better knowledge of the temperature gradient within the winding, arrangements were made with a manufacturer to have thermocouples embedded in the winding of transformer 100-7 at the manufacturing stage. Eight thermocouples were installed on a single plane, half-way down the winding.

Figure 1 shows the temperature difference between the measured temperature and the average oil temperature for a test current of 1.27 p.u.. By comparing with test results at 2.0 p.u., it can be extrapolated that at rated current the gradient between the hottest spot (under the center line of the core) and the "nose" of the winding would be 1.8°C for the inner LV winding, 1.2°C for the HV winding and 0.7°C for the outer LV winding. A uniform and conservative value of 3°C was used in the report to allow for design variations.

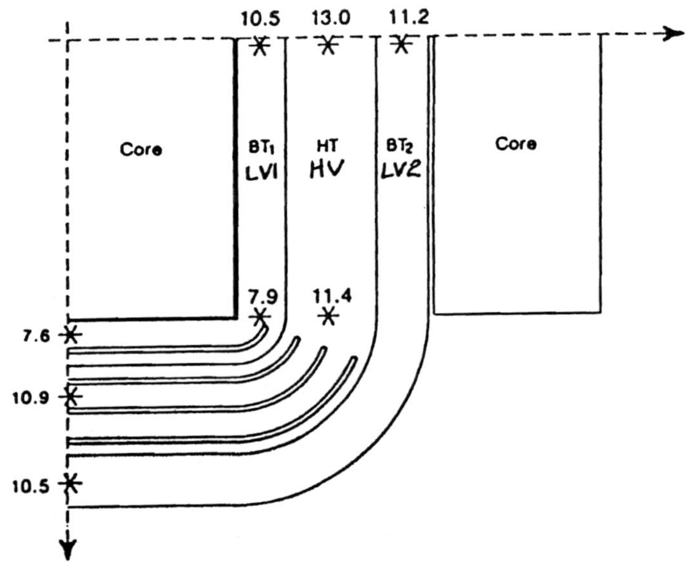

Fig. 1. Temperature Differences Measured in the Winding of Transformer 100-7.

MARS : AN AID FOR NETWORK RESTORATION AFTER A LOCAL DISTURBANCE

E. Mondon(*) - B. Heilbronn(*) - Y. Harmand(*) - O. Paillet(**) - H. Fargier(**)

(*)EDF - Direction des Etudes et Recherches
Service Etude de Réseaux
1 avenue du Général de Gaulle
92141 - CLAMART CEDEX - FRANCE

(**)Laboratoires de Marcoussis
Centre de Recherche de la CGE
Route de Nozay
91460 - MARCOUSSIS - FRANCE

ABSTRACT

Power transmission networks are large and complex systems, and are subject to local failures. They are protected by automatic devices that disconnect faulted areas after a failure to suppress short circuits or overloads on lines. The objectives of the operators are to restore the power supply of loads in the disconnected areas and to consolidate the network by the restoration of non-faulty network components. Restoration is a reconfiguration process: a restoration plan is a sequence of discrete actions on the network (opening or closing circuit-breakers, disconnecting of automatons, ...). To construct these plans, the operators must take into account general restoration guidelines or local constraints, but most of the procedure relies on their experience. The aim of the MARS project is to provide the operator with an aid during these delicate phases of network monitoring. MARS is a second generation expert system which takes into account the man-machine cooperation problematic. MARS constructs restoration plans after a local failure on a power transmission network. Organized around a blackboard architecture, it integrates planning knowledge sources containing the restoration expertise, a qualitative model to predict the results of the considered actions and a quantitative model to verify the correctness of the plans towards numerical constraints (load flow).

Keywords : Control Centre - Network restoration - Expert system

1. INTRODUCTION

Electric Power Utilities must face an increasing demand regarding the quality and availability of electric power supply. However, transmission networks remain complex systems, including physical components subject to failure. It is therefore more and more important to minimize the impact of such failures on the customers.

A great part of this objective is achieved through network redundancy, action of protective relays and automatons. Nevertheless, the monitoring function performed by the operators is a prime necessity in a large number of cases. When an incident occurs, the operators must analyze the system state, localize the failure and find its origine in order to do suitable actions for network restoration.

Papers presented at the Seventeenth PICA
Conference at the Hyatt Regency Baltimore
Hotel, Baltimore, Maryland, May 7 - 10, 1991
Sponsored by the IEEE Power Engineering Society

A new generation of operator aid tools is under development at *Electricité De France* (EDF) in connection with the renewal of the regional control centers. The main purposes are to enable the operators to assess the Power System state more quickly, to devise corrective or restorative strategies in the case of incidents and to restore the conditions for a secure operation. The paper describes in this context, the MARS sytem which has been designed to generate restoration plans in the case of a local disturbance, leading to the loss of power supply on a limited number of HV or EHV substations. By speeding-up the restoration process, it will then minimize the impact of such incidents and increase the availability of electric power supply.

2. THE MARS PROJECT ENVIRONMENT

The operation of EDF generation-transmission system is currently co-ordinated by a national control center in charge of the power exchanges, the global unit commitment and the 400 kV system security, and by seven regional control centers which are in charge of the regional transmission networks (225 kV to 63 kV). The orders decided on by the operators are transmitted by phone to the staff operating in the thermal plants, the main substations and the hydro-electric generation control centers.

The regional networks are partly meshed, and each of them includes about one thousand electric nodes. They will be shortly monitored at the regional level by a new generation of control centers named Regional Control Center. The purpose of the RCC project is to group the function presently performed in each regional control center with the remote control functions which are presently provided locally. Generalized remote control of the substations from the regional control centers will make it possible to adapt the network topology permanently and as optimally as possible, and to facilitate and speed-up the network restoration after an incident.

At the same time, this new conception of the RCC will considerably increase the number of informations flowing to the control system, as well as the orders to be sent, especially in disturbed conditions. Therefore, the RCC will support two innovative functions allowing the operator to manage this situation :

- an event synthesis and fault-diagnosis function based on an expert system [1],
- a function allowing to send "functional orders" grouping several elementary orders to the network substations, and then making the remote operation easier.

In a further stage, the MARS system will be articulated with these functions, as shown on figure 1. It will use the event synthesis function to know the state of the electric system components, and cooperate with the operator to generate restoration plans, which will be directly applied by the remote operation function.

Reprinted from *IEEE Transactions on Power Systems,* Vol. 7, No. 2, pp. 850–855, May 1992.

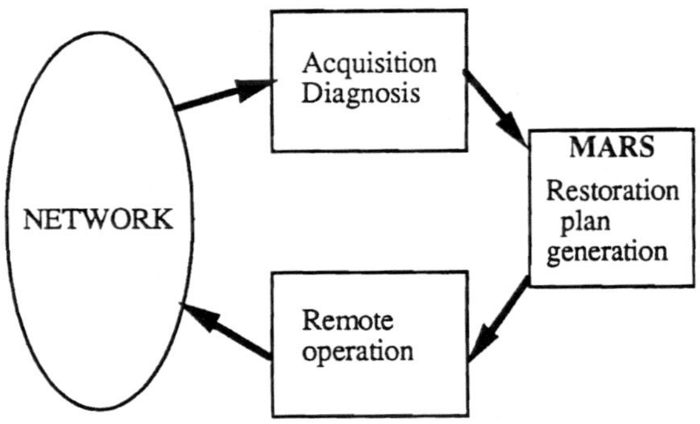

Figure 1
Possible insertion of MARS in the future Regional Control Centers environment

MARS is representative of a new generation of intelligent control functions whose conception, ergonomy and man-machine cooperation level will be particularly suited to system operation in disturbed conditions. At the present stage, MARS is a prototype running in its own environment and allowing to generate restoration plans with various man-machine cooperation levels.

3. THE RESTORATION PROBLEM

The Regional transmission networks are protected by automatic devices that disconnect faulted elements (lines, busbars) after a failure in order to suppress short-circuits or overloads. Automatons divide the networks into sections after a lack of voltage and in some cases manage the network restoration automatically. In the final state, the faulty components are isolated, and a part of the network is, in some cases, out of voltage.

The objectives of the operators are then to restore the power supply of loads in the disconnected area and to consolidate the network by a restoration of non-faulted components. Restoration is a reconfiguration process : a restoration plan is a sequence of discrete actions on the network (opening or closing of circuit-breakers, disconnecting of automatons, ...). The restoration problem can then be stated as a plan-construction problem, related to the general class of "planning problems" in the field of Artificial Intelligence. This "planning problem" has however some specific characteristics :

- The effect of actions depends on the state of the network. For example, the closing of a circuit-breaker does not only change its position, but may also restore the power supply of a part of the network, then provoke the operation of automatons that can themselves open or close other circuit-breakers, and so on.

- The state of the network is sometimes incompletely known. In some case the operator must take into account incomplete information while he conceives the restoration plan. Practically, when operators have a doubt on the state of a component that would be useful in the restoration process, they can include in the plan special trial operations to determine its actual state.

- Errors may be detected at the execution stage of the restoration plan, if the network does not behave as foreseen (for example, if a component which has been considered as sane reveals faulted). The initial plan must then be modified.

Constructing a restoration plan is a heuristic process, and no available algorithmic technique seems to be able to find directly an acceptable solution. An expert system approach has been therefore adopted for the conception of MARS. Naturally, such an approach relied a lot on the available expertise about the restoration process.

Presently, only some general rules concerning the authorized operations on the network components (breakers, switches, ...) are really available in a formal way, as most of the restoration procedure relies on the operator experience and skillness.

An important step in the design of MARS was then to derive, from that implicit knowledge, a systematic and explicit expertise on the restoration procedure. This task was performed through the careful analysis, with a group of experts, of real cases of network incidents. Such a formalization of the restoration expertise, which revealed possible in spite of the diversity of regional practices, is in itself a very interesting result of the MARS project.

4. MAIN DESIGN PRINCIPLES

The goal of MARS is to generate restoration plans after a local failure, using diagnostic information on the nature and location of the failure, and the information on the state of the electrical and control devices. To manage the restoration problem, two kinds of knowledge are required :

- The knowledge of the electrical system, which includes the description of the network and its components, and the behaviour of this network towards events ; this knowledge is handled by a qualitative simulator linked with a quantitative model (load-flow),

- The restoration expertise, which includes the operator skillness and the operation rules to manage network restoration.

4.1 Network representation

4.1.1 The object hierarchy

The use of an object oriented programming environment allows to describe the network as an object hierarchy, these objects being characterized by their structure, the link which interconnect them, and their behavior towards events. At a high level of abstraction, the network may be considered as a set of elements divided into three types :

- indivisible electrical devices which correspond to physical objects from various types : transformers, transmission lines, busbars, ...

- divisible electrical components named "breakers" : circuit-breakers, switches, ... A breaker is a link between two electrical components. It is characterized by a state variable : "open" or "closed". The values of these variables define the connex components (Islands) of the network. These components are handled by the operator or by automatons.

- automatons which react to the change of the network state, mainly when the devices on which they are connected are re-energized or de-energized. Automatons behave as follow : they can open or close breakers ; they can also trigger other automatons.

The figure below describes the object hierarchy model which has been used to represent the network and its components. The full model takes into account, as subclasses of these object classes, various types of automatons or devices. It also gathers some of these components into logical objects (for example, substation as a set of busbars, ...) and includes the description of the existing links between the different objects of the tree hierarchy.

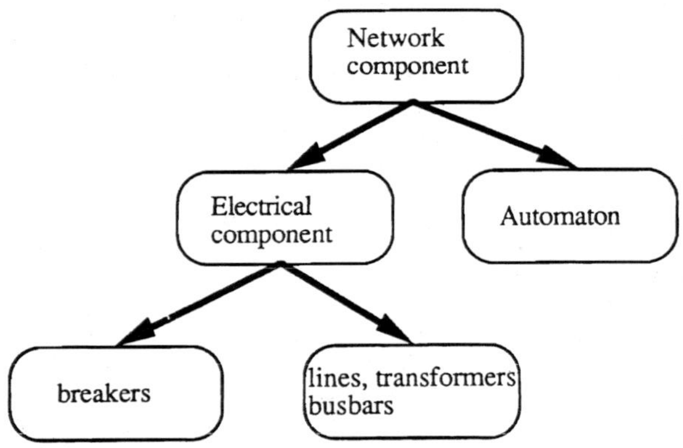

Figure 2
Hierarchichal representation of the network

4.1.2 The qualitative simulator

The purpose of the qualitative simulator is to predict the state of the network resulting of elementary restoration actions : energizing of line, transformer, power supply of loads, ... This model uses the network representation previously described. Such a model is required to manage the restoration planning process for the following reasons :

- First, a restoration action may not change the value of only one variable, but it may modify several variables. This is due, for example, to the voltage propagation in the network : the closing of a breaker between a not powered area and a powered area energizes the not powered area.

- Then, the network is equiped with automatons which react to the modification of the network state and which can themselves open or close breakers.

The solution adopted to solve this problem has been to conceive a qualitative simulator to compute the stable state of the network resulting of a restoration action. This simulator allows to determine the state resulting of planning actions . It presents the main following features :

- In the qualitative simulator, the network is described by a qualitative model, containing elements whose states are determined by a set of discrete variables (like "powered" or "not powered" for an electrical device, "opened" or "closed" for a breaker or a switch, "awake" or "asleep" for an automaton). The behaviour of these components is determined by transfer functions that respond to events. All events are collected in a central agenda and sorted by priority. The discrete event simulation performs a loop on the agenda, activating each time the event with the highest priority.

- Normally, the state of the network components have to be known to perform effective simulations : the behavior of a component can be defined in the "in operation" or "out of operation" states, but not in an unknown state. However, restoration planning has to deal with uncertainness and incomplete information. This problem has been handled by an *alarm generation mechanism* : when the simulation process is facing a situation where it is unable to determine the result, it generates an alarm signal meaning "the simulation of this operation concerns a doubtful equipment", that will be returned to the planning process.

4.1.3 The quantitative model

The discrete event simulation determines which parts of the network are energized after a restoration action, but cannot verify if the restoration plan is operationnally acceptable and does not violate constraints such as voltage or current limits. For this purpose, a load flow model is added to the system, able to perform conventional load-flow calculations. These calculations are done on a stable state of the network (after the functionning of protective-relays and automatons). Dynamical and transient phenomena are not taken into account, the quantitative model being used to characterize only the stable state of the network after restoration operations. In the present stage, the load-flow model is an active (DC) one.

The interfacing of this model to the qualitative simulator offers possibilities to generate alarms if numerical constraints are violated. These alarms are transmitted to the planning system. The plan can be modified to prevent the occurence of these problems.

4.2 Restoration expertise and planning system

The analysis of the operator expertise in the field of network restoration revealed various kinds of independant knowledge elements. This main characteristic of the restoration problem leads us to propose a *blackboard* architecture that we describe in the following paragraph. Then we present the various kinds of knowledge of MARS and the way they have been introduced in the blackboard architecture [2].

4.2.1 The blackboard architecture

The restoration process looks like a planning problem from an initial state of the network including out of power areas and disconnected loads. The final state which has to be reached is the same network in which most of the load is supplied and most of the electrical devices are re-energized. According to the main characteristics of the restoration problem, a *blackboard* architecture was proposed.

This resolution method is based on the decomposition of knowledge into several *knowledge sources* (KS) and the cooperation of these KSs under a *control mechanism* in order to solve the restoration planning problem [3]. The blackboard is a common memory storing the state of the resolution process.

Each KS contains the knowledge required to solve a specific problem. For instance the know-how and the operator skillness have been modelized into restoration KSs named "connecting of loads", "energizing of busbars", "energizing of lines ", ... Knowledge is represented as rules and heuristics that permit to solve the specific problem.

The control mechanism manages the cooperation between the different KSs. At each stage of the planning process, some of the KSs are candidates to interfere with the planning process by using Knowledge Source Activation Records (KSARS). A KSAR is defined as a candidature of a KS on a particular problem. For instance, the KS "energizing of busbar" can be triggered in several substations in which there is at least one energized line ; then, there will be in the blackboard as many KSARs of this KS as busbars which can be energized. The purpose of the control mechanism is to solve conflicts between KSARs by choosing at each stage of the planning process the "best" one, according to criteria such as efficiency (for instance volume and priority of the reenergized load), progress towards a solution, risk related to the new situation,...

The action of a KS can be to solve a partial problem by expanding the plan, but sometimes also to create new tasks which can be lately solved by other KSs. Then, the restoration problem can be solved by a decomposition of goals into subgoals. The overall control strategy of the planning process is then simply represented by a standard loop, until a solution is found and accepted (no more KSARs in the blackboard or all the disconnected load supplied) :

(1) detect KSARS
(2) select best KSAR
(3) execute the associated KS. Go to (1)

4.2.2 Planning tasks and knowledge representation

The analysis of the current restoration practice revealed three kinds of planning knowledge :

- *Strategic knowledge* is used to determine priority between restoration zones, and the order in which substations have to be re-energized according to criteria such as the priority and the level of disconnected loads.

- *Tactic knowledge* is used to opportunistically choose a restoration operation, taking into account the result of previous operations. For instance, the reenergizing of a transmission line may be chosen, even if these operation is not emergent from the initial strategical point of view, but allows to reenergize an important substation at this point of the restoration plan.

- *Procedural knowledge* is used to define how a restoration operation will be performed, by a recursive decomposition of goals into subgoals until elementary actions are reached. For instance, the KS "trial of a doubtful busbar" includes the knowledge to decompose the trial by choosing the line which must be used and the operations which have to be done to disconnect the other components from the busbar.

Taking advantage of this natural decomposition, the planning task is decomposed in a set of subtasks, having their particular knowledge representation and control structure, and integrated in a blackboard architecture :

- Procedural knowledge is encoded in Knowledge Sources (KSs). Various kind of procedural KSs exist in MARS :

  - *planning KSs* that try to expand a partial plan and solve a typical subproblem. "energizing of a busbar" and "connecting of load" are such KSs. Twelve planning KSs have been so encoded.

  - *correction KSs* that correct plans the execution of which can raise problems. These problems are the energizing of a faulty component and overloads. They are stressed by alarms generated in the discrete event simulation.

- Tactic knowledge is encoded in two tasks :

  - a *trigerring task.* Its goal is to determine the KSARs.
  - a *control task* , responsible for the characterization, comparison and selection between KSARs. Control knowledge is represented by criteria that evaluate the pertinent KSARs according to preference rules related to the choice of the network element on which a given action can be done, the choice among different types of actions, the choice among situations The result of the control task is a permanent updating of a "best KSAR" that should be activated next.

- Strategic knowledge is encoded in a specialized KS, which is only activated once at the beginning of the planning process.

4.2.3 The planning process

The solution, restoration plans, can be defined as a fully ordered set of *operations* possibly recursively decomposed into sub-operations or elementary actions. We call the state resulting of the execution of a partial plan a *situation.* A situation is characterized by the values of the state variables at that point.

As mentioned earlier, the state of some elements is sometimes unknown. Practically, when operators have a doubt on the state of an element that would be useful in the restoration process, they can include in the plan special trial operations to determine its actual state. So, the plans elaborated by MARS contain test operations, which have two resulting situations, depending on the *hypothesis* that the tested equipment is sane or not. Plans in MARS are then *conditional plans.* The figure below shows a typical plan :

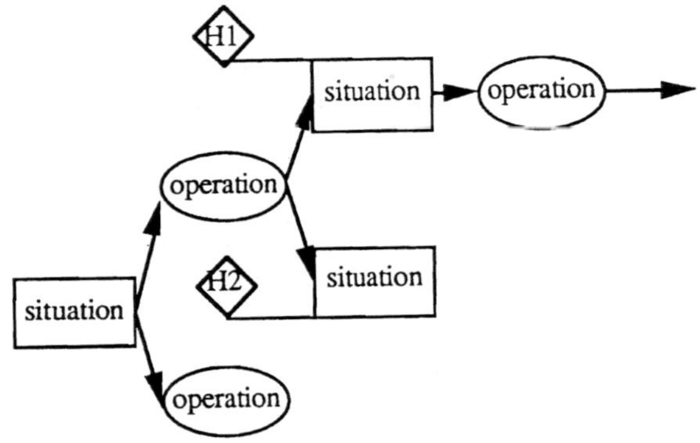

Figure 3
A restoration plan
H1 and H2 are the hypothesis related to an operation on a doubtful element

The blackboard is divided in two areas : domain and control. The domain area describes the restoration plan at two levels of abstraction : strategy, defining a general guide-line to the restoration process in terms of restoration areas and priorities, and procedure defining the restoration plan : the different situations, operations and hypothesis.

We also included *management KSs* to characterize the plan. In particular , a specific management KS is responsible for calling the discrete event simulator to evaluate a new situation in the plan. The overall architecture of MARS is presented on Figure 4.

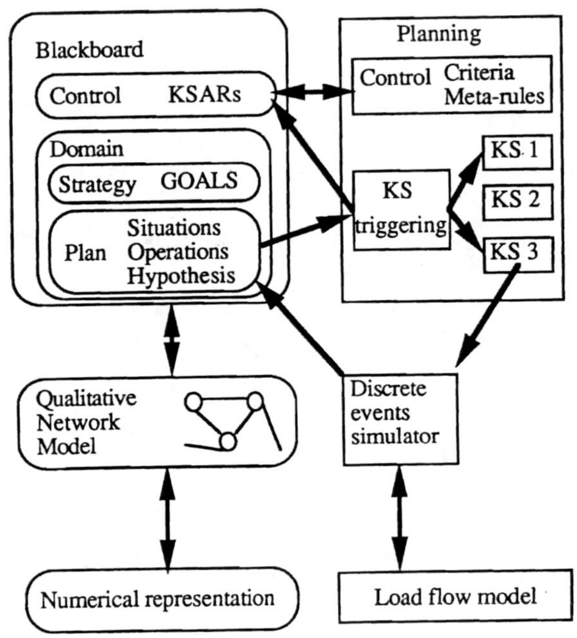

Figure 4
The architecture of MARS

5. THE MARS PROTOTYPE

5.1 An object-oriented implementation

Except the load-flow in the quantitative model which is implemented in FORTRAN, the whole MARS prototype is implemented in SPOKE , an object-oriented environment written in C, including a forward chaining system for first-order rules management and a X11 graphic interface. In its current state of development, MARS includes about fourty thousands code lines. The knowledge base contains twelve KSs and approximatively one hundred and twenty rules related to the restoration expertise.

The MARS prototype runs on Unix Workstations DEC or SUN. The time required to add a new step in a restoration plan is a few seconds, and a whole restoration plan is produced in one to five minutes, depending on the complexity of the initial situation of the network.

5.2 A flexible man-machine interface

In the current MARS prototype, the man-machine interface has been designed to allow a detailed validation of the resolution principles on real cases, and to point out different cooperation levels between the system and the operator, who can interfere with the planning process to modify the choices of the system.

Therefore, various interaction modes can be chosen during a planning session, containing different levels of interaction. The highest level allows the operator to modify only the order of the substations in which MARS suggests to do the restoration. At the lowest level, the operator can choose the next elementary restoration action. In order to limit the backtracking process, MARS also provides the operator with the possibility to validate or invalidate parts of the plan already constructed during the planning process.

Naturally, the MARS prototype provides the operator with several display functions concerning the state of the resolution of the problem and the solution obtained. They are mainly the following ones :

- a graphic view of the restoration plan in a tree representation,
- a graphic view of the network in a particular situation (see figure 5),
- the possibility to simulate one or several operations of the restoration plan,
- the possibility to list the possible actions at any step.

5.3 A prototype able to generate correct restoration plans in real cases

Several cases taken out from real operation have been used to create the knowledge base of the MARS prototype and a lot of knowledge is derived from paper-and-pen simulations and analysis on these cases.

The first simulations with the MARS prototype itself have been completed on simple cases as well as on more complicated ones, and reveal the good performance of MARS :

- MARS is able to construct restoration plans in short delays, typically a few minutes, this time being related to the size of the network to be restored,
- the plans proposed by the model are similar to those predicted by the experts.

This validation stage will shortly be extended to several other cases.

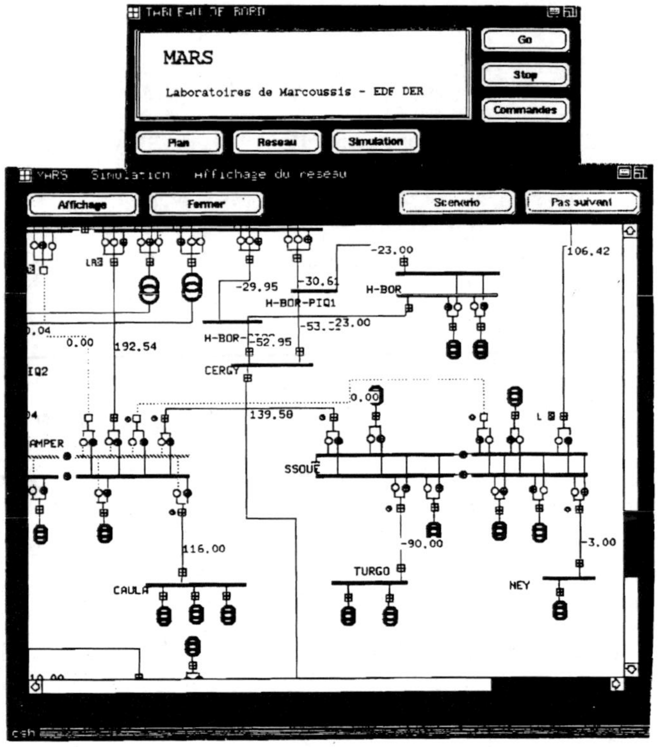

Figure 5
A view of the network during the restoration process

6. FUTURE DEVELOPMENTS AND APPLICATIONS

In its current state, the MARS prototype will be used to evaluate and define more precisely the future applications of a restoration function in different environments. The future developments of MARS include the following :

- the extension of the restoration expertise ; new expertise fields corresponding to new KS will be added to the MARS prototype : a voltage problem expertise (linked with an AC load-flow which will be added to MARS), expertise on circuit-breakers failures, ...
- the definition and realisation of an operator training tool derived from the MARS prototype : its high degree of interactivity is particularly well suited to an operator training tool. However, the environment of such a tool, the way it will be used by the operators, its ability to give explanations or to guide the operator in the training session must be precisely defined,
- the precise definition and realisation of a real time restoration function running in the Regional Control Centers environment ; this point is the final aim of the MARS project and needs a very precise definition of the real-time objectives, the restoration function interface, the desired man-machine cooperation level ; the MARS prototype has proved that such a function is possible, and it will be a very precise tool to define and experiment the right man-machine cooperation level required for a restoration function in a real-time environment.

7. CONCLUSION

At the present stage of development, the benefits of the MARS project already look considerable.

At first, a precise expertise on the restoration procedures has been made explicit, and clearly represented in a modular way in the MARS prototype.

Secondly, the MARS prototype has shown that advanced Artificial Intelligence techniques, combining object-oriented approaches, rule-based representations and simulation techniques, can solve such a difficult problem in a satisfactory way.

Finally, the defined expertise as well as the MARS prototype will constitute a very strong basis for the precise definition and realization of a restoration function, with the right man-machine cooperation level, in an operator training application as well as in the future real-time environment.

In that way, we can say that the MARS project seems to be a very significant step towards the availability of well-adapted monitoring and control tools for real-time operation in disturbed situations.

REFERENCES

[1] G. De Montravel - D. Arrieta
"An Alarm Synthesis in the Future Power System Control Centres"
Revue Générale de l'Electricité - Vol. 1 - January 1989

[2] O. Paillet - H. Fargier - Y. Harmand - E. Mondon
"Heuristics and models for reactive planning : the MARS experiment"
10th International Workshop on Expert Systems and their applications, AVIGNON 1990

[3] P. Nii
Blackboard Systems
AI magazine - Summer 86

AN ANALYTICAL APPROACH FOR STEP-BY-STEP RESTORATION OF DISTRIBUTION SYSTEMS FOLLOWING EXTENDED OUTAGES

Canbolat Uçak
Student Member

Anil Pahwa
Senior Member

EECE Department, Kansas State University
Manhattan, Kansas 66506-5105

Abstract— **Loss of diversity among loads may cause cold load pickup problems after an extended outage in distribution systems. Thermostatically-controlled devices such as air-conditioners, heaters, heat pumps, water heaters etc. become the largest portion of load during restoration.**

In this paper, an analytical approach for restoration of distribution systems during cold load pickup is presented. A delayed exponential model is used to represent the load during cold load pickup. A thermodynamic model of transformer is used to determine the loading limits for this type of load. Whenever loading limits are reached, sections are restored step-by-step. In step-by-step restoration, the sequence of sections restored plays an important role in total restoration time. An optimal restoration sequence of different sections of the distribution system is determined for minimum total restoration time as well as minimum customer interruption duration.

Keywords : **Distribution automation, distribution restoration, cold load pickup, step-by-step restoration, transformer loading.**

I. Introduction

Individual loads on a residential feeder can be categorized into two different groups, namely, thermostatically-controlled and manually-controlled. Generally, thermostatically-controlled devices such as air-conditioners, heaters, heat pumps etc. provide largest contribution to the total load in a typical house. Manually-controlled loads are switched on and off by occupants of the house in a random fashion. Life-style of the occupants of the house have a significant influence on the contribution of these loads to the total load of the house.

During normal condition, diversity amongst loads is present, and therefore, the aggregate load of a number of houses is less than the connected load. The average load for each thermostatically-controlled device is proportional to its duty cycle. If an abnormal condition, such as an extended outage occurs in a distribution system, some or all thermostatically-controlled devices will be on as soon as the power is restored. Similarly, the aggregate load of manually-controlled devices will be higher than normal upon restoration because more people may want to use different devices. Such loss of diversity may cause load higher than the normal by 2 to 5 times [1-4].

Cold Load PickUp (CLPU) appeared first in literature as a problem due to high inrush currents which last a few seconds and prevent the circuit from being re-energized after extended outages. Usage of very inverse characteristic relays or sectionalizing the system were some of the solutions used to overcome this problem. Since then, thermostatically-controlled devices in distribution circuit have increased [1, 5]. These types of loads may cause restoration problems during CLPU before they cause serious overload problems in the steady-state operation. Therefore, sustained (undiversified) load after restoration becomes an important issue for thermal and loading limitations of transformers.

Ideally, a power transformer can be loaded up to its nameplate rating 24 hours a day with normal loss of life. If the distribution system experiences a peak load, for example as in a CLPU situation, overloading of transformers may be necessary for a short duration. Limitations for short term loading of 65 °C winding rise transformers are given in ANSI/IEEE C57.92-1981 [6] and they are summarized as

- 4% loss of life in any one emergency operation
- 180 °C hottest-spot temperature
- 110 °C top-oil temperature
- maximum short term loading of 2 p.u.

Based on these limitations, Wilde [7] investigated the effects of CLPU at the substation transformer by using a CLPU model in which post-outage load is constant for some time and then decreases linearly from undiversified load to diversified load. Similar study using the same CLPU model was done by J. Aubin *et al.* [8] for the distribution transformers.

One or more of the loading limits of a transformer may be reached when only a fraction of the total load is restored after an extended outage. However, due to the characteristics of CLPU, loads will gain diversity with time after restoration and the aggregate load will decrease gradually to allow remaining load to be energized in steps. Such step-by-step restoration will be necessary for those systems where the substation transformers operate near rated capacity under normal conditions.

In the existing distribution systems, many substation transformers are loaded to 40% to 60% of their rating under normal operation. Hence, CLPU during restoration is not a significant problem. However, in the future, due to constraints on capital expenses for system expansion and upgrade, and due to increased reliance on distribution automation, the transformers may be more heavily loaded. Therefore, it is important to investigate impact of CLPU on distribution system restoration. In fact, some utilities have experienced restoration problems because of sustained high load due to CLPU [9].

To accomplish step-by-step restoration, the distribution system should be divided into sections using remotely-controlled sectionalizers. The order in which the sections are restored has

94 WM 072-9 PWRD A paper recommended and approved by the IEEE Transmission and Distribution Committee of the IEEE Power Engineering Society for presentation at the IEEE/PES 1994 Winter Meeting, New York, New York, January 30 - February 3, 1994. Manuscript submitted August 2, 1993; made available for printing November 29, 1993.

Reprinted from *IEEE Transactions on Power Delivery,* Vol. 9, No. 3, pp. 1717–1723, July 1994.

a significant impact on the total time of restoration. An order of sections can be selected either to minimize the total restoration time, which is the time when the last section is restored, or to minimize customer interruption duration.

In this paper, an analytical approach for restoration of a distribution system based on the delayed exponential CLPU model is investigated. Transformer loading limits are determined using this model. Expressions for restoration times of different sections in a distribution system are found. Restoration order of sections is searched to minimize the total restoration time and the customer interruption duration while keeping the transformer loading within limits. Determination of the optimal order is done using a global search as well as a reduced order search technique called Adjacent Pairwise Interchange Method (APIM).

II. CLPU model

An aggregate load model is required to study the impact of CLPU. Importance of CLPU models during distribution system restoration has been discussed in a recent paper [10]. Several models available in literature range from very simple piece-wise linear model [7, 8] to somewhat more complicated physically-based models [11-17]. Some oscillatory behavior in the aggregated load for similar type of houses is noticeable when physically-based models are used. However, parameter dispersion and increase in system noise result in increased variance and decreased oscillatory behavior. The resultant characteristics appear to match closely a delayed exponential model. A delayed exponential model for CLPU has also been suggested by Lang [5]. Physically-based models can predict the CLPU behavior of thermostatically-controlled loads quite accurately; but it is difficult to use them to model manually-controlled loads. However, based on engineering judgment, the aggregate load of manually-controlled loads upon restoration is also expected to be higher than normal and after a time delay it will gradually approach a reduced level. Thus, a delayed exponential model is a good analytical model for representation of aggregate load in distribution system during CLPU.

In the delayed exponential model undiversified load S_{U_i} remains constant from restoration time T_i to t_i where diversity starts and after t_i load decreases exponentially to diversified load level S_{D_i} for i^{th} section as shown in Fig.1.

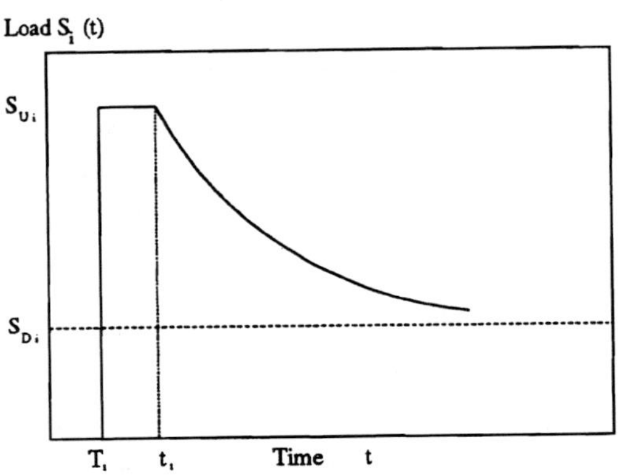

Fig. 1. CLPU behavior of a distribution section.

A mathematical expression for this model can be written as

$$\begin{aligned} S_i(t) &= [S_{D_i} + (S_{U_i} - S_{D_i})e^{-\alpha_i(t-t_i)}]u(t-t_i) + \\ & \quad S_{U_i}[1 - u(t-t_i)]u(t-T_i) \end{aligned} \tag{1}$$

where α_i is the rate of decay of load on i^{th} section, and $S_i(t)$ is load of ith section. $u(\tau)$ is a unit step function.

III. Transformer thermal characteristics

Temperatures inside a transformer determine its loading capability. The highest temperatures occur in the top section of transformer. Winding outlet oil temperature which is also called top-oil temperature is the hottest part of oil and average temperature of the uppermost disc is the hottest section of winding. The hottest-spot temperature is temperature of the conductor which has the maximum temperature in the uppermost disc and it is generally a few degrees higher than the average temperature of the uppermost disc. The highest deterioration of insulation material will be in the location of the hottest-spot temperature. Therefore, this temperature is used to calculate the life expectancy of a transformer.

Loss of life is a cumulative process. Thus, when the hottest-spot temperature is a function of time, the total loss of life for a period of T can be determined by

$$\%(\text{loss of life}) = 100 \int_0^T \frac{dt}{10^{\frac{A}{\theta_{hs}(t)}+B}}. \tag{2}$$

Where A and B are constants from desired life expectancy curve. For 65 °C winding rise transformer, $A = 6972.15$ and $B = -13.391$ [6]. This equation shows that the transformer can be loaded in excess of its normal hottest-spot temperature to give a normal loss of life if it has been loaded below its rating previously. For example, if there is an outage in the system and the transformer lost its load partially or completely, the hottest-spot temperature will reduce to a lower value. Thus, upon restoration, load higher than the transformer rating could be allowed for a short duration. Short term load level and duration can be determined based on thermal characteristic and loading limitations of the transformer.

In what follows, we will briefly state the equations to determine the top-oil and the hottest-spot temperatures. A list of symbols used in this section is given at the end of the paper. More information on thermal characteristic of transformers can be found in references [6, 18, 19].

Transformer hottest-spot temperature (θ_{hs}) is given by

$$\theta_{hs} = \theta_a + \theta_o + \theta_g. \tag{3}$$

The ambient temperature (θ_a), oil temperature rise over ambient (θ_o) and the hottest-spot winding rise over the top-oil temperature (θ_g) are required to find θ_{hs}. The top-oil temperature rise can be expressed by a first order differential equation [18] as

$$P_{loss} = k\theta_o + C\frac{d\theta_o}{dt}. \tag{4}$$

Here k is given as $k = \Delta P_{loss}/(\theta_u - \theta_i)$. Final top-oil temperature rise is not directly proportional to the losses in transformer. It can be more accurately represented by a relation

$$\theta_u = \theta_{fl}\left(\frac{P_{loss}}{P_{loss_N}}\right)^n. \tag{5}$$

Where n changes between 0.8 and 1.0 depending on the transformer type. Similar to the top-oil temperature change, temperature change of the hottest-spot conductor over the top-oil is given by a first order differential equation

$$\tau_g\frac{d\theta_g}{dt} + \theta_g = \theta_{gu}. \tag{6}$$

Time constant of the conductor (τ_g) is small compared to the time constant of transformer oil. Ultimate hottest-spot winding rise over the top-oil for a loading ratio of K is given by

$$\theta_{gu} = \theta_{g(fl)}K^{2m}. \tag{7}$$

Since temperature rises for any load must be added to the ambient temperature θ_a to determine operating temperatures, the ambient temperature is important in determining load carrying capability of the transformer. In general, 30 °C is used as a standard ambient temperature. If the ambient temperature changes are large during the period T, a weighted ambient temperature should be used to calculate loss of life in the transformer.

IV. Loading limits of forced air-cooled (133% or less) transformers for a delayed exponential load

In this section, loading limits of a forced air-cooled (133% or less) transformer for the delayed exponential CLPU model using 30 °C ambient temperature are presented. Since the top-oil temperature rise is not directly proportional to losses in the transformer, the differential equation is solved numerically as suggested in reference [6]. Fig.2 shows both the top-oil and the hottest-spot temperatures during a cold load pickup. The values used for CLPU model are S_U = 1.7 p.u., S_D = 1 p.u., α = 0.5 hr^{-1}, and Δt = 30 min. (this value is consistent with that reported in literature). An outage duration of 1 hour was considered. It was assumed that the pre-outage and the post-outage diversified load is the same. As it can be seen from Fig.2, the top-oil temperature reaches the limit before the hottest-spot temperature does. Based on the limitations given by standards on oil-immersed transformers in ANSI/IEEE C57.92-1981, maximum undiversified load for different diversified loads and outage durations was determined and is shown in Fig.3. This figure does not show thermal limits for S_D = 0.5 p.u. or less because the transformer can be loaded up to 2 p.u. for all outage durations. These are the cases where the transformer thermal limits are not reached but recommended maximum loading limit of 2 p.u. becomes the limiting factor. Outage duration also affects the loading limit of a transformer, i.e. the longer the outage the more load can be restored during CLPU.

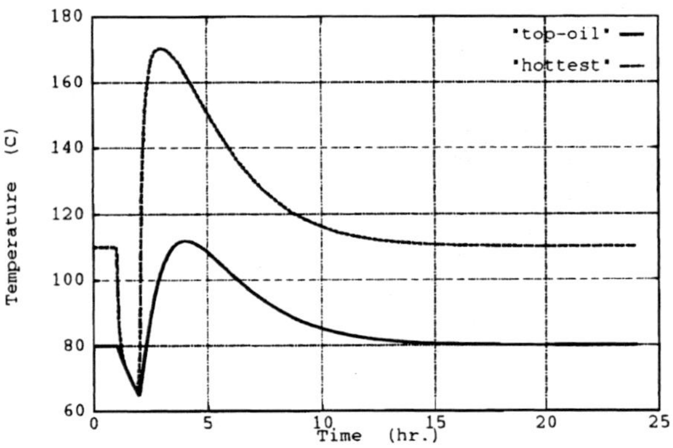

Fig. 2. The top-oil and the hottest-spot temperature during a CLPU

In an emergency situation, a substation transformer may need to supply power to sections which had been connected to different transformers in normal configuration. As a result, the post-outage diversified load may be higher than the pre-outage diversified load. Fig.4 shows loading limits of the transformer for this case. Loading limits in Fig.4 are determined for one hour outage since it is the most severe case as seen in Fig.3. For outages more than one hour, loading limits will be higher.

Although Fig.3 and Fig.4 show the maximum loading of the transformer based on thermal limitations, there may be other limitations. Some of these limitations are bushing ratings, oil expansion, tap changers and stray flux heating as well as the thermal capability of related equipment such as cables, reactors, circuit breakers, disconnecting switches and current transformer. These limitations could also be included in determining the maximum loading of a transformer if necessary.

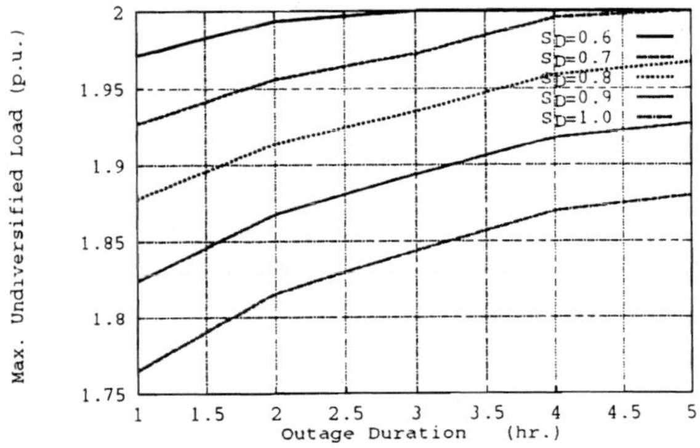

Fig. 3. Maximum undiversified load that can be restored after an outage as a function of outage duration (Δt = 0.5 hr. and α = 0.5 hr^{-1}.)

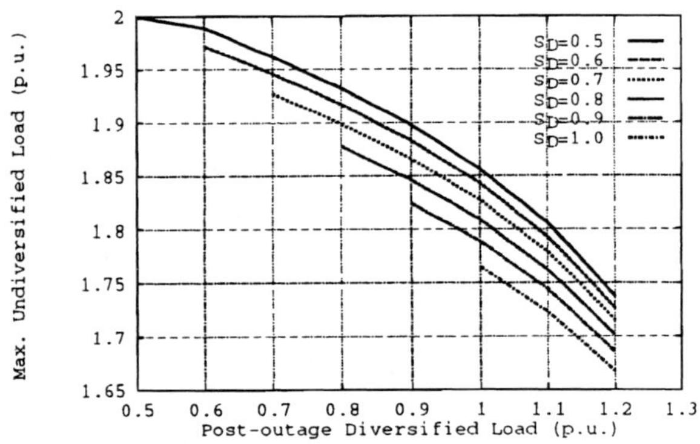

Fig. 4. Loading limits when the post-outage diversified load level is higher than the pre-outage diversified load for an outage of one hour.

V. Model for step-by-step restoration

The limitations on transformer loading may not allow restoration of the whole system in one step due to high undiversified load resulting from an outage. In such cases, sectionalizers will be required and step-by-step restoration must be implemented. A one-line diagram of a typical distribution system with automatic sectionalizers is shown in Fig.5. If the total undiversified load is represented by S_U, and loading limit by S_{MT}, then step-by-step restoration of sections is necessary when $S_U > S_{MT}$. The order in which the sections are restored has a significant impact on the total time needed to restore the system. Moreover, the customer interruption duration should also be considered in determining the order of restoration. Therefore, determination of the best sequence of sections for restoration is important. Some sections could have important loads such as hospital, fire station, police station etc. and should be given priority during restoration. Also, if a section that needs to be supplied power is at the end of a feeder then all sections upstream must already have power. Thus, this becomes a precedence constraints in the restoration procedure. These two constraints, priority of sections and precedence constraints, however, reduce the combinatorial search space to allow faster solution to the problem.

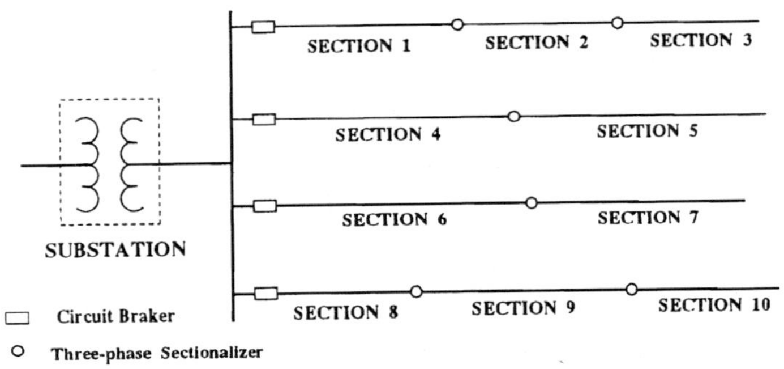

Fig. 5. One-line diagram of a distribution system.

Let there be n sections to be supplied power from the substation transformer. Each section is supplied power with the maximum transformer capacity constraint given by

$$\sum_{i=1}^{n} S_i(t) \leq S_{MT} \qquad t \geq 0. \tag{8}$$

Restoration times of sections will depend on the restoration order. For mathematical representation of restoration times, each section will be represented with a number such as 1st, 2nd, ...,n^{th} section and their restoration order will be enclosed with brackets. For Example, if i^{th} section is restored as the first section in the restoration sequence then the first element of the index will be i, i.e. [1] gives the section number which is restored first.

There will be m sections which could be restored in the first step of restoration without violating Eq.8. In that case restoration times of sections [1], [2],...,[m] will be equal to T_0, which is the time when restoration started. m will change depending on undiversified load of sections, loading capacity S_{MT}, and priority and precedence constraints. Remaining $n-m$ sections will be restored in steps at times $T_{[m+1]}$, $T_{[m+2]}$,..., $T_{[n]}$. Thus, the transformer load as a function of time during step-by-step restoration is obtained using Eq.1,

$$\begin{aligned} S(t) &= \sum_{i=1}^{n}\{[S_{D_{[i]}} + (S_{U_{[i]}} - S_{D_{[i]}})e^{-\alpha_{[i]}(t-t_{[i]})}]u(t-t_{[i]}) + \\ &\quad S_{U_{[i]}}u(t-T_{[i]})[1-u(t-t_{[i]})]\}. \end{aligned} \tag{9}$$

If $\alpha_{[i]}$'s are the same for all the sections, an analytical expression for restoration time $T_{[i]}$ can be derived. Otherwise a closed form solution for restoration times is not possible and numerical techniques would be required. Using $\alpha_{[i]} = \alpha$ and with a condition that $T_{[i]} - T_{[i-1]} \geq \Delta t_{[i-1]}$ for all $i \leq k$, the solution for restoration time of section $[k]$ is

$$T_{[k]} = -\frac{1}{\alpha}\ln\frac{S_{MT} - S_{U_{[k]}} - \sum_{i=1}^{k-1} S_{D_{[i]}}}{\sum_{i=1}^{k-1}(S_{U_{[i]}} - S_{D_{[i]}})e^{\alpha(T_{[i]}+\Delta t_{[i]})}}, \tag{10}$$

where $\Delta t_{[i]}$ is delay part of the load of section $[i]$, and defined as $\Delta t_{[i]} = t_{[i]} - T_{[i]}$. The case where $T_{[i]} - T_{[i-1]} < \Delta t_{[i-1]}$ is not considered because during analysis it was found that the load does not decay sufficiently to allow restoration of an additional section with this condition. Eq.10 is valid for $m < k \leq n$.

Note that restoration time $T_{[k]}$ depends on all the previous restoration times.

Upon substitution of expressions of $T_{[1]}$, $T_{[2]}$..., $T_{[k-1]}$ into Eq.10, a general equation where restoration time of kth section is a function of $S_{U_{[i]}}$, $S_{D_{[i]}}$, $\Delta t_{[i]}$, α, and S_{MT} is obtained

$$T_{[k]} = -\frac{1}{\alpha}\ln\frac{\prod_{i=m+1}^{k} S_{\eta_{[i]}}}{\sum_{i=1}^{m} S_{\mu_{[i]}} \prod_{i=m+1}^{k-1}(S_{\eta_{[i]}} + S_{\mu_{[i]}})} \tag{11}$$

$S_{\eta_{[i]}}$ and $S_{\mu_{[i]}}$ are defined as

$$S_{\eta_{[i]}} = S_{MT} - S_{U_{[i]}} - \sum_{j=1}^{i-1} S_{D_{[j]}} \tag{12}$$

$$S_{\mu_{[i]}} = (S_{U_{[i]}} - S_{D_{[i]}})e^{\alpha\Delta t_{[i]}} \tag{13}$$

Total restoration time T for n sections can be obtained from Eq.11 by replacing k with n.

VI. Optimization of restoration order

The main goal is to restore power to the distribution system as fast as possible without exceeding loading limits of the transformer. Moreover, the customer interruption duration should be given consideration because it affects the reliability of the system. Thus, there could be two different approaches to this problem; 1) to minimize the total restoration time, and 2) to minimize the customer interruption duration. To meet the first objective, restoration of sections should be done in such a way that $T_{[n]}$ is minimized. However, for high reliability levels, the customer interruption duration should be minimized. But minimization of one does not necessarily lead to minimization of the other.

A. Minimization of total restoration time

This is a single-machine scheduling problem where scheduling times are sequence dependent [20]. Total number of different permutations of n sections is given by $n!$. But some of the permutations will give the same total restoration time since the number of sections restored in the first step of restoration is more than one. We seek a restoration order of sections which will minimize total restoration time, i.e. find the order $\{[1], [2], ..., [n]\}$ so that $T_{[n]}$ is the minimum.

Let us assume that a restoration sequence is selected. The improvement on total restoration time $T_{[n]}$ by interchanging two (a and b) adjacent sections can be investigated using Eq.11. Now, define the total restoration times $T^1_{[n]}$ and $T^2_{[n]}$ for two different ordering,

1. $T^1_{[n]}$ for restoration order $(a \rightarrow b)$
2. $T^2_{[n]}$ for restoration order $(b \rightarrow a)$

If $T^1_{[n]} < T^2_{[n]}$ then restoration order $(a \rightarrow b)$ gives smaller total restoration time. This method can be applied between two adjacent sections in the restoration sequence until there is no reduction in the total restoration time. In other words, a sequence is sought for which all adjacent pairwise interchanges lead to increase in $T_{[n]}$ [20]. This is called adjacent pairwise interchange method (APIM). However this method does not guarantee a global minimum solution for $T_{[n]}$. The method may get stuck in local minima depending on initial sequence used. An ordering of sections in descending order of undiversified load provides a good starting sequence [21].

B. Minimization of customer interruption duration

Minimization of the customer interruption duration is important because it affects both system average interruption duration index (SAIDI) and customer average interruption duration index (CAIDI) [22]. Both of these indices are the indicator of reliability levels and are very sensitive to the duration and the number of outages. System and customer reliability levels can be improved with remotely controlled switches and by carefully selected restoration sequence of sections during CLPU.

In this section, a CLPU restoration procedure which minimizes SAIDI and CAIDI is considered. Assume that outage duration is given by T_{out} and restoration is initiated immediately after outage is over. Customer interruption duration (CID) for i^{th} section is given by

$$CID_{[i]} = C_{[i]}(T_{[i]} + T_{out}). \tag{14}$$

Where $C_{[i]}$ is the number of customers on section $[i]$. Total CID (TCID) for the distribution system can be written as

$$TCID = \sum_{i=1}^{n} CID_{[i]} = \sum_{i=1}^{n} C_{[i]}(T_{[i]} + T_{out}) \tag{15}$$

or

$$TCID = \sum_{i=1}^{n} C_{[i]}T_{[i]} + \sum_{i=1}^{n} C_{[i]}T_{out}. \tag{16}$$

Last part of equation does not depend on the restoration order. Therefore, it can be dropped from minimization process. Minimization of TCID involves only the first term in Eq.16 and can be written as

$$\text{minimize } \{\sum_{i=1}^{n} C_{[i]}T_{[i]}\} \tag{17}$$

If the number of customers $C_{[i]}$ is unknown, undiversified load in sections may be used to approximate the number of customers on each section as done in the example presented in this paper.

VII. Numerical example of restoration

A numerical example to illustrate restoration procedure for the distribution system shown in Fig.5 is presented in this section. It is considered that before outage, all the sections except 3rd and 10th are supplied power from the substation transformer under consideration. They are normally supplied power from another substation transformer. But at the time of restoration due to emergency situation, such as failure of a transformer elsewhere, they will be tied to the transformer under consideration.

Table I gives values of undiversified and diversified load levels of each section in per unit for an outage duration of one hour. Again, $\Delta t = 30$ min. and $\alpha = 0.5$ hr^{-1} are considered for all the sections. Diversity factors for these sections are also given in the table. These values lie between 1.5833 and 3.333. The problem can be easily reformulated for a case where all the diversity factors are equal.

The transformer is loaded to 90% of its nameplate rating (30 MVA) before outage. With addition of section 3 and 10, diversified load after restoration increases by 10% over the rating. Since restoration times depend on the transformer short term loading capacity, search for the best restoration order requires S_{MT}. The value of S_{MT} for a load with 30 minutes delay and $\alpha = 0.5$ hr^{-1} can be found from Fig.4. For 0.9 p.u. pre-outage and 1.1 p.u. post-outage diversified load, S_{MT} is given as 1.75

TABLE I

Undiversified and diversified loads of the sections for one hour outage duration (S_{base} = 30 MVA.)

Section Number	S_{U_i} (p.u.)	S_{D_i} (p.u.)	S_{U_i}/S_{D_i}
1	0.3000	0.1000	3.0000
2	0.1833	0.1000	1.833
3 (boxed)	0.2000	0.0833	2.4
4	0.1667	0.0500	3.333
5	0.1667	0.1000	1.6667
6	0.3167	0.2000	1.5833
7	0.2000	0.0833	2.4
8	0.2167	0.1167	1.8571
9	0.3667	0.1500	2.444
10 (boxed)	0.2667	0.1167	2.2857
Total	2.3835 (post)	1.1000 (post) 0.9000 (pre)	

p.u. But the total undiversified load of the distribution system considered is 2.3835 p.u. Thus, step-by-step restoration is required. However, 1.75 p.u. is the limit for a load which is restored in one step and it decays to diversified load. Since in step-by-step restoration more load is added later, transformer constraints during restoration must be checked for violations. If there is a violation, S_{MT} is decreased and the problem is run again. If it is found that S_{MT} has been decreased too much, it is increased by a smaller value. After a few iterations a proper value of S_{MT} can be found. For the example under consideration the value of S_{MT} was found to be 1.45 p.u.

It is assumed that there are no priority loads and therefore no priority constraints (although they could be included easily) are considered. Precedence constraints for the system are $1^{st} \rightarrow 2^{nd} \rightarrow 3^{rd}$, $4^{th} \rightarrow 5^{th}$, $6^{th} \rightarrow 7^{th}$, and $8^{th} \rightarrow 9^{th} \rightarrow 10^{th}$. In other words, for example, 10th section cannot be restored if either 8th or 9th section has not been energized. Any restoration order which does not account for the precedence constraints is not feasible. Thus, during the search strategy for the best restoration order, sequences which violate these constraints are discarded by the search algorithm.

Table II shows results of both the APIM algorithm and the global search. The global search basically searches for all permutations and finds the minimum. The sequences that minimize the total restoration time and the customer interruption duration are given in this table. Note that optimum restoration order for the total restoration time and for the customer interruption duration is the same. In general, this may not be true. The APIM algorithm found a local minimum which is off from the global solution by 7.5 min.

The minimum total restoration time was found to be 275.1 min. This is 51 minutes lesser than the maximum total restoration time of 326.1 min. The minimum customer interruption duration per customer is 64.7 min. whereas the maximum customer interruption duration is 83.2 min. per customer. Thus, for a large system it is an important factor to be considered during restoration. These values are without inclusion of the outage duration (T_{out}).

The hottest-spot and the top-oil temperature changes of the transformer, when sections are restored in order as given in Table II, are shown in Fig.6. As seen from the figure, the maximum hottest-spot temperature reached is 159.3 °C and the maximum top-oil temperature reached is 108.3 °C. Both temperatures are below the limits suggested by ANSI/IEEE

TABLE II
Restoration orders and restoration times for APIM and GLOBAL search.

Order	APIM				GLOBAL			
	Restoration Time		CID		Restoration Time		CID	
	Sec. No	$T_{[i]}$ (min)	Sec. No	$T_{[i]}$ (min)	Sec. No	$T_{[i]}$ (min)	Sec. No	$T_{[i]}$ (min)
1	1	0.0	8	0.0	1	0.0	1	0.0
2	8	0.0	1	0.0	2	0.0	2	0.0
3	6	0.0	4	0.0	8	0.0	8	0.0
4	9	0.0	6	0.0	9	0.0	9	0.0
5	4	0.0	9	0.0	3	0.0	3	0.0
6	10	63.6	10	63.6	4	0.0	4	0.0
7	7	109.8	7	109.8	10	72.8	10	72.8
8	2	157.0	2	157.0	6	147.4	6	147.4
9	3	218.6	3	218.6	7	211.0	7	211.0
10	5	282.6	5	282.6	5	275.1	5	275.1
Minimum Total Restoration Time = 275.1 min.								
Maximum Total Restoration Time = 326.1 min.								
Minimum Customer Interruption Duration = 64.7 min. per customer								
Maximum Customer Interruption Duration = 83.2 min. per customer								

57.92-1981 standard. However, top-oil temperature is the limiting factor. Loss of life of transformer is 0.664% during this restoration for a period of 24 hours.

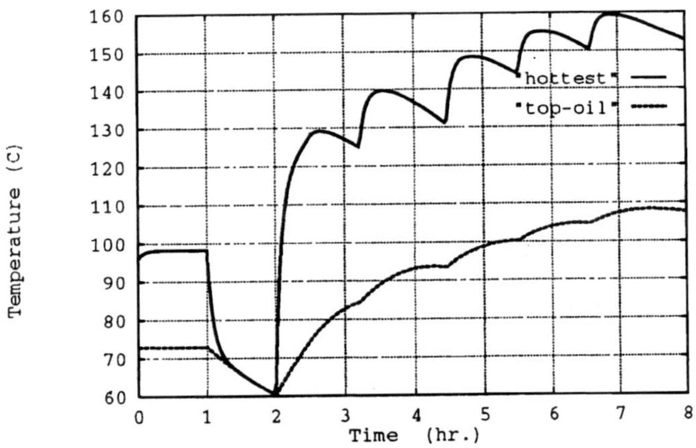

Fig. 6. The hottest-spot and the top-oil temperature change of transformer when sections are restored in steps.

VIII. Conclusion

A step-by step procedure for restoration of distribution systems following extended outages is investigated for those cases where the undiversified load exceeds the available transformer capacity. A delayed exponential model is used to model the aggregate load during cold load pickup. Restoration time of sections as a function of restoration order and parameters of CLPU model are derived. Using these formulas, the total restoration time and the customer interruption duration are minimized. Optimization of restoration procedure is done while taking into consideration the transformer loading limits and precedence constraints.

The example considered showed that a considerable reduction in the total restoration time and the customer interruption duration can be achieved with proper restoration. The amount of reduction depends on CLPU parameters, transformer thermal characteristics and loading capacity, number of sections, and other constraints such as precedence and priority.

Basic ideas of this paper could be extended to include other constraints such as thermal and the maximum load limit of feeders, the maximum voltage drop at the end of feeders during sustained current, and limits of protection equipment etc. Since CLPU problem is one of the most severe situation that the distribution system experiences when thermal limits are considered, restoration capabilities of the system and procedures to return the system to normal operation as fast as possible will benefit operation as well as design engineers. However, determination of parameters of cold load pickup model for a given system is very important for implementation of this method. Furthermore, in the future availability of high resolution data from field during actual restoration can provide validation or improvement of CLPU model used.

The conceptual framework of this paper in conjunction with distribution automation will allow the distribution systems to be operated with lesser spare transformer capacity. Hence, utilities will be able to accrue financial savings by deferring upgrade of existing transformers and installation of new transformers.

IX. List of symbols

P_{loss} : total losses at load S_L (W)
ΔP_{loss} : total loss difference between initial and final load(W)
P_{loss_N} : total losses at rated load (W)
θ_{fl} : top-oil rise at rated load (°C)
θ_{hs} : transformer hottest-spot temperature (°C)
θ_a : ambient temperature (°C)
θ_o : top-oil rise over ambient temperature (°C)
θ_u : ultimate top-oil rise for load S_L (°C)
θ_i : initial top-oil rise for t=0 (°C)
K : loading ratio
C : thermal capacity of transformer (Whr./°C)
k : heat transfer coefficient (W/°C)
n : exponential power of total loss versus top-oil temperature rise above ambient
m : exponential power of winding loss versus winding temperature rise above oil
θ_g : hottest-spot winding rise over top-oil (°C)
$\theta_{g(fl)}$: hottest-spot winding rise over top-oil temperature at rated load
θ_{gu} : ultimate hottest-spot winding rise over top-oil (°C)
τ_g : winding thermal time constant at rated load (hr.)

REFERENCES

[1] James E. McDonald, Armin M. Bruning, and William R. Mahieu, "Cold load pickup", *IEEE Transactions on Power Apparatus and Systems*, vol. PAS-98, pp. 1384–1386, July/August 1979.

[2] C. E. Hartay and C. J. Couy, "Diversity: A new problem in feeder pickup", *Electric Light and Power*, pp. 142–146, October 1952.

[3] R. S. Smithley, "Normal relay settings handle cold load", *Electrical World*, pp. 52–54, June 15, 1959.

[4] Oliver Ramsaur, "A new approach to cold load restoration", *Electrical World*, pp. 101–103, October 6, 1952.

[5] Walter W. Lang, Max D. Anderson, and David R. Fannin, "An analytical method for quantifying the electrical space heating component of a cold load pick up", *IEEE Transactions on Power Apparatus and Systems*, vol. PAS-101, pp. 924–932, April 1982.

[6] ANSI/IEEE C57.92-1981, *Guide for loading mineral-oil-immersed power transformers up to and including 100 MVA with 55 °C or 65 °C winding rise*, The Institute of Electrical and Electronics Engineers, Inc., 1981.

[7] R. L. Wilde, "Effects of cold load pickup at the distribution substation transformer", *IEEE Transactions on Power Apparatus and Systems*, vol. PAS-104, pp. 704–710, March 1985.

[8] J. Aubin, R. Bergeron, and R. Morin, "Distribution transformer overloading capability under cold-load pickup conditions", *IEEE Transactions on Power Delivery*, vol. 5, pp. 1883–1891, November 1990.

[9] D. W. Butts, "Winter cold load pickup study", Technical report, Planing Department- llinois Power Company, Decatur, Illinois, Feb. 1979.

[10] M. M. Adibi, "Special consideration in power system restoration faults: The second working group report", *in IEEE PES Winter Meeting*, number 93 WM 202-2 PWRS, Columbus, Ohio, Jan 1993.

[11] Satoru Ihara and Fred C. Schweppe, "Physically based modeling of cold load pickup", *IEEE Transactions on Power Apparatus and Systems*, vol. PAS-100, pp. 4142–4150, September 1981.

[12] Chee-Yee Chong and Roland P. Malhamé, "Statistical synthesis of physically based load models with applications to cold load pickup", *IEEE Transactions on Power Apparatus and Systems*, vol. PAS-103, pp. 1612–1628, July 1984.

[13] Roland Malhamé and Chee-Yee Chong, "Electric load model synthesis by diffusion approximation of a high-order hybrid-state stochastic system", *IEEE Transactions on Automatic Control*, vol. AC-30, pp. 854–860, September 1985.

[14] R. E. Mortensen and K. P. Haggerty, "A stochastic computer model for heating and cooling loads", *IEEE Transactions on Power Systems*, vol. 3, pp. 1213–1219, August 1988.

[15] C. Alvarez, R. Malhamé, and A. Gabaldón, "A class of models for load management application and evaluation revisited", *IEEE Transactions on Power Systems*, vol. 7, pp. 1435–1443, November 1992.

[16] R. E. Mortensen and K. P. Haggerty, "Dynamics of heating and cooling loads: Models, simulation, and actual utility data", *IEEE Transactions on Power Systems*, vol. 5, pp. 243–249, February 1990.

[17] J. C. Laurent and R. P. Malhamé, "A physically-based computer model of aggregate electic water heating loads", *in IEEE/PES Summer Meeting*, Vancouver, B.C., Canada, 1993, paper no. 93 SM 496-0 PWRS.

[18] A. C. Monteith and C. F. Wagner, editors, *Electrical Transmission and Distribution Reference Book*, Westinghouse Electric Corporation, East Pittsburgh, PA., 1950.

[19] K. Karsai, D. Kerényi, and L. Kiss, *Large Power Transformers*, Elsevier Science Publishing Company, Inc., 52 Vanderbilt Avenue, New York , NY 10017, U.S.A., 1987.

[20] Kenneth R. Baker, *Introduction to Sequencing and Scheduling*, John Wiley & Sons, Inc., 1974.

[21] A. Pahwa and C. Uçak, "Restoration of power distribution systems in the wake of storms", *in The Proceedings of the 23th Annual North American Power Symposium*, pp. 54–63, Carbondale, Illinois, October 7-8 1991.

[22] Don O. Koval, "Impact of rural design, operating, and restoration practices on the costs of service interruptions", *IEEE Transactions on Industry Applications*, vol. 25, pp. 186–194, January/February 1989.

Knowledge-Based Behavior Interface: Its Application to Power Network Restoration Support System

Jun-ichi Shinohara, Member, IEEE
Toshiba Corporation

Isao Kozakai
Chubu Electric Power Company

Masahiko Kunugi, Member, IEEE
Toshiba Corporation

Jun-ichi Nagata
Toshiba Corporation

Hideki Saito
Toshiba Corporation

***Abstract*–The characteristics of power system restoration (its combinational aspects, use of knowledge from a wide variety of origins and of different types and number of criteria to satisfy) make it a difficult problem. As a solution, a number of research teams are considering the use of expert systems to generate restoration plans. Promising results have already been obtained.**

Yet most expert systems remain in the prototype stage.[1] One reason seems to be that the study of support systems for knowledge-based behavior (unable to describe with any rule) is insufficient.

This paper begins by analyzing power system restoration based on the human performance model [6] and discusses knowledge-based behavior as a high conceptual level human performance to solve this problem with combinational aspects. The paper then reports on its applications to the trunk line power network dispatching center.

Finally the paper verifies the relation between knowledge-based behavior and designed human interfaces with a power network restoration case-study.

I. INTRODUCTION

Recent years have seen our society become increasingly dependent upon electricity, and this has created a need for greater reliability in the power system. The demand for more reliability is particularly high in the upper voltage system (trunk system). Quickly restoring a power system that is experiencing power failure as a result of an accident has become an key issue in power system operation.

The planning of the power system restoration is one of combinational problems. A solution requires extensive knowledge. Moreover, the issue also involves a number of restrictions and conditions that require the judgement of operators, rendering the issue more complex. A system to provide answers to the issue at a speed sufficient to enable practical application has not been achieved with such conventional techniques as optimization algorithms.

This paper was presented at the 1995 IEEE Power Industry Computer Applications Conference held in Salt Lake City, Utah, May 7–12, 1995.

In recent years, research has begun on the application of expert systems to power system restoration issue. Although promising results have been obtained, most are still limited to the stage of prototype development [1].

The following issues must be solved in order to develop a system that can be put to an actual use. These issues have made power system restoration more complex:

(1) It is important to conduct a preliminary examination of the restart of thermal and hydraulic generators, their connection to the power system, the startup characteristics of these generators, output capacity and the division of systems used for power system restoration, and then plan for power system restoration in a way that matches the characteristics of each power system.[2]

(2) The load fluctuates according to the time zone, seasons and weather, and also fluctuates thermostatically to correspond to the duration of the power blackout. Power system restoration must be conducted with an understanding of such elements.[3]

(3) Economy, reliability and other aspects are objective coefficients related to power system restoration. However, the evaluation of these have generally different criteria and these have a trade-off relationship. Power system restoration is a multi-objective optimization issue in which improving one objective deteriorates another. [4]

Automation is made difficult by the existence of such diverse, complex or uncertain elements as those listed above. Consequently, creating a system to support operators is more realistic than building a completely automated system. The following are proposed as dialog functions with operators under the support system:[5]

- Function to indicate current system status and progress of restoration operations.
- Function to enable the operator to nominate usable facilities.
- Function to build a targeted restoration system and indicate the system to the operator.
- Function to generate restoration procedures leading to the targeted restoration system and indicate such procedures to the operator.

Even with the existence of these research results, restoration systems are not in widespread use. The authors believe it is partly because of the lack of research to develop a system that will to some extent substitute for human operations in power system restoration operations and of research on information processing with the aim of acceptance by humans.

This paper begins by describing the flow of power system restoration operations streamlined through interviews with skilled human operators. Next, we state the ideas behind the

Reprinted from *IEEE Transactions on Power Systems,* Vol. 11, No. 1, pp. 383–389, February 1996.

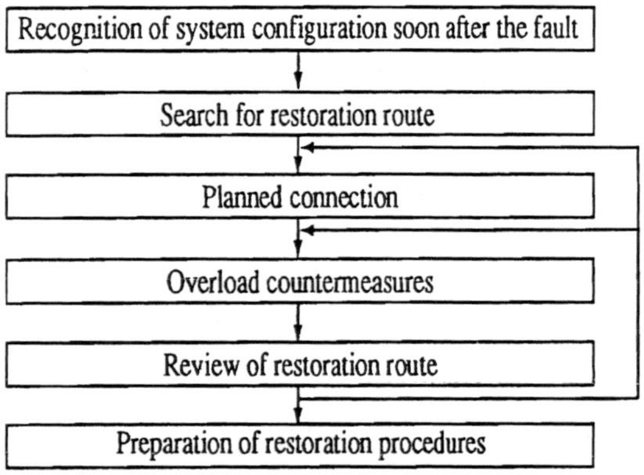

Fig. 1 Processing Flow for Restoration

"knowledge-based behavior interface," which is designed to clarify that part which has been difficult to express in the form of rules and procedures, based on past experience. This is where generalizations are required. Then, we will provide examples in which the central load dispatch/control system has been installed. We will conclude with an verification using a case study of an example of an power failure.

II. FAULT RESTORATION FLOW

Restoration refers to a series of operations in which the facility that has experienced a fault is repaired and the power system is restored to the status that prevailed before the power failure occurred. However, restoration often refers to a series of operations in a power system operation, including the separation of the facility that has experienced the fault from the system to minimize interruption to power supply. This paper will adopt system operation concepts and describe the flow from the time the determination of which facility has experienced the fault is made to the production of restoration procedures.

Fig. 1 outlines the flow of restoration when an a fault occurs in the power system. This is a typical flow created based on the experience of skilled human operators and was produced with the cooperation of operators with substantial experience in power dispatch operations. Explanations are based on this flow.

1) Recognition of system configuration soon after the fault: When a fault takes place in the power system, the circuit breaker trips and the facility experiencing the fault is separated from the system which is sound. Furthermore, the area in which the power was supplied through the facility encountering the fault will experience a power blackout. Concerning this range, even though the facility experiencing the power failure is not included, preset operations known as independent restoration operations will be conducted in order to configuration a system status suited to receiving power. Here, the circuit breakers on the receiving side will generally be placed in ON status, while the circuit breakers on the load side will be placed in OFF status. The group of more than one facility connected by circuit breakers maintained in ON status is called a block and the circuit breakers between the blocks which are maintained in OFF status are called linkage points. In the first stage of fault restoration, the system experiencing the power blackout caused by the power failure and the system that is sound will be recognized. The block, the foundation for the system application, will also be acknowledged.

2) Search for the restoration route: A search is made for the restoration route to the system that has been made to be an independent system through separation from the overall power system, the generators which came to a standstill, the load which is experiencing power blackout. The following four points comprise the basic guidelines for this. Here, the connection status of the system before the occurrence of the fault is called the original route and the other status is called the route to avoid the fault point:

- Get as close as possible to the original route.
- Supply power to the generators targeted for restoration.
- Make maximum use of the restoration power source and recover as much load as possible
- Execute restoration in as few operation procedures as possible

Examples of the knowledge required of operators to implement these basic guidelines are given below.

- If there is no facility experiencing the fault along the original route, use the original route as the restoration route.
- When there is more than one possible route to avoid the fault point, give priority to the route with the fewest linkage points.
- Give priority to the route with the higher voltage linkage point for connection to the sound system.
- Select the route with the smallest increase in voltage in relation to the sound system.
- Give priority to the route with the shorter electric distance (smaller impedance)
- Give priority to the route with the greater margin of facility capacity.

3) Planned connection: Temporary power generation output and load values will be connected while taking into consideration the supply and demand balance of the system looked for during the search for a restoration route. We call this a planned connection. This action creates a hypothetical system on the assumption that the connection is made depending on the priority among all the routes including the generator, load and restoration routes.

4) Overload countermeasures: A power flow calculation for the hypothetical system will be conducted when the planned connections conducted. When an overload is anticipated, countermeasures are adopted by conducting examinations in the following order.

(1) Actions against overload by adjusting generator output.
For example, when the emergency power system is located in the up stream of the facility expected to experience the overload, increase the generator output located on the downstream system. In this case, be sure that the increase in the targeted generator output will not trigger an overload in another facility and be sure that no overload occurs in facilities other than targeted facilities.

(2) System switchover
Ensure that the switch that will be closed and the switch that will be open at the time of system switchover belong to an identical voltage class. The operations target will be a linkage point. Give priority to switches located within the same substation.

5) Review of the restoration route: When the system will be switched as part of actions taken against overload, this may call for revisions in the restoration route for the generators or the load. In this case, with the status after the system has been switched (division of blocks) as the basic status, search for the restoration route again, and carry out the planned connection, take actions against overloads and produce a targeted system after the restoration.

6) Preparation of restoration procedures: Produce the restoration procedures up to the production of the targeted system based on the current system. Examples of the knowledge required is given below.

- Knowledge of procedure production

Parallel operations for two switches simultaneously should be possible. However, operations conducted within the same substations and operations along identical restoration routes should be for one switch.

- Production of power source restoration procedures

Restore generators in the order in which higher priority for restoration is given. When the load targeted for restoration is included (connected) on the block located along the restoration route, transit power to the load at the same time. The output value of the restored generator will be determined in accordance with the startup characteristics inherent to the generator.

- Production of load restoration procedures

Restore the load in the order of restoration priority. The transmission of power to the load will be conducted within the excess capacity of the generator output.

III. USER INTERFACE ISSUES IN FAULT RESTORATION SUPPORT SYSTEM AND IDEAS BEHIND THE KNOWLEDGE-BASED BEHAVIOR INTERFACE

When human beings think, particularly when they are unable to form thoughts based on simple rules, it is said that they make generalizations. Many issues in fault restoration operations cannot be described using given rules. The operators involved in such operations therefore make generalizations. Furthermore, when restoration policies are prepared, the judgement that the plans are generally correct is made via a comparison to other plans. Then, actual restoration procedures are determined.

In this section, based on a human performance model that indicates the behavior of skilled human operators, we begin by trying to understand what is a generalization. Then, in the restoration flow described in Section II, we clarify the relationship of items that require generalization with other items. Finally, we explain about the ideas behind user interface, knowledge-based behavior interface (KBI) which supports the behavior known as operators' knowledge-based behavior in fault restoration operations.

A. Human Performance Model

The human performance model [6] was proposed as a means to analyze the behavior of the operator. As Fig. 2 shows, the human performance model is classified into three levels.

(1) Skill-based behavior

Skill-based behavior is behavior that operators can carry out smoothly and unconsciously even under a stressful conditions. Generally, human activity is thought to comprise a series of behavior required of the person in each case. When such behavior is based on skill, it is conducted unconsciously. Moreover, the operator is able to give answers to questions about what the operator is controlling and what information is used to conduct that control

(2) Rule-based behavior

Rule-based behavior is behavior which enables operators to act on rules and procedures prepared in advance based on past experience.

(3) Knowledge-based behavior

Knowledge-based behavior is behavior that the operator will adopt when the operator faces a situation where relevant expertise or rules do not exist. In this situation, the goal is explicitly formulated, based on an analysis of the environment and the overall aims of the operator. The operator will then select an effective plan through trial and error. The generalization which human beings carry out in tackling multipurpose optimization issue thus refers to knowledge base behavior.

In Fig. 2, when the operator receives a sensory input, the operator conducts an examination by classifying the information input into that which can be processed unconsciously, that which can be processed based on a certain format using rules and procedures, and that which requires judgement through trial and error. The operator acts after reaching the results of judgement. Here, for each item of input, the operator examines appropriateness by repeating several times knowledge-based behavior, rule-based behavior and skill-based behavior until he or she finally decides on the definite action that should be taken.

An expert system is generally called a knowledge-based system. This is based on the idea that the specialized knowledge of skilled human operators is valued.

This special knowledge is believed to support the knowledge-based behavior and rule-based behavior shown in Fig. 2. However, what can actually be provided as rules and procedures to the expert system follows rule-based behavior.

B. Skill-Based Behavior, Rule-Based Behavior and Knowledge-Based Behavior in Fault Restoration.

Section II described the fault restoration flow. In that section, for example in searching for the restoration route, the operator prepares a route which avoids the point of the fault by using information acquired through daily operation and training, such as attributes including each facility's voltage class and capacity and the restoration priority of generators and loads. In the fault restoration flow, for example in searching for the restoration route, the operator prepares a route which avoids the point of the fault.

When we try to understand it in terms of human performance model;

- Recognition of power system connection status before the occurrence of the fault, namely the recognition of the original route is a skill-based behavior.
- Recognition of block structure and the locational relationship between the system which is sound and the system which is experiencing the power blackout is a rule-based behavior.
- Operation of a route to avoid the point of the fault that is carried out when the original route will not be recovered after following the above procedures is a knowledge-based behavior.

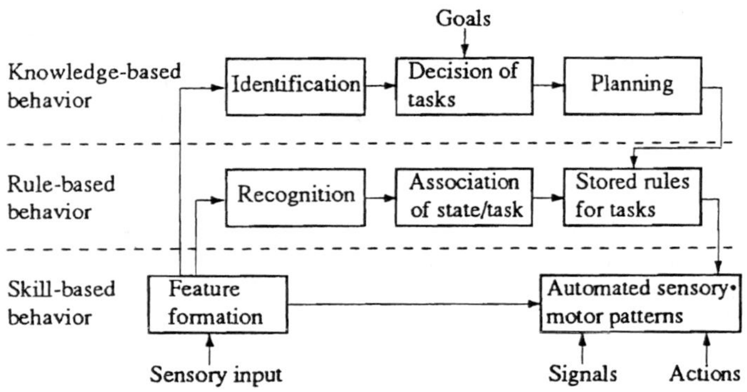

Fig. 2 Simplified illustration of three levels of performance of skilled human operators

Table 1 shows the results which classify all the situations based on the above model. As already mentioned, human generalizations correspond to knowledge-based behavior. Consequently, Table 1 shows the part considered to be the part where computer-based support has been considered difficult, which is one of the major issues of conventional computer systems.

C. Ideas Behind KBI

As mentioned above, knowledge-based behavior deals with the multi-objective optimization issue that contains more than one evaluation criteria and where such criteria have a trade-off relationship. To work out knowledge-based behavior using a computer requires that the goal and the criteria, which the operator has based on his or her own judgement, be input into the computer.

The computer must provide output in such a manner that the computation results and the goal or criteria can be compared. The goal and criteria of the operator for fault restoration operations can be described as follows:

- Quickly reduce the degradation of supply.
- Quickly restore the important load.
- Appropriately select the restoration and the route to avoid the point of fault.

The precondition for a restoration support system is that it can indicate an outline of policies and detailed procedures for restoration operations in a manner that enables the operator to assess the above mentioned items. Therefore, based on this precondition, a function that allows the operator to obtain solutions through dialogue with the computer system while varying the goals and evaluation criteria is also necessary. These are considered the requisites for KBI.

Fig. 3 shows the KBI conceptual chart.

IV. FAULT RESTORATION SYSTEM INSTALLED AT THE CENTRAL LOAD DISPATCH/CONTROL SYSTEM

Fig. 4 shows the Configuration of the Dispatching System of the part related to the power failure restoration incorporated in the system of Chubu Electric Power's central load dispatch/control system.

The function to calculate total expected demand is used to forecast power demand value at each hour of the relevant day. Fault restoration procedures are prepared with the value calculated here set as the target value for the load after restoration.

The function to prepare restoration guidance is designed to show the operator the planned outline of the fault evasion route and the restoration route. The name of the substation which are located on the restored power source side, the name of the lines used for power transmission, the name of the power station and the substation that will be connected to the system, will be indicated on CRT.

The function to prepare restoration procedures is designed to show the operator the restoration procedures while taking into consideration the restoration priority of the generators and load. The expected time when the operation will be implemented, the name of the power station and the substation which will become the target of operations, the name of the generator, the name of the line, the operation and other detailed procedures will be displayed on the CRT.

Table 1 Analysis based on the human performance model

Restoration situation	Sensory input	Skill-based behavior	Rule-based behavior	Knowledge-based behavior
Recognition of the power system configuration soon after the fault	CB in symbol form Operation messages of protective relay	Recognition of • CB ON/OFF status • Power system energized/deenergized	Recognition of • Block • Sound power system	
Search for restoration root	CB in symbol form Attributes of each facility • Capacity • Restoration priority	Recognition of the original route	Recognition of • Block • Sound power system	Preparation for the route to avoid the fault point
Planned connection	Predicted demand Status of generator Restoration priority	Recognition of the power supplied from others	Determination of restoration target • Generator • Load	Preparation of the outline of the restoration route
Overload countermeasure	Facility capacity		Judgement of the need for countermeasures	Preparation of method for overload countermeasures
Preparation of restoration procedures	Restoration priority of each facility Startup characters of generators Restoration unit of the load	Recognition of prohibited matters in the operations procedures	Preparation of operations procedures Understanding of the remaining generator output capacity	Evaluation of restoration priority Forecast trend in the interruption of the supply of electrical power

The function to produce a power supply interruption diagram is designed to show in polygonal graph form the changes in the amount of supply problem anticipated as a result of implementing the restoration procedures described previously.

Apart from the above, the system also incorporates a user interface which allows the operator to change the restoration priority in certain areas.

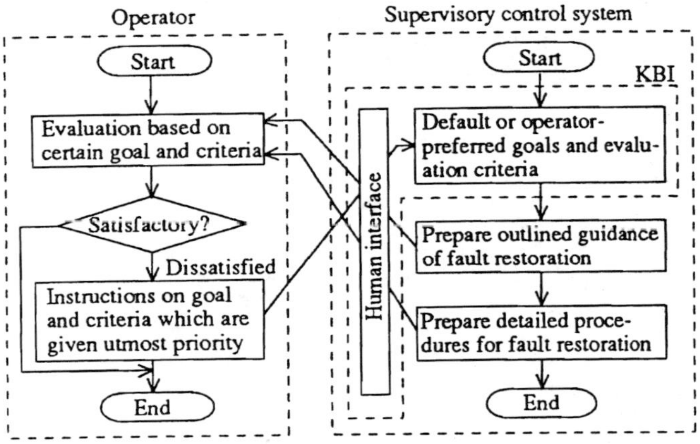

Fig.3 KBI Conceptual Chart

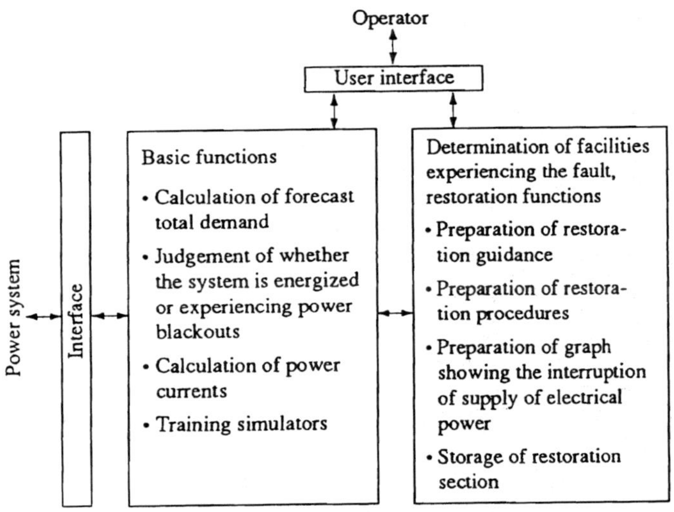

Fig. 4 Configuration of the Dispatching System

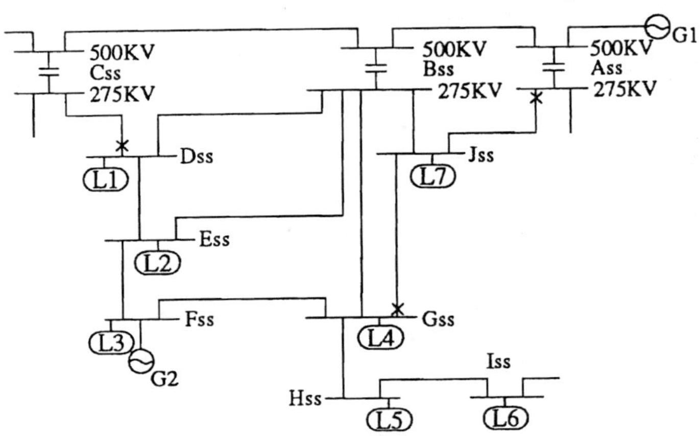

Fig. 5 System Configuration Before a Fault Occurs

V. VERIFICATION OF THE KBI

Section IV gave specific examples of screens displayed by KBI, developed by the authors. Here, we use a fault restoration case study to examine whether KBI will support knowledge-based behavior.

Fig. 5 shows the power system under normal operation before the fault occurs. Ass, Bss and Css show the substations with a 500 kV and 275 kV bus. More than one substation operated at 275 kV are connected to Bss in a looped configuration. In addition, L1 - L7 are loads, G1 and G2 are generators and X shows whether the switcher is cut.

Based on this system, when a power failure occurs on the 275 kV bus, the 275 Kv substations shown in Fig. 5 will all experience a power blackout. Fig. 6 shows the system configuration after the fault where independent restoration operations have been carried out.

In Fig. 6, there are two routes to avoid the point of fault: the route from Css to Dss and one from Ass to Jss. There are also several combinations of restoration routes to restore Ess, Fss, Gss, Hss and Iss. From these several possibilities, the expert system function is used to select a set of routes to avoid the point of fault and the restoration route. The expert system mentioned here indicates a system equipped with the function to express the operator's knowledge as clear rules. Until the expert system obtains one definite solution, the existence of overload will be checked and the restoration route will be searched to avoid the over load.

Fig. 7 shows the route to avoid the point of power failure and the restoration route marked with an arrow. These are output in message format, namely restoration guidance, as shown in Fig. 8.

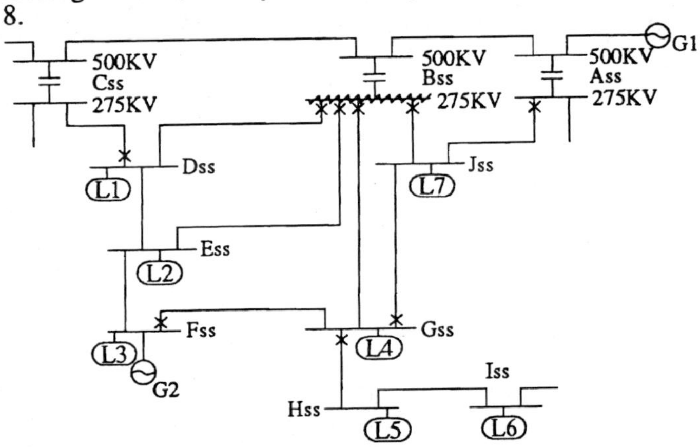

Fig. 6 System Configuration After a Fault Occurs

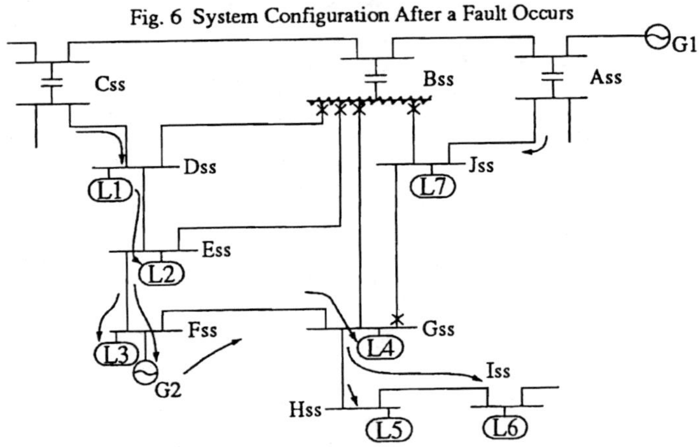

Fig. 7 System Configuration After Restoration

From Css, using the Css - Dss line, Dss, Ess, Fss, G2, Gss, Hss and Iss will be restored.
From Ass, using the Ass - Jss line, Jss will be restored.

Fig. 8 Restoration guidance

The restoration guidance shown in Fig. 8 expresses the following: With the Css-Dss line set as the route to avoid the point of the fault, the power source for restoration will be sent from Css to Dss and subsequent portions. That power source goes through Dss, Ess and Fss and reaches generator G2, and is then sent to Gss, Hss and Iss. The restoration guidance outlines the route to avoid the point of the fault and the restoration route. This enables the operator to understand the outline of the restoration guidance and corresponds to what was stated in Item C of Section III as a prerequisite of KBI.

The order of load restoration in temporal terms is not indicated. For example, whether load L4 of Gss will be supplied after restoration of G2 or whether load L4 of Gss will be directly supplied from Css will depend on the time characteristics relating to the G2's start up as well as L4's restoration priority.

As Fig. 7 shows, the expert system will prepare the restoration procedures depending on the restoration priority of the generators and the loads through the routes established once the route to avoid the point of the fault and restoration route are finalized. For example, when the load value, restoration priority and facility capacity are given as shown in Table 2, the restoration procedures shown in Fig. 9 is prepared. In Fig. 9, "0:03:00 Dss Css-Dss line IN" signifies that the Css-Dss line will be connected at the Dss substation at 0 hours 03 minutes 00 seconds, the relative time lapsed since the fault occurred. The power transmitted to load L1 (200 MW) will transmit the power of 200 MW to load L1 is also indicated. Therefore, Fig. 9 helps the operator understand the order of load restoration and the time prepared by the expert system.

Table 2 Restoration Procedures Based on Standard Priorities

Name of load	Load value (MW)	Priority	Name of facility	Capacity (MW)
L1	200	1	Css-Dss line	500
L2	100	3	Dss-Ess line	400
L3	200	2	Ess-Fss line	300
L4	200	4	Fss-Gss line	300
L5	30	5	Ass-Jss line	400
L6	20	6	Jss-Gss line	300
L7	200	7		

0:03:00 Dss Css - Dss line IN
L1 Power transmitted to load (200MW)
Ass Ass - Jss line IN
L7 Power transmitted to load (200 MW)
0:03:30 Ess Dss - Ess line IN
G2 IN
L3 Power transmitted to load (200 MW)
L2 Power transmitted to load (100 MW)
0:04:00 Fss Fss - Gss line IN
0:23:00 L4 Power transmitted to load
L5 Power transmitted to load (30 MW)
L6 Power transmitted to load (20 MW)

Fig. 9 Restoration procedures based on the standard priorities

The priority of loads tend to change in accordance with the organization of the events in each region. For example, the operator becomes dissatisfied with the procedures shown in Fig. 9 and wants to place more priority on restoring L4. When the operator indicates that intention to the expert system, Figs 10, 11 and 12 are shown as alternative plans as the power systems after the restoration, restoration guidance and restoration procedures, respectively. The alternative plans are prepared in order to supply power from Ass to L4 by separating the Gss bus for L4, which is given a higher priority of restoration. Fig. 13 compares the original plans with the alternative plans using a graph of the interruption of the supply of electrical power. When the above is put together, the operator makes a comparative evaluation of the following points and selects the restoration guidance.

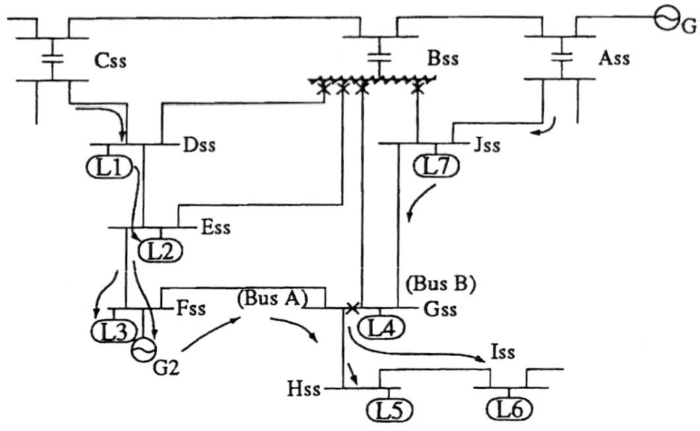

Fig. 10 Another system configuration after restoration

From Css, using the Css - Dss line, Dss, Ess, Fss, G2, Gss A, Hss and Iss will be restored.
From Ass, using the Ass - Jss line, Jss and Gss B will be restored.

Fig. 11 Another Alternative restoration guidance

0:03:00 Ass Ass - Iss IN
Gss Bus sect OFF
0:03:30 Gss Jss - Gss IN
L4 Power transmitted to L4 (200 MW)
L7 Power transmitted to L7 (200 MW)
Dss Css - Dss IN
L1 Power transmitted to L1 (200 MW)
0:04:00 Ess Dss - Ess IN
G2 IN
L3 Power transmitted to L3 (200 MW)
L2 Power transmitted to L2 (200 MW)
0:04:30 Fss Fss - Gss IN
0:23:30 L5 Power transmitted to L5 (30 MW)
L6 Power transmitted to L6 (20 MW)

Fig. 12 Another restoration procedure

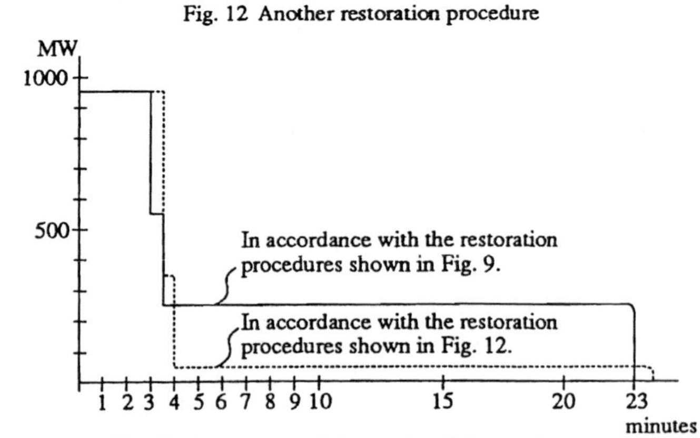

Fig. 13 Interruption of the supply of electrical power

- In the alternative plan, the time of restoration of L4 is 20 minutes earlier than the original plan.
- The alternative plan uses the Jss-Gss line as the restoration route and separates the Gss bus.
- The alternative plan eliminates the interruption of the supply faster than the original plan.

As mentioned above, by analyzing the fault restoration behavior, which has almost always been conducted by skilled human operators using human performance models, we were able to clearly indicate the following: (a) The part where the computer is able to provide answers; (b) The part where trial and error is conducted through a dialog between the computer and the operator. Furthermore, as understood from the case study, in the fault restoration, the answers of rule-based behavior (the answers obtained from the computer as a part of (a)) are evaluated based on knowledge-based behavior (to examine the plan provided by computer as a part of (b) while the operator uses it as a reference.) Therefore, the system which was developed by the authors and which was shown in Fig. 3 can be deemed to support the knowledge-based behavior.

VI. AFTERWORD

Since 1987, Chubu Electric Power Company and Toshiba Corporation have been jointly developing a fault restoration assistance expert system targeting the trunk line power system. The central load dispatch/control system which incorporated the results of this joint work began operating in January 1994. In this paper, we used a small and simple power system chart to simplify our explanations. However, the relevant power system targets a part of the large-scale 154 kV system as well as all of the 500 kV and 275 kV systems of Chubu Electric Power. The fault restoration expert support system is constructed targeting these large-scale power systems.

Section IV included restoration guidance and examples of the display of a graph of the interruption in the supply of electrical power as examples of our development. These functions have been prepared on the urging of skilled human operators involved in power supply operations and correspond to the analysis using human performance models.

In shown in Fig. 4, the system incorporates a training simulator. It is possible to simulate fault using any system configuration and examine the restoration guidance. When the system configuration are changed owing to maintenance of the electricity facilities or implementation of new facilities, the operator is supposed to prepare restoration policies each time the maintenance or implementation takes place and should also prepare restoration procedures. These are the fundamentals of operators' tasks. The system which assists these operators is not only effective during power failure but also during normal operations.

REFERENCES

[1] CIGRE SC 8 WG 38.08 TF04: The Use of Expert System for Power System Restoration (Final Report (Draft) Feb 1994)

[2] M. M. Adibi et al., "Power System Restoration–A Task Force Report" IEEE Transaction on Power System, Vol. PWRS-2, No. 2 May 1987 pp 271-277.

[3] M. M. Adibi et al., "Special Consideration in Power System Restoration, The Second Working Group Report", 93 WM 202-2 PWRS, 1993.

[4] No. 354 of Electric Association's Technical Report (Part 2), "Restoration Operations During Power System Accidents," Special Committee to Investigate Restoration Operations at During Power System Faults

[5] Y. Kojima et al., "Development of A Guidance Method for Power System Restoration" IEEE Transaction on Power Systems, Vol. 4, No. 3, August 1989 pp1219-1227

[6] J. Rasmusen, "Skill, Rules and Knowledge: Signals, Signs, Symbols and Other Distinctions in Human Performance Models," IEEE SMC-13-3, No. 3 May/June 1983 pp257-266

[7] I.Takeyasu et al., "An Expert System for Fault Analysis and Restoration of Trunk Line Power Systems" Symposium on Expert Systems Application to Power Systems Stockholm - Helsinki, August 22-26, 1988

An Object-Oriented Approach for Implementing Power System Restoration Package

Z. L. Gaing C. N. Lu
Department of Electrical Engineering
National Sun Yat-Sen University
Kaohsiung, Taiwan

B. S. Chang C. L. Cheng
System Operations Department
Taiwan Power Company
Taipei, Taiwan

Abstract- Due to many unforeseen circumstances that could happen in the today's bulk power system, there is a possibility of a system wide outage. In order to provide aids to the power system dispatchers following a complete collapse of the power system, a prototype package has been developed. Through interactive and friendly graphic interface, the package suggests a guideline for the dispatcher to restore the power system. With an aim to increase the ease of maintenance, object-oriented technique was adopted to implement the package. The development of the software involved three stages: 1) Object oriented analysis, 2) Object oriented design, and 3) Integration and testing. In this paper, the structure and the development procedure of the prototype system are presented.

Keywords - power system restoration, object-oriented technique

1. INTRODUCTION

Many electric companies have developed system restoration schemes to meet their particular needs in the blackout period. Most of these schemes contain general guidelines which serve to define the required restoration plans [1-5]. Due to voluminous data and complex operation procedure involved in the system restoration process, a tool that can manage all pertinent data and coordinate related aspects of the problem will certainly minimize the blackout period. In recent years, expert systems have shown potential since heuristic knowledge plays an important role in the power system restoration [6-7]. Interactive graphic environment used to visualize and control the restoration process can also enhance the effectiveness of the process[5].

Taiwan Power Company (Taipower) is the sole electricity supplier in the island. Table 1 shows a summary of generation and transmission facilities of the current Taipower system. Since the island wide blackout occurred in 1976, reviews of system operation procedure for handling blackout situation have been conducted at frequent intervals by the engineers of various departments of Taipower.

In order to provide aids to the power system dispatchers following a complete collapse of the power system, a research project was conducted to develop a computer package for this purpose. Through interactive and friendly graphic interface, the prototype package can suggest guidelines for the dispatchers to restore the system. Practical operation rules obtained from dispatcher operation documents are utilized in this package to provide a feasible and practical procedure for the system restoration. Analytical tools such as load flow and transient voltage calculation are integrated into the system to assess the system state of a tentative measure. Figure 1 shows part of a restoration plan suggested by the system for an assumed blackout case in the southern part of the Taipower system. On average, the system takes 3.5 seconds to come out with a tentative measure in each stage of the restoration process. Figure 2 shows a network display along with a transient voltage analysis result of closing a transmission line.

To overcome the inflexibility in the traditional software development, in recent years, object-oriented technique has gained widespread acceptance to implement complex problems as its advantages of flexibility and ease of maintenance have been recognized[8-13]. In order to exploit these advantages, an object-oriented approach was adopted in the development of the system restoration assistant program.

Three stages were involved in the development of the prototype system, 1) Object-oriented analysis (OOA), 2) Object-oriented design (OOD), and 3) Integration and testing [11]. In the OOA stage, the following three system models were built: i) Object model which was described by object diagram to depict the static structure of the system in terms of objects and relationships corresponding to real-world entities, ii) Dynamic model which utilized state diagram to describe the control structure of the system in terms of events and states, iii) Functional model which utilized data flow diagram to describe the computational structure of a system in terms of values and functions.

Two design phases were included in the OOD stage: i) System design phase, in this phase the overall structure and style of system architecture were decided, including its breakdown into subsystems, its inherent concurrence, and boundary conditions, etc., ii) Object design phase which determined the full definitions of the classes and associations used in the implementation as well as the interface and algorithms of the methods used to implement operations. Each module of the system can be implemented independently based on the style of the design[11].

In the final integration and testing stage, objects that have been implemented were integrated to became a complete software program. The developed system was

This paper was presented at the 1995 IEEE Power Industry Computer Applications Conference held in Salt Lake City, Utah, May 7–12, 1995.

Reprinted from *IEEE Transactions on Power Systems,* Vol. 11, No. 1, pp. 483–489, February 1996.

then tested to validate its satisfaction of the system requirements. The structure of the prototype system and the development procedure are described in the following Sections.

Table 1 : A Summary of Generation and Transmission Facilities of Current Taipower System.

Generation				Substation			Transmission Line		
Type	Nuclear	thermal	hydro	Type	UHV (345kv)	EHV (161kv)	Type	UHV (345kv)	EHV (161kv)
No. of Stations	3	6	34	No. of Stations	12	78	No. of Lines	52	217
No. of Units	6	31	165	-	-	-	Length (km)	2524	4004

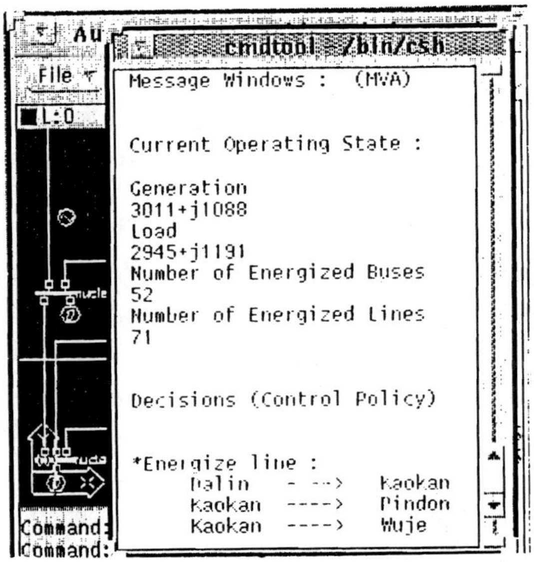

Figure 1: A View of a Restoration Plan

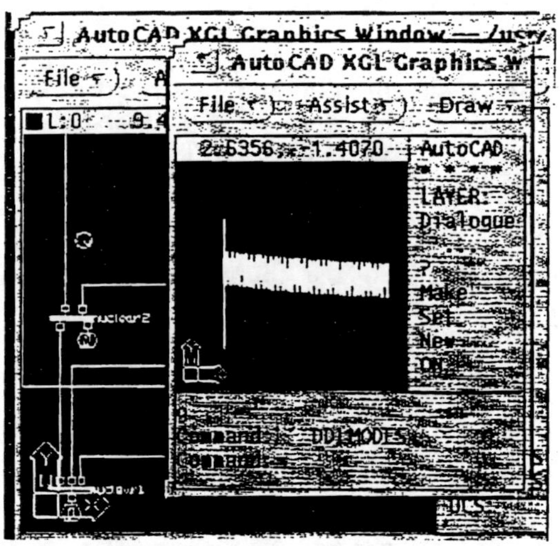

Figure 2: A View of Transient Voltage Analysis Result

2. PROBLEM DESCRIPTION

Power system restoration procedures are normally developed to meet specific requirements of an individual company so that the service interruption period can be minimized. A restoration procedure is developed based on certain operating principles, past experiences, characteristics of their power plants as well as other apparatuses. In order to handle the blackout situation, Taipower has developed system restoration rules which were accumulated from their operational experiences for several decades. Despite the complexity of the restoration procedure, the procedure can be broken down into a sequence of tasks. The tasks and issues of particular concern addressed in this package are as follows [1]:

1. Identify the status of the collapsed system, components and equipments. After a system wide outage, the system should be divided into three areas: North, Central and South areas. Restart of each area should be conducted independently. Breaker should be changed to the pre-specified status.
2. Restart and supply station service to plants, and substations. Hydro and black-start (gas turbine) units in each area are used to provide cranking power for the thermal units.
3. All sources of reactive power should be removed from the system and automatic load shedding should be deactivated. The circuit breaker on the primary side of transformers should be closed.
4. Coordinate power plant start-up timings with load pick-ups to bring generators to their stable minimum levels.
5. Energize sections of transmission lines with acceptable transient and sustained overvoltages. Restoration of 161 KV system should be completed before the 345 KV system.
6. Pick up loads in large increments without the risk of frequency decline.
7. Reintegrate the skeleton of the entire power system with consideration of the requisite time-consuming switching operations. The breaker status and the sequences of transmission line energizing operations have been pre-defined and documented for the restoration period, they should be followed in the restoration process.
8. Reduce the standing phase angles when closing loops to firm up transmission paths. Area synchronizations are conducted only at substations with synchronization equipments.

Figure 3 shows the roughly divided stages of the restoration process. Network analytical algorithms and graph theory are used in the system to analyze transient and steady state system behavior and to conceive a restoration path. In the first stage, a connection analysis algorithm is used to identify network status. Tasks 1 and 3 of the above mentioned issues are taken care in this stage. In the second stage, a shortest path algorithm is used to find the shortest path between any two split groups. In this stage, it performs tasks 2, 4, 5, and 8. In the third stage, a minimum spanning tree algorithm is used to expand the live area and tasks 4, 5, and 7 are carried out. In the fourth stage, tasks 4 and 6 are executed.

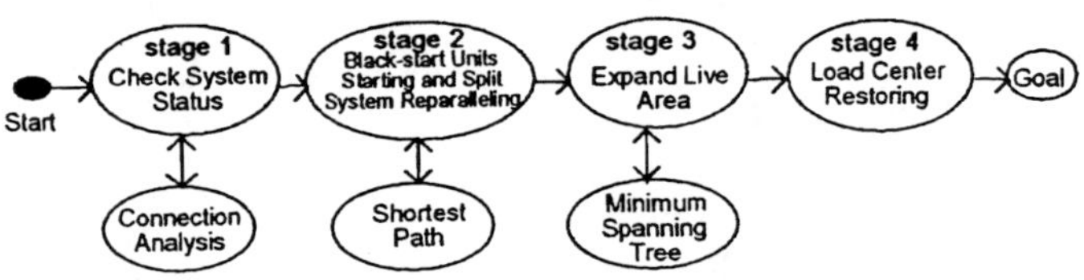

Figure 3 : Stages of Restoration Procedure

3. HARDWARE AND SOFTWARE REQUIREMENTS

The system was implemented on a SUN Sparc 10 workstation with 32 MB main memory and a 2 Giga byte hard disk. The main software requirements of the present system are (1) the general purpose graphics software AutoCAD, (2) conventional power system simulation software, and (3) restoration assistant program. Figure 4 shows the software sub-modules of the environment.

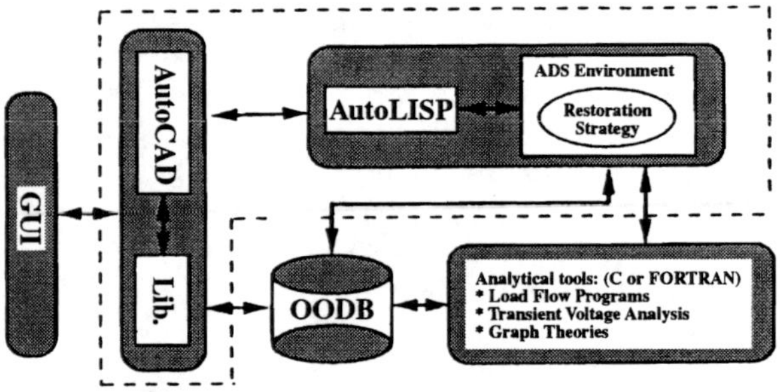

Figure 4 : Software Sub-modules

Graphics User Interface (GUI)

The graphics module allows users to interface with external programs, to display simulation results as well as the following data:

- * Summary tables and system diagram of each subsystem as well as detailed substation one-line diagram.
- * Physical characteristic data, such as parameters and ratings.
- * Operational characteristic data, such as system load, generation, and operating reserve in the system.

The restoration assistant program can be activated by keying in the "restore" command on the command window or choosing the item in the pull-down menu on the screen. With the graphics environment the user can visualize the energization process at its various stages. It also provides the dispatchers with the sequence of switching operations required to energize the system.

Object-Oriented Database (OODB)

Object-oriented technique is used to manage all pertinent data[10-12]. In addition to the necessary data for graphical representations of the power system, the following data are classified and stored in the database:

Generation stations :
- max. and min. output and rate of loading under normal and emergency conditions.
- max. time intervals for hot restart of boilers.
- black start capabilities.
- P(I,J,t), Q(I,J,t) I: boiler turbine generator conditions, J: start-up stage, t: time.
- substation structures and equipments.

Transmission lines :
- line parameters
- line chargings
- thermal limits
- status of the line

Substations :
- switchgear status
- transformer parameters
- transformer rating
- transmission line numbers and destination

Loads :
- number and size of low voltage AC network loads under peak and light load conditions.
- load priority

Analytical Program Module

Analytical programs include load flow calculation and transient voltage calculation. The environment permits the addition of other simulation programs. Power flow program is used for maintaining the real and reactive power balance to avoid sustained overvoltages and thermal overloads. Switching transient program is used to determine transient voltages in energizing high voltage transmission lines.

Integration of AutoCAD and External Programs

AutoCAD release 12 package is used as the graphics development environment. In addition to the graphics display capability, AutoCAD allows the inclusion of many application programs. Refer to Figure 4, AutoLISP lets users write marco programs, particular drawing instructions or calculations in a high-level language. ADS (AutoCAD Development System) is a C-language programming environment for AutoCAD applications. An ADS application is written as a set of external functions loaded by, and called from the AutoLISP interpreter.

4. SYSTEM IMPLEMENTATION[11]

In the object-oriented approach, the decomposition of the problem is based upon the concept of an object. An object is an entity whose behavior is characterized by the actions that it suffers and that it requires of other objects. Each module in the system denotes an object or class of objects from the problem space [12].

Objects are the software modules which incorporate instance variables (data) and a set of methods (operations) to operate on the data. Abstract and information hiding (encapsulation) form the foundation of object-oriented development. As Shaw reports, " an abstraction is a simplified description of a system that emphasizes some of the system's details or properties while suppressing others" [13]. If the parts of the requirements analysis effort that are most volatile are encapsulated, then the changing of requirements becomes less of a threat to the overall effort [13].

Object-oriented programming technique is very flexible because changes to one object will not affect other objects of the program. Thus, it allows engineers to focus on applying the knowledge of their field rather than on computer related details such as keeping the variables and data structures of various functions consistent[8].

The development of the system restoration assistant program was divided into three major steps - OOA, OOD, and integration and testing. The description of each step is given in the following. Due to limited space, the description of OOA and OOD will be concentrated on the implementation of power system restoration strategy. Same technique was applied to other functions of the system such as GUI. The notations and development techniques in [11] are used in this paper.

OOA[11]

The purpose of OOA is to state and understand the problem and application domain so that a current design can be constructed. We first consulted with dispatcher to write an initial problem statement. The problem statement may be incomplete, and ambiguous, but it was just a starting point. We then looked into the operation procedure documents pertinent to the blackout situation to gain more information. After the data and dispatchers' experiences were gathered, they were used to build three system models. These models are object model, dynamic model, and functional model.

Object Modeling

The object model shows the static structure of the real world. First we identified object classes, then we identified associations between objects. Object attributes and links were also identified. Inheritance were used to organize and simplify the class structure. Inheritance refers to the property that a class inherits the declaration of object and functions defined in its ancestors. Tightly-coupled classes and associations were organized into modules. Information in object models were supplemented

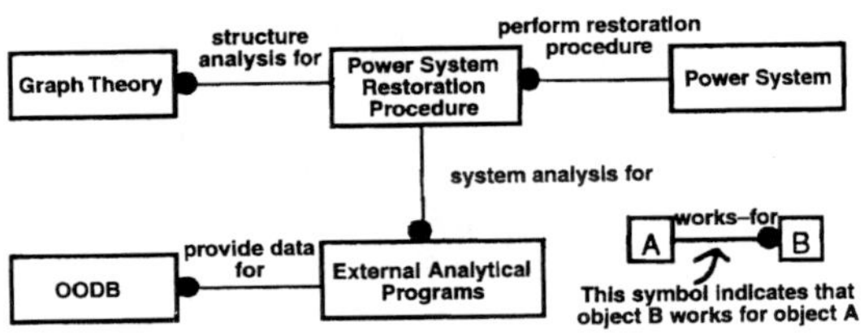

Figure 5 : A Coarse Object Diagram of System Restoration

with brief textual descriptions, including the purpose and scope of each entity. Object diagram was used to describe the object model. Figure 5 shows a coarse object diagram of the package based on problem description mentioned in Sections 2 and 3. The pertinent data and restoration procedure of the problem were divided into five modules. Each module was treated as a special object that consisted of other objects. Figure 6 shows a more detailed object diagram of the problem.

Dynamic Modeling

The dynamic model shows the behavior of the system, especially the sequence of interactions. Typical and exceptional scenarios of the restoration process were obtained from the dispatchers. We then identified events between object and the outside world. A state diagram was built for each active object showing the patterns of events it receives and sends, together with actions that it performs. Events could be represented as operations on the object model. After matching events between state diagrams to verify consistency, the resulting set of state diagrams constitute the dynamic model. The dynamic model specifies allowable sequences of changes to objects from the object model. A state diagram describes all or part of behavior of one object of a given class. States were equivalence classes of attribute and link values for the object.

Figure 7 shows the event flow diagram of power system restoration procedure module. Figure 8, for example, show the detail state flow diagram of stage 1 of the procedure.

Based on Figures 7-8 and other state diagrams, we can add the attribute and event (function) data to the objects shown in Figure 6, and the result (Figure 9) provides us a more detailed description of the problem. Notice that in Figure 9 only data and functions pertinent to the power system restoration procedure module are added.

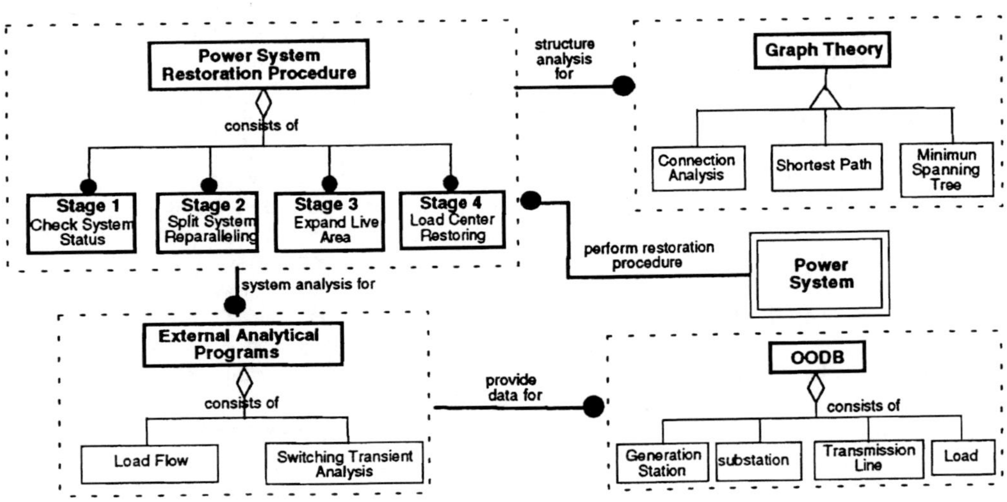

Figure 6 : Object Diagram of System Restoration Procedure

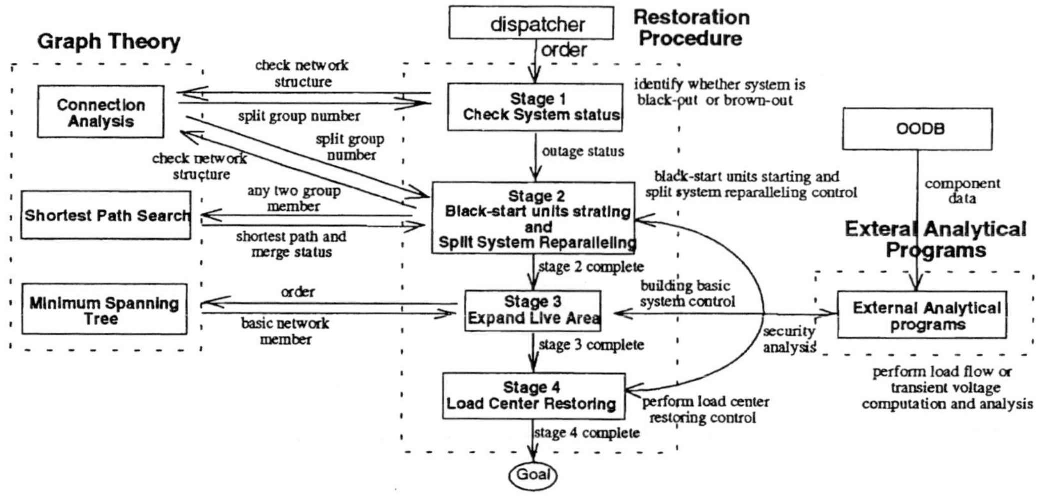

Figure 7 : Event Flow Diagram of Restoration Procedure Module

Functional Modeling

The functional model describes the data transformations within a system. The functional model shows how output values in a computation are derived from input values, without regard for the order in which the values are computed. These functions of computation are expressed in various ways, including mathematical equations, and natural language. First we identified input and output values of objects as parameters of external events, then constructed data flow diagrams to show the computation of each output value from other values and ultimately input values.

The processes on a data flow diagram correspond to activities or actions in the state diagrams of the classes. The data flows on a data flow diagram correspond to objects or attribute values in an object diagram.

Figure 10 shows an example of the data flow diagram of the power system restoration procedure module. It shows the relationship between values in a computation that is expressed as natural language.

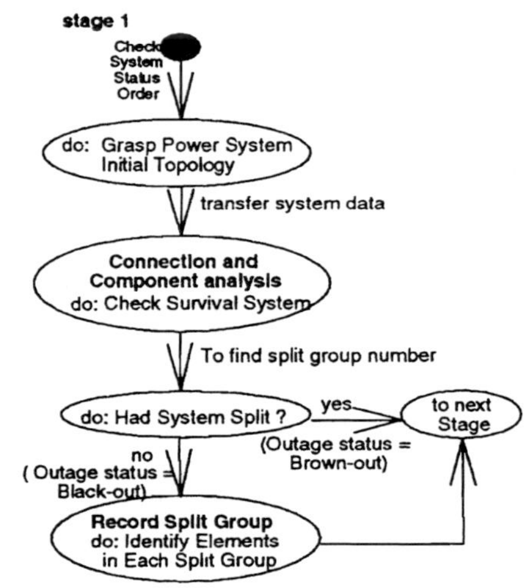

Figure 8 : State Flow Diagram of Stage 1

OOD[11]

After the problem was analyzed, we must decide how to approach the design. During design stage, decisions were made about how the problem will be solved, first at a high level, then at increasingly detailed levels. In the OOD stage, it consisted of two design phases: system design, and object design.

System Design

During the system design, the overall structure and style were decided. The organization of the system implementation was divided into components called subsystems. In general, the following decisions need to be made in the system design:

* Organize the system into subsystems,
* Identify concurrence inherent in the problem,
* Allocate subsystems to processors and tasks,
* Choose an approach for management of data stores.

The implementation of the system was divided into subsystems as shown in Figure 11.

Object Design

Object design did not begin from scratch but rather elaborated on the previous analysis and system design. Object design extended the analysis models with specific implementation decisions and additional internal classes, attributes, associations, and operations. In the object design phase implementation details were added, such as restructuring class for efficiency, internal data structures and algorithms to implement operations, implementation of control flow, implementation of associations, and packaging into physical modules.

Operations from the functional and dynamic models were transferred onto the object model (Figure 9) for implementation. A process from the functional model becomes an operation on the object. An event from the dynamic model may also become an operation on an object, depending on the implementation of system control flow.

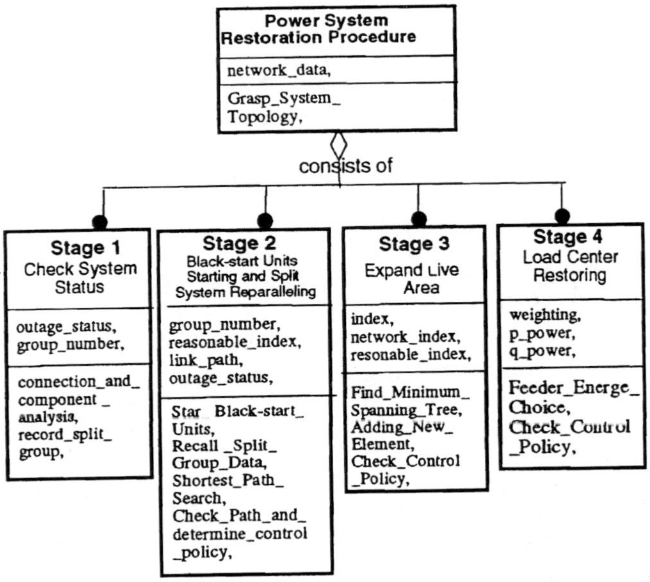

Figure 9 : A More Detailed Object Diagram of Power System Restoration Procedure

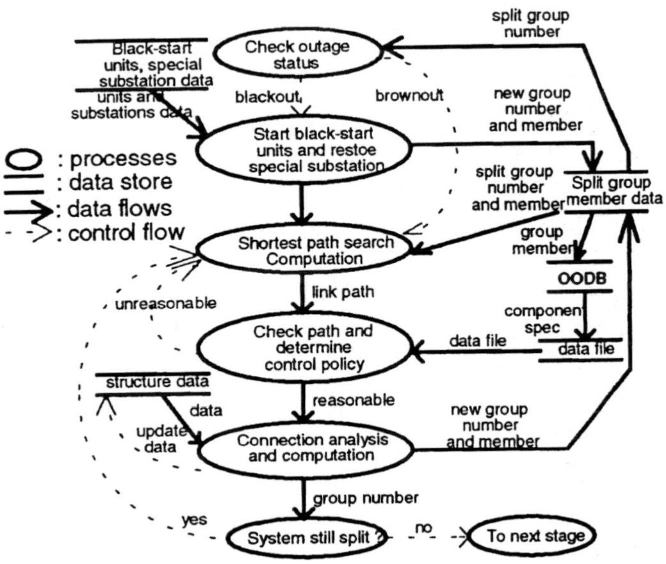

(a) Data Flow Diagram of Stage 2

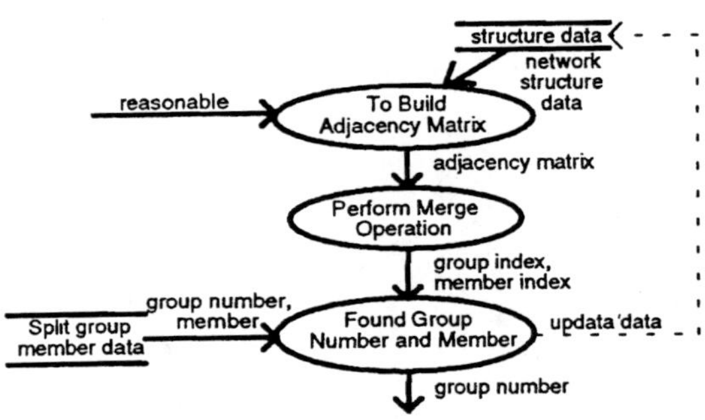

(b) Data Flow Diagram of "Connection analysis and computation" Process in (a)

Figure 10 : Data Flow Diagram of Restoration Procedure of stage 2

Figure 12 shows the implementation of the stages 1 and 2 of power system restoration procedure module with C++ language. The functions and data within a class were broken down into three (3) different types : private, protected, and public. Public functions and data can be accessed and modified by functions in other classes. Private data can only be accessed by the functions defined in their associated class. Protected data can be accessed by functions defined in the class and its subclass. Each class in Figure 12 inherits data and functions from other classes of objects.

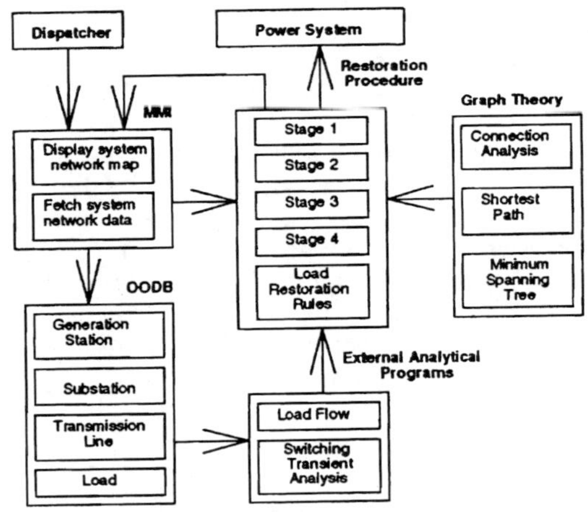

Figure 11 : Implementation Subsystem

```
class stage_1: public connection_analysis{
    / *Stage 1 class inherits data from class
                        connection_analysis.*/
private:
  int group_number;            /* attributes abstraction */
  int outage_index,
private:
  int connection_and_component_analysis(int [dim][dim] );
  void record_split_group( int );  /* methods declaration */
  };
class stage_2 :public connection_analysis,
               public shortest_path,
               public load_restoration_rules{
                              / *multiple inheritance */
  private:
   int group_number;           /* attributes abstraction */
   int reasonale_index;
   int link_path[dim][2];
  public:
   void recall_split_group_data();
   int start_blackstrat_units();  /* methods declaration */
   int *shortestness_path_search( int );
   int check_path_and_determine_control_policy( int[dim][2]);
   int connection_and_component_analysis(int [dim][dim] );
   };
      < methods implementation >
```

Figure 12 : Class Realization of Stages 1 and 2 of System Restoration Procedure Module

Integration and Testing

In this phase all parts of the program were integrated and tested. Since every object's functions have been realized they can be integrated based on the system architecture defined in the system design stage. Experience has demonstrated that the majority of the problems commonly encountered during the system integration are easily identified and resolved as a result of the module style of the object-oriented approach.

5. DISCUSSIONS

Currently, the prototype system is under off-line testing. Since dispatchers' experience is included in the system, in several cases, the system gives similar plans as those suggested by the user. In certain cases, the plans suggested by the user may not be feasible due to overvoltage problem in energizing high voltage transmission lines, while the prototype system is able to detect the problem and provide alternate routes. Until now, 2 man-years have been spent in implementing this prototype system. All the developing work was conducted at the University. Pertinent system data and dispatching experience for the work were provided by the staffs of Taipower.

Experience has shown that in the OOA, three models are not equally important in every problem. Almost all problems have useful object models derived from real-world entities. In the problems concerning interactions and times, such as interfaces and procedure process, the dynamic models are important. For the problems containing significant computation or data process, such as engineering calculation and database management, then the functional models are important. In our package the object model and dynamic model are stressed.

6. CONCLUSION

A prototype package for providing aids to the dispatchers for power system restoration has been presented in this paper. This package provides a convenient tool for coordinating pertinent data and complex operation procedure during system wide outage. The package was implemented with object-oriented approach and programming language. The greatest advantage of the object-oriented approach is that it provides a mechanism that captures a model of the real world. This leads to an improvement in the maintainability and understandability of a complex system.

7. ACKNOWLEDGMENT

The support provided by the National Science Council of the ROC is gratefully acknowledged. The first two authors would like to thank the system operators of Taipower Central Dispatching Office for their time and support.

8. REFERENCES

1 Taiwan Power Company, " Power System Operation Manual," July 1991.
2 M. M. Adibi et al. :Power System Restoration - A Task Force Report," IEEE Transactions on Power Systems, Vol. PWRS-2, No. 2, pp. 271-277, May 1987.
3 M. M. Adibi et al. "Power System Restoration - The Second Task force Report," IEEE Transactions on Power systems, vol. PWRS-2, No. 4, pp. 927-933, November 1987.
4 M. M. Adibi, D. Scheurer, "System Operations Challenges: Issues and Problems in Power system Restoration," IEEE Transactions on Power Systems, Vol. PWRS-3, No. 1, pp. 123-124, February 1988.
5 R. J. Kafka, D. R. Penders, S.H. Bouchey, M.M. Adibi, "Role of Interactive and Control Computers in the Development of a System Restoration Plan," IEEE Transactions on Power Apparatus and Systems, Vol. PAS-101, No.1, pp. 43-52, January 1982.
6 C. C. Liu, S.J. Lee, S.S. Venkata, "An Expert System Operational Aid for Restoration and Loss Reduction of Distribution Systems, " IEEE Transactions on Power Systems, Vol. 3, No. 2, pp. 619-626, May 1988.
7 D. S. Kirschen, T. L. Volkmann, " Guiding a Power System Restoration with an Expert System," IEEE Trans. on Power Systems, Vol. 2, No. 2, pp. 558-566, May 1991.
8 A. F. Neyer, F. F. Wu and K. Imhof, "Object-Oriented Programming for Flexible Software, Example of a Load Flow," IEEE Transactions on Power Systems, Vol. 5, No. 3, Aug. 1990, pp. 689-696.
9 Mike Foley, Anjan Bose, " Object-Oriented on-line Network Analysis," IEEE PWRS Winter Meeting, New York, pp 216-2, Jan., 1994.
10 S. J. Pollinger, C. C. Liu, and M. J. Damborg, "Design Guidelines for Object-Oriented Software with an Ems Man-Machine Interface Application," Electrical Power & Energy Systems, Vol. 14, No. 213, April/June 1992, pp. 122-130.
11 J. Rumbaugh, M. Blaha, W. Premerlani, F. Eddy, W. Lorensen, Object-Oriented Modeling and Design, Prentice Hall, 1991.
12 Grady Booch, " Object-Oriented Development," IEEE Trans. Software Eng., vol. SE-12, No. 2, February 1986.
13 M. Shaw, "Abstraction techniques in modern programming languages," IEEE Software, vol. 1, No. 4 ,p 10, Oct. 1984.

Distribution System Design Optimization for Cold Load Pickup

Jiryes J. Wakileh   Anil Pahwa
Department of Electrical and Computer Engineering
Kansas State University
Manhattan, KS 66506

Abstract— **This paper presents a study on the optimal design of power distribution systems. Distribution substation transformer size and feeder sectionalizing switches are selected such that a cold load pickup situation can be handled while minimizing an annual cost function. The cases of a distribution substation with one or more transformer(s) are considered. In the latter case, the optimization problem is formulated based on the worst-case scenario of having one of the transformers out of service.**

Keywords— **Power distribution systems, Optimal design, Cold load pickup.**

I. Introduction

Cold load pickup is a term given to the event of restoring power to a distribution system that has experienced a long-duration outage involving a large number of customers, and resulting in the distribution substation load being several times the rating of its transformer(s). This situation is encountered because during normal conditions, diversity exists among the loads served by a substation, so that the maximum coincident demand is less than the sum of the individual maximum demands. However, following an outage, a distribution substation has to supply power for the simultaneous starting of all ON connected loads, of which the thermostatically-controlled ones (such as air conditioners, water heaters and refrigerators) constitute the largest portion.

Sectionalizing a distribution system, in other words dividing it into sections, and restoring power to these sections in steps represents one of the solutions used to overcome excessive substation overloading due to cold load pickup.

The issue of cold load pickup has been dealt with over the years. For example, researchers studied and analyzed cold load pickup situations and further developed models to characterize cold load pickup [1]. Others investigated the effects of cold load pickup on the distribution substation transformer [2]. Also, individuals developed algorithms for the automation of distribution systems [3] as well as the quick identification of system faults [4], thus reducing the severity of power outages. Later, several IEEE papers focused on the development of search techniques, heuristic methods and knowledge-based approaches [5], [6], [7] that seek quick service restoration in distribution systems. Finally, Uçak and Pahwa [8] developed a numerical technique to calculate the restoration times of outaged-sections and to find the restoration sequence that minimizes customer average interruption duration or total restoration time.

This paper deals with the problem of cold load pickup from a design point of view. Transformer size and feeder sectionalizing switches are selected to handle a cold load pickup situation while minimizing an annual cost function that is composed of the cost of components, the net cost of energy interruption (cost of energy interruption less the revenue of energy sold during restoration), and the cost of transformer overloading.

96 WM 246-9 PWRS A paper recommended and approved by the IEEE Power System Engineering Committee of the IEEE Power Engineering Society for presentation at the 1996 IEEE/PES Winter Meeting, January 21-25, 1996, Baltimore, MD. Manuscript submitted June 29, 1995; made available for printing December 5, 1995,

II. Optimal System Design

It is desired to find the optimal size of substation transformer(s) and number of sectionalizing switches that result in minimizing a total annual cost that is broken down into the following components:

- Cost of transformers and sectionalizing switches
- Cost of energy interruption
- Revenue of energy sold during power restoration
- Cost of transformer loss of life due to overloading.

Thus, our objective function can be formulated as

$$\min Z = x\, S_T\, f(S_T) + K_s N_s + \rho(K_c\,\mathrm{TEI} - K_e\, E(t) + x\, K_{lol})$$

where

Z	is the total cost function, \$/year
x	is the number of substation transformers
S_T	is the transformer rating, MVA
$f(S_T)$	is the transformer cost function, \$/MVA.year
K_s	is the cost of a sectionalizing switch, \$/switch.year
N_s	is the number of sectionalizing switches
ρ	is the average number of extended outages per year, 1/2
K_c	is the cost of service interruption to customers, \$/kWh
TEI	is the total energy interruption, kWh/outage
K_e	is the cost of energy to customers, 0.05 \$/kWh
$E(t)$	is the energy sold during restoration, kWh/outage
K_{lol}	is the cost of transformer loss of life, \$/outage.

The different components of the objective function will be discussed thereafter. Moreover, since this paper addresses a design problem, all sections will be considered identical.

III. Transformer Cost

A preliminary step involves obtaining the transformer cost function. The data given in Table I is for a 115 kV Δ/12.47 kV Y OA/FA/FA 65 °C transformer and was obtained from a utility-type transformer manufacturer. For a 12 % interest rate and a 30-year life, Burke [9] calculates the total levelized annual carrying charges factor as 20.19 %. The annual cost/MVA, shown in the fourth column of the table, has been calculated using this factor.

A C program was written to perform least-squares approximation on the data. Both linear and quadratic functions were fit to the data. These in addition to the original data points are plotted in Figure 1. Transformer cost can, now, be approximated as

$$f(S_T) = K_{T0} + K_{T1} S_T + K_{T2} S_T^2 = \begin{cases} 5283.52 - 70.28 S_T & \text{, linearly} \\ 7199.263 - 235.015 S_T + 2.728 S_T^2 & \text{, quadratically.} \end{cases} \quad (1)$$

Reprinted from *IEEE Transactions on Power Systems,* Vol. 11, No. 4, pp. 1879–1884, November 1996.

TABLE I

TRANSFORMER COSTS AND CALCULATED ANNUAL CHARGES.

S_T, MVA	Cost		
	\$	\$/MVA	\$/MVA.year
7.5/8.5/9.5	246000	25894.7	5228.15
12/16/20	350000	17500	3533.25
15/20/25	380000	15200	3068.88
24/32/40	430000	10750	2170.42
30/40/50	560000	11200	2261.28

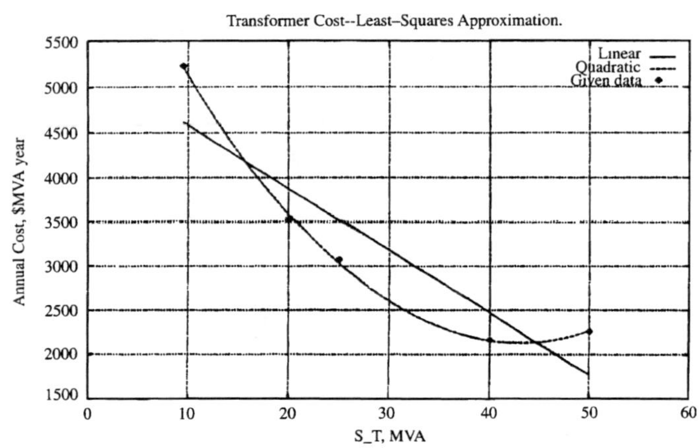

Fig. 1. Transformer least-squares approximated data.

Also, taking the cost of a sectionalizing switch to be \$ 11000, and using the 0.2019 factor, one finds $K_s = 2220.9$.

IV. BACKGROUND

Consider a distribution substation with x transformers, that has experienced an extended outage with a duration of T_{out}. If we shift our reference so that the instant power restoration starts corresponds to zero time, and with only one section being restored, then the transformer load before and after the outage will be as shown in Figure 2, where Δt is the undiversified load duration [10].

When n sections are to be restored by x transformers, each transformer would be responsible for restoring n/x of the sections. Transformer load would then be the sum of the loads of n/x sections. Based on the work of Uçak [10] and considering identical sections, the load of any section i following an extended outage would be [11]

$$S_i(t) = \frac{1}{n}\{S_U\,[u(t-T_i) - u(t-t_i)] + [S_D + (S_U - S_D)e^{-\alpha(t-t_i)}]\,u(t-t_i)\} \quad (2)$$

where

$S_i(t)$ is the load of the ith section

T_i is the restoration time of the ith section, minutes

t_i is the time at which the load of the restored i^{th} section begins to gain diversity

α is the load rate of decay

$1/\alpha$ is the load time constant

S_U is the substation total undiversified load

S_D is the substation total diversified load

n is the total number of sections

$u(t)$ is a unit step function, $u(t) = \begin{cases} 1 & \text{for } t \geq 0 \\ 0 & \text{for } t < 0. \end{cases}$

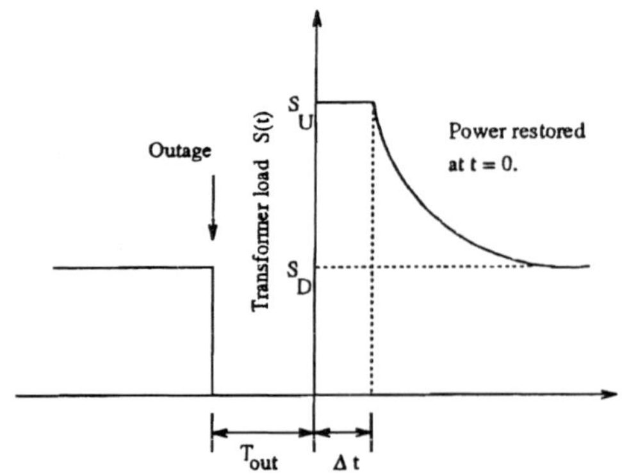

Fig. 2. Transformer load pre and post an outage.

If the transformer maximum capacity S_{Tm} is not to be exceeded, then the total number of sections restored in the first step would be [11]

$$s = x\,Floor(\frac{nS_{Tm}}{S_U}) \quad , \quad S_{Tm} = K_p\,S_T \quad (3)$$

where

K_p is the transformer peak loading factor

S_T is the transformer nameplate rating.

Section k of n identical sections is restored at time $t = T_k$, called the restoration time of section k, when [11]

$$T_k = \begin{cases} T_0 = 0 & , k \leq \frac{s}{x} \\ -\frac{1}{\alpha}\ln\dfrac{nS_{Tm} - S_U - (k-1)S_D}{(S_U - S_D)\sum_{i=1}^{k-1} e^{\alpha(T_i+\Delta t)}} & , \frac{s}{x} < k \leq \frac{n}{x} \end{cases} \quad (4)$$

Total customer interruption duration and customer average interruption duration index are, respectively, given by [11]

$$\text{TCID} = \frac{x\,C}{n}\sum_{i=1}^{n/x} T_i \quad , \quad \text{CAIDI} = \frac{x}{n}\sum_{i=1}^{n/x} T_i \quad (5)$$

where C is the total number of customers in the system.

Total energy interruption is calculated as [11]

$$\text{TEI} = \frac{1000\,x\,S_D\,pf}{60\,n}\sum_{i=1}^{n/x} T_i = 1000\,S_D\,pf\,\frac{\text{CAIDI}}{60} \quad \text{kWh} \quad (6)$$

where

pf is the power factor as measured at the distribution s/s

CAIDI is the customer average interruption duration, minutes.

Assuming a constant power factor, the total energy supplied while n sections are being restored by x transformers is [11]

$$E(t) = pf\,[(S_U - S_D)\Delta t + S_D(t - \frac{x}{n}\sum_{i=1}^{n/x} T_i) + \frac{S_U - S_D}{\alpha}(1 - \frac{x}{n}\sum_{i=1}^{n/x} e^{-\alpha(t-T_i-\Delta t)})] \quad (7)$$

Finally, Substation transformer loss of life due to overloading is taken into consideration by numerically solving the differential equations for top-oil temperature rise over ambient and hottest-spot temperature rise over top-oil [11]. Transformer loss of life is then calculated as [11]

$$\% \text{ LOL in } (t_f - t_0) \text{ hrs} = 100 \sum_{i=0}^{\text{pts}} 10^{-\left(\frac{A}{T_{h,s}(i)+273}+B\right)} \cdot h \qquad (8)$$

where h is the step size when solving using (pts) points in the time interval $[t_0, t_f]$, $h = (t_f - t_0)/\text{pts}$.

The cost of loss of life would therefore be

$$K_{lol} = \frac{S_T\, f(S_T)}{\text{ACCF}} \cdot \frac{\text{LOL}}{100} \qquad (9)$$

where ACCF is the annual carrying charges factor.

V. Cost of Service Interruption to Customers

According to Chen et al. [12], the cost of energy interruption to customers depends on outage duration. Taken from [12], residential customer interruption costs in 1988 \$ are shown in Table II for various outage durations. The corresponding costs in 1995 actual (current) and real (constant) \$ are calculated using a combined interest rate of 12 % and a general price inflation rate of 8 %.

TABLE II

Cost of Service Interruption to Residential Customers.

Outage duration hours	Interruption cost per kWh		
	1988 \$	1995 A\$	1995 R\$
1	0.48	1.06	0.619
2	0.72	1.59	0.929
4	1.20	2.65	1.548
8	1.98	4.38	2.554

VI. Problem Formulation

For a distribution substation with one transformer, as depicted in Figure 3, we need that transformer to be capable of supplying the substation load in the steady state, namely the substation total diversified load S_D. To account for a cold load pickup situation, an upper bound on the transformer size is S_U, the substation total undiversified load. Thus, we have the constraints

$$S_D \leq S_T \leq S_U$$

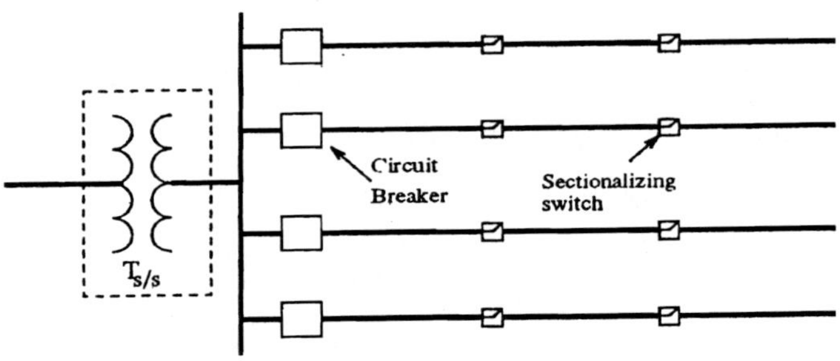

Fig. 3. Single-line diagram of a distribution system.

A third constraint comes into effect from Equation 4, where for the restoration times T_k to be calculated for all sections $k = 1, \ldots, n$, we need the argument of the natural log function to be positive. Since the denominator is for sure positive, we are left with the requirement that the numerator be positive, that is

$$nS_{Tm} - S_U - (k-1)S_D > 0$$

$$\Rightarrow \quad S_{Tm} > \frac{S_U}{n} + (k-1)\frac{S_D}{n} \qquad , s < k \leq n$$

However, it is seen that satisfying the above constraint for $k = n$ ensures its satisfaction for $k < n$. Therefore, all what we need is

$$S_{Tm} > \frac{S_U}{n} + (n-1)\frac{S_D}{n}$$

Inspecting this constraint carefully, one can see that it ensures all n sections are restored. The constraint actually asks for a transformer whose size is capable of restoring the undiversified load of the nth section following the restoration of the diversified load of $(n-1)$ identical sections.

With the number of sections n being related to the number of sectionalizing switches N_s as $n = N_s + f$, where f is the number of feeders, and $S_{Tm} = K_p S_T$, the above constraint becomes

$$(N_s + f)(K_p S_T - S_D) > S_U - S_D$$

For a distribution substation with x transformers, our design will be based on the worst case of having a cold load pickup situation while one of the substation transformers is out of service. Each remaining-in-service transformer would then be responsible for restoring $n/(x-1)$ sections of which $s/(x-1)$ will be restored in the first step. The total number of sections restored in the first step would then be given by Equation 3 with x replaced by $(x-1)$.

Reformulating our problem so that the constraints take care of the outage of one transformer, we need the remaining $(x-1)$ transformers to pick up the load of the original x transformers. [1] Based on the above discussion, we have the constraints

$$S_D \leq (x-1)S_T \leq S_U$$

and similar to the case of a substation with one transformer, the third constraint is

$$S_{Tm} > \frac{S_U}{n} + \left(\frac{n}{x-1} - 1\right)\frac{S_D}{n}$$

which, with $n = N_s + f$, becomes

$$(N_s + f)(K_p S_T - \frac{S_D}{x-1}) > S_U - S_D$$

This constraint says that a transformer should be of a size to enable it restore the diversified load of $(n/(x-1) - 1)$ of the sections plus the undiversified load of an additional section, thus restoring a total of $n/(x-1)$ sections.

Denoting the number of in-service transformers by x_o, where

$$x_o = \begin{cases} x & \text{for } x = 1 \\ x-1 & \text{for } x \geq 2, \end{cases}$$

and aiming at minimizing the total annual cost, our problem can be formulated as

$$\begin{aligned} \min Z = \; & x S_T\, f(S_T) + K_s N_s + \rho(K_c \,\text{TEI} - K_e E(t) + x_o K_{lol}) \\ s.t. \quad & x_o S_T \leq S_U \\ & x_o S_T \geq S_D \\ & (N_s + f)(K_p S_T - \frac{S_D}{x_o}) > S_U - S_D \\ & 1.0 \leq Kp \leq 2.0 \\ & S_T, N_s \geq 0. \end{aligned}$$

[1] In a steady state situation, and for the case of a distribution substation with x transformers, this will result in each transformer being $1 - 1/x$ loaded (50 % for $x = 2$ and 66.7 % for $x = 3$). The outage of one transformer results in its load being carried by the other transformers, each of which becomes 100 % loaded.

where

T_k are the restoration times for identical sections, as given by Equation 4 and with x replaced by x_o

TCID is the total customer interruption duration as given by Equation 5 with x replaced by x_o

CAIDI is the average interruption duration per customer as given by Equation 5 with x replaced by x_o

TEI is the total energy interruption as given by Equation 6 with x replaced by x_o

$E(t)$ is the energy sold during the restoration process of n identical sections, as given by Equation 7 and with x replaced by x_o

K_{lol} is the cost of transformer loss of life due to overloading as given by Equation 9.

VII. A Simplified Cost Function

In the following, we will study a simplified form of the cost function consisting of the cost of substation transformer and sectionalizing switches only. That is, our objective function is

$$\min Z = f(S_T)\,S_T + K_s N_s = (K_{T0} + K_{T1} S_T + K_{T2} S_T^2) S_T + K_s N_s$$

Specifically, five cases will be considered here:

- Z continuous and differentiable, that is both $f(S_T)$ and N_s are continuous with $f(S_T)$ quadratic
- Z continuous and differentiable, that is both $f(S_T)$ and N_s are continuous with $f(S_T)$ linear
- $f(S_T)$ continuous quadratic and N_s discrete
- $f(S_T)$ continuous linear and N_s discrete
- Both $f(S_T)$ and N_s discrete.

The last case is the one which represents the real situation where the transformer cost function is discrete and corresponds to the standard transformer sizes available in the market, and the number of sectionalizing switches is integer-valued.

A. $f(S_T)$ and N_s continuous, $f(S_T)$ quadratic

For this case, we have

$$\min Z = K_{T0} S_T + K_{T1} S_T^2 + K_{T2} S_T^3 + K_s N_s$$

with K_{T0}, K_{T1} and K_{T2} as given by Equation 1. Taking $\mathbf{X}$ to be

$$\mathbf{X} = \begin{bmatrix} S_T \\ N_s \end{bmatrix}$$

the gradient would be given by

$$\nabla_{\mathbf{X}} Z = \begin{bmatrix} \dfrac{\partial Z}{\partial S_T} \\[2ex] \dfrac{\partial Z}{\partial N_s} \end{bmatrix} = \begin{bmatrix} K_{T0} + 2K_{T1} S_T + 3K_{T2} S_T^2 \\[2ex] K_s \end{bmatrix}$$

For a minimum point, we need $\dfrac{\partial^2 Z}{\partial S_T^2} \geq 0$ and $\dfrac{\partial^2 Z}{\partial N_s^2} \geq 0$, so that

$$2K_{T1} + 6K_{T2} S_T \geq 0 \qquad \text{or} \qquad S_T \geq \frac{-K_{T1}}{3K_{T2}}$$

Alternatively, one can form the Hessian as

$$\mathbf{H} = \begin{bmatrix} \dfrac{\partial^2 Z}{\partial S_T^2} & \dfrac{\partial^2 Z}{\partial S_T \partial N_s} \\[2ex] \dfrac{\partial^2 Z}{\partial N_s \partial S_T} & \dfrac{\partial^2 Z}{\partial N_s^2} \end{bmatrix} = \begin{bmatrix} 2K_{T1} + 6K_{T2} S_T & 0 \\[2ex] 0 & 0 \end{bmatrix}$$

so that,

$$\mathbf{X}^T \mathbf{H}\, \mathbf{X} = 2K_{T1} S_T^2 + 6K_{T2} S_T^3 = 2S_T^2 (K_{T1} + 3K_{T2} S_T) \begin{cases} < 0 & , S_T < -K_{T1}/3K_{T2} \\ = 0 & , S_T = -K_{T1}/3K_{T2} \\ > 0 & , S_T > -K_{T1}/3K_{T2} \end{cases}$$

which shows that $\mathbf{H}$ is indefinite and $S_T = -K_{T1}/3K_{T2}$ ($= 28.72$ MVA for our data) is an inflection point. Bearing in mind that a necessary condition for a stationary point in a minimization problem is a semi-positive definite $\mathbf{H}$, or $\mathbf{X}^T \mathbf{H}\, \mathbf{X} \geq 0$, we can say that looking over the range $S_T \geq -K_{T1}/3K_{T2}$, the obtained minimum would be a relative not a global one. The cost function, plotted in Figure 4, validates the above discussion.

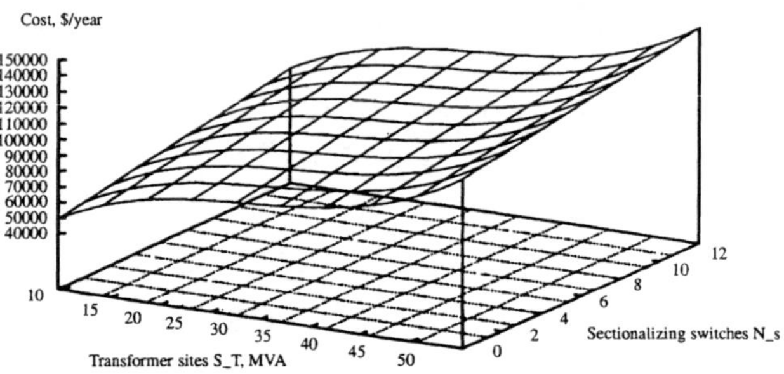

Fig. 4. Cost function: $f(S_T)$ continuous and quadratic, N_s continuous.

B. $f(S_T)$ and N_s continuous, $f(S_T)$ linear

This time $f(S_T)$ is linear, so that $K_{T2} = 0$ and

$$\mathbf{X}^T \mathbf{H}\, \mathbf{X} = 2K_{T1} S_T^2 < 0 \qquad \forall\, S_T \neq 0$$

This means that $\mathbf{H}$ is negative definite, which is a necessary condition for a maximization problem. Hence, using a linear least-squares function to approximate the transformer cost is not a good choice for our purpose. This can also be made clear by the concavity of the total cost plot, shown in Figure 5.

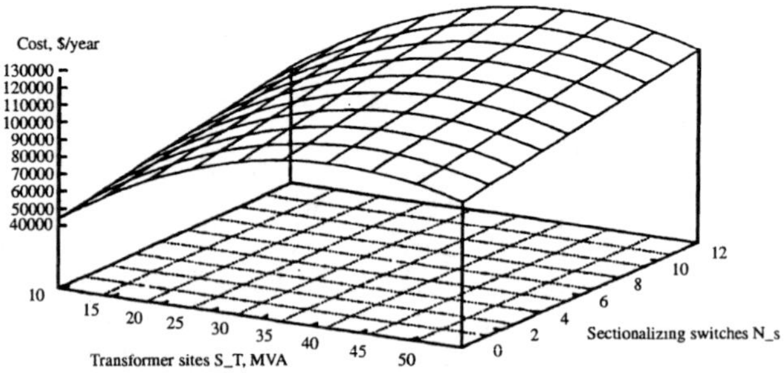

Fig. 5. Cost function: $f(S_T)$ continuous and linear, N_s continuous.

C. $f(S_T)$ and N_s discrete

This time the cost surface, plotted in Figure 6 consists of points that correspond to discrete values of S_T and N_s.

The cases of $f(S_T)$ continuous, N_s discrete, with $f(S_T)$ being quadratic or linear, are similar to the first two cases, except that the cost function is no longer a surface but consists of strings.

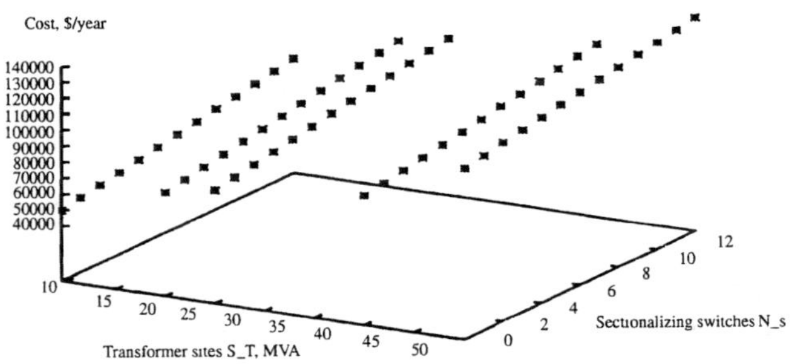

Fig. 6. Cost function: $f(S_T)$ and N_s discrete.

It is worth mentioning here that when other components are included in the cost function, its characteristics will change. Thus, a numerical technique will be required to find the optimal solution.

VIII. Analysis of an Example System

As a preliminary step, a C program whose flowchart is shown in Figure 7, was written to solve the optimization problem through **enumerating all possible combinations**. The developed program was used to run several studies on an example system with $S_D = 17$ MVA, $S_U = 49$ MVA, $1/\alpha = 2$ hours and $\Delta t = 1/2$ hour. The other system characteristics as well as C program menu and sample output will not be included for lack of space.

With the cost of transformer loss of life due to overloading not being considered, the results for different numbers of transformers and feeders are summarized in Table III for a quadratic transformer cost. Inspecting the output data reveals that:

- The minimum cost corresponds to the smallest possible transformer being overloaded to the maximum possible limit (2.0 pu), and the maximum cost corresponds to the largest possible transformer without being overloaded.

With the transformer loss of life due to overloading considered, the corresponding results are summarized in Table IV. These show that the optimal solution is now represented by a larger size transformer overloaded to a reduced level with almost no change on the number of sectionalizing switches. The reason for that can be one of two:

- The cost associated with the smaller size transformer overloaded to 2.0 pu did not favor its choice as optimal when the cost of loss of life was considered
- The smaller size transformer overloaded to 2.0 pu no longer represents a feasible solution for violating ANSI/IEEE C57.92-1981 [13] recommended limits.

Also, for a certain number of transformers, the minimum cost is associated with an increased number of feeders with no sectionalizing switches, meaning that the feeder circuit breakers would do the job.

A. Effect of Outage Duration

According to Table II, outage durations of 1, 2 and 4 hours result in service interruption costs of 1.06, 1.59 and 2.65 $/kWh, respectively. Keeping ρ constant at 1/2, the effect of reducing K_c (corresponding to a shorter outage duration) can be seen by comparing Table V to the base case of Table IV, and is one of three depending on the combination of x and f:

- No effect at all on S_T, K_p, N_s or the cost, meaning that all sections are restored at $T_0 = 0$ so that TEI=0
- No effect on S_T, K_p or N_s but the cost is lower meaning that not all sections are restored at $T_0 = 0$ so that TEI $\neq 0$
- A smaller size transformer overloaded to a reduced level with no effect on N_s. The cost is of course lower.

B. Effect of the Number of Extended Outages

Keeping K_c constant at 1.59 $/kWh, the effect of reducing ρ to 1/5 can be seen by comparing Table VI to the base case of Table IV, and is one of two depending on the combination of x and f:

- No effect on the transformer size or peak loading factor but the cost is lower meaning that not all sections are restored at $T_0 = 0$ so that TEI $\neq 0$
- A smaller size transformer overloaded to a reduced level. The cost is of course lower.

The number of sectionalizing switches was not affected at all.

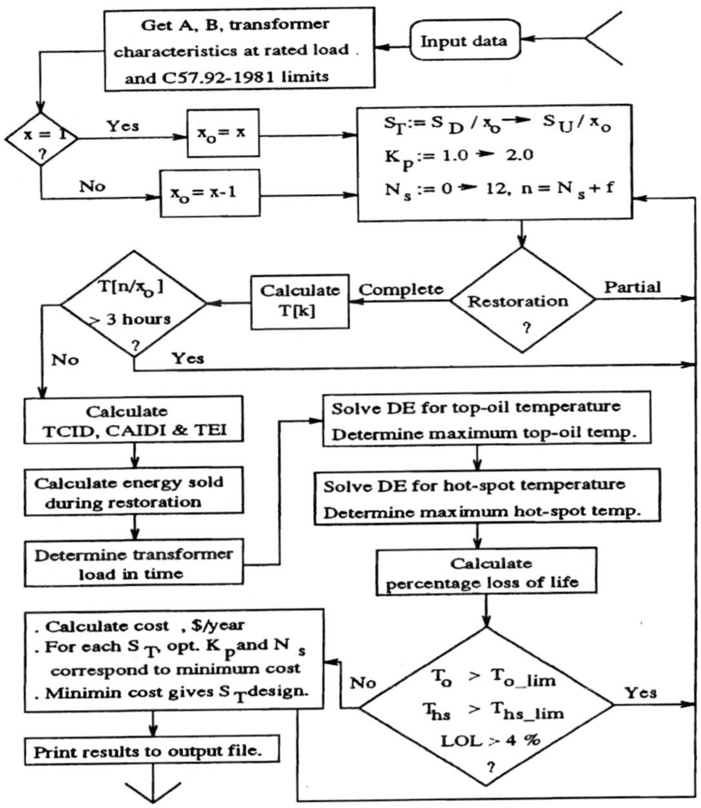

Fig. 7. Flowchart of the developed C program.

IX. Further Research

The following is an overview of the work to be pursued:

- Solving the optimization problem through a technique that takes care of the special nature of the problem.
- Studying and analyzing a real-life system. System restoration following an outage will be considered subject to voltage drop and current capacity constraints. The system will be compared against the output of our optimal design.

X. Conclusions

This paper presented a formulation of an optimization problem for the design of power distribution systems, taking cold load pickup into consideration. Solution of the optimization problem results in the distribution substation transformer size and number of sectionalizing switches that minimize an annual cost function representing the cost of substation transformer(s), the cost of sectionalizing switches, the net cost of energy interruption and the cost of transformer overloading.

Acknowledgment

This research is funded by the National Science Foundation, grant # ECS-9311257.

References

[1] J. Law, D. Minford, L. Elliott and M. Storms, "Measured and Predicted Cold Load Pickup and Feeder Parameter Determination Using the Harmonic Model Algorithm", IEEE Paper 94 SM 546-2 PWRS, IEEE/PES 1994 Summer Meeting, San Francisco, CA, July 1994.

[2] J. Aubin, R. Bergeron, and R. Morin, "Distribution Transformer Overloading Capability under Cold Load Pickup Conditions", *IEEE Transactions on Power Delivery*, Vol. 5, pp. 1883–1891, November 1990.

[3] H. Kim, Y. Ko and K. Jung, "Algorithm of Transferring the Load of the Faulted Substation Transformer Using the Best-First Search Method", IEEE Paper 91 SM 403-6 PWRD, IEEE/PES 1991 Summer Meeting, San Diego, CA, July/August 1991.

[4] A. A. Girgis and M. B. Johns, "A Hybrid Expert System for Faulted Section Identification, Fault Type Classification, and Selection of Fault Location Algorithms", *IEEE Transactions on Power Delivery*, Vol. PWRD-4, pp. 978–985, April 1989.

[5] K. Aoki et al., "A New Algorithm for Service Restoration in Distribution Systems", *IEEE Transactions on Power Delivery*, Vol. 4, pp. 1832–1839, July 1989.

[6] A. L. Morelato and A. Monticelli, "Heuristic Search Approach to Distribution System Restoration", *IEEE Transactions on Power Delivery*, Vol. 4, pp. 2235–2241, October 1989.

[7] Chen-Ching Liu, Seung Jae Lee, and S. S. Venkata, "An Expert System Operational Aid for Restoration and Loss Reduction of Distribution Systems", *IEEE Transactions on Power Systems*, Vol. 3, pp. 619–626, May 1988.

[8] C. Uçak and A. Pahwa, "An Analytical Approach for Step-by-Step Restoration of Distribution Systems Following Extended Outages.", *IEEE Transactions on Power Delivery*, Vol. 9, pp. 1717-1723, July 1994.

[9] J. Burke, *Power Distribution Engineering: Fundamentals and Applications*, Marcel Dekker, Inc., New York, NY, 1994.

[10] C. Uçak, *Restoration of Distribution Systems Following Extended Outages*, Ph.D. Dissertation, Kansas State University, 1994.

[11] J. J. Wakileh, "Substation Transformer Loading Following Extended Outages", submitted to the *IEEE PES Winter Meeting*.

[12] Rong-Liang Chen, K. Allen and R. Billinton, "Value-based Distribution Reliability Assessment and Planning", *IEEE PES T&D Conference and Exposition*, Chicago, IL, April 1994.

[13] IEEE Guide for Loading Mineral-Oil-Immersed Power Transformers up to and Including 100 MVA with 55 or 65 °C Average Winding Rise, ANSI/IEEE Standard C57.92-1981.

TABLE III

OPTIMAL VALUES OF S_T, K_p AND N_s, $f(S_T)$ QUADRATIC.

x	f = 1	2	3	6
1	$S_T = 24.6$ $K_p = 2.0$ $N_s = 0$ $/y 72240.9	24.5 2.0 0 72181.6	24.5 2.0 0 72181.6	20.5 2.0 0 71785.8
2		$S_T = 18.4$ $K_p = 2.0$ $N_s = 2$ $/y 147441		17.0 2.0 0 140921
3			$S_T = 8.5$ $K_p = 2.0$ $N_s = 3$ $/y 154301	8.5 2.0 0 147639

TABLE IV

OPTIMAL VALUES OF S_T, K_p AND N_s, OVERLOADING CONSIDERED, $f(S_T)$ QUADRATIC, $K_c = 1.59$, $\rho = 1/2$.

x	f = 1	2	3	6
1	$S_T = 24.6$ $K_p = 2.0$ $N_s = 0$ $/y 72847.5	24.5 2.0 0 72835.6	24.5 2.0 0 72835.6	24.5 2.0 0 72835.6
2		$S_T = 24.5$ $K_p = 2.0$ $N_s = 0$ $/y 148268		22.0 1.88 0 148077
3			$S_T = 9.4$ $K_p = 1.8$ $N_s = 3$ $/y 165521	9.4 1.8 0 158858

TABLE V

OPTIMAL VALUES OF S_T, K_p AND N_s, OVERLOADING CONSIDERED, $f(S_T)$ QUADRATIC, $K_c = 1.06$, $\rho = 1/2$.

x	f = 1	2	3	6
1	$S_T = 24.6$ $K_p = 2.0$ $N_s = 0$ $/y 72847.5	24.5 2.0 0 72835.6	24.5 2.0 0 72835.6	24.5 2.0 0 72835.6
2		$S_T = 24.5$ $K_p = 2.0$ $N_s = 0$ $/y 148268		19.6 1.67 0 146808
3			$S_T = 9.4$ $K_p = 1.8$ $N_s = 3$ $/y 161269	9.4 1.8 0 154607

TABLE VI

OPTIMAL VALUES OF S_T, K_p AND N_s, OVERLOADING CONSIDERED, $f(S_T)$ QUADRATIC, $K_c \doteq 1.59$, $\rho = 1/5$.

x	f = 1	2	3	6
1	$S_T = 24.6$ $K_p = 2.0$ $N_s = 0$ $/y 74434.1	24.5 2.0 0 74393.7	24.5 2.0 0 74393.7	22.0 1.88 0 73967.7
2		$S_T = 19.0$ $K_p = 1.98$ $N_s = 0$ $/y 148551		19.6 1.67 0 145565
3			$S_T = 9.4$ $K_p = 1.8$ $N_s = 3$ $/y 158717	9.4 1.8 0 152055

Distribution Feeder Reconfiguration For Service Restoration And Load Balancing

Qin Zhou
Consultant
San Francisco, California

Dariush Shirmohammadi
Shir Consultants
San Ramon, California

W.-H. Edwin Liu
Pacific Gas and Electric Company
San Francisco, California

Abstract - This paper describes two feeder reconfiguration algorithms for the purpose of service restoration and load balancing in a real-time operation environment. The developed methodologies combine optimization techniques with heuristic rules and fuzzy logic for efficiency and robust performance. Many of practical operating concerns of feeder reconfiguration and the coordination with other distribution automation applications are also addressed. The developed algorithms have been implemented as a production grade software. Test results on PG&E distribution feeders show that the performance is efficient and robust.

Keywords: distribution automation, distribution planning and operation, feeder reconfiguration, service restoration, load balancing, optimization, heuristic rules, fuzzy logic.

I. INTRODUCTION

Feeder reconfiguration entails altering the topological structure of distribution feeders by changing the open/close status of the switches under both normal and abnormal operating conditions. In this paper we will present the algorithms for the following functions:

1. Service restoration in respond to a fault by 1) isolating the faulted area(s), 2) supplying power to the non-faulted out-of-service area(s) , and 3) minimizing the inevitable load shedding.
2. Load balancing in order to relieve the overloads in the distribution system.

Most of the algorithms presented so far in the literature are basically planning tools and they are heuristic based methods. The algorithms presented in this paper provide a real-time operation decision support tool using a combination of optimization techniques with heuristic rules and the fuzzy logic approach to obtain efficient and robust performance. The practical concerns such as operating constraints, proper selection of a sequence for implementing switching operations, and the coordination with other distribution automation functions are also addressed.

96 SM 488-7 PWRS A paper recomemended and approved by the IEEE Power System Engineering Committee of the IEEE Power Engineering Society for presentation at the 1996 IEEE/PES Summer Meeting, July 28 - August 1,1996, in Denver, Colorado. Manuscript submitted January 2, 1996; made available for printing April 23, 1996.

We developed a production grade software, named FRECON, based on the algorithms presented in this paper and have implemented it for use with PG&E distribution systems.

II. STATE-OF-THE-ART TECHNIQUES

Most of the algorithms in the literature are heuristic search techniques based on a series of switching operations [1]-[4]. The major advantage of these algorithms is that they are very fast. Other algorithms use a blend of optimization and heuristics to efficiently determine the switching operations for service restoration [5]. The advantage of this method is that its solution is independent of the initial system configuration. The main problem for this kind of method is that if a feasible solution does not exist, it cannot provide a "best possible solution" which can minimize the inevitable load drop. Hence, It is not suitable for the real-time application since currently only less than 5% of the switches in the distribution systems are remote-controlled.

Definitions:

In the follows we define a number of terms commonly used in this paper in order to make the presentation of this paper clearer:

Tie switch: A normally open switch connecting two adjacent feeders.

Faulted area: A portion of the feeder which is experiencing a permanent fault.

Unfaulted out-of-service area: A portion of the feeder downstream from the faulted area which can be separated from the faulted area by opening a normally closed switch.

Adjacent feeder: A feeder with tie switches to the faulted or overloaded feeder.

Secondary feeder: A feeder with tie switches to the adjacent feeder(s) but no tie switch to the faulted or overloaded feeder.

Boundary switch: A normally open switch that connects the out-of-service area(s) or the area(s) with overloaded line sections to the adjacent feeders or an unfaulted section of the faulted or overloaded feeder.

Equivalent load: The summation of all the loads on the line sections enclosed by switches (open or closed) (see Appendix A).

Reprinted from *IEEE Transactions on Power Systems,* Vol. 12, No. 2, pp. 724–729, May 1997.

III. METHODOLOGY

In order to account for all concerns that prevail in a real-time operating environment, we pursued the following strategy in developing the proposed reconfiguration algorithms:

- Combine the optimization techniques with heuristic rules and fuzzy logic approach to find a solution which could restore service or relieve the overloads in the shortest period of time with minimum number of switching operations possible.
- Account for all operating constraints.
- Consider remote-controlled switches first and seek a solution using remote-controlled switches only. If remote-controlled switches alone cannot fully restore the service or relieve the overloads, consider all the switches.
- Relieve overload in line sections farther away from the substation first (for load balancing function). Relieving the overloads on these line sections may automatically relieve some or even all the overloads in line sections close to the substation.
- If the adjacent feeders cannot pick up the loads to restore the service or relieve the overloads, try to transfer load to secondary feeders in order to increase the capacity of the adjacent feeders to pick up the loads.
- The solution should be verified for protection requirements by a relay coordination program.
- If the feasible solution cannot be found, provide the solution that will minimize the inevitable load shedding.

Service Restoration

We developed an Integer Linear Programming (ILP) technique which guarantees a final solution for service restoration problem with the objective of minimizing the number of switching operations. This method can give multiple solutions allowing the operator to pick the best one based on his/her knowledge of the distribution system. We used a constant current load model in this model.

The ILP method for the service restoration problem can be considered as an assignment problem where the problem is how to assign loads in the unfaulted out-of-service area(s) to boundary switches.

Let

$$x_{ij} = \begin{cases} 1 & \text{load } j \text{ being assigned to boundary switch } i \\ 0 & \text{otherwise} \end{cases}$$

$$x_{i0} = \begin{cases} 1 & \text{boundary switch } i \text{ is closed} \\ 0 & \text{otherwise} \end{cases} \qquad (1)$$

The formulation of optimization problem to find the minimum number of switching operations for service restoration, can be written as,

Objective:

$$Min \sum_{i=1}^{n} x_{i0} \qquad (2)$$

Subject to:

1. Radial system constraints: Each load must only be assigned to one boundary switch.
2. Connectivity constraints: Entails that if load j is assigned to boundary switch i, then all the loads along the path from load j to boundary switch i are also assigned to boundary switch i.
3. Boundary switch capacity constraints:
   - The total loads being assigned to a boundary switch must be less than the capacity margin of that switch.
   - The loads assigned to a set of correlated boundary switches must be less than the simultaneous capacity margin of these switches.

When using the assignment formulation, if there are N boundary switches and M closed switches, then the total number of control variables will be N*M. Thus, the numbers of control variables will increase dramatically as the numbers of boundary or closed switches increases. In order to avoid the so called "dimension curse" the switch status is used as the control variables and the following modifications are made to form the mathematical formulations and speed up the solution search.

Let

$$y_i = \begin{cases} 1 & \text{closed switch } i \text{ is closed} \\ 0 & \text{otherwise} \end{cases} \qquad i=1,2,...,m$$

$$z_j = \begin{cases} 1 & \text{boundary switch } j \text{ is closed} \\ 0 & \text{otherwise} \end{cases} \qquad j=1,2,...,n$$

The formulation of optimal solution, which finds the minimum number of switching operations, can be written as,

Objective:

$$Min \ \{ \sum_{i=1}^{m} (1-y_i) + \sum_{j=1}^{n} z_j \} \qquad (3)$$

Subject to:

1. Topology constraints as described below:

$$(1-y_i) + \sum_{k\in\Omega_i} (1-y_k) - \sum_{l\in\Psi_i} z_l \le 0 \qquad i=1,2,...,m \qquad (4)$$

Where,

Ω_i: set of closed switches downstream from closed switch i

Ψ_i: set of boundary switches downstream from closed switch i

$$\sum_{k \in loop_{ij}} (1-y_k)-(z_i+z_j)+1 \geq 0$$

$$i=1,2,...,n \quad j=i+1,i+2,...,n \quad (5)$$

Where,

$loop_{ij}$: set of the closed switches in the path from boundary switch i to boundary switch j.

Equations (4) and (5) guarantee the radial configuration and also no isolation for any switch.

2. Boundary switch capacity constraints:
   For each solution candidate the following constraints shall be checked by using the load assignment variable.
   - The total loads being assigned to a boundary switch is less than the capacity margin of that switch:

     $$\sum_{j=1}^{m} x_{ij} I_j \leq C_i \qquad i=1,2,...,n \qquad (6)$$

     Where,

     m: the number of equivalent loads

     C_i: the capacity margin of boundary switch i

     I_j: equivalent load j
   - The loads assigned to a set of correlated boundary switches are less than the simultaneous capacity margin of these switches:

     $$\sum_{i \in \omega_k} \sum_{j=1}^{m} x_{ij} I_j \leq D_k \qquad (7)$$

     Where,

     k: node number of the node upstream from a number of boundary switches

     D_k: the minimum capacity of line sections upstream from node k

     ω_k: set of boundary switches downstream from node k

The solution candidate that satisfies the constraints (4) - (7) and also has the minimum numbers of switching operations will be the optimal solution.

An implicit enumeration technique with breath first search logic is used to search the minimum number of switching operations. A set of heuristic rules has been developed to speed up the search. Some key heuristic rules are,

- Use the total loads in the unfaulted out-of-service area to estimate the minimum number of boundary switches needed.
- Ignore all closed switches that do not have downstream boundary switches and all boundary switches that cannot pick up any equivalent load due to the capacity limitation.
- Ignore all closed switches that can separate an isolated out-of-service area into two sub-areas where one of the sub-areas cannot be picked up by its adjacent boundary switch(es).

Solution Framework for Service Restoration:

The framework of service restoration module consists of the following steps:

1. Identify and isolate faulted area(s) using remote-controlled switches only.
2. Considering remote-controlled switches only, use the ILP method to search for solution(s) with minimum number of switching operations. If no feasible solution exists, then find the "best possible solution" to pick up as much load as possible.
3. Include all switches and, if possible, further reduce the size of isolated faulted area(s).
4. Consider all switches and search for solution(s) with minimum number of switching operations.
5. If no feasible solution has been found, use heuristic method to transfer loads from adjacent feeders to the secondary feeders in order to increase the load pick up capability of the adjacent feeders.
6. If the feasible solution is still not available after step 5, then pick up the solution that can restore the maximum amount of loads possible so that the inevitable load shedding will be minimum.

Load Balancing

The algorithm we have developed for load balancing can be summarized as follows:

1. Assign layers to each line section per reference [5].
2. Start with line sections (or distribution equipment) in the farthest layers from the substation and identify the overloaded line sections.
3. Determine switching operations that relieve overloads in all line sections in the same layer using the ILP method described earlier. Several solutions may be identified at this step.
4. Use Fuzzy Logic approach to select the best solution of relieving the overloads at each layer. The best solution should lead to the lowest overload for the rest of the overloaded line sections in the upper layers and meanwhile result in the minimum number of switching operations.

The mathematical formulas of ILP method have been presented in the previous section. For load balancing function the only difference is that now the objective is to remove the overload instead of restoring the service.

Fuzzy Logic Reasoning Approach:

In step 3 of the heuristic algorithm mentioned above, many solutions could be found to relieve overloads in line sections in a specific layer. Moving to line sections in the upper layers requires that one of the solutions be retained and others be discarded. We used the Fuzzy Logic method with the following criteria for selecting the best solution to carry over to the next step: 1) the number of switching operations, 2) the amount of overloads being relieved, and 3) the capacity margin of boundary switches. Thus, three membership functions were developed with the following variables:

1. Number of switching operations N (in pairs): The membership function U(N) is shown in Figure 1,

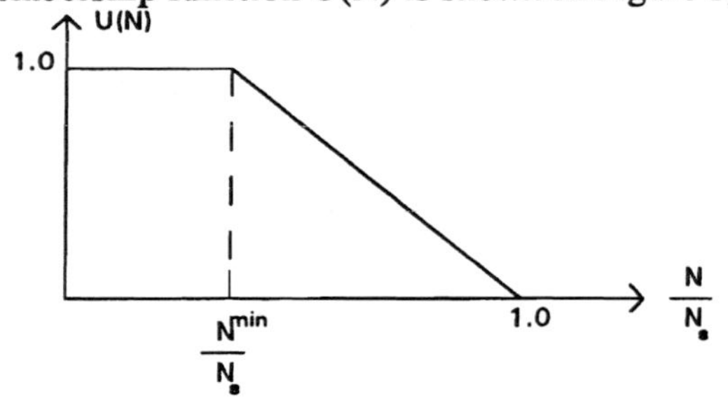

Figure 1 membership function U(N)

Where, N is the number of pairs of switching operations, N_s is the total number of available boundary switches, and N^{min} is the threshold of minimum number of pairs of switching operations defined by the operator.

2. Total amount of overloads OL: The membership function U(OL) is shown in Figure 2,

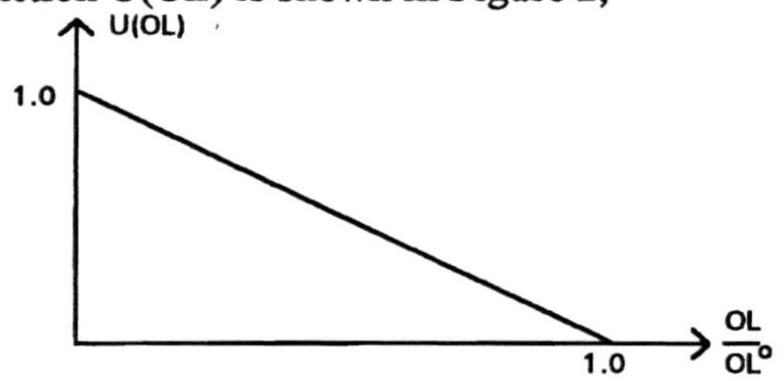

Figure 2 membership function U(OL)

Where, OL is the "total amount of overloads" defined as,

$$OL = \sum \left| \frac{J_i}{J_i^R} \right|^2 \qquad i = 1, 2, ..., n$$

where J_i is the line section current of overloaded line i, J_i^R is the line ampacity, and OL° is the "total amount of overloads" at the initial operating point.

3. Total amount of boundary switches' capacity margin CM: The membership function U(CM) is shown in Figure 3,

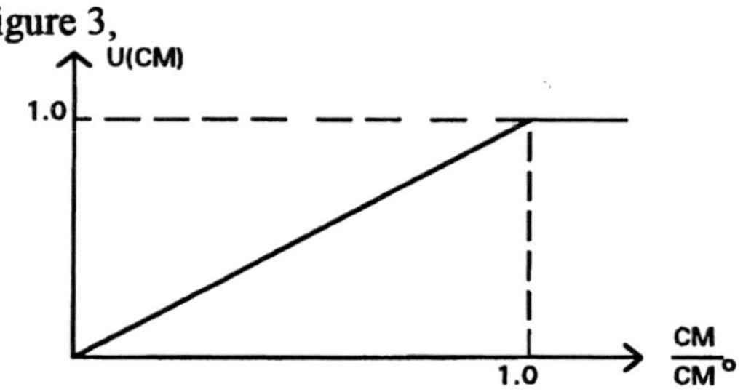

Figure 3 membership function U(CM)

Where, CM is the "total amount of boundary switches' capacity margin", and CM° is the "total amount of boundary switches' capacity margin" at initial operating condition.

The steps of applying the Fuzzy Logic approach to select the best solution of relieving the overloads are as follows,

- For each layer that has the overloads, find all the solutions that can relieve the overloads on this layer.
- Calculate U(N), U(OL), and U(CM) for each solution.
- Identify the membership value U_S for each solution. The membership value is the minimum value of U(N), U(OL), and U(CM). Thus, the membership value of solution i will be,

  $$U_{si} = \mathrm{Min}\{U(N), U(OL), U(CM)\} \qquad i = 1, 2, ..., n$$
- Select the best solution of relieving the overloads in this layer. The best solution is the one with maximum membership value. That is,

  $$U_{smsx} = \mathrm{Max}\{U_{s1}, U_{s2}, ..., U_{sn}\}$$

IV. NUMERICAL RESULTS

We developed the software package FRECON based on the service restoration and load balancing algorithms presented here. It was successfully implemented and tested on many of PG&E's distribution feeders. The following are two examples of the test results.

Case 1 service restoration:

The test system consists of two 12kv feeders with total load 4.85MVA with 920 line sections and 55 switches. The faulted feeder (feeder 1) has 430 branches and 28 switches. The total load of this feeder is 2.56MVA (354.9 amps). The other feeder (feeder 2), which is adjacent to feeder 1, has 490 branches and 32 switches. There are 5 tie switches between two feeders. Two fault locations are selected for simulation.

a) Fault close to the substation on feeder 1: In this case the entire feeder is out of service. The following are the test system characteristics for this scenario.

Simulated fault scenarios:

No. of boundary switches available for load transfer	:	5
No. of closed switches in the out-of-service area	:	20
No. of line sections in the out-of-the-service area	:	430
Total load to be restored	:	354.9 amp

Under this fault situation the entire feeder load was picked up by closing one tie switch.

b) fault at the middle of the feeder 1: In this case more than half of the feeder branches are out of service. The following are the test results.

Simulated fault scenarios:

No. of boundary switches available

for load transfer	:	4
No. of closed switches in the out-of-service area	:	8
No. of line sections in the out-of-the-service area	:	277
Total load to be restored	:	245.6 amp

The service restoration program selected two switch pairs to restore the service. The majority of the loads were picked up by one tie switch and the rest of the loads was picked up by a sectionalizing switch. It is a normall open intra-feeder switch in feeder 1.

Case 2 load balancing:
For the same test system and under another operating condition, initially there are six overloaded branches at different layers in feeder 1. The maximum overload is 89.6% above the normal line section ampacity. The first three overloads are close to each other and they are located near to the substation. The rest of overloaded line sections are located close to the end of the feeder. The load balancing program starts from line sections in layer farthest from the substation and finds the solution to relieve the overloads in that layer. In the first intermediate solution 55 amps of load is transferred to feeder 2 by a pair of switching operations.

After implementing the first switching pair and running the power flow for the new configuration three line sections remain overloaded with the maximum overload being 27.1% above the normal line section ampacity.

We noticed that solution relieves three overloads in the farthest layers and it also helps to reduce the overload values of the other three line sections. In the second intermediate solution 165 amps of load is transferred to feeder 2 by another pair of switching operations.

After implementing the second switching pair and running the power flow for the new configuration, all the remaining overloads are relieved.

V. IMPLEMENTATION ISSUES

Sequence of Implementing the Switching Operations

After service restoration or load balancing program determines the optimal solution, which is a set of open/close switching operations, it is necessary to find a sequence for implementing the switching operations. This is necessary since at any intermediate step of switching implementation we must guarantee that there are no major operating constraints violations. The answer is definitely not trivial.

We developed an algorithm to deal with this problem. The objective of this algorithm is to find the sequence of switching operations for implementing the optimal solution of service restoration or load balancing program. Following are the major steps of the algorithms for service restoration and load balancing functions:

Service Restoration:

1. Open the selected closed switches that would isolate the faulted area(s).
2. Open all the selected closed switches that are to be opened to transfer the load.
3. Close all the selected open switches that are to be closed one by one. Attention must be paid to situations where extensive changes in load levels may have occurred due to prolong outages.

Load Balancing:

1. Start the sequence of switching operations with the switch pairs that eliminate the overloads on line sections in the farthest layer from the substation.
2. Simulate the switch closing for the selected switch pair and check the following constraints violations.
   - Overloads in the loop formed by the switch closing;
     Load break capability of the switch that should be opened.
3. Identify the bypass switch candidates to deal with constraints violations, if any, identified in step 3. The selected bypass switch should be able to:
   - relieve the violations generated in step 2;
   - ensure no new constraints violations are generated due to the switching operations of bypass switch.
4. If no bypass switch is available in step 3 then find the alternative switching operations that may have to first temporarily interrupt the service for some customers then do the load transfer to relieve the overloads.
5. Go to step 1 or finish the sequence implementation.

Protection Considerations

From system protection point of view the viability of new network configuration recommended by service restoration or load balancing program must be verified to assure that the new configuration can be fully protected by relay devices. The following are the major steps in the coordination of service restoration or load balancing program with relay coordination program [9]:

P1. Send the following data to relay coordination program:
   - List of recommended switching operations.
   - Power flow outputs of the recommended new system configuration.

P2. Request from the relay coordination program whether the recommended switching operations are acceptable from protection point of view:
   - If the recommended switching operations are acceptable, the service restoration or load balancing program has achieved the final solution and it shall pass the outputs to operator for his action.
   - If the recommended switching operations are not acceptable, the solution is still passed to the

operator along with relay coordination warnings. Operator will either override the warnings and implement the selected switching operations or reject the solution. If the solution is rejected based on relay coordination consideration, the feeder reconfiguration program shall discard the selected solution and find the next available solution. In case there is no available solution, it must pass the information to operator.

VI. CONCLUSIONS

Two new feeder reconfiguration algorithms for the purpose of service restoration and load balancing have been developed and tested as part of the distribution automation system at PG&E. The methodologies developed combine optimization techniques with heuristic rules and fuzzy logic. The practical operating concerns of feeder reconfiguration and the coordination with other distribution automation applications are addressed. The developed algorithms have been implemented into a production grade software program. Test results on PG&E distribution feeders show that the performance is efficient and robust.

VII. ACKNOWLEDGMENT

The authors would like to thank Dr. Carol Cheng and Mr. Ken Lau of PG&E for their useful comments in the development and implementation of this software package. PG&E's R&D department funded the development of the proposed algorithms and the FRECON software program.

VIII. REFERENCES

[1] K. Aoki, H. Kuwabara, etc. "An efficient algorithm for load balancing of transformers and feeders by switch operation in large scale distribution systems," IEEE Trans. on Power Delivery, Vol. 3, No. 4, October 1988.

[2] K. Aoki, K. Nara, etc. "A new algorithm for service restoration in distribution systems," IEEE Trans. on Power Delivery, Vol. 4, No. 3, July 1989.

[3] J.S. Wu, K.L. Tonsovic, C.S. Chen, "A heuristic search approach to feeder switching operation for overload, faults, unbalance flow and maintenance," IEEE Trans. on Power Delivery, Vol. No. 4, Oct. 1991.

[4] Yuan-Yih Hsu, H. Kuo, "A heuristic based fuzzy reasoning approach for distribution system service restoration," IEEE Trans. on Power Delivery, Vol. 9, April 1994, pp. 1950-1957.

[5] D. Shirmohammadi, "Service restoration in distribution networks via network reconfiguration," IEEE Transaction on Power Delivery, April 1992, pp. 952-958.

[6] Mesut E. Baran, F. F. Wu, "Network reconfiguration in distribution systems for loss reduction and load balancing," IEEE Trans. Power Delivery, 4 (1989) 1401-1407

[7] B. Naga Raj, K. S. P. Rao, "A new fuzzy reasoning approach for load balancing in distribution system," IEEE 1995 Winter Meeting, 85 WM 126-3 PWRS.

[8] J.C. Wang, H. D. Chiang, G. R. Darling, "An efficient algorithm for real-time network reconfiguration in large scale unbalanced distribution systems," PICA 1995, pp. 510-516.

[9] F.Soudi and J. Yee, "How distribution automation and protection systems can complement each other," proceeding of the 21st annual western protective relay conference, Spokane, WA, Oct. 1994.

Appendix A Summation of Loads into Equivalent Load

In Figure A1 loads I_1, I_2, I_3, and I_4 are surrounded by closed switches SW_1, SW_2, and SW_3. In order to find the optimal solution by solving an assignment problem and also simplify the calculation we need to sum them into an equivalent load as shown in Figure A2.

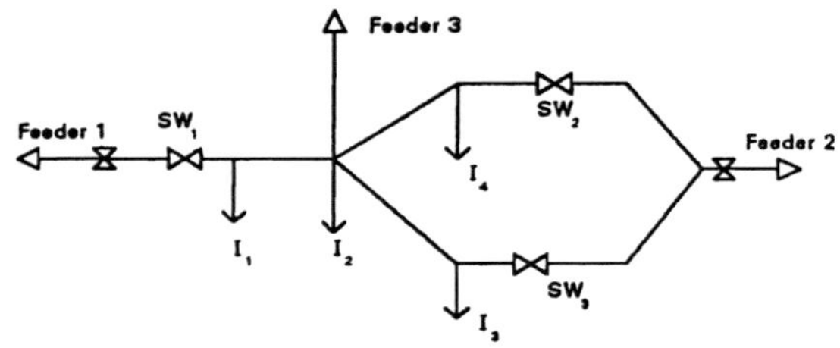

Figure A1

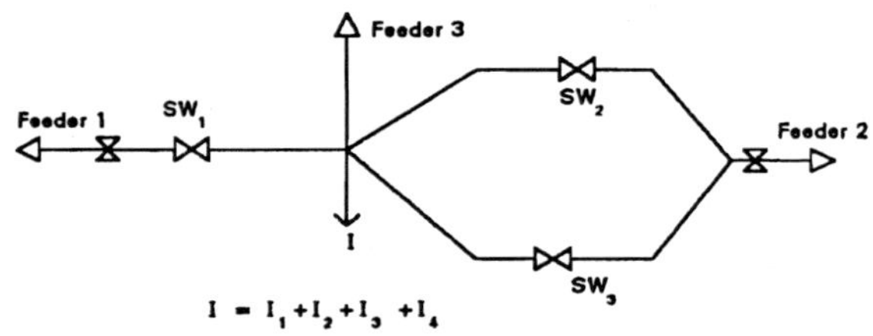

Figure A2

Application of Expert System to Power System Restoration in Sub-control Center

Young-Moon Park, Fellow, IEEE
Department of Electrical Engineering
Seoul National University
Seoul 151-742 Korea

Kwang-Ho Lee, Member, IEEE
Department of Electrical Engineering
Dankook University
Seoul 140-714 Korea

Abstract — An expert system is developed for emergency operation of power system restoration in a sub-control center. The proposed method restores the blackout area in power system by using sensitivity analysis and heuristic rules. The rules are categorized into flow-control rules, energization rules, line switching rules, load shedding rules, and voltage correction rules. The switching (opening, closing) of transmission lines, transformers and shunt capacitors, and the load shedding are used for the emergency operations in this paper. For the real time recognition of the network topology, the power system components are represented in hierarchical data structure.

I. INTRODUCTION

Power systems are getting more complicated and larger. Under these circumstances, a fast and reliable restoration algorithm is required for energizing the unserved power. The studies on the blackout restoration started in the early 1980s. The restorative state is classified into two : the state of complete collapse and the state of partial outage [1,2]. The restoration for complete collapse deals with a number of complicated and difficult problems such as : appropriate sectionalization of blackout area, energizing transmission lines, and considering the dynamic characteristics and restart conditions of generators, etc.. The restoration of partial outage, whereas, usually deals with the reconfiguration of the networks from the restorative state to the normal state. Practically, blackout areas are restored by the power system operators using off-line data and their experiences [3,4].

96 WM 346-7 PWRS A paper recommended and approved by the IEEE Power System Engineering Committee of the IEEE Power Engineering Society for presentation at the 1996 IEEE/PES Winter Meeting, January 21-25, 1996, Baltimore, MD. Manuscript submitted July 25, 1995; made available for printing January 5, 1996.

Some papers [5,6] classify the power balance into three, viz. power surplus, deficit and blackout of the isolated area. The line power flow in these papers are checked by load flow calculation. However, since the load flow calculation requires excess of time, these methods do not seem that appropriate for emergency operations.

This paper proposes a method to restore the blackout areas of partial outage, preventing at the same time the line overloads and voltage violations by : network switching, load shedding and shunt capacitor switching. Load shedding has been one of the topics of recent interests [7,8]. The load in these cases is usually modeled as continuous load, i.e. the load power is continuously changeable. But in this paper, since the load denotes the feeder power in a substation, the load shedding decreases the load power discretely by opening the feeder switches connected to the low voltage buses in substations. For the estimation of line overloads and voltage violations, and for the determination of the lines, feeders and shunt capacitors to be switched, the switching sensitivities are used for calculating the line power flows and the bus voltages. Such approximate calculations contribute to the speedy restoration planning and lead to an efficient performance compared to the approaches with no line power calculations but only heuristics.

The research on the power system restoration was activated by applications of expert system in the middle of 1980s. A number of papers [5,11,12] using expert systems for restoration problems based on the heuristics of operators and the patterns of restorative operations were presented. However, since the possible heuristics and patterns are predefined, they are not always appropriate in all cases. And also the effective selection of switching depends on the state of power system. The proposed scheme has the expert system structure which is suitable to program the operational knowledge for handling the difficulties of analytical approach and the requirements of operator's heuristics. In this system, the numerical data of switching sensitivities are incorporated with the inference and the restoration knowledge.

A prototype is developed for emergency operation at a sub-control center in the metropolitan area of Korea. It is confirmed that it can give appropriate restoration plan to the operators in the sub-control center.

Reprinted from *IEEE Transactions on Power Systems,* Vol. 12, No. 2, pp. 629–635, May 1997.

II. RESTORATION PROBLEM

Problem Formulation

The operating conditions of a power system can be described mathematically in terms of two sets of equations [10] : the power balance (or equality) constraints, and the operating (or inequality) constraints. The power balance constraints impose the requirement that the customer load demand be met at all times, while the operating constraints reflect the fact that the system variables (e.g., power flow and bus voltage) must always be kept within limits representing the physical limitations of the power system equipment. Power system operation can be described as being in one of three operating states : normal, emergency, or restorative.

The restoration problem is a very complex process which typically involves all of the components of the power system, including generation, transmission and distribution. This paper concerns only with the restoration of a local transmission system, primarily conducted in a sub-control center. Therefore the issues in a bulk power transmission such as Ferranti rise, and resonance are not considered in this paper. As stated, the restoration control problem can be formulated as a multi-objective function, composed of at least the following two separate objectives : (i) to minimize the unserved load power including the load shedding, (ii) to minimize the number of the switching of lines, control actions for the speedy restoration.

Network Reduction

In the on-line control and operation of a sub-control system, the system is usually interconnected to other systems. Therefore the difficulty in predicting the impact of a contingency arises from the fact that the external network is not monitored as carefully as the internal network. While the internal system variables are known on-line by monitoring, as for the external system, on-line information is normally restricted to items such as interlink power flows and status of switch connections. Since the state of the external network is not fully known, some approximations are required.

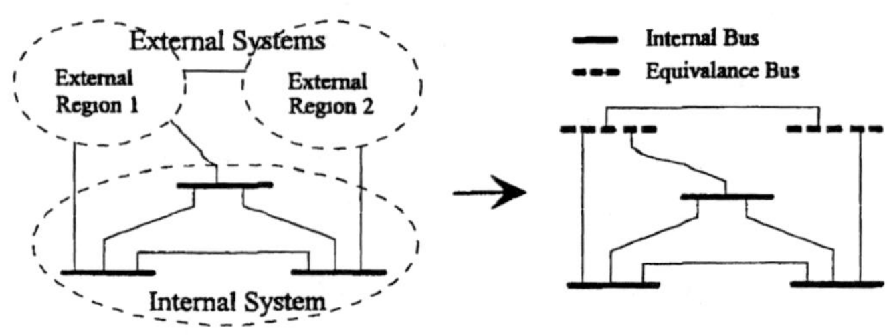

Fig.1. Sample Diagram of Network Reduction.

The external system can be decomposed into several regions as shown in Fig.1. In a decomposed region, all the bus voltages have a tendency to change in the same rate. This phenomenon is called voltage coherency. Assuming voltage coherency, a decomposed region is equivalenced by a bus which has an average voltage among the buses in the region. The injection power of the equivalence bus is assumed to be equal to the aggregated injection power in the region.

These two assumptions make it possible to predict the interconnections between internals and externals approximately. The key point of this network reduction technique lies in keeping the voltages in voltage coherency. In this paper, the interconnections between internals and externals can also be predicted by the sensitivity analysis.

Sensitivity of Switching

The prediction of the line power flows and bus voltages for the reconfigured network is required after the energization process. The restoration plan must include overload alleviation and/or voltage correction, if any line is overloaded and/or any bus voltage is violated, respectively. The predicted power flows and bus voltages are required, also when the planning for the network modified by switching control is carried out. The sensitivities of switching indicate the changes of the line power flow and bus voltage.

The power balance equation of complex form in the i-th bus is as follows :

$$S_{gi} - S_{di} - f_i(\mathbf{y},\mathbf{v},\theta) = 0, \tag{1}$$

where f_i, S_{gi}, S_{di} are the injection, generation and load power including shunt capacitor at the i-th bus respectively, $\mathbf{y}$ is the admittance vector of lines and transformers, $\mathbf{v}$ and θ are the vectors of bus voltage magnitude and angle respectively.

The line switching and the load shedding including shunt capacitor switching at the i-th bus are assumed.

$$S_{gi} - S_{di} - \Delta S_{di} - f_i(\mathbf{y}+\Delta\mathbf{y},\mathbf{v}+\Delta\mathbf{v},\theta+\Delta\theta) = 0. \tag{2}$$

The Eqn. (2) is approximated as follows :

$$S_{gi} - S_{di} - f_i(\mathbf{y},\mathbf{v},\theta) - \Delta S_{di} - \Delta f_i(\Delta\mathbf{y},\Delta\mathbf{v},\Delta\theta) = 0, \tag{3}$$

where ΔS_{di} is the load power to be shed, or changed by shunt capacitor switching,

$$\begin{aligned}\Delta f_i(\Delta\mathbf{y},\Delta\mathbf{v},\Delta\theta) &= (\partial f_i/\partial\mathbf{y})^t\Delta\mathbf{y} + (\partial f_i/\partial\mathbf{v})^t\Delta\mathbf{v} + (\partial f_i/\partial\theta)^t\Delta\theta \\ &+ \Delta\mathbf{y}^t(\partial^2 f_i/\partial\mathbf{y}\partial\mathbf{v})\Delta\mathbf{v} + \Delta\mathbf{y}^t(\partial^2 f_i/\partial\mathbf{y}\partial\theta)\Delta\theta\end{aligned} \tag{4}$$

In the Eqn.(4) of the second order Taylor expansion, the terms $\Delta\theta$-$\Delta\theta$, $\Delta\mathbf{v}$-$\Delta\mathbf{v}$ and $\Delta\mathbf{v}$-$\Delta\theta$ are neglected, and the term of $\Delta\mathbf{y}$-$\Delta\mathbf{y}$ is zero because of the linear relation between the bus power equation and the line admittance. From (1), (3) and (4),

$$\begin{aligned}&\Delta S_{di} + (\partial f_i/\partial\mathbf{y})^t\Delta\mathbf{y} + (\partial f_i/\partial\mathbf{v})^t\Delta\mathbf{v} + (\partial f_i/\partial\theta)^t\Delta\theta \\ &+ \Delta\mathbf{y}^t(\partial^2 f_i/\partial\mathbf{y}\partial\mathbf{v})\Delta\mathbf{v} + \Delta\mathbf{y}^t(\partial^2 f_i/\partial\mathbf{y}\partial\theta)\Delta\theta = 0\end{aligned} \tag{5}$$

The real terms and reactive terms in (5) are decomposed

and expanded to all the buses using the decoupling features of the Jacobian in the power equations. This gives the following equations :

$$\Delta P_d + Re\{f_y' \Delta \mathbf{y}\} + [H + \Delta H]\Delta\theta = 0, \tag{6a}$$

$$\Delta Q_d + Im\{f_y' \Delta \mathbf{y}\} + [L + \Delta L]\Delta \mathbf{v} = 0, \tag{6b}$$

where H, L are the Jacobian of $\partial f / \partial \theta$, $\partial f / \partial \mathbf{v}$ respectively, and ΔH, ΔL corresponds to $\Delta \mathbf{y}'(\partial^2 f / \partial \mathbf{y} \partial \theta)$, $\Delta \mathbf{y}'(\partial^2 f / \partial \mathbf{y} \partial \mathbf{v})$ respectively.

The value of the elements in the vector $\Delta \mathbf{y}$ is $+\mathbf{y}$ in closing, $-\mathbf{y}$ in opening, so the term can be substituted by $diag(\mathbf{y})\Delta u_s$, where Δu_s takes one of the values -1, 0, +1. Similarly, the real load shedding term, ΔP_d is substituted as $D_r \Delta u_d$, and the reactive load shedding and shunt capacitor term, ΔQ_d is substituted as $C\Delta u_c + D_i \Delta u_d$. The values in the matrix D_r and D_i are the real and reactive powers of feeder loads respectively, where the row index corresponds to the buses, column index corresponds to all the feeders of all the buses. The value of the element in the vector Δu_d takes one of the values, -1, 0. The matrix C indicates the ratings of shunt capacitors. The term of shunt capacitor switching, Δu_c also takes one of the values, -1, 0, +1.

From Eqn.(6), $\Delta\theta$ and $\Delta \mathbf{v}$ are obtained by the following equations.

$$\Delta\theta = A_r' \Delta u_d + B_r' \Delta u_s, \tag{7a}$$

$$\Delta \mathbf{v} = A_i \Delta u_d + A_c \Delta u_c + B_i \Delta u_s, \tag{7b}$$

where $A_i = -[L + \Delta L]^{-1} D_i$, $A_c = -[L + \Delta L]^{-1} D_c$, $B_i = -[L + \Delta L]^{-1} Im\{f_y'\, diag(\mathbf{y})\}$, $A_r' = -[H + \Delta H]^{-1} D_r$, $B_r' = -[H + \Delta H]^{-1} Re\{f_y'\, diag(\mathbf{y})\}$.

In Eqn. (7b), A_i, A_c and B_i are voltage sensitivities with respect to load shedding, shunt capacitor switching and line switching respectively.

The power flow of the line between m-th and n-th buses is

$$P_{mn} = -v_m^2 Re\{y_{mn}\} + v_m v_n (Re\{y_{mn}\}\cos\theta_{mn} + Im\{y_{mn}\}\sin\theta_{mn}). \tag{8}$$

The variation of Eqn.(8) is

$$\Delta P_{mn} = v_m v_n (-Re\{y_{mn}\}\sin\theta_{mn} + Im\{y_{mn}\}\cos\theta_{mn})\Delta\theta_{mn}. \tag{9}$$

The equation of line power sensitivity is derived as follows by expanding (9) to all the lines and by combining Eqn.(7).

$$\Delta P_l = A_r \Delta u_d + B_r \Delta u_s, \tag{10}$$

where $A_r = -H_l A_r'$, $B_r = -H_l B_r'$. The matrix H_l indicates the Jacobian value between both buses of each line. In the matrix B_r, the row index corresponds to the lines to be checked for overloads, and the column index corresponds to the lines to be switched. In the matrix A_r, the row index corresponds to the lines to be checked, and column index corresponds to the buses where the load shedding is considered.

III. DESCRIPTION OF SYSTEM COMPONENTS

The hierarchical knowledge-base is constructed by subsystems for the fast restoration. The cells and subsystems describe the power systems and provide the expert systems with the data structure which ensures the efficient search.

Cells and Subsystems

A cell is a conceptual object which corresponds to the element that can be isolated by opening all the switches. A cell is represented by the data that can describe a certain element such as bus, line, generator, load, transformer, etc.. A subsystem is defined as a group of cells or the other subsystems in one level lower in the hierarchical structure. Subsystem describes some part of power systems. Therefore the network reconfiguration is recognized by rebuilding the subsystem hierarchies.

Switch and Its Representation

The electrical network connection of the power systems depends on the on/off status of the switches(circuit breakers, line switches). In this paper, the restoration plan is determined by the simulation of many cases of switching. Therefore during restoration planning, rapid recognition of the reconfigured networks is required. This recognition is done by rebuilding the subsystems with respect to the on/off status of the switches. A CB is represented as :

CB(index, $Cell_1$, $Cell_2$, status),

where the index gives the identification of the CB. The $Cell_1$ and the $Cell_2$ indicate the cells which are linked by the CB.

Representation of Subsystems

The data object, SUB, which describes a subsystem, is stored in the data base with the following format.

SUB(index, type, [internal SUB list], [external CB list], [data]),

where the index is the identification of the subsystem, and the other components are as follows :

1) Type : It describes the type of the subsystem. It is one of the equipment : bus, line, generator, load, transformer, and SUB. The subsystem of SUB-type is formed by grouping the other subsystems, which locate in one-level lower in the hierarchical structures.

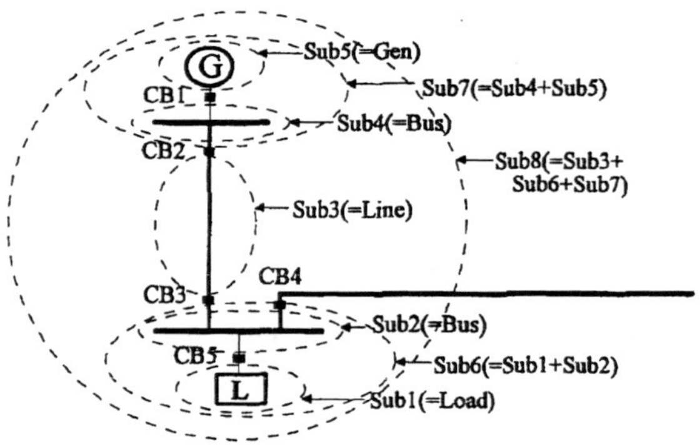

Fig.2. Diagram of Sample Subsystems.

2) Internal SUB List (ISL) : The internal SUB of the subsystem is a subsystem which locates in one-level lower inside the subsystem. The ISL of the subsystem lists all the internal SUBs of the subsystem, then the ISL of a cell-type subsystem is empty. In Fig.2, the ISL of SUB 6 is {1, 2}, and the ISL of SUB 8 is {3, 6, 7}.

3) External CB List (ECL) : ECL lists all the external CBs which connect the subsystem in consideration with the other external subsystems. Since the internal CB list is not needed to rebuild the hierarchical structures, the recognition of the network reconfiguration from the switching is done only by the ISL and the ECL. In Fig.2, the ECL of SUB 6 is {3, 4}, the ECL of SUB 4 is {1, 2}.

4) Data : According to the types of the subsystems, each has information of state variables and their limits allowed such as voltage, power, line capacity, etc.. The SUB-type subsystem has the data of the total generation power and the total load power inside itself.

A subsystem is classified into energized and deenergized subsystems. A subsystem with the power generation source in the data field is defined as an energized subsystem, while a subsystem without the source power in the data field is defined as a deenergized subsystem. After the recognition of the network modified by switch operations, the state of the separated partial systems are to be identified according to the power balance. The identification of the power state such as blackout, surplus and deficit, is done by the information in the data field of the subsystem which belongs to the highest level in the hierarchical structures.

Sequence of Reconfiguration

In case of opening a switch, the separation of a subsystem occurs, and in case of closing, the merge of subsystems occurs. When a switch operates to connect, the hierarchical structures are modified only if the changed switch is the external CB of the highest subsystem. This is obvious since the other subsystems lower than the highest one already are merged to build the hierarchical structure.

IV. KNOWLEDGE-BASED SYSTEM

The knowledge base consists of following types described by production rules : energization, overload alleviation, voltage correction, and flow-control knowledge. The flow-control knowledge carries out the reasoning control for other types of knowledge. The overload alleviation knowledge decomposed into line switching and load shedding knowledge refers to the switching sensitivities in off-line data. The knowledge base constitution is explained in Fig. 3.

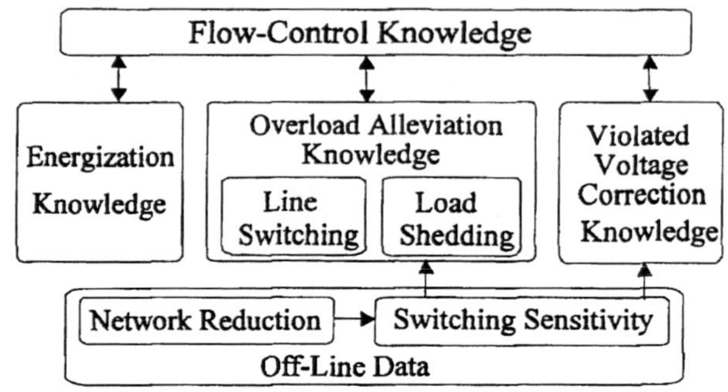

Fig.3. Knowledge Configuration.

Flow-Control Knowledge

The reasoning tree of flow-control knowledge is depicted in Fig.4. Each node has the truth value "true" or "false". The task of restoration is ended when the root node 'System Restoration' has the value "true". As the inference engine traces the AND/OR tree, each node is evaluated to have one of the truth values.

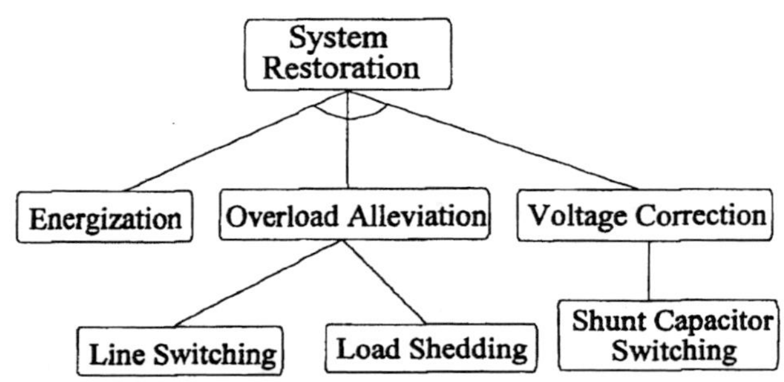

Fig.4. Flow-Control Knowledge Tree.

The node 'Energization' has the value "true" when the plan of energizing blackout areas is determined, or when blackout areas do not arise. After the plan of line switching is determined, if it makes all the overloads alleviated, then the node 'Overload Alleviation' has the value "true" without executing the load shedding knowledge, otherwise, the load shedding is executed. After the overload alleviation, if any bus voltage violation is predicted, the correction control by shunt capacitor switching is considered.

Energization Knowledge

After a fault occurs and the protection devices operate, the subsystems in highest level belong to one of three states, viz. blackout, deficit, surplus according to power balance. A blackout area corresponds to the highest subsystem in which the total generation power equals to zero, and the total load power does not. To recognize the blackout area, the only thing to do is to check the data field of the highest subsystems. Once the blackout area is recognized, it is connected to the surplus area having the maximum surplus power with a line of maximum line power capacity

Line Switching Knowledge

After the line switching for energizing blackout areas, the power flows of the changed networks are predicted for checking the line overloads by the switching sensitivities of the lines (including transformers). If the overloads are predicted, the supplementary line switching is considered to alleviate the overloads. For speedy restoration, the number of lines to be switched must be minimized satisfying the operating constraints at the same time.

There are a lot of combinations of lines to be switched. This means the searching space to find the best switching is vast. But information of line switching sensitivity, the matrix B_r in Eqn.(10), confines the searching space by selecting the effective lines which have negative values of sensitivities, and by avoiding the combination of switching lines which increase the overloads. Since the searching proceeds from the cases of single line switching to those of more lines, the solution switching that is first found in the searching process satisfying the line power constraints has the minimum number of switching. These heuristics help reduce the time required to determine the switching lines.

Load Shedding Knowledge

When severe fault occurs, it is possible that the line switching is insufficient to alleviate all the overloads. In such case, after the overloads are decreased maximally by the line switching, the load shedding is considered to alleviate the remaining overloads. The load power of a bus is the load sum of the feeders at the bus, and it decreases discretely according to the load shedding which is carried out by opening the feeder switches. Since the minimization of the unserved power is the objective of the restoration plan, it is efficient to minimize the amount of load shedding power satisfying the operating constraints at the same time. Information of load shedding sensitivity, the matrix A_r in Eqn.(10), also confines the searching space, and reduces the amount of the shedding power.

Voltage Correction Knowledge

After the energization and overload alleviation, if any bus voltage is violated, the shunt capacitors at the bus are switched for voltage correction. The amount of voltage violation is calculated approximately by the sensitivities in Eqn.(7b). If a bus voltage is low, a shunt capacitor is switched on, and if high, vice versa. The number of switching capacitors is determined by the sensitivity, A_c in Eqn.(7b). If the reactive power compensation devices at the voltage violated bus is not sufficient, another bus which is effective to correct the violated voltage is determined. If any reactive power compensation is not satisfactory, the restoration planning is terminated with an emergency message of voltage violation.

V. CASE STUDY

This restoration guidance system is installed at a sub-control center (SCC) equipped with Minor-SCADA in Seoul, Korea. As shown in system configuration figure 5, the power system data are acquired through the connection and communication modules from RTUs. The sensitivities are updated each several minutes, and its calculation time is about 1 sec. by 486 PC.

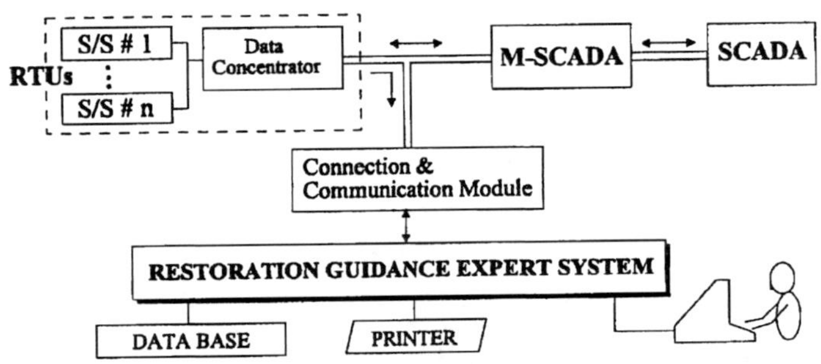

Fig.5. System Configuration.

Table 1. Feeder Load Data [MW]

Load Cell Number	*	1st feeder	2nd	3rd	4th	5th	6th	7th
c64	3	3.0	1.1	0.8	-	-	-	-
c65	6	12.5	17.5	33.7	16.3	16.4	13.6	-
c66	7	7.4	12.5	9.5	8.52	11.0	12.7	7.3
c67	6	7.7	3.8	16.5	8.48	10.0	9.0	-
c68	5	5.2	3.8	16.0	8.9	13.1	-	-
c69	4	10.0	25.0	20.0	22.0	-	-	-
c70	6	5.0	13.0	10.5	7.5	10.1	15.7	-
c71	5	17.1	10.1	24.1	18.1	16.0	-	-
c72	4	11.7	14.1	6.25	15.6	-	-	-
c73	6	21.9	8.2	18.9	11.2	27.3	3.2	-
c74	5	12.7	5.3	5.27	11.5	5.0	-	-
c75	4	2.8	5.6	3.6	5.0	-	-	-
c76	4	7.6	4.2	7.1	3.1	-	-	-
c77	4	3.2	1.8	4.9	4.1	-	-	-
c78	4	9.7	3.3	2.8	3.2	-	-	-
c79	4	0.8	3.2	0.75	2.25	-	-	-
c80	3	0.453	0.28	0.72	-	-	-	-
c81	1	1094.	-	-	-	-	-	-

(* : Number of feeders)

The power system has 18 lumped loads in heavy loads as shown in Fig.6. The equivalence buses and lines outside the SCC. are shown in dotted lines. The nominal voltages are 154kv in the high-side of the transformers 44, 45 and 46, and 66kv in the low-side. Each lumped load consists of several feeders in a substation and total number of feeders in the SCC. is 81, as shown in Table 1. In normal operation of the network, the lines 49 and 55 in Fig.6 are open to suppress single phase short circuit capacity.

Two faults have occurred at the line 29 and at the transformer 45. These faults cause the blackout area **B1**, as shown in the Fig.6, and the overload at the line 30. If no countermeasures are performed, the overload lasts to trip the CBs for protecting the line 30. Then another blackout area **B2** arises, as shown in Fig.6. The result of the energization routine shows connecting the line 49 for **B1** and the line 30 for **B2**. But the overloads at the line 30 and 49 are predicted by the line switching sensitivities. The predicted line powers at the line 30 and 49 are 157.14 and 38.92[MW] respectively, so the overloads are 4.76 and 14.45% relative to the power flow capacities of the lines, 150 and 34[MW] respectively.

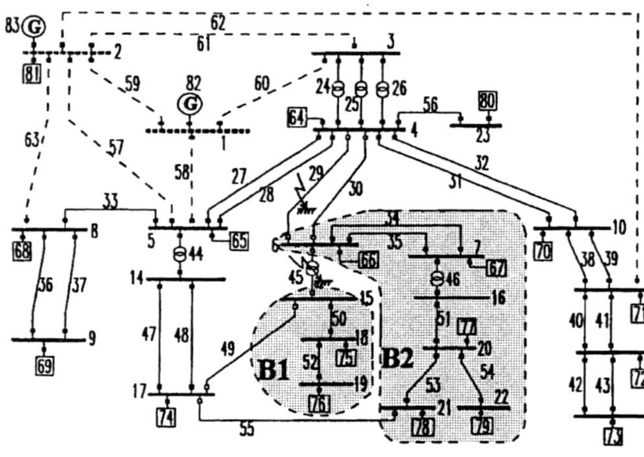

Fig.6. Network in case of a fault at line 29.

The overload alleviation by line switching shows closing the line 55, and opening the line 53. This indicates that the lumped load 78 must be transferred from the bus 20 to the bus 17 for alleviating the overload at the line 30. Closing the lines 30 and 55 does not relieve the overload at the line 30, and it is predicted by switching sensitivities. The overload at the line 49 can not be reduced by line switching, so the load shedding routines are activated. The feeders to be opened are as follows :

Bus 18, feeder 3, 3.6[MW]
Bus 19, feeder 4, 3.1[MW]

The Fig.7 shows the line power changes of all the lines. In the figure, the blanked bars denotes the line powers before the routine of overload alleviation and load shedding, and the gray bars denotes those after the routine. The error analysis shows that the line powers are predicted by sensitivities with accuracy of 95%.

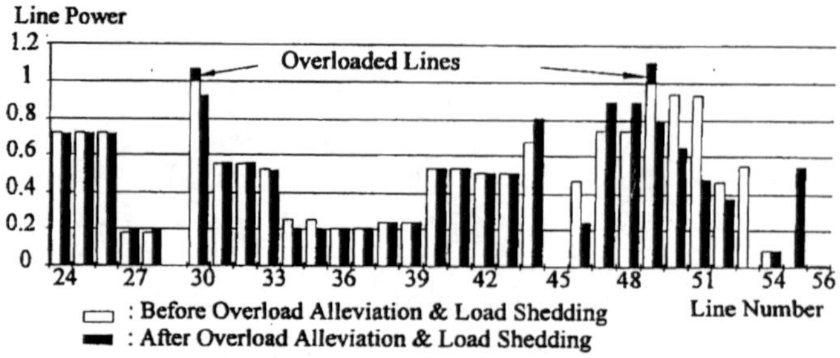

Fig.7. Comparisons of Line Powers in Case 1.

In the voltage correction routine, the voltage violation is predicted by the sensitivities of line switching and load shedding. The predicted voltage at bus 19, 0.941 is below the limit 0.95. The result of voltage correction shows that the switch of a shunt capacitor, 5.0[MVAR] is to be closed. After the restoration planning of overall switching strategy, following operations are presented to the operators.

- open the 3rd feeder at bus 18
- open the 4th feeder at bus 19
- open the line 53
- close the lines 55, 49, 30
- close the shunt capacitor, 5.0[MVAR] at bus 19

VI. CONCLUSION

This paper proposed the expert systems which can not only restore the blackout areas but also prevent the line overloads and voltage violations. A blackout area is restored by a line switching that interconnects the blackout area and the energized areas. The overload alleviation is done by the line switching and the feeder switching which means the load shedding in a substation. The voltage violation is corrected by the shunt capacitor switching. During restoration plan, the numerous cases of switching are generated, but the restricted cases are considered by using the heuristic rules. Since the line switching changes the network topology, the recognition of the network reconfiguration is needed. The suggested method of the data representation is appropriate to recognize the changed topology rapidly. The line powers and the bus voltages of post-switching play a key role in determining the restorative switching. These states of power system are predicted easily by switching sensitivities derived in this paper.

This restoration guidance system is installed at a sub-control center(SCC) with M-SCADA in Seoul. This expert system gives the appropriate solutions that are applicable to the power system operators. However for more realistic assistance to operators, dynamics of loads, length of blackout and cold load pickup are to be considered in extending this system.

Reference

[1] TASK FORCE Report, "Power System Restoration", IEEE Trans. on Power Systems, Vol. PWRS-2, No.2, May 1987

[2] R.J.Kafka, "Role of Interactive And Control Computers in The Development of A System Restoration Plan", Vol.PAS-101, No.1, Jan. 1982

[3] M.Kato, "The Development of Power System Restoration Method for A Bulk Power System by Applying Knowledge Engineering Techniques", IEEE Trans. on Power Systems, Vol.4, No.3, pp.1228-1235, Aug. 1989

[4] K.Hotta, "Implementation of A Real-Time Expert System for A Restoration Guide in A Dispatching Center", IEEE Trans. on Power Systems, Vol.5, No.3, Aug. 1990

[5] T.Sakaguchi, "Development of A Knowledge Based System for Power System Restoration", IEEE Trans. on PAS. Vol.PAS-102, No.2, Feb. 1983

[6] Shi-Mo Wang, "An Expert System for Bulk Power System Restoration", Second Symposium on ESAP, Seattle, USA, July 17-20, 1989

[7] F.C.Schweppe, "A Generation Reallocation And Load Shedding Algorithm", IEEE Trans. on PAS, Vol.PAS-98, No.1, Jan/Feb., 1979

[8] M.M.Adibi, "Local Load Shedding", Trans. on Power Systems, Vol.3, No.3, Aug. 1988

[9] C.C.Liu, "An Expert system Assistaing Decision-Making of Reactive Power/Voltage Control", IEEE Trans. on Power Systems, Vol.1, No.3, Aug. 1986

[10] Dy Liacco, "A Hierarchical Interactive Approach to Electric Power System Restoration", IEEE Trans. on Power Systems, Vol.7, No.3, Aug. 1992

[11] M.Matsumoto, "Development of A Guidance Method for Power System Restoration", IEEE Trans. on Power Systems, Vol.4, No.3, Aug. 1989

[12] S.Fukui, "A Knowledge-Based Method for Making Restoration Plan of Bulk Power System", IEEE Trans. on Power Systems, Vol.7, No.2, May. 1992

[13] Atif.S. Debs, Modern Power Systems Control and Operation, Kluwer Academic Press, 1988.

Author Index

Subject Index

NOTE: Italic f indicates pages with illustrations; italic t indicates pages with tables.

A

B

C

E

L

M

T

U

V

W

Z

About the Editor

M. M. Adibi earned the B.Sc.E.E. with honors in 1950 from the University of Birmingham, U.K., and the M.E.E. in 1960 from Polytechnic Institute of Brooklyn, New York City. He has spent over 49 years in the service of electric utilities, assuming various responsibilities, including manufacturing and testing of heavy electrical equipment at General Electric Company, U.K. (1950–1951); operation and maintenance of large thermal power plants at British Petroleum, Iran (1952–1955), where he was promoted to power plant superintendent; and power system planning at Ebasco International, New York (1956–1959), where he was in charge of developing 10-year construction programs for several South and Central American Utilities.

In 1960, Mr. Adibi joined IBM Corporation, developing computer applications in the engineering and operation of electric utilities. As the industry consultant, he conducted and participated in many R&D projects related to power system and power plant operation, and advanced network functions. He was the recipient of the IBM Industry Grant to study the computer-based fuel model for the power industry.

Since 1979 he has assumed industry leadership by founding and chairing the IEEE System Restoration Working Group. He has addressed many restoration issues in the 23 IEEE Transaction papers in which he is the principal author, and he has organized many paper and panel sessions at the meetings of the IEEE Power Engineering Society. His effort culminated in 1993 with the successful publication of the IEEE special publication on Power System Restoration, 93 THO 605-6 PWR, for which he received the 1995 Power Engineering Society's Outstanding Technical Report Award. Mr. Adibi is also the recipient of the IEEE Power Engineering Society's prize for his paper on "Power System Restoration Planning." He is a life fellow of the IEEE.

Mr. Adibi is currently the president of IRD Corporation, Chartered Engineer, Institute of Electrical Engineers (IEE), U.K., and a registered professional engineer in the state of Maryland.

About the Authors

Jerry Ancona earned his B.S., in electrical engineering from Clarkson University in 1969 and his M.B.A. from Syracuse University in 1982. He is a registered professional engineer in the state of New York, a senior member of the IEEE, and a member of the Power Engineering Society's System Economics Subcommittee and the Power System Restoration Working Group.

Since joining Niagara Mohawk Power Corporation in 1969, Mr. Ancona has held positions in Power Control Operations, Transmission Planning, Generation Planning, and Distribution Planning. Within Power Control, he developed more detailed power system restoration procedures, guidelines, and training sessions to improve Niagara Mohawk's preparedness and to help coordinate its restoration with the New York Power Pool Restoration Plan.

Currently, Mr. Ancona is the Independent System Operator (ISO) coordinator for Niagara Mohawk and chairman of the Pricing Working Group–Generation Issues Resolution Team within the New York Power Pool. In this role, he has been heavily involved in the development and dissemination of market restructuring rules and operating procedures for the New York ISO, including its tariff filing. He is also helping establish procedures and training programs to assist Niagara Mohawk's transition to the new environment.

Lester H. Fink retired in 1997 as executive vice president of KEMA Consulting. Earlier in his career, he conducted and managed plant and system control research at the Philadelphia Electric Company (1950–1974), developed and managed the National Systems Engineering for Power Research Program at the U.S. Energy Research and Development Administration, engaged in contract research and consulting in the private sector (1979–1997), and served as adjunct professor at the University of Pennsylvania, Drexel University, and the University of Maryland. His publications include three U.S. patents (two on power systems control) and more than 50 papers. He is a life fellow of IEEE and the Instrument Society of America (ISA), Attwood Associate of the U.S. National Committee of Conference Internationale das Grandes Researex Electriques a Haute Tension (CIGRE), and received the Meritorious Service Award of the U.S. Department of Energy.

R. J. Kafka received the B.S. degree in physics from Regis College, Denver, Colorado, in 1970 and the M.S. degree from Purdue University, West Lafayette, Indiana, in 1972. He has been employed by the Potomac Electric Power Company (PEPCO) since 1973. He was assigned to PEPCO's system restoration study team in 1979, which produced one of the first formal power flow studies for a system restoration plan. Mr. Kafka is principal engineer, Transmission System Operations of PEPCO, a charter member of the Power System Restoration Working Group, chair of the Transmission Operation Subcommittee, and a registered professional engineer in the state of Maryland. He is a fellow of the IEEE.

Chen-Ching Liu received his Ph.D. degree from the University of California, Berkeley. He is presently professor of Electrical Engineering and director of Advanced Power Technologies

(APT) Center at the University of Washington. Dr. Liu is chairing the Subcommittee on Intelligent System Applications, Technical Committee on Power System Analysis, Computing and Economics, and the IEEE Power Engineering Society. He is also one of the founders of the International Symposia on Intelligent System Applications to Power Systems (ISAPs). Professor Liu is a fellow of the IEEE.

A. J. Monticelli received his B.S. degree in electronic engineering from the Institute Technologicode Aeronautica (ITA) in 1970, the M.S. degree from Universidade Federal da Paraiba (UFPB) in 1972, and the Ph.D. degree from Universidade Estadual de Campinas (Unicamp) in 1975, all in Brazil. From 1982 to 1985, he was a visiting professor at the University of California, Berkeley, and from 1991 to 1992 with Mitsubishi Electric Corporation, Japan. Currently, he is a professor of electrical engineering at Unicamp. He is a fellow of the IEEE and a member of the Brazilian Academy of Sciences.

M. Rafian received the M.S. and Ph.D. in electrical engineering from the University of London. Currently, he is the team leader and developer of Operator Training Simulator applications at Siemens Power Systems Control.

Printed in the USA/Agawam, MA
September 21, 2026

910661.041